ちゃ ゆ ー と　　す ゐ ら う は ゝ ん う ゜ ら

ちゃゆーと

Algebra
and Trigonometry

Structure and Method   *Book 2*

Algebra
and Trigonometry
Structure and Method
Book 2

Richard G. Brown
Mary P. Dolciani
Robert H. Sorgenfrey
Robert B. Kane

Contributing Authors
Sandra K. Dawson
Barbara Nunn

Teacher Consultants
Keith Bangs
Doris Holland
Ouidasue Nash
Al Varness

McDougal Littell
A HOUGHTON MIFFLIN COMPANY
Evanston, Illinois ◆ Boston ◆ Dallas

Authors

Richard G. Brown, Mathematics Teacher, Phillips Exeter Academy, Exeter, New Hampshire

Mary P. Dolciani, formerly Professor of Mathematical Sciences, Hunter College of the City University of New York

Robert H. Sorgenfrey, Professor Emeritus of Mathematics, University of California, Los Angeles

Robert B. Kane, Professor of Mathematics Education, Purdue University, Lafayette, Indiana

Contributing Authors

Sandra K. Dawson, Mathematics Teacher, Glenbrook South High School, Glenview, Illinois

Barbara Nunn, Mathematics Department Chairperson, Marjory Stoneman Douglas High School, Parkland, Florida

Teacher Consultants

Keith Bangs, Mathematics Teacher, Owatonna High School, Owatonna, Minnesota

Doris Holland, Mathematics Teacher, O.D. Wyatt High School, International Baccalaureate Magnet Program, Fort Worth, Texas

Ouidasue Nash, Mathematics Teacher, Columbia High School, Columbia, South Carolina

Al Varness, former Mathematics Teacher, Lake Washington High School, Kirkland, Washington

The authors wish to thank **David L. Myers**, Computer Coordinator and Mathematics Teacher, Winsor School, Boston, Massachusetts, for writing the Portfolio Projects.

Printed in U.S.A.

ISBN: 0-395-77118-8

456789-DC-00 99 98

Contents

1 Basic Concepts of Algebra

The Language of Algebra

1-1	Real Numbers and Their Graphs	1
1-2	Simplifying Expressions	6

Operating with Real Numbers

1-3	Basic Properties of Real Numbers	14
1-4	Sums and Differences	21
1-5	Products	27
1-6	Quotients	33

Solving Equations and Solving Problems

1-7	Solving Equations in One Variable	37
1-8	Words into Symbols	43
1-9	Problem Solving with Equations	49

• *Explorations* Exploring Irrational Numbers 832

• *Technology* Computer Exercises 5, 32 Calculator Key-In 12

• *Special Topics* Reading Algebra xiv, 26 Logical Symbols: Quantifiers 20 Biographical Note 32 Historical Note 42 Challenge 48

• *Reviews and Tests* Mixed Review Exercises 5, 20, 31, 42, 54 Self-Tests 13, 36, 54 Chapter Summary 55 Chapter Review 55 Chapter Test 57

2 Inequalities and Proof

Working with Inequalities

2-1	Solving Inequalities in One Variable	59
2-2	Solving Combined Inequalities	65
2-3	Problem Solving Using Inequalities	69

Working with Absolute Value

2-4	Absolute Value in Open Sentences	73
2-5	Solving Absolute Value Sentences Graphically	76

Proving Theorems

2-6	Theorems and Proofs	81
2-7	Theorems about Order and Absolute Value	88

- *Explorations*   Exploring Inequalities 833

- *Technology*   Computer Exercises 64, 75   Computer Key-In 80

- *Special Topics*   Career Note / Automotive Engineer 64   Historical Note 68   Symbolic Logic: Boolean Algebra 95

- *Reviews and Tests*   Mixed Review Exercises 63, 72, 79, 91 Self-Tests 72, 79, 92   Chapter Summary 92   Chapter Review 93 Chapter Test 94   Mixed Problem Solving 98   Preparing for College Entrance Exams 99

3 Linear Equations and Functions

Linear Equations and Their Graphs

3-1	Open Sentences in Two Variables	101
3-2	Graphs of Linear Equations in Two Variables	107
3-3	The Slope of a Line	112
3-4	Finding an Equation of a Line	118

Linear Systems

3-5	Systems of Linear Equations in Two Variables	124
3-6	Problem Solving: Using Systems	131
3-7	Linear Inequalities in Two Variables	135

Functions and Relations

3-8	Functions	141
3-9	Linear Functions	146
3-10	Relations	153

- *Explorations*   Exploring Functions 834

- *Technology*   Computer Graphing Ideas 114, 119, 160 Computer Exercises 122

- *Special Topics*   Challenge 140, 158   Historical Note 152 Application / Linear Programming 159

- *Reviews and Tests*   Mixed Review Exercises 106, 117, 130, 139, 152 Self-Tests 123, 140, 158   Chapter Summary 161   Chapter Review 162 Chapter Test 164   Cumulative Review 165

4 Products and Factors of Polynomials

Working with Polynomials

4-1	Polynomials	167
4-2	Using Laws of Exponents	171
4-3	Multiplying Polynomials	174

Factors of Polynomials

4-4	Using Prime Factorization	179
4-5	Factoring Polynomials	183
4-6	Factoring Quadratic Polynomials	188

Applications of Factoring

4-7	Solving Polynomial Equations	194
4-8	Problem Solving Using Polynomial Equations	198
4-9	Solving Polynomial Inequalities	202

• *Explorations*   Exploring Polynomial Factors 835

• *Technology*   Computer Exercises 187   Computer Key-In 193

• *Special Topics*   Biographical Note 177   Reading Algebra 178
Historical Note 182

• *Reviews and Tests*   Mixed Review Exercises 170, 176, 187, 197, 205
Self-Tests 177, 192, 205   Chapter Summary 206   Chapter Review 207
Chapter Test 208   Preparing for College Entrance Exams 209

5 Rational Expressions

Using the Laws of Exponents

5-1	Quotients of Monomials	211
5-2	Zero and Negative Exponents	216
5-3	Scientific Notation and Significant Digits	221

Rational Expressions

5-4	Rational Algebraic Expressions	227
5-5	Products and Quotients of Rational Expressions	232
5-6	Sums and Differences of Rational Expressions	235
5-7	Complex Fractions	238

Problem Solving Using Fractional Equations

5-8	Fractional Coefficients	242
5-9	Fractional Equations	247

• *Explorations*   Exploring Continued Fractions 836

• *Technology*   Computer Key-In 215   Calculator Key-In 225

• *Special Topics*   Biographical Note 220   Graphing Rational Functions 230
Application / Electrical Circuits 253

• *Reviews and Tests*   Mixed Review Exercises 215, 225, 234, 241, 252
Self-Tests 226, 241, 252   Chapter Summary 254   Chapter Review 255
Chapter Test 257

6 Irrational and Complex Numbers

Roots and Radicals

6-1	Roots of Real Numbers	259
6-2	Properties of Radicals	264
6-3	Sums of Radicals	270
6-4	Binomials Containing Radicals	274
6-5	Equations Containing Radicals	277

Real Numbers and Complex Numbers

6-6	Rational and Irrational Numbers	283
6-7	The Imaginary Number i	288
6-8	The Complex Numbers	292

- *Explorations* Exploring Radicals 837

- *Technology* Calculator Key-In 263, 269

- *Special Topics* The Irrationality of $\sqrt{2}$ 273   Historical Note 297
  Conjugates and Absolute Value 298

- *Reviews and Tests* Mixed Review Exercises 263, 273, 282, 291
  Self-Tests 282, 297   Chapter Summary 301   Chapter Review 302
  Chapter Test 303   Mixed Problem Solving 304   Preparing for College
  Entrance Exams 305

7 Quadratic Equations and Functions

Solving Quadratic Equations

7-1	Completing the Square	307
7-2	The Quadratic Formula	311

Roots of Quadratic Equations

7-3	The Discriminant	317
7-4	Equations in Quadratic Form	322

Quadratic Functions and Their Graphs

7-5	Graphing $y - k = a(x - h)^2$	326
7-6	Quadratic Functions	333
7-7	Writing Quadratic Equations and Functions	338

- *Explorations* Exploring Quadratic Equations 838

- *Technology* Computer Graphing Ideas 327, 838   Computer Exercises 321

- *Special Topics* Biographical Note 316   Challenge 325
  Career Note / Statistician 337

- *Reviews and Tests* Mixed Review Exercises 310, 321, 332, 345
  Self-Tests 316, 325, 345   Chapter Summary 346   Chapter Review 346
  Chapter Test 348   Cumulative Review 348

8 Variation and Polynomial Equations

Variation and Proportion

8-1	Direct Variation and Proportion	351
8-2	Inverse and Joint Variation	358

Polynomial Equations

8-3	Dividing Polynomials	364
8-4	Synthetic Division	368
8-5	The Remainder and Factor Theorems	372
8-6	Some Useful Theorems	377

Solving Polynomial Equations

8-7	Finding Rational Roots	382
8-8	Approximating Irrational Roots	386
8-9	Linear Interpolation	391

- *Explorations*   Exploring Direct Variation 839
- *Technology*   Computer Graphing Ideas 386, 387   Computer Key-In 389   Computer Exercises 395
- *Special Topics*   Challenge 363   Reading Algebra 371   Historical Note 385
- *Reviews and Tests*   Mixed Review Exercises 357, 367, 376, 385   Self-Tests 363, 381, 396   Chapter Summary 396   Chapter Review 397   Chapter Test 398   Preparing for College Entrance Exams 399

9 Analytic Geometry

Conic Sections: Circles and Parabolas

9-1	Distance and Midpoint Formulas	401
9-2	Circles	407
9-3	Parabolas	412

Conic Sections: Ellipses and Hyperbolas

9-4	Ellipses	418
9-5	Hyperbolas	426
9-6	More on Central Conics	432

Systems of Equations

9-7	The Geometry of Quadratic Systems	436
9-8	Solving Quadratic Systems	439
9-9	Systems of Linear Equations in Three Variables	444

- *Explorations*   Exploring Circles and Ellipses 840
- *Technology*   Computer Graphing Ideas 436, 840   Computer Exercises 406
- *Special Topics*   Challenge 411   Career Note / Computer Graphics Artist 423   Application / Planetary Orbits 424   Historical Note 431

• *Reviews and Tests* Mixed Review Exercises 406, 417, 431, 438, 449
Self-Tests 417, 435, 450 Chapter Summary 450 Chapter Review 451
Chapter Test 453

10 Exponential and Logarithmic Functions

Exponential Functions

10-1	Rational Exponents	455
10-2	Real Number Exponents	459
10-3	Composition and Inverses of Functions	463

Logarithmic Functions

10-4	Definition of Logarithms	468
10-5	Laws of Logarithms	473

Applications

10-6	Applications of Logarithms	478
10-7	Problem Solving: Exponential Growth and Decay	483
10-8	The Natural Logarithm Function	489

• *Explorations* Exploring Powers and Roots 841

• *Technology* Computer Graphing 460, 464, 841 Calculator Key-In 492

• *Special Topics* Growth of Functions 462 Challenge 492
Application / Radiocarbon Dating 493 Historical Note 472

• *Reviews and Tests* Mixed Review Exercises 458, 467, 477, 488
Self-Tests 467, 477, 493 Chapter Summary 495 Chapter Review 496
Chapter Test 497 Mixed Problem Solving 498 Preparing for College
Entrance Exams 499

11 Sequences and Series

Sequences

11-1	Types of Sequences	501
11-2	Arithmetic Sequences	507
11-3	Geometric Sequences	510

Series

11-4	Series and Sigma Notation	518
11-5	Sums of Arithmetic and Geometric Series	525
11-6	Infinite Geometric Series	531

Binomial Expansions

11-7	Powers of Binomials	537
11-8	The General Binomial Expansion	540

• *Explorations* Exploring Pascal's Triangle 842

• *Technology* Computer Exercises 530 Calculator Key-In 543

- *Special Topics*   Challenge 516   Graphing Sequences 517
  Career Note / Marine Biologist 522   Induction 523   Biographical Note 544
- *Reviews and Tests*   Mixed Review Exercises 506, 515, 530, 539
  Self-Tests 516, 536, 543   Chapter Summary 544   Chapter Review 545
  Chapter Test 546   Cumulative Review 547

12 Triangle Trigonometry

Trigonometric Functions

12-1	Angles and Degree Measure	549
12-2	Trigonometric Functions of Acute Angles	555
12-3	Trigonometric Functions of General Angles	561
12-4	Values of Trigonometric Functions	568

Triangle Trigonometry

12-5	Solving Right Triangles	574
12-6	The Law of Cosines	580
12-7	The Law of Sines	586
12-8	Solving General Triangles	591
12-9	Areas of Triangles	597

- *Explorations*   Exploring Trigonometric Ratios 843
- *Technology*   Calculator Key-In 554   Computer Exercises 584
- *Special Topics*   Career Note / Flight Engineer 567   Historical Note 573
  Reading Algebra 585   Biographical Note 601   Challenge 601
- *Reviews and Tests*   Mixed Review Exercises 554, 567, 579, 590, 600
  Self-Tests 572, 600   Chapter Summary 602   Chapter Review 602
  Chapter Test 604   Preparing for College Entrance Exams 605

13 Trigonometric Graphs; Identities

Circular Functions and Their Graphs

13-1	Radian Measure	607
13-2	Circular Functions	613
13-3	Periodicity and Symmetry	619
13-4	Graphs of the Sine and Cosine	624
13-5	Graphs of the Other Functions	630

Trigonometric Identities

13-6	The Fundamental Identities	636
13-7	Trigonometric Addition Formulas	641
13-8	Double-Angle and Half-Angle Formulas	646
13-9	Formulas for the Tangent	650

- *Explorations*   Exploring Sine Curves 844
- *Technology*   Computer Graphing 624, 625, 631   Computer Exercises 645

- *Special Topics*   Biographical Note 618   Historical Note 618
  Challenge 623   Application / Frequencies in Music 634

- *Reviews and Tests*   Mixed Review Exercises 612, 623, 633, 645, 653
  Self-Tests 634, 654   Chapter Summary 654   Chapter Review 655
  Chapter Test 657

14 Trigonometric Applications

Vectors

14-1	Vector Operations	659
14-2	Vectors in the Plane	666

Polar Coordinates and Complex Numbers

14-3	Polar Coordinates	675
14-4	The Geometry of Complex Numbers	680
14-5	De Moivre's Theorem	685

Inverse Functions

14-6	The Inverse Cosine and Inverse Sine	689
14-7	Other Inverse Functions	693
14-8	Trigonometric Equations	697

- *Explorations*   Exploring Polar Coordinate Equations 845

- *Technology*   Computer Exercises 670   Computer Key-In 701
  Computer Graphing Ideas 845

- *Special Topics*   Application / Force, Work, and Energy 672
  Spirals 679   Biographical Note 684

- *Reviews and Tests*   Mixed Review Exercises 665, 679, 688, 696
  Self-Tests 671, 688, 700   Chapter Summary 702   Chapter Review 703
  Chapter Test 704   Mixed Problem Solving 705   Preparing for College
  Entrance Exams 706   Cumulative Review 707

15 Statistics and Probability

Statistics

15-1	Presenting Statistical Data	709
15-2	Analyzing Statistical Data	713
15-3	The Normal Distribution	719
15-4	Correlation	724

Counting

15-5	Fundamental Counting Principles	730
15-6	Permutations	734
15-7	Combinations	738

Probability

15-8	Sample Spaces and Events	743
15-9	Probability	745
15-10	Mutually Exclusive and Independent Events	754

- *Explorations*   Exploring Probability with Experiments 846

- *Technology*   Computer Exercises 742, 750   Computer Key-In 762

- *Special Topics*   Career Note / Electrician 718   Challenge 733   Reading Algebra 751   Application / Sampling 752   Random Numbers 753

- *Reviews and Tests*   Mixed Review Exercises 712, 723, 733, 741, 750 Self-Tests 729, 742, 761   Chapter Summary 763   Chapter Review 764 Chapter Test 765

16 Matrices and Determinants

Matrices

16-1	Definition of Terms	767
16-2	Addition and Scalar Multiplication	770
16-3	Matrix Multiplication	774
16-4	Applications of Matrices	779

Inverses of Matrices

16-5	Determinants	787
16-6	Inverses of Matrices	790

Working with Determinants

16-7	Expansion of Determinants by Minors	794
16-8	Properties of Determinants	798
16-9	Cramer's Rule	801

- *Explorations*   Exploring Matrices in Geometry 847

- *Special Topics*   Augmented Matrices 785   Historical Note 793

- *Reviews and Tests*   Mixed Review Exercises 769, 778, 789, 797, 805 Self-Tests 784, 793, 805   Chapter Summary 806   Chapter Review 806 Chapter Test 808   Preparing for College Entrance Exams 809

Tables 810   **Explorations** 832   **Portfolio Projects** 848
Appendix
    A-1 Common Logarithms: Notation and Interpolation 856
    A-2 Common Logarithms: Computation 858
    A-3 Discrete Mathematics 862
    A-4 Preparing for College Entrance Exams 864
Glossary 869   **Index** 879   **Answers to Selected Exercises**

Reading Your Algebra Book

An algebra book requires a different type of reading than a novel or a short story. Every sentence in a math book is full of information and logically linked to the surrounding sentences. You should read the sentences carefully and think about their meaning. As you read, remember that algebra builds upon itself; for example, the method of factoring that you'll study on page 188 will be useful to you on page 697. Be sure to read with a pencil in your hand: Do calculations, draw sketches, and take notes.

Vocabulary

You'll learn many new words in algebra. Some, such as *polynomial* and *parabola,* are mathematical in nature. Others, such as *power* and *proof,* are used in everyday speech but have different meanings when used in algebra. Important words whose meanings you'll learn are printed in **heavy type.** Also, they are listed at the beginning of each Self-Test. If you don't recall the meaning of a word, you can look it up in the Glossary or the Index at the back of the book. The Glossary will give you a definition, and the Index will give you page references for more information.

Symbols

Algebra, and mathematics in general, has its own symbolic language. You must be able to read these symbols in order to understand algebra. For example, $|x| > 2$ means "the absolute value of x is greater than 2." If you aren't sure what a symbol means, check the list of symbols on page xvi.

Diagrams

Throughout this book you'll find many diagrams. They contain information that will help you understand the concepts under discussion. Study the diagrams carefully as you read the text that accompanies them.

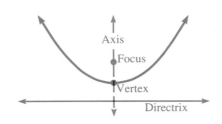

Displayed Material

Throughout this book important information is displayed in gray boxes.
This information includes properties, definitions, methods, and summaries.
Be sure to read and understand the material in these boxes. You should find
these boxes useful when reviewing for tests and exams.

> If a is a real number and m and n are positive integers, then $a^m \cdot a^n = a^{m+n}$.

This book also contains worked-out examples. They will help you in doing
many of the exercises and problems.

Example   Simplify $x^3 \cdot x^5$.

Solution $x^3 \cdot x^5 = x^{3+5} = x^8$ **Answer**

Reading Aids

Throughout this book you will find
sections called Reading Algebra.
These sections deal with such topics
as independent study and problem
solving strategies. They will help
you become a more effective reader
and problem solver.

Exercises, Tests, and Reviews

Each lesson in this book is followed by Oral and Written Exercises. Lessons
may also include Problems, Mixed Review Exercises, and optional Computer
Exercises. Answers for all Mixed Review Exercises and for selected Written
Exercises, Problems, and Computer Exercises are given at the back of this book.

Within each chapter you will find Self-Tests that you can use to check your
progress. Answers for all Self-Tests are also given at the back of this book.

Each chapter concludes with a Chapter Summary that lists important ideas
from the chapter, a Chapter Review in
multiple-choice format, and a Chapter
Test. Lesson numbers in the margins
of the Review and Test indicate which
lesson a group of questions covers.

Reading Algebra/Symbols

		Page			Page
{ }	set	1	$\log_b N$	logarithm of N to the base b	469
$\|a\|$	absolute value of a	3	Σ	summation sign	518
$=$	equals *or* is equal to	6	!	factorial	540
$\neq$	does not equal	6	$^\circ$	degree	549
$>$	is greater than	6	$'$	minute	550
$<$	is less than	6	$''$	second	550
$\leq$	is less than or equal to	6	$\overrightarrow{AB}$	vector AB	659
$\geq$	is greater than or equal to	6	$\|\mathbf{u}\|$	norm of the vector $\mathbf{u}$	661
$\therefore$	therefore	6	Cos^{-1}	inverse cosine *or* arc cosine	689
a^n	the nth power of a	7	σ	standard deviation	715
$\in$	is an element of	9	r	correlation coefficient	725
$-a$	additive inverse of a *or* the opposite of a	15	$_nP_r$	number of permutations of n elements taken r at a time	735
$\emptyset$	empty set or null set	38	$_nC_r$	number of combinations of n elements taken r at a time	738
$f(x)$	f of x *or* the value of f at x	142	$P(E)$	probability of event E	745
$\approx$	is approximately equal to	222	$\cap$	intersection	754
$\pm$	plus-or-minus sign	259	$\cup$	union	754
$\sqrt[n]{b}$	nth root of b	260	$\overline{E}$	complement of event E	757
i	imaginary unit ($i^2 = -1$)	288	det	determinant	787
$\overline{z}$	conjugate of the complex number z	298	A^{-1}	inverse of matrix A	790
$a^{p/q}$	qth root of pth power of a	455			
f^{-1}	inverse function of f	464			

Greek letters: $\alpha, \beta, \gamma, \theta, \pi, \sigma, \phi, \omega$ *alpha, beta, gamma, theta, pi, sigma, phi, omega*

Reading Algebra/*Table of Measures*

Metric Units

Length	10 millimeters (mm)	= 1 centimeter (cm)
	100 centimeters $\rbrace$ 1000 millimeters	= 1 meter (m)
	1000 meters	= 1 kilometer (km)
Area	100 square millimeters (mm^2)	= 1 square centimeter (cm^2)
	10,000 square centimeters	= 1 square meter (m^2)
Volume	1000 cubic millimeters (mm^3)	= 1 cubic centimeter (cm^3)
	1,000,000 cubic centimeters	= 1 cubic meter (m^3)
Liquid Capacity	1000 milliliters (mL)	= 1 liter (L)
	1000 cubic centimeters	= 1 liter
Mass	1000 milligrams (mg)	= 1 gram (g)
	1000 grams	= 1 kilogram (kg)
Temperature in degrees Celsius (°C)	0°C	= freezing point of water
	100°C	= boiling point of water

United States Customary Units

Length	12 inches (in.)	= 1 foot (ft)
	36 inches $\rbrace$ 3 feet	= 1 yard (yd)
	5280 feet $\rbrace$ 1760 yards	= 1 mile (mi)
Area	144 square inches (in.^{2})	= 1 square foot (ft^2)
	9 square feet	= 1 square yard (yd^2)
Volume	1728 cubic inches (in.^{3})	= 1 cubic foot (ft^3)
	27 cubic feet	= 1 cubic yard (yd^3)
Liquid Capacity	16 fluid ounces (fl oz)	= 1 pint (pt)
	2 pints	= 1 quart (qt)
	4 quarts	= 1 gallon (gal)
Weight	16 ounces (oz)	= 1 pound (lb)
Temperature in degrees Fahrenheit (°F)	32°F	= freezing point of water
	212°F	= boiling point of water

Time

60 seconds (s)	=	1 minute (min)
60 minutes	=	1 hour (h)

Facts and Formulas from Geometry

Angle: A figure formed by two rays that have the same end-point. To indicate that the measure of $\angle A$ is 35°, we write $\angle A = 35°$.

Complementary angles: two angles whose measures have the sum 90°.

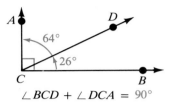

$\angle BCD + \angle DCA = 90°$

Supplementary angles: two angles whose measures have the sum 180°.

$\angle BCD + \angle DCE = 180°$

Isosceles triangle: a triangle having at least two equal sides and two equal angles. The two sides are called *legs* and the included angle is called the *vertex angle.*

Equilateral triangle: a triangle having three equal sides. Each angle measures 60°, so the triangle is also called *equiangular.*

Parallelogram: a quadrilateral (a four-sided polygon) whose opposite sides are equal and parallel. The opposite angles are also equal.

The sum of the measures of the angles of a polygon with n sides is $180(n - 2)°$. For example, the sum of the measures of the angles of a triangle is 180° and the sum of the measures of the angles of a quadrilateral is 360°.

Let: A = area, P = perimeter, C = circumference, S = lateral surface area, T = total surface area, V = volume, $\pi \approx 3.1416$

Triangle	Rectangle	Parallelogram
$P = a + b + c$; $A = \frac{1}{2}bh$	$P = 2l + 2w$; $A = lw$	$P = 2b + 2s$; $A = bh$

Facts and Formulas from Geometry

Trapezoid

$$A = \frac{1}{2}h(b_1 + b_2)$$

Circle

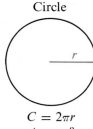

$$C = 2\pi r$$
$$A = \pi r^2$$

Rectangular Box

$$V = lwh$$
$$T = 2lw + 2lh + 2wh$$

Sphere

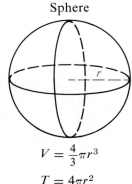

$$V = \frac{4}{3}\pi r^3$$
$$T = 4\pi r^2$$

Right Circular Cylinder

$$V = \pi r^2 h$$
$$S = 2\pi rh$$
$$T = 2\pi rh + 2\pi r^2$$

Right Circular Cone

$$V = \frac{1}{3}\pi r^2 h$$
$$S = \pi rs$$
$$T = \pi rs + \pi r^2$$

Using Technology with This Course

There are three types of optional computer material in this text: Computer Key-In features, Computer Exercises, and suggestions for using computer graphing techniques to explore concepts and confirm results.

The Computer Key-In features can be used by students without previous programming experience. These features teach some programming in BASIC and usually include a program that students can run to explore an algebra topic covered in the chapter.

The optional Computer Exercises are designed for students who have some familiarity with programming in BASIC. Students are usually asked to write one or more programs related to the lesson just presented.

The suggestions for applying computer graphing techniques are appropriate for use with a graphing calculator or graphing software such as *Algebra Plotter Plus*.

Calculator Key-In features and certain exercise sets also suggest appropriate use of scientific calculators with this course.

1 Basic Concepts of Algebra

The ability to understand numerical relationships and to use them in making decisions is important in many careers, as well as in daily life. Studying algebra can help develop this ability.

The Language of Algebra

1-1 Real Numbers and Their Graphs

Objective To graph real numbers on a number line, to compare numbers, and to find their absolute values.

In mathematics you have met such numbers as

$$4, -3, -\frac{5}{3}, \sqrt{3}, -4.16, \pi, \text{ and } 0.$$

These are members of the set of **real numbers.**

Some important subsets of the real numbers are shown in the chart below. Remember that when you list members of a set within braces, { }, you use three dots to show that a list continues without end.

Real Numbers

Natural numbers	{1, 2, 3, . . .}
Whole numbers	{0, 1, 2, 3, . . .}
Integers	{. . . , −3, −2, −1, 0, 1, 2, 3, . . .}
Rational numbers	numbers, such as $-\frac{5}{3}, \frac{7}{9}, 0.5$, and $\frac{4}{1}$, that result when an integer is divided by a nonzero integer
Irrational numbers	numbers, such as $\sqrt{3}, -\sqrt{7}$, and π, that are not rational

You have probably worked with number lines like the one below.

When using a number line, we take the following facts for granted.

1. Each point on a number line is paired with exactly one real number, called the **coordinate** of the point.
2. Each real number is paired with exactly one point on the line, called the **graph** of the number.

For example, on the number line above, the coordinate of point Q is 4, and the graph of −3 is point P. The graph of 0 (zero) is the **origin.**

Example 1 Find the coordinate of the point one third of the way from *C* to *D* for the number line shown.

Solution The distance from *C* to *D* is 6 units. Since $\frac{1}{3} \cdot 6 = 2$, move two units to the right from *C* to find the required point. The coordinate of this point is -2. **Answer**

On a horizontal number line, you can see that the larger a number is, the farther to the right its graph is. In particular, positive numbers (numbers greater than 0) are graphed to the right of the origin and negative numbers (numbers less than 0) to the left.

Notice that a number line shows the *order* of the numbers. The fact that the graph of -3 is to the left of the graph of 1 corresponds to these two equivalent statements:

In Words	In Symbols
Negative three is less than one.	$-3 < 1$
One is greater than negative three.	$1 > -3$

Look at the number line to see that the following statements are true.

$$-4 < -1 \qquad -4 < 0 \qquad -1 < 3$$
$$-1 > -4 \qquad 0 > -4 \qquad 3 > -1$$

Example 2 Graph the numbers -2 and -6 on a number line. Then write an inequality statement comparing them.

Solution

$-6 < -2$, or $-2 > -6$ **Answer**

The diagram below illustrates that every real number has an **opposite**.

In Words	In Symbols
The opposite of three is negative three.	$-(3) = -3$
The opposite of negative five is five.	$-(-5) = 5$
The opposite of zero is zero.	$-(0) = 0$

On a number line, the graphs of opposites are on opposite sides of the origin but are the same distance from the origin. For example, the graphs of 5 and −5 are both 5 units from 0. The **absolute value** of a number is the distance between the graph of the number and the origin on a number line. Vertical bars | | are used to denote absolute value. Thus:

$$|5| = 5 \qquad\qquad |-5| = 5$$

Read: The absolute value of   Read: The absolute value of
five is five.   negative five is five.

In general:

1. The absolute value of a positive number is the number itself.
2. The absolute value of a negative number is the opposite of the number.
3. The absolute value of zero is zero.

Examples

$|7| = 7$

$|-1| = -(-1) = 1$

$|0| = 0$

Example 3   Find the value of each expression.
          a. $-|-4|$ **b.** $|-6| - |6|$

Solution   **a.** $-|-4| = -4$ **b.** $|-6| - |6| = 6 - 6 = 0$

Oral Exercises

Estimate the coordinate of each point shown in red to the nearest half unit.

1.

2.

Give the opposite of each number.

3. 2   **4.** 0   **5.** $-\sqrt{2}$ **6.** $-\dfrac{3}{2}$

Find the value of each expression.

7. $|-7|$ **8.** $|4 - 4|$ **9.** $|3 - 2|$ **10.** $|3| - |2|$

Tell whether each statement is true or false.

11. $-4 < 1$ **12.** $-6 > 0$ **13.** $-2 > -7$ **14.** $-3 < 3$

15. $|-4| < 1$ **16.** $|-6| > 0$ **17.** $|-2| > |-7|$ **18.** $|-3| < |3|$

19. Is the absolute value of the sum of two numbers always equal to the sum of their absolute values? Explain.

Written Exercises

Find the coordinate of each point described. Use the number line below.

A 1. *B*   2. *M*   3. *H*   4. *G*

5. The point halfway between *D* and *J*
6. The point halfway between *C* and *F*
7. The point one third of the way from *B* to *H*
8. The point two thirds of the way from *C* to *I*
9. The point one fourth of the way from *F* to *L*
10. The point three fourths of the way from *B* to *J*

Write each statement using symbols.

| **Sample** | Negative four is less than one. | **Solution** | $-4 < 1$ |

11. Zero is greater than negative six.
12. Negative two is less than two.
13. Negative three is less than negative one.
14. One is greater than negative ten.
15. Six is greater than negative five.
16. Negative six is less than negative five eighths.

Graph each pair of numbers on a separate number line. Then write an inequality statement comparing the numbers.

17. 0 and -4 18. 3 and -3 19. $\frac{1}{2}$ and $-\frac{3}{2}$

20. $-\frac{5}{4}$ and $-\frac{3}{4}$ 21. 0.5 and -1.5 22. -2.5 and -0.75

Find the value of each expression.

23. $-|-5|$ 24. $2 \cdot |-5|$ 25. $|5| - |-2|$ 26. $|-5| - |2|$

Arrange each list of numbers in order from least to greatest.

B 27. $4, -5, 0, -1, 2$ 28. $1, -1, -5, 3, -3$

29. $-\frac{1}{2}, 2, -\frac{3}{2}, \frac{5}{2}, -2$ 30. $-\frac{3}{4}, -\frac{1}{4}, \frac{1}{2}, -\frac{1}{2}, \frac{3}{4}$

31. $-1.6, -2.6, -0.6, -1.8$ 32. $-2.4, -1.7, -1.9, -2.0$

On a number line, point A has coordinate -5 and point B has coordinate 1. Find the coordinate of each point described.

33. The point 2 units to the left of B

34. The point 3 units to the right of A

35. The point 1.4 units to the right of A

36. The point 2.6 units to the left of B

37. The point $5\frac{1}{3}$ units from A and $\frac{2}{3}$ units from B

38. The point $4\frac{1}{2}$ units from A and $1\frac{1}{2}$ units from B

C **39.** Each point that is twice as far from A as from B

40. Each point that is half as far from A as from B

Mixed Review Exercises

Tell whether each statement is true or false.

1. $\frac{1}{2} + \frac{2}{3} = \frac{3}{5}$

2. $3\frac{2}{5} - 2\frac{3}{10} < 1$

3. $\frac{4}{7} \times \frac{4}{7} > \frac{4}{7}$

4. $6\frac{3}{8} \div 2\frac{1}{4} = 2\frac{5}{6}$

5. $5.31 + 24.7 = 29.01$

6. $0.1 - 0.09 < 0.001$

7. $1.3 \times 0.1 = 0.13$

8. $1 \div 0.99 > 1$

9. $|-6| > -6$

10. $|-1| = -(-1)$

11. $|0| = 0$

12. $|-3| < 0$

Computer Exercises

The optional computer exercises in this book require familiarity with the fundamental concepts and terminology of programming in BASIC.

1. Write a program in which the user enters two numbers, and the program then prints a statement of comparison using $>$, $=$, or $<$. For example, if the user enters -3 and -5, the program should print $-3 > -5$.

2. Run the program in Exercise 1 for the following pairs of numbers.

    a. -8, -7 **b.** 0.1, 0.01   **c.** 0, 0

    d. 3.65, 3.56   **e.** 4, -4 **f.** 800, 8000

3. Write a program in which the user enters the coordinates of two points on a number line, and the program then prints the coordinates of the point halfway between the entered points.

4. Run the program in Exercise 3 to find the coordinate of the point halfway between the two points with the given coordinates.

    a. 3, 8   **b.** -1.7, 1.7   **c.** -2.73, 5.86

    d. -7.92, -2.40 **e.** 9.1, 9.01   **f.** 102, -100

1-2 Simplifying Expressions

Objective   To review the methods used to simplify numerical expressions and to evaluate algebraic expressions.

The words and symbols reviewed in the following charts are used throughout this book. Study the examples to be sure you understand the meaning of the words shown in **heavy type**.

Definition	Examples
Numerical expression, or **numeral:** a symbol or group of symbols used to represent a number	3×4 $5 + 5 + 2$ $24 \div 2$ 12
Value of a numerical expression: the number represented by the expression	Twelve is the value of 3×4.
Equation: a sentence formed by placing an *equals sign* = (read "equals" or "is equal to") between two expressions, called the *sides* of the equation. The equation is a true statement if both sides have the same value.	$3 \times 4 = 12$ the sides $36 \div 4 = 1 \times 9$
Inequality symbol: one of the symbols $<$, $>$, $\neq$ (read "does not equal" or "is not equal to"), $\leq$ (read "is less than or equal to"), $\geq$ (read "is greater than or equal to")	$3 + 9 < 20$ $3 \times 9 > 20$ $3 + 9 \neq 3 \times 9$ $7 \leq 1 + 6$ $8 \geq 0$
Inequality: a sentence formed by placing an inequality symbol between two expressions, called the *sides* of the inequality	$-3 > -5$ the sides $-3 \leq -0.3$

Example 1   Use one of the symbols $<$, $=$, or $>$ to make a true statement.
   a. $10 - 2 \underline{\ \ ?\ \ } 2 \times 4$ **b.** $27 \div 3 \underline{\ \ ?\ \ } 3^2 + 2$

Solution   Find the value of each side.

   a. $10 - 2 = 8$ and   **b.** $27 \div 3 = 9$ and
      $2 \times 4 = 8$ $3^2 + 2 = 9 + 2 = 11$
      $\therefore 10 - 2 = 2 \cdot 4$ $\therefore 27 \div 3 < 3^2 + 2$

   (read "therefore")

As you can see in Example 1, expressions may involve sums, differences, products, quotients, and powers. The following chart reviews the meaning of these words as well as the various grouping symbols that are used to enclose parts of expressions.

Definition	Examples
Sum: the result of adding numbers, called the **terms** of the sum	$6 + 15 = 21$ terms sum
Difference: the result of subtracting one number from another	$8 - 6 = 2$ difference
Product: the result of multiplying numbers, called the **factors** of the product. Multiplication is sometimes shown by a raised dot · instead of a times sign ×.	$3 \times 15 = 45$ factors product $6 \cdot 7 = 42$
Quotient: the result of dividing one number by another. Division is indicated by the symbol ÷ or by a fraction bar.	$35 \div 5 = 7 = \dfrac{35}{5}$ quotient
Power, base, and **exponent:** A *power* is a product of equal factors. The repeated factor is the *base*. A positive *exponent* tells the number of times the base occurs as a factor.	Let the base be 3. exponent ⌐ base ⌐ First power: $3 = 3^1$ Second power: $3 \times 3 = 3^2$ 3^2 is also called "three squared." Third power: $3 \times 3 \times 3 = 3^3$ 3^3 is also called "three cubed." Fourth power: $3 \times 3 \times 3 \times 3 = 3^4$
Grouping symbols: pairs of parentheses ( ), brackets [ ], or a bar — used to enclose part of an expression that represents a single number	$3 + (2 \times 6) = 3 + 12$ $8 \times [5 + 1] = 8 \times 6$ $\dfrac{4 + 18}{11} = \dfrac{22}{11}$

To **simplify** an expression you replace it by the simplest or most common symbol having the same value. For example,

$$4 \cdot (7 - 5) = 4 \cdot 2 = 8$$

and $\dfrac{5 + 3}{5 - 3} = \dfrac{8}{2} = 4.$

The following principle is used in simplifying expressions.

Substitution Principle

An expression may be replaced by another expression that has the same value.

When simplifying expressions, we agree to perform operations in the following order.

Order of Operations

1. Simplify the expression within each grouping symbol, working outward from the innermost grouping.
2. Simplify powers.
3. Perform multiplications and divisions in order from left to right.
4. Perform additions and subtractions in order from left to right.

Recall that the symbols $\times$ and $\cdot$ often are omitted in expressions with grouping symbols. For example, $3(5 + 2)$ means $3 \times (5 + 2)$, and $(2^3 - 1)5$ means $(2^3 - 1) \cdot 5$.

Example 2 Simplify: **a.** $2 \cdot 5 - 6 \div 3$

   b. $2[4 + (2^3 - 1)]$

   c. $5^2 - \dfrac{1 + 5}{2}$

Solution **a.** $\quad 2 \cdot 5 - 6 \div 3 = 10 - \underbrace{6 \div 3}$

$= 10 - \quad 2$

$= 8 \quad \textit{Answer}$

b. $2[4 + (2^3 - 1)] = 2[4 + \underbrace{(8 - 1)}]$

$= 2[\underbrace{4 + \quad 7}]$

$= \quad 2[11]$

$= 22 \quad \textit{Answer}$

c. $\quad 5^2 - \dfrac{1 + 5}{2} = 5^2 - \underbrace{\dfrac{6}{2}}$

$= 25 - 3$

$= 22 \quad \textit{Answer}$

The chart at the top of the next page reviews the meanings of *variable* and some related terms.

Definition	Examples
Variable: a symbol, usually a letter, used to represent any member of a given set, called the **domain** or **replacement set,** of the variable	a, x, or y If the domain of x is {0, 1, 2, 3}, we write $x \in$ {0, 1, 2, 3} and read "x is a member of (or belongs to) the set whose members are 0, 1, 2, and 3."
Values of a variable: the members of the domain of the variable	If the domain of a is the set of positive integers, then a can have these values: 1, 2, 3, 4, . . .
Algebraic expression: a numerical expression; a variable; or a sum, difference, product, or quotient that contains one or more variables	$2^4 + 3 \qquad x \qquad y^2 - 2y + 6$ $\dfrac{a+b}{c} \qquad\qquad 2c^2d - \dfrac{4}{d}$ (Note that $2c^2d$ stands for $2 \cdot c^2 \cdot d$.)

The process of replacing each variable in an expression by a given value and simplifying the result is known as **evaluating** the expression, or **finding the value of** the expression.

Example 3 Evaluate each expression if $x = 4$ and $y = 3$.

a. $\dfrac{3x}{x^2 - 8}$ 　　　　　　　　　**b.** $x^2y - xy^2$

Solution 　**a.** $\dfrac{3x}{x^2 - 8} = \dfrac{3 \cdot 4}{4^2 - 8}$ 　　　　**b.** $x^2y - xy^2 = 4^2 \cdot 3 - 4 \cdot 3^2$

$\qquad\qquad = \dfrac{12}{16 - 8}$ 　　　　　　　　$= 16 \cdot 3 - 4 \cdot 9$

$\qquad\qquad = \dfrac{12}{8}$ 　　　　　　　　　　$= 48 - 36$

$\qquad\qquad = \dfrac{3}{2}$, or $1\dfrac{1}{2}$ 　　　　　　　$= 12$

To evaluate some expressions, you may need to find the absolute value of one or more of the terms. You learned earlier that the absolute value of a number is the distance between the graph of the number and the origin on a number line. Here is a formal definition using a variable.

Definition of Absolute Value

For each real number a,

$$|a| = \begin{cases} a & \text{if } a > 0 \\ 0 & \text{if } a = 0 \\ -a & \text{if } a < 0 \end{cases}$$

Example 4 Evaluate $10|x| + |y|$ if $x = 3$ and $y = -2$.

Solution $\quad 10|x| + |y| = 10|3| + |-2|$
$$= 30 + 2$$
$$= 32 \quad \textit{Answer}$$

You may wish to use a calculator to evaluate some expressions. If you do, be sure to read the Calculator Key-In on page 12 first.

Oral Exercises

Use one of the symbols $=$ or $\neq$ to make a true statement.

1. 3^2 _?_ $3 \cdot 3$

2. 2^2 _?_ $2 + 2$

3. $2^2 + 1$ _?_ $3 \cdot 2$

4. $\dfrac{2+3}{5}$ _?_ $\dfrac{2}{5} + \dfrac{3}{5}$

5. $\dfrac{1}{3+4}$ _?_ $\dfrac{1}{3} + \dfrac{1}{4}$

6. $\dfrac{1}{3 \cdot 2}$ _?_ $\dfrac{1}{3} \cdot \dfrac{1}{2}$

Tell which operation to perform first. Then simplify the expression.

7. $8 - 2 \cdot 3$

8. $(8 - 2) \cdot 3$

9. $2(3 + 2)^2$

10. $2(3 + 2^2)$

11. $2(3 + 2 \cdot 4)$

12. $14 \div [(2 + 3) - 4]$

Evaluate each expression (a) if $x = 2$ and (b) if $x = 3$.

13. $3x + 2$

14. $3x - 1$

15. $\dfrac{x+1}{6}$

16. $\dfrac{12}{x}$

17. $3(x - 1)$

18. $2(2x - 1)$

19. $\dfrac{1}{2}x^3 + 1$

20. $2x^2 - 3$

Evaluate each expression if $x = 3$ and $y = -2$.

21. $|x| + |y|$

22. $|x| - |y|$

23. $|x|^2 + |y|^2$

24. $|x|^2 - |y|^2$

Written Exercises

Use one of the symbols $<$, $=$, or $>$ to make a true statement.

A **1.** $5 \cdot 1$ _?_ $5 \div 1$

2. $1 \cdot 5$ _?_ $1 \div 5$

3. $\dfrac{3+2}{3-2}$ _?_ $\dfrac{4+2}{4-2}$

4. $\dfrac{3+2}{3-2}$ _?_ $\dfrac{6+4}{6-4}$

5. $3^2 \cdot 4^2$ _?_ $(3 \cdot 4)^2$

6. $3^2 + 4^2$ _?_ $(3 + 4)^2$

7. $(9 - 3) - 2$ _?_ $9 - (3 - 2)$

8. $(9 + 3) + 2$ _?_ $9 + (3 + 2)$

9. $(9 \cdot 3) \cdot 2$ _?_ $9 \cdot (3 \cdot 2)$

10. $(12 \div 6) \div 2$ _?_ $12 \div (6 \div 2)$

Simplify.

11. **a.** $11 - 3 + 5 - 2$ **b.** $11 - (3 + 5) - 2$ **c.** $11 - (3 + 5 - 2)$

12. **a.** $12 - 5 - 2 + 3$ **b.** $12 - (5 - 2) + 3$ **c.** $12 - (5 - 2 + 3)$

13. **a.** $3 \cdot 8 + 4 \cdot 5$ **b.** $3 \cdot (8 + 4) \cdot 5$ **c.** $3 \cdot (8 + 4 \cdot 5)$

14. **a.** $4^2 - 6 \div 2 + 3$ **b.** $(4^2 - 6) \div 2 + 3$ **c.** $(4^2 - 6) \div (2 + 3)$

15. $6 - [7 - (5 - 2)]$

16. $14 - 2[9 - 2(5 - 3)]$

17. $\dfrac{2^3 + 1}{2^2 - 1}$

18. $\dfrac{3^2}{5 - (3 - 1)}$

19. $\dfrac{1}{3}\left|\dfrac{1 + 7^2}{5^2}\right|$

20. $\dfrac{2^2(3^2 + 4^2)}{10^2}$

21. $64 \div 4^2 + 3(3^2 - 1)$

22. $2^2 \cdot 3^2 - (5^2 - 4^2)$

23. $[3^3 - (2^3 + 2^2)] \div 5$

24. $\dfrac{1}{10}[2(3 + 4) - 3^2]$

Evaluate each expression if $x = 3$, $y = 2$, and $z = 5$.

25. $2x^2 + x - 2$

26. $3y^2 - y - 5$

27. $(yz - x)^3$

28. $(xz - zy)^3$

29. $\dfrac{4z^3}{x^2 - y^2}$

30. $\dfrac{4xyz}{z^2 - x^2}$

31. $\dfrac{x + z}{y} - \dfrac{x + y}{2z}$

32. $\dfrac{z^2}{x + y} - \dfrac{y^2}{z - x}$

33. $\left(\dfrac{xyz}{x - y + z}\right)^4$

34. $\left(\dfrac{z^2 - y^2 - x^2}{xy}\right)^5$

35. $\dfrac{z^2 - (x^2 - y^2)}{3y^2z}$

36. $\dfrac{z^2 - y^2}{xz - 2y(z - x)}$

Evaluate each expression if $a = 6$ and $b = -2$.

37. $2|a| + |b|$

38. $|a| - 3|b|$

39. $|a|^2 - |b|^2$

40. $|a^2 - b^2|$
6   (2)(2)
36 - 4

Evaluate each expression for the given values of the variables.

B

41. $\dfrac{3u^2 - 2(v - 3)^2}{2(u^2 - 1) - v^2}$; $u = 4$, $v = 5$

42. $\dfrac{3(r^2 - s^2) + 5(r - s)^2}{(r + s)^2}$; $r = 7$, $s = 3$

43. $\dfrac{(2r^2 + s^2)(13r - 3s)^2}{5r^2 + 4rs + s^2}$; $r = 3$, $s = 8$

44. $\dfrac{(g^2 + h^2)(g^2 - h)^2}{(g + h)^2(3h^2 - 2gh - g^2)}$; $g = 6$, $h = 8$

Insert grouping symbols in each expression to make a true equation.

Sample $\qquad 3 + 2^2 - 10 \div 5 = 3 \qquad$ **Solution** $\qquad [(3 + 2)^2 - 10] \div 5 = 3$

45. $18 \div 2 - 3 \cdot 2 + 1 = 0$ $\qquad\qquad$ **46.** $7 - 2 \cdot 5 - 3 + 2 = 12$

47. $6 - 5 - 3 \cdot 2 = 8$ $\qquad\qquad\qquad$ **48.** $24 \div 3 + 1 - 2 \cdot 2 + 3 = 60$

49. $3^2 - 2^2 - 4 \cdot 3 + 3 = 25$ $\qquad\quad$ **50.** $6 + 5 \cdot 2 - 7 \div 2 \cdot 2 + 1 = 3$

51. a. Find the value of $\frac{1}{4}n^2(n + 1)^2$ when $n = 2$, $n = 3$, $n = 4$, and $n = 5$.

$\qquad$ **b.** Evaluate $1^3 + 2^3$, $1^3 + 2^3 + 3^3$, $1^3 + 2^3 + 3^3 + 4^3$, and $1^3 + 2^3 + 3^3 + 4^3 + 5^3$.

In Exercises 52–54, you may use any operation signs and any grouping symbols.

C **52.** Use five 2's to write an expression whose value is 7.

$\quad$ **53.** Use five 3's to write an expression whose value is 11.

$\quad$ **54.** Use four 2's to write expressions for 0, 1, 2, 3, 4, 5, and 6.

 Calculator Key-In

Different calculators perform a sequence of operations in different ways. You can test your calculator by entering the expression $4 + 3 \times 2$ and pressing the equals ($=$) key. If you get a result of 10, your calculator performs multiplication before addition. Your calculator is therefore following the order of operations as stated in Lesson 1-2. If you get 14, however, your calculator performs operations in the order in which they are entered. To get the correct result, you must enter the expression as $3 \times 2 + 4$ or, if your calculator has parentheses, as $4 + (3 \times 2)$.

$\qquad$ Be especially careful when simplifying fractional expressions. You must remember that the fraction bar acts as a grouping symbol. For example, $\frac{4 + 6}{2}$ should be entered as $(4 + 6) \div 2$ to produce 5, the correct result. If you simply enter $\frac{4 + 6}{2}$ as $4 + 6 \div 2$, you may get 7 instead. Likewise, the expression $\frac{10}{2 + 3}$ should be entered as $10 \div (2 + 3)$ and not as $10 \div 2 + 3$.

Simplify each expression on your calculator.

1. $3 + 6 \times 7$ $\qquad\qquad$ **2.** $12 + 84 \div 7$ $\qquad\qquad$ **3.** $12 \div 3 \times 4 - 1$

4. $12(10 - 5.5) - 4$ $\qquad$ **5.** $20 - 3.5 \cdot 4.5$ $\qquad\quad$ **6.** $3 \cdot 12 - 5 \cdot 2.5$

7. $\dfrac{3.8 + 7.7}{4.6 - 1.7}$ $\qquad\qquad$ **8.** $(3 + 2.5)(4.5 + 5.5)$ $\qquad$ **9.** $\dfrac{10.1 + 13.7}{48.5 - 17.9}$

10. $\dfrac{2.8 \times 7.3}{6.9 \times 2.5}$ $\qquad\qquad$ **11.** $26.54 - (3.9)^2 \div 4$ $\qquad$ **12.** $\dfrac{6.1(15.8 - 9.9)}{4.3 \times 15}$

Self-Test 1

Vocabulary   natural numbers (p. 1)
whole numbers (p. 1)
integers (p. 1)
rational numbers (p. 1)
irrational numbers (p. 1)
real numbers (p. 1)
coordinate (p. 1)
graph (p. 1)
origin (p. 1)
is less than (p. 2)
is greater than (p. 2)
absolute value (p. 3 and p. 9)
numerical expression (p. 6)
numeral (p. 6)
value of an expression (p. 6)
equation (p. 6)

inequality (p. 6)
sum (p. 7)
term (p. 7)
difference (p. 7)
product (p. 7)
factor (p. 7)
quotient (p. 7)
power (p. 7)
base (p. 7)
exponent (p. 7)
simplifying an expression (p. 7)
evaluating an expression (p. 9)
variable (p. 9)
domain (p. 9)
values of a variable (p. 9)
algebraic expression (p. 9)

1. Graph the following numbers on a number line:

$$4, \ -2\tfrac{2}{3}, \ 0, \ -1, \ 2.5$$

Obj. 1-1, p. 1

2. Use symbols to write the following statement:

"Negative two is less than negative one half."

Tell whether the statement is true or false.

3. $-3 < -5$

4. $1 > -2$

5. $|-4| = 4$

6. $-|7| = -7$

Simplify.

Obj. 1-2, p. 6

7. $3[5(6 - 2) - 4^2]$

8. $\dfrac{9 + 3 \cdot 5}{18 \div 6 + 3}$

Evaluate each expression if $r = 5$, $s = 3$, and $t = 1$.

9. $r^2 + 2s - rst$

10. $\dfrac{2s^2 + 3(r - t)}{r + s + 2t}$

Check your answers with those at the back of the book.

Operating with Real Numbers

1-3 Basic Properties of Real Numbers

Objective   To review properties of equality of real numbers and properties for adding and multiplying real numbers.

When you play softball, there are definite rules that you must follow. In fact, every game has its own set of rules, and this is certainly true of the "game" of algebra. In algebra, the properties of real numbers form an important part of the rules.

    The five properties below concern equality of real numbers.

Properties of Equality

For all real numbers a, b, and c:

Reflexive Property	$a = a$
Symmetric Property	If $a = b$, then $b = a$.
Transitive Property	If $a = b$ and $b = c$, then $a = c$.
Addition Property	If $a = b$, then $a + c = b + c$ and $c + a = c + b$.
Multiplication Property	If $a = b$, then $ac = bc$ and $ca = cb$.

Example 1   Name the property of equality illustrated in each statement.

    a. If $x + 4 = 3$, then $2(x + 4) = 6$.

    b. If $z = a + 2$ and $a + 2 = 3$, then $z = 3$.

    c. If $x = y + z$, then $y + z = x$.

Solution   **a.** Multiplication property

    b. Transitive property

    c. Symmetric property

    A set of numbers that has all the properties listed in the chart on the next page is called a **field.** The set of real numbers is a field. Note that when we say a **unique** real number has a certain field property we mean that *one and only one* real number has this property.

Field Properties of Real Numbers For all real numbers a, b, and c:	Examples Let $a = 12$, $b = 6$, and $c = 5$.
Closure Properties $a + b$ and ab are unique real numbers.	The sum of 12 and 6 is the real number 18; the product of 12 and 6 is the real number 72.

Commutative Properties
   $a + b = b + a$
   $ab = ba$
	$12 + 6 = 6 + 12$
	$12 \cdot 6 = 6 \cdot 12$

Associative Properties
   $(a + b) + c = a + (b + c)$
   $(ab)c = a(bc)$
	$(12 + 6) + 5 = 12 + (6 + 5)$
	$(12 \cdot 6) \cdot 5 = 12 \cdot (6 \cdot 5)$

Identity Properties
There are unique real numbers 0 and 1
$(1 \neq 0)$ such that:
   $a + 0 = a$ and $0 + a = a$
   $a \cdot 1 = a$ and $1 \cdot a = a$
	$12 + 0 = 12$ and $0 + 12 = 12$
	$12 \cdot 1 = 12$ and $1 \cdot 12 = 12$

Inverse Properties
   Property of Opposites: For each a, there is a unique real number $-a$ such that:
      $a + (-a) = 0$ and $(-a) + a = 0$.
   ($-a$ is called the *opposite* or *additive inverse* of a.)
   Property of Reciprocals: For each a except 0, there is a unique real number $\frac{1}{a}$ such that:
      $a \cdot \frac{1}{a} = 1$ and $\frac{1}{a} \cdot a = 1$.
   $\left(\frac{1}{a}$ is called the *reciprocal* or *multiplicative inverse* of a.$\right)$
	$12 + (-12) = 0$ and
	   $-12 + 12 = 0$
	
	$12 \cdot \frac{1}{12} = 1$ and
	
	$\frac{1}{12} \cdot 12 = 1$

Distributive Property (of multiplication with respect to addition)
   $a(b + c) = ab + ac$ and
   $(b + c)a = ba + ca$
	$12(6 + 5) = 12 \cdot 6 + 12 \cdot 5$
	$(6 + 5)12 = 6 \cdot 12 + 5 \cdot 12$

   The next two examples illustrate that you can use the commutative and associative properties to add or to multiply numbers in any convenient order and in any groups of two.

Example 2 Simplify: **a.** $7 + 11 + 13 + 19$ **b.** $26 + 8y + 14$

Solution   **a.** $7 + 11 + 13 + 19 = (7 + 13) + (11 + 19) = 20 + 30 = 50$ *Answer*

b. $26 + 8y + 14 = (26 + 14) + 8y = 40 + 8y$ *Answer*

Example 3 Simplify $16(4xy^2)(5z^3)$.

Solution $16(4xy^2)(5z^3) = 16 \cdot (4 \cdot 5)xy^2z^3 = 16 \cdot 20xy^2z^3 = 320xy^2z^3$ *Answer*

The real number properties can be used to change a given expression into a simpler *equivalent expression*. Two expressions are **equivalent** when they are equal for every value of each variable they contain. *In this book, unless otherwise stated, the domain of every variable is the set of real numbers.*

Example 4 Name the property used in each step of simplifying $\frac{1}{3}(1 + 3t)$.

Solution

1. $\frac{1}{3}(1 + 3t) = \frac{1}{3} \cdot 1 + \frac{1}{3}(3t)$ 1. Distributive property

2. $= \frac{1}{3} \cdot 1 + \left(\frac{1}{3} \cdot 3\right)t$ 2. Associative property of multiplication

3. $= \frac{1}{3} \cdot 1 + 1 \cdot t$ 3. Property of reciprocals

4. $= \frac{1}{3} + t$ 4. Identity property of multiplication

5. $\therefore \frac{1}{3}(1 + 3t) = \frac{1}{3} + t$ 5. Transitive property of equality

---

Note that in Step 3 the property of reciprocals was used to justify the substitution of 1 for $\frac{1}{3} \cdot 3$. Of course, the substitution principle (page 8) was also used in this step. In this book we will follow the common practice of omitting many references to the substitution principle.

Oral Exercises

Give (a) the opposite and (b) the reciprocal of each number. If a number has no reciprocal, say so.

1. -2 **2.** $\frac{1}{2}$ **3.** 1 **4.** 0

5. $-\frac{4}{3}$ **6.** -1 **7.** 0.1 **8.** -0.25

Name the property illustrated in each statement.

Sample $7(2y) = (7 \cdot 2)y$ **Solution**   Associative property of multiplication

9. $3 + t = t + 3$

10. $5 \cdot \frac{1}{5} = 1$

11. $1 \cdot (abc) = abc$

12. $-2y + 0 = -2y$

13. $2\left(x + \frac{1}{2}\right) = 2x + 2 \cdot \frac{1}{2}$

14. If $x - 2 = 3$, then $(x - 2) + 2 = 3 + 2$.

15. $-1.5 + 1.5 = 0$

16. $(a + 1)(a - 1) = (a - 1)(a + 1)$

17. If $x + 2 = 8$, then $8 = x + 2$.

18. $\frac{1}{3}(3z^2) = \left(\frac{1}{3} \cdot 3\right)z^2$

19. $\left(-\frac{2}{3}\right)\left(-\frac{3}{2}\right) = 1$

20. $(2 + x) + 3 = (x + 2) + 3$

21. $2000 + 20$ is a unique real number.

22. If $3x = 6$ and $6 = 2y$, then $3x = 2y$.

23. Does $a(bc) = (ab)(ac)$ for all real numbers a, b, and c? Explain.

24. Explain why 0 has no reciprocal.

Written Exercises

Simplify.

A

1. $96 + 13 + 4 + 37$

2. $-6 + x + 6$

3. $\frac{1}{3}(1 \cdot 3) + (-1)$

4. $\frac{1}{3}(1 \cdot 3) + (-3 + 3)$

5. $4\left(z + \frac{1}{4}\right)$

6. $\frac{1}{4}(z + 4)$

7. $1 \cdot (-t + t)$

8. $\left(\frac{2}{3}a\right)\left(\frac{3}{2}b\right)$

9. $2(a + 4) + (-8)$

10. $\left(\frac{1}{3} \cdot 3\right)[p + (-1)] + 1$

Determine whether each simplification is true or false.

Sample 1 $7\left(2 + \frac{1}{7}x\right) - 14 = x$

Solution   True, because $7\left(2 + \frac{1}{7}x\right) - 14 = 14 + x - 14 = x$.

11. $(-x + 6) + (-6 + x) = 0$

12. $2(x + 3) + (-3) = 2x$

13. $\frac{1}{2}(2t + 2) + (-2) = t$

14. $(7p)\frac{1}{7} + (-p) = 0$

15. $5(2n + 1) + (-5) = 10n$

16. $5(xy) = (5x)(5y)$

In Exercises 17–22, name the property used in each step of the simplification.

17. $\frac{1}{2}(1 + 2t) = \frac{1}{2} \cdot 1 + \frac{1}{2}(2t)$ **a.** __?__

$\quad\quad\quad = \frac{1}{2} \cdot 1 + \left(\frac{1}{2} \cdot 2\right)t$ **b.** __?__

$\quad\quad\quad = \frac{1}{2} \cdot 1 + 1 \cdot t$ **c.** __?__

$\quad\quad\quad = \frac{1}{2} + t$ **d.** __?__

18. $(2 + a) + (-2) = (a + 2) + (-2)$ **a.** __?__

$\quad\quad\quad\quad\quad = a + [2 + (-2)]$ **b.** __?__

$\quad\quad\quad\quad\quad = a + 0$ **c.** __?__

$\quad\quad\quad\quad\quad = a$ **d.** __?__

19. $\frac{1}{5}[(n + 5) + (-n)] = \frac{1}{5}[(5 + n) + (-n)]$ **a.** __?__

$\quad\quad\quad\quad\quad = \frac{1}{5}[5 + (n + (-n))]$ **b.** __?__

$\quad\quad\quad\quad\quad = \frac{1}{5}[5 + 0]$ **c.** __?__

$\quad\quad\quad\quad\quad = \frac{1}{5} \cdot 5$ **d.** __?__

$\quad\quad\quad\quad\quad = 1$ **e.** __?__

20. $3 + 4(x + 1) = 3 + (4x + 4 \cdot 1)$ **a.** __?__

$\quad\quad\quad\quad = 3 + (4x + 4)$ **b.** __?__

$\quad\quad\quad\quad = (4x + 4) + 3$ **c.** __?__

$\quad\quad\quad\quad = 4x + (4 + 3)$ **d.** __?__

$\quad\quad\quad\quad = 4x + 7$ Substitution

21. $k + (k + 2) = (k + k) + 2$ **a.** __?__

$\quad\quad\quad\quad = (1 \cdot k + 1 \cdot k) + 2$ **b.** __?__

$\quad\quad\quad\quad = (1 + 1)k + 2$ **c.** __?__

$\quad\quad\quad\quad = 2k + 2$ Substitution

$\quad\quad\quad\quad = 2k + 2 \cdot 1$ **d.** __?__

$\quad\quad\quad\quad = 2(k + 1)$ **e.** __?__

22. $x(y + 1) + (-1)x = x(y + 1) + x(-1)$ **a.** __?__

$\quad\quad\quad\quad = x[(y + 1) + (-1)]$ **b.** __?__

$\quad\quad\quad\quad = x[y + (1 + (-1))]$ **c.** __?__

$\quad\quad\quad\quad = x[y + 0]$ **d.** __?__

$\quad\quad\quad\quad = xy$ **e.** __?__

B **23.** Show that if $3x + (-12) = 0$, then $x = 4$ by justifying each indicated step.

$$3x + (-12) = 0 \qquad \text{Given}$$
$$[3x + (-12)] + 12 = 0 + 12 \qquad \textbf{a.} \quad \underline{\;?\;}$$
$$3x + [(-12) + 12] = 0 + 12 \qquad \textbf{b.} \quad \underline{\;?\;}$$
$$3x + 0 = 0 + 12 \qquad \textbf{c.} \quad \underline{\;?\;}$$
$$3x = 12 \qquad \textbf{d.} \quad \underline{\;?\;}$$
$$\frac{1}{3}(3x) = \frac{1}{3} \cdot 12 \qquad \textbf{e.} \quad \underline{\;?\;}$$
$$\frac{1}{3}(3x) = 4 \qquad \text{Substitution}$$
$$\left(\frac{1}{3} \cdot 3\right)x = 4 \qquad \textbf{f.} \quad \underline{\;?\;}$$
$$1 \cdot x = 4 \qquad \textbf{g.} \quad \underline{\;?\;}$$
$$x = 4 \qquad \textbf{h.} \quad \underline{\;?\;}$$

24. Use steps similar to those in Exercise 23 to show that if $\frac{1}{2}x + (-3) = 0$, then $x = 6$.

A set S of real numbers is *closed under addition* if the sum of any two members of S is in S. A set S is *closed under multiplication* if the product of any two members of S is in S. For each set given below tell whether the set is closed under (a) addition and (b) multiplication. If the set is not closed, give an example to show this.

Sample 2 {0, 1}

Solution **a.** Not closed under addition, since $1 + 1 = 2$ and 2 is not in the set. (Notice that ''any two members'' includes the same number used twice.)

b. Closed under multiplication, since every possible product ($0 \cdot 0$, $1 \cdot 1$, and $0 \cdot 1$) is in the set.

25. {0}   + sula closed   **26.** {−1, 0, 1}   **27.** {3, 6, 9, 12, . . .}

28. {1, 3, 5, . . .} x หนู closed.   **29.** {2, 4, 6, ...}   **30.** $\left\{1, \frac{1}{2}, \frac{1}{3}, \frac{1}{4}, \ldots\right\}$

31. The rational numbers   **32.** The irrational numbers   **33.** The real numbers greater than −1 and less than 1

Decide whether each set is a field under the operations of addition and multiplication. If the set is not a field, name at least one field property that does not hold.

C **34.** The natural numbers   **35.** The integers

36. The rational numbers   **37.** The negative rational numbers

Mixed Review Exercises

Use <, =, or > to make a true statement.

1. $0.02 \underline{\quad ? \quad} 0.002$

2. $3(8 - 5) \underline{\quad ? \quad} 3 \cdot 8 - 5$

3. $17 - (8 + 5) \underline{\quad ? \quad} 17 - 8 + 5$

4. $(3 + 4)^2 \underline{\quad ? \quad} 3^2 + 2 \cdot 3 \cdot 4 + 4^2$

5. $\dfrac{7 + 5}{8 - 2} \underline{\quad ? \quad} \dfrac{8 + 2}{7 - 5}$

6. $\dfrac{1 + 1}{2 + 2} \underline{\quad ? \quad} \dfrac{1}{2} + \dfrac{1}{2}$

Extra / *Logical Symbols: Quantifiers*

Statements about the properties of real numbers often include phrases like "for all real numbers a," "for each b," or "there exists a unique real number c," that express the idea of "how many" or of quantity. Such a phrase is called a **quantifier.** In mathematical logic, quantifiers are represented by symbols, as shown in the table at the right.

Symbol	Meaning
$\forall_x$	"for all x," "for every x," "for each x," and so on
$\exists_x$	"there exists an x," "for some x," and so on
$\exists!_x$	"there is a unique x"

For examples of the use of symbolic quantifiers, study the following sentences. All variables here and in the exercises represent real numbers.

For every x, $x + 2x = 3x$.	$\forall_x \, x + 2x = 3x$
For all x, $x + 7 = 7 + x$.	$\forall_x \, x + 7 = 7 + x$
For some x, $x + 5 = 10$.	$\exists_x \, x + 5 = 10$
For all a and b, $(a + b) + 2 = a + (b + 2)$.	$\forall_a \forall_b \, (a + b) + 2 = a + (b + 2)$
For every a, there is a b for which $b = 2a$.	$\forall_a \exists_b \, b = 2a$
For every x there is a unique y such that $x + y = 0$.	$\forall_x \exists!_y \, x + y = 0$

Exercises

Express each statement using a symbol for the quantifier.

1. For all x, $2x = x + x$.

2. For some y, $y^2 = 17$.

3. For all real numbers x and all real numbers y, $x + y = y + x$.

4. For all real numbers r, there exists a unique real number x such that $r + x = 0$.

Tell whether each statement is true or false.

5. $\forall_a \exists_x \, x^2 = a$

6. $\forall_a \exists_x \, x = a^2$

7. $\forall_x \exists_a \, x > a$

8. $\forall_a \forall_b \forall_c \exists!_x \, ax + b = c$

1-4 Sums and Differences

Objective   To review the rules for adding and subtracting real numbers.

In arithmetic you learned to add positive numbers $(4 + 7 = 11)$ and to subtract a smaller positive number from a larger one $(12 - 7 = 5)$. In this lesson you will review how to add and subtract *any* real numbers, positive or not.

The identity property of addition tells you how to add zero and any given number: *The sum is the given number.*

$$-6 + 0 = -6 \qquad\qquad 0 + \left(-\frac{1}{2}\right) = -\frac{1}{2}$$

The property of opposites tells you that the sum of opposites is zero.

$$-1.2 + 1.2 = 0 \qquad\qquad 4 + (-4) = 0$$

The rules covering the remaining cases can be stated in terms of absolute value. Study the examples to be sure you understand the rules.

Rules for Addition For real numbers a and b:	Examples
1. If a and b are negative numbers, then $a + b$ is negative and $a + b = -(\lvert a \rvert + \lvert b \rvert).$	$-5 + (-9) = -(\lvert -5 \rvert + \lvert -9 \rvert)$ $= -(5 + 9)$ $= -14$
2. If a is a positive number, b is a negative number, and $\lvert a \rvert$ is greater than $\lvert b \rvert$, then $a + b$ is a positive number and $a + b = \lvert a \rvert - \lvert b \rvert.$	$9 + (-5) = \lvert 9 \rvert - \lvert -5 \rvert$ $= 9 - 5$ $= 4$
3. If a is a positive number, b is a negative number, and $\lvert a \rvert$ is less than $\lvert b \rvert$, then $a + b$ is a negative number and $a + b = -(\lvert b \rvert - \lvert a \rvert).$	$5 + (-9) = -(\lvert -9 \rvert - \lvert 5 \rvert)$ $= -(9 - 5)$ $= -4$

Since addition of real numbers is commutative, these rules also cover sums like $-4 + 9 = 5$ and $-6 + 2 = -4$.

We say that the sign of a positive number is *plus* and the sign of a negative number is *minus,* and that a positive and a negative number have *opposite* signs. Thus, the rules for adding two nonzero numbers can be restated as follows:

If two numbers have the *same* sign, add their absolute values and then prefix their common sign.

If two numbers have *opposite* signs, subtract the lesser absolute value from the greater and then prefix the sign of the number having the greater absolute value.

Example 1 Simplify $-14 + 9 + (-12) + 20 + (-8)$.

Solution 1 Add the terms in order from left to right.
$$-14 + 9 + (-12) + 20 + (-8) = \underbrace{-5 + (-12)} + 20 + (-8)$$
$$= \underbrace{-17 \quad + 20} + (-8)$$
$$= \underbrace{3 \quad + (-8)}$$
$$= -5 \quad \textbf{\textit{Answer}}$$

Solution 2 Use the commutative and associative properties to order and group the terms conveniently. For example, you may choose to group the positive terms and the negative terms.
$$-14 + 9 + (-12) + 20 + (-8) = [-14 + (-12) + (-8)] + [9 + 20]$$
$$= -34 + 29$$
$$= -5 \quad \textbf{\textit{Answer}}$$

You may also use the rules shown on page 21 to subtract because subtraction of real numbers is defined in terms of addition.

Definition of Subtraction

For all real numbers a and b, $\qquad a - b = a + (-b)$.

To subtract any real number, add its opposite.

Example 2 Simplify.
    a. $-7 - (-8)$ **b.** $-7 - 8$ **c.** $7 - (-8)$ **d.** $7 - 8$

Solution   **a.** $-7 - (-8) = -7 + 8 = 1$ **b.** $-7 - 8 = -7 + (-8) = -15$
    c. $7 - (-8) = 7 + 8 = 15$ **d.** $7 - 8 = 7 + (-8) = -1$

Example 3 Simplify $10 - 8 - 15 + 6$.

Solution 1 Add the terms in order from left to right.
$$10 - 8 - 15 + 6 = 2 - 15 + 6$$
$$= -13 + 6$$
$$= -7 \quad \textbf{\textit{Answer}}$$

Solution 2 Group the positive terms and the negative terms.
$$10 - 8 - 15 + 6 = 10 + (-8) + (-15) + 6$$
$$= [10 + 6] + [(-8) + (-15)]$$
$$= 16 + (-23)$$
$$= -7 \quad \textbf{\textit{Answer}}$$

Example 4 Simplify $|14 - 5| - |5 - 19|$.

Solution

$$|14 - 5| - |5 - 19| = |9| - |-14|$$
$$= 9 - 14$$
$$= -5 \quad \textbf{\textit{Answer}}$$

You know that multiplication is distributive with respect to addition (page 15). Since subtraction is just the addition of an opposite, multiplication is also distributive with respect to subtraction.

Distributive Property (of multiplication with respect to subtraction)
For all real numbers a, b, and c,
$$a(b - c) = ab - ac$$
and $\quad (b - c)a = ba - ca.$

Example 5 Multiply $3(4x^2 - 9)$.

Solution $3(4x^2 - 9) = 3(4x^2) - 3(9) = 12x^2 - 27 \quad \textbf{\textit{Answer}}$

Example 6 Simplify $4y + 2(6y - 5)$.

Solution

$$4y + 2(6y - 5) = 4y + 12y - 10$$
$$= (4 + 12)y - 10$$
$$= 16y - 10 \quad \textbf{\textit{Answer}}$$

In Example 6, we used the distributive property to combine the *similar terms* $4y$ and $12y$. Remember that **similar terms** contain the same variable factors. For example, $8ab$ and $2ab$ are similar terms, but $3x^2$ and $3x$ are not. Similar terms are sometimes referred to as **like terms**.

Oral Exercises

In Exercises 1–9, (a) add the numbers and (b) subtract the lower number from the upper one.

1. 3
   7

2. −2
   5

3. −9
   4

4. 4
   −7

5. 8
   −3

6. −5
   5

7. -5
$\underline{-5}$

8. -8
$\underline{-3}$

9. 0
$\underline{-6}$

Simplify.

10. $12 + (-7)$

11. $12 - 7$

12. $5 - 13$

13. $5 + (-13)$

14. $-5 - (-13)$

15. $5 - (-13)$

16. $0 - 6$

17. $0 - (-6)$

18. $-7 - (-7)$

19. $-7 - 7$

20. $6 - 4 + 3$

21. $8 - 3 - 2$

22. $8 - (3 - 2)$

23. $2 - (7 + 3) - 4$

24. $3 - (8 - 2 + 7)$

25. $|5| - |-6|$

26. $|5 - (-6)|$

27. $|-5 + (-6)|$

28. Is the sum of a positive number and a negative number always, sometimes, or never a positive number?

29. Give a numerical example that shows $|a + b| \neq |a| + |b|$.

Written Exercises

Simplify. You may wish to check your answers on a calculator.

A **1.** $32 - 53$

2. $-16 - 42$

3. $-23 + 57$

4. $17 - (-26)$

5. $-15 - (-40)$

6. $16 + (-50)$

7. $-10.2 + 17.6$

8. $21.2 - 32.3$

9. $-9.6 - 13.4$

10. $26.2 - (-26.2)$

11. $0 - (-30.7)$

12. $-15.85 - (-20.15)$

13. $57 - 13 - 46$

14. $33 - 72 + 14$

15. $68 + (-42) + (-35) + 17$

16. $-87 + 16 - 22 + 61$

17. $5 - (2 - 9) - (4 - 11)$

18. $-16 - [2 - (-61)] - [2 + (-6)]$

19. $(-4 + 7 - 10) - [-6 - (-8)]$

20. $(3 - 6 - 9) - [8 + (-4) - (-7)]$

21. $|6 - 13| - |22 - (-6)|$

22. $|-22 - 33| - |16 - 7|$

Multiply.

23. $4(p - 2q)$

24. $5(3y - 2x)$

25. $\frac{1}{2}(2a - 4b + 6)$

26. $\frac{2}{3}(6r - 9s - 27)$

Simplify by combining similar terms.

27. $8c + 2(c + 3)$

28. $5(d + 2) - 3d$

29. $7(x + 2) + 4(x - 4)$

30. $4(3 - y) + 2(1 - y)$

31. $7p - 4q + 3q - 10p$

32. $6m - 4n + (-7)m - (-5)n$

33. $(-2r + s + 5) + 2(r - 3s + 2)$

34. $(6x - 5y + 4) + 2(-2x + 3y - 2)$

Find the number that makes the equation true.

B 35. $-7 + \underline{\quad ? \quad} = 2$ 36. $\underline{\quad ? \quad} - 8 = -3$

37. $\underline{\quad ? \quad} - 4 = -7$ 38. $-6 - \underline{\quad ? \quad} = -1$

39. $-6 + \underline{\quad ? \quad} + 2 = 0$ 40. $-6 + \underline{\quad ? \quad} + 11 = -4$

41. $-6.4 + \underline{\quad ? \quad} = 3.7$ 42. $2.6 + \underline{\quad ? \quad} - 8.7 = 3.2$

Solve.

43. Temperatures on the planet Mars range from $-122.2°$ C to $30.5°$ C. Find the difference between these two temperatures.

44. The highest point in the United States is Mt. McKinley, Alaska, at 20,320 ft above sea level, and the lowest is Bad Water, California, at 282 ft below sea level. Find the difference between these elevations.

45. Lake Baikal, in Siberia, is 5315 ft deep, and its surface is 1493 ft above sea level. How many feet below sea level is the bottom of the lake at its deepest point?

46. A subway train was carrying 103 passengers. At the next three stops, 15 people got on and 9 got off, 27 got on and 13 got off, and 8 got on and 53 got off. How many passengers were then on the train?

47. A newspaper lists the closing price of a stock at $72\frac{3}{4}$ following a drop of $1\frac{7}{8}$ that day. What was the stock's opening price for the day?

48. At 7:38:16 A.M. (38 minutes and 16 seconds after 7 o'clock) the count-down on a rocket launch stood at $-14:42$ (14 minutes and 42 seconds before launch). There was a 37-second delay in resuming the count. At what time did the rocket take off?

49. Give a numerical example to show that subtraction is not commutative.

50. Give a numerical example to show that subtraction is not associative.

Tell whether each set is closed (see page 19) under subtraction. If the set is not closed, give an example to show this.

C 51. $\{0, 1\}$ 52. $\{-1, 0, 1\}$

53. The whole numbers   54. The negative integers

55. The integers   56. The irrational numbers

Reading Algebra / Independent Study

Your algebra textbook has been designed and written to make reading and learning algebra as easy as possible. To take full advantage of your book, begin your study of each lesson by getting an idea of the goal of the lesson. Read the lesson title and the stated objective. You may find it useful to skim the lesson first.

Heavy type indicates new words or phrases that you must understand in order to understand the lesson. It is usually very helpful to learn the definitions of these words or phrases before you begin a slow and careful reading of the lesson. When you do start to read the lesson, pay particular attention to the important information set off in boxes. If you come across any words whose meaning you do not understand, look them up in the glossary at the back of the book or in a dictionary. You may also find additional information by checking references from the index.

Be sure to work through all the examples that are included in the lesson discussions. Try doing these examples on your own before reading the solutions that are given following the examples. If you do not understand a concept, and rereading does not seem to help, make a note of the concept so that you can discuss it with your teacher.

When you have finished reading a lesson and feel that you understand what you have read, try some of the Oral and Written Exercises. They will help you determine whether you have achieved the goal set forth in the objective. Note that the answers to most odd-numbered exercises are given in the Selected Answers at the back of your book. Doing the Self-Tests and checking your answers with those at the back of the book will help you to see whether you have understood a group of lessons. The Chapter Reviews, Chapter Tests, Cumulative Reviews, and Mixed Reviews will also give you a good idea of your progress.

Exercises

Skim through Lesson 1–7 and answer the following questions.

1. What should you be able to do after studying Lesson 1–7?

2. What new words or phrases are introduced in Lesson 1–7?

3. What is an open sentence? What are equivalent equations? Find the definitions in your book.

4. Describe a transformation that produces equivalent equations.

5. Look at Example 1 on page 38. Then solve $3(4x - 9) = 5x - 6$.

6. Suppose that you have forgotten the definition of the word *identity*. Where could you look it up? On what page is this word first used?

7. Lesson 1–7 also shows you how to rewrite formulas. Where in your book could you find a list of geometric formulas?

1-5 Products

Objective   To review rules for multiplying real numbers.

The following facts are useful in finding products of real numbers.

Multiplicative Property of 0

For every real number a,

$$a \cdot 0 = 0 \quad \text{and} \quad 0 \cdot a = 0.$$

Multiplicative Property of −1

For every real number a,

$$a(-1) = -a \quad \text{and} \quad (-1)a = -a.$$

The multiplicative property of 0 tells you that the product of any real number and 0 is 0. For example,

$$-5 \cdot 0 = 0 \quad \text{and} \quad 0\left(\frac{1}{2}\right) = 0.$$

The multiplicative property of −1 tells you that the product of any real number and −1 is the opposite of the number. For example,

$$6(-1) = -6, \quad (-1)4 = -4, \quad \text{and} \quad (-1)(-1) = -(-1) = 1.$$

You can use the multiplicative property of −1 along with the facts for positive numbers to find the product of *any* two nonzero real numbers. For example:

$$(5)(7) = 35$$
$$(-5)(-7) = (-1)(5)(-1)(7) = (-1)(-1)(5)(7) = 1(35) = 35$$
$$(5)(-7) = (5)(-1)(7) = (-1)(5)(7) = (-1)(35) = -35$$

All three products have the same absolute value, 35. When the two factors have the same sign, the product is positive; when they have opposite signs, the product is negative. Examples like these suggest the following rules.

Rules for Multiplication

1. The product of two positive numbers or two negative numbers is a positive number.
2. The product of a positive number and a negative number is a negative number.
3. The absolute value of the product of two or more numbers is the product of their absolute values.

Example 1 Simplify.

    a. $(-3)(-2)(-1)(4)(-5)$

    b. $(-4) \cdot 3 \cdot (-2)\left(-\frac{1}{3}\right)$

    c. $24(-15)(0)(13)$

Solution

    a. $(-3)(-2)(-1)(4)(-5) = \underline{(6)(-1)}\underline{(4)(-5)}$

                                  $= \ \ (-6) \cdot (-20)$

                                  $= 120$ *Answer*

    b. $(-4) \cdot 3 \cdot (-2)\left(-\frac{1}{3}\right) = (-12)(-2)\left(-\frac{1}{3}\right)$

                                  $= 24\left(-\frac{1}{3}\right)$

                                  $= -8$ *Answer*

    c. $24(-15)(0)(13) = 0$ *Answer*

Example 1 illustrates the following facts.

1. A product of nonzero numbers is *positive* if the number of negative factors is *even*.
2. The product is *negative* if the number of negative factors is *odd*.
3. The product is *zero* if any one of the factors is *0*.

Example 2 Name the property used in each step.

Solution

Step	Property
1. $-ab = (-1)(ab)$	1. Multiplicative property of -1
2. $= [(-1)a]b$	2. Associative property of multiplication
3. $= (-a)b$	3. Multiplicative property of -1
4. $\therefore \ -ab = (-a)b$	4. Transitive property of equality

The last step in Example 2 states one case of the following useful fact:
The opposite of a product of real numbers is the product of the opposite of one factor and the other factors.

Property of the Opposite of a Product

For all real numbers a and b,

$$-ab = (-a)b \quad \text{and} \quad -ab = a(-b).$$

Example 3 Simplify.

    a. $2(-3x)(-5y)$

    b. $(-p)(2q) + (3p)(-2q)$

Solution

    a. $2(-3x)(-5y) = 2(-3)(-5)xy = 30xy$ *Answer*

    b. $(-p)(2q) + (3p)(-2q) = -2pq + (-6pq) = -8pq$ *Answer*

Example 4 Simplify.

    a. $-2\left(3t - \dfrac{1}{2}\right)$

    b. $16d - 5(3d - 4)$

Solution

    a. $-2\left(3t - \dfrac{1}{2}\right) = (-2)3t - (-2)\dfrac{1}{2}$

                        $= -6t - (-1)$

                        $= -6t + 1,$ or $1 - 6t$ *Answer*

    b. $16d - 5(3d - 4) = 16d - 15d + 20$

                          $= d + 20$ *Answer*

Example 5 Name the property used in each step.

Solution

Step	Property
1. $-(a + b) = (-1)(a + b)$	1. Multiplicative property of -1
2. $= (-1)a + (-1)b$	2. Distributive property
3. $= (-a) + (-b)$	3. Multiplicative property of -1
4. $\therefore\ -(a + b) = (-a) + (-b)$	4. Transitive property of equality

      Example 5 shows that *the opposite of a sum of real numbers is equal to the sum of the opposites of the numbers.*

Property of the Opposite of a Sum

For all real numbers a and b,

$$-(a + b) = (-a) + (-b).$$

Example 6 Give the opposite of each expression.

    a. $3y + 5$ **b.** $-6x + 4$ **c.** $-xy - 5z$ **d.** $4y^3 - 8y^2 + 3y - 7$

Solution

    a. $-3y - 5$ **b.** $6x - 4$ **c.** $xy + 5z$ **d.** $-4y^3 + 8y^2 - 3y + 7$

Oral Exercises

Tell whether the given expression represents a positive number, a negative number, or zero. Then simplify the expression.

1. $4(-3)$

2. $(-2)(-5)$

3. $-2 \cdot 7$

4. $-[2(-3)]$

5. $(-2) \cdot 3 \cdot (-5)$

6. $(-6)(-1)(0)(-2)$

Without doing the computation, tell whether the value of each expression is positive, negative, or zero.

Sample $\quad (-2)(3)\left(-\frac{1}{4}\right)\left(-\frac{1}{8}\right)(-12)$

Solution $\quad$ Positive $\quad$ (The expression has an even number of negative factors.)

7. $2(-6)(-13)(-37)$

8. $(-1)(-2)(-3)(-4)(-5)(-6)$

9. $2(-2)(2 + 2)(2 - 2)$

10. $7\left(-\frac{1}{3}\right)\left(-\frac{2}{5}\right)(9)\left(\frac{1}{9}\right)$

11. $-(-1)^4$

12. $(-2)^{101}$

Give the opposite of each expression.

13. $5e - f$

14. $-1 + 3t$

15. y^3

16. $-3y^3 + 2y - 1$

Tell whether each expression always, sometimes, or never represents a negative number.

17. $-y$

18. $|-y|$

19. $y \cdot y$

20. $-|-(-y)|$

Written Exercises

Simplify. You may wish to check your answers on a calculator.

A

1. $5(-2)(-7)(-3)$

2. $7(-2)(6)(-1)$

3. $9\left(-\frac{1}{7}\right)\left(\frac{1}{3}\right)(-28)$

4. $\left(\frac{3}{4}\right)(-10)(-8)\left(\frac{1}{5}\right)$

5. $(0.5)(-6)(-4)(-0.2)$

6. $(1.5)(-3)(0.2)(-2)(-1)$

7. $2(-5x)(-6y)$

8. $(-3)(-u)(-7v)$

9. $(-a)(-2b)(-3c)$

10. $\left(-\frac{1}{2}\right)(4r)(-s)$

11. $(-6 - 4)(-6 + 5)$

12. $17(-13) + 17(-7)$

13. $12(-1)^7(-2)^3$

14. $(-5)^3\left(-\frac{1}{5}\right)^2$

15. $(-1)^3(-2 - 2)(-6)(-4)$

16. $(-9)^2(-2 + 2)(-5)$

17. $(-2)(1 - 2x - 3x^2)$

18. $(-3)\left(2a - \frac{2}{3}\right)$

19. $\left(-\dfrac{1}{2}\right)(6z^2 - 4z + 2)$ **20.** $(-1)[(-t)^3 - 1]$ **21.** $(-x)(3y) + (2x)(-3y)$

22. $(-a)(-b) + 6(ab)(-1)$ **23.** $7k - 4(3k + 6)$ **24.** $-3(p - 5) - 7p$

25. $5(x - y) - 3(x - y)$ **26.** $-4(a - b) + 2(a - b)$ **27.** $5(-3a^3 - 2) + 3(-2 - a^3)$

28. $4(2y^2 - 3) - 2(4 + y^2)$ **29.** $t(2w - 9) - 2(2t + 7)$ **30.** $-c(d + 5) + 6(2 - cd)$

Evaluate each expression for the given values of the variable.

Sample $(x + 1)(x + 3)(x - 3)$ **a.** $x = 1$ **b.** $x = -2$

Solution   **a.** $(x + 1)(x + 3)(x - 3) = (1 + 1)(1 + 3)(1 - 3) = (2)(4)(-2) = -16$

            b. $(x + 1)(x + 3)(x - 3) = (-2 + 1)(-2 + 3)(-2 - 3) = (-1)(1)(-5) = 5$

B   **31.** $x(x - 2)(x - 4)$ **a.** $x = 1$ **b.** $x = 3$ **c.** $x = 2$

     32. $t(t + 1)(t - 3)$ **a.** $t = 2$ **b.** $t = -1$ **c.** $t = 1$

     33. $y^3 - 3y^2 + 4$ **a.** $y = 2$ **b.** $y = -1$ **c.** $y = 3$

     34. $2b^3 + 3b^2 - b + 3$ **a.** $b = 2$ **b.** $b = -2$ **c.** $b = 0$

     35. Name the property used in each step.

         1. $-a = -(a \cdot 1)$ 1. __?__

         2. $= a(-1)$ 2. __?__

         3. $= (-1)a$ 3. __?__

         4. $-a = (-1)a$ 4. __?__

         5. and $(-1)a = -a$ 5. __?__

     (This exercise is a proof of the multiplicative property of -1.)

C   **36.** Find a value of y for which the expression $(1 - y)(3 - y)(6 - y)$ has the given value.

      a. 18   **b.** 0   **c.** -70 **d.** 120   **e.** $3\dfrac{1}{8}$

Mixed Review Exercises

Evaluate each expression if $x = 3$ and $y = -2$.

   1. $x + y - 1$ **2.** $|x| - |y|$ **3.** $x^2 + y$

   4. $-|x + y|$ **5.** $y - (x - 1)$ **6.** $4x - y$

Name the property illustrated in each statement.

   7. $3 + 5 = 5 + 3$ **8.** $-3 + 0 = -3$ **9.** If $x = y + 1$, then $y + 1 = x$.

   10. $4 \cdot \dfrac{1}{4} = 1$ **11.** $4(5 \cdot 8) = (4 \cdot 5)8$ **12.** $5(2 + 7) = 5 \cdot 2 + 5 \cdot 7$

Computer Exercises

For students with some programming experience.

1. Write a program that will evaluate the expression $5x^3 - 7x^2 + 3x + 2$ for any value of x. Use an INPUT statement to enter the values of x.

2. Run the program in Exercise 1 to find the value of the expression for each value of x.

   a. $x = 1$ **b.** $x = 4.7$ **c.** $x = -2.5$ **d.** $x = -3.04$

3. Modify the program of Exercise 1 so that it will evaluate *any* expression of the form $ax^3 + bx^2 + cx + d$ for given values of x. That is, modify the program so that the user can enter the constants a, b, c, and d in the expression as well as the value to be substituted for x.

4. Run the program in Exercise 3 to evaluate $-3x^3 + 8x^2 - 6x + 11$ for each value of x.

   a. $x = 2.1$ **b.** $x = 4.95$ **c.** $x = -1.06$ **d.** $x = -4.17$

 Biographical Note / *Sonya Kovalevski*

When the Russian mathematician Sonya Kovalevski (1850–1891) went to Sweden to teach, a Stockholm newspaper proclaimed her "the Princess of Science." She did not reach such a position of acclaim easily, however. European universities resisted accepting women as students or faculty members during the nineteenth century.

To obtain her education, Kovalevski became the private student of Karl Weierstrass, an eminent German mathematician whose own university in Berlin would not grant her admission. Her papers on partial differential equations, Abelian integrals, and the rings of Saturn eventually qualified her for a degree from the University of Göttingen.

At the end of her five-year teaching appointment in Stockholm, Kovalevski became the first woman in modern times to hold a tenured position in mathematics. Also during this time, she won the Prix Bordin from the French Academy of Science for a paper on the rotation of solid bodies around fixed points. Her work grew out of previous work on ultraelliptic integrals, but her extension of the concept was so remarkable that the prize money was nearly doubled. In 1889 she was elected a corresponding member of the Russian Academy of Sciences.

In addition to her mathematical studies, Kovalevski published poetry, novels, plays, and essays on politics and education.

1-6 Quotients

Objective   To review rules for dividing real numbers.

You know that subtraction is defined in terms of addition and opposites. Similarly, division is defined in terms of multiplication and reciprocals.

Definition of Division

The quotient a divided by b is written $\frac{a}{b}$ or $a \div b$. For every real number a and nonzero real number b,

$$\frac{a}{b} = a \cdot \frac{1}{b}, \qquad \text{or} \qquad a \div b = a \cdot \frac{1}{b}.$$

To divide by any nonzero number, multiply by its reciprocal. Since 0 has no reciprocal, division by 0 is not defined.

Example 1   Simplify: **a.** $12 \div 4$ **b.** $\frac{-18}{-3}$ **c.** $15 \div \left(-\frac{3}{2}\right)$

Solution   **a.** $12 \div 4 = 12 \cdot \frac{1}{4} = 3$ ***Answer***

   b. $\frac{-18}{-3} = (-18)\left(-\frac{1}{3}\right) = 6$ ***Answer***

   c. $15 \div \left(-\frac{3}{2}\right) = 15 \cdot \left(-\frac{2}{3}\right) = -10$ ***Answer***

A number and its reciprocal have the same sign. Therefore, the sign of a quotient of real numbers follows the same rules as the sign of a product.

Rules for Division

1. The quotient of two positive numbers or two negative numbers is a positive number.
2. The quotient of two numbers when one is positive and the other negative is a negative number.

Example 2   Simplify $\dfrac{-9 \div (-3)}{(-1)^2(-3)}$.

Solution $\dfrac{-9 \div (-3)}{(-1)^2(-3)} = \dfrac{-9 \cdot \left(-\frac{1}{3}\right)}{(1)(-3)} = \dfrac{3}{-3} = -1$ ***Answer***

Basic Concepts of Algebra   **33**

Notice that $\frac{a + b}{c} = (a + b) \cdot \frac{1}{c} = a \cdot \frac{1}{c} + b \cdot \frac{1}{c} = \frac{a}{c} + \frac{b}{c}$. Division, like multiplication, has these distributive properties.

For all real numbers a and b and *nonzero* real numbers c,

$$\frac{a + b}{c} = \frac{a}{c} + \frac{b}{c} \qquad \text{and} \qquad \frac{a - b}{c} = \frac{a}{c} - \frac{b}{c}.$$

Caution: $\frac{c}{a + b} \neq \frac{c}{a} + \frac{c}{b}$. For example, $\frac{12}{2 + 4} \neq \frac{12}{2} + \frac{12}{4}$, since $2 \neq 6 + 3$.

Example 3 Simplify $\frac{48 - 12x^2}{-3}$.

Solution
$$\frac{48 - 12x^2}{-3} = \frac{48}{-3} - \frac{12x^2}{-3}$$
$$= -16 - (-4x^2)$$
$$= -16 + 4x^2, \text{ or } 4x^2 - 16 \quad \textit{Answer}$$

Oral Exercises

Tell whether the expression represents a positive number, a negative number, or zero. Then simplify the expression.

1. $\frac{-40}{8}$
2. $\frac{-70}{-10}$
3. $\frac{101}{-101}$

4. $\frac{0}{-5}$
5. $-20 \div (-4)$
6. $-8 \div (-24)$

7. $0 \div (-8)$
8. $19 \div (-1)$
9. $\frac{2(-6)}{-3}$

10. $\frac{6(-8)}{(-3)(-4)}$
11. $\frac{8 \div (-4)}{-12 \div 3}$
12. $(-30) \div (15) \div (-2)$

Without doing the computation, tell whether the value of the expression is positive, negative, or zero.

13. $\frac{(-23) \cdot 17 \cdot (-13)}{(41)(-31)(-29)}$
14. $\frac{(-16)(-7)}{(-3)(-1)(-9)}$

15. $\frac{-14 + (-26)}{(-8)(-3)(-5)}$
16. $\frac{(-16)(-7)}{(-3) + (-9)}$

What number makes the statement true?

17. $-12 \div (\underline{\quad ? \quad}) = 4$
18. $-12 \div (\underline{\quad ? \quad}) = -4$
19. $12 \div (\underline{\quad ? \quad}) = -4$

Written Exercises

Simplify. You may wish to check your answers on a calculator.

A **1.** $-21 \div (-7)$

2. $-7 \div (-21)$

3. $64 \div (-4) \div (-2)$

4. $64 \div [(-4) \div (-2)]$

5. $-6 \div \left(-\frac{1}{3}\right) \div (-1)$

6. $-\frac{1}{2} \div \left(\frac{1}{4}\right) \div (-4)$

7. $\dfrac{8(-18)}{3(-12)}$

8. $\dfrac{(-3)(-4)(-2)}{(-6)(-2)}$

9. $[60 \div (-5)][8 \div (-2)]$

10. $[27(-2)] \div (-3)^4$

11. $\dfrac{-16 \cdot 3 \div 2}{(-2)^4}$

12. $\dfrac{24 \div (-3)}{4(-5) \div 2}$

13. $\dfrac{3^2 - 5^2}{3 + (-5)}$

14. $\dfrac{4^2 - 5^2}{(-4) + (-5)}$

15. $\dfrac{144 \div (-24)}{(-10) \div \left(-\frac{2}{3}\right)}$

16. $\dfrac{(-12)\left(-\frac{3}{4} - \frac{1}{2}\right)}{\frac{5}{9} \div (-10)}$

17. $\dfrac{-24\left[-18 \div \left(-\frac{2}{3}\right)\right]}{(-18)\left(-\frac{2}{3}\right)}$

18. $\dfrac{\left[\frac{4}{9} - \left(-\frac{2}{9}\right)\right]\left[\frac{2}{3} - \left(-\frac{2}{3}\right)\right]^2}{\frac{5}{9} \div \left(-\frac{10}{3}\right)}$

19. $\dfrac{9x^2 + 27}{-3}$

20. $\dfrac{24 - 6t^2}{2}$

21. $\dfrac{1 - (-n)^2}{-1}$

22. $\dfrac{2n^2 - 2^2}{-2}$

23. $\dfrac{36c^2 - 24c - 6}{6}$

24. $\dfrac{-15r^3 - 5r - 5}{-5}$

Evaluate each expression for the given values of the variable.

B **25.** $\dfrac{y(y - 3)}{y - 2}$ **a.** $y = 1$ **b.** $y = 4$ **c.** $y = 3$

26. $\dfrac{(x^2 - 4)(x - 3)}{x + 1}$ **a.** $x = 1$ **b.** $x = 2$ **c.** $x = -2$

27. $\dfrac{(t - 1)(t + 1)(t - 3)}{\frac{1}{2}t + 2}$ **a.** $t = 2$ **b.** $t = 0$ **c.** $t = -2$

28. $\dfrac{(r - 3)(r)(r + 3)}{(r - 2)(r + 2)}$ **a.** $r = \frac{1}{2}$ **b.** $r = -1$ **c.** $r = 0$

29. Give a numerical example to show that division is not commutative.

30. Give a numerical example to show that division is not associative.

Tell whether the given set is closed (see page 19) under division. If the set is not closed, give an example to show this.

C **31.** $\{1, -1\}$

32. The rational numbers

33. The rational numbers without 0

34. $\left\{ \ldots, \frac{1}{8}, \frac{1}{4}, \frac{1}{2}, 1, 2, 4, 8, \ldots \right\}$

35. The squares of the rational numbers without 0

36. The irrational numbers

Self-Test 2

Vocabulary   field (p. 14)   reciprocal (p. 15)
                unique (p. 14)   multiplicative inverse (p. 15)
                opposite (p. 15)   equivalent expressions (p. 16)
                additive inverse (p. 15)   similar terms (p. 23)

1. Name the property that justifies each step in the simplification below.   **Obj. 1-3, p. 14**

$$\tfrac{1}{2}[(x + 2) + (-x)] = \tfrac{1}{2}[-x + (x + 2)] \qquad \textbf{a.} \ \underline{\ ?\ }$$

$$= \tfrac{1}{2}[(-x + x) + 2] \qquad \textbf{b.} \ \underline{\ ?\ }$$

$$= \tfrac{1}{2}(0 + 2) \qquad \textbf{c.} \ \underline{\ ?\ }$$

$$= \tfrac{1}{2} \cdot 2 \qquad \textbf{d.} \ \underline{\ ?\ }$$

$$= 1 \qquad \textbf{e.} \ \underline{\ ?\ }$$

Simplify.

2. $-6 + 13 + (-5)$ **Obj. 1-4, p. 21**

3. $-11 + (-8) + 9$

4. $7 + (-4) - (-5)$

5. $-3.2 - (4.6 - 5)$

6. $-3(2 - 7) + 4(-6)$ **Obj. 1-5, p. 27**

7. $(-6c)(4d)\left(-\dfrac{1}{8}\right)$

8. Evaluate the expression $2x^2 - 5x - 8$ if $x = -3$.

Simplify.

9. $24 \div (-6) \div (-2)$ **Obj. 1-6, p. 33**

10. $\dfrac{4(-8) - 3(-4)}{(-5)(-2)}$

11. $\dfrac{-6x^2 + 10x - 2}{-2}$

12. Evaluate the expression $\dfrac{y + 9}{(y - 1)(3 - y)}$ if $y = -1$.

Check your answers with those at the back of the book.

Solving Equations and Solving Problems

1-7 Solving Equations in One Variable

Objective To solve certain equations in one variable.

An equation or inequality that contains a variable, such as

$$2t - 1 = 5 \quad \text{and} \quad x + 3 > 0,$$

is called an **open sentence.** Any value of the variable that makes an open sentence a true statement is called a **solution,** or **root,** of the open sentence and is said to **satisfy** it. For example,

3 *is* a solution of $2t - 1 = 5$, because $2 \cdot 3 - 1 = 5$ is true,

but 2 is *not* a solution, because $2 \cdot 2 - 1 = 5$ is false. The set of all solutions of an open sentence that belong to a given domain of the variable is the **solution set** of the sentence *over* that domain. *Unless otherwise stated, open sentences in this book are to be solved over the set of real numbers.*

Study the following two sequences of equations.

$$
\begin{array}{ll}
7x + 12 = 47 & x = 5 \\
7x + 12 - 12 = 47 - 12 & 7x = 7 \cdot 5 \\
7x = 35 & 7x = 35 \\
\dfrac{7x}{7} = \dfrac{35}{7} & 7x + 12 = 35 + 12 \\
x = 5 & 7x + 12 = 47
\end{array}
$$

The properties of real numbers guarantee that if the first statement in either sequence is true for some value of x, then the last statement in the sequence is true for that value of x. Therefore, $7x + 12 = 47$ and $x = 5$ have the same solution set, namely $\{5\}$. Equations having the same solution set over a given domain are called **equivalent equations.** To **solve** an equation you usually change, or *transform,* it into a simple equivalent equation whose solution set is easy to see.

Transformations That Produce Equivalent Equations

1. Simplifying either side of an equation.
2. Adding to (or subtracting from) each side of an equation the same number or the same expression.
3. Multiplying (or dividing) each side of an equation by the same *nonzero* number.

Example 1 Solve $5(2z - 9) = 7z - 60$.

Solution

$5(2z - 9) = 7z - 60$	Copy the equation.
$10z - 45 = 7z - 60$	Simplify the left side.
$10z - 45 + 45 = 7z - 60 + 45$	Add 45 to each side.
$10z = 7z - 15$	
$10z - 7z = 7z - 15 - 7z$	Subtract $7z$ from each side.
$3z = -15$	
$\dfrac{3z}{3} = \dfrac{-15}{3}$	Divide each side by 3
$z = -5$	

$$\text{Check:}\quad 5(2z - 9) = 7z - 60$$
$$5[2(-5) - 9] \stackrel{?}{=} 7(-5) - 60 \qquad\qquad \text{Substitute } -5 \text{ for } z.$$
$$5(-10 - 9) \stackrel{?}{=} -35 - 60$$
$$-95 = -95 \;\; \checkmark$$

$\therefore$ the solution set is $\{-5\}$. **Answer**

Example 1 illustrates how to check your work for mistakes by showing that the solution actually satisfies the *given* equation. A calculator is often helpful in checking solutions.

When solving an equation with one variable, you cannot assume that it has exactly one root. Example 2(a) shows that an equation may have *no* solutions, so that its solution set has no members. The set with no members is called the **empty set,** or the **null set,** and is denoted by $\emptyset$.

Example 2(b) shows an equation that is satisfied by *all* values of the variable. Such an equation is called an **identity.** The solution set of an identity is the set of all real numbers.

Example 2 Solve.

a. $3(2s - 3) = 6(s + 1) - 10$ **b.** $3(2s - 3) = 6(s + 1) - 15$

Solution

a. $3(2s - 3) = 6(s + 1) - 10$
$$6s - 9 = 6s + 6 - 10$$
$$6s - 9 = 6s - 4$$
$$0 = 5 \quad \text{False!}$$

Since the given equation is equivalent to the *false* statement $0 = 5$, it has *no* root.

$\therefore$ the solution set is $\emptyset$. **Answer**

b. $3(2s - 3) = 6(s + 1) - 15$
$$6s - 9 = 6s + 6 - 15$$
$$6s - 9 = 6s - 9$$

Since the given equation is equivalent to $6s - 9 = 6s - 9$, which is *true* for all values of s, it is an identity.

$\therefore$ the solution set is the set of all real numbers. **Answer**

A **formula** is an equation that states a relationship between two or more variables. The variables usually represent physical or geometric quantities. For example, the formula $h = -16t^2 + vt$ gives the height h (in feet) of a launched object t seconds after firing with initial velocity v (in ft/s). Given values for all but one of the variables in a formula, you can find the value of the remaining variable.

Example 3 A model rocket launched with initial velocity v reaches a height of 40 ft after 2.5 s. Find v.

Solution Substitute the given values of the variables.

$$h = -16t^2 + vt$$
$$40 = -16(2.5)^2 + v(2.5)$$
$$40 = -100 + 2.5v$$
$$140 = 2.5v$$
$$v = 56 \text{ (ft/s)} \quad \textbf{\textit{Answer}}$$

When you solve a formula or equation for a certain variable, you can think of all the other variables as **constants,** that is, as fixed numbers. Then solve by the usual methods.

Example 4 The volume V of a pyramid with height h and square base with sides s is given by the formula $V = \frac{1}{3}s^2h$. Solve this formula for h.

Solution You need to express h in terms of the other variables. That is, you need to get h alone on one side of the equation.

$$V = \frac{1}{3}s^2h$$
$$3(V) = 3\left(\frac{1}{3}s^2h\right)$$
$$3V = s^2h$$
$$\frac{3V}{s^2} = \frac{s^2h}{s^2} \quad \text{(Assume } s \neq 0.)$$
$$h = \frac{3V}{s^2} \quad \textbf{\textit{Answer}}$$

Oral Exercises

Explain how you transform the first equation to obtain the second equation.

a. $2t + 5 = 11$; $2t = 6$ **b.** $3x - 2 - x = 16$; $2x - 2 = 16$

a. Subtract 5 from both sides.   **b.** Simplify the left side of the equation.

1. $6z = 42$; $z = 7$ **2.** $4x - 3 = 7$; $4x = 10$

3. $r = 2(r - 1) + 5$; $r = 2r + 3$ **4.** $\frac{1}{2}(t - 5) = 3$; $t - 5 = 6$

5. $2x - 7 = x + 1$; $2x = x + 8$ **6.** $5y - 1 = 3y$; $2y - 1 = 0$

7. $\frac{t - 5}{3} = 2t$; $t - 5 = 6t$ **8.** $3x - 9 = 12$; $x - 3 = 4$

Each formula below has been transformed into an equivalent one. Name the transformation used at each step.

9. Area of a triangle:

$$A = \frac{1}{2}bh$$

   a. $2A = bh$ **a.** __?__

   b. $\frac{2A}{h} = b$ **b.** __?__

10. Kinetic energy:

$$K = \frac{1}{2}mv^2$$

   a. $2K = mv^2$ **a.** __?__

   b. $\frac{2K}{v^2} = m$ **b.** __?__

11. Explain how you would solve the equation $A = \dfrac{x + y}{2}$ for y.

12. a. Solve $2x = 6$.   **b.** Solve $0 \cdot 2x = 0 \cdot 6$.

   c. If you multiply each side of an equation by 0, do you get an equivalent equation?

   d. Explain why it is not possible to divide each side of an equation by 0.

Written Exercises

Solve. Check your work when there is a single solution.

A

1. $3x - 4 = 5$ **2.** $4z + 11 = 3$

3. $\frac{2}{3}t - 8 = 0$ **4.** $15 - \frac{1}{8}d = -1$

5. $5r = 18 + 2r$ **6.** $24 - 2y = 6y$

7. $3(t - 1) = -(t - 5)$ **8.** $2(x - 3) = x + 3$

9. $2k - 1 = k - 5 + 3k$

10. $3(2z - 1) = 6z + 5$

11. $-(5 - x) = x + 3$

12. $\frac{1}{3}(s - 2) = s + 4$

13. $1.5(u + 2) = 7.5$

14. $1.6(2t - 1) = 14.4$

15. $0.2(x - 5) = x + 5$

16. $0.3(2r - 3) = 0.2r + 0.9$

17. $3(x - 2) - x = 2(2x + 1)$

18. $\frac{2}{5}(x - 2) = x + 4$

19. $2z - (1 - z) = 11 - z$

20. $3(1 - t) + 5 = 3(1 + t) - 7$

21. $2(5t - 3) - t = 3(3t - 2)$

22. $3(5z - 1) + 5(3z + 2) = 7$

23. $\dfrac{6x - 2(x - 4)}{3} = 8$

24. $\dfrac{3y - 2(y - 1)}{6} = -1$

Tell whether each number at the right of the given equation is a solution of the equation.

25. $x(x - 3)(x + 2) = 0$; $-2, -3$

26. $z(z + 1)(z - 2) = 0$; $0, -1$

27. $z^3 - 4z^2 + z + 6 = 0$; $2, 3$

28. $u^3 - 7u - 6 = 0$; $1, -2$

29. $\dfrac{x + 12}{x - 4} = x - 3$; $-2, 0$

30. $\dfrac{2y}{2y - 1} = \dfrac{y + 2}{y + 1}$; $0, 2$

Solve the equation for the given variable.

31. $2x - 5y = 10$ for x

32. $A = \frac{1}{2}bh$ for h

33. $I = prt$ for p

34. $C = 2\pi r$ for r

35. $y = mx + b$ for x

36. $ax + by = c$ for y

37. $P = 2l + 2w$ for w

38. $P = 2(l + w)$ for l

B   **39.** $a(x - b) = c + ab$ for x

40. $5cy - d = 4d - cy$ for y

41. $S = -\frac{1}{2}gt^2 + vt$ for v

42. $C = \frac{5}{9}(F - 32)$ for F

In each formula, substitute the given values of the variables. Then find the value of the remaining variable, which is printed in red.

43. Volume of a cylinder: $V = \pi r^2 h$; $V = 128$, $r = 8$

44. Volume of a cone: $V = \frac{1}{3}\pi r^2 h$; $V = 48$, $r = 4$

45. Amount at simple interest: $A = P(1 + rt)$; $A = 168$, $P = 150$, $r = 0.08$

46. Distance an object falls: $d = \dfrac{v^2}{2g}$; $d = 1000$, $v = 140$

47. Area of a trapezoid: $A = \dfrac{h}{2}(b_1 + b_2)$; $A = 100$, $h = 5$, $b_2 = 12$

48. Total area of a cylinder: $A = 2\pi r(r + h)$; $A = 80\pi$, $r = 5$

49. Total area of a cone: $A = \pi r(s + r)$; $A = 100\pi$, $r = 5$

50. Sum of a geometric series: $S = \dfrac{a(1 - r^n)}{1 - r}$; $S = 80$, $r = 3$, $n = 4$

C 51–53. Solve the formulas in Exercises 48–50 for the variables printed in red. (Ignore the given values of the variables.)

Mixed Review Exercises

Simplify.

1. $7 - (3 - 4)$

2. $-2(-5 + 8)$

3. $(4 - 9)^2$

4. $|-6 + 2|$

5. $(-8)\left(\dfrac{1}{2}\right)(-3)(-1)$

6. $\dfrac{-3 + 11}{-1 - 3}$

7. $3x - 2(x + 4)$

8. $(-4a)(5b) + (-3a)(-8b)$

9. $(-x)^2(-y)^3$

10. $2(c - 3d) + 5(2d - c)$

11. $-3 + \dfrac{1}{2}(6 - 10x) + 5x$

12. $\dfrac{1 - m}{-1}$

Historical Note / Word Problems

Consider the following problems.

1. A number added to $\dfrac{1}{7}$ of the number is 19. What is the number?

2. A dog chasing a rabbit, which has a head start of 150 feet, jumps 9 feet every time the rabbit jumps 7 feet. In how many jumps will the dog catch up with the rabbit?

3. If B gives A 7 denars, then A will have 5 times as much money as B. If instead A gives B 5 denars, then B will have 7 times as much as A. How much money has each?

    Students of mathematics have been solving problems similar to these for thousands of years. Problem 1 above appears in the Rhind Papyrus, written about 1650 B.C.; Problem 2 appears in a Latin problem collection compiled about A.D. 775; and Problem 3 is from a book by the Italian mathematician Fibonacci dating from A.D. 1202.

    The mathematical symbols we use to represent quantities and relationships in word problems have their history too. For example, the plus sign (+) was used for addition as early as 1514; the equals sign (=) we know today originated with Robert Recorde in 1557; and the dot (·) was adopted as a symbol for multiplication by Leibniz in the seventeenth century. With these symbols you can write Problem 1 as

$$\text{Number} + \frac{1}{7} \cdot \text{Number} = 19.$$

1-8 Words into Symbols

Objective To translate word phrases into algebraic expressions and word sentences into equations.

To use algebra as a problem-solving tool, you often must translate word phrases into algebraic expressions.

Example 1 Represent each word phrase by an algebraic expression. Use n for the variable.

    a. A number decreased by 2

    b. Five more than three times a number

    c. The difference between a number and its square

    d. The sum of twice a number and 6

    e. Twice the sum of a number and 6

Solution **a.** $n - 2$ **b.** $3n + 5$ **c.** $n - n^2$

            d. $2n + 6$ **e.** $2(n + 6)$

Notice that the answer to part (c) of Example 1 is $n - n^2$ and *not* $n^2 - n$. In this book, when we say "the difference between x and y," we mean $x - y$. Similarly, "the quotient of x and y" means $\frac{x}{y}$, or $x \div y$.

Example 2 Ann is biking at r mi/h. Use the variable r to represent each word phrase by an algebraic expression.

    a. Ann's speed if she bikes 5 mi/h slower

    b. Ann's speed if she bikes 3 mi/h faster

    c. The average of Ann's and Juan's speeds if Juan bikes at 10 mi/h

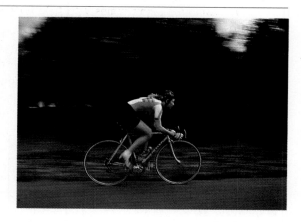

Solution **a.** $r - 5$ **b.** $r + 3$ **c.** $\dfrac{r + 10}{2}$

In certain applications, like the one in Example 3, facts and formulas from geometry are needed.

Example 3 The base of an isosceles triangle has length b cm and each base angle measures $a°$.

  a. Find the measure of the vertex angle in terms of a.

  b. If the perimeter is 120 cm, find the length of one of the legs in terms of b.

Solution   **a.** The sum of the measures of the angles of a triangle is 180°.

  ∴ the measure of the vertex angle is $180° - (a + a)°$, or $(180 - 2a)°$.   **Answer**

  b. The legs of an isosceles triangle are equal in length. Call this length s. Then,

$$s + s + b = 120$$
$$2s + b = 120$$
$$s = \frac{1}{2}(120 - b)$$

  ∴ the length of each leg is $\frac{1}{2}(120 - b)$ cm.   **Answer**

Formulas from science and technology are used in many applications. For example, to describe **uniform motion,** that is, motion at a constant speed, you use the formula

$$\text{distance} = \text{rate} \times \text{time, or } d = rt.$$

Example 4 A helicopter left Midcity airport at noon and flew east at 110 km/h. One hour later a light plane left Midcity flying west at 320 km/h. How far apart were the aircraft x hours after noon? Express your answer in terms of x.

Solution   The table and the diagram show the given information. Note that the time for the plane is represented by $x - 1$ because the plane left one hour *later* than the helicopter.

	Rate	× Time	= Distance
	r (km/h)	t (h)	d (km)
Helicopter	110	x	$110x$
Plane	320	$x - 1$	$320(x - 1)$

total distance $= 320(x - 1) + 110x = 430x - 320$

  ∴ the aircraft are $(430x - 320)$ km apart.   **Answer**

Use the chart below to review the meaning of consecutive numbers.

Consecutive Numbers

Integers	$\ldots, -3, -2, -1, 0, 1, 2, 3, \ldots$	$n - 1, n, n + 1, n + 2$ are four consecutive integers if n is an integer.
Even Integers (multiples of 2)	$\ldots, -6, -4, -2, 0, 2, 4, 6, \ldots$	$n - 2, n, n + 2, n + 4$ are four consecutive even integers if n is even
Odd Integers	$\ldots, -5, -3, -1, 1, 3, 5, \ldots$	and four consecutive odd integers if n is odd.

Example 5 What is the sum of five consecutive odd integers if:

a. the middle one is m?   **b.** the next to largest is x?

Solution **a.** $(m - 4) + (m - 2) + m + (m + 2) + (m + 4)$, or $5m$

b. $(x - 6) + (x - 4) + (x - 2) + x + (x + 2)$, or $5x - 10$

The next example shows how you can choose a variable to represent an unknown number and then write an equation to describe a given situation.

Example 6 A state legislature has 45 people. The number of men is six less than twice the number of women.

a. Choose a variable to represent the number of women.
b. Write an expression for the number of men in terms of that variable.
c. Write an equation that describes the situation.

Solution **a.** Let w = the number of women.

b. Then $2w - 6$ = the number of men.

c. The sum of the number of women and the number of men is 45.
$\therefore w + (2w - 6) = 45$, or $3w - 6 = 45$ *Answer*

Oral Exercises

Represent each word phrase by an algebraic expression.

1. Five more than a number

2. Ten less than a number

3. One less than twice a number

4. The difference between a number and six

5. Seven more than half a number

6. The sum of a number and its reciprocal

7. One more than the square of a number

8. The square of one more than a number

9. The reciprocal of two more than a number

10. Two more than the reciprocal of a number

In Exercises 11–20, express each answer in simplest form in terms of the given variable.

11. What is the area of a rectangular desk top that is w cm wide and twice as long as it is wide?

12. How long is a rug that is w meters wide and covers an area of 20 m²?

13. How far will a car traveling at 90 km/h go in:

   a. h hours?   **b.** $h + 2$ hours?

14. How long will it take a car to go 200 km traveling at:

   a. v km/h?   **b.** $v - 5$ km/h?

15. The vertex angle of an isosceles triangle measures $v°$. What is the measure of each base angle?

16. What is the sum of the measures of a supplement and a complement of an angle with measure $x°$?

17. What is the area of an isosceles right triangle if one of its legs is r cm long?

18. Yoneko is y inches tall. Her height and the heights of her two taller sisters are consecutive odd numbers. What is the average of their heights?

19. What is the cost *in cents* of one baseball if baseballs are priced at d dollars per dozen?

20. What is the cost *in dollars* of 20 pounds of potatoes priced at c cents per pound?

Written Exercises

Express each answer in simplest form in terms of the given variable.

A 1. A rectangular garden that is w ft wide is enclosed by 120 ft of fencing. How long is the garden?

2. The perimeter of an isosceles triangle is 300 cm, and its base is b cm long. How long is each leg?

3. In a basketball game, one team's score is two points less than half the other team's score, which is x. What is the difference in the scores?

4. Dan has y cassette tapes. If he had 15 more tapes, he would have half the number his brother has. How many tapes does his brother have?

5. The length and the width of a rectangle are consecutive even integers, and the length is l cm. Find **(a)** the area and **(b)** the perimeter of the rectangle.

6. The length, width, and height of a rectangular box are consecutive integers, and the largest dimension is k cm. Find the volume V of the box. (*Hint:* $V = lwh$.)

7. Two jets leave an airport at noon, one flying north at r mi/h, and the other flying south at twice that speed. After 3 h, how far apart are the planes?

8. A bus traveled for 2 h at r mi/h, then decreased the speed by 10 mi/h and traveled for 1 more hour. How far did the bus go?

9. One angle of a quadrilateral has measure $a°$. Find the average of the measures of the other three angles. (*Hint:* The sum of the measures of the angles of a quadrilateral is $360°$.)

10. An angle has measure $x°$. Find the average of the measures of a complement and a supplement of the angle.

In Exercises 11 and 12, use the fact that the volume V of a pyramid is given by $V = \frac{1}{3}Bh$, where B is the area of the base and h is the height.

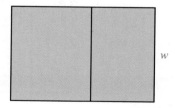

11. A pyramid has height x cm and a *square base* whose edges are 3 cm less than twice the height. Find the volume.

12. A pyramid has a *rectangular base*. Find the volume if the length and width of the base and the height are three consecutive odd integers and x is the largest integer.

13. The Drama Club sold t students' tickets at $1.50 each and 100 fewer adults' tickets at $2.50 each. How much money did the club collect?

14. Jorge bought s 40-cent stamps and three times as many 25-cent stamps. How many dollars did he spend?

15. Jessica's bank contains 18 quarters and dimes, of which q are quarters. Find the total value of the coins in dollars.

16. The length of a rectangular field is 45 ft greater than its width, w ft. How much fencing is needed to enclose the field and divide it into two parts as shown?

In Exercises 17–29, choose a variable to represent an unknown number, and then write an equation to describe the given situation.

17. A quadrilateral has perimeter 60 cm, and the lengths of its sides (in centimeters) are consecutive odd numbers.

18. The numerator of a fraction is 6 less than its denominator, and the value of the fraction is $\frac{3}{5}$.

B **19.** Adam purchased a shirt at regular price. Later, when the shirts were on sale, he purchased two more at $2 off the regular price. He spent a total of $41 for the three shirts.

20. Lupe swims two fewer laps than Mary. If both added seven laps to their daily swims, the sum of their laps would be three times as many as Mary now swims.

21. Max has twice as much money as Katy, who has $12 more than Greg. All three together have $124.

22. Paula's purse contains twice as many dimes as quarters and three fewer nickels than dimes. The total of these coins is $3.15.

23. A car and a truck left Elton at noon and traveled in opposite directions. The truck's speed is two thirds of the car's speed, and the vehicles are 140 mi apart at 2 P.M.

24. Tom and Tina set out on their bikes at noon and travel toward each other, meeting at 2:30 P.M. Tina's speed is 4 mi/h faster than Tom's speed, and their starting points are 50 mi apart.

25. In quadrilateral $ABCD$, the measure of $\angle A$ exceeds the measure of $\angle B$ by 20°. Also, the measure of $\angle D$ is twice the measure of $\angle B$ and half the measure of $\angle C$.

26. In an equilateral triangle, the length of one side is 20 cm more than one third the length of another.

27. A grocer mixed cashews and almonds to produce 20 kg of mixed nuts worth $7.80/kg. Cashews are worth $7/kg, and almonds are worth $9/kg.

28. When the Kims went to the ball game, they bought two adults' tickets and three children's tickets. A child's ticket cost $1.50 less than an adult's, and the family's average price per ticket was $3.35.

C **29.** Kevin drove 320 mi to a mountain resort. His return trip took 20 min longer because his speed returning was 4 mi/h slower than his speed going.

Challenge

Every positive integer can be written as a sum of cubes of positive integers. For example, 10 can be written as a sum of three cubes, and 44 can be written as a sum of four cubes: $10 = 2^3 + 1^3 + 1^3$ and $44 = 3^3 + 2^3 + 2^3 + 1^3$. Find a positive integer that can be written as a sum of nine cubes, but that cannot be written as the sum of any smaller number of cubes.

1-9 Problem Solving with Equations

Objective   To solve word problems by using an equation in one variable.

"Word problems" describe relationships among numbers. If you can translate the relationship described in a problem into an equation, then you can solve the problem by solving the equation. Example 1 illustrates a step-by-step method for solving word problems.

Example 1   Tickets to the Civic Center Auditorium for a rock concert were $19 for seats near the front and $14 for rear seats. There were 525 more rear seats sold than front seats, and sales for all tickets totaled $31,770. How many of each kind of ticket were sold?

Solution

Step 1 The problem asks for the number of each type of ticket sold.

Step 2 Let x = the number of front-seat tickets sold.
Then $x + 525$ = the number of rear-seat tickets sold.

	Price × Number =		Sales
Front	19	x	$19x$
Rear	14	$x + 525$	$14(x + 525)$
	Total Ticket Sales		31,770

Step 3 Front-seat Sales + Rear-seat Sales = Total Sales
$$19x \quad + \quad 14(x + 525) \quad = \quad 31{,}700$$

Step 4
$$19x + 14x + 7350 = 31{,}770$$
$$33x + 7350 = 31{,}770$$
$$33x = 24{,}420$$
$$x = 740 \quad \text{(Front)}$$
$$x + 525 = 1265 \quad \text{(Rear)}$$

Step 5 Is the number of rear-seat tickets 525 more than the number of front-seat tickets?
$$1265 \stackrel{?}{=} 740 + 525$$
$$1265 = 1265 \quad \checkmark$$

Is the sum of the ticket sales of front seats and rear seats $31,770?
$$740(19) + 1265(14) \stackrel{?}{=} 31{,}770$$
$$14{,}060 + 17{,}710 \stackrel{?}{=} 31{,}770$$
$$31{,}770 = 31{,}770 \quad \checkmark$$

∴ 740 front-seat tickets and 1265 rear-seat tickets were sold.   ***Answer***

The step-by-step method of Example 1 is part of this general strategy.

Plan for Solving a Word Problem

Step 1 Read the problem carefully a few times. Decide what numbers are asked for and what information is given. Making a sketch may be helpful.

Step 2 Choose a variable and use it with the given facts to represent the number(s) described in the problem. Labeling your sketch or arranging the given information in a chart may help.

Step 3 Reread the problem. Then write an equation that represents relationships among the numbers in the problem.

Step 4 Solve the equation and find the required numbers.

Step 5 Check your results with the original statement of the problem. Give the answer.

Example 2 Melissa, a city planner, had two pieces of wire of equal length. She shaped one piece into a square to represent a new building site and the other into an isosceles triangle to represent a nearby park. The base of the isosceles triangle is 4 cm shorter than a side of the square, and each leg is 9 cm longer than a side of the square. How long was each piece of wire?

Solution

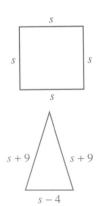

Step 1 The problem asks for the length of each piece of wire. This length is the perimeter of the square and the perimeter of the triangle. Sketch each figure.

Step 2 Let s = the length of a side of the square in centimeters. (Note that the variable does not have to represent a number that is asked for.) Then the base of the triangle has length $s - 4$, and each leg has length $s + 9$.

Perimeter of square = $4s$
Perimeter of triangle = $(s - 4) + (s + 9) + (s + 9)$

Step 3 The pieces of wire were equally long.
Perimeter of square = Perimeter of triangle
$$4s = (s - 4) + (s + 9) + (s + 9)$$

Step 4
$$4s = 3s + 14$$
$$s = 14$$
Perimeter of square = $4s = 56$ (cm)

Step 5 ∴ each piece of wire was 56 cm long. (The check is left for you.) **Answer**

Example 3 Find the length of each piece of wire in Example 2 if the base of the isosceles triangle is 6 cm shorter than a side of the square, and each leg is 3 cm longer than a side of the square.

Solution Modify Step 2 of Example 2 to obtain the following equation in Step 3:

Step 3 Perimeter of square = Perimeter of triangle
$$4s = (s - 6) + (s + 3) + (s + 3)$$

Step 4
$$4s = (s - 6) + (s + 3) + (s + 3)$$
$$4s = 3s$$
$$s = 0$$

Step 5 ∴ the side of the square is 0.

But there is *no* square with sides 0 units long. Therefore, the problem has no solution. ***Answer***

As Example 3 shows, *not every word problem has a solution.* Some problems without solutions have contradictory facts like the one in Example 3; others do not give enough facts to provide a solution. Of course, *sometimes a problem will give you more information than you need to find the solution.*

Example 4 At noon a cargo plane leaves McHare Airport and heads east at 180 mi/h. Its destination is Jamesville, 500 mi away. At 1:00 P.M. a jet takes off from McHare and flies east after the cargo plane at 450 mi/h. At what time will the jet overtake the cargo plane?

Solution

Step 1 The problem asks for the time at which the jet overtakes the cargo plane.

Step 2 Let t = the number of hours *after noon* when the jet overtakes the cargo plane.

Make a chart and a sketch.

(Solution continues on the next page.)

	Rate × Time = Distance		
	(mi/h)	(h)	(mi)
Cargo plane	180	t	$180t$
Jet	450	$t - 1$	$450(t - 1)$

McHare airport

Step 3 The jet will overtake the cargo plane when:

Cargo plane's distance = Jet's distance

$$180t = 450(t - 1)$$

Step 4

$$180t = 450t - 450$$
$$450 = 270t$$
$$t = 1\frac{2}{3} \quad \text{(hours after noon)}$$

Step 5 ∴ the jet overtakes the cargo plane $1\frac{2}{3}$ h after noon, or 1:40 P.M.

The check is left for you. ***Answer***

The information concerning the cargo plane's destination 500 mi away was not needed in the solution of this example.

Problems

A
1. Amy has $8 less than Maria. Together they have $30. How much money does each girl have?

2. Jim's weekly pay is two thirds of Alicia's. Together they earn $600 per week. What is each person's weekly pay?

3. A music dealer ran a sale of records and tapes. Records were reduced to $7 each and tapes to $7.50 each. The dealer sold 60 more records than tapes for a total sale of $2160. How many records did the dealer sell?

4. At the homecoming football game, the Senior Class officers sold slices of pizza for $.75 each and hamburgers for $1.35 each. They sold 40 more slices of pizza than hamburgers, and sales totaled $292.50. How many slices of pizza did they sell?

5. The perimeter of a certain basketball court is 266 ft, and its length is 35 ft more than its width. Find the dimensions of the court.

6. If one side of a square is increased by 8 cm and an adjacent side decreased by 2 cm, a rectangle is formed whose perimeter is 40 cm. Find the length of a side of the square.

7. The measure of a supplement of an angle is 12° greater than three times the measure of a complement. Find the measure of the angle.

8. The degree measures of the angles of a pentagon are consecutive even integers. Find the measure of the largest angle. (*Hint:* The sum of the measures of the angles of a pentagon is 540°.)

9. At 10:30 A.M. two planes leave Houston, one flying east at 560 km/h and the other flying west at 640 km/h. At what time will they be 2100 km apart?

10. In a walkathon to raise money for a charity, Elisa walked a certain distance at 5 mi/h and then jogged twice that distance at 8 mi/h. Her total time walking and jogging was 2 h and 15 min. How many miles long was the walkathon? $\left(\text{Hint: time} = \frac{\text{distance}}{\text{rate}} . \right)$

11. A jar contains 40 coins consisting of dimes and quarters and having a total value of $4.90. How many of each kind of coin are there?

12. Larry has an annual return of $213 from $3000 invested at simple interest, some at 5% and the rest at 8%. How much is invested at each rate? (*Hint:* Interest earned = Amount invested × Rate of interest.)

In Problems 13–18, find the solution if possible. If there is not enough information to solve the problem or if it has no solution, say so. If extra information is given, identify it.

13. Two planes leave Wichita at noon. One plane flies east 30 mi/h faster than the other plane, which is flying west. At what time will they be 1200 mi apart?

14. A collection of 30 coins worth $5.50 consists of nickels, dimes, and quarters. There are twice as many dimes as nickels. How many quarters are there?

15. John drove part of a 260 km trip at 80 km/h and the rest at 100 km/h. Find the distance he traveled at 80 km/h if his total driving time was 2 h and 30 min.

16. A triangle has perimeter 29 cm. The sides have lengths, in centimeters, that are consecutive odd integers. What is the length of the longest side?

B 17. Jan invested $1200 at a certain simple interest rate and $2200 at a rate 3% higher. Her annual earnings were $253. Find the two interest rates if she earned $121 more on the larger investment than on the smaller.

18. A school cafeteria sells milk at 25 cents per carton and salads at 45 cents each. One week the total sales for these items were $132.50. How many salads were sold that week?

Solve.

19. At noon a train leaves Bridgton heading east at 90 mi/h to Cogsville, 450 mi away. At 12:15 P.M. a train leaves Cogsville heading west to Bridgton at 100 mi/h. At what time will they pass each other?

20. In a bicycle race, Lionel gives Robert a 500 m advantage. Also, Lionel agrees to start 15 min after Robert. If Lionel bikes at 17 km/h and Robert at 14 km/h, how long will it take Lionel after he starts biking to overtake Robert?

21. A pollution control device reduces the rate of emission of an air pollutant in a car's exhaust by 56 ppm (parts per million) per hour. When the device is installed, it will take the car 10 h to emit the same amount of pollutant that it formerly did in 3 h. What was the original rate of emission of the pollutant in the car's exhaust?

22. A grocer wants to mix peanuts and cashews to produce 20 lb of mixed nuts worth $6.20/lb. How many pounds of each kind of nut should she use if peanuts cost $4.80/lb and cashews cost $8/lb?

C **23.** In a two-candidate election 1401 votes were cast. If 30 voters had switched their votes from the winner to the loser, the loser would have won by 5 votes. How many votes did each candidate actually receive?

Mixed Review Exercises

Solve.

1. $3x + 2 = -4$ **2.** $4(2 - y) = y - 7$ **3.** $1 - 5t = t + 7$

Evaluate each expression if $c = -4$ and $d = 6$.

4. $c^2 - d^2$ **5.** $(c - d)^2$ **6.** $|2c - 3d|$

7. $\dfrac{c + d}{-2}$ **8.** $\dfrac{cd}{8}$ **9.** $\dfrac{d - c}{5}$

Self-Test 3

Vocabulary   open sentence (p. 37)   equivalent equations (p. 37)
                   solution (p. 37)   empty set (p. 38)
                   root (p. 37)   identity (p. 38)
                   solve (p. 37)   formula (p. 39)
                   solution set (p. 37)   constant (p. 39)

Solve.

1. $3x - 8 = 7$ **2.** $4(1 - x) = 2(x - 4)$ **Obj. 1-7, p. 37**

3. Solve the formula $m = \dfrac{1}{2}(a + b)$ for b.

4. Using n for the variable, translate this word phrase into an algebraic expression: Twice the sum of a number and its square.   **Obj. 1-8, p. 43**

5. Express your answer in terms of x: What is the perimeter of a rectangle that is x cm wide and 5 cm longer than it is wide?

6. Two cars, heading toward each other on a divided highway, are 250 mi apart. If one car travels 45 mi/h and the other 10 mi/h faster, in how many hours will the cars pass each other?   **Obj. 1-9, p. 49**

Chapter Summary

1. The set of real numbers has several familiar subsets, which are shown in the chart on page 1.

2. Each point on a number line can be paired with exactly one real number, which is the coordinate of the point. The point is called the graph of the number.

3. The words and symbols used in algebra have specific mathematical definitions. (See the charts on pages 6, 7, and 9.)

4. A numerical expression can be simplified by following the rules for order of operations (page 8).

5. An algebraic expression can be simplified by using the field properties of real numbers (page 15) along with the definitions of addition, subtraction, multiplication, and division.

6. An algebraic expression can be evaluated by replacing each variable by a given value and simplifying the result.

7. An equation can be solved by applying the transformations on page 37. The solution set of an equation is the set of values of the variable that make the equation a true statement.

8. Word problems can be solved algebraically after translating the given information into an equation. Refer to the five-step problem-solving method presented on page 50.

Chapter Review

Write the letter of the correct answer.

1. On a number line, the coordinate of point A is -3, and the coordinate of point B is 7. Find the coordinate of the point halfway between A and B.   1-1
   - **a.** 4
   - **b.** 3
   - **c.** 2
   - **d.** 1

2. Simplify $|-5| - |5|$.
   - **a.** -10
   - **b.** 0
   - **c.** 10
   - **d.** -5

3. Simplify $\dfrac{2 \cdot 5^2 + 1}{2 \cdot 3^2 - 1}$.   1-2
   - **a.** 2
   - **b.** $\dfrac{13}{4}$
   - **c.** $\dfrac{101}{35}$
   - **d.** 3

4. Evaluate $x(y - 1)^2$ if $x = 4$ and $y = 3$.
   - **a.** 16
   - **b.** 12
   - **c.** 64
   - **d.** 32

5. Simplify $\frac{1}{2}(2a + 1)$.

    a. $a + \frac{1}{2}$ **b.** $\frac{1}{2}a + 1$ **c.** $\frac{1}{2}a + \frac{1}{2}$ **d.** $a + 1$

1-3

6. Name the property that is illustrated by this statement:
$$2a + (3b + c) = (3b + c) + 2a.$$

    a. Symmetric property of equality   **b.** Associative property of addition

    c. Addition property of equality   **d.** Commutative property of addition

7. Simplify $(-14 + 8) - (4 - 11)$.

    a. -13 **b.** 13 **c.** 1 **d.** -1

1-4

8. Simplify $8m - 6n - m + 2n$.

    a. $3mn$ **b.** $7m - 4n$ **c.** $4n - 7m$ **d.** $8 - 4n$

9. Simplify $3(2x - y) - 4(x - 2y)$.

    a. $2x - 11y$ **b.** $10x - 11y$ **c.** $2x - 5y$ **d.** $2x + 5y$

1-5

10. Simplify $(-a)^2 + a^2$.

    a. $-2a^2$ **b.** $2a$ **c.** 0 **d.** $2a^2$

11. Simplify $6(3 - 7) \div (-2)^3 \div (-1)$.

    a. 3 **b.** 4 **c.** -3 **d.** -4

1-6

12. Simplify $\frac{24 - 20x}{-4}$.

    a. $5x - 6$ **b.** $5x + 6$ **c.** $-5x - 6$ **d.** $-5x + 6$

13. Solve $2(3 - x) = 3x + 1$.

    a. $\frac{5}{4}$ **b.** 1 **c.** -1 **d.** 5

1-7

14. Solve $A = 2\pi r(r + h)$ for h.

    a. $\frac{A - r}{2\pi r}$ **b.** $\frac{2\pi r}{A} - r$ **c.** $\frac{2\pi r}{A - r}$ **d.** $\frac{A}{2\pi r} - r$

15. The hourly wages of three workers are consecutive even integers. If the highest-paid worker's hourly wage is x, what is the sum of all three workers' hourly wages?

    a. $3x$ **b.** $3x + 6$ **c.** $3x - 6$ **d.** $3x - 3$

1-8

16. The perimeter of a rectangle is 24 cm. If the length is y cm, what is the width?

    a. $12 - y$ **b.** $24 - y$ **c.** $24 - \frac{y}{2}$ **d.** $12 - \frac{y}{2}$

17. The school book store sold 8 more pencils than pens one day. The cost of a pencil is $.05, and the cost of a pen is $.20. If the day's sales of pens and pencils totaled $8.90, how many pencils were sold?

    a. 42 **b.** 34 **c.** 26 **d.** 50

1-9

18. A woman drove part of a 185 mi trip at 50 mi/h and the rest at 55 mi/h. Find the distance she traveled at 50 mi/h if her total driving time was 3 h and 30 min.

    a. 110 mi   **b.** 100 mi   **c.** 75 mi   **d.** 150 mi

Chapter Test

1. On a number line, point A has coordinate -4, and point B has coordinate 5. Find the coordinate of the point two thirds of the way from A to B.

1-1

2. List 0.6, -0.8, 1.4, -1, and -0.2 in order from least to greatest.

3. Simplify $5 \cdot 3^2 - 2(4 + 6)$.

1-2

4. Evaluate $\dfrac{(x + y)^2}{x^2 + y^2}$ if $x = 2$ and $y = 4$.

5. Name the property that justifies each lettered step.

1-3

$$\frac{1}{2}(2x + 2) = \frac{1}{2}(2x) + \frac{1}{2} \cdot 2 \qquad \textbf{a.} \ \underline{\ ?\ }$$
$$= \left(\frac{1}{2} \cdot 2\right)x + \frac{1}{2} \cdot 2 \qquad \textbf{b.} \ \underline{\ ?\ }$$
$$= 1x + 1 \qquad \textbf{c.} \ \underline{\ ?\ }$$
$$= x + 1 \qquad \textbf{d.} \ \underline{\ ?\ }$$

Simplify.

6. $|8 - 13| - |-7 - (-3) + 1|$

7. $8m - 4 - 9m + 7$

1-4

8. $-(-1)^3(-2)^2$

9. $5(3x - 2y) - 4(6y - x)$

1-5

Evaluate if $x = -3$.

10. $x(4 - x)(x + 1)$

11. $7 - 2x - x^2$

Simplify.

12. $\left|\dfrac{(-6)^2 - 3(-4)}{-8 + 2}\right|$

13. $\dfrac{9 - 12x^2}{-3}$

1-6

Solve.

14. $\dfrac{3}{2}x - 7 = 2x + 3$

15. $6(5x - 4) = 7(4x + 5) - 19$

1-7

Express your answer in terms of the given variable.

16. A car's gas tank held x gallons of gas before a trip. The trip consumed three quarters of the gas in the tank, but 10 gallons of gas were added after the trip. How much gas is now in the tank?

1-8

Solve.

17. A man invested $6000, part of it at 5% simple interest and the rest at 7% simple interest. If his annual interest income is $372, how much did he invest at each rate?

1-9

Range of Temperatures (°C)

950 1000 1050 1100 1150 1200 t

$1000 \leq t \leq 1150$

Earthenware pottery is a mixture of clay, flint, and feldspar. Depending on how these elements are mixed, pottery must be heated to a temperature between 1000°C and 1150°C.

Working with Inequalities

2-1 Solving Inequalities in One Variable

Objective   To solve simple inequalities in one variable.

The inequality

$$x > -2$$

is satisfied by every real number greater than -2. The graph of this inequality is shown in red on the number line below. Note the use of the open circle to show that -2 is not a solution.

To find solutions of more complicated inequalities like

$$5x + 17 < 2$$

and $$5(3 - t) < 7 - t$$

you use methods similar to those used to solve equations. These methods are based on the properties of order for real numbers stated below. Recall that a real number c is called *positive* if $c > 0$ and *negative* if $c < 0$.

Properties of Order

Let a, b, and c be any real numbers.

Comparison Property

Exactly one of the following statements is true:

$$a < b, \quad a = b, \quad \text{or } a > b.$$

Transitive Property

If $a < b$ and $b < c$, then $a < c$.

Addition Property

If $a < b$, then $a + c < b + c$.

Multiplication Property

1. If $a < b$ and c is *positive*, then $ac < bc$.
2. If $a < b$ and c is *negative*, then $ac > bc$.

Since subtraction is defined in terms of addition, the addition property of order applies to subtraction. Similarly, the multiplication properties apply to division. Furthermore, since the statement $b > a$ has the same meaning as $a < b$, the properties of order hold true if $<$ and $>$ are interchanged throughout.

When you multiply or divide both sides of an inequality by a *negative* number, you must *reverse* the direction of the inequality. For example:

$$5 < 8, \quad \text{but} \quad (-2)5 > (-2)8 \quad \text{(that is, } -10 > -16)$$

$$1 > -4, \quad \text{but} \quad \frac{1}{-2} < \frac{-4}{-2} \quad \left(\text{that is, } -\frac{1}{2} < 2\right)$$

You solve an inequality by transforming it into an inequality whose solution set is easy to see. Transformations that produce **equivalent inequalities,** that is, inequalities with the same solution set, are listed in the chart below.

Transformations that Produce Equivalent Inequalities

1. Simplifying either side of an inequality.

2. Adding to (or subtracting from) each side of an inequality the same number or the same expression.

3. Multiplying (or dividing) each side of an inequality by the same *positive* number.

4. Multiplying (or dividing) each side of an inequality by the same *negative* number and *reversing* the direction of the inequality.

Notice how these transformations are used in the following example.

Example 1 Solve each inequality and graph its solution set.

a. $\quad 5x + 17 < 2$
b. $\quad 5(3 - t) < 7 - t$

Solution

a. $5x + 17 - 17 < 2 - 17$

$5x < -15$

$\dfrac{5x}{5} < \dfrac{-15}{5}$

$x < -3$

∴ the solution set consists of all real numbers less than -3.

b. $15 - 5t < 7 - t$

$15 - 5t + t < 7 - t + t$

$15 - 4t < 7$

$-4t < -8$

$\dfrac{-4t}{-4} > \dfrac{-8}{-4}$

$t > 2$

∴ the solution set consists of all real numbers greater than 2.

To quickly check whether the direction of the inequality in your answer is correct, choose a test point and see if it satisfies the given inequality. For example, in part (a) of Example 1, substituting 0 for x gives $17 < 2$, which is false. So 0 is not in the solution set and the direction of the inequality $x < -3$ is correct.

The solution set of part (a) of Example 1 can be written as $\{x: x < -3\}$, which is read, "the set of all x such that x is less than -3." Similarly, the solution set of part (b) can be written $\{t: t > 2\}$. This notation will be used throughout the rest of this book.

Some inequalities are true for all real numbers, and others have no solution, as Example 2 illustrates.

Example 2 Solve each inequality and graph its solution set.

    a. $4x > 2(3 + 2x)$ **b.** $2t < -\frac{1}{2}(-6 - 4t)$

Solution   **a.** $4x > 2(3 + 2x)$ **b.** $2t < -\frac{1}{2}(-6 - 4t)$

$\qquad\qquad 4x > 6 + 4x \qquad\qquad\qquad 2t < 3 + 2t$

$\qquad\qquad\quad 0 > 6 \qquad\qquad\qquad\qquad\quad 0 < 3$

Since the equivalent inequality $0 > 6$ is *false,* the given inequality is false and has no solution.

$\therefore$ the solution set is $\emptyset$, and there is no graph.

Since $0 < 3$ is *true,* the given inequality is true for all values of t.

$\therefore$ the solution set is {real numbers}.

Oral Exercises

Explain how to transform the first inequality into the second inequality.

Sample $-3t > 15; t < -5$

Solution   Divide each side by -3 and reverse the direction of the inequality.

1. $x - 3 < 4; x < 7$ **2.** $3s > 6; s > 2$

3. $-\frac{u}{4} > 3; u < -12$ **4.** $k + 2 < 1; k < -1$

5. $2(y + 1) + 3 < y; 2y + 5 < y$ **6.** $10 > -5y; -2 < y$

7. $\frac{3x + 1}{2} < 5; 3x + 1 < 10$ **8.** $5t < 3(t - 2) + 1; 5t < 3t - 5$

9. $z - 3 > 3z - 7; -3 > 2z - 7$ **10.** $4x - 7 > x - 2; 4x > x + 5$

Match each inequality with the graph of its solution set.

11. $-6x > 3$

12. $4 - x > 2$

13. $1 - x < x$

14. $2(x - 1) < 2x - 3$

15. $2(x + 1) + 1 > 0$

16. $2(x - 1) > 2x - 3$

a. ← —|——|——|——|——|——|——⊕—|→
    -2 -1 0 1 2

b. ← —|⊕—|——|——|——|——|——|——|→
    -2 -1 0 1 2

c. ← —|——|——|——|——|——|——|——|→
    -2 -1 0 1 2

d. ← —|——|——⊕—|——|——|——|——|→
    -2 -1 0 1 2

e. ← —|——|——|——|——⊕—|——|——|→
    -2 -1 0 1 2

f. ← —|——|——|——|——|——|——|——|→
    -2 -1 0 1 2

g. ← —|——|—⊕—|——|——|——|——|→
    -2 -1 0 1 2

h. ← —|——|——|——|——|⊕—|——|——|→
    -2 -1 0 1 2

17. a. Explain the meaning of this statement: *If $a < 0$, then $a^2 > 0$.*
   b. Decide whether the statement in (a) is true or false. Give a convincing argument to support your answer.

Written Exercises

Solve each inequality and graph each solution set that is not empty.

> **Sample** $1 - 2t > -5$
>
> **Solution** $\{t: t < 3\}$

    -3 -2 -1 0 1 2 3 4 5

A **1.** $x - 7 > -5$ **2.** $y + 4 < 3$

3. $2t < 6$ **4.** $3u > -6$

5. $-5x < 10$ **6.** $-12 > -4y$

7. $-\dfrac{t}{2} > \dfrac{3}{2}$ **8.** $-\dfrac{3}{4}k < -6$

9. $3s - 1 > -4$ **10.** $2r + 5 < -1$

11. $y < 7y - 24$ **12.** $3t > 6t + 12$

13. $2 - h < 4 + h$ **14.** $1 + 2x < 2(x - 1)$

15. $5(2u + 3) > 2(u - 3) + u$ **16.** $3(x - 2) - 2 < x - 5$

17. $5(x - 7) + 2(1 - x) > 3(x - 11)$ **18.** $4s + 3(2 - 3s) < 5(2 - s)$

19. $7y - 2(y - 4) > 6 - (2 - y)$ **20.** $4(2 - x) - 3(1 + x) < 5(1 - x)$

Solve.

B   **21.** $k - 3(2 - 4k) < 7 - (8k - 9 + k)$

22. $\frac{2}{3}t - (2 - 3t) < 5t + 2(1 - t)$

23. $4(y + 2) - 9y > y - 3(2y + 1) - 1$

24. $4[5x - (3x - 7)] < 2(4x - 5)$

Tell whether each statement is true for all real numbers. If you think it is not, give a numerical example to support your answer.

25. If $a < b$, then $a - c < b - c$.

26. If $a < b$, then $a - b < 0$.

27. If $a < b$, then $a^2 < b^2$.

28. If $a < b$, then $a^3 < b^3$.

29. If $a < b$ and $c < d$, then $a + c < b + d$.

30. If $a < b$ and $c < d$, then $a - c < b - d$.

C   **31.** If $a \neq b$, then $a^2 + b^2 > 2ab$.

32. If $a > 0$ and $a \neq 1$, then $a + \frac{1}{a} > 2$.

33. The following statement is true:

$$\text{If } a > 0 \text{ and } a < 1, \text{ then } a^2 < a.$$

Write a short paragraph explaining the meaning of this statement to some-one who has studied arithmetic but not algebra.

Mixed Review Exercises

Simplify.

1. $(8t - 5) - (5 - 8t)$ **2.** $(-4)(5)(-1)(-3)$

3. $(-2)^4(-p)^3$ **4.** $\frac{5^2 - 7^2}{5 - 7}$

5. $\frac{8cd - 6}{-2}$ **6.** $\frac{6(2 + 3)}{6 \cdot 2 + 3}$

7. $|3(-4) - 2|$ **8.** $-2(5 - 8)^3$

9. $2a^2 - 5a - (a^2 - 7)$ **10.** $6x - y + 2x - 4y$

11. $|-4| - |-9|$ **12.** $4(3 - m) - (2m + 1)$

Computer Exercises

For students with some programming experience.

1. Write a computer program that will print all solutions of the inequality $ax + b < c$, where a, b, and c are entered by the user. In this program the domain of x will be a set of consecutive integers, where the smallest and largest members of the set will also be entered by the user.
   (*Hint:* Use a FOR . . . NEXT loop to test each integer from smallest to largest.)

2. Run the program in Exercise 1 to find solutions of the inequality $3x - 7 < 17$ for the following domains.
   a. {1, 2, . . . , 6}   **b.** {5, 6, . . . , 12}   **c.** {10, 11, . . . , 20}

3. Modify the program in Exercise 1 so that if there are no solutions in the given domain, the computer will print a message to this effect.

4. Run the program in Exercise 3 to find solutions of the inequality $5x + 3 < 12$ for the following domains.
   a. {6, 7, . . . , 12}   **b.** {−10, −9, . . . , 6}   **c.** {1, 2, . . . , 6}

 Career Note / *Automotive Engineer*

When you read that automobile manufacturers have increased the fuel efficiency of cars, you are actually reading about the work of automotive engineers. Automotive engineering is a specialized field of mechanical engineering, which is concerned with the use, production, and transmission of mechanical power.

The role of an automotive engineer, as with all engineers, is to apply the theories and principles of science and mathematics to the solution of practical problems. Automotive engineers in particular are responsible for designing automobiles that deliver efficient and economical performance. Evaluating the overall cost, reliability, and safety of automobiles is also part of an automotive engineer's work.

To assist them in their work, automotive engineers rely upon calculators, computer simulations, and other engineers.

Computer-aided design systems have become especially important for the production and analysis of automotive engine-and-body designs.

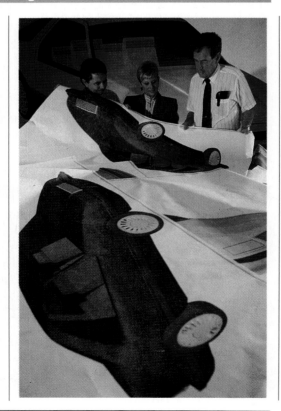

2-2 Solving Combined Inequalities

Objective   To solve conjunctions and disjunctions.

The general admission price at the Cinema V theater is $4.50. Children 12 years of age or under and adults who are at least 65 are charged only half price. Therefore, you must pay the full price of $4.50 if your age, a, satisfies both of the inequalities $a > 12$ and $a < 65$. The combined inequality "$a > 12$ and $a < 65$" is an example of a *conjunction*.

A sentence formed by joining two sentences with the word *and* is called a **conjunction.** A conjunction is true when *both* sentences are true.

Example 1   Graph the solution set of the conjunction $x > -2$ *and* $x < 3$.

Solution   The conjunction is true for all values of x between -2 and 3. The graph is shown below. Notice that the numbers *between* -2 and 3 include neither -2 nor 3.

The conjunction "$x > a$ and $x < b$" is usually written as

$$a < x < b,$$

which is read "x is greater than a and less than b." Using this notation, the solution set of Example 1 is written $\{x: -2 < x < 3\}$.

A sentence formed by joining two sentences with the word *or* is called a **disjunction.** A disjunction is true when *at least one* of the sentences is true.

You learned earlier that "$x \leq 2$" means "x is less than or equal to 2" and therefore represents the disjunction

$$x < 2 \quad or \quad x = 2.$$

In the graph of this disjunction, shown below, a solid red dot has been used to show that 2 is included in the solution set.

The chart below summarizes how to solve conjunctions and disjunctions of open sentences in one variable.

Conjunction: Find the values of the variable for which *both* sentences are true.

Disjunction: Find the values of the variable for which *at least one* of the sentences is true.

Example 2 Solve $3 < 2x + 5 \le 15$ and graph its solution set.

Solution 1 You can first rewrite the inequality using *and*. Then solve both inequalities.

$$3 < 2x + 5 \quad \text{and} \quad 2x + 5 \le 15$$
$$-2 < 2x \qquad\qquad\qquad 2x \le 10$$
$$-1 < x \quad \text{and} \qquad x \le 5$$

∴ the solution set is $\{x: -1 < x \le 5\}$. **Answer**

Solution 2 You can solve the inequality using this shortened method that involves operating on all three parts of the inequality at the same time.

$$3 < 2x + 5 \le 15 \qquad \text{Subtract 5 from all three parts.}$$
$$-2 < \quad 2x \quad \le 10 \qquad \text{Divide all three parts by 2.}$$
$$-1 < \quad x \quad \le 5$$

∴ the solution set is $\{x: -1 < x \le 5\}$. **Answer**

Example 3 Solve the disjunction $7 - 2y \le 1$ *or* $3y + 10 < 4 - y$ and graph its solution set.

Solution

$$7 - 2y \le 1 \qquad \text{or} \qquad 3y + 10 < 4 - y$$
$$-2y \le -6 \qquad\qquad\qquad 4y < -6$$
$$y \ge 3 \qquad \text{or} \qquad y < -\frac{3}{2}$$

∴ the solution set is $\left\{y: y \ge 3 \text{ or } y < -\frac{3}{2}\right\}$. **Answer**

Oral Exercises

Tell whether each conjunction or disjunction is true or false.

1. $-5 < -3$ and $3 < 5$

2. $-1 < 2$ and $-1 < -2$

3. $-6 > -2$ or $-6 > 2$

4. $-6 > -3$ or $-3 < 6$

5. $-7 < -5$ or $7 < -5$

6. $-4 < -6$ and $-4 < 6$

Match each graph with one of the open sentences in a–h.

7.

8.

9.

10.

11.

12.

a. $-2 \le x \le 3$

b. $-2 < x < 3$

c. $x < 0$ or $x > 0$

d. $x \ge -2$ or $x \le 3$

e. $x > 3$ or $x \le -2$

f. $-2 \le x < 3$

g. $x < -2$ or $x \ge 3$

h. $x \le -2$ or $x > 3$

13. The inequality $x \ne 2$ is equivalent to the combined inequality $x < 2 \underline{\ ?\ } x > 2$.

14. Explain why it would be incorrect to write $-2 < x < -5$.

Written Exercises

Solve each conjunction or disjunction and graph each solution set that is not empty.

A

1. $3 \le x < 5$

2. $z > -1$ and $z < 3$

3. $t < 1$ or $t \ge 3$

4. $p > 1$ or $p < 1$

5. $y \ge -1$ and $y \ge 3$

6. $y \ge -1$ or $y \ge 3$

7. $t > 0$ or $t < 2$

8. $w < 0$ and $w \ge 4$

9. $0 \le x - 2 < 3$

10. $2 > y + 2 \ge 0$

11. $-1 > 2r - 5 > -9$

12. $-1 \le 3z + 2 \le 8$

13. $2z - 1 \le 5$ or $3z - 5 > 10$

14. $3k + 7 < 1$ or $2k - 3 > 1$

15. $2t + 7 \geq 13$ or $5t - 4 < 6$

16. $2x + 3 > 1$ or $5x - 9 \leq 6$

17. $2t + 7 \geq 13$ and $5t - 4 < 6$

18. $2x + 3 > 1$ and $5x - 9 \leq 6$

19. $-5 < 1 - 2k < 3$

20. $-6 \leq 2 - 3m \leq 7$

21. $-3 < 2 - \dfrac{d}{3} \leq -1$

22. $3 \geq 1 - \dfrac{n}{2} > -2$

B **23.** $7q - 1 > q + 11$ or $-11q > -33$

24. $5n - 1 > 0$ and $4n + 2 < 0$

25. $x - 7 < 3x - 5 < x + 11$

26. $3y + 5 \geq 2y + 1 > y - 1$

27. $-\dfrac{3}{4}m \geq m - 1$ or $-\dfrac{3}{4}m < m + 1$

28. $3z + 7 \leq 4z$ and $3z + 7 > -4z$

Solve.

29. $-3 \leq -2(t - 3) < 6$

30. $-5 < 2(2 - s) + 1 \leq 9$

31. $\dfrac{t}{4} + 2 < t + 3$ and $t - 3 > \dfrac{t}{2} - 4$

32. $\dfrac{r - 3}{6} \leq r - 1$ or $\dfrac{r - 6}{3} \leq r + 4$

C **33.** $0 < 1 - x \leq 3$ or $-1 \leq 2x - 3 \leq 5$

34. $1 < -(2s + 1) < 5$ or $1 < 2s - 1 < 5$

35. $2 < \dfrac{y + 6}{2} < 5$ and $(4 - y > 5$ or $4 + y > 7)$

36. $\left(x \leq \dfrac{x + 4}{3} + 2 \text{ or } x \geq 2x - 1\right)$ and $1 \leq \dfrac{x - 1}{2} \leq 3$

Historical Note / *Linkages*

The problem of constructing a linkage made of hinged rods that will draw a straight line has practical as well as theoretical interest. James Watt, the inventor of the steam engine, tried to construct such a device to guide the motion of the engine's piston. Watt's linkage, however, drew only an approximation of a straight line.

In 1864 a French army officer named Peaucellier solved the problem with the linkage shown at the right, in which one pivot moves in a perfectly straight line. Peaucellier was awarded the mechanical prize of the Institute of France in 1873.

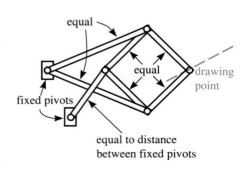

2-3 Problem Solving Using Inequalities

Objective   To solve word problems by using inequalities in one variable.

Sometimes solving a word problem involves using an inequality.

Example 1   A bus is to be chartered for the senior class trip. The basic fare is $9.50 per passenger. If more than 20 people go, everyone's fare is reduced by $.30 for each passenger over this number (20). *At least* how many people must go to make the fare less than $7.50 per passenger?

Solution

Step 1   The problem asks for the least number of passengers needed to make the fare for each less than $7.50.

Step 2   Let n = the number of passengers.
Then $n - 20$ = the number of passengers over 20;
$0.30(n - 20)$ = the amount each passenger's fare is reduced; and
$9.50 - 0.30(n - 20)$ = the reduced fare per passenger.

Step 3   Reduced fare per passenger   is less than $7.50.
$$9.50 - 0.30(n - 20) \qquad < \qquad 7.50$$

Step 4   Multiplying both sides of the inequality in Step 3 by 10 clears the decimals and gives this equivalent inequality to solve:
$$95 - 3(n - 20) < 75$$
$$95 - 3n + 60 < 75$$
$$-3n < -80$$
$$n > \frac{-80}{-3}$$
$$n > 26\frac{2}{3}$$

Interpret the result: Since the number of passengers must be an integer, $n \geq 27$.

Step 5   *Check:*   Is the reduced fare less than $7.50? If at least 27 people go, the fare per passenger is reduced by $0.30(27 - 20) = 2.10$.
Then the reduced fare is $9.50 - 2.10 = 7.40 < 7.50$.

Is 27 the least number of passengers? Try 26. If 26 people go, the fare per passenger is reduced by $0.30(26 - 20) = 1.80$.
Then the reduced fare is $9.50 - 1.80 = 7.70 > 7.50$.

∴ at least 27 passengers must go.   *Answer*

Certain phrases can be translated into mathematical terms using inequalities. Here is a list of the more common ones.

Phrase	Translation
x is at least a. x is no less than a.	$x \geq a$
x is at most b. x is no greater than b.	$x \leq b$
x is between a and b. x is between a and b, inclusive.	$a < x < b$ $a \leq x \leq b$

Example 2 Find all sets of 4 consecutive integers whose sum is between 10 and 20.

Solution

Step 1 The problem asks for 4 consecutive integers; their sum must be greater than 10 and less than 20.

Step 2 Let $n =$ the first of these integers.
Then the other three are $n + 1$, $n + 2$, and $n + 3$.

Step 3 $10 <$ the sum < 20
$10 < n + (n + 1) + (n + 2) + (n + 3) < 20$

Step 4 $10 < 4n + 6 < 20$
$\phantom{10 <} 4 < 4n < 14$
$\phantom{10 <} 1 < n < \dfrac{14}{4}$

Interpret the result: Since n is an integer, there are only two values possible for n: 2 and 3. There are two sets of consecutive integers that fulfill the requirements of the problem:

$$\{2, 3, 4, 5\} \text{ and } \{3, 4, 5, 6\}$$

Step 5 *Check:* Is the sum between 10 and 20?

 For $\{2, 3, 4, 5\}$: $10 < 2 + 3 + 4 + 5 < 20$
$ 10 < 14 < 20 \checkmark$
 For $\{3, 4, 5, 6\}$: $10 < 3 + 4 + 5 + 6 < 20$
$ 10 < 18 < 20 \checkmark$

To complete the check, you must show that any other set of four consecutive integers will not satisfy the requirements. In fact, you need only eliminate the set of "next greater" integers, $\{4, 5, 6, 7\}$, and the set of "next smaller" integers, $\{1, 2, 3, 4\}$. That work is left for you to do.

$\therefore$ the required sets are $\{2, 3, 4, 5\}$ and $\{3, 4, 5, 6\}$. *Answer*

Problems

Solve.

A **1.** For the Hawks' 80 basketball games next year, you can buy separate tickets for each game at $9 each, or you can buy a season ticket for $580. At most how many games could you attend at the $9 price before spending more than the cost of a season ticket?

2. The usual toll charge to use the Bingham tunnel is 50 cents. If you purchase a special sticker for $5.50, the toll is only 35 cents. At least how many trips through the tunnel are needed before the sticker costs less than paying for each trip separately? $.50x > 5.50 + .35x$ $.15x > 5.50$

3. The length of a rectangle is 5 cm more than twice its width. Find the largest possible width if the perimeter is at most 64 cm. $2(2w+5)+2w \le 64$ $x > 36\frac{2}{3}$

4. The lengths of the legs of an isosceles triangle are integers. The base is half as long as each leg. What are the possible lengths of the legs if the perimeter is between 6 units and 16 units?

5. Find all sets of three consecutive *odd* integers whose sum is between 20 and 30.

6. Find all sets of three consecutive *even* integers whose sum is between 25 and 45.

7. Jeannie's scores on her first four tests were 80, 65, 87, and 75. What will she have to score on her next test to obtain an average of at least 80 for the term?

8. Jim's second test score was 8 points higher than his first score. His third score was 88. He had a B average (between 80 and 89, inclusive) for the three tests. What can you conclude about his first test score?

9. The sides $\overline{AB}$ and $\overline{AD}$ of a square are extended 10 cm and 6 cm, respectively, to form sides $\overline{AE}$ and $\overline{AF}$ of a rectangle. At most how long is the side of the square if the perimeter of the rectangle is at least twice the perimeter of the square?

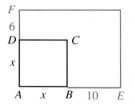

10. The three sides of an equilateral triangle are increased by 20 cm, 30 cm, and 40 cm, respectively. The perimeter of the resulting triangle is between twice and three times the perimeter of the original triangle. What can you conclude about the length of a side of the original triangle?

B **11.** The telephone company offers two types of service. With Plan A, you can make an unlimited number of local calls per month for $18.50. With Plan B, you pay $6.50 monthly, plus 10 cents for each min of calls after the first 40 min. At least how many min would you have to use the telephone each month to make Plan A the better option?

12. During the first 20 mi of a 50 mi bicycle race, Roger's average speed was 16 mi/h. What must his average speed be during the remainder of the race if he is to finish the race in less than 2.5h?

13. A subway train makes six stops of equal length during its 21 km run. The train is actually moving for 20 min of the trip. At most how long can the train remain at each station if the average speed for the trip, including stops, is to be at least 36 km/h?

14. The length of a rectangular sheet of paper was twice its width. After 1 cm was trimmed from each edge of the sheet, the perimeter was at most 1 m. Find the largest possible dimensions of the trimmed sheet.

C **15.** A swimming pool is 5 m longer than it is wide and is surrounded by a deck 2 m wide. The area of the pool and deck together is at least 140 m^2 greater than the area of the pool alone. What can you conclude about the dimensions of the pool?

16. Find all triples of consecutive integers such that 11 times the largest of the integers is at least 46 more than the product of the other two.

Mixed Review Exercises

Solve each open sentence and graph each solution set that is not empty.

1. $3x - 2 \geq -8$

2. $\frac{1}{2}t \leq -2$ or $t - 4 \geq -3$

3. $2(m - 2) > 4 - 3(1 - m)$

4. $-2y < -6$ and $y + 3 \leq 1$

5. $7 - 4d < 3$

6. $-1 < 5 - 2p < 5$

Evaluate if $x = -7$ and $y = 3$.

7. $|x + y|$ **8.** $|x| + |y|$ **9.** $|xy|$ **10.** $|x| \cdot |y|$

Self-Test 1

Vocabulary   equivalent inequalities (p. 60)   disjunction (p. 65)
              conjunction (p. 65)

Solve each inequality and graph the solution set.

1. $6m - 13 \leq 5$ **2.** $2y + 9 < 5(y + 3)$ **3.** $8 - 3x > -7$ Obj. 2-1, p. 59

4. $\frac{n - 3}{2} \geq n - 1$ **5.** $1 < 4a + 5 < 9$ **6.** $c - 3 > 2c$ or $\frac{c}{3} \geq 1$ Obj. 2-2, p. 65

7. Bill's scores on his first two tests were 75 and 82. What will he have to score on his next test to obtain an average of at least 80?   Obj. 2-3, p. 69

Working with Absolute Value

2-4 Absolute Value in Open Sentences

Objective   To solve open sentences involving absolute value.

If you think of the absolute value of a real number x (see page 3) as the distance between the graph of x and the origin on a number line, you can see why the sentences below are equivalent.

Sentence	Equivalent Sentence	Graph
$\|x\| = 1$ Distance between x and 0 equals 1.	$x = -1$ or $x = 1$	
$\|x\| > 1$ Distance between x and 0 is greater than 1.	$x < -1$ or $x > 1$	
$\|x\| < 1$ Distance between x and 0 is less than 1.	$-1 < x < 1$	

You can often solve an open sentence involving absolute value by first writing an equivalent disjunction or conjunction.

Example 1   Solve $|3x - 2| = 8$.

Solution $|3x - 2| = 8$ is equivalent to this disjunction:

$$3x - 2 = -8 \quad \text{or} \quad 3x - 2 = 8$$
$$3x = -6 \qquad\qquad 3x = 10$$
$$x = -2 \quad \text{or} \quad x = \frac{10}{3}$$

∴ the solution set is $\left\{-2, \frac{10}{3}\right\}$.   **Answer**

Example 2   Solve $|3 - 2t| < 5$.

Solution $|3 - 2t| < 5$ is equivalent to this conjunction:

$$-5 < 3 - 2t < 5$$
$$-8 < \quad -2t \quad < 2$$
$$4 > \quad t \quad > -1$$

∴ the solution set is $\{t: -1 < t < 4\}$.   **Answer**

Example 3 Solve $|2z - 1| + 3 \geq 8$ and graph its solution set.

Solution First transform the inequality to an equivalent inequality in which the expression involving absolute value is alone on one side.

$$|2z - 1| + 3 \geq 8$$
$$|2z - 1| + 3 - 3 \geq 8 - 3$$
$$|2z - 1| \geq 5$$

The last inequality is equivalent to this disjunction:

$$
\begin{array}{ccc}
2z - 1 \leq -5 & \text{or} & 2z - 1 \geq 5 \\
2z \leq -4 & \downarrow & 2z \geq 6 \\
z \leq -2 & \text{or} & z \geq 3
\end{array}
$$

∴ the solution set is $\{z: z \leq -2 \text{ or } z \geq 3\}$. *Answer*

You can help guard against errors by testing one value from each region of the graph. Substitute values in the original inequality $|2z - 1| + 3 \geq 8$.

Try $z = -4$: $|2(-4) - 1| + 3 = |-9| + 3 = 12 \geq 8$ True ✓
Try $z = 0$: $|2 \cdot 0 - 1| + 3 = |-1| + 3 = 4 \geq 8$ False ✓
Try $z = 4$: $|2 \cdot 4 - 1| + 3 = |7| + 3 = 10 \geq 8$ True ✓

You can tell at a glance that an inequality such as $|x - 3| \geq -2$ is true for all real numbers x, because the absolute value of every real number is nonnegative. On the other hand, an inequality such as $|t + 5| < -1$ has ∅ as its solution set (why?).

Oral Exercises

Express each open sentence as an equivalent conjunction or disjunction without absolute value.

Sample 1 $|3t - 1| > 2$ **Solution** $3t - 1 < -2 \text{ or } 3t - 1 > 2$

1. $|x| \leq 3$
2. $|t| = 2$
3. $|z| > 0$
4. $|y - 3| \leq 2$
5. $|s + 3| = 3$
6. $|2x - 3| \geq 1$
7. $|3t - 1| \leq 2$
8. $|5 - 2z| < 3$

Express each conjunction or disjunction as an equivalent open sentence involving absolute value.

Sample 2 $-1 \leq x - 2 \leq 1$ **Solution** $|x - 2| \leq 1$

9. $u = -3 \text{ or } u = 3$
10. $t \geq -3 \text{ and } t \leq 3$
11. $3 > 4(x - 1) > -3$

Written Exercises

A **1–8.** Graph the solution set of each open sentence in Oral Exercises 1–8.

Solve and graph the solution set.

9. $|2t + 5| < 3$

10. $|3x + 2| > 4$

11. $|2u - 5| = 0$

12. $8 = |5y + 2|$

13. $\left|1 - \dfrac{x}{3}\right| \geq \dfrac{2}{3}$

14. $\left|1 - \dfrac{p}{2}\right| \leq 2$

15. $0 \leq |4u - 7|$

16. $|3r - 12| > 0$

17. $\left|\dfrac{t - 2}{4}\right| \leq \dfrac{1}{2}$

18. $1 > |2 - 0.8n|$

Solve.

B **19.** $|x + 5| - 3 = 1$

20. $|2t - 3| + 2 = 5$

21. $|2u - 1| + 3 \leq 6$

22. $4 - |3k + 1| < 2$

23. $7 - 3|4d - 7| \geq 4$

24. $6 + 5|2r - 3| \geq 4$

25. $4 + 2\left|\dfrac{3t - 5}{2}\right| > 5$

26. $2\left|\dfrac{2t - 5}{3}\right| - 3 \geq 5$

27. $7 + 5|c| \leq 1 - 3|c|$

28. $\dfrac{1}{2}|d| + 5 \geq 2|d| - 13$

Graph the solution set of each open sentence.

C **29.** $2 < |w| < 4$

30. $1 \leq |s - 2| \leq 3$

31. $1 \leq |2x + 1| < 3$

32. $0 < |2 - r| \leq 2$

Solve.

33. $|2x| \leq |x - 3|$

34. $|t| > |2t - 6|$

Computer Exercises

For students with some programming experience.

1. Write a program to list all integers x in the interval $-50 \leq x \leq 50$ that are solutions of an open sentence of the form $a < |cx + d| < b$. The values of a, b, c, and d are to be entered by the user. If no integers in the given interval satisfy the inequality, have the output state this. You will need to use the BASIC function ABS in your program.

2. Use the program in Exercise 1 to find the integer solutions of each open sentence.

   a. $17 < |3x - 25| < 35$ **b.** $1 < |18x + 120| < 100$

2-5 Solving Absolute Value Sentences Graphically

Objective   To use number lines to obtain quick solutions to certain equations and inequalities involving absolute value.

You know that on a number line the distance between the graph of a real number x and the origin is $|x|$. *The distance on the number line between the graphs of real numbers a and b is $|a - b|$, or equivalently, $|b - a|$.* You can use this fact to solve many open sentences almost at sight.

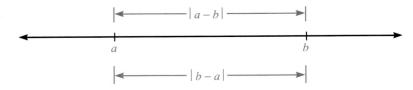

Example 1   Find the distance between the graphs of each pair of numbers.

        a. 8 and 17   **b.** -11 and -6 **c.** -9 and 12

Solution   **a.** $|8 - 17| = |-9| = 9$ *Answer*
          b. $|-11 - (-6)| = |-11 + 6| = |-5| = 5$ *Answer*
          c. $|-9 - 12| = |-21| = 21$ *Answer*

Example 2   Solve $|x - 3| = 2$.

Solution   To satisfy $|x - 3| = 2$, x must be a number whose distance from 3 is 2 units. So, to find x, start at 3 and move 2 units in each direction on a number line.

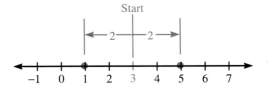

You arrive at 1 and 5 as the values of x.

$\therefore$ the solution set is $\{1, 5\}$.   *Answer*

Example 3   Solve $|y + 2| \le 3$.

Solution $y + 2 = y - (-2)$
          Therefore, $|y + 2| \le 3$ is equivalent to $|y - (-2)| \le 3$.

So the distance between y and -2 must be 3 units or less. To find y, start at -2 and move 3 units in each direction on a number line.

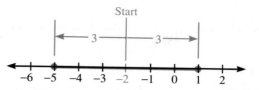

The numbers up to and including 1 and the numbers down to and including -5 will satisfy the inequality.

∴ the solution set is $\{y: -5 \le y \le 1\}$. **Answer**

Certain equations and inequalities, such as the ones in Examples 1–3, lend themselves more easily to a graphic solution. With these types, the expression involving absolute value is of the form $|x - \text{constant}|$. When the expression is more complicated, as in Example 4, a graphic method can also be applied, though not as easily. For these types, you might prefer to use the algebraic method learned in the previous lesson.

Example 4 Solve $|5 - 2t| > 3$.

Solution Use the facts that $|a - b| = |b - a|$ and $|ab| = |a| \cdot |b|$ (page 27) to rewrite $|5 - 2t|$ this way:

$$|5 - 2t| = |2t - 5| = \left|2 \cdot \left(t - \frac{5}{2}\right)\right|$$

$$= |2| \cdot \left|t - \frac{5}{2}\right|$$

$$= 2\left|t - \frac{5}{2}\right|$$

Therefore, the given inequality

$$|5 - 2t| > 3$$

is equivalent to

$$2\left|t - \frac{5}{2}\right| > 3, \text{ or } \left|t - \frac{5}{2}\right| > \frac{3}{2}$$

To find t, start at $\frac{5}{2}$ and move *more than* $\frac{3}{2}$ units to the right and to the left.

∴ the solution set is $\{t: t < 1 \text{ or } t > 4\}$. **Answer**

Oral Exercises

Find the distance between the graphs of each pair of numbers.

1. 3 and -4
2. -6 and -1
3. 12 and 5
4. -7 and 1
5. 11 and 8
6. -11 and 8
7. 11 and -8
8. -11 and -8
9. t and $4t$ if (a) t is positive, and (b) t is negative.
10. $-2k$ and $2k$ if (a) k is positive, and (b) k is negative.

Translate each open sentence into a statement involving distance.

Sample $|z + 2| \geq 3$ **Solution** The distance between z and -2 is at least 3.

11. $|t| \leq 5$
12. $|r - 2| > 1$
13. $|3 - s| = 4$
14. $|x + 1| \leq 4$
15. $|u + 5| \geq 3$
16. $|2 - y| < 2$

Translate each statement into an open sentence using absolute value.

Sample The numbers whose distance from -3 is at most 2   **Solution** $|x - (-3)| \leq 2$, or $|x + 3| \leq 2$

17. The numbers whose distance from 2 is less than 5
18. The numbers whose distance from -3 is at least 3
19. The numbers whose distance from -1 is not more than 3
20. The numbers whose distance from 4 is equal to 4
21. The numbers whose distance from $-\frac{2}{3}$ is not less than 5
22. The numbers whose distance from 0 is at most 6
23. The numbers whose distance from -1.2 is greater than 4.8
24. The numbers whose distance from a is not more than h

Written Exercises

Solve each open sentence graphically.

A

1. $|w| = 4$
2. $|x - 3| = 2$
3. $|t| < 3$
4. $|y| > 4$
5. $|u| \geq 2$
6. $|p| \leq 5$
7. $|y - 2| < 3$
8. $|k - 4| > 1$
9. $\left| t - \frac{3}{2} \right| \leq \frac{5}{2}$
10. $\left| x - \frac{1}{2} \right| > \frac{3}{2}$
11. $|r + 2| > 5$
12. $|w + 2| \leq 2$

Solve each open sentence. Use whichever method you prefer.

B 13. $|2x - 1| = 3$

14. $\left|\frac{1}{4}y + 1\right| = \frac{1}{2}$

15. $|2p + 5| \geq 3$

16. $|3k - 2| < 4$

17. $\left|\frac{1}{3}t - 1\right| \leq \frac{2}{3}$

18. $\left|\frac{y}{4} + \frac{1}{2}\right| > \frac{3}{4}$

19. $3 - |2x - 3| > 1$

20. $7 - |3y - 2| \leq 1$

21. $|9 + 3f| < 4$

22. $|4 + 2y| \geq 3$

23. $|1.2 + 0.4t| < 2$

24. $|1 - 0.3x| \geq 1.5$

Solve for x in terms of the other variables. Assume that a, b, and c are positive numbers.

C 25. $|x - a| \geq c$

26. $|x + a| \leq c$

27. $|x - c| < c$

28. $|x + c| > c$

29. $|a + bx| \geq c$

30. $|a - bx| \leq c$

31. $a - |bx| < c$
   (Assume $a - c \geq 0$.)

32. $a + |bx| > c$
   (Assume $c - a \geq 0$.)

33. $b|x + a| < c$

Mixed Review Exercises

Solve each open sentence.

1. $1 \leq 3x + 4 \leq 13$

2. $|5 - 2n| = 3$

3. $4(2c - 3) > 7c - 9$

4. $|6y + 6| > 0$

5. $|w| + 4 \geq 6$

6. $p + 2 < -1$ or $-4p \leq -8$

7. $5z + 11 \leq 1$

8. $|t + 4| = 0$

9. $\frac{1}{2}m > -1$ and $5 - m > 1$

Self-Test 2

1. Express the following conjunction as an equivalent open sentence involving absolute value: $3x + 2 \geq -4$ and $3x + 2 \leq 4$.

   Obj. 2-4, p. 73

Solve and graph each solution set.

2. $|2x + 1| > 3$

3. $5 - |3 - y| \geq 4$

4. Translate the following statement into an open sentence using absolute value and the variable x: The numbers whose distance from -2 is at least 5 units.

   Obj. 2-5, p. 76

Solve each open sentence graphically.

5. $|m - 4| \leq 2$

6. $\left|n + \frac{3}{2}\right| > \frac{1}{2}$

Check your answers with those at the back of the book.

In Lesson 2-4 you solved inequalities involving absolute value. The program below solves inequalities of the form $|ax + b| < c$ or $|ax + b| > c$ when the user enters the values of a, b, and c, and the inequality sign. The program uses the following four basic solutions.

For $|ax + b| < c$, where $c > 0$:

If $a > 0$, $\dfrac{-c - b}{a} < x$ and $x < \dfrac{c - b}{a}$.   If $a < 0$, $\dfrac{c - b}{a} < x$ and $x < \dfrac{-c - b}{a}$.

For $|ax + b| > c$, where $c \geq 0$:

If $a > 0$, $x < \dfrac{-c - b}{a}$ or $x > \dfrac{c - b}{a}$.   If $a < 0$, $x < \dfrac{c - b}{a}$ or $x > \dfrac{-c - b}{a}$.

The program also handles two special cases, which you will be asked about in Exercises 7 and 8 below.

```
10 PRINT "THIS PROGRAM WILL SOLVE AN INEQUALITY"
20 PRINT "OF THE FORM ABS (A * X + B) SIGN C,"
30 PRINT "WITH A NOT ZERO. THE SIGN IS < OR >."
40 INPUT "ENTER A,B,C, AND THE SIGN: ";A,B,C,S$
50 IF S$ = "<" AND C < = 0 THEN 60
55 PRINT "SOLUTION SET CONSISTS OF X SUCH THAT"
60 IF S$ = "<" AND A > 0 AND C > 0 THEN PRINT
   (-C - B) / A;" < X AND X < "; (C - B) / A
70 IF S$ = "<" AND A < 0 AND C > 0 THEN PRINT
   (C - B) / A;" < X AND X < "; (-C - B) / A
80 IF S$ = ">" AND A > 0 AND C > = 0 THEN PRINT
   "X < ";(-C - B) / A;" OR X > ";(C - B) / A
90 IF S$ = ">" AND A < 0 AND C > = 0 THEN PRINT
   "X < "; (C - B) / A;" OR X > ";(-C - B) / A
100 IF S$ = ">" AND C < 0 THEN PRINT
   "X IS ANY REAL NUMBER."
110 IF S$ = "<" AND C < = 0 THEN PRINT
   "THERE IS NO SOLUTION."
120 END
```

Exercises

Run the program to solve each inequality. Be sure each inequality is in one of the forms $|ax + b| < c$ or $|ax + b| > c$.

1. $|-4x + 5| < 11$ **2.** $|2x + 9| > 1$ **3.** $|5x - 1| + 4 > 6$

4. $3|5 - x| > 12$ **5.** $|2x + 3| - 6 < 7$ **6.** $-4|x - 1| > -8$

Run the program to solve each inequality. Explain the result.

7. $|3x + 4| < -1$ **8.** $|-2x + 5| > -3$

Proving Theorems

2-6 Theorems and Proofs

Objective   To use axioms, definitions, and theorems to prove some properties of real numbers.

You have used many properties of real numbers in the previous lessons of this book. It can be shown that if just a few of these properties are accepted as true statements, then all the other properties will necessarily follow.

Statements that we assume to be true are called **axioms** (or **postulates**). The axioms that we accept include:

The substitution principle (page 8)
The properties of equality (page 14)
The field properties of real numbers (page 15)
The properties of order (page 59)

At this time, you should review all the properties and the definitions included up to this point.

Example 1 shows how you can reason from a **hypothesis** (a statement that is given or assumed to be true) to a **conclusion** (a statement that follows logically from the assumptions). In the example, each step of reasoning from the hypothesis "a, b, and c are real numbers and $a + c = b + c$" to the conclusion "$a = b$" is justified by a given fact or an axiom.

Example 1   Show that for all real numbers a, b, and c, if $a + c = b + c$, then $a = b$.

Solution

Statements	Reasons
1. a, b, and c are real numbers; $a + c = b + c$.	1. Hypothesis (or Given)
2. $-c$ is a real number.	2. Property of opposites
3. $(a + c) + (-c) = (b + c) + (-c)$	3. Addition property of equality
4. $a + [c + (-c)] = b + [c + (-c)]$	4. Associative property of addition
5. $\quad\quad a + 0 = b + 0$	5. Property of opposites
6. $\quad\quad \therefore a = b$	6. Identity property of addition

This form of logical reasoning from hypothesis to conclusion is called a **proof.** A statement that can be proved is called a **theorem.**

A theorem that can be proved easily from another is called a **corollary.** You can use the commutative property of addition to prove the following corollary of the theorem proved in Example 1.

Corollary: If $c + a = c + b$, then $a = b$.

The theorem of Example 1 and its corollary make up the *cancellation property of addition*.

Cancellation Property of Addition

For all real numbers a, b, and c:

$$\text{If } a + c = b + c, \text{ then } a = b.$$
$$\text{If } c + a = c + b, \text{ then } a = b.$$

A proof of the following property is outlined in Exercise 14.

Cancellation Property of Multiplication

For all real numbers a and b and *nonzero* real numbers c:

$$\text{If } ac = bc, \text{ then } a = b.$$
$$\text{If } ca = cb, \text{ then } a = b.$$

If you interchange the hypothesis and conclusion of an if-then statement, you get the **converse** of the statement. The converse of a true statement is not necessarily true. For example, the converse of "If $a = 1$, then $a^2 = a$" is "If $a^2 = a$, then $a = 1$." This converse is false, as shown by the *counterexample* $0^2 = 0$, but $0 \neq 1$. (In algebra, a **counterexample** is a single numerical example that makes a statement false.)

After a theorem has been proved, it may be used along with axioms and definitions in other proofs. For example, the cancellation property of addition is used in the following proof.

Example 2 Prove the *multiplicative property of 0* (page 27):
For every real number a, $a \cdot 0 = 0$ and $0 \cdot a = 0$.

Proof

Statements	Reasons
1. a is a real number.	1. Given
2. $\quad 0 + 0 = 0$	2. Identity property of addition
3. $\quad a \cdot (0 + 0) = a \cdot 0$	3. Multiplication property of equality
4. $a \cdot 0 + a \cdot 0 = a \cdot 0$	4. Distributive property
5. $a \cdot 0 + a \cdot 0 = 0 + a \cdot 0$	5. Identity property of addition
6. $\quad \therefore a \cdot 0 = 0$	6. Cancellation property of addition
7. $\quad$ and $0 \cdot a = 0$	7. Commutative property of multiplication

The multiplicative property of 0 can be used to prove an important property, shown at the top of the next page.

Zero-Product Property

For all real numbers a and b:

$$ab = 0 \text{ if and only if } a = 0 \text{ or } b = 0.$$

An "if and only if" statement such as the one above is equivalent to two "if-then" statements that are converses of each other:

> If $ab = 0$, then $a = 0$ or $b = 0$.
>
> If $a = 0$ or $b = 0$, then $ab = 0$.

The second statement follows directly as a corollary to the multiplicative property of 0. A proof of the first statement is outlined in Exercise 11.

The property of opposites (page 15) states that the opposite of a real number *exists* and that it is *unique* (exactly one exists). These facts are used in the two following proofs.

Example 3 Prove: For every real number a, $-(-a) = a$.

Proof

Statements	Reasons
1. a is a real number.	1. Given
2. $\quad -a + a = 0$	2. Property of opposites (existence of $-a$)
3. $-a + [-(-a)] = 0$	3. Property of opposites (existence of $-(-a)$)
4. $\quad \therefore -(-a) = a$	4. Steps 2 and 3 and property of opposites (uniqueness)

Example 4 Prove the property of the *opposite of a product* (page 28): For all real numbers a and b, $-ab = (-a)b$ and $-ab = a(-b)$.

Proof

Statements	Reasons
1. a and b are real numbers.	1. Given
2. $ab + (-a)b = [a + (-a)]b$	2. Distributive property
3. $ab + (-a)b = 0 \cdot b$	3. Property of opposites
4. $ab + (-a)b = 0$	4. Multiplicative property of 0
5. $ab + (-ab) = 0$	5. Property of opposites (existence of $-ab$)
6. $\therefore -ab = (-a)b$	6. Steps 4 and 5 and property of opposites (uniqueness)

(The proof of the second part is outlined in Exercise 5.)

Inequalities and Proof **83**

A corollary to the theorem of Example 4 is given in the next example.

Example 5 Prove the *multiplicative property of* -1 (page 27):
For every real number a, $a(-1) = -a$ and $(-1)a = -a$.

Proof

Statements	Reasons
1. a is a real number.	1. Given
2. $-a$ is a real number.	2. Property of opposites
3. $\quad -a = -(a \cdot 1)$	3. Identity property of multiplication
4. $\quad -a = a(-1)$	4. Property of the opposite of a product
5. $\therefore a(-1) = -a$	5. Symmetric property of equality
6. $\quad a(-1) = (-1)a$	6. Commutative property of multiplication
7. $\therefore (-1)a = -a$	7. Substitution

Example 6 Prove the property of the *opposite of a sum* (page 29):
For all real numbers a and b, $-(a + b) = (-a) + (-b)$.

Proof

Statements	Reasons
1. a and b are real numbers.	1. Given
2. $\quad -(a + b) = (-1)(a + b)$	2. Multiplicative property of -1
3. $\quad -(a + b) = (-1)a + (-1)b$	3. Distributive property
4. $\therefore -(a + b) = (-a) + (-b)$	4. Multiplicative property of -1

Oral Exercises

For each statement, (a) identify the hypothesis and the conclusion, (b) give
the converse of the statement, and (c) tell if the converse is true or false,
and if false, give a counterexample.

1. If $x = 1$, then $|x| = 1$.

2. If $|x - 1| = 0$, then $x = 1$.

3. An integer is even if it is divisible by 2.

4. $y^2 > 0$ if $y > 0$.

5. If $0 < x < 1$, then $|x| < 1$.

6. If a triangle is isosceles, then two of its angles are congruent.

Give reasons for the steps shown in each proof. The domain of each variable is the set of real numbers unless otherwise stated.

7. If $a + a = a$, then $a = 0$.

Proof

1. $a + a = a$
2. $-a$ is a real number.
3. $(a + a) + (-a) = a + (-a)$
4. $a + [a + (-a)] = a + (-a)$
5. $a + 0 = 0$
6. $\therefore a = 0$

8. If $b \neq 0$, then $\dfrac{ab}{b} = a$.

Proof

1. $b \neq 0$
2. $\dfrac{1}{b}$ is a real number.
3. $\dfrac{ab}{b} = (ab) \cdot \dfrac{1}{b}$
4. $\dfrac{ab}{b} = a\left(b \cdot \dfrac{1}{b}\right)$
5. $\dfrac{ab}{b} = a \cdot 1$
6. $\therefore \dfrac{ab}{b} = a$

Written Exercises

Give a counterexample to show that each statement is false. The domain of each variable is the set of real numbers.

A   **1.** If $a^2 = b^2$, then $a = b$.   **2.** If $b < a$, then $a - b < 0$.

     3. $|a - b| = |a| - |b|$ **4.** $|-a| < 0$

Give reasons for the steps shown in each proof. You may use the axioms listed at the top of page 81, definitions, theorems proved in the examples of this lesson, and the results of earlier exercises as reasons. The domain of each variable is the set of real numbers unless otherwise stated.

5. The opposite of a product:
$-ab = a(-b)$.

Proof

1. $-ab = -ba$
2. $-ab = (-b)a$
    (See Example 4.)
3. $\therefore -ab = a(-b)$

7. $(a + b) - b = a$.

Proof

1. $(a + b) - b = (a + b) + (-b)$
2. $(a + b) - b = a + [b + (-b)]$
3. $(a + b) - b = a + 0$
4. $\therefore (a + b) - b = a$

6. The product of opposites:
$(-a)(-b) = ab$.

Proof

1. $(-a)(-b) = -[a(-b)]$
2. $(-a)(-b) = -[-(ab)]$
3. $\therefore (-a)(-b) = ab$
    (See Example 3.)

8. If $x + c = 0$, then $x = -c$.

Proof

1. $x + c = 0$
2. $(x + c) + (-c) = 0 + (-c)$
3. $x + [c + (-c)] = 0 + (-c)$
4. $x + 0 = 0 + (-c)$
5. $\therefore x = -c$

9. If $u \neq 0$ and $u^2 = u$, then $u = 1$.

Proof

1. $u \neq 0$; $u^2 = u$
2. $\quad u \cdot u = u$
3. $\dfrac{1}{u}$ is a real number.
4. $(u \cdot u) \cdot \dfrac{1}{u} = u \cdot \dfrac{1}{u}$
5. $u \cdot \left(u \cdot \dfrac{1}{u}\right) = u \cdot \dfrac{1}{u}$
6. $\quad u \cdot 1 = 1$
7. $\quad \therefore u = 1$

10. If $c \neq 0$ and $cx = 1$, then $x = \dfrac{1}{c}$.

Proof

1. $c \neq 0$; $cx = 1$
2. $\dfrac{1}{c}$ is a real number.
3. $\quad \dfrac{1}{c}(cx) = \dfrac{1}{c} \cdot 1$
4. $\left(\dfrac{1}{c} \cdot c\right)x = \dfrac{1}{c} \cdot 1$
5. $\quad 1 \cdot x = \dfrac{1}{c} \cdot 1$
6. $\quad \therefore x = \dfrac{1}{c}$

11. Zero-product property: If $ab = 0$, then $a = 0$ or $b = 0$.

Proof

1. $\quad ab = 0$
2. Suppose $a \neq 0$. Then $\dfrac{1}{a}$ is a real number. (Note that if $a = 0$, the given statement is obviously true.)
3. $\quad \dfrac{1}{a}(ab) = \dfrac{1}{a} \cdot 0$
4. $\quad \dfrac{1}{a}(ab) = 0$
5. $\left(\dfrac{1}{a} \cdot a\right)b = 0$
6. $\quad 1 \cdot b = 0$
7. $\quad b = 0$
8. $\therefore$ if $a \neq 0$, then $b = 0$.

12. Multiplication is distributive with respect to subtraction: $a(b - c) = ab - ac$.

Proof

1. $\quad a(b - c) = a[b + (-c)]$
2. $\quad a(b - c) = ab + a(-c)$
3. $\quad a(b - c) = ab + (-ac)$ (*Hint:* Use Exercise 5.)
4. $\therefore a(b - c) = ab - ac$

13. Division is distributive with respect to subtraction: If $c \neq 0$, then $\dfrac{a - b}{c} = \dfrac{a}{c} - \dfrac{b}{c}$.

Proof

1. $c \neq 0$; $\dfrac{1}{c}$ is a real number.
2. $\quad \dfrac{a - b}{c} = (a - b) \cdot \dfrac{1}{c}$
3. $\quad \dfrac{a - b}{c} = a \cdot \dfrac{1}{c} - b \cdot \dfrac{1}{c}$ (*Hint:* Use Exercise 12.)
4. $\therefore \dfrac{a - b}{c} = \dfrac{a}{c} - \dfrac{b}{c}$

14. Cancellation property of multiplication: If $c \neq 0$ and $ca = cb$, then $a = b$.

Proof

1. $c \neq 0$; $ca = cb$
2. $\dfrac{1}{c}$ is a real number.
3. $\quad \dfrac{1}{c}(ca) = \dfrac{1}{c}(cb)$
4. $\left(\dfrac{1}{c} \cdot c \right) a = \left(\dfrac{1}{c} \cdot c \right) b$
5. $\quad\quad 1 \cdot a = 1 \cdot b$
6. $\quad\quad \therefore a = b$

B **15.** Follow the steps outlined in Example 3 (with multiplication replacing addition) to prove that if $a \neq 0$, then $\dfrac{1}{\frac{1}{a}} = a$.

16. Prove: If $b \neq 0$, then $\dfrac{-a}{b} = -\dfrac{a}{b}$.
(*Hint:* Use the definition of division and Example 4.)

17. Prove: If $b \neq 0$, then $\dfrac{1}{-b} = -\dfrac{1}{b}$.
$\left(\textit{Hint: Show that } -\dfrac{1}{b} \text{ is the reciprocal of } -b. \right)$

18. Prove: If $b \neq 0$, then $\dfrac{a}{-b} = -\dfrac{a}{b}$. (*Hint:* Use Exercises 17 and 5.)

19. Prove: If $b \neq 0$, then $\dfrac{-a}{-b} = \dfrac{a}{b}$. (*Hint:* Use Exercises 17 and 6.)

20. Prove: If $a \neq 0$ and $b \neq 0$, then $\dfrac{1}{a} \cdot \dfrac{1}{b} = \dfrac{1}{ab}$.
$\left(\textit{Hint: Show that } \dfrac{1}{a} \cdot \dfrac{1}{b} \text{ is the reciprocal of } ab. \right)$

21. Prove: If $b \neq 0$ and $d \neq 0$, then $\dfrac{a}{b} \cdot \dfrac{c}{d} = \dfrac{ac}{bd}$.
(*Hint:* Use Exercise 20.)

C **22.** Prove: If $c \neq 0$ and $d \neq 0$, then $\dfrac{1}{\frac{c}{d}} = \dfrac{d}{c}$.

23. Prove: If $c \neq 0$ and $d \neq 0$, then $\dfrac{a}{b} \div \dfrac{c}{d} = \dfrac{ad}{bc}$.

24. Use the zero-product property to give a convincing argument why 0 has no reciprocal.

25. Use the zero-product property to give a convincing argument why the product of two nonzero real numbers is never equal to 0.

2-7 Theorems about Order and Absolute Value

Objective   To prove theorems about inequalities and absolute value.

The properties of order (page 59) play an important part in proving the theorems of this section.

Example 1   Prove: If a is a real number and $a \neq 0$, then $a^2 > 0$.

Proof

Statements	Reasons
1. a is a real number; $a \neq 0$	1. Given
2. Either $a > 0$ or $a < 0$.	2. Comparison property of order
3. *Case 1:* $a > 0$ Multiply this inequality by the *positive* number a: $\underbrace{a \cdot a} > \underbrace{a \cdot 0}$	3. First multiplication property of order (with $<$ replaced by $>$)
4. $\therefore \ \ a^2 \ > \ \ 0$	4. Definition of a^2 (page 7) and multiplicative property of 0
5. *Case 2:* $a < 0$ Multiply this inequality by the *negative* number a: $\underbrace{a \cdot a} > \underbrace{a \cdot 0}$	5. Second multiplication property of order
6. $\therefore \ \ a^2 \ > \ \ 0$	6. See the reasons in Step 4.

The rules for multiplication are proved as theorems in Examples 2 and 3.

Example 2   Prove for all real numbers a and b:
   a. If $a > 0$ and $b > 0$, then $ab > 0$.
   b. If $a < 0$ and $b < 0$, then $ab > 0$.
   c. If $a > 0$ and $b < 0$, then $ab < 0$.

Proof   We will prove parts (**b**) and (**c**). The proof of (**a**) is left as Exercise 9.

Proof of part (**b**)

Statements	Reasons
1. $a < 0$ and $b < 0$	1. Given
2. Multiply both sides of $b < 0$ by the *negative* number a: $ab > a \cdot 0$	2. Second multiplication property of order
3. $\therefore ab > 0$	3. Multiplicative property of 0

Proof of part (c)

Statements	Reasons
1. $a > 0$ and $b < 0$	1. Given
2. Multiply both sides of $b < 0$ by the *positive* number a: $$ab < a \cdot 0$$	2. First multiplication property of order
3. $\therefore ab < 0$	3. Multiplicative property of 0

Recall that the **absolute value** of a real number x is defined as follows:

$$|x| = x \quad \text{if } x \geq 0$$
$$|x| = -x \text{ if } x < 0$$

Example 3 Prove: For all real numbers a and b, $|ab| = |a| \cdot |b|$.

Proof If either a or b is 0, $|ab| = |a| \cdot |b|$ because both products are 0. There are three remaining cases to prove. Case 1: $a > 0$ and $b > 0$; Case 2: $a < 0$ and $b < 0$; and Case 3: $a > 0$ and $b < 0$. Cases 2 and 3 follow. Case 1 is left as Exercise 10.

Case 2: $a < 0$ and $b < 0$

Statements	Reasons
1. $a < 0$; $b < 0$	1. Given
2. $ab > 0$	2. Example 2, part (b)
3. $|ab| = ab$	3. Definition of absolute value
4. $|a| = -a,\ |b| = -b$	4. Definition of absolute value
5. $|a| \cdot |b| = (-a)(-b)$	5. Multiplication property of equality
6. $|a| \cdot |b| = ab$	6. Exercise 6, page 85
7. $\therefore |ab| = |a| \cdot |b|$	7. Substitution principle (Steps 3 and 6)

Case 3: $a > 0$ and $b < 0$

Statements	Reasons
1. $a > 0$; $b < 0$	1. Given
2. $ab < 0$	2. Example 2, part (c)
3. $|ab| = -ab$	3. Definition of absolute value
4. $|a| = a,\ |b| = -b$	4. Definition of absolute value
5. $|a| \cdot |b| = a(-b)$	5. Multiplication property of equality
6. $|a| \cdot |b| = -ab$	6. Property of the opposite of a product
7. $\therefore |ab| = |a| \cdot |b|$	7. Substitution principle (Steps 3 and 6)

Oral Exercises

Give reasons for the steps shown in each proof.

1. Prove: For all real numbers a and b, if $a < b$, then $-a > -b$.

Proof

1. $a < b$
2. $(-1)a > (-1)b$
3. $\therefore -a > -b$

2. Prove: If a is a real number and $a < 0$, then $-a > 0$.

Proof

1. $a < 0$
2. $a + (-a) < 0 + (-a)$
3. $0 < 0 + (-a)$
4. $\therefore 0 < -a$, or $-a > 0$

3. In Example 3, why is there no need to consider the case $a < 0$ and $b > 0$?

Written Exercises

Give reasons for the steps shown in the proof of each theorem. The domain of each variable is the set of real numbers unless otherwise stated.

A

1. If $a > b$, then $-a < -b$.

Proof

1. $a > b$
2. $(-1)a < (-1)b$
3. $\therefore -a < -b$

2. If $a > b$, then $a - c > b - c$.

Proof

1. $a > b$
2. $a + (-c) > b + (-c)$
3. $\therefore a - c > b - c$

3. If $a < b$ and $c < d$, then $a + c < b + d$.

Proof

1. $a < b$
2. $a + c < b + c$
3. $c < d$
4. $b + c < b + d$
5. $\therefore a + c < b + d$

4. If $0 < a < b$, then $a^2 < b^2$.

Proof

1. $0 < a < b$
2. $a \cdot a < a \cdot b$
3. $a \cdot b < b \cdot b$
4. $a \cdot a < b \cdot b$
5. $\therefore a^2 < b^2$

5. If $a < b < 0$, then $a^2 > b^2$.

Proof

1. $a < b$ and $a < 0$
2. $a \cdot a > a \cdot b$
3. $a < b$ and $b < 0$
4. $a \cdot b > b \cdot b$
5. $a \cdot a > b \cdot b$
6. $\therefore a^2 > b^2$

6. If $0 < a < b$ and $0 < c < d$, then $ac < bd$.

Proof

1. $a < b$ and $c > 0$
2. $ac < bc$
3. $c < d$ and $b > 0$
4. $bc < bd$
5. $\therefore ac < bd$

7. Prove: If $a > 0$, then $-a < 0$. (*Hint:* See Oral Exercise 2.)

8. Prove: If $0 < a < 1$, then $a^2 < a$.

9. Prove: If $a > 0$ and $b > 0$, then $ab > 0$. (Part (a), Example 2)

10. Prove: If $a > 0$ and $b > 0$, then $|ab| = |a|\,|b|$. (Case 1, Example 3)

In Exercises 11–16, assume that if $a > 0$, then $\frac{1}{a} > 0$, and if $a < 0$, then $\frac{1}{a} < 0$.

B 11. Prove: If $a > 0$ and $b > 0$, then $\frac{a}{b} > 0$.

12. Prove: If $a < 0$ and $b < 0$, then $\frac{a}{b} > 0$.

13. Prove: If $a > 0$ and $b < 0$, then $\frac{a}{b} < 0$.

14. Prove: If $a < 0$ and $b > 0$, then $\frac{a}{b} < 0$.

15. Prove: If $a > 0$, $b > 0$, and $\frac{1}{a} > \frac{1}{b}$, then $a < b$.

16. Prove: If $0 < a < b$, then $\frac{1}{a} > \frac{1}{b}$. $\left(\textit{Hint: } \text{Multiply by } \frac{1}{ab}.\right)$

C 17. Prove: If $a < b$, then $a < \dfrac{a + b}{2} < b$. This proof shows that the average of two numbers lies between them.

18. Prove: For every $a > 0$, if $a > \frac{1}{a}$, then $a > 1$.
$\Big($*Hint:* Give an *indirect proof*. That is, show that the assumptions $a = 1$ and $a < 1$ each lead to a contradiction of the hypothesis that $a > \frac{1}{a}.\Big)$

Mixed Review Exercises

Tell whether each statement is true for all real numbers.

1. If $a > b$, then $b < a$.

2. If $a > b$ and $c < 0$, then $ac < bc$.

3. If $ac = bc$ and $c \neq 0$, then $a = b$.

Solve each open sentence and graph each solution set that is not empty.

4. $|x - 3| = 1$
5. $4d + 5 \geq 1$
6. $-1 < 2 - y < 3$
7. $6r + 13 = 25$
8. $|n| + 7 < 5$
9. $-2k > 8$ or $k - 4 \geq 0$

Self-Test 3

Vocabulary   axiom (p. 81)   theorem (p. 81)
                hypothesis (p. 81)   corollary (p. 81)
                conclusion (p. 81)   converse (p. 82)
                proof (p. 81)   counterexample (p. 82)

1. Give reasons for the steps shown in the following proof. If   Obj. 2-6, p. 81
   $a + b = c$, then $a = c - b$.

   Proof

   1. $a + b = c$
   2. $-b$ is a real number.
   3. $(a + b) + (-b) = c + (-b)$
   4. $a + [b + (-b)] = c + (-b)$
   5. $a + 0 = c + (-b)$
   6. $a = c + (-b)$
   7. $\therefore a = c - b$

2. Prove: If $ab = c$ and $b \neq 0$, then $a = \dfrac{c}{b}$.

3. Prove: If $a > 1$, then $a^2 > a$.   Obj. 2-7, p. 88

Chapter Summary

1. Inequalities can be solved by using the properties of order on page 59 and the transformations on page 60.

2. If an inequality is a conjunction, it can be solved by finding the values of the variable for which both sentences are true. If an inequality is a disjunction, it can be solved by finding the values of the variable for which at least one of the sentences is true.

3. Some word problems can be solved algebraically by translating the given information into an inequality and then solving the inequality.

4. Equations and inequalities that involve absolute value can be solved algebraically by writing an equivalent conjunction or disjunction.

5. Equations and inequalities that involve absolute value can be solved geometrically by using this fact: On a number line the distance between the graphs of two numbers is the absolute value of the difference between the numbers.

6. Axioms are statements assumed to be true. Using the axioms for real numbers (referred to on page 81), other properties of real numbers can be proved as theorems. Every step in the proof of a theorem can be justified by either an axiom, a definition, a given fact, or a theorem.

Chapter Review

Give the letter of the correct answer.

Solve.

1. $-\dfrac{m}{2} < -2$ $\dfrac{m}{2} > 2$ $m > 4$ 2-1

   a. $m > 1$ **b.** $m > 4$ **c.** $m < 4$ **d.** $m > -4$

2. $3(n - 1) > 5n + 7$ $3(n - 1) < 5n + 7$

   a. $n > -5$ **b.** $n < -4$ **c.** $n > -4$ **d.** $n < -5$

   $3n - 3 < 5n + 7$ $3n < 5n + 10$

3. $-3 < 4c + 5 \le 1$ 2-2

   a. $-2 < c \le -1$ **b.** $-2 < c \le 1$

   c. $2 < c \le -1$ **d.** $2 < c \le 1$

4. Graph the solution set of the following disjunction:

   $4 - w \le 3$ or $w + 5 < 3$ $w < -2$

   $w \ge 1$

   a.

   b.

   c.

   d.

Solve.

5. At the video store, you can rent tapes for $1.50 each per day, or you can get unlimited rentals for a monthly fee of $20. At most how many tapes could you rent at $1.50 before spending more than the cost for a month's unlimited rentals?   2-3

   a. 12   **b.** 13   **c.** 14   **d.** 15

6. $\left|5 - \dfrac{x}{3}\right| = 7$ 2-4

   a. $\{-6, 36\}$ **b.** $\{-6, -36\}$ **c.** $\{6, -36\}$ **d.** $\{6, 36\}$

7. $|2y + 9| < 13$

   a. $-2 < y < 11$ **b.** $y > 2$ **c.** $-11 < y < 2$ **d.** $y < -11$ or $y > 2$

8. $|4 - h| \ge 5$

   a. $h \le -1$ or $h \ge 9$ **b.** $h \le -1$

   c. $h \le -9$ or $h \ge 1$ **d.** $-1 \le h \le 9$

9. Describe the graph of the solution set of $|k + 3| < 2$.   2-5

   a. Points at least 2 units from 3.   **b.** Points less than 2 units from -3.

   c. Points more than 3 units from 2.   **d.** Points at most 2 units from -3.

10. A theorem that can be proved easily from another theorem is a(n) __?__.   2-6

    a. axiom   **b.** corollary   **c.** property   **d.** theorem

11. The converse of a true statement is __?__.

    a. a corollary   **b.** never true   **c.** always true   **d.** sometimes true

Supply the conclusion that makes each statement true.

12. For real numbers a and b, $ab = 0$ if and only if ___?___ .

    a. $a = 0$ **b.** $b = 0$ **c.** $a = 0$ or $b = 0$ **d.** $a = 0$ and $b = 0$

13. For real numbers a, b, and c, if $a > b$ and $c < 0$, then ___?___ .   2-7

    a. $\dfrac{a}{c} > \dfrac{b}{c}$ **b.** $ac > bc$ **c.** $\dfrac{a}{c} < \dfrac{b}{c}$ **d.** $a + c < b + c$

Chapter Test

Solve.

1. $5x - 9 > 6x$ **2.** $3(2y + 1) < 2(y - 3) + 1$ 2-1

3. $-3 < 4 - m < 6$ **4.** $-2n \ge 8$ or $n + 3 < 7$ 2-2

5. The dimensions of a rectangle are consecutive odd integers. Find the small-   2-3
est such rectangle with a perimeter of at least 35 cm.

Solve.

6. $|9 - 2k| = 5$ **7.** $\left|\dfrac{c}{3} + 1\right| > 2$ **8.** $|4f + 3| \le 5$ 2-4

Solve each sentence graphically.

9. $|w - 4| \ge 1$ **10.** $|h + 2| < 3$ 2-5

11. Write the converse of the following statement and tell whether the converse   2-6
is true or false: "If $x > 1$, then $x^2 > x$."

12. Supply reasons for the steps in the proof of the following statement: For all
real numbers a and nonzero real numbers b, if $ab = b$, then $a = 1$.

    1. a is a real number; b is a   1. ___?___
       nonzero real number; $ab = b$

    2. $\dfrac{1}{b}$ is a real number.   2. ___?___

    3. $(ab) \cdot \dfrac{1}{b} = b \cdot \dfrac{1}{b}$ 3. ___?___

    4. $a\left(b \cdot \dfrac{1}{b}\right) = b \cdot \dfrac{1}{b}$ 4. ___?___

    5. $a(1) = 1$ 5. ___?___
    6. $a = 1$ 6. ___?___

**Tell whether the statement is true for all real numbers. If it is not, give a
counterexample.**

13. If $a > b$, then $|a| > |b|$.   **14.** If $a < b$ and $c > d$, then $ad < bc$.   2-7

15. If $a < b$ and $c < 0$, then $ac < bc$.   **16.** If $a < b < 0$, then $a^2 < b^2$.

Extra / Symbolic Logic: Boolean Algebra

Most people think of algebra as the study of operations with numbers and variables. Another kind of algebra that is used in the design of electronic digital computers involves operations with logical statements. In fact, complex electrical circuits can be analyzed using logical statements similar to those that you will work with in Exercises 1–18. This "algebra of logic" is called **Boolean algebra** in honor of its originator, George Boole (British, 1815–1864).

In Boolean algebra you use letters such as p, q, r, s, and so on, to stand for statements. For example, you might let p represent the statement "3 is an odd integer," and q, the statement "5 is less than 3." In this case, the statement p has the truth value T (the statement is true), whereas q has the truth value F (the statement is false).

The table below shows the operations that are used in Boolean algebra to produce compound statements from any given statements p and q.

Operation	In words	In symbols
Conjunction	p and q	$p \wedge q$
Disjunction	p or q	$p \vee q$
Conditional	If p, then q	$p \rightarrow q$
Equivalence	p if and only if q	$p \leftrightarrow q$
Negation	not p	$\sim p$

Note that in ordinary English the word *or* is sometimes used in the exclusive sense, to mean that just one of two alternatives occurs. In mathematical usage, however, the word *or* generally is used in the inclusive sense. As you can see in the table below, the disjunction p or q is true if only p is true, if only q is true, or if both are true. The rules for assigning truth values to compound statements are shown in the following five *truth tables*.

The conjunction $p \wedge q$ is true whenever *both* p and q are true.

The disjunction $p \vee q$ is true whenever *at least one* of the statements is true.

Conjunction

p	q	$p \wedge q$
T	T	T
T	F	F
F	T	F
F	F	F

Disjunction

p	q	$p \vee q$
T	T	T
T	F	T
F	T	T
F	F	F

The conditional $p \rightarrow q$ is false only when p is true and q is false.

The equivalence $p \leftrightarrow q$ is true whenever p and q have the same truth value.

Conditional

p	q	$p \rightarrow q$
T	T	T
T	F	F
F	T	T
F	F	T

Equivalence

p	q	$p \leftrightarrow q$
T	T	T
T	F	F
F	T	F
F	F	T

The negation $\sim p$ is the denial of p. Therefore, it is reasonable to agree that $\sim p$ is false when p is true, and true when p is false.

Negation

p	$\sim p$
T	F
F	T

Example 1 Let r stand for "$1 > 2$," and s for "$2 < 4$." Read each of the following statements. Then by referring to the truth tables above and on page 95, give the truth value of the statement and a reason for your answer.

  a. $r \wedge s$ **b.** $r \vee s$ **c.** $r \rightarrow s$ **d.** $r \leftrightarrow s$ **e.** $\sim r$

Solution   **a.** $r \wedge s$: $1 > 2$ and $2 < 4$. F; the truth value of r is F.

  b. $r \vee s$: $1 > 2$ or $2 < 4$. T; the truth value of s is T.

  c. $r \rightarrow s$: If $1 > 2$, then $2 < 4$. T; the truth value of r is F, and that of s is T.

  d. $r \leftrightarrow s$: $1 > 2$ if and only if $2 < 4$. F; r and s have different truth values.

  e. $\sim r$: not ($1 > 2$), that is $1 \leq 2$. T; the truth value of r is F.

Two compound statements are *logically equivalent* if they have the same truth value for each combination of truth values of the individual statements.

Example 2 Construct a truth table for $q \vee \sim p$ and $p \rightarrow q$ for all possible truth values of p and q. Show that $q \vee \sim p$ and $p \rightarrow q$ are logically equivalent.

Solution

p	q	$\sim p$	$q \vee \sim p$	$p \rightarrow q$
T	T	F	T	T
T	F	F	F	F
F	T	T	T	T
F	F	T	T	T

Column 4 was constructed by using disjunction with columns 2 and 3.

Column 5 was constructed by using the conditional with columns 1 and 2.

Since columns 4 and 5 match, $q \vee \sim p$ and $p \rightarrow q$ are logically equivalent.

A compound statement that is true for all truth values of its component statements is called a **tautology.** For example, $p \leftrightarrow p$ is a tautology.

Exercises

In Exercises 1–9 assume that r and p are true statements and that q is a false statement. Determine the truth value of each statement.

1. $q \rightarrow r$ **2.** $\sim r \wedge p$

3. $p \vee \sim q$ **4.** $r \wedge (p \vee q)$

5. $p \vee (q \wedge r)$ **6.** $(p \vee r) \rightarrow q$

7. $\sim p \rightarrow (q \vee \sim r)$ **8.** $r \rightarrow (q \rightarrow r)$

9. $[r \wedge (p \vee q)] \rightarrow [(r \wedge p) \vee (r \wedge q)]$

Given that $p \rightarrow q$ is false, show that each statement in Exercises 10–12 is true.

10. $q \rightarrow p$ **11.** $p \wedge \sim q$ **12.** $(p \vee q) \wedge p$

Construct a truth table and determine if the two statements in each exercise are logically equivalent.

13. $p \vee q, \ q \vee p$ **14.** $q \vee \sim q, \ q$

Construct a truth table for each statement and determine whether the statement is a tautology.

15. $(p \vee q) \rightarrow p$ **16.** $(q \wedge \sim q) \rightarrow p$

17. $r \wedge (p \vee q) \leftrightarrow (r \wedge p) \vee (r \wedge q)$ **18.** $[(p \rightarrow q) \wedge (q \rightarrow r)] \rightarrow (p \rightarrow r)$

Mixed Problem Solving

Solve each problem that has a solution. If a problem has no solution, explain why.

A **1.** A child has 40 coins worth more than $6.00. If the coins are dimes and quarters only, what can you conclude about the number of quarters?

2. The width and length of a rectangle are consecutive odd integers. If the width is w, find the perimeter, in simplest form, in terms of w.

3. Percy works 8 hours more per week than Selena. Selena, however, earns $2 more per hour than Percy, who earns $6 per hour. If their weekly pay is the same, how many hours does Percy work per week?

4. In Sioux Falls, South Dakota, the average low temperatures in January and July are $-16°$ C and $17°$ C, respectively. Find the difference between these temperatures.

5. Marcus invested $4000 at simple annual interest, some at 5% and the rest at 8%. If he received $272 in interest for one year, how much did he invest at each rate?

6. Megan bought four times as many pencils as erasers. Pencils cost $.04 each, and erasers cost $.19 each. If Megan spent $2.10 in all, how many pencils did she buy?

7. Four tennis balls cost d dollars. How many tennis balls can you buy for $8? Give your answer in terms of d.

8. Drew's second test score was 6 points higher than his first score. If his average was less than 80, what can you conclude about his first score?

9. The measure of a supplement of an angle is four times the measure of a complement. Find the measure of the angle.

10. Helen has $57 more in her checking account than in her savings account. If she has $239 altogether, how much is in savings?

B **11.** At a health food store, dried apple slices that cost $1.80/lb are mixed with dried banana slices that cost $2.10/lb. If the mixture weighs 5 lb and costs $1.92/lb, how many pounds of apple slices were used?

12. The Petersons drove to a friend's house at 80 km/h. On the return trip, rain reduced their speed by 10 km/h. If their total driving time was 3 h, find the total distance driven.

13. The width of a rectangle is 1 cm more than half the length. When both the length and the width are increased by 1 cm, the area increases by 20 cm^2. Find the original dimensions of the rectangle.

14. At 1:00 P.M. Sue left her home and began walking at 6 km/h toward Sandy's house. Fifteen minutes later, Sandy left her home and walked at 8 km/h toward Sue's house. If Sue lives 5 km from Sandy, at what time did they meet? Who walked farther?

Preparing for College Entrance Exams

Strategy for Success

If there is a penalty for incorrect answers, random guessing may not be to your advantage. However, if you can eliminate several answers in a given question and then take a guess, your chance of selecting the correct answer is improved and it may be worthwhile to guess.

Decide which is the best of the choices given and write the corresponding letter on your answer sheet.

1. If $a > b$ and $c < 0$, which of the following *must* be true:

   I. $ab > bc$ II. $a^2 > b^2$ III. $\dfrac{a+c}{c} < \dfrac{b+c}{c}$

   (A) I only   **(B)** II only   **(C)** III only   **(D)** I, II, and III   **(E)** II and III only

2. Which of the following open sentences is true for all real numbers x?

   (A) $3[x - 5(2 - x)] = 2(9x + 15)$ **(B)** $|7 - x| > 0$

   (C) $(7x - 1) \div \dfrac{1}{2} = 2 - 14x$ **(D)** $-\dfrac{5}{2}(-6 + 4x) = 10\left(\dfrac{3}{2} - x\right)$

   (E) $x + 3 < 3 - x$

3. Lucy spent \$2.36 on 15¢ and 22¢ stamps. If she bought twice as many 22¢ stamps as 15¢ stamps, how many 22¢ stamps did she buy?

   (A) 2   **(B)** 4   **(C)** 8   **(D)** 5   **(E)** 10

4. Evaluate $x^3 - 3x^2 \div (4 + x)$ if $x = -2$.

   (A) 2   **(B)** -2 **(C)** -10 **(D)** -14 **(E)** -6

5. Solve $A = \dfrac{h}{2}(b_1 + b_2)$ for b_2.

   (A) $\dfrac{A}{2h} - b_1$ **(B)** $A - \dfrac{hb_1}{2}$ **(C)** $\dfrac{2A - b_1}{h}$ **(D)** $\dfrac{2A}{h} - b_1$ **(E)** $\dfrac{2A}{hb_1}$

6. How many sets of four consecutive integers are there such that half the sum of the first three is greater than the fourth, and the sum of the last three is at least four times the first?

   (A) none   **(B)** one   **(C)** two   **(D)** three   **(E)** four

7. The graph at the right shows the solution set to which inequality?

   (A) $|2t + 3| < 1$ **(B)** $|2t + 3| > 1$

   (C) $|3t + 2| < 1$ **(D)** $|3t + 2| \geq -1$

   (E) $\dfrac{1}{2} \geq \left|t + \dfrac{3}{2}\right|$

3 Linear Equations and Functions

The ratio of a roof's rise to its run is referred to as "pitch" by a builder. In mathematics, this ratio is called "slope."

Linear Equations and Their Graphs

3-1 Open Sentences in Two Variables

Objective   To find solutions of open sentences in two variables and to solve problems involving open sentences in two variables.

Equations and inequalities such as

$$9x + 2y = 15, \quad y = x^2 - 4, \quad \text{and} \quad 2x - y \geq 6$$

are called **open sentences in two variables.**

A **solution** of an open sentence in the two variables x and y is a pair of numbers, one a value of x and the other a value of y, that together make the sentence a true statement. We usually write such a solution as an **ordered pair** in which the value of x is listed first and value of y is listed second. For example,

(1, 3)   is a solution of $9x + 2y = 15$ because $9(1) + 2(3) = 15$.

However, (3, 1) is *not* a solution because $9(3) + 2(1) \neq 15$. A solution of an open sentence is said to *satisfy* the sentence. The set of all solutions is called the **solution set** of the open sentence. Finding the solution set is called **solving** the open sentence.

Example 1   Solve the equation $9x + 2y = 15$ if the domain of x is $\{-1, 0, 1, 2, 3\}$.

Solution 1   Solve the equation for y.

$$y = \frac{15 - 9x}{2}$$

Then replace x with each value in its domain and find the corresponding value of y. The last column of the table lists the five solutions. Solving the equation over other domains will produce different solutions.

x	$\dfrac{15 - 9x}{2}$	y	Solution
-1	$\dfrac{15 - 9(-1)}{2}$	12	$(-1, 12)$
0	$\dfrac{15 - 9(0)}{2}$	$\dfrac{15}{2}$	$\left(0, \dfrac{15}{2}\right)$
1	$\dfrac{15 - 9(1)}{2}$	3	$(1, 3)$
2	$\dfrac{15 - 9(2)}{2}$	$-\dfrac{3}{2}$	$\left(2, -\dfrac{3}{2}\right)$
3	$\dfrac{15 - 9(3)}{2}$	-6	$(3, -6)$

$\therefore$ the solution set is $\left\{(-1, 12), \left(0, \dfrac{15}{2}\right), (1, 3), \left(2, -\dfrac{3}{2}\right), (3, -6)\right\}$.   **Answer**

Solution 2 Substitute each value in the domain of x in the given equation $9x + 2y = 15$. Then solve for y.

$$x = -1: 9(-1) + 2y = 15$$
$$2y = 24$$
$$y = 12$$

$$x = 0: 9(0) + 2y = 15$$
$$2y = 15$$
$$y = \frac{15}{2}$$

Solution $(-1, 12)$ *Solution* $\left(0, \frac{15}{2}\right)$

When you substitute the other values in the domain of x in the given equation, you will find the complete solution set:

$$\left\{(-1, 12), \left(0, \frac{15}{2}\right), (1, 3), \left(2, -\frac{3}{2}\right), (3, -6)\right\}. \quad \textbf{\textit{Answer}}$$

Open sentences in two variables can be used to solve certain word problems.

Example 2 A customer asks a bank teller for $390 in traveler's checks, some worth $50 and some worth $20. Find all possibilities for the number of each type of check the customer could receive.

Solution

Step 1 The problem asks for the number of $50 checks and the number of $20 checks whose total value is $390.

Step 2 Let f = the number of $50 checks and let t = the number of $20 checks.

Step 3 The total value of the $50 checks is $50f$ dollars and the total value of the $20 checks is $20t$ dollars. Write an equation that expresses the total value of the checks in dollars.

$$50f + 20t = 390$$

Step 4 Solve the equation for one variable, say t, in terms of the other. First divide both sides by 10 to obtain this simpler equation:

$$5f + 2t = 39$$
$$t = \frac{39 - 5f}{2}$$

Remember: The number of each type of check *must* be a whole number. If f is even, then $5f$ is even and $39 - 5f$ is odd. If $39 - 5f$ is odd, t is not a whole number. Also, if f exceeds 7, t is negative. Therefore, replace f with 1, 3, 5, and 7 as shown in the table.

f	t	(f, t)
1	17	(1, 17)
3	12	(3, 12)
5	7	(5, 7)
7	2	(7, 2)

$\therefore$ the solution set of $50f + 20t = 390$ over the domain of the whole numbers is $\{(1, 17), (3, 12), (5, 7), (7, 2)\}$.

Step 5 The check is left for you.

∴ the customer can receive: 1 $50 check and 17 $20 checks
or 3 $50 checks and 12 $20 checks
or 5 $50 checks and 7 $20 checks
or 7 $50 checks and 2 $20 checks. ***Answer***

Example 3 Find all positive two-digit odd numbers with this property: When the digits are interchanged, the result exceeds the original number by more than 36.

Solution

Step 1 The problem asks for all positive two-digit odd numbers such that when the digits are interchanged, the new number is greater than the original number plus 36. Recall that every positive integer can be written in expanded form. For example:

$$73 = 7 \cdot 10 + 3$$

units' digit
tens' digit

Step 2 Let u = the units' digit and let t = the tens' digit. Then the original number is $10t + u$. The number with the digits interchanged is $10u + t$.

Step 3 Write an inequality relating t and u.

new number	is greater than	original number	plus	36
$10u + t$	$>$	$(10t + u)$	$+$	36

Step 4 Solve the inequality:

$$10u + t > 10t + u + 36$$
$$9u > 9t + 36$$

Divide both sides by 9: $u > t + 4$

Since the original number has two digits, the tens' digit is not zero, so $t \neq 0$ and the replacement set for t is {1, 2, 3, 4, 5, 6, 7, 8, 9}. Since the number is odd, the replacement set for u is {1, 3, 5, 7, 9}.

t	$t + 4$	$u > t + 4$	(t, u)
1	5	7, 9	(1, 7), (1, 9)
2	6	7, 9	(2, 7), (2, 9)
3	7	9	(3, 9)
4	8	9	(4, 9)

If $t > 4$, then $u > 9$ and there are no more possibilities.

Step 5 The check is left for you.

∴ the numbers are 17, 19, 27, 29, 39, and 49. ***Answer***

Oral Exercises

Tell whether each ordered pair is a solution of the open sentence.

1. $3x + y = 1$ $(1, 0), (0, 1), (1, -2), (1, -4)$
2. $2x - 3y = 5$ $(-1, 1), (1, -1), (2, -3), (-5, -5)$
3. $7x - 2y = 8$ $(1, -1), (2, 3), (0, -4), (4, 10)$
4. $3x + 5y = 4$ $(-2, -2), (-2, 2), (8, 4), (7, -3)$
5. $5x - 2y > 6$ $(2, 0), (0, -3), (1, -1), (1, 1)$
6. $4x - 3y \le 2$ $(1, 2), (2, 1), (1, -1), (0, -1)$

Give three ordered pairs of integers that satisfy the open sentence.

7. $x - y = 2$ 8. $5x + y = 6$ 9. $2x + 3y > 10$ 10. $3x + 5y \le 8$

11. Make up a word problem that could be solved using this equation: $10d + 25q = 160$.

Written Exercises

Solve each equation if the domain of x is $\{-1, 0, 2\}$.

A
1. $2x + 3y = 7$ 2. $3x + 6y = 9$ 3. $-x - 2y = 0$
4. $-2x + y = -3$ 5. $4x - 9y = 5$ 6. $6x - \frac{1}{2}y = 3$

7–12. Solve each equation in Exercises 1–6 if the domain of x is $\{-2, 1, 3\}$.

Complete each ordered pair to form a solution of the equation.

Sample $x + 2y = 8$; $(0, \underline{\,?\,}), (\underline{\,?\,}, 0)$ **Solution** $(0, 4), (8, 0)$

13. $3x + 2y = 12$ $(0, \underline{\,?\,}), (\underline{\,?\,}, 0), (2, \underline{\,?\,})$
14. $4x + 3y = 8$ $(0, \underline{\,?\,}), (\underline{\,?\,}, 0), (5, \underline{\,?\,})$
15. $5x - 2y = 7$ $(0, \underline{\,?\,}), (\underline{\,?\,}, 0), (-3, \underline{\,?\,})$
16. $x + 6y = -9$ $(0, \underline{\,?\,}), (\underline{\,?\,}, 0), (-3, \underline{\,?\,})$
17. $2x - 2y = 3$ $(1, \underline{\,?\,}), \left(\frac{1}{2}, \underline{\,?\,}\right), \left(\underline{\,?\,}, \frac{1}{2}\right)$
18. $3x + 5y = 3$ $(1, \underline{\,?\,}), \left(-\frac{2}{3}, \underline{\,?\,}\right), \left(\underline{\,?\,}, \frac{7}{5}\right)$
19. $\frac{1}{2}x - 2y = 1$ $(1, \underline{\,?\,}), (6, \underline{\,?\,}), (\underline{\,?\,}, 0)$
20. $x + \frac{1}{3}y = 2$ $(1, \underline{\,?\,}), (\underline{\,?\,}, 6), \left(\frac{1}{3}, \underline{\,?\,}\right)$

Find the value of k so that the ordered pair satisfies the equation.

$x + 2y = k$; (3, 1) $x + 2y = k$
$3 + 2(1) = k$
$k = 5$

21. $2x + y = k$; (2, 1)   22. $3x - y = k$; (1, −3)
23. $3x - ky = 4$; (2, −1)   24. $kx + 3y = 7$; (−1, 3)
25. $kx + 2y = k$; (3, 3)   26. $6x - ky = k$; (2, 2)

Solve each equation if each variable represents a *whole number*.

B 27. $x + y = 4$ 28. $2x + y = 6$
   29. $4x + y = 15$ 30. $x + 5y = 24$
   31. $2x + 3y = 18$ 32. $5x + 2y = 30$

Solve each open sentence if each variable represents a *positive integer*.

33. $x + y < 5$ 34. $2x + y < 6$
35. $2x + 3y \leq 12$ 36. $3x + 5y \leq 19$
37. $x + y^2 = 10$ 38. $x^2 + 2y < 11$

In Exercises 39–40, the digits of a positive two-digit integer N are interchanged to form an integer K.

C 39. Show that $N - K$ is an integral multiple of 9.
   40. Show that $N + K$ is an integral multiple of 11.

Problems

In each problem (a) choose two variables to represent the numbers asked for, (b) write an open sentence relating the variables, and (c) solve the open sentence and give the answer to the problem. (Include solutions in which one of the variables is zero.)

A 1. A bank teller needs to pay out $75 using $5 bills and $20 bills. Find all possibilities for the number of each type of bill the teller could use.

   2. Bruce, an appliance salesman, earns a commission of $50 for each washing machine he sells and $100 for each refrigerator. Last month he earned $500 in commissions. Find all possibilities for the number of each kind of appliance he could have sold.

   3. Luis has 95 cents in dimes and quarters. Find all possibilities for the number of each type of coin he could have.

   4. Kimberly has $1.95 in dimes and quarters. Find all possibilities for the number of each type of coin she could have.

5. A certain quadrilateral has three sides of equal length and its perimeter is 19 cm. Find all integral possibilities for the lengths of the sides in centimeters. (*Hint:* The sum of the lengths of any three sides of a quadrilateral must exceed the length of the fourth side.)

6. An isosceles triangle has perimeter 15m. Find all integral possibilities for the lengths of the sides in meters. (*Hint:* The sum of the lengths of any two sides of a triangle must exceed the third side.)

B **7.** A box contains nickels, dimes, and quarters worth $2.00. Find all possibilities for the number of each coin if there are three more dimes than quarters.

8. A bag contains twice as many pennies as nickels and four more dimes than quarters. Find all possibilities for the number of each coin if their total value is $2.01.

In Exercises 9–12, the digits of a positive two-digit integer *N* are interchanged to form an integer *K*. Find all possibilities for *N* under the conditions described.

9. *N* is odd and exceeds *K* by more than 18.

10. The average of *N*, *K*, and 35 is 30.

C **11.** The sum of *K* and twice *N* is less than 60.

12. *N* is even and exceeds *K* by more than 50.

Solve.

13. A 15-member special committee met three times. Twice as many members were present at the third meeting as at the first, and the average attendance was 9 people. Find all possibilities for the number of people present at each meeting.

14. A stick of wood is to be cut into three unequal pieces. The first piece is shorter than the second piece, and the second piece is shorter than the third. If the stick is 24 cm in length and the length of each piece is an even integer, what are the possibilities for the lengths of the pieces?

Mixed Review Exercises

Evaluate each expression if $x = -3$ and $y = 4$.

1. $2x - 5y$ **2.** $-x^2y$ **3.** $|x - y|$ **4.** $(x - 2)(y + 1)$

5. $\dfrac{3x + 1}{y}$ **6.** $|xy|$ **7.** $x + 3y$ **8.** $\dfrac{x - y}{x + y}$

Solve each open sentence and graph each solution set that is not empty.

9. $-7 < 2y - 5 \le 3$ **10.** $|3 - m| > 1$ **11.** $3n + 7 \le 8n - 13$

3-2 Graphs of Linear Equations in Two Variables

Objective   To graph a linear equation in two variables.

Solutions of open sentences in x and y can be graphed in an **xy-coordinate plane.** To set up a **plane rectangular coordinate system,** draw two number lines, or **axes,** meeting at right angles at a point O, the **origin.** The horizontal axis is called the **x-axis,** and the vertical axis, the **y-axis.** The axes divide the plane into four **quadrants.**

With each ordered pair of numbers, you can associate a unique point in the plane. To associate a point with the pair $(-4, -3)$, imagine drawing a vertical line through -4 on the x-axis and a horizontal line through -3 on the y-axis, as shown below. These lines intersect at a point P, the **graph** of $(-4, -3)$. Locating a point in this way is called **graphing** the ordered pair, or **plotting** the point.

By reversing the process just described, you can associate with each point P in the plane a unique ordered pair (a, b) of real numbers, called the **coordinates** of P (see the diagram above). The first coordinate, a, is called the **x-coordinate,** or **abscissa,** of P; the second coordinate, b, is called the **y-coordinate,** or **ordinate** of P.

This **one-to-one correspondence** between ordered pairs of real numbers and points of the plane can be summarized as follows:

1. There is exactly one point in the coordinate plane associated with each ordered pair of real numbers.

2. There is exactly one ordered pair of real numbers associated with each point in the coordinate plane.

A rectangular coordinate system is sometimes called a **Cartesian coordinate system** in honor of René Descartes (1596–1650), the French mathematician who introduced coordinates.

Example 1 Graph the ordered pairs (4, 1), (0, 2), (−2, 4), (2, −4), and (−4, 0) in the same coordinate plane.

Solution

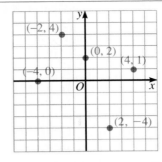

Example 2 Find and graph five solutions of $3x + 2y = 4$.

Solution Solve for one variable, say y, $\qquad y = \dfrac{4 - 3x}{2}$

in terms of the other. Choose convenient values of x and find the corresponding values of y. Then graph the resulting ordered pairs.

x	$\dfrac{4 - 3x}{2}$	y
-2	$\dfrac{4 - 3(-2)}{2}$	5
-1	$\dfrac{4 - 3(-1)}{2}$	$\dfrac{7}{2}$
0	$\dfrac{4 - 3(0)}{2}$	2
1	$\dfrac{4 - 3(1)}{2}$	$\dfrac{1}{2}$
2	$\dfrac{4 - 3(2)}{2}$	-1

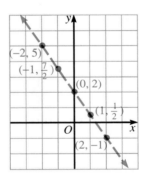

The **graph** of an open sentence in two variables is the set of *all* points in the coordinate plane whose coordinates satisfy the open sentence. Example 2 suggests that if we were to graph *all* the ordered pairs satisfying $3x + 2y = 4$, we would obtain the line shown in blue. The following theorem, which is proved in more advanced courses, tells us that the graph of the equation $3x + 2y = 4$ is in fact a line.

Theorem

The graph of every equation of the form

$$Ax + By = C \quad (A \text{ and } B \text{ not both } 0)$$

is a line. Conversely, every line in the coordinate plane is the graph of an equation of this form.

Because of this property, any equation that can be expressed in the form $Ax + By = C$ (A and B not both 0) is called a **linear equation in two variables.** The equation $3y + 8 = -5x$ is linear since it can be written as $5x + 3y = -8$. The following equations are *not* linear:

$$2x + 3y^2 = 4, \quad xy = 2, \quad 2x + \frac{3}{y} = 5$$

Although you need only two points to determine the graph of a linear equation, it is a good practice to plot a third point as a check. Points where the graph crosses the axes are often easy to find and plot.

Example 3 Graph $2x - 3y = -9$.

Solution The graph crosses the y-axis at a point whose x-coordinate is 0. The graph crosses the x-axis at a point whose y-coordinate is 0.

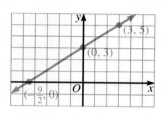

Let $x = 0$.	Let $y = 0$.
$2(0) - 3y = -9$	$2x - 3(0) = -9$
$-3y = -9$	$2x = -9$
$y = 3$	$x = -\dfrac{9}{2}$
Solution $(0, 3)$	*Solution* $\left(-\dfrac{9}{2}, 0\right)$

The graph is the line through the points with coordinates $(0, 3)$ and $\left(-\dfrac{9}{2}, 0\right)$. As a check, note that $(3, 5)$ is a solution of $2x - 3y = -9$ and its graph lies on the line.

Example 4 Graph in a coordinate plane.

a. $y = 4$

b. $x = -\dfrac{5}{2}$

Solution **a.** The equation $y = 4$ can be written as

$$0 \cdot x + 1 \cdot y = 4.$$

The graph consists of all points having y-coordinate 4 and is therefore a horizontal line.

b. The equation $x = -\dfrac{5}{2}$ can be written as

$$1 \cdot x + 0 \cdot y = -\dfrac{5}{2}.$$

The graph consists of all points having x-coordinate $-\dfrac{5}{2}$ and is therefore a vertical line.

Example 4 illustrates the fact that the graph of $Ax + By = C$ is a horizontal line if $A = 0$, and is a vertical line if $B = 0$.

From now on, we will follow the common practice of using the simpler phrase "the point (a, b)" instead of "the point whose coordinates are (a, b)." Also, the phrase "the line $Ax + By = C$" will mean "the line whose equation is $Ax + By = C$."

Oral Exercises

1–12. Give the coordinates of each of the points A through L, and name the quadrant or axis that contains the point.

Name the quadrant of the point described.

13. Its coordinates are both negative.

14. Its coordinates are both positive.

15. Its x-coordinate is negative and its y-coordinate is positive.

16. Its x-coordinate is positive and its y-coordinate is negative.

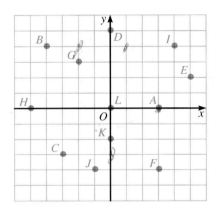

In Exercises 17–24, (a) tell whether or not the equation is linear, and (b) if it is linear, tell whether its graph is a horizontal line, a vertical line, or neither.

17. $x + y = 2$

18. $2x^2 + y = 3$

19. $x - 3 = 0$

20. $2x + 3y + 4 = 0$

21. $x^2 - y^2 = 4$

22. $2y + 3 = 0$

23. $\dfrac{x}{2} - \dfrac{y}{3} = 1$

24. $\dfrac{2}{x} - \dfrac{3}{y} = 1$

Give the coordinates of the points where each line crosses the coordinate axes.

Sample $2x + 3y = 6$ **Solution** $(0, 2)$ and $(3, 0)$

25. $x + 2y = 4$

26. $x + y = 3$

27. $x - y = 2$

28. $2x - y = 6$

29. $\dfrac{x}{2} + \dfrac{y}{3} = 1$

30. $\dfrac{x}{5} + \dfrac{y}{-2} = 1$

31. a. Give the coordinates of the point where the line $x = -2$ crosses the x-axis.

b. Explain why the line $x = -2$ does not cross the y-axis.

Written Exercises

For each exercise, graph the ordered pairs in the same coordinate plane.

A
1. $(1, 2), (0, 3), (3, -2), (-3, 2), (-4, 0)$
2. $(-3, -3), (0, -3), (-2, 4), (3, 0), (1, 4)$
3. $(6, -4), (3, -2), (0, 0), (-3, 2), \left(\frac{3}{2}, -1\right)$
4. $(3, 3), (1, 1), \left(-\frac{3}{2}, -\frac{3}{2}\right), (0, 0), (-2, -2)$

Graph each equation. You may wish to verify your graphs on a computer or a graphing calculator.

5. $x - y = 4$
6. $x + y = 3$
7. $2y - 3 = 0$
8. $2x + 1 = 0$
9. $x + 2y = 4$
10. $2x - y = 6$
11. $3x - 2y + 18 = 0$
12. $2x + 3y + 12 = 0$
13. $x - 3y = 0$
14. $x = 2 - 2y$
15. $2x + 5y = 15$
16. $2x - 3y = 0$
17. $3x - 2y = 7$
18. $3x + 2y - 9 = 0$

B
19. $y = \frac{1}{2}x - 2$
20. $\frac{x}{3} + y = 1$
21. $y - 2 = x - y + 2$
22. $x + y = 2(x + y + 3)$

Find the value of k so that the point P lies on the line L.

23. $P(2, 1)$, L: $3x + ky = 8$
24. $P(2, 3)$, L: $kx - 2y + k = 0$
25. $P(2, 2)$, L: $kx + (k + 1)y = 2$
26. $P(k, -2)$, L: $3x + 2y = k$

Graph each pair of equations in the same coordinate plane. Find the coordinates of the point where the graphs intersect. Then show by substitution that the coordinates satisfy both equations.

27. $2x + 5y = 0$
    $2x + \ \ y = 8$
28. $\ \ x - 2y = -4$
    $3x + 2y = 12$
29. $3x + 2y - \ \ 1 = 0$
    $\ \ x - 2y + 13 = 0$
30. $3x + \ \ y = -6$
    $3x - 5y = 12$

Graph each equation.

31. $|x| = 2$
32. $|y| = 3$
33. $y = |x|$

C
34. $|y| = |x|$
35. $y = |x - 1|$
36. $y = |x| - 1$
37. $y = |x| - x$
38. $y = x - |x|$
39. $y = |x| + x$

3-3 The Slope of a Line

Objective   To find the slope of a line and to graph a line given its slope and a point on it.

A jet plane can take off at an angle that will lessen the noise on the ground. The path of the plane shown is a line that rises 150 ft for each 500 ft traveled horizontally. The steepness, or *slope*, of the path is

$$\frac{\text{rise}}{\text{run}} = \frac{150}{500} = \frac{3}{10}, \text{ or } 0.30.$$

You can use this same method to measure the steepness, or *slope*, of a nonvertical line L in a coordinate plane. First choose any two distinct points (x_1, y_1) and (x_2, y_2) on L. (See Figure 1. You read (x_1, y_1) as "*x* one, *y* one" or "*x* sub one, *y* sub one.") Then,

$$\textbf{slope of line } L = \frac{\text{rise}}{\text{run}} = \frac{y_2 - y_1}{x_2 - x_1} \quad (x_2 \neq x_1).$$

Figure 1

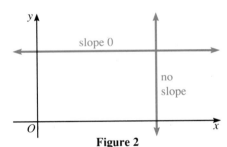

Figure 2

Look at Figure 2. If line L is vertical, then x_2 equals x_1, $x_2 - x_1 = 0$, and $\frac{y_2 - y_1}{x_2 - x_1}$ does not represent a real number. *Vertical lines have no slope.* If L is horizontal, then $y_2 = y_1$ and $y_2 - y_1 = 0$. *Horizontal lines have slope 0.*

Example 1   Find the slope of the line containing the given points.

   a. $(-2, 3)$ and $(4, -1)$ **b.** $(3, -4)$ and $(3, 5)$

Solution   **a.** slope $= \dfrac{-1 - 3}{4 - (-2)} = \dfrac{-4}{6} = -\dfrac{2}{3}$ *Answer*

   b. slope $= \dfrac{5 - (-4)}{3 - 3} = \dfrac{9}{0}$; the line is vertical and has no slope.   *Answer*

When you know an equation of a line, an easy way to find the slope is to use the following theorem.

Theorem

The slope of the line $Ax + By = C$ $(B \neq 0)$ is $-\dfrac{A}{B}$.

Proof   Let (x_1, y_1) and (x_2, y_2) be any two different points of the line $Ax + By = C$. Since $B \neq 0$ (given), the line is nonvertical and $x_2 \neq x_1$. Therefore, $x_2 - x_1 \neq 0$ and the slope of the line is $\dfrac{y_2 - y_1}{x_2 - x_1}$.

Since (x_1, y_1) and (x_2, y_2) are on the line, the coordinates of these points must satisfy $Ax + By = C$.

$$Ax_1 + By_1 = C$$
$$Ax_2 + By_2 = C$$

Subtracting the first equation from the second gives:

$$(Ax_2 + By_2) - (Ax_1 + By_1) = C - C$$
$$Ax_2 - Ax_1 + By_2 - By_1 = 0$$
$$A(x_2 - x_1) + B(y_2 - y_1) = 0$$
$$B(y_2 - y_1) = -A(x_2 - x_1)$$

Dividing both sides by $B(x_2 - x_1)$ completes the proof by showing that the slope is the constant $-\dfrac{A}{B}$:

$$\frac{y_2 - y_1}{x_2 - x_1} = -\frac{A}{B}. \quad \text{(Recall that } B \neq 0 \text{ and } x_2 - x_1 \neq 0.)$$

When you solve $Ax + By = C$ for y, you obtain $y = -\dfrac{A}{B}x + \dfrac{C}{B}$. Thus, *the slope of a line is the numerical factor, or **coefficient**, of the x-term when the equation of the line is solved for y.*

Example 2   Find the slope of the line $5x - 2y = 20$.

Solution 1   Solve the given equation for y.

$$-2y = -5x + 20$$
$$y = \frac{5}{2}x - 10$$

The slope is the coefficient of x, which is $\dfrac{5}{2}$.   ***Answer***

Solution 2   Since $A = 5$ and $B = -2$, slope $= -\dfrac{A}{B} = -\dfrac{5}{-2} = \dfrac{5}{2}$.   ***Answer***

The following theorem states that there is exactly one line passing through a given point and having a given slope. A proof of the theorem is outlined in Exercises 49–54.

Theorem

Let $P(x_1, y_1)$ be a point and m a real number. There is one and only one line L through P having slope m. An equation of L is

$$y - y_1 = m(x - x_1).$$

Example 3 Graph the line through the point $P(2, 1)$ having slope $m = -\dfrac{3}{4}$.

Solution   Since

$$m = \frac{\text{rise}}{\text{run}} = -\frac{3}{4} = \frac{-3}{4},$$

start at $P(2, 1)$ and move 3 units down and 4 units to the right to reach the point $Q(6, -2)$. Draw the line through P and Q. As a check, you should also find a third point on the line, such as $(-2, 4)$.

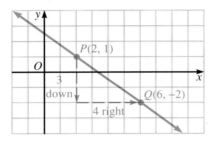

Answer

You can explore the meaning of slope by drawing graphs such as those shown at the right. A computer or a graphing calculator may be helpful. The diagram shows the graphs of the equation

$$y - y_1 = m(x - x_1)$$

when $(x_1, y_1) = (0, 0)$ for various values of m.

Notice that as you move from left to right along a line having slope m, the line:

rises if m is positive;
is horizontal if $m = 0$;
falls if m is negative.

The larger $|m|$ is, the steeper the line is.

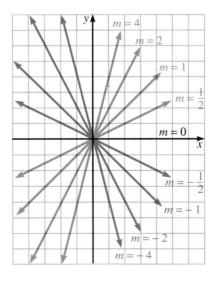

Oral Exercises

Give the slope of each line shown.

1.
2.
3.

4.
5.
6.

Give the slope of the line containing the given points.

7. $(0, 0), (-2, 6)$ **8.** $(4, -2), (0, 0)$ **9.** $(3, 0), (0, 9)$

10. $(3, 2), (-1, 2)$ **11.** $(3, 1), (2, 2)$ **12.** $(1, 1), (-1, -1)$

Give the slope of each line.

13. $y = 2x + 3$ **14.** $y = 3x - 1$ **15.** $y = -x + 2$

16. $y + 2x = 4$ **17.** $y = 0$ **18.** $y = \frac{1}{2}x$

Determine whether the points whose coordinates are given in the table lie on a line. If they do, give the slope of the line.

Sample

x	y
-2	5
1	3
4	1
10	-3

$3\{\ \ \ \}-2$
$3\{\ \ \ \}-2$
$6\{\ \ \ \}-4$

Solution Each change of 3 in x produces a change of -2 in y.

$\therefore$ the points lie on a line because the slope between any two points is a constant, $-\frac{2}{3}$.

19.

x	y
1	9
2	7
3	5
4	3

20.

x	y
0	3
2	0
6	3
8	6

21.

x	y
-4	2
1	4
6	6
11	8

22.

x	y
0	1
1	3
3	7
7	15

What do you know about the value(s) of A and B if the line $Ax + By = C$ has the property described?

23. The line is horizontal.

24. The line is vertical.

25. The line rises from left to right.

26. The line falls from left to right.

Written Exercises

Find the slope of the line containing the given points. If the line has no slope, write "vertical."

A
1. $(3, 1)$, $(5, 5)$
2. $(4, 3)$, $(0, 1)$
3. $(4, -1)$, $(-2, 3)$

4. $(-5, -2)$, $(5, -2)$
5. $(3, -4)$, $(3, -2)$
6. $(3, -1)$, $(-3, 1)$

7. $\left(\frac{3}{2}, -3\right)$, $\left(\frac{1}{2}, -7\right)$
8. $\left(\frac{1}{2}, -2\right)$, $(0, -4)$
9. $(6, -5)$, $(-4, 3)$

10. $(0.5, 2.4)$, $(1.5, -1.6)$
11. (a, b), (b, a) $(a \ne b)$
12. (a, b), $(-b, -a)$ $(a \ne -b)$

Find the slope of each line.

13. $x + y = 7$
14. $x - y + 1 = 0$
15. $2x + 4y = 5$

16. $4x - 3y = 3$
17. $3x - 3y = 5$
18. $4y - 5 = 6x$

19. $x = 3y + 2$
20. $2(1 - y) = x$
21. $\frac{1}{2}x + \frac{1}{3}y = 1$

22. $\frac{1}{4}x - \frac{1}{2}y = 1$
23. $\frac{x}{-1} + \frac{y}{6} = 1$
24. $\frac{x}{3} - \frac{y}{-5} = 1$

Graph the line through point P having slope m. Find the coordinates of two other points on the line.

25. $P(0, 2)$, $m = 1$
26. $P(1, 0)$, $m = -1$
27. $P(3, -1)$, $m = -2$

28. $P(-2, -1)$, $m = 3$
29. $P(2, -3)$, $m = \frac{1}{2}$
30. $P(0, 3)$, $m = -\frac{3}{2}$

B
31. $P(-2, 3)$, $m = -\frac{2}{3}$
32. $P(-1, -4)$, $m = \frac{5}{3}$
33. $P(0, 0)$, $m = 0.25$

34. $P(-3, -1)$, $m = 0.75$
35. $P(2, 1)$, $m = 0$
36. $P(2, 1)$, no slope

Solve.

37. A ramp to provide handicapped people access to a certain building is to be constructed with a slope of 5%. If the entrance to the building is 3 ft above ground level, how long should the base of the ramp be?

Solve.

38. A jetliner covered a horizontal distance of 5 mi while following a flight path with slope 0.25. How much altitude did it gain?

Find the value of k so that the given line has slope m.

39. $kx - 3y = 7$, $m = 2$ **40.** $6x + ky = 10$, $m = -2$

41. $(k + 3)x - 3y = 1$, $m = k$ **42.** $(k + 1)x + 2y = 6$, $m = k - 2$

Find the value of k so that the line through the given points has slope m.

43. $(2k, 3)$, $(1, k)$; $m = 2$ **44.** $(k, k + 1)$, $(3, 2)$; $m = 3$

45. $(k + 1, k - 1)$, $(k, -k)$; $m = k + 1$ **46.** $(k + 1, 3 + 2k)$, $(k - 1, 1 - k)$; $m = k$

C   **47.** $(3, k)$, $(-1, |k|)$; $m = -2$ **48.** $(1, k)$, $(5, |k|)$; $m = 3$

Exercises 49–54 outline a proof of the theorem stated on page 114.

49. Show that the graph of the equation $y - y_1 = m(x - x_1)$ is a line L by expressing the equation in the form $Ax + By = C$.

50. Explain how you know L passes through $P(x_1, y_1)$.

51. Explain how you know L has slope m.

Exercises 49–51 showed that there is *at least one* line through P having slope m. Now let L' be *any* line through P having slope m.

52. Let $Q(x', y')$ be any point of L' different from $P(x_1, y_1)$. Explain why
$$\frac{y' - y_1}{x' - x_1} = m.$$

53. From Exercise 52, $y' - y_1 = m(x' - x_1)$. Explain why this shows that $Q(x', y')$ is on L.

54. Explain why L' and L must be the same line.

Therefore, there is *only one* line through P having slope m.

Mixed Review Exercises

Complete the ordered pair to form a solution of the given equation.

1. $2x + y = 5$; $(4, \underline{\ ?\ })$ **2.** $x - 3y = 7$; $(\underline{\ ?\ }, -2)$

3. $-x + 4y = 9$; $(-1, \underline{\ ?\ })$ **4.** $5x + 2y = -8$; $(\underline{\ ?\ }, 1)$

Graph each equation.

5. $-x + y = 2$ **6.** $y = -4$ **7.** $2x + 3y = 6$

8. $y = -x$ **9.** $x + 3 = 0$ **10.** $y = \frac{1}{2}x - 3$

3-4 Finding an Equation of a Line

Objective   To find an equation of a line given its slope and a point on the line, or two points, or its slope and the y-intercept.

You know that the graph of $Ax + By = C$ is a line in an xy-plane. Equations of lines are usually given in this **standard form,** with A, B, and C integers. However, when you are asked to *find* the equation of a certain line, the two other forms discussed in this lesson may be more useful.

　　By the theorem on page 114, the line containing the point (x_1, y_1) and having slope m has the equation $y - y_1 = m(x - x_1)$.

The equation
$$y - y_1 = m(x - x_1)$$
is called the **point-slope form** of the equation of a line.

Example 1   (a point and a slope given) Find an equation in standard form of the line containing the point $(4, -3)$ and having slope $-\dfrac{2}{5}$.

Solution   Use the point-slope form with $(x_1, y_1) = (4, -3)$ and $m = -\dfrac{2}{5}$.

$$y - y_1 = m(x - x_1)$$

$$y - (-3) = -\frac{2}{5}(x - 4) \qquad \text{Multiply both sides by 5.}$$

$$5(y + 3) = -2(x - 4)$$

$$5y + 15 = -2x + 8$$

$$2x + 5y = -7 \quad \textbf{\textit{Answer}}$$

Check:　$2(4) + 5(-3) = -7$; $m = -\dfrac{A}{B} = -\dfrac{2}{5}$　✓

Example 2   (two points given) Find an equation in standard form of the line containing $(1, -2)$ and $(5, 1)$.

Solution   First find the slope of the line:　$m = \dfrac{1 - (-2)}{5 - 1} = \dfrac{3}{4}$.

Then use the point-slope form with either given point.

$$y - y_1 = m(x - x_1) \qquad\qquad y - y_1 = m(x - x_1)$$

$$y - (-2) = \frac{3}{4}(x - 1) \qquad\qquad y - 1 = \frac{3}{4}(x - 5)$$

$$4y + 8 = 3x - 3 \qquad\qquad 4y - 4 = 3x - 15$$

$$3x - 4y = 11 \qquad\qquad 3x - 4y = 11 \quad \textbf{\textit{Answer}}$$

You should check the answer in Example 2 by verifying that $(1, -2)$ and $(5, 1)$ are both solutions of $3x - 4y = 11$.

A line (or any curve) has **y-intercept** b if it intersects the y-axis at the point $(0, b)$. A line has **x-intercept** a if it intersects the x-axis at the point $(a, 0)$. If you let $(x, y) = (0, b)$ in the point-slope form, then you get $y - b = m(x - 0)$, or $y = mx + b$.

If b is the y-intercept of a line, the equation

$$y = mx + b$$

is called the **slope-intercept** form of the equation of the line.

Example 3 (slope and y-intercept given) Find an equation in standard form of the line having slope -2 and y-intercept $\frac{4}{3}$.

Solution Use $y = mx + b$ with $m = -2$ and $b = \frac{4}{3}$.

$$y = -2x + \frac{4}{3} \qquad \text{Multiply both sides by 3.}$$

$$3y = -6x + 4$$

$$6x + 3y = 4 \quad \textbf{\textit{Answer}}$$

You can explore the relationships between the slopes and the graphs of two equations by graphing $y = mx + b$ for various values of m and b. A computer or a graphing calculator may be helpful. If you graph $y = 3x + b$ and $y = -\frac{1}{3}x + b$ for various values of b on the same axes, your results should suggest the following theorem. This theorem tells how slopes indicate when graphs of equations are parallel or perpendicular lines. (A proof of part (1) is outlined in Exercises 42 and 43, page 130, and of part (2) in Problem 14, page 406.)

Theorem

Let L_1 and L_2 be two different lines, with slopes m_1 and m_2 respectively.
1. L_1 and L_2 are *parallel* if and only if $m_1 = m_2$.
2. L_1 and L_2 are *perpendicular* if and only if $m_1 m_2 = -1$.

By this theorem, the lines $y = 3x - 2$ and $3x - y = 5$ are parallel because the slope of each is 3. The lines $y = 3x - 2$ and $-3y = x + 1$ are perpendicular because their slopes are 3 and $-\frac{1}{3}$, respectively, and $(3)(-\frac{1}{3}) = -1$.

Because the equation $m_1m_2 = -1$ can be written as $m_2 = -\dfrac{1}{m_1}$ and also as $m_1 = -\dfrac{1}{m_2}$, we say that nonvertical lines are perpendicular if and only if their slopes are *negative reciprocals* of each other.

Example 4 Let P be the point $(4, 1)$ and L the line $2x + 5y = -10$.

a. Find an equation in standard form of the line L_1 through P and parallel to L.

b. Find an equation in standard form of the line L_2 through P and perpendicular to L.

Solution Solve the equation of line L for y:

$$2x + 5y = -10; \; y = -\frac{2}{5}x - 2$$

$\therefore$ the slope of L is $-\dfrac{2}{5}$.

a. To find an equation of L_1, use the point-slope form with $(x_1, y_1) = (4, 1)$ and $m = -\dfrac{2}{5}$:

$$y - 1 = -\frac{2}{5}(x - 4)$$

$$5y - 5 = -2x + 8$$

$$L_1\!: 2x + 5y = 13$$

b. The slope of L_2 is the negative reciprocal of $-\dfrac{2}{5}$, which is $\dfrac{5}{2}$. Use the point-slope form with $(x_1, y_1) = (4, 1)$ and $m = \dfrac{5}{2}$:

$$y - 1 = \frac{5}{2}(x - 4)$$

$$2y - 2 = 5x - 20$$

$$L_2\!: 5x - 2y = 18$$

Oral Exercises

Give the slope and y-intercept of each line.

1. $y = 3x + 2$
2. $y = x - 1$
3. $y = 3 - x$
4. $y = 2(x + 1)$
5. $2y = x + 6$
6. $x + 2y = 4$
7. $6x - 2y = 1$
8. $2x + y + 5 = 0$

Give an equation of the line described in point-slope or slope-intercept form.

9. Has slope 2 and y-intercept 3
10. Passes through $(2, 1)$ with slope 1
11. Contains $(1, 3)$ and has slope 2
12. Has y-intercept -2 and slope -1
13. Has slope -2 and x-intercept 3
14. Has slope $\dfrac{2}{3}$ and passes through $(0, 0)$

Give the slope of any line that is (a) parallel to, and (b) perpendicular to, the graph of the given equation.

15. $y = -2x + 6$
16. $y = 5(x + 1)$
17. $x = 2y + 5$
18. $x - 3y = 3$

Tell whether the given lines are parallel, perpendicular, or neither.

19. $y = x + 2$

$x - y = 3$

20. $y = 2x - 4$

$y = 3 - \dfrac{1}{2}x$

21. $2x + y = 1$

$x + 2y = 3$

22. $x - 2y = 1$

$2x = 4y + 3$

23. $2x + 3y = 6$

$3x - 2y = 4$

24. $3x - y = 6$

$x - 3y = 6$

25. Explain why the lines $Ax + By = C_1$ and $Ax + By = C_2$ $(C_1 \neq C_2)$ are parallel.

Written Exercises   2-26 EVEN

Find an equation in standard form of the line containing point P and having slope m.

A

1. $P(2, 3)$, $m = 1$

2. $P(2, 1)$, $m = -1$

3. $P(5, 0)$, $m = -2$

4. $P(-1, 4)$, $m = 0$

5. $P(-3, -2)$, $m = \dfrac{1}{2}$

6. $P(2, 1)$, $m = \dfrac{2}{3}$

7. $P(4, -3)$, $m = \dfrac{1}{5}$

8. $P(0, 6)$, $m = -\dfrac{3}{2}$

9. $P(-2, -1)$, $m = 0$

10. $P(-3, 3)$, $m = -\dfrac{4}{3}$

11. $P(-2, 4)$, $m = 0.4$

12. $P(4, 0)$, $m = -0.6$

Find an equation in standard form of the line having slope m and y-intercept b.

13. $m = -1$, $b = 2$

14. $m = 1$, $b = -3$

15. $m = \dfrac{1}{2}$, $b = \dfrac{3}{2}$

16. $m = -\dfrac{3}{4}$, $b = -\dfrac{5}{4}$

17. $m = 1.2$, $b = -0.6$

18. $m = -0.8$, $b = 1.4$

Find an equation in standard form of the line containing the given points.

19. $(0, 0)$, $(5, -2)$

20. $(0, 0)$, $(-3, 1)$

21. $(3, -2)$, $(-2, 3)$

22. $(3, -2)$, $(2, -3)$

23. $(3, -2)$, $(-3, 2)$

24. $(3, -2)$, $(-3, -2)$

25. $(-2, 3)$, $(-2, -3)$

26. $(4, -5)$, $(1, -4)$

27. $\left(-3, \dfrac{1}{2}\right)$, $\left(3, \dfrac{1}{2}\right)$

28. $\left(\dfrac{3}{2}, -\dfrac{1}{2}\right)$, $\left(-\dfrac{1}{2}, \dfrac{5}{2}\right)$

29. $\left(\dfrac{2}{3}, -\dfrac{1}{2}\right)$, $\left(\dfrac{1}{6}, -1\right)$

30. $\left(\dfrac{3}{4}, \dfrac{5}{4}\right)$, $\left(-\dfrac{1}{4}, \dfrac{1}{2}\right)$

Find equations in standard form of the lines through point P that are (a) parallel to, and (b) perpendicular to, line L.   Ex: $m = -2$ $/ m = \dfrac{1}{2}$.

31. $P(0, 3)$; $L: x + y = 5$

32. $P(0, -2)$; $L: y = x - 3$

33. $P(0, -4)$; $L: 2y = x$

34. $P(0, 1)$; $L: 3x + 2y = 1$

35. $P(2, 0)$; $L: x + 2y = 3$

36. $P(-1, 2)$; $L: x - 3y = -2$

37. $P(-4, 1)$; $L: y + 2 = 0$

38. $P(-1, -2)$; $L: x - 3 = 0$

Find an equation in standard form for the line described.

B **39.** Passing through the points $(1, 4)$ and $(-3, 4)$

40. Passing through the points $(-2, 3)$ and $(-2, 6)$

41. Passing through the origin and with no slope

42. Passing through the origin and with slope 0

43. Having y-intercept 6 and parallel to the x-axis

44. Having x-intercept -4 and parallel to the y-axis

45. Having x-intercept 4 and y-intercept 3

46. Having x-intercept -3 and y-intercept -1

47. Through $P(-2, 1)$ and parallel to the line containing $(1, 4)$ and $(2, 3)$

48. Through $Q(-3, 2)$ and parallel to the line containing $(2, 3)$ and $(1, -2)$

In Exercises 49–54 the vertices of a quadrilateral $ABCD$ are given. Determine whether or not $ABCD$ is a parallelogram, and if it is, whether or not it is a rectangle. (*Hint:* Check to see if the lines containing opposite sides are parallel. If they are, check to see if the lines containing two adjacent sides are perpendicular.)

49. $A(2, 0)$, $B(-2, 2)$, $C(2, 8)$, $D(6, 6)$ **50.** $A(-2, 3)$, $B(1, 4)$, $C(3, -1)$, $D(0, -2)$

51. $A(0, -2)$, $B(6, 0)$, $C(5, 3)$, $D(-1, 1)$ **52.** $A(4, -2)$, $B(5, 1)$, $C(0, 3)$, $D(-2, 0)$

53. $A(7, 1)$, $B(1, 4)$, $C(-2, -1)$, $D(6, -5)$ **54.** $A(5, 1)$, $B(-1, 5)$, $C(-3, 2)$, $D(3, -2)$

C **55.** Consider two different nonvertical lines.
$$L_1: A_1x + B_1y = C_1 \quad \text{and} \quad L_2: A_2x + B_2y = C_2$$
   a. Show that L_1 and L_2 are parallel if and only if $A_1B_2 = A_2B_1$.
   b. Show that L_1 and L_2 are perpendicular if and only if $A_1A_2 + B_1B_2 = 0$.

56. Show that an equation of the line containing the points (x_1, y_1) and (x_2, y_2) when $x_1 \neq x_2$ is $y - y_1 = \dfrac{y_2 - y_1}{x_2 - x_1}(x - x_1)$. This is called the **two-point form.**

57. Show that an equation of the line having x-intercept a and y-intercept b is $\dfrac{x}{a} + \dfrac{y}{b} = 1$. This is called the **intercept form.**

Computer Exercises

For students with some programming experience.

1. Write a program to find the slope of a line, given two distinct points on the line. If the line has no slope, then the program should print that the line is vertical.

2. Run the program in Exercise 1 to find the slope of the line through each pair of points or to identify the line as vertical.
   a. $(3, -1)$ and $(5, 8)$ **b.** $(-5, 7)$ and $(2, -9)$ **c.** $(7, -3)$ and $(7, 10)$

3. Write a program to find the slope-intercept equation of a line, given the slope and a point on the line.

4. Run the program in Exercise 3 to find the slope-intercept equation of a line with slope 1.72 and containing the point $(-5, 8.6)$.

5. Write a program based on the programs in Exercises 1 and 3 to find the slope-intercept equation of a line given two distinct points on the line. Allow for the possibility that the line may be vertical.

6. Run the program in Exercise 5 to find the slope-intercept equation of the line containing each pair of points in Exercise 2.

Self-Test 1

Vocabulary   open sentence in two variables (p. 101)
ordered pair (p. 101)
solution set (p. 101)
xy-coordinate plane (p. 107)
plane rectangular coordinate system (p. 107)
axes (p. 107)
origin (p. 107)
quadrant (p. 107)
graphing an ordered pair (p. 107)
plotting a point (p. 107)
coordinates (p. 107)

x-coordinate (p. 107)
abscissa (p. 107)
y-coordinate (p. 107)
ordinate (p. 107)
Cartesian coordinate system (p. 107)
graph of an open sentence (p. 108)
linear equation (p. 109)
slope (p. 112)
standard form (p. 118)
point-slope form (p. 118)
y-intercept (p. 119)
x-intercept (p. 119)
slope-intercept form (p. 119)

1. Solve $4x - 2y = 9$ if the domain of x is $\{-3, -1, 2\}$.   **Obj. 3-1, p. 101**

2. At a yard sale, Toni bought some jazz records at $4 each and some classical records at $6 each. If her purchases totaled $46, find all possibilities for the number of each type of record that she bought.

3. Graph $2x - 5y = 10$.   4. Graph $x = -3$.   **Obj. 3-2, p. 107**

5. Find the slope of line $9x - 6y = 5$.   **Obj. 3-3, p. 112**

6. Find the slope of the line containing the points $(-2, 6)$ and $(2, 1)$.

7. Find the slope of the line $2y - 12 = 0$.

8. Find an equation in point-slope form of the line containing $(2, 4)$ and $(-4, 3)$.   **Obj. 3-4, p. 118**

9. Find an equation in standard form of the line through $(-3, 2)$ and parallel to the line $2x - y = 1$.

10. Find an equation in standard form of the line through $(5, 1)$ that is perpendicular to the line $x = 4$.

Check your answers with those at the back of the book.

Linear Systems

3-5 Systems of Linear Equations in Two Variables

Objective To solve systems of linear equations in two variables.

A set of linear equations in the same two variables is called a **system of linear equations,** or a **linear system.** Any ordered pair of numbers that is a solution of each equation in the system is called a **solution** of the system, or a **simultaneous solution** of the equations.

When you graph a linear system with two equations in the same coordinate plane, the result is one of three types of graphs, as illustrated below.

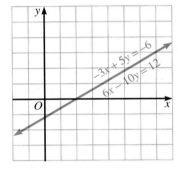

(a) Intersecting lines

$$x - 2y = 5$$
$$4x + 3y = 9$$

(b) Parallel lines

$$x + 2y = 1$$
$$y = -\frac{1}{2}x + 4$$

(c) Coinciding lines

$$-3x + 5y = -6$$
$$6x - 10y = 12$$

Algebraically, these geometric relationships mean:

(a) When the graphs intersect in only one point, the system has one solution.

(b) When the graphs are parallel, the system has no solution.

(c) When the graphs coincide (that is, are the same line), every ordered pair that satisfies one equation also satisfies the other equation. The solution set for the system contains infinitely many ordered pairs.

We can extend the definition of equivalent equations to systems. **Equivalent systems** are systems that have the same solution set. To solve a system of equations, you transform the system into an equivalent system whose solution is easily seen. Transformations that produce equivalent systems are listed in the chart at the top of the next page.

Transformations that Produce Equivalent Systems

1. Replacing an equation by an equivalent equation. (Recall that multiplying each side of an equation by the same nonzero number produces an equivalent equation.)

2. Substituting for one variable in any equation an equivalent expression for that variable obtained from another equation in the system. (This expression may be the value of the variable, if known.)

3. Replacing any equation by the sum of that equation and another equation in the system. (Recall that to add two equations, you add their left sides, add their right sides, and equate the results.)

Example 1 Solve this system: $4x + 3y = 5$
$\qquad\qquad\qquad\qquad\quad 2x - 5y = -17$

Solution At each step replace the system by an equivalent one, using one of the transformations listed in the chart.

1. Multiply the second given equation by -2. (Transformation 1)

$$\left.\begin{array}{r} 4x + 3y = 5 \\ -4x + 10y = 34 \end{array}\right\} (1)$$

2. Replace the second equation in (1) by the sum of the two equations. (Transformation 3)

$$\left.\begin{array}{r} 4x + 3y = 5 \\ 13y = 39 \end{array}\right\} (2)$$

3. Solve the second equation in (2) for y. (Transformation 1)

$$\left.\begin{array}{r} 4x + 3y = 5 \\ y = 3 \end{array}\right\} (3)$$

4. Substitute 3 for y in the *first* equation in (3) (Transformation 2)

$$\left.\begin{array}{r} 4x + 3(3) = 5 \\ y = 3 \end{array}\right\} (4)$$

5. Solve the first equation in (4) for x. (Transformation 1)

$$\left.\begin{array}{r} x = -1 \\ y = 3 \end{array}\right\} (5)$$

Check that the solution $(-1, 3)$ satisfies *both of the original equations:*

$$4x + 3y = 5 \qquad\qquad\qquad 2x - 5y = -17$$
$$4(-1) + 3(3) \stackrel{?}{=} 5 \qquad\qquad 2(-1) - 5(3) \stackrel{?}{=} -17$$
$$5 = 5 \;\checkmark \qquad\qquad\qquad -17 = -17 \;\checkmark$$

$\therefore$ the solution of the system is $(-1, 3)$. **Answer**

When you add two equations, the result is a **linear combination** of the equations. To solve a system of equations, a useful goal is to obtain a linear combination that has fewer variables than the given system.

Solution 1 of Example 2 illustrates the linear-combination method for solving a linear system and shows you an efficient way to set up your work. You do not need to write out each system in the chain of equivalent systems as was done in Example 1.

Example 2 Solve this system: $\quad x - 2y = 5$
$\quad\quad\quad\quad\quad\quad\quad\quad\quad\quad 4x + 3y = 9$

Solution 1 (Linear-Combination Method)

Find a linear combination that eliminates x.

1. Multiply the first equation by -4 so that the coefficients of x will be *opposites*.

$$\begin{array}{r} -4x + 8y = -20 \\ \underline{4x + 3y = 9} \end{array}$$

2. Add the equations in Step 1 and solve the resulting equation for y.

$$\begin{array}{r} 0x + 11y = -11 \\ 11y = -11 \\ y = -1 \end{array}$$

3. Substitute -1 for y in either of the original equations to find x.

$$\begin{array}{r} x - 2(-1) = 5 \\ x = 3 \end{array}$$

4. Check that $(3, -1)$ satisfies both equations.

$$\begin{array}{r} 3 - 2(-1) = 5 \quad \checkmark \\ 4(3) + 3(-1) = 9 \quad \checkmark \end{array}$$

$\therefore$ the solution of the system is $(3, -1)$. ***Answer***

This algebraic solution of the system verifies the graphic solution shown in diagram (a) on page 124.

Solution 2 (Substitution Method)

1. Express x in terms of y in the first equation (since the coefficient of x is 1).

$$\begin{array}{r} x - 2y = 5 \\ x = 2y + 5 \end{array}$$

2. Substitute $2y + 5$ for x in the second equation. Solve for y.

$$\begin{array}{r} 4x + 3y = 9 \\ 4(2y + 5) + 3y = 9 \\ 8y + 20 + 3y = 9 \\ 11y = -11 \\ y = -1 \end{array}$$

3. Substitute -1 for y in either of the original equations to find x.

$$\begin{array}{r} 4x + 3(-1) = 9 \\ 4x = 12 \\ x = 3 \end{array}$$

4. The check is the same as Step 4 above.

$\therefore$ the solution of the system is $(3, -1)$. ***Answer***

Example 3 Solve this system: $\quad 3x - 4y = 14$
$\quad\quad\quad\quad\quad\quad\quad\quad\quad\quad 4x + 10y = 11$

Solution (Linear-Combination Method)

Find a linear combination that eliminates y.

1. Multiply the first equation by 10 and the second equation by 4 so that the coefficients of y will be *opposites*.

$$\begin{array}{r} 30x - 40y = 140 \\ \underline{16x + 40y = 44} \end{array}$$

2. Add the equations in Step 1 and solve the resulting equation for x.

$$46x + 0y = 184$$
$$46x = 184$$
$$x = 4$$

3. Substitute 4 for x in either of the original equations to find y.

$$3(4) - 4y = 14$$
$$-4y = 2$$
$$y = -\frac{1}{2}$$

4. Check that $\left(4, -\frac{1}{2}\right)$ satisfies both equations.

$$3(4) - 4\left(-\frac{1}{2}\right) = 14 \quad \checkmark$$
$$4(4) + 10\left(-\frac{1}{2}\right) = 11 \quad \checkmark$$

$\therefore$ the solution is $\left(4, -\frac{1}{2}\right)$.   **Answer**

When the graphs of the equations of a system are parallel (see diagram (b) on page 124), the system has no solution. If you try to solve such a system algebraically, both variables are eliminated and the result is a false statement.

Example 4   Solve this system: $x + 2y = 1$
$$y = -\frac{1}{2}x + 4$$

Solution   Substitute the expression for y from the second equation into the first equation:

$$x + 2y = 1$$
$$x + 2\left(-\frac{1}{2}x + 4\right) = 1$$
$$x - x + 8 = 1$$
$$8 = 1 \qquad \leftarrow \text{False!}$$

Since $8 = 1$ is false, the system has no solution.   **Answer**

When the graphs of the equations of a system coincide (see diagram (c) on page 124), the system has infinitely many solutions. If you try to solve such a system algebraically, you obtain an identity and all solutions of one equation are solutions of the other.

Example 5   Solve this system: $-3x + 5y = -6$
$$6x - 10y = 12$$

Solution   Multiply the first equation by 2, and then add the equations:

$$-6x + 10y = -12$$
$$\underline{6x - 10y = 12}$$
$$0 = 0$$

(Solution continues on the next page.)

Since $0 = 0$ is an identity, the solution set of the system is the same as the solution set of either of the equations, namely, the infinite set $\{(x, y): -3x + 5y = -6\}$. **Answer**

Three of the solutions are $\left(0, -\frac{6}{5}\right)$, $(2,0)$, and $(7, 3)$.

If a system of equations has at least one solution, the equations in the system are called **consistent.** This was the case in Examples 1, 2, 3, and 5. If the system has no solution, the equations are called **inconsistent** and their graphs are parallel lines. This was the case in Example 4. If a consistent system has infinitely many solutions, then the equations are called **dependent** and their graphs are coinciding lines. This was the case in Example 5.

Another method for solving linear systems, called *Cramer's Rule,* will be presented in Chapter 16. The method is especially useful when using a computer to solve linear systems.

Oral Exercises

Tell whether the given ordered pair is a solution of the system.

1. $(2, 1)$
$x + 2y = 4$
$3x - 2y = 4$

2. $(4, 3)$
$3x - 2y = 6$
$4x - 3y = 7$

3. $(-3, 2)$
$x + 3y = 3$
$3x - 2y = 13$

4. $(3, -5)$
$3x + 2y = -1$
$2x + y = 1$

5. $(-4, -2)$
$x - 3y = 2$
$2x - 5y = 2$

6. $(-3, 4)$
$2x + 3y = 6$
$3x - 2y = -1$

Explain how you would form a linear combination of the given equations to eliminate the variable printed in red.

Sample $5x + 2y = 2$
$2x + 3y = 7$; y

Solution   Multiply the first equation by 3 and the second by -2; then add.

7. $2x + y = 1$
$2x + 3y = 7$; x

8. $2x - 3y = 7$
$3x + y = 5$; y

9. $5x - 6y = 9$
$2x - 3y = 3$; y

10. $4x - 3y = 6$
$2x - 5y = -4$; x

11. $2x - 3y = 3$
$5x + 2y = 17$; y

12. $4x + 5y = 3$
$3x + 2y = 4$; x

Use *intersect, are parallel,* or *coincide* to make a true statement.

13. If two lines have the same slope and different y-intercepts, then the lines __?__.

14. If one equation can be obtained from another equation by multiplying both sides by the same nonzero number, then the graphs __?__.

Use *intersect*, *are parallel*, or *coincide* to make a true statement.

15. If two lines have different slopes and the same x-intercept, then the lines' __?__ .

16. If two lines have more than one point in common, then the lines __?__ .

17. Why do you think the linear-combination method is more efficient than substitution in solving the system in Example 3?

Written Exercises

Solve each system.

A   **1–6.** The systems in Oral Exercises 7–12.

7. $2x - 7y = 10$
  $5x - 6y = 2$

8. $5x + 6y + 8 = 0$
  $3x - 2y + 16 = 0$

9. $3x - 2y + 2 = 0$
  $x + 3y = 14$

10. $8x - 3y = 3$
  $3x - 2y + 5 = 0$

11. $6u + 5v = -2$
  $2u + 3v = 6$

12. $3p + 2q = -2$
  $9p - q = -6$

Graph both equations of each system in the same coordinate plane. Then estimate the solution (that is, the coordinates of the point of intersection) to the nearest half unit.

13. $y = -x + 3$
  $y = x - 4$

14. $2x + y = -2$
  $2x - 3y = 15$

15. $y = 5x$
  $x + y = 10$

16. $3x + 5y = 15$
  $x - y = 4$

Solve each system. If the system has an infinite solution set, specify it and give three solutions. If the system has no solution, say so.

17. $3x + 3y = 6$
  $5x - 6y = 15$

18. $3x - 2y = 6$
  $5x + 3y + 9 = 0$

19. $x - y = 2x - 2$
  $x + y = 2y - 2$

20. $6x = 4y + 5$
  $6y = 9x - 5$

21. $2p - 5q = 14$
  $p + \frac{3}{2}q = 5$

22. $d = 2 - 6c$
  $\frac{1}{2}d - c = 1$

23. $x + y = 3x - 1$
  $x - y = 1 - x$

24. $2x + y = 2 - x$
  $x + 2y = 2 + y$

25. $3y = x - 2y - 1$
  $3x = 2x - y - 1$

26. $x + y = 4(y + 2)$
  $x - y = 2(y + 4)$

27. $2x - 3y = 2 - x$
  $3x - 2y = -2 + y$

28. $2(y - x) = 5 + 2x$
  $2(y + x) = 5 - 2y$

Write each equation of the system in slope-intercept form. By comparing slopes and y-intercepts, determine whether the equations are consistent or inconsistent.

B   **29.** $3x = 4y - 4$
  $4y = 3x - 3$

30. $3x = 4y + 8$
  $3y = 4x + 8$

31. $3x - 4y = 12$
  $4x - 3y = 12$

32. $3x - 6y = 9$
  $4x - 3y = 12$

Solve each system. Note that the equations in Exercises 33–38 are not linear in the original variables. $\left(\textit{Hint: Let } x = \dfrac{1}{u} \text{ and } y = \dfrac{1}{v}. \text{ Rewrite the}\right.$ system in terms of x and y and solve. Then use x and y to find u and $v.\big)$

33. $\dfrac{6}{u} + \dfrac{3}{v} = 2$

$\dfrac{2}{u} - \dfrac{9}{v} = 4$

34. $\dfrac{6}{u} + \dfrac{5}{v} = 1$

$\dfrac{3}{u} - \dfrac{10}{v} = 3$

35. $\dfrac{2}{u} - \dfrac{3}{v} + 2 = 0$

$\dfrac{4}{u} + \dfrac{3}{v} + 1 = 0$

36. $\dfrac{3}{u} + \dfrac{4}{v} = 1$

$\dfrac{6}{u} - \dfrac{2}{v} = 1$

37. $\dfrac{3}{u} - \dfrac{4}{v} = 4$

$\dfrac{5}{u} - \dfrac{6}{v} = 7$

38. $\dfrac{4}{u} + \dfrac{3}{v} = 3$

$\dfrac{6}{u} + \dfrac{5}{v} = 4$

In Exercises 39–41, a, b, c, d, e, and f represent constants. Solve each system. What conditions must be placed on the constants in each system to be certain that the system has a unique solution?

C **39.** $ax + y = b$

$bx + y = a$

40. $ax - by = c$

$bx + ay = d$

41. $ax + by = e$

$cx + dy = f$

Exercises 42 and 43 outline a proof of the first part of the theorem stated on page 119.

42. Proof of the "if" part: If $m_1 = m_2 = m$, then L_1 and L_2 have equations of the form L_1: $y = mx + b_1$ and L_2: $y = mx + b_2$ where $b_1 \neq b_2$. Show that these equations have no common solution.

43. The "only if" part is equivalent to: If $m_1 \neq m_2$, then L_1 and L_2 are not parallel. Show that if $m_1 \neq m_2$, then the equations L_1: $y = m_1x + b_1$ and L_2: $y = m_2x + b_2$ have a common solution.

Mixed Review Exercises

For the line containing the given points, find (a) the slope and (b) an equation in standard form.

1. $(-2, 0)$, $(0, 1)$

2. $(1, -2)$, $(3, 4)$

3. $(-6, -4)$, $(3, 2)$

4. $(1, -1)$, $(5, -1)$

5. $(-3, 2)$, $(1, -2)$

6. $(0, 0)$, $(-3, 4)$

Find the slope and y-intercept of each line.

7. $x - y = 4$

8. $5x + 3y = 6$

9. $-2x + y = 1$

10. $y - 5 = 0$

11. $y = 3x + 2$

12. $x = 4 - 6y$

3-6 Problem Solving: Using Systems

Objective   To use systems of equations to solve problems.

Sometimes you can solve a problem involving two unknown quantities most easily by using a system of equations. The problem solving plan given in Lesson 1-9 is useful here. However, you must set up and solve a system of equations rather than a single equation.

Example 1   To use a certain computer data base, the charge is $30/h during the day and $10.50/h at night. If a research company paid $411 for 28 h of use, find the number of hours charged at the daytime rate and at the nighttime rate.

Solution

Step 1 The problem asks for the number of hours charged at each rate.

Step 2 Let d = the number of hours of use at the daytime rate.
Let n = the number of hours of use at the nighttime rate.

Step 3 Set up a system of two equations.

Total number of hours charged is 28. $\longrightarrow$ $d + n = 28$
Total amount charged is 411. $\longrightarrow$ $30d + 10.50n = 411$

Step 4 Solve the system. Express d in terms of n in the first equation and substitute in the second equation.

$$d = 28 - n$$
$$30(28 - n) + 10.50n = 411$$
$$840 - 30n + 10.50n = 411$$
$$-19.50n = -429$$
$$n = 22$$

Since $n = 22$, $d = 28 - 22 = 6$. The solution is (6, 22).

Step 5 The check is left for you.
∴ there were 6 h charged at the daytime rate and 22 h at the nighttime rate.   *Answer*

The following terms are used in connection with aircraft flight.

air speed	the speed of an aircraft in still air
wind speed	the speed of the wind
tail wind	a wind blowing in the same direction as the path of the aircraft
head wind	a wind blowing in the direction opposite to the path of the aircraft
ground speed	the speed of the aircraft relative to the ground

With a tail wind: ground speed = air speed + wind speed.
With a head wind: ground speed = air speed − wind speed.

Example 2 To measure the speed of the jet stream (a high-speed, high-altitude, west-to-east wind), a weather-service plane flew 1800 km with the jet stream as a tail wind and then back again. The eastbound flight took 2 h, and the westbound one took 3 h 20 min. Find the speed of the jet stream and the air speed of the plane.

Solution

Step 1 The problem asks for the plane's air speed and the speed of the wind.

Step 2 Let p = the air speed in km/h and w = the wind speed in km/h.

Step 3 Using the fact that rate × time = distance, construct a table.

	Ground speed km/h	Time h	Distance km
Eastbound	$p + w$	2	$2(p + w)$
Westbound	$p - w$	$3\frac{1}{3}$, or $\frac{10}{3}$	$\frac{10}{3}(p - w)$

The distance east is 1800 km.
$$2(p + w) = 1800$$

The distance west is 1800 km.
$$\frac{10}{3}(p - w) = 1800$$

Step 4 Solve the system.

$2(p + w) = 1800$ is equivalent to $p + w = 900$

$\frac{10}{3}(p - w) = 1800$ is equivalent to $p - w = 540$

Using the linear-combination method, you will find that $p = 720$ and $w = 900 - 720 = 180$. The solution is (720, 180).

Step 5 The check is left for you.

∴ the air speed of the plane is 720 km/h and the speed of the jet stream is 180 km/h. **Answer**

Problems

A

1. Kerry asked a bank teller to cash a $390 check using $20 bills and $50 bills. If the teller gave her a total of 15 bills, how many of each type of bill did she receive?

2. Tickets for the homecoming dance cost $20 for a single ticket or $35 for a couple. Ticket sales totaled $2280, and 128 people attended. How many tickets of each type were sold?

3. Find the measures of the angles of an isosceles triangle if the measure of the vertex angle is 40° less than the sum of the measures of the base angles.

4. Two isosceles triangles have the same base length. The legs of one of the triangles are twice as long as the legs of the other. Find the lengths of the sides of the triangles if their perimeters are 23 cm and 41 cm.

5. On Friday, the With-It Clothiers sold some jeans at $25 a pair and some shirts at $18 each. Receipts for the day totaled $441. On Saturday the store priced both items at $20, sold exactly the same number of each item, and had receipts of $420. How many pairs of jeans and how many shirts were sold each day?

6. A grain-storage warehouse has a total of 30 bins. Some hold 20 tons of grain each, and the rest hold 15 tons each. How many of each type of bin are there if the capacity of the warehouse is 510 tons?

7. With a tail wind, a helicopter traveled 300 mi in 1 h 40 min. The return trip against the same wind took 20 min longer. Find the wind speed and also the air speed of the helicopter.

8. With a head wind, a plane traveled 1000 mi in 4 h. With the same wind as a tail wind, the return trip took 3 h 20 min. Find the plane's air speed and the wind speed.

9. An overseas phone call is charged at one rate (a fixed amount) for the first minute and at a different rate for each additional minute. If a 7 min call costs $10, and a 4 min call costs $6.40, find each rate.

10. A caterer's total cost for catering a party includes a fixed cost, which is the same for every party. In addition the caterer charges a certain amount for each guest. If it costs $300 to serve 25 guests and $420 to serve 40 guests, find the fixed cost and the cost per guest.

11. A financial planner wants to invest $8000, some in stocks earning 15% annually and the rest in bonds earning 6% annually. How much should be invested at each rate to get a return of $930 annually from the two investments?

12. For a recent job, a plumber earned $28/h, and the plumber's apprentice earned $15/h. The plumber worked 3 hours more than the apprentice. If together they were paid $213, how much did each earn? (*Hint:* First write an expression for the number of hours each worked on the job.)

B 13. Marcia flew her ultralight plane to a nearby town against a head wind of 15 km/h in 2 h 20 min. The return trip under the same wind conditions took 1 h 24 min. Find the plane's air speed and the distance to the nearby town.

14. A plane whose air speed is 150 mi/h flew from Abbot to Blair in 2 h with a tail wind. On the return trip against the same wind, the plane was still 60 mi from Abbot after two hours. Find the wind speed and the distance between Abbot and Blair.

15. If a particle starting with initial speed v_0 has constant acceleration a, then its speed after t seconds is given by

$$v = v_0 + at.$$

Find v_0 and a if $v = 28$ m/s when $t = 4$ s and $v = 43$ m/s when $t = 7$ s.

16. While training for a biathlon race, Kevin covered a total distance of 9 km by swimming for 45 min and running for 20 min. The next day he swam for 30 min and ran for 40 min, covering 14 km. Find his rates (in km/h) for swimming and running. (Assume that these rates are constant.)

17. Davis Rent-A-Car charges a fixed amount per weekly rental plus a charge for each mile driven. A one-week trip of 520 miles costs $250, and a two-week trip of 800 miles costs $440. Find the weekly charge and the charge for each mile driven.

18. In a certain mill the cost C in thousands of dollars of producing x tons of steel is given by

$$C = 0.3x + 3.5.$$

The revenue R in thousands of dollars from selling x tons is given by

$$R = 0.5x.$$

a. Graph each equation in the same coordinate plane. (Let the vertical axis represent cost or revenue in thousands of dollars.)
b. Find the point at which $R = C$ (the *break-even point*).
c. What does the difference $R - C$ measure for all x greater than the x-coordinate of the break-even point?

19. The *supply S* of a commodity (in thousands of units), if the price is p dollars per unit, is given by the equation

$$S = 0.3p + 3.$$

The *demand D* for that commodity (in thousands of units), if the price is p dollars per unit, is given by the equation

$$D = -0.5p + 9.$$

The value of p for which $S = D$ is called the *equilibrium price*.

a. Graph both equations in the same coordinate plane. (Let the horizontal axis represent price and the vertical axis represent supply or demand in thousands of units.)
b. Determine the equilibrium price.
c. What will happen at prices greater than this price?

C **20.** Two miles upstream from his starting point, a canoeist passed a log floating in the river's current. After paddling upstream for one more hour, he paddled back and reached his starting point just as the log arrived. Find the speed of the current.

3-7 Linear Inequalities in Two Variables

Objective   To graph linear inequalities in two variables and systems of such inequalities.

When you replace the equals sign in a linear equation in two variables by one of the symbols $>$, $<$, $\geq$, or $\leq$, you obtain a **linear inequality in two variables.** For example,

$$y > \frac{1}{2}x + 1, \qquad 2x - 3y < 12, \qquad \text{and} \qquad y \geq x$$

are linear inequalities in the two variables x and y.

A **solution** of an inequality in two variables is an ordered pair of numbers that satisfies the inequality. For example, the pair (4, 5) is a solution of $y > \frac{1}{2}x + 1$ since $5 > \frac{1}{2}(4) + 1$. There are infinitely many ordered pairs that satisfy this inequality, and so its graph contains infinitely many points.

When graphing a linear inequality, it is helpful first to graph the **associated equation** of the inequality. The associated equation of $y > \frac{1}{2}x + 1$ is $y = \frac{1}{2}x + 1$. Its graph is shown at the right. Every point $P(x,y)$ on the line $y = \frac{1}{2}x + 1$ has y-coordinate equal to $\frac{1}{2}x + 1$. It is not difficult to see that if a point Q is above P, its y-coordinate is greater than $\frac{1}{2}x + 1$. Similarly, if a point R is below P, its y-coordinate is less than $\frac{1}{2}x + 1$. Therefore, the graph of $y > \frac{1}{2}x + 1$ is the set of all points *above* the line $y = \frac{1}{2}x + 1$, and the graph of $y < \frac{1}{2}x + 1$ is the set of all points *below* the line $y = \frac{1}{2}x + 1$.

The graphs of $y > \frac{1}{2}x + 1$ and $y < \frac{1}{2}x + 1$ are shown as shaded regions in Figure 1. Each graph is a **half-plane** with the line $y = \frac{1}{2}x + 1$ as its **boundary.** We show each boundary as a dashed line because it is not included in the graph. We call each graph an **open half-plane.**

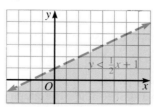

Figure 1

The graphs of $y \geq \frac{1}{2}x + 1$ and $y \leq \frac{1}{2}x + 1$ are shown in Figure 2. Since the boundary line $y = \frac{1}{2}x + 1$ is included in each graph, we show it as a solid line. We call a half-plane together with its boundary a **closed half-plane.**

 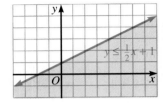

Figure 2

Example 1 Graph $3x + 2y + 4 \geq 0$.

Solution Transform the inequality into an equivalent inequality with y alone on one side.

$$3x + 2y + 4 \geq 0$$
$$2y \geq -3x - 4$$
$$y \geq -\frac{3}{2}x - 2$$

Graph the associated equation

$$y = -\frac{3}{2}x - 2$$

as a solid line. The graph of the inequality includes this boundary line and the half-plane above it.

Example 2 Graph $5x - 4y > 10$.

Solution
$$5x - 4y > 10$$
$$-4y > -5x + 10$$
$$y < \frac{5}{4}x - \frac{5}{2}$$

Graph the associated equation

$$y = \frac{5}{4}x - \frac{5}{2}$$

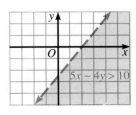

as a dashed line. The graph of the inequality is the open half-plane below this boundary line.

You can tell whether you have shaded the correct half-plane by testing a point that is not on the boundary. The origin is a convenient point to use. In Example 1, the ordered pair (0, 0) *satisfies* the given inequality ($4 \geq 0$ is true), and the point (0, 0) is *in* the shaded region. In Example 2, (0, 0) *does not satisfy* the given inequality ($0 > 10$ is false), and the point (0, 0) is *not in* the shaded region.

Example 3 Graph $x < 2$ in a coordinate plane.

Solution The graph of $x = 2$ is a vertical line. A point $P(x, y)$ belongs to the graph of the inequality if and only if the x-coordinate of P is less than 2 (that is, if and only if P is to the left of the boundary line $x = 2$).

The graph of a **system of inequalities** consists of points satisfying all of the inequalities in the system. This is the region where the graphs of the individual inequalities overlap.

Example 4 Graph this system:

$$x + y \leq 6$$
$$2y > x$$

Solution Graph $x + y \leq 6$, showing the boundary as a solid line. In the same coordinate plane graph $2y > x$, showing the boundary as a dashed line. The graph of the system is the doubly-shaded region in the figure at the right.

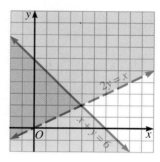

Example 5 Graph this system:

$$x > 0$$
$$y > 0$$
$$3x + 2y \leq 12$$
$$x - 2y \geq -4$$

Solution The inequalities $x > 0$ and $y > 0$ tell you that the graph of the system is contained in the first quadrant. The graph of the system is the shaded region in the first quadrant below both of the lines $3x + 2y = 12$ and $x - 2y = -4$.

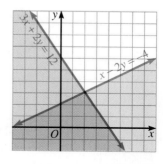

Oral Exercises

Tell whether each ordered pair satisfies the given inequality.

1. $(3, -4)$, $(-3, 4)$; $x + y > 0$

2. $(3, -1)$, $(1, -1)$; $x - y < 2$

3. $(2, 1)$, $(1, 2)$; $2x - y \le 3$

4. $(-2, -1)$, $(-1, -2)$; $3x - y \ge -1$

5. $(3, -2)$, $(2, -3)$; $3x + 2y < 0$

6. $(5, 1)$, $(-3, -1)$; $x - 4y + 1 > 0$

Tell whether point P lies above, on, or below the line L.

7. $P(0, 2)$; L: $y = 2x + 1$

8. $P(2, 2)$; L: $y = \frac{3}{2}x - 1$

9. $P(-1, 2)$; L: $y = -\frac{1}{2}x + 2$

10. $P(-3, 2)$; L: $y = \frac{1}{2}x + 3$

11. $P(4, 5)$; L: $y = \frac{1}{2}x + 3$

12. $P(-5, -3)$; L: $y = -\frac{1}{2}x - 1$

Give the letter of the inequality whose graph is shown.

13. a. $x + 1 \ge 0$
 b. $y \ge 1$
 c. $x - 1 \ge 0$
 d. $x > 1$

14. a. $y \le 2$
 b. $y - 2 < 0$
 c. $y + 2 < 0$
 d. $x < 2$

15. a. $y > -x$
 b. $x - y > 0$
 c. $x + y \ge 0$
 d. $y > x$

16. a. $x + y \ge 0$
 b. $x + y > 0$
 c. $x - y \ge 0$
 d. $y - x \ge 0$

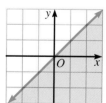

Written Exercises

Graph each inequality in a coordinate plane. You may wish to check your graphs on a computer or a graphing calculator.

A

1. $y - 2 \ge 0$

2. $x - 1 < 0$

3. $x + 1 < 0$

4. $y + 2 \ge 0$

5. $x < y$

6. $x + y < 0$

7. $x \le 2y$

8. $x + 2y \ge 0$

9. $x + y > 1$

10. $x - y < 1$

11. $2x - y \le 2$

12. $x + 2y \le 2$

13. $3x + 2y < 6$

14. $2x - 3y < 6$

15. $3x \ge 2(y - 1)$

16. $y \le \frac{1}{2}(4 - x)$

17. $y \ge \frac{1}{3}(2x + 3)$

18. $2(y - 1) > 3(x + 1)$

Graph each system of inequalities.

19. $y \le 2$
    $y \ge x$

20. $x + y > 0$
    $y + 2 > 0$

21. $x + y > 0$
    $x - y \le 0$

22. $x - y \le 0$
    $x + y < 0$

23. $y < x + 2$
    $1 - y < 0$

24. $y + x \le 2$
    $y - x \le 2$

25. $x - y \le 1$
    $y - x \le 1$

26. $x + y > 0$
    $x + y < 2$

27. $x + 2y < 2$
    $x - 2y > 2$

28. $x - 2y + 2 \ge 0$
    $x + 2y - 2 \ge 0$

29. $3x + 2y \le 6$
    $2x + 3y \ge 6$

30. $y - 3x < 3$
    $3y - x > 3$

B **31.** $x \ge 0$
    $y \ge 0$
    $x + 2y \le 4$

32. $y \ge 0$
    $y \le x$
    $x + y \le 3$

33. $y - x > 0$
    $y + x > 0$
    $y - 2 < 0$

34. $y - x < 3$
    $y + x < 3$
    $y - 1 > 0$

35. $x \ge 0$
    $y \ge 0$
    $x + y \le 6$
    $x - y \le 2$

36. $x \ge 0$
    $y \ge 0$
    $x - y + 2 > 0$
    $2x - y < 2$

Graph each inequality or system of inequalities.

Sample $|x + 1| < 2$

Solution   The given inequality is equivalent to
$$-2 < x + 1 < 2$$
so $x > -3$ and $x < 1$.
The graph is the shaded region between the vertical lines $x = -3$ and $x = 1$.

37. $0 \le y \le 3$

38. $1 < x < 3$

39. $|x - 2| < 2$

40. $|y - 2| \le 1$

41. $|x| \le 2$
    $|y| \le 3$

42. $|x - 1| < 1$
    $|y - 1| < 1$

C **43.** $|x| < |y|$

44. $|y| \le |x|$

45. $|x| + |y| \le 2$

Mixed Review Exercises

Each system has a unique solution; find it.

1. $2x + y = -3$
    $x + 3y = 1$

2. $3x - y = 4$
    $x + 2y = -8$

3. $5x + 2y = 7$
    $x - y = 7$

Find an equation in standard form of the line through P having the given slope. Then graph the line.

4. $P(-1, 3)$, $m = 1$

5. $P(2, 3)$, vertical

6. $P(0, 0)$, $m = -4$

7. $P(-6, -2)$, $m = \dfrac{2}{3}$

8. $P(-5, 7)$, $m = 0$

9. $P(0, -3)$, $m = -\dfrac{1}{2}$

Self-Test 2

Vocabulary   system of linear equations (p. 124)   linear inequality in two variables
          linear system (p. 124)   (p. 135)
          solution of a system (p. 124)   associated equation (p. 135)
          simultaneous solution (p. 124)   open half-plane (p. 135)
          equivalent systems (p. 124)   boundary (p. 135)
          linear combination (p. 125)   closed half-plane (p. 136)
          consistent (p. 128)   system of inequalities in two
          inconsistent (p. 128)   variables (p. 137)
          dependent (p. 128)

Solve each system. If the system has an infinite solution set, specify it and give three solutions. If the system has no solution, say so.

1. $4x - y = 12$
   $3x + 4y = 66$

2. $6x + 3y = 5$
   $4x + 2y = 9$

Obj. 3-5, p. 124

3. $4x = 10 - 3y$
   $12x = 30 - 9y$

4. $8x + 3y = 5$
   $5x + 4y = 18$

5. If it takes an airplane 3 h to fly 360 mi with the wind and 4 h to make the return trip against the wind, find the speed of the wind and the air speed of the airplane.

Obj. 3-6, p. 131

6. Last summer Bret earned $32 a day and Sandra earned $36 a day. Together they earned a total of $3560. How much did each earn, if Bret worked 5 days more than Sandra?

7. Graph $2x + 3y < 12$.

8. Graph $5x - 2y \geq -10$.

9. Graph $y > 0$ in a coordinate plane.

Obj. 3-7, p. 135

10. Graph this system: $x + 2y \leq 6$
                        $x - 3y \geq 4$

Check your answers with those at the back of the book.

Challenge

Suppose that you want to rank four people (Alicia, Beth, Carlos, and Dave) in order from youngest to oldest. No two of them have the same age. They won't tell you their ages, but they will answer questions like "Dave, are you older than Beth?" What is the least number of such questions with which you can be *sure* of finding out the order of their ages? Describe your method for finding the order.

Functions and Relations

3-8 Functions

Objective   To find values of functions and to graph functions.

The post office provides an everyday example of a *function:* The cost of mailing a letter is a function of its weight. Here there is a correspondence, cost to weight, given by a definite rule. The mathematical concept of function involves such a correspondence between two sets.

The **mapping diagram** at the right pictures a correspondence between two sets, *D* and *R*. A rule that defines this particular correspondence is:

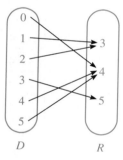

> To each whole number, assign
> the number of letters in its name.

For example, 5 is assigned to 3 because there are 5 letters in the word "three." Such a pairing is called a *mapping,* or a *function.*

In Lesson 3-10, you will learn a general definition of function stated in terms of ordered pairs. Here you will work with a definition that emphasizes a correspondence between two sets.

> A **function** is a correspondence between two sets, *D* and *R*, that assigns to each member of *D* exactly one member of *R*.

The set *D* is the **domain** of the function. The **range** of the function is the set of all members of *R* assigned to at least one member of *D*. *In this book, domains and ranges of functions will be sets of real numbers, unless otherwise stated.*

To define a function, you can give a rule describing the correspondence and then specify the domain. For example, if the function with the rule

> "double the number"

has the set of integers as its domain, it assigns 4 to 2, 2 to 1, 0 to 0, −2 to −1, −4 to −2, and so on, as illustrated at the right. The range of this function is the set of even integers.

The letters f, g, F, G, and ϕ (Greek phi) are often used to name functions. If the rule of a function is given by an algebraic expression, you can describe it using "arrow notation." For example, if g is the name of the doubling function, you can write

$$g: x \rightarrow 2x.$$

This is read "g, the function that assigns to x the number $2x$."

The **graph of a function** f is the set of all points (x, y) such that x is in the domain of f and the rule of the function assigns y to x.

Example 1 Given $f: x \rightarrow 4x - x^2$ with domain $D = \{1, 2, 3, 4, 5\}$.
 a. Find the range of f.   **b.** Graph f.

Solution

a. Make a table showing the numbers that the rule of f assigns to each member of the domain.

x	$4x - x^2$	(x, y)
1	$4 \cdot 1 - 1^2 = 3$	$(1, 3)$
2	$4 \cdot 2 - 2^2 = 4$	$(2, 4)$
3	$4 \cdot 3 - 3^2 = 3$	$(3, 3)$
4	$4 \cdot 4 - 4^2 = 0$	$(4, 0)$
5	$4 \cdot 5 - 5^2 = -5$	$(5, -5)$

From the second column,
$R = \{-5, 0, 3, 4\}$. **Answer**

b. Graph the ordered pairs from the table.

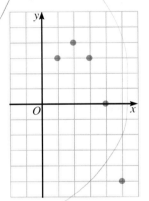

The **values** of a function are the members of its range. Thus, in Example 1, f assigns to 2 the *value* 4. To indicate this you use **functional notation** and write $f(2) = 4$, read "the value of f at 2 is 4" or "f of 2 is 4." In general, for any function f,

$$f(x) \text{ denotes the value of } f \text{ at } x.$$

Caution: $f(2)$ is *not* the product of f and 2.

Example 2 Given $f: x \rightarrow 2x - 1$ and $g: x \rightarrow x^2 + 1$, find the indicated values.
 a. $f(2)$ **b.** $g(2)$ **c.** $f(a)$ **d.** $f(g(2))$ **e.** $g(f(2))$

Solution   **a.** $f(2) = 2 \cdot 2 - 1 = 3$
   b. $g(2) = 2^2 + 1 = 5$
   c. $f(a) = 2a - 1$
   d. Use (b): $f(g(2)) = f(5) = 2 \cdot 5 - 1 = 9$
   e. Use (a): $g(f(2)) = g(3) = 3^2 + 1 = 10$

Often a function f is defined simply by giving an equation for $f(x)$ and specifying the domain, as in Example 3.

Example 3 Given $f(x) = |x - 1|$ with domain $D = \{-2, -1, 0, 1, 2\}$.
        a. Find the range of f.   **b.** Graph f.

Solution

a.

x	$|x - 1| = f(x)$	(x, y)
-2	$|-2 - 1| = 3$	$(-2, 3)$
-1	$|-1 - 1| = 2$	$(-1, 2)$
0	$|0 - 1| = 1$	$(0, 1)$
1	$|1 - 1| = 0$	$(1, 0)$
2	$|2 - 1| = 1$	$(2, 1)$

b.

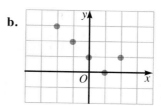

From the second column, $R = \{3, 2, 1, 0\}$.   *Answer*

Sometimes a function is defined by an equation and its domain is not specified. *For such functions, take as the domain the set of those real numbers for which the equation produces real numbers.*

Example 4 Give the domain of each function.

    a. $f(x) = \dfrac{1}{x^2 + 4}$ **b.** $g(x) = \dfrac{1}{x(x + 2)}$ **c.** $h(x) = \sqrt{2 - x}$

Solution   **a.** $D = \{\text{all real numbers}\}$

    b. The expression $\dfrac{1}{x(x + 2)}$ is not defined when $x = 0$ and $x = -2$.
    $\therefore D = \{\text{all real numbers except 0 and } -2\}$
    c. The expression $\sqrt{2 - x}$ is not defined when $2 - x < 0$, or $x > 2$.
    $\therefore D = \{x: x \le 2\}$

Oral Exercises

For each function, find the indicated values.

1. $g: x \to x + 2$ **a.** $g(0)$ **b.** $g(-2)$ **c.** $g(-1)$ **d.** $g(2)$
2. $f: x \to 1 - x$ **a.** $f(-2)$ **b.** $f(1)$ **c.** $f(0)$ **d.** $f(2)$
3. $G: x \to 1 - 2x$ **a.** $G(-2)$ **b.** $G(-1)$ **c.** $G(0)$ **d.** $G(3)$
4. $F: x \to 3x - 2$ **a.** $F(1)$ **b.** $F(0)$ **c.** $F(-1)$ **d.** $F(2)$
5. $h: x \to 1 - x^2$ **a.** $h(1)$ **b.** $h(-1)$ **c.** $h(2)$ **d.** $h(-2)$

For each function, find the indicated values.

6. $\phi: x \to x - x^2$ **a.** $\phi(0)$ **b.** $\phi(-1)$ **c.** $\phi(2)$ **d.** $\phi(-2)$

7. $k: x \to |2 - x|$ **a.** $k(0)$ **b.** $k(-2)$ **c.** $k(2)$ **d.** $k(3)$

8. $H: x \to 2 - |x|$ **a.** $H(0)$ **b.** $H(-2)$ **c.** $H(2)$ **d.** $H(-3)$

9. $s: x \to \dfrac{10}{x^2 + 1}$ **a.** $s(0)$ **b.** $s(1)$ **c.** $s(2)$ **d.** $s(3)$

10. $h: x \to \dfrac{1 - x}{x^3}$ **a.** $h(-1)$ **b.** $h(0)$ **c.** $h(1)$ **d.** $h(2)$

Give the domain and range of the function whose complete graph is shown in red.

11.

12.

13.

14.

15.

16.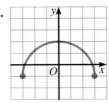

17. Suppose that the domain of a function has n members.

   a. What is the least number of members in its range?

   b. What is the greatest number of members in its range?

Written Exercises

In Exercises 1–12, the domains D and rules of some functions are given. Find the range of each function.

A
1. $F: x \to 3 - 2x$; $D = \{0, 1, 2, 3\}$

2. $\phi: x \to 3x - 5$; $D = \{0, 1, 2, 3\}$

3. $g: x \to x^2 - 2$; $D = \{-2, 0, 2\}$

4. $f: x \to 1 - x^2$; $D = \{-1, 0, 1\}$

5. $f: x \to x^2 - 3x$; $D = \{-1, 0, 1, 2, 3\}$

6. $h: x \to 4x - x^2$; $D = \{-1, -2, 0\}$

7. $G: x \to x^2 - 4x + 4$; $D = \{0, 1, 2, 3\}$

8. $k: t \to t^2 + t - 2$; $D = \{-2, -1, 0, 1\}$

9. $g: x \to x^4 - x^2$; $D = \{-2, 0, 2\}$

10. $H: z \to z^2 - z^3$; $D = \{-1, 0, 1, 2\}$

11. $m: z \to 1 - |z|$; $D = \{-2, -1, 0, 2\}$

12. $r: t \to |1 - t|$; $D = \{-2, -1, 0, 1, 2\}$

13–24. Graph the functions in Exercises 1–12.

Complete the equation for the function pictured in each mapping diagram.

B 25. $f(x) = \underline{\ ?\ } \cdot x + 1$ **26.** $g(x) = x \cdot \underline{\ ?\ }$ **27.** $h(x) = \underline{\ ?\ }^{\,x}$ **28.** $F(x) = \underline{\ ?\ } \cdot x + 5$

Give the domain of each function.

29. $f(x) = x^3 - 2$

30. $g(x) = \dfrac{2}{x + 3}$

31. $h(x) = \dfrac{2}{x^2 + 3}$

32. $F(x) = \sqrt{2x - 1}$

33. $f(x) = \dfrac{1}{x^2 + 5x + 6}$

34. $G(x) = \dfrac{2}{(x - 1)(x + 2)}$

Let $f(x) = x^2 - 1$ and $g(x) = 1 - 2x$. Find the indicated values.

35. $f(g(1))$ and $g(f(1))$

36. $f(g(-1))$ and $g(f(-1))$

37. $f(g(2))$ and $g(f(2))$

38. $f(g(-2))$ and $g(f(-2))$

39. $f(f(2))$ and $f(2f(1))$

40. $g(g(2))$ and $g(2g(1))$

C 41. $f\left(\dfrac{1}{g(3)}\right)$

42. $g\left(\dfrac{1}{f(3)}\right)$

43. $g(a + 1) - g(a)$

44. $f(a + 1) - f(a)$

45. $\dfrac{f(x + h) - f(x)}{h}$

46. $\dfrac{g(x + h) - g(x)}{h}$

47. If f is a function such that $f(a + b) = f(a) + f(b)$, for all real numbers a and b, then show that the following statements are true.

 a. $f(0) = 0$ (*Hint:* Let $a = b = 0$.)
 b. $f(2a) = 2f(a)$ (*Hint:* Let $b = a$.)
 c. $f(-a) = -f(a)$

48. The **greatest-integer function** is defined for all real numbers x as follows:

$$g(x) = \text{the greatest integer } not\ exceeding\ x$$

For example: $g(2) = 2$, $g\left(2\dfrac{1}{2}\right) = 2$, $g(3) = 3$, $g(\pi) = 3$, and $g\left(-2\dfrac{1}{2}\right) = -3$. Graph the greatest-integer function for $-3 \le x \le 3$.

49. The **signum function** is defined for all real numbers x as follows:

$$s(x) = \begin{cases} 1 & \text{if } x > 0 \\ 0 & \text{if } x = 0 \\ -1 & \text{if } x < 0 \end{cases}$$

Graph the signum function for $3 \le x \le 3$. How does the signum function differ from the function $f(x) = \dfrac{x}{|x|}$?

3-9 Linear Functions

Objective   To find equations of linear functions and to apply properties of linear functions.

The graph of the function $f(x) = 2x - 3$ is the set of all points (x, y) such that $y = 2x - 3$. This graph is the line with slope 2 and y-intercept -3, as shown at the right. A function such as f, which has a straight line for its graph, is called a *linear function*.

A **linear function** is a function f that can be defined by

$$f(x) = mx + b$$

where x, m, and b are any real numbers. The graph of f is the graph of $y = mx + b$, a line with slope m and y-intercept b.

If $f(x) = mx + b$ and $m = 0$, then $f(x) = b$ for all x, and f is a **constant function.** Its graph is the horizontal line $y = b$.

For any linear function f that is not a constant function, a change in the value of x produces a change in the value of $f(x)$. For a linear function $f(x) = mx + b$, the *rate of change of $f(x)$ with respect to x* is the constant m. (See Exercise 32.) The tables below illustrate this fact.

$f(x) = 2x - 3$, $m = 2$

x	$f(x)$
0	-3
1	-1
2	1
3	3
4	5

$f(x) = -5x + 7$, $m = -5$

x	$f(x)$
0	7
1	2
2	-3
3	-8
4	-13

When x increases by 1, $f(x)$ changes by $2 \cdot 1$, or 2. When x increases by 3, $f(x)$ changes by $2 \cdot 3$, or 6.

When x increases by 1, $f(x)$ changes by $-5 \cdot 1$, or -5. When x increases by 3, $f(x)$ changes by $-5 \cdot 3$, or -15.

Notice that changing x by an amount a changes $f(x)$ by a corresponding amount $m \cdot a$. In general, to find the change in $f(x)$ you multiply the change in x by the rate of change m. If m is positive, the change in $f(x)$ is an increase; if m is negative, the change in $f(x)$ is a decrease.

When you know the change in $f(x)$ produced by some change in x, you can find the rate of change m by using this formula:

$$\text{rate of change } m = \frac{\text{change in } f(x)}{\text{change in } x}$$

Example 1 Find equations of the linear functions f and g using the given information.

  a. $f(4) = 1$ and $f(8) = 7$

  b. $g(0) = 5$ and an increase of 4 units in x causes a decrease of 12 units in $g(x)$.

Solution **a.** Since f is linear, let $f(x) = mx + b$. First find the value of m.

$$\text{rate of change } m = \frac{\text{change in } f(x)}{\text{change in } x}$$

$$m = \frac{f(8) - f(4)}{8 - 4} = \frac{7 - 1}{4} = \frac{3}{2}$$

Thus, $m = \frac{3}{2}$ and $f(x) = \frac{3}{2}x + b$. Now find the value of b.

Since $f(4) = 1$, $1 = \frac{3}{2}(4) + b$

$1 = 6 + b$

$-5 = b$

$\therefore f(x) = \frac{3}{2}x - 5$ **Answer**

  b. Since g is linear, let $g(x) = mx + b$. First find m.

$$\text{rate of change } m = \frac{\text{change in } g(x)}{\text{change in } x} = \frac{-12}{4} = -3$$

Thus, $m = -3$ and $g(x) = -3x + b$. Now find b.

Since $g(0) = 5$, $5 = -3(0) + b$

$5 = b$

$\therefore g(x) = -3x + 5$ **Answer**

Many real-life situations can be described by linear functions. Here are some examples:

1. A manufacturer of television sets spends $6000 per day in fixed costs and $250 for each set produced. The total cost C for producing x sets a day is given by

$$C(x) = 6000 + 250x.$$

Notice that the domain of the linear function C is restricted to positive integers, since x represents a number of television sets.

2. A computer purchased new for $3600 loses $600 in value each year (straight-line depreciation). Its remaining worth, or book value B, y years after purchase is given by

$$B(y) = 3600 - 600y.$$

Notice that the domain of B is restricted to $\{y: 0 \leq y \leq 6\}$, since after 6 years the book value has reached $0.

3. A hiker climbing Mount Jefferson left the 4000-foot level at 6 A.M. and gained elevation at the rate of 800 ft/h. The hiker's elevation E at any time t hours after 6 A.M. is given by

$$E(t) = 4000 + 800t.$$

Here t is restricted also, because the hiker's elevation cannot exceed the height of the mountain.

It is customary to call functions such as C, B, and E linear functions even though the domain of each is not the set of all real numbers.

In Example 2, an equation of a linear function is found using a method different from that shown in Example 1.

Example 2 At noon a pump started emptying an oil storage tank at a constant rate. At 2 P.M. there were 1680 barrels left in the tank. At 5 P.M. there were 600 barrels left.

    a. Write a linear function to describe this situation.
    b. How many barrels were in the tank at noon?
    c. What time will the tank be empty?

Solution **a.** Let $f(t) =$ the number of barrels of petroleum in the tank t hours after noon. Then, since f is linear, $f(t) = mt + b$.

    To find the values of m and b, use the facts that $f(2) = 1680$ and $f(5) = 600$ to write a system of linear equations:

$$f(2) = 2m + b = 1680$$
$$f(5) = 5m + b = 600$$

Solve this system to obtain $m = -360$ and $b = 2400$. Therefore,

$$f(t) = -360t + 2400. \quad \textit{Answer}$$

    b. At noon, $t = 0$, and

$$f(0) = -360(0) + 2400 = 2400.$$

$\therefore$ there were 2400 barrels in the tank at noon.   *Answer*

    c. The tank will be empty when $f(t) = 0$.

$$0 = -360t + 2400$$
$$360t = 2400$$
$$t = \frac{2400}{360} = 6\frac{2}{3} \text{ (hours after noon)}$$

$\therefore$ the tank will be empty at 6:40 P.M.   *Answer*

Oral Exercises

Each table is a table of values of a *linear* function. Supply the missing entries.

Sample

x	f(x)
0	3
3	5
6	?
12	?

3 { (0→3), 3 { (3→6) } 2, 6 { (6→12)

Solution

The rate of change $m = \frac{2}{3}$. This means an increase of 3 in x (from 3 to 6) causes an increase of $3 \cdot \frac{2}{3}$, or 2, in $f(x)$: from 5 to 7. An increase of 6 in x (from 6 to 12) causes an increase of $6 \cdot \frac{2}{3}$, or 4, in $f(x)$: from 7 to 11.

∴ the missing entries are 7 and 11.

1.

x	f(x)
1	1
2	4
3	?
4	?

2.

x	f(x)
0	3
1	1
2	?
3	?

3.

x	f(x)
0	5
2	4
4	?
6	?

4.

x	f(x)
2	−1
4	1
6	?
8	?

5.

x	f(x)
0	?
1	2
3	?
6	2

6.

x	f(x)
0	?
−1	−1
−2	2
−3	?

7.

x	f(x)
2	−5
4	?
6	?
8	4

8.

x	f(x)
1	6
−1	?
−3	?
−5	−6

Written Exercises

149–150 # 2–30 EVEN

$f(x) = mx + b$

Find an equation of the linear function f using the given information.

A　**1.** $m = 2, b = 3$　$f(x) = 2x + 3$

2. $m = -1, b = \frac{1}{2}$

3. $m = 3, f(0) = 1$

4. $m = -\frac{3}{2}, f(0) = -\frac{1}{2}$

5. $f(0) = -2$, slope of graph $= \frac{1}{2}$

6. $f(0) = 1$, slope of graph $= -1$

7. $f(0) = 1$; $f(x)$ increases by 6 when x increases by 3.

8. $f(0) = -1$; $f(x)$ decreases by 3 when x increases by 1.　$f(x) = x - 4$

9. $m = 2, f(1) = 5$

10. $m = -1, f(2) = 3$

11. $m = -\frac{3}{2}, f(4) = -1$

12. $m = \frac{2}{3}, f(6) = -2$

13. $f(0) = 1, f(3) = 7$

14. $f(0) = -2, f(2) = 4$

15. $f(0) = -3, f(3) = -3$

16. $f(0) = 0, f(-2) = 4$

17. $f(1) = 2, f(2) = 5$

18. $f(2) = 6, f(4) = 0$

19. $f(-2) = 3, f(2) = -3$

20. $f(-1) = 2, f(2) = 2$

21. $f(-1) = 1, f(1) = -2$

22. $f(3) = 4, f(6) = -3$

Complete the table, given that g is a linear function. (*Hint:* See the Sample on page 149.)

B

23.

x	$g(x)$
3	4
1	-2
0	?
?	-8

24.

x	$g(x)$
2	1
4	4
?	-2
-2	?

25.

x	$g(x)$
-4	4
?	2
2	1
3	?

26.

x	$g(x)$
?	-3
2	-1
1	?
-1	1

In Exercises 27–30, f is a linear function.

27. If $f(6) = 7$ and $f(3) = 2$, find $f(-3)$ and $f(10)$.

28. If $f(2) = 10$ and $f(10) = -2$, find $f(-10)$ and $f(100)$.

29. If $f(1) = 2$ and $f(7) = -6$, find $f(10)$ and $f(20)$.

30. If $f(5) = -5$ and $f(-25) = -25$, find $f(2)$ and $f(50)$.

C 31. If $f(x) = mx + b$ and $k \neq 0$, find $\dfrac{f(x + k) - f(x)}{k}$.

32. If $f(x) = mx + b$ and $x_1 \neq x_2$, show that $\dfrac{f(x_2) - f(x_1)}{x_2 - x_1} = m$.

33. A function f is *increasing* if $f(x_2) > f(x_1)$ whenever $x_2 > x_1$. A function f is *decreasing* if $f(x_2) < f(x_1)$ whenever $x_2 > x_1$. Show that if $f(x) = mx + b$, then (**a**) f is increasing if $m > 0$, and (**b**) f is decreasing if $m < 0$.

Problems

Solve Problems 1–11. Assume that each situation can be described by a linear function.

A 1. A photocopying machine purchased new for $4500 loses $900 in value each year.
   a. Find the book value of the machine after 18 months.
   b. When will the book value be $1200?

2. The charge for a one-day rental of a car from Shurtz Car Rental Agency is $24 plus 15 cents for each mile driven.
   a. If Jill drives 85 mi in one day, how much is the charge?
   b. If a one-day rental cost Jill $42, how far did she drive?

3. A plumber charged $110 for a three-hour job and $160 for a five-hour job. At this rate, how much would he charge for an eight-hour job?

4. It costs Ace Electronics Company $1900 to manufacture 10 VCRs and $2200 to manufacture 16 VCRs. At this rate, what would be the cost of manufacturing 25 VCRs?

5. Allied Airlines charges $90 for a ticket to fly between two cities 260 mi apart and $150 for a ticket to fly between two cities 500 mi apart. At this rate, what would it cost for a trip between two cities 1000 mi apart?

6. A load of 8 kg attached to the bottom of a coil spring stretches the spring to a length of 76 cm, and a load of 14 kg stretches it to a length of 85 cm. Find the natural (unstretched) length of the spring.

7. Fifteen days after Alan began a diet he weighed 176 lb. After 45 days he weighed 170 lb.
   a. How much did he weigh at the beginning of the diet?
   b. At this rate, when will he weigh 165 lb?

8. A climber left base camp at 5 A.M. to ascend a 7400-meter peak. The climber gained altitude at the rate of 240 m/h and at 8 A.M. was at the 6500-meter level.
   a. Find the elevation of the base camp.
   b. At what time did the climber reach the summit?

B 9. The table below shows the freezing and boiling points of water in degrees Celsius (°C) and degrees Fahrenheit (°F).

	Celsius	Fahrenheit
freezing	0	32
boiling	100	212

   a. Find a linear function that gives °C in terms of °F.
   b. What °C temperature corresponds to 98.6° F?
   c. At what temperature do the two scales give the same reading?

10. On a 50 question true-false test, a student received 3 points for each correct answer. For each incorrect answer, the student lost 1 point. For an unanswered question, no points were added or deducted.
   a. Find the best score possible.   **b.** Find the worst score possible.
   c. Find the linear function that converts the raw score described above to a score from 0 to 100.
   d. What score from 0 to 100 corresponds to a test result of 38 correct, 4 incorrect, and 8 omitted?

C 11. South Coast Utilities makes the following monthly charges for residential electrical services:

Service Charge $4.50
Energy Charge:
First 800 kW · h $\longrightarrow$ $0.062 per kW · h
All over 800 kW · h $\longrightarrow$ $0.055 per kW · h

Express the monthly charge C in terms of the number n of kilowatt-hours used, for **(a)** $0 \leq n \leq 800$ and **(b)** $n > 800$. Then **(c)** draw a graph of $C(n)$ for $0 \leq c \leq 1500$.

Mixed Review Exercises

Graph each open sentence.

1. $2x + 3y = 9$

2. $x - y < 4$

3. $y = x$

4. $y + 2 \geq 0$

5. $x + 2 < 0$

6. $y \geq x$

7. $y \geq \frac{1}{2}x - 1$

8. $2x - y = 5$

9. $3x + 4y < 8$

Find the range of each function with domain D.

10. $f: x \rightarrow 3x + 2$, $D = \{-2, -1, 0\}$

11. $g: x \rightarrow 1 - x^2$, $D = \{-1, 0, 1\}$

12. $F: x \rightarrow |x - 4|$, $D = \{-3, -2, -1\}$

13. $G: x \rightarrow 8 - 5x$, $D = \{1, 2, 3\}$

Historical Note / Functions

In the seventeenth century, two different concepts of function developed. One was *geometric:* Mathematicians studied curves defined by geometric conditions (like the conic sections you will study in Chapter 9) and explored geometric properties of graphs such as their slopes and maximum and minimum points.

Another approach dealt with functions that could be defined by a formula, such as a polynomial in the variable x. The methods of calculus were used to analyze these and other basic functions, such as the exponential, logarithmic, and trigonometric functions you will meet in Chapters 10 and 12–14. It was Leibniz, a pioneer of this *analytic* approach, who introduced the term *function*.

As mathematicians became interested in a greater variety of functions, a more formal definition was needed. In 1837 the German mathematician Dirichlet proposed the following definition: y is a single-valued function of the variable x when a definite value of y corresponds to each value of x, no matter in what form this correspondence is specified.

Most modern treatments of functions use a definition based on ordered pairs, like the one stated on page 153.

3-10 Relations

Objective   To graph relations and to determine when relations are also functions.

The mapping diagram, the table, and the graph below illustrate a correspondence between the members of the set $D = \{1, 2, 3\}$ and those of the set $R = \{0, 1, 2, 3, 4\}$.

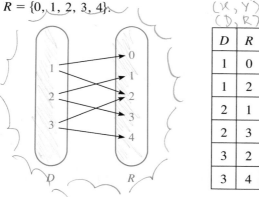

D	R
1	0
1	2
2	1
2	3
3	2
3	4

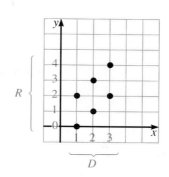

A rule that defines this particular correspondence is: "to each x in D assign a number y if and only if $|x - y| = 1$." Another way of describing this correspondence is to list the ordered pairs determined by the rule:

$$\{(1, 0), (1, 2), (2, 1), (2, 3), (3, 2), (3, 4)\}$$

A more efficient way of describing this set is to use *set notation:*

$$\{(x, y): |x - y| = 1 \text{ and } x \in D\}$$

This is read "the set of all ordered pairs (x, y) such that $|x - y| = 1$ and x is in D."

    The correspondence just discussed does *not* assign exactly one member of R to each member of D. For example, both 0 and 2 are assigned to 1. Therefore, this correspondence is *not* a function. It is, however, a *relation*.

A **relation** is any set of ordered pairs. The set of first coordinates in the ordered pairs is the **domain** of the relation, and the set of second coordinates is the **range**.

A function can now be defined as a special kind of relation.

A **function** is a relation in which different ordered pairs have *different first coordinates*.

Note that the definition just given does not refer to any rule. The set of ordered pairs {(1, 5), (3, 2), (5, 2)} is a function even though no rule is given. The set of ordered pairs {(1, 5), (2, 3), (2, 5)} is a relation but *not* a function, since (2, 3) and (2, 5) are different pairs with the *same* first coordinate.

When a relation is defined by an open sentence in two variables and a domain and range are not specified, *include in the domain and range those real numbers and only those real numbers for which the open sentence is true.* The relation is the solution set of the open sentence. The graph of the relation is the graph of the open sentence.

Example 1 Graph each relation, and then tell whether it is a function.

a. $\left\{(1, 1), (2, 1), \left(2, \frac{1}{2}\right), \left(3, \frac{1}{3}\right), \left(4, \frac{1}{4}\right)\right\}$

b. $\{(x, y): |x| + y = 3\}$

c. $\{(x, y): y > |x|\}$

Solution

a. This relation is *not* a function. The ordered pairs (2, 1) and $\left(2, \frac{1}{2}\right)$ have the same first coordinate. **Answer**

b. The open sentence $|x| + y = 3$ is equivalent to:

$$y = 3 - |x| = \begin{cases} 3 - x & \text{if } x \geq 0 \\ 3 + x & \text{if } x < 0 \end{cases}$$

This relation *is* a function, since each value of x is paired with exactly one value of y. **Answer**

c. $y = |x| = \begin{cases} x & \text{if } x \geq 0 \\ -x & \text{if } x < 0 \end{cases}$

The boundary of $y > |x|$ is the V-shaped region shown at the right. The relation is *not* a function, since it includes different pairs with the same first coordinates, for example, (1, 2) and (1, 3). **Answer**

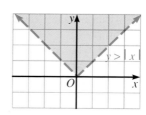

The graphs in Example 1 suggest a convenient way of telling whether a relation is a function. It is called the *vertical-line test.*

Vertical-Line Test

A relation is a function if and only if no vertical line intersects its graph more than once.

Example 2 In the relation $\{(x, y): |x| + |y| = 3 \text{ and } |x| \leq 2\}$, x and y are integers. Find the domain D of the relation and draw its graph. Is the relation a function?

Solution Since x is an integer and $|x| \leq 2$, the domain $D = \{-2, -1, 0, 1, 2\}$.

The relation $$|x| + |y| = 3$$
is equivalent to $$|y| = 3 - |x|.$$

Here is a table of values.

Here is a graph.

x	$|y| = 3 - |x|$	y
-2	1	1 or -1
-1	2	2 or -2
0	3	3 or -3
1	2	2 or -2
2	1	1 or -1

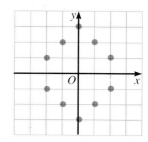

Since the vertical line $x = 1$ intersects the graph in the two points $(1, 2)$ and $(1, -2)$, the relation is *not* a function. **Answer**

Oral Exercises

Give the domain and range of each relation. Is the relation a function?

Sample $\{(3, 2), (1, 4), (5, 2), (7, 6)\}$

Solution $D = \{1, 3, 5, 7\}$, $R = \{2, 4, 6\}$; yes

1. $\{(1, 3), (2, 5), (3, 5), (4, 3)\}$
2. $\{(2, 4), (1, 3), (0, 2), (1, 1)\}$
3. $\{(-1, 0), (1, 0), (0, -1), (-1, 1)\}$
4. $\{(4, 1), (3, 2), (2, 3), (1, 4)\}$
5. $\{(a, b), (b, a)\}$
6. $\{(a, a + 1), (a, a - 1)\}$

List the ordered pairs in the relation pictured in each diagram. Is the relation a function?

7.
8.
9.
10.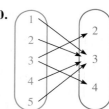

List the ordered pairs in the relation pictured in each graph. Is the relation a function?

11.

12.

13.

14.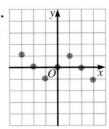

Give the domain and range of each relation graphed below. Is the relation a function?

15.

16.

17.

18.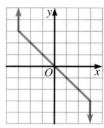

Give the domain of each relation.

19. $\left\{(x, y): y = \dfrac{3}{x - 5}\right\}$

20. $\left\{(x, y): y < \dfrac{1}{x^2 - 4}\right\}$

21. $\left\{(x, y): y = \dfrac{x}{|x - 7|}\right\}$

22. $\{(x, y): |x| + |y| = 1\}$

True or false?

23. All relations are functions.

24. All functions are relations.

Written Exercises

Graph each relation. Then tell whether it is a function. If it is not a function, draw a vertical line that intersects the graph more than once.

A

1. $\{(1, 2), (2, 0), (1, 1)\}$

2. $\{(-1, 2), (0, 1), (1, 2)\}$

3. $\{(-2, 1), (|-2|, 0), (-1, |-2|), (|-1|, -2)\}$

4. $\{(2, 1), (1, -1), (0, 2), (2, 0)\}$

5. $\{(|-3|, 2), (3, -2), (-2, -1), (2, |-1|)\}$

6. $\{(1, 2), (2, -1), (-1, 1), (1, -1), (0, 1)\}$

In each relation in Exercises 7–20, x and y are *integers*. **Find the domain of the relation and draw its graph. Is the relation a function?**

7. $\{(x, y): |x| = |y| \text{ and } |x| \leq 1\}$

8. $\{(x, y): |x| = |y| \text{ and } |y| \leq 2\}$

9. $\{(x, y): |y| = x \text{ and } x \leq 3\}$

10. $\{(x, y): y + |x| = 0 \text{ and } y > -3\}$

11. $\{(x, y): |x| + |y| = 2\}$

12. $\{(x, y): |x| + |y| \leq 1\}$

13. $\{(x, y): |x + y| = 0 \text{ and } |y| \leq 2\}$

14. $\{(x, y): |x| \leq y \text{ and } y \leq 2\}$

15. $\{(x, y): |y| = 2 \text{ and } |x| \leq 2\}$

16. $\{(x, y): |x| = 2 \text{ and } |y| \leq 2\}$

B **17.** $\{(x, y): |xy| = 2\}$

18. $\{(x, y): x|y| = 2\}$

19. $\{(x, y): |x|y = 2\}$

20. $\{(x, y): |xy| \leq 2\}$

Two functions are *equal* if they consist of the same ordered pairs. In Exercises 21–24, f and g have the same domain D and the same range. **Determine whether f and g are equal.**

21. $D = \{-1, 0, 1, 2, 3\}$ $f: x \rightarrow 2x + 1$ $g: x \rightarrow 5 - 2x$

22. $D = \{x: x \geq -1\}$ $f: x \rightarrow x^2 + x$ $g: x \rightarrow x|1 + x|$

23. $D = \{-2, -1, 1, 2\}$ $f: x \rightarrow 4 - x^2$ $g: x \rightarrow 6 - 3|x|$

24. $D = \{-2, -1, 0, 1, 2\}$ $f: x \rightarrow |x| + 2x$ $g: x \rightarrow |x| - 2x$

In Exercises 25–36, each open sentence defines a relation, and x and y are real numbers. **Graph each relation and tell whether it is a function.**

25. $\{(x, y): y = -|x|\}$

26. $\{(x, y): |y| = 2\}$

27. $\{(x, y): y = 1 - |x|\}$

28. $\{(x, y): y = 1 + |x|\}$

29. $\{(x, y): y = |1 - x|\}$

30. $\{(x, y): y = |1 + x|\}$

C **31.** $\{(x, y): |y| = 1 - x\}$

32. $\{(x, y): |y| = 1 + x\}$

33. $\{(x, y): |y| \leq 1 - |x|\}$

34. $\{(x, y): |y| \geq 1 + |x|\}$

35. $\{(x, y): x^2 = 4y^2\}$

36. $\{(x, y): x^2 = 9 - y^2\}$

37. A *definition of "ordered pair"*: The **pair** "a, b" is the set $\{a, b\}$ (which is the same as $\{b, a\}$). We can define **ordered pair** in terms of the simpler concept of pair: The ordered pair "a, b," written (a, b), is the pair having as its two members a and the pair "a, b." Thus,

$$(a, b) = \{a, \{a, b\}\} \quad \text{or} \quad (a, b) = \{\{b, a\}, a\}.$$

For example, in $\{\{3, 5\}, 5\}$, the pair member $\{3, 5\}$ tells you that the pair is either $(3, 5)$ or $(5, 3)$, while the lone member 5 tells you that the first coordinate is 5, so that $\{\{3, 5\}, 5\} = (5, 3)$.

a. Express the ordered pair $(5, 3)$ in terms of pairs in three other ways.

b. What ordered pair is $\{4, \{1, 4\}\}$?

c. Write a definition of the ordered triple (a, b, c) in terms of **pairs**.

Self-Test 3

Vocabulary mapping diagram (p. 141)   functional notation (p. 142)
function (p. 141, 153)   linear function (p. 146)
domain (p. 141)   constant function (p. 146)
range (p. 141)   relation (p. 153)
graph of a function (p. 142)   domain of a relation (p. 153)
values of a function (p. 142)   range of a relation (p. 153)

1. If $f: x \to x^2 - x + 1$ has domain $D = \{-2, -1, 0, 1, 2\}$, what is the range of f?   **Obj. 3-8, p. 141**

2. Give the domain of $g(x) = \dfrac{3x + 4}{2x - 5}$.

3. Graph $f(x) = |x - 2|$ if the domain $D = \{0, 1, 2, 3, 4\}$.

4. Find an equation of the linear function f if $f(3) = -1$ and $f(x)$ decreases by 4 when x increases by 6.   **Obj. 3-9, p. 146**

5. Find an equation of a linear function f having slope -2 if $f(1) = 3$.

6. Write an equation of a linear function f that describes this situation: A bank's total monthly charge for maintaining a checking account is \$.20 for each check written plus an \$8.50 service charge.

7. The cost of printing the school newspaper is \$500 for 800 copies and \$620 for 1200 copies. If the printing cost is a linear function of the copies printed, find the cost of printing 1500 copies.

8. Graph the relation $\{(-2, 0), (-1, -1), (0, -2), (2, 0), (1, 1), (0, 2)\}$. Is the relation a function?   **Obj. 3-10, p. 153**

9. If x and y are integers, find the domain of the relation $\{(x, y): y = 2 - |x| \text{ and } y \geq 0\}$ and draw its graph. Is the relation a function?

Check your answers with those at the back of the book.

Challenge

Using three different weights, you want to be able to balance an object with any integer weight from 1 gram to 13 grams on an equal-arm balance. You can use one, two, or all three of your weights, and one or two of them can go in the same pan with the object you are balancing. What should your three weights weigh?

A farmer wants to maximize the revenue resulting from the production and harvesting of corn and wheat. For every 100 bushels of corn he produces he receives $265 and for every 100 bushels of wheat he produces he receives $365. However, his production is restricted by the availability of land, capital, and labor. For example, he needs one acre to produce 100 bushels of corn and three acres to produce 100 bushels of wheat, and he can plant a total of 100 acres with these crops. The following table shows the availability of resources and the restrictions placed upon them.

	Input requirements per 100 bushels of		Available Material
	Corn	Wheat	
Land (acre)	1	3	100
Capital ($)	120	90	9000
Aug. labor (h)	1	2	200
Sept. labor (h)	1	6	160
Value of output of 100 bushels ($)	265	365	

The farmer can formulate the problem of maximizing revenue mathematically.

Let x = the number of hundreds of bushels of corn produced.
Let y = the number of hundreds of bushels of wheat produced.

If R denotes the total return in dollars, then $R = 265x + 365y$.

The complete problem facing the farmer is to maximize $R = 265x + 365y$ subject to

$$x + 3y \leq 100$$
$$120x + 90y \leq 9000$$
$$x + 2y \leq 200$$
$$x + 6y \leq 160$$
$$x \geq 0$$
$$y \geq 0$$

Notice that the set of restrictions forms a system of linear inequalities. Each linear inequality is called a **constraint,** and the expression to be maximized is called the **objective function.** The problem of maximizing (or minimizing) a linear function subject to a set of linear constraints is called a problem in

linear programming. Decision makers in many organizations formulate and solve linear programming problems as part of their management activities.

The graph of the solution set of the system of constraints for a linear programming problem is called the **feasible region.** The figure below shows the feasible region for the farmer's problem. A computer may be helpful in finding the feasible region, or you can obtain it by hand.

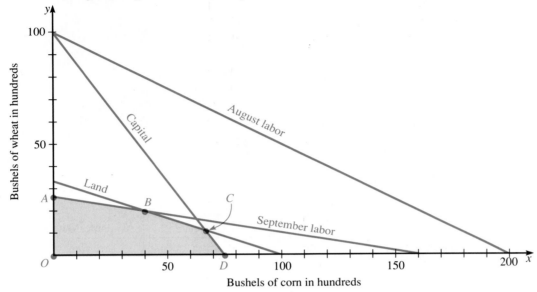

Bushels of corn in hundreds

The farmer needs to determine which point in the feasible region gives the maximum value of R. Although there are infinitely many points in this region, the following theorem from linear programming (which is proved in more advanced courses) makes the list of points to check manageable.

The maximum and minimum values of the objective function occur at vertices of the feasible region.

By solving $x + 3y = 100$ and $x + 6y = 160$ simultaneously, point B, (40, 20), is obtained. Each of the remaining vertices O, A, C, and D is obtained by solving a system of two equations associated with a pair of constraints. The table shows the vertices and the corresponding values of R.

Vertex	$265x + 365y = R$
$O(0, 0)$	$265(0) + 365(0) = 0$
$A\left(0, 26\frac{2}{3}\right)$	$265(0) + 365\left(26\frac{2}{3}\right) = 9733.33$
$B(40, 20)$	$265(40) + 365(20) = 17900$
$C\left(66\frac{2}{3}, 11\frac{1}{9}\right)$	$265\left(66\frac{2}{3}\right) + 365\left(11\frac{1}{9}\right) = 21722.22$
$D(75, 0)$	$265(75) + 365(0) = 19875$

A maximum revenue of $21,722.22 is achieved if the farmer produces $66\frac{2}{3} \times 100 = 6667$ bushels of corn and $11\frac{1}{9} \times 100 = 1111$ bushels of wheat.

Thus he should plant $1 \times 66\frac{2}{3}$, or $66\frac{2}{3}$, acres of corn and $3 \times 11\frac{1}{9}$, or $33\frac{1}{3}$ acres of wheat.

Exercises (A computer may be helpful in Exercises 3–5.)

1. A linear programming problem gives $(0, 0)$, $(0, 90)$, $(50, 80)$, $(70, 20)$, and $(70, 0)$ as consecutive vertices of the feasible region. Find the maximum and minimum values of the objective function $P = 3x + 4y$ in the feasible region.

2. A linear programming problem gives $(10, 0)$, $(10, 40)$, $(30, 50)$, and $(40, 20)$ as the consecutive vertices of the feasible region. Find the maximum and minimum values of the objective function $P = 13x - 5y$ in the feasible region.

3. Find the maximum and minimum values of the objective function $S = 10x + 3y$ in the region which is the solution set of the system $0 \le x \le 10$, $y \ge 5$, and $y \le -0.3x + 10$.

4. Find the maximum and minimum values of the objective function $S = 6x - 2y$ in the region which is the solution set of the system $0 \le y \le 10$, $y \ge -2x + 4$, and $y \le -2x + 12$.

5. A nutrition center sells health food to mountain-climbing teams. The Trailblazer mix package contains one pound of corn cereal mixed with four pounds of wheat cereal and sells for $9.75. The Frontier mix package contains two pounds of corn cereal mixed with three pounds of wheat cereal and sells for $9.50. The center has available 60 pounds of corn cereal and 120 pounds of wheat cereal. How many packages of each mix should the center sell to maximize its income?

Chapter Summary

1. An equation that can be expressed in the form
$$Ax + By = C$$
(A and B not both zero) is called a *linear equation* because its graph in the coordinate plane is a straight line. The line is horizontal if $A = 0$ and vertical if $B = 0$.

2. The slope, m, of a nonvertical line L is given by
$$m = \frac{y_2 - y_1}{x_2 - x_1}$$
where (x_1, y_1) and (x_2, y_2) are two points of L. An equation of L is
$$y - y_1 = m(x - x_1) \qquad \text{(point-slope form)},$$
and if L has y-intercept b, it has the equation
$$y = mx + b \qquad \text{(slope-intercept form)}.$$

3. If the slopes of the nonvertical lines L_1 and L_2 are m_1 and m_2, then:
   (a) L_1 and L_2 are parallel if and only if $m_1 = m_2$.
   (b) L_1 and L_2 are perpendicular if and only if $m_1 m_2 = -1$.

4. A solution of a system of linear equations in two variables is an ordered pair that satisfies each of the equations. The number of solutions of such a system is one, none, or infinitely many according to whether the graphs of the equations intersect in one point, are parallel, or coincide.

5. Systems of equations can be solved algebraically by the *substitution method* or the *linear-combination method*.

6. The graph of a linear inequality in two variables is an open or a closed half-plane. The graph of the solution set of a system of such inequalities is the intersection of the solution sets of the individual inequalities.

7. A *function* is a correspondence that assigns to each member of its *domain* exactly one member of its *range*. The *value* assigned by the function f to the number x is denoted by $f(x)$. The *graph* of a function f is the set of all points having coordinates (x, y).

8. A *linear function* is a function f that can be defined by $f(x) = mx + b$ where x, m, and b are any real numbers. The graph of f is the graph of $y = mx + b$, a line with slope m and y-intercept b.

9. A set of ordered pairs is called a *relation*. A function is a relation in which different ordered pairs have different first coordinates.

Chapter Review

Give the letter of the correct answer.

1. Find the solution set of $x + 2y = 4$ if each variable represents a whole number.   3-1

    a. $\{(2, 1)\}$ **b.** $\{(2, 0), (1, 2), (0, 4)\}$

    c. $\{(0, 2), (2, 1), (4, 0)\}$ **d.** $\{(0, 2), \left(1, \frac{3}{2}\right), (2, 1), \left(3, \frac{1}{2}\right), (4, 0)\}$

2. Frank has \$1.85 in dimes and quarters. Find all possibilities for the number of dimes that he has.

    a. 1, 3, 5, or 7   **b.** 7 or fewer
    c. 1, 6, 11, or 16   **d.** 18 or fewer

3. Graph the equation $2x - y = 1$.   3-2

    a.   **b.**   **c.**   **d.**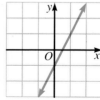

4. Determine k so that the line $3x + ky = 7$ will pass through the point $P(-3, -2)$.

    a. 1   **b.** 8   **c.** -8 **d.** $-\frac{13}{3}$

5. Find the slope of the line containing the points $(-4, 3)$ and $(2, -5)$.

a. $\dfrac{3}{4}$ **b.** $-\dfrac{4}{3}$ **c.** $\dfrac{4}{3}$ **d.** $-\dfrac{3}{4}$

6. Find the slope of the line $2x - 5y = 8$.

a. $-\dfrac{8}{5}$ **b.** $\dfrac{8}{5}$ **c.** $\dfrac{2}{5}$ **d.** $-\dfrac{2}{5}$

Find an equation in standard form for each line.

7. The line through $(-1, 2)$ having slope $-\dfrac{3}{4}$.

3-4

     a. $3x + 4y = 5$ **b.** $-3x + 4y = 11$ **c.** $4x + 3y = 2$ **d.** $4x + 3y = 5$

8. The line through $(3, -2)$ and parallel to $4x - y = 5$.
     a. $x + 4y = -5$ **b.** $4x - y = 14$ **c.** $4x + y = 10$ **d.** $x - 4y = 11$

9. The line through $(4, 1)$ and perpendicular to $3x + y = 5$.
     a. $x + 3y = 7$ **b.** $3x + y = 13$ **c.** $x - 3y = 1$ **d.** $3x - y = 11$

10. Solve this system: $4x + 3y = 7$
                         $-2x + y = 9$

3-5

     a. $(1, 1)$ **b.** $(-4, 1)$ **c.** $(-2, 5)$ **d.** $(5, -2)$

11. Find an equation of the form $y = ax^2 + b$ whose graph passes through the points $(-1, 1)$ and $(2, 7)$.
     a. $y = 2x^2 + 1$ **b.** $y = -2x^2 - 1$ **c.** $y = -2x^2 + 1$ **d.** $y = 2x^2 - 1$

12. Two shirts and one tie cost \$42, but one shirt and two ties cost \$39. What does one shirt cost?

3-6

     a. \$12   **b.** \$13   **c.** \$14   **d.** \$15

13. Find the ordered pair that does *not* satisfy this system: $2x + y \le 3$
                                                     $x - 3y > 4$

3-7

     a. $(-1, -3)$ **b.** $(1, 1)$ **c.** $(0, -2)$ **d.** $(2, -1)$

14. If $f(x) = 3 - x^2$ and $g(x) = x + 2$, find $f(g(1))$.

3-8

     a. -6 **b.** 0 **c.** 2 **d.** 4

15. The function $g: x \rightarrow x + |x|$ has domain $\{-2, -1, 0, 1, 2\}$. Find the range of g.
     a. $\{0, 1, 2\}$ **b.** $\{-4, -2, 0\}$ **c.** $\{-2, 0, 2, 4\}$ **d.** $\{0, 2, 4\}$

16. The function f is linear. If $f(2) = 2$ and $f(-6) = 6$, find $f(6)$.

3-9

     a. 0 **b.** -6 **c.** 6 **d.** 1

17. A student's test score was 684 on a scale of 200–800. Convert this score to a scale of 0–100.

     a. 80 **b.** $80\dfrac{1}{3}$ **c.** $80\dfrac{2}{3}$ **d.** 81

18. Which relation graphed below is a function?

3-10

     a.   **b.**   **c.**   **d.**

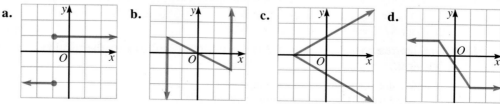

Chapter Test

1. Find the value of k so that $(-2, 1)$ satisfies $kx + 6y = k$.

2. Find the solution set of $3x + 2y = 12$ if each variable represents a positive integer.

3. Graph each equation.
   a. $y - 2 = 0$
   b. $x = -3$
   c. $4x - 3y = 6$
   d. $y = -\frac{1}{2}x + 2$

4. Find the slope of each line described.
   a. Passes through $(-7, 1)$ and $(3, -4)$.
   b. Has equation $y = -\frac{1}{3}x + 7$.
   c. Has equation $2x - 5y = 3$.

5. Find the value of k so that $(k + 2)x - 3y = 1$ has slope 2.

6. Find an equation in standard form for each line described.
   a. Through $(-2, 4)$ and slope -3.
   b. Through $(-5, -1)$ and $(3, 3)$.
   c. Having x-intercept 4 and y-intercept -6.

7. Find equations in standard form for the lines through $(3, -1)$ that are (a) parallel to and (b) perpendicular to the line $4x - 3y = 7$.

8. Solve this system: $5x - 4y = 13$
   $2x + 3y = -4$

9. Two cement blocks and three bricks weigh 102 lb, as do one cement block and ten bricks. What does one brick weigh?

10. Graph this system: $2x + y \geq 2$
    $x - y < 1$

11. The domain of the function $f: x \rightarrow (x - 2)^2$ is $D = \{0, 1, 2, 3, 4\}$. Find the range of f.

12. If $f(x) = |2x + 1|$ and $g(x) = x - 4$, find $f(g(1))$ and $g(f(1))$.

13. Find an equation of the linear function f if $f(1) = -1$ and $f(4) = 8$.

14. A refrigerator repairman charges a fixed amount for a service call in addition to his hourly rate. If a two-hour repair costs \$50 and a four-hour repair costs \$74, what is the repairman's hourly rate?

15. Graph the relation $\{(-2, 2), (-1, 1), (0, 1), (1, 1), (2, 0)\}$. Is the relation a function?

16. If x and y are integers, draw the graph of the relation $\{(x, y): |x| + |y| = 1\}$. Is the relation a function?

Cumulative Review (Chapters 1–3)

Simplify.

1. $-1(4 - 7)^2 + 6$
2. $2 - |(-3)(4) + (-1)^2|$
3. $2(x + y) - 3(y - x)$

4. Evaluate $\dfrac{|a| + |b|}{|a + b|}$ if $a = -6$ and $b = 2$.

5. On a 5-hour trip, Thomas drove part of the time at 30 mi/h and the rest at 55 mi/h. If he traveled a distance of 250 mi, how much time did he spend driving at 55 mi/h?

Solve each open sentence and graph each solution set.

6. $3(2x - 1) = 4x + 7$
7. $13 - 7y \geq 34$
8. $5 < 3w + 8 < 14$

9. $|2m + 3| = 5$
10. $\left|\dfrac{c - 3}{2}\right| < 1$
11. $6 - |n| \geq 4$

12. Prove: If a and b are real numbers, then $a + (b - a) = b$.
13. Solve $3x - 2y = 1$ if the domain of x is $\{-1, 0, 1\}$.
14. Find the slope of the line $8x - 6y = 5$.
15. Find the slope of the line $3y - 9 = 0$.

Find an equation in standard form for the line described.

16. Passes through $(4, -3)$ with slope $-\dfrac{1}{2}$

17. Passes through $(-2, 4)$ and $\left(-3, \dfrac{1}{2}\right)$

18. Has x-intercept 6 and y-intercept 4
19. Parallel to the line $5x - 4y = 7$ and contains $(-1, -2)$
20. Perpendicular to the line $2x + y = 5$ and contains $(3, 2)$

Solve each system. If the system has no solution, say so.

21. $3x - 4y = 10$
    $2x + 3y = 1$
22. $y = \dfrac{1}{2}x - 1$
    $x - 2y = 4$
23. $x = 7y - 4$
    $y = 7x + 4$
24. Graph this system: $2x - y < 3$
    $x + y \geq 0$

25. Admission to a county fair costs \$20 for two adults and one child and \$16 for one adult and two children. What is the cost for one adult?

26. If $f(x) = 4 - x^2$ and $g(x) = |2x - 7|$, find $f(g(2))$ and $g(f(2))$.

27. If $h(x)$ is a linear function such that $h(-3) = 2$ and $h(1) = -4$, find $h(7)$.

28. Graph the relation $\{(-2, 0), (-1, 1), (0, 2), (2, 0), (1, -1), (0, -2)\}$. Is the relation a function?

4 Products and Factors of Polynomials

Just as a prism separates white light into its component colors, you can often factor a polynomial into a product of "simpler" polynomials.

Working with Polynomials

4-1 Polynomials

Objective   To simplify, add, and subtract polynomials.

Here is a review of some terms you'll use when working with polynomials.

Definition	Examples
Constant: a number	-2, $\frac{3}{5}$, 0
Monomial: a constant, a variable, or a product of a constant and one or more variables	-7, u, $\frac{1}{3}m^2$, $-s^2t^3$, $6xy^3$
Coefficient (or **numerical coefficient**): the constant (or numerical) factor in a monomial	The coefficient of $3m^2$ is 3. The coefficient of u is 1. The coefficient of $-s^2t^3$ is -1.
Degree of a variable in a monomial: the number of times the variable occurs as a factor in the monomial	The degree of x is 1. The degree of y is 3.
Degree of a monomial: the sum of the degrees of the variables in the monomial. A nonzero constant has degree 0. The constant 0 has *no degree*.	$6xy^3$ has degree $1 + 3$, or 4. $-s^2t^3$ has degree $2 + 3$, or 5. u has degree 1. -7 has degree 0.
Similar (or **like**) **monomials:** monomials that are identical or that differ only in their coefficients	$-s^2t^3$ and $2s^2t^3$ are similar. $6xy^3$ and $6x^3y$ are *not* similar.
Polynomial: a monomial or a sum of monomials. The monomials in a polynomial are called the **terms** of the polynomial.	$x^2 + (-4)x + 5$, or $x^2 - 4x + 5$ The terms are x^2, $-4x$, and 5.
Simplified polynomial: a polynomial in which no two terms are similar. The terms are usually arranged in order of decreasing degree of one of the variables.	$2x^3 - 5 + 4x + x^3$ is *not* simplified, but $3x^3 + 4x - 5$ is simplified.
Degree of a polynomial: the greatest of the degrees of its terms after it has been simplified	The degrees of the terms of $x^4 - 2x^2y^3 + 6y - 11$ are, in order, 4, 5, 1, and 0. The polynomial has degree 5.

When simplifying a polynomial, you use many of the properties of real numbers discussed in Chapter 1. By using the commutative and associative properties of addition, you can order and group the terms of a polynomial in any way. Similar terms are usually grouped together and then combined. For example,

$$4s - 3s^2t - s + 5s^2t - 7 = (-3s^2t + 5s^2t) + (4s - s) - 7$$
$$= 2s^2t + 3s - 7$$

Example 1 Simplify, arranging terms in order of decreasing degree of x. Then write the degree of the polynomial.
  a. $x - 3x^2 + 8 + x^2 - 2 + 4x$
  b. $x^3y^3 - 6xy^4 + 2x^3y - x^3y^3 + 3xy^4 - 4x^2y$

Solution **a.** $x - 3x^2 + 8 + x^2 - 2 + 4x = (-3x^2 + x^2) + (x + 4x) + (8 - 2)$
$$= -2x^2 + 5x + 6$$
The degrees of the terms are, in order, 2, 1, and 0.
∴ the degree of the polynomial is 2. *Answer*

  b. $x^3y^3 - 6xy^4 + 2x^3y - x^3y^3 + 3xy^4 - 4x^2y =$
$(x^3y^3 - x^3y^3) + 2x^3y - 4x^2y + (-6xy^4 + 3xy^4) =$
$0 + 2x^3y - 4x^2y - 3xy^4 =$
$2x^3y - 4x^2y - 3xy^4$
The degrees of the terms are, in order, 4, 3, and 5.
∴ the degree of the polynomial is 5. *Answer*

Adding and Subtracting Polynomials

To add two or more polynomials, write their sum and then simplify by combining similar terms.

To subtract one polynomial from another, add the opposite of each term of the polynomial you're subtracting.

Example 2 Add $2x^2 - 3x + 5$ and $x^3 - 5x^2 + 2x - 5$.

Solution 1 $(2x^2 - 3x + 5) + (x^3 - 5x^2 + 2x - 5) =$
$x^3 + [2x^2 + (-5x^2)] + (-3x + 2x) + [5 + (-5)] =$
$x^3 - 3x^2 - x$ *Answer*

Solution 2 You can also add vertically.
$$\begin{array}{r} 2x^2 - 3x + 5 \\ x^3 - 5x^2 + 2x - 5 \\ \hline x^3 - 3x^2 - x \qquad \end{array}$$ *Answer*

Example 3 Subtract $2x^2 - 3x + 5$ from $x^3 - 5x^2 + 2x - 5$.

Solution
$(x^3 - 5x^2 + 2x - 5) - (2x^2 - 3x + 5) =$
$(x^3 - 5x^2 + 2x - 5) + (-2x^2 + 3x - 5) =$
$x^3 + [-5x^2 + (-2x^2)] + (2x + 3x) + [-5 + (-5)] =$
$x^3 - 7x^2 + 5x - 10$
Answer

Example 4 Simplify $x(2y - 3) + 4(x + 2y) - 3y(x - 1)$.

Solution Use the distributive property first. Then combine similar terms.

$$x(2y - 3) + 4(x + 2y) - 3y(x - 1) = 2xy - 3x + 4x + 8y - 3xy + 3y$$
$$= x - xy + 11y$$
Answer

Oral Exercises

Give (a) the degree and (b) the coefficient of each monomial.

1. $3x^4$ **2.** $-t$ **3.** xy^2z^2 **4.** 2^3

Name the similar monomials.

5. $3t,\quad -t^4,\quad 4t^2,\quad -\dfrac{1}{2}t,\quad 6,\quad 3t^3,\quad -1,\quad \dfrac{3}{4}t^2,\quad -\dfrac{1}{2}t^4,\quad 16t^2$

6. $uv^2,\quad -3u^2v,\quad -u^3,\quad \dfrac{3}{2}uv^2,\quad uv,\quad 4,\quad 2vu,\quad v^3,\quad u^2v,\quad -7$

7. $6xy,\quad -4xz,\quad yz,\quad 4zx,\quad -zxy,\quad -xy,\quad 2yzx,\quad \dfrac{xz}{2},\quad \dfrac{2yx}{3}$

8. $p^2q,\quad -3pq^3,\quad 2qp,\quad 6qp^2,\quad \dfrac{pq}{2},\quad 4pq^3,\quad -8p^2q,\quad p^2q^3,\quad q^3p$

Read each polynomial with terms in order of decreasing degree of x. Then give the degree of the polynomial.

9. $11x - 7 + 2x^2$ **10.** $1 - x^3 + 2x - 4x^2$

11. $1 - 3x^2 + 5x^4 - 7x^6$ **12.** $7xz - 5z^2 + 3x^2$

13. $ux - u^3x^2 + ux^3 - u^2$ **14.** $xy^3 - 2x^3y + 4x^2y^2 - y^4$

In Exercises 15–18, (a) add the polynomials and (b) subtract the second polynomial from the first.

15. $x + 7,\ 3x + 2$ **16.** $3t + 5,\ 4t - 7$

17. $3r - 5,\ r^2 - 2r + 6$ **18.** $u^2 - 6u + 7,\ u^2 + 6u - 8$

Written Exercises

Simplify, arranging terms in order of decreasing degree of x. Then write the degree of the polynomial.

A

1. $2 - x^2 + 3x + 2x^2 - 5x$

2. $x^3 - 4x + 7x^2 + 3 + 2x$

3. $-x^2 + 3x^3 - 3x + x^2 + 2x$

4. $2x^3 - 7 + 5x^2 - x^3 + 3x - x^3$

5. $x^2y^2 - x^2 + 8x^2y^2 + 5xy^2 - 2x^2$

6. $4x^2y^3 - xy^2 + 2x^3y - 2xy^2$

7. $4x^2yz^3 - xyz + 2x^2yz^3 + 5x^3y^2z^2$

8. $7xy^2z^3 - 4xy^2z^3 + 2x^2yz^2 - 3xy^2z^3$

In Exercises 9–16, (a) add the polynomials and (b) subtract the second polynomial from the first.

9. $5m - 4, \ 2m + 3$

10. $3u + 7, \ u - 8$

11. $t^2 - 8t - 7, \ t^2 + 5t - 6$

12. $2n^2 - n + 5, \ n^2 + 1$

13. $5v^3 - 2v + 1, \ v^2 + 2v - 2$

14. $w^3 - w^2 + w - 1, \ 1 - w - w^2 - w^3$

15. $3x^2 - 2xy + 4y^2, \ 2x^2 + 3y^2$

16. $4a^2 + 3ab - b^2, \ b^2 - 2ab$

Simplify.

B

17. $3(x^2 - 2x + 4) + 2(5x^2 - 7)$

18. $4(3y^2 - 2y) + 3(y^2 + 5y - 1)$

19. $2(4m^2 + 3) - 7(m^2 - 2) + 1$

20. $5(2n^2 - 3) - 2(5n^2 + 2) - 6$

21. $4a(x - y) + 3a(x + y) + ay$

22. $2d(3m + n) - 5d(m - 4n) - 10dm$

23. $3[2p^2 - q(3p + 4q)] - 2[4q^2 - 3p(p - 2q)]$

24. $4[2a(3a - b) + 3ab] + 5[3b(a + 2b) - 4ab]$

Find values of a, b, c, and d that make the equation true.

C

25. $(4t^3 - at^2 - 2bt + 5) - (ct^3 + 2t^2 - 6t + 3) = t^3 - 2t + d$

26. $(ax^3 - 3x^2 + 2bx - 2) - (2x^3 - cx^2 - 5x - 4d) = x^2 + x - 6$

27. $(x^2 + ax + 2b) + (x^2 - 2bx + 3a) = cx^2 - 7x + 3$

Mixed Review Exercises

Find an equation of the form $f(x) = mx + b$ for the linear function f.

1. $m = -2; \ f(4) = -6$

2. $f(0) = 5; \ f(2) = 7$

3. $f(-1) = 3; \ f(3) = 1$

For the line containing the given points, find (a) the slope and (b) an equation in standard form.

4. $(0, 4), (4, 0)$

5. $(-3, 1), (5, 3)$

6. $(0, 0), (-3, 2)$

7. $(6, -5), (-2, 1)$

8. $(-4, 7), (1, 9)$

9. $(-2, 5), (3, 5)$

4-2 Using Laws of Exponents

Objective   To use laws of exponents to multiply a polynomial by a monomial.

In this lesson you'll review the laws of exponents that are used when multiplying polynomials.

Laws of Exponents

Let a and b be real numbers and m and n be positive integers. Then:

1. $a^m \cdot a^n = a^{m+n}$ **2.** $(ab)^m = a^m b^m$ **3.** $(a^m)^n = a^{mn}$

Without the first law you would have to simplify a product of powers, such as $c^5 \cdot c^3$, by counting factors:

$$c^5 \cdot c^3 = \overbrace{(c \cdot c \cdot c \cdot c \cdot c)}^{\text{5 factors}}\overbrace{(c \cdot c \cdot c)}^{\text{3 factors}} = c^8.$$
$$\underbrace{}_{\text{8 factors}}$$

But using the first law you can write

$$c^5 \cdot c^3 = c^{5+3} = c^8.$$

Using the second law you can simplify a power of a product:

$$(2x)^3 = 2^3 x^3 = 8x^3,$$

and using the third law you can simplify a power of a power:

$$(10^3)^2 = 10^{3 \cdot 2} = 10^6.$$

We'll prove the law $(ab)^m = a^m b^m$ by counting factors; the proofs of the other laws are left as Exercises 39 and 40.

Statements	Reasons
1. $(ab)^m = \overbrace{(ab)(ab) \ldots (ab)}^{m \text{ factors of } ab}$	1. Definition of a power
2. $\quad = \underbrace{(a \cdot a \cdot \ldots \cdot a)}_{m \text{ factors of } a}\underbrace{(b \cdot b \cdot \ldots \cdot b)}_{m \text{ factors of } b}$	2. Commutative and associative properties of multiplication
3. $\quad = a^m b^m$	3. Definition of a power

You can use the laws of exponents along with the commutative and associative properties of multiplication to simplify products and powers like the ones in Examples 1 and 2 on the next page.

Example 1 Simplify: **a.** $(-3x^2y^3)(4xy^2)$ **b.** $(st^4)^3$ **c.** $(-x^3)^2$

Solution **a.** Use the first law of exponents and the fact that $x = x^1$.

$$(-3x^2y^3)(4xy^2) = (-3 \cdot 4)(x^2 \cdot x)(y^3 \cdot y^2)$$
$$= -12x^3y^5 \quad \textit{Answer}$$

b. Use the second law of exponents. Then apply the third law.

$$(st^4)^3 = s^3(t^4)^3 = s^3t^{12} \quad \textit{Answer}$$

c. Use the fact that $-x^3 = (-1)x^3$ and the second law of exponents.

$$(-x^3)^2 = [(-1)x^3]^2$$
$$= (-1)^2(x^3)^2 = x^6 \quad \textit{Answer}$$

Example 2 Simplify: **a.** $u \cdot (u^2)^3 \cdot u^5$ **b.** $(3xy^2z^3)^3$

Solution **a.** $u(u^2)^3u^5 = u^1 \cdot u^6 \cdot u^5$

$$= u^{1+6+5} = u^{12} \quad \textit{Answer}$$

b. $(3xy^2z^3)^3 = 3^3x^3(y^2)^3(z^3)^3 = 27x^3y^6z^9 \quad \textit{Answer}$

To multiply a polynomial by a monomial, use the distributive property.

Example 3 Simplify $3t^2(t^3 - 2t^2 + t - 4)$.

Solution $3t^2(t^3 - 2t^2 + t - 4) = (3t^2)t^3 - (3t^2)(2t^2) + (3t^2)t - (3t^2)4$

$$= 3t^5 - 6t^4 + 3t^3 - 12t^2 \quad \textit{Answer}$$

Example 4 Simplify. Assume that variable exponents represent positive integers.
a. $(a^2)^k(a^k)^3$ **b.** $x^{m-n}(x^{m+n} + x^n)$

Solution **a.** $(a^2)^k(a^k)^3 = a^{2k} \cdot a^{3k} = a^{5k} \quad \textit{Answer}$

b. $x^{m-n}(x^{m+n} + x^n) = x^{m-n} \cdot x^{m+n} + x^{m-n} \cdot x^n$

$$= x^{(m-n)+(m+n)} + x^{(m-n)+n}$$
$$= x^{2m} + x^m \quad \textit{Answer}$$

Oral Exercises

Simplify. Assume that variable exponents represent positive integers.

1. $y^3 \cdot y^2$ **2.** $c \cdot c^3$ **3.** $(a^2)^3$ **4.** $(2k)^4$

5. $(-x^3)^2$ **6.** $(-x^2)^3$ **7.** $z^3 \cdot z^n$ **8.** $(z^3)^n$

9. $(p^3q)^2$ **10.** $b^3 \cdot b^2 \cdot b$ **11.** $r^k \cdot r^k$ **12.** $(r^k)^k$

If the statement is true, give the law of exponents that justifies it. If the statement is false, give a counterexample.

13. $(ab)^2 = ab^2$ **14.** $a^3 \cdot a^2 = a^6$ **15.** $a^3 \cdot a^3 = a^6$

16. $(a^3)^3 = a^9$ **17.** $(a^3)^4 = a^7$ **18.** $(a^2b)^2 = a^2b^2$

Simplify. Assume that variable exponents represent positive integers.

19. $t^2(t^2 - 3)$ **20.** $uv(u + v)$ **21.** $p^2(p - 2q)$

22. $4y^2z(y^2 - z^2)$ **23.** $-3s(s^4 + 2s^2t^3)$ **24.** $-2xy^2(x^3 - xy^2)$

25. $u^3 \cdot u^{m-3}$ **26.** $a^{n-1} \cdot a^{n+1}$ **27.** $x^4(x^{k-2})^2$

Written Exercises

Simplify. Assume that variable exponents represent positive integers.

A **1.** $3z^2 \cdot 2z^3$ **2.** $5r^2 \cdot r^4$

$x \cdot x^n = x^{m+n}$

3. $(-t^4)^3$ **4.** $(-t^3)^4$

5. $(3x^2y)(xy^2)$ **6.** $(4p^2q)(p^2q^3)$

$(xy)^m = x^m y^m$

7. $(-2u^2)(uv^3)(-u^2v^2)$ **8.** $(r^2s)(-3rs^3)(2rs)$

9. $(4a^3b^2)^2$ **10.** $(2c^2d^3)^3$

$(m^m)^n = m^{m \cdot n}$

11. $(-3pq^4r^2)^3$ **12.** $(-x^2yz^3)^4$

13. $(-z^3)(-z)^3$ **14.** $(-c)^2(-c^4)$

15. $(s^2t)^3(st^3)^2$ **16.** $(2x^2y^3)^3(3x^3y)^2$

17. $3y(y^3 - 2y^2 + 3)$ **18.** $x^2(x - 2x^2 + 3x^3)$

19. $rs^2(r^2 - 2rs - s^2)$ **20.** $p^2q^3(p^2 - 4q)$

21. $z^{n-2} \cdot z^{n+2}$ $p^{n+m-n+1}$ **22.** $t^4 \cdot t^{k-4}$

23. $x^{m-1} \cdot x \cdot x^m$ $p^{m+1} + p^{n+m-h}$ **24.** $y^{p+2} \cdot y^p \cdot y^{p-2}$

25. $r^{h-2}(r^{h+1})^2$ **26.** $s^3(s^{2k-1})^3$

$p^{m+1} + p^m$

B **27.** $t(t^{n-1} + t^n + t^{n+1})$ **28.** $x^2(x^k - x^{k-1} + x^{k-2})$

29. $p^n(p^{m-n+1} + p^{m-n})$ **30.** $s^{2n}(s^{2m-n} - s^{m-2n})$

31. $z^{m-n}(z^{n+m} - z^{n-m} + z^n)$ **32.** $x^{h+k}(x^{2h-k} - x^{h-2k} + x^k)$

33. $(t^m)^n(t^n)^{n-m}$ **34.** $(y^{h-k})^h(y^{h+k})^k$

In Exercises 35–38, solve for n.

35. $3^{5n} = 3^5(3^{2n})^2$ **36.** $(2^{3n})^2 = (2^n)^3 \cdot 2^{n+6}$

37. $3 \cdot 9^{2n} = (3^{n+1})^3$ **38.** $4^{n+3} \cdot 16^n = 8^{3n}$

C **39.** Prove the first law of exponents. **40.** Prove the third law of exponents.

41. Prove that for positive integers m, n, and r, $((a^m)^n)^r = a^{mnr}$.

42. Prove that for positive integers m and n, $(a^m)^n = (a^n)^m$.

4-3 Multiplying Polynomials

Objective   To multiply polynomials.

To multiply two polynomials, you use the distributive property: Multiply each term of one polynomial by each term of the other and add the resulting monomials.

Example 1 Multiply: $(2x + 3)(x^2 + 4x - 5)$

Solution 1 You can arrange the work horizontally.

$$(2x + 3)(x^2 + 4x - 5) = 2x(x^2 + 4x - 5) + 3(x^2 + 4x - 5)$$
$$= 2x^3 + 8x^2 - 10x + 3x^2 + 12x - 15$$
$$= 2x^3 + 11x^2 + 2x - 15 \quad \textbf{\textit{Answer}}$$

Solution 2 You can arrange the work vertically.

$$
\begin{array}{r}
x^2 + 4x - 5 \\
2x + 3 \\
\hline
\end{array}
$$

$2x^3 + 8x^2 - 10x$ ⟵——————— This is $2x(x^2 + 4x - 5)$.

$\underline{3x^2 + 12x - 15}$ ⟵——— This is $3(x^2 + 4x - 5)$.

$2x^3 + 11x^2 + 2x - 15$ **Answer**

In Example 1, the factors of the product $(2x + 3)(x^2 + 4x - 5)$ are a *binomial* and a *trinomial*, respectively. A **binomial** is a polynomial that has two terms. A **trinomial** is a polynomial that has three terms.

When multiplying two binomials, you can save time if you learn to find the product mentally. Study the method illustrated in Example 2.

Example 2 Multiply: $(2a - b)(3a + 5b)$

Solution $(2a - b)(3a + 5b) = 6a^2 + 7ab - 5b^2$

1. Multiply the *first* terms of the binomials.

    $(2a)(3a)$ ————

2. Multiply the *outer* terms of the binomials.
   Multiply the *inner* terms of the binomials.
   Then add the products.

    $(2a)(5b) + (3a)(-b)$ ————

3. Multiply the *last* terms of the binomials.

    $(-b)(5b)$ ————

The method used in Example 2 is sometimes called the FOIL method.

The word FOIL reminds you to multiply the *First*, *Outer*, *Inner*, and *Last* terms when multiplying two binomials.

Certain special products occur so frequently that their patterns should be memorized.

Special Product	Pattern and Example
$(a + b)^2 = a^2 + 2ab + b^2$	$(\text{first} + \text{second})^2 = (\text{first})^2 + 2(\text{first})(\text{second}) + (\text{second})^2$ $(4s + 3t)^2 = (4s)^2 + 2(4s)(3t) + (3t)^2$ $= 16s^2 + 24st + 9t^2$
$(a - b)^2 = a^2 - 2ab + b^2$	$(\text{first} - \text{second})^2 = (\text{first})^2 - 2(\text{first})(\text{second}) + (\text{second})^2$ $(3x - 5)^2 = (3x)^2 - 2(3x)(5) + 5^2$ $= 9x^2 - 30x + 25$
$(a + b)(a - b) = a^2 - b^2$	$(\text{first} + \text{second})(\text{first} - \text{second}) = (\text{first})^2 - (\text{second})^2$ $(2p + 3q)(2p - 3q) = (2p)^2 - (3q)^2$ $= 4p^2 - 9q^2$

Oral Exercises

Use the FOIL method to find each product mentally.

1. $(x + 6)(x + 4)$

2. $(y + 7)(y - 3)$

3. $(3m + 1)(2m + 5)$

4. $(2n - 3)(n + 8)$

5. $(2 - 3p)(3 + 2p)$

6. $(1 - 4w)(3 - 2w)$

7. $(3x + y)(x + 2y)$

8. $(c + d)(2c - 5d)$

9. $(2p - q)(3p - q)$

Find each special product.

10. $(x + 3)^2$

11. $(2y + 5)^2$

12. $(3m + n)^2$

13. $(a - 2)^2$

14. $(4z - 3)^2$

15. $(p - 2q)^2$

16. $(k + 1)(k - 1)$

17. $(3r + 5)(3r - 5)$

18. $(4c + d)(4c - d)$

Written Exercises

Multiply.

A

1. $(3v + 1)(2v - 5)$

2. $(2x - 3)(3x + 2)$

3. $(4z + 3)(3z - 4)$

4. $(r - 4)(3r - 2)$

5. $(3x + 10)^2$

6. $(4k - 5)^2$

7. $(5y - 2)(5y + 2)$

8. $(2s + 7)(2s - 7)$

9. $(7t + 2)(2t - 1)$

10. $(5z + 6)(6z - 5)$

11. $(9t + 1)(1 - 9t)$

12. $(9 - 5t)(5t - 9)$

Multiply. Assume that variable exponents represent positive integers.

13. $(x - 2y)(3x + 4y)$

14. $(5h - 3k)(h - 2k)$

15. $(2p + 3q)(3p - 2q)$

16. $(10r - 3s)(r + 2s)$

17. $(x^2 - 3)(x^2 + 3)$

18. $(p^2 - 2q^2)(p^2 + 2q^2)$

19. $(s^3 + t^3)^2$

20. $(2z^2 - 5)^2$

21. $t(t - 2)(t + 1)$

22. $x^2(x - 3)(x + 3)$

23. $xy(x - y)^2$

24. $mn(m - n)(m - 2n)$

25. $(2c + 1)(c^2 - 3c + 2)$

26. $(t - 3)(2t^2 - t + 2)$

27. $(x^2 + 3x - 5)(x + 2)$

28. $(z^2 - 2z + 4)(z + 3)$

29. $(y^4 - 3y^2 + 1)(y^2 - 2)$

30. $(3 - k^2)(2 - k^2 - k^4)$

B 31. $(x^2 - x + 2)(x^2 + x - 1)$

32. $(y^2 - 2y + 1)(y^2 + y + 1)$

33. $(a + 2b)(a^3 - 2a^2b - b^3)$

34. $(3s + 2t)(s^3 - 3st^2 + 2t^3)$

35. $(p^n - 1)^2$

36. $(x^{2n} - y^n)^2$

37. $(r^n - s^n)(r^n + 2s^n)$

38. $(x^n + 1)(x^n - 1)$

39. $(a - b)^3$

40. $(a + b)^3$

41. $(a + b)(a^2 - ab + b^2)$

42. $(a - b)(a^2 + ab + b^2)$

43. $(a - b)(a^3 + a^2b + ab^2 + b^3)$

44. $(a + b)(a^3 - a^2b + ab^2 - b^3)$

45. $(x + y)(x - y)(x^2 + y^2)$

46. $(x + y)^2(x - y)^2$

47. $(x^2 + 2x + 2)(x^2 - 2x + 2)$

48. $(x^2 - 4x + 8)(x^2 + 4x + 8)$

Without actually finding the product, determine how many terms the simplified product has.

C 49. $(u + v)(x + y)(u - v)(x - y)$

50. $(u + v + w)(x + y + z)$

51. $(u + v + w)(u + v - w)$

52. $(x - y + z)(x + y + z)$

Find the value or values of k that make the equation true.

53. $(x + 2k)(x - 3k) = x^2 + 2x - 24$

54. $(2x + k)(x - 2k) = 2x^2 + 9x - 18$

55. $(2x - k)(3x + 2k) = 6x^2 + kx - 32$

56. $(3kx + 2)^2 = 81x^2 + 12kx + 4$

Mixed Review Exercises

Simplify.

1. $(3x^2 - 7x + 9) - (x^2 + 4x - 1)$

2. $(a^2b^3)^3$

3. $(4m^2n)(-3mn^3)$

4. $5(2y - 1) - 3(y + 2)$

5. $2c(d - 3) + 3d(c + 4)$

6. $4p(2p^2 - p + 5)$

7. $(-u^2)^4(-u)^3$

8. $(9z^3 - 4z) + (5z^2 - 8)$

9. $(y^2 - 5y + 9) - (9 + 5y + y^2)$

10. $\left(-\frac{1}{2}x^2\right)(-4x^4)$

Self-Test 1

Vocabulary   constant (p. 167)
                monomial (p. 167)
                coefficient (p. 167)
                degree of a variable (p. 167)
                degree of a monomial (p. 167)
                similar monomials (p. 167)

polynomial (p. 167)
simplified polynomial (p. 167)
degree of a polynomial (p. 167)
binomial (p. 174)
trinomial (p. 174)

Simplify. Assume that variable exponents represent positive integers.

1. $(x^2 + 3x - 2) + (1 - 2x)$

2. $3(t^2 - 2) - t(3t - 1)$

Obj. 4-1, p. 167

3. $(3p^2)(-2pq^2)^2$

4. $t^2 \cdot t^{n+1} \cdot t^{n-3}$

Obj. 4-2, p. 171

5. $(4a - 5b)(2a + 3b)$

6. $(2y - 1)(y^2 + 2y + 3)$

Obj. 4-3, p. 174

7. $(3x - 2)^2$

8. $(4c - 3)(4c + 3)$

Check your answers with those at the back of the book.

Biographical Note / *Ernest Everett Just*

How the biologist Ernest Everett Just (1883–1941) felt about life affected how he studied it. He saw living things as more than a mass of cells and a chain of chemical interactions. He was therefore very careful about how he conducted his laboratory work.

Isolated cells, unless handled properly, can easily die or grow unnaturally. At the Marine Biology Laboratories in Woods Hole, Massachusetts, Just showed how to keep cells normal and healthy in a laboratory.

Just was particularly interested in the function of the cell wall. Describing the cell wall as more than just a "dam against the outside world," he saw it as a living, interacting part of the cell. One result of his work with cells was an important original book, *The Biology of the Cell Surface*.

Besides spending the summers between 1912 and 1929 working at Woods Hole, Just taught and performed research

at Howard University in Washington, D.C. By applying new experimental techniques to the study of cell development, Just became an important force in the advancement of genetics and embryology.

Reading Algebra / Symbols

At times you may think that the use of symbols only complicates algebra and that it's not worth the effort to master the skill of reading symbols. To see why this is not the case, try to write in simple words an understandable version of the law $a^m \cdot a^n = a^{m+n}$. The task is difficult even if you are allowed to use variables. The use of symbols permits the presentation of ideas in a clear, concise, and easy-to-remember fashion.

When you come across an unfamiliar symbol, be sure you learn what it means. If necessary, check the list on page xvi, which includes each symbol in your textbook, its meaning, and a reference to the page on which it is introduced. Practice translating back and forth between expressions using the symbol and equivalent expressions in words.

The exercises below deal with two symbols related to divisibility. Let a, b, and n be integers with $n \neq 0$. We say that *a is congruent to b (mod n)* if n divides the difference $a - b$. (Recall that one integer is said to *divide* another if and only if the second integer is an integral multiple of the first.) The symbol $\equiv$ stands for "is congruent to," and the symbol $|$ stands for "divides." Therefore, $a \equiv b$ (mod n) if $n|(a - b)$. For example, $18 \equiv 3$ (mod 5) since $5|(18 - 3)$.

Exercises

1. Rewrite each expression in words and tell whether it is true or false.
   a. $2|8$ **b.** $4|4$ **c.** $2|1$ **d.** $2|0$ **e.** $0|3$

2. What does the symbol $\neq$ mean? What do you think the symbol $\nmid$ means?

3. Tell whether each statement is true or false.
   a. $2 \nmid 5$ **b.** $n \nmid (n + 1)$ if $n \geq 2$ **c.** $5 \nmid 0$

4. If $a|b$ and $b|c$, is it necessarily true that $a|c$?

5. If $m|n$ and $n|m$, is it necessarily true that $m = n$?

6. Write the statement that 17 is congruent to 2 (mod 5) using the congruency symbol $\equiv$.

7. Write $19 \equiv 4$ (mod 5) using the divisibility symbol $|$.

8. If $k|s$ and $k|t$, is it necessarily true that $s \equiv t$ (mod k)?

9. Rewrite each expression in words and tell whether it is true or false.
   a. $100 \equiv 25$ (mod 5)   **b.** $-3 \equiv 3$ (mod 5)
   c. $6 \equiv 1$ (mod 5)   **d.** $0 \equiv 5$ (mod 5)

10. What do we call the set of integers n that satisfy $n \equiv 0$ (mod 2)?

11. What do we call the set of integers n that satisfy $n \equiv 1$ (mod 2)?

12. Find a value of n that makes the statement true.
    a. $11 \equiv 27$ (mod n)   **b.** $6 \equiv n$ (mod 2)

Factors of Polynomials

4-4 Using Prime Factorization

Objective   To find the GCF and LCM of integers and monomials.

To **factor** a number over a set of numbers, you write it as a product of numbers chosen from that set, called the **factor set.** For example, the number 14 can be factored over the integers in the following ways:

$$(1)(14), \quad (-1)(-14), \quad (2)(7), \quad \text{and} \quad (-2)(-7).$$

In this book, integers will be factored over the set of integers unless some other set is specified.

The adjective corresponding to the word "integer" is *integral*. Because $2 \cdot 7 = 14$, you say that 2 and 7 are *integral factors* of 14 and that 14 is an *integral multiple* of 2 and of 7.

A **prime number,** or **prime,** is an integer greater than 1 whose only positive integral factors are itself and 1. The first ten primes are

$$2, 3, 5, 7, 11, 13, 17, 19, 23, \text{ and } 29.$$

If the factor set is restricted to the set of primes, then 14 has only one factorization, $2 \cdot 7$. This is called the *prime factorization* of 14. To find the **prime factorization** of a positive integer, you write the integer as a product of primes. If a prime factor occurs more than once, use an exponent.

Example 1   Find the prime factorization of 936.

Solution   Here is a systematic way to find the prime factorization of a large number: Try the primes, in order, as factors. Use each repeatedly until it is no longer a factor. Then try the next prime.

$$936 = 2 \cdot 468$$
$$= 2 \cdot 2 \cdot 234$$
$$= 2 \cdot 2 \cdot 2 \cdot 117$$
$$= 2 \cdot 2 \cdot 2 \cdot 3 \cdot 39$$
$$= 2 \cdot 2 \cdot 2 \cdot 3 \cdot 3 \cdot 13$$
$$\therefore 936 = 2^3 \cdot 3^2 \cdot 13 \quad \textit{Answer}$$

The **greatest common factor (GCF)** of two or more integers is the greatest integer that is a factor of each. The **least common multiple (LCM)** of two or more integers is the least positive integer having each as a factor. When given two or more integers, you can use their prime factorizations to find their GCF and LCM.

Example 2 Find **(a)** the GCF and **(b)** the LCM of the following numbers:

$$72, \; 108, \text{ and } 126.$$

Solution The prime factorizations are

$$72 = 2^3 \cdot 3^2, \qquad 108 = 2^2 \cdot 3^3, \qquad \text{and} \qquad 126 = 2 \cdot 3^2 \cdot 7.$$

a. To find the GCF, take the *least* power of each *common* prime factor.

$$\therefore \quad \text{GCF} = 2 \cdot 3^2 = 18$$

b. To find the LCM, take the *greatest* power of *each* prime factor.

$$\therefore \quad \text{LCM} = 2^3 \cdot 3^3 \cdot 7 = 1512$$

Factoring monomials is very similar to factoring integers. For example, $6xy^2$ is a *factor* of $18x^3y^2$ because $18x^3y^2 = 6xy^2 \cdot 3x^2$. Moreover, $18x^3y^2$ is a *multiple* of $6xy^2$ and of $3x^2$.

In this book, monomials with integral coefficients will be factored over the set of monomials with integral coefficients unless some other factor set is specified.

The **greatest common factor (GCF)** of two or more monomials is the common factor that has the greatest degree and the greatest numerical coefficient. The **least common multiple (LCM)** of two or more monomials is the common multiple that has the least degree and the least positive numerical coefficient. Here is an example.

The GCF of $9a^2x$ and $-6ax^3$ is $3ax$.

The LCM of $9a^2x$ and $-6ax^3$ is $18a^2x^3$.

Although numerical coefficients of monomials may be negative, the numerical coefficients of the GCF and LCM of monomials are always positive.

Example 3 Find **(a)** the GCF and **(b)** the LCM of the following monomials:

$$48u^2v^2 \quad \text{and} \quad 60uv^3w.$$

Solution The prime factorizations of the coefficients are

$$48 = 2^4 \cdot 3 \qquad \text{and} \qquad 60 = 2^2 \cdot 3 \cdot 5.$$

a. The GCF of 48 and 60 is
$$2^2 \cdot 3 = 12.$$

Compare the powers of each variable occurring in *both* monomials. Use the power with the *least* exponent.

Compare u^2 and u. Use u
Compare v^2 and v^3. Use v^2.

$$\therefore \quad \text{GCF} = 12uv^2$$

b. The LCM of 48 and 60 is
$$2^4 \cdot 3 \cdot 5 = 240.$$

Compare the powers of each variable occurring in *either* monomial. Use the power with the *greatest* exponent.

Compare u^2 and u. Use u^2.
Compare v^3 and v^2. Use v^3.
Use w.

$$\therefore \quad \text{LCM} = 240u^2v^3w$$

Oral Exercises

Find all positive factors of each integer.

1. 12   **2.** 15   **3.** 26   **4.** 35

Find the prime factorization of each integer.

5. 10   **6.** 33   **7.** 25   **8.** 50

9. 20   **10.** 49   **11.** 98   **12.** 18

Find the GCF and LCM of each pair of integers.

13. 8 and 12   **14.** 15 and 20   **15.** 9 and 11   **16.** 13 and 26

Find the GCF and LCM of the given monomials.

17. a^2b and ab^2 **18.** $2x^3$ and x^2y

19. $4st^2$ and $6rs^2t$ **20.** $4uv^2w$ and $-2u^2v^2$

21. $15x^2y$, $-25xyz^2$, and $20xy^2$ **22.** $12ar^2s^3$, $8a^2r^2s$, and ar^3

23. There are fifteen primes less than 50. Ten of them are given on page 179. What are the other five?

24. Give a convincing argument that justifies the following statement:
If a is a factor of b and b is a factor of c, then a is a factor of c.

Written Exercises

Find the prime factorization of each integer.

A

1. 140   **2.** 198   **3.** 89   **4.** 756

5. 441   **6.** 203   **7.** 2548   **8.** 3861

Find (a) the GCF and (b) the LCM of the following monomials.

9. 20, 35   **10.** 45, 75

11. -48, 108   **12.** 315, -525

13. 84, -56, 140   **14.** 168, 280, 196

15. 3, 5, 7, 9   **16.** 30, 35, 36, 42

17. $9p^3q$, $15p^2$ **18.** $49x^3$, $35x^2y$

19. $68xy^2z$, $51y^2z^2$ **20.** $52r^2s$, $78rs^2t$

21. $110h^3k^2r$, $-88h^2k^2r^2$ **22.** $98a^2b^2c$, $-70abc^2$

23. $14ab$, $14bc$, $21ac$ **24.** $22xy^2z^2$, $33x^2yz^2$, $44x^2yz$

25. $26p^3q^2r^2$, $39p^2q^3r^2$, $78p^2q^2r^3$ **26.** $200a^3b^2c$, $300a^2bc^3$, $400ab^3c^2$

B **27.** The GCF of 84 and another integer is 42, and their LCM is 252. Find the other integer.

28. The GCF of $12x^2y^3$ and another monomial is $6xy^3$, and their LCM is $36x^2y^4$. Find the other monomial.

A positive integer n is *perfect* if it is the sum of all its positive factors except n itself.

29. Show that 496 is perfect.

30. There are two perfect integers less than 30. Find them.

In Exercises 31 and 32, p and q represent different prime numbers.

31. List all the positive factors of each product.
   a. pq **b.** p^2q^2 **c.** p^3q^3

32. How many positive factors does p^mq^m have?

Two integers are *relatively prime* if their greatest common factor is 1. Give a convincing argument to justify each statement.

33. The LCM of two relatively prime integers is their product.

34. If each of two integers is divided by their GCF, the two quotients are relatively prime.

C **35.** Prove or disprove: If a and b are relatively prime and b and c are relatively prime, then a and c are relatively prime.

36. Prove that m and n are relatively prime if there are integers h and k such that $hm - kn = 1$. (The converse of this statement is also true. See the Euclidean Algorithm, page 215.)

Historical Note / *Prime Numbers*

The Greek mathematician Euclid, author of the *Elements,* is best known for his work in geometry. In fact his famous book still provides the basis for most modern geometry courses. The book, however, is not entirely devoted to geometry. Euclid also stated and proved many algebraic theorems. One of the most important theorems states that there are *infinitely many* primes. Euclid's reasoning, considered today a model of elegant simplicity, was as follows:

> Suppose there were only finitely many prime numbers: 2, 3, 5, . . . , p. Let $Q = 2 \cdot 3 \cdot 5 \cdot \ \ldots \ \cdot p$, the product of all the existing primes. Then $Q + 1$ could *not* be prime, since $Q + 1 > p$ and p is the largest in the finite list of primes. But if $Q + 1$ were composite (that is, not a prime), it would have to be divisible by one of the primes 2, 3, 5, . . . , p. This is impossible, since each of these primes divides Q and therefore leaves a remainder of 1 when divided into $Q + 1$. Therefore, there must exist infinitely many primes, since the assumption that there are only a finite number of them leads to a contradiction.

4-5 Factoring Polynomials

Objective   To factor polynomials by using the GCF, by recognizing special products, and by grouping terms.

Many real life situations can be described by using polynomials. For example, the height of a baseball hit directly upward with initial speed 96 ft/s will be about $96t - 16t^2$ feet after t seconds. This polynomial can be written as $16t(6 - t)$ by factoring, as you'll soon see in this lesson. The factored polynomial can then be used to determine how long the ball is in the air and what its maximum height is, as you'll see later in this chapter.

To **factor** a polynomial you express it as a product of other polynomials taken from a specified factor set. *In this chapter the factor set is the set of polynomials having integral coefficients.*

The first step in factoring a polynomial is to find its **greatest monomial factor,** that is, the GCF of its terms. If this factor is other than 1, "factor it out." In other words, write the given polynomial as the product of its greatest monomial factor and a polynomial whose greatest monomial factor is 1.

Example 1   Factor each polynomial.

      a. $2x^4 - 4x^3 + 8x^2$ **b.** $10ab^3 - 15a^2b^2$

Solution   **a.** The GCF of $2x^4$, $-4x^3$, and $8x^2$ is $2x^2$.

              $\therefore 2x^4 - 4x^3 + 8x^2 = 2x^2(x^2 - 2x + 4)$ *Answer*

      b. The GCF of $10ab^3$ and $-15a^2b^2$ is $5ab^2$.

              $\therefore 10ab^3 - 15a^2b^2 = 5ab^2(2b - 3a)$ *Answer*

When a polynomial has 1 as its greatest monomial factor, you may still be able to factor it by recognizing it as one of the special products listed on page 175. The polynomials $a^2 + 2ab + b^2$ and $a^2 - 2ab + b^2$, which are the result of squaring $a + b$ and $a - b$, respectively, are called **perfect square trinomials.** Also, the polynomial $a^2 - b^2$, which is the product of $a + b$ and $a - b$, is called a **difference of squares.**

Perfect Square Trinomials	**Difference of Squares**
$a^2 + 2ab + b^2 = (a + b)^2$	$a^2 - b^2 = (a + b)(a - b)$
$a^2 - 2ab + b^2 = (a - b)^2$	

Example 2 Factor each polynomial.

       a. $z^2 + 6z + 9$ **b.** $4s^2 - 4st + t^2$ **c.** $25x^2 - 16a^2$

Solution   **a.** $z^2 + 6z + 9 = z^2 + 2 \cdot z \cdot 3 + 3^2$ ⟵——————— perfect square trinomial

                         $= (z + 3)^2$ *Answer*

      b. $4s^2 - 4st + t^2 = (2s)^2 - 2(2s)t + t^2$ ⟵——————— perfect square trinomial

                         $= (2s - t)^2$ *Answer*

      c. $25x^2 - 16a^2 = (5x)^2 - (4a)^2$ ⟵——————— difference of squares

                         $= (5x + 4a)(5x - 4a)$ *Answer*

Sometimes you need to use more than one method to factor a given polynomial. Remember that the first step is to look for and factor out the GCF of the terms.

Example 3 Factor $3x^5 - 48x$.

Solution   The GCF of the terms is $3x$.

$$3x^5 - 48x = 3x(x^4 - 16)$$
$$= 3x[(x^2)^2 - 4^2] \quad\longleftarrow\quad \text{difference of squares}$$
$$= 3x(x^2 + 4)(x^2 - 4)$$
$$= 3x(x^2 + 4)(x^2 - 2^2) \quad\longleftarrow\quad \text{difference of squares}$$
$$= 3x(x^2 + 4)(x + 2)(x - 2)$$

The following patterns are useful when factoring polynomials that are sums or differences of cubes.

Sum and Difference of Cubes

$$a^3 + b^3 = (a + b)(a^2 - ab + b^2)$$
$$a^3 - b^3 = (a - b)(a^2 + ab + b^2)$$

Example 4 Factor each polynomial.

       a. $y^3 - 1$ **b.** $8u^3 + v^3$

Solution   **a.** $y^3 - 1 = y^3 - 1^3 = (y - 1)(y^2 + y + 1)$ *Answer*

      b. $8u^3 + v^3 = (2u)^3 + v^3 = (2u + v)(4u^2 - 2uv + v^2)$ *Answer*

When a polynomial is not a special product, you may be able to factor it by rearranging and grouping its terms. This method can be particularly helpful when there are four or more terms, as Example 5 on the next page shows.

Example 5 Factor each polynomial.

      a. $3xy - 4 - 6x + 2y$ **b.** $s^2 - 4t^2 - 4s + 4$

Solution **a.** The first and third terms have a common factor of $3x$, and the second and fourth terms have a common factor of 2.

$$3xy - 4 - 6x + 2y = (3xy - 6x) + (2y - 4)$$
$$= 3x(y - 2) + 2(y - 2)$$
$$= (3x + 2)(y - 2) \quad \textbf{\textit{Answer}}$$

      b. The first, third, and fourth terms form a perfect square trinomial.

$$s^2 - 4t^2 - 4s + 4 = (s^2 - 4s + 4) - 4t^2$$
$$= (s - 2)^2 - 4t^2$$
$$= (s - 2)^2 - (2t)^2 \longleftarrow \text{difference of squares}$$
$$= [(s - 2) + 2t][(s - 2) - 2t]$$
$$= (s + 2t - 2)(s - 2t - 2) \quad \textbf{\textit{Answer}}$$

Don't be discouraged if your first attempt at factoring doesn't work. In Example 5(b) you might write $s^2 - 4t^2 - 4s + 4 = (s + 2t)(s - 2t) - 4(s - 1)$, but that would not lead to a factorization, so try a different grouping.

Oral Exercises

Factor each polynomial.

1. $3z^2 + 6z$ **2.** $11x^2 - 33x^3$ **3.** $9ay^2 - 15a^2y$

4. $18r^2s^3 + 12r^4s$ **5.** $x^2 - 16$ **6.** $9t^2 - 1$

7. $4a^2 - z^2$ **8.** $x^2 - 8x + 16$ **9.** $h^2 + 6h + 9$

10. $u^4 - 4v^2$ **11.** $25y^2 - 10y + 1$ **12.** $4s^2 + 4s + 1$

13. $x^3 - x$ **14.** $4hk^2 + 16h^2k$ **15.** $t^3 - 12t^2 + 36t$

16. $xy^3 - 2xy^2 + xy$ **17.** $ax + bx + a + b$ **18.** $ax - bx + a - b$

For what value(s) of k will the polynomial be a perfect square trinomial?

19. $x^2 - 10x + k$ **20.** $y^2 + ky + 9$ **21.** $100t^2 + kt + 1$

Written Exercises

Factor each polynomial.

A **1.** $16x^3 - 64x^2$ **2.** $6x^2y^2 + 8x^3y$

    3. $t^2 + 18t + 81$ **4.** $z^2 - 12z + 36$

Factor each polynomial.

5. $16k^2 - 1$

6. $121x^2 - 1$

7. $4y^2 + 20y + 25$

8. $9s^2 - 24s + 16$

9. $16x^2 - 25$

10. $4h^2 - 81$

11. $121s^2 - 66st + 9t^2$

12. $16x^2 + 40xy + 25y^2$

13. $36p^2 - 49q^2$

14. $9x^4 - 16z^2$

15. $st^2 - s$

16. $p^3q - pq$

17. $t^3 - 27$

18. $8p^3 + 1$

19. $16r^4s + 2rs^4$

20. $3x^2y^4 - 81x^2y$

21. $x(y - 3) + 2(y - 3)$

22. $u(v - 1) - 2(v - 1)$

23. $x(y - 3) + 2(3 - y)$

24. $u(v - 1) - 2(1 - v)$

25. $pq - 2q + 2p - 4$

26. $xy - 2y - x + 2$

27. $ab - 2 - 2b + a$

28. $4ab + 1 - 2a - 2b$

29. $x^2 - 6x + 9 - 4y^2$

30. $z^2 + 2z + 1 - w^2$

31. $u^2 - v^2 + 2v - 1$

32. $x^2 - y^2 - 4y - 4$

B 33. $x^4 - 2x^2y + y^2$

34. $4u^4v^2 + 4u^2v + 1$

35. $a^6 + b^3$

36. $250x^2 - 2x^5$

37. $16s^4 - 81$

38. $p^4 - q^4$

39. $x^6 - y^6$

40. $64 - z^6$

41. $u^2 - v^2 - 2u - 2v$

42. $a^2 - b^2 + a - b$

43. $(p + q)^3 - (p - q)^3$

44. $(x + y)^3 + (x - y)^3$

45. $s^3 + t^3 + s^2t + st^2$

46. $u^3 - v^3 - u^2v + uv^2$

C 47. $(a + b)^6 - (a - b)^6$

48. $(a + b)^4 - (a - b)^4$

Factor each polynomial. In Exercises 49–54, assume that n represents a positive integer.

49. $x^{2n} - 1$

50. $x^{2n} - 2x^n + 1$

51. $x^{2n} + 2x^ny^n + y^{2n}$

52. $x^{3n} + y^{3n}$

53. $x^{4n} - x^{2n}y^{2n}$

54. $x^{4n} - 2x^{3n}y^n + x^{2n}y^{2n}$

55. $x^4 + x^2 + 1$ (*Hint:* Add and subtract x^2.)

56. $x^4 + 4$ (*Hint:* Add and subtract $4x^2$.)

57. $x^4 + x^2y^2 + y^4$ (*Hint:* See Exercise 55.)

58. $x^4 + 4y^4$ (*Hint:* See Exercise 56.)

59. **a.** Show that $x^4 - y^4 = (x - y)(x + y)(x^2 + y^2)$.

   b. Show that $x^8 - y^8 = (x - y)(x + y)(x^2 + y^2)(x^4 + y^4)$.

   c. Write a general pattern for the factorization of $x^n - y^n$ where n is a power of 2.

Mixed Review Exercises

Write as a simplified polynomial.

1. $(x - 3)^2$

2. $(2y + 3)(y - 4)$

3. $m^2n(2m - 3n)$

4. $(3c - 4) - (5 - c)$

5. $(a^2 + 2)(a - 6)$

6. $(-w)^2(2w)^3$

Find (a) the GCF and (b) the LCM of the following monomials.

7. $18x$, $24x^3$

8. $30a^2b^4$, $75a^3b^3$

9. $63mp^2$, $42mp^3$

Computer Exercises

For students with some programming experience.

How can a computer determine whether a given positive integer is a perfect square? One way is to use a loop like this:

```
LET R=0
FOR I=1 TO N
IF I*I=N THEN LET R=I
NEXT I
```

If the given integer N is a perfect square of some integer I, then the value of I will be stored in R by the time the computer completes the FOR . . . NEXT loop; otherwise the value of R will be 0.

You can use loops like the one above to determine whether the polynomial $Ax^2 + Bxy + Cy^2$ with $A > 0$ and $C > 0$ is a perfect square trinomial.

Exercises

1. Write a program to determine whether $Ax^2 + Bxy + Cy^2$ is a perfect square trinomial for given integers A, B, and C. (Assume $A > 0$ and $C > 0$.) If the polynomial is a perfect square trinomial, the program should print its factorization.

2. Use the program in Exercise 1 to test each polynomial and, if the polynomial is a perfect square trinomial, print its factorization.

   a. $x^2 - 56xy + 784y^2$

   b. $9x^2 + 144xy + 576y^2$

   c. $64x^2 + 56xy + 49y^2$

   d. $625x^2 - 1050xy + 441y^2$

3. Modify the program in Exercise 1 to determine whether the polynomial $Ax^2 + Bxy + Cy^2$ is a perfect square trinomial or a difference of squares. (Assume $A > 0$.) If the polynomial is factorable, print the factorization.

4. Use the program in Exercise 3 to test each polynomial and, if it is factorable, print the factorization.

   a. $144x^2 - 48xy + 4y^2$

   b. $144x^2 - 4y^2$

   c. $144x^2 + 48xy + 4y^2$

   d. $144x^2 + 4y^2$

4-6 Factoring Quadratic Polynomials

Objective   To factor quadratic polynomials.

Polynomials of the form
$$ax^2 + bx + c \quad (a \neq 0)$$
are called **quadratic** or **second-degree polynomials.** The term ax^2 is the **quadratic term,** bx is the **linear term,** and c is the **constant term.**

A **quadratic trinomial** is a quadratic polynomial for which a, b, and c are all nonzero integers. In Lesson 4-5 you factored a special type of quadratic trinomial, the perfect square trinomial. In this lesson you'll review how to factor quadratic trinomials that are not necessarily perfect squares.

If the quadratic trinomial $ax^2 + bx + c$ can be factored into the product $(px + q)(rx + s)$ where p, q, r, and s are integers, then
$$ax^2 + bx + c = (px + q)(rx + s)$$
$$= prx^2 + (ps + qr)x + qs$$

Setting corresponding coefficients equal gives
$$a = pr, \qquad b = ps + qr, \qquad \text{and} \qquad c = qs.$$
These equations suggest a way to find p, q, r, and s; it is illustrated in the following examples.

Example 1   Factor $x^2 + 2x - 15$.

Solution   1. Since the coefficient of x^2 is 1, $pr = 1$. Therefore, let both p and r equal 1 so that the factorization begins $(x \quad)(x \quad)$.

2. Since the constant term is -15, $qs = -15$. Therefore, q and s must have opposite signs. There are four possibilities:

$(-1)(15), \qquad (1)(-15), \qquad (-3)(5), \qquad$ and $\qquad (3)(-5)$.

3. Since the coefficient of x is 2, $ps + qr = 2$. But p and r are both 1, so $ps + qr = s + q = 2$. Therefore, choose the pair of factors having sum 2: -3 and 5.

$\therefore x^2 + 2x - 15 = (x - 3)(x + 5)$ *Answer*

Example 2   Factor $15t^2 - 16t + 4$.

Solution   1. $pr = 15$. The positive factors of 15 are 1, 3, 5, and 15, so the factorization begins in one of these ways:

$(t \quad)(15t \quad) \qquad$ or $\qquad (3t \quad)(5t \quad)$.

2. $qs = 4$. Since the product qs is positive, its factors are both positive or both negative. But the coefficient of x is negative, so choose negative factors of 4. Therefore, the factorization ends as

$(\quad -1)(\quad -4), \quad (\quad -2)(\quad -2), \quad$ or $\quad (\quad -4)(\quad -1)$.

3. There are six possible factorizations to check:

Trial Factorization	Linear Term of Product
$(t - 1)(15t - 4)$	$-19t$
$(t - 2)(15t - 2)$	$-32t$
$(t - 4)(15t - 1)$	$-61t$
$(3t - 1)(5t - 4)$	$-17t$
$(3t - 2)(5t - 2)$	$-16t$
$(3t - 4)(5t - 1)$	$-23t$

Since the required linear term is $-16t$, the fifth factorization is the correct one.

$\therefore 15t^2 - 16t + 4 = (3t - 2)(5t - 2)$ **Answer**

Example 3   Factor $3 - 2z - z^2$.

Solution $3 - 2z - z^2 = -z^2 - 2z + 3$ Write the terms in order of decreasing degree of z.

$\qquad = -1(z^2 + 2z - 3)$ Factor out -1.

$\qquad = -(z + 3)(z - 1)$ Factor $z^2 + 2z - 3$.

$\therefore 3 - 2z - z^2 = -(z + 3)(z - 1)$ **Answer**

There are quadratic polynomials with integral coefficients that cannot be factored over the set of polynomials with integral coefficients.

Example 4   Factor $x^2 + 4x - 3$.

Solution   Since 1 is the coefficient of the quadratic term, you need factors of -3 whose sum is 4. The only factorizations of -3 are $(-1)(3)$ and $(1)(-3)$. Neither pair of factors has sum 4. Therefore $x^2 + 4x - 3$ cannot be factored over the set of polynomials with integral coefficients.

A polynomial that has more than one term and cannot be expressed as a product of polynomials of lower degree taken from a given factor set is said to be **irreducible** over that set. Example 1 shows that $x^2 + 2x - 15$ is reducible, while Example 4 shows that $x^2 + 4x - 3$ is irreducible. An irreducible polynomial with integral coefficients is **prime** if the greatest common factor of its coefficients is 1. Therefore $x^2 + 4x - 3$ is prime, but $2x^2 + 8x - 6$ is *not* prime since 2 can be factored out.

A polynomial is **factored completely** when it is written as a product of factors and each factor is either a monomial, a prime polynomial, or a power of a prime polynomial.

Example 5 Factor $3x^6 - 48x^2$ completely.

Solution

$$3x^6 - 48x^2 = 3x^2(x^4 - 16) \qquad \text{Factor out } 3x^2.$$
$$= 3x^2(x^2 + 4)(x^2 - 4) \qquad \text{Factor } x^4 - 16.$$
$$= 3x^2(x^2 + 4)(x + 2)(x - 2) \qquad \text{Factor } x^2 - 4.$$

The polynomial is now factored completely since $x^2 + 4$, $x + 2$, and $x - 2$ are prime polynomials.

$$\therefore 3x^6 - 48x^2 = 3x^2(x^2 + 4)(x + 2)(x - 2) \quad \textit{Answer}$$

The **greatest common factor (GCF)** of two or more polynomials is the common factor having the greatest degree and the greatest constant factor. The **least common multiple (LCM)** of two or more polynomials is the common multiple having least degree and least positive constant factor. To find the GCF and LCM of two or more polynomials, first write the complete factorization of each.

Example 6 Find **(a)** the GCF and **(b)** the LCM of the following polynomials:
$$12p^3q + 12p^2q^2 + 3pq^3 \text{ and } 12p^4 - 6p^3q - 6p^2q^2.$$

Solution Factor the polynomials completely.

$$12p^3q + 12p^2q^2 + 3pq^3 = 3pq(4p^2 + 4pq + q^2)$$
$$= 3pq(2p + q)^2$$
$$12p^4 - 6p^3q - 6p^2q^2 = 6p^2(2p^2 - pq - q^2)$$
$$= 6p^2(2p + q)(p - q)$$

a. The GCF of the monomial factors is $3p$.
The GCF of the binomial factors is $2p + q$, because each binomial factor is taken the *least* number of times that it occurs in *both* factorizations.

$$\therefore \text{ the GCF of the polynomials is } 3p(2p + q). \quad \textit{Answer}$$

b. The LCM of the monomial factors is $6p^2q$.
The LCM of the binomial factors is $(2p + q)^2(p - q)$, because each binomial factor is taken the *greatest* number of times that it occurs in *either* factorization.

$$\therefore \text{ the LCM of the polynomials is } 6p^2q(2p + q)^2(p - q). \quad \textit{Answer}$$

Oral Exercises

Tell which factorization is correct.

1. $x^2 - 8x - 48$
   $(x - 4)(x + 12)$ or $(x + 4)(x - 12)$

2. $4x^2 - 4x - 15$
   $(2x - 5)(2x + 3)$ or $(4x + 5)(x - 3)$

Factor each polynomial.

3. $x^2 + 3x + 2$

4. $t^2 + 11t + 10$

5. $u^2 + 7u - 8$

6. $y^2 + 4y - 5$

7. $z^2 + z - 12$

8. $r^2 - r - 2$

State whether or not each polynomial is prime.

9. $4x + 9$

10. $x^2 + 3x$

11. $x^2 - 2x + 1$

12. $x^2 + 1$

13. $x^4 + x^2$

14. $x^3 + 1$

15. One factor of $x^2 + 3x - 10$ is $x - 2$. What is the other?

16. One factor of $2x^2 + 5x - 12$ is $x + 4$. What is the other?

Written Exercises

Factor completely. If the polynomial is prime, say so.

A

1. $x^2 - 9x + 8$

2. $t^2 + 9t + 14$

3. $z^2 - 11z + 18$

4. $u^2 - 10u + 9$

5. $r^2 + 12r + 20$

6. $y^2 - 5y + 6$

7. $p^2 - 8p + 9$

8. $h^2 - 10h + 24$

9. $s^2 - 20s + 36$

10. $z^2 - 9z + 12$

11. $x^2 + x - 12$

12. $t^2 + 2t - 15$

13. $t^2 - 2t - 35$

14. $s^2 - 6s - 27$

15. $3z^2 + 4z + 1$

16. $5v^2 + 4v - 1$

17. $8 + 2s - s^2$

18. $21 - 4x - x^2$

19. $x^2 - xy - 30y^2$

20. $p^2 + 2pq - 24q^2$

21. $u^2 - 8uv - 12v^2$

22. $h^2 - 8hk - 15k^2$

23. $2t^2 + 5t - 3$

24. $3x^2 - 8x + 5$

25. $3p^2 - 7p - 6$

26. $4r^2 + 8r + 3$

27. $6x^2 - 7xy - 3y^2$

28. $6s^2 + st - 5t^2$

29. $2h^2 + 7hk - 15k^2$

30. $2u^2 + uv - 21v^2$

B

31. $6x^2 + 7x - 10$

32. $4y^2 - 17y + 15$

33. $4t^2 - 9t + 6$

34. $25u^2 - 20u + 4$

35. $12p^2 - 32pq - 5q^2$

36. $4r^2 + 16rs - 10s^2$

37. $4x^3 + 8x^2y - 5xy^2$

38. $4x^2 + 3xy - 15y^2$

39. $4pq^4 - 32pq$

40. $81uv^3 + 3u^4$

41. $r^4 - 16s^4$

42. $x^6 - 64y^6$

43. $x^4 - 3x^2 - 4$

44. $z^4 - 10z^2 + 9$

Find (a) the GCF and (b) the LCM of the given polynomials.

45. $x^2 - 3x + 2$; $x^2 - 4x + 4$

46. $x^2 - 16$; $x^2 + 2x - 8$

47. $t^3 - 2t^2 - 3t$; $t^3 + 5t^2 + 4t$

48. $y^3 - 4y$; $y^4 - 8y$

49. $p - q$; $p^3 - q^3$

50. $x^3 + y^3$; $x^2 + y^2$; $x + y$

C **51.** Show that $x^2 + x + k$ is prime for all positive integers k.

52. Show that $x^2 + (k + 1)x + k$ is reducible for all positive integers k.

Factor completely. Assume that n represents a positive integer.

53. $a^{4n} - b^{4n}$

54. $a^{6n} - b^{6n}$

55. $x^{4n} - 5x^{2n} + 4$

56. $x^{8n} - 2x^{4n} + 1$

57. $(x^2 - 2x + 3)^2 - (x^2 - 4x - 1)^2$

58. $(x^2 + 2x - 4)^2 - (x^2 - 2x - 4)^2$

59. One factor of $x^3 - 3x^2 + 3x - 2$ is $x - 2$. What is the other?

Self-Test 2

Vocabulary

factor, factor set (p. 179)
prime number, prime (p. 179)
prime factorization (p. 179)
greatest common factor (GCF) of
  integers (p. 179)
least common multiple (LCM) of
  integers (p. 179)
greatest common factor (GCF) of
  monomials (p. 180)
least common multiple (LCM) of
  monomials (p. 180)
greatest monomial factor (p. 183)
perfect square trinomial (p. 183)
difference of squares (p. 183)

sum of cubes (p. 184)
difference of cubes (p. 184)
quadratic polynomial (p. 188)
quadratic term (p. 188)
linear term (p. 188)
constant term (p. 188)
quadratic trinomial (p. 188)
irreducible polynomial (p. 189)
prime polynomial (p. 189)
factor completely (p. 189)
greatest common factor (GCF) of
  polynomials (p. 190)
least common multiple (LCM) of
  polynomials (p. 190)

Find the prime factorization of each integer.

1. 990

2. 3000

Obj. 4-4, p. 179

Find (a) the GCF and (b) the LCM of the following.

3. $36a^2b^3$, $54abc^2$

4. $6p^3q^2r$, $8p^2q^2r^2$, $4p^3q^3$

Factor completely.

5. $27x^3 - 12x$

6. $9t^2 + 6t + 1$

Obj. 4-5, p. 183

7. $125s^3 - 8t^3$

8. $ab - a + b - 1$

9. $6t^2 + 4t - 2$

10. $4z^2 + z - 14$

Obj. 4-6, p. 188

In Lesson 4–6 you learned how to factor $ax^2 + bx + c$ where a, b, and c are integers and $a \neq 0$. The following program uses a similar method to factor $Ax^2 + Bx + C$ where $A \neq 0$ and $C \neq 0$. Although the process of complete factorization includes factoring out the GCF of the terms of a polynomial, this program does *not* factor out the GCF of A, B, and C. (See the Computer Key-In in Chapter 5 for a program that will find the GCF of two integers.)

```
10 PRINT "THIS PROGRAM FINDS FACTORS OF"
20 PRINT "A*X/\2+B*X+C."
30 INPUT "ENTER A, B, AND C (A, C NONZERO): "; A, B, C
40 IF A=0 OR C=0 THEN 30
50 LET FLAG=0
60 FOR M=1 TO SQR (ABS (A))
70 IF INT (A/M)<>A/M OR FLAG<>0 THEN 170
80 LET P=M
90 LET R=A/M
100 FOR N=1 TO ABS (C)
110 IF INT (C/N)<>C/N OR FLAG<>0 THEN 160
120 LET Q=N
130 LET S=C/N
140 IF P*S+Q*R=B THEN LET FLAG=1
150 IF P*S+Q*R=-B THEN LET FLAG=2
160 NEXT N
170 NEXT M
180 IF FLAG=0 THEN
       PRINT A; "*X/\2+ "; B; " *X+ "; C; " IS IRREDUCIBLE."
190 IF FLAG=1 THEN
       PRINT "("; P; " *X+ "; Q; ")*("; R; " *X+ "; S; ")"
200 IF FLAG=2 THEN
       PRINT "("; P; " *X+ ";-Q; ")*("; R; " *X+ "; -S; ")"
210 END
```

For each possible pair of factors P and R of A, the program tests each possible pair of factors Q and S of C. If $PS + QR = B$ in line 140, the program will print the factorization $(Px + Q)(Rx + S)$. If $PS + QR = -B$ in line 150, then reversing the signs of Q and S will yield a factorization. If all pairs of factors of A and C are checked and $PS + QR$ is neither B nor $-B$, then the computer will print that the given trinomial is irreducible.

Exercises

Run the program to display a factorization, if possible, for each quadratic trinomial.

1. $12x^2 - 13x - 120$

2. $-15x^2 + 22x + 48$

3. $6x^2 - 55x + 56$

4. $x^2 + 5x - 4$

5. $2x^2 + 8x - 10$

6. $4x^2 - 25$

Applications of Factoring

4-7 Solving Polynomial Equations

Objective   To solve polynomial equations.

The factoring that you have learned to do is a necessary step in solving polynomial equations. A **polynomial equation** is an equation that is equivalent to one with a polynomial as one side and 0 as the other. Both $x^2 - 5x - 24 = 0$ and $x^2 = 5x + 24$ are polynomial equations.

A **root,** or **solution,** of a polynomial equation is a value of the variable that satisfies the equation. The roots of $x^2 - 5x - 24 = 0$ are -3 and 8, because $(-3)^2 - 5(-3) - 24 = 0$ and $8^2 - 5(8) - 24 = 0$.

When a polynomial equation in x is written with 0 as one side, you can solve the equation if you can factor the polynomial on the other side into *linear factors* of the form $ax + b$ $(a \neq 0)$. The basis of this method is the *zero-product property* (page 83):

$$ab = 0 \text{ if and only if } a = 0 \text{ or } b = 0.$$

The zero-product property can be extended to any number of factors. Therefore, a product of real numbers is zero if and only if at least one of its factors is zero.

To use the *zero-product property* to solve a polynomial equation, you need to

1. write the equation with 0 as one side,
2. factor the other side of the equation, and
3. solve the equation obtained by setting each factor equal to 0.

Example 1   Solve $(x - 5)(x + 2) = 0$.

Solution   The given equation already has 0 as one side, and the other side is already factored. Simply set each factor equal to 0 and solve:

$$
\begin{array}{ccc}
x - 5 = 0 & \text{or} & x + 2 = 0 \\
x = 5 & \text{or} & x = -2
\end{array}
$$

$$
\begin{array}{cc}
\textit{Check } x = 5\text{:} & \textit{Check } x = -2\text{:} \\
(5 - 5)(5 + 2) \stackrel{?}{=} 0 & (-2 - 5)(-2 + 2) \stackrel{?}{=} 0 \\
0 \cdot 7 \stackrel{?}{=} 0 & -7 \cdot 0 \stackrel{?}{=} 0 \\
0 = 0 \ \checkmark & 0 = 0 \ \checkmark
\end{array}
$$

$\therefore$ the solution set is $\{5, -2\}$.   *Answer*

Example 2 Solve $x^2 = x + 30$.

Solution
1. $x^2 - x - 30 = 0$ Make one side 0.
2. $(x + 5)(x - 6) = 0$ Factor the polynomial.
3. $x + 5 = 0$ or $x - 6 = 0$ Use the zero-product property.
       $x = -5$ or $x = 6$

Check: Substitute -5 and 6 into $x^2 = x + 30$ to verify that they are
           solutions.

$\therefore$ the solution set is $\{-5, 6\}$.   ***Answer***

Example 3 Solve $3x^3 = 4x(2x - 1)$.

Solution
1. $3x^3 = 8x^2 - 4x$
    $3x^3 - 8x^2 + 4x = 0$ Make one side 0.
2. $x(3x^2 - 8x + 4) = 0$ Factor the
    $x(3x - 2)(x - 2) = 0$ polynomial.
3. $x = 0$ or $3x - 2 = 0$ or $x - 2 = 0$ Use the zero-product
    $x = 0$ or $x = \dfrac{2}{3}$ or $x = 2$ property.

$\therefore$ the solution set is $\left\{0, \dfrac{2}{3}, 2\right\}$. The check is left for you.   ***Answer***

      As you have seen, the zero-product property is used to solve polynomial
equations. You can also use it to find *zeros* of polynomial functions. A number
r is a **zero** of a function f if $f(r) = 0$. For example, 3 and -3 are zeros of
$g(x) = x^2 - 9$ because $g(3) = 3^2 - 9 = 0$ and $g(-3) = (-3)^2 - 9 = 0$.

Example 4 Find the zeros of $f(x) = (x - 4)^3 - 4(3x - 16)$.

Solution
1. $f(x) = (x - 4)^3 - 4(3x - 16)$ To find the zeros of f,
    $(x - 4)^3 - 4(3x - 16) = 0$ write $f(x) = 0$.
2. $(x^3 - 12x^2 + 48x - 64) - 12x + 64 = 0$
                  $x^3 - 12x^2 + 36x = 0$ Simplify.
                $x(x^2 - 12x + 36) = 0$
                     $x(x - 6)^2 = 0$ Factor the polynomial.
3. $x = 0$ or $x - 6 = 0$ or $x - 6 = 0$ Use the zero-product
    $x = 0$ or $x = 6$ or $x = 6$ property.

Check: $f(0) = (0 - 4)^3 - 4(3 \cdot 0 - 16) = -64 + 64 = 0$ $\checkmark$
          $f(6) = (6 - 4)^3 - 4(3 \cdot 6 - 16) = 8 - 8 = 0$ $\checkmark$

$\therefore$ the zeros of f are 0 and 6.   ***Answer***

From Example 4 you can see that $f(x) = x(x - 6)^2$. Since $x - 6$ occurs as a factor of f twice, the number 6 is called a **double zero** of the function f and a **double root** of the equation $f(x) = 0$. In general, the zeros and roots arising from repeated factors are called **multiple zeros** of functions and **multiple roots** of equations.

Oral Exercises

Solve.

1. $(x - 3)(x - 7) = 0$ **2.** $(y + 2)(y - 5) = 0$

3. $(u - 1)(u + 2)(u - 3) = 0$ **4.** $t(t - 3)(t + 4) = 0$

5. $y^2 - 2y = 0$ **6.** $s^2 - 5s = 0$

7. $x^2 - 4x + 4 = 0$ **8.** $z^2 + 6z + 9 = 0$

9. $t^2 - 25 = 0$ **10.** $4x^2 - 1 = 0$

Find all zeros of the given function. Identify all double zeros.

11. $f(x) = 3x - 12$ **12.** $g(y) = y^2 - 16$

13. $h(t) = (2t - 3)(t + 2)$ **14.** $k(z) = z(z - 3)(z + 1)$

15. $f(s) = s(s - 2)^2$ **16.** $g(u) = u^2(u - 2)$

17. $g(x) = x(x^2 - 2x + 1)$ **18.** $f(t) = (t - 2)^2(t - 3)^2$

Written Exercises

Solve. Identify all double roots.

A   **1.** $(x - 1)(x - 4) = 0$ **2.** $(t + 2)(t - 5) = 0$

3. $t(t + 1)(t - 2) = 0$ **4.** $z^2(2z - 1) = 0$

5. $(s - 1)^2(s - 3)^2 = 0$ **6.** $y(y + 1)^2(y - 2) = 0$

7. $z^2 + 3 = 4z$ **8.** $x^2 - 12 = 4x$

9. $t^3 - t = 0$ **10.** $s^3 - s^2 = 0$

11. $x^3 + 4x = 4x^2$ **12.** $y^3 + 6y^2 = 27y$

13. $3r^2 = 4r - 1$ **14.** $6x^2 = 1 - x$

15. $2y^2 + y = 6$ **16.** $10t^2 - 9t = 1$

17. $6 - 7u = 3u^2$ **18.** $5s - 1 = 6s^2$

19. $6(x + 12) = x^2$ **20.** $(u + 3)(u - 3) = 8u$

21. $(y - 4)^2 = 2y$ **22.** $x = (x - 6)^2$

23. $3t(t + 1) = 4(t + 1)$ **24.** $2(r^2 + 1) = 5r$

25. $(x - 1)(x^2 + x - 2) = 0$ **26.** $(x + 2)(x^2 - 4) = 0$

27. $y^2(y - 3)(y^2 - 9) = 0$ **28.** $(x^2 - 1)(x^2 + 3x + 2) = 0$

Solve. Identify all double roots.

B **29.** $x^4 - 2x^2 + 1 = 0$

31. $t^6 + 9t^2 = 10t^4$

33. $(u - 2)^3 - 3u + 8 = 0$

30. $y^4 - 5y^2 + 4 = 0$

32. $x^2(x^4 + 16) = 8x^4$

34. $(s - 6)^3 = 27(s - 8)$

Multiply both sides of the given equation by a number that will eliminate decimals or fractions. Then solve.

35. $0.3x^2 + 0.2x - 0.1 = 0$

37. $\frac{1}{6}x^2 + \frac{1}{2}x - \frac{2}{3} = 0$

36. $0.2x^2 - 1.1x + 0.5 = 0$

38. $\frac{1}{3}x^2 - \frac{1}{2}x - \frac{1}{3} = 0$

Solve. Identify all multiple roots.

39. $(x - 1)^3 - (x - 1)^2 = 0$

41. $(x^2 - 4)^3 = 0$

43. $(x^2 + 1)^2 - 4(x^2 + 1) + 4 = 0$

40. $(t - 2)^3 - (t - 2) = 0$

42. $(x^2 - 3x + 2)^3 = 0$

44. $(x^2 - 3)^2 + (x^2 - 3) - 2 = 0$

Find all zeros of f. Identify all multiple zeros.

45. $f(x) = (x - 4)^3 - 4(x - 4)$

47. $f(z) = 9z^4 - 12z^3 + 4z^2$

49. $f(x) = (x - 1)^4 - 4(x - 1)^3$

46. $f(t) = (t - 2)^3 - (t - 2)$

48. $f(s) = 4s^4 - 17s^2 + 4$

50. $f(y) = (y^2 - 9)^3$

In Exercises 51–54, a and b are constants. Solve each equation for x.

C **51.** $a^2x^2 - b^2 = 0$ $(a \neq 0)$

53. $x^3 + ax^2 = x + a$

52. $x^2 + ax - bx - ab = 0$

54. $x^3 - bx^2 - 4x + 4b = 0$

55. Under what conditions placed on x and y is $(x + y)^2 = x^2 + y^2$ true?

56. Under what conditions placed on x and y is $(x + y)^3 = x^3 + y^3$ true?

Mixed Review Exercises

Factor completely. If the polynomial is prime, say so.

1. $x^2 - 6x + 9$

3. $8mn - 10n + 12m - 15$

5. $12x + 24x^3 - 34x^2$

2. $3y^2 - 5y + 8$

4. $16a^2 - 25b^2$

6. $8u^3 + 1$

Express as a simplified polynomial.

7. $z^2(z - 1)$

9. $(3m - 7) + 2(m + 4)$

11. $x - (x + 1)$

8. $(4a + 5)^2$

10. $(p^2 + 2)(p^2 - 3)$

12. $(-3c^2d)^3$

4-8 Problem Solving Using Polynomial Equations

Objective   To solve problems using polynomial equations.

An equation that represents a real life problem is called a **mathematical model.** The plan for solving word problems outlined in Lesson 1-9 can be used to develop a mathematical model for a given problem. After solving the equation, you should always check your answer against the conditions stated in the original problem. Checking your answer just in the equation is not sufficient, because an acceptable answer in mathematics may not be reasonable in real life.

Example 1   A graphic artist is designing a poster that consists of a rectangular print with a uniform border. The print is to be twice as tall as it is wide, and the border is to be 3 in. wide. If the area of the poster is to be 680 in.^{2}, find the dimensions of the print.

Solution

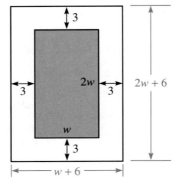

Step 1 You are asked to find the width and height of the print. A border 3 in. wide surrounds the print. Draw a diagram.

Step 2 Let w = width of the print. Then $2w$ = height of the print. Since the dimensions of the poster are 6 in. greater than the dimensions of the print, $w + 6$ = width of the poster and $2w + 6$ = height of the poster. Label your diagram.

Step 3 Write an equation that shows the area of the poster is 680 in.^{2}:

$$\text{Length} \times \text{Width} = \text{Area}$$
$$(2w + 6)(w + 6) = 680$$

Step 4 Solve the equation.

$2(w + 3)(w + 6) = 680$	Factor $2w + 6$.
$(w + 3)(w + 6) = 340$	Divide both sides by 2.
$w^2 + 9w + 18 = 340$	
$w^2 + 9w - 322 = 0$	Make one side 0.
$(w - 14)(w + 23) = 0$	Factor the polynomial.
$w - 14 = 0$ or $w + 23 = 0$	Use the zero-product property.
$w = 14$ or $w = -23$	

Step 5 Although $w = -23$ is a solution of $(2w + 6)(w + 6) = 680$, it must be rejected because width cannot be negative. The other solution, $w = 14$, results in a print that is 14 in. wide and 28 in. tall. Check these two dimensions:

$$\text{Print's height} = 28 = 2 \cdot 14 = \text{twice print's width} \quad \checkmark$$
$$\text{Poster's area} = (14 + 6)(28 + 6) = 20 \cdot 34 = 680 \quad \checkmark$$

$\therefore$ the print is 14 in. by 28 in. **Answer**

Vertical motion affected only by gravity leads to a mathematical model that is a polynomial equation. When a projectile is launched vertically upward with an initial speed v, its height h above the launch point t seconds later is given approximately by:

$$h = vt - 4.9t^2 \text{ if the distance is measured in meters,}$$

and $\quad h = vt - 16t^2$ if the distance is measured in feet.

Example 2 A batter hits a baseball directly upward with speed 96 ft/s.

a. How long is the ball in the air before being caught by the catcher?

b. How high does the ball go?

Solution **a.** Use the formula $h = 96t - 16t^2$. When the ball returns to the level at which it is hit, the height h is 0.

$$0 = 96t - 16t^2$$
$$0 = 16t(6 - t)$$
$$t = 0 \quad \text{or} \quad t = 6$$

$\therefore$ the ball is in the air 6 seconds. **Answer**

b. The ball reaches its maximum height halfway between the time when it is hit ($t = 0$) and the time when it is caught ($t = 6$).

$$\text{If } t = \frac{6}{2} = 3, \ h = 96(3) - 16(3)^2 = 144.$$

$\therefore$ the maximum height of the ball above the level of the bat is 144 ft. **Answer**

Problems   HW   # 1, 3, 5, 6, 9, and 15   Feb 15$\frac{th}{}$, 2000

Solve each problem. If there are two correct answers, give both of them.

A **1.** The sum of a number and its square is 72. Find the number.

2. Find a number that is 56 less than its square.

3. Find two consecutive odd integers whose product is 143.

4. Find two consecutive odd integers the sum of whose squares is 130.

5. A rectangle is 4 cm longer than it is wide, and its area is 117 cm². Find its dimensions. $P = 2l + 2w$

6. A rectangular garden has perimeter 66 ft and area 216 ft². Find the dimensions of the garden. $A = l \cdot w$ Ans   9ft by 24 ft

7. The area of a right triangle is 44 m². Find the lengths of its legs if one of the legs is 3 m longer than the other.

8. Two ships leave port, one sailing east and the other south. Some time later they are 17 mi apart, with the eastbound ship 7 mi farther from port than the southbound ship. How far is each from port?

9. The top of a 15-foot ladder is 3 ft farther up a wall than the foot of the ladder is from the bottom of the wall. How far is the foot of the ladder from the bottom of the wall?

10. The height of a triangle is 7 cm greater than the length of its base, and its area is 15 cm². Find the height.

11. The hypotenuse of a right triangle is 25 m long. The length of one leg is 10 m less than twice the other. Find the length of each leg.

12. The side of a large tent is in the shape of an isosceles triangle whose area is 54 ft² and whose base is 6 ft shorter than twice its height. Find the height and the base of the side of the tent.

13. A rectangle is 15 cm wide and 18 cm long. If both dimensions are decreased by the same amount, the area of the new rectangle formed is 116 cm² less than the area of the original. Find the dimensions of the new rectangle.

14. A rectangle is twice as long as it is wide. If its length is increased by 4 cm and its width is decreased by 3 cm, the new rectangle formed has an area of 100 cm². Find the dimensions of the original rectangle. $V_0 = ?$

15. A projectile is launched upward from ground level with an initial speed of 98 m/s. How high will it go? When will it return to the ground?

16. A ball is thrown directly upward from ground level with an initial speed of 80 ft/s. How high will it go? When will it return to the ground?

B **17.** A signal flare is fired upward from ground level with an initial speed of 294 m/s. A balloonist cruising at a height of 2450 m sees it pass on the way up. How long will it be before the flare passes the balloonist again on the way down?

เมตร ใช้ 9.8
เมตร ft ใช้ 32

$0 = V_0 t - \frac{1}{2} g t^2$

$t = time$ (เวลาตลอด ตอนที่ อยู่ตรงกลาง S_0 หาร ตัว 2)

$V_0 = ความเร็ว$

18. Luis wanted to throw an apple to Kim, who was on a balcony 40 ft above him, so he tossed it upward with an initial speed of 56 ft/s. Kim missed it on the way up, but then caught it on the way down. How long was the apple in the air?

19. A rocket is moving vertically upward with speed 245 m/s when its fuel runs out. How much farther will it travel upward before starting to fall back to the ground?

20. A ball is thrown upward from the top of a 98-meter tower with an initial speed of 39.2 m/s. How much later will it hit the ground? (*Hint:* If h is the height of the ball above the top of the tower, then $h = -98$ when the ball hits the ground.)

21. The distance d, in feet, required for a vehicle traveling at r mi/h to come to a stop is given approximately by $d = 0.05r^2 + r$. If an automobile in a 55 mi/h speed zone required 240 ft to stop, was its speed within the legal limit?

22. The cost C of manufacturing n calculators per day at a certain plant is given by $C = n(20 - 0.01n) + 100$. The size of the plant limits the maximum output to 500 calculators per day. If the company plans to invest $5200 per day in manufacturing costs, how many calculators per day can it manufacture?

23. When 0.5 cm was planed off each of the six faces of a wooden cube, its volume decreased by 169 cm^3. Find its new volume.

24. A farmer plans to use 21 m of fencing to enclose a rectangular pen having area 55 m^2. Only three sides of the pen need fencing because part of an existing wall will form the fourth side. Find the dimensions of the pen.

25. The width, length, and diagonal of a rectangle are consecutive even integers. Find the integers.

26. Give a convincing argument why the width, length, and diagonal of a rectangle cannot be consecutive odd integers.

27. A rectangular corner lot, originally twice as long as it was wide, lost a 2-meter strip along two adjacent sides due to street widening. Its new area is 684 m^2. Find its new dimensions.

28. A decorator plans to place a rug in a room 9 m by 12 m in such a way that a uniform strip of flooring around the rug will remain uncovered. If the rug is to cover half the floor space, what should the dimensions of the rug be?

29. A garden plot 5 m by 15 m has one of its longer sides next to a wall. The area of the plot is to be doubled by digging up a strip of uniform width along the other three sides. How wide should the border be?

30. A rancher plans to use 160 yd of fencing to enclose a rectangular corral and to divide it into two parts by a fence parallel to the shorter sides of the corral. Find the dimensions of the corral if its area is 1000 yd^2.

C 31. The general formula for the height h after t seconds of a projectile launched upward from ground level with initial speed v is $h = vt - \frac{1}{2}gt^2$, where g is the gravitational constant (approximately 9.8 m/s^2 or 32 ft/s^2).

Show that the greatest height of the projectile is $\dfrac{v^2}{2g}$.

4-9 Solving Polynomial Inequalities

Objective   To solve polynomial inequalities.

A **polynomial inequality** is an inequality that is equivalent to an inequality with a polynomial as one side and 0 as the other side. Both $x^2 - x - 6 > 0$ and $x^2 > x + 6$ are polynomial inequalities.

    You can often solve a polynomial inequality that has 0 as one side by factoring the polynomial into linear factors and applying one of the following facts:

$$ab > 0 \text{ if and only if } a \text{ and } b \text{ have the same sign;}$$
$$ab < 0 \text{ if and only if } a \text{ and } b \text{ have opposite signs.}$$

Example 1   Find and graph the solution set of $x^2 - 1 > x + 5$.

Solution
$$x^2 - 1 > x + 5$$
$$x^2 - x - 6 > 0 \qquad \text{Make one side 0.}$$
$$(x + 2)(x - 3) > 0 \qquad \text{Factor the polynomial.}$$

The product is positive if and only if both factors have the *same sign*.

Both factors positive	or	**Both factors negative**

$x + 2 > 0 \quad$ and $\quad x - 3 > 0 \qquad\qquad\qquad x + 2 < 0 \quad$ and $\quad x - 3 < 0$

$\quad x > -2 \quad$ and $\qquad x > 3 \qquad\qquad\qquad\quad x < -2 \quad$ and $\qquad x < 3$

The solution set of this conjunction is $\{x: x > 3\}$.   The solution set of this conjunction is $\{x: x < -2\}$.

$\therefore$ the solution set of the given inequality is $\{x: x > 3 \text{ or } x < -2\}$.
The graph of the solution set is given below.

Example 2   Find and graph the solution set of $3t < 4 - t^2$.

Solution
$$3t < 4 - t^2$$
$$t^2 + 3t - 4 < 0 \qquad \text{Make one side 0.}$$
$$(t - 1)(t + 4) < 0 \qquad \text{Factor the polynomial.}$$

The product is negative if and only if the factors have opposite signs.

$t - 1 > 0 \quad$ and $\quad t + 4 < 0 \qquad\qquad\text{or}\qquad\qquad t - 1 < 0 \quad$ and $\quad t + 4 > 0$

$\quad t > 1 \quad$ and $\qquad t < -4 \qquad\qquad\qquad\qquad\qquad t < 1 \quad$ and $\qquad t > -4$

$\qquad\qquad$ no solution $\qquad\qquad\qquad\qquad\qquad\qquad\qquad \{t: -4 < t < 1\}$

∴ the solution set of the given inequality is $\{t: -4 < t < 1\}$.
The graph of the solution set is shown below.

After solving a polynomial inequality, you can make a partial check of your solution by choosing *test points* on the number line. For instance, with the answer in Example 2, first check a number to the left of the endpoint -4, say -5: -5 does *not* satisfy the inequality. Then check a number between -4 and 1, say 0: 0 *does* satisfy the inequality. Finally, check a number to the right of the endpoint 1, say 2: 2 does *not* satisfy the inequality.

In Example 3 a **sign graph** is used to help find and graph the solution set of a polynomial inequality. This visual method is particularly helpful when the polynomial has three or more factors.

Example 3   Find and graph the solution set of $9x \le x^3$.

Solution

$$9x \le x^3$$
$$x^3 - 9x \ge 0 \qquad \text{Make one side 0.}$$
$$x(x^2 - 9) \ge 0 \qquad \text{Factor the polynomial.}$$
$$x(x + 3)(x - 3) \ge 0$$

To draw a sign graph, draw a number line for each factor and label the part where the factor is positive, where it is zero, and where it is negative.

The sign of the product $x(x + 3)(x - 3)$ is determined by the number of negative factors. When exactly one or exactly three factors are negative, the product is negative. When no factors or exactly two factors are negative, the product is positive.

∴ the solution set of the given inequality is $\{x: -3 \le x \le 0 \text{ or } x \ge 3\}$.
The graph of the solution set is shown below.

Oral Exercises

Match each inequality with the graph of its solution set.

1. $(x - 2)(x + 3) < 0$

2. $(x - 2)(x + 3) > 0$

3. $(x + 2)(x - 3) < 0$

4. $(x + 2)(x - 3) \geq 0$

5. $(x + 2)(x - 3) > 0$

6. $(x - 2)(x + 3) \leq 0$

a.

b.

c.

d.

e.

f.

Justify the following statements.

7. The solution set of $x^2 \leq 0$ is $\{0\}$.

8. The solution set of $x^2 > 0$ is {real numbers except 0}.

9. The solution set of $x^2 + 4 \leq 0$ is $\emptyset$.

10. The solution set of $x^2 + 1 > 0$ is {real numbers}.

11. The solution set of $x(x^2 + 1) > 0$ is $\{x: x > 0\}$.

12. The solution set of $(x - 2)(x^2 + 1) < 0$ is $\{x: x < 2\}$.

Give the solution set of each inequality.

13. $x(x + 1) < 0$ **14.** $x(x - 3) > 0$ **15.** $x(x + 2) \geq 0$ **16.** $x(x - 4) \leq 0$

Written Exercises Feb 17th 2000
1-18

Find and graph the solution set of each inequality.

A **1.** $(x - 2)(x - 5) < 0$

3. $x^2 - 4 \leq 0$

5. $z^2 - 4z > 0$

7. $4y^2 \geq 36$

9. $x^2 - 5x + 4 < 0$

11. $t^2 > 9(t - 2)$

13. $4z(z - 1) \leq 15$

2. $(t + 1)(t - 2) > 0$

4. $x^2 \leq 9$

6. $x^2 + 2x \geq 0$

8. $3s^2 < 48$

10. $z^2 - z - 6 > 0$

12. $r^2 \leq 2(r + 4)$

14. $4x(x + 1) \geq 3$

15. $t^2 + 16 \geq 8t$

16. $x^2 + 9 \leq 6x$

17. $12 + s - s^2 \geq 0$

18. $9 - 3t - 2t^2 > 0$

B **19.** $x^3 - 16x > 0$

20. $t^3 < 9t^2$

21. $y^3 + y^2 < 6y$

22. $z^3 + 7z^2 + 10z > 0$

Find and graph the solution set of each inequality.

23. $(x^2 - 4x)(x^2 - 4) < 0$

24. $(x^2 + 4)(x^2 + 4x) > 0$

25. $x^4 > 3x^2 + 4$

26. $x^4 - 18 < 7x^2$

C **27.** $x^4 + 9 \leq 10x^2$

28. $x^4 + 100 \geq 29x^2$

29. $(x^2 - x - 6)(x^2 - 2x + 1) > 0$

30. $(x^2 + x - 2)(x^2 - 4x + 4) \leq 0$

31. $x^2(x^2 + 4) \leq 4x^3$

32. $x^2(x^2 + 9) > 6x^3$

33. Under what conditions placed on the real numbers p and q will $(x + p)(x + q) < 0$ have a nonempty solution set?

Mixed Review Exercises

Solve.

1. $(2x + 3)(x - 1) = 0$

2. $3p - 7 = 8$

3. $5 - |b| = 1$

4. $z(z - 1) = z(z + 1)$

5. $|2u - 5| = 3$

6. $2a^2 = 7a + 4$

7. $|8 - t| = 0$

8. $(n - 1)^2 = 1$

9. $(3c + 1)^2 = 0$

10. $m^2 = m$

11. $y^2(4 - y) = 0$

12. $4(3d - 5) = 2(5d + 1)$

Self-Test 3

Vocabulary   polynomial equation (p. 194)   multiple zero (p. 196)
root (p. 194)   multiple root (p. 196)
zero of a function (p. 195)   mathematical model (p. 198)
double zero (p. 196)   polynomial inequality (p. 202)
double root (p. 196)   sign graph (p. 203)

Solve each equation.

1. $x^2 - 11x - 42 = 0$ **2.** $3t^2 = 2 - t$ **Obj. 4-7, p. 194**

3. A 10 m by 20 m pool is surrounded by a deck of uniform width.   **Obj. 4-8, p. 198**
The area of the deck is 216 m². How wide is the deck?

Find and graph the solution set of each inequality.

4. $2x^2 + 7x - 15 > 0$ **5.** $(x - 2)^2 \leq x$ **Obj. 4-9, p. 202**

Check your answers with those at the back of the book.

Chapter Summary

1. A *monomial* is a constant, a variable, or a product of a constant and one or more variables. A *polynomial* is a monomial or a sum of monomials. A *simplified* polynomial has no two terms similar. The terms of a simplified polynomial are usually arranged in decreasing degree of one of the variables.

2. If a and b are real numbers and m and n are positive integers, then:

$$\text{Law 1} \quad a^m \cdot a^n = a^{m+n}$$
$$\text{Law 2} \quad (ab)^m = a^m b^m$$
$$\text{Law 3} \quad (a^m)^n = a^{mn}$$

These *laws of exponents* are useful in simplifying polynomials.

3. To find the product of two polynomials, multiply each term of one polynomial by each term of the other and simplify the result.

4. A *prime* is an integer greater than 1 whose only positive integral factors are itself and 1. To find the *prime factorization* of an integer, write it as the product of primes.

5. Use prime factorization to determine the *greatest common factor* (GCF) or *least common multiple* (LCM) of two or more integers. To find the GCF, take the least power of each common prime factor in the factorization. To find the LCM, take the greatest power of each prime factor in the factorizations.

6. The GCF of two or more monomials is the factor of each that has the greatest degree and greatest coefficient. The *greatest monomial factor* of a polynomial is the GCF of its terms.

7. The LCM of two or more monomials is the multiple of each that has the least degree and least positive coefficient.

8. The following strategies are useful in factoring polynomials:

   Factor out the greatest monomial factor.

   Look for special products such as a perfect square trinomial, a difference of squares, and a sum or difference of cubes.

   Rearrange and group terms.

   Factor quadratic trinomials into products of linear factors.

9. Many *polynomial equations* can be solved by factoring and using the *zero-product property*. Many *polynomial inequalities* can be solved by factoring and determining the signs of the factors. If the polynomial has three or more factors, it may be helpful to use a *sign graph* to determine the solution.

Chapter Review

Write the letter of the correct answer.

1. Simplify $(x^3 - 3x^2 - 2x + 5) - (x^2 - 2x + 2)$.

    a. $-x^2 + 7$ **b.** $x^3 - 2x^2 - 4x + 3$

    c. $x^3 - 4x^2 + 3$ **d.** $x^3 - 4x^2 - 4x + 3$

2. Simplify $p(p - 2q) + 3(pq - q^2)$.

    a. $p^2 + pq - q^2$ **b.** $p^2 + pq - 3q^2$

    c. $p^2 - 2q + 3pq - q^2$ **d.** $p^2 - 5pq - 3q^2$

3. Simplify $(-x^3)^2 x^4$.

    a. $-x^{10}$ **b.** x^{24} **c.** $-x^9$ **d.** x^{10}

4. Simplify $xy^2(x + y) - x^2y(y - x)$.

    a. $xy^3 - x^3y$ **b.** 0 **c.** $xy^3 + x^3y$ **d.** $2x^2y^2$

5. Express $(3t - 4)(2t + 3)$ as a simplified polynomial.

    a. $6t^2 + t - 12$ **b.** $6t^2 - t + 12$ **c.** $6t^2 + 11t - 12$ **d.** $6t^2 - 12$

6. Express $ax(x - a)(x + a)$ as a simplified polynomial.

    a. $ax^3 - 2a^2x^2 + a^3x$ **b.** $ax^3 + 2a^2x^2 + a^3x$

    c. $ax^3 - a^3x$ **d.** $-a^4x^4$

7. Find the prime factorization of 1350.

    a. $2^3 \cdot 3 \cdot 5^2$ **b.** $2^2 \cdot 3^4 \cdot 5$ **c.** $2 \cdot 3^3 \cdot 5^2$ **d.** $2^2 \cdot 3^2 \cdot 5^2$

8. Find the GCF of $54s^2t^3$, $90s^3t^2$, and $108s^4t$.

    a. $18s^2t$ **b.** $9s^2t$ **c.** $540s^4t^3$ **d.** $36st$

9. Find the LCM of $9a^2b^2c$, $15a^2b^3$, and $6a^3b^2$.

    a. $45abc$ **b.** $90a^2b^2$ **c.** $180a^3b^3c$ **d.** $90a^3b^3c$

10. Factor $4t^2 + 4t + 1$ completely.

    a. $4t(t + 1) + 1$ **b.** $4(t^2 + t + 1)$ **c.** $(2t + 1)^2$ **d.** $(2t - 1)^2$

11. Factor $3x^3 - 27x$ completely.

    a. $3x(x - 3)^2$ **b.** $3x(x + 3)(x - 3)$

    c. $3(x - 3)(x^2 + 3x + 9)$ **d.** $3(x + 3)(x^2 - 3x + 9)$

12. Factor $x^2 - x - a - a^2$ completely.

    a. $(x + a)(x - a - 1)$ **b.** $x(x - 1) - a(1 + a)$

    c. $(x - a)(x + a - 1)$ **d.** $(x - a)(x + a + 1)$

13. Factor $3x^2 - 4x - 4$ completely.

    a. $(3x - 2)(x + 2)$ **b.** $(3x + 2)(x - 2)$

    c. $(3x - 4)(x + 1)$ **d.** $(3x - 1)(x + 4)$

14. Factor $x^4 - 3x^2 - 4$ completely.

    a. $(x^2 + 4)(x + 1)(x - 1)$ **b.** $(x + 1)(x - 1)(x + 2)(x - 2)$

    c. $(x^2 + 1)(x - 2)(x + 2)$ **d.** $(x^2 + 1)(x^2 + 4)$

15. Solve the equation $6x^2 = 5x - 1$.

    a. $\{2, 3\}$ **b.** $\left\{0, \frac{1}{5}\right\}$ **c.** $\left\{\frac{1}{2}, \frac{1}{3}\right\}$ **d.** $\{2, 3\}$

4-1

4-2

4-3

4-4

4-5

4-6

4-7

16. Find the zeros of the function $f(x) = x^3 - 4x$.

    a. 0 and 2   **b.** 2 and -2 **c.** 0, 2, and -2 **d.** 0 and 4

17. A farmer plans to use 25 m of fencing to enclose a rectangular pen that has   4-8
area 63 m². Only three sides of the pen need fencing because part of an
existing wall will be used for one of the longer sides. Find the dimensions
of the pen.

    a. 3.5 m by 18 m   **b.** 4.5 m by 16 m   **c.** 7 m by 9 m   **d.** 6.3 m by 10 m

18. Find the solution set of $x^2 + 2x - 8 > 0$.   4-9

    a. $\{x: -4 < x < 2\}$ **b.** $\{x: x < -4 \text{ or } x > 2\}$

    c. $\{x: -2 < x < 4\}$ **d.** $\{x: x < -2 \text{ or } x > 4\}$

19. Graph the solution set of $x^3 + x^2 < 2x$.

    a.   **b.**

    c.   **d.**

Chapter Test

Simplify.

1. $-3(x^2 - 2) + 2(3 - 2x^2)$ 4-1

2. $(2xy^2 + 2x^3 - x^2y) - (-2x^2y + 2xy^2 - y^3)$

3. $(-2p^2q^3)^2(-p^2q)^3$ **4.** $3xy^2(x^2 - 2xy)$ 4-2

5. $(2x - 3a)(x + 2a)$ **6.** $rs(r + 2s)(2s - r)$ 4-3

Find (a) the GCF and (b) the LCM of the following.

7. 315 and 882   **8.** $12a^2b^2$, $8ab^3$, and $16b^4$ 4-4

Factor each polynomial completely.

9. $x^4 - 8x^2 + 16$ **10.** $xy + 2 - 2x - y$ 4-5

11. $5s^2 - 16s + 12$ **12.** $6a^3 - 7a^2b - 2ab^2$ 4-6

Solve. Identify double roots.

13. $3t^2 - 10t + 3 = 0$ **14.** $x^4 = 8(x^2 - 2)$ 4-7

15. The minute hand of a clock is 3 cm longer than the hour hand. If the   4-8
distance between the tips of the hands at nine o'clock is 15 cm, how
long is the minute hand? (*Hint:* Draw a diagram.)

Find and graph the solution set of each inequality.

16. $x^2 \le x + 12$ **17.** $x^3 > 4x$ 4-9

Preparing for College Entrance Exams

Strategy for Success

Use your time wisely. Solve the easy problems first and skip the more difficult ones. Do not spend too much time on one problem. When you come to the end of the test, return to the problems that were skipped. If time remains, check your work.

Decide which is the best of the choices given and write the corresponding letter on your answer sheet.

1. The lengths of the three sides of a right triangle are consecutive multiples of three. What is the area of the triangle?
   (A) 108   (B) 54   (C) 36   (D) 90   (E) 45

2. A line contains the points $(2, -7)$ and $(-4, 2)$. Which statement is *not* true?
   (A) The line has y-intercept -4.

   (B) The line contains $\left(1, -\dfrac{11}{2}\right)$.

   (C) The line is parallel to the line $3x - 2y = 6$.

   (D) The line is perpendicular to the line $y = \dfrac{2}{3}x + 4$.

   (E) No portion of the graph of the line lies in the first quadrant.

3. The graph of the solution set of the system $2(x + y) = 3(x - 1)$
   $$x - 3y = 3 - y$$
   is best described as:
   (A) a point   (B) no point   (C) a line
   (D) a pair of parallel lines   (E) a pair of intersecting lines

4. The Simon brothers rowed 18 km upstream in 3.75 h. The return trip with the same current took only 2.5 h. What was the speed of the current?
   (A) 1 km/h   (B) 1.2 km/h   (C) 4.5 km/h   (D) 6 km/h   (E) 6.5 km/h

5. Which expression is equivalent to $x^{n+5} \cdot x^{n-5}$?
   (A) x^{2n} (B) x^{10} (C) x^{n^2-25} (D) $(x^2)^{n^2-25}$ (E) x^{n^2}

6. Which polynomial is prime?
   (A) $4x^2 - 17x + 15$ (B) $4x^4 + 13x$ (C) $8x^6 + 27$
   (D) $8x^2 + 7x - 16$ (E) $x^4 - 4$

7. Which inequality is false for every real number?
   (A) $r^2 - 4r \geq 5$ (B) $s^3 < 1$ (C) $t^3 + 9t \leq 0$
   (D) $9u^2 + 49 < 42u$ (E) $(3z - 1)(z - 2) \leq 6$

$E = I$

$E = \frac{1}{4} I$

$E = \frac{1}{9} I$

$d = 1\,\text{ft}$

$d = 2\,\text{ft}$

$d = 3\,\text{ft}$

The illumination (E) on a surface depends directly on the intensity (I) of the light source, but varies according to the square of the distance (d) between the surface and the light source.

Using the Laws of Exponents

5-1 Quotients of Monomials

Objective   To simplify quotients using the laws of exponents.

When you multiply fractions, you use the *multiplication rule for fractions*. For example,

$$\frac{3}{5} \cdot \frac{4}{7} = \frac{3 \cdot 4}{5 \cdot 7} = \frac{12}{35}.$$

Multiplication Rule for Fractions

Let p, q, r, and s be real numbers with $q \neq 0$ and $s \neq 0$. Then

$$\frac{p}{q} \cdot \frac{r}{s} = \frac{pr}{qs}.$$

A proof of this rule is asked for in Exercise 33.

Because equality is symmetric, the rule can be rewritten as $\frac{pr}{qs} = \frac{p}{q} \cdot \frac{r}{s}$. If $r = s$, you can replace s by r, obtaining $\frac{pr}{qr} = \frac{p}{q} \cdot \frac{r}{r} = \frac{p}{q} \cdot 1 = \frac{p}{q}$. This proves the following *rule for simplifying fractions*.

Let p, q, and r be real numbers with $q \neq 0$ and $r \neq 0$. Then

$$\frac{pr}{qr} = \frac{p}{q}.$$

Example 1 Simplify: **a.** $\dfrac{30}{48}$ **b.** $\dfrac{9xy^3}{15x^2y^2}$

Solution   Find the GCF of the numerator and denominator. Then use the rule $\dfrac{pr}{qr} = \dfrac{p}{q}$ to simplify the fractions.

a. $\dfrac{30}{48} = \dfrac{5 \cdot 6}{8 \cdot 6} = \dfrac{5}{8}$
              ↑
            GCF

b. $\dfrac{9xy^3}{15x^2y^2} = \dfrac{3y \cdot 3xy^2}{5x \cdot 3xy^2} = \dfrac{3y}{5x}$
                   ↑
                GCF

In Lesson 4-2, you learned the first three laws of exponents listed below. The additional laws, Laws 4 and 5, cover quotients involving exponents.

The Laws of Exponents

Let m and n be positive integers and a and b be real numbers, with $a \neq 0$ and $b \neq 0$ when they are divisors. Then:

1. $a^m \cdot a^n = a^{m+n}$ **2.** $(ab)^m = a^m b^m$ **3.** $(a^m)^n = a^{mn}$

4a. If $m > n$, $\dfrac{a^m}{a^n} = a^{m-n}$ **4b.** If $n > m$, $\dfrac{a^m}{a^n} = \dfrac{1}{a^{n-m}}$

5. $\left(\dfrac{a}{b}\right)^m = \dfrac{a^m}{b^m}$

Here is a concise proof of Law 4(b). We omit many steps that use only the basic properties of real numbers. For example, no reasons are given for the fact that $n = (n - m) + m$.

Statement	Reason
1. $a^n = a^{(n-m)+m} = a^{n-m} \cdot a^m$	Law 1 of Exponents
2. $\dfrac{a^m}{a^n} = \dfrac{1 \cdot a^m}{a^{n-m} \cdot a^m}$	Substitution
3. $\dfrac{a^m}{a^n} = \dfrac{1}{a^{n-m}}$	Rule for simplifying fractions

Proofs of Laws 4(a) and 5 are left as Exercises 34 and 35.

Notice how Laws 4(a), 4(b), and 5 are used in Example 2.

Example 2 Simplify: **a.** $\dfrac{2^8}{2^5}$ **b.** $\dfrac{5x^3}{x^7}$ **c.** $\left(\dfrac{t^3}{3}\right)^2$

Solution **a.** $\dfrac{2^8}{2^5} = 2^{8-5}$ **b.** $\dfrac{5x^3}{x^7} = \dfrac{5}{x^{7-3}}$ **c.** $\left(\dfrac{t^3}{3}\right)^2 = \dfrac{(t^3)^2}{3^2}$

$\qquad\qquad = 2^3$ $\qquad\qquad = \dfrac{5}{x^4}$ *Answer* $\qquad\qquad = \dfrac{t^6}{9}$ *Answer*

$\qquad\qquad = 8$ *Answer*

A quotient of monomials having integral coefficients is *simplified* when:

1. the integral coefficients are relatively prime, that is, have no common factor except 1 and -1;
2. each base appears only once; and
3. there are no "powers of powers" [such as $(a^3)^4$].

Example 3 Simplify: **a.** $\dfrac{24s^4t^3}{32st^5}$　　　　　**b.** $\dfrac{5x}{4y^2}\left(\dfrac{2y}{x^2}\right)^3$

Solution **a.** *Method 1* Think of each variable separately and use Law 4(a) or 4(b).

$$\frac{24s^4t^3}{32st^5} = \frac{24}{32}\cdot\frac{s^4}{s^1}\cdot\frac{t^3}{t^5} = \frac{3}{4}\cdot\frac{s^3}{1}\cdot\frac{1}{t^2} = \frac{3s^3}{4t^2}$$

Method 2 As in Example 1, find the GCF of the numerator and the denominator. Then simplify the fraction.

$$\frac{24s^4t^3}{32st^5} = \frac{3s^3\cdot 8st^3}{4t^2\cdot 8st^3} = \frac{3s^3}{4t^2}$$

b. $\dfrac{5x}{4y^2}\left(\dfrac{2y}{x^2}\right)^3 = \dfrac{5x}{4y^2}\cdot\dfrac{8y^3}{x^6} = \dfrac{5\cdot 8}{4}\cdot\dfrac{x}{x^6}\cdot\dfrac{y^3}{y^2} = \dfrac{10y}{x^5}$

(This solution used Laws 5 and 3 and then Method 1.)

Oral Exercises

Simplify. Assume that no denominator equals 0 and that variables in exponents represent positive integers.

1. $\dfrac{6^5}{6^3}$　　　　**2.** $\dfrac{3^5}{3^2}$　　　　**3.** $\dfrac{(-2)^7}{(-2)^4}$　　　　**4.** $\dfrac{(-5)^2}{(-5)^4}$

5. $\dfrac{t^8}{t^5}$　　　　**6.** $\dfrac{x^4}{x^7}$　　　　**7.** $\dfrac{y^n}{y^{2n}}$　　　　**8.** $\dfrac{s^{k+2}}{s^k}$

9. $\dfrac{6x^3y^2}{3x^2y}$　　　　**10.** $\dfrac{4ab^4}{12ab^2}$　　　　**11.** $\dfrac{(x^2)^3}{x^5}$　　　　**12.** $\dfrac{(pq)^2}{pq^2}$

Written Exercises

Simplify. Assume that no denominator equals 0.

A **1.** $\dfrac{18x^3}{6x}$　　　　**2.** $\dfrac{5t^3}{15t^5}$　　　　**3.** $\dfrac{-12p^3q}{4p^2q^2}$　　　　**4.** $\dfrac{30x^2y^3}{-6x^3y^2}$

5. $\dfrac{-15u^5v^3}{-25u^4v^2}$　　　　**6.** $\dfrac{48x^5y^5}{32x^4y^6}$　　　　**7.** $\left(\dfrac{3r}{s^2}\right)^3$　　　　**8.** $\left(\dfrac{2x^2}{-y}\right)^4$

9. $\dfrac{3s}{t^2}\cdot\dfrac{s^2}{t}$　　　　**10.** $\dfrac{2u}{v^2}\cdot\dfrac{3u}{2v^2}$　　　　**11.** $\dfrac{3x^2}{y^2}\cdot\dfrac{3y}{6x}$　　　　**12.** $\dfrac{xy^2}{2}\cdot\dfrac{6x}{y^2}$

13. $\dfrac{rs^2t^3}{r^3s^2t}$　　　　**14.** $\dfrac{a^2b^3c}{a^3bc^2}$　　　　**15.** $\dfrac{u^2}{v}\left(\dfrac{3v}{u^2}\right)^2$　　　　**16.** $\dfrac{2x^2}{y^3}\left(\dfrac{-y^3}{2x^2}\right)^2$

17. $\dfrac{(4r^2s^2)^2}{(4r^2s)^2}$　　　　**18.** $\dfrac{(2hk^3)^3}{(-h^2k^2)^2}$　　　　**19.** $\dfrac{(xyz^2)^2}{(x^2yz)^2}$　　　　**20.** $\dfrac{(pq^2r^3)^3}{(p^3qr^2)^2}$

B **21.** $\left(\dfrac{2y^2}{3}\right)^2 \cdot \dfrac{3x}{y^4}$ **22.** $\dfrac{4x^2}{yz^2}\left(\dfrac{z}{2x}\right)^3$ **23.** $\left(\dfrac{c^3}{d^4}\right)^2\left(\dfrac{-cd}{h}\right)^3$ **24.** $\left(\dfrac{-4a^2}{3b}\right)^2\left(\dfrac{-b}{2a}\right)^3$

In Exercises 25–30, assume that no denominator equals 0 and that m and n are integers greater than 1.

25. $\dfrac{a^{2m}b^{2m+1}}{(a^2b^2)^m}$ **26.** $\dfrac{x^{n+1}y^n}{x^n y^{n-1}}$ **27.** $\dfrac{(pq)^n}{pq^n}$

28. $\dfrac{(z^n)^3}{z^n z^3}$ **29.** $\dfrac{t^{n+1}t^{n-1}}{t^n}$ **30.** $\dfrac{a^{n-1}b^{2n}}{a^{n+1}(b^2)^{n-1}}$

31. The multiplication rule for fractions can be rewritten in the form
$(p \div q) \cdot (r \div s) = (p \cdot r) \div (q \cdot s)$. Replace $\div$ by $-$ and $\cdot$ by $+$. Is the resulting statement true?

32. a. Rewrite the rule $\dfrac{pr}{qr} = \dfrac{p}{q}$ using $\cdot$ for multiplication and $\div$ for division.

   b. In your answer to (a) replace $\cdot$ by $+$ and $\div$ by $-$. Is the resulting statement true?

33. Supply reasons for the steps in the following proof of the multiplication rule for fractions.

1. $\dfrac{p}{q} \cdot \dfrac{r}{s} = \left(p \cdot \dfrac{1}{q}\right)\left(r \cdot \dfrac{1}{s}\right)$ _?_

2. $= pr\left(\dfrac{1}{q} \cdot \dfrac{1}{s}\right)$ _?_ and _?_

3. $= pr \cdot \dfrac{1}{qs}$ _?_

4. $= \dfrac{pr}{qs}$ _?_

$\therefore \dfrac{p}{q} \cdot \dfrac{r}{s} = \dfrac{pr}{qs}$

34. Supply reasons for the steps in the following proof of Exponent Law 5.

1. $\left(\dfrac{a}{b}\right)^m = \underbrace{\dfrac{a}{b} \cdot \dfrac{a}{b} \cdot \ldots \cdot \dfrac{a}{b}}_{m \text{ factors}}$ _?_

2. $= \dfrac{\overbrace{a \cdot a \cdot \ldots \cdot a}}{\underbrace{b \cdot b \cdot \ldots \cdot b}_{m \text{ factors}}}$ _?_

$\therefore \left(\dfrac{a}{b}\right)^m = \dfrac{a^m}{b^m}$

C **35.** Give a concise proof of Exponent Law 4a.

36. In each of the laws of exponents, replace $\cdot$ by $+$, $\div$ by $-$, and, for example, a^m by ma. Is the resulting statement true? (*Example:* $a^m \cdot a^n = a^{m+n}$ becomes $ma + na = (m + n)a$, which is true.)

Mixed Review Exercises

Find the solution set of each inequality.

1. $x^2 + 2x - 8 > 0$

2. $4y + 9 \leq 1$

3. $w^2 < 4$

4. $5 - 3p > -7$

5. $|2t + 5| \leq 3$

6. $2a^2 < a + 3$

7. $-3 < 2c - 1 < 5$

8. $m(m - 1) \geq 2$

9. $|d + 2| > 1$

10. $g \leq g^2$

11. $n^2 + 2 \leq 3n$

12. $-2u < -6$ or $u - 1 \leq 1$

Computer Key-In

If you divide a by b, you obtain a quotient q and a remainder r less than b.
For example, $192 \div 84$ gives

$$192 = 84 \cdot 2 + 24$$
$$a = b \cdot q + r.$$

The *Euclidean Algorithm* uses repeated division to find the GCF of two positive integers a and b. To find the GCF of 192 and 84 by the Euclidean Algorithm you would proceed as follows.

$$192 = 84 \cdot 2 + 24$$
$$84 = 24 \cdot 3 + 12$$
$$24 = 12 \cdot 2 + 0$$

The last nonzero remainder, 12, is the GCF of 192 and 84.

Here is a program that uses the Euclidean Algorithm to find GCF's.

```
10  PRINT "FIND THE GCF OF TWO POSITIVE INTEGERS."
20  INPUT "ENTER THE NUMBERS (LARGER FIRST): "; A, B
30  PRINT "THE GCF OF "; A; " AND "; B; " IS ";
40  IF INT (A/B)=A/B THEN 90
50  LET R=A−B*INT (A/B)
60  LET A=B
70  LET B=R
80  GOTO 40
90  PRINT B
95  END
```

Exercises

1. Run the program to find the GCF of each pair of numbers.

a. (432, 528)   **b.** (2592, 4682)   **c.** (999, 259)   **d.** (30,030, 13,122)

2. Run the program to find the GCF of the numerator and the denominator of each fraction. Write each fraction in simplest form.

a. $\dfrac{87}{1653}$ **b.** $\dfrac{11,776}{15,872}$ **c.** $\dfrac{2772}{5940}$ **d.** $\dfrac{2048}{40,387}$

5-2 Zero and Negative Exponents

Objective   To simplify expressions involving the exponent zero and negative integral exponents.

The definition of *power* can be extended so that any integer may be used as an exponent, not just a positive integer. The power a^n when n is 0 or n is a negative integer is defined in such a way that the laws of exponents stated on page 212 continue to hold.

If Law 1, $a^m \cdot a^n = a^{m+n}$, is to hold when $n = 0$, then this statement must be true:

$$a^m \cdot a^0 = a^{m+0} = a^m$$

If $a \neq 0$, you can divide both sides of the resulting equation $a^m \cdot a^0 = a^m$ by a^m to obtain

$$a^0 = \frac{a^m}{a^m} = 1.$$

If Law 1 is to hold for negative exponents, then this statement must be true: If $-n$ is a negative integer and $a \neq 0$,

$$a^{-n} \cdot a^n = a^{-n+n} = a^0 = 1.$$

Since $a^{-n} \cdot a^n = 1$, a^{-n} must be the reciprocal of a^n:

$$a^{-n} = \frac{1}{a^n}$$

The discussion above leads to these definitions.

If n is a positive integer and $a \neq 0$:

$$a^0 = 1 \qquad\qquad a^{-n} = \frac{1}{a^n}$$

The expression 0^0 is not defined.

Example 1   Write in simplest form without negative or zero exponents. Assume that the variables are nonzero.

   a. 10^{-2} **b.** $-3c^{-2}$ **c.** $2^{-1}a^0b^{-3}$ **d.** $(3x)^{-2}$

Solution   **a.** $10^{-2} = \dfrac{1}{10^2} = \dfrac{1}{100}$ **b.** $-3c^{-2} = -3 \cdot \dfrac{1}{c^2} = -\dfrac{3}{c^2}$

   c. $2^{-1}a^0b^{-3} = \dfrac{1}{2} \cdot 1 \cdot \dfrac{1}{b^3}$ **d.** $(3x)^{-2} = \dfrac{1}{(3x)^2}$

   $\qquad\qquad\quad = \dfrac{1}{2b^3}$ $\qquad\qquad\quad = \dfrac{1}{9x^2}$

Since $a^{-n} = \dfrac{1}{a^n}$ and division by zero is not defined, bases of powers with negative exponents must not be zero. To simplify matters we make the following agreement, to hold throughout the rest of this book.

The domains of all variables in any algebraic expression are automatically restricted so that denominators of fractions and bases of powers with negative or zero exponents will not be zero.

For example, in the expression

$$\frac{3a^0b^{-2}}{x^2 - 4},$$

you may assume that $a \neq 0$, $b \neq 0$, $x \neq 2$, and $x \neq -2$ even though these restrictions are not stated.

It can be shown that all the laws of exponents hold even if some of the exponents are negative or zero. For example, Law 3 now implies that

$$(a^3)^{-2} = a^{3(-2)} = a^{-6}.$$

This statement can be justified with the help of Law 3 for *positive* exponents:

$$(a^3)^{-2} = \frac{1}{(a^3)^2} = \frac{1}{a^6} = a^{-6}$$

You can now see that Laws 4(a) and 4(b) give the same results. For example,

by Law 4(a): $\quad \dfrac{a^2}{a^7} = a^{2-7} = a^{-5} = \dfrac{1}{a^5}$

by Law 4(b): $\quad \dfrac{a^2}{a^7} = \dfrac{1}{a^{7-2}} = \dfrac{1}{a^5}.$

Example 2 Write in simplest form without negative or zero exponents.

a. $10^{-2} \times 10^{-3}$ **b.** $(-3 \cdot 5^{-1})^{-2}$ **c.** $\dfrac{3^0 x^{-3} y}{2x^{-1} y^{-2}}$

Solution **a.** $10^{-2} \times 10^{-3} = 10^{-2+(-3)} = 10^{-5} = \dfrac{1}{10^5} = \dfrac{1}{100,000}$ *Answer*

b. $(-3 \cdot 5^{-1})^{-2} = (-3)^{-2}(5^{-1})^{-2}$

$\qquad = \dfrac{1}{(-3)^2} \cdot 5^2$

$\qquad = \dfrac{25}{9}$ *Answer*

c. $\dfrac{3^0 x^{-3} y}{2x^{-1} y^{-2}} = \dfrac{1 \cdot x^{-3-(-1)} y^{1-(-2)}}{2}$

$\qquad = \dfrac{1x^{-2} y^3}{2}$

$\qquad = \dfrac{y^3}{2x^2}$ *Answer*

Since a^{-n} and a^n are reciprocals, $\frac{1}{a^{-n}} = a^n$. Example 3 uses this fact.

Example 3 Write in simplest form without negative or zero exponents.

    a. $\left(\frac{2}{3}\right)^{-3}$ **b.** $\left(\frac{2x^{-2}}{5y^3}\right)^{-1}$

Solution **a.** $\left(\frac{2}{3}\right)^{-3} = \frac{2^{-3}}{3^{-3}} = 2^{-3} \cdot \frac{1}{3^{-3}} = \frac{1}{2^3} \cdot 3^3 = \frac{3^3}{2^3} = \frac{27}{8}$ *Answer*

                b. $\left(\frac{2x^{-2}}{5y^3}\right)^{-1} = \frac{(2x^{-2})^{-1}}{(5y^3)^{-1}} = \frac{2^{-1}x^2}{5^{-1}y^{-3}} = \frac{5x^2y^3}{2}$ *Answer*

Negative exponents are introduced in order to write expressions without using fractions. The form of the answer to Example 4(a) below is called *scientific notation*. This notation will be discussed in Lesson 5-3.

Example 4 Write without using fractions: **a.** $\frac{3}{10,000}$ **b.** $\frac{5x^2}{yz^3}$

Solution **a.** $\frac{3}{10,000} = \frac{3}{10^4} = 3 \times 10^{-4}$ **b.** $\frac{5x^2}{yz^3} = 5x^2y^{-1}z^{-3}$

Oral Exercises

Express in simplest form without negative or zero exponents.

1. $\frac{1}{2^{-3}}$ 2. $\frac{5}{3^{-1}}$ 3. $10^7 \cdot 10^{-5}$ 4. $(3^{-1})^{-2}$

5. $c^{-3} \cdot c^5$ 6. $a^{-4} \cdot a^3$ 7. $(y^{-1})^{-2}$ 8. $y^{-1} \cdot y^{-2}$

9. $(5x)^0$ 10. $5x^0$ 11. p^2q^{-2} 12. $(p^2q)^{-2}$

13. $\left(\frac{a}{b}\right)^{-2}$ 14. $\frac{a}{2^{-1}b^{-2}}$ 15. $\frac{x^{-3}}{y^{-3}}$ 16. $\left(\frac{r^{-1}}{s^{-1}}\right)^{-1}$

Written Exercises

Write in simplest form without negative or zero exponents.

A 1. $3 \cdot 5^{-1}$ 2. $(3 \cdot 5)^{-1}$ 3. $(-3^{-1})^{-2}$ 4. $(-2^{-2})^{-1}$

     5. $(2^{-2} \cdot 3^{-1} \cdot 5^0)^{-1}$ 6. $5^{-1}(3^{-2} \cdot 2^{-3})^0$ 7. $2\left(\frac{2}{5}\right)^{-2}$ 8. $\left(\frac{3}{4}\right)^{-1}\left(\frac{4}{3}\right)^{-2}$

Write without using fractions.

9. $\dfrac{7}{10,000}$ **10.** $\dfrac{3}{1000}$ **11.** $\dfrac{6x^2}{y^3}$ **12.** $\dfrac{x^2}{yz^4}$

Express as a decimal numeral. A calculator may be helpful.

Sample 1 $(-5 \cdot 3^{-2})^{-2} = (-5)^{-2} \cdot 3^4 = \dfrac{81}{25} = 3.24$

13. 596×10^{-2} **14.** 238×10^{-3} **15.** 7.2×10^{-3} **16.** 1.45×10^{-2}

17. $\left(-\dfrac{5}{3}\right)^{-3}$ **18.** $\left(-\dfrac{2}{5}\right)^{-2}$ **19.** $(3^{-1} \cdot 2^2)^{-2}$ **20.** $(-5^{-1} \cdot 2^{-2})^2$

Write in simplest form without negative or zero exponents.

21. $\dfrac{3x^{-2}}{y^{-1}}$ **22.** $\dfrac{p^{-1}q^{-2}}{p^{-3}}$ **23.** $\dfrac{s^{-2}t^{-3}}{s^{-1}t^0}$

24. $\dfrac{6xy^{-1}}{-2x^{-2}y^{-1}}$ **25.** $\left(\dfrac{u^{-2}}{v}\right)^{-1}$ **26.** $\left(\dfrac{2}{h^2k^{-3}}\right)^{-2}$

27. $(2x^{-2}y^2)^{-2}$ **28.** $\dfrac{(3x^{-2}y)^{-1}}{(2xy^{-2})^0}$ **29.** $3x^2(3xy^{-1})^{-2}$

30. $5t(s^{-1}t^{-2})^{-2}$ **31.** $\dfrac{(2x^{-1})^{-2}}{2(y^{-1})^{-2}}$ **32.** $\left(\dfrac{2pq^{-1}}{4q^2}\right)^{-1}$

33. $\left(\dfrac{x}{y^2}\right)^{-1}\left(\dfrac{x^{-2}}{y}\right)^2$ **34.** $\left(\dfrac{3}{t^2}\right)^{-1}\left(\dfrac{t}{3}\right)^{-2}$ **35.** $\left(\dfrac{p^{-2}q^{-1}}{p^{-1}q^{-2}}\right)^{-1}$

36. $\dfrac{(ax^2)^{-1}}{a^{-2}x^{-2}}$ **37.** $\left(\dfrac{x^2}{y^{-1}}\right)^{-2}\left(\dfrac{y^2}{x^{-1}}\right)^2$ **38.** $\dfrac{r^{-2}}{s^2}\left(\dfrac{1}{rs}\right)^{-2}$

39. $\left(\dfrac{u}{v^{-1}}\right)^0\left(\dfrac{u^{-1}}{v^2}\right)^2(uv^2)^{-1}$ **40.** $\left(\dfrac{a^0}{b}\right)^{-2}\left(\dfrac{a}{b^{-2}}\right)^{-2}$

41. $4x^3y^{-6} + (x^{-1}y^2)^{-3}$ **42.** $\left(\dfrac{u^2}{v}\right)^2 + (-u^{-2}v)^{-2}$

Show by counterexample that the given expressions are *not* equivalent.

Sample 2 $(x + y)^{-2}$ and $x^{-2} + y^{-2}$

Solution Let $x = 1$ and $y = 2$. Then:

$$(x + y)^{-2} = (1 + 2)^{-2} \qquad x^{-2} + y^{-2} = 1^{-2} + 2^{-2}$$

$$= 3^{-2} = \dfrac{1}{9} \qquad\qquad = 1 + \dfrac{1}{4} = \dfrac{5}{4} \qquad \left(\dfrac{1}{9} \neq \dfrac{5}{4}\right)$$

B **43.** $(x + y)^{-1}$ and $x^{-1} + y^{-1}$ **44.** $(xy)^{-1}$ and $\dfrac{x}{y}$

45. xy^{-1} and $\dfrac{1}{xy}$ **46.** $(1 - x)^{-2}$ and $1 - x^{-2}$

In Exercises 47–52, replace the (__?__) by a polynomial to make a true statement.

Sample 3 $x^{-3} + 2x^{-2} - 3x^{-1} = x^{-3}(\underline{\quad?\quad})$

Solution $x^{-3} + 2x^{-2} - 3x^{-1} = x^{-3}(1 + 2x - 3x^2)$

47. $x^{-1} - 4x^{-2} + 2x^{-3} = x^{-3}(\underline{\quad?\quad})$ 48. $2x^{-2} + x^{-1} - 3 = x^{-2}(\underline{\quad?\quad})$

49. $4 - 5x^{-1} + x^{-2} = x^{-2}(\underline{\quad?\quad})$ 50. $x^{-1} - 9x^{-3} = x^{-3}(\underline{\quad?\quad})$

C 51. $x^2(x-1)^{-2} - 4(x-1)^{-1} = (x-1)^{-2}(\underline{\quad?\quad})$

52. $(x^2+4)^{-1} - 5(x^2+4)^{-2} = (x^2+4)^{-2}(\underline{\quad?\quad})$

In Exercises 53–58, use the exponent laws for positive exponents to prove the forms of the laws given below. In each case, m and n are positive integers with $m > n$.

53. $a^m a^{-n} = a^{m+(-n)}$ 54. $(a^m)^{-n} = a^{m(-n)}$ 55. $a^{-m}a^n = a^{(-m)+n}$

56. $\dfrac{a^{-m}}{a^n} = a^{(-m)-n}$ 57. $\dfrac{a^m}{a^{-n}} = a^{m-(-n)}$ 58. $\dfrac{a^{-m}}{a^{-n}} = a^{(-m)-(-n)}$

Biographical Note / Ch'in Chiu-Shao

Ch'in Chiu-Shao (ca. 1202–1261) has been called one of the greatest mathematicians of his time. This achievement is particularly remarkable because Ch'in did not devote his life to mathematics. He was accomplished in many other fields and held a series of bureaucratic positions in Chinese provinces.

Ch'in's mathematical reputation rests on one celebrated treatise, *Shu-shu chiu-chang* (''Mathematical Treatise in Nine Sections''), which appeared in 1247. The treatise covers topics ranging from indeterminate analysis to military matters and surveying. Ch'in included a version of the Chinese remainder theorem, which used algorithms to solve problems.

His interest in indeterminate analysis led Ch'in into related fields. He wrote down the earliest explanation of how Chinese calendar experts calculated astronomical data according to the timing of the winter solstice. He also introduced techniques for

solving equations, finding sums of arithmetic series, and solving linear systems. His use of the zero symbol is a milestone in Chinese mathematics.

5-3 Scientific Notation and Significant Digits

Objective   To use scientific notation and significant digits.

In scientific work, you meet very large and very small numbers. For example, the distance that light travels in a year is about 5,680,000,000,000 mi. The time that it takes a signal to travel from one component of a supercomputer to another might be 0.0000000024 s. The wavelength of ultraviolet light is about 0.00000136 cm. To make such numbers easier to work with, you can write them in *scientific notation*:

$$5,680,000,000,000 = 5.68 \times 10^{12}$$
$$0.0000000024 = 2.4 \times 10^{-9}$$
$$0.00000136 = 1.36 \times 10^{-6}$$

In **scientific notation,** a number is expressed in the form

$$m \times 10^n$$

where $1 \le m < 10$, and n is an integer.

The digits in the factor m should all be *significant*. A **significant digit** of a number written in decimal form is any nonzero digit or any zero that has a purpose other than placing the decimal point. In the following examples the significant digits are printed in red.

4006	0.4050	320.0	0.00203

In scientific notation

$$4006 = 4.006 \times 10^3, \qquad 0.4050 = 4.050 \times 10^{-1},$$
$$320.0 = 3.200 \times 10^2, \qquad 0.00203 = 2.03 \times 10^{-3}.$$

For a number such as 2300 it is not clear which, if any, of the zeros are significant. Scientific notation eliminates this problem. Writing 2300 as

$$2.3 \times 10^3, \qquad 2.30 \times 10^3, \qquad \text{or } 2.300 \times 10^3$$

indicates, respectively, that none, one, or two of the zeros are significant.

Most measurements are approximate, and the more significant digits that are given in such an approximation, the more *accurate* it is. For example, if a length is given as 2.30×10^3 cm, the measurement has three significant digits. Since the 0 is significant, you know that the length is *not* 2.31×10^3 cm or 2.29×10^3 cm, but it might be 2.302×10^3 cm.

When you multiply or divide measurements, their accuracy limits the accuracy of the answer. *When approximations are multiplied or divided, the product or quotient should have the same number of digits as the least accurate factor.*

Example 1 Write each number in scientific notation. If the number is an integer and ends in zeros, assume that the zeros are not significant.

    a. 75,040   **b.** 0.0000702   **c.** 3,050,000

Solution   **a.** $75,040 = 7.504 \times 10^4$ decimal point moved 4 places to the left
               b. $0.0000702 = 7.02 \times 10^{-5}$ decimal point moved 5 places to the right
               c. $3,050,000 = 3.05 \times 10^6$ decimal point moved 6 places to the left

---

Numbers expressed in scientific notation are easy to compare.

Example 2   **a.** $6.84 \times 10^{-2} > 4.96 \times 10^{-2}$ because $6.84 > 4.96$.
               b. $8.7 \times 10^4 < 5.3 \times 10^5$ because $4 < 5$.
               c. $5.25 \times 10^{-2} > 7.50 \times 10^{-3}$ because $-2 > -3$.

Example 3 Find a one-significant-digit estimate of x, where

$$x = \frac{5260 \times 0.0682 \times 86.1}{0.420}.$$

Solution $x = \dfrac{5.26 \times 10^3 \times 6.82 \times 10^{-2} \times 8.61 \times 10^1}{4.2 \times 10^{-1}}$ Write each number in scientific notation.

            $\approx \dfrac{5 \times 10^3 \times 7 \times 10^{-2} \times 9 \times 10^1}{4 \times 10^{-1}}$ Round each decimal to a whole number.

            $= \dfrac{5 \times 7 \times 9}{4} \times 10^{3+(-2)+1-(-1)}$ Compute, and give the result to one significant digit.

            $= \dfrac{315}{4} \times 10^3$

            $\approx 80 \times 10^3 = 8 \times 10^4$

            $\therefore x \approx 8 \times 10^4$, or 80,000

Round off errors may cause the digit obtained in the process shown in Example 3 to be incorrect. (In the example, $x = 73,500$ to three significant digits.) Nevertheless, such estimates do describe the "size" of the answer.

Before using a calculator in the exercises, see the Calculator Key-In on page 225 to determine how your calculator handles numbers that are too large or small or too long for the display.

Tell how many significant digits each number has.

1. 40.05   **2.** 10,608   **3.** 0.050   **4.** 0.0250

5. 4.0×10^3 **6.** 7.06×10^{-2} **7.** 3.600×10^2 **8.** 4.8×10^6

Replace the __?__ with $<$ or $>$ to make a true statement.

9. 4.60×10^3 __?__ 3.75×10^3 **10.** 2.60×10^3 __?__ 6.75×10^4

11. 7.73×10^{-3} __?__ 2.50×10^{-2} **12.** 2.78×10^{-4} __?__ 2.78×10^{-6}

Written Exercises   # 3 - 27   ODD

Write each number in scientific notation. If the number is an integer and ends in zeros, assume that the zeros are not significant.

A

1. 7500   **2.** 106,000   **3.** 0.608   **4.** 0.0038

5. 10.05   **6.** 762.20   **7.** 0.0320   **8.** 0.0000460

9. 655×10^3 **10.** 0.025×10^{-2} **11.** 0.560×10^{-3} **12.** 4775×10^{-2}

Write each number in decimal form.

13. 5×10^3 **14.** 10^{-4} **15.** 4.3×10^{-3} **16.** 10^7

17. 6.75×10^4 **18.** 6.20×10^{-2} **19.** 7.50×10^{-3} **20.** 4.000×10^2

Replace the __?__ with $<$ or $>$ to make a true statement.

21. $(2 \times 10^3)^2$ __?__ 5×10^5 **22.** 1.6×10^{18} __?__ $(1.2 \times 10^9)^2$

23. $(0.005)^3$ __?__ 2×10^{-8} **24.** $(2 \times 10^3)^{-2}$ __?__ $(2 \times 10^{-3})^2$

Find a one-significant-digit estimate of each of the following.

B

25. $\dfrac{478 \times 0.230}{0.0281 \times 32.4}$ **26.** $\dfrac{0.525 \times 7820}{22.5 \times 0.00475}$

27. $\dfrac{73.1 \times (0.493)^2}{0.620 \times (32.6)^2}$ **28.** $\dfrac{0.000212 \times (588)^2}{57.7 \times (0.0620)^2}$

Simplify, assuming that the factors are approximations. Give answers in scientific notation with the same number of significant digits as in the least accurate factor.

29. $\dfrac{(8 \times 10^6)(2 \times 10^{-2})}{4 \times 10^2}$ **30.** $\dfrac{(7.5 \times 10^6)(5.0 \times 10^{-1})}{1.5 \times 10^8}$ **31.** $\dfrac{(8.4 \times 10^{15})(1.5 \times 10^{-5})}{(4.0 \times 10^4)(1.2 \times 10^3)}$

Problems

In these problems, all given data are approximate. In Problems 1–5, give answers reflecting the situation in 1987 using the following table. A calculator may be helpful.

	1987 Population	Land Area (km^2)
U.S.A.	2.44×10^8	9.36×10^6
China	1.06×10^9	9.60×10^6
Italy	5.74×10^7	3.01×10^5
World	5.03×10^9	1.49×10^8

A

1. Write the table above using decimal notation.

2. What percent of the world's land area has (a) the United States, (b) China, and (c) Italy?

3. What percent of the world's population has (a) the United States, (b) China, and (c) Italy?

4. Find the population density (persons per km^2) of (a) the United States, (b) China, (c) Italy, and (d) the world.

5. In 1987 the national debt of the United States was about 2.50×10^{12} dollars. How much debt is this per person?

6. How many nanoseconds (1 nanosecond = 10^{-9} s) does it take a computer signal to travel 60 cm at a rate of 2.4×10^{10} cm/s?

7. The designer of a minicomputer wants to limit to 4.0 nanoseconds (see Problem 6) the time that it takes signals to travel from one component to another. What restriction does this put on the distance between components?

8. The estimated masses of a proton and an electron are 1.67×10^{-24} g and 9.11×10^{-28} g, respectively. Find the ratio of the mass of the proton to the mass of the electron.

B

9. The astronomical unit (AU), the light year, and the parsec are units of distance used in astronomy. Copy and complete the following table showing how these units are related.

	AU	parsec	light year
One AU equals	1	?	1.58×10^{-5}
One parsec equals	?	1	?
One light year equals	?	0.307	1

10. Given that 1 light year = 9.46×10^{12} km, express 1 AU and 1 parsec in kilometers. (See Problem 9.)

11. The mean distances from the sun to the planets Mercury and Neptune are 5.79×10^7 km and 4.51×10^9 km, respectively. Express these distances in astronomical units. (See Problems 9 and 10.)

12. A steady current of 1 ampere flowing through a solution of silver nitrate will deposit 1.12×10^{-3} g of silver in one second. How much silver will be deposited by a current of 12.0 amperes in one hour?

13. An electric current of one ampere corresponds to a flow of 6.2×10^{18} electrons per second past any point of the circuit. How many electrons flow through the filament of a 100 watt, 115 volt light bulb during one hour? (*Note:* Watts = amperes × volts)

14. If you were to receive as wages 1 cent for the first day, 2 cents for the second day, 4 cents for the third day, and so on, with each day's wages twice the previous day's, about how many dollars would you receive on the 31st day? Use the approximation $2^{10} \approx 10^3$ and express your answer in decimal form.

C 15. The masses of the sun and Earth are 2.0×10^{30} kg and 6.0×10^{24} kg, respectively, and their radii are 7.0×10^8 m and 6.4×10^6 m. Show that the average density of Earth is about four times that of the sun. (*Note:* Density = mass ÷ volume)

Mixed Review Exercises

Simplify.

1. $\dfrac{18x^3y}{27x^2y^4}$

2. $\left(\dfrac{3a}{2b}\right)^2$

3. $\dfrac{4m}{n} \cdot \dfrac{m^2}{5n}$

4. $\dfrac{(c^2d)^3}{(cd^2)^3}$

5. $\dfrac{2p^2}{q} \cdot \dfrac{q^3}{2p^3}$

6. $\dfrac{(-w)^2}{-w^2}$

7. $\dfrac{14z^5}{10z^5}$

8. $\dfrac{u^4}{v}\left(\dfrac{v^2}{u^3}\right)^2$

Express in simplest form without negative or zero exponents.

9. $(x^{-3}y^2)^{-1}$

10. $\left(\dfrac{c^{-1}}{d}\right)^{-2}$

11. $(6m^2)(2m)^{-2}$

12. $\dfrac{a^{-4}b}{a^2b^{-3}}$

Calculator Key-In

Some calculators will indicate an error when the result of a computation has too many digits to be shown in full on the display. Others will display such a number in scientific notation, with the exponent of 10 shown at the right. Most scientific calculators have a key that allows the user to select scientific notation for all numbers displayed and to enter numbers in scientific notation.

When the result of a computation has more significant digits than the display can show, the number may be *truncated* (the extra digits are discarded) or *rounded* (the last place shown is rounded to the correct value).

You can find answers to some of the exercises below without a calculator, but doing them on your calculator will help you find out how your calculator handles numbers that are too large or small or too long for the display.

Evaluate on your calculator.

1. $20,000 \times 50,000$ **2.** $352,129 \times 416,187$ **3.** $(0.00001)^2$

Find the value of each expression to four significant digits. Give your answers in scientific notation.

4. $(4.325 \times 10^5)(9.817 \times 10^{12})$ **5.** $(1.589 \times 10^3)(6.805 \times 10^{-8})$

6. $2.563 \div 493,412$ **7.** $\dfrac{(4.303 \times 10^6)(2.115 \times 10^3)}{9.563 \times 10^8}$

8. a. Find $\frac{2}{3}$ as a decimal on your calculator.

    b. From part (a), tell whether your calculator truncates or rounds.

Self-Test 1

Vocabulary   scientific notation (p. 221)   significant digit (p. 221)

Simplify.

1. $\dfrac{24s^3t^2}{16st^3}$ **2.** $\left(\dfrac{2x^2y}{w^2}\right)^2\left(\dfrac{y^2w}{x^2}\right)^3$ Obj. 5-1, p. 211

Write in simplest form without negative or zero exponents.

3. $\dfrac{p^{-2}qr^{-1}}{p^0q^{-2}r^{-3}}$ **4.** $\left(\dfrac{x^2y^{-1}}{z}\right)^{-2}\left(\dfrac{x^2y^2}{z^{-3}}\right)$ Obj. 5-2, p. 216

Replace the __?__ by a polynomial to make a true statement.

5. $x^{-1} - 4x^{-3} = x^{-3}\,(\underline{\ ?\ })$

Write each number in scientific notation.

6. 482.60 **7.** 0.000210 Obj. 5-3, p. 221

Write each number in decimal form.

8. 3.609×10^3 **9.** 5.400×10^{-2}

10. Give a one-significant-digit estimate of $\dfrac{683 \times 0.536}{0.00392}$.

Rational Expressions

5-4 Rational Algebraic Expressions

Objective   To simplify rational algebraic expressions.

You know that a *rational number* is one that can be expressed as a quotient of integers. Similarly, a **rational algebraic expression,** or **rational expression,** is one that can be expressed as a quotient of polynomials. Some examples of rational expressions are

$$\frac{4xy^2}{7y}, \qquad \frac{x^2 - 3x - 4}{x^2 - 1}, \qquad \text{and} \qquad x(x^2 - 4)^{-1} = \frac{x}{x^2 - 4}.$$

A rational expression is *simplified*, or in *simplest form*, when it is expressed as a quotient of polynomials whose greatest common factor is 1. (Recall Lesson 4-6.) To simplify a rational expression you factor the numerator and denominator and then look for common factors.

Example 1   Simplify $\dfrac{x^2 - 2x}{x^2 - 4}$.

Solution $\dfrac{x^2 - 2x}{x^2 - 4} = \dfrac{x(x - 2)}{(x + 2)(x - 2)}$ Factor the numerator
and the denominator.

$\qquad\qquad = \dfrac{x}{x + 2}$ Simplify.

Answer

Factors of the numerator and denominator may be opposites of each other, as in Example 2.

Example 2   Simplify $(3x - 5x^2 - 2x^3)(6x^2 - 5x + 1)^{-1}$.

Solution

$(3x - 5x^2 - 2x^3)(6x^2 - 5x + 1)^{-1} = \dfrac{3x - 5x^2 - 2x^3}{6x^2 - 5x + 1}$ Write as
a fraction.

$\qquad\qquad = \dfrac{x(3 + x)(1 - 2x)}{(2x - 1)(3x - 1)}$ Factor.

$\qquad\qquad = \dfrac{x(3 + x)(-1)(2x - 1)}{(2x - 1)(3x - 1)}$ Replace $1 - 2x$
by $(-1)(2x - 1)$.

$\qquad\qquad = \dfrac{-x(3 + x)}{3x - 1}, \text{ or } -\dfrac{x(x + 3)}{3x - 1}$

Answer

A function that is defined by a simplified rational expression in one variable is called a **rational function.**

Example 3 Let $f(x) = \dfrac{2x^2 - 7x + 3}{x^3 + x^2 - 2x}$.

a. Find the domain of f. b. Find the zeros of f, if any.

Solution
$$f(x) = \frac{2x^2 - 7x + 3}{x^3 + x^2 - 2x} = \frac{(2x - 1)(x - 3)}{x(x - 1)(x + 2)}$$

The zero-product property is used for both (a) and (b).

a. According to the agreement made on page 217, the domain of f consists of all real numbers except 0, 1, and -2.

b. $f(x) = 0$ if and only if $(2x - 1)(x - 3) = 0$. Therefore, the zeros of f are $\dfrac{1}{2}$ and 3.

Oral Exercises

Simplify. # 1-26 every 3rd problem

1. $\dfrac{5x^2 y^2}{xy^3}$

2. $\dfrac{5x^2 y^3}{3xy^2}$

3. $\dfrac{x(x - 2)}{x(x + 2)}$

4. $\dfrac{(x - 1)(x + 2)}{2 + x}$

5. $\dfrac{x^2 - x}{x - 1}$

6. $\dfrac{2 - x}{x - 2}$

Find (a) the domain of each rational function f and (b) the zeros of f, if any.

7. $f(x) = \dfrac{x - 2}{x}$

8. $f(x) = \dfrac{x}{x - 2}$

9. $f(x) = \dfrac{x + 1}{x - 1}$

10. $f(x) = \dfrac{x^2 + 1}{x + 1}$

11. $f(x) = \dfrac{x + 3}{x^2 + 9}$

12. $f(x) = \dfrac{(x - 1)(x - 3)}{x - 2}$

13. $f(x) = \dfrac{x^2 + 1}{x^2 - 1}$

14. $f(x) = \dfrac{x^2 - 4}{x^2 + 4}$

15. $f(x) = \dfrac{x - 4}{x^2 - 4}$

Written Exercises

Simplify.

A 1. $\dfrac{5x^2 - 15x}{10x^2}$

2. $\dfrac{3t^4 - 9t^3}{6t^2}$

3. $\dfrac{u^2 - u - 2}{u^2 + u}$

4. $\dfrac{z^3 - 4z}{z^2 - 4z + 4}$

5. $(p - q)(q - p)^{-1}$

6. $(r^2 - rs)(r^2 - s^2)^{-1}$

7. $\dfrac{s^2 - t^2}{(t - s)^2}$

8. $\dfrac{(a - x)^2}{x^2 - a^2}$

9. $\dfrac{x^2 - 5x + 6}{x^2 - 7x + 12}$

10. $\dfrac{2t^2 + 5t - 3}{2t^2 + 7t + 3}$

11. $\dfrac{6y^2 - 5y + 1}{1 - y - 6y^2}$

12. $\dfrac{9 - 4z^2}{6z^2 - 5z - 6}$

13. $(r^2 - 5r + 4)(r - 4)^{-2}$

14. $(p^2 + 4p - 5)(p - 1)^{-2}$

15. $\dfrac{x^2 + 2x - 8}{(2 - x)(4 + x)}$

16. $\dfrac{(t + 1)(t^2 - 1)}{(t - 1)(t + 1)^2}$

17. $(z^4 - 1)(z^4 - z^2)^{-1}$

18. $(1 - r^3)(1 - r)^{-3}$

19. $\dfrac{(x^2 - a^2)^2}{(x - a)^2}$

20. $\dfrac{t^4 - c^4}{(t + c)^2(t^2 + c^2)}$

Find (a) the domain of each function and (b) its zeros, if any.

21. $f(t) = \dfrac{t^2 - 9}{t^2 - 9t}$

22. $g(x) = \dfrac{x^3 + 2x}{x^2 - 4}$

23. $F(x) = (x^4 - 16)(x^3 - 1)^{-1}$

24. $h(y) = (y^3 - 8)(y + 2)^{-3}$

25. $g(t) = \dfrac{2t^2 + 3t - 9}{t^3 - 4t}$

26. $G(s) = \dfrac{4s^2 + 15s - 4}{(2s - 1)^2}$

B **27.** $f(x) = \dfrac{x^3 - 2x^2 + x - 2}{x^4 + x^2 - 2}$

28. $h(t) = \dfrac{t^3 + 4t^2 - t - 4}{t^3 - t^2 + t - 1}$

Simplify.

29. $\dfrac{x^3 + x^2 - x - 1}{x^3 - x^2 - x + 1}$

30. $\dfrac{t^4 - 1}{t^3 + t^2 + t + 1}$

31. $\dfrac{x^3 - x^2y + xy^2 - y^3}{x^4 - y^4}$

32. $\dfrac{x^4 - 2x^2y^2 + y^4}{x^4 - x^3y - xy^3 + y^4}$

33. $\dfrac{s^4 - t^4}{s^4 - 2s^2t^2 + t^4}$

34. $\dfrac{u^4 - v^4}{u^4 + 2u^2v^2 + v^4}$

35. $\dfrac{x^4 + x^3y - xy^3 - y^4}{x^4 - y^4}$

36. $\dfrac{x^2 - y^2 - 4x + 4y}{x^2 - y^2 + 4x - 4y}$

37. $\dfrac{x^2 - y^2 - 4y - 4}{x^2 - y^2 - 4x + 4}$

38. $\dfrac{ax + by - bx - ay}{ax - by + bx - ay}$

C **39.** $\dfrac{x^2 - y^2 - z^2 - 2yz}{x^2 - y^2 + z^2 - 2xz}$

40. $\dfrac{x^2 + y^2 - z^2 - 2xy}{x^2 - y^2 + z^2 - 2xz}$

41. $\dfrac{x^4 + x^2y^2 + y^4}{x^3 + y^3}$ (*Hint:* See Exercise 57, p. 186.)

42. $\dfrac{x^{2n} + 2x^ny^n - 3y^{2n}}{x^{2n} + 5x^ny^n + 6y^{2n}}$

43. $\dfrac{x^{2n} - 2x^ny^n + y^{2n}}{x^{2n} + 3x^ny^n - 4y^{2n}}$

Graphing Rational Functions

To graph a rational function, $f(x) = \dfrac{x}{x-2}$ for example, you need to plot points (x, y) for which $y = \dfrac{x}{x-2}$. As you will see, the graph of a rational function can approach, without intersecting, certain lines called *asymptotes*.

Example 1 Graph $f(x) = \dfrac{x}{x-2}$.

Solution The domain of f consists of all real numbers except 2. Using a calculator, prepare two tables of x-y values: one for $x < 2$ and one for $x > 2$.

$x < 2$

x	$y = \dfrac{x}{x-2}$
-2	$\dfrac{-2}{-2-2} = 0.5$
-1	$\dfrac{-1}{-1-2} \approx 0.33$
0	$\dfrac{0}{0-2} = 0$
1	$\dfrac{1}{1-2} = -1$
1.5	$\dfrac{1.5}{1.5-2} = -3$
1.9	$\dfrac{1.9}{1.9-2} = -19$
1.99	$\dfrac{1.99}{1.99-2} = -199$

$x > 2$

x	$y = \dfrac{x}{x-2}$
6	$\dfrac{6}{6-2} = 1.5$
5	$\dfrac{5}{5-2} \approx 1.67$
4	$\dfrac{4}{4-2} = 2$
3	$\dfrac{3}{3-2} = 3$
2.5	$\dfrac{2.5}{2.5-2} = 5$
2.1	$\dfrac{2.1}{2.1-2} = 21$
2.01	$\dfrac{2.01}{2.01-2} = 201$

As x approaches 2 from the left, y becomes increasingly small.

As x approaches 2 from the right, y becomes increasingly large.

Some of the points from the two tables are plotted on the grid at the right. Notice that for values of x near 2, the points on the graph of f "follow" the vertical line $x = 2$. The line $x = 2$ is an asymptote of the graph of f.

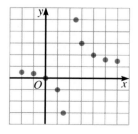

The graph of f has another asymptote. As $|x|$ gets increasingly large, $f(x)$ approaches 1:

$$f(12) = 1.2, \qquad f(102) = 1.02, \qquad f(1002) = 1.002;$$
$$f(-8) = 0.8, \qquad f(-98) = 0.98, \qquad f(-998) = 0.998.$$

Therefore the horizontal line $y = 1$ is also an asymptote of the graph of f.

To obtain the complete graph of f, draw the asymptotes $x = 2$ and $y = 1$ as dashed lines and connect the points already plotted with a smooth curve, as shown at the right.

Example 2 Graph $f(x) = \dfrac{1}{(x + 2)^2}$.

Solution The domain of f consists of all real numbers except -2. Because $f(x)$ gets increasingly large as x approaches -2, the line $x = -2$ is a vertical asymptote of the graph of f.

As $|x|$ gets increasingly large, $f(x)$ approaches 0. Therefore the x-axis is a horizontal asymptote of the graph of f.

A table of x-y values and the graph of f are shown below.

x	$y = \dfrac{1}{(x + 2)^2}$
-4	0.25
-3	1
-2.5	4
-1.5	4
-1	1
0	0.25

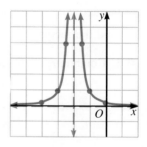

Exercises

Graph each function. Show any asymptotes as dashed lines. You may wish to check your graphs on a computer or a graphing calculator.

A 1. $f(x) = \dfrac{1}{x}$

2. $g(x) = -\dfrac{1}{x}$

3. $h(x) = \dfrac{x + 1}{x - 1}$

4. $f(x) = \dfrac{x - 1}{x + 1}$

5. $g(x) = \dfrac{2x}{x - 3}$

6. $h(x) = \dfrac{3x}{x + 2}$

B 7. $f(x) = \dfrac{1}{x^2}$

8. $g(x) = \dfrac{1}{(x - 1)^2}$

9. $h(x) = \dfrac{x^2 + 1}{x^2}$

10. $f(x) = \dfrac{1 - x^2}{x^2}$

11. $g(x) = \dfrac{10}{x^2 + 1}$

12. $h(x) = \dfrac{10x^2}{x^2 + 1}$

5-5 Products and Quotients of Rational Expressions

Objective   To multiply and divide rational expressions.

To find the product of two or more rational expressions, you use the multiplication rule for fractions: $\frac{a}{b} \cdot \frac{c}{d} = \frac{ac}{bd}$. Products of rational expressions should always be expressed in simplest form.

Example 1   Simplify $\dfrac{3x^2 - 6x}{x^2 - 6x + 9} \cdot \dfrac{x^2 - x - 6}{x^2 - 4}$.

Solution
$$\frac{3x^2 - 6x}{x^2 - 6x + 9} \cdot \frac{x^2 - x - 6}{x^2 - 4} = \frac{3x(x - 2)}{(x - 3)(x - 3)} \cdot \frac{(x + 2)(x - 3)}{(x + 2)(x - 2)}$$

$$= \frac{3x(x - 2)(x - 3)}{(x - 3)(x - 3)(x - 2)}$$

$$= \frac{3x}{x - 3} \quad \textit{Answer}$$

By the definition of division (page 33), a quotient is the product of the dividend and the reciprocal of the divisor. The reciprocal of $\frac{r}{s}$ is $\frac{s}{r}$ because $\frac{r}{s} \cdot \frac{s}{r} = 1$. Combining these facts gives the following rule.

Division Rule for Fractions

Let p, q, r, and s be real numbers with $q \neq 0$, $r \neq 0$, and $s \neq 0$. Then

$$\frac{p}{q} \div \frac{r}{s} = \frac{p}{q} \cdot \frac{s}{r}.$$

Example 2   Simplify.

    a. $\dfrac{14}{15} \div \dfrac{7}{5}$ **b.** $\dfrac{6xy}{a^2} \div \dfrac{3y}{a^3x}$

Solution
    a. $\dfrac{14}{15} \div \dfrac{7}{5} = \dfrac{14}{15} \cdot \dfrac{5}{7} = \dfrac{14 \cdot 5}{15 \cdot 7} = \dfrac{2}{3}$ *Answer*

    b. $\dfrac{6xy}{a^2} \div \dfrac{3y}{a^3x} = \dfrac{6xy}{a^2} \cdot \dfrac{a^3x}{3y} = \dfrac{6a^3x^2y}{3a^2y} = 2ax^2$ *Answer*

When multiplying rational expressions, you can divide out factors common to a numerator and a denominator *before* you write the product as a single fraction. For instance, for the product in Example 2(b),

$$\frac{\overset{2}{6x\cancel{y}}}{\cancel{a^2}} \cdot \frac{\overset{a}{a^{\cancel{3}}x}}{\cancel{3y}} = 2ax^2.$$

Sometimes you may find it helpful to rewrite a quotient using the $\div$ sign.

Example 3 Simplify $\dfrac{\dfrac{a^2 - 4ab + 3b^2}{a + 2b}}{a^2 - ab - 6b^2}$.

Solution

$$\frac{\dfrac{a^2 - 4ab + 3b^2}{a + 2b}}{a^2 - ab - 6b^2} = \frac{a^2 - 4ab + 3b^2}{a + 2b} \div \frac{a^2 - ab - 6b^2}{1}$$

$$= \frac{a^2 - 4ab + 3b^2}{a + 2b} \cdot \frac{1}{a^2 - ab - 6b^2}$$

$$= \frac{(a - b)(a - 3b)}{a + 2b} \cdot \frac{1}{(a + 2b)(a - 3b)}$$

$$= \frac{a - b}{(a + 2b)^2} \quad \textbf{Answer}$$

When you simplify expressions, perform multiplication and division in order from left to right, as in Example 4.

Example 4 Simplify $\dfrac{6p^2q}{r} \div \dfrac{3pq^2}{r} \cdot \dfrac{2q^2}{pr}$.

Solution

$$\frac{6p^2q}{r} \div \frac{3pq^2}{r} \cdot \frac{2q^2}{pr} = \left(\frac{6p^2q}{r} \div \frac{3pq^2}{r} \right) \cdot \frac{2q^2}{pr}$$

$$= \frac{\overset{2}{6p^2q}}{r} \cdot \frac{r}{3pq^2} \cdot \frac{2q^2}{pr}$$

$$= \frac{4q}{r} \quad \textbf{Answer}$$

Oral Exercises

Simplify.

1. $\dfrac{4}{5} \cdot 2$

2. $\dfrac{4}{5} \div 2$

3. $\dfrac{4}{5} \div \dfrac{1}{2}$

4. $\dfrac{4}{5} \cdot \dfrac{1}{2}$

5. $\dfrac{a^2}{b} \div \dfrac{a}{b^2}$

6. $\dfrac{a^2}{b} \cdot \dfrac{a}{b^2}$

7. $\dfrac{3s}{4t} \cdot \dfrac{2t}{s^2}$

8. $\dfrac{3s}{4t} \div \dfrac{2t}{s^2}$

I'M LAZY !!

Written Exercises

Simplify. Write answers without negative or zero exponents.

HW

March 31 2000

A

1. $\dfrac{-10}{21} \div \dfrac{15}{28}$

2. $-\dfrac{26}{25} \div \dfrac{39}{20}$

3. $\dfrac{5x^3}{-3} \cdot \dfrac{-6}{10x^2}$

4. $\dfrac{22z^2}{15} \cdot \dfrac{-5}{11z^3}$

5. $\dfrac{8t^2}{3} \div \dfrac{2t}{9}$

6. $\dfrac{28x}{25} \div \dfrac{21x^3}{15}$

7. $\dfrac{x^2}{4} \cdot \left(\dfrac{xy}{6}\right)^{-1} \cdot \dfrac{2y^2}{x}$

8. $2uv \div \dfrac{2u^2}{v} \div \dfrac{2v^2}{u}$

9. $\dfrac{4rs^2}{45} \div \dfrac{8s}{27r} \div \dfrac{9rs}{10}$

10. $\dfrac{7x^2}{9y} \div \dfrac{4x}{15y^2} \cdot \left(\dfrac{35}{6xy}\right)^{-1}$

11. $\dfrac{x(x-1)}{(x-2)^2} \div \dfrac{(x-1)^2}{x-2}$

12. $\dfrac{t-2}{t+3} \cdot \dfrac{t^2+2t-3}{t^2-t-2}$

13. $\dfrac{4u^2-1}{u^2-4} \cdot \dfrac{u-2}{2u-1}$

14. $\dfrac{x^2}{x-1} \cdot \dfrac{x+1}{x+2} \div \dfrac{x}{(x-1)(x+2)}$

15. $\dfrac{x^2-4}{2x^2-5x+2} \div \dfrac{2x^2-3x-2}{4x^2-1}$

16. $\dfrac{3x^2-8x+4}{9x^2-4} \div \dfrac{3x^2-5x-2}{9x^2-3x-2}$

B

17. $\dfrac{\dfrac{p^4-q^4}{(p+q)^2}}{p^2+q^2}$

18. $\dfrac{\dfrac{x^2-y^2}{x+y}}{x^4-y^4}$

19. $\dfrac{u^2v}{u+v} \div (u+v) \cdot \dfrac{u^2+2uv+v^2}{uv^2-u^2v}$

20. $\dfrac{x^2+3ax}{3a-x} \cdot \dfrac{x^2-4ax+3a^2}{a^2-x^2} \div \dfrac{x+3a}{x+a}$

21. $\dfrac{3x^2+xy-2y^2}{3x^2-xy-2y^2} \div \dfrac{3x^2+7xy-6y^2}{3x^2-2xy-y^2} \div \dfrac{3x+y}{3x+2y}$

22. $\dfrac{r^2+4rs+3s^2}{r^2+5rs+6s^2} \cdot (r+2s)^{-1} \div \dfrac{r+s}{r^2+4rs+4s^2}$

23. $(a^4+2a^2b^2+b^4) \div (a^4-b^4) \cdot (a-b)$

C

24. $(u^4+u^2v^2+v^4) \div (u^6-v^6) \cdot (u^2-v^2)$

Mixed Review Exercises

Express in scientific notation.

1. 0.00054

2. $63{,}400{,}000$

3. 0.1

4. 3281

Simplify.

5. $\dfrac{x^2+x-6}{x^2-x-2}$

6. $\dfrac{15y^2}{15y^3-10y^2}$

7. $\dfrac{16c^3d^2}{24cd^5}$

8. $\dfrac{a^2-1}{a^2+2a+1}$

9. $\dfrac{3m+12}{2m+8}$

10. $\dfrac{(6z^2)^2}{(4z^3)^3}$

11. $\dfrac{4t^2+4}{8t^2-8}$

12. $(u^3+1)(u+1)^{-1}$

WW.

5-6 Sums and Differences of Rational Expressions

Objective   To add and subtract rational expressions.

The definition of division and the distributive property can be used to prove the following rules for adding and subtracting fractions with *equal* denominators.

$$\frac{a}{c} + \frac{b}{c} = \frac{a+b}{c} \qquad\qquad \frac{a}{c} - \frac{b}{c} = \frac{a-b}{c}$$

These rules can be extended to more than two terms.

Example 1   Simplify: **a.** $\dfrac{4}{15} + \dfrac{13}{15} - \dfrac{7}{15}$ **b.** $\dfrac{3x^2 - 8}{x + 2} - \dfrac{x^2 - 9}{x + 2}$

Solution   **a.** $\dfrac{4}{15} + \dfrac{13}{15} - \dfrac{7}{15} = \dfrac{4 + 13 - 7}{15} = \dfrac{10}{15} = \dfrac{2}{3}$ ***Answer***

b. $\dfrac{3x^2 - 8}{x + 2} - \dfrac{x^2 - 9}{x + 2} = \dfrac{3x^2 - 8 - x^2 + 9}{x + 2}$

$$= \dfrac{2x^2 + 1}{x + 2} \quad \textit{Answer}$$

To add or subtract fractions with *different* denominators, use this rule.

Rule for Adding and Subtracting Fractions

1. Find the **least common denominator (LCD)** of the fractions (that is, the least common multiple (LCM) of their denominators).

2. Express each fraction as an equivalent fraction with the LCD as denominator.

3. Add or subtract and then simplify the result.

Example 2   Simplify $\dfrac{25}{42} + \dfrac{11}{18} - 2$.

Solution   The LCM of $42 = 2 \cdot 3 \cdot 7$ and $18 = 2 \cdot 3^2$ is $2 \cdot 3^2 \cdot 7 = 126$.
So the LCD is 126.

$$\frac{25}{42} + \frac{11}{18} - 2 = \frac{25 \cdot 3}{42 \cdot 3} + \frac{11 \cdot 7}{18 \cdot 7} - \frac{2 \cdot 126}{1 \cdot 126}$$

$$= \frac{75}{126} + \frac{77}{126} - \frac{252}{126}$$

$$= -\frac{100}{126} = -\frac{50}{63} \quad \textit{Answer}$$

Recall (page 190) that the least common multiple (LCM) of two or more polynomials is the common multiple having the least degree and least positive common factor.

Example 3 Simplify $\dfrac{1}{6a^2} - \dfrac{1}{2ab} + \dfrac{3}{8b^2}$.

Solution The LCM of $6a^2$, $2ab$, and $8b^2$ is $24a^2b^2$. So the LCD is $24a^2b^2$.

$$\frac{1}{6a^2} - \frac{1}{2ab} + \frac{3}{8b^2} = \frac{1 \cdot 4b^2}{6a^2 \cdot 4b^2} - \frac{1 \cdot 12ab}{2ab \cdot 12ab} + \frac{3 \cdot 3a^2}{8b^2 \cdot 3a^2}$$

$$= \frac{4b^2}{24a^2b^2} - \frac{12ab}{24a^2b^2} + \frac{9a^2}{24a^2b^2}$$

$$= \frac{4b^2 - 12ab + 9a^2}{24a^2b^2}$$

$$= \frac{(2b - 3a)^2}{24a^2b^2} \quad \textbf{\textit{Answer}}$$

When you add fractions in which the denominators are trinomials, factoring these trinomials will often shorten the work.

Example 4 Simplify $\dfrac{3}{x^2 + x - 6} - \dfrac{2}{x^2 - 3x + 2}$.

Solution $x^2 + x - 6 = (x - 2)(x + 3)$; $x^2 - 3x + 2 = (x - 1)(x - 2)$
So the LCD is $(x - 1)(x - 2)(x + 3)$.

$$\frac{3}{x^2 + x - 6} - \frac{2}{x^2 - 3x + 2} = \frac{3}{(x - 2)(x + 3)} - \frac{2}{(x - 1)(x - 2)}$$

$$= \frac{3(x - 1)}{(x - 1)(x - 2)(x + 3)} - \frac{2(x + 3)}{(x - 1)(x - 2)(x + 3)}$$

$$= \frac{3x - 3 - 2x - 6}{(x - 1)(x - 2)(x + 3)}$$

$$= \frac{x - 9}{(x - 1)(x - 2)(x + 3)} \quad \textbf{\textit{Answer}}$$

Oral Exercises

Simplify.

1. $\dfrac{3}{5} + \dfrac{1}{5}$

2. $\dfrac{5}{7} - \dfrac{2}{7}$

3. $\dfrac{5}{6} - \dfrac{1}{6}$

4. $\dfrac{3}{8} + \dfrac{1}{4}$

5. $\dfrac{1}{3} + \dfrac{1}{3a}$

6. $\dfrac{2}{x} - \dfrac{1}{2x}$

7. $\dfrac{1}{x} + \dfrac{1}{y}$

8. $z + \dfrac{1}{z}$

Written Exercises

H.W. March April 1ST, 2000

Simplify.

A 1. $\dfrac{5}{16} - \dfrac{12}{16} + \dfrac{3}{16}$

2. $\dfrac{7}{10} + \dfrac{11}{20} - \dfrac{9}{20}$

3. $\dfrac{1}{2} - \dfrac{3}{7} + \dfrac{5}{14}$

4. $\dfrac{2}{3} - \dfrac{3}{5} + \dfrac{4}{15}$

5. $\dfrac{5}{12} + \dfrac{5}{18} - \dfrac{5}{36}$

6. $\dfrac{7}{12} - 1 + \dfrac{19}{20}$

7. $\dfrac{t+2}{3} + \dfrac{t-4}{6}$

8. $\dfrac{x+3}{5} - \dfrac{2x+1}{10}$

9. $\dfrac{z-1}{z} + \dfrac{z+1}{z^2}$

10. $\dfrac{x+2}{x^2} + \dfrac{x-2}{2x}$

11. $\dfrac{t-4}{2t} - \dfrac{t-6}{3t}$

12. $\dfrac{2x+5}{4x^2} + \dfrac{2x-5}{10x}$

13. $\dfrac{1}{2pq^4} + \dfrac{2}{p^3q^2}$

14. $\dfrac{1}{xy^3} - \dfrac{1}{x^3y}$

15. $ab^{-1} - a^{-1}b$

16. $4s^{-2} - (4t)^{-2}$

17. $\dfrac{a}{bc} + \dfrac{b}{ac} + \dfrac{c}{ab}$

18. $\dfrac{1}{yz} + \dfrac{1}{zx} + \dfrac{1}{xy}$

19. $\dfrac{y-z}{yz} - \dfrac{z-x}{zx} - \dfrac{x-y}{xy}$

20. $\dfrac{1}{x^2} + \dfrac{1}{xy} + \dfrac{1}{4y^2}$

21. $\dfrac{1}{z-4} - \dfrac{1}{z+4}$

22. $\dfrac{x}{x-1} - \dfrac{1}{x+1}$

23. $\dfrac{1}{t^2+t} + \dfrac{1}{t^2-t}$

24. $\dfrac{1}{u^2-2u} - \dfrac{1}{u^2-4}$

25. $\dfrac{1}{x^2-1} - \dfrac{1}{(x-1)^2}$

26. $\dfrac{1}{y^2-y-2} + \dfrac{1}{y^2+y}$

27. $\dfrac{1}{s^2+2s+1} - \dfrac{1}{s^2-1}$

28. $\dfrac{1}{p^2-2p+1} - \dfrac{1}{p^2+p-2}$

B 29. $\dfrac{a+b}{a-b} + \dfrac{a-b}{a+b} + \dfrac{b-a}{a-b} + \dfrac{b-a}{a+b}$

30. $\dfrac{a+b}{a-b} + \dfrac{a-b}{a+b} - \dfrac{b-a}{a-b} + \dfrac{b-a}{a+b}$

31. $(x-y)^{-1} - (x+y)^{-1}$

32. $(x-y)^{-2} - (x+y)^{-2}$

33. $\dfrac{3}{x^2-5x+6} + \dfrac{2}{x^2-4}$

34. $\dfrac{1}{4t^2-4t+1} + \dfrac{1}{4t^2-1}$

35. $\dfrac{1}{2u^2-3uv+v^2} + \dfrac{1}{4u^2-v^2}$

36. $\dfrac{3}{4x^2-12xy+9y^2} + \dfrac{1}{2xy-3y^2}$

37. $\dfrac{x}{x-a} - \dfrac{x^2+a^2}{x^2-a^2} + \dfrac{a}{x+a}$

38. $\dfrac{3u}{2u-v} - \dfrac{2u}{2u+v} + \dfrac{2v^2}{4u^2-v^2}$

39. Prove: **a.** $\dfrac{a}{c} + \dfrac{b}{c} = \dfrac{a+b}{c}$ **b.** $\dfrac{a}{c} - \dfrac{b}{c} = \dfrac{a-b}{c}$

Find constants A and B that make the equation true.

C 40. $\dfrac{2x-9}{x^2-x-6} = \dfrac{A}{x-3} + \dfrac{B}{x+2}$

41. $\dfrac{x-7}{x^2+x-6} = \dfrac{A}{x+3} + \dfrac{B}{x-2}$

5-7 Complex Fractions

Objective   To simplify complex fractions.

A fraction is a **complex fraction** if its numerator or denominator (or both) has one or more fractions or powers with negative exponents. For example,

$$\frac{\dfrac{7}{12} - \dfrac{1}{6}}{2 + \dfrac{4}{9}} \qquad \text{and} \qquad \frac{x^{-1} - y^{-1}}{x^{-2} - y^{-2}}$$

are complex fractions. Here are two methods for simplifying such fractions.

Method 1   Simplify the numerator and denominator separately; then divide.

Method 2   Multiply the numerator and denominator by the LCD of all the fractions appearing in the numerator and denominator.

Example 1   Simplify $\dfrac{\dfrac{7}{12} - \dfrac{1}{6}}{2 + \dfrac{4}{9}}$.

Solution

Method 1

$$\frac{\dfrac{7}{12} - \dfrac{1}{6}}{2 + \dfrac{4}{9}} = \frac{\dfrac{7-2}{12}}{\dfrac{18+4}{9}} = \frac{\dfrac{5}{12}}{\dfrac{22}{9}} = \frac{5}{12} \cdot \frac{9}{22} = \frac{15}{88}$$

Method 2

$$\frac{\dfrac{7}{12} - \dfrac{1}{6}}{2 + \dfrac{4}{9}} = \frac{\left(\dfrac{7}{12} - \dfrac{1}{6}\right) \cdot 36}{\left(2 + \dfrac{4}{9}\right) \cdot 36} = \frac{21 - 6}{72 + 16} = \frac{15}{88}$$

Example 2   Simplify $\dfrac{z - \dfrac{1}{z}}{1 - \dfrac{1}{z}}$.

Solution

Method 1

$$\frac{z - \dfrac{1}{z}}{1 - \dfrac{1}{z}} = \frac{\dfrac{z^2 - 1}{z}}{\dfrac{z - 1}{z}}$$

$$= \frac{z^2 - 1}{z} \cdot \frac{z}{z - 1}$$

$$= \frac{z(\cancel{z - 1})(z + 1)}{\cancel{z}(\cancel{z - 1})}$$

$$= z + 1 \quad \textbf{Answer}$$

Method 2

$$\frac{z - \dfrac{1}{z}}{1 - \dfrac{1}{z}} = \frac{\left(z - \dfrac{1}{z}\right) \cdot z}{\left(1 - \dfrac{1}{z}\right) \cdot z}$$

$$= \frac{z^2 - 1}{z - 1}$$

$$= \frac{(\cancel{z - 1})(z + 1)}{\cancel{z - 1}}$$

$$= z + 1 \quad \textbf{Answer}$$

When the numerator or denominator of a complex fraction has powers with negative exponents, you should first rewrite the powers using positive exponents. Then simplify the fraction using either of the methods shown in Example 3.

Example 3 Simplify $\dfrac{a^{-1} - x^{-1}}{a^{-2} - x^{-2}}$.

Solution

Method 1

$$\frac{a^{-1} - x^{-1}}{a^{-2} - x^{-2}} = \left(\frac{1}{a} - \frac{1}{x}\right) \div \left(\frac{1}{a^2} - \frac{1}{x^2}\right)$$

$$= \frac{x - a}{ax} \div \frac{x^2 - a^2}{a^2x^2}$$

$$= \frac{x - a}{ax} \cdot \frac{a^2x^2}{x^2 - a^2}$$

$$= \frac{a^2x^2(x - a)}{ax(x + a)(x - a)}$$

$$= \frac{ax}{x + a}$$
Answer

Method 2

$$\frac{a^{-1} - x^{-1}}{a^{-2} - x^{-2}} = \frac{\left(\dfrac{1}{a} - \dfrac{1}{x}\right) \cdot a^2x^2}{\left(\dfrac{1}{a^2} - \dfrac{1}{x^2}\right) \cdot a^2x^2}$$

$$= \frac{ax^2 - a^2x}{x^2 - a^2}$$

$$= \frac{ax(x - a)}{(x + a)(x - a)}$$

$$= \frac{ax}{x + a}$$
Answer

Written Exercises

Simplify.

A **1.** $\dfrac{1 - \dfrac{1}{3}}{\dfrac{1}{2} - \dfrac{1}{6}}$

2. $\dfrac{\dfrac{1}{2} + \dfrac{1}{3}}{1 - \dfrac{1}{6}}$

3. $\dfrac{1 - \dfrac{4}{5}}{\dfrac{1}{4} - \dfrac{1}{5}}$

4. $\dfrac{\dfrac{2}{3} - \dfrac{5}{6}}{\dfrac{1}{3} + \dfrac{2}{9}}$

5. $\dfrac{x + 1}{1 + \dfrac{1}{x}}$

6. $\dfrac{z - \dfrac{1}{z}}{1 - \dfrac{1}{z}}$

7. $\dfrac{a - b}{a^{-1} - b^{-1}}$

8. $\dfrac{1 - xy^{-1}}{x^{-1} - y^{-1}}$

9. $\dfrac{u^{-2} - v^{-2}}{u^{-1} - v^{-1}}$

10. $\dfrac{a^{-2} - b^{-2}}{a^{-1} + b^{-1}}$

11. $\dfrac{\dfrac{1}{x^2} - \dfrac{1}{y^2}}{\dfrac{1}{x^2} + \dfrac{2}{xy} + \dfrac{1}{y^2}}$

12. $\dfrac{\dfrac{1}{p^2} - \dfrac{1}{q^2}}{\dfrac{2}{p^2} - \dfrac{1}{pq} - \dfrac{1}{q^2}}$

13. $\dfrac{h + h^{-2}}{1 + h^{-1}}$

14. $\dfrac{x^{-2} - x^2}{x^{-1} - x}$

15. $\dfrac{s^2 - t^{-2}}{s - t^{-1}}$

Simplify.

16. $\dfrac{\dfrac{1}{x} - \dfrac{1}{y}}{\dfrac{y}{x} - \dfrac{x}{y}}$

17. $\dfrac{\dfrac{2}{y+2} - 1}{\dfrac{1}{y+2} + 1}$

18. $\dfrac{1 + \dfrac{1}{t-1}}{1 - \dfrac{1}{t+1}}$

B 19. $\dfrac{\dfrac{1}{a+1} + \dfrac{1}{a-1}}{\dfrac{1}{a+1} - \dfrac{1}{a-1}}$

20. $\dfrac{\dfrac{1}{x} + \dfrac{1}{x+1}}{\dfrac{1}{x} - \dfrac{1}{x+1}}$

21. $\dfrac{1 + \dfrac{1}{x-1}}{1 + \dfrac{1}{x^2-1}}$

22. $\dfrac{\dfrac{1}{1-t} - \dfrac{1}{t}}{\dfrac{1}{1+t} - \dfrac{1}{t}}$

23. $\dfrac{\dfrac{a}{b} - \dfrac{a-b}{a+b}}{\dfrac{a}{b} + \dfrac{a+b}{a-b}}$

24. $\dfrac{\dfrac{u+v}{u-v} - \dfrac{u-v}{u+v}}{\dfrac{u+v}{u-v} + \dfrac{u-v}{u+v}}$

25. $\dfrac{1 - \dfrac{2 - \dfrac{1}{x}}{x}}{1 - \dfrac{1}{x}}$

26. $\dfrac{u + \dfrac{1}{1 + \dfrac{1}{u}}}{\dfrac{1}{u+1}}$

27. Evaluate to three decimal places. A calculator may be helpful.

    a. $1 + \dfrac{1}{2}$

    b. $1 + \dfrac{1}{2 + \dfrac{1}{2}}$

    c. $1 + \dfrac{1}{2 + \dfrac{1}{2 + \dfrac{1}{2}}}$

    d. $1 + \dfrac{1}{2 + \dfrac{1}{2 + \dfrac{1}{2 + \dfrac{1}{2}}}}$

    (The farther this process is carried out, the closer the results will be to $\sqrt{2} = 1.41421 \ldots .$)

In Exercises 28–31, express $\dfrac{f(x+h) - f(x)}{h}$ as a single simplified fraction. (These exercises might be met in calculus.)

Sample $f(x) = \dfrac{1}{1-x}$

Solution $\dfrac{f(x+h) - f(x)}{h} = \dfrac{\dfrac{1}{1-(x+h)} - \dfrac{1}{1-x}}{h}$

$$= \dfrac{(1-x) - (1-x-h)}{h(1-x-h)(1-x)}$$

$$= \dfrac{1}{(1-x-h)(1-x)}$$

C 28. $f(x) = \dfrac{1}{x}$

     30. $f(x) = \dfrac{1-x}{x}$

     29. $f(x) = \dfrac{1}{x+1}$

     31. $f(x) = \dfrac{1}{x^2}$

In Exercises 32–35, express $f(f(x))$ as a single simplified fraction.

32. $f(x) = \dfrac{1}{x+1}$

33. $f(x) = \dfrac{x}{x+1}$

34. $f(x) = \dfrac{1+x}{1-x}$

35. $f(x) = (1-x)^{-1}$

Mixed Review Exercises

Simplify.

1. $\dfrac{1}{x^2 y} + \dfrac{1}{xy^2}$

2. $\dfrac{8}{a^2 - 3a} \cdot \dfrac{a^2 - 9}{6a}$

3. $\dfrac{36u^2}{25v} \div \dfrac{27u^4}{10v^3}$

4. $\dfrac{t}{t-2} - \dfrac{1}{t+2}$

5. $-\dfrac{8a^2 b^3}{16a^3 b^2}$

6. $\dfrac{x}{2y} + \dfrac{y}{2x}$

7. $\dfrac{x-2}{4} - \dfrac{2-x}{8}$

8. $\dfrac{2y-x}{x-2y}$

9. $\dfrac{(a-2)^2}{b^2} \div \dfrac{6a-12}{4b}$

Find the unique solution of each system. Check your answer by using substitution.

10. $4x - y = 14$
$5x + 3y = 9$

11. $2x + 5y = 17$
$x - 2y = -5$

12. $x - 4y = -4$
$2x + 10y = 1$

Self-Test 2

Vocabulary   rational algebraic expression (p. 227)   rational function (p. 228)
rational expression (p. 227)   least common denominator (p. 235)
complex fraction (p. 238)

Simplify.

1. $\dfrac{t^3 - t}{t^3 - t^2}$

2. $\dfrac{x^3 - x^2 - 4x + 4}{x^3 - 4x}$ Obj. 5-4, p. 227

3. $\dfrac{2p^4 r^3}{3s^2} \cdot \dfrac{p^3 r^2}{6s^3}$

4. $\dfrac{2u^2 + 3uv - 2v^2}{2u^2 - 3uv - 2v^2} \cdot \dfrac{2u+v}{2u-v}$ Obj. 5-5, p. 232

5. $\dfrac{1}{a^2} - \dfrac{2}{ab} + \dfrac{1}{b^2}$

6. $\dfrac{1}{4x^2 - 4x + 1} - \dfrac{1}{4x^2 - 1}$ Obj. 5-6, p. 235

7. $\dfrac{\dfrac{2}{3} - \dfrac{1}{4}}{1 - \dfrac{1}{6}}$

8. $\dfrac{1 - hk^{-1}}{h^{-1} - k^{-1}}$ Obj. 5-7, p. 238

Check your answers with those at the back of the book.

Problem Solving Using Fractional Equations

5-8 *Fractional Coefficients*

Objective   To solve equations and inequalities having fractional coefficients.

To solve an equation or an inequality having fractional coefficients, it is helpful to multiply both sides of the open sentence by the least common denominator of the fractions.

Example 1 Solve $\dfrac{x^2}{2} = \dfrac{2x}{15} + \dfrac{1}{10}$.

Solution

$$\frac{x^2}{2} = \frac{2x}{15} + \frac{1}{10}$$

$$30 \cdot \frac{x^2}{2} = 30\left(\frac{2x}{15} + \frac{1}{10}\right)$$ Multiply both sides by 30, the LCM of 2, 15, and 10.

$$15x^2 = 4x + 3$$ Use the distributive property.

$$15x^2 - 4x - 3 = 0$$

$$(3x + 1)(5x - 3) = 0$$

$$x = -\frac{1}{3} \quad \text{or} \quad x = \frac{3}{5}$$

$\therefore$ the solution set is $\left\{-\dfrac{1}{3}, \dfrac{3}{5}\right\}$.   ***Answer***

Example 2 Solve $\dfrac{x}{8} - \dfrac{x-2}{3} \geq \dfrac{x+1}{6} - 1$.

Solution

$$\frac{x}{8} - \frac{x-2}{3} \geq \frac{x+1}{6} - 1$$

$$24\left(\frac{x}{8} - \frac{x-2}{3}\right) \geq 24\left(\frac{x+1}{6} - 1\right)$$ Multiply both sides by 24, the LCM of 8, 3, and 6.

$$3x - 8x + 16 \geq 4x + 4 - 24$$ Use the distributive property.

$$-9x \geq -36$$

$$x \leq 4$$ Reverse the inequality.

$\therefore$ the solution set is $\{x: x \leq 4\}$.   ***Answer***

Example 3 Crane A can unload the container ship in 10 h, and crane B can unload it in 14 h. Crane A started to unload the ship at noon and was joined by crane B at 2 P.M. At what time was the unloading job of the ship completed?

Solution

Step 1 Essentially, the problem asks for the number of hours crane A worked.

Step 2 Let t = the number of hours crane A worked.

Then $t - 2$ = the number of hours crane B worked.

The part of the job crane A can do in 1 hour = $\dfrac{1}{10}$.

Thus, the part of the job crane A can do in t hours is $t \cdot \dfrac{1}{10} = \dfrac{t}{10}$.

Similarly, the part of the job crane B can do in $t - 2$ hours is $(t - 2) \cdot \dfrac{1}{14} = \dfrac{t - 2}{14}$.

Step 3

$$\underbrace{\text{Part done by crane A}}_{\dfrac{t}{10}} \quad \text{plus} \quad \underbrace{\text{part done by crane B}}_{\dfrac{t-2}{14}} \quad \text{is} \quad \underbrace{\text{the whole job.}}_{1}$$

Step 4 $\dfrac{t}{10} + \dfrac{t-2}{14} = 1$ Multiply both sides by the LCM, 70.

$$7t + 5t - 10 = 70$$
$$12t = 80$$
$$t = \frac{80}{12}$$
$$t = \frac{20}{3} = 6\frac{2}{3}$$

Step 5 The check is left for you.

$\therefore$ the unloading job was completed $6\frac{2}{3}$ h after noon, or at 6:40 P.M.

Recall that *percent* means hundredths. For example, 60% of 400 means 0.60×400, or 240. Fractions often enter problems through the use of percents, as in Example 4 at the top of the next page.

Example 4 A nurse wishes to obtain 800 mL of a 7% solution of boric acid by mixing 4% and 12% solutions. How much of each should be used?

Solution

Step 1 The problem asks for a number of mL of 4% and 12% solutions of boric acid to be used.

Step 2 Let x = number of mL of 4% solution.

Then $800 - x$ = number of mL of 12% solution.

Show the known facts in a table.

	mL of solution	× % boric acid	= mL of boric acid
4% solution	x	4%	$0.04x$
12% solution	$800 - x$	12%	$0.12(800 - x)$
Final mixture	800	7%	$0.07(800)$

Step 3 From the last column of the table, write an equation relating the 4% and 12% solutions to the final mixture.

$$0.04x + 0.12(800 - x) = 0.07(800)$$
or $$4x + 12(800 - x) = 7(800)$$

To clear decimals, multiply both sides by 100.

Step 4
$$4x + 9600 - 12x = 5600$$
$$-8x = -4000$$
$$x = 500 \quad \text{(4\% solution)}$$
$$800 - x = 300 \quad \text{(12\% solution)}$$

Step 5 The check is left for you.

∴ the nurse should mix 500 mL of the 4% solution and 300 mL of the 12% solution. **Answer**

Oral Exercises

Find an open sentence with integers as coefficients that is equivalent to the given open sentence.

1. $\frac{x}{2} - 1 = \frac{5}{6}$

2. $\frac{y}{6} + 2 = \frac{2}{3}$

3. $\frac{t}{3} - \frac{1}{2} > 1$

4. $\frac{x}{3} - \frac{2}{5} \leq 2$

5. $u = \frac{2}{5} - \frac{u}{2}$

6. $\frac{t}{3} + \frac{5}{6} = \frac{3}{2}$

7. $2.7 + 0.5s = 1.2$

8. $0.14x - 0.06 = 0.15$

9. $\frac{z}{3} - \frac{3}{4} = \frac{5}{6}$

Written Exercises

Homework : April 6$^{\#}$, 2000

Solve each open sentence.

A 1. $\frac{x}{9} + \frac{1}{6} = \frac{2}{3}$

2. $\frac{3u}{5} - \frac{5}{6} = \frac{u}{10}$

3. $\frac{2t-1}{6} = \frac{t+2}{4} + \frac{1}{3}$

4. $\frac{s-2}{2} - \frac{s-1}{5} = \frac{1}{4}$

5. $\frac{z}{4} - \frac{z-1}{6} \leq \frac{5}{12}$

6. $\frac{t}{4} - \frac{t+2}{3} + \frac{1}{6} \geq 0$

7. $\frac{r-2}{8} < \frac{3r+1}{6} + \frac{1}{3}$

8. $\frac{x-5}{9} \leq \frac{2x}{15} - \frac{2}{5}$

9. $\frac{y^2}{4} - \frac{3y}{2} + 2 = 0$

10. $\frac{t^2}{6} - \frac{t}{2} - \frac{2}{3} = 0$

11. $\frac{z^2}{3} - \frac{z}{6} = 1$

12. $\frac{v^2}{6} + \frac{1}{8} = \frac{v}{3}$

13. $\frac{x^2}{9} + \frac{x-1}{10} = 0$

14. $\frac{t(t-1)}{3} = \frac{t+1}{2}$

15. $\frac{y(2y-1)}{2} = \frac{1-y}{3}$

16. $\frac{w(w-1)}{3} + \frac{1}{2} = \frac{w+1}{4}$

B 17. $\frac{x(x+1)}{5} - \frac{x+1}{6} = \frac{1}{3}$

18. $\frac{2t(3t+1)}{5} - \frac{t+1}{2} = \frac{1}{10}$

19. $\frac{y(2y-1)}{4} + \frac{3}{10} = \frac{y(y+2)}{5}$

20. $\frac{u(u-1)}{2} + \frac{1}{3} = \frac{u(2-u)}{4}$

21. $\frac{y^2+4}{6} + \frac{y+1}{3} < \frac{3}{2}$

22. $\frac{z^2+1}{6} \geq \frac{z+2}{3}$

23. $\frac{t^2}{6} + \frac{t-2}{4} \geq \frac{t+1}{3}$

24. $\frac{u^2}{15} \leq \frac{u+2}{6} - \frac{2}{5}$

Problems

In Problems 1–4, find the number described.

A 1. $\frac{5}{16}$ of $\frac{4}{5}$ of the number is 15.

2. 12 is $\frac{3}{5}$ of $\frac{10}{21}$ of the number.

3. 30 is 20% of 30% of the number.

4. 75% of 60% of the number is 36.

5. Pump A can unload the *Lunar Petro* in 30 h and pump B can unload it in 24 h. Because of an approaching storm, both pumps were used. How long did they take to empty the ship? How much Lunar Petro in the ship.?

6. An old conveyor belt takes 21 h to move one day's coal output from the mine to a rail line. A new belt can do it in 15 h. How long does it take when both are used at the same time?

7. How much pure antifreeze must be added to 12 L of a 40% solution of antifreeze to obtain a 60% solution?

8. How much water must be evaporated from a 300 L tank of a 2% salt solution to obtain a 5% solution?

9. The river boat *Delta Duchess* paddled upstream at 12 km/h, stopped for 2 h of sightseeing, and paddled back at 18 km/h. How far upstream did the boat travel if the total time for the trip, including the stop, was 7 h?

10. Pam jogged up a hill at 6 km/h and then jogged back down at 10 km/h. How many kilometers did she travel in all if her total jogging time was 1 h 20 min?

11. The Computer Club invested $2200, part at 4.5% interest and the rest at 7%. The total annual interest earned was $144. How much was invested at each rate?

12. Lina Chen invested $24,000, part at 8% and the rest at 7.2%. How much did she invest at each rate if her income from the 8% investment is two thirds that of the 7.2% investment?

B 13. A pharmacist wishes to make 1.8 L of a 10% solution of boric acid by mixing 7.5% and 12% solutions. How much of each type of solution should be used?

14. How much of an 18% solution of sulfuric acid should be added to 360 mL of a 10% solution to obtain a 15% solution?

15. The county's new asphalt paving machine can surface 1 km of highway in 10 h. A much older machine can surface 1 km in 18 h. How long will it take them to surface 21 km of highway if they start at opposite ends and work day and night?

16. Pipes A and B can fill a storage tank in 8 h and 12 h, respectively. With the tank empty, pipe A was turned on at noon, and then pipe B was turned on at 1:30 P.M. At what time was the tank full?

17. Sharon drove for part of a 150 km trip at 45 km/h and the rest of the trip at 75 km/h. How far did she drive at each speed if the entire trip took her 2 h 40 min?

18. An elevator went from the bottom to the top of a tower at an average speed of 4 m/s, remained at the top for 90 s, and then returned to the bottom at 5 m/s. If the total elapsed time was $4\frac{1}{2}$ min, how high is the tower?

C 19. A commercial jet can fly from San Francisco to Dallas in 3 h. A private jet can make the same trip in $3\frac{1}{2}$ h. If the two planes leave San Francisco at noon, after how many hours is the private jet twice as far from Dallas as the commercial jet?

20. A car radiator is filled with 5 L of a 25% antifreeze solution. How many liters must be drawn off and replaced by a 75% antifreeze solution to leave the radiator filled with a 55% antifreeze solution?

21. The rail line between two cities consists of two segments, one 96 km longer than the other. A passenger train averages 60 km/h over the shorter segment, 120 km/h over the longer, and 100 km/h for the entire trip. How far apart are the cities?

5-9 Fractional Equations

Objective   To solve and use fractional equations.

An equation in which a variable occurs in a denominator, such as

$$\frac{3}{x^2 - 7x + 10} + 2 = \frac{x - 4}{x - 5},$$

is called a **fractional equation.** When you solve a fractional equation by multiplying both sides by the LCD of the fractions, the resulting equation is *not* always equivalent to the original one.

Example 1   Solve $\dfrac{3}{x^2 - 7x + 10} + 2 = \dfrac{x - 4}{x - 5}.$

Solution $x^2 - 7x + 10 = (x - 2)(x - 5)$. So the LCD is $(x - 2)(x - 5)$.

$$(x - 2)(x - 5)\left[\frac{3}{x^2 - 7x + 10} + 2\right] = (x - 2)(x - 5)\left[\frac{x - 4}{x - 5}\right]$$

$$3 + 2(x - 2)(x - 5) = (x - 2)(x - 4)$$

$$3 + 2x^2 - 14x + 20 = (x - 2)(x - 4)$$

$$2x^2 - 14x + 23 = x^2 - 6x + 8$$

$$x^2 - 8x + 15 = 0$$

$$(x - 3)(x - 5) = 0$$

$$x = 3 \quad \text{or} \quad x = 5$$

Check: When x equals 2 or 5, a denominator in the original equation has a value of 0. Therefore, 2 and 5 are not in the domain of x and cannot be solutions. Since 5 is not permissible, check 3 in the original equation:

$$\frac{3}{3^2 - 7 \cdot 3 + 10} + 2 \stackrel{?}{=} \frac{3 - 4}{3 - 5}$$

$$-\frac{3}{2} + 2 \stackrel{?}{=} \frac{-1}{-2}$$

$$\frac{1}{2} = \frac{1}{2} \quad \checkmark$$

$\therefore$ the solution set is $\{3\}$.   *Answer*

In Example 1, the equation obtained by multiplying the given equation by $(x - 2)(x - 5)$ has the *extraneous root* 5. An **extraneous root** is a root of the transformed equation that is *not* a root of the original equation.

Caution: If you transform an equation by multiplying by a polynomial, always check each root of the new equation in the *original* one.

Example 2 One pump can empty the town swimming pool in 7 h less time than a smaller second pump can. Together they can empty the pool in 12 h. How much time would it take the larger pump alone to empty it?

Solution

Step 1 The problem asks for the number of hours for the larger pump to empty the pool.

Step 2 Let t = the number of hours for the larger pump to empty the pool.

Then $t + 7$ = the number of hours for the smaller pump to empty the pool.

Now write an expression for the part of the pool that is emptied by each pump in 1 h and the part emptied by each in 12 h.

$\dfrac{1}{t}$ = part of pool emptied by larger pump in 1 h

$\dfrac{12}{t}$ = part of pool emptied by larger pump in 12 h

$\dfrac{1}{t + 7}$ = part of pool emptied by smaller pump in 1 h

$\dfrac{12}{t + 7}$ = part of pool emptied by smaller pump in 12 h

Step 3 The sum of the parts emptied by each pump in 12 h is 1.

$$\frac{12}{t} + \frac{12}{t + 7} = 1$$

Step 4
$$12(t + 7) + 12t = t(t + 7)$$
$$t^2 - 17t - 84 = 0$$
$$(t + 4)(t - 21) = 0$$
$$t = -4 \quad \text{or} \quad t = 21$$

Since the time t cannot be negative, reject $t = -4$ as an answer.

Step 5 *Check $t = 21$:* If the larger pump can empty the pool in 21 h, then the smaller can empty it in 21 + 7, or 28 h. Therefore, the parts of the pool emptied by the pumps in 12 h are $\dfrac{12}{21} = \dfrac{4}{7}$ and $\dfrac{12}{28} = \dfrac{3}{7}$.

$$\frac{4}{7} + \frac{3}{7} = 1 \quad \checkmark$$

∴ the larger pump can empty the pool in 21 h.
Answer

Example 3 Because of strong headwinds, an airplane's ground speed (see page 131) for the first half of a 2000 km trip averaged only 600 km/h. What must its ground speed be for the rest of the trip if it is to average 720 km/h for the entire trip?

Solution Let r = the ground speed in km/h that the airplane must travel for the second half of its trip. Use the fact that $time = \dfrac{distance}{rate}$ and make a table.

	Distance (km)	Rate (km/h)	Time (h)
First half	1000	600	$\dfrac{1000}{600} = \dfrac{5}{3}$
Second half	1000	r	$\dfrac{1000}{r}$
Entire trip	2000	720	$\dfrac{2000}{720} = \dfrac{25}{9}$

Time for first half of trip	plus	Time for second half of trip	equals	Time for entire trip
$\dfrac{5}{3}$	$+$	$\dfrac{1000}{r}$	$=$	$\dfrac{25}{9}$

Multiply both sides by $9r$: $15r + 9000 = 25r$

$$r = 900$$

The check is left for you.

∴ the ground speed for the second half must be 900 km/h.   **Answer**

Written Exercises

Solve and check. If an equation has no solution, say so.

A **1.** $\dfrac{3}{t} - \dfrac{1}{3t} = \dfrac{2}{3}$

2. $\dfrac{2}{x} + \dfrac{1}{4} = \dfrac{1}{x}$

3. $\dfrac{1}{x} = \dfrac{2}{x-3}$

4. $\dfrac{3}{u+2} = \dfrac{1}{u-2}$

5. $\dfrac{2}{s+3} - \dfrac{1}{s-3} = 0$

6. $\dfrac{5}{2r+1} - \dfrac{3}{2r-1} = 0$

7. $\dfrac{t}{t-1} = \dfrac{t+2}{t}$

8. $\dfrac{x-3}{x} = \dfrac{x-4}{x-2}$

9. $\dfrac{6t^2 - t - 1}{3(t^2+1)} = 2$

10. $\dfrac{(y+1)^2}{(y-3)^2} = 1$

11. $\dfrac{x}{x+3} + \dfrac{1}{x-3} = 1$

12. $\dfrac{1}{s} + \dfrac{s}{s+2} = 1$

13. $\dfrac{1}{y-2} + \dfrac{1}{y+2} = \dfrac{4}{y^2-4}$

14. $\dfrac{k}{k+1} + \dfrac{k}{k-2} = 2$

15. $\dfrac{3}{x+1} - \dfrac{1}{x-2} = \dfrac{1}{x^2-x-2}$

16. $\dfrac{1}{t-1} + \dfrac{1}{t+2} = \dfrac{3}{t^2+t-2}$

17. $\dfrac{6}{t} - \dfrac{2}{t-1} = 1$

18. $\dfrac{1}{x+1} - \dfrac{1}{x+2} = \dfrac{1}{2}$

B 19. $\dfrac{u}{u-2} + \dfrac{30}{u+2} = 9$ 20. $\dfrac{1}{s+3} + \dfrac{1}{s-5} = \dfrac{1-s}{s+3}$

21. $\dfrac{2}{x-1} - \dfrac{x}{x+3} = \dfrac{6}{x^2+2x-3}$ 22. $\dfrac{x}{x+3} + \dfrac{1}{x-1} = \dfrac{4}{x^2+2x-3}$

23. $\dfrac{5}{u^2+u-6} = 2 - \dfrac{u-3}{u-2}$ 24. $\dfrac{y}{y-2} - \dfrac{2}{y+3} = \dfrac{10}{y^2+y-6}$

25. $\dfrac{t}{t-1} = \dfrac{1}{t+2} + \dfrac{3}{t^2+t-2}$ 26. $\dfrac{9}{t^2-2t-8} + \dfrac{t}{t+2} = 2$

27. $\left(\dfrac{x-3}{x+1}\right)^2 = 2 \cdot \dfrac{x-3}{x+1} + 3$ 28. $\left(\dfrac{t+3}{t-1}\right)^2 = 2 + \dfrac{t+3}{t-1}$

C 29. $\dfrac{\dfrac{1}{x^2} - x^2}{\dfrac{1}{x} + x} = \dfrac{3}{2}$ 30. $\left(\dfrac{1}{x} - \dfrac{x}{2}\right)\left(\dfrac{2}{x} - x\right) = \dfrac{1}{2}$

Problems

Solve.

A 1. Find two positive numbers that differ by 8 and whose reciprocals differ by $\dfrac{1}{6}$.

2. Find two numbers whose sum is 25 and the sum of whose reciprocals is $\dfrac{1}{6}$.

3. The reciprocal of half a number increased by half the reciprocal of the number is $\dfrac{1}{2}$. Find the number.

4. The reciprocal of one third of a number decreased by one third of the reciprocal of the number is $\dfrac{1}{3}$. Find the number.

5. A town's old street sweeper can clean the streets in 60 h. The old sweeper together with a new sweeper can clean the streets in 15 h. How long would it take the new sweeper to do the job alone?

6. The intake pipe can fill a certain tank in 6 h when the outlet pipe is closed, but with the outlet pipe open it takes 9 h. How long would it take the outlet pipe to empty a full tank?

7. During 60 mi of city driving, Jenna averaged 15 mi/gal. She then drove 140 mi on an expressway and averaged 25 mi/gal for the entire 200 mi. Find the average fuel consumption on the expressway.

8. Helped by a strong jet stream, a Los Angeles-to-Boston plane flew 10% faster than usual and made the 4400 km trip in 30 min less time than usual. At what speed does the plane usually fly?

9. The excursion boat *Holiday* travels 35 km upstream and then back again in 4 h 48 min. If the speed of the *Holiday* in still water is 15 km/h, what is the speed of the current?

10. Tim paddled his kayak 12 km upstream against a 3 km/h current and back again in 5 h 20 min. In that time how far could he have paddled in still water?

11. Members of the Computer Club were assessed equal amounts to raise $1200 to buy some software. When 8 new members joined, the per-member assessment was reduced by $7.50. What was the new size of the club?

12. Members of the Ski Club contributed equally to obtain $1800 for a holiday trip. When 6 members found that they could not go, their contributions were refunded and each remaining member then had to pay $10 more to raise the $1800. How many went on the trip?

13. To measure the speed of the jet stream, a weather plane left its base at noon and flew 800 km directly against the stream with an air speed of 750 km/h. It then returned directly to its base, arriving at 2:24 P.M. What was the speed of the jet stream?

14. When Ace Airlines changed to planes that flew 100 km/h faster than its old ones, the time of its 2800 km Dallas-Seattle flight was reduced by 30 min. Find the speed of the new planes.

B 15. Elvin drove halfway from Ashton to Dover at 40 mi/h and the rest of the way at 60 mi/h. What was his average speed for the whole trip? (*Hint:* Let the distance for the whole trip be, say, 100 mi.)

16. Elizabeth drove the first half of a trip at 36 mi/h. At what speed should she cover the remaining half in order to average 45 mi/h for the whole trip? (See the hint in Exercise 15.)

17. A train averaged 120 km/h for the first two thirds of a trip and 100 km/h for the whole trip. Find its average speed for the last third of the trip.

18. Because of traffic Maria could average only 40 km/h for the first 20% of her trip, but she averaged 75 km/h for the whole trip. What was her average speed for the last 80% of her trip?

19. Pipe A can fill a tank in 5 h. Pipe B can fill it in 2 h less time than it takes pipe C, a drainpipe, to empty the tank. With all three pipes open, it takes 3 h to fill the tank. How long would it take pipe C to empty it?

20. An elevator went from the bottom to the top of a 240 m tower, remained there for 12 s, and returned to the bottom in an elapsed time of 2 min. If the elevator traveled 1 m/s faster on the way down, find its speed going up.

A number x is the **harmonic mean** of a and b if $\frac{1}{x}$ is the average of $\frac{1}{a}$ and $\frac{1}{b}$.

C **21.** The harmonic mean of a number and 5 is 8. Find the number.

22. Find two positive numbers that differ by 12 and have harmonic mean 5.

23. Suppose that a vehicle averages u km/h for the first half of a trip and v km/h for the second half. Show that its average speed for the whole trip is the harmonic mean of u and v.

Mixed Review Exercises

Simplify.

1. $\dfrac{x^2 - 1}{1 - x}$

2. $\dfrac{8t^2 - 2}{8t + 4}$

3. $\dfrac{32m^5n}{24m^2n^2}$

4. $\dfrac{(4z^3)^{-2}}{(2z^4)^{-1}}$

5. $\dfrac{a^{-1} + b^{-2}}{a^{-2} + b^{-1}}$

6. $\dfrac{c^2 - c - 12}{c^2 - 2c - 8}$

Solve.

7. $\dfrac{2x - 1}{3} = \dfrac{3x + 2}{6}$

8. $3 - \dfrac{1}{2}x \le 4$

9. $\dfrac{x^2}{6} = \dfrac{x}{3} + \dfrac{5}{2}$

10. $\dfrac{3x}{4} + 1 > \dfrac{x - 1}{2}$

11. $\dfrac{3}{2}x - 1 = 5$

12. $\dfrac{1}{2}x^2 + \dfrac{7}{3}x - 4 < 0$

Self-Test 3

Vocabulary   fractional equation (p. 247)   extraneous root (p. 247)

Solve.

1. $\dfrac{3t - 2}{8} = \dfrac{t - 1}{12} + 1$

2. $\dfrac{y^2}{3} + \dfrac{y}{15} - \dfrac{2}{5} = 0$ Obj. 5-8, p. 242

3. One computer can process a payroll in 6 h. An older computer can do the same job in 7 h 30 min. How long would it take both computers to do the processing?

Solve and check. If an equation has no solution, say so.

4. $\dfrac{x}{x - 1} = \dfrac{x - 2}{x - 3}$

5. $\dfrac{z}{z - 2} - \dfrac{1}{z + 3} = \dfrac{10}{z^2 + z - 6}$ Obj. 5-9, p. 247

6. A train averaged 80 km/h for the first half of its trip. How fast must it travel for the second half of the trip in order to average 96 km/h for the whole trip?

Compare your answers with those at the back of the book.

Application / *Electrical Circuits*

Figure (a) pictures an electrical circuit containing an E-volt (V) battery, an R-ohm (Ω) resistance, and an ammeter for measuring the current I in amperes (A). Ohm's law states that $I = \dfrac{E}{R}$.

$$R_c = R_1 + R_2$$

$$\frac{1}{R_c} = \frac{1}{R_1} + \frac{1}{R_2}$$

(a)   **(b)**   **(c)**

Figures (b) and (c) show resistances R_1 and R_2 in series and in parallel, respectively, and also formulas for finding the total resistance R_c of the combination.

Example   Find the quantity labeled X in the following circuit.

Solution   The resistance of the 10 ohm–X ohm combination is R_c, where

$$\frac{1}{R_c} = \frac{1}{10} + \frac{1}{X} = \frac{X + 10}{10X}.$$

$$\therefore R_c = \frac{10X}{X + 10}.$$

The total resistance is

$$R = 5 + R_c$$

$$= 5 + \frac{10X}{X + 10}.$$

From $I = \dfrac{E}{R}$, or $IR = E$, you have

$$1.2\left(5 + \frac{10X}{X + 10}\right) = 14$$

Solving the equation, you find that $X = 20\ \Omega$.   *Answer*

Exercises

Find the quantity labeled _X_ in each of the following circuits.

1.

2.

3.

4.

Chapter Summary

1. The definitions

$$a^0 = 1 \quad \text{and} \quad a^{-n} = \frac{1}{a^n}$$

permit zero and negative integers as exponents. The laws of exponents stated on page 212 continue to hold.

2. In *scientific notation* a positive number is expressed in the form

$$m \times 10^n$$

where $1 \le m < 10$ and n is an integer. The digits of m should all be *significant*.

3. A quotient of polynomials is called a *rational expression*. Such an expression is simplified if its numerator and denominator have no common factors other than 1.

4. To multiply and divide rational expressions, you can use the product and quotient rules for fractions:

$$\frac{p}{q} \cdot \frac{r}{s} = \frac{pr}{qs}$$

and

$$\frac{p}{q} \div \frac{r}{s} = \frac{p}{q} \cdot \frac{s}{r}$$

5. To add or subtract rational expressions, find their *least common denominator* (LCD), express each as an equivalent fraction having the LCD as denominator, and then add or subtract.

6. To simplify a *complex fraction,* use either of these methods:

   Simplify the numerator and denominator separately; then divide.

   Multiply the numerator and denominator by the LCD of all the fractions appearing in the numerator and the denominator.

7. To solve an equation or an inequality having fractional coefficients, multiply both sides by the LCD of the fractions. To solve a *fractional equation* (one having a variable in the denominator), use the same method, but always check the results for any extraneous roots that may have been introduced.

Chapter Review

Give the letter of the correct answer.

1. Simplify $\dfrac{18x^3y}{12x^2y^4}$.

5-1

  a. $\dfrac{2}{3xy^3}$ **b.** $\dfrac{3x}{2y^3}$ **c.** $\dfrac{2xy^3}{3}$ **d.** $\dfrac{3y^3}{2x}$

2. Simplify $\dfrac{(a^2bc^3)^3}{(-ab^3c^2)^2}$.

  a. $-\dfrac{a^4c^5}{b^3}$ **b.** $a^4b^3c^5$ **c.** $-\dfrac{ac}{b^2}$ **d.** $\dfrac{a^4c^5}{b^3}$

3. Simplify $\dfrac{(p^2q)^{-1}}{p^2q^{-1}}$.

5-2

  a. $\dfrac{1}{p^4}$ **b.** q^2 **c.** $\dfrac{q^2}{p^4}$ **d.** 1

4. Simplify $\left(\dfrac{m^2}{n}\right)^{-2} \cdot \left(\dfrac{m^0}{n^{-1}}\right)^3$.

  a. $\dfrac{n^5}{m}$ **b.** m^4n **c.** $\dfrac{1}{m^4n}$ **d.** $\dfrac{n^5}{m^4}$

5. Express 0.000543 in scientific notation.

5-3

  a. 5.43×10^{-3} **b.** 5.43×10^{-4} **c.** 54.3×10^{-5} **d.** 543×10^{-6}

6. Find a one-significant-digit estimate for $(0.0593)(2045) \div 0.32$.

  a. 5×10^3 **b.** 4×10^2 **c.** 2×10^{-1} **d.** 3×10

7. Simplify $\dfrac{y^2 - 4}{y^2 + y - 6}$.

5-4

  a. $\dfrac{2}{y + 3}$ **b.** $\dfrac{y - 2}{y - 3}$ **c.** $\dfrac{y - 4}{y - 6}$ **d.** $\dfrac{y + 2}{y + 3}$

8. Find the zeros of $f(x) = \dfrac{2x^2 - x - 1}{2x^2 + 3x - 2}$.

   a. $-\dfrac{1}{2}$, 1   **b.** $\dfrac{1}{2}$, -2 **c.** $\dfrac{1}{2}$, -1 **d.** $-\dfrac{1}{2}$, 2

9. Simplify $\dfrac{u^2 + 3u}{u^2 + 2u - 3} \cdot \dfrac{u^3 - u}{u^2 - u - 2}$.   5-5

   a. $\dfrac{u^2}{u - 2}$ **b.** $u - 2$ **c.** $\dfrac{1}{u - 2}$ **d.** $\dfrac{u}{u - 2}$

10. Simplify $\dfrac{m^2}{n^3} \cdot \dfrac{n}{m^4} \div \dfrac{1}{mn}$.

   a. mn **b.** $m^3 n^3$ **c.** $\dfrac{1}{mn}$ **d.** $\dfrac{1}{m^3 n^3}$

11. Simplify $\dfrac{a + 2}{2a} + \dfrac{1 - a}{a^2}$.   5-6

   a. $\dfrac{a^2 + 2}{2a^3}$ **b.** $\dfrac{a^2 + 2}{2a^2}$ **c.** $\dfrac{a + 2}{2a}$ **d.** $\dfrac{a + 1}{a}$

12. Simplify $\dfrac{x - 1}{x + 1} - \dfrac{x^2 - x - 6}{x^2 + 4x + 3}$.

   a. $\dfrac{1}{x + 1}$ **b.** $\dfrac{2x^2 + x - 9}{(x + 1)(x + 3)}$

   c. $\dfrac{3}{x + 3}$ **d.** $\dfrac{x - 9}{(x + 1)(x + 3)}$

13. Simplify $\dfrac{y - \dfrac{2}{y + 1}}{1 - \dfrac{2}{y + 1}}$.   5-7

   a. $y + 1$ **b.** $\dfrac{y}{y - 1}$ **c.** $\dfrac{y - 2}{y + 1}$ **d.** $y + 2$

14. Solve $\dfrac{x + 1}{6} < x - \dfrac{3x - 2}{4}$.   5-8

   a. $\{x: x > -4\}$ **b.** $\{x: x < 4\}$ **c.** $\{x: x < -4\}$ **d.** $\{x: x > 4\}$

15. Solve $\dfrac{x(x - 2)}{6} = \dfrac{x + 1}{8}$.

   a. $\{3\}$ **b.** $\left\{-3, \dfrac{1}{4}\right\}$ **c.** $\left\{-\dfrac{1}{4}, 3\right\}$ **d.** $\{-1, 0, 2\}$

16. How much of an 8% saline solution should be added to 600 mL of a 3% solution to produce a 5% solution?

   a. 200 mL   **b.** 300 mL   **c.** 400 mL   **d.** 500 mL

17. Solve $\dfrac{2}{x + 2} + \dfrac{x^2}{x^2 - 4} = \dfrac{1}{x - 2}$.   5-9

   a. $\{-3, 2\}$ **b.** $\{-3\}$ **c.** $\{-2, 3\}$ **d.** $\{3\}$

18. Bill rowed 30 km upstream against a 2 km/h current and back again in a total of 8 h. How fast can he row in still water?

   a. 7 km/h   **b.** 8 km/h   **c.** 9 km/h   **d.** 10 km/h

Chapter Test

Simplify. Use only positive exponents in the answer.

1. $\dfrac{(3xy^3)^2}{6x^5y^4}$

2. $\dfrac{m^4}{4n^3}\left(\dfrac{2n}{m^3}\right)^3$

5-1

3. $pq^4(p^{-1}q^2)^{-2}$

4. $\left(\dfrac{u^2}{v}\right)^{-2}\left(\dfrac{v^3}{u}\right)^{-1}$

5-2

Evaluate. Give each answer in scientific notation with two significant digits.

5. $(2.1 \times 10^3)(1.6 \times 10^{-5})$

6. $\dfrac{5.2 \times 10^2}{1.2 \times 10^{-4}}$

5-3

Find (a) the domain and (b) the zeros of each function.

7. $f(x) = \dfrac{3x - 2}{x^2 - x - 12}$

8. $f(x) = \dfrac{x^2 - x - 12}{3x - 2}$

5-4

Simplify.

9. $\dfrac{6x^3}{y^2} \div \dfrac{3xy}{2}$

10. $\dfrac{u^2 - 9}{u + 2} \div \dfrac{u + 3}{u^2 - 4}$

5-5

11. $\dfrac{p - 2}{2p} + \dfrac{p + 3}{3p}$

12. $\dfrac{1}{s - 1} - \dfrac{2}{s^2 - 1}$

5-6

13. $\dfrac{ab^{-1} + 1}{a^{-1}b + 1}$

14. $\dfrac{\dfrac{x + y}{x - y}}{\dfrac{1}{x} - \dfrac{1}{y}}$

5-7

Solve.

15. $\dfrac{x + 7}{8} - \dfrac{1}{2} \geq \dfrac{x}{4}$

16. $\dfrac{x^2}{9} = \dfrac{x + 2}{2}$

5-8

17. One crew can detassel a field of corn in 6 days. Another crew can do the same job in 4 days. If the slower crew works alone for the first two days of detasseling and then is joined by the faster crew, how long will it take the two crews to finish?

18. Solve $\dfrac{3}{x^2 - 2x - 8} + \dfrac{1}{x + 2} = \dfrac{4}{x^2 - 16}$.

5-9

19. The members of a computer club were to be assessed equally to raise $400 for the purchase of software. When four more persons joined the club, the per-member assessment was reduced by $5.00. What was the new size of the club membership?

6 Irrational and Complex Numbers

$$E = E_0 e^{iwt}$$

The flow of an electrical current through a circuit can be described using a formula that includes the imaginary number i, defined as $\sqrt{-1}$.

Roots and Radicals

$$\sqrt{\frac{1}{9}} = \sqrt{\frac{1}{3}}$$

6-1 Roots of Real Numbers

Objective   To find roots of real numbers.

A **square root** of a number b is a solution of the equation $x^2 = b$.
Every positive number b has two square roots, denoted $\sqrt{b}$ and $-\sqrt{b}$.

For example, the square roots of 25 are $\sqrt{25} = 5$ and $-\sqrt{25} = -5$, since
$5^2 = 25$ and $(-5)^2 = 25$. The *positive* square root, $\sqrt{b}$, is called the
principal square root of b. The principal square root of 25 is $\sqrt{25} = 5$.

Example 1 Simplify: **a.** $\sqrt{9}$ **b.** $-\sqrt{9}$ **c.** $\sqrt{\frac{1}{9}}$ **d.** $\sqrt{0.09}$

Solution   **a.** 3   **b.** -3 **c.** $\frac{1}{3}$ **d.** 0.3

$1^0 = \cancel{1}\ \sqrt{-1}$
$1^2 = 1$
$1^3 = -1^p$
$1^4 = 1$

   Since the square of a real number is never negative, the equation $x^2 = b$
has no real-number solution if $b < 0$. Therefore, *a negative number b has no
real square roots.*

Root ไม่ สามารถ $\sqrt{-x}$ ได้   ถ้าเอา = no / root.

Example 2 Find the real roots of each equation. If there are none, say so.
    a. $x^2 = 9$ **b.** $x^2 + 4 = 0$ **c.** $5x^2 = 15$

Solution   **a.** $x^2 = 9$ **b.** $x^2 + 4 = 0$ **c.** $5x^2 = 15$
         $x = \pm\sqrt{9} = \pm 3$ $x^2 = -4$ $x^2 = 3$
         $\therefore$ the roots are $\therefore$ there are no $x = \pm\sqrt{3}$
            3 and -3.   real roots. $\therefore$ the roots are
                                                      $\sqrt{3}$ and $-\sqrt{3}$.

   In part (a) of Example 2, notice that the two solutions are written $\pm\sqrt{9}$,
or ± 3. You read $\pm\sqrt{9}$ as "plus or minus the square root of 9."

Caution: Do not confuse solving $x^2 = 9$, as in Example 2(a), with simplifying $\sqrt{9}$, as
        in Example 1(a). It is correct to write $x = \pm\sqrt{9} = \pm 3$ in solving $x^2 = 9$.
        However, it is *incorrect* to write $\sqrt{9} = \pm 3$ in simplifying $\sqrt{9}$, since $\sqrt{9}$
        represents only the *positive* square root of 9.

A **cube root** of a number b is a solution of the equation $x^3 = b$.

Every number b, whether positive, negative, or zero, has exactly one real cube root, denoted $\sqrt[3]{b}$.

Example 3 illustrates that the cube root of a positive number is positive and the cube root of a negative number is negative.

Example 3 Simplify.

    a. $\sqrt[3]{8}$ **b.** $\sqrt[3]{-27}$ **c.** $\sqrt[3]{10^6}$ **d.** $\sqrt[3]{a^9}$

Solution **a.** Since $2^3 = 8$, $\sqrt[3]{8} = 2$.
        b. Since $(-3)^3 = -27$, $\sqrt[3]{-27} = -3$.
        c. Since $(10^2)^3 = 10^6$, $\sqrt[3]{10^6} = 10^2$, or 100.
        d. Since $(a^3)^3 = a^9$, $\sqrt[3]{a^9} = a^3$.

Instead of defining the fourth, fifth, and higher roots of a number b separately, we will give a general definition of the *n*th root of b (where n is a positive integer).

1. An ***n*th root** of b is a solution of the equation $x^n = b$.

2. a. If n is even and $b > 0$, there are two real *n*th roots of b.
   The **principal** (or positive) ***n*th root** of b is denoted $\sqrt[n]{b}$.
   The other *n*th root of b is denoted $-\sqrt[n]{b}$.

   b. If n is even and $b = 0$, there is one *n*th root: $\sqrt[n]{0} = 0$.

   c. If n is even and $b < 0$, there is no real *n*th root of b.

3. If n is odd, there is exactly one real *n*th root of b, whether b is positive, negative, or zero.

Example 4 Simplify.

    a. $\sqrt[4]{81}$ **b.** $\sqrt[5]{32}$ **c.** $\sqrt[5]{-32}$ **d.** $\sqrt[6]{-1}$

Solution **a.** Since $3^4 = 81$, $\sqrt[4]{81} = 3$.
        b. Since $2^5 = 32$, $\sqrt[5]{32} = 2$.
        c. Since $(-2)^5 = -32$, $\sqrt[5]{-32} = -2$.
        d. Since $x^6 = -1$ has no solution, $\sqrt[6]{-1}$ does not represent a real number. (See part 2(c) in the chart above.)

The symbol $\sqrt[n]{b}$ is called a **radical.** Each part of a radical is given a name, as indicated below.

The index must be a positive integer. For a square root, the index 2 is usually omitted: $\sqrt{b}$ has the same meaning as $\sqrt[2]{b}$.

Here are some properties that will help you in simplifying radicals.

Properties of Radicals	Examples
1. $(\sqrt[n]{b})^n = b$, because $\sqrt[n]{b}$ satisfies the equation $x^n = b$.	$(\sqrt{7})^2 = 7$ $(\sqrt[3]{-5})^3 = -5$
2. $\sqrt[n]{b^n} = b$ if n is odd.	$\sqrt[3]{10^3} = 10$ $\sqrt[5]{x^5} = x$
3. $\sqrt[n]{b^n} = \|b\|$ if n is even, because the principal nth root is always nonnegative for even values of n.	$\sqrt{(-3)^2} = \|-3\| = 3$ $\sqrt{(x-1)^2} = \|x - 1\| = \begin{cases} x - 1 & \text{if } x \geq 1 \\ 1 - x & \text{if } x < 1 \end{cases}$

Oral Exercises

Simplify each expression. If the expression does not represent a real number, say so.

1. $\sqrt{36}$

2. $\sqrt[3]{-8}$

3. $\sqrt{-\frac{1}{4}}$

4. $-\sqrt{\frac{1}{25}}$

5. $-\sqrt[4]{16}$

6. $(\sqrt[3]{5})^3$

7. $(\sqrt{2})^2$

8. $(-\sqrt{11})^2$

9. $(-\sqrt[3]{13})^3$

10. $\sqrt[4]{(-7)^4}$

11. $\sqrt[5]{(-6)^5}$

12. $\sqrt[6]{0}$

Give the real roots of each equation. If an equation has no solution, say so.

13. $x^2 = 49$

14. $x^2 = 1$

15. $x^2 = -1$

16. $x^2 - 6 = 0$

17. $3x^2 = 27$

18. $5x^2 = 20$

Tell whether each equation is true for all real values of the variable.

19. $\sqrt{w^2} = \|w\|$

20. $\sqrt{y^4} = y^2$

21. $(\sqrt[4]{y})^4 = \|y\|$

22. $\sqrt[3]{x^3} = \|x\|$

23. $(\sqrt[4]{\|y\|})^4 = y$

24. $(\sqrt[5]{z})^5 = z$

Tell whether each statement is true for all real values of the variable.

25. $\sqrt[3]{x}$ is always a real number.

26. $\sqrt[6]{x}$ is always a real number.

27. $\sqrt{x + 1}$ is always a real number.

28. $\sqrt[3]{x + 1}$ is always a real number.

29. $\sqrt{x^2 + 1}$ is always a real number.

30. $\sqrt[4]{x^2 + 1}$ is always a real number.

$= 7\sqrt{2}$

31. For every real number x, $\sqrt[4]{x^4} = |x|$ and $\sqrt[5]{x^5} = x$. Using a numerical example, explain why the absolute value symbol must be used in the first case but not in the second.

$3\sqrt{25}$, $\sqrt[3]{10}$ $\quad$ $\sqrt[3]{250}$ $= 5\sqrt{2}$

Written Exercises

Simplify each expression. If the expression does not represent a real number, say so.

A

1. a. $\sqrt{16}$ **b.** $-\sqrt{16}$ **c.** $\sqrt{-16}$ **d.** $\sqrt[4]{16}$

2. a. $\sqrt{64}$ $= 8$ **b.** $\sqrt{-64}$ $= not$ **c.** $\sqrt[3]{64}$ $= 4$ **d.** $\sqrt[3]{-64}$ $= -4$

3. a. $\sqrt{81}$ **b.** $-\sqrt{81}$ **c.** $\sqrt{-81}$ **d.** $\sqrt[4]{81}$

4. a. $\sqrt{144}$ **b.** $\sqrt{-144}$ **c.** $-\sqrt{144}$ **d.** $\sqrt[4]{-144}$

5. a. $\sqrt{0.01}$ $= .1$ **b.** $\sqrt{-0.01}$ $= not$ **c.** $\sqrt[3]{0.001}$ $= .1$ **d.** $\sqrt[3]{-0.001}$ $= -.1$

6. a. $\sqrt{0.04}$ **b.** $-\sqrt{0.04}$ **c.** $\sqrt{0.0004}$ **d.** $\sqrt{-0.0004}$

7. a. $\sqrt{7^2}$ $= 7$ **b.** $\sqrt[3]{7^3}$ $= 7$ **c.** $\sqrt[4]{(-7)^4}$ $= 7$ **d.** $\sqrt[5]{(-7)^5}$ $= -7$

8. a. $\sqrt{5^2}$ **b.** $\sqrt{-5^2}$ **c.** $\sqrt{(-5)^3}$ **d.** $\sqrt[3]{-5^3}$

9. a. $\sqrt{\dfrac{1}{64}}$ **b.** $\dfrac{1}{\sqrt{64}}$ **c.** $\sqrt[3]{-\dfrac{1}{64}}$ **d.** $-\dfrac{1}{\sqrt[3]{64}}$

10. a. $\sqrt{\dfrac{1}{16}}$ **b.** $\sqrt{\dfrac{81}{16}}$ **c.** $\sqrt[4]{\dfrac{1}{16}}$ **d.** $\sqrt[4]{\dfrac{81}{16}}$

11. a. $\sqrt{10^2}$ **b.** $\sqrt{10^4}$ **c.** $\sqrt{10^6}$ **d.** $\sqrt{10^{20}}$

12. a. $\sqrt[3]{10^{-3}}$ **b.** $\sqrt[3]{10^{-6}}$ **c.** $\sqrt[3]{10^{-9}}$ **d.** $\sqrt[3]{10^{-30}}$

13. a. $\sqrt{a^2}$ a **b.** $\sqrt{a^4}$ a **c.** $\sqrt[3]{a^6}$ $\sqrt{a}$ **d.** $\sqrt[6]{a^6}$ a

14. a. $\sqrt{-a^2}$ **b.** $\sqrt{(-a)^2}$ $= a$ **c.** $\sqrt[4]{a^4}$ $= a$ **d.** $\sqrt[6]{a^6}$ $= a^3$

Find the real roots of each equation. If there are none, say so.

15. $x^2 = 144$ ± 12 **16.** $y^2 = 0$ **17.** $x^2 + 9 = 0$

18. $y^2 - 7 = 0$ **19.** $9x^2 = 4$ **20.** $25y^2 = -16$

21. $16y^2 = 25$ **22.** $9x^2 - 81 = 0$ **23.** $4 - 16x^2 = 0$

24. $0 = 4 + 16x^2$ no **25.** $81 - 9x^2 = 0$ **26.** $25y^2 + 16 = 17$

For what values of the variable is each equation true?

Sample $\sqrt{(x-3)^2} = x - 3$

Solution $\sqrt{(x-3)^2} = |x-3| = x - 3$ if $x \geq 3$

(Note that if $x < 3$, $|x - 3| = -(x-3) = 3 - x$.)

B 27. $\sqrt{(x+5)^2} = x + 5$

28. $\sqrt{(a-2)^2} = 2 - a$

29. $\sqrt[3]{(n+1)^3} = n + 1$

30. $\sqrt{y^4} = y^2$

31. $\sqrt{c^2 + 4c + 4} = |c + 2|$

32. $\sqrt{b^2 - 2b + 1} = b - 1$

For what values of x does each expression represent a real number?

33. **a.** $\sqrt{x+1}$ **b.** $\sqrt{x-1}$ **c.** $\sqrt[3]{x-1}$ **d.** $\sqrt{x^2-1}$

34. **a.** $\sqrt{4-x}$ **b.** $\sqrt{4-x^2}$ **c.** $\sqrt[3]{4-x^2}$ **d.** $\sqrt{4+x^2}$

C 35. $\sqrt{x^3 - 9x}$ 36. $\sqrt{16x - x^2}$ 37. $\sqrt{\sqrt{x} - x}$ 38. $\sqrt{x - \sqrt{x}}$

Mixed Review Exercises

Solve. If an equation has no solution, say so.

1. $\dfrac{x}{x-1} = \dfrac{2x}{x+2}$

2. $-3y = y^2$

3. $\dfrac{3n-1}{4} + \dfrac{n+1}{2} = 4$

4. $|5 - 2w| = 1$

5. $\dfrac{d}{d+2} + \dfrac{3}{d} = 2$

6. $\dfrac{u^2}{3} = \dfrac{u+1}{6}$

7. $3(2m - 1) = 4m + 7$

8. $2p^2 = 12 - 5p$

9. $3k + 6 = 3(k + 4)$

10. $\dfrac{1}{f} + \dfrac{1}{f-1} = \dfrac{1}{f^2-f}$

Calculator Key-In

Use the square root key, usually labeled $\sqrt{x}$, to simplify each radical. (In Exercises 5 and 6, use scientific notation.)

1. $\sqrt{11{,}025}$

2. $\sqrt{1{,}010{,}025}$

3. $\sqrt{0.9801}$

4. $\sqrt{0.998001}$

5. $\sqrt{40{,}000{,}000{,}000}$

6. $\sqrt{0.00000000025}$

7. The expressions $\sqrt{2} - \sqrt{3}$ and $\dfrac{\sqrt{6} - \sqrt{2}}{2}$ represent the same real number. Use your calculator to confirm this fact.

6-2 Properties of Radicals

Objective   To simplify expressions involving radicals.

Since $\sqrt{4 \cdot 9} = \sqrt{36} = 6$ and $\sqrt{4} \cdot \sqrt{9} = 2 \cdot 3 = 6$, you can see that

$$\sqrt{4 \cdot 9} = 6 = \sqrt{4} \cdot \sqrt{9}.$$

Similarly, $$\sqrt{\frac{4}{9}} = \frac{2}{3} = \frac{\sqrt{4}}{\sqrt{9}}.$$

These examples illustrate two important properties of radicals, stated in the chart below.

Product and Quotient Properties of Radicals

If $\sqrt[n]{a}$ and $\sqrt[n]{b}$ are real numbers, then:

$$\textbf{1. } \sqrt[n]{ab} = \sqrt[n]{a} \cdot \sqrt[n]{b}$$

$$\textbf{2. } \sqrt[n]{\frac{a}{b}} = \frac{\sqrt[n]{a}}{\sqrt[n]{b}}$$

Caution: Do *not* assume that $\sqrt[n]{a + b} = \sqrt[n]{a} + \sqrt[n]{b}$ or $\sqrt[n]{a - b} = \sqrt[n]{a} - \sqrt[n]{b}$.
For example, $\sqrt{9 + 16} \neq \sqrt{9} + \sqrt{16}$, and $\sqrt{4 - x^2}$ is not equivalent to $2 - x$.

You are asked to prove the product and quotient properties of radicals in Exercises 67 and 68. As the following examples illustrate, these properties can be used to simplify expressions involving radicals.

Example 1 Simplify.

    a. $\sqrt{98}$ **b.** $\sqrt[3]{25} \cdot \sqrt[3]{10}$ **c.** $\sqrt[3]{\frac{81}{8}}$ **d.** $\frac{\sqrt{60}}{\sqrt{5}}$

Solution **a.** $\sqrt{98} = \sqrt{49 \cdot 2}$ **b.** $\sqrt[3]{25} \cdot \sqrt[3]{10} = \sqrt[3]{250} = \sqrt[3]{125 \cdot 2}$
$$= \sqrt{49} \cdot \sqrt{2} \qquad\qquad\qquad\qquad = \sqrt[3]{125} \cdot \sqrt[3]{2}$$
$$= 7\sqrt{2} \quad \textit{Answer} \qquad\qquad\qquad = 5\sqrt[3]{2} \quad \textit{Answer}$$

c. $\sqrt[3]{\frac{81}{8}} = \frac{\sqrt[3]{81}}{\sqrt[3]{8}}$ **d.** $\frac{\sqrt{60}}{\sqrt{5}} = \sqrt{\frac{60}{5}}$

$$= \frac{\sqrt[3]{27 \cdot 3}}{2} \qquad\qquad\qquad = \sqrt{12}$$

$$= \frac{\sqrt[3]{27} \cdot \sqrt[3]{3}}{2} \qquad\qquad\quad = \sqrt{4 \cdot 3}$$

$$= \frac{3\sqrt[3]{3}}{2} \quad \textit{Answer} \qquad\quad = \sqrt{4} \cdot \sqrt{3}$$

$$\qquad\qquad\qquad\qquad\qquad\qquad = 2\sqrt{3} \quad \textit{Answer}$$

Example 2 Simplify. Assume that each radical represents a real number.

     a. $\sqrt{2a^2b}$ **b.** $\sqrt{36w^3}$ **c.** $\sqrt[3]{\dfrac{x^5}{y^3}}$ **d.** $\sqrt{4a^2 + 4b^2}$

Solution

    a. $\sqrt{2a^2b} = \sqrt{a^2} \cdot \sqrt{2b}$

                $= |a|\sqrt{2b}$ *Answer*

    b. $\sqrt{36w^3} = \sqrt{36w^2 \cdot w}$

                $= \sqrt{36w^2} \cdot \sqrt{w}$

                $= 6w\sqrt{w}$ *Answer*

       (Since $\sqrt{36w^3}$ is real, w is nonnegative, and
       you can write $6w\sqrt{w}$ rather than $6|w|\sqrt{w}$.)

    c. $\sqrt[3]{\dfrac{x^5}{y^3}} = \dfrac{\sqrt[3]{x^5}}{\sqrt[3]{y^3}} = \dfrac{\sqrt[3]{x^3 \cdot x^2}}{y} = \dfrac{\sqrt[3]{x^3} \cdot \sqrt[3]{x^2}}{y} = \dfrac{x\sqrt[3]{x^2}}{y}$ *Answer*

    d. $\sqrt{4a^2 + 4b^2} = \sqrt{4(a^2 + b^2)}$

                    $= \sqrt{4} \cdot \sqrt{a^2 + b^2}$

                    $= 2\sqrt{a^2 + b^2}$ *Answer*

The next example illustrates a process known as **rationalizing the denominator,** in which a perfect square, cube, or other power is created in the denominator so that the radical expression can be written without a fraction in the radicand or a radical in the denominator.

Example 3 Simplify: **a.** $\sqrt{\dfrac{5}{3}}$ **b.** $\dfrac{4}{\sqrt[3]{c}}$

Solution

    a. $\sqrt{\dfrac{5}{3}} = \sqrt{\dfrac{5}{3} \cdot \dfrac{3}{3}}$ $\longleftarrow$ Numerator and denominator are multiplied by 3 so that the denominator will be a perfect square.

              $= \sqrt{\dfrac{15}{9}}$

              $= \dfrac{\sqrt{15}}{\sqrt{9}}$

              $= \dfrac{\sqrt{15}}{3}$ *Answer*

    b. $\dfrac{4}{\sqrt[3]{c}} = \dfrac{4}{\sqrt[3]{c}} \cdot \dfrac{\sqrt[3]{c^2}}{\sqrt[3]{c^2}}$ $\longleftarrow$ Numerator and denominator are multiplied by $\sqrt[3]{c^2}$ so that the radicand in the denominator will be a perfect cube.

            $= \dfrac{4\sqrt[3]{c^2}}{\sqrt[3]{c^3}}$

            $= \dfrac{4\sqrt[3]{c^2}}{c}$ *Answer*

The process of rationalizing denominators was very important before the invention of calculators. You can see why if you try to evaluate $\sqrt{\frac{5}{3}}$ using a table of square roots (see Table 1, page 810) instead of a calculator: $\frac{\sqrt{15}}{3}$ is easier to evaluate than $\sqrt{\frac{5}{3}}$ $\left(\text{or even } \frac{\sqrt{5}}{\sqrt{3}}\right)$.

Although rationalizing denominators is now less important for numerical work than it used to be, it is often used in mathematics classes as a convenient way to compare answers. This is also the reason that answers involving radicals are given in *simplest radical form*.

An expression containing nth roots is in **simplest radical form** if:

1. no radicand contains a factor (other than 1) that is a perfect nth power, and

2. every denominator has been rationalized, so that no radicand is a fraction and no radical is in a denominator.

All the answers in Examples 1, 2, and 3 are in simplest form.

You can use a calculator or a table of roots to obtain a decimal approximation of some radicals. Table 1 and Table 2 on pages 810–811 are provided for this purpose.

Example 4 Give a decimal approximation to the nearest hundredth for each radical.

    a. $\sqrt{53}$ **b.** $\sqrt[3]{7100}$

Solution   **a.** Use Table 1. Find 5.3 and look under $\sqrt{10N}$.

        $\sqrt{53} \approx 7.28$ ***Answer***

    b. Use Table 2.

        $\sqrt[3]{7100} = \sqrt[3]{1000} \cdot \sqrt[3]{7.1} \approx 10(1.922) = 19.22$ ***Answer***

The following two theorems are often useful in numerical work with radicals. You are asked to prove these theorems in Exercises 69 and 70.

Theorem 1. If each radical represents a real number, then $\sqrt[nq]{b} = \sqrt[n]{\sqrt[q]{b}}$.

Theorem 2. If $\sqrt[n]{b}$ represents a real number, then $\sqrt[n]{b^m} = (\sqrt[n]{b})^m$.

The first of these theorems allows you to transform a radical with a large index into two radicals with smaller indexes. The second allows you to transform a root of a power into a power of a root.

Example 5 Give a decimal approximation to the nearest hundredth for each radical.

  a. $\sqrt[4]{100}$ **b.** $\sqrt[3]{170^2}$

Solution   **a.** $\sqrt[4]{100} = \sqrt{\sqrt{100}} = \sqrt{10} \approx 3.16$

  b. $\sqrt[3]{170^2} = (\sqrt[3]{170})^2 \approx (5.540)^2 \approx 30.69$

To investigate how you can use your calculator to find roots of numbers, try the Calculator Key-Ins on pages 263 and 269.

Oral Exercises

Tell whether each radical is in simplest form. Explain your answer.

1. $\sqrt{\dfrac{1}{3}}$ **2.** $\dfrac{\sqrt{6}}{5}$ **3.** $\sqrt{72}$ **4.** $\dfrac{2}{\sqrt[3]{4}}$

Simplify.

5. $\sqrt{75}$ **6.** $\sqrt{24}$ **7.** $\sqrt{600}$ **8.** $\sqrt{98}$

9. $\sqrt{\dfrac{3}{16}}$ **10.** $\dfrac{1}{\sqrt{5}}$ **11.** $\dfrac{2}{\sqrt{2}}$ **12.** $\sqrt{\dfrac{2}{3}}$

13. $\sqrt{2} \cdot \sqrt{6}$ **14.** $\sqrt{10} \cdot \sqrt{15}$ **15.** $\dfrac{\sqrt{27}}{\sqrt{3}}$ **16.** $\dfrac{\sqrt{56}}{\sqrt{7}}$

17. $\sqrt[3]{40}$ **18.** $\sqrt[3]{\dfrac{16}{27}}$ **19.** $\sqrt[3]{8} \cdot \sqrt[3]{-8}$ **20.** $\dfrac{\sqrt[3]{48}}{\sqrt[3]{2}}$

21. Can you apply the product property of radicals to $\sqrt{6} \cdot \sqrt[3]{5}$? Explain your answer.

22. Does $\sqrt{x^2 + y^2} = x + y$? Explain.

23. Does $\sqrt{4 - x^2} = 2 - x$? Explain.

Written Exercises

Simplify.

A   **1.** $\sqrt{52}$ **2.** $\sqrt{125}$ **3.** $\sqrt{162}$

    4. $\sqrt{363}$ **5.** $\sqrt{196}$ **6.** $\sqrt{324}$

    7. $\sqrt{\dfrac{8}{9}}$ **8.** $\sqrt{\dfrac{50}{49}}$ **9.** $\sqrt{\dfrac{4}{3}}$

Simplify.

10. $\sqrt{\dfrac{9}{5}}$

11. $\dfrac{4}{\sqrt{2}}$

12. $\dfrac{6}{\sqrt{3}}$

13. $\dfrac{\sqrt{270}}{\sqrt{6}}$

14. $\dfrac{\sqrt{96}}{\sqrt{3}}$

15. $\sqrt{30} \cdot \sqrt{42}$

16. $\sqrt{35} \cdot \sqrt{21}$

17. $\sqrt{6} \cdot \sqrt{\dfrac{2}{3}}$

18. $\sqrt{15} \cdot \sqrt{\dfrac{3}{5}}$

19. $\sqrt[3]{250}$

20. $\sqrt[3]{135}$

21. $\sqrt[3]{\dfrac{5}{4}}$

22. $\sqrt[3]{\dfrac{2}{9}}$

23. $\dfrac{9\sqrt{2}}{\sqrt{18}}$

24. $\dfrac{4\sqrt{3}}{\sqrt{12}}$

25. $(2\sqrt{7})^2$

26. $(3\sqrt{6})^2$

27. $\sqrt[3]{45} \cdot \sqrt[3]{12}$

28. $\sqrt[3]{20} \cdot \sqrt[3]{14}$

29. $\dfrac{\sqrt[3]{60}}{\sqrt[3]{36}}$

30. $\dfrac{\sqrt[3]{175}}{\sqrt[3]{50}}$

31. a. $\sqrt{32}$ **b.** $\sqrt[3]{32}$ **c.** $\sqrt[4]{32}$ **d.** $\sqrt[5]{32}$

32. a. $\sqrt{\dfrac{3}{8}}$ **b.** $\sqrt[3]{\dfrac{3}{8}}$ **c.** $\sqrt[4]{\dfrac{3}{8}}$ **d.** $\sqrt[5]{\dfrac{3}{8}}$

Give a decimal approximation to the nearest hundredth for each radical. Use a calculator or Tables 1 and 2 on pages 810–811.

33. $\sqrt{39}$

34. $\sqrt{870}$

35. $\sqrt[3]{450}$

36. $\sqrt[3]{7300}$

37. $\sqrt[4]{144}$

38. $\sqrt[4]{900}$

Simplify. Assume that each radical represents a real number.

39. $\sqrt{18x^2}$

40. $\sqrt{12x^5}$

41. $\sqrt[3]{375a^5}$

42. $\sqrt[3]{16c^4}$

43. $\sqrt{\dfrac{x^2}{y^3}}$

44. $\sqrt{\dfrac{y^2}{x^5}}$

45. $\sqrt[3]{\dfrac{27a}{4b^4}}$

46. $\sqrt[3]{\dfrac{8c}{9d^5}}$

B 47. $\sqrt{16a + 16b}$

48. $\sqrt{9a^2 - 9b^2}$

49. $\sqrt{2a^2 + 4a + 2}$

50. $\sqrt{3x^2 - 12x + 12}$

Evaluate the following radicals if $x = 4$, $y = 3$, and $z = 8$.

51. $\sqrt{x^{-1}y^{-2}}$

52. $\sqrt[3]{(xz)^{-1}}$

53. $\sqrt[3]{x^{-1} + z^{-1}}$

54. $\sqrt{x^{-2} + y^{-2}}$

55. $\sqrt[3]{(y - x)^{-1}}$

56. $\sqrt{(x + z)^{-1}}$

57. $\sqrt[6]{(xy)^z}$

58. $\sqrt[4]{xy^z}$

Simplify, giving answers with no negative exponents. Assume that each radical represents a real number.

59. $\sqrt{27x^3y^{-2}}$

60. $\sqrt[3]{32a^{-3}b^4}$

61. $\sqrt{3a} \cdot \sqrt{15a^{-2}}$

62. $\sqrt{2x^{-3}} \cdot \sqrt{6x}$

63. $\sqrt[3]{6y^{-4}} \cdot \sqrt[3]{9y^2}$

64. $\sqrt[3]{6c} \cdot \sqrt[3]{8c^{-5}}$

65. $\dfrac{\sqrt{2x^{-1}}}{\sqrt{8x}}$

66. $\dfrac{\sqrt{27x}}{\sqrt{3x^{-1}}}$

C **67.** The purpose of this exercise is to prove the product property of radicals.

    a. Using the exponent law $(xy)^n = x^n \cdot y^n$, show that $(\sqrt[n]{a} \cdot \sqrt[n]{b})^n = ab$.

    b. Part (a) shows that $\sqrt[n]{a} \cdot \sqrt[n]{b}$ is a solution of $x^n = ab$ and is therefore an nth root of ab. Prove that it is the principal nth root of ab.

68. Prove the quotient property of radicals. Follow the method suggested in Exercise 67.

69. Prove the first theorem on page 266. (*Hint:* Show that $\sqrt[n]{\sqrt[q]{b}}$ is an nqth root of b by justifying these statements:

$$(\sqrt[n]{\sqrt[q]{b}})^{nq} = [(\sqrt[n]{\sqrt[q]{b}})^n]^q = (\sqrt[q]{b})^q = b$$

Then show that $\sqrt[n]{\sqrt[q]{b}}$ is the principal nqth root by considering the cases $b \geq 0$ and $b < 0$. Note that if $b < 0$, then q and n are both odd.)

70. Prove the second theorem on page 266. (*Hint:* Show that $(\sqrt[n]{b})^m$ is an nth root of b^m by justifying these statements:

$$[(\sqrt[n]{b})^m]^n = [(\sqrt[n]{b})^n]^m = b^m$$

Then show that it is the principal nth root by considering the cases n even and n odd.)

Calculator Key-In

Example 5(a) in Lesson 6-2 illustrates a method for finding the fourth root of a number using square roots. Use your calculator's square root key to find a decimal approximation for each radical.

1. $\sqrt[4]{897}$ **2.** $\sqrt[4]{523.72}$ **3.** $\sqrt[8]{0.002}$ **4.** $\sqrt[16]{8.3 \times 10^{-80}}$

5. Can you use the square root key to find $\sqrt[10]{1024}$? Explain.

You can evaluate the nth root of a positive number for any n if your calculator has an $\sqrt[x]{y}$ key. (Some calculators instead have an $x^{1/y}$ key or a combination of INV and y^x keys.) Investigate how your calculator works by finding $\sqrt[3]{8}$, which you know is 2. Then find a decimal approximation to the nearest hundredth for each radical.

6. $\sqrt[3]{2463}$ **7.** $\sqrt[3]{479.552}$ **8.** $\sqrt[6]{0.518}$ **9.** $\sqrt[20]{4.5 \times 10^{40}}$

10. What value does $\sqrt[n]{n}$ approach as n gets very large?

6-3 Sums of Radicals

Objective   To simplify expressions involving sums of radicals.

Two radicals with the same index and radicand are called *like radicals*. You can apply the distributive property to add or subtract like radicals in the same way as like terms.

Combine like terms:

$$5x + y - 4x = (5x - 4x) + y$$
$$= x + y$$

Combine like radicals:

$$5\sqrt{2} + \sqrt[3]{2} - 4\sqrt{2} = (5\sqrt{2} - 4\sqrt{2}) + \sqrt[3]{2}$$
$$= \sqrt{2} + \sqrt[3]{2}$$

You cannot combine $\sqrt{2}$ and $\sqrt[3]{2}$ because they are *not* like radicals; they have different indexes. Example 1 shows that sometimes you can combine unlike radicals if you first simplify them so that they become like radicals.

Example 1   Simplify.

    a. $\sqrt{8} + \sqrt{98}$ **b.** $\sqrt[3]{81} - \sqrt[3]{24}$ **c.** $\sqrt{\dfrac{32}{3}} + \sqrt{\dfrac{2}{3}}$

Solution   **a.** $\sqrt{8} + \sqrt{98} = \sqrt{4 \cdot 2} + \sqrt{49 \cdot 2}$
$$= 2\sqrt{2} + 7\sqrt{2}$$
$$= 9\sqrt{2} \quad \textit{Answer}$$

    b. $\sqrt[3]{81} - \sqrt[3]{24} = \sqrt[3]{27 \cdot 3} - \sqrt[3]{8 \cdot 3}$
$$= 3\sqrt[3]{3} - 2\sqrt[3]{3}$$
$$= \sqrt[3]{3} \quad \textit{Answer}$$

    c. $\sqrt{\dfrac{32}{3}} + \sqrt{\dfrac{2}{3}} = \sqrt{\dfrac{32 \cdot 3}{3 \cdot 3}} + \sqrt{\dfrac{2 \cdot 3}{3 \cdot 3}}$
$$= \sqrt{\dfrac{96}{9}} + \sqrt{\dfrac{6}{9}}$$
$$= \dfrac{\sqrt{96}}{\sqrt{9}} + \dfrac{\sqrt{6}}{\sqrt{9}}$$
$$= \dfrac{\sqrt{16 \cdot 6}}{3} + \dfrac{\sqrt{6}}{3}$$
$$= \dfrac{4\sqrt{6}}{3} + \dfrac{\sqrt{6}}{3}$$
$$= \dfrac{5\sqrt{6}}{3} \quad \textit{Answer}$$

Example 2 Simplify.

a. $\sqrt{6}(\sqrt{2} + \sqrt{3})$ **b.** $\dfrac{\sqrt{21} + \sqrt{15}}{\sqrt{3}}$

Solution **a.** $\sqrt{6}(\sqrt{2} + \sqrt{3}) = \sqrt{6} \cdot \sqrt{2} + \sqrt{6} \cdot \sqrt{3}$
$= \sqrt{12} + \sqrt{18}$
$= \sqrt{4 \cdot 3} + \sqrt{9 \cdot 2}$
$= 2\sqrt{3} + 3\sqrt{2}$ *Answer*

b. $\dfrac{\sqrt{21} + \sqrt{15}}{\sqrt{3}} = \dfrac{\sqrt{21}}{\sqrt{3}} + \dfrac{\sqrt{15}}{\sqrt{3}}$
$= \sqrt{\dfrac{21}{3}} + \sqrt{\dfrac{15}{3}}$
$= \sqrt{7} + \sqrt{5}$ *Answer*

Example 3 Simplify. Assume that each radical represents a real number.

a. $\sqrt{12x^5} - x\sqrt{3x^3} + 5x^2\sqrt{3x}$ **b.** $\sqrt{6y} - \dfrac{\sqrt{3y}}{\sqrt{2}}$

Solution **a.** $\sqrt{12x^5} - x\sqrt{3x^3} + 5x^2\sqrt{3x} = \sqrt{4x^4 \cdot 3x} - x\sqrt{x^2 \cdot 3x} + 5x^2\sqrt{3x}$
$= 2x^2\sqrt{3x} - x^2\sqrt{3x} + 5x^2\sqrt{3x}$
$= 6x^2\sqrt{3x}$ *Answer*

b. $\sqrt{6y} - \dfrac{\sqrt{3y}}{\sqrt{2}} = \sqrt{6y} - \dfrac{\sqrt{3y}}{\sqrt{2}} \cdot \dfrac{\sqrt{2}}{\sqrt{2}}$
$= \sqrt{6y} - \dfrac{\sqrt{6y}}{2}$
$= \dfrac{2\sqrt{6y} - \sqrt{6y}}{2}$
$= \dfrac{\sqrt{6y}}{2}$ *Answer*

Oral Exercises

Simplify. If no simplification is possible, say so.

1. $\sqrt{5} + \sqrt{10}$ **2.** $\sqrt{7} + \sqrt[3]{7}$ No **3.** $\sqrt{5} + \sqrt{2} + 2\sqrt{5}$

4. $\sqrt{18} + 2\sqrt{3}$ **5.** $\sqrt[3]{16} - \sqrt[3]{2}$ **6.** $\sqrt{2} + \sqrt{\dfrac{1}{2}}$

7. $\sqrt[3]{3} - \sqrt[3]{2}$ **8.** $\sqrt{12} - \sqrt{27}$ **9.** $\sqrt{3} + \sqrt{30} + \sqrt{300}$

10. $\sqrt{3}(\sqrt{3} + \sqrt{6})$ **11.** $\sqrt{2}(2 - \sqrt{2})$ **12.** $\sqrt{\dfrac{3}{4}} + \sqrt{\dfrac{27}{4}}$

Simplify.

13. $\dfrac{\sqrt{8} + \sqrt{18}}{\sqrt{2}}$

14. $\dfrac{\sqrt{15} + \sqrt{35}}{\sqrt{5}}$

15. $\dfrac{8\sqrt{6} - 2\sqrt{3}}{2\sqrt{3}}$

16. Is $\sqrt{x} + \sqrt{x}$ equivalent to $\sqrt{2x}$? Explain.

Written Exercises

Simplify. If no simplification is possible, say so.

A 1. $\sqrt{50} + \sqrt{18}$

2. $\sqrt{45} - \sqrt{20}$

3. $3\sqrt{12} - \sqrt{48}$

4. $\sqrt{27} + 2\sqrt{75}$

5. $5\sqrt{2} - 2\sqrt{5}$

6. $7\sqrt{3} - 3\sqrt{7}$

7. $\sqrt{6} + \sqrt{36} + \sqrt{216}$

8. $\sqrt{5} + \sqrt{25} + \sqrt{125}$

9. $\sqrt{50} + \sqrt{63} - \sqrt{32}$

10. $\sqrt{18} + \sqrt{24} - \sqrt{54}$

11. $\sqrt[3]{54} + \sqrt[3]{40} + \sqrt[3]{16}$

12. $\sqrt[3]{24} - \sqrt[3]{56} + \sqrt[3]{81}$

13. $\sqrt{\dfrac{27}{5}} - \sqrt{\dfrac{3}{5}}$

14. $\sqrt{\dfrac{75}{2}} - \sqrt{\dfrac{3}{2}}$

15. $\sqrt{\dfrac{2}{3}} + \sqrt{\dfrac{3}{2}}$

16. $\sqrt{\dfrac{5}{2}} + \sqrt{\dfrac{2}{5}}$

17. $\sqrt[3]{4} + \sqrt[3]{\dfrac{1}{2}}$

18. $\sqrt[3]{16} - \sqrt[3]{\dfrac{1}{4}}$

19. $\sqrt{2}(\sqrt{8} + \sqrt{10})$

20. $\sqrt{3}(\sqrt{12} - \sqrt{24})$

21. $\sqrt{15}(\sqrt{3} + 2\sqrt{5})$

22. $\sqrt{7}(3\sqrt{14} - \sqrt{21})$

23. $2\sqrt{3}(\sqrt{48} - 5\sqrt{12})$

24. $3\sqrt{5}(\sqrt{5} + 2\sqrt{75})$

25. $\dfrac{\sqrt{6} - \sqrt{24}}{\sqrt{2}}$

26. $\dfrac{\sqrt{18} - \sqrt{6}}{\sqrt{3}}$

27. $\dfrac{4\sqrt{300} - \sqrt{108}}{\sqrt{12}}$

28. $\dfrac{\sqrt{40} - 2\sqrt{5}}{\sqrt{10}}$

29. $\sqrt{\dfrac{2}{3}}\left(\sqrt{\dfrac{27}{2}} - \dfrac{3}{\sqrt{2}}\right)$

30. $\sqrt{\dfrac{3}{8}}\left(\sqrt{\dfrac{3}{4}} + \dfrac{2}{\sqrt{3}}\right)$

B 31. $\sqrt[3]{5}(\sqrt[3]{200} - \sqrt[3]{16})$

32. $\sqrt[3]{40}(\sqrt[3]{25} + 2\sqrt[3]{5})$

33. $\dfrac{\sqrt[3]{18} + 3\sqrt[3]{54}}{\sqrt[3]{3}}$

34. $\dfrac{\sqrt[3]{320} + \sqrt[3]{1250}}{2\sqrt[3]{5}}$

Simplify. Assume that each radical represents a real number.

35. $\sqrt{8x^3} - x\sqrt{18x}$

36. $y^2\sqrt{45y} + 2y\sqrt{5y^3}$

37. $\sqrt{p^3r} + \sqrt{pr^3}$

38. $\sqrt{2a^2b^4} + \sqrt{8a^2b^4}$

39. $\sqrt{10a} - \dfrac{\sqrt{5a}}{\sqrt{2}} + \sqrt{\dfrac{2a}{5}}$

40. $\sqrt{6x} + \dfrac{\sqrt{2x}}{\sqrt{3}} - \sqrt{\dfrac{3x}{2}}$

41. $\sqrt{6w}(\sqrt{3w} + \sqrt{2w^3})$

42. $\sqrt{10t}(\sqrt{2t^5} - \sqrt{5t})$

Use the Pythagorean theorem to find x in each right triangle shown.

43.

44.

![bar]

Mixed Review Exercises

Simplify.

1. $\sqrt{28x^3}$

2. $\sqrt{\dfrac{81}{49}}$

3. $\dfrac{a^3x^2y}{a^2x^3y^2}$

4. $\sqrt{\dfrac{45x}{32y^4}}$

5. $\dfrac{x^2 - 5x + 6}{2 - x}$

6. $\sqrt[3]{-135}$

7. $(3x + 1)(3x - 1)$

8. $\sqrt{\dfrac{1}{x}}$

9. $\sqrt[3]{8x^4y^6}$

Extra / The Irrationality of $\sqrt{2}$

The proof that the number $\sqrt{2}$ is irrational, outlined below, is usually attributed to Pythagoras. The proof is *indirect;* that is, we first assume that $\sqrt{2}$ is rational and then show that this assumption leads to a contradiction. Here is the proof:

Suppose that $\sqrt{2}$ is rational. Then we can write $\sqrt{2}$ as a fraction and reduce this fraction to lowest terms. That is,

$$\sqrt{2} = \frac{a}{b}$$

where a and b are integers having no common factor except 1. Then $\dfrac{a^2}{b^2} = 2$ and $a^2 = 2b^2$, so a^2 is even. Since a^2 is even, a is even, and we can write $a = 2c$ for some integer c. Then $a^2 = 4c^2$, so $2b^2 = 4c^2$ and $b^2 = 2c^2$. Therefore b^2 is even and b is even. Since a and b are both even, they have the common factor 2, which contradicts our assumption that $\dfrac{a}{b}$ is in lowest terms. Therefore $\sqrt{2}$ *cannot* be written as a quotient of integers; that is, $\sqrt{2}$ is irrational.

Exercises

1. Prove that $\sqrt{3}$ is irrational.

2. Prove that $\sqrt{p}$, where p is a prime number, is irrational.

Irrational and Complex Numbers **273**

6-4 Binomials Containing Radicals

Objective   To simplify products and quotients of binomials that contain radicals.

You can multiply binomials containing radicals just as you would multiply any binomials.

Example 1   Simplify.

a. $(4 + \sqrt{7})(3 + 2\sqrt{7})$ **b.** $(2\sqrt{3} - \sqrt{6})^2$

c. $(4\sqrt{5} + 3\sqrt{2})(4\sqrt{5} - 3\sqrt{2})$

Solution   **a.** Recall how you simplify $(4 + x)(3 + 2x)$:

$$(4 + x)(3 + 2x) = 12 + 8x + 3x + 2x^2$$
$$= 12 + 11x + 2x^2$$
$$\therefore \quad (4 + \sqrt{7})(3 + 2\sqrt{7}) = 12 + 8\sqrt{7} + 3\sqrt{7} + 2(\sqrt{7})^2$$
$$= 12 + 11\sqrt{7} + 14$$
$$= 26 + 11\sqrt{7} \quad \textit{Answer}$$

b. Recall this pattern:

$$(a - b)^2 = a^2 - 2ab + b^2$$
$$\therefore \quad (2\sqrt{3} - \sqrt{6})^2 = (2\sqrt{3})^2 - 2 \cdot 2\sqrt{3} \cdot \sqrt{6} + (\sqrt{6})^2$$
$$= 12 - 4\sqrt{18} + 6$$
$$= 12 - 12\sqrt{2} + 6$$
$$= 18 - 12\sqrt{2} \quad \textit{Answer}$$

c. Recall this pattern:

$$(a + b)(a - b) = a^2 - b^2$$
$$\therefore \quad (4\sqrt{5} + 3\sqrt{2})(4\sqrt{5} - 3\sqrt{2}) = (4\sqrt{5})^2 - (3\sqrt{2})^2$$
$$= 80 - 18 = 62 \quad \textit{Answer}$$

In part (c) of Example 1, notice that the product of $4\sqrt{5} + 3\sqrt{2}$ and $4\sqrt{5} - 3\sqrt{2}$ is an integer. Expressions of the form $a\sqrt{b} + c\sqrt{d}$ and $a\sqrt{b} - c\sqrt{d}$ are called **conjugates.** Since

$$(a\sqrt{b} + c\sqrt{d})(a\sqrt{b} - c\sqrt{d}) = (a\sqrt{b})^2 - (c\sqrt{d})^2$$
$$= a^2b - c^2d,$$

the product of conjugates is always an integer when a, b, c, and d are integers.

Conjugates can be used to rationalize any denominator containing a binomial radical expression, as Examples 2 and 3 illustrate.

Example 2 Simplify $\dfrac{3 + \sqrt{5}}{3 - \sqrt{5}}$.

Solution

$$\dfrac{3 + \sqrt{5}}{3 - \sqrt{5}} = \dfrac{3 + \sqrt{5}}{3 - \sqrt{5}} \cdot \dfrac{3 + \sqrt{5}}{3 + \sqrt{5}} \qquad \longleftarrow \text{Multiply numerator and denomi-}$$
$$\text{nator by the conjugate of the}$$
$$= \dfrac{9 + 6\sqrt{5} + 5}{9 - 5} \qquad\qquad \text{denominator.}$$
$$= \dfrac{14 + 6\sqrt{5}}{4} = \dfrac{7 + 3\sqrt{5}}{2} \quad \textit{Answer}$$

Example 3 If $f(x) = \dfrac{x + 1}{x}$, find $f(\sqrt{3} + 2)$.

Solution

$$f(\sqrt{3} + 2) = \dfrac{(\sqrt{3} + 2) + 1}{\sqrt{3} + 2}$$
$$= \dfrac{\sqrt{3} + 3}{\sqrt{3} + 2} \cdot \dfrac{\sqrt{3} - 2}{\sqrt{3} - 2}$$
$$= \dfrac{3 + 3\sqrt{3} - 2\sqrt{3} - 6}{3 - 4} = \dfrac{-3 + \sqrt{3}}{-1} = 3 - \sqrt{3} \quad \textit{Answer}$$

Oral Exercises

Simplify. Assume that each radical represents a real number.

1. a. $(x + y)(x - y)$ **b.** $(\sqrt{5} + \sqrt{2})(\sqrt{5} - \sqrt{2})$ **c.** $(2 + \sqrt{3})(2 - \sqrt{3})$

2. a. $(x + y)^2$ **b.** $(\sqrt{5} + \sqrt{2})^2$ **c.** $(2 + \sqrt{3})^2$

3. a. $(x - y)^2$ **b.** $(\sqrt{5} - \sqrt{2})^2$ **c.** $(2 - \sqrt{3})^2$

4. a. $\dfrac{1}{2 + \sqrt{3}} \cdot \dfrac{2 - \sqrt{3}}{2 - \sqrt{3}}$ **b.** $\dfrac{1}{x + \sqrt{y}} \cdot \dfrac{x - \sqrt{y}}{x - \sqrt{y}}$

5. a. $\dfrac{1}{\sqrt{5} - \sqrt{2}} \cdot \dfrac{\sqrt{5} + \sqrt{2}}{\sqrt{5} + \sqrt{2}}$ **b.** $\dfrac{1}{\sqrt{x} - \sqrt{y}} \cdot \dfrac{\sqrt{x} + \sqrt{y}}{\sqrt{x} + \sqrt{y}}$

Written Exercises

Simplify.

A

1. $(3 + \sqrt{7})(3 - \sqrt{7})$ **2.** $(5 + \sqrt{2})(5 - \sqrt{2})$ **3.** $(\sqrt{7} + 1)^2$

4. $(\sqrt{5} + 2)^2$ **5.** $(1 + \sqrt{2})(3 + \sqrt{2})$ **6.** $(6 - \sqrt{3})(4 + \sqrt{3})$

7. $\dfrac{1}{4 - \sqrt{3}}$ **8.** $\dfrac{1}{6 + \sqrt{3}}$ **9.** $(\sqrt{7} - \sqrt{2})^2$

10. $(3\sqrt{11} - \sqrt{10})^2$ **11.** $(3 + 4\sqrt{3})(2 - \sqrt{3})$ **12.** $(5 - \sqrt{2})(3 - 2\sqrt{2})$

13. $\dfrac{3}{\sqrt{5} + \sqrt{2}}$

14. $\dfrac{10}{2\sqrt{3} - \sqrt{7}}$

15. $(\sqrt{11} - \sqrt{7})(\sqrt{11} + \sqrt{7})$

16. $(\sqrt{13} - \sqrt{3})(\sqrt{13} + \sqrt{3})$

17. $(5 + \sqrt{3})(8 - 2\sqrt{3})$

18. $(3 + 2\sqrt{6})(4 - 5\sqrt{6})$

19. $\dfrac{\sqrt{15}}{\sqrt{3} + \sqrt{5}}$

20. $\dfrac{\sqrt{6}}{\sqrt{2} + \sqrt{3}}$

21. $(2\sqrt{5} + \sqrt{7})^2$

22. $(3\sqrt{2} + \sqrt{6})^2$

23. $(2\sqrt{3} + \sqrt{5})(2\sqrt{3} - \sqrt{5})$

24. $(3\sqrt{7} - 2\sqrt{5})(3\sqrt{7} + 2\sqrt{5})$

25. $(\sqrt{6} - \sqrt{15})^2$

26. $(2\sqrt{5} - \sqrt{10})^2$

27. $\dfrac{\sqrt{5} + \sqrt{3}}{2} \cdot \dfrac{\sqrt{5} - \sqrt{3}}{2}$

28. $\dfrac{2\sqrt{7} + 1}{3} \cdot \dfrac{2\sqrt{7} - 1}{3}$

29. $(5\sqrt{6} + 3\sqrt{2})(2\sqrt{6} - 4\sqrt{3})$

30. $(3\sqrt{5} + 2\sqrt{15})(4\sqrt{3} - 3\sqrt{15})$

31. $\dfrac{\sqrt{5} + 1}{\sqrt{5} - 3}$

32. $\dfrac{2\sqrt{7} - \sqrt{3}}{\sqrt{7} + \sqrt{3}}$

B **33.** If $f(x) = \dfrac{x}{x + 1}$, find $f(1 - \sqrt{2})$.

34. If $g(x) = \dfrac{x^2}{x - 1}$, find $g(1 + \sqrt{2})$.

35. Show by substitution that $3 + \sqrt{5}$ and $3 - \sqrt{5}$ are roots of $x^2 - 6x + 4 = 0$.

36. Show by substitution that $1 + \dfrac{\sqrt{2}}{2}$ and $1 - \dfrac{\sqrt{2}}{2}$ are roots of $2x^2 - 4x + 1 = 0$.

37. **a.** What is the conjugate of $2\sqrt{5} - 3\sqrt{2}$?
 b. What is the reciprocal of the conjugate of $2\sqrt{5} - 3\sqrt{2}$?
 c. What is the conjugate of the reciprocal of $2\sqrt{5} - 3\sqrt{2}$?

38. Show that the reciprocal of $\dfrac{\sqrt{5} + 1}{2}$ is also the conjugate of $\dfrac{\sqrt{5} + 1}{2}$.

Use the Pythagorean theorem to find x.

39.

40.

Simplify. Assume that each radical represents a real number.

41. $(\sqrt{n + 1} + \sqrt{n})(\sqrt{n + 1} - \sqrt{n})$

42. $(b + \sqrt{b})^2 - (b - \sqrt{b})^2$

43. $\dfrac{\sqrt{w}}{\sqrt{w} + 1} + \dfrac{\sqrt{w}}{\sqrt{w} - 1}$

44. $\sqrt{1 - y^2} + \dfrac{y^2}{\sqrt{1 - y^2}}$

C **45.** $\dfrac{\sqrt{a} - \sqrt{a} \cdot \sqrt{a} + \sqrt{a}}{\sqrt{a} - 1}$

46. $\dfrac{\sqrt{x}}{\sqrt{x} + \sqrt{y}} + \dfrac{\sqrt{y}}{\sqrt{x} - \sqrt{y}}$

47. Prove that if a and b are rational numbers, $(a + \sqrt{b})^3 + (a - \sqrt{b})^3$ is also rational.

6-5 Equations Containing Radicals

Objective   To solve equations containing radicals.

By measuring a car's skid marks, investigators can determine how fast the car was traveling when the brakes were applied. On dry pavement, the speed s in miles per hour is given approximately by

$$s = \sqrt{22d},$$

where d is the distance in feet that the car travels as it brakes to a complete stop. For example, if the length of the skid marks is $d = 130$ ft, then the car's speed was

$$s = \sqrt{22 \cdot 130} \approx 53.5 \text{ (mi/h)}.$$

The formula $s = \sqrt{22d}$ can also be used to find a car's braking distance for a given speed. For example, at a speed s of 40 mi/h,

$$
\begin{array}{ll}
40 = \sqrt{22d} & \text{Substitute 40 for } s. \\
40^2 = (\sqrt{22d})^2 & \text{Square both sides.} \\
1600 = 22d & \text{Solve for } d. \\
d = \dfrac{1600}{22} \approx 73 \text{ (ft)} &
\end{array}
$$

Therefore, a speed of 40 mi/h requires a braking distance of 73 ft.

An equation like $40 = \sqrt{22d}$, which contains a radical with a variable in the radicand, is called a **radical equation.** To solve a radical equation involving square roots, start by isolating the radical term on one side of the equation. Then square both sides of the equation. If cube roots are involved instead of square roots, then cube both sides of the equation.

Example 1 Solve: **a.** $\sqrt{2x - 1} = 3$ **b.** $2\sqrt[3]{x} - 1 = 3$

Solution

a.
$$
\begin{array}{ll}
\sqrt{2x - 1} = 3 & \text{The radical term is isolated.} \\
(\sqrt{2x - 1})^2 = 3^2 & \text{Square both sides.} \\
2x - 1 = 9 & \text{Solve for } x. \\
x = 5 &
\end{array}
$$

Check:
$$\sqrt{2 \cdot 5 - 1} \overset{?}{=} 3$$
$$3 = 3 \quad \checkmark$$
∴ the solution set is $\{5\}$.

b.
$$
\begin{array}{ll}
2\sqrt[3]{x} - 1 = 3 & \text{Isolate the radical term.} \\
2\sqrt[3]{x} = 4 & \\
(2\sqrt[3]{x})^3 = 4^3 & \text{Cube both sides.} \\
8x = 64 & \text{Solve for } x. \\
x = 8 &
\end{array}
$$

Check:
$$2\sqrt[3]{8} - 1 \overset{?}{=} 3$$
$$3 = 3 \quad \checkmark$$
∴ the solution set is $\{8\}$.

Example 2   Solve $3x - 5\sqrt{x} = 2$.

Solution

$$3x - 2 = 5\sqrt{x} \qquad \text{Isolate the radical term.}$$
$$9x^2 - 12x + 4 = 25x \qquad \text{Square both sides.}$$
$$9x^2 - 37x + 4 = 0 \qquad \text{Solve for } x.$$
$$(x - 4)(9x - 1) = 0$$
$$x = 4 \quad \text{or} \quad x = \frac{1}{9}$$

Check: Check each possible solution in the *original* equation.

$$3x - 5\sqrt{x} = 2 \qquad\qquad\qquad 3x - 5\sqrt{x} = 2$$
$$3 \cdot 4 - 5\sqrt{4} \overset{?}{=} 2 \qquad\qquad 3 \cdot \frac{1}{9} - 5\sqrt{\frac{1}{9}} \overset{?}{=} 2$$
$$12 - 10 \overset{?}{=} 2 \qquad\qquad\qquad \frac{1}{3} - \frac{5}{3} \overset{?}{=} 2$$
$$2 = 2 \quad \checkmark \qquad\qquad\qquad\qquad -\frac{4}{3} \neq 2$$

The root 4 checks. $\qquad\qquad\qquad$ The root $\frac{1}{9}$ does not check.

$\therefore$ the solution set is $\{4\}$.   **Answer**

Example 2 shows that when you square both sides of a radical equation, the resulting equation may have a solution that is not a solution of the original equation. Such a solution is called an **extraneous root.** When you square the equation in the first step of this example, you get $(3x - 2)^2 = (5\sqrt{x})^2$, which is equivalent to $(3x - 2) = \pm 5\sqrt{x}$. The root 4 satisfies $3x - 2 = +5\sqrt{x}$ and the extraneous root $\frac{1}{9}$ satisfies $3x - 2 = -5\sqrt{x}$.

Caution: Since you may get an extraneous root when you solve a radical equation, you must not forget to check your answer in the *original* equation.

If a radical equation has more than one term with a variable in a radical, it may be necessary to square both sides of the equation more than once.

Example 3   Solve $\sqrt{2x + 5} = 2\sqrt{2x} + 1$.

Solution

$$(\sqrt{2x + 5})^2 = (2\sqrt{2x} + 1)^2 \qquad \text{Square both sides.}$$
$$2x + 5 = 8x + 4\sqrt{2x} + 1 \qquad \text{Isolate the radical term.}$$
$$4 - 6x = 4\sqrt{2x} \qquad \text{Square both sides.}$$
$$16 - 48x + 36x^2 = 32x$$
$$36x^2 - 80x + 16 = 0 \qquad \text{Divide both sides by 4.}$$
$$9x^2 - 20x + 4 = 0 \qquad \text{Solve for } x.$$
$$(9x - 2)(x - 2) = 0$$
$$x = \frac{2}{9} \quad \text{or} \quad x = 2$$

Check:

$$\sqrt{2 \cdot \frac{2}{9} + 5} \overset{?}{=} 2\sqrt{2 \cdot \frac{2}{9}} + 1$$

$$\sqrt{\frac{49}{9}} \overset{?}{=} 2 \cdot \frac{2}{3} + 1$$

$$\frac{7}{3} = \frac{7}{3} \ \checkmark$$

$\therefore$ the solution set is $\left\{\frac{2}{9}\right\}$. *Answer*

Check:

$$\sqrt{2 \cdot 2 + 5} \overset{?}{=} 2\sqrt{2 \cdot 2} + 1$$

$$\sqrt{9} \overset{?}{=} 2\sqrt{4} + 1$$

$$3 \neq 5$$

Consider these two similar-looking equations:

(a) $3x = 2 + 5\sqrt{x}$ (b) $3x = 2 + x\sqrt{5}$

Although both equations contain radicals, only equation (a) is a radical equation. Equation (b) is a *linear* equation having the form $ax = c + bx$, where the coefficient b in this case is a radical. Such linear equations can be solved without squaring both sides, as Example 4 illustrates.

Example 4   Solve $3x = 2 + x\sqrt{5}$ without squaring both sides.

Solution

$$3x = 2 + x\sqrt{5}$$

$$3x - x\sqrt{5} = 2$$

$$(3 - \sqrt{5})x = 2$$

$$x = \frac{2}{3 - \sqrt{5}}$$

$$= \frac{2}{3 - \sqrt{5}} \cdot \frac{3 + \sqrt{5}}{3 + \sqrt{5}}$$

$$= \frac{2(3 + \sqrt{5})}{4} = \frac{3 + \sqrt{5}}{2}$$

$\therefore$ the solution set is $\left\{\dfrac{3 + \sqrt{5}}{2}\right\}$. *Answer*

Oral Exercises

Solve. If an equation has no real solution, say so.

1. $\sqrt{x} = 5$ **2.** $\sqrt{2x} = 4$ **3.** $\sqrt{x - 1} = 3$

4. $\sqrt{x} - 7 = 0$ **5.** $\sqrt{x} + 6 = 0$ **6.** $2\sqrt{x} - 3 = 1$

7. $\sqrt[3]{x - 3} = 2$ **8.** $\sqrt{x} + \sqrt{x + 2} = 0$ **9.** $\sqrt[3]{x} + 5 = 3$

10. Tell whether each equation is a radical equation or a linear equation.

     a. $x\sqrt{2} = 3$ **b.** $2\sqrt{x} = 3$ **c.** $x\sqrt{3} + x\sqrt{2} = 1$

11. Solve each of the equations in Exercise 10.

12. Explain why $\sqrt{x}$ should be isolated in an equation like $\sqrt{x} + 4 = 7$ *before* both sides are squared.

Written Exercises

HW
1-18 All

Solve. If an equation has no real solution, say so.

A **1.** $\sqrt{4x - 3} = 5$ **2.** $\sqrt{3n + 1} = 7$ **3.** $3\sqrt{t} - 5 = 13$

4. $7 + 4\sqrt{a} = 3$ **5.** $\sqrt{2x^2 - 7} = 5$ **6.** $\sqrt{5y^2 + 1} = 9$

7. $\sqrt[3]{3m + 1} = 4$ **8.** $\sqrt[3]{2w - 5} = 3$ **9.** $\sqrt[3]{2d} + 5 = 3$

10. $7 - \sqrt[3]{9c} = 4$ **11.** $2\sqrt[3]{x} = \sqrt[3]{x^2}$ **12.** $\dfrac{\sqrt[3]{x}}{2} = \sqrt[3]{x - 7}$

13. $\sqrt{x + 2} = x$ **14.** $\sqrt{2n + 3} = n$ **15.** $\sqrt{t - 2} + t = 4$

16. $5 + \sqrt{a + 7} = a$ **17.** $\sqrt{2x + 5} - 1 = x$ **18.** $\sqrt{3n + 10} - 4 = n$

In Exercises 19–24, a radical equation and a linear equation are given. Solve the radical equation by squaring both sides. Solve the linear equation without squaring both sides.

19. a. $5\sqrt{x} = 10$ **20. a.** $3\sqrt{x} = 12$ **21. a.** $5 + 2\sqrt{x} = 7$

    b. $x\sqrt{5} = 10$ **b.** $x\sqrt{3} = 12$ **b.** $5 + x\sqrt{2} = 7$

22. a. $2 + 3\sqrt{x} = 8$ **23. a.** $x = 3 + 2\sqrt{x}$ **24. a.** $3x = 7\sqrt{x} - 2$

    b. $2 + x\sqrt{3} = 8$ **b.** $x = 3 + x\sqrt{2}$ **b.** $3x = x\sqrt{7} - 2$

Solve. If an equation has no real solution, say so.

B **25.** $\sqrt{y} + \sqrt{y + 5} = 5$ **26.** $\sqrt{x - 7} + \sqrt{x} = 7$

27. $\sqrt{2n - 5} - \sqrt{3n + 4} = 2$ **28.** $\sqrt{3a - 2} - \sqrt{2a - 3} = 1$

29. $\sqrt{3b - 2} - \sqrt{2b + 5} = 1$ **30.** $\sqrt{5y - 1} - \sqrt{7y + 9} = 2$

31. $\sqrt{x} + \sqrt{3} = \sqrt{x + 3}$ **32.** $\sqrt{n + 6} - \sqrt{n} = \sqrt{6}$

33. If you are near the top of a tall building on a clear day, how far can you see? If a building is h ft high, then the distance d (in miles) to the earth's horizon is approximately

$$d = \sqrt{\frac{3}{2}h}.$$

    a. The observatory of a tall building in Chicago is 607 ft high. What is the distance to the horizon from this observatory?

    b. Solve the formula for h.

34. If a pendulum is l cm long, then the time T (in seconds) that it takes the pendulum to swing back and forth once is given by $T = 2\pi\sqrt{\dfrac{l}{g}}$, where $\pi \approx 3.14$ and $g \approx 980$.

   a. Find the value of T if the pendulum is 20 cm long.

   b. Solve the formula for l in terms of T, g, and π.

35. The diagram shows an isosceles right triangle.

   a. If its perimeter is 10, find x.

   b. If its area is 12, find x.

36. The diagram shows a 30°-60°-90° right triangle.

   a. If the perimeter of the triangle is 18, find x.

   b. If the area of the triangle is 24, show that $x = 4\sqrt[4]{3}$.

37. PQR is an isosceles triangle with base $\overline{QR}$ 10 cm long. The length of altitude $\overline{PN}$ is x cm.

   a. Show that each leg of the triangle has length $\sqrt{25 + x^2}$.

   b. If the perimeter of the triangle is three times the length of $\overline{PN}$, find x.

Ex. 37

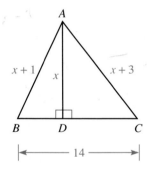

Ex. 38

C **38.** Use the dimensions given in the figure to find x, the length of altitude $\overline{AD}$.

39. The eight edges of the square pyramid shown each have length x.

   a. Show that the volume of the pyramid is $\dfrac{x^3\sqrt{2}}{6}$.

$$\left(\text{Volume} = \frac{1}{3} \times \text{base area} \times \text{height}\right)$$

   b. If the volume is 9, find x.

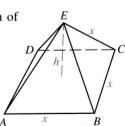

Mixed Review Exercises

Simplify.

1. $\dfrac{\sqrt{35} + \sqrt{14}}{\sqrt{7}}$

2. $\sqrt{6}(2\sqrt{3} - 3\sqrt{2})$

3. $\dfrac{a^3 - b^3}{a - b}$

4. $\dfrac{2}{4 - \sqrt{6}}$

5. $(m^2n)^3(-mn^3)^2$

6. $(\sqrt{5} + \sqrt{10})^2$

7. $(1 - x)^{-2}(x - 1)$

8. $\sqrt{75} - \sqrt{12} + \sqrt{27}$

9. $(3\sqrt{6} + \sqrt{2})(\sqrt{6} - 4\sqrt{3})$

10. $\dfrac{3\sqrt{32}}{2\sqrt{27}}$

11. $(5y + 2)(y^2 - 3y + 4)$

12. $\sqrt[3]{40} + \sqrt[3]{625} - \sqrt[3]{135}$

Self-Test 1

Vocabulary   square root (p. 259)
principal square root (p. 259)
cube root (p. 260)
nth root (p. 260)
principal nth root (p. 260)
radical (p. 261)
radicand (p. 261)

radical sign (p. 261)
index of a radical (p. 261)
rationalizing the denominator (p. 265)
simplest radical form (p. 266)
conjugates (p. 274)
radical equation (p. 277)
extraneous root (p. 278)

Simplify.

1. $\sqrt{\dfrac{49}{121}}$

2. $\sqrt[3]{-0.125}$

3. $\sqrt[4]{(-13)^4}$ Obj. 6-1, p. 259

4. For what values of x is $\sqrt{(x - 4)^2} = 4 - x$ true?

Simplify.

5. $\sqrt{10} \cdot \sqrt{15}$

6. $\sqrt{\dfrac{4a^4}{9b^3}}$

7. $\sqrt{2x^2 - 12x + 18}$ Obj. 6-2, p. 264

8. $\sqrt[3]{250} - \sqrt[3]{128} + \sqrt[3]{16}$

9. $\sqrt{\dfrac{3}{5}}\left(\sqrt{\dfrac{20}{3}} - \dfrac{\sqrt{15}}{6}\right)$ Obj. 6-3, p. 270

10. $(2\sqrt{3} + 5)(4\sqrt{3} - 3)$

11. $\dfrac{4}{3\sqrt{2} - 4}$ Obj. 6-4, p. 274

Solve.

12. $2\sqrt{x} + 3 = x$

13. $1 + \sqrt{x - 2} = \sqrt{x + 3}$ Obj. 6-5, p. 277

Check your answers with those at the back of the book.

Real Numbers and Complex Numbers

6-6 Rational and Irrational Numbers

Objective   To find and use decimal representations of real numbers.

The set of real numbers includes both rational numbers, such as $\frac{37}{16}$ and $\frac{19}{22}$, and irrational numbers, such as $\sqrt{2}$ and π. One way to learn about real numbers is to look at their decimal representations. We accept as an axiom the following property of real numbers.

Completeness Property of Real Numbers

Every real number has a decimal representation, and every decimal represents a real number.

Recall that a **rational number** is any number that can be expressed as the ratio, or quotient, of two integers. To find the decimal representation of a rational number, you can use the division process.

Example 1   Find a decimal representation for each rational number.

        a. $\dfrac{37}{16}$ **b.** $\dfrac{19}{22}$

Solution   **a.** $\dfrac{37}{16} = 37 \div 16$ **b.** $\dfrac{19}{22} = 19 \div 22$

```
        2.3125                          0.86363
   16)37.0000                      22)19.00000
      32                              17 6
       5 0                            1 40
       4 8                            1 32
         20                             80
         16                             66
         40                            140
         32                            132
          80                            80
          80                            66
           0                            14
```

Division terminates   Recurring remainders

$\therefore \dfrac{37}{16} = 2.3125$ ***Answer*** $\therefore \dfrac{19}{22} = 0.8636363 \ldots$ ***Answer***

Notice that the representation for $\frac{37}{16}$ is a **terminating decimal.** The representation for $\frac{19}{22}$, however, is a nonterminating **repeating decimal.** Terminating decimals are also called *finite decimals*. Nonterminating decimals are called *infinite decimals*. When you write a repeating decimal, you can use a bar to indicate the block of digits that repeats. For example,

$$2.\overline{45} = 2.454545 \ldots$$

and

$$0.8\overline{63} = 0.8636363 \ldots$$

The conversion of $\frac{37}{16}$ and $\frac{19}{22}$ to decimals illustrates the first fact stated in the chart below.

1. The decimal representation of any rational number is either terminating or repeating.
2. Every terminating or repeating decimal represents a rational number.

   (In other words, every terminating or repeating decimal can be written in

   the form $\frac{p}{q}$, where p and q are integers and $q \neq 0$.)

Exercise 42 asks you to prove the second fact stated above.

Example 2 Write each terminating decimal as a common fraction in lowest terms.

   a. 2.571   **b.** 0.0036

Solution **a.** $2.571 = 2\dfrac{571}{1000}$ **b.** $0.0036 = \dfrac{36}{10,000}$

$$= \dfrac{2571}{1000} \quad \textit{Answer}$$ $$= \dfrac{9}{2500} \quad \textit{Answer}$$

Example 3 Write each repeating decimal as a common fraction in lowest terms.

   a. $0.3\overline{27}$ **b.** $1.89\overline{189}$

Solution In each case let N be the number. Multiply the given number by 10^n where n is the number of digits in the block of repeating digits. Then subtract N.

   a. Let $N = 0.3\overline{27}$

$$100N = 32.7\overline{27}$$

$$\text{Subtract } N = 0.3\overline{27}$$

$$99N = 32.4$$

$$N = \frac{32.4}{99} = \frac{324}{990} = \frac{18}{55}$$

$$\therefore \quad 0.3\overline{27} = \frac{18}{55} \quad \textit{Answer}$$

   b. Let $N = 1.89\overline{189}$

$$1000N = 1891.89\overline{189}$$

$$\text{Subtract } N = 1.89\overline{189}$$

$$999N = 1890$$

$$N = \frac{1890}{999} = \frac{70}{37}$$

$$\therefore \quad 1.89\overline{189} = \frac{70}{37} \quad \textit{Answer}$$

An **irrational number** is a real number that is not rational. Therefore, the decimal representation of an irrational number is neither terminating nor repeating. For example,

$$\pi = 3.14159265\ldots \quad \text{and} \quad \sqrt{3} = 1.7320508\ldots$$

are irrational.

1. The decimal representation of any irrational number is infinite and nonrepeating.
2. Every infinite and nonrepeating decimal represents an irrational number.

This fact is significant when you use a calculator to obtain the decimal representation of an irrational number. Because the calculator's display is finite, you can get only a *rational approximation* of the irrational number.

Example 4 Classify each real number as either rational or irrational.

    a. $\sqrt{2}$ **b.** $\sqrt{\dfrac{4}{9}}$

    c. $2.030303\ldots$ **d.** $2.030030003\ldots$

Solution   **a.** Since $\sqrt{2} = 1.41421356\ldots$ (an infinite, nonrepeating decimal), $\sqrt{2}$ is irrational. (For a proof of the irrationality of $\sqrt{2}$, see the Extra on page 273.)

    b. Since $\sqrt{\dfrac{4}{9}} = \dfrac{2}{3}$, $\sqrt{\dfrac{4}{9}}$ is rational.

    c. Since $2.030303\ldots$ (or $2.\overline{03}$) is a repeating decimal, $2.030303\ldots$ is rational.

    d. Since $2.030030003\ldots$ is infinite and nonrepeating, $2.030030003\ldots$ is irrational.

A set S of real numbers is **dense** if between any two numbers in the set there is a member of S. Both the set of rational numbers and the set of irrational numbers are dense.

Example 5 Find a rational number r and an irrational number s between 1.51287 and 1.51288.

Solution   There are an infinite number of possible answers for r and s.

For example, $r = 1.5128711111111111111\ldots$

and $s = 1.5128710110111011110\ldots$

Notice that r is repeating and is therefore rational and that s is nonrepeating and is therefore irrational.

Oral Exercises

Tell whether each number is rational or irrational.

1. $\frac{15}{34}$ 2. -7 3. 9.26 4. $\sqrt{0.01}$

5. $\sqrt[3]{8}$ 6. $\sqrt[3]{9}$ 7. $\sqrt{\frac{16}{25}}$ 8. $\frac{\sqrt{2}}{2}$

9. π 10. $\frac{\pi}{\pi + 1}$ 11. $(2\sqrt{3})^2$ 12. $(2\sqrt{3})^3$

Tell whether or not each number is real. If it is real, tell whether it is rational or irrational.

13. $\sqrt{1000}$ 14. $\sqrt{-1000}$ 15. $\sqrt{(-1000)^2}$ 16. $\sqrt[3]{-1000}$

Written Exercises

Classify each real number or expression as either rational or irrational.

A 1. a. $\sqrt{49}$ b. $\sqrt{50}$ 2. a. π b. $\frac{22}{7}$

3. a. $\pi + \frac{1}{\pi}$ b. $\pi \cdot \frac{1}{\pi}$ 4. a. $\sqrt{\frac{1}{2}} + \sqrt{\frac{1}{8}}$ b. $\sqrt{\frac{1}{2}} \cdot \sqrt{\frac{1}{8}}$

Classify each real number as either rational or irrational.

5. a. 1.23 b. $1.\overline{23}$ c. $1.2345678910111213 \ldots$

6. a. 75.4682 b. $75.4\overline{682}$ c. $75.468244682444682 \ldots$

Write each fraction as a terminating or repeating decimal.

7. $\frac{5}{8}$ 8. $\frac{5}{11}$ 9. $\frac{13}{7}$ 10. $\frac{13}{4}$

Write each decimal as a common fraction in lowest terms.

11. 5.06 12. 3.004 13. 4.72 14. 0.1375
15. $0.\overline{4}$ 16. $0.\overline{5}$ 17. $0.8\overline{3}$ 18. $0.08\overline{3}$
19. $2.\overline{36}$ 20. $1.\overline{27}$ 21. $3.0\overline{33}$ 22. $1.\overline{101}$

Find (a) a rational, and (b) an irrational, number between each given pair.

23. 0.1 and 0.2 24. 0.3725 and 0.3726 25. $\sqrt{6}$ and $\sqrt{7}$
26. 3 and π 27. $\sqrt{15}$ and 4 28. 10^{-9} and 10^{-8}
29. $\frac{7}{8}$ and $\frac{8}{9}$ 30. $3\frac{1}{7}$ and $3\frac{1}{6}$ 31. $0.\overline{4}$ and $0.\overline{45}$

B **32. a.** The *arithmetic mean* of two numbers x and y is the number $\frac{x + y}{2}$. If x and y are rational numbers, what can you conclude about their arithmetic mean?

b. The *geometric mean* of two positive numbers x and y is the number $\sqrt{xy}$. If x and y are positive rational numbers, can you conclude that their geometric mean is also rational? Explain.

33. a. Find two irrational numbers x and y such that $\frac{x + y}{2}$ is rational.

b. Find two irrational numbers a and b such that $\frac{a + b}{2}$ is irrational.

34. If $x < y$, then $x < \frac{x + y}{2} < y$. Use this fact to give a convincing argument for the following: Between any two rational numbers there are infinitely many rational numbers.

35. Each of these numbers is between $\sqrt{2}$ and $\sqrt{3}$. Which are irrational?
(1) $\sqrt{2\frac{1}{2}}$ **(2)** $\sqrt{2\frac{1}{4}}$ **(3)** $1.\overline{4}$ **(4)** $1.456789101112\ldots$

36. a. A student who used long division to find the repeating decimal for $\frac{1}{17}$ claimed that there were 20 digits in the repeating block of digits. Explain why this is incorrect.

b. Find the repeating decimal for $\frac{1}{17}$. (Be patient!)

37. The following unit fractions all have finite decimals: $\frac{1}{2}, \frac{1}{4}, \frac{1}{5}, \frac{1}{8}, \frac{1}{25}, \frac{1}{40}$. Name four other unit fractions with finite decimals.

38. Suppose x and y are rational numbers. Prove that $x + y$ is also rational. $\left(Hint:\ \text{Let } x = \frac{a}{b} \text{ and } y = \frac{c}{d} \text{ where } a, b, c, \text{ and } d \text{ are integers. Show that } x + y \text{ can be written as a quotient of integers.}\right)$

39. a. Prove that the product of two rational numbers is rational. (See the hint for Exercise 38.)

b. What can you say about the product of two irrational numbers?

C **40.** Suppose x is rational and z is irrational. Prove that $x + z$ is irrational. $\left(Hint:\ \text{Let } x = \frac{a}{b} \text{ where } a \text{ and } b \text{ are integers, and use an indirect proof by assuming that } x + z \text{ is a rational number } \frac{c}{d}.\right)$

41. What can you say about the product of a nonzero rational number and an irrational number? Prove your answer by using an indirect proof.

42. Show that each repeating decimal represents a rational number as follows: Let N = the number. Let p = the number of digits in the repeating block of N. Justify the following statements.

a. $N \cdot 10^p$ has the same repeating block as N.

b. $N \cdot 10^p - N$ is a terminating decimal.

c. $N(10^p - 1)$ is rational. **d.** $\frac{N(10^p - 1)}{10^p - 1}$ is rational.

6-7 The Imaginary Number i

Objective   To use the number i to simplify square roots of negative numbers.

Throughout the history of mathematics, new numbers have been invented to extend the existing number system. For example, the invention of negative numbers made it possible to solve an equation like $x + 3 = 2$. Similarly, the invention of irrational numbers made an equation like $x^2 = 2$ solvable.

Now consider the equation $x^2 = -1$. This equation has no real solution, because the square of a real number is never negative. To solve this equation the *imaginary number i* was invented.

Definition of i
$$i = \sqrt{-1} \quad \text{and} \quad i^2 = -1.$$

The equation $x^2 = -1$ can now be solved: $x = \pm\sqrt{-1} = \pm i$. Because i and $-i$ are solutions of $x^2 = -1$, both i and $-i$ are square roots of -1.

The number i can be used to find square roots of other negative numbers. For example, since
$$(2i)^2 = 2^2 \cdot i^2 = 4(-1) = -4,$$
$$2i = \sqrt{-4}.$$
Similarly,
$$3i = \sqrt{-9}.$$

If r is a positive real number, then $\sqrt{-r} = i\sqrt{r}$.

Example 1 Simplify: **a.** $\sqrt{-5}$ **b.** $\sqrt{-25}$ **c.** $\sqrt{-50}$

Solution   **a.** $\sqrt{-5} = i\sqrt{5}$ **b.** $\sqrt{-25} = 5i$ **c.** $\sqrt{-50} = i\sqrt{50} = 5i\sqrt{2}$

Numbers like $i\sqrt{5}$ and $5i$, which have the form bi (where b is a nonzero real number), are called **pure imaginary numbers.** Example 2 illustrates how you can add and subtract these numbers.

Example 2 Simplify.

   a. $\sqrt{-16} - \sqrt{-49}$ **b.** $i\sqrt{2} + 3i\sqrt{2}$

Solution   **a.** $\sqrt{-16} - \sqrt{-49} = 4i - 7i$ **b.** $i\sqrt{2} + 3i\sqrt{2} = (1 + 3)i\sqrt{2}$
                 $= -3i$ *Answer* $= 4i\sqrt{2}$ *Answer*

Examples 3 and 4 illustrate how you can multiply and divide pure imaginary numbers.

Example 3 Simplify.

    a. $\sqrt{-4} \cdot \sqrt{-25}$ **b.** $i\sqrt{2} \cdot i\sqrt{3}$

Solution

$$\begin{aligned}\textbf{a.} \quad \sqrt{-4} \cdot \sqrt{-25} &= i\sqrt{4} \cdot i\sqrt{25}\\ &= 2i \cdot 5i\\ &= 10i^2 = -10 \quad \textit{Answer}\end{aligned}$$

$$\begin{aligned}\textbf{b.} \quad i\sqrt{2} \cdot i\sqrt{3} &= i^2\sqrt{6}\\ &= (-1)\sqrt{6}\\ &= -\sqrt{6} \quad \textit{Answer}\end{aligned}$$

Caution: The rule $\sqrt{a} \cdot \sqrt{b} = \sqrt{ab}$ does *not* hold if a and b are both negative. Notice that if you tried to use this rule in part (a) in Example 3 you would *incorrectly* get $\sqrt{-4} \cdot \sqrt{-25} = \sqrt{100} = 10$. Therefore, whenever you are simplifying expressions involving square roots of negative numbers, you must first express the radicals as pure imaginary numbers.

Example 4 Simplify.

    a. $\dfrac{2}{3i}$ **b.** $\dfrac{6}{\sqrt{-2}}$

Solution

$$\begin{aligned}\textbf{a.} \quad \frac{2}{3i} &= \frac{2}{3i} \cdot \frac{i}{i}\\[4pt] &= \frac{2i}{3i^2}\\[4pt] &= \frac{2i}{3(-1)}\\[4pt] &= -\frac{2}{3}i \quad \textit{Answer}\end{aligned}$$

$$\begin{aligned}\textbf{b.} \quad \frac{6}{\sqrt{-2}} &= \frac{6}{i\sqrt{2}}\\[4pt] &= \frac{6}{i\sqrt{2}} \cdot \frac{i\sqrt{2}}{i\sqrt{2}}\\[4pt] &= \frac{6i\sqrt{2}}{2i^2}\\[4pt] &= \frac{6i\sqrt{2}}{2(-1)}\\[4pt] &= -3i\sqrt{2} \quad \textit{Answer}\end{aligned}$$

Example 5 Simplify. Assume that each variable represents a positive real number.

    a. $\sqrt{-9x^2} + \sqrt{-x^2}$ **b.** $\sqrt{-6y} \cdot \sqrt{-2y}$

Solution

$$\begin{aligned}\textbf{a.} \quad \sqrt{-9x^2} + \sqrt{-x^2} &= 3xi + xi\\ &= 4xi \quad \textit{Answer}\end{aligned}$$

$$\begin{aligned}\textbf{b.} \quad \sqrt{-6y} \cdot \sqrt{-2y} &= i\sqrt{6y} \cdot i\sqrt{2y}\\ &= i^2\sqrt{12y^2}\\ &= (-1)\sqrt{4y^2 \cdot 3}\\ &= -2y\sqrt{3} \quad \textit{Answer}\end{aligned}$$

Using pure imaginary numbers, you can solve simple quadratic equations that have no real solutions.

Example 6 Solve $2x^2 + 19 = 3$.

Solution

$$2x^2 + 19 = 3$$
$$2x^2 = -16$$
$$x^2 = -8$$
$$x = \pm\sqrt{-8} = \pm 2i\sqrt{2}$$

$\therefore$ the solution set is $\{2i\sqrt{2}, -2i\sqrt{2}\}$.   *Answer*

Oral Exercises

Simplify.

1. $\sqrt{-64}$
2. $\sqrt{-100}$
3. $\sqrt{-11}$
4. $-\sqrt{-9}$
5. $\sqrt{-12}$
6. $-\sqrt{-18}$
7. $3i^2$
8. $(3i)^2$
9. $(-4i)^2$
10. $2i \cdot 6i$
11. $-i \cdot 8i$
12. $(i\sqrt{5})^2$
13. $\sqrt{-3} \cdot \sqrt{-3}$
14. $\sqrt{11} \cdot \sqrt{-11}$
15. $-\sqrt{2} \cdot \sqrt{-2}$

Solve.

16. $x^2 = -16$
17. $-y^2 = 11$
18. $3w^2 = -12$
19. $-2n^2 = 14$
20. $6 - t^2 = 7$
21. $8 + z^2 = 3$

22. What is the reciprocal of i (in simplified form)?

Written Exercises

$(-25$ all

Simplify.

A

1. $\sqrt{-81}$
2. $\sqrt{-121}$
3. $-4\sqrt{-36}$
4. $-2\sqrt{-144}$
5. $\sqrt{-20}$
6. $\sqrt{-75}$
7. $3\sqrt{-8}$
8. $5\sqrt{-27}$
9. $2i \cdot 3i$
10. $5i \cdot 3i$
11. $\sqrt{7} \cdot \sqrt{-7}$
12. $\sqrt{-6} \cdot \sqrt{2}$
13. $\sqrt{-5} \cdot \sqrt{-10}$
14. $\sqrt{-3} \cdot \sqrt{-6}$
15. $(7i)^2$
16. $(8i)^2$
17. $(-i)^2$
18. $(-5i)^2$
19. $(i\sqrt{2})^2$
20. $(3i\sqrt{5})^2$
21. $(-i\sqrt{3})^2$
22. $(-3i\sqrt{6})^2$
23. $-\dfrac{2}{i}$
24. $\dfrac{8}{3i}$

Simplify.

25. $\dfrac{1}{\sqrt{-5}}$

26. $\dfrac{4}{\sqrt{-4}}$

27. $\dfrac{\sqrt{18}}{2i\sqrt{6}}$

28. $\dfrac{\sqrt{28}}{4i\sqrt{7}}$

29. $\dfrac{\sqrt{60}}{\sqrt{-15}}$

30. $-\dfrac{\sqrt{12}}{\sqrt{-18}}$

Solve.

31. $x^2 + 144 = 0$

32. $y^2 + 400 = 0$

33. $2w^2 = -98$

34. $5t^2 = -20$

35. $3u^2 + 40 = 4$

36. $4z^2 + 39 = 7$

Simplify.

37. a. $\sqrt{-25} + \sqrt{-36}$ **b.** $\sqrt{-25} \cdot \sqrt{-36}$

38. a. $\sqrt{-3} + \sqrt{-27}$ **b.** $\sqrt{-3} \cdot \sqrt{-27}$

39. a. $3\sqrt{-2} - \sqrt{-50}$ **b.** $3\sqrt{-2} \cdot (-\sqrt{-50})$

40. a. $2\sqrt{-24} - \sqrt{-54}$ **b.** $2\sqrt{-24} \cdot (-\sqrt{-54})$

41. a. $i\sqrt{18} + \sqrt{-8}$ **b.** $i\sqrt{18} \cdot \sqrt{-8}$

42. a. $i\sqrt{-98} - \sqrt{98}$ **b.** $i\sqrt{-98} \cdot (-\sqrt{98})$

Simplify. Assume that each variable represents a positive number.

B   **43.** $\sqrt{-12a} \cdot \sqrt{-3a}$

44. $-\sqrt{18c} \cdot \sqrt{-2c^3}$

45. $\sqrt{-\dfrac{r}{5}} \cdot \sqrt{-\dfrac{20}{r}}$

46. $\sqrt{-\dfrac{t^5}{2}} \cdot \sqrt{-\dfrac{2}{t^3}}$

47. $\sqrt{-3c^2} + \sqrt{-27c^2} - \sqrt{-45c^2}$

48. $\sqrt{-2t^5} + \sqrt{-8t^5} - \sqrt{-18t^5}$

49. $\sqrt{-4r^3} + \sqrt{-64r^3} - 4r\sqrt{-16r}$

50. $\sqrt{-25a^3} - \sqrt{-225a^3} + 20a\sqrt{-a}$

51. $\sqrt{-x^5} + x\sqrt{-25x^2} - x^2\sqrt{-25x}$

52. $yi\sqrt{-16y^2} + \sqrt{16y^4} - y^2i\sqrt{-9}$

53. Simplify i^n for $n = 2, 3, 4, \ldots, 12$. What pattern do you see?

54. Simplify: **a.** i^{100} **b.** i^{101} **c.** i^{102} **d.** i^{103}

Mixed Review Exercises

Solve. If an equation has no real solutions, say so.

1. $\sqrt{2x - 3} = 5$

2. $15 - 2n = n^2$

3. $\sqrt{y^2 + 12} = 2y$

4. $\dfrac{3y - 4}{5} = \dfrac{y + 1}{2}$

5. $\dfrac{1}{n} + \dfrac{2}{n - 2} = \dfrac{4}{n(n - 2)}$

6. $2\sqrt[3]{x} + 9 = 5$

7. $y = \sqrt{5y - 6}$

8. $5|n| - 7 = 3$

9. $x = 2 + \sqrt{x + 4}$

Classify each real number as either rational or irrational.

10. $\sqrt[3]{-125}$ **11.** $3.7\overline{82}$ **12.** $\sqrt[5]{12}$ **13.** $0.121121112 \ldots$

6-8 The Complex Numbers

Objective   To add, subtract, multiply, and divide complex numbers.

In this lesson, you will see how the real numbers can be extended to form a number system that includes the imaginary number i. This extended system, called the *complex numbers,* has become an important mathematical tool for physicists and engineers working in fields such as electricity and magnetism, optics, and hydrodynamics.

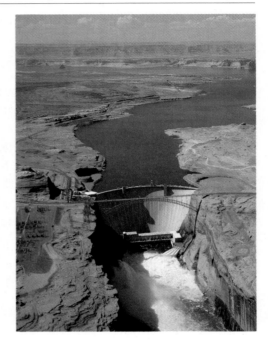

In the preceding lesson you worked with pure imaginary numbers, which have the form bi $(b \neq 0)$. When pure imaginary and real numbers are combined, **imaginary numbers** of the form $a + bi$ $(b \neq 0)$ are the result. The real numbers and the imaginary numbers together form the set of *complex numbers.*

A **complex number** is a number of the form $a + bi$, where a and b are real numbers. The chart below shows the relationship between the various sets of numbers that make up the complex numbers. In the chart, a and b are real numbers.

Complex numbers: $a + bi$
$3 + 2i,\ 5 + 0i,\ -i\sqrt{2}$

Real numbers: $a + bi,\ b = 0$
$5,\ \sqrt[3]{5},\ \pi,\ -\frac{2}{7},\ 5.\overline{67}$

Imaginary numbers: $a + bi,\ b \neq 0$
$3 + 2i,\ \frac{2}{3} - \frac{1}{2}i,\ 7i,\ -i\sqrt{2}$

Rational numbers
$5,\ -\frac{2}{7},\ 5.\overline{67}$

Irrational numbers
$\sqrt[3]{5},\ \pi$

Pure imaginary numbers:
$a + bi,\ a = 0,\ b \neq 0$
$7i,\ -i\sqrt{2}$

The number a is called the **real part** of the complex number $a + bi$, and b (not bi) is called the **imaginary part.**

Two complex numbers are equal if and only if they have equal real parts and equal imaginary parts. For example, if $x + yi = 5 - i$, then $x = 5$ and $y = -1$.

Equality of Complex Numbers

$$a + bi = c + di \quad \text{if and only if} \quad a = c \text{ and } b = d.$$

The sum and the product of two complex numbers are defined as follows.

Sum of Complex Numbers

$$(a + bi) + (c + di) = (a + c) + (b + d)i$$

Product of Complex Numbers

$$(a + bi)(c + di) = (ac - bd) + (ad + bc)i$$

Notice that you add or subtract complex numbers by combining their real parts and their imaginary parts separately.

Example 1 Simplify.

    a. $(3 + 6i) + (4 - 2i)$ **b.** $(3 + 6i) - (4 - 2i)$

Solution   **a.** $(3 + 6i) + (4 - 2i) = (3 + 4) + (6 - 2)i = 7 + 4i$

          b. $(3 + 6i) - (4 - 2i) = (3 - 4) + (6 + 2)i = -1 + 8i$

The definition of multiplication given above is equivalent to the following rule for multiplying two complex numbers: *Multiply two complex numbers as you would multiply any two binomials and use the fact that $i^2 = -1$.*

Example 2 Simplify.

    a. $(3 + 4i)(5 + 2i)$ **b.** $(3 + 4i)^2$ **c.** $(3 + 4i)(3 - 4i)$

Solution   **a.** $(3 + 4i)(5 + 2i) = 15 + 20i + 6i + 8i^2$

$$= 15 + 26i + 8(-1)$$

$$= 7 + 26i \quad \textbf{\textit{Answer}}$$

        b. $(3 + 4i)^2 = 9 + 24i + 16i^2$

$$= 9 + 24i - 16$$

$$= -7 + 24i \quad \textbf{\textit{Answer}}$$

        c. $(3 + 4i)(3 - 4i) = 9 - 16i^2$

$$= 9 + 16 = 25 \quad \textbf{\textit{Answer}}$$

The numbers $3 + 4i$ and $3 - 4i$ in part (c) of Example 2 are examples of *conjugates*. In general, the complex numbers $a + bi$ and $a - bi$ are **complex conjugates,** and their product is the real number $a^2 + b^2$. This fact can be used to simplify the quotient of two imaginary numbers.

Example 3 Simplify $\dfrac{5 - i}{2 + 3i}$.

Solution $\dfrac{5 - i}{2 + 3i} = \dfrac{5 - i}{2 + 3i} \cdot \dfrac{2 - 3i}{2 - 3i}$ ⟵ Multiply numerator and denominator by the conjugate of the denominator.

$$= \frac{10 - 2i - 15i + 3i^2}{4 - 9i^2}$$

$$= \frac{10 - 17i + 3(-1)}{4 - 9(-1)}$$

$$= \frac{7 - 17i}{13} = \frac{7}{13} - \frac{17}{13}i \quad \textbf{\textit{Answer}}$$

You usually express the result of a computation with complex numbers in the form $a + bi$, as was done in Example 3. This form emphasizes that the answer is another complex number.

The examples of this lesson show that the sum, difference, product, or quotient of two complex numbers is another complex number. A power or root of a complex number is also a complex number. In short, the complex numbers form a closed number system.

Example 4 illustrates that the reciprocal of a complex number is another complex number.

Example 4 Find the reciprocal of $3 - i$.

Solution $\dfrac{1}{3 - i} = \dfrac{1}{3 - i} \cdot \dfrac{3 + i}{3 + i}$

$$= \frac{3 + i}{9 - i^2}$$

$$= \frac{3 + i}{10} = \frac{3}{10} + \frac{1}{10}i \quad \textbf{\textit{Answer}}$$

In addition to being closed for all operations, the complex number system has many other properties that the real number system has: Addition and multiplication are commutative and associative, and the distributive property holds. (See Exercises 53–56.) However, the complex numbers cannot be ordered along a number line like the real numbers can. Although you cannot say that one complex number is greater than another, you can measure the "size" of a complex number by using absolute value. See the Extra on page 298 for more on absolute values of complex numbers and conjugates.

Oral Exercises

Simplify.

1. $(3 + 2i) + (5 + 7i)$

2. $(9 + 4i) + (7 - 5i)$

3. $(-4 + 3i) - (6 - i)$

4. $(9 - 2i) - (6 - 3i)$

5. $2i(5 - i)$

6. $3i(7 + 2i)$

7. $(6 + i)(6 - i)$

8. $(3 + 2i)(3 - 2i)$

9. $(2 + i)(2 - i)$

10. $(-1 + 3i)(-1 - 3i)$

11. $(1 + i)^2$

12. $(4 - i)^2$

Study the chart on page 292 and tell in which category each of the following numbers belongs.

13. $5 - 9i$

14. $-3i$

15. $5.\overline{67}$

16. $\sqrt{27}$

17. $\sqrt[3]{27}$

18. 3^{-2}

19. $7 + i\sqrt{3}$

20. $0.121121112\ldots$

Written Exercises

Simplify.

A

1. $(9 + 2i) + (1 - 7i)$

2. $(3 - 4i) + (-5 - 2i)$

3. $(5 - 7i) - (8 + 2i)$

4. $(6 - 4i) - (-4 + i)$

5. $3(-2 + i) - 4(3 - 2i)$

6. $2(-1 + 6i) - 3(2 + 5i)$

7. $i(3 + 4i)$

8. $3i(5 - 6i)$

9. $-4i(-2 + i)$

10. $-2i(1 - 3i)$

11. $(3 - i)(3 + i)$

12. $(4 + i)(4 - i)$

13. $(3 + 7i)(3 - 7i)$

14. $(-5 + 3i)(2 - 3i)$

15. $(-4 + i)(8 + 5i)$

16. $(3 - 7i)(2 + 4i)$

17. $(-2 + 5i)(1 + 3i)$

18. $(3 + 4i)(-2 + 3i)$

19. $(4 + i\sqrt{5})(4 - i\sqrt{5})$

20. $(2 - i\sqrt{3})(2 + i\sqrt{3})$

21. $(2 - 4i)^2$

22. $(6 - 7i)^2$

23. $(-1 + i\sqrt{3})^2$

24. $(3 + i\sqrt{5})^2$

25. $(3 + 2i)^2(3 - 2i)^2$

26. $(2 - 3i)^2(2 + 3i)^2$

27. $(\sqrt{2} - \sqrt{-5})(\sqrt{2} + \sqrt{-5})$

28. $(\sqrt{3} + \sqrt{-7})(\sqrt{3} - \sqrt{-7})$

Simplify each quotient.

29. $\dfrac{5}{3 + 4i}$

30. $\dfrac{15}{2 - i}$

31. $\dfrac{2}{3 - i}$

32. $\dfrac{10}{1 + 7i}$

33. $\dfrac{-1 - 2i}{-1 + 2i}$

34. $\dfrac{5 + i}{5 - i}$

35. $\dfrac{6 - i\sqrt{2}}{6 + i\sqrt{2}}$

36. $\dfrac{-3 + i\sqrt{5}}{-3 - i\sqrt{5}}$

Find the reciprocal of each complex number.

37. $2 + 3i$ **38.** $1 - 4i$ **39.** $-\sqrt{3} + i\sqrt{6}$ **40.** $-\sqrt{5} - i\sqrt{2}$

41. If $f(x) = x + \dfrac{1}{x}$, find $f(1 + 3i)$.   **42.** If $g(x) = x - \dfrac{1}{x}$, find $g(2 - i)$.

43. If $g(x) = \dfrac{x - 1}{x + 1}$, find $g(1 + i\sqrt{3})$.   **44.** If $f(x) = \dfrac{2 + x}{2 - x}$, find $f(1 - i)$.

B **45.** Show by substitution that $2 + i$ is a solution of the equation $x^2 - 4x + 5 = 0$.

46. Show by substitution that $1 - 3i$ is a solution of the equation $x^2 - 2x + 10 = 0$.

47. a. Show that $\dfrac{\sqrt{2}}{2} + \dfrac{\sqrt{2}}{2}i$ is a square root of i.

$\left(\textit{Hint: Square } \dfrac{\sqrt{2}}{2} + \dfrac{\sqrt{2}}{2}i.\right)$

     b. Can you find another square root of i?

48. a. Show that $-1 - i\sqrt{3}$ is a cube root of 8. (*Hint:* Cube $-1 - i\sqrt{3}$.)
     b. Decide whether or not $-1 + i\sqrt{3}$ is a cube root of 8.

49. Show that the conjugates $\dfrac{3}{5} + \dfrac{4}{5}i$ and $\dfrac{3}{5} - \dfrac{4}{5}i$ are also reciprocals.

50. Show that the conjugates $\dfrac{1}{2} + \dfrac{\sqrt{3}}{2}i$ and $\dfrac{1}{2} - \dfrac{\sqrt{3}}{2}i$ are also reciprocals.

C **51.** The numbers in Exercises 49 and 50 are both conjugates and reciprocals. Find another pair of such numbers.

52. What condition must be placed on a and b if the reciprocal of $a + bi$ is to be $a - bi$?

In Exercises 53–56, prove each property of complex numbers.

Sample   The addition of complex numbers is commutative.

Solution   Let $a + bi$ and $c + di$ be any two complex numbers where a, b, c, and d are real numbers.

$$
\begin{aligned}
(a + bi) + (c + di) &= (a + c) + (b + d)i &&\text{Def. of complex add.}\\
&= (c + a) + (d + b)i &&\text{Comm. prop. of real add.}\\
&= (c + di) + (a + bi) &&\text{Def. of complex add.}
\end{aligned}
$$

53. The addition of complex numbers is associative.

54. The multiplication of complex numbers is commutative.

55. The multiplication of complex numbers is associative.

56. In the complex number system, multiplication is distributive over addition.

Self-Test 2

Vocabulary   rational number (p. 283)   imaginary number (p. 292)
terminating decimal (p. 284)   complex number (p. 292)
repeating decimal (p. 284)   real part (p. 292)
irrational number (p. 285)   imaginary part (p. 292)
dense (p. 285)   complex conjugate (p. 294)
pure imaginary number (p. 288)

1. Find the decimal representation of $\frac{5}{13}$.   **Obj.6-6, p. 283**

2. Express $0.1\overline{72}$ as a common fraction in lowest terms.

3. Find a rational number and an irrational number between 10 and $\sqrt{101}$.

Simplify.

4. $(-3i)^2$ **5.** $\sqrt{-28} + \sqrt{-63}$ **Obj. 6-7, p. 288**

6. $\sqrt{-6} \cdot \sqrt{-10}$ **7.** $-\dfrac{1}{2i}$

8. Solve $2y^2 + 90 = 0$.

Simplify.

9. $(5 + 3i) - (2 - i)$ **10.** $(1 - 2i)^2$ **Obj. 6-8, p. 292**

11. $(4 - i\sqrt{3})(-5 + 2i\sqrt{3})$ **12.** $\dfrac{2 + 3i}{1 - 4i}$

Historical Note / Complex Numbers

As early as the mid-sixteenth century, mathematicians like Girolamo Cardano and Rafael Bombelli began working with square roots of negative numbers. They used these non-real numbers in a formula that produced *real* solutions of cubic equations. Although the Swiss mathematician Leonhard Euler introduced the symbol i to represent $\sqrt{-1}$ about 1750, most mathematicians shunned "imaginary" numbers. Attitudes changed 50 years later when the great German mathematician Karl Friedrich Gauss used complex numbers in his doctoral thesis to prove this geometric fact: A regular polygon can be constructed with straightedge and compass if its number of sides is a prime number of the form $2^m + 1$, where m is a positive integer. (The numbers 3, 5, and 17 are the first three such primes, so regular triangles, pentagons, and 17-gons can be constructed with straightedge and compass.)

Gauss' accomplishment finally made complex numbers acceptable and paved the way for their widespread use in modern mathematics and physics.

Extra / Conjugates and Absolute Value

When complex numbers are being discussed, they are often represented by single letters. For example, let

$$z = x + yi.$$

The *complex conjugate* of z is denoted by $\bar{z}$ (read "z bar"). That is,

$$\bar{z} = x - yi.$$

The following theorem summarizes the properties of conjugates.

Theorem 1

Let w and z be complex numbers.

a. $\overline{w + z} = \bar{w} + \bar{z}$ **b.** $\overline{w - z} = \bar{w} - \bar{z}$

c. $\overline{w \cdot z} = \bar{w} \cdot \bar{z}$ **d.** $\overline{\left(\dfrac{w}{z}\right)} = \dfrac{\bar{w}}{\bar{z}}$ $(z \neq 0)$

Proof of part (c):   Let $w = u + vi$ and $z = x + yi$. Then:

$$\overline{w \cdot z} = \overline{(u + vi)(x + yi)}$$
$$= \overline{(ux - vy) + (uy + vx)i}$$
$$= (ux - vy) - (uy + vx)i$$

$$\bar{w} \cdot \bar{z} = (u - vi)(x - yi)$$
$$= [ux - (-v)(-y)] + [u(-y) + (-v)x]i$$
$$= (ux - vy) - (uy + vx)i$$

$$\therefore \overline{w \cdot z} = \bar{w} \cdot \bar{z}$$

Proofs of the other parts of Theorem 1 are left as Exercises 16 and 17.

To measure the "size" of the complex number $z = x + yi$, you can use its **absolute value,** defined by

$$|z| = \sqrt{x^2 + y^2}.$$

Example   Find $|z|$ for each complex number z.

a. $z = 4 - 3i$ **b.** $z = -6i$ **c.** $z = -2$

Solution   **a.** $|z| = \sqrt{4^2 + (-3)^2} = \sqrt{25} = 5$

b. $|z| = \sqrt{0^2 + (-6)^2} = \sqrt{36} = 6$

c. $|z| = \sqrt{(-2)^2 + 0^2} = \sqrt{4} = 2$

Notice that if z is real, as in part (c) of Example 1, the definition of absolute value for *complex* numbers agrees with the definition of absolute value for *real* numbers found in Chapter 1.

The following theorem states a useful relationship between absolute value and conjugates.

Theorem 2

If z is a complex number, then $|z|^2 = z \cdot \bar{z}$.

Proof: Let $z = x + yi$. Then

$$|z|^2 = (\sqrt{x^2 + y^2})^2 = x^2 + y^2,$$
$$\text{and} \quad z \cdot \bar{z} = (x + yi)(x - yi) = x^2 + y^2.$$
$$\therefore |z|^2 = z \cdot \bar{z}$$

If you divide both sides of $|z|^2 = z \cdot \bar{z}$ by $|z|^2 \cdot z$, you obtain this convenient expression for the reciprocal of a nonzero complex number:

$$\frac{1}{z} = \frac{\bar{z}}{|z|^2}$$

Example 2 Use the fact that $\dfrac{1}{z} = \dfrac{\bar{z}}{|z|^2}$ to find $\dfrac{1}{z}$ for each z given in Example 1.

Solution **a.** Since $|z|^2 = 25$, $\dfrac{1}{z} = \dfrac{\bar{z}}{|z|^2} = \dfrac{4 + 3i}{25} = \dfrac{4}{25} + \dfrac{3}{25}i$.

b. Since $|z|^2 = 36$, $\dfrac{1}{z} = \dfrac{\bar{z}}{|z|^2} = \dfrac{6i}{36} = \dfrac{1}{6}i$.

c. Since $|z|^2 = 4$, $\dfrac{1}{z} = \dfrac{\bar{z}}{|z|^2} = \dfrac{-2}{4} = -\dfrac{1}{2}$.

Some properties of absolute value are stated in the following theorem.

Theorem 3

Let w and z be complex numbers.

a. $|w \cdot z| = |w| \cdot |z|$ **b.** $\left| \dfrac{w}{z} \right| = \dfrac{|w|}{|z|}$ $(z \neq 0)$

c. $|w + z| \leq |w| + |z|$

The proof of part (a) is given on the next page. Proofs of parts (b) and (c) are left as Exercises 20–22.

Proof of part (a):   By Theorems 1 and 2 and the commutative and associative properties:

$$|w \cdot z|^2 = (w \cdot z)(\overline{w \cdot z}) = w \cdot \overline{w} \cdot z \cdot \overline{z} = |w|^2 \cdot |z|^2$$

Since $|wz|$, $|w|$, and $|z|$ are all nonnegative, when you take the principal square root of each side of $|w \cdot z|^2 = |w|^2 \cdot |z|^2$, you obtain

$$|w \cdot z| = |w| \cdot |z|.$$

Exercises

Find $|z|$ for each complex number z.

A   **1.** $z = -4 + 3i$ 　　　　　　 **2.** $z = 12 - 5i$ 　　　　　　 **3.** $z = 2i$

　　4. $z = -3$ 　　　　　　 **5.** $z = \dfrac{\sqrt{2}}{2} + \dfrac{\sqrt{2}}{2}i$ 　　　　 **6.** $z = \dfrac{\sqrt{3}}{2} - \dfrac{1}{2}i$

　　7–12. Use the fact that $\dfrac{1}{z} = \dfrac{\overline{z}}{|z|^2}$ to find $\dfrac{1}{z}$ for each complex number z in Exercises 1–6.

13. Show that $|\overline{z}| = |z|$.

14. Show that $z + \overline{z}$ is twice the real part of z.

15. a. Express the imaginary part of z in terms of z and $\overline{z}$.

　　b. Show that z is real if and only if $z = \overline{z}$.

16. Prove part (a) of Theorem 1. 　　　　 **17.** Prove part (b) of Theorem 1.

B   **18.** Prove that $\overline{\left(\dfrac{1}{z}\right)} = \dfrac{1}{\overline{z}}$.

19. Prove part (d) of Theorem 1 by combining part (c) and Exercise 18.
$\left(\textit{Hint:}\ \dfrac{w}{z} = w \cdot \dfrac{1}{z}.\right)$

20. Prove part (b) of Theorem 3. $\left(\textit{Hint:}\ \text{Apply Theorem 2 to}\ \left|\dfrac{w}{z}\right|^2.\ \text{Then use} \right.$ part (d) of Theorem 1.$\Big)$

Exercises 21–22 outline a proof of part (c) of Theorem 3.

C   **21.** Show that $w \cdot \overline{z} + \overline{w} \cdot z \le 2|w||z|$.
　　　(*Hint:* Show that $w \cdot \overline{z} + \overline{w} \cdot z = w \cdot \overline{z} + \overline{w \cdot \overline{z}}$ and therefore is twice the real part of $w \cdot \overline{z}$. How does the real part of a complex number compare in size with the absolute value of the number?)

22. Show that $|w + z| \le |w| + |z|$ by using the theorems of this Extra and the preceding exercises to justify these steps:

$$\begin{aligned}
|w + z|^2 &= (w + z)(\overline{w + z}) \\
&= (w + z)(\overline{w} + \overline{z}) \\
&= w \cdot \overline{w} + w \cdot \overline{z} + \overline{w} \cdot z + z \cdot \overline{z} \\
&\le |w|^2 + 2|w||z| + |z|^2 \\
&\le (|w| + |z|)^2 \\
\therefore \quad |w + z| &\le |w| + |z|
\end{aligned}$$

Chapter Summary

1. A solution of the equation $x^n = b$ is called an *n*th root of *b*.
    The radical $\sqrt[n]{b}$ denotes the *principal n*th root of *b*.

    $$(\sqrt[n]{b})^n = b \qquad \sqrt[n]{b^n} = \begin{cases} b & \text{if } n \text{ is odd} \\ |b| & \text{if } n \text{ is even} \end{cases}$$

2. In working with radicals you can use the following properties when the radicals involved represent real numbers.

    $$\sqrt[n]{ab} = \sqrt[n]{a} \cdot \sqrt[n]{b} \qquad\qquad \sqrt[n]{\frac{a}{b}} = \frac{\sqrt[n]{a}}{\sqrt[n]{b}}$$

    $$\sqrt[nq]{b} = \sqrt[n]{\sqrt[q]{b}} \qquad\qquad \sqrt[n]{b^m} = (\sqrt[n]{b})^m$$

3. An expression containing *n*th roots is in *simplest radical form* if:
    (a) no radicand contains a factor (other than 1) that is a perfect *n*th power, and
    (b) every denominator has been rationalized, so that no radicand is a fraction and no radical is in a denominator.

4. Radical expressions with the same index and radicand can be added or subtracted in the same way as like terms.

5. To solve a *radical equation* involving square roots, first isolate the radical term and then square both sides of the equation. Repeat this process if necessary. Be sure to check all possible solutions in the original equation to determine if any extraneous roots were introduced.

6. A real number that can be expressed as a quotient of two integers is *rational*. Otherwise it is *irrational*.

7. (a) Every terminating or repeating decimal represents a rational number, and any rational number can be expressed as a terminating or repeating decimal.
    (b) Every infinite nonrepeating decimal represents an irrational number, and any irrational number can be expressed as a nonrepeating decimal.

8. A *complex number* is a number of the form $a + bi$, where *a* and *b* are real numbers and $i = \sqrt{-1}$. The number *a* is called the *real part* of the complex number, and *b* is called the *imaginary part*. The *complex conjugate* of $a + bi$ is $a - bi$. The product of these complex conjugates is the real number $a^2 + b^2$.

9. To simplify an expression containing square roots of negative numbers, first rewrite each radical using *i*.

10. Complex numbers can be added or subtracted by combining their real parts and their imaginary parts separately. Multiply two complex numbers as you would multiply two binomials, and use the fact that $i^2 = -1$. To simplify a quotient, multiply both the numerator and denominator by the conjugate of the denominator.

Chapter Review

Write the letter of the correct answer.

6-1

1. Simplify $\sqrt{x^2}$.
   a. $-x$ **b.** x **c.** $|x|$ **d.** $-|x|$

2. Find the real roots of the equation $2 - y^2 = 0$.
   a. $\sqrt{2}$ **b.** $\pm i\sqrt{2}$ **c.** $\pm\sqrt{2}$ **d.** no real roots

6-2

3. Simplify $\sqrt{12x^4y^7}$. Assume that the radical represents a real number.
   a. $x^2y^3\sqrt{12y}$ **b.** $2x^2y^3\sqrt{3y}$
   c. $6x^2y^3\sqrt{y}$ **d.** $x^2\sqrt{12y^7}$

4. Simplify $\sqrt{18} \cdot \sqrt{\frac{3}{10}}$.

   a. $\dfrac{3\sqrt{15}}{5}$ **b.** $15\sqrt{3}$

   c. $\dfrac{\sqrt{15}}{5}$ **d.** $\dfrac{\sqrt{15}}{5}$

6-3

5. Simplify $\sqrt{6}\left(\dfrac{\sqrt{2}}{2} + \sqrt{3}\right) - \sqrt{8}$.
   a. $\sqrt{3} + \sqrt{2}$ **b.** $\sqrt{3} + 5\sqrt{2}$
   c. $\sqrt{5}$ **d.** $2\sqrt{3} + \sqrt{2}$

6. Simplify $\sqrt{72x^3} - 5x\sqrt{2x}$. Assume that each radical represents a real number.
   a. $\sqrt{2x}$ **b.** $2x\sqrt{x}$
   c. $x^2\sqrt{2x}$ **d.** $x\sqrt{2x}$

6-4

7. Simplify $(2\sqrt{6} - \sqrt{3})^2$.
   a. 21 **b.** $27 - 6\sqrt{2}$
   c. $27 - 12\sqrt{2}$ **d.** $15 - 12\sqrt{2}$

8. Simplify $\dfrac{6}{3 + 2\sqrt{3}}$.
   a. $6 - 4\sqrt{3}$ **b.** $6 + 4\sqrt{3}$ **c.** $-6 + 4\sqrt{3}$ **d.** $-6 - 4\sqrt{3}$

6-5

9. Solve $3 - 2\sqrt{x} = 7$.
   a. $\{-2\}$ **b.** $\{4\}$ **c.** $\{-4\}$ **d.** $\varnothing$

10. Solve $2 - y = \sqrt{y + 4}$.
    a. $\{0\}$ **b.** $\{0, 5\}$ **c.** $\{5\}$ **d.** $\varnothing$

6-6

11. Express $0.\overline{675}$ as a common fraction in lowest terms.
    a. $\dfrac{27}{40}$ **b.** $\dfrac{25}{37}$ **c.** $\dfrac{67}{99}$ **d.** $\dfrac{2}{3}$

12. Which of the following is *not* a rational number?
    a. $\dfrac{407}{528}$ **b.** $\sqrt[3]{-8}$ **c.** $0.\overline{34}$ **d.** $1.626626662 \ldots$

13. Simplify $3i\sqrt{2} \cdot \sqrt{-12}$.

    a. $-6\sqrt{6}$ **b.** $6i\sqrt{6}$ **c.** $-6i\sqrt{6}$ **d.** $6\sqrt{6}$

6-7

14. Solve $3x^2 + 14 = 8$.

    a. $\{\pm 2i\}$ **b.** $\{\pm\sqrt{2}\}$ **c.** $\{\pm i\sqrt{2}\}$ **d.** $\{\pm 2\}$

15. Simplify $(2 + 3i)(1 - i)$.

    a. $-1 + i$ **b.** $5 + i$ **c.** $-1 + 5i$ **d.** -1

6-8

16. Simplify $\dfrac{3 - 4i}{-2 + i}$.

    a. $-\dfrac{2}{3} + \dfrac{5}{3}i$ **b.** -2 **c.** $-\dfrac{2}{3} + \dfrac{11}{3}i$ **d.** $-2 + i$

Chapter Test

Simplify.

1. $\sqrt[3]{(-8)^3}$ **2.** $(\sqrt{49})^2$

6-1

3. $\sqrt[3]{-64}$ **4.** $\sqrt[4]{81}$

5. Find the real roots of the equation $16x^2 = 81$.

6. Find the real roots of the equation $2x^3 = -16$.

Simplify. Assume that each radical represents a real number.

7. $\dfrac{\sqrt{15} \cdot \sqrt{42}}{\sqrt{35}}$ **8.** $\sqrt{\dfrac{75a^6}{32b^3}}$

6-2

9. $13\sqrt{5} - \sqrt{10}(3\sqrt{2} + 4\sqrt{5})$ **10.** $\dfrac{\sqrt{5x}}{\sqrt{3}} + \sqrt{\dfrac{3x}{5}}$

6-3

11. $(5 + 2\sqrt{6})(8 - 3\sqrt{6})$ **12.** $\dfrac{2}{3 - \sqrt{5}}$

6-4

Solve.

13. $\sqrt[3]{4x - 1} = 3$ **14.** $\sqrt{3y + 4} = 2 + \sqrt{y + 2}$

6-5

15. Express $0.7\overline{45}$ as a common fraction in lowest terms.

6-6

16. Find a rational number and an irrational number between 1.999 and 2.

Simplify.

17. $\sqrt{-45} \cdot \sqrt{-20}$ **18.** $2i\sqrt{3} - \sqrt{-75}$

6-7

19. $(-3 + 4i) + (5 - i)$ **20.** $(7 - 3i) - (-2 + 5i)$

6-8

21. $(4 - 6i)(3 + 8i)$ **22.** $\dfrac{6 + i}{3 - 2i}$

Mixed Problem Solving

Solve each problem that has a solution. If a problem has no solution, explain why.

A 1. For $1 you can buy either five lemons or four limes. If you pay $2.10 for some lemons and limes, find all possibilities for the number of each fruit bought.

2. Cheryl has 20 coins worth $3.15. If the coins are dimes and quarters only, how many of each type of coin does she have?

3. A season pass to the city pool costs $34. If daily passes cost $1.25 each, at most how many daily passes can be purchased before their cost exceeds that of the season pass?

4. A micron is a millionth of a meter. Some bacteria are about 0.2 micron in diameter. Use scientific notation to express this diameter in meters.

5. The Khos invested $7000, part at 6% and the rest at 10% simple annual interest. If they had reversed the amounts invested at each rate, they would have earned $120 less in annual interest. How much was invested at each rate?

6. As a salesman, George earns a monthly base salary of $1500 plus a 2% commission on his sales. If he wants an income of at least $2000 each month, what must his minimum monthly sales total be?

7. The numerator of a fraction is three more than twice the denominator. If the fraction is equal to $\frac{9}{4}$, find the fraction.

B 8. A boat traveled 36 km downstream in 2 h. The return trip upstream took 3 h. Find the rate of the current.

9. Find the dimensions of a rectangle whose perimeter is 18 cm and whose area is 18 cm^2.

10. A company's monthly sales of $50,000 result in $6000 profit, while monthly sales of $70,000 result in $9000 profit. If profit is considered as a linear function of sales, what minimum amount of sales is needed to avoid a loss?

11. Marie can prepare the salads at the Lunch Inn in 2 h. Bob needs 3 h to prepare the salads. At 10:00 A.M. Marie began the job. At 11:00 A.M. Bob joined her. At what time did they finish preparing the salads?

12. The lengths of the sides of a right triangle are consecutive multiples of 3. Find the length of the hypotenuse.

Preparing for College Entrance Exams

Decide which is the best of the choices given and write the corresponding letter on your answer sheet.

1. If $y^2 - y - 6 = 0$, then $y =$

   (A) 1, -6 (B) -2, 3   (C) -1, 6   (D) 2, 3   (E) -3, 2

2. If $\dfrac{1}{1-x} + \dfrac{1}{1+x} = 4$, then $x^4 =$

   (A) $\dfrac{\sqrt{2}}{2}$ (B) $\dfrac{1}{4}$ (C) $\dfrac{1}{2}$ (D) $\pm\dfrac{\sqrt{2}}{2}$ (E) $\pm\dfrac{1}{2}$

3. $\sqrt[3]{-54} - \sqrt[3]{250} + \sqrt[3]{32} =$

   (A) $-6\sqrt[3]{2}$ (B) $-12\sqrt[3]{2}$ (C) $2\sqrt[3]{2}(\sqrt[3]{2} + 4)$ (D) $2(\sqrt[3]{4} - 4\sqrt[3]{2})$ (E) $2(\sqrt[3]{4} - \sqrt[3]{2})$

4. What are the zeros of $\dfrac{r^4 - 13r^2 + 36}{r^3 + r^2 - 6r}$?

   (A) 0, ± 2, 3   (B) 0, 3   (C) 2, ± 3 (D) -2, 3   (E) 0, ± 2, ± 3

5. How many solutions does $\sqrt{\dfrac{1}{2}x + 5} = \dfrac{1}{2}x - 1$ have?

   (A) 0   (B) 1   (C) 2   (D) 3   (E) cannot be determined

6. $\sqrt{-36}\left(\sqrt{-81} - \dfrac{\sqrt{8}}{\sqrt{-18}}\right) =$

   (A) -58 (B) 58   (C) $58i$ (D) -50 (E) $-50i$

7. $\dfrac{\left(\frac{1}{3}\right)^{-2} - (-3)^{-1}}{\left(\frac{1}{2}\right)^{-1} - (-2)^{-2}} =$

   (A) $\dfrac{112}{27}$ (B) $\dfrac{16}{9}$ (C) $\dfrac{16}{3}$ (D) $\dfrac{104}{21}$ (E) $-\dfrac{8}{9}$

8. What is the reciprocal of $\sqrt{2} + \sqrt{3} - \sqrt{5}$?

   (A) $\sqrt{2} + \sqrt{3} + \sqrt{5}$ (B) $\dfrac{2\sqrt{3} + 3\sqrt{2} + \sqrt{30}}{12}$ (C) $\dfrac{15\sqrt{2} + 10\sqrt{3} - 6\sqrt{5}}{30}$

   (D) $\sqrt{5} - \sqrt{2} - \sqrt{3}$ (E) $\dfrac{2\sqrt{3} + 3\sqrt{2} + \sqrt{30}}{6}$

7 Quadratic Equations and Functions

Light at Focus

Headlights and flashlights use a curved mirror to direct light from a single bulb into parallel rays. The curve of the mirror, in cross section, is a parabola.

Solving Quadratic Equations

7-1 Completing the Square

Objective   To solve quadratic equations by completing the square.

Here are three general types of polynomial equations in one variable ($a \neq 0$):

$$\textit{linear: } ax + b = 0$$
$$\textbf{\textit{quadratic: } } ax^2 + bx + c = 0$$
$$\textit{cubic: } ax^3 + bx^2 + cx + d = 0$$

Earlier in this book you learned how to solve all linear equations and some quadratic equations. In this chapter you'll learn methods for solving *all* types of quadratic equations. Cubic equations will be considered in the next chapter.

In Chapter 6 you learned to solve a simple quadratic equation $x^2 = r$ by writing the equivalent equation $x = \pm\sqrt{r}$. Example 1 illustrates how a similar method can be used to solve quadratic equations of the form $(px + q)^2 = r$.

Example 1 Solve.

   a. $(x - 3)^2 = 7$ **b.** $(2x - 3)^2 = 7$ **c.** $(x + 5)^2 = -4$

Solution   **a.** $(x - 3)^2 = 7$

$$x - 3 = \pm\sqrt{7}$$
$$x = 3 \pm \sqrt{7}$$

∴ the solution set is $\{3 + \sqrt{7},\ 3 - \sqrt{7}\}$.   *Answer*

   b. $(2x - 3)^2 = 7$

$$2x - 3 = \pm\sqrt{7}$$
$$2x = 3 \pm \sqrt{7}$$
$$x = \frac{3 \pm \sqrt{7}}{2}$$

∴ the solution set is $\left\{\dfrac{3 + \sqrt{7}}{2},\ \dfrac{3 - \sqrt{7}}{2}\right\}$.   *Answer*

   c. $(x + 5)^2 = -4$

$$x + 5 = \pm\sqrt{-4} = \pm 2i$$
$$x = -5 \pm 2i$$

∴ the solution set is $\{-5 + 2i,\ -5 - 2i\}$.   *Answer*

To solve a quadratic equation of the form $ax^2 + bx + c = 0$, you can transform the equation into the form $(x + q)^2 = r$, which can be solved by the method of Example 1.

The method of changing the form of a quadratic equation so that the left side is a trinomial square is called **completing the square.**

Example 2 Solve $x^2 - 6x - 3 = 0$.

Solution

$$x^2 - 6x - 3 = 0$$

Step 1 Add 3 to both sides to get 3 alone on one side.

$$x^2 - 6x \quad = 3$$

Step 2 Check the coefficient of x^2. Since it is 1, you are ready to complete the square.

$$1x^2 - 6x \quad = 3$$

Step 3 Add the square of half the coefficient of x to both sides. Thus, add $\left(\frac{-6}{2}\right)^2$, or 9, to both sides.

$$x^2 - 6x + 9 = 3 + 9$$

Step 4 Factor the left side as the square of a binomial.

$$(x - 3)^2 = 12$$

Step 5 Solve as in Example 1.

$$x - 3 = \pm\sqrt{12}$$
$$x = 3 \pm 2\sqrt{3}$$

∴ the solution set is $\{3 + 2\sqrt{3},\ 3 - 2\sqrt{3}\}$.   **Answer**

Solving $ax^2 + bx + c = 0$ by Completing the Square

Step 1 Transform the equation so that the constant term c is alone on the right side.

Step 2 If a, the coefficient of the second-degree term, is not equal to 1, then divide both sides by a.

Step 3 Add the square of half the coefficient of the first-degree term, $\left(\frac{b}{2a}\right)^2$, to both sides. (Completing the square)

Step 4 Factor the left side as the square of a binomial.

Step 5 Complete the solution using the fact that $(x + q)^2 = r$ is equivalent to $x + q = \pm\sqrt{r}$.

Example 3 Solve $2y^2 + 2y + 5 = 0$.

Solution $2y^2 + 2y + 5 = 0$

Step 1 $2y^2 + 2y \quad = -5$ $\Big\{$ Since $a = 2$, divide

Step 2 $y^2 + y \quad = -\dfrac{5}{2}$ $\Big\{$ both sides by 2.

Step 3 $y^2 + y + \left(\frac{1}{2}\right)^2 = -\frac{5}{2} + \left(\frac{1}{2}\right)^2$ ← Complete the square.

$\qquad y^2 + y + \frac{1}{4} = -\frac{5}{2} + \frac{1}{4}$

Step 4 $\qquad \left(y + \frac{1}{2}\right)^2 = -\frac{9}{4}$

Step 5 $\qquad y + \frac{1}{2} = \pm\sqrt{-\frac{9}{4}} = \pm\frac{3}{2}i$

$\qquad\qquad y = -\frac{1}{2} \pm \frac{3}{2}i$

$\therefore$ the solution set is $\left\{-\frac{1}{2} + \frac{3}{2}i, \ -\frac{1}{2} - \frac{3}{2}i\right\}$. *Answer*

Oral Exercises

What two numbers are represented by each of the following?

1. 4 ± 1 **2.** -3 ± 5 **3.** $\frac{1}{2} \pm \frac{3}{2}$ **4.** $-2 \pm \frac{1}{4}$

5. $11 \pm \sqrt{2}$ **6.** $-6 \pm \sqrt{6}$ **7.** $6 \pm 3i$ **8.** $\frac{4}{3} \pm \frac{i\sqrt{2}}{3}$

Give the first step in the solution of each equation.

9. a. $x^2 = 5$ **b.** $(x - 3)^2 = 5$ **c.** $(x + 3)^2 = 5$ **d.** $(2x + 3)^2 = 5$

10. a. $y^2 = -1$ **b.** $(y + 2)^2 = -1$ **c.** $(y - 2)^2 = -1$ **d.** $(6y - 2)^2 = -1$

What must be added to make the given expression a trinomial square?

11. $x^2 + 2x + \underline{\ ?\ }$ **12.** $m^2 - 4m + \underline{\ ?\ }$ **13.** $y^2 + 8y + \underline{\ ?\ }$

14. $a^2 + 12a + \underline{\ ?\ }$ **15.** $t^2 - t + \underline{\ ?\ }$ **16.** $y^2 - 3y + \underline{\ ?\ }$

17. $z^2 - 7z + \underline{\ ?\ }$ **18.** $y^2 + \frac{2}{5}y + \underline{\ ?\ }$ **19.** $r^2 - \frac{1}{2}r + \underline{\ ?\ }$

20. Solve $t^2 - 6t = 0$ by completing the square and by factoring. Which method seems easier?

Written Exercises

Solve.

A

1. a. $x^2 = 3$ **b.** $(x - 1)^2 = 3$ **c.** $(2x - 1)^2 = 3$

2. a. $x^2 = 6$ **b.** $(x + 4)^2 = 6$ **c.** $(3x + 4)^2 = 6$

3. a. $y^2 = 16$ **b.** $(y + 7)^2 = 16$ **c.** $(3y + 7)^2 = 16$

4. a. $y^2 = 49$ **b.** $(y - 8)^2 = 49$ **c.** $(5y - 8)^2 = 49$

Solve.

5. a. $x^2 = -4$ **b.** $(x + 7)^2 = -4$ **c.** $(2x + 7)^2 = -4$

6. a. $z^2 = -5$ **b.** $(z - 3)^2 = -5$ **c.** $(5z - 3)^2 = -5$

7. $(y - 7)^2 = 12$ **8.** $(t - 3)^2 = 8$ **9.** $\left(\frac{1}{3}n + 2\right)^2 = 18$

10. $\left(\frac{1}{2}t - 12\right)^2 = 50$ **11.** $3(y - 7)^2 = -12$ **12.** $\frac{(x - 5)^2}{3} = -8$

Solve by completing the square.

13. $x^2 - 2x - 5 = 0$ **14.** $x^2 - 4x + 2 = 0$ **15.** $y^2 + 6y - 2 = 0$

16. $y^2 + 8y + 6 = 0$ **17.** $p^2 + 20p + 200 = 0$ **18.** $k^2 - 10k + 30 = 0$

19. $x^2 - 1 = 4x$ **20.** $t^2 + 8 = 4t$ **21.** $2t^2 + 4t + 1 = 0$

22. $3n^2 + 12n + 1 = 0$ **23.** $5n^2 + 100 = 30n$ **24.** $2n^2 - 8n - 3 = 0$

25. $x^2 - x - 1 = 0$ **26.** $y^2 - 3y - 5 = 0$ **27.** $3k^2 + 5k + 2 = 0$

B **28.** $\frac{1}{2}x^2 - 3x = 2$ **29.** $\frac{y^2}{4} - \frac{y}{2} + 1 = 0$ **30.** $0.1x^2 - 0.6x + 9 = 0$

31. $0.6x^2 + 2 = 2.4x$ **32.** $7x(1 - x) = 5(x - 2)$ **33.** $2x(x - 4) = 3(1 - x)$

Solve each equation. Be sure to check for extraneous roots by substituting your answers into the original equation.

34. $\dfrac{1}{x + 1} + \dfrac{1}{x - 1} = 1$ **35.** $\dfrac{1}{y + 2} + \dfrac{1}{y + 6} = 1$

36. $\sqrt{x + 3} = 2x$ **37.** $\dfrac{x}{x - 1} - \dfrac{x}{x + 1} = 3 + \dfrac{2x^2}{1 - x^2}$

38. $\dfrac{x + 2}{x - 2} + \dfrac{x - 2}{x + 2} = \dfrac{8 - 4x}{x^2 - 4}$ **39.** $\sqrt{x - 4} - \dfrac{2}{\sqrt{x - 4}} = 1$

Solve by completing the square. Assume that a, b, and c are nonzero constants.

C **40.** $x^2 + x + c = 0$ **41.** $x^2 + bx + c = 0$ **42.** $ax^2 + bx + c = 0$

Mixed Review Exercises

Simplify.

1. $\sqrt{-50}$ **2.** $(3 - 2i)(4 + i)$ **3.** $(3\sqrt{2})^3$

4. $\sqrt{98} - \sqrt{8}$ **5.** $(2i\sqrt{6})^2$ **6.** $2(8 - 3i) + (-4 + 5i)$

7. $\sqrt{\dfrac{2}{5}} + \sqrt{\dfrac{5}{2}}$ **8.** $\dfrac{5x^2y^3}{-15xy^4}$ **9.** $\dfrac{x^2 - 5x + 6}{x^2 - 9}$

7-2 The Quadratic Formula

Objective To solve quadratic equations by using the quadratic formula.

Quadratic equations are used in many applications, so it is useful to have a formula that gives their solutions directly from the coefficients. You can derive this formula by applying the method of completing the square to the general quadratic equation:

$$ax^2 + bx + c = 0$$

$$x^2 + \frac{b}{a}x + \frac{c}{a} = 0 \text{ (Recall that } a \neq 0.)$$

$$x^2 + \frac{b}{a}x = -\frac{c}{a}$$

$$x^2 + \frac{b}{a}x + \left(\frac{b}{2a}\right)^2 = -\frac{c}{a} + \left(\frac{b}{2a}\right)^2 \qquad \text{Complete the square.}$$

$$\left(x + \frac{b}{2a}\right)^2 = \frac{-4ac + b^2}{4a^2} \qquad \text{Factor the left side.}$$

$$x + \frac{b}{2a} = \pm\sqrt{\frac{b^2 - 4ac}{4a^2}} = \pm\frac{\sqrt{b^2 - 4ac}}{2a}$$

$$x = -\frac{b}{2a} \pm \frac{\sqrt{b^2 - 4ac}}{2a} = \frac{-b \pm \sqrt{b^2 - 4ac}}{2a}$$

The Quadratic Formula

The solutions of the quadratic equation $ax^2 + bx + c = 0$ ($a \neq 0$) are given by the formula

$$x = \frac{-b \pm \sqrt{b^2 - 4ac}}{2a}.$$

Example 1 Solve $3x^2 + x - 1 = 0$.

Solution For the equation $3x^2 + 1x - 1 = 0$,

$$a = 3, \ b = 1, \text{ and } c = -1.$$

Substitute these values in the quadratic formula.

$$x = \frac{-b \pm \sqrt{b^2 - 4ac}}{2a}$$

$$x = \frac{-(1) \pm \sqrt{(1)^2 - 4(3)(-1)}}{2(3)}$$

$$= \frac{-1 \pm \sqrt{1 - (-12)}}{6} = \frac{-1 \pm \sqrt{13}}{6}$$

$\therefore$ the solution set is $\left\{\dfrac{-1 + \sqrt{13}}{6}, \dfrac{-1 - \sqrt{13}}{6}\right\}$. *Answer*

The solutions just obtained are exact and expressed in simplest radical form. In applications you may want approximate solutions to the nearest hundredth. Since $\sqrt{13} \approx 3.606$ (from a calculator or Table 1, page 810),

$$x \approx \frac{-1 + 3.606}{6} \quad \text{or} \quad x \approx \frac{-1 - 3.606}{6}$$

$$x \approx 0.43 \quad \text{or} \quad x \approx -0.77$$

$\therefore$ the solution set is $\{0.43, -0.77\}$. **Answer**

Example 2 Solve $5y^2 = 6y - 3$.

Solution First rewrite the equation in the form $ax^2 + bx + c = 0$:
$$5y^2 - 6y + 3 = 0$$
Then substitute 5 for a, -6 for b, and 3 for c in the quadratic formula.

$$y = \frac{-b \pm \sqrt{b^2 - 4ac}}{2a}$$

$$y = \frac{-(-6) \pm \sqrt{(-6)^2 - 4(5)(3)}}{2(5)}$$

$$= \frac{6 \pm \sqrt{-24}}{10}$$

$$= \frac{6 \pm 2i\sqrt{6}}{10}$$

$$= \frac{3 \pm i\sqrt{6}}{5}$$

$\therefore$ the solution set is $\left\{ \dfrac{3 + i\sqrt{6}}{5}, \dfrac{3 - i\sqrt{6}}{5} \right\}$. **Answer**

In Examples 1 and 2, the coefficients a, b, and c were integers. Keep in mind, however, that the quadratic formula can be used to solve *any* quadratic equation, whether the coefficients are fractions, decimals, irrational numbers, or imaginary numbers. (See Exercises 29–40, page 314.)

Example 3 A swimming pool 6 m wide and 10 m long is to be surrounded by a walk of uniform width. The area of the walk happens to equal the area of the pool. What is the width of the walk?

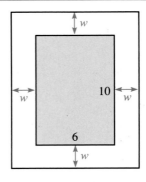

Solution 1. Make a sketch.
   Let w = the width of the walk.
   Then the dimensions of the pool plus the walkway are $10 + 2w$ by $6 + 2w$.

2. The area of the pool is $10 \cdot 6 = 60$ (m²).
The area of the walk can be determined by subtracting the pool's area from the total area covered by both the pool and the walk.

$$\text{Area of walk} = \text{Total area} - \text{Area of pool}$$
$$= (10 + 2w)(6 + 2w) - 60$$

3.
$$\text{Area of walk} = \text{Area of pool}$$
$$(10 + 2w)(6 + 2w) - 60 = 60$$
$$(60 + 32w + 4w^2) - 60 = 60$$
$$4w^2 + 32w = 60$$
$$4w^2 + 32w - 60 = 0 \longleftarrow \text{Divide both sides by 4.}$$
$$w^2 + 8w - 15 = 0$$

4. By completing the square or using the quadratic formula you will find that
$$w = -4 \pm \sqrt{31}.$$
Since $-4 - \sqrt{31} \approx -9.57$, you must reject this root because the width of the walk cannot be negative.

∴ the width of the walk is $(-4 + \sqrt{31})$ m, or approximately 1.57 m. *Answer*

5. *Check:* A calculator is helpful in checking approximate solutions.
$$\text{Area of walk} = \text{Area of pool}$$
$$(10 + 2w)(6 + 2w) - 60 \overset{?}{\approx} 60$$
$$(10 + 2 \cdot 1.57)(6 + 2 \cdot 1.57) - 60 \overset{?}{\approx} 60$$
$$(13.14)(9.14) - 60 \overset{?}{\approx} 60$$
$$120.10 - 60 \overset{?}{\approx} 60 \quad \checkmark$$

Oral Exercises

Give the values that you would substitute for *a*, *b*, and *c* in the quadratic formula.

1. $2x^2 - 3x + 7 = 0$ **2.** $3x^2 + 7x - 2 = 0$ **3.** $5 - 7x - 4x^2 = 0$
4. $x^2 = 4 - 2x$ **5.** $x^2 - x\sqrt{5} + 1 = 0$ **6.** $x(x - 2) = 9$

Written Exercises   # 5-17 odd �-ǂ 25, 27

Solve each equation. Give answers involving radicals in simplest radical form.

A   **1.** $x^2 + 6x + 4 = 0$ **2.** $v^2 + 3v - 5 = 0$ **3.** $y^2 - 4y + 13 = 0$
4. $t^2 + 6t + 6 = 0$ **5.** $5k^2 + 3k - 2 = 0$ **6.** $2p^2 - 3p - 2 = 0$

Solve each equation after rewriting it in the form $ax^2 + bx + c = 0$. Give answers involving radicals in simplest radical form.

7. $5r^2 + 8 = -12r$

8. $2w^2 + 4w = -3$

9. $3y^2 = 1 - y$

10. $8x = 1 - x^2$

11. $2x(x + 1) = 7$

12. $5 = 4r(2r + 3)$

13. $(3n - 5)(2n - 2) = 6$

14. $(2x + 1)(2x - 1) = 4x$

15. $\dfrac{w^2}{2} - w = \dfrac{3}{4}$

16. $\dfrac{t^2}{2} + 1 = \dfrac{t}{5}$

17. $\dfrac{2m^2 + 16}{5} = 2m$

18. $\dfrac{4 - 2y^2}{7} = 2y$

Solve each equation and approximate solutions to the nearest hundredth. A calculator may be helpful.

19. $2n^2 - 4n = 8$

20. $2x^2 - 3x = 7$

21. $3t^2 - 6t - 7 = 0$

22. $4x(x + 1) = 2.75$

23. $3x(x + 2) = -2.5$

24. $2t(t - 4) = -3$

Solve each equation (a) by factoring and (b) by using the quadratic formula.

25. $5x^2 - 45 = 0$

26. $3y^2 - 48 = 0$

27. $3x^2 - 6x + 3 = 0$

28. $4y^2 + 4y - 15 = 0$

Solve each equation. Give answers involving radicals in simplest radical form.

B **29.** $x^2 - x\sqrt{2} - 1 = 0$

30. $x^2 - x\sqrt{5} - 1 = 0$

31. $t^2 - 2t\sqrt{2} + 1 = 0$

32. $u^2 + 2u\sqrt{3} - 3 = 0$

33. $\sqrt{2}x^2 + 5x + 2\sqrt{2} = 0$

34. $\sqrt{3}x^2 - 2x + 2\sqrt{3} = 0$

35. $z^2 + iz + 2 = 0$

36. $z^2 + 2iz - 1 = 0$

37. $z^2 - (3 + 2i)z + (1 + 3i) = 0$

38. $iz^2 + (2 - 3i)z - (3 + i) = 0$

39. $\dfrac{2w + i}{w - i} = \dfrac{3w + 4i}{w + 3i}$

40. $\dfrac{1}{2z + i} + \dfrac{1}{2z - i} = \dfrac{4}{z + 2i}$

41. Show that the solutions of $3x^2 - 2x + 3 = 0$ are reciprocals.

C **42.** Prove that if the roots of $ax^2 + bx + c = 0$ $(a \neq 0)$ are reciprocals, then $a = c$.

Problems

Solve each problem. Approximate any answers involving radicals to the nearest hundredth. A calculator may be helpful.

A **1.** Each side of a square is 4 m long. When each side is increased by x m, the area is doubled. Find the value of x.

2. A rectangle is 6 cm long and 5 cm wide. When each dimension is increased by x cm, the area is tripled. Find the value of x.

3. A positive real number is 1 more than its reciprocal. Find the number.

4. Two positive real numbers have a sum of 5 and product of 5. Find the numbers.

5. A rectangular field with area 5000 m² is enclosed by 300 m of fencing. Find the dimensions of the field.

6. A rectangular animal pen with area 1200 m² has one side along a barn. The other three sides are enclosed by 100 m of fencing. Find the dimensions of the pen.

7. A walkway of uniform width has area 72 m² and surrounds a swimming pool that is 8 m wide and 10 m long. Find the width of the walkway.

8. A 5 in. by 7 in. photograph is surrounded by a frame of uniform width. The area of the frame equals the area of the photograph. Find the width of the frame.

9. When mineral deposits formed a coating 1 mm thick on the inside of a pipe, the area through which fluid can flow was reduced by 20%. Find the original inside diameter of the pipe.
(*Remember:* Area of circle = πr^2 and diameter = $2r$.)

10. The area of the trapezoid shown below is 90 square units. Find the value of x.

Ex. 10   **Ex. 11**

B 11. The total surface area of the rectangular solid shown is 36 m². Find the value of x.

12. In a *golden rectangle* the ratio of the length to the width equals the ratio of the length plus width to the length. Find the value of this *golden ratio*. (Do not approximate the answer.)

13. A box with height $(x + 5)$ cm has a square base with side x cm. A second box with height $(x + 2)$ cm has a square base with side $(x + 1)$ cm. If the two boxes have the same volume, find the value of x.

14. A box with a square base and no lid is to be made from a square piece of metal by cutting squares from the corners and folding up the sides. The cut-off squares are 5 cm on a side. If the volume of the box is 100 cm³, find the dimensions of the original piece of metal.

15. A hydrofoil made a round trip of 144 km in 4 h. Because of head winds, the average speed on returning was 15 km/h less than the average speed going out. Find the two speeds.

Self-Test 1

Vocabulary quadratic equation (p. 307) quadratic formula (p. 311)
completing the square (p. 308)

Solve by completing the square.

1. $x^2 - 6x + 2 = 0$

2. $3y^2 + 9y = -2$

Obj. 7-1, p. 307

Solve by using the quadratic formula.

3. $w^2 - 5w = 3$

4. $4z^2 + 2z + 1 = 0$

Obj. 7-2, p. 311

Solve each problem. Approximate answers involving radicals to the nearest hundredth. A calculator may be helpful.

5. Find the dimensions of a rectangle whose perimeter is 10 cm and whose area is 3 cm².

6. A sidewalk of uniform width has area 180 ft² and surrounds a flower bed that is 11 ft wide and 13 ft long. Find the width of the sidewalk.

Check your answers with those at the back of the book.

Biographical Note / Charles Steinmetz

Charles Proteus Steinmetz (1865–1923) was an electrical engineer and mathematician who combined technical skill with theoretical insight. He showed that sophisticated mathematics could help to solve problems in the design of motors and transformers, and he pioneered the use of complex numbers in the analysis of alternating current circuits.

Born in Germany, Steinmetz immigrated to the United States at the age of 24 and was soon working for an electric company, where he spent most of his career. Even though he suffered from a deformed spine, Steinmetz had a tremendous capacity for work. In his laboratory he devised and improved designs for arc-lamp electrodes and generators and studied the effects of transient currents like those produced by lightning. Steinmetz also researched solar energy, electrical networks, electrification of railways, synthetic production of protein, and electric cars. His 195 patents are proof of the range of his invention.

Roots of Quadratic Equations

7-3 The Discriminant

Objective To determine the nature of the roots of a quadratic equation by using its discriminant.

Using the quadratic formula, you can write the roots of the quadratic equation

$$ax^2 + bx + c = 0$$

in the form

$$r_1 = \frac{-b + \sqrt{D}}{2a} \quad \text{and} \quad r_2 = \frac{-b - \sqrt{D}}{2a},$$

where

$$D = b^2 - 4ac.$$

The following examples illustrate how D is related to the nature of the roots.

Example 1 Solve.

    a. $x^2 + 6x - 2 = 0$ **b.** $3x^2 - 4\sqrt{3}x + 4 = 0$ **c.** $x^2 - 6x + 10 = 0$

Solution **a.** $x^2 + 6x - 2 = 0$

$$D = 6^2 - 4(1)(-2) = 36 + 8 = 44 \longrightarrow \text{D is positive.}$$

$$r_1 = \frac{-6 + \sqrt{44}}{2} = -3 + \sqrt{11}$$

The roots are real and unequal.

$$r_2 = \frac{-6 - \sqrt{44}}{2} = -3 - \sqrt{11}$$

b. $3x^2 - 4\sqrt{3}x + 4 = 0$

$$D = (-4\sqrt{3})^2 - 4(3)(4) = 48 - 48 = 0 \longrightarrow \text{D is zero.}$$

$$r_1 = \frac{-(-4\sqrt{3}) + \sqrt{0}}{2 \cdot 3} = \frac{2\sqrt{3}}{3}$$

The roots are real and equal. We say there is a *double root*.

$$r_2 = \frac{-(-4\sqrt{3}) - \sqrt{0}}{2 \cdot 3} = \frac{2\sqrt{3}}{3}$$

c. $x^2 - 6x + 10 = 0$

$$D = (-6)^2 - 4(1)(10) = 36 - 40 = -4 \longrightarrow \text{D is negative.}$$

$$r_1 = \frac{-(-6) + \sqrt{-4}}{2} = 3 + i$$

The roots are imaginary conjugates.

$$r_2 = \frac{-(-6) - \sqrt{-4}}{2} = 3 - i$$

The Nature of the Roots of a Quadratic Equation

Let $ax^2 + bx + c = 0$ be a quadratic equation with *real* coefficients.

1. If $b^2 - 4ac > 0$, then there are two unequal real roots.
2. If $b^2 - 4ac = 0$, then there is a real double root.
3. If $b^2 - 4ac < 0$, then there are two conjugate imaginary roots.

Since the value $D = b^2 - 4ac$ "discriminates" among the three cases, D is called the **discriminant** of the quadratic equation.

The discriminant also shows you whether a quadratic equation with integral coefficients has rational roots. If $b^2 - 4ac$ is a perfect square, then

$$x = \frac{-b \pm \sqrt{\text{perfect square}}}{2a}$$

$$x = \frac{-b \pm \text{integer}}{2a}$$

Since $\dfrac{-b \pm \text{integer}}{2a}$ is always a quotient of integers, the roots are rational.

Test for Rational Roots

If a quadratic equation has integral coefficients and its discriminant is a perfect square, then the equation has rational roots.

If a quadratic equation can be transformed into an equivalent equation that meets this test, then it has rational roots.

Example 2 Without solving each equation, determine the nature of its roots.

a. $3x^2 - 7x + 5 = 0$ **b.** $2x^2 - 13x + 15 = 0$

c. $x^2 + 2\sqrt{3}x - 1 = 0$ **d.** $\dfrac{x^2}{4} - \dfrac{5}{2}x + \dfrac{25}{4} = 0$

Solution   **a.** $D = (-7)^2 - 4(3)(5) = 49 - 60 = -11$ (negative)
∴ the roots are imaginary conjugates.   *Answer*

b. $D = (-13)^2 - 4(2)(15) = 169 - 120 = 49$ (positive, perfect square)
The given equation has integral coefficients.
∴ the roots are real, unequal, and rational.   *Answer*

c. $D = (2\sqrt{3})^2 - 4(1)(-1) = 12 + 4 = 16$ (positive, perfect square)
The given equation does *not* have integral coefficients.
∴ the roots are real and unequal.   *Answer*

d. The given equation is equivalent to $x^2 - 10x + 25 = 0$, which has integral coefficients.
$D = (-10)^2 - 4(1)(25) = 100 - 100 = 0$ (perfect square)
∴ there is a rational double root.   *Answer*

You now have three methods of solving quadratic equations:

1. Using the quadratic formula
2. Completing the square
3. Factoring

Using the quadratic formula: For real-life applications, computing the quadratic formula on a calculator to find exact rational roots or to approximate irrational roots may be the most efficient method.

      If a calculator is not available, the quadratic formula can be computed by hand to solve *any* quadratic equation.

Completing the square: If the equation has the form

$$x^2 + (\text{even number}) \cdot x + \text{constant} = 0,$$

then completing the square may be the easiest method.

Factoring: If $b^2 - 4ac$ is a perfect square, then you can solve by factoring.

Example 3   The equation $kx^2 + 12x + 9k = 0$ has different roots for different values of k. Find the values of k for which the equation has the following:

   a. a real double root   **b.** two different real roots   **c.** imaginary roots

Solution   The discriminant $D = 12^2 - 4(k)(9k) = 144 - 36k^2$.

a. Equation has a double root if $D = 0$:

$$144 - 36k^2 = 0$$
$$-36k^2 = -144$$
$$k^2 = 4$$
$$k = \pm 2$$

b. Equation has real roots if $D > 0$:

$$144 - 36k^2 > 0$$
$$-36k^2 > -144$$
$$k^2 < 4$$
$$|k| < 2, \text{ or}$$
$$-2 < k < 2$$

c. Equation has imaginary roots if $D < 0$:

$$144 - 36k^2 < 0$$
$$-36k^2 < -144$$
$$k^2 > 4$$
$$|k| > 2, \text{ or}$$
$$k > 2 \text{ or } k < -2$$

Caution: When solving $k^2 = 4$, it is correct to write $k = 2$ or $k = -2$. However, when solving $k^2 < 4$, it is *incorrect* to write $k < 2$ or $k < -2$. To see why, consider $k = -3$. This value satisfies $k < -2$, but *not* $k^2 < 4$.

Oral Exercises

Give the value of the discriminant and determine the nature of the roots of each equation. If the equation is factorable, say so.

1. $x^2 + 6x + 3 = 0$ **2.** $x^2 + 6x + 5 = 0$ **3.** $x^2 + 8x + 16 = 0$

4. $x^2 + 6x + 10 = 0$ **5.** $t^2 - 5t - 5 = 0$ **6.** $3y^2 - 4y + 2 = 0$

7. $5x^2 + \sqrt{5}x - 1 = 0$ **8.** $k^2 - 4\sqrt{2}k + 4 = 0$ **9.** $\sqrt{5}x^2 - 6x + \sqrt{5} = 0$

10. Explain why the roots of $ax^2 + bx + c = 0$ are unequal when $b^2 - 4ac > 0$.

Written Exercises   # 1-11 odd   13, 14, 15, 17, 20, 23, 25

Without solving each equation, determine the nature of its roots. A calculator may be helpful.

A

1. $x^2 + 3x - 9 = 0$ **2.** $x^2 - 4x - 5 = 0$ **3.** $t^2 + 8t + 20 = 0$

4. $3m^2 - 8m - 5 = 0$ **5.** $2y^2 - 9y + 3 = 0$ **6.** $5t^2 - 4t + 3 = 0$

7. $z^2 + \dfrac{5}{4} = z$ **8.** $\dfrac{r^2}{4} + 1 = r$ **9.** $d^2 + \dfrac{7}{3}d = 2$

10. $\sqrt{3}x^2 - 4\sqrt{3} = 0$ **11.** $2u^2 - 5u + \sqrt{8} = 0$ **12.** $7y^2 + 2 = 2\sqrt{14}y$

Solve each equation using whichever method seems easiest to you.

13. $x^2 - 6x + 5 = 0$ **14.** $y^2 + 2y - 24 = 0$ **15.** $9x^2 - 12x + 4 = 0$

16. $4y^2 + 12y + 9 = 0$ **17.** $5(x + 7)^2 = 0$ **18.** $5(x + 7)^2 = 25$

19. $y^2 - 2y = 99$ **20.** $10 = 6t - t^2$ **21.** $3(x - 2)^2 = 18$

22. $x^2 + 4x - 396 = 0$ **23.** $(2x + 5)(x - 3) = 0$ **24.** $(2x + 5)(x - 3) = 6$

B

25. $2(y - 1)^2 = y^2$ **26.** $2w^2 = 3(w - 2)^2$ **27.** $\dfrac{x}{3} + \dfrac{3}{x} = 1$

28. $\dfrac{x + 1}{x} - \dfrac{x}{x + 1} = 2$ **29.** $\dfrac{y + 1}{y - 1} - \dfrac{y}{3} = \dfrac{2}{y - 1}$ **30.** $\dfrac{t}{t - 2} + \dfrac{2t}{t - 1} = 6$

Find the value(s) of k for which each equation has the following:

    a. a real double root   **b. two different real roots**   **c. imaginary roots**

31. $2x^2 + 4x + k = 0$ **32.** $3x^2 - 6x - k = 0$ **33.** $k^2x^2 - 8x + 4 = 0$

34. $9x^2 - 6x + k^2 = 0$ **35.** $kx^2 - 4x + k = 0$ **36.** $3x^2 - 6kx + 12 = 0$

37. Find the value(s) of k for which the expression $16x^2 + 8x + 2k$ is a perfect square.

38. Find three values of k for which the expression $3x^2 - 6x + k$ is factorable.

39. Cliff solved the equation $2x^2 - 5x + 7 = 0$ and noticed that the sum of the roots was $\dfrac{5}{2}$ and the product of the roots was $\dfrac{7}{2}$.

    a. Solve the equation and verify Cliff's results.

    b. Find the sum and product of the roots of $2x^2 + 13x + 11 = 0$.

    c. On the basis of your answers to parts (a) and (b), predict the sum and the product of the roots of the equation $ax^2 + bx + c = 0$.

40. a. Find the roots of $9x^2 - 6x + 2 = 0$.

    b. Verify that the sum of the roots is $\dfrac{6}{9}$ and the product of the roots is $\dfrac{2}{9}$.

    c. Predict the sum and the product of the roots of the equation $3x^2 + 7x + 8 = 0$.

C **41.** Show that the roots of $\dfrac{1}{x+r} = \dfrac{1}{x} + \dfrac{1}{r}$ are always imaginary if r is a non-zero real number.

42. Show that the roots of $x^2 + kx + k = 1$ are always rational if k is an integer.

43. Consider the equation $iz^2 - 3z - 2i = 0$.
  a. Show that the discriminant is 1.
  b. The roots of the given equation are imaginary. Find these roots.
  c. Explain why parts (a) and (b) do not contradict the rule that a quadratic with positive discriminant has real roots.

Mixed Review Exercises

Solve each equation over the complex numbers.

1. $(x - 2)^2 = 18$ **2.** $y^2 = 3y + 4$ **3.** $2w(w - 3) = -1$

4. $u^2 + 5 = u$ **5.** $6v^2 - 13v - 28 = 0$ **6.** $3t^2 = 1 - 4t$

Express in simplest form without negative exponents. Assume that all radicals represent real numbers.

7. $(x^3)^2$ **8.** $\sqrt{72x^3}$ **9.** $(x^{-1})^2$ **10.** $(x^2y)(xy^3)$

11. $(\sqrt{x})^{-3}$ **12.** $\dfrac{x^{-2}y}{(xy)^{-1}}$ **13.** $(\sqrt{5x})^2$ **14.** $\left(\dfrac{x^2}{y^3}\right)^{-2}$

Computer Exercises   *For students with some programming experience.*

1. Write a program in which the user enters the integer coefficients a, b, and c of any quadratic equation $ax^2 + bx + c = 0$, and the program then prints the nature of the roots (imaginary roots, a double real root, rational real roots, or irrational real roots). Use $X * X$ to represent x^2.

2. Run the program in (1) to find the nature of the roots of each equation.
  a. $2x^2 + 3x - 1 = 0$ **b.** $x^2 - 6x = 16$ **c.** $5x^2 - 7x + 12 = 0$

3. Modify the program in (1) so that the computer finds and prints the values of the roots when they are real.

4. Run the program in (3) to print the roots of each equation.
  a. $3x^2 - 8x - 35 = 0$ **b.** $x^2 + 5x - 1 = 0$ **c.** $x^2 = x + 21$

5. Modify the program in (3) so that the computer finds the roots of any quadratic, whether the roots are real or imaginary. Have the computer print imaginary roots in the form $a + bi$. You will need to have the computer print the symbol "*i*."

6. Run the program in (5) to print the roots of each equation.
  a. $x^2 + 2x + 8 = 0$ **b.** $4x^2 - 5x - 11 = 0$ **c.** $2x^2 - 6x + 7 = 0$

7-4 Equations in Quadratic Form

Objective   To recognize and solve equations in quadratic form.

Study the three equations below and note their similarity.

$$x^2 - 5x - 6 = 0 \qquad \text{This equation is quadratic in } x.$$
$$(3x - 2)^2 - 5(3x - 2) - 6 = 0 \qquad \text{This equation is quadratic in } 3x - 2.$$
$$\left(\frac{1}{2x}\right)^2 - 5\left(\frac{1}{2x}\right) - 6 = 0 \qquad \text{This equation is quadratic in } \frac{1}{2x}.$$

Each of the equations above has the same *quadratic form:*

$$(\text{function of } x)^2 - 5 \cdot (\text{function of } x) - 6 = 0$$

In general, an equation is in **quadratic form** if it can be written as

$$a[f(x)]^2 + b[f(x)] + c = 0,$$

where $a \neq 0$ and $f(x)$ is some function of x. When solving an equation in quadratic form, it is often helpful to replace $f(x)$ by a single variable.

Example 1   Solve $(3x - 2)^2 - 5(3x - 2) - 6 = 0$.

Solution   Let $z = 3x - 2$. Then $z^2 = (3x - 2)^2$ and the equation becomes quadratic in z.

$$z^2 - 5z - 6 = 0$$
$$(z + 1)(z - 6) = 0$$

$$z = -1 \qquad \text{or} \qquad z = 6$$
$$3x - 2 = -1 \qquad\qquad 3x - 2 = 6 \longleftarrow \text{Remember } z = 3x - 2$$
$$x = \frac{1}{3} \qquad\qquad\qquad x = \frac{8}{3}$$

∴ the solution set is $\left\{\frac{1}{3}, \frac{8}{3}\right\}$.   ***Answer***

The check is left for you.

Example 2   Solve $\left(\frac{1}{2x}\right)^2 - 5\left(\frac{1}{2x}\right) - 6 = 0$.

Solution   Let $z = \frac{1}{2x}$. Then $z^2 = \left(\frac{1}{2x}\right)^2$ and the equation becomes quadratic in z.

$$z^2 - 5z - 6 = 0$$
$$(z + 1)(z - 6) = 0$$

$$z = -1 \qquad \text{or} \qquad z = 6$$
$$\frac{1}{2x} = -1 \qquad\qquad \frac{1}{2x} = 6 \longleftarrow \text{Remember } z = \frac{1}{2x}.$$
$$x = -\frac{1}{2} \qquad\qquad x = \frac{1}{12}$$

∴ the solution set is $\left\{-\frac{1}{2}, \frac{1}{12}\right\}$.   ***Answer***

Example 3 illustrates that the quadratic form of an equation can sometimes be less obvious than in Examples 1 and 2.

Example 3 Solve $3x + 4\sqrt{x} - 2 = 0$. Approximate real solutions to the nearest hundredth.

Solution Let $z = \sqrt{x}$. Then $z^2 = x$ and the equation becomes

$$3z^2 + 4z - 2 = 0.$$

By using the quadratic formula, you will find that

$$z = \frac{-2 \pm \sqrt{10}}{3} \approx \frac{-2 \pm 3.162}{3}.$$

$$z = \sqrt{x} \approx \frac{-2 + 3.162}{3} \qquad \text{or} \qquad z = \sqrt{x} \approx \frac{-2 - 3.162}{3}$$

$$\sqrt{x} \approx 0.387 \qquad\qquad\qquad \sqrt{x} \approx -1.72$$

$$x \approx (0.387)^2 \qquad\qquad \text{No solution, because the}$$

$$x \approx 0.15 \qquad\qquad\quad \text{principal square root is}$$

always nonnegative.

$\therefore$ the solution set is $\{0.15\}$. *Answer*

Example 4 Solve $x^4 + 7x^2 - 18 = 0$.

Solution The given equation is quadratic in x^2. You can either do all your work in terms of x or let $z = x^2$. Both methods are shown below.

Method 1

$$(x^2)^2 + 7(x^2) - 18 = 0$$
$$(x^2 + 9)(x^2 - 2) = 0$$
$$x^2 = -9 \quad \text{or} \quad x^2 = 2$$
$$x = \pm 3i \quad \text{or} \quad x = \pm \sqrt{2}$$

Method 2

Let $z = x^2$. Then:

$$z^2 + 7z - 18 = 0$$
$$(z + 9)(z - 2) = 0$$
$$z = -9 \quad \text{or} \quad z = 2$$
$$x^2 = -9 \quad \text{or} \quad x^2 = 2$$
$$x = \pm 3i \quad \text{or} \quad x = \pm \sqrt{2}$$

$\therefore$ the solution set is $\{\sqrt{2}, -\sqrt{2}, 3i, -3i\}$. *Answer*

Oral Exercises

For each equation, substitute z for some expression in x to make the equation easier to solve. Then give the resulting quadratic equation in z.

1. $(x + 2)^2 - 5(x + 2) - 14 = 0$

2. $(3x + 4)^2 + 6(3x + 4) - 16 = 0$

3. $\left(\frac{1}{3x}\right)^2 - 4\left(\frac{1}{3x}\right) - 5 = 0$

4. $\left(\frac{2}{x}\right)^2 + 5\left(\frac{2}{x}\right) - 24 = 0$

5. $x^4 - 10x^2 + 9 = 0$

6. $x^6 - 9x^3 + 8 = 0$

7. $2x - 3\sqrt{x} + 1 = 0$

8. $7|x|^2 - 20|x| - 3 = 0$

9. $(x^2 - 5)^2 - (x^2 - 5) - 2 = 0$

10. $5\left(\dfrac{x+2}{x}\right)^2 - 3\left(\dfrac{x+2}{x}\right) = 2$

11. $2x^{-2} + 3x^{-1} - 2 = 0$

12. $3x^{-2} - 4x^{-1} + 1 = 0$

Written Exercises

Solve each equation.

A 1. a. $(x + 3)^2 - 5(x + 3) + 4 = 0$

   b. $(2x - 1)^2 - 5(2x - 1) + 4 = 0$

   c. $x^4 - 5x^2 + 4 = 0$

2. a. $(x - 7)^2 - 13(x - 7) + 36 = 0$

   b. $(1 - 3x)^2 - 13(1 - 3x) + 36 = 0$

   c. $x^4 - 13x^2 + 36 = 0$

3. a. $2\left(\dfrac{1}{2y}\right)^2 + 5\left(\dfrac{1}{2y}\right) - 3 = 0$

   b. $2(y^2 - 4)^2 + 5(y^2 - 4) - 3 = 0$

   c. $2y^{-2} + 5y^{-1} - 3 = 0$

4. a. $3\left(\dfrac{w}{6}\right)^2 - 8\left(\dfrac{w}{6}\right) + 4 = 0$

   b. $3(w^2 - 2)^2 - 8(w^2 - 2) + 4 = 0$

   c. $3w^{-2} - 8w^{-1} + 4 = 0$

5. a. $x^4 - 3x^2 - 4 = 0$

   b. $x - 3\sqrt{x} - 4 = 0$

6. a. $x^4 + 7x^2 - 8 = 0$

   b. $x + 7\sqrt{x} - 8 = 0$

7. a. $x^4 + 5x^2 - 36 = 0$

   b. $x^{-4} + 5x^{-2} - 36 = 0$

8. a. $2x^4 + 7x^2 - 4 = 0$

   b. $2(x - 1)^4 + 7(x - 1)^2 - 4 = 0$

9. $x - 11\sqrt{x} + 30 = 0$

10. $x + 4\sqrt{x} - 21 = 0$

11. $(x^2 - 1)^2 - 11(x^2 - 1) + 24 = 0$

12. $(x^2 + 3)^2 - 6(x^2 + 3) = 7$

B 13. $\left(\dfrac{1+x}{2}\right)^2 - 3\left(\dfrac{1+x}{2}\right) = 18$

14. $\left(\dfrac{1}{x-1}\right)^2 - \dfrac{1}{x-1} = 2$

15. $\dfrac{2}{y} + \dfrac{1}{\sqrt{y}} = 1$

16. $\dfrac{3}{n} - \dfrac{7}{\sqrt{n}} - 6 = 0$

Solve each equation (a) by squaring both sides of the equation and then (b) by using a quadratic form. Be sure to check whether either method gives an extraneous root.

17. $x - 10 = 3\sqrt{x}$

18. $x + 3 = 4\sqrt{x}$

19. $2y - 2 = 3\sqrt{y}$

20. $2t + 3 = 5\sqrt{t}$

Solve each equation. Approximate real solutions to the nearest hundredth. A calculator may be helpful.

21. $3|x|^2 = 7|x| + 5$

22. $|x| \cdot (|x| - 3) = 1$

23. $3x + 6\sqrt{x} - 2 = 0$

24. $\dfrac{2}{x} - \dfrac{5}{\sqrt{x}} = 1$

25. $(x^2 - 3x)^2 - 3(x^2 - 3x) = 10$

26. $(x^2 + 6x)^2 + 9(x^2 + 6x) + 20 = 0$

27. Suppose you know the fourth-degree equation $4x^4 - 73x^2 + 144 = 0$ has roots $\pm\frac{3}{2}$ and ± 4. Write a fourth-degree equation that has roots $\pm\frac{2}{3}$ and $\pm\frac{1}{4}$. (*Hint:* See Exercise 7.)

28. Suppose you know the fourth-degree equation $4x^4 - 55x^2 + 39 = 0$ has roots $\pm\sqrt{13}$ and $\pm\frac{\sqrt{3}}{2}$. Write a fourth-degree equation that has roots $1 \pm \sqrt{13}$ and $1 \pm \frac{\sqrt{3}}{2}$. (*Hint:* See Exercise 8.)

Solve.

C 29. $\sqrt{x+6} - 6\sqrt[4]{x+6} + 8 = 0$ **30.** $\sqrt{x-2} + \sqrt{\sqrt{x-2}} = 2$

31. $(s^2 - 9)(s^4 - 3s^2 - 2) = 2(s^2 - 9)$ **32.** $(\sqrt{y} - 3)(y - \sqrt{y} - 1) = \sqrt{y} - 3$

Self-Test 2

Vocabulary   discriminant (p. 318)   quadratic form (p. 322)

Without solving each equation, determine the nature of its roots.

1. $2t^2 - 6t + 5 = 0$ **2.** $7y^2 - 9y + 2 = 0$ Obj. 7-3, p. 317

3. $2x^2 - 6x - 5 = 0$ **4.** $3y^2 - 9y = 0$

Solve each equation using whichever method seems easiest to you.

5. $4(x - 2)^2 = 20$ **6.** $w^2 - 4 = 3w$

7. $(x - 3)(x - 5) = 9$

8. Find all real values of k for which $3t^2 - 4t + k = 0$ has imaginary roots.

Solve.

9. $x^4 + x^2 - 12 = 0$ **10.** $3x + 2\sqrt{3x} - 8 = 0$ Obj. 7-4, p. 322

Challenge

Find three different positive integers a, b, and c, such that the sum

$$\frac{1}{a} + \frac{1}{b} + \frac{1}{c}$$

is an integer.

Quadratic Functions and Their Graphs

7-5 Graphing $y - k = a(x - h)^2$

Objective   To graph parabolas whose equations have the form
$y - k = a(x - h)^2$ and to find the vertices and axes
of symmetry.

To graph the equation

$$y = x^2$$

first make a table of pairs of values of x and y that satisfy the equation. Then
plot the graph of each ordered pair of coordinates, as shown in Figure 1. If you
plotted more points, you would see that they all lie on the smooth curve shown
in Figure 2. This curve, called a **parabola,** is the graph of $y = x^2$.

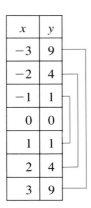

x	y
-3	9
-2	4
-1	1
0	0
1	1
2	4
3	9

Figure 1

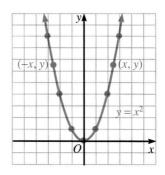

Figure 2

In Figure 2 notice that if the point (x, y) is on the parabola,
then $(-x, y)$, its "mirror image" across the y-axis, is also on
the parabola. This property can also be seen in the table where
the coordinate pairs connected by red lines are mirror images.
Because of this property, the y-axis is called the **axis of
symmetry,** or simply the **axis,** of the parabola. The **vertex** of
a parabola is the point where the parabola crosses its axis. In
the case of $y = x^2$, the vertex is the origin.

Figure 3

The graph of $y = -x^2$, shown in Figure 3 at the right, is
a mirror image or *congruent copy* of the graph of $y = x^2$. If
the graph of $y = x^2$ is reflected across the x-axis, then the result is the graph of $y = -x^2$.

Figures 4 and 5 on the next page show the effect of the
value of a on the graph of an equation of the form $y = ax^2$.

The graph of $y = ax^2$ opens upward if $a > 0$ and downward if $a < 0$.
The larger $|a|$ is, the "narrower" the graph is.

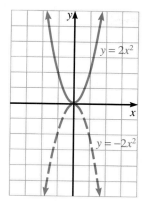

Figure 4

Figure 5

By graphing pairs of quadratic equations on the same axes, such as those in Figures 6–11 that follow, you may investigate the methods for graphing parabolas described in this lesson. A computer or a graphing calculator may be helpful.

Figures 6 and 7 below illustrate the following method for graphing a parabola whose equation has the form $y = a(x - h)^2$.

To graph $y = a(x - h)^2$, slide the graph of $y = ax^2$ horizontally h units.

If $h > 0$, slide it to the right; if $h < 0$, slide it to the left.

The graph has vertex $(h, 0)$ and its axis is the line $x = h$.

Slide right 3.

Slide left 3.

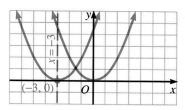

$y = \frac{1}{2}(x - 0)^2$, or $y = \frac{1}{2}x^2$

$y = \frac{1}{2}(x - 3)^2$

$y = \frac{1}{2}x^2$

$y = \frac{1}{2}(x - (-3))^2$, or $y = \frac{1}{2}(x + 3)^2$

Figure 6

Figure 7

Figures 8 and 9 on the next page illustrate a method for graphing a parabola whose equation has the form $y - k = ax^2$.

To graph $y - k = ax^2$, slide the graph of $y = ax^2$ vertically k units.

If $k > 0$, slide it upward; if $k < 0$, slide it downward.

The graph has vertex $(0, k)$ and its axis is the line $x = 0$ (the y-axis).

Slide up 3.

Slide down 3.

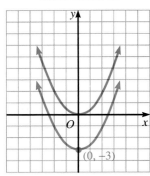

$y - 0 = \frac{1}{2}x^2$, or $y = \frac{1}{2}x^2$

$y - 3 = \frac{1}{2}x^2$

$y = \frac{1}{2}x^2$

$y - (-3) = \frac{1}{2}x^2$, or $y + 3 = \frac{1}{2}x^2$

Figure 8

Figure 9

Figures 10 and 11 below illustrate the following method for graphing a parabola whose equation has the form $y - k = a(x - h)^2$.

To graph $y - k = a(x - h)^2$,

slide the graph of $y = ax^2$ horizontally h units and vertically k units.

The graph has vertex (h, k) and its axis is the line $x = h$.

Slide right 4
and up 2.

Slide left 2
and up 3.

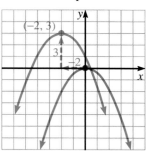

$y = x^2$ $y - 2 = (x - 4)^2$

$y = -\frac{1}{2}x^2$ $y - 3 = -\frac{1}{2}(x + 2)^2$

Figure 10

Figure 11

Example 1 Graph $y - 5 = -(x + 2)^2$. Label the vertex and axis.

Solution Since $a = -1$, the parabola opens downward.
Since $h = -2$ and $k = 5$, the vertex is $(-2, 5)$.
The axis of symmetry is the line $x = -2$.

Calculate a few convenient ordered pairs and plot the corresponding points. Also plot their images by reflection across the axis $x = -2$. Now draw the parabola by connecting the points with a smooth curve.

x	y
0	1
−1	4
−2	5

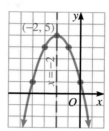

When graphing an equation in the coordinate plane, it is usually helpful to know the *intercepts* of the graph. The y-coordinate of a point where a graph crosses the y-axis is called the *y-intercept*. The x-coordinate of a point where a graph crosses the x-axis is called an *x-intercept*.

A parabola may have no x-intercepts, one x-intercept, or two x-intercepts, as illustrated below.

no x-intercepts
y-intercept 2

x-intercept −2
y-intercept −1

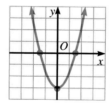

x-intercepts 1.5 and −1.5
y-intercept −3

To find the y-intercept of a parabola, set x equal to zero in the equation of the parabola and solve for y. To find the x-intercepts of a parabola, set y equal to zero in the equation and solve the resulting quadratic equation for x. If the roots are real they are the x-intercepts. If the roots are imaginary, the graph has no x-intercepts.

Example 2 Graph $y + 6 = 2(x + 1)^2$. Label the vertex and axis. Find all intercepts.

Solution 1. Since $a = 2$, the parabola opens upward.
Since $h = -1$ and $k = -6$ the vertex is $(-1, -6)$
The axis of symmetry is the line $x = -1$.

(Solution continues on the next page.)

2. To find the y-intercept, set $x = 0$ and solve for y.
$$y + 6 = 2(0 + 1)^2$$
$$y + 6 = 2$$
$$y = -4 \longleftarrow y\text{-intercept}$$

Therefore, the graph crosses the y-axis at $(0, -4)$. Since $(0, -4)$ is on the graph, so is its mirror image across the axis of symmetry, $(-2, -4)$.

3. To find any x-intercepts, set $y = 0$.
$$0 + 6 = 2(x + 1)^2$$
$$3 = (x + 1)^2$$
$$\pm\sqrt{3} = x + 1$$
$$x = -1 \pm \sqrt{3}$$

To the nearest tenth, $\sqrt{3} \approx 1.7$.
Therefore, the x-intercepts are:

$x \approx -1 - 1.7$ or $x \approx -1 + 1.7$
$x \approx -2.7$ or $x \approx 0.7$

4. Plot the vertex $(-1, -6)$ and the intercepts. Then complete the curve using symmetry.

Example 3   Find an equation $y - k = a(x - h)^2$ of the parabola having vertex $(1, -2)$ and containing the point $(3, 6)$.

Solution   Substitute $(1, -2)$ for (h, k) in the equation $y - k = a(x - h)^2$.
$$y - (-2) = a(x - 1)^2$$
$$y + 2 = a(x - 1)^2$$

Since the parabola contains the point $(3, 6)$, the coordinates of this point must satisfy the equation.
$$6 + 2 = a(3 - 1)^2$$
$$8 = 4a$$
$$a = 2$$

$\therefore$ an equation of the parabola is $y + 2 = 2(x - 1)^2$.   *Answer*

Oral Exercises

Match each equation with one of the graphs at the top of the next page.

1. $y = 3x^2$

2. $y = -\dfrac{1}{3}x^2$

3. $y = 3(x + 2)^2$

4. $y + 1 = 3(x + 2)^2$

5. $y - 1 = \dfrac{1}{3}(x + 2)^2$

6. $y + 1 = -3(x - 2)^2$

7. $y = -\dfrac{1}{3}(x - 2)^2$

8. $y - 1 = -\dfrac{1}{3}(x - 2)^2$

(a) (b) (c) (d)

(e) (f) (g) (h)

Tell: (a) whether the parabola opens upward or downward, (b) the coordinates of the vertex, and (c) the equation of the axis of symmetry.

9. $y = -3x^2$

10. $y + 1 = x^2$

11. $y = -3(x - 1)^2$

12. $y - 4 = 2(x - 5)^2$

13. $y + 3 = -(x - 2)^2$

14. $y + 2 = 5(x + 1)^2$

Tell how to shift the graph of $y = 2x^2$ to obtain the graph of each equation.

15. $y = 2(x - 1)^2$

16. $y + 4 = 2x^2$

17. $y + 4 = 2(x - 1)^2$

Written Exercises

$1-18$ all

Graph each equation. Label the vertex and axis of symmetry. Find all intercepts. You may wish to check your graphs on a computer or a graphing calculator.

A

1. $y = -3x^2$

2. $y = \frac{1}{3}x^2$

3. $y - 4 = -x^2$

4. $y + 1 = \frac{1}{2}x^2$

5. $y = (x - 3)^2$

6. $y = -(x + 3)^2$

7. $y + 2 = (x - 1)^2$

8. $y - 5 = -(x + 2)^2$

9. $y - 3 = 2(x + 1)^2$

10. $y - 8 = -2(x - 3)^2$

11. $y + 8 = \frac{1}{2}(x + 1)^2$

12. $y + 3 = \frac{1}{3}(x - 6)^2$

In Exercises 13–18, graph the given equations on the same axes.

13. $y = x^2$
$y = (x - 2)^2$
$y = (x + 3)^2$

14. $y = 2x^2$
$y = 2(x - 1)^2$
$y = 2(x + 3)^2$

15. $y = -x^2$
$y + 3 = -x^2$
$y - 4 = -x^2$

16. $y = \frac{1}{2}x^2$

$y - 2 = \frac{1}{2}(x - 1)^2$

17. $y = -2x^2$
$y + 1 = -2(x - 2)^2$

18. $y = -\frac{1}{3}x^2$

$y - 3 = -\frac{1}{3}(x + 6)^2$

Quadratic Equations and Functions **331**

Find an equation $y - k = a(x - h)^2$ for each parabola described.

19. Vertex $(4, -3)$; contains $(2, -1)$ **20.** Vertex $(4, 5)$; contains $(5, 3)$

21. Vertex $(0, 0)$; contains $(-3, 3)$ **22.** Vertex $(-3, 6)$; contains the origin

B   **23.** Vertex $(3, 5)$; y-intercept 2   **24.** Vertex $(-2, 6)$; y-intercept -2

25. Vertex $(4, 2)$; one x-intercept is 3   **26.** Vertex $(-3, 4)$; one x-intercept is -1

27. If the parabola $y - 8 = 3(x - h)^2$ has the line $x = 5$ as its axis, find h.

28. If the parabola $y - k = -3(x - 1)^2$ passes through the origin, find k.

29. If the parabola $y - 3 = a(x - 1)^2$ passes through the point $(2, 5)$, find a.

30. If the parabola $y + 5 = a(x + 2)^2$ has y-intercept 4, find a.

Find values of a and k for which each parabola contains the given points.

31. $y - k = a(x - 1)^2$; $(3, 11)$, $(1, 3)$ **32.** $y - k = a(x + 3)^2$; $(-5, 1)$, $(1, 7)$

C   **33.** Graph $y + 3 = |x + 5|$.   **34.** Graph $y - 1 = |x + 4|$.

35. Prove algebraically that if the point $(h + r, s)$ is on the parabola $y - k = (x - h)^2$, then so is the point $(h - r, s)$. What fact about a parabola is established by this proof?

36. a. Using integral values of y from -3 to 3, make a table of values for the relation $x = y^2$.
     b. Use the table in part (a) to draw the graph of $x = y^2$.
     c. Is $x = y^2$ a function? Explain.

37. Make a table of values for each relation using integer values of y in the given range. Then draw the graph of the relation over that range.
     a. $x = -y^2$; -3 to 3   **b.** $x = y^2 + 2$; -3 to 3
     c. $x = (y + 1)^2$; -4 to 2   **d.** $x = (y + 1)^2 + 2$; -4 to 2
     e. $x = -\frac{1}{2}(y - 2)^2 + 3$; -1 to 5

38. Find the coordinates of the vertex and the equation of the axis of symmetry for each parabola in Exercise 37.

39. a. Find a general equation that represents all parabolas that have a horizontal axis.
     b. How many x-intercepts can a parabola with a horizontal axis have? How many y-intercepts?

Mixed Review Exercises

Without solving each equation, determine the nature of its roots.

1. $5x^2 - 3x + 2 = 0$ **2.** $16y^2 - 24y + 9 = 0$ **3.** $t^2 + t - 1 = 0$

Solve each equation over the complex numbers.

4. $t^4 - 5t^2 + 4 = 0$ **5.** $4w^2 - 4w + 5 = 0$ **6.** $x + \sqrt{x} - 6 = 0$

7-6 Quadratic Functions

Objective   To analyze a quadratic function, draw its graph, and find its minimum or maximum value.

A **quadratic function** is a function that can be written in either of two forms.

$$\textit{General form:} \quad f(x) = ax^2 + bx + c \quad (a \neq 0)$$
$$\textit{Completed-square form:} \quad f(x) = a(x - h)^2 + k \quad (a \neq 0)$$

The graph of a quadratic function in either form is a parabola.

Example 1   Graph $f(x) = 2(x - 3)^2 + 1$.

Solution   When graphing a function, it is often helpful to replace $f(x)$ with y.

$$f(x) = 2(x - 3)^2 + 1$$
$$y = 2(x - 3)^2 + 1$$

Then, $y - 1 = 2(x - 3)^2$.

This last equation is the equation of a parabola with vertex (3, 1) and axis $x = 3$.   **Answer**

Example 2   **a.** Show that the graph of $f(x) = 3x^2 - 6x + 1$ is a parabola.
b. Find the vertex and graph the parabola.

Solution   **a.** You want to rewrite the equation in the form $y - k = a(x - h)^2$.

Step 1   Replace $f(x)$ with y. $y = 3x^2 - 6x + 1$

Step 2   Subtract the constant $y - 1 = 3x^2 - 6x$
term from both sides.

Step 3   Factor so that the coefficient $y - 1 = 3(x^2 - 2x)$
of x^2 becomes 1.

Step 4   Complete the square in x. $y - 1 + 3 \cdot 1 = 3(x^2 - 2x + 1)$
$$y + 2 = 3(x - 1)^2$$

∴ since the last equation is in the form
$y - k = a(x - h)^2$, the graph is a parabola.
                                        Answer

b. Since $h = 1$ and $k = -2$, the vertex is $(1, -2)$.
Since $a = 3$, a is positive and the parabola
opens upward. The graph is shown at the right.
                                        Answer

The completing-the-square method illustrated in Example 2 can be used (see Exercises 41 and 42) to show that the quadratic function $y = ax^2 + bx + c$ is equivalent to

$$y - \left(-\frac{b^2 - 4ac}{4a}\right) = a\left(x - \frac{-b}{2a}\right)^2.$$

$$\downarrow \qquad\qquad\qquad \downarrow$$

$$y - k \quad = \quad a(x - h)^2$$

Therefore, the graph of $y = ax^2 + bx + c$ is a parabola having the vertex $(h, k) = \left(-\dfrac{b}{2a}, -\dfrac{b^2 - 4ac}{4a}\right)$. You do not have to remember the expression for k, because you can easily find it by substituting $-\dfrac{b}{2a}$ for x in the equation of the function. Example 3 illustrates.

Example 3 Given $g(x) = 6 + 6x - 3x^2$, find the following.

    a. The vertex of the graph of g.   **b.** The domain of g.

    c. The range of g.   **d.** The zeros of g.

Solution **a.** Since $a = -3$ and $b = 6$, the x-coordinate of the vertex is

$$x = -\frac{b}{2a} = -\frac{6}{2(-3)} = 1.$$

Then the corresponding y-coordinate is

$$g(1) = 6 + 6(1) - 3(1)^2 = 9.$$

∴ the vertex is $(1, 9)$. **Answer**

b. g is defined for all real values of x.

∴ the domain $D = \{\text{real numbers}\}$. **Answer**

c. Since a is negative, the parabola opens downward and the vertex $(1, 9)$ is the highest point of the graph. This means 9 is the *maximum value* of the function.

∴ the range $R = \{y: y \le 9\}$.
Answer

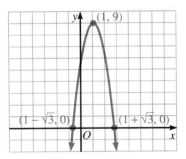

d. Recall (page 195) that the zeros of g are the roots of $g(x) = 0$.

$$g(x) = 6 + 6x - 3x^2$$
$$0 = -3x^2 + 6x + 6$$
$$x = \frac{-6 \pm \sqrt{6^2 - 4 \cdot (-3) \cdot 6}}{2 \cdot (-3)}$$
$$x = 1 \pm \sqrt{3}$$

∴ the zeros of g are $1 + \sqrt{3}$ and $1 - \sqrt{3}$. **Answer**

Note that the zeros of g are the x-intercepts of the graph of g.

Example 3, part (c), illustrated one of the following important facts about a quadratic function.

Let $f(x) = ax^2 + bx + c, a \neq 0$.

If $a < 0$, f has a **maximum value.**

If $a > 0$, f has a **minimum value.**

The graph of f is a parabola. This maximum or minimum value of f is the y-coordinate when $x = -\dfrac{b}{2a}$, at the vertex of the graph.

Example 4 Given $f(x) = \frac{1}{2}x^2 + 3x - \frac{7}{4}$, find the following.

      a. The maximum or minimum value of f.

      b. The vertex of the graph of f.

Solution **a.** Since $a = \frac{1}{2}$, $a > 0$ and f has a minimum value.

$$\text{This minimum occurs when } x = -\frac{b}{2a} = -\frac{3}{2 \cdot \frac{1}{2}} = -3.$$

$$f(-3) = \frac{1}{2}(-3)^2 + 3(-3) - \frac{7}{4}$$

$$= \frac{9}{2} - 9 - \frac{7}{4}$$

$$= -\frac{25}{4}$$

$$\therefore \text{ the minimum value of } f \text{ is } -\frac{25}{4}. \quad \textit{Answer}$$

      b. The vertex is $\left(-3, -\frac{25}{4}\right)$. *Answer*

Oral Exercises

Tell whether each function has a maximum value or a minimum value. Give the value of x associated with this maximum or minimum.

1. $f(x) = x^2 - 5$ **2.** $f(x) = 5 - x^2$ **3.** $f(x) = x^2 - 2x$

4. $f(x) = 2x - x^2$ **5.** $f(x) = (x - 5)^2$ **6.** $f(x) = (5 - x)^2$

7. Mentally draw the graph of $f(x) = (x - 1)^2 + 5$.

      a. Does the graph have 0, 1, or 2 x-intercepts?

      b. How many zeros does f have?

Written Exercises

Graph each function. Follow the method of Example 1. You may wish to check your graphs on a computer or a graphing calculator.

A
1. $f(x) = x^2 - 1$

2. $f(x) = 4 - x^2$

3. $f(x) = (x - 1)^2 + 3$

4. $f(x) = 2(x + 3)^2 + 5$

5. $f(x) = 4 - 2(x - 1)^2$

6. $f(x) = 6 - \frac{1}{2}(x + 4)^2$

Graph each function. Follow the method of Example 2. You may wish to check your graphs on a computer or a graphing calculator.

7. $h(x) = x^2 + 2x - 3$

8. $g(x) = x^2 + 4x + 1$

9. $f(x) = 2x^2 - 4x + 1$

10. $g(x) = 3x^2 - 6x + 5$

11. $f(x) = 5 + 2x - x^2$

12. $h(x) = 6 - 6x - x^2$

13. $f(x) = 3x^2 - 12x$

14. $h(x) = 4x - 2x^2$

15. $g(x) = 2x^2 - 12x + 18$

16. $h(x) = 4x^2 + 4x - 5$

17. $g(x) = \frac{1}{2}x^2 + x + \frac{1}{2}$

18. $f(x) = \frac{1}{4}x^2 - 2x + \frac{5}{2}$

Find the maximum value or minimum value of each function. Then give the coordinates of the vertex of its graph.

19. $g(x) = 2x^2 + 8x$

20. $f(x) = x^2 + 8x + 6$

21. $g(x) = 2x^2 - 6x + 2$

22. $f(x) = 3x^2 - 5x - 2$

23. $h(x) = (2 - x)(5 + x)$

24. $h(x) = (2x - 5)(2x + 3)$

For each function find (a) the vertex of its graph, (b) its domain, (c) its range, and (d) its zeros.

25. $f(x) = x^2 - 4x - 3$

26. $g(x) = 6x - 3x^2$

27. $f(x) = 8 - 2x - x^2$

28. $F(x) = 9 - 8x - x^2$

29. $f(x) = x^2 - 2x - 5$

30. $G(x) = 2x^2 + 8x + 5$

B
31. $h(x) = 2(x - 7)(x + 5)$

32. $g(x) = \frac{1}{2}(6 + x)(4 + x)$

33. $f(x) = 9 - (x + 6)^2$

34. $F(x) = 2(x - 5)^2 - 8$

35. $G(x) = 2 - \frac{1}{2}(x + 1)^2$

36. $H(x) = \frac{1}{2}(3x - 2)(3x - 4)$

In Exercises 37–39, a parabola represented by an equation $y = ax^2 + bx + c$ is given. Study each graph and then tell whether the value of each expression that follows is *positive*, *negative*, or *zero*:

a. $b^2 - 4ac$
   b. c
   c. a
   d. b

37.

38.

39.

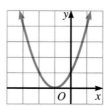

40. Tell how many times the parabola $y = x^2 - 6x + 10$ intersects:

    a. the x-axis

    b. the line $y = 1$

    c. the line $y = 2$

41. By completing the square, rewrite the equation $y = x^2 + bx + c$ in the form $y - k = a(x - h)^2$.

42. Repeat Exercise 41 using the equation $y = ax^2 + bx + c$.

C **43.** A function f is an **even function**

$$\text{if } f(-x) = f(x) \text{ for every } x.$$

Show that $f(x) = ax^2 + bx + c$ is an even function if and only if $b = 0$.

44. A function f is an **odd function**

$$\text{if } f(-x) = -f(x) \text{ for every } x.$$

Show that there are no odd quadratic functions. That is, show that if $f(x) = ax^2 + bx + c$, then $f(-x) \neq -f(x)$.

Career Note / *Statistician*

Every day, people's opinions are being polled by means of surveys—opinions about political candidates, television favorites, consumer goods, and taste preferences, for example. Statisticians are involved in designing such surveys and experiments, implementing them, and interpreting their numerical results. Statistical methods are applied to a wide variety of situations, from predicting human behavior to controlling the quality of a manufactured product.

One of the most important characteristics of statistical work is the way in which accurate information about a large group of people can be obtained by surveying a small part of the group. Television rating services, for example, can determine the size of a national network's audience by asking only a few thousand households, rather than all viewers, what programs are watched. To achieve accurate results, however, statisticians must carefully choose where and how to get their data and determine the type and size of the sample group.

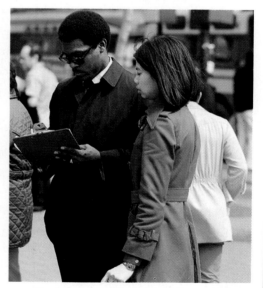

Because statistics are used so widely, statisticians generally combine their knowledge of mathematics with another field, such as business or psychology.

Statisticians also find a background in computer science helpful, since computers are frequently used as tools in statistical applications.

7-7 Writing Quadratic Equations and Functions

Objective	To learn the relationship between the roots and coefficients of a quadratic equation. To write a quadratic equation or function using information about the roots or the graph.

So far in this chapter, you have learned to find roots of quadratic equations and to graph quadratic functions. In this lesson, you will reverse the process and learn how to *write* a quadratic equation or function using given information about the roots or the graph. These new problems are closely related to ones you've already studied. For example, consider the familiar problem of solving the quadratic equation $2x^2 - 6x - 8 = 0$:

$$2x^2 - 6x - 8 = 0$$
$$2(x^2 - 3x - 4) = 0$$
$$2(x + 1)(x - 4) = 0$$
$$x = -1 \quad \text{or} \quad x = 4$$

Now suppose you are given that -1 and 4 are the roots of a quadratic equation. You can reverse the steps above and reason that the equation is

$$(x + 1)(x - 4) = 0$$
$$x^2 - 3x - 4 = 0$$
$$\text{or} \qquad a(x^2 - 3x - 4) = 0 \qquad (a \neq 0).$$

This idea can be generalized as follows: If r_1 and r_2 are roots of a quadratic equation, then

$$(x - r_1)(x - r_2) = 0.$$

Multiplying the binomials on the left side gives

$$x^2 - (r_1 + r_2)x + r_1r_2 = 0.$$

Theorem

A quadratic equation with roots r_1 and r_2 is

$$x^2 - (r_1 + r_2)x + r_1r_2 = 0$$
$$\text{or} \qquad a[x^2 - (r_1 + r_2)x + r_1r_2] = 0.$$

The equation just given is equivalent to

$$a[x^2 - (\text{sum of roots})x + (\text{product of roots})] = 0.$$

The theorem above provides an easy method of writing a quadratic equation from the sum and product of its roots, as Example 1 illustrates.

Example 1 Find a quadratic equation with roots $\dfrac{2 + i}{3}$ and $\dfrac{2 - i}{3}$.

Solution Sum of roots $= \dfrac{2 + i}{3} + \dfrac{2 - i}{3} = \dfrac{4}{3}$

Product of roots $= \dfrac{2 + i}{3} \cdot \dfrac{2 - i}{3} = \dfrac{4 - i^2}{9} = \dfrac{5}{9}$

$x^2 - (\text{sum of roots})x + (\text{product of roots}) = 0$

$$x^2 - \tfrac{4}{3}x + \tfrac{5}{9} = 0.$$

To clear fractions, multiply both sides of the equation by 9.

$\therefore$ the equation is $9x^2 - 12x + 5 = 0$. **Answer**

The next theorem gives you an easy way to find the sum and product of the roots of a quadratic equation of the form $ax^2 + bx + c = 0$.

Theorem

If r_1 and r_2 are the roots of a quadratic equation $ax^2 + bx + c = 0$, then

$$r_1 + r_2 = \text{sum of roots} = -\frac{b}{a} \qquad \text{and} \qquad r_1 r_2 = \text{product of roots} = \frac{c}{a}.$$

Proof By the theorem on page 338, an equation with roots r_1 and r_2 is

$$a[x^2 - (r_1 + r_2)x + r_1 r_2] = 0.$$

Factoring a from the left side of the equation $ax^2 + bx + c = 0$ gives this equivalent equation:

$$a\left[x^2 - \left(-\frac{b}{a}\right)x + \frac{c}{a}\right] = 0$$

Setting the corresponding coefficients equal to each other gives

$$r_1 + r_2 = -\frac{b}{a} \qquad \text{and} \qquad r_1 r_2 = \frac{c}{a}.$$

This theorem is often used to check solutions of a quadratic equation.

Example 2 Find the roots of $2x^2 + 9x + 5 = 0$. Check your answer by using the theorem about the sum and product of the roots.

Solution $2x^2 + 9x + 5 = 0$

$$x = \frac{-9 \pm \sqrt{9^2 - 4(2)(5)}}{2(2)} = \frac{-9 \pm \sqrt{41}}{4}$$

Thus, $r_1 = \dfrac{-9 + \sqrt{41}}{4}$ and $r_2 = \dfrac{-9 - \sqrt{41}}{4}$.

(Solution continues on the next page.)

$$\text{Check: } \quad r_1 + r_2 = \frac{-9 + \sqrt{41}}{4} + \frac{-9 - \sqrt{41}}{4} = -\frac{9}{2} = -\frac{b}{a} \quad \checkmark$$

$$r_1 r_2 = \frac{-9 + \sqrt{41}}{4} \cdot \frac{-9 - \sqrt{41}}{4} = \frac{81 - 41}{16} = \frac{5}{2} = \frac{c}{a} \quad \checkmark$$

$$\therefore \frac{-9 + \sqrt{41}}{4} \text{ and } \frac{-9 - \sqrt{41}}{4} \text{ are the roots. } \quad \textbf{\textit{Answer}}$$

Example 3 shows how to write the equation of a quadratic function using information about its graph.

Example 3 Find a quadratic function $f(x) = ax^2 + bx + c$ such that the maximum value of f is 8 and the graph of f has x-intercepts 3 and 7.

Solution Recall that the x-intercepts of the graph of f are also the roots of the related quadratic equation $f(x) = 0$. Therefore, you can use the theorem on page 338 and write this equation:

$$f(x) = a(x^2 - 10x + 21)$$

Since the maximum value of f is 8, the y-coordinate of the vertex of the graph is 8. To find the x-coordinate of the vertex, note that the axis of symmetry intersects the x-axis at a point midway between the x-intercepts. This means the x-coordinate of the vertex is simply the *average* of the x-intercepts:

$$x = \frac{3 + 7}{2} = 5$$

Therefore, the vertex is (5, 8). Since the vertex lies on the graph of f, you can substitute its coordinates in the equation of f to find a.

$$f(x) = a(x^2 - 10x + 21)$$
$$8 = a(25 - 50 + 21)$$
$$8 = -4a$$
$$a = -2$$

$\therefore$ the equation is $f(x) = -2(x^2 - 10x + 21)$.   ***Answer***

There are many situations in which knowing when maximum or minimum values occur is useful. For example, a business may want to maximize a profit or minimize a loss. If the profit (or loss) can be described as a quadratic function, then you can find the maximum value (or minimum value). Example 4 illustrates this idea.

Example 4 An automobile dealership sells only 35 new cars per month when the markup over factory price is $3000. Marketing research indicates that for each $200 decrease in the markup, the dealership can expect to sell an additional 5 cars per month. What should the markup be per car in order to maximize total monthly profits? What will the total profit be?

Solution

Step 1 The problem asks for the markup needed per car to produce maximum monthly profits. It also asks for the maximum monthly profit.

Step 2 Let n = the number of $200 decreases in markup per car. Making a chart like the one below is often helpful.

Number of $200 decreases	Markup per car	Number of car sales	Total monthly profit
1	$3000 - 200(1) = 2800$	$35 + 5(1) = 40$	$(2800)(40) = 112,000$
2	$3000 - 200(2) = 2600$	$35 + 5(2) = 45$	$(2600)(45) = 117,000$
$\vdots$	$\vdots$	$\vdots$	$\vdots$
n	$3000 - 200n$	$35 + 5n$	$(3000 - 200n)(35 + 5n)$

Step 3 Express the total profit P as a function of n.

Total profit = (Markup per car)(Number of car sales)
$$P(n) = (3000 - 200n)(35 + 5n)$$
$$P(n) = 105,000 + 8000n - 1000n^2$$

Step 4 This function has its maximum value when
$$n = -\frac{b}{2a} = -\frac{8000}{2(-1000)} = 4.$$

Then the markup per car is $3000 - 200(4) = 2200$, and the maximum profit is
$$P(4) = 105,000 + 8000(4) - 1000(4)^2$$
$$= 121,000$$

Step 5 The check is left for you.

$\therefore$ the markup per car should be $2200 to obtain maximum monthly profit of $121,000. **Answer**

Oral Exercises

Give the sum and product of the roots of each equation.

1. $x^2 - 3x + 6 = 0$ **2.** $x^2 - 6x + 8 = 0$ **3.** $2y^2 - y - 6 = 0$

4. $2t^2 - 6t + 1 = 0$ **5.** $2x^2 + 2 = 3x$ **6.** $3u^2 - 4 = 0$

Find a quadratic equation whose roots have the given sum and product.

7. $r_1 + r_2 = 3$, $r_1 r_2 = 4$

8. $r_1 + r_2 = -2$, $r_1 r_2 = -2$

9. $r_1 + r_2 = 0$, $r_1 r_2 = -1$

10. $r_1 + r_2 = -4$, $r_1 r_2 = 0$

Find a quadratic equation with the given roots.

11. $3, -2$ **12.** $1, 9$ **13.** $0, 7$ **14.** $\sqrt{5}, -\sqrt{5}$

15. Explain why 2 and $\frac{3}{2}$ cannot be the roots of $2y^2 - 7y + 3 = 0$.

Written Exercises

A **1–6.** Solve the equations in Oral Exercises 1–6. Check each answer by using the theorem about the sum and product of the roots.

Find a quadratic equation with *integral coefficients* having the given roots.

7. $2, 5$

8. $-3, 1$

9. $-2, \dfrac{5}{2}$

10. $\dfrac{3}{2}, -\dfrac{1}{2}$

11. $-\sqrt{3}, \sqrt{3}$

12. $-\dfrac{\sqrt{5}}{2}, \dfrac{\sqrt{5}}{2}$

13. $1 + \sqrt{3}, 1 - \sqrt{3}$

14. $2 + \sqrt{7}, 2 - \sqrt{7}$

15. $\dfrac{1 + \sqrt{2}}{3}, \dfrac{1 - \sqrt{2}}{3}$

16. $\dfrac{-2 + \sqrt{5}}{4}, \dfrac{-2 - \sqrt{5}}{4}$

17. $i\sqrt{5}, -i\sqrt{5}$

18. $2i\sqrt{2}, -2i\sqrt{2}$

19. $3 + i, 3 - i$

20. $4 + 2i, 4 - 2i$

21. $5 + i\sqrt{2}, 5 - i\sqrt{2}$

22. $-2 + i\sqrt{7}, -2 - i\sqrt{7}$

23. $\dfrac{1 - i\sqrt{5}}{4}, \dfrac{1 + i\sqrt{5}}{4}$

24. $\dfrac{2 + i\sqrt{3}}{2}, \dfrac{2 - i\sqrt{3}}{2}$

Find a quadratic function $f(x) = ax^2 + bx + c$ for each parabola described.

25. maximum value 10
x-intercepts 1 and 3

26. maximum value 6
x-intercepts -2 and 4

27. minimum value -8
x-intercepts 0 and 8

28. minimum value -5
x-intercepts -2 and 3

B **29.** vertex $(2, 12)$; x-intercepts -4 and 8

30. vertex $(-1, -10)$; x-intercepts -6 and 4

31. minimum value -6; zeros of f are -1 and 5
(*Hint:* Recall that the zeros of f are the *roots* of $f(x) = 0$.)

32. maximum value 9; zeros of f are -6 and 0

33. maximum value 6 when $x = -2$; one zero is 1

34. minimum value -4 when $x = 3$; one zero is 6

35. y-intercept 2; x-intercepts 1 and 5

36. range: $\{y: y \le 9\}$; x-intercepts -2 and 4

37. a. Show that $3 + i\sqrt{2}$ is a root of $x^2 - 6x + 11 = 0$.

    b. Find another root.

38. a. Show that $\dfrac{5 - i\sqrt{2}}{4}$ is a root of $16x^2 - 40x + 27 = 0$.

    b. Find another root.

C **39.** Show that the roots of $ax^2 + bx + c = 0$ are reciprocals of each other if and only if $a = c$.

    40. Show that the sum of the reciprocals of the roots of $ax^2 + bx + c = 0$ is $-\dfrac{b}{c}$.

Problems

A **1.** The sum of two numbers is 20. If one number is x, then the other number is __?__ . Their product is $p(x) = $ __?__ . Find the maximum value of p.

    2. The difference of two numbers is 8. If the smaller number is x, then the other number is __?__ . Their product is $p(x) = $ __?__ . Find the minimum value of p.

    3. The sum of two numbers is 40. Find their greatest possible product.

    4. Find two numbers such that their sum is 20 and the sum of their squares is as small as possible.

    5. A rectangle has a perimeter of 100 cm. Find the greatest possible area for the rectangle. (*Hint:* Draw a diagram and label the length and width.)

    6. A rectangular pen is made with 100 m of fencing on three sides. The fourth side is a stone wall. Find the greatest possible area of such an enclosure.

    7. There are currently 20 members in a school's Ski Club and the dues are $8 per member. To encourage the recruiting of new members, the club treasurer suggests that for each *new* member recruited, the dues for *all* members be reduced by 10 cents.

        a. How much membership money will the club have if it recruits 10 new members? 15 new members? n new members?

        b. For what value of n will the club's total membership money be a maximum?

    8. In Problem 7, suppose that the Ski Club has 24 members instead of 20. How many new members should be recruited in order to maximize the total membership money? What will this total be?

    9. A charter company will provide a plane for a fare of $60 each for 20 or fewer passengers. For each passenger in excess of 20, the fare is decreased $2 per person for everyone. What number of passengers will produce the greatest revenue for the company?

B **10.** A ferry service transporting passengers to an island charges a fare of $10 and carries 300 persons per day. The manager estimates that the company will lose 15 passengers for each increase of $1 in the fare. Find the fare that yields the greatest income.

11. In a 120 volt electrical circuit having a resistance of 16 ohms, the available power P in watts is a function of I, the amount of current flowing in amperes. If $P = 120I - 16I^2$, how many amperes will produce the maximum power in the circuit? What will this maximum power be?

12. A ball is thrown vertically upward with an initial speed of 80 ft/s. Its height after t seconds is given by $h = 80t - 16t^2$.

a. How high does the ball go?

b. When does the ball hit the ground?

13. A rectangular field is to be enclosed by a fence and divided into two parts by another fence. Find the maximum area that can be enclosed and separated in this way with 800 m of fencing.

14. A parking lot is to be formed by fencing in a rectangular plot of land except for an entrance 12 m wide along one of the sides. Find the dimensions of the lot of greatest area if 300 m of fencing is to be used.

15. A football kicker tries to make a field goal with the ball on the 25 yard line (105 ft from the goal posts). If you imagine a coordinate system with the ball being kicked at $x = 0$ and the goal post at $x = 105$, then the equation of the ball's path is $y = \frac{4}{3}x - \frac{x^2}{90}$, where y is the ball's height in feet.

a. What is the maximum height of the ball?

b. Will the ball clear the goal-post crossbar that is 10 ft above the ground?

C **16.** A rectangle is to be inscribed in an isosceles triangle of height 8 and base 10, as shown. Find the greatest area of such a rectangle. (*Hint:* $\triangle APQ$ is similar to $\triangle ABC$. Use this fact to express y in terms of x.)

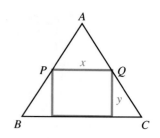

Mixed Review Exercises

Sketch the graph of each equation.

1. $y + 2 = (x - 3)^2$

2. $y = \frac{3}{4}x - 1$

3. $y = 3 - 2x - x^2$

4. $2x + 3y = 12$

5. $y = 2x^2 + 3$

6. $y = -3$

Find the domain and zeros of each function. If there are no zeros, say so.

7. $f(x) = 9 - x^2$

8. $g(x) = 3x - 8$

9. $h(x) = x^2 + 6x - 2$

10. $F(x) = 2$

11. $G(x) = 2x^2 - 3x + 4$

12. $H(x) = \frac{4x + 1}{2x - 5}$

Self-Test 3   ODD

Vocabulary   parabola (p. 326)   vertex (p. 326)
              axis of symmetry (p. 326)   quadratic function (p. 333)

1. Find the vertex and axis of the parabola $y - 4 = -2(x + 3)^2$.   **Obj. 7-5, p. 326**

2. Find all intercepts of the parabola $y - 2 = -\frac{1}{2}(x - 4)^2$.

3. Find an equation $y - k = a(x - h)^2$ of the parabola having vertex $(-2, 1)$ and containing the point $(-3, 4)$.

4. Graph the function $f(x) = 2x^2 + 12x + 17$.   **Obj. 7-6, p. 333**

5. Find the domain, range, and zeros of the function $g(x) = x^2 - 6x + 3$.

6. Find a quadratic equation with integral coefficients having   **Obj. 7-7, p. 338**
roots $\frac{1 + \sqrt{3}}{2}$ and $\frac{1 - \sqrt{3}}{2}$.

7. A ball is thrown vertically upward with an initial speed of 48 ft/s. Its height, in feet, after t seconds is given by $h = 48t - 16t^2$. How high does the ball go?

Check your answers with those at the back of the book.

Quadratic Equations and Functions   **345**

Chapter Summary

1. The general quadratic equation, $ax^2 + bx + c = 0$ $(a \neq 0)$, can be solved either by *completing the square* or by using the *quadratic formula:*

$$x = \frac{-b \pm \sqrt{b^2 - 4ac}}{2a}$$

2. The *discriminant,* $b^2 - 4ac$, can be used to determine the nature and number of the roots of a quadratic equation: real or imaginary; rational or irrational; one root or two roots.

3. An equation in *quadratic form,* $a[f(x)]^2 + b[f(x)] + c = 0$, can usually be solved by first replacing the expression for $f(x)$ by a single variable such as z and solving the resulting equation for z. Then to find x, substitute for z the expression for $f(x)$.

4. A parabola with vertex (h, k) and a vertical axis of symmetry has an equation of the form

$$y - k = a(x - h)^2.$$

If $a > 0$, the parabola opens upward; if $a < 0$, the parabola opens downward. The axis has the equation $x = h$.

5. The graph of the *quadratic function* $f(x) = ax^2 + bx + c$ (or $f(x) = a(x - h)^2 + k$) is a parabola. The *minimum* or *maximum* value of a quadratic function is the y-coordinate of its vertex. This value is a minimum if $a > 0$, a maximum if $a < 0$.

6. The sum of the roots of a quadratic equation is $-\frac{b}{a}$; the product of the roots is $\frac{c}{a}$.

Chapter Review

Write the letter of the correct answer.

1. Solve $2(x - 1)^2 = -16$.   7-1
   a. $\{1 + 2i,\ 1 - 2i\}$ **b.** $\{1 + 2\sqrt{2},\ 1 - 2\sqrt{2}\}$
   c. $\{1 + 2i\sqrt{2},\ 1 - 2i\sqrt{2}\}$ **d.** $\{1 - i\sqrt{2},\ 1 + i\sqrt{2}\}$

2. If $3y^2 - 18y + d$ is a trinomial square, find d.
   a. 81   **b.** 27   **c.** 6   **d.** 108

3. Find the roots of $2x^2 = 3x - 1$.   7-2
   a. $\frac{1}{2}, 1$ **b.** $\frac{-3 \pm \sqrt{41}}{4}$ **c.** $-\frac{1}{2}, -1$ **d.** $\frac{3 \pm \sqrt{41}}{4}$

4. A vegetable garden measures 8 ft by 12 ft. By what equal amount must each dimension be increased if the area is to be doubled?
   a. 2 ft   **b.** 4 ft   **c.** 6 ft   **d.** 8 ft

5. Describe the roots of $9y^2 - 6y + 5 = 0$. 7-3
    a. imaginary conjugates   **b.** real, equal, rational
    c. real, unequal, rational   **d.** real, unequal, irrational

6. Find the real values of k for which the equation $3y^2 - 2y + k = 0$ has two real roots.
    a. $k > \frac{1}{3}$ **b.** $k > -\frac{1}{3}$ **c.** $k = \frac{1}{3}$ **d.** $k < \frac{1}{3}$

7. Solve $(2x + 1)^2 - 5(2x + 1) + 6 = 0$. 7-4
    a. $\left\{1, \frac{1}{2}\right\}$ **b.** $\left\{-1, \frac{5}{2}\right\}$ **c.** $\left\{-2, -\frac{3}{2}\right\}$ **d.** $\left\{0, -\frac{7}{2}\right\}$

8. Find the roots of $y^{-2} - 2y^{-1} + 2 = 0$.
    a. $\frac{1 \pm i}{2}$ **b.** $1 \pm i$ **c.** $-1 \pm i$ **d.** $\frac{-1 \pm i}{2}$

9. Find the vertex of the parabola $y + 3 = -(x - 2)^2$. 7-5
    a. $(-3, 2)$ **b.** $(2, 3)$ **c.** $(2, -3)$ **d.** $(3, -2)$

10. Find an equation of the parabola having vertex $(-1, 4)$ and containing the point $(1, 2)$.
    a. $y + 4 = \frac{1}{2}(x - 1)^2$ **b.** $y - 4 = \frac{1}{2}(x + 1)^2$

    c. $y + 4 = -\frac{1}{2}(x - 1)^2$ **d.** $y - 4 = -\frac{1}{2}(x + 1)^2$

11. Find the minimum value of $f(x) = 2x^2 - 6x + 5$. 7-6
    a. $\frac{3}{2}$ **b.** $\frac{1}{2}$ **c.** $-\frac{3}{2}$ **d.** $\frac{37}{2}$

12. Which curve is the graph of $g(x) = 4x - x^2$?

    a.   **b.**   **c.**   **d.**

13. Find the range of the function $f(x) = 1 - 6x - 3x^2$.
    a. $\{y: y \le 4\}$ **b.** $\{y: y \ge 4\}$ **c.** $\{y: y \le -8\}$ **d.** $\{y: y \ge -8\}$

14. Find a quadratic equation with integral coefficients having the roots 7-7
    $\frac{1 + i\sqrt{2}}{2}$ and $\frac{1 - i\sqrt{2}}{2}$.
    a. $4x^2 + 4x + 3 = 0$ **b.** $4x^2 - 4x - 3 = 0$
    c. $4x^2 - 4x + 3 = 0$ **d.** $4x^2 + 4x - 3 = 0$

15. Find a quadratic function for the parabola having minimum value -2 and x-intercepts 1 and -3.
    a. $f(x) = 2x^2 + 4x - 6$ **b.** $f(x) = 6 - 4x - 2x^2$
    c. $f(x) = \frac{3}{2} - x - \frac{1}{2}x^2$ **d.** $f(x) = \frac{1}{2}x^2 + x - \frac{3}{2}$

Chapter Test

1. Solve $(3x + 1)^2 = 8$.

7-1

2. Solve by completing the square: $2x^2 + 6x + 3 = 0$.

3. Use the quadratic formula to solve $4x^2 - 3x + 2 = 0$.

7-2

4. Two positive real numbers have a sum of 7 and a product of 11. Find the numbers.

5. Without solving the equation, determine the nature of its roots.

7-3
   a. $5x^2 + 7x + 2 = 0$ **b.** $3x^2 - 4x + 2 = 0$

6. Find all real values of k for which $2x^2 + kx + 3 = 0$ has a double root.

7. Solve each equation over the complex numbers.

7-4
   a. $x^4 + x^2 - 12 = 0$ **b.** $x^{-2} - 2x^{-1} - 1 = 0$

8. Graph the parabola $y + 3 = -\frac{1}{2}(x - 2)^2$. Label the vertex and axis of

7-5
   symmetry.

9. Find an equation in the form $y - k = a(x - h)^2$ for the parabola having vertex $(-2, 5)$ and containing the point $(2, 9)$.

10. Graph the function $f(x) = 2x^2 - 4x + 1$ after rewriting its equation in

7-6
    the form $y - k = a(x - h)^2$.

11. Find the domain, range, and zeros of $g(x) = x^2 - 6x + 4$.

12. Find a quadratic equation with *integral* coefficients having roots $\dfrac{1 + \sqrt{3}}{4}$

7-7
    and $\dfrac{1 - \sqrt{3}}{4}$.

13. Find a quadratic function $f(x) = ax^2 + bx + c$ having minimum value -9 and zeros $\dfrac{1}{2}$ and $-\dfrac{5}{2}$.

14. Find the dimensions of the rectangle of greatest area whose perimeter is 20 cm.

Cumulative Review *(Chapters 4–7)*

Simplify.

1. $2(3x^2 - 4x + 1) - (5x^2 + x - 6)$ 2. $(4a^2b)(-3a^3b^2)$
3. $(-x^2yz^3)^4$ 4. $(4y - 3)(2y + 5)$

Factor completely.

5. $2x^2 - 11x + 12$ 6. $3x^2 - 2x + 6xy - 4y$
7. $16a^2 - 9b^2$ 8. $3y^4 - 6y^3 - 9y^2$

Solve.

9. $(x - 4)(x + 1) = 0$

10. $x^2 - 10x + 25 = 0$

11. $x^3 = 9x$

12. $y^2 + y < 20$

13. Find two consecutive odd integers whose product is 99.

14. One leg of a right triangle is 8 cm less than the hypotenuse, while the other leg is only 1 cm less than the hypotenuse. Find the length of the hypotenuse.

Simplify.

15. $\dfrac{12x^3y}{(4xy^2)^2}$

16. $\left(\dfrac{x^3}{y^2}\right)^{-1} \cdot \left(\dfrac{x^2}{y}\right)^{-2}$

17. $\dfrac{a^2 - a}{a^2 + a - 2}$

18. $\dfrac{18x^2}{x^2 - 2x - 8} \cdot \dfrac{x^2 - x - 6}{12x} \div \dfrac{3x - 9}{x - 4}$

19. $\dfrac{1}{2y + 4} + \dfrac{1}{y^2 + 2y}$

20. $\dfrac{1 - m^{-2}}{m^{-1} - 1}$

Solve.

21. $\dfrac{x}{x + 2} - \dfrac{1}{x + 1} = \dfrac{2}{x^2 + 3x + 2}$

22. $y + \sqrt{y - 1} = 7$

23. Mr. Ramirez invested \$4000, part at 5% simple annual interest and the rest at 8%. If his total interest income for a year is \$284, how much did he invest at each rate?

Simplify.

24. $\sqrt{80}$

25. $\sqrt{24} - \sqrt{\dfrac{3}{2}}$

26. $(7 - \sqrt{5})(3 + 4\sqrt{5})$

27. $\dfrac{2}{\sqrt{7} - \sqrt{3}}$

28. $\sqrt{-8} \cdot \sqrt{-6}$

29. $\dfrac{3 - 2i}{3 + 2i}$

30. Write $0.1\overline{54}$ as a common fraction in lowest terms.

31. Solve by completing the square: $4x^2 - 12x + 7 = 0$.

32. Solve by using the quadratic formula: $9y^2 + 12y + 5 = 0$.

33. Find the real values of k for which $kx^2 - 6x + 3 = 0$ has two imaginary roots.

34. Solve: $(4x - 3)^2 - 6(4x - 3) + 5 = 0$.

35. Graph $y - 1 = 2(x - 3)^2$. Label the vertex and axis of symmetry.

36. Find the domain, range, and zeros of $f(x) = x^2 - 8x + 9$.

37. If two numbers differ by 6, then what is the least possible value of their product?

8 Variation and Polynomial Equations

Kepler's third law of planetary motion states that the square of the period (P) of a planet's orbit is directly proportional to the cube of its mean distance (D) from the Sun: $P^2 = kD^3$ for some constant k.

Period (Length of Orbit)

Mean Distance from Sun

Planet

Variation and Proportion

8-1 Direct Variation and Proportion

Objective   To solve problems involving direct variation.

The water pressure, y, on a scuba diver increases with the diver's depth x meters beneath the surface, as shown in the table at the right. (Here pressure is measured in kilopascals (kPa).) Notice that for each ordered pair (x, y), the ratio of pressure to depth is constant: $\frac{y}{x} = 9.8$; and therefore, $y = 9.8x$. Because of this, we say that the pressure *varies directly* as the depth.

Depth x (m)	Pressure y (kPa)	$\frac{y}{x}$ (kPa/m)
3	29.4	9.8
6	58.8	9.8
9	88.2	9.8
12	117.6	9.8

A linear function defined by an equation of the form
$$y = mx \qquad (m \neq 0)$$
is called a **direct variation,** and we say that y *varies directly as x.* The constant m is called the **constant of variation.**

Example 1   The stretch in a loaded spring varies directly as the load it supports (within the spring's elastic limit). A load of 8 kg stretches a certain spring 9.6 cm.
   a. Find the constant of variation (the stiffness constant) and the equation of the direct variation.
   b. What load would stretch the spring 6 cm?

Unloaded spring

Same spring with load

Solution   Let x = the load in kilograms and y = the resulting stretch in centimeters. Since y varies directly as x, $y = mx$ where m is the constant of variation.
   a. Find m. Since $y = 9.6$ when $x = 8$, substitute these values of y and x in $y = mx$ and solve for m:
$$9.6 = m(8)$$
$$m = 9.6 \div 8 = 1.2$$
   ∴ the constant of variation is 1.2, and the equation is $y = 1.2x$.

(Solution continues on the next page.)

b. Use the equation $y = 1.2x$ to find x when $y = 6$:

$$6 = 1.2x$$
$$x = 6 \div 1.2 = 5$$

∴ the load is 5 kg.   **Answer**

The graph of $y = mx$ is a line through the origin with slope m. In most applications of direct variation, the variables are positive, so the graph of the variation is a half-line, as shown at the right. If (x_1, y_1) and (x_2, y_2) are ordered pairs of the variation $y = mx$, and neither x_1 nor x_2 is zero, then $\frac{y_1}{x_1} = m$ and $\frac{y_2}{x_2} = m$. Therefore,

$$\frac{y_1}{x_1} = \frac{y_2}{x_2}.$$

Such an equality of ratios is called a **proportion.** For this reason, in a direct variation y is often said to be **directly proportional** to x, and m is called the **constant of proportionality.** The proportion is sometimes written

$$y_1 : x_1 = y_2 : x_2$$

and is read "y_1 is to x_1 as y_2 is to x_2." The numbers x_1 and y_2 are called the **means,** and y_1 and x_2 the **extremes** of the proportion.

$$\begin{array}{c} \text{means} \\ \downarrow \quad \downarrow \\ y_1 : x_1 = y_2 : x_2 \\ \uparrow \text{extremes} \uparrow \end{array}$$

Multiplying both sides of $\frac{y_1}{x_1} = \frac{y_2}{x_2}$ by $x_1 x_2$ gives the result

$$y_1 x_2 = x_1 y_2.$$

Therefore, in any proportion, *the product of the extremes equals the product of the means.*

Example 2   If y varies directly as x, and $y = 15$ when $x = 24$, find x when $y = 25$.

Solution 1   First find m and write an equation of the direct variation.

$$y = mx$$
$$15 = m(24)$$
$$m = \frac{15}{24} = \frac{5}{8}$$

∴ an equation of the direct variation is $y = \frac{5}{8}x$.

Then, to find x when $y = 25$, substitute in $y = \frac{5}{8}x$.

$$25 = \frac{5}{8}x$$
$$x = 40 \quad \textbf{Answer}$$

Solution 2 Since "y varies directly as x" means that y is directly proportional to x, a proportion can be used.

$$\frac{y_1}{x_1} = \frac{y_2}{x_2}$$

$$\frac{15}{24} = \frac{25}{x}$$

$$15x = 24 \cdot 25$$

$$x = \frac{24 \cdot 25}{15} = 40 \quad \textbf{\textit{Answer}}$$

Example 3 The electrical resistance in ohms (Ω) of a wire varies directly as its length. If a wire 110 cm long has a resistance of 7.5 Ω, what length wire will have a resistance of 12 Ω?

Solution Let l be the required length in centimeters. Since the ratio $\frac{\text{resistance}}{\text{length}}$ is constant, you can write this proportion:

$$\frac{7.5}{110} = \frac{12}{l}$$

$$7.5l = 12(110)$$

$$l = 176$$

$\therefore$ the wire's length is 176 cm. **_Answer_**

In many applications direct variations are nonlinear functions. One quantity may vary directly as a power of another (that is, $y = kx^n$ for some constant k) or as a root of another (that is, $y = k\sqrt[n]{x}$ for some constant k).

Example 4 The period of a pendulum (the time for a complete back-and-forth trip) is directly proportional to the square root of the length of the pendulum. If a pendulum 64 cm long has period 1.6 seconds, what is the period of a pendulum 1 m long?

Solution Let p = the period of the pendulum, l = the length of the pendulum, and k = the constant of proportionality. Then $p = k\sqrt{l}$, or $\frac{p}{\sqrt{l}} = k$. Therefore, you can use this proportion:

$$\frac{p_1}{\sqrt{l_1}} = \frac{p_2}{\sqrt{l_2}}$$

$$\frac{1.6}{\sqrt{64}} = \frac{p_2}{\sqrt{100}} \qquad \text{Use 100 cm for 1 m.}$$

$$\frac{1.6}{8} = \frac{p_2}{10}$$

$$8p_2 = 16, \text{ so } p_2 = 2$$

$\therefore$ the period of the pendulum is 2 s. **_Answer_**

Oral Exercises

In Exercises 1–4, several ordered pairs of a function are given in a table. Tell whether the function is a direct variation. If it is, give the constant of variation.

1.

x	y
0.5	3
2	12
7	42

2.

w	z
8	6
12	9
15	12

3.

s	t
4	10
10	25
12	30

4.

t	d
$\frac{3}{2}$	$\frac{3}{4}$
6	3
$\frac{32}{3}$	$\frac{16}{3}$

For each direct variation, give an equation that relates the two variables and includes the value of the constant of variation.

5. The circumference (C) of a circle varies directly as the radius (r).

6. The perimeter (P) of a square is directly proportional to the length (s) of a side.

7. The area (A) of a square is directly proportional to the square of the length (s) of a side.

8. The area (A) of a circle varies directly as the square of the radius (r).

9. The momentum (M) of a moving 1-kg mass is directly proportional to the speed (v) of the mass in m/s. (Constant of proportionality = 1)

10. The kinetic energy (E) of a moving 1-kg mass varies directly as the square of the speed (v) of the mass in m/s. (Constant of variation = $\frac{1}{2}$)

11. The distance (d) a car travels at 90 km/h is directly proportional to the time (t) traveled.

12. A shop is selling all its merchandise at a 20% discount. The sale price (s) varies directly as the original price (p).

Written Exercises

Solve.

A

1. If y varies directly as x, and $y = 6$ when $x = 15$, find y when $x = 25$.

2. If s is directly proportional to t, and $s = 40$ when $t = 15$, find t when $s = 64$.

3. If p is directly proportional to q, and $p = 9$ when $q = 7.5$, find q when $p = 24$.

4. If a varies directly as b, and $a = 75$ when $b = 40$, find a when $b = 12$.

5. If s varies directly as r^2, and $s = 12$ when $r = 2$, find s when $r = 5$.

6. If y is directly proportional to $\sqrt{x}$, and $y = 25$ when $x = 3$, find x when $y = 100$.

7. If p is directly proportional to $r - 2$, and $p = 20$ when $r = 6$, find p when $r = 12$.

8. If w varies directly as $2x - 1$, and $w = 9$ when $x = 2$, find x when $w = 15$.

If a, b, and c are positive, and $\dfrac{a}{b} = \dfrac{b}{c}$, then b is called the *mean proportional*, or **geometric mean**, between a and c. **Find the mean proportional between each pair of numbers.**

9. 2 and 18   **10.** 3 and 27

11. 8 and 9   **12.** 5 and 15

Let a, b, c and d be positive. **Prove the following properties of the proportion** $\dfrac{a}{b} = \dfrac{c}{d}$. $\left(\textit{Hint:}\text{ For Exercises 13--16, multiply both sides of } \dfrac{a}{b} = \dfrac{c}{d} \text{ by an expression that will produce the given equation.}\right)$

B   **13.** $ad = bc$ **14.** $\dfrac{a}{c} = \dfrac{b}{d}$ **15.** $\dfrac{d}{b} = \dfrac{c}{a}$ **16.** $\dfrac{d}{c} = \dfrac{b}{a}$

17. $\dfrac{a + b}{b} = \dfrac{c + d}{d}$ $\left(\textit{Hint:}\text{ Add 1 to each side of } \dfrac{a}{b} = \dfrac{c}{d}.\right)$

18. $\dfrac{a - b}{b} = \dfrac{c - d}{d}$ $\left(\textit{Hint:}\text{ Add } -1 \text{ to each side of } \dfrac{a}{b} = \dfrac{c}{d}.\right)$

19. If $c \neq d$, then $\dfrac{a - b}{c - d} = \dfrac{a}{c}$.   (*Hint:* Use Exercises 14 and 18.)

20. If $a \neq b$ and $c \neq d$, then $\dfrac{a + b}{a - b} = \dfrac{c + d}{c - d}$.
(*Hint:* Use Exercises 17 and 18.)

C   **21.** Prove that if g is a direct variation over the real numbers, then for every pair of real numbers a and c,
$$g(a + c) = g(a) + g(c).$$

22. Prove that if f is a linear function but not a direct variation and not the constant function 0, then for every pair of real numbers a and c
$$f(a + c) \neq f(a) + f(c).$$

In Exercises 23–25, x, y, and z are positive numbers such that y varies directly as x, and z varies directly as x.

23. Show that $y + z$ varies directly as x.

24. Show that $\sqrt{yz}$ varies directly as x.

25. Show that yz varies directly as $y^2 + z^2$.

Problems

Solve.

A 1. Refer to the table on page 351 that shows water pressure at various depths. At what depth would the pressure be 147 kPa?

2. If the sales tax on a $60 purchase is $3.90, what would it be on a $280 purchase?

3. A real estate agent made a commission of $5400 on a house that sold at $120,000. At this rate, what commission will the agent make on a house that sells for $145,000?

4. The acceleration of an object varies directly as the force acting on it. If a force of 240 newtons causes an acceleration of 150 m/s^2, what force will cause an acceleration of 100 m/s^2?

5. On a certain map, a field 280 ft long is represented by a 5 in. by 8 in. rectangle. How wide is the field?

6. At the Gourmet Grocery Mart, a 575 g can of green beans costs 39¢ and a 810 g can of the same product costs 52¢. Which is the better buy?

7. To estimate the number of flamingos in the Lake Manyara flock, the Kenyan Wildlife Service banded 600 birds. At a later time, they caught 1000 birds and found that 48 were banded. About how large is the entire flock?

8. A public-opinion poll found that of a sample of 450 voters, 252 favored a school bond measure. If 20,000 persons vote, about how many are likely to vote for the bond measure?

9. The stretch in a loaded spring varies directly as the load it supports. A load of 15 kg stretches a certain spring 3.6 cm. What load would stretch the spring 6 cm?

10. The speed of an object falling from rest in a vacuum is directly proportional to the time it has fallen. After an object has fallen for 1.5 s, its speed is 14.7 m/s. What is its speed after it has fallen 5 s?

B 11. *Absolute temperature* is measured in degrees Kelvin (° K). A degree on the Kelvin scale is the same size as a Celsius degree, but 0° K is −273° C. The volume of a fixed amount of gas kept at constant pressure varies directly as its absolute temperature. If the gas occupies 100 L at −13° C, what is its volume at 26° C? (*Hint:* First convert the Celsius temperatures to absolute temperatures.)

12. *Newton's law of cooling* states that the rate at which an object cools varies directly as the difference between its temperature and the temperature of the surrounding air. At the moment a steel plate at 270° C is placed in air that is at 20° C, its rate of cooling is 50° C per minute. How fast is it cooling when its temperature is 100° C?

13. A load of 12 kg stretches a spring to a total length of 15 cm, and a load of 30 kg stretches it to a length of 18 cm. Find the natural (unstretched) length of the spring.

14. The centrifugal force acting on an object moving in a circle is directly proportional to the square of the speed of the object. If the force is 2240 newtons when the object is moving at 8 m/s, what is the force when the object is moving at 12 m/s?

15. The distance an object falls from rest varies directly as the square of the time it has fallen. If the object fell 4 ft during the first half second, how far did it fall during the next two seconds?

16. The speed of an object falling from rest is directly proportional to the square root of the distance the object has fallen. When an object has fallen 36 ft, its speed is 48 ft/s. How much farther must it fall before its speed is 80 ft/s?

17. The power developed by an electric current varies directly as the square of the magnitude of the current. If a current of 0.5 amperes develops 100 watts of power, what current will develop 1.6 kilowatts of power?

18. The average number of red cells in a milliliter (mL) of human blood is 5×10^6. For a certain test, a laboratory technician diluted one part of a blood sample with 199 parts of salt water. How many red cells would the technician expect to find in 6 mL of the diluted solution?

19. The volume of a sphere is directly proportional to the cube of its diameter and is 288π cm^3 when the diameter is 12 cm. Find the constant of proportionality in terms of π. Then write the formula for the volume of a sphere in terms of its diameter.

C 20. The kinetic energy of a moving object varies directly as the square of its speed. By how much would a car traveling at v_0 km/h need to increase its speed in order to double its kinetic energy?

Mixed Review Exercises

Solve each equation over the real numbers.

1. $3x^2 - 5x + 1 = 0$

2. $\dfrac{y - 1}{y + 2} = 1 - \dfrac{2}{y}$

3. $2m^2 = 5m$

4. $\left| \dfrac{a}{2} + 1 \right| = 3$

5. $3(2 - c) = c + 4$

6. $\dfrac{w^2}{2} + \dfrac{7w}{4} = 1$

7. $p^{-2} - 2p^{-1} - 1 = 0$

8. $\sqrt{5n + 6} = n$

9. $(3k - 1)^2 = 12$

8-2 Inverse and Joint Variation

Objective   To solve problems involving inverse and joint variation.

You know that the distance d traveled in time t at speed r is given by

$$rt = d.$$

If the distance d is held constant and r increases, the time t decreases. For example, the greater the average speed of a sprinter, the less time it takes the sprinter to run 100 m. We say that the time *varies inversely* as the speed. In the case of the sprinter, the formula above becomes

$$rt = 100 \quad \text{or} \quad t = \frac{100}{r}$$

where t is measured in seconds and r in meters per second.

A function defined by an equation of the form

$$xy = k \quad \text{or} \quad y = \frac{k}{x} \quad (x \neq 0, \ k \neq 0)$$

is called an **inverse variation,** and we say that y *varies inversely as x,* or y is *inversely proportional to x.* The constant k is called the **constant of variation,** or the **constant of proportionality.**

In most applications of inverse variation, x and y are positive. In such cases the larger x is, the smaller y is, as shown by the graph of $xy = k$ in the diagram at the right. (You will study this type of curve in Lesson 9-5.)

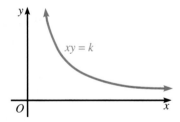

Example 1   If y is inversely proportional to x, and $y = 6$ when $x = 5$, find x when $y = 12$.

Solution   First find k and write an equation of the inverse variation.

$$xy = k$$
$$(5)(6) = k$$
$$k = 30$$

∴ an equation of the inverse variation is $xy = 30$.

Then, to find x when $y = 12$, substitute in $xy = 30$.

$$x(12) = 30$$
$$x = 2.5 \quad \textit{Answer}$$

When a quantity varies directly as the product of two or more other quantities, the variation is called **joint variation**. For example, if z varies jointly as x and the square of y, then $z = kxy^2$ for some nonzero constant k. In this case, we also say that z is **jointly proportional** to x and y^2.

Example 2 If z varies jointly as x and the square root of y, and $z = 6$ when $x = 3$ and $y = 16$, find z when $x = 7$ and $y = 4$.

Solution First find k and write an equation of the joint variation.

$$z = kx\sqrt{y}$$
$$6 = k(3)\sqrt{16}$$
$$6 = 12k$$
$$k = \frac{1}{2}$$

∴ an equation of the joint variation is $z = \frac{1}{2}x\sqrt{y}$.

Then, to find z when $x = 7$ and $y = 4$, substitute in $z = \frac{1}{2}x\sqrt{y}$.

$$z = \frac{1}{2}(7)\sqrt{4}$$
$$z = 7 \quad \textbf{\textit{Answer}}$$

Several important physical laws combine joint and inverse variation. For example, Newton's law of gravitation states that the force of attraction F between two spherical bodies varies jointly as their masses, m_1 and m_2, and inversely as the square of the distance r between their centers. That is,

$$F = k\frac{m_1 m_2}{r^2}.$$

Example 3 The electrical resistance of a wire varies directly as its length and inversely as the square of its diameter. One hundred meters of a wire with diameter 6 mm has resistance 12 ohms (Ω). Eighty meters of a second wire of the same material has resistance 15 Ω. Find the diameter of the second wire.

Solution Let R = the resistance of the wire in ohms, d = the diameter of the wire in millimeters, and l = its length in meters. Then R varies directly as l and inversely as d^2, that is,

$$R = k\frac{l}{d^2}.$$

For the first wire, $R = 12$, $d = 6$, and $l = 100$.

$$12 = k \cdot \frac{100}{6^2}$$
$$k = \frac{12 \cdot 6^2}{100}$$

(Solution continues on the next page.)

For the second wire, $R = 15$ and $l = 80$.

$$15 = k \cdot \frac{80}{d^2}$$

$$15 = \frac{12 \cdot 6^2}{100} \cdot \frac{80}{d^2} \qquad \text{Use the value of } k$$
$$\text{obtained for the first wire.}$$

$$d^2 = \frac{12 \cdot 6^2 \cdot 80}{100 \cdot 15} = 23.04$$

$$d = 4.8$$

∴ the diameter of the second wire is 4.8 mm. *Answer*

Oral Exercises

In Exercises 1–4, k is a constant and the other letters are variables. Express in words each equation as a variation.

Sample $E = k\dfrac{p}{r}$ **Solution** E varies directly as p and inversely as r.

1. $P = kst$ **2.** $z = kqr^2$ **3.** $h = k\dfrac{s}{t^2}$ **4.** $M = k\dfrac{jm^2}{r}$

In Exercises 5–10, give an equation of the variation described. Use k for the constant of variation.

5. V varies directly as the cube of r.

6. i is inversely proportional to the square of j.

7. h is directly proportional to t and inversely proportional to z.

8. A is jointly proportional to l and w.

9. w varies jointly as x and y and inversely as z.

10. m varies directly as n and inversely as z.

In Exercises 11–14, z varies jointly as x and y. Tell what happens to z in each of the situations described.

11. x and y are both doubled.   **12.** x and y are both halved.

13. x is halved and y is quadrupled.   **14.** x is doubled and y is quadrupled.

Written Exercises

A **1.** If y varies inversely as x, and $y = 3$ when $x = 6$, find x when $y = 18$.

2. If z is inversely proportional to r, and $z = 32$ when $r = 1.5$, find r when $z = 8$.

3. If w is inversely proportional to the square of v, and $w = 3$ when $v = 6$, find w when $v = 3$.

4. If p varies inversely as the square root of q, and $p = 12$ when $q = 36$, find p when $q = 16$.

5. If z is jointly proportional to x and y, and $z = 18$ when $x = 0.4$ and $y = 3$, find z when $x = 1.2$ and $y = 2$.

6. If w is jointly proportional to u and v, and $w = 24$ when $u = 0.8$ and $v = 5$, for what value of u will $w = 18$ when $v = 2$?

B 7. If s varies directly as r and inversely as t, and $s = 10$ when $r = 5$ and $t = 3$, for what value of t will $s = 3$ when $r = 4$?

8. Suppose that r varies directly as p and inversely as q^2, and that $r = 27$ when $p = 3$ and $q = 2$. Find r when $p = 2$ and $q = 3$.

9. Suppose that z varies jointly as u and v and inversely as w, and that $z = 0.8$ when $u = 8$, $v = 6$ and $w = 5$. Find z when $u = 3$, $v = 10$ and $w = 5$.

10. Suppose that w varies directly as z^2 and inversely as xy, and that $w = 10$ when $x = 15$, $y = 2$, and $z = 5$. Find z when $w = 2$, $x = 8$, and $y = 27$.

Problems

A 1. The frequency of a radio signal varies inversely as the wave length. A signal of frequency 1200 kilohertz (kHz), which might be the frequency of an AM radio station, has wave length 250 m. What frequency has a signal of wave length 400 m?

2. By *Ohm's law,* the current flowing in a wire is inversely proportional to the resistance of the wire. If the current is 5 amperes (A) when the resistance is 24 ohms (Ω), for what resistance will the current be 8 A?

3. The heat loss through a glass window varies jointly as the area of the window and the difference between the inside and outside temperatures. If the loss through a window with area 3 m^2 is 720 BTU when the temperature difference is 15° C, what is the heat loss through a window with area 4.5 m^2 when the temperature difference is 12° C?

4. The conductance of a wire varies directly as the square of the wire's diameter and inversely as its length. Fifty meters of wire with diameter 2 mm has conductance 0.12 mho (*mho,* which is "ohm" spelled backwards, is a unit of conductance). If a wire of the same material has length 75 m and diameter 2.5 mm, what is its conductance?

In Problems 5 and 6, use the fact that the intensity of light, measured in *lux*, is inversely proportional to the square of the distance between the light source and the object illuminated.

5. A light meter 7.5 m from a light source registers 24 lux. What intensity would it register 15 m from the light source?

6. A light hangs 4.8 ft above the center of a circular table 7.2 ft in diameter. If the illumination is 25 lux at the center of the table, what is it at the edge of the table?

7. A sprocket gear 8 in. in diameter is connected to a gear 3 in. in diameter. How fast does the smaller gear rotate when the larger one rotates at 216 r/min (revolutions per minute)? If the smaller gear is attached to the 28-inch-diameter rear wheel of a bicycle, how fast does the wheel rotate? (*Hint:* r/min · circumference of gear = k.)

8. The volume of a cone varies jointly as the height and the square of the radius of the base. A cone of height 8 cm and base diameter 9 cm has volume 54π. Find the constant of variation and a general formula for the volume of a cone.

B 9. The stretch in a wire under a given tension varies directly as the length of the wire and inversely as the square of its diameter. A wire having length 2 m and diameter 1.5 mm stretches 1.2 mm. If a second wire of the same material (and under the same tension) has length 3 m and diameter 2.0 mm, find the amount of stretch.

10. The volume of a given mass of a gas varies directly as its absolute temperature and inversely as its pressure. (See Problem 11, page 356.) At 2° C and 99 kPa, 4 g of helium occupies 23.1 L. What is the volume of the helium at 27° C and 121 kPa?

In Problems 11 and 12, use the fact that the load a beam with a rectangular cross section can support is jointly proportional to the beam's width and the square of its depth and inversely proportional to its length.

11. A beam 3 cm wide and 5 cm deep can support a load of 630 kg. What load can it support when turned on its side?

12. One beam is half as wide, twice as long, and three times as deep as a second beam. Find the ratio of the loads they can support.

C 13. By one of Kepler's laws (proved by Newton), the square of the period of a satellite in circular orbit is directly proportional to the cube of the radius of the orbit. A satellite in orbit 9600 km above the surface of the Earth has period 5.6 h. How high is a satellite in a ''stationary'' orbit that has period 24 h? (Use 6400 km as the radius of Earth.)

14. Part of a monument consists of two iron spheres whose centers are 3 m apart. Each sphere has radius 1 m and mass 3200 kg. How many times greater is the gravitational force of Earth on each of them than between them? (*Hint:* Use Newton's law of gravitation (see page 359). Also use 6400 km as the radius of Earth and 6×10^{24} kg as the mass of Earth.)

Self-Test 1

Vocabulary
direct variation (p. 351)
  y varies directly as x (p. 351)
  constant of (direct) variation
    (p. 351)
  proportion (p. 352)
  directly proportional (p. 352)
  constant of proportionality (p. 352)
  means (p. 352)

extremes (p. 352)
inverse variation (p. 358)
y varies inversely as x (p. 358)
inversely proportional (p. 358)
constant of (inverse) variation
  (p. 358)
joint variation (p. 359)
jointly proportional (p. 359)

1. Suppose that y varies directly as x, and $y = 8$ when $x = 56$.   **Obj. 8-1, p. 351**
   a. Find the constant of variation.
   b. Find y when $x = 21$.

2. If w is directly proportional to z, and $z = 81$ when $w = 45$,
   find z when $w = 120$.

3. The water pressure on a diver is directly proportional to the
   diver's depth below the surface. If the pressure at 21 m is
   205.8 kPa, what is the pressure at 7 m?

4. Suppose that y varies inversely as x, and $y = 45$ when $x = \frac{1}{3}$.   **Obj. 8-2, p. 358**

   a. Find the constant of variation.
   b. Find x when $y = 30$.

5. If w is inversely proportional to the square of t, and $w = 1$
   when $t = 6$, find w when $t = 2$.

6. Suppose y varies jointly as r and s^2 and inversely as t, and
   $y = 120$ when $r = 5$, $s = 3$, and $t = 2$. Find r when $y = 80$,
   $s = 1$, and $t = 6$.

7. Two meshed gears have 32 and 50 teeth, respectively. If the
   speeds of the gears are inversely proportional to the numbers
   of teeth, at what speed should the second gear be driven so
   that the first gear will run at 1250 revolutions per minute?

Check your answers with those at the back of the book.

Challenge

A group of friends went to a restaurant for lunch. They had agreed to split the
bill equally. However, when the bill arrived two of them discovered that they
had left their money at home. The others in the group then agreed to make up
the difference, which resulted in each one having to pay an extra $1.30. If the
total bill was $78.00, how many people were in the group?

Polynomial Equations

8-3 Dividing Polynomials

Objective   To divide one polynomial by another polynomial.

Polynomial division is similar to the division process you used in arithmetic to write a fraction as a mixed number. For example:

$$\frac{19}{5} = 3 + \frac{4}{5}, \quad \text{or} \quad 3\frac{4}{5}$$

$$\frac{3x + 4}{x} = 3 + \frac{4}{x}$$

These two equations illustrate the **division algorithm:**

$$\frac{\text{Dividend}}{\text{Divisor}} = \text{Quotient} + \frac{\text{Remainder}}{\text{Divisor}}$$

or   Dividend = Quotient × Divisor + Remainder

When you cannot perform a division mentally, you can use long division, illustrated below for $\frac{873}{14}$.

$$
\begin{array}{r}
62 \\
14\overline{)873} \\
\underline{84} \quad \leftarrow \text{subtract } 6 \times 14 \\
33 \\
\underline{28} \quad \leftarrow \text{subtract } 2 \times 14 \\
5
\end{array}
$$

$$\therefore \frac{873}{14} = 62\frac{5}{14}$$

Check: 873 $\overset{?}{=}$ 62 × 14 + 5

873 $\overset{?}{=}$ 868 + 5

873 = 873   √

You can use a similar long-division process to divide one polynomial by another, as Example 1 illustrates.

Example 1   Divide: $\dfrac{x^3 - 5x^2 + 4x - 2}{x - 2}$

Solution

$$
\begin{array}{r}
x^2 - 3x \ - 2 \quad \longleftarrow \text{quotient} \\
\text{divisor} \longrightarrow x - 2\overline{)x^3 - 5x^2 + 4x - 2} \quad \longleftarrow \text{dividend} \\
\underline{x^3 - 2x^2} \quad \longleftarrow \text{subtract } x^2(x - 2) \\
-3x^2 + 4x \\
\underline{-3x^2 + 6x} \quad \longleftarrow \text{subtract } -3x(x - 2) \\
-2x - 2 \\
\underline{-2x + 4} \quad \longleftarrow \text{subtract } -2(x - 2) \\
-6 \quad \longleftarrow \text{remainder}
\end{array}
$$

$$\therefore \frac{x^3 - 5x^2 + 4x - 2}{x - 2} = x^2 - 3x - 2 + \frac{-6}{x - 2}$$

$$\underbrace{\frac{\text{Dividend}}{\text{Divisor}}}_{} = \underbrace{\text{Quotient}}_{} + \underbrace{\frac{\text{Remainder}}{\text{Divisor}}}_{}$$

Check: To check the result, use this form of the division algorithm:

$$\text{Quotient} \times \text{Divisor} + \text{Remainder} = \text{Dividend}$$

$$(x^2 - 3x - 2)(x - 2) + (-6) \overset{?}{=} x^3 - 5x^2 + 4x - 2$$

$$x^3 - 3x^2 - 2x - 2x^2 + 6x + 4 - 6 \overset{?}{=} x^3 - 5x^2 + 4x - 2$$

$$x^3 - 5x^2 + 4x - 2 = x^3 - 5x^2 + 4x - 2 \quad \checkmark$$

Notice that the division process ends when the remainder is 0 or is of lower degree than the divisor.

The next example illustrates the two points mentioned in the following caution.

Caution: Before using long division, always arrange the terms of both dividend and divisor in order of decreasing degree of the same variable. Be sure to insert any ''missing'' terms by using 0 as a coefficient.

Example 2 Divide: $\dfrac{3x^4 - 9a^2x^2 + 4a^4 - 2ax^3}{ax - 2a^2 + 3x^2}$

Solution Arrange the terms of the dividend and divisor in decreasing degree of x.

$$
\begin{array}{r}
x^2 - ax - 2a^2 \\
3x^2 + ax - 2a^2 \overline{\smash{)}\,3x^4 - 2ax^3 - 9a^2x^2 + 0a^3x + 4a^4} \\
\underline{3x^4 + ax^3 - 2a^2x^2} \\
-3ax^3 - 7a^2x^2 + 0a^3x \\
\underline{-3ax^3 - a^2x^2 + 2a^3x} \\
-6a^2x^2 - 2a^3x + 4a^4 \\
\underline{-6a^2x^2 - 2a^3x + 4a^4} \\
0
\end{array}
$$

$\therefore$ the quotient is $x^2 - ax - 2a^2$. ***Answer***

Check: Since the remainder is 0, you only have to show that

$$\text{Quotient} \times \text{Divisor} = \text{Dividend}.$$

$$
\begin{array}{r}
x^2 - ax - 2a^2 \\
\underline{3x^2 + ax - 2a^2} \\
3x^4 - 3ax^3 - 6a^2x^2 \\
ax^3 - a^2x^2 - 2a^3x \\
\underline{- 2a^2x^2 + 2a^3x + 4a^4} \\
3x^4 - 2ax^3 - 9a^2x^2 + 4a^4 \quad \checkmark
\end{array}
$$

Oral Exercises

Give the quotient and remainder in each division.

1. $\dfrac{3x + 5}{x}$

2. $\dfrac{x^2 - 2}{x}$

3. $\dfrac{2x^2 + 1}{x^2}$

4. $\dfrac{x^2 + 2x}{x}$

5. $\dfrac{x^2 + 3x + 5}{x}$

6. $\dfrac{2x^2 - x - 3}{x}$

Sample $\dfrac{x + 1}{x + 4} = \dfrac{(x + 4) - 3}{x + 4} = \dfrac{x + 4}{x + 4} + \dfrac{-3}{x + 4} = 1 + \dfrac{-3}{x + 4}$

$\therefore$ the quotient is 1 and the remainder is -3. *Answer*

7. $\dfrac{x + 4}{x + 1}$

8. $\dfrac{x - 2}{x - 3}$

9. $\dfrac{x}{x + 2}$

10. $\dfrac{x - 3}{x + 2}$

11. $\dfrac{x^2 + 5x + 6}{x + 5}$

12. $\dfrac{x^2 - 3x + 2}{x - 3}$

Written Exercises

Divide.

A

1. $\dfrac{x^2 + 3x - 4}{x + 2}$

2. $\dfrac{x^2 - x + 3}{x + 1}$

3. $\dfrac{6 - 2x - x^2}{2 - x}$

4. $\dfrac{9z - z^2}{z - 3}$

5. $\dfrac{4t^2 - 4t + 1}{2t + 1}$

6. $\dfrac{6u^2 + 7u + 5}{3u - 1}$

7. $\dfrac{x^3 - x^2 - 10x + 10}{x - 3}$

8. $\dfrac{z^3 + 3z^2 - 13z + 6}{z - 2}$

9. $\dfrac{2s^3 - 29s + 13}{s + 4}$

10. $\dfrac{15y^3 + y^2 - 21y}{5y - 3}$

11. $\dfrac{6t^3 + t^2 + 7t + 10}{3t + 2}$

12. $\dfrac{6x^3 - 5x^2 + 15x - 5}{2x^2 - x + 3}$

13. $\dfrac{15z^3 - z^2 - 11z - 3}{3z^2 - 2z - 1}$

14. $\dfrac{4u^4 - 4u^3 - 5u^2 - 9u - 1}{2u - 1}$

15. $\dfrac{6x^4 - 3x^3 + 5x^2 + 2x - 6}{3x^2 - 2}$

16. $\dfrac{4t^4 - 2t^3 - 3t - 9}{2t^2 - t - 3}$

17. $\dfrac{9u^4 + 6u^3 + 4u + 4}{3u^2 + 2u + 2}$

18. $\dfrac{2x^4 - 3x^2 + 7x - 8}{x^2 + x - 3}$

B

19. $\dfrac{x^3 + a^3 + 4a^2x + 4ax^2}{x + 2a}$

20. $\dfrac{2p^3 - 2q^3 + pq^2 - p^2q}{2p + q}$

21. $\dfrac{6t^4 + ct^3 - c^3t + c^4}{2t^2 + ct + c^2}$

22. $\dfrac{2x^4 - x^3y + x^2y^2 + 4xy^3 - 4y^4}{2x^2 + xy - 2y^2}$

23. $\dfrac{x^6 + x^5 + x^3 + x + 1}{x^4 - x^2 + 1}$

24. $\dfrac{x^5 - x^4 + x^3 - x^2 + x - 1}{x^2 - x + 1}$

25. $\dfrac{x^4 + a^4}{x^2 + a^2}$

26. $\dfrac{x^5 - a^5}{x - a}$

27. $\dfrac{x^6 - a^6}{x^2 + ax + a^2}$

28. $\dfrac{x^8 - a^8}{x^3 - ax^2 + a^2x - a^3}$

29. When the polynomial $P(x)$ is divided by $x - 3$, the quotient is $x^2 + 2x + 6$ and the remainder is 8. Find $P(x)$.

C 30. When $x^3 - 7x + 4$ is divided by the polynomial $D(x)$, the quotient is $x^2 - 3x + 2$ and the remainder is -2. Find $D(x)$.

31. Find k so that when $x^3 + kx^2 - kx + 1$ is divided by $x - 2$, the remainder is 0.

32. Find k so that when $x^3 + kx^2 + k^2x + 14$ is divided by $x + 2$, the remainder is 0.

33. When $3x^2 - 5x + c$ is divided by $x + k$, the quotient is $3x + 1$ and the remainder is 3. Find c and k.

34. Here is a complete statement of the division algorithm for polynomials:

> Let $P(x)$ and $D(x)$ be polynomials with $D(x)$ not equal to the constant 0. Then there are *unique* polynomials $Q(x)$ and $R(x)$ such that
> $$\frac{P(x)}{D(x)} = Q(x) + \frac{R(x)}{D(x)}$$
> and either R is the constant 0, or the degree of R is less than the degree of D.

Verify that

$$\frac{x^3 + 2x^2 - 5x - 1}{x - 2} = x^2 + 4x + 3 + \frac{5}{x - 2}$$

and that

$$\frac{x^3 + 2x^2 - 5x - 1}{x - 2} = x^2 + 4x + 2 + \frac{x + 3}{x - 2}.$$

Explain why this does not violate the uniqueness part of the algorithm.

Mixed Review Exercises

In Exercises 1–4, assume that y varies directly as x.

1. If $y = 8$ when $x = 6$, find y when $x = 9$.

2. If $y = 4$ when $x = \dfrac{1}{2}$, find y when $x = 2$.

3. If $y = 3$ when $x = \sqrt{2}$, find y when $x = \sqrt{6}$.

4. If $y = 3.2$ when $x = 0.2$, find y when $x = 1.6$.

5–8. Solve Exercises 1–4 assuming that y varies inversely as x.

8-4 Synthetic Division

Objective   To use synthetic division to divide a polynomial by a first-degree binomial.

Synthetic division is an efficient way to divide a polynomial by a binomial of the form $x - c$. The process is illustrated below with the division of $2x^3 - 10x^2 + 9x + 15$ by $x - 3$. In this example, $c = 3$.

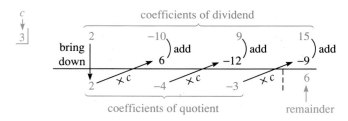

Therefore, $\dfrac{2x^3 - 10x^2 + 9x + 15}{x - 3} = 2x^2 - 4x - 3 + \dfrac{6}{x - 3}$

or $2x^3 - 10x^2 + 9x + 15 = (2x^2 - 4x - 3)(x - 3) + 6.$

Synthetic division is derived from ordinary long division by using only the coefficients of the polynomials involved. Compare the two methods:

$$
\begin{array}{r}
2x^2 - 4x - 3 \\
x - 3\overline{)2x^3 - 10x^2 + 9x + 15} \\
\underline{2x^3 - 6x^2} \\
-4x^2 + 9x \\
\underline{-4x^2 + 12x} \\
-3x + 15 \\
\underline{-3x + 9} \\
6
\end{array}
$$

$$
\begin{array}{r|rrrr}
3 & 2 & -10 & 9 & 15 \\
  &   & 6 & -12 & -9 \\
\hline
  & 2 & -4 & -3 & 6
\end{array}
$$

Example 1   Use synthetic division to divide $x^4 - 2x^3 + 13x - 6$ by $x + 2$.

Solution   Since $x + 2 = x - (-2)$, you use -2 for c. Insert the missing term, $0x^2$, as you would in long division. Note that the coefficient of x^4 is 1.

$$
\begin{array}{r|rrrrr}
-2 & 1 & -2 & 0 & 13 & -6 \\
   &   & -2 & 8 & -16 & 6 \\
\hline
   & 1 & -4 & 8 & -3 & 0
\end{array}
$$

The quotient is $x^3 - 4x^2 + 8x - 3$, and the remainder is 0.

$\therefore \dfrac{x^4 - 2x^3 + 13x - 6}{x + 2} = x^3 - 4x^2 + 8x - 3$ **Answer**

Example 1 illustrates the use of synthetic division when the coefficient of x in the divisor is 1. In fact, synthetic division can be used when the divisor is *any* first-degree binomial.

Example 2 Use synthetic division to divide $6x^3 + 7x^2 + x + 1$ by $2x + 3$.

Solution $2x + 3 = 2\left(x + \dfrac{3}{2}\right)$

$$\therefore \frac{6x^3 + 7x^2 + x + 1}{2x + 3} = \frac{1}{2}\left(\frac{6x^3 + 7x^2 + x + 1}{x + \frac{3}{2}}\right)$$

Now use synthetic division to divide $6x^3 + 7x^2 + x + 1$ by $x + \dfrac{3}{2}$.

$$
\begin{array}{r|rrrr}
-\frac{3}{2} & 6 & 7 & 1 & 1 \\
 & & -9 & 3 & -6 \\
\hline
 & 6 & -2 & 4 & -5
\end{array}
$$

$$\therefore \frac{6x^3 + 7x^2 + x + 1}{2x + 3} = \frac{1}{2}\left(6x^2 - 2x + 4 + \frac{-5}{x + \frac{3}{2}}\right)$$

$$= 3x^2 - x + 2 + \frac{-5}{2x + 3} \quad \textbf{\textit{Answer}}$$

Oral Exercises

For each synthetic division shown below, express as a polynomial (a) the divisor, (b) the dividend, (c) the quotient, and (d) the remainder. Use x as the variable.

1.
$$
\begin{array}{r|rrrr}
2 & 1 & -3 & 5 & -4 \\
 & & 2 & -2 & 6 \\
\hline
 & 1 & -1 & 3 & 2
\end{array}
$$

2.
$$
\begin{array}{r|rrrr}
-3 & 1 & 1 & -4 & 3 \\
 & & -3 & 6 & -6 \\
\hline
 & 1 & -2 & 2 & -3
\end{array}
$$

3.
$$
\begin{array}{r|rrrrr}
-1 & 5 & 1 & 0 & 6 & 2 \\
 & & -5 & 4 & -4 & -2 \\
\hline
 & 5 & -4 & 4 & 2 & 0
\end{array}
$$

4.
$$
\begin{array}{r|rrrrr}
-2 & 3 & 5 & -2 & -1 & 1 \\
 & & -6 & 2 & 0 & 2 \\
\hline
 & 3 & -1 & 0 & -1 & 3
\end{array}
$$

5.
$$
\begin{array}{r|rrrrr}
3 & 1 & -3 & -1 & 5 & -3 \\
 & & 3 & 0 & -3 & 6 \\
\hline
 & 1 & 0 & -1 & 2 & 3
\end{array}
$$

6.
$$
\begin{array}{r|rrrrr}
1 & 2 & -1 & 3 & 0 & -4 \\
 & & 2 & 1 & 4 & 4 \\
\hline
 & 2 & 1 & 4 & 4 & 0
\end{array}
$$

Written Exercises

In Exercises 1–18, divide using synthetic division.

A 1. $\dfrac{3x^3 - 5x^2 + x - 2}{x - 2}$

2. $\dfrac{2x^3 - 4x^2 - 7x + 5}{x - 3}$

3. $\dfrac{x^3 + 3x^2 - 2x - 6}{x + 3}$

4. $\dfrac{3x^3 - 2x^2 + x + 4}{x + 1}$

5. $\dfrac{t^4 + 5t^3 - 2t - 7}{t + 5}$

6. $\dfrac{2u^4 - 5u^3 - 12u^2 + 2u - 8}{u - 4}$

7. $\dfrac{2s^4 - 7s^3 + 7s + 6}{s - 3}$

8. $\dfrac{y^4 - 4y^2 + y + 4}{y + 2}$

9. $\dfrac{x^5 - 1}{x - 1}$

10. $\dfrac{x^6 - 1}{x + 1}$

11. $\dfrac{2x^4 + x^3 - x - 2}{x + 1}$

12. $\dfrac{3x^6 - 2x^5 - x^4 - x^2 - 2x + 3}{x - 1}$

B 13. $\dfrac{2x^3 - 3x^2 + 4x - 2}{2x + 1}$

14. $\dfrac{4x^3 + 2x^2 - 4x + 3}{2x + 3}$

15. $\dfrac{6t^4 + 5t^3 - 10t + 4}{3t - 2}$

16. $\dfrac{5s^4 - 3s^3 + 10s + 2}{5s - 3}$

Sample $\dfrac{2z^3 - z^2 + 8z - 4}{z + 2i}$

Solution

$$
\begin{array}{r|rrrr}
-2i & 2 & -1 & 8 & -4 \\
    &   & -4i & -8 + 2i & 4 \\
\hline
    & 2 & -1 - 4i & 2i & 0 \\
\end{array}
$$

∴ the quotient is $2z^2 - (1 + 4i)z + 2i$, and the remainder is 0.

17. $\dfrac{z^3 - 2z^2 + 4z - 5}{z - 2i}$

18. $\dfrac{z^3 + 3z^2 - 2z + 3}{z - i}$

Find the polynomial $Q(x)$ and the constant R.

19. $2x^3 + 5x^2 + 4 = (x + 3)Q(x) + R$

20. $x^4 - 5x^3 + 2x - 5 = (x - 5)Q(x) + R$

21. $2z^3 - 3z^2 + 8z - 10 = (z + 2i)Q(z) + R$

22. $z^3 - 2z^2 - 3z + 10 = (z - 2 + i)Q(z) + R$

Determine k so that the first polynomial is a factor of the second.

C 23. $x + 2;\ 2x^3 + 3x^2 + k$

24. $x - 2;\ x^4 - 2x^3 + kx + 6$

Reading Algebra / Problem Solving

Problem solving is not an automatic process. Aside from the calculations that produce the solution, solving a problem involves reading, thinking, planning, and, often, rethinking.

The first step in solving any problem is knowing what is given and what you must find. The key to making that first step is reading. Before you start to solve a problem, you should read the related lesson in your textbook and work through the examples.

The next step is finding a solution method. Don't assume that there is a single correct approach that will give you the solution on the first try; problem solving often involves false starts. How can you decide if you are at least on the right track? One way is to estimate a solution using your method, rounding numbers to make the computations easier. If the answer is unreasonable, reconsider your strategy. Perhaps you have chosen the wrong approach, or perhaps you have made an error in setting up an equation. Check for both possibilities. Once you are certain that you have chosen a suitable method, complete your solution to find the exact answer.

The last step in solving a problem is to make sure that your answer is reasonable and that your calculations are correct.

Exercises

1. Describe two methods for solving a quadratic equation. What are some of the advantages and disadvantages of using each method?

2. Read Problem 6 on page 362. Suppose you obtained

$$\frac{I}{(6.0)^2} = \frac{25}{(4.8)^2}$$

as the equation for the intensity I at the edge of the table. According to this equation, will I be greater than 25 or less than 25? Is this answer reasonable; that is, should the intensity at the edge of the table be greater than or less than the intensity at the center? Is the equation correct?

3. Read Problem 11 on page 246. If an *equal* amount of money was invested at each rate, what would the total annual yield be? Since the actual yield was $144, which account has more money?

4. Read Problem 20 on page 134. Is the current greater than 2 mi/h or less than 2 mi/h?

5. Working together, Tom and Sue can complete a task in 3 h. Working alone, Sue takes 5 h.
   a. If they worked at the *same* rate, how long would each person take working alone?
   b. Is Sue's actual time greater than or less than your answer to (a)? What, then, can you conclude about Tom's time working alone?
   c. Find Tom's time working alone, and check it against your conclusion from (b).

8-5 The Remainder and Factor Theorems

Objective To use the remainder and factor theorems to find factors of polynomials and to solve polynomial equations.

You can evaluate the polynomial $P(x) = 2x^3 - 7x^2 + 5x - 1$ when $x = 3$ by substituting 3 for x:

$$P(3) = 2 \cdot 3^3 - 7 \cdot 3^2 + 5 \cdot 3 - 1 = 54 - 63 + 15 - 1 = 5$$

To find the remainder when $P(x)$ is divided by $x - 3$, use synthetic division:

$$
\begin{array}{r|rrrr}
3 & 2 & -7 & 5 & -1 \\
  &   & 6 & -3 & 6 \\
\hline
  & 2 & -1 & 2 & 5
\end{array}
$$

Notice that the remainder when $P(x)$ is divided by $x - 3$ is equal to $P(3)$. To see that this did not just happen by chance, rewrite $P(x)$ using the division algorithm with $x - 3$ as divisor:

$$\text{Dividend} = \text{Quotient} \times \text{Divisor} + \text{Remainder}$$
$$P(x) = (2x^2 - x + 2)(x - 3) + 5$$
$$P(3) = (2 \cdot 3^2 - 3 + 2)(3 - 3) + 5$$
$$= (2 \cdot 3^2 - 3 + 2)(0) + 5$$
$$= 0 + 5$$
$$= 5$$

In general, if $Q(x)$ is the quotient and R is the remainder when the polynomial $P(x)$ is divided by $x - c$, then $P(c) = R$. You can reason as follows:

$$P(x) = Q(x) \cdot (x - c) + R$$
$$P(c) = Q(c) \cdot (c - c) + R$$
$$= Q(c) \cdot (0) + R$$
$$= 0 + R$$
$$= R$$

We have just proved the *remainder theorem*.

Remainder Theorem

Let $P(x)$ be a polynomial of positive degree n. Then for any number c,
$$P(x) = Q(x) \cdot (x - c) + P(c),$$
where $Q(x)$ is a polynomial of degree $n - 1$.

The remainder theorem is *not* restricted to polynomials with real coefficients. It holds for any polynomial $P(x)$ with complex coefficients and any complex number c.

Because synthetic division provides a convenient way to find values of polynomials, the process is sometimes called **synthetic substitution.**

Example 1 Use synthetic substitution to find the value $P(-4)$ for the polynomial $P(x) = x^4 - 14x^2 + 5x - 3$.

Solution By the remainder theorem, $P(-4)$ is equal to the remainder when $P(x)$ is divided by $x + 4$. Since $x + 4 = x - (-4)$, $c = -4$.

$$
\begin{array}{r|rrrrr}
c \rightarrow -4 & 1 & 0 & -14 & 5 & -3 \\
 & & -4 & 16 & -8 & 12 \\
\hline
 & 1 & -4 & 2 & -3 & 9 \leftarrow P(c)
\end{array}
$$

$\therefore P(-4) = 9$ *Answer*

Check: You can check the answer by directly calculating $P(-4)$.

$$P(-4) = (-4)^4 - 14(-4)^2 + 5(-4) - 3$$
$$= 256 - 224 - 20 - 3 = 9 \ \checkmark$$

The following theorem is a corollary of the remainder theorem.

Factor Theorem

The polynomial $P(x)$ has $x - r$ as a factor if and only if r is a root of the equation $P(x) = 0$.

Proof If r is a root of $P(x) = 0$, then by the definition of root, $P(r) = 0$. By the remainder theorem,

$$P(x) = Q(x) \cdot (x - r) + P(r) = Q(x) \cdot (x - r) + 0 = Q(x) \cdot (x - r).$$

Therefore, $x - r$ is a factor of $P(x)$.

Conversely, if $x - r$ is a factor of $P(x)$, then $P(x) = Q(x) \cdot (x - r)$ for some polynomial $Q(x)$, and thus

$$P(r) = Q(r) \cdot (r - r) = Q(r) \cdot 0 = 0.$$

Therefore r is a root of $P(x) = 0$.

Example 2 Determine whether $x + 1$ is a factor of $P(x) = x^{12} - 3x^8 - 4x - 2$.

Solution By the factor theorem, if $P(-1) = 0$, then $x + 1$ is a factor.

$$P(-1) = (-1)^{12} - 3(-1)^8 - 4(-1) - 2 = 1 - 3 + 4 - 2 = 0$$

$\therefore x + 1$ is a factor of $P(x)$.   *Answer*

Example 3 Find a polynomial equation with integral coefficients that has 1, -2, and $\frac{3}{2}$ as roots.

Solution By the factor theorem, the required polynomial must have factors $(x - 1)$, $(x - (-2))$, and $\left(x - \frac{3}{2}\right)$. To obtain integers, you can use $(2x - 3)$ in place of $\left(x - \frac{3}{2}\right)$, since both produce the root $\frac{3}{2}$. Therefore, a solution is

$$(x - 1)(x + 2)(2x - 3) = 0, \quad \text{or}$$
$$2x^3 - x^2 - 7x + 6 = 0 \quad \textit{Answer}$$

In Example 3 many polynomial equations with the given roots are possible. If a is any nonzero constant, then $a(x - 1)(x + 2)(2x - 3) = 0$ has the given roots 1, -2, and $\frac{3}{2}$.

When a given root is a multiple root, the corresponding factor of the polynomial will be to a power greater than 1. For example, an equation with the root 0 and the double root 3 is $x(x - 3)^2 = 0$, or $x^3 - 6x^2 + 9x = 0$.

Suppose that you are given one root r of the nth degree polynomial equation $P(x) = 0$. By the theorems of this lesson,

$$P(x) = Q(x) \cdot (x - r),$$

where $Q(x)$ is a polynomial of degree $n - 1$. You can find the remaining roots of $P(x) = 0$ by solving the **depressed equation**

$$Q(x) = 0$$

because any root of $Q(x)$ is also a root of $P(x)$.

Example 4 Solve $x^3 + x + 10 = 0$ given that -2 is a root.

Solution To find the depressed equation, divide $P(x) = x^3 + x + 10$ by $x - (-2)$, or $x + 2$.

$$
\begin{array}{r|rrrr}
-2 & 1 & 0 & 1 & 10 \\
   &   & -2 & 4 & -10 \\
\hline
   & 1 & -2 & 5 & 0 \\
\end{array}
$$

So $x^3 + x + 10 = (x + 2)(x^2 - 2x + 5)$ and the *depressed equation* is
$$x^2 - 2x + 5 = 0.$$

To solve the depressed equation, use the quadratic formula:
$$x = \frac{2 \pm \sqrt{4 - 20}}{2} = 1 \pm 2i$$

$\therefore$ the solution set is $\{-2, 1 + 2i, 1 - 2i\}$. *Answer*

Oral Exercises

Use the remainder theorem to find the remainder when $P(x)$ is divided by each of the two given binomials.

Sample $P(x) = x^3 + x^2 + 3x + 3$ **a.** $x - 1$ **b.** $x + 2$

Solution   **a.** $P(1) = 1^3 + 1^2 + 3(1) + 3 = 8$; the remainder is 8.
b. $P(-2) = (-2)^3 + (-2)^2 + 3(-2) + 3 = -7$; the remainder is -7.

1. $P(x) = x^3 - 3x^2 - 2x + 2$
   a. $x - 1$ **b.** $x + 1$
3. $P(x) = x^3 - 2x^2 - x + 2$
   a. $x - 1$ **b.** $x - 2$
5. $P(x) = x^6 - x^4 - x^2 + 1$
   a. $x + 1$ **b.** $x - 1$

2. $P(x) = x^3 - x^2 - x - 2$
   a. $x - 2$ **b.** $x + 1$
4. $P(x) = x^4 - x^3 + x^2 - x + 1$
   a. $x - 1$ **b.** $x + 1$
6. $P(x) = x^4 - 4x^2 + 2x - 4$
   a. $x + 2$ **b.** $x - 2$

Written Exercises

Use synthetic substitution to find $P(c)$ for the given polynomial $P(x)$ and the given number c.

A   1. $P(x) = x^3 - 2x^2 - 5x - 7$; $c = 4$ 2. $P(x) = x^3 + 4x^2 - 8x - 6$; $c = -5$
   3. $P(x) = 2 - 5x + 3x^2 + 2x^3$; $c = -3$ 4. $P(x) = 1 - 7x - 4x^2 + x^3$; $c = 6$
   5. $P(x) = 4x^3 - 4x^2 + 5x + 1$; $c = \dfrac{3}{2}$ 6. $P(x) = 6x^3 - x^2 + 4x + 3$; $c = -\dfrac{1}{3}$
   7. $P(x) = 2x^4 - x^3 + x - 2$; $c = -\dfrac{3}{2}$ 8. $P(x) = 2x^4 - 3x^3 + 3x^2 + 1$; $c = -\dfrac{1}{2}$

Use the factor theorem to determine whether the binomial is a factor of the given polynomial.

9. $x + 1$; $P(x) = x^7 - x^5 + x^3 - x$ 10. $t + 1$; $P(t) = t^5 + t^4 + t^3 + t^2 + t + 1$
11. $s + 1$; $P(s) = s^6 - s^5 - s + 1$ 12. $z + 2$; $P(z) = z^5 + 2z^4 + z^3 + 2z^2 + z + 2$
13. $x - \sqrt{3}$; $P(x) = x^3 - 2x^2 - 3x + 6$ 14. $t + \sqrt{2}$; $P(t) = t^5 + t^4 + 4t + 4$
15. $z - i$; $P(z) = z^7 + z^6 + z^5 + z^4 + z^3 + z^2 + z + 1$
16. $z + 2i$; $P(z) = z^3 + z^2 + 4z + 4$

In Exercises 17–20, a root of the equation is given. Solve the equation.

17. $x^3 + 3x^2 - 3x - 9 = 0$; -3 18. $2x^3 + 9x^2 + 7x - 6 = 0$; -2

19. $t^3 - 11t + 20 = 0$; -4 20. $2z^3 + z^2 - 8z + 3 = 0$; $\dfrac{3}{2}$

Find a polynomial equation with integral coefficients that has the given numbers as roots.

21. $1, 2, -3$ **22.** $-2, 2, -3$ **23.** $0, -1, 2, -3$ **24.** $-2, -i, i$

B   **25.** $\frac{1}{2}, -2, 3$ **26.** $1, -\frac{3}{2}, 2$ **27.** -2 (double)$, 1, 4$ **28.** $3, 1 + 2i, 1 - 2i$

Solve each equation given the two indicated roots. (*Hint:* Perform the process illustrated in Example 4 twice.)

29. $x^4 - 3x^3 - 8x^2 + 12x + 16 = 0; \quad -1, 4$ **30.** $2x^4 - 5x^3 - 11x^2 + 20x + 12; \quad -2, 3$

31. $3x^4 + 5x^3 - 7x^2 - 3x + 2; \quad 1, -\frac{2}{3}$ **32.** $2x^4 - 3x^3 - 3x - 2 = 0; \quad 2, -\frac{1}{2}$

Show that r is a double root of $P(x) = 0$ by verifying that $x - r$ is a factor of both $P(x)$ and $P(x) \div (x - r)$.

33. $P(x) = x^4 + 2x^2 + 8x + 5; \quad r = -1$

34. $P(x) = 4x^4 - 12x^3 + 13x^2 - 12x + 9; \quad r = \frac{3}{2}$

Find the number of times r is a root of $P(x) = 0$.

35. $P(x) = x^4 + 4x^3 - 16x - 16; \quad r = -2$
36. $P(x) = x^5 + 3x^4 + 2x^3 - 2x^2 - 3x - 1; \quad r = -1$
37. When $x^3 - 5x^2 + 4x + 5$ is divided by $x - c$, the quotient is $x^2 - 3x - 2$ and the remainder is 1. Find c.

C   **38.** Show that $x - a - b$ is a factor of $x^3 - a^3 - b^3 - 3ab(a + b)$.
     39. Show that for every nonzero number a and integer $n > 1$, $x(x - a)$ is a factor of $ax^n - a^nx$.

Mixed Review Exercises

Divide.

1. $\dfrac{x^2 + 3x - 1}{x + 2}$ **2.** $\dfrac{y^3 - 1}{y^2 + y + 1}$ **3.** $\dfrac{4u^2 - 8u + 3}{2u - 1}$

4. $\dfrac{3w^4 - 4w^2 - 5}{w^2 - 2}$ **5.** $\dfrac{a^5 - 4a^3 + 3a}{a + 1}$ **6.** $\dfrac{3c^3 + 8c^2 - 8}{3c^2 + 2c - 4}$

Without solving, determine whether the roots of each equation are rational, irrational, or imaginary.

7. $2x^2 - 3x + 2 = 0$ **8.** $2x^2 - 3x + 1 = 0$ **9.** $2x^2 - 3x - 1 = 0$

8-6 Some Useful Theorems

Objective   To find or solve a polynomial equation with real coefficients and positive degree n by using these facts:
1. There are exactly n roots.
2. The imaginary roots occur in conjugate pairs.
3. Descartes' rule of signs gives information about the numbers of positive and negative real roots.

The following equations can be solved using methods you already know. Each is a polynomial equation of the form $P(x) = 0$, and the **degree of the polynomial equation** is the degree of $P(x)$.

Equation	Roots	Degree of equation Number of roots
$x + 1 = 0$	-1	1
$x^2 - 4x + 13 = 0$	$2 + 3i, 2 - 3i$	2
$x^3 + 4x^2 + 4x = 0$	$0, -2, -2$	3
$x^4 - 10x^2 + 9 = 0$	$1, -1, 3, -3$	4
$x^5 - 16x = 0$	$0, 2, -2, 2i, -2i$	5

Notice that in each case the degree of the equation and the number of roots of the equation are equal, provided that you count multiple roots as many times as each appears. This relationship is true for any polynomial equation.

Theorem

Every polynomial equation with complex coefficients and positive degree n has exactly n roots.

The theorem above follows from the *Fundamental Theorem of Algebra,* proved by the German mathematician Karl Friedrich Gauss (1777–1855):

> Every polynomial equation with complex coefficients and positive degree has at least one complex root.

In the examples given above, the imaginary roots occurred in conjugate pairs. For example, $x^2 - 4x + 13 = 0$ has roots $2 + 3i$ and $2 - 3i$. This is true of every polynomial equation with *real* coefficients.

Conjugate Root Theorem

If a polynomial equation with *real* coefficients has $a + bi$ as a root (a and b real, $b \neq 0$), then $a - bi$ is also a root.

The proof of this theorem is given in Exercise 29, page 381.

Example 1 Find a cubic equation with integral coefficients that has 2 and $3 - i$ as roots.

Solution The third root must be $3 + i$, the conjugate of $3 - i$.
An equation with these three roots is:
$$(x - 2)[x - (3 - i)][x - (3 + i)] = 0$$
$$(x - 2)[x^2 - (3 + i)x - (3 - i)x + (9 - i^2)] = 0$$
$$(x - 2)(x^2 - 6x + 10) = 0$$
$$x^3 - 8x^2 + 22x - 20 = 0 \quad \textit{Answer}$$

Example 2 Solve $x^4 - 12x - 5 = 0$ given that $-1 + 2i$ is a root.

Solution Let $P(x) = x^4 - 12x - 5$.

1. If $-1 + 2i$ is a root of $P(x) = 0$, then $-1 - 2i$ is also a root. Therefore, $x - (-1 + 2i)$ and $x - (-1 - 2i)$ are factors of $P(x)$. Their product,
$$[x - (-1 + 2i)][x - (-1 - 2i)] = x^2 + 2x + 5,$$
is also a factor of $P(x)$; that is, $P(x) = (x^2 + 2x + 5) \cdot Q(x)$.

2. Using long division, $Q(x) = P(x) \div (x^2 + 2x + 5) = x^2 - 2x - 1$.

3. The resulting equation is $x^2 - 2x - 1 = 0$, and its roots are $1 + \sqrt{2}$ and $1 - \sqrt{2}$.

$\therefore$ the solution set of the given equation is
$$\{-1 + 2i, -1 - 2i, 1 + \sqrt{2}, 1 - \sqrt{2}\}. \quad \textit{Answer}$$

The next theorem, **Descartes' rule of signs,** gives you information about the number of real roots of a polynomial equation with *real* coefficients. Consider a simplified polynomial, $P(x)$, with terms arranged in decreasing degree of x. Whenever the coefficients of two adjacent terms have opposite signs, we say that $P(x)$ has a **variation in sign.** For example,

$$x^6 - 2x^4 - 5x^2 + 3x - 6$$
$$\quad 1 \qquad\quad 2 \qquad 3$$

has three variations in sign. (Notice that you ignore any "missing" terms.)

Descartes' Rule of Signs

Let $P(x)$ be a simplified polynomial with real coefficients and terms arranged in decreasing degree of x.

1. The number of positive real roots of $P(x) = 0$ equals the number of variations of sign of $P(x)$ or is fewer than this number by an even integer.

2. The number of negative real roots of $P(x) = 0$ equals the number of variations of sign of $P(-x)$ or is fewer than this number by an even integer.

Sometimes Descartes' rule gives complete information about the number of roots of various kinds. For example, $P(x) = x^4 - 12x - 5$ has one variation in sign, so that $P(x) = 0$ has exactly one positive real root. Since the polynomial $P(-x) = (-x)^4 - 12(-x) - 5 = x^4 + 12x - 5$ also has one variation in sign, $P(x) = 0$ has exactly one negative real root. Because $P(x) = 0$ has four roots in all, two of them must be imaginary. This is what was found in Example 2.

Usually, instead of giving complete information, Descartes' rule leaves us with several possibilities.

Example 3 List the possibilities for the nature of the roots (positive real, negative real, and imaginary) for the equation $P(x) = 0$, where

$$P(x) = x^5 + x^4 - 3x^2 + 4x + 6.$$

Solution

1. $P(x)$ has two variations in sign, so by part (1) of Descartes' rule the number of positive real roots of $P(x) = 0$ is 2 or 0.

2. $P(-x) = (-x)^5 + (-x)^4 - 3(-x)^2 + 4(-x) + 6$
   $= -x^5 + x^4 - 3x^2 - 4x + 6$

   $P(-x)$ has three variations in sign, so by part (2) of Descartes' rule the number of negative real roots of $P(x) = 0$ is 3 or 1.

3. Since $P(x) = 0$ has five roots in all, there are only four possibilities.

Number of positive real roots	Number of negative real roots	Number of imaginary roots
2	3	0
2	1	2
0	3	2
0	1	4

Oral Exercises

In Exercises 1–6, a polynomial equation with real coefficients has the given root(s). What other root(s) must it have?

1. $2 - 3i$ **2.** $2i - 1$ **3.** $-i, 2i$

4. $3i, -\frac{1}{3}i$ **5.** $i + 1, i - 1$ **6.** $2 - i, -2 + i$

In Exercises 7 and 8, explain why each statement is true.

7. Every third-degree polynomial equation with real coefficients has at least one real root.

8. Every fifth-degree polynomial equation with real coefficients has at least one real root.

9. What general statement is suggested by Exercises 7 and 8?

Use Descartes' rule of signs to give the possibilities for the numbers of positive and negative real roots.

10. $x^3 - 3x^2 - 6 = 0$

11. $x^3 + x^2 + 2 = 0$

12. $x^4 - 2x^3 - 4 = 0$

13. $x^4 - 6x + 3 = 0$

14. $x^5 - x^2 + 3x - 1 = 0$

15. $x^6 - 10 = 0$

Written Exercises

Find a cubic equation with integral coefficients that has the given numbers as roots.

A **1.** $-1, 5i$ **2.** $3, i\sqrt{2}$ **3.** $-2, -1 + i$ **4.** $1, 2 - 3i$

In Exercises 5–8, all but one of the equation's roots are given. Find the remaining root. Check your answer by substituting it for x in the equation.

5. $x^3 - 3x^2 + 4x - 12 = 0$; 3 and $2i$

6. $x^3 - 2x + 4 = 0$; -2 and $1 - i$

7. $x^4 - 2x^3 + 4x^2 + 2x - 5 = 0$; 1, -1, and $1 - 2i$

8. $x^4 - 3x^3 + 4x^2 - 6x + 4 = 0$; 1, 2, and $i\sqrt{2}$

In Exercises 9–12, a root of the equation is given. Solve the equation.

9. $x^3 + x - 10 = 0$; $-1 + 2i$

10. $2x^3 - x^2 + 10x - 5 = 0$; $i\sqrt{5}$

11. $x^4 - 6x^3 + 60x - 100 = 0$; $3 + i$

12. $x^4 - 5x^2 - 10x - 6 = 0$; $-1 + i$

List the possibilities for the nature of the roots of each equation.

B **13.** $x^4 + 3x^2 - 4 = 0$

14. $x^4 - x + 3 = 0$

15. $x^4 + 2x^3 + x^2 + 1 = 0$

16. $x^4 - 3x^3 + 5x^2 - 2x + 5 = 0$

17. $x^5 - x^3 - x - 2 = 0$

18. $x^5 - x^4 + 2x - 3 = 0$

19. $x^5 - x^3 - x^2 + x - 2 = 0$

20. $x^6 + x^5 + x^4 + 3x - 2 = 0$

Find a fourth-degree polynomial equation with integral coefficients that has the given numbers as roots.

21. $2i, 1 - i$

22. $1 + i, 2 + i$

For the equations in Exercises 23 and 24, show that (a) the given number is a root, and (b) its conjugate is *not* a root.

23. $x^3 + 2x^2 + x - 1 + i = 0$; $-1 + i$

24. $x^3 - 2x^2 + 2x + 5i = 0$; $2 - i$

25. Explain why Exercises 23 and 24 do *not* contradict the conjugate root theorem.

26. There can be six possibilities for the nature of the roots of a cubic equation with real coefficients. List them in a table.

In Exercises 27 and 28, find one real root of the equation by inspection. Then use Descartes' rule to show that there are no other real roots.

C **27.** $x^5 - x^4 + 2x^3 - 2x^2 + 3x - 3 = 0$ **28.** $x^5 + x^4 + x^3 + x^2 + 2x + 2 = 0$

29. A proof of the conjugate root theorem is outlined below. Let $P(x)$ be a polynomial with real coefficients, and let $a + bi$ $(b \neq 0)$ be an imaginary root of $P(x) = 0$. Justify each of the following statements.

    1. If $S(x) = [x - (a + bi)][x - (a - bi)] = x^2 - 2ax + (a^2 + b^2)$, then $S(a + bi) = 0$ and $S(a - bi) = 0$.

    2. There are real numbers c and d and a polynomial $Q(x)$ for which $P(x) = Q(x)S(x) + cx + d$.

    3. Setting $x = a + bi$, you obtain $0 = Q(a + bi) \cdot 0 + c(a + bi) + d$.

    4. $c(a + bi) + d = 0$, or $(ac + d) + bci = 0$

    5. $bc = 0$, and since $b \neq 0$, $c = 0$

    6. $ac + d = 0$, and since $c = 0$, $d = 0$

    7. $P(x) = Q(x)S(x)$

    8. $P(a - bi) = Q(a - bi)S(a - bi) = Q(a - bi) \cdot 0 = 0$

Self-Test 2

Vocabulary   division algorithm (p. 364)   depressed equation (p. 374)
                           synthetic division (p. 368)   degree of a polynomial equation
                           remainder theorem (p. 372)   (p. 377)
                           synthetic substitution (p. 373)   conjugate-root theorem (p. 377)
                         factor theorem (p. 373)   Descartes' rule of signs (p. 378)
                                                                 variation in sign (p. 378)

1. Divide: $\dfrac{x^4 + x^3 - 5x^2 + 13x - 6}{x^2 + 3x - 2}$ Obj. 8-3, p. 364

2. Use synthetic division to divide $2x^4 - 3x^3 - 5x + 4$ Obj. 8-4, p. 368
    by $x - 2$.

3. If $P(x) = 2x^3 - 4x^2 + x - 5$, use synthetic substitution to find $P(3)$.   Obj. 8-5, p. 372

4. Determine whether $x + 1$ is a factor of $x^7 + x^4 + x + 1$.

5. Find a cubic equation with integral coefficients that has $\frac{1}{2}$, -1, and 3 as roots.

6. Consider the equation $2x^4 - 9x^3 + 13x^2 - x - 5 = 0$.   Obj. 8-6, p. 377
    a. What are the possibilities for the numbers of positive and negative real roots?
    b. Given that one root is $2 + i$, solve the equation.

7. Find a cubic equation with integral coefficients that has -2 and $1 + 3i$ as roots.

Check your answers with those at the back of the book.

Solving Polynomial Equations

8-7 Finding Rational Roots

Objective   To find rational roots of polynomial equations with integral coefficients.

The theorems in Lesson 8-6 give you information about the nature of the roots of polynomial equations but not about how to find them. The following theorem (proved in Exercises 27 and 28, page 385) tells you how to find rational roots of any polynomial equation with integral coefficients. Recall that every rational number can be written as the quotient of relatively prime integers, that is, integers whose GCF is 1.

Rational Root Theorem

Suppose that a polynomial equation with integral coefficients has the root $\frac{h}{k}$, where h and k are relatively prime integers. Then h must be a factor of the constant term of the polynomial and k must be a factor of the coefficient of the highest-degree term.

For example, any rational root of

$$6x^3 + 8x^2 - 7x - 3 = 0$$

must have a numerator that is a factor of -3 (namely, ± 1 or ± 3) and a denominator that is a factor of 6 (namely, ± 1, ± 2, ± 3, or ± 6). There are, therefore, only twelve *possible* rational roots:

$$\pm \frac{1}{1}, \ \pm \frac{3}{1}, \ \pm \frac{1}{2}, \ \pm \frac{3}{2}, \ \pm \frac{1}{3}, \ \pm \frac{1}{6}$$

You can verify that the only root among the twelve possibilities is $-\frac{1}{3}$.

Example 1   Solve the equation $2x^4 + 3x^3 - 7x^2 + 3x - 9 = 0$ by first finding any rational roots.

Solution   The only possible rational roots are

$$\pm 1, \ \pm 3, \ \pm 9, \ \pm \frac{1}{2}, \ \pm \frac{3}{2}, \text{ and } \pm \frac{9}{2}.$$

Since computing with integers is easier, try the integral possibilities first. When you substitute 1 and -1 in the equation, you find that neither is a root. To try other possibilities, it is best to use synthetic substitution.

$$
\begin{array}{r|rrrrr}
3 & 2 & 3 & -7 & 3 & -9 \\
  &   & 6 & 27 & 60 & 189 \\ \hline
  & 2 & 9 & 20 & 63 & 180
\end{array}
\;\leftarrow 3 \text{ is } \textit{not} \text{ a root.}
$$

$$
\begin{array}{r|rrrrr}
-3 & 2 & 3 & -7 & 3 & -9 \\
   &   & -6 & 9 & -6 & 9 \\ \hline
   & 2 & -3 & 2 & -3 & 0
\end{array}
\;\leftarrow -3 \textit{ is } \text{a root.}
$$

The depressed equation $2x^3 - 3x^2 + 2x - 3 = 0$ has

$$\pm 1, \ \pm 3, \ \pm\frac{1}{2}, \text{ and } \pm\frac{3}{2}$$

as possible rational roots. The possibilities 1, -1, and 3 were eliminated using the original equation. Since -3 may be a multiple root, it remains a possibility, along with $\pm\frac{1}{2}$ and $\pm\frac{3}{2}$. Try each of these until another root is found.

$$
\begin{array}{r|rrrr}
-3 & 2 & -3 & 2 & -3 \\
   &   & -6 & 27 & -87 \\ \hline
   & 2 & -9 & 29 & -90
\end{array}
\qquad
\begin{array}{r|rrrr}
\frac{1}{2} & 2 & -3 & 2 & -3 \\
            &   & 1 & -1 & \frac{1}{2} \\ \hline
            & 2 & -2 & 1 & -\frac{5}{2}
\end{array}
$$

$$
\begin{array}{r|rrrr}
-\frac{1}{2} & 2 & -3 & 2 & -3 \\
             &   & -1 & 2 & -2 \\ \hline
             & 2 & -4 & 4 & -5
\end{array}
\qquad
\begin{array}{r|rrrr}
\frac{3}{2} & 2 & -3 & 2 & -3 \\
            &   & 3 & 0 & 3 \\ \hline
            & 2 & 0 & 2 & 0
\end{array}
$$

The synthetic substitution shows that $\frac{3}{2}$ is another root. The second depressed equation is $2x^2 + 2 = 0$ (or $x^2 + 1 = 0$). From your study of quadratic equations you know that it has roots $\pm i$.

$\therefore$ the original equation has solution set $\{-3, \frac{3}{2}, i, -i\}$.   *Answer*

Sometimes you can eliminate several possibilities by using Descartes' rule of signs. In Example 1, the given equation $2x^4 + 3x^3 - 7x^2 + 3x - 9 = 0$ has only one negative root. Therefore, after finding that -3 is a root, you could have eliminated the other negative possibilities, leaving only $\frac{1}{2}$ and $\frac{3}{2}$ to try.

The rational root theorem can be used to prove that certain numbers are irrational.

Example 2   Show that $\sqrt[3]{9}$ is irrational.

Solution $\sqrt[3]{9}$ is a root of the equation $x^3 - 9 = 0$. The only possible rational roots of this equation are ± 1, ± 3, and ± 9. None of these six numbers satisfies the equation. Since $x^3 - 9 = 0$ has no rational roots, $\sqrt[3]{9}$ is irrational.

Oral Exercises

List the possible rational roots of each equation.

1. $x^3 - 7x + 6 = 0$

2. $x^3 + x^2 - 4x + 4 = 0$

3. $x^3 - 3x^2 + 2x - 8 = 0$

4. $x^4 + 3x^3 - x^2 - 9x - 6 = 0$

5. $x^4 - 2x^2 - 16x - 15 = 0$

6. $x^4 + 3x^2 - 8x + 10 = 0$

7. $2x^3 + 7x^2 + 6x - 5 = 0$

8. $2x^3 - 5x^2 - 11x - 4 = 0$

9. $2x^3 - 11x^2 + 16x - 6 = 0$

10. $4x^4 + 4x^3 + 17x^2 + 16x + 4 = 0$

11. $3x^4 + 4x^3 - x^2 + 4x - 4 = 0$

12. $6x^4 - 7x^3 + 8x^2 - 7x + 2 = 0$

Tell how you know that the following equations have no rational roots.

13. $x^3 - x + 1 = 0$

14. $x^3 + x^2 - x + 1 = 0$

15. $x^4 + 2x + 2 = 0$

16. $x^3 + 3x - 3 = 0$

Written Exercises

A **1–12.** For each equation in Oral Exercises 1–12, find any rational roots. If the equation has at least one rational root, solve it completely.

Use the method of Example 2 to show that the following numbers are irrational.

13. $\sqrt{3}$

14. $\sqrt{6}$

15. $\sqrt[3]{-4}$

16. $\sqrt[3]{2}$

17. $\sqrt[4]{8}$

18. $\sqrt[5]{-9}$

Verify that the given number is a root of the equation and use this fact to show that the number is irrational.

19. $\sqrt{3} + \sqrt{2}$; $x^4 - 10x^2 + 1 = 0$

20. $\sqrt{5} - \sqrt{3}$; $x^4 - 16x^2 + 4 = 0$

B **21.** Explain how to use the rational root theorem with an equation that has rational coefficients.

Use the method you described in Exercise 21 to solve the following equations.

22. $\frac{1}{3}x^3 - \frac{1}{2}x^2 + \frac{1}{3}x + \frac{1}{3} = 0$

23. $\frac{1}{3}x^3 + \frac{1}{2}x^2 + \frac{2}{3}x + 1 = 0$

24. $2.0x^3 - 0.8x^2 + 0.5x - 0.2 = 0$

25. $0.2x^3 - 0.5x^2 + 0.8x - 2.0 = 0$

C **26.** Show that if k is an integer and $\sqrt[n]{k}$ is rational, then k is the nth power of an integer. (*Hint:* Consider the possible rational roots of $x^n - k = 0$.)

The proof of the rational root theorem is essentially the same whatever the degree of the polynomial. The proof for degree 3 is given in Exercises 27 and 28. Let $\frac{h}{k}$ (h and k relatively prime integers) be a root of the polynomial equation $ax^3 + bx^2 + cx + d = 0$, where a, b, c, and d are integers. We must show that h is a factor of d and k is a factor of a.

27. Justify the following statements.

1. $a\left(\dfrac{h}{k}\right)^3 + b\left(\dfrac{h}{k}\right)^2 + c\left(\dfrac{h}{k}\right) + d = 0$
2. $ah^3 + bh^2k + chk^2 + dk^3 = 0$
3. $dk^3 = -ah^3 - bh^2k - chk^2$
4. $\dfrac{dk^3}{h} = -ah^2 - bhk - ck^2$
5. $\dfrac{dk^3}{h}$ is an integer.

The proof is completed as follows: Since $\dfrac{dk^3}{h}$ is an integer, all the prime factors of h must divide out with the prime factors of dk^3. But h and k have no common prime factor. Therefore h must be a factor of d.

28. Modify the proof in Exercise 27 to show that k is a factor of a.

Mixed Review Exercises

Solve each equation completely. In Exercises 7–10, one root is given.

1. $4(2 - m) + 5 = 3m - 8$

2. $3u^2 - 4u + 2 = 0$

3. $\sqrt{2r + 1} = r - 1$

4. $\dfrac{1}{w} + \dfrac{w}{w - 1} = \dfrac{1}{w - w^2}$

5. $2v^{-2} - 5v^{-1} - 12 = 0$

6. $n^4 - 2n^2 - 8 = 0$

7. $x^3 - 2x^2 + x - 2 = 0$; i

8. $c^3 + c^2 - 7c - 3 = 0$; -3

9. $a^3 - 3a^2 + a + 5 = 0$; $2 - i$

10. $y^4 - 3y^3 - 2y^2 + 10y - 12 = 0$; $1 + i$

Historical Note / Roots of Polynomial Equations

Although the Fundamental Theorem of Algebra guarantees the existence of roots for any polynomial equation, it does not provide a method for finding them. Of course, the quadratic formula gives roots of any quadratic equation, and there are similar but more complicated formulas for the roots of cubic and fourth-degree equations. But in 1824 the Norwegian mathematician Niels Henrik Abel (1802–1829) showed that for equations of degree greater than four no such general formulas exist. Numerical methods like those you are studying in this chapter, therefore, are often used to find approximate values for the roots of higher-degree polynomial equations.

8-8 Approximating Irrational Roots

Objective   To approximate the real roots of a polynomial equation $P(x) = 0$ by using the graph of $y = P(x)$.

Using methods discussed in previous lessons, you can show that the equation

$$x^3 - 2x^2 - 4x + 2 = 0$$

has either one or three real roots, none of which are rational. In this lesson you will learn a method for approximating the irrational roots of such equations. Because the real *roots* of the given equation are the same as the *x-intercepts* of the graph of $y = x^3 - 2x^2 - 4x + 2$, the graph can give us more information about these roots.

Example 1   **a.** Graph the equation $y = x^3 - 2x^2 - 4x + 2$.

   b. Use the graph to estimate to the nearest half unit the roots of the equation $x^3 - 2x^2 - 4x + 2 = 0$.

Solution   **a.** You may wish to use a computer or a graphing calculator if one is available. Otherwise make a table of values, plot the corresponding points, and join the points with a smooth unbroken curve. (Use a smaller scale on the y-axis than on the x-axis to accommodate the greater range of y-values.)

$$y = x^3 - 2x^2 - 4x + 2$$

x	y
−2	−6
−1	3
0	2
1	−3
2	−6
3	−1
4	18

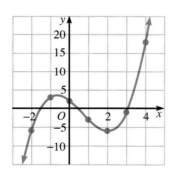

   b. The roots of the equation are the x-coordinates of the points where the graph intersects the x-axis. To the nearest half unit, the roots are −1.5, 0.5, and 3.   ***Answer***

   The method of Example 1 can be described in terms of the polynomial *function* $P(x) = x^3 - 2x^2 - 4x + 2$. The *zeros* of P are the *roots* of the polynomial *equation* $P(x) = 0$. When you draw the graph of the function P as an unbroken curve, you assume that as x goes from a to b, P takes on all values between $P(a)$ and $P(b)$. For example, since $P(-2) = -6$ and $P(-1) = 3$, and 0 is between −6 and 3, $P(x)$ must equal 0 for some x between −2 and −1.

This property, known as the *intermediate-value theorem*, is stated formally below.

Intermediate-Value Theorem

If P is a polynomial function with real coefficients, and m is any number between $P(a)$ and $P(b)$, then there is at least one number c between a and b for which $P(c) = m$.

Example 2 Approximate to the nearest tenth the real zero of the function $P(x) = x^3 - 2x^2 + x - 5$.

Solution By Descartes' rule $P(x) = 0$ has no negative roots, so you can make a table using nonnegative values of x. The table at the left below shows that $P(2) < 0$ and $P(3) > 0$. Therefore, for some r between 2 and 3, $P(r) = 0$.

	x	$P(x)$		x	$P(x)$
	0	-5		2.1	-2.459
	1	-5		2.2	-1.832
$r \rightarrow$	2	-3 $\leftarrow$ 0		2.3	-1.113
	3	7	$r \rightarrow$	2.4	-0.296 $\leftarrow$ 0
				2.5	0.625

The table of values at the right above, which is more easily generated using a computer or a calculator, shows that r is between 2.4 and 2.5. Since $P(2.4)$ is closer to 0 than $P(2.5)$, you may assume that r is closer to 2.4. Therefore, to the nearest tenth, the real zero of P is 2.4. *Answer*

The graphs below show the "magnification" process used in Example 2. You can obtain the same result on a computer or graphing calculator by changing the scaling of the graph.

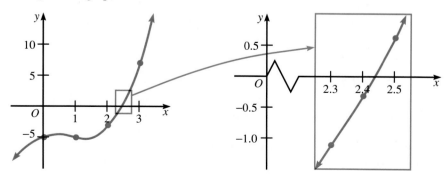

The "magnification" process shown on the previous page can be repeated if a better approximation is desired. For example, another "magnification" shows that the zero asked for in Example 2 is about 2.43.

Oral Exercises

In Exercises 1 and 2, a table of values for a polynomial function P is given. In each case find one or more pairs of numbers between which roots of $P(x) = 0$ must lie.

1.

x	-3	-2	-1	0	1	2	3	4
$P(x)$	6	-4	3	5	4	1	-3	-10

2.

x	-0.1	-0.2	-0.3	-0.4	-0.5	-0.6	-0.7	-0.8	-0.9
$P(x)$	-2.64	-1.25	-0.72	0.27	1.31	0.98	0.14	-0.55	-1.28

Suppose P is a polynomial function for which $P(1) = 6$, $P(2) = 1$, and $P(3) = -5$. Tell whether each statement is true or false.

3. P has no zeros between 1 and 2.   **4.** P has exactly one zero between 2 and 3.

Locate between consecutive integers each real root of each equation.

5. $x^3 - x^2 - 5 = 0$ **6.** $x^3 + x + 3 = 0$

7. $x^3 - x + 1 = 0$ **8.** $x^4 - 2x^3 - x + 1 = 0$

Written Exercises

For each polynomial P, draw a graph to approximate to the nearest half unit the real root(s) of the equation $P(x) = 0$. You may wish to use a computer or a graphing calculator.

A **1.** $P(x) = x^3 - 16$ **2.** $P(x) = x^3 + 11$

3. $P(x) = x^3 + x^2 + 1$ **4.** $P(x) = x^3 + 2x - 8$

5. $P(x) = x^3 - x^2 - x - 3$ **6.** $P(x) = x^3 - 3x^2 + 3x + 2$

7. $P(x) = x^3 - 3x - 1$ **8.** $P(x) = x^3 - 4x + 1$

9. $P(x) = x^4 - 3x^3 + 5$ **10.** $P(x) = x^4 + x^3 - 10x^2 - 4x + 10$

11–16. In Exercises 1–6 above, each polynomial equation $P(x) = 0$ has one real root. Approximate it to the nearest tenth.

B **17–20.** In Exercises 7–10 above, each polynomial equation $P(x) = 0$ has several real roots. Approximate each root to the nearest tenth.

For each polynomial P, (a) verify that m is between $P(a)$ and $P(b)$, and (b) find a number c between a and b such that $P(c) = m$.

21. $P(x) = x^2 - 2x + 3$; $a = 1$; $b = 4$; $m = 6$
22. $P(x) = x^2 + x - 5$; $a = -5$; $b = -1$; $m = 7$
23. $P(x) = 2x^2 + 5x - 4$; $a = -2$; $b = 2$; $m = -1$
24. $P(x) = x^3 + x^2 - 2x - 5$; $a = -3$; $b = 2$; $m = -5$

C 25. Show that $P(x) = x^3 - 3x + 1$ has a zero between 1.5 and 1.6 and approximate it to the nearest hundredth.

26. Show that $P(x) = x^3 + x^2 - 3x + 2$ has a zero between -2.6 and -2.5 and approximate it to the nearest hundredth.

27. This exercise develops a method for finding an *upper bound* for the positive roots of the polynomial equation $P(x) = 0$. Assume that the coefficient of the highest-degree term, called the *leading coefficient*, of $P(x)$ is positive. Let $P(x) = (x - m)Q(x) + P(m)$, where m is a positive number. Show that if $P(m)$ and the coefficients of $Q(x)$ are all nonnegative, then $P(x) = 0$ has no roots greater than m. Therefore m is an upper bound for the roots. (*Hint:* For all $x > m$, $x - m > 0$ and $Q(x) > 0$.)

28. Explain how to use Exercise 27 to find a *lower bound* for the negative roots of $P(x) = 0$. (*Hint:* Both $P(-x) = 0$ and $-P(-x) = 0$ have the absolute values of the negative roots of $P(x) = 0$ as their positive roots, and one of them has a positive leading coefficient.)

Use Exercises 27 and 28 to find the *least* integral upper bound and *greatest* integral lower bound for the roots of the given equations.

29. $2x^3 - 5x^2 - 8x - 9 = 0$

30. $2x^4 - 6x^2 - 3x - 9 = 0$

 Computer Key-In

In Example 2 of Lesson 8-8 you found that a zero of $P(x) = x^3 - 2x^2 + x - 5$ was between 2 and 3 by observing that $P(2) = -3 < 0$ and $P(3) = 7 > 0$ and by using the intermediate-value theorem. Then, dividing the interval from 2 to 3 into tenths and observing that $P(2.4) < 0$ and $P(2.5) > 0$, you found that the zero was between 2.4 and 2.5.

By continuing to subdivide intervals containing the zero and by using the intermediate-value theorem, you could approximate the zero more accurately.

The program given on the next page can be used to search for real zeros of a polynomial function. The program is based on subdividing an interval and using the intermediate-value theorem. Lines 20–100 accept data about the given polynomial and the interval to be searched. In lines 140–170 the polynomial is evaluated by means of synthetic substitution.

The program then tests whether the current x-value is a zero (line 180) or whether a zero lies between the current x-value and its predecessor (line 190). The intermediate-value theorem is used in line 190, where the program checks whether the corresponding y-values have opposite sign. If the product of the *y*-values is negative, then they have opposite signs, and there must be a zero between the two subdivision points that produced those *y*-values.

```
10  PRINT "THIS PROGRAM WILL SEARCH FOR REAL ZEROS OF A
      POLYNOMIAL FUNCTION."
20  INPUT "ENTER THE DEGREE OF THE POLYNOMIAL: "; D
30  DIM C(D)
40  FOR J=D TO 1 STEP−1
50  PRINT "ENTER THE COEFFICIENT OF X^"; J; ": ";
60  INPUT C(J)
70  NEXT J
80  INPUT "ENTER THE CONSTANT TERM: "; C(0)
90  INPUT "ENTER THE ENDPOINTS OF THE INTERVAL TO BE
      SEARCHED: "; X1, X2
100 INPUT "ENTER THE SUBDIVISION SIZE: "; I
110 LET F=0
120 LET Y1=0
130 FOR X=X1 TO X2 STEP I
140 LET Y=C(D)
150 FOR J=D−1 TO 0 STEP −1
160 LET Y=Y*X+C(J)
170 NEXT J
180 IF Y=0 THEN PRINT X; " IS A ZERO."
190 IF Y*Y1<0 THEN PRINT "THERE IS A ZERO BETWEEN ";
      X−I; " AND "; X; "."
200 IF Y=0 OR Y*Y1 < 0 THEN LET F=1
210 LET Y1=Y
220 NEXT X
230 IF F=0 THEN PRINT "NO ZEROS FOUND."
240 END
```

Exercises

1. The function $P(x) = x^3 - 4x^2 + 2x + 4$ has three real zeros.
   a. Approximate these zeros to the nearest tenth. (*Hint:* Use -3 and 3 as the left and right endpoints. Use a subdivision size of 0.01.)
   b. Approximate the largest of the zeros to the nearest hundredth.
2. The function $P(x) = x^5 - 14x^4 + 38x^3 + 32x^2 - 118x - 20$ has five real zeros. Approximate each zero to the nearest tenth as follows:
   a. Run the program for the interval from $x = -3$ to $x = 3$. What zeros did you locate?
   b. Run the program again with an interval large enough to locate the remaining zeros.
3. Find $\sqrt[5]{2}$ to three decimal places. (*Hint:* Use $P(x) = x^5 - 2$.)
4. The function $P(x) = 42x^2 - 13x + 1$ has two real zeros between $x = 0$ and $x = 1$. What happens when you use a subdivision size of 0.1 to locate these zeros? Explain why this happens.

8-9 Linear Interpolation

Objective   To use linear interpolation to find values not listed in a given table of data.

Newspapers, magazines, and journals often display numerical information in tables. Look, for instance, at the Census Bureau data shown in the table at the right.

To approximate values not given in the table, you can use **linear interpolation**. For example, to find the population of the United States in 1953, you reason as follows: 1953 is $\frac{3}{10}$ of the way from 1950 to 1960; therefore the number p that is $\frac{3}{10}$ of the way from 151 to 179 is an acceptable approximation of the 1953 population (in millions). To find p, add $\frac{3}{10}$ of the difference between 179 and 151 to 151:

$$p = 151 + \frac{3}{10}(179 - 151) \approx 151 + 8 = 159$$

Therefore, the 1953 population was about 159 million.

Example 1 shows how the work done above can be arranged in a way that makes it easy to set up a proportion.

Year	U.S. population (millions)
1900	76
1910	92
1920	106
1930	123
1940	132
1950	151
1960	179
1970	203
1980	227
1990	243

Example 1   Find the approximate population of the United States in 1953.

Solution

Year	Population

$$10 \begin{bmatrix} 3 \begin{bmatrix} 1950 \\ 1953 \\ \\ 1960 \end{bmatrix} & \begin{matrix} 151 \\ p \\ \\ 179 \end{matrix} \end{bmatrix} d \Big] 28$$

$\dfrac{d}{28} = \dfrac{3}{10}$; $d = \dfrac{3}{10} \cdot 28 \approx 8$

$p = 151 + d \approx 151 + 8 = 159$

∴ the population in 1953 was about 159 million.   *Answer*

Given a population size, the process used above can be reversed to find the corresponding year. This process is called **inverse interpolation.**

Example 2   Refer to the table above. Use inverse interpolation to find approximately in which year the population was 140 million.

Solution

Year	Population

$$10 \begin{bmatrix} c \begin{bmatrix} 1940 \\ y \\ \\ 1950 \end{bmatrix} & \begin{matrix} 132 \\ 140 \\ \\ 151 \end{matrix} \Big] 8 \end{bmatrix} 19$$

$\dfrac{c}{10} = \dfrac{8}{19}$; $c = \dfrac{80}{19} \approx 4$

$y = 1940 + c \approx 1940 + 4 = 1944$

∴ the population was 140 million in about 1944.   *Answer*

In the table used in Examples 1 and 2, the population values increased as the years increased. In some tables the entries in one column (or row) decrease as the entries in the other increase.

Example 3 The table below gives the density of dry air at various altitudes.

Altitude (m)	0	500	1000	1500	2000	2500	3000	3500
Density (kg/m³)	1.225	1.167	1.112	1.058	1.007	0.957	0.909	0.863

a. Approximate the density at 2300 m.

b. At about what altitude is the density 1.025 kg/m³?

Solution **a.**

Altitude	Density
2000	1.007
2300	y
2500	0.957

$$\frac{d}{0.050} = \frac{300}{500} = \frac{3}{5};$$

$$d = \frac{3}{5} \cdot 0.050 = 0.030$$

$$y = 1.007 - d$$

$$\approx 1.007 - 0.030 = 0.977$$

∴ the density at 2300 m is about 0.977 kg/m³. *Answer*

b.

Altitude	Density
1500	1.058
x	1.025
2000	1.007

$$\frac{c}{500} = \frac{0.033}{0.051} = \frac{11}{17};$$

$$c = \frac{11}{17} \cdot 500 \approx 324$$

$$x = 1500 + c$$

$$\approx 1500 + 324 = 1824$$

∴ the density is 1.025 km/m³ at about 1824 m. *Answer*

You can use linear interpolation to approximate the zeros of functions.

Example 4 Approximate to the nearest hundredth the real zero of the function $P(x) = x^3 - 2x^2 + x - 5$.

Solution First use the method of Example 2, page 387, to locate the zero between consecutive tenths. In that example you found that $P(2.4) = -0.296$ and $P(2.5) = 0.625$. Now use linear interpolation.

x	$P(x)$
2.4	-0.296
r	0
2.5	0.625

$$\frac{c}{0.1} = \frac{0.296}{0.921} \approx 0.32;$$

$$c \approx 0.1 \cdot 0.32 \approx 0.03$$

$$r = 2.4 + c$$

$$\approx 2.4 + 0.03 = 2.43$$

∴ the zero of P is about 2.43. *Answer*

To illustrate how linear interpolation works, suppose $P_1(x_1, f(x_1))$ and $P_2(x_2, f(x_2))$ are two points on the graph of a function f. Linear interpolation essentially replaces the graph of f between these points by the line segment $\overline{P_1 P_2}$. (See the figure below, where the graph of f is in red and the approximation line in blue.)

For any x between x_1 and x_2, linear interpolation approximates $f(x)$ by $f(x_1) + d$. The amount of error in this approximation depends upon the vertical distance between the graph of f and the approximation line. The less the graph of f "strays" from the approximation line, the smaller the error.

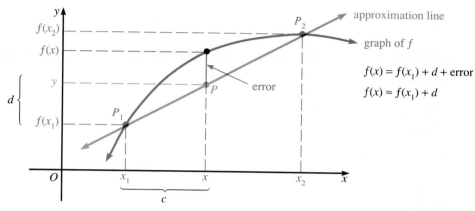

Using P_1 and P you can see that the slope of the approximation line is $\dfrac{d}{c}$; using P_1 and P_2 the slope is $\dfrac{f(x_2) - f(x_1)}{x_2 - x_1}$. Since the slope of a line is constant, you can write $\dfrac{f(x_2) - f(x_1)}{x_2 - x_1} = \dfrac{d}{c}$, which is equivalent to

$$\frac{c}{x_2 - x_1} = \frac{d}{f(x_2) - f(x_1)}.$$

This proportion is the one used in linear interpolation.

	x	$f(x)$	
	x_1	$f(x_1)$	
$x_2 - x_1\left[c\left[\begin{matrix} \\ x \\ \end{matrix}\right.\right.$	x	y	$\left.\left.\begin{matrix} \\ \\ \end{matrix}\right]d\right] f(x_2) - f(x_1)$
	x_2	$f(x_2)$	

Oral Exercises

1. Explain why, in part (a) of Example 3, d is *subtracted* from 1.007 in order to approximate the density at 2300 m.
2. Use linear interpolation to estimate the indicated function value.
   a. $f(1) = 0$, $f(5) = 100$; $f(2) = $ ___?___
   b. $g(-1) = 10$, $g(1) = -10$; $g(0.5) = $ ___?___

3. This exercise shows that the error in using linear interpolation may be large: Use linear interpolation to approximate $f(x) = x^2$ at $x = 1$ with $x_1 = 0$ and $x_2 = 2$, as follows.

a. $f(0) = \underline{\ ?\ }$ and $f(2) = \underline{\ ?\ }$.

b. What linear function g has the same values as f at 0 and 2?

c. What is the value of this linear function at $x = 1$?

d. By how much does the functional value found in part (c) differ from $f(1)$?

e. How could you choose x_1 and x_2 to make the approximation for $f(1)$ closer to the true value?

Written Exercises

Throughout these exercises, use linear interpolation. In Exercises 1–8, use the table on page 391.

Approximate the population of the United States in each year.

A **1.** 1915 **2.** 1963 **3.** 1968 **4.** 1976

Approximate the year in which the United States had each population (in millions).

5. 100 **6.** 115 **7.** 170 **8.** 220

Use the table in Example 3 on page 392 to find an approximation of the density of air at each altitude (in meters).

9. 1200 **10.** 3200 **11.** 400 **12.** 320

Use the table in Example 3 on page 392 to find an approximation of the altitude at which air has the given density (in kg/m³).

13. 1.200 **14.** 1.030 **15.** 1.000 **16.** 0.930

In Exercises 17–26, use Tables 1 and 2 on pages 810 and 811 to find an approximation of each square root and cube root.

Sample $\sqrt[3]{726}$

Solution   Use the column headed $\sqrt[3]{100N}$ with $N = 7.2$ and $N = 7.3$.

x	$\sqrt[3]{x}$
720	8.963
726	$\sqrt[3]{726}$
730	9.004

$$10\left[6\left[\begin{array}{c}720 \\ 726\end{array}\right. \quad \begin{array}{c}8.963 \\ \sqrt[3]{726}\end{array}\left.\right]d\right]0.041$$

$d = \dfrac{6}{10} \cdot 0.041 \approx 0.024$

$\sqrt[3]{726} \approx 8.963 + 0.024$

$\qquad\quad = 8.987$ *Answer*

B   **17.** $\sqrt{7.63}$ **18.** $\sqrt{4.87}$ **19.** $\sqrt{67.8}$ **20.** $\sqrt{24.2}$

     21. $\sqrt[3]{5.26}$ **22.** $\sqrt[3]{526}$ **23.** $\sqrt[3]{26.8}$ **24.** $\sqrt[3]{84.3}$

     25. $\sqrt{5280}$ (*Hint:* $5280 = 10 \times 5.28 \times 10^2$)

     26. $\sqrt[3]{62300}$ (*Hint:* $62300 = 10 \times 6.23 \times 10^3$)

In Exercises 27–30, (a) locate the real zero of each polynomial function between consecutive tenths, and (b) approximate the zero to the nearest hundredth using linear interpolation.

27. $P(x) = x^3 + 2x - 5$ **28.** $P(x) = x^3 - 2x^2 - 3$

29. $P(x) = x^3 + x^2 + 9$ **30.** $P(x) = x^3 + 2x + 8$

Computer Exercises

For students with some programming experience.

1. Write a program that will carry out linear interpolation between two given functional values, $y_1 = f(x_1)$ and $y_2 = f(x_2)$. You will enter the numbers for x_1, y_1, x_2, y_2, and x (where $x_1 < x < x_2$). Then have the program compute an estimate for $y = f(x)$ using linear interpolation.

For Exercises 2 and 3, refer to the population table on page 391.

2. **a.** Run the program in Exercise 1 to estimate the population (in millions) for the years 1916, 1949, and 1972.
   b. Run the program with $x_1 = 1910$ and $x_2 = 1930$ to estimate the population in the year 1920. How does this approximate value compare to the exact value for the 1920 population given in the table?

Instead of estimating a value *between* two known values, it is possible to use the same process involved in linear interpolation to estimate a value that is either *less than* or *greater than* the two known values. This process is called **linear extrapolation.**

3. Run the program in Exercise 1 to estimate the population (in millions) for the years 1890, 1895, 1995, and 2000.

4. Modify the program in Exercise 1 so that you can enter a rule for the function f. You may want to use the DEF FN statement so that you can change the function by retyping one line. After finding an estimate of $y = f(x)$, have the program compute the exact value of $y = f(x)$ and print it for comparison.

5. Run the program in Exercise 4 for each given function and each given value of x_1, x_2, and x.

   a. $f(x) = x^3$; $x_1 = 2$; $x_2 = 3$, and $x = 2.8$

   b. $f(x) = 3x - 7$; $x_1 = 5$, $x_2 = 10$, and $x = 7$

   c. $f(x) = \dfrac{1}{x}$; $x_1 = 10$, $x_2 = 20$, and $x = 13.71$

   d. $f(x) = x^5 + 0.75x^3 - x^2 + 8$; $x_1 = 14$, $x_2 = 15$, and $x = 14.3$

Self-Test 3

Vocabulary   rational root theorem (p. 382)   linear interpolation (p. 391)
             intermediate-value theorem (p. 387)   inverse interpolation (p. 391)

1. Explain why $x^3 + 2x^2 - x + 3 = 0$ has no rational roots.   Obj. 8-7, p. 382

2. Solve the equation $2x^4 - x^3 + 7x^2 - 4x - 4 = 0$ completely.

3. Approximate to the nearest tenth the real zero of the polynomial   Obj. 8-8, p. 386
   function $P(x) = x^3 + 5$.

4. Given that $\sqrt{5.3} \approx 2.302$ and $\sqrt{5.4} \approx 2.324$, use linear interpo-   Obj. 8-9, p. 391
   lation to find an approximation of $\sqrt{5.36}$.

5. The temperature of a beaker of water is read every 0.1 min as it is
   heated, with the results shown in the following table.

Time in minutes	0.0	0.1	0.2	0.3	0.4
Temperature (° C)	30.2	31.5	32.7	34.0	35.1

   a. Estimate the temperature of the water 0.24 min after the
      heating began.
   b. Estimate how long it took the temperature to rise to 32.1° C.

Check your answers with those at the back of the book.

Chapter Summary

1. If $y = mx$, then y is said to *vary directly* as x or to be *directly proportional*
   to x. If $y_1 = mx_1$ and $y_2 = mx_2$, then the *proportion* $\frac{y_1}{x_1} = \frac{y_2}{x_2}$ holds. If $y = \frac{k}{x}$,
   then y *varies inversely* as x, and if $z = kxy$, then z *varies jointly* as x and y.

2. To divide one polynomial by another, find the quotient and remainder using
   the *division algorithm:*

$$\frac{\text{Dividend}}{\text{Divisor}} = \text{Quotient} + \frac{\text{Remainder}}{\text{Divisor}}$$

   Synthetic division can be used instead of long division if the divisor is a
   first-degree binomial.

3. The *remainder theorem* states that when a polynomial $P(x)$ is divided by
   $x - c$, the remainder is $P(c)$, that is,

$$P(x) = Q(x) \cdot (x - c) + P(c).$$

   A corollary is the *factor theorem:* The polynomial $P(x)$ has $x - r$ as a factor
   if and only if r is a root of the equation $P(x) = 0$.

4. Every polynomial equation $P(x) = 0$ with complex coefficients and of positive degree n has exactly n roots. By the *conjugate root theorem*, if $P(x)$ has *real* coefficients and $a + bi$ (a and b real, $b \neq 0$) is a root of $P(x) = 0$, then $a - bi$ is also a root. *Descartes' rule of signs* gives information about the numbers of positive and negative roots.

5. By the *rational root theorem*, if a polynomial equation $P(x) = 0$ with integral coefficients has the *rational* root $\frac{h}{k}$ (h and k relatively prime), then h is a factor of the constant term of $P(x)$ and k is a factor of the coefficient of the highest-degree term.

6. The *irrational* roots of a polynomial equation $P(x) = 0$ (P having real coefficients) can be approximated using the *intermediate-value theorem*. If P is a polynomial function with real coefficients, and m is any number between $P(a)$ and $P(b)$, then there is at least one number c between a and b for which $P(c) = m$.

7. *Linear interpolation* can be used to approximate values not given in a table.

Chapter Review

Write the letter of the correct answer.

1. If t varies directly as s and $t = 21$ when $s = 12$, find t when $s = 28$.   8-1
   a. 9   **b.** 16   **c.** 49   **d.** 54

2. In a pre-election poll of 480 voters, 260 favored candidate Harrison. How many of an anticipated 12,000 votes is Harrison likely to get?
   a. 6000   **b.** 6500   **c.** 7000   **d.** 7500

3. Suppose z varies directly as the square of x and inversely as y. If $z = 8$ when $x = 4$ and $y = 6$, find z when $x = 6$ and $y = 12$.   8-2
   a. $\frac{3}{2}$ **b.** 6   **c.** $\frac{64}{9}$ **d.** 9

4. The kinetic energy of an object varies jointly as the mass and the square of the speed. The kinetic energy of an object with mass 3 kg and speed 4 m/s is 24 joules. Find the kinetic energy of an object with mass 4 kg and speed 3 m/s.
   a. 30 joules   **b.** 24 joules   **c.** 18 joules   **d.** 12 joules

5. Divide $3x^3 - x^2 - 4x + 1$ by $x^2 - 2$.   8-3
   a. $3x - 1 + \dfrac{2x + 1}{x^2 - 2}$ **b.** $3x - 1 + \dfrac{2x - 1}{x^2 - 2}$
   c. $3x + 1 + \dfrac{2x - 1}{x^2 - 2}$ **d.** $3x + 1 + \dfrac{2x + 1}{x^2 - 2}$

6. Use synthetic division to find the remainder when $x^4 + 4x^3 - 5x + 3$ is divided by $x + 2$.   8-4
   a. 21   **b.** 41   **c.** -3 **d.** 17

7. For what value of k will $x + 1$ be a factor of $x^{13} - 2x^7 + 3x + k$?
   a. -6 **b.** -2 **c.** 2 **d.** 6

8-5

8. Find $P(2)$ if $P(x) = x^5 - 2x^4 + 3x^2 - 1$.
   a. 11 **b.** 5 **c.** -23 **d.** -55

8-6

9. Solve $x^4 - 4x^3 + 4x^2 - 9 = 0$ given that $1 + i\sqrt{2}$ is a root.
   a. $\{1 + i\sqrt{2}, 1 - i\sqrt{2}\}$
   b. $\{1 + i\sqrt{2}, 1 - i\sqrt{2}, -3, -1\}$
   c. $\{1 + i\sqrt{2}, 1 - i\sqrt{2}, -1, 3\}$
   d. $\{1 + i\sqrt{2}, 1 - i\sqrt{2}, 1, -3\}$

10. Find a cubic equation with integral coefficients that has -2 and $3 - i$ as roots.
    a. $x^3 + 8x^2 + 22x + 20 = 0$
    b. $x^3 - 8x^2 + 22x - 20 = 0$
    c. $x^3 + 4x^2 - 2x - 20 = 0$
    d. $x^3 - 4x^2 - 2x + 20 = 0$

8-7

11. Use the rational root theorem to determine which of the following could *not* be a root of $6x^3 + 23x^2 - 6x - 8 = 0$.
    a. -4 **b.** $\frac{2}{3}$ **c.** $\frac{3}{4}$ **d.** $-\frac{1}{2}$

8-8

12. Approximate to the nearest tenth the real zero of $P(x) = x^3 - 4$.
    a. 1.5 **b.** 1.6 **c.** 1.7 **d.** 1.8

8-9

13. Use linear interpolation to find an approximation of $f(2.64)$ to the nearest hundredth given that $f(2.6) = 4.97$ and $f(2.7) = 5.12$.
    a. 5.01 **b.** 5.02 **c.** 5.03 **d.** 5.04

Chapter Test

1. The distance an object falls from rest is proportional to the square of the length of time it has fallen. If an object falls 64 ft in 2 s, how far will it fall in 3 s?

8-1

2. The volume of a cone varies jointly as the height and the square of the base radius. A cone with height 6 cm and base radius 4 cm has volume 32π cm^3. What is the volume of a cone with height 4 cm and base radius 6 cm?

8-2

3. Divide $x^4 - 3x^3 + 6x - 5$ by $x^2 - 2x + 2$.

8-3

4. Use synthetic division to divide $x^4 + 2x^3 + 4x + 5$ by $x + 2$.

8-4

5. Find $P\left(\frac{1}{2}\right)$ given that $P(x) = 6x^3 - x^2 + 3x + 5$.

8-5

6. Solve $x^4 - 2x^3 + 6x - 9 = 0$ completely given that $1 - i\sqrt{2}$ is a root.

8-6

7. Solve $2x^4 + 3x^3 - 11x^2 + 2x + 4 = 0$ completely by first finding any rational roots.

8-7

8. Graph $P(x) = x^3 - x^2 - 2$ and estimate the real root of $x^3 - x^2 - 2 = 0$ to the nearest half unit.

8-8

9. Use linear interpolation to find an approximation of $f(-1.76)$ to the nearest hundredth given that $f(-1.7) = 4.63$ and $f(-1.8) = 4.45$.

8-9

Preparing for College Entrance Exams

Strategy for Success

You are not expected to solve the test problems mentally. Do your calculations in the test booklet and cross out inappropriate choices. If you decide to skip a particular problem, mark it in your test booklet so you can easily find it later. Do not put any stray marks on your answer sheet.

Decide which is the best of the choices given and write the corresponding letter on your answer sheet.

1. How many real roots does $2(x^2 + 1)^2 + (x^2 + 1) - 3 = 0$ have?
   (A) 0   (B) 1   (C) 2   (D) 3   (E) 4

2. The roots of $1.5y^2 - 5y + 2 = 0$:
   (A) are imaginary   (B) are rational   (C) are irrational
   (D) have the sum $-\dfrac{10}{3}$ (E) have the product $-\dfrac{4}{3}$

3. Find $P(1 - i)$ given that $P(x) = 2x^3 - x^2 + 3x + 1$.
   (A) $2 + 9i$ (B) 0   (C) $1 + i$ (D) $-1 + i$ (E) $-5i$

4. Given that $\sqrt{44} \approx 6.633$ and $\sqrt{45} \approx 6.708$, use linear interpolation to find an approximate value for $\sqrt{44.2}$.
   (A) 6.652   (B) 6.618   (C) 6.693   (D) 6.648   (E) 6.723

5. Suppose z varies directly as x and inversely as y. If $z = 12$ when $x = 4$ and $y = 5$, find z when $x = 6$ and $y = 45$.
   (A) $\dfrac{1}{9}$ (B) 2   (C) 112.5   (D) 15   (E) 72

6. Use the rational root theorem to determine which number is *not* a possible root of $6t^3 + t^2 - 31t + 10 = 0$.
   (A) $-\dfrac{2}{3}$ (B) $\dfrac{5}{2}$ (C) 1   (D) -2 (E) 3

7. Find a quadratic equation whose roots have the sum -2 and the product $-\dfrac{3}{2}$.
   (A) $2x^2 + 3x - 4 = 0$ (B) $2x^2 - 4x + 3 = 0$ (C) $2x^2 + 4x - 3 = 0$
   (D) $2x^2 - 4x - 3 = 0$ (E) $2x^2 - 3x + 4 = 0$

8. Find the maximum value of g if $g(x) = 7 - 8x - 2x^2$.
   (A) 7   (B) -2 (C) 15   (D) -17 (E) 13

9. The volume of a cylinder varies jointly as the height and the square of the base radius. If a cylinder's height is doubled and its base radius is halved, then its volume:
   (A) is quadrupled   (B) is tripled   (C) is doubled
   (D) remains the same   (E) is halved

Circles, ellipses, parabolas, and hyperbolas are called "conics" because each of these curves is the intersection of a plane and a double cone.

Conic Sections:
Circles and Parabolas

9-1 Distance and Midpoint Formulas

Objective   To find the distance between any two points and the midpoint of the line segment joining them.

The distance between any two points P and Q is written as PQ. If P and Q are points on a number line, PQ is the absolute value of the difference between their coordinates. Since $|a - b| = |b - a|$, you can subtract the coordinates in either order. For the points P and Q shown,

$$PQ = |2 - (-3)| = |5| = 5$$
or $$PQ = |-3 - 2| = |-5| = 5.$$

A similar method can be used to find the distance between two points *in a coordinate plane* if the points lie on the same horizontal or vertical line. For example, the distances between the points shown in the diagram at the right are

$$AB = |4 - (-2)| = 6$$
$$CD = |-1 - 6| = 7$$
$$QR = |y_2 - y_1| = |y_1 - y_2|$$
$$PQ = |x_2 - x_1| = |x_1 - x_2|.$$

You can use the Pythagorean theorem to get a general formula for finding the distance between *any* two points in a coordinate plane.

Pythagorean Theorem

If the length of the hypotenuse of a right triangle is c, and the lengths of the other two sides (legs) are a and b, then

$$c^2 = a^2 + b^2.$$

The converse of this theorem is true as well: If the lengths a, b, and c of the sides of a triangle satisfy the equation $c^2 = a^2 + b^2$, then the triangle is a right triangle and c is the length of the hypotenuse.

To find a general formula for distance, let $P_1(x_1, y_1)$ and $P_2(x_2, y_2)$ be any two points not on the same horizontal or vertical line. Then P_1P_2 is the length of the hypotenuse of the right triangle having vertices P_1, P_2, and $Q(x_2, y_1)$. The lengths of the legs are

$$P_1Q = |x_2 - x_1| \quad \text{and} \quad P_2Q = |y_2 - y_1|.$$

By the Pythagorean theorem,

$$(P_1\,P_2)^2 = (P_1\,Q)^2 + (P_2Q)^2$$
$$= |x_2 - x_1|^2 + |y_2 - y_1|^2$$
$$= (x_2 - x_1)^2 + (y_2 - y_1)^2 \quad \longleftarrow \text{Remember, } |a|^2 = a^2.$$
$$P_1P_2 = \sqrt{(x_2 - x_1)^2 + (y_2 - y_1)^2} \quad \longleftarrow \text{Taking the principal square root of each side.}$$

The Distance Formula

The distance between the points $P_1(x_1, y_1)$ and $P_2(x_2, y_2)$ is
$$P_1P_2 = \sqrt{(x_2 - x_1)^2 + (y_2 - y_1)^2}.$$

Example 1 Find the distance between $P_1(-2, -1)$ and $P_2(-4, 3)$.

Solution
$$P_1P_2 = \sqrt{(-4 - (-2))^2 + (3 - (-1))^2}$$
$$= \sqrt{(-2)^2 + 4^2}$$
$$= \sqrt{4 + 16}$$
$$= \sqrt{20}$$
$$= 2\sqrt{5} \quad \textbf{\textit{Answer}}$$

The distance formula can be used to prove the next result (see Problem 11).

The Midpoint Formula

The midpoint of the line segment joining $P_1(x_1, y_1)$ and $P_2(x_2, y_2)$ is
$$M\left(\frac{x_1 + x_2}{2}, \frac{y_1 + y_2}{2}\right).$$

Example 2 Find the midpoint of the line segment joining $(4, -6)$ and $(-3, 2)$.

Solution $\left(\dfrac{x_1 + x_2}{2}, \dfrac{y_1 + y_2}{2}\right) = \left(\dfrac{4 + (-3)}{2}, \dfrac{-6 + 2}{2}\right)$, or $\left(\dfrac{1}{2}, -2\right)$.

$\qquad\qquad\qquad\qquad\qquad\qquad\qquad\qquad\qquad\qquad\qquad\qquad$ **_Answer_**

Example 3 Prove that the midpoint of the hypotenuse of any right triangle is equidistant from the three vertices.

Solution Any right triangle can be placed so that its vertices are $O(0, 0)$, $P(a, 0)$, and $Q(0, b)$, as shown. By the distance formula or the Pythagorean theorem, $PQ = \sqrt{a^2 + b^2}$.

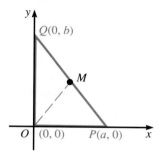

Let point M be the midpoint of $\overline{PQ}$. Then $MP = MQ = \frac{1}{2}\sqrt{a^2 + b^2}$. The coordinates of M are $\left(\frac{a + 0}{2}, \frac{0 + b}{2}\right)$, or $\left(\frac{a}{2}, \frac{b}{2}\right)$.

By the distance formula,

$$MO = \sqrt{\left(\frac{a}{2} - 0\right)^2 + \left(\frac{b}{2} - 0\right)^2}$$
$$= \sqrt{\frac{a^2}{4} + \frac{b^2}{4}} = \sqrt{\frac{a^2 + b^2}{4}} = \frac{1}{2}\sqrt{a^2 + b^2}$$

$\therefore MO = MP = MQ$

Example 4 Find an equation of the perpendicular bisector of $\overline{AB}$ for the points $A(-2, 1)$ and $B(1, -3)$.

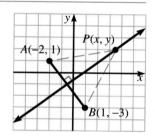

Solution 1 Make a sketch of the points $A(-2, 1)$ and $B(1, -3)$, as shown. Let $P(x, y)$ be any point on the perpendicular bisector. Since any point on the perpendicular bisector is equidistant from A and B,

$$AP = BP$$
$$\sqrt{(x + 2)^2 + (y - 1)^2} = \sqrt{(x - 1)^2 + (y + 3)^2}$$

Square both sides and multiply out the binomials.
$$x^2 + 4x + 4 + y^2 - 2y + 1 = x^2 - 2x + 1 + y^2 + 6y + 9$$
$$6x - 8y - 5 = 0 \quad \textbf{\textit{Answer}}$$

Solution 2 The midpoint of $\overline{AB}$ is $\left(-\frac{1}{2}, -1\right)$, and the slope of the line containing A and B is $-\frac{4}{3}$. Then the perpendicular bisector is the line passing through $\left(-\frac{1}{2}, -1\right)$ having slope which is the negative reciprocal of $-\frac{4}{3}$, namely $\frac{3}{4}$. The point-slope form of the equation of this line is $y + 1 = \frac{3}{4}\left(x + \frac{1}{2}\right)$, which is equivalent to $6x - 8y - 5 = 0$.

Oral Exercises

Find the distance between each pair of points.

1. $(8, 4)$, $(8, -2)$
2. $(6, -5)$, $(0, -5)$
3. $(4, -1)$, $(-2, -1)$
4. $(2, -7)$, $(2, -1)$
5. $(0, 0)$, $(4, 3)$
6. $(0, -8)$, $(6, 0)$
7. $(2, 3)$, $(0, 0)$
8. $(7, 0)$, $(0, -2)$
9. $(3, 1)$, $(2, 4)$

Find the midpoint of the line segment joining each pair of points.

10. $(7, 5)$, $(-1, -3)$
11. $(3, -8)$, $(5, -9)$
12. $(-3, 6)$, $(2, 2)$
13. $(5, -1)$, $(-6, -2)$
14. $(2.5, 0.5)$, $(6.5, 3)$
15. $(6, -4.5)$, $(-3, -1.5)$

Written Exercises

Find (a) the distance between each pair of points and (b) the midpoint of the line segment joining the points. Express all radicals in simplest form.

A
1. $(13, 6)$, $(0, 6)$
2. $(0, 8)$, $(-6, 0)$
3. $(0, 6)$, $(-5, -1)$
4. $(9, 1)$, $(2, -1)$
5. $(3, 2)$, $(5, 6)$
6. $(-4, -3)$, $(2, 1)$
7. $(2, 2)$, $\left(\frac{1}{3}, -2\right)$
8. $\left(\frac{1}{2}, -1\right)$, $(-1, 1)$
9. $(0, 0)$, $(11, 11)$
10. $(0, 0)$, $(5, 5)$
11. $(\sqrt{2}, 1)$, $(-\sqrt{2}, 0)$
12. $(5, \sqrt{5})$, $(3, -\sqrt{5})$
13. $(1 + \sqrt{5}, 2 + \sqrt{3})$, $(1 - \sqrt{5}, -2 + \sqrt{3})$
14. $(\sqrt{6} + 1, \sqrt{3} - \sqrt{2})$, $(\sqrt{6} - 1, \sqrt{3} + \sqrt{2})$
15. (a, b), $(0, b)$
16. $(-a, b)$, $(2a, 4b)$
17. $(a + b, a - b)$, $(b - a, b + a)$
18. $(a, \sqrt{ab})$, $(b, -\sqrt{ab})$

Find the coordinates of Q given that M is the midpoint of $\overline{PQ}$.

Sample $P(-6, 2)$, $M(-1, 1)$

Solution   Let Q be the point (x, y). Then use the midpoint formula.

$$\frac{x_1 + x_2}{2} = -1 \quad \text{and} \quad \frac{y_1 + y_2}{2} = 1$$

$$\frac{-6 + x}{2} = -1 \qquad \frac{2 + y}{2} = 1$$

$$-6 + x = -2 \qquad 2 + y = 2$$

$$x = 4 \qquad y = 0$$

$\therefore Q$ is the point $(4, 0)$.   *Answer*

B
19. $P(0, 0)$, $M(3, 5)$
20. $P(-4, 3)$, $M(0, 0)$
21. $P(-4, 0)$, $M(3, 3)$
22. $P(6, -2)$, $M(0, 5)$
23. $P(h, k)$, $M(0, 0)$
24. $P(0, 0)$, $M(h, k)$

The vertices A, B, and C of a triangle are given. For each triangle, determine (a) whether it is isosceles, and (b) whether it is a right triangle. If it is a right triangle, find its area.

25. $A(-2, 2)$, $B(2, 1)$, $C(1, -3)$

26. $A(4, -5)$, $B(-2, -8)$, $C(-8, 4)$

27. $A(5, -4)$, $B(-3, 4)$, $C(-2, -3)$

28. $A(2, -3)$, $B(2, 3)$, $C(5, 0)$

Use the distance formula to determine whether the given points are collinear. (*Hint:* Find the distance between each pair of points. If one of these distances is the sum of the other two, the three points are collinear.)

29. $(1, 2)$, $(7, 4)$, $(-2, 1)$

30. $(5, 0)$, $(-7, 3)$, $(1, 1)$

31. $(-5, -2)$, $(-2, 1)$, $(1, 3)$

32. $(1, 5)$, $(2, 0)$, $(4, -10)$

Problems

Find an equation of the perpendicular bisector of $\overline{AB}$.

A

1. $A(-3, 0)$, $B(0, 5)$

2. $A(2, 1)$, $B(-2, 3)$

3. $A(8, -3)$, $B(-2, 5)$

4. $A(-9, -3)$, $B(1, -7)$

5. Find the points on the coordinate axes that are equidistant from the points A and B in Problem 1.

6. Which of the following represents the distance from any point (x, y) to the origin $(0, 0)$?

a. $\sqrt{x^2 + y^2}$ *b.* $\sqrt{x} + \sqrt{y}$ *c.* $x + y$

7. The sum of the squares of the lengths of the sides of any parallelogram equals the sum of the squares of the lengths of its diagonals. Verify this for the parallelogram with vertices $C(0, 0)$, $D(5, 0)$, $E(8, 5)$, and $F(3, 5)$.

8. The midpoints of the sides of any quadrilateral are the vertices of a parallelogram. Verify this for the quadrilateral with vertices $A(-2, 3)$, $B(2, 7)$, $C(6, 1)$, and $D(0, -3)$.

B

9. Prove: The segment joining the midpoints of two sides of a triangle is parallel to, and half as long as, the third side. [*Hint:* Any triangle can be placed so that its vertices are $A(0, 0)$, $B(a, 0)$, and $C(b, c)$.]

10. Suppose that the midpoints of the sides of a triangle are given. Explain how to draw the triangle.

11. Use the distance formula to prove the midpoint formula. (That is, show that $MP_1 = MP_2 = \frac{1}{2}P_1P_2$.)

C

12. Prove the fact stated in Problem 7. [*Hint:* Any parallelogram can be placed so that its vertices are $A(0, 0)$, $B(a, 0)$, $C(a + b, c)$, and $D(b, c)$.]

13. Prove the fact stated in Problem 8. [*Hint:* Any quadrilateral can be placed so that its vertices are $A(0, 0)$, $B(2a, 0)$, $C(2b, 2c)$, and $D(2d, 2e)$.]

14. This problem outlines a proof of part (2) of the theorem on page 119.

Let $P(r, s)$ be the point of intersection of

$\quad L_1\colon y = m_1x + b_1 \quad$ and $\quad L_2\colon y = m_2x + b_2,$

and let T_1 and T_2 be the points where the line $x = r + 1$ intersects L_1 and L_2, respectively.

a. Show that $T_1 = (r + 1, s + m_1)$ and $T_2 = (r + 1, s + m_2)$.

b. Find the distances T_1T_2, PT_1, and PT_2.

c. Proof of the "if" part: You are given that $m_1m_2 = -1$. Use this fact and part (b) to show that $(T_1T_2)^2 = (PT_1)^2 + (PT_2)^2$. It follows from the converse of the Pythagorean theorem that triangle T_1PT_2 has a right angle at P. Therefore, L_1 and L_2 are perpendicular.

d. Proof of the "only if" part: You are given that L_1 and L_2 are perpendicular, so that triangle T_1PT_2 has a right angle at P. Apply the Pythagorean theorem and part (b). Then simplify to obtain $m_1m_2 = -1$.

Mixed Review Exercises

Graph each equation.

1. $x - 2y = 4$ **2.** $y = 3 - x^2$ **3.** $y = x^2 + 4x + 5$

Find the value of each function if $x = -3$.

4. $f(x) = -x + 5$ **5.** $g(x) = \dfrac{x - 2}{3x - 1}$ **6.** $h(x) = x^5 - 7x^3 - 13x + 11$

Computer Exercises   *For students with some programming experience.*

1. Write a program to compute the distance between two given points.

2. Run your program to find the distance between the given points.

     a. $(0, 0)$, $(5, 5)$ **b.** $(-4, 7)$, $(8, -7)$ **c.** $(100, 85)$, $(242, -173)$

3. Modify the program to accept the coordinates of 3 distinct points and compute the distance between each pair of points. If the 3 points are collinear, the program should print a message stating this; and if they are not collinear, the program should print the perimeter of the resulting triangle. (*Hint:* Since the computer approximates the square roots used in finding distances, consider A, B, and C collinear if $|AB - (BC + AC)|$, $|AC - (BC + AB)|$, or $|BC - (AC + AB)| < 0.00001$.)

4. Run the program in Exercise 3 for each set of points.

     a. $(4, 7)$, $(5, 9)$, $(6, 18)$ **b.** $(74, -18)$, $(100, 36)$, $(120, 0)$

9-2 Circles

Objective   To learn the relationship between the center and radius of a circle and the equation of the circle.

Analytic geometry uses algebra to investigate geometric figures. In Lessons 9-2 through 9-6 you will study plane curves having second-degree equations. These curves are called **conic sections,** or simply **conics,** because they can be obtained by slicing a double cone with a plane (see page 400).

Of these conics you are probably most familiar with the circle. A **circle** is the set of all points in a plane that are a fixed distance, called the **radius,** from a fixed point, called the **center.**

The diagram shows a circle having center $C(2, -3)$ and radius 6. To find the equation of the circle, use the distance formula. A point $P(x, y)$ is on the circle if and only if the distance between P and the center $C(2, -3)$ equals 6.

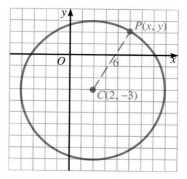

$$CP = 6$$
$$\sqrt{(x - 2)^2 + (y + 3)^2} = 6$$
Thus $(x - 2)^2 + (y + 3)^2 = 36$

Every circle has an equation of this form.

Equation of a Circle

The circle with center (h, k) and radius r has the equation

$$(x - h)^2 + (y - k)^2 = r^2.$$

Example 1   Find an equation of the circle with center $(-2, 5)$ and radius 3.

Solution   Substitute $h = -2$, $k = 5$, and $r = 3$ in this equation:
$$(x - h)^2 + (y - k)^2 = r^2.$$
$$[x - (-2)]^2 + (y - 5)^2 = 3^2$$
$$(x + 2)^2 + (y - 5)^2 = 9 \quad \textbf{\textit{Answer}}$$

If the center of a circle is at the origin, then (h, k) is $(0, 0)$ and the equation of the circle is

$$(x - 0)^2 + (y - 0)^2 = r^2$$
or
$$x^2 + y^2 = r^2.$$

The graph of the equation $x^2 + y^2 = 9$ is a circle with center at the origin and radius 3. If you slide every point of this circle to the right 2 units and up 5 units, the equation of the *translated* circle is

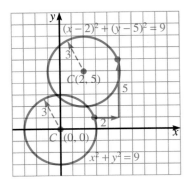

$$(x - 2)^2 + (y - 5)^2 = 9.$$

Sliding a graph to a new position in the coordinate plane without changing its shape is called a **translation.**

In general, replacing x by $x - h$ and y by $y - k$ in an equation slides the corresponding graph h units horizontally and k units vertically. Slide:

right if h is positive,   left if h is negative,
up if k is positive,   down if k is negative.

Example 2   Graph $(x - 2)^2 + (y + 6)^2 = 4$.

Solution 1   The graph of the equation $x^2 + y^2 = 4$ is a circle with center at the origin and radius 2. Slide this circle 2 units to the right and then 6 units down to get the graph of $(x - 2)^2 + [y - (-6)]^2 = 4$, or $(x - 2)^2 + (y + 6)^2 = 4$.

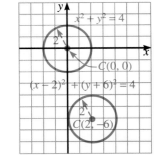

Solution 2   Rewrite the given equation in the form $(x - h)^2 + (y - k)^2 = r^2$:

$$(x - 2)^2 + [y - (-6)]^2 = 2^2$$

The graph is a circle. The center (h, k) is $(2, -6)$ and the radius $r = 2$.

Example 3   If the graph of the given equation is a circle, find its center and radius. If the equation has no graph, say so.
   a. $x^2 + y^2 + 10x - 4y + 21 = 0$ **b.** $x^2 + y^2 - 8x + 6y + 30 = 0$

Solution   **a.** If the graph is a circle, the equation can be written in the form $(x - h)^2 + (y - k)^2 = r^2$. Complete the square twice, once using the terms in x, and once using the terms in y.

$$x^2 + y^2 + 10x - 4y + 21 = 0$$

$$(x^2 + 10x + \underline{\ ?\ }) + (y^2 - 4y + \underline{\ ?\ }) = -21 \qquad \text{Rearrange terms.}$$

$$(x^2 + 10x + 25) + (y^2 - 4y + 4) = -21 + 25 + 4 \qquad \text{Add 25 and 4.}$$

$$(x + 5)^2 + (y - 2)^2 = 8$$

∴ the center is $(-5, 2)$ and the radius is $2\sqrt{2}$.   *Answer*

b.
$$x^2 + y^2 - 8x + 6y + 30 = 0$$
$$(x^2 - 8x + 16) + (y^2 + 6y + 9) = -30 + 16 + 9$$
$$(x - 4)^2 + (y + 3)^2 = -5$$

Since the square of any number is positive and the sum of two positive numbers is positive, *no* ordered pair satisfies the equation. This equation has no graph.
Answer

Recall this fact from geometry: Let L be the line tangent to a given circle at a point P. Then the line perpendicular to L at P passes through the center of the circle.

Example 4 Find an equation of a circle of radius 3 that has its center in the first quadrant and is tangent to the y-axis at $(0, 2)$.

Solution First make a sketch using the given information. From the geometric fact stated above, the center of the circle lies on the line $y = 2$ and must be 3 units from $(0, 2)$, which is on the circle. Thus the center is $(3, 2)$.

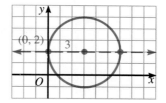

The equation is

$$(x - 3)^2 + (y - 2)^2 = 3^2.$$
Answer

Oral Exercises

Give the center and radius of each circle.

1. $x^2 + y^2 = 49$ **2.** $x^2 + y^2 = 1$

3. $(x - 2)^2 + (y - 4)^2 = 36$ **4.** $(x - 5)^2 + y^2 = 25$

5. $x^2 + (y + 5)^2 = \frac{1}{9}$ **6.** $(x + 4)^2 + (y - 3)^2 = 7$

Give an equation of the circle with the indicated center and radius.

7. $(0, 0)$; 4   **8.** $(0, 0)$; $\sqrt{2}$ **9.** $(1, 1)$; 3

10. $(3, 2)$; 5   **11.** $(4, 0)$; 1   **12.** $(-1, 0)$; 4

13. $(-3, -3)$; 3   **14.** $(-3, 4)$; $\sqrt{8}$ **15.** $(1, -1)$; $\frac{1}{2}$

Written Exercises

Find an equation of the circle with the given center and radius.

A **1.** $(3, 0)$; 3

2. $(0, -1)$; 1

3. $(2, -5)$; 8

4. $(-3, 1)$; 5

5. $(0, 0)$; 12

6. $(-4, -2)$; 10

7. $(6, 1)$; $\sqrt{2}$

8. $(-5, 3)$; $\frac{1}{6}$

Graph each equation. You may wish to check your graphs on a computer or a graphing calculator.

9. $x^2 + y^2 = 25$

10. $x^2 + y^2 = 4$

11. $(x - 4)^2 + (y - 5)^2 = 1$

12. $(x + 2)^2 + (y + 3)^2 = 81$

13. $(x - 3)^2 + y^2 = 36$

14. $x^2 + (y + 6)^2 = 4$

If the graph of the given equation is a circle, find its center and radius. If the equation has no graph, say so.

15. $x^2 + y^2 - 16 = 0$

16. $x^2 + y^2 - 81 = 0$

17. $x^2 + y^2 = -8y$

18. $x^2 + y^2 - 6x = 0$

19. $x^2 + y^2 - 4x + 2y - 4 = 0$

20. $x^2 + y^2 + 10x - 4y + 20 = 0$

21. $x^2 + y^2 + 8x + 2y + 18 = 0$

22. $x^2 + y^2 + 12x - 6y = 0$

23. $x^2 + y^2 + 3x - 4y = 0$

24. $x^2 + y^2 - 5y + 4 = 0$

Graph each inequality.

Sample $x^2 + y^2 - 8x + 2y + 1 < 0$

Solution Rewrite the inequality by completing the squares in x and y. The inequality becomes $(x - 4)^2 + (y + 1)^2 < 16$.

The graph of the equation $(x - 4)^2 + (y + 1)^2 = 16$ is the circle with center $(4, -1)$ and radius 4. Sketch this circle using dashed lines to indicate that points on the circle do not satisfy the inequality. To determine whether the graph is the set of points inside or outside the circle, substitute a test point, say $(3, 0)$, in the inequality. Since $(3, 0)$ satisfies the inequality, shade the region inside the circle for the graph.

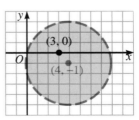

B **25.** $x^2 + y^2 \geq 1$

26. $x^2 + y^2 < 4$

27. $x^2 + y^2 - 4y > 0$

28. $x^2 + y^2 \geq 2y$

29. $x^2 + y^2 + 6x - 6y + 9 \leq 0$

30. $x^2 + y^2 + 4x - 10y < 7$

Find the center and radius of each circle. (*Hint:* First divide both sides by the coefficient of the second-degree terms.)

31. $4x^2 + 4y^2 - 16x - 24y + 36 = 0$

32. $9x^2 + 9y^2 + 6x + 18y + 9 = 0$

33. $16x^2 + 16y^2 - 32x + 8y = 0$

34. $3x^2 + 3y^2 - 6x + 24y + 24 = 0$

Find an equation of the circle described. (A sketch may be helpful.)

35. Center $(0, 5)$; passes through $(0, 0)$.

36. Center $(-2, 0)$; passes through $(2, 0)$.

37. A diameter has endpoints $(2, 5)$ and $(0, 3)$.

38. Center in quadrant two; radius 3; tangent to y-axis at $(0, 4)$.

39. Center on line $y - 4 = 0$; tangent to x-axis at $(-2, 0)$.

40. Center on line $x + y = 4$; tangent to both coordinate axes.

41. Center in quadrant four; tangent to the lines $x = 1$, $x = 9$, and $y = 0$.

42. Tangent to both coordinate axes and the line $x = -8$. (Two answers)

Graph each semicircle. Recall that $\sqrt{a} \geq 0$.

C 43. $y = \sqrt{25 - x^2}$

44. $y = -\sqrt{25 - x^2}$

45. $y = \sqrt{4x - x^2}$

46. $x = \sqrt{2y - y^2}$

47. A boat sails so that it is always twice as far from one buoy as from a second buoy 3 mi from the first one. Describe the path of the boat. (*Hint:* Introduce a coordinate system so that one buoy is at $(0, 0)$ and the other is at $(3, 0)$.)

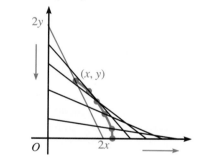

48. A ladder 6 m long leaning against a wall slips to the ground. Describe the path followed by the midpoint of the ladder. (*Hint:* See the figure at the right.)

49. Find an equation of the circle of radius 4 that is tangent to both branches of the graph of $y = |x|$.

50. Use analytic geometry to prove that an angle inscribed in a semicircle is a right angle. (*Hint:* Let the semicircle be the top half of the circle $x^2 + y^2 = r^2$ and $P(x, y)$ be any point on the semicircle. Use the slopes of $\overline{PA}$ and $\overline{PB}$ to show that they are perpendicular.)

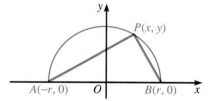

Challenge

Points with integer coordinates are called *lattice points*. For example, $(0, 0)$ and $(-2, 5)$ are lattice points, but $\left(\frac{1}{2}, 3\right)$ is not. The circle $x^2 + y^2 = 1$ passes through the 4 lattice points $(1, 0)$, $(-1, 0)$, $(0, 1)$, and $(0, -1)$. Find an equation of a circle that passes through exactly 3 lattice points.

9-3 Parabolas

Objective   To learn the relationship between the focus, directrix, vertex, and axis of a parabola and the equation of the parabola.

Suppose that you draw the line $y = 4$ and plot the point $F(3, 0)$. Then plot several points P that appear to be the same distance from the line $y = 4$ as they are from the point F.

In the diagram, the distance from point P to the line is measured along the perpendicular $\overline{PD}$. To find an equation of the path of P, you use the distance formula.

$$PD = PF$$
$$\sqrt{(x - x)^2 + (y - 4)^2} = \sqrt{(x - 3)^2 + (y - 0)^2}$$
$$(y - 4)^2 = (x - 3)^2 + y^2$$
$$-8y + 16 = (x - 3)^2$$
$$-8(y - 2) = (x - 3)^2$$
$$y - 2 = -\frac{1}{8}(x - 3)^2$$

The last equation is of the form $y - k = a(x - h)^2$. In Lesson 7-5, you learned that the graph of such an equation is a parabola with vertex (h, k) and axis $x = h$. Therefore, the graph of the set of points P is a parabola with vertex $(3, 2)$ and axis $x = 3$. The following general definition of a parabola is stated in terms of distance.

A **parabola** is the set of all points equidistant from a fixed line, called the **directrix,** and a fixed point not on the line, called the **focus.**

The important features of a parabola are shown in the diagram below. Notice that the vertex is midway between the focus and the directrix.

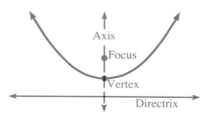

Example 1   The vertex of a parabola is $(-5, 1)$ and the directrix is the line $y = -2$. Find the focus of the parabola.

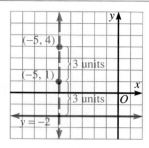

Solution   It is helpful to make a sketch. The vertex is 3 units above the directrix. Since the vertex is midway between the focus and the directrix, the focus is $(-5, 4)$.   **Answer**

Example 2   Find an equation of the parabola having the point $F(0, -2)$ as focus and the line $x = 3$ as directrix. Draw the curve and label the vertex V, the focus F, the directrix, and the axis of symmetry.

Solution   From the definition, $P(x, y)$ is on the parabola if and only if $PD = PF$, where PD is the perpendicular distance from P to the directrix.

$$PD = PF$$
$$\sqrt{(x - 3)^2 + (y - y)^2} = \sqrt{(x - 0)^2 + (y - (-2))^2}$$
$$(x - 3)^2 = x^2 + (y + 2)^2$$
$$x^2 - 6x + 9 - x^2 = (y + 2)^2$$
$$-6\left(x - \frac{3}{2}\right) = (y + 2)^2$$
$$x - \frac{3}{2} = -\frac{1}{6}(y + 2)^2$$

To plot a few points, choose convenient values of y and compute the corresponding values of x.

x	y
$-\dfrac{9}{2}$	-8
0	-5
$\dfrac{3}{2}$	-2
0	1
$-\dfrac{9}{2}$	4

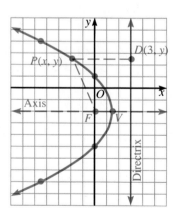

Notice that the parabola in Example 2 has a horizontal axis and an equation of the form $x - h = a(y - k)^2$ where the point (h, k) is the vertex. This is similar to the equation of a parabola that has a vertical axis, except that the roles of x and y are reversed.

If the distance between the vertex and the focus of a parabola is $|c|$, then it can be shown that $a = \dfrac{1}{4c}$ in the equation of the parabola (see Exercise 36).

The parabola whose equation is

$$y - k = a(x - h)^2, \text{ where } a = \frac{1}{4c},$$

opens upward if $a > 0$, downward if $a < 0$;
has   vertex $V(h, k)$,
    focus $F(h, k + c)$,
    directrix $y = k - c$,
and   axis of symmetry $x = h$.

The parabola whose equation is

$$x - h = a(y - k)^2, \text{ where } a = \frac{1}{4c},$$

opens to the right if $a > 0$, to the left if $a < 0$;
has   vertex $V(h, k)$,
    focus $F(h + c, k)$,
    directrix $x = h - c$,
and   axis of symmetry $y = k$.

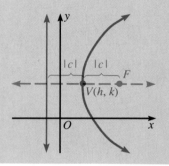

Example 3 Find the vertex, focus, directrix, and axis of symmetry of the parabola $y^2 - 12x - 2y + 25 = 0$. Then graph the parabola.

Solution Complete the square using the terms in y:

$$y^2 - 12x - 2y + 25 = 0$$
$$y^2 - 2y = 12x - 25$$
$$y^2 - 2y + 1 = 12x - 25 + 1$$
$$(y - 1)^2 = 12(x - 2)$$
$$x - 2 = \frac{1}{12}(y - 1)^2$$

Comparing this equation with $x - h = a(y - k)^2$, you can see that $a = \dfrac{1}{12}$, $h = 2$, and $k = 1$.

Since $a = \dfrac{1}{12} = \dfrac{1}{4c}$, $c = 3$.

Thus, the parabola opens to the right (since $a > 0$).

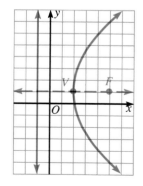

The vertex is $V(2, 1)$,   the focus is $F(5, 1)$,   the directrix is $x = -1$, and the axis of symmetry is $y = 1$. The graph is shown above. *Answer*

Example 4 Find an equation of the parabola that has vertex (4, 2) and directrix $y = 5$.

Solution The distance from the vertex to the directrix is 3, so $|c| = 3$. Since the directrix is above the vertex, the parabola opens downward. Therefore the squared term is the term with x, and c is negative. If $c = -3$, then $a = -\dfrac{1}{12}$. Thus the equation is

$$y - 2 = -\frac{1}{12}(x - 4)^2. \quad \textbf{\textit{Answer}}$$

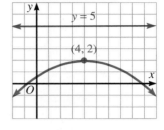

Oral Exercises

In the following exercises, V denotes the vertex of a parabola, F the focus, and D the directrix. Two of these are given. Find the third.

1. $V(0, 0)$, $F(-3, 0)$　　　　**2.** $V(0, 0)$, $D: x = -2$　　　　**3.** $F(4, 6)$, $D: x = 0$

4. $D: y = -1$, $F(4, -4)$　　**5.** $D: y = -2$, $V(3, -1)$　　**6.** $F(-2, 5)$, $V(-2, 1)$

For each parabola, find (a) $V(h, k)$, (b) $|c|$, the distance between its vertex and focus, and (c) the direction in which it opens (up, down, left, right).

7. $y = \dfrac{1}{8}x^2$　　　　**8.** $x = -\dfrac{1}{12}y^2$　　　　**9.** $x = -(y - 2)^2$

10. $y = (x + 5)^2$　　　**11.** $x + 3 = \dfrac{1}{4}(y - 1)^2$　　　**12.** $y + 4 = -\dfrac{1}{36}(x - 4)^2$

Written Exercises

In the following exercises, V denotes the vertex of a parabola, F the focus, and D the directrix. Two of these are given. Find the third.

A　**1.** $V(4, 2)$, $D: y = -3$　　　　　　**2.** $D: y = 2$, $V(2, 4)$

　　3. $V(0, 2)$, $F(0, 0)$　　　　　　　　**4.** $F(-3, -1)$, $V(1, -1)$

　　5. $F(1, -2)$, $V(1, -5)$　　　　　　**6.** $D: x = -2$, $F(2, 0)$

Find an equation of the parabola described. Then graph the parabola.

7. Focus $(0, 0)$; directrix $y = 4$　　　　　**8.** Focus $(0, 0)$; directrix $x = 4$

9. Vertex $(0, 0)$; focus $(0, -4)$　　　　　**10.** Vertex $(0, 0)$; focus $(3, 0)$

11. Vertex $(0, 0)$; directrix $x + 1 = 0$　　**12.** Vertex $(0, 0)$; directrix $y = -4$

13. Focus $(0, 2)$; directrix $x = 2$　　　　**14.** Focus $(-2, 0)$; directrix $y = 3$

15. Focus $(3, 4)$; vertex $(3, 2)$　　　　　**16.** Focus $(-2, 1)$; vertex $(-3, 1)$

Find the vertex, focus, directrix, and axis of symmetry of each parabola. Then graph the parabola. You may wish to check your graphs on a computer or a graphing calculator.

17. $y = \frac{1}{6}x^2$

18. $6x + y^2 = 0$

19. $4x = y^2 - 4y$

20. $x^2 = y + 2x$

B 21. $x^2 + 8y + 4x - 4 = 0$

22. $y^2 + 6y + 8x - 7 = 0$

23. $y^2 - 8x - 6y - 3 = 0$

24. $x^2 - 6x + 10y - 1 = 0$

25. $x^2 + 10x - 2y + 21 = 0$

26. $y^2 + 3x - 2y - 11 = 0$

Graph each inequality.

27. $y < (x - 3)^2$

28. $y \geq 2(x + 1)^2$

29. $y + 4 \geq (x + 2)^2$

30. $x - 11 < y^2 + 6y$

Graph each equation. Each graph is half of a parabola, since $\sqrt{a} \geq 0$.

31. $y = \sqrt{x}$

32. $y = \sqrt{x - 2}$

33. $y = -\sqrt{1 - x}$

34. $y = -\sqrt{3 - x}$

C 35. The line through the focus of a parabola perpendicular to its axis intersects the parabola in two points, P and Q. ($\overline{PQ}$ is called the *latus rectum* of the parabola.) Explain why the length of $\overline{PQ}$ is twice the distance from the focus to the directrix.

36. A parabola having vertex (h, k) and a vertical axis has as focus the point $(h, k + c)$ and as directrix the line $y = k - c$. Show that an equation of the parabola is $y - k = \frac{1}{4c}(x - h)^2$.

37. F is the focus of the parabolic arc shown below. Line L is parallel to the directrix, D. Explain why the sum $FP + PQ$ is the same for all positions of P. (*Hint:* Extend $\overline{PQ}$ so that it intersects the directrix.)

 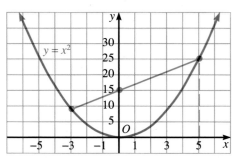

38. A *multiplication machine:* Draw the parabola $y = x^2$. Given any two positive numbers a and b, locate the points of the parabola having the x-coordinates $-a$ and b. Explain why the line joining these points has y-intercept ab. (*Hint:* Find an equation of the line through $(-a, a^2)$ and (b, b^2). Use the diagram above as a model.)

Mixed Review Exercises

For the given points P **and** Q, **find (a) the slope of** $\overline{PQ}$, **(b) the distance** PQ, **and (c) the midpoint of** $\overline{PQ}$. **Express radicals in simplest form.**

1. $P(-3, 2)$, $Q(5, 2)$ **2.** $P(1, -6)$, $Q(3, 2)$ **3.** $P(4, 3)$, $Q(-3, 4)$

Find an equation for each figure described.

4. The line having slope $\frac{2}{3}$ and passing through $(-2, -3)$.

5. The perpendicular bisector of the segment with endpoints $(7, -2)$ and $(1, -8)$.

6. The circle with center $(-1, 3)$ and radius 4.

Self-Test 1

Vocabulary   Pythagorean theorem (p. 401)   center (p. 407)
                            distance formula (p. 402)   translation (p. 408)
                            midpoint formula (p. 402)   parabola (p. 412)
                            conic section (p. 407)   directrix (p. 412)
                            conic (p. 407)   focus (of a parabola) (p. 412)
                            circle (p. 407)   vertex (p. 412)
                            radius (p. 407)   axis of symmetry (p. 412)

1. Find the distance between $(3, -6)$ and $(4, 2)$.   **Obj. 9-1, p. 401**

2. Find the midpoint of the line segment having endpoints $(-3, 5)$ and $(8, 11)$.

3. Determine whether the three points $A(2, 2)$, $B(8, 6)$, and $C(11, 10)$ are collinear by comparing the lengths of $\overline{AB}$, $\overline{BC}$, and $\overline{AC}$.

4. Find an equation of the circle with center $(5, -4)$ and radius 7.   **Obj. 9-2, p. 407**

5. Find the center and radius of the circle whose equation is $x^2 + y^2 + 12x - 4y + 32 = 0$.

6. Find an equation of the parabola having focus $(0, 3)$ and vertex $(-4, 3)$.   **Obj. 9-3, p. 412**

7. Find the vertex, focus, directrix, and axis of the parabola $x^2 + 10x + 16y - 7 = 0$.

8. Graph $(x + 1) = \frac{1}{8}(y + 3)^2$. Label the vertex, focus, directrix, and axis.

Check your answers with those at the back of the book.

Conic Sections: Ellipses and Hyperbolas

9-4 Ellipses

Objective   To learn the relationship between the center, foci, and intercepts of an ellipse and the equation of the ellipse.

The diagram at the right shows a piece of string whose ends are fastened at the points F_1 and F_2. If you use a pencil to keep the string tight, the point P of the pencil will move along a path that is an oval curve called an *ellipse*. Because the sum of the distances from the pencil to the points F_1 and F_2 is the length of the string, this sum, $PF_1 + PF_2$, is the same for all positions of P.

An **ellipse** is the set of all points P in the plane such that the sum of the distances from P to two fixed points is a given constant.

Each of the fixed points is called a **focus** (plural: *foci*) of the ellipse. If the foci are F_1 and F_2, then for each point P of the ellipse, PF_1 and PF_2 are the **focal radii** of P.

   To find an equation of the ellipse having foci $F_1(-4, 0)$ and $F_2(4, 0)$ and sum of focal radii 10, use the distance formula. A point $P(x, y)$ is on the ellipse if and only if the following is true:

$$PF_1 \quad + \quad PF_2 \quad = 10$$
$$\sqrt{(x + 4)^2 + y^2} + \sqrt{(x - 4)^2 + y^2} = 10$$
$$\sqrt{(x + 4)^2 + y^2} = 10 - \sqrt{(x - 4)^2 + y^2}$$

Square both sides and then simplify:

$$(x + 4)^2 + y^2 = 100 - 20\sqrt{(x - 4)^2 + y^2} + (x - 4)^2 + y^2$$
$$20\sqrt{(x - 4)^2 + y^2} = 100 - 16x$$
$$5\sqrt{(x - 4)^2 + y^2} = 25 - 4x$$

Square both sides again:

$$25(x^2 - 8x + 16 + y^2) = 625 - 200x + 16x^2$$
$$9x^2 + 25y^2 = 225$$

Divide both sides by 225:

$$\frac{x^2}{25} + \frac{y^2}{9} = 1$$

The following analysis will help you graph the equation $\dfrac{x^2}{25} + \dfrac{y^2}{9} = 1$.

1. The graph has four *intercepts*.
   If $y = 0$, then

   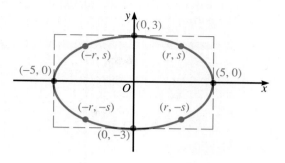

   $$\frac{x^2}{25} + \frac{0^2}{9} = 1.$$

   $\therefore x^2 = 25$, or $x = \pm 5$.

   If $x = 0$, then

   $$\frac{0^2}{25} + \frac{y^2}{9} = 1.$$

   $\therefore y^2 = 9$, or $y = \pm 3$.

2. The *extent* of the graph is limited.
   Solving the equation for each variable in terms of the other gives

   $$y = \pm \frac{3}{5}\sqrt{25 - x^2} \quad \text{and} \quad x = \pm \frac{5}{3}\sqrt{9 - y^2}.$$

   In the first equation, y is a real number if and only if $|x| \le 5$. In the second equation, x is a real number if and only if $|y| \le 3$. Thus the extent of the graph is limited to the dashed rectangle shown above.

3. The graph is *symmetric about the x-axis*: If (r, s) is on the graph, so is $(r, -s)$. The graph is *symmetric about the y-axis*: If (r, s) is on the graph, so is $(-r, s)$. Thus the graph is also *symmetric about the origin*: If (r, s) is on the graph, then so is $(-r, -s)$. See the diagram above.

Because of the symmetry described, you need to choose only values of x in the first quadrant from $x = 0$ to $x = 5$. A calculator may be helpful to set up a table of values. You can then draw the complete graph.

x	$y = 0.6\sqrt{25 - x^2}$
0	3
1	2.9
2	2.7
3	2.4
4	1.8
5	0

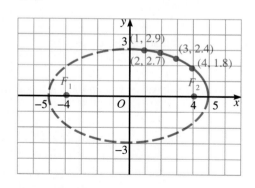

The **center** of an ellipse is the midpoint of the line segment joining its foci. The **major axis** of an ellipse is the chord passing through its foci. The **minor axis** is the chord containing the center and perpendicular to the major axis.

An ellipse has a *quadratic equation in two variables*. A general form for the equation of an ellipse with foci on the *x*-axis and equidistant from the origin is shown at the left in the chart. An ellipse may of course have its foci on the *y*-axis. The corresponding equation is shown at the right.

The ellipse having center $(0, 0)$, foci $(-c, 0)$ and $(c, 0)$, and sum of focal radii $2a$ has the equation

$$\frac{x^2}{a^2} + \frac{y^2}{b^2} = 1$$

where $b^2 = a^2 - c^2$.

The ellipse having center $(0, 0)$, foci $(0, -c)$ and $(0, c)$, and sum of focal radii $2a$ has the equation

$$\frac{x^2}{b^2} + \frac{y^2}{a^2} = 1$$

where $b^2 = a^2 - c^2$.

For an ellipse having an equation with one of the forms above, the major axis is a segment of the *x*-axis or the *y*-axis. Note that the major axis is horizontal if the x^2-term has the larger denominator and vertical if the y^2-term has the larger denominator. Since the larger of the two denominators is a^2, the length of the major axis is always $2a$. The length of the minor axis is $2b$.

The distance from the center to either focus is $|c|$. If you know the values of a and b and the coordinates of the center, you can find the coordinates of each focus by solving the equation $b^2 = a^2 - c^2$ for c (see Exercise 34).

Example 1 Graph the ellipse $4x^2 + y^2 = 64$ and find its foci.

Solution $4x^2 + y^2 = 64$ ⟵ Divide both sides by 64.

$$\frac{x^2}{16} + \frac{y^2}{64} = 1 \qquad\qquad y = \pm 2\sqrt{16 - x^2}$$

Since the denominator of the y^2-term is the larger, the major axis is vertical. The *y*-intercepts are 8 and -8. The *x*-intercepts are 4 and -4. Solve the equation for *y* and make a short table of first-quadrant points.

x	y
0	8
1	7.7
2	6.9
3	5.3
4	0

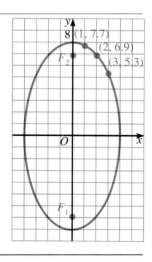

Use the relationship $c^2 = a^2 - b^2$ to find the foci. Since $a^2 = 64$ and $b^2 = 16$, $c^2 = 48$ and $c = \sqrt{48} = 4\sqrt{3} \approx 6.9$. Thus the foci of the ellipse are $F_1(0, -4\sqrt{3})$ and $F_2(0, 4\sqrt{3})$. **Answer**

Example 2   Find an equation of an ellipse having x-intercepts $\sqrt{2}$ and $-\sqrt{2}$ and y-intercepts 3 and -3.

Solution   Since $3^2 > (\sqrt{2})^2$, the y^2-term has the larger denominator, and so the major axis is vertical. The center is $(0, 0)$, $a^2 = 9$, and $b^2 = 2$.

$\therefore$ an equation is $\dfrac{x^2}{2} + \dfrac{y^2}{9} = 1$.   **Answer**

Example 3   Find an equation of an ellipse having foci $(-3, 0)$ and $(3, 0)$ and sum of focal radii equal to 12.

Solution   The center of the ellipse is $(0, 0)$. The distance from each focus to the center is 3, so $c = 3$. The sum of the focal radii is $2a$. Since $2a = 12$, $a = 6$ and $a^2 = 36$. So $b^2 = a^2 - c^2 = 36 - 9 = 27$. Since the foci are on the x-axis, the major axis is horizontal.

$\therefore$ an equation is $\dfrac{x^2}{36} + \dfrac{y^2}{27} = 1$.   **Answer**

Oral Exercises

Give the x- and y-intercepts of each ellipse and tell on which of the coordinate axes its foci lie.

1. $\dfrac{x^2}{9} + \dfrac{y^2}{4} = 1$ **2.** $\dfrac{x^2}{16} + \dfrac{y^2}{25} = 1$ **3.** $x^2 + 9y^2 = 36$

4. $x^2 + 4y^2 = 16$ **5.** $3x^2 + y^2 = 9$ **6.** $2x^2 + 3y^2 = 6$

Written Exercises

Graph each ellipse and find its foci. You may wish to check your graphs on a computer or a graphing calculator.

A   **1–6.** Use the equations in Oral Exercises 1–6.

7. $5x^2 + y^2 = 25$ **8.** $x^2 + 25y^2 = 100$

9. $3x^2 + y^2 = 12$ **10.** $2x^2 + y^2 = 8$

11. $4x^2 + 3y^2 = 48$ **12.** $5x^2 + 9y^2 = 45$

13. $x^2 + 9y^2 = 1$ **14.** $9x^2 + 4y^2 = 9$

Find an equation of an ellipse having the given intercepts.

15. x-intercepts: ± 5
   y-intercepts: ± 2

16. x-intercepts: ± 3
   y-intercepts: ± 4

17. x-intercepts: ± 2
   y-intercepts: $\pm \sqrt{2}$

18. x-intercepts: $\pm \sqrt{6}$
   y-intercepts: $\pm 2\sqrt{3}$

Find an equation of the ellipse having the given points as foci and the given number as sum of focal radii.

B **19.** $(-6, 0)$, $(6, 0)$; 18

20. $(0, -5)$, $(0, 5)$; 20

21. $(0, -4)$, $(0, 4)$; 24

22. $(-9, 0)$, $(9, 0)$; 30

23. Describe a way of constructing the foci of a given ellipse using a compass. Give a convincing argument to justify the method. (*Hint:* See the figure at the left below.)

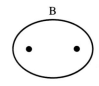

24. An indicator of the shape of the ellipse $\dfrac{x^2}{a^2} + \dfrac{y^2}{b^2} = 1$ is its *eccentricity e*,

defined by $e = \dfrac{c}{a}$, where $c = \sqrt{a^2 - b^2}$.

   a. Explain why is e a number between 0 and 1.
   b. Which of the ellipses shown above has the greater eccentricity?

25. Suppose that $a = b$ in the equation $\dfrac{x^2}{a^2} + \dfrac{y^2}{b^2} = 1$.

   a. What is the value of c?
   b. What is the value of e?
   c. Where are the two foci?
   d. What special ellipse will the graph be?

26. Describe the graph of $\dfrac{x^2}{a^2} + \dfrac{y^2}{b^2} = 0$.

27. The arch of a bridge is in the form of half an ellipse, with the major axis horizontal. The span of the bridge (the length of the major axis) is 12 m and the height of the arch above the water at the center is 4 m. How high above the water is the arch at a point on the water 2 m from one of the ends of the arch?

Graph each inequality.

28. $4x^2 + 9y^2 < 36$

29. $4x^2 + y^2 \geq 16$

30. $5x^2 + 4y^2 \geq 20$

31. $3x^2 + 16y^2 \leq 48$

Graph each equation. Each graph is a semi-ellipse, since $\sqrt{a}$ is always nonnegative.

32. $y = 3\sqrt{1 - x^2}$

33. $y = \frac{1}{2}\sqrt{4 - x^2}$

C 34. For the ellipse $\dfrac{x^2}{a^2} + \dfrac{y^2}{b^2} = 1$ with foci at $(-c, 0)$ and $(c, 0)$, show that $b^2 = a^2 - c^2$.

35. Use the definition to find an equation of the ellipse with foci at $(-c, 0)$ and $(c, 0)$ and sum of focal radii equal to $2a$. At the appropriate point let $b^2 = a^2 - c^2$ to simplify the equation.

Career Note / Computer Graphics Artist

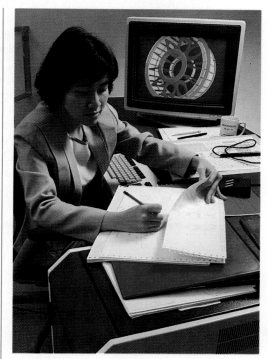

In movies and on television you have probably seen believable pictures of things that do not exist and cannot be photographed. These pictures are the product of computer graphics. The computer graphics artist gives the computer a complete description of each object that is to be in a picture. The object's shape may be built up from flat polygons or from geometric solids such as spheres. The location of each object in the scene is given in three-dimensional coordinates so that objects farther from the viewer will appear smaller.

A computer program compares the boundaries of objects and decides which surfaces are visible to the viewer. In order to give realistic texture to images, the graphics artist uses formulas describing how different materials reflect light. Even using powerful computers, it may take the computer graphics artist hours to make one second of movie film.

The computer has become a valuable tool for graphic artists in many other fields. For example, computer-generated shapes and color patterns are now used in the design of magazine illustrations, posters, packaging, and textiles.

Application / *Planetary Orbits*

In 1609 the German astronomer Johann Kepler proposed two of his three famous laws of planetary motion (all of which were later confirmed by Newton's laws). In one of these, Kepler stated that the orbits of the planets in our solar system are ellipses with the sun at one focus. This discovery was all the more remarkable because it represented the first practical application of the geometry of the ellipse, which had been extensively studied by Greek mathematicians about 2000 years earlier.

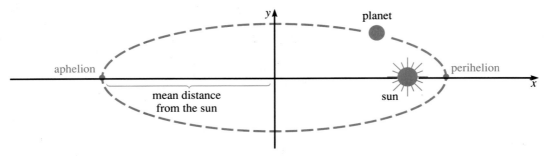

The greater the eccentricity of an ellipse, the more elongated it is (see Exercise 24 on page 422). The eccentricity of the orbit shown above has been exaggerated for clarity. The planetary orbits are all more nearly circular.

As the diagram above indicates, a planet's *mean distance from the sun*, a figure given in many astronomical tables, is defined to be half the length of the major axis of the orbit. This is the mean, or average, of the planet's maximum and minimum distances from the sun. The positions at which a planet achieves these distances are called *aphelion* and *perihelion*, respectively.

Much of the computation involving distances within the solar system is carried out using the *astronomical unit* (AU), defined to be Earth's mean distance from the sun, or about 150 million km, as a unit of measure. The table below gives the eccentricities of the orbits of the planets and their mean distances from the sun.

Planet	Mean Distance (in AU)	Eccentricity
Mercury	0.387	0.206
Venus	0.723	0.007
Earth	1.000	0.017
Mars	1.524	0.093
Jupiter	5.203	0.049
Saturn	9.539	0.056
Uranus	19.182	0.047
Neptune	30.058	0.009
Pluto	39.44	0.250

Suppose that you choose a coordinate system so that the major axis of the orbit lies on the x-axis and the center of the ellipse is at the origin. Then using the information from the table, you can find the Cartesian equation of the orbit of a planet, such as Mercury, in the standard form

$$\frac{x^2}{a^2} + \frac{y^2}{b^2} = 1.$$

First note that for Mercury a = mean distance from the sun = 0.387. Then use the fact that eccentricity = $\frac{c}{a}$ = 0.206. To find the value of c:

$$\frac{c}{0.387} = 0.206, \text{ so } c = 0.080.$$

Now use the relationship $b^2 = a^2 - c^2$, which holds for any ellipse, to find the value of b:

$$b = \text{half the length of the minor axis}$$
$$= 0.379.$$

Thus the equation for the orbit of Mercury is

$$\frac{x^2}{(0.387)^2} + \frac{y^2}{(0.379)^2} = 1.$$

Exercises

1. Find the maximum and minimum distances from Earth to the sun in AU and in km.

2. What is the planet with the least circular orbit? Find the lengths of the major and minor axes for this orbit. What is the ratio of these lengths? Sketch the graph.

3. Find the ratio of the length of the major axis of Earth's orbit to the length of the minor axis.

4. Using the equation for the orbit of Mercury, find how far the planet is from the sun in AU when Mercury is on the y-axis of the coordinate system.

5. The asteroid Icarus has an elliptical orbit around the sun. Its mean distance from the sun is 1.078 AU and the eccentricity of its orbit is 0.827. Find the ratio of the length of the major axis of this orbit to the length of the minor axis. Contrast this ratio with the same ratio for Earth (Exercise 3).

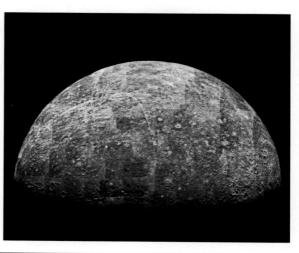

9-5 Hyperbolas

Objective   To learn the relationship between the foci, intercepts, and
asymptotes of a hyperbola and the equation of the hyperbola.

A ship's navigator can plot the location of a ship by using a radio navigation system called LORAN (LOng RAnge
Navigation). In this system, a pair of radio sending stations
continually transmit signals at different time intervals. The
navigator measures the difference in time between the two
signals received and then uses this information to plot the
path of the ship. Its curved path is called a *hyperbola*. The
signals from another pair of stations are used to plot a second hyperbolic path. The intersection of the two hyperbolas determines the
exact position of the ship. In the diagram above, the sending stations are represented by two fixed points, F_1 and F_2, and the ship by the point P. The
ship, or point P, moves along a path that maintains a constant difference between the distances from point P to the fixed points. In other words, the difference $PF_1 - PF_2$ is the same for all positions of P.

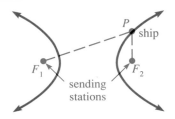

A **hyperbola** is the set of all points P in the plane such that the difference between the distances from P to two fixed points is a given constant.

Each of the fixed points is a **focus.** If P is a point on the hyperbola and F_1 and
F_2 are the foci, then PF_1 and PF_2 are the **focal radii.** As you will see, hyperbolas are curves with two pieces, or *branches*.

To find an equation of the hyperbola having foci $F_1(-5, 0)$ and $F_2(5, 0)$
and difference of focal radii 6, use the definition and write:

$$|PF_1 - PF_2| = 6, \quad \text{or} \quad PF_1 - PF_2 = \pm 6.$$

Now apply the distance formula:

$$\sqrt{(x + 5)^2 + y^2} - \sqrt{(x - 5)^2 + y^2} = \pm 6.$$

Next eliminate the radicals and simplify.

$$\sqrt{(x + 5)^2 + y^2} = \sqrt{(x - 5)^2 + y^2} \pm 6$$
$$(x + 5)^2 + y^2 = (x - 5)^2 + y^2 \pm 12\sqrt{(x - 5)^2 + y^2} + 36$$
$$20x - 36 = \pm 12\sqrt{(x - 5)^2 + y^2}$$
$$5x - 9 = \pm 3\sqrt{(x - 5)^2 + y^2}$$
$$25x^2 - 90x + 81 = 9[(x - 5)^2 + y^2]$$
$$16x^2 - 9y^2 = 144$$

Divide both sides by 144:

$$\frac{x^2}{9} - \frac{y^2}{16} = 1$$

Before graphing this equation, you analyze it using similar steps as for ellipses (page 419).

1. The x-intercepts, 3 and -3, are the values of x for which $y = 0$. When $x = 0$, the equation becomes $-\dfrac{y^2}{16} = 1$. Since no real value of y satisfies this equation, there are no y-intercepts.

2. Determine the extent of the graph.

$$y = \pm\frac{4}{3}\sqrt{x^2 - 9} \qquad x = \pm\frac{3}{4}\sqrt{y^2 + 16}$$

In the first equation, y is real if and only if $|x| \geq 3$. Thus, no part of the graph lies between the lines $x = -3$ and $x = 3$. In the second equation, x is real for all values of y, so no values of y are excluded.

3. It is easy to see that the graph is symmetric with respect to the x-axis, the y-axis, and the origin.

4. Notice that $y = \pm\dfrac{4}{3}\sqrt{x^2 - 9} = \pm\dfrac{4}{3}x\sqrt{1 - \dfrac{9}{x^2}}$.

When $|x|$ is very large, $\sqrt{1 - \dfrac{9}{x^2}}$ is very near 1, and therefore $y \approx \pm\dfrac{4}{3}x$. Thus as $|x|$ becomes large the graph approaches the lines

$$y = \frac{4}{3}x \quad \text{and} \quad y = -\frac{4}{3}x.$$

These lines are the *asymptotes* of the hyperbola $\dfrac{x^2}{3^2} - \dfrac{y^2}{4^2} = 1$. These asymptotes are also the diagonals of the rectangle formed by the lines $x = 3$, $x = -3$, $y = 4$, and $y = -4$. Asymptotes are useful guides in drawing a hyperbola, as you can see in the diagram below.

Using the facts above, you can construct a table of first-quadrant points and then draw the complete graph. A calculator may be helpful.

Since $\dfrac{x^2}{9} - \dfrac{y^2}{16} = 1$,

$$y = \pm\frac{4}{3}\sqrt{x^2 - 9}.$$

x	y
3	0
4	3.5
5	5.3
7	8.4

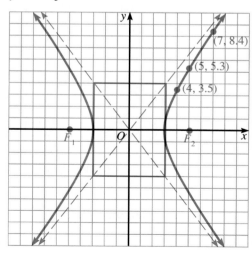

The graph is shown at the right. Notice that the graph consists of two *branches*.

The **center** of a hyperbola is the midpoint of the line segment joining its foci. A hyperbola with foci on the x-axis or y-axis and equidistant from the origin has a *quadratic equation in two variables* with one of the two forms shown below. Notice that in each case a^2 is the denominator of the *positive* squared term.

The hyperbola with center $(0, 0)$, foci $(-c, 0)$ and $(c, 0)$, and difference of focal radii $2a$ has the equation

$$\frac{x^2}{a^2} - \frac{y^2}{b^2} = 1, \text{ where } b^2 = c^2 - a^2.$$

The equations of the asymptotes are $y = \frac{b}{a}x$ and $y = -\frac{b}{a}x$.

If the foci are on the y-axis at $(0, -c)$ and $(0, c)$, then the equation is

$$\frac{y^2}{a^2} - \frac{x^2}{b^2} = 1, \text{ where } b^2 = c^2 - a^2.$$

In this case the equations of the asymptotes are $y = \frac{a}{b}x$ and $y = -\frac{a}{b}x$.

For a hyperbola having an equation with one of the forms shown above, the foci are either on the x-axis or the y-axis. If the x^2-term is positive, the line containing the center and foci is horizontal. If the y^2-term is positive, the line containing the center and foci is vertical.

The denominator of the positive squared term is a^2. The difference of the focal radii is $2a$. If you know the values of a and b and the coordinates of the center, you can find the coordinates of the foci by solving $b^2 = c^2 - a^2$ for c.

Example 1 Graph the hyperbola $y^2 = 3x^2 + 12$, showing its asymptotes as dashed lines. Find its foci.

Solution
$$y^2 = 3x^2 + 12$$
$$y^2 - 3x^2 = 12$$
$$\frac{y^2}{12} - \frac{x^2}{4} = 1$$

The y-intercepts are $\pm\sqrt{12} \approx 3.5$. There are no x-intercepts.
Determine the extent of the graph:
$$y = \pm\sqrt{3x^2 + 12} \quad \text{and} \quad x = \pm\frac{\sqrt{3}}{3}\sqrt{y^2 - 12}$$

$\therefore$ *x is real if and only if* $|y| \geq \sqrt{12}$.
The asymptotes have equations
$$y = \pm\frac{\sqrt{12}}{\sqrt{4}}x, \text{ or } y = \sqrt{3}x \text{ and } y = -\sqrt{3}x.$$

Make a short table of first-quadrant points and sketch the complete graph using symmetry.

$$y = \pm\sqrt{3x^2 + 12}$$

x	y
0	3.5
1	3.9
2	4.9
3	6.2

The graph is shown at the right.
Since $c^2 = a^2 + b^2 = 12 + 4 = 16$, $c = 4$.
∴ the foci are $F_1(0, -4)$ and $F_2(0, 4)$. *Answer*

Example 2 Find an equation of the hyperbola having foci at $(3, 0)$ and $(-3, 0)$ and difference of focal radii equal to 4.

Solution The center of the hyperbola is $(0, 0)$. The distance from each focus to the center is 3, so $c = 3$. The difference of the focal radii is $2a = 4$, and so $a = 2$. Since $b^2 = c^2 - a^2$, $b^2 = 3^2 - 2^2 = 5$. Since the foci are on the x-axis, an equation is $\dfrac{x^2}{4} - \dfrac{y^2}{5} = 1$. *Answer*

Example 3 Find an equation of the hyperbola with asymptotes $y = \dfrac{3}{4}x$ and $y = -\dfrac{3}{4}x$ and foci $(5, 0)$ and $(-5, 0)$.

Solution The center of the hyperbola is $(0, 0)$. The x^2-term is positive because the foci are on the x-axis. Since the equations of the asymptotes are $y = \pm\dfrac{3}{4}x$, $b = 3$ and $a = 4$. Thus $b^2 = 9$ and $a^2 = 16$. An equation of the hyperbola is $\dfrac{x^2}{16} - \dfrac{y^2}{9} = 1$. *Answer*

In Lesson 8-2, you learned that an inverse variation has an equation of the form

$$xy = k \quad (k \neq 0).$$

It can be shown that the graph of such an equation is a hyperbola (see Exercises 34 and 35 on page 431). The coordinate axes are the asymptotes. If $k > 0$, the branches of the hyperbola are in the first and third quadrants. If $k < 0$, the branches are in the second and fourth quadrants. The graph of $xy = 9$ is shown at the right.

Oral Exercises

Give the intercepts of the hyperbola and the equations of its asymptotes.
Tell on which of the coordinate axes its foci lie.

1. $\dfrac{x^2}{25} - \dfrac{y^2}{16} = 1$
2. $\dfrac{y^2}{1} - \dfrac{x^2}{9} = 1$
3. $x^2 - 25y^2 + 25 = 0$

4. $4x^2 - y^2 = 16$
5. $25x^2 - 4y^2 = 100$
6. $4x^2 - 9y^2 + 36 = 0$

Describe the graph of each equation.

7. $xy = k$, when $k = 0$
8. $y = \dfrac{1}{x}$
9. $\dfrac{x^2}{a^2} - \dfrac{y^2}{b^2} = 0$

Written Exercises

In Exercises 1–12, (a) graph each hyperbola, showing its asymptotes as
dashed lines; (b) find the coordinates of the foci. You may wish to check
your graphs on a computer or a graphing calculator.

A 1–6. Use the equations in Oral Exercises 1–6.

7. $x^2 = 9y^2 - 81$
8. $y^2 = 5x^2 + 25$
9. $75x^2 - 100y^2 = 7500$

10. $25x^2 - 144y^2 = 3600$
11. $4x^2 - y^2 + 1 = 0$
12. $16x^2 - 4y^2 + 64 = 0$

Find an equation of the hyperbola described.

13. Foci $(0, -8)$ and $(0, 8)$; difference of focal radii 10.

14. Foci $(-4, 0)$ and $(4, 0)$; difference of focal radii 4.

15. Asymptotes $y = \dfrac{3}{2}x$ and $y = -\dfrac{3}{2}x$; foci $(0, -\sqrt{13})$ and $(0, \sqrt{13})$.

16. Asymptotes $y = \dfrac{\sqrt{2}}{2}x$ and $y = -\dfrac{\sqrt{2}}{2}x$; foci $(0, -\sqrt{6})$ and $(0, \sqrt{6})$.

B **17.** Asymptotes $y = 3x$ and $y = -3x$; y-intercepts 3 and -3.

18. Asymptotes $y = x$ and $y = -x$; foci $(-4, 0)$ and $(4, 0)$.

Graph each inequality.

19. $y^2 - x^2 > 4$
20. $y^2 \le x^2 - 4$
21. $4x^2 \le y^2 + 16$
22. $9x^2 > 4y^2 - 36$

The hyperbolas $\dfrac{x^2}{p^2} - \dfrac{y^2}{q^2} = 1$ and $\dfrac{y^2}{q^2} - \dfrac{x^2}{p^2} = 1$ are *conjugates* of each other.
Graph the following conjugate hyperbolas on the same coordinate axes.

23. $\dfrac{x^2}{9} - \dfrac{y^2}{4} = 1$; $\dfrac{y^2}{4} - \dfrac{x^2}{9} = 1$
24. $\dfrac{x^2}{1} - \dfrac{y^2}{4} = 1$; $\dfrac{y^2}{4} - \dfrac{x^2}{1} = 1$

Graph each equation. Each graph is half of a hyperbola, since $\sqrt{a}$ is nonnegative.

25. $y = \sqrt{x^2 + 16}$

26. $y = \sqrt{x^2 - 1}$

27. $y = \sqrt{x^2 - 16}$

28. $y = \sqrt{x^2 + 1}$

29. The statement "traveling 200 miles at x mi/h for y hours" can be described by the equation $xy = 200$. Consider the restrictions on x and y and then graph this equation. What does the graph tell you about the relationship between x and y?

Use the definition to find an equation of the hyperbola having the given points as foci and the given number as difference of focal radii.

30. $(0, -5)$, $(0, 5)$; 4

31. $(-3, 0)$, $(3, 0)$; 2

C 32. $(-c, 0)$, $(c, 0)$; $2a$

33. $(0, -c)$, $(0, c)$; $2a$

34. (a, a), $(-a, -a)$; $2a$

35. $(-a, a)$, $(a, -a)$; $2a$

Mixed Review Exercises

Graph each equation.

1. $x^2 + 4y^2 = 16$

2. $x^2 + y^2 - 2x + 4y + 1 = 0$

3. $3x - 4y = 6$

4. $2x^2 - 4x + y + 5 = 0$

Draw the graph of each relation and tell whether or not it is a function.

5. $\{(x, y): y = 2x\}$

6. $\{(x, y): x = 2y\}$

7. $\{(x, y): x = y^2\}$

8. $\{(x, y): y = x^2\}$

9. $\{(x, y): y = |x|\}$

10. $\{(x, y): x = |y|\}$

Historical Note / The Area of a Parabolic Section

About 240 B.C. the great Greek mathematician Archimedes discovered a formula for the area of any section of a parabola. By summing the areas of a sequence of smaller and smaller triangles that fill the section more and more completely, Archimedes was able to show that the area of any section is $\frac{2}{3}bh$, where b is the length of the base (the chord that cuts off the section) and h is the height (the length of a perpendicular segment to the base from the point on the parabola where a tangent parallel to the base touches the parabola). Therefore, the area of the section is $\frac{4}{3}$ as large as the area of the largest triangle that can be inscribed in it.

9-6 More on Central Conics

Objective   To find an equation of a conic section with center not at the origin and to identify a conic as a circle, ellipse, or hyperbola.

Circles, ellipses, and hyperbolas are called **central conics** because they have centers. Now you'll study central conics with centers not at the origin.

As you learned in Lesson 9-2, you can translate a graph centered at the origin by sliding every point of the graph h units horizontally and k units vertically. Replacing x by $x - h$ and y by $y - k$ in the equation of a central conic with center at the origin gives the equation of the same conic with center now at (h, k). Using this fact, the formulas of Lessons 9-4 and 9-5 can be rewritten in the following more general forms. The constants a and b have the same meaning here as in those lessons.

Ellipses with Center (h, k)	*Hyperbolas with Center (h, k)*
Horizontal major axis:	*Horizontal major axis:*
$$\frac{(x - h)^2}{a^2} + \frac{(y - k)^2}{b^2} = 1$$	$$\frac{(x - h)^2}{a^2} - \frac{(y - k)^2}{b^2} = 1$$
Foci at $(h - c, k)$ and $(h + c, k)$, where $c^2 = a^2 - b^2$.	Foci at $(h - c, k)$ and $(h + c, k)$, where $c^2 = a^2 + b^2$.
Vertical major axis:	*Vertical major axis:*
$$\frac{(x - h)^2}{b^2} + \frac{(y - k)^2}{a^2} = 1$$	$$\frac{(y - k)^2}{a^2} - \frac{(x - h)^2}{b^2} = 1$$
Foci at $(h, k - c)$ and $(h, k + c)$, where $c^2 = a^2 - b^2$.	Foci at $(h, k - c)$ and $(h, k + c)$, where $c^2 = a^2 + b^2$.

Example 1   Find an equation of the ellipse having foci $(-3, 4)$ and $(9, 4)$ and sum of focal radii 14.

Solution   The sum of the focal radii, $2a$, is 14. So $a = 7$. The center is halfway between the foci, at $(3, 4)$. The distance from the center to each focus is 6, so $c = 6$. Substituting 7 for a and 6 for c in the equation $b^2 = a^2 - c^2$ (from Lesson 9-4) gives $b^2 = 49 - 36 = 13$.

∴ an equation of the ellipse is $\dfrac{(x - 3)^2}{49} + \dfrac{(y - 4)^2}{13} = 1$.   ***Answer***

Example 2   Find an equation of the hyperbola having foci $(-3, -2)$ and $(-3, 8)$ and difference of focal radii 8.

Solution The difference of the focal radii is $8 = 2a$, so $a = 4$. The center is halfway between the foci, at $(-3, 3)$. The distance from the center to each focus is 5, so $c = 5$. Substituting 4 for a and 5 for c in the equation $b^2 = c^2 - a^2$ gives $b^2 = 25 - 16 = 9$. The y^2-term is positive because the line containing the foci is vertical.

∴ an equation of the hyperbola is $\dfrac{(y - 3)^2}{16} - \dfrac{(x + 3)^2}{9} = 1$.   *Answer*

When the equation of a central conic is given in the form
$$Ax^2 + By^2 + Cx + Dy + E = 0,$$
you can complete the square in x and y to identify the conic and find its center and its foci, as shown in Example 3.

Example 3 Identify the conic $x^2 - 4y^2 - 2x - 16y - 11 = 0$. Find its center and foci. Then draw its graph.

Solution Complete the square using the x-terms and then using the y-terms.
$$(x^2 - 2x + \underline{\ ?\ }) + (-4y^2 - 16y + \underline{\ ?\ }) = 11 + \underline{\ ?\ } + \underline{\ ?\ }$$
$$(x^2 - 2x + 1) - 4(y^2 + 4y + 4) = 11 + 1 - 16$$
$$(x - 1)^2 - 4(y + 2)^2 = -4$$

Divide both sides by -4.

$$\dfrac{(y + 2)^2}{1} - \dfrac{(x - 1)^2}{4} = 1$$

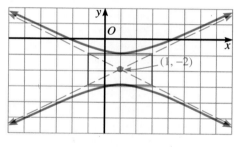

Comparing this equation with the general forms, you can see that the conic is a hyperbola with center $(1, -2)$. The line containing the center and foci is vertical. Since $a^2 = 1$ and $b^2 = 4$, $c^2 = a^2 + b^2 = 5$, and $c = \sqrt{5}$. Thus, the foci are the points $(1, -2 + \sqrt{5})$ and $(1, -2 - \sqrt{5})$, or approximately $(1, 0.2)$ and $(1, -4.2)$. The graph is shown above. *Answer*

If you apply the vertical-line test (see page 154) to the graph of the hyperbola in Example 3, you will see that the graph does *not* represent a function. *No central conic can be a function.*

Oral Exercises

Identify each conic and give its center and foci.

1. $\dfrac{(x + 3)^2}{16} + \dfrac{(y - 5)^2}{12} = 1$

2. $\dfrac{(x + 7)^2}{9} - \dfrac{(y + 1)^2}{16} = 1$

The given conic is to be translated so that its new center is at the given point. What will its new equation be?

3. $\dfrac{x^2}{25} + \dfrac{y^2}{4} = 1$; $(0, -5)$

4. $\dfrac{x^2}{16} - \dfrac{y^2}{1} = 1$; $(-5, 0)$

5. $x^2 - y^2 = 49$; $(-4, 3)$

6. $x^2 + y^2 = 9$; $(-1, 1)$

7. $4x^2 + y^2 = 16$; $(1, -4)$

8. $4x^2 - 9y^2 = 36$; $(3, -2)$

Match each equation with its graph.

9. $\dfrac{(x + 1)^2}{1} + \dfrac{(y - 2)^2}{4} = 1$

10. $\dfrac{(x + 1)^2}{1} - \dfrac{(y - 2)^2}{4} = 1$

11. $x + 1 = (y - 2)^2$

12. $(x + 1)^2 + (y - 2)^2 = 1$

13. $\dfrac{(y - 2)^2}{1} - \dfrac{(x + 1)^2}{4} = 1$

14. $\dfrac{(x + 1)^2}{4} + \dfrac{(y - 2)^2}{1} = 1$

(a)

(b)

(c)

(d)

(e)

(f)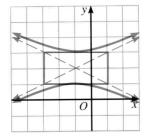

Written Exercises

Find an equation of the ellipse having the given foci and sum of focal radii.

A **1.** $(6, 0)$, $(6, 6)$; 10

2. $(0, 0)$, $(0, 8)$; 12

3. $(-3, -3)$, $(-3, 3)$; 8

4. $(-5, 1)$, $(3, 1)$; 16

5. $(-2, -3)$, $(6, -3)$; 10

6. $(-10, 2)$, $(-2, 2)$; 14

Find an equation of the hyperbola having the given foci and difference of focal radii.

7. $(0, -2)$, $(8, -2)$; 2

8. $(0, 4)$, $(0, 10)$; 4

9. $(3, -8)$, $(3, -2)$; 4

10. $(-5, 3)$, $(9, 3)$; 6

11. $(5, -9)$, $(5, -1)$; 6

12. $(-4, -4)$, $(4, -4)$; 6

Identify each conic. Find its center and its foci (if any). Then draw its graph. You may wish to check your graphs on a computer or graphing calculator.

13. $x^2 - 4y^2 - 2x - 24y - 39 = 0$

14. $x^2 + 9y^2 + 2x - 18y + 1 = 0$

15. $x^2 + y^2 - 6x - 16y + 57 = 0$

16. $9x^2 - y^2 - 18x - 6y - 9 = 0$

17. $9x^2 + 25y^2 + 36x - 150y + 36 = 0$

18. $16x^2 - 9y^2 + 64x + 18y + 199 = 0$

B   19. Use the definition of an ellipse to find an equation of the ellipse having foci $(1, 1)$ and $(-1, -1)$ and sum of focal radii 3.

  20. Use the definition of a hyperbola to find an equation of the hyperbola having foci $(-1, 1)$ and $(1, -1)$ and difference of focal radii 2.

C   21. Every conic section has an equation of the form

$$Ax^2 + By^2 + Cx + Dy + E = 0$$

where A and B are not both zero. Let $A = 1$, $C = 2$, $D = -8$, and $E = 1$. Graph the resulting quadratic equation in two variables for each given value of B. Then identify the graph.

  a. $B = 0$ **b.** $B = 1$ **c.** $B = 4$ **d.** $B = -4$

  e. Analyze the different equations you graphed in parts (a)–(d). What is the relationship between the coefficients A and B for which the general equation gives a circle? a parabola? an ellipse? a hyperbola?

Self-Test 2

Vocabulary   ellipse (p. 418)
               focus (of an ellipse)(p. 418)
               focal radii (of an ellipse) (p. 418)
               symmetric about the x-axis (p. 419)
               symmetric about the y-axis (p. 419)
               symmetric about the origin (p. 419)

            center (of an ellipse) (p. 419)
            major axis (p. 419)
            minor axis (p. 419)
            hyperbola (p. 426)
            focal radii (of a hyperbola) (p. 426)
            asymptotes (p. 427)
            central conic (p. 432)

1. Graph $16x^2 + 9y^2 = 144$.   **Obj. 9-4, p. 418**

2. Find an equation of the ellipse having foci $(3, 0)$ and $(-3, 0)$ and sum of focal radii 8.

3. Graph $25y^2 - x^2 = 25$, showing the asymptotes as dashed lines.   **Obj. 9-5, p. 426**

4. Find an equation of the hyperbola having foci $(0, 4)$ and $(0, -4)$ and difference of focal radii 4.

5. Identify and graph $4x^2 - 25y^2 - 24x + 50y - 89 = 0$.   **Obj. 9-6, p. 432**

6. Find an equation of the ellipse having foci $(5, 2)$ and $(-5, 2)$ and sum of focal radii 12.

Check your answers with those at the back of the book.

Systems of Equations

9-7 The Geometry of Quadratic Systems

Objective   To use graphs to determine the number of real solutions of a quadratic system and to estimate the solutions.

You know that a solution of a linear system can be found geometrically by locating the point where the graphs of the system's equations intersect. You can also use graphs to determine the real solutions of a *quadratic system*. A system containing only quadratic equations or a combination of linear and quadratic equations in the same two variables is called a **quadratic system.**

    If you have a computer or a graphing calculator, you may wish to explore the geometry of some quadratic systems. For example, a system of one quadratic and one linear equation may have 2, 1, or 0 real solutions, as illustrated below.

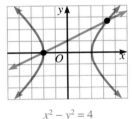

$$x^2 - y^2 = 4$$
$$x - 2y + 2 = 0$$
Two solutions

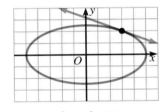

$$x^2 + 4y^2 = 25$$
$$3x + 8y = 25$$
One solution

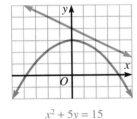

$$x^2 + 5y = 15$$
$$x + 2y = 8$$
No solution

    A system of two quadratic equations may have 4, 3, 2, 1, or 0 real solutions. For example, a circle and an ellipse can have four points of intersection and their equations would have four common real solutions (see Oral Exercise 9).

    Finding the points of intersection of quadratic graphs can be useful in astronomy. Like planetary orbits, the orbits of many comets are elliptical. To observe a comet closely, astronomers are naturally interested in where its orbit crosses that of the earth's.

    You can use graphs not only to find the number of real solutions of a quadratic system, but also to estimate these solutions. Example 1 at the top of the next page illustrates.

Graph each system and estimate the real solutions.

a. $25x^2 + 4y^2 = 100$ **b.** $x^2 + 6y = 30$
  $5x = y^2 - 15$ $xy + 6 = 0$

Graph the equations of each system and estimate the coordinates of their points of intersection.

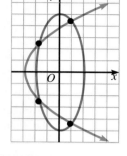

a. $25x^2 + 4y^2 = 100$
  $5x = y^2 - 15$

There are four real solutions. Estimates of the solutions are $(0.9, 4.4)$, $(0.9, -4.4)$, $(-1.7, 2.5)$, and $(-1.7, -2.5)$.

b. $x^2 + 6y = 30$
  $xy + 6 = 0$

There are three real solutions. Estimates of the solutions are $(6, -1)$, $(-1.3, 4.7)$, and $(-4.7, 1.3)$.

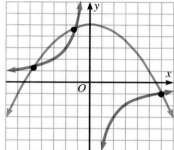

Actually, $(6, -1)$ is an exact solution of the system in part (b) of the example, as the following check shows.

$(6)^2 + 6(-1) = 30$ ✓
$(6)(-1) + 6 = 0$ ✓

The other solutions in the example are approximations.
If you have a calculator, you might check to see how accurate they are.

Oral Exercises

Mentally sketch the graph of each equation and give the *number* of real solutions the system has.

1. $x^2 + y^2 = 4$ **2.** $x^2 + y^2 = 4$ **3.** $x^2 + y^2 = 4$ **4.** $y = x^2$
  $y = x$ $x + y = 8$ $y = 2$ $y = 2x$

5. $x^2 + y^2 = 1$ **6.** $x^2 - y^2 = 1$ **7.** $x^2 - y^2 = 4$ **8.** $x^2 + y^2 = 9$
  $y = x^2$ $x^2 + y^2 = 9$ $x + y = 2$ $\dfrac{x^2}{3} + \dfrac{y^2}{4} = 1$

9. By drawing rough sketches, illustrate that a circle and an ellipse can have the following possible points of intersection.
  a. four   **b.** three   **c.** two   **d.** one   **e.** zero

Can a line intersect a conic of the specified type in a single point without being tangent to it? If so, explain how.

10. A parabola   **11.** A hyperbola   **12.** An ellipse

Written Exercises

By sketching graphs, find the *number* of real solutions the system has.

1. $4x^2 + 9y^2 = 36$
   $2x + 3y = 6$

2. $16x^2 + 4y^2 = 64$
   $3x + 4y = 12$

3. $x^2 - 9y = 0$
   $x - 2y = 2$

4. $4x^2 + 25y^2 = 100$
   $x^2 + y^2 = 64$

5. $x^2 - y^2 = 4$
   $x + y = 4$

6. $y^2 - x^2 = 4$
   $y = x^2 - 3$

7. $x^2 + y^2 = 25$
   $xy = 10$

8. $9x^2 + 25y^2 = 225$
   $y^2 = 2x + 4$

9. $x^2 - 4y^2 = 4$
   $6y^2 - x = 2$

10–15. Estimate the real solutions (if any) of the systems of Exercises 4–9 to the nearest half unit.

Graph each system of inequalities.

Sample	$x^2 + 4y^2 \leq 4$	**Solution**
	$x^2 \geq y^2 + 1$	

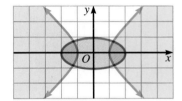

B **16.** $y \geq x + 3$
   $y \leq x^2$

17. $x^2 + 4y^2 \leq 16$
   $x^2 \leq y^2 + 4$

18. $4x^2 + y^2 \leq 16$
   $x^2 + 4y^2 \leq 16$

19. $x \geq -y^2$
   $x^2 + y^2 \leq 49$

C **20.** Use parts (a), (b), and (c) below to show that no isosceles triangle has perimeter 4 and area 1.

   a. Write equations for the area and perimeter.
   (*Hint:* Let base $= 2x$ and height $= y$.)
   b. Simplify the perimeter equation to eliminate radicals.
   c. Graph the resulting system and use the graph to reach the required conclusion.

Mixed Review Exercises

Find an equation for each figure described.

1. Parabola with focus (2, 3) and directrix $y = -1$.
2. Ellipse with x-intercepts $\pm \sqrt{6}$ and y-intercepts ± 2.
3. Hyperbola with foci (0, -3) and (0, 3) and difference of focal radii 4.
4. The circle having center (-3, 4) and passing through the origin.

Find the unique solution for each system.

5. $y = 3x + 2$
   $2x - y = -3$

6. $x - 3y = 5$
   $4x + y = 7$

7. $5x + 2y = 9$
   $4x - 3y = 21$

9-8 Solving Quadratic Systems

Objective To use algebraic methods to find exact solutions of quadratic systems.

The substitution and linear-combination methods that you used in Chapter 3 to solve linear systems can also be used to solve quadratic systems. Although it is possible for the solutions of quadratic systems to be complex numbers, in this lesson we will consider only real solutions.

Example 1 Solve this system: $4x^2 + y^2 = 25$
$$2x + y = -1$$

Solution (Substitution Method) Solve the linear equation for one of the variables. If you solve for y, you can avoid fractions.
$$y = -2x - 1$$

Substitute $-2x - 1$ for y in the quadratic equation and solve the resulting equation.

$$4x^2 + (-2x - 1)^2 = 25$$
$$4x^2 + 4x^2 + 4x + 1 = 25$$
$$8x^2 + 4x - 24 = 0$$
$$2x^2 + x - 6 = 0$$
$$(2x - 3)(x + 2) = 0$$
$$x = \frac{3}{2} \quad \text{or} \quad x = -2$$

Substitute $\frac{3}{2}$ and -2 for x in $y = -2x - 1$ to find the y-values.

$$y = -2\left(\frac{3}{2}\right) - 1 = -4$$
$$\therefore \left(\frac{3}{2}, -4\right) \text{ is a solution.}$$

$$y = -2(-2) - 1 = 3$$
$$\therefore (-2, 3) \text{ is a solution.}$$

Checking the ordered pairs $\left(\frac{3}{2}, -4\right)$ and $(-2, 3)$ in both *given* equations is left for you.

The solution set is $\left\{\left(\frac{3}{2}, -4\right), (-2, 3)\right\}$.

Answer

The equations are graphed in the diagram at the right. As you can see, the graphs have two points of intersection.

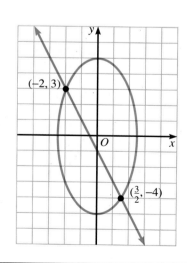

Example 2 Solve this system: $x^2 - y^2 = 12$
$$xy = 8$$

Solution (Substitution Method) **Solve the second equation for** y.

$$y = \frac{8}{x} \qquad (x \neq 0)$$

Substitute $\frac{8}{x}$ for y in the first equation and solve the resulting equation.

$$x^2 - \left(\frac{8}{x}\right)^2 = 12$$

$$x^2 - \frac{64}{x^2} = 12$$

$$x^4 - 64 = 12x^2$$

$$x^4 - 12x^2 - 64 = 0 \qquad \text{This equation is quadratic in } x^2.$$

$$(x^2 - 16)(x^2 + 4) = 0$$

$$x^2 - 16 = 0 \qquad x^2 + 4 \text{ has no real solutions.}$$

$$x = 4 \quad \text{or} \quad x = -4$$

Substitute the x-values in $y = \frac{8}{x}$

If $x = 4$, $y = \frac{8}{4} = 2$.

If $x = -4$, $y = \frac{8}{-4} = -2$.

The solution set is $\{(4, 2), (-4, -2)\}$.

Answer

The graph is shown at the right.

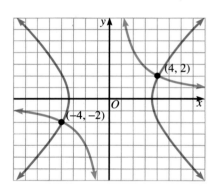

Example 3 Solve this system: $x^2 + 2y^2 = 23$
$$2x^2 - y^2 = 1$$

Solution (Linear-Combination Method) **Multiply the second equation by 2 and add the two equations. Then solve the resulting equation.**

$$x^2 + 2y^2 = 23$$

$$4x^2 - 2y^2 = 2$$

$$5x^2 = 25$$

$$x^2 = 5$$

$$x = \sqrt{5} \qquad \text{or} \qquad x = -\sqrt{5}$$

Substitute $\sqrt{5}$ and $-\sqrt{5}$ for x in $x^2 + 2y^2 = 23$ to find the corresponding values of y.

	If $x = \sqrt{5}$:	If $x = -\sqrt{5}$:

$$\begin{aligned}(\sqrt{5})^2 + 2y^2 &= 23\\ 2y^2 &= 18\\ y^2 &= 9\\ y = 3 \quad \text{or} \quad y &= -3\end{aligned}$$

$$\begin{aligned}(-\sqrt{5})^2 + 2y^2 &= 23\\ 2y^2 &= 18\\ y^2 &= 9\\ y = 3 \quad \text{or} \quad y &= -3\end{aligned}$$

$\therefore$ the solution set is $\{(\sqrt{5}, 3), (\sqrt{5}, -3), (-\sqrt{5}, 3), (-\sqrt{5}, -3)\}$.

As the preceding examples show, substitution is usually the more appropriate method for solving a system consisting of a linear and a quadratic equation. When a system's equations are both quadratic, either the substitution or the linear-combination method may be used.

Oral Exercises

Which of the given ordered pairs are solutions of the system?

1. $xy + 5 = 0$
$2x + y + 3 = 0$
    $(5, -1), (-5, 1), (1, -5), (-1, 5)$

2. $x^2 + y^2 = 20$
$x + y + 2 = 0$
    $(4, 2), (-4, 2), (2, 4), (-2, 4)$

3. $x^2 + 4y^2 = 25$
$y^2 - x = 1$
    $(3, 2), (-3, 2), (-3, -2), (3, -2)$

Written Exercises

Find the real solutions, if any, of each system. You may wish to check your answers visually on a computer or a graphing calculator.

A

1. $x^2 - y = 5$
$2x + y = 3$

2. $x = y^2 - 9$
$x - 4y - 12 = 0$

3. $y = x^2$
$x^2 + y^2 = 12$

4. $y^2 = 2x$
$x^2 + y^2 = 8$

5. $x^2 - y^2 = 15$
$x + y = 1$

6. $x^2 + y^2 = 61$
$x - 2y + 7 = 0$

7. $xy = 8$
$x + y = 6$

8. $xy + 6 = 0$
$x - y = 5$

9. $2y^2 + 3x = 33$
$x + 4y + 7 = 0$

10. $4x^2 - y^2 + 12 = 0$
$x + y = 3$

11. $x^2 + y^2 = 25$
$x^2 - y^2 = 7$

12. $8x^2 + y^2 = 25$
$8x^2 - y^2 = 39$

B

13. $2x^2 - 3y^2 = 30$
$x^2 + y^2 = 25$

14. $x^2 + 2y^2 = 12$
$3x^2 - y^2 = 8$

15. $5x^2 + 3y^2 = 7$
$3x^2 - 7y^2 = 13$

16. $9x^2 + 9y^2 = 1$
$x = y^2 + 1$

17. $x^2 + y^2 = 13$
$xy + 6 = 0$

18. $2x^2 - y^2 = 7$
$xy = 3$

Exercises 19 and 20 lead to cubic equations. Use the methods of Lesson 8-7 to find their real roots.

19. $y^2 = x + 7$
    $xy = 6$

20. $y = x^2 - 1$
    $xy + 6 = 0$

Find the square roots of each complex number.

Sample $12 + 16i$

Solution   Let a square root of $12 + 16i$ be $x + yi$, where x and y are real. Then $(x + yi)^2 = 12 + 16i$, or $(x^2 - y^2) + 2xyi = 12 + 16i$. Equating the real and imaginary parts gives this system:

$$x^2 - y^2 = 12$$
$$2xy = 16$$

This system is equivalent to the one in Example 2, page 440, and thus has the solutions $(4, 2)$ and $(-4, -2)$. Therefore, the square roots of $12 + 16i$ are $4 + 2i$ and $-4 - 2i$.

C   21. $3 + 4i$

22. $7 - 24i$

23. $5 + 12i$

24. $-5 - 12i$

25. $-7 + 24i$

26. $12 - 16i$

Problems

Solve.

A   1. The sum of two numbers is 16, and the sum of their squares is 146. Find the numbers.

2. The product of two numbers is 1, and the difference of their squares is $\frac{15}{4}$. Find the numbers.

3. The fence around a rectangular piece of property is 156 m long. If the area of the property is 1505 m^2, find the dimensions of the property.

4. An ellipse with center at the origin and horizontal major axis is to fit snugly inside a rectangle that has its longer sides horizontal. The area of the rectangle is 12 square units, and the perimeter is 14 units. Find an equation for the ellipse.

5. Find the dimensions of a rectangle having perimeter 34 ft and a diagonal of length 13 ft.

6. Find the length of the legs of a right triangle having perimeter 56 m if the hypotenuse is 25 m.

7. The product of a two-digit number and its tens digit is 285. The units digit is two more than the tens digit. Find the original number.

8. Find the dimensions of a rectangle that has area 10 and a diagonal of length 5. Leave your answer in terms of radicals.

B **9.** A rectangular plot of land having area 1350 m² is to be enclosed and divided into two parts, as shown. Find the dimensions of the plot if the total length of fencing used is 180 m.

Ex. 9

Ex. 10

10. Four squares, each with sides 4 cm long, are cut from the corners of a rectangular piece of cardboard having area 560 cm². The flaps are then bent up to form an open-topped box having volume 960 cm³. Find the dimensions of the original piece of cardboard.

11. Two people part company and walk along perpendicular paths. One person walks 1 km/h faster than the other. They are 6 km apart after one hour. Find the rate at which each person walks. Give your answers to the nearest tenth.

12. A 20 m ladder and a 15 m ladder were leaned against a building. The bottom of the longer ladder was 7 m farther from the building than the bottom of the shorter ladder, but both ladders reached the same distance up the building. Find the distance.

C **13.** Find the point on the circle with equation $x^2 + y^2 = 1$ that is closest to the point (4, 3).

14. From the top of a vertical canyon wall 300 m high, a person throws a rock with a speed of 20 m/s toward the top of the opposite side of the canyon, which is at the same elevation, but is 510 m away. According to the laws of physics, the rock follows a path described approximately by the equation

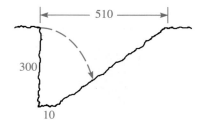

$$y = 300 - \frac{x^2}{80}$$

where y is the rock's height in meters above the canyon floor and x is the rock's horizontal distance in meters to the right of the vertical canyon wall. The opposite side of the canyon is steeply sloping and becomes level at a point 10 m from the base of the vertical wall. When the rock strikes the sloping side of the canyon, how far is it from where it was thrown? Express your answer rounded to the nearest tenth of a meter. (*Hint:* Introduce a coordinate system whose origin is the base of the vertical wall. Write an equation for the sloping side, and find a common solution of this equation and the equation of the rock's path.)

9-9 Systems of Linear Equations in Three Variables

Objective   To solve systems of linear equations in three variables.

Just as the ordered pair $(x, y) = (2, -4)$ is a solution of the equation $3x + 2y = -2$, so the **ordered triple** $(x, y, z) = (2, -4, 3)$ is a solution of the equation

$$3x + 2y - z = -5$$

because $3(2) + 2(-4) - 3 = -5$ is a true statement.

Any equation of the form $Ax + By + Cz = D$, where A, B, C, and D are real numbers with A, B, and C not all zero, is called a **linear equation in three variables.** Although such equations are called "linear," their graphs are planes, not lines.

Solutions of equations in x, y, and z can be graphed in a three-dimensional coordinate system having an x-axis, a y-axis, and a z-axis that meet at right angles at a common point O, the origin. The graph of the *ordered triple* $(2, -4, 3)$ is shown in the diagram above.

To graph a *system* of three linear equations in three variables, you draw the three graphs in the same coordinate system. Although any two of the planes may intersect, coincide, or be parallel, the graph of the system's solution consists only of the points that *all three* planes have in common. Some of the ways that three planes can intersect are shown below.

(a)

(b)

(c)

(d)

Algebraically, these geometrical relationships mean:

(a) When the three planes intersect in only *one* point, the system has *one solution*.

(b)–(c) When the three planes do *not* contain a common point, the system has *no solution*. (This is also true when the three planes are parallel.)

(d) When the three planes intersect in a common line, the system has *infinitely many solutions*. (This is also true when the three planes coincide.)

You can solve linear systems algebraically by substitution or linear combinations, as you did in Lesson 3-5.

Example 1 Solve this system:
$$2x + 3y + 2z = 13$$
$$2y + z = 1$$
$$z = 3$$

Solution (Substitution Method) The third equation gives the value of z: $z = 3$. Substitute 3 for z in the second equation and solve for y.

$$2y + z = 1$$
$$2y + 3 = 1$$
$$2y = -2$$
$$y = -1$$

Substitute 3 for z and -1 for y in the first equation and solve for x.

$$2x + 3y + 2z = 13$$
$$2x + 3(-1) + 2(3) = 13$$
$$2x - 3 + 6 = 13$$
$$2x = 10$$
$$x = 5$$

The check is left for you.

∴ the solution of the system is $(5, -1, 3)$. **Answer**

Because its shape suggests a triangle, the system in Example 1 is called a **triangular system.** Since triangular systems are easy to solve, a good way to solve any linear system is to transform it into an equivalent triangular system and then to proceed as in Example 1. This method is called *Gaussian elimination* in honor of Karl Friedrich Gauss (1777–1855), a very famous mathematician.

$$2x + 3y + 2z = 13$$
$$2y + z = 1$$
$$z = 3$$

Example 2 Solve this system:
$$\left. \begin{array}{l} x + y - 2z = 7 \\ -x + 4y + 3z = 2 \\ 2x - 3y + 2z = -2 \end{array} \right\} \quad (1)$$

Solution (Triangular-System Method)

In (1), add the first equation to the second. Then multiply the first equation by -2 and add the result to the third equation.

$$\left. \begin{aligned} x + y - 2z &= 7 \\ 5y + z &= 9 \\ -5y + 6z &= -16 \end{aligned} \right\} \quad (2)$$

In (2), add the second and third equations together.

$$\left. \begin{aligned} x + y - 2z &= 7 \\ 5y + z &= 9 \\ 7z &= -7 \end{aligned} \right\} \quad (3)$$

When you solve the triangular system in (3) you get $x = 3$, $y = 2$, and $z = -1$. The solution is $(3, 2, -1)$. **Answer**

The system in Example 2 has a unique solution. Example 3 shows a system that has infinitely many solutions.

Example 3 Solve the system:
$$\left. \begin{aligned} x - 2y - z &= -1 \\ 2x - y + z &= 1 \\ x + 4y + 5z &= 5 \end{aligned} \right\} \quad (1)$$

Solution (Triangular-System Method)

In (1), multiply the first equation by -2 and add it to the second equation. Then subtract the third equation from the first. Here is the resulting system:

$$\left. \begin{aligned} x + 2y + 2z &= +1 \\ 3y + 3z &= 3 \\ -6y - 6z &= -6 \end{aligned} \right\} \quad (2)$$

In (2), add twice the second equation to the third.

$$\left. \begin{aligned} x - 2y - z &= -1 \\ 3y + 3z &= 3 \\ 0 &= 0 \end{aligned} \right\} \quad (3)$$

Solve $3y + 3z = 3$ for y and substitute for y in the first equation in (3).

$$3y + 3z = 3$$
$$y + z = 1$$
$$y = 1 - z$$
$$x - 2(1 - z) - z = -1$$
$$x - 2 + 2z - z = -1$$
$$x = 1 - z$$

You can check that, for any number z, the triple $(1 - z, 1 - z, z)$ is a solution of the given system. If, in this general expression, you assign z the values 0, 2, and -3, you obtain the particular solutions $(1, 1, 0)$, $(-1, -1, 2)$, and $(4, 4, -3)$, respectively. **Answer**

Recall from Chapter 3 that a linear system having infinitely many solutions, such as the one in Example 3, is said to be *dependent*. If the second equation in Example 3 had been $x - 2y - z = 2$, then the system would have had no solution, since the first two equations would have had no common solution. Such a system is said to be *inconsistent*.

Other methods can be used to solve systems of equations. One such method involves using numbers in a rectangular array called a **matrix.** You will learn about matrices in Chapter 16.

Oral Exercises

Which of the ordered triples, if any, is a solution of the equation?

1. $x + y - z = 7$; $(4, -2, 5)$, $(3, 0, -4)$

2. $2x + y - z = 0$; $(-1, 2, 3)$, $(0, 2, 2)$

3. $4x - y + 2z = 8$; $(1, 2, -3)$, $(2, -3, -1)$

4. $x - 3y + 4z = 6$; $(-2, 0, 2)$, $(2, -2, 0)$

Find two ordered triples that satisfy the equation.

5. $x + y + z = 7$ **6.** $x + y + z = -5$ **7.** $y + z = 6$ **8.** $3x + z = 0$

Tell how to combine the given equations to obtain an equation in which the coefficient of x is 0. What is the new coefficient of y?

9. $x + 3y - 2z = 5$
    $-x + y + 4z = 6$

10. $x + 2y - z = 2$
     $2x + 5y - 4z = 20$

11. $3x + y - 2z = 6$
     $2x + 2y + z = 4$

12. $7x - y + 4z = 3$
     $-2x + 3y - z = 5$

Written Exercises

Solve each system.

A   **1.** $x + y - 3z = 10$
       $y + z = 12$
          $z = -2$

2. $x + 2y + 3z = -1$
     $3y + 2z = -1$
         $z = -2$

3. $x + 2y - z = 3$
     $3y + z = -10$
       $-2z = 8$

4. $5x - 3y + 4z = 4$
     $4y - 3z = 10$
        $3z = -6$

5. $2z + 3y - x = 0$
     $4y - x = -4$
      $-2x = 8$

6. $z + 3x + y = 3$
     $2x + 3y = 10$
        $2y = 8$

7. $2x - y - z = 2$
     $x - 5y + 3z = -13$
   $-2x - 2y + z = -2$

8. $x + 3y + 2z = -1$
   $-3x - 2y + z = 3$
    $2x - y + 3z = -8$

9. $2x + y + 3z = 10$
     $x - 2y + z = 10$
  $-4x + 3y + 2z = 5$

10. $2x + 5y + 2z = -5$
   $-3x + 3y + 5z = 2$
    $x + 4y - z = 3$

11. $2x + 3y = 6 + z$
    $x - 2y = -1 - z$
   $3x + y = -1 + 3z$

12. $2x - y = 3z - 3$
    $3x + 2y = z - 1$
    $x + 3y = z - 10$

B **13.** $3a - 2b + 2c = 1$
$2a + 5b - 5c = 7$
$4a - 3b + c = -3$

14. $2x + 2y + z = 3$
$3x + 2y - 2z = -1$
$5x - 2y - 6z = 17$

15. $3u + 2v + w = 4$
$5u + 3v - w = -2$
$2u + w = 1$

16. $a + 3b + 4c = 6$
$a - 2b + c = 10$
$2a + 3b - c = 0$

17. $\dfrac{1}{x} - \dfrac{2}{y} + \dfrac{3}{z} = -3$
$\dfrac{2}{x} - \dfrac{3}{y} - \dfrac{1}{z} = 7$
$\dfrac{3}{x} + \dfrac{1}{y} - \dfrac{2}{z} = 6$

18. $\dfrac{1}{r} + \dfrac{3}{s} - \dfrac{2}{t} = 1$
$\dfrac{2}{r} + \dfrac{3}{s} - \dfrac{4}{t} = 1$
$\dfrac{1}{r} - \dfrac{6}{s} - \dfrac{6}{t} = 0$

Show that each of the following systems is dependent by (a) finding a general expression for all of its solutions and (b) using this expression to find three particular solutions.

19. $x - 2y + 3z = 1$
$x + y - 3z = 7$
$3x - 4y + 5z = 7$

20. $x + 2y - 2z = 3$
$x + 3y - 4z = 6$
$4x + 5y - 2z = 3$

Show that each of the following systems is inconsistent.

C **21.** $x - 2y - z = -1$
$2x - y + z = 3$
$x + 4y + 5z = 5$

22. $x + 2y - 2z = 3$
$x + 3y - 4z = 6$
$4x + 5y - 2z = 6$

The following systems are said to be *homogeneous* because the constant terms are all 0. Each of them has $(0, 0, 0)$ as an obvious solution. Determine whether there are any other solutions, and, if there are, find a general expression for them.

23. $x - 2z = 0$
$2x - y + z = 0$
$x - y + 3z = 0$

24. $x + 2y - 2z = 0$
$2x + 5y + 2z = 0$
$3x + 4y - 2z = 0$

25. $x - 2y + 4z = 0$
$3x - y + 2z = 0$
$x + 3y - 6z = 0$

Problems

Solve.

A **1.** Barbara has nickels, dimes, and quarters worth $2.35 in her purse. The number of dimes is three less than the sum of the number of nickels and quarters. How many of each type of coin does she have if there are 19 coins in all?

2. David paid $9.50 for some 15¢, 25¢, and 45¢ stamps. He bought 38 stamps in all. The number of 25¢ stamps was 8 more than twice the number of 45¢ stamps. How many stamps of each type did David buy?

3. When three large diamonds are weighed in pairs, the masses of the pairs are found to be 6 carats, 10 carats, and 12 carats. Find the mass of each diamond.

4. A chemist has three samples of unknown mass. Each pair of the samples is balanced against the third sample and one of three known masses. Find the unknown masses if the known ones are 9 g, 15 g, and 27 g.

5. In a certain triangle, the measure of the largest angle is 20° more than twice the measure of the smallest angle. Five times the measure of the smallest angle equals the sum of the measures of the other two angles. Find the measures of the three angles. (*Hint:* Remember that the sum of the measures of the angles of a triangle is 180°.)

6. The sum of the length, width, and height of a rectangular box is 75 cm. The length is twice the sum of the width and height, and twice the width exceeds the height by 5 cm. Find the dimensions of the box.

B **7.** The Charity Ball Committee sold $10 patron tickets, $5 sponsor tickets, and $2.50 donor tickets. The number of donor tickets sold was 24 more than the number of sponsor and patron tickets combined. There were 326 tickets sold, and the receipts totaled $1432.50. How many tickets of each type were sold?

8. The sum of the digits of a 3-digit number is 21. The sum of the ten's and hundred's digits is 3 less than twice the unit's digit. The number is increased by 198 if the digits are reversed. Find the original number.

A parabola with vertical axis passes through the given points. Find its equation. (*Hint:* Substitute the coordinates of the given points in turn in $y = ax^2 + bx + c$. Then solve the resulting system for a, b, and c.)

9. $(-1, 4), (2, 7), (1, 0)$ **10.** $(1, 3), (2, 10), (-2, -6)$

A circle passes through the given points. Find its equation. (*Hint:* Use the equation $x^2 + y^2 + Cx + Dy + E = 0$.)

11. $(2, -1), (4, -3), (0, -3)$ **12.** $(-6, -3), (1, 4), (2, 3)$

Mixed Review Exercises

Find the real solutions of each system.

1. $x^2 - y^2 = 1$
$x^2 + y^2 = 3$

2. $4x - y = 3$
$x + 3y = 17$

3. $y^2 = x + 1$
$2x + y = 4$

Simplify.

4. $(3x^{-1}y^2)^{-2}$ **5.** $3\sqrt{16x^3}$ **6.** $(4x^3y)(-3x^2y^4)$ **7.** $(2x\sqrt{3})^2$

8. $\dfrac{10xy^{-2}}{6x^{-1}y^3}$ **9.** $\dfrac{18x^3y}{30x^2y^4}$ **10.** $\sqrt{\dfrac{27x^4}{4}}$ **11.** $\left(\dfrac{x}{y^2}\right)^{-1} \cdot \left(\dfrac{y}{x^2}\right)^2$

Self-Test 3

Vocabulary   quadratic system (p. 436)   linear equation in three variables
            ordered triple (p. 444)   (p. 444)
            triangular system (p. 445)   matrix (p. 447)

For each system, (a) graph the equations, (b) determine the number of real solutions, and (c) estimate each real solution to the nearest half-unit.

1. $4x^2 - 9y^2 = 36$
$\quad\ x^2 + y^2 = 25$

2. $x^2 = 4y$
$\quad\ xy = 8$

3. $9y^2 + 16x^2 = 144$ Obj. 9-7, p. 436
$\quad\ x^2 + y^2 = 9$

Solve each system algebraically.

4. $25x^2 + 16y^2 = 400$
$\quad\ x^2 + y^2 = 25$

5. $3x^2 - 2y^2 = 1$ Obj. 9-8, p. 439
$\quad\ y = x - 2$

6. $3x + 2y - z = 1$
$\quad\quad\ 4y + 3z = 5$
$\quad\quad\quad\quad\ z = 3$

7. $2x - y + z = 11$ Obj. 9-9, p. 444
$\quad\ x + 2y + 3z = 8$
$\quad\ 3x - 4y - 5z = -2$

Check your answers with those at the back of the book.

Chapter Summary

1. The *distance* between the points $P_1(x_1, y_1)$ and $P_2(x_2, y_2)$ is given by
$$P_1P_2 = \sqrt{(x_2 - x_1)^2 + (y_2 - y_1)^2}.$$
   The *midpoint* of the line segment P_1P_2 is
$$\left(\frac{x_1 + x_2}{2}, \frac{y_1 + y_2}{2} \right).$$

2. The *circle* with center (h, k) and radius r has the equation
$$(x - h)^2 + (y - k)^2 = r^2.$$

3. A *parabola* with vertex (h, k) has an equation of the form
$$y - k = a(x - h)^2 \text{ if its axis is vertical,}$$
$$x - h = a(y - k)^2 \text{ if its axis is horizontal,}$$
   where $a = \frac{1}{4c}$. The distance between its vertex and focus is $|c|$.

4. The *ellipse* with center $(0, 0)$, foci $(-c, 0)$ and $(c, 0)$, and sum of focal radii $2a$ has the equation
$$\frac{x^2}{a^2} + \frac{y^2}{b^2} = 1, \quad \text{where } b^2 = a^2 - c^2.$$
   If the foci are $(0, -c)$ and $(0, c)$, then the equation has the form
$$\frac{x^2}{b^2} + \frac{y^2}{a^2} = 1, \quad \text{where } b^2 = a^2 - c^2.$$

5. The *hyperbola* with center $(0, 0)$, foci $(-c, 0)$ and $(c, 0)$, and difference of focal radii $2a$ has the equation

$$\frac{x^2}{a^2} - \frac{y^2}{b^2} = 1, \quad \text{where } b^2 = c^2 - a^2.$$

If the foci are $(0, c)$ and $(0, -c)$, then the equation has the form

$$\frac{y^2}{a^2} - \frac{x^2}{b^2} = 1, \quad \text{where } b^2 = c^2 - a^2.$$

6. If an ellipse (or hyperbola) having foci on a horizontal or vertical line has its center at (h, k), then its equation can be obtained from the one given in 4 (or 5) by replacing x by $x - h$ and y by $y - k$.

7. When the equation of a central conic is given in the form
$$Ax^2 + By^2 + Cx + Dy + E = 0,$$
you can complete the squares in x and y to identify the conic and find its center and its foci.

8. Graphical methods can be used to determine the number of real solutions of a quadratic system and to estimate these solutions. To obtain exact solutions, the *substitution method* or the *linear-combination method* often can be used.

9. A good way to solve any system of linear equations in three variables algebraically is to transform it into an equivalent *triangular system* by using linear combinations and then to solve the triangular system by substitution.

Chapter Review

Write the letter of the correct answer.

Exercises 1–3 refer to the points $P(-4, 3)$ and $Q(-6, -1)$.

1. Find the distance PQ.   9-1
   a. 12   **b.** 6   **c.** $2\sqrt{26}$ **d.** $2\sqrt{5}$

2. Find the midpoint of $\overline{PQ}$.
   a. $(5, 1)$ **b.** $(-5, 1)$ **c.** $(-1, 1)$ **d.** $(-5, -1)$

3. If Q is the midpoint of $\overline{PM}$, find M.
   a. $(-5, 1)$ **b.** $(-2, -4)$ **c.** $(-8, -5)$ **d.** $(2, 4)$

4. Find an equation of the circle having center $(4, -3)$ and radius 5.   9-2
   a. $x^2 + y^2 - 8x + 6y - 25 = 0$ **b.** $x^2 + y^2 - 8x + 6y = 0$
   c. $x^2 + y^2 + 8x - 6y = 0$ **d.** $x^2 + y^2 - 4x + 3y = 0$

5. Find the center and radius of the circle $x^2 + y^2 - 8x + 4y + 12 = 0$.
   a. $(4, -2)$; $2\sqrt{2}$ **b.** $(-4, 2)$; $2\sqrt{2}$
   c. $(4, -2)$; $2\sqrt{3}$ **d.** $(-4, 2)$; $2\sqrt{3}$

6. Find an equation of the circle shown at the right.
   a. $x^2 + y^2 - 8x + 8y + 57 = 0$
   b. $x^2 + y^2 - 8x + 8y + 7 = 0$
   c. $x^2 + y^2 + 8x - 8y + 57 = 0$
   d. $x^2 + y^2 + 8x - 8y + 7 = 0$

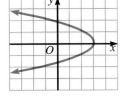

9-3

7. Find an equation of the parabola having vertex $(0, 0)$ and directrix $y = 2$.
   a. $y^2 = -4x$ b. $x^2 = -4y$
   c. $x^2 = -8y$ d. $x^2 = 4 - 4y$

8. Find the vertex of the parabola $4x = y^2 - 4y$.
   a. $(1, -2)$ b. $(-1, 2)$ c. $(0, -2)$ d. $(0, 4)$

9. Find an equation of the parabola shown at the right.
   a. $x^2 = -y + 3$ b. $y^2 = -x - 3$
   c. $y^2 = -x + 3$ d. $x^2 = -y - 3$

9-4

10. Find an equation of the ellipse having foci $(0, -2)$ and $(0, 2)$ and sum of focal radii 8.
   a. $\dfrac{x^2}{4} + \dfrac{y^2}{16} = 1$ b. $\dfrac{x^2}{16} + \dfrac{y^2}{20} = 1$
   c. $\dfrac{x^2}{12} + \dfrac{y^2}{16} = 1$ d. $\dfrac{x^2}{16} + \dfrac{y^2}{12} = 1$

11. Find a focus of the ellipse $\dfrac{x^2}{4} + \dfrac{y^2}{20} = 1$.
   a. $(0, 4)$ b. $(4, 0)$ c. $(0, 2\sqrt{6})$ d. $(2\sqrt{6}, 0)$

12. Find a focus of the hyperbola $8x^2 - 3y^2 = 48$.
   a. $(0, \sqrt{22})$ b. $(\sqrt{22}, 0)$ c. $(0, \sqrt{10})$ d. $(\sqrt{10}, 0)$

9-5

13. Find an equation of the hyperbola shown at the right.
   a. $xy = 4$ b. $xy = 2$
   c. $x^2 - y^2 = 4$ d. $y^2 - x^2 = 4$

14. Find the center of the conic $9x^2 - y^2 - 18x + 4y - 31 = 0$.
   a. $(1, 2)$ b. $(3, 2)$
   c. $(1, -2)$ d. $(-1, -2)$

9-6

15. Find an equation of the ellipse having foci $(1, 0)$ and $(3, 0)$ and sum of focal radii 4.
   a. $x^2 + 9y^2 = 9$ b. $4x^2 + 9y^2 = 36$
   c. $4x^2 + 3y^2 = 12y$ d. $3x^2 + 4y^2 = 12x$

16. How many solutions does this system have? $x^2 - 4y^2 = 4$
                                                 $x + y^2 = 1$
   a. 0   b. 1   c. 2   d. 3   e. 4

9-7

17. Find the real solutions of the system: $x^2 + y^2 = 17$
                                              $x^2 - 2y = 9$

   a. $\{(1, -4), (-1, -4)\}$ b. $\{(-\sqrt{13}, 2), (\sqrt{13}, 2)\}$
   c. $\{(\sqrt{13}, -2), (-\sqrt{13}, -2)\}$ d. $\{(1, -4), (-1, -4), (\sqrt{13}, 2), (-\sqrt{13}, 2)\}$

9-8

18. Find the dimensions of a rectangle having perimeter 14 m and a diagonal of length 5 m.

    a. 3 m by 5 m   **b.** 4 m by 5 m   **c.** 3 m by 4 m   **d.** 2 m by 3 m

19. Solve this system: $x + y + z = 2$ 9-9

                             $x - 2y - z = 2$

                          $3x + 2y + z = 2$

    a. $\{(-1, -2, 5)\}$ **b.** $\{(1, 2, -1)\}$ **c.** $\{(1, -2, 3)\}$ **d.** $\{(1, -3, 4)\}$

20. Sara has $36 in $1, $5, and $10 bills. She has the same number of $5 bills as $10 bills, and she has 10 bills in all. How many bills of each denomination does she have?

    a. 8 $1 bills, 1 $5 bill, 1 $10 bill   **b.** 6 $1 bills, 2 $5 bills, 2 $10 bills

    c. 4 $1 bills, 3 $5 bills, 3 $10 bills   **d.** 2 $1 bills, 4 $5 bills, 4 $10 bills

Chapter Test

1. Given the points $P(3, -4)$ and $Q(-5, 6)$, find **(a)** the length PQ, **(b)** the   9-1
midpoint of $\overline{PQ}$, and **(c)** the point X such that P is the midpoint of $\overline{QX}$.

2. Find an equation of the circle with center $(6, -3)$ that passes through $(-4, 1)$.   9-2

3. Find the center and radius of the circle $x^2 + y^2 - 8x + 2y + 11 = 0$.

4. Find an equation for the set of all points equidistant from the point $(2, 0)$ 9-3
and the line $y = -4$.

5. Find the vertex, focus, and directrix of the parabola $x - 1 = \frac{1}{8}(y + 3)^2$.

6. Find an equation of the ellipse having foci $(\sqrt{5}, 0)$ and $(-\sqrt{5}, 0)$ and sum   9-4
of focal radii 6.

7. Graph the ellipse $x^2 + 4y^2 = 16$.

8. Graph the hyperbola $x^2 - y^2 + 4 = 0$, showing the asymptotes as dashed lines.   9-5

9. Find an equation of the ellipse having foci $(0, 0)$ and $(4, 0)$ and sum of   9-6
focal radii 8.

10. Identify the conic $x^2 - 4y^2 - 4x - 8y - 4 = 0$ and find its center and foci.

By sketching graphs, find the *number* of real solutions of each system.

11. $4x^2 + 9y^2 = 36$ **12.** $4x^2 - y^2 = 4$ 9-7

     $x^2 = y + 2$ $xy = -9$

13. Solve this system: $x^2 - 4y^2 = 4$ 9-8

                       $y^2 = x + 2$

14. Find the dimensions of a rectangle whose area is 36 m² and whose diagonal is $5\sqrt{3}$ m long.

15. Solve this system: $2x - y - 3z = -1$ 9-9

                       $2x - y + z = -9$

                      $x + 2y - 4z = 17$

10 Exponential and Logarithmic Functions

In a Petri dish, a colony of bacteria can grow exponentially. The graph shows that a colony may double in size every 20 minutes.

Exponential Functions

10-1 Rational Exponents

Objective   To extend the meaning of exponents to include rational numbers.

You know the meaning of the power b^x when x is an integer. For example,

$$5^2 = 25, \quad 5^{-2} = \frac{1}{25}, \quad \text{and} \quad 5^0 = 1.$$

Now you'll learn the meaning of b^x when x is any *rational number*.

Definition of $b^{1/2}$:   (read as ''b to the one-half power'' or as ''b to the one half'')
If the laws of exponents on page 212 are to hold, then

$$(b^{1/2})^2 = b^{(1/2)2} = b^1 = b.$$

Since the square of $b^{1/2}$ is b, $b^{1/2}$ is defined to be $\sqrt{b}$.

Definition of $b^{1/3}$: $b^{1/3}$ is defined to be $\sqrt[3]{b}$, since its cube is b:

$$(b^{1/3})^3 = b^{(1/3)3} = b^1 = b$$

Definition of $b^{2/3}$:   Using the law $(a^m)^n = a^{mn}$, you can express $b^{2/3}$ in either of two ways:

$$b^{2/3} = (b^{1/3})^2 = (\sqrt[3]{b})^2 \quad \text{or} \quad b^{2/3} = (b^2)^{1/3} = \sqrt[3]{b^2}$$

Therefore, $b^{2/3}$ is defined to be either of the equivalent expressions $(\sqrt[3]{b})^2$ or $\sqrt[3]{b^2}$.

Definition of $b^{p/q}$

If p and q are integers with $q > 0$, and b is a positive real number, then

$$b^{p/q} = (\sqrt[q]{b})^p = \sqrt[q]{b^p}.$$

Example 1   Simplify $16^{3/4}$.

Solution 1 $16^{3/4} = (\sqrt[4]{16})^3 = 2^3 = 8$ *Answer*

Solution 2 $16^{3/4} = \sqrt[4]{16^3} = \sqrt[4]{4096} = 8$ *Answer*

Notice that computing the root first is easier. This is usually the case.

Example 2 Simplify: **a.** $25^{-3/2}$ **b.** $9^{2.5}$

Solution   **a.** $25^{-3/2} = \dfrac{1}{25^{3/2}}$

$\qquad\qquad = \dfrac{1}{(\sqrt{25})^3}$

$\qquad\qquad = \dfrac{1}{125}$ *Answer*

b. $9^{2.5} = 9^{5/2}$

$\qquad\quad = (\sqrt{9})^5$

$\qquad\quad = 3^5$

$\qquad\quad = 243$ *Answer*

The laws of exponents on page 212 also hold for powers with rational exponents. These laws can be used with the definition of rational exponents to simplify many radical expressions. The first step is to write the expression in **exponential form,** that is, as a power or a product of powers.

Example 3   **a.** Write $\sqrt[3]{\dfrac{a^5 b^3}{c^2}}$ in exponential form.   **b.** Simplify $\left(\dfrac{1}{\sqrt[3]{4}}\right)^{-3/2}$.

Solution   **a.** $\sqrt[3]{\dfrac{a^5 b^3}{c^2}} = \left(\dfrac{a^5 b^3}{c^2}\right)^{1/3}$

$\qquad\qquad = \dfrac{a^{5/3} b^{3/3}}{c^{2/3}}$

$\qquad\qquad = a^{5/3} b c^{-2/3}$ *Answer*

b. $\left(\dfrac{1}{\sqrt[3]{4}}\right)^{-3/2} = (4^{-1/3})^{-3/2}$

$\qquad\qquad\quad = 4^{(-1/3)(-3/2)}$

$\qquad\qquad\quad = 4^{1/2}$

$\qquad\qquad\quad = 2$ *Answer*

Example 4 Write $\sqrt{8} \cdot \sqrt[3]{4}$ in exponential form and in simplest radical form.

Solution $\sqrt{8} \cdot \sqrt[3]{4} = \sqrt{2^3} \cdot \sqrt[3]{2^2}$

$\qquad\qquad = 2^{3/2} \cdot 2^{2/3}$

$\qquad\qquad = 2^{13/6} \longleftarrow$ in exponential form   *Answer*

$\qquad\qquad = 2^2 \cdot 2^{1/6}$

$\qquad\qquad = 4\sqrt[6]{2} \longleftarrow$ in simplest radical form   *Answer*

Example 5 Solve: **a.** $(x-1)^{3/2} = 8$ **b.** $5x^{-1/3} = 20$

Solution

a. $(x-1)^{3/2} = 8$

Raise both sides to the $\frac{2}{3}$ power. $[(x-1)^{3/2}]^{2/3} = 8^{2/3}$

$\qquad\qquad\qquad x - 1 = 4$

$\qquad\qquad\qquad\quad x = 5$

$\therefore$ the solution set is $\{5\}$.

Check: $(5-1)^{3/2} = 4^{3/2} = 8$ $\checkmark$

b. $5x^{-1/3} = 20$

$\qquad x^{-1/3} = 4$

$\qquad (x^{-1/3})^{-3} = 4^{-3}$ Raise both sides to the -3 power.

$\qquad\qquad x = \dfrac{1}{64}$

$\therefore$ the solution set is $\left\{\dfrac{1}{64}\right\}$.

Check: $5\left(\dfrac{1}{64}\right)^{-1/3} = 5 \cdot 4 = 20$ $\checkmark$

Oral Exercises

Simplify.

1. a. $9^{1/2}$ b. $9^{-1/2}$
2. a. $8^{1/3}$ b. $8^{-1/3}$
3. a. $8^{2/3}$ b. $8^{-2/3}$
4. a. $16^{1/4}$ b. $16^{-1/4}$
5. $27^{-1/3}$
6. $27^{2/3}$
7. $4^{3/2}$
8. $25^{-1/2}$

Give the letter of the correct answer. Assume x and y are positive numbers.

9. $x^{1/2}$ equals: a. $\dfrac{x}{2}$ b. $\dfrac{2}{x}$ c. $\sqrt{x}$

10. $x^{2/3}$ equals: a. $(\sqrt[3]{x})^2$ b. $\sqrt{x^3}$ c. $\dfrac{2\sqrt{x}}{3}$

11. $2x^{-1/2}$ equals: a. $\dfrac{1}{\sqrt{2x}}$ b. $\dfrac{1}{2\sqrt{x}}$ c. $\dfrac{2}{\sqrt{x}}$

12. $(2x)^{-1/2}$ equals: a. $\dfrac{1}{\sqrt{2x}}$ b. $\dfrac{1}{2\sqrt{x}}$ c. $\dfrac{\sqrt{2}}{x}$

13. $-8x^{-1/3}$ equals: a. $\dfrac{-2}{\sqrt[3]{x}}$ b. $\dfrac{-8}{\sqrt[3]{x}}$ c. $\dfrac{-1}{2\sqrt[3]{x}}$

14. $\sqrt[4]{x^3 y^{-4}}$ equals: a. $\dfrac{x^{3/4}}{y}$ b. $\dfrac{x^{3/4}}{y^{-1}}$ c. $\dfrac{3x}{4y}$

15. $\sqrt[3]{\dfrac{x^2}{8y^{-1}}}$ equals: a. $\dfrac{x^{2/3}\, y}{8}$ b. $\dfrac{x^{2/3} y^{1/3}}{2}$ c. $\dfrac{x^{2/3}}{2y}$

16. $8 \cdot 2^{1/2}$ equals: a. $2^{7/2}$ b. $16^{1/2}$ c. $16\sqrt{2}$

Tell whether each equation is true or false.

17. a. $9^{1/2} + 4^{1/2} = (9 + 4)^{1/2}$
    b. $9^{1/2} \cdot 4^{1/2} = (9 \cdot 4)^{1/2}$

18. a. $2^{1/3} + 4^{1/3} = (2 + 4)^{1/3}$
    b. $2^{1/3} \cdot 4^{1/3} = (2 \cdot 4)^{1/3}$

19. a. $\left(\dfrac{1}{a} + \dfrac{1}{b}\right)^{-1} = a + b$
    b. $\left(\dfrac{1}{a} \cdot \dfrac{1}{b}\right)^{-1} = ab$

20. a. $(\sqrt{a} + \sqrt{b})^2 = a + b$
    b. $(\sqrt{a} \cdot \sqrt{b})^2 = ab$

21. a. $(a^{-1} + b^{-1})^{-2} = a^2 + b^2$
    b. $(a^{-1} \cdot b^{-1})^{-2} = a^2 b^2$

22. a. $(x^{1/3} + y^{1/3})^6 = x^2 + y^2$
    b. $(x^{1/3} \cdot y^{1/3})^6 = x^2 y^2$

Give the power to which you would raise both sides of each equation in order to solve the equation.

23. $x^{1/2} = 9$
24. $x^{2/3} = 4$
25. $x^{-1/3} = 2$
26. $x^{-3/4} = 8^{-1}$

Tell what steps you would use to solve each equation.

27. $3x^{1/4} = 6$
28. $5x^{-3/2} = 40$
29. $(x - 3)^{-2} = \dfrac{1}{4}$
30. $(5x)^{-1/2} = 3$

Written Exercises

Simplify.

A
1. $81^{1/2}$ 2. $27^{1/3}$ 3. $49^{-1/2}$ 4. $32^{-1/5}$

5. $4^{3/2}$ 6. $27^{2/3}$ 7. $16^{3/4}$ 8. $25^{-3/2}$

9. $(-125)^{-1/3}$ 10. $(-32)^{-3/5}$ 11. $4^{-0.5}$ 12. $\left(\frac{4}{9}\right)^{-1.5}$

13. $-8^{2/3}$ 14. $-9^{3/2}$ 15. $(5^{1/3})^{-3}$ 16. $(7^{-2/3})^3$

17. $(16^{-5})^{1/20}$ 18. $(27^4)^{-1/12}$ 19. $(9^{1/2} + 16^{1/2})^2$ 20. $(8^{2/3} - 8^{1/3})^3$

Write in exponential form.

21. $\sqrt{x^3 y^5}$ 22. $\sqrt[3]{p^4 q}$ 23. $\sqrt{a^{-2} b^3}$ 24. $\sqrt[3]{x^6 y^{-4}}$

25. $(\sqrt{a^{-2} b})^5$ 26. $\sqrt[3]{8 b^6 c^{-4}}$ 27. $\sqrt[4]{\dfrac{16^3 \cdot a^{-2}}{b^6}}$ 28. $\dfrac{1}{\sqrt[4]{p^4 q^{-8}}}$

Express in simplest radical form.

29. $\sqrt[3]{4} \cdot \sqrt[3]{4}$ 30. $\sqrt{8} \cdot \sqrt[6]{8}$ 31. $\dfrac{\sqrt[3]{4}}{\sqrt[6]{2}}$ 32. $\dfrac{\sqrt[5]{27^3}}{\sqrt[5]{9^2}}$

33. $\sqrt[10]{32} \div \sqrt[8]{4}$ 34. $\sqrt[6]{8^3} \div \sqrt[6]{4^2}$ 35. $\sqrt[4]{27} \cdot \sqrt[8]{9}$ 36. $\sqrt[4]{128} \cdot \sqrt[8]{256}$

Simplify each expression. Give answers in exponential form.

B
37. $\sqrt{x} \cdot \sqrt[3]{x} \cdot \sqrt[6]{x}$ 38. $\sqrt[3]{a^2} \cdot \sqrt[3]{a^4}$ 39. $\sqrt[4]{x} \cdot \sqrt[6]{x} \div \sqrt[3]{x}$

40. $((b^{1/2})^{-2/3})^{3/4}$ 41. $a^{1/2}(a^{3/2} - 2a^{1/2})$ 42. $(x^{3/2} - 2x^{5/2}) \div x^{1/2}$

Solve each equation.

43. **a.** $a^{3/4} = 8$
    b. $(3x + 1)^{3/4} = 8$

44. **a.** $y^{-1/2} = 6$
    b. $(3y)^{-1/2} = 6$

45. **a.** $2y^{-1/2} = 10$
    b. $(2y)^{-1/2} = 10$

46. **a.** $(9t)^{-2/3} = 4$
    b. $9t^{-2/3} = 4$

47. $(8 - y)^{1/3} = 4$ 48. $(3n - 1)^{-2/3} = \dfrac{1}{4}$ 49. $(x^2 + 4)^{2/3} = 25$ 50. $(x^2 + 9)^{1/2} = 5$

Mixed Review Exercises

Express in simplest form without negative exponents.

1. $\dfrac{x}{x - 1} - \dfrac{2}{x^2 - 1}$ 2. $\sqrt{\dfrac{8x}{y^3}}$ 3. $(a^2 + 5) - (4 + a - a^2)$

4. $(-2c^2)(-2c)^2$ 5. $\left(\dfrac{a}{b^2}\right)^{-2} \cdot \left(\dfrac{b}{a^3}\right)^{-1}$ 6. $(2n + 3)(n - 4)$

10-2 Real Number Exponents

Objective To extend the meaning of exponents to include irrational numbers and to define exponential functions.

You have learned the meaning of the power b^x when x is a rational number. Now let's consider the meaning of b^x when x is an irrational number, as in the expression $2^{\sqrt{3}}$.

The graph of $y = 2^x$ for the rational values of x in the table below is shown in Figure 1. In the table irrational values of y are rounded to the nearest tenth.

x	-3	-2	-1	-0.5	0	0.5	1	1.5	2
$y = 2^x$	0.125	0.25	0.5	0.7	1	1.4	2	2.8	4

Figure 1

Figure 2

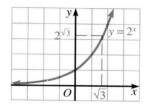

Figure 3

Figure 2 illustrates what happens when you plot many more points of $y = 2^x$ for rational values of x. If the complete graph is to be the smooth curve shown in Figure 3, there must be a point on the curve for every real value of x. In order for there to be no "breaks" in the curve, there must be a point where $x = \sqrt{3}$ and $y = 2^{\sqrt{3}}$.

To find a decimal approximation of $2^{\sqrt{3}}$, look at a sequence of rational numbers x that more and more closely approximate $\sqrt{3}$. Then consider the corresponding sequence of powers 2^x.

sequence of exponents x:	1,	1.7,	1.73,	1.732
sequence of powers 2^x:	2^1,	$2^{1.7}$,	$2^{1.73}$,	$2^{1.732}$
	$\downarrow$	$\downarrow$	$\downarrow$	$\downarrow$
approximations:	2,	3.2490,	3.3173,	3.3219

As the value of the exponent x gets closer and closer to $\sqrt{3}$, the power 2^x gets closer and closer to a definite positive real number, which is defined to be $2^{\sqrt{3}}$. A scientific calculator can approximate $2^{\sqrt{3}}$ to eight or more significant digits: $2^{\sqrt{3}} \approx 3.3219971$.

The method just discussed can be used to define b^x where $b > 0$ and x is *any real number*. It can be shown in more advanced courses that the laws of exponents also hold for these powers.

Exponential and Logarithmic Functions **459**

Example 1 Simplify: **a.** $\dfrac{6^{\sqrt{2}}}{6^{-\sqrt{2}}}$ **b.** $4^{\pi} \cdot 2^{3-2\pi}$ **c.** $(3^{2\sqrt{2}})^{\sqrt{2}}$

Solution **a.** $\dfrac{6^{\sqrt{2}}}{6^{-\sqrt{2}}} = 6^{\sqrt{2}-(-\sqrt{2})}$ $\qquad$ **b.** $4^{\pi} \cdot 2^{3-2\pi} = (2^2)^{\pi} \cdot 2^{3-2\pi}$

$\qquad\qquad\qquad = 6^{2\sqrt{2}}$, or $36^{\sqrt{2}}$ $\qquad\qquad\qquad\qquad = 2^{2\pi} \cdot 2^{3-2\pi}$

$\qquad\qquad\qquad\qquad$ ***Answer*** $\qquad\qquad\qquad\qquad\qquad = 2^{2\pi+(3-2\pi)}$

$\qquad\qquad\qquad\qquad\qquad\qquad\qquad\qquad\qquad\qquad = 2^3 = 8$ ***Answer***

c. $(3^{2\sqrt{2}})^{\sqrt{2}} = 3^{2\sqrt{2}\cdot\sqrt{2}} = 3^4 = 81$ ***Answer***

Now that the meaning of exponents has been extended to include all real numbers, *exponential functions* can be defined.

If $b > 0$ and $b \neq 1$, the function defined by $y = b^x$ is called the **exponential function** with base b.

A computer or a graphing calculator may be helpful in exploring the graphs of $y = b^x$ for various values of b.

If $b > 1$, the graph has the general shape shown in red in the diagram. It rises continuously as x increases.

If $0 < b < 1$, the graph has the general shape shown in blue. It falls continuously as x increases.

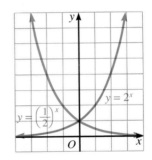

The graph of every exponential function has 1 for its y-intercept (since $b^0 = 1$) and the x-axis as an asymptote.

For any exponential function,

$$b^p = b^q \text{ if and only if } p = q.$$

This condition expresses the fact that every exponential function is *one-to-one*. A function f is called **one-to-one** if for every p and q in the domain of f, $f(p) = f(q)$ if and only if $p = q$. This is equivalent to saying that any horizontal line intersects the graph of a one-to-one function in at most one point.

In the diagram you can see that any horizontal line intersects the graph of $y = 2^x$ in at most one point. This is also true of the graph of $y = (\frac{1}{2})^x$.

The fact that every exponential function is one-to-one allows you to solve certain kinds of *exponential equations*. An **exponential equation** is an equation in which a variable appears in an exponent.

Solving Exponential Equations

Step 1 Express each side of the equation as a power of the same base.

Step 2 Set the exponents equal and then solve.

Step 3 Check the answer.

Example 2 Solve: **a.** $8^x = \dfrac{1}{4}$ **b.** $5^{4-t} = 25^{t-1}$

Solution

a. *Step 1* $(2^3)^x = 2^{-2}$
$$2^{3x} = 2^{-2}$$

    Step 2 $3x = -2$
$$x = -\dfrac{2}{3}$$

    Step 3   *Check:* $8^{-2/3} = \dfrac{1}{8^{2/3}} = \dfrac{1}{4}$ ✓

                $\therefore$ the solution set is $\left\{-\dfrac{2}{3}\right\}$.   **Answer**

b. *Step 1* $5^{4-t} = (5^2)^{t-1}$
$$5^{4-t} = 5^{2t-2}$$

    Step 2 $4 - t = 2t - 2$
$$-3t = -6$$
$$t = 2$$

    Step 3   The check is left to you.
                $\therefore$ the solution set is $\{2\}$.
                                Answer

Oral Exercises

Which number of the given pair is greater?

1. $7^{\sqrt{2}}, 7^{1.5}$ **2.** $5^{\pi}, 5^{3.14}$ **3.** $4^{2\pi}, 16^3$ **4.** $6^{-\sqrt{2}}, 6^{-\sqrt{3}}$

Simplify.

5. $3^{1+\pi} \cdot 3^{1-\pi}$ **6.** $8^{\sqrt{2}} \cdot 8^{-\sqrt{2}}$ **7.** $\dfrac{7^{3\sqrt{2}}}{7^{2\sqrt{2}}}$ **8.** $\dfrac{9^{\sqrt{10}}}{3^{\sqrt{10}}}$

Solve.

9. $3^x = \dfrac{1}{9}$ **10.** $25^x = 125$ **11.** $4^x = \dfrac{1}{8}$ **12.** $36^x = \sqrt{6}$

Written Exercises

Simplify.

A   **1. a.** $3^{\sqrt{2}} \cdot 3^{\sqrt{2}}$ **b.** $(3^{\sqrt{2}})^2$ **c.** $(3^{\sqrt{2}})^{\sqrt{2}}$ **d.** $\dfrac{3^{\sqrt{2}+2}}{3^{\sqrt{2}-2}}$

     2. a. $7^{\sqrt{3}} \cdot 7^{\sqrt{2}}$ **b.** $(7^{\sqrt{3}})^2$ **c.** $(7^{\sqrt{3}})^{\sqrt{2}}$ **d.** $\dfrac{7^{\sqrt{3}+2}}{49}$

Simplify.

3. $(10^{\pi})^2$ **4.** $(5^{-\pi})^{-1}$ **5.** $\sqrt{6^{2\pi}}$ **6.** $\sqrt[3]{4^{6\pi}}$

7. $\dfrac{10^{\sqrt{3}-2}}{10^{\sqrt{3}+2}}$ **8.** $\dfrac{6^{\sqrt{2}} \cdot 6^{\sqrt{8}}}{6^{3\sqrt{2}}}$ **9.** $[(\sqrt{2})^{\pi}]^0$ **10.** $(\sqrt{3})^{\sqrt{2}}(\sqrt{3})^{-\sqrt{2}}$

Simplify.

11. $(2^{\sqrt{2}})^{-1/\sqrt{2}}$ **12.** $(\sqrt{2}^{\sqrt{2}})^{\sqrt{2}}$ **13.** $8^{1.2} \cdot 2^{-3.6}$ **14.** $\dfrac{25^{2.4}}{5^{5.8}}$

15. $\dfrac{(1 + \sqrt{3})^{\pi - 1}}{(1 + \sqrt{3})^{\pi + 1}}$ **16.** $\dfrac{(\sqrt{2} - 1)^{2 + \pi}}{(\sqrt{2} - 1)^{\pi}}$ **17.** $\sqrt[4]{\dfrac{9^{1 - \pi}}{9^{1 + \pi}}}$ **18.** $\sqrt{\dfrac{2^{\sqrt{3} + 3}}{8}}$

Solve. If an equation has no solution, say so.

19. $3^x = \dfrac{1}{27}$ **20.** $5^x = \sqrt{125}$ **21.** $8^{2 + x} = 2$ **22.** $4^{1 - x} = 8$

23. $27^{2x - 1} = 3$ **24.** $49^{x - 2} = 7\sqrt{7}$ **25.** $4^{2x + 5} = 16^{x + 1}$ **26.** $3^{-(x + 5)} = 9^{4x}$

27. $25^{2x} = 5^{x + 6}$ **28.** $6^{x + 1} = 36^{x - 1}$ **29.** $10^{x - 1} = 100^{4 - x}$

B   **30.** $3^{2x} - 6 \cdot 3^x + 9 = 0$ **31.** $4^{2x} - 63 \cdot 4^x - 64 = 0$ **32.** $3^{2x} - 10 \cdot 3^x + 9 = 0$

Graph each pair of equations in the same coordinate system. You may wish to verify your graphs on a computer or a graphing calculator.

33. $y = 3^x$ and $y = 3^{x - 2}$ **34.** $y = 3^x - 2$ and $y = 3^x + 2$

35. $y = 2^x$ and $y - 3 = 2^{x - 4}$ **36.** $y = 2^{-x}$ and $y + 3 = 2^{-(x + 1)}$

Solve each equation.

C   **37.** $2^{(2/3)x + 1} - 3 \cdot 2^{(1/3)x} - 20 = 0$

    38. $2^{2x - 1} - 3 \cdot 2^{x - 1} + 1 = 0$

    39. $2^x - 3^x = 0$

Extra / Growth of Functions

A function f is *increasing* for positive x if $f(x_2) > f(x_1)$ whenever $x_2 > x_1 > 0$. To compare the rate at which two increasing functions grow with x, you look at the ratio $\dfrac{f(x)}{g(x)}$. The function f will *grow faster* than g if you can make $\dfrac{f(x)}{g(x)}$ larger than any positive number by taking x large enough.

     An important fact about exponential functions is that the exponential function b^x with $b > 1$ grows faster than any polynomial. For example, 2^x grows faster than the polynomial function x^{100}. When $x = 10$ the ratio $\dfrac{2^x}{x^{100}}$ is very small (approximately 10^{-97}!), but as x increases, the numerator catches up. For $x = 996$ the ratio is about 1, for $x = 1000$ it is about 10, and by taking x still larger we can make the ratio as large as we want. Even an exponential function with a base very close to 1, such as 1.0001^x, will eventually overtake any power x^n, no matter how large n is.

10-3 Composition and Inverses of Functions

Objective   To find the composite of two given functions and to find the inverse of a given function.

Consider the squaring function $f(x) = x^2$ and the doubling function $g(x) = 2x$. As the diagram below shows, these two functions can be combined to produce a new function whose value at x is $f(g(x))$, read "f of g of x."

  g is the doubling function f is the squaring function

3 $\longrightarrow$ $g(3) = 6$ $\longrightarrow$ $f(g(3)) = f(6) = 36$

5 $\longrightarrow$ $g(5) = 10$ $\longrightarrow$ $f(g(5)) = f(10) = 100$

x $\longrightarrow$ $g(x) = 2x$ $\longrightarrow$ $f(g(x)) = f(2x) = (2x)^2$

Notice that $f(g(x))$ is evaluated by working from the innermost parentheses to the outside. You begin with x, then g doubles x, and then f squares the result.

$$f(g(x)) = f(2x) = (2x)^2$$

The function whose value at x is $f(g(x))$ is called the **composite** of the functions f and g. The operation that combines f and g to produce their composite is called **composition.**

Example 1   If $f(x) = 3x - 5$ and $g(x) = \sqrt{x}$, find the following.

   a. $f(g(4))$ **b.** $g(f(4))$
   c. $f(g(x))$ **d.** $g(f(x))$

Solution   **a.** Since $g(4) = \sqrt{4} = 2$, $f(g(4)) = 3 \cdot 2 - 5 = 1$.
   b. Since $f(4) = 3 \cdot 4 - 5 = 7$, $g(f(4)) = \sqrt{7}$.
   c. $f(g(x)) = f(\sqrt{x}) = 3\sqrt{x} - 5$
   d. $g(f(x)) = g(3x - 5) = \sqrt{3x - 5}$

Notice in Example 1 that $f(g(x)) \neq g(f(x))$.

The function $I(x) = x$ is called the **identity function.** It behaves like the multiplicative identity 1.

$$a \cdot 1 = a \qquad \text{for all numbers } a$$
$$f(I(x)) = f(x) \quad \text{for all functions } f$$

For the two functions f and g defined in the example at the top of the next page, the composites $f(g(x))$ and $g(f(x))$ are both equal to the identity function.

Example 2 If $f(x) = \dfrac{x+4}{2}$ and $g(x) = 2x - 4$, find the following.

    a. $g(1)$ and $f(g(1))$ **b.** $f(-3)$ and $g(f(-3))$ **c.** $f(g(x))$ **d.** $g(f(x))$

Solution

a. $g(1) = -2$ and $f(g(1)) = f(-2) = 1$.   ***Answer***
Notice that $g: 1 \to -2$ and $f: -2 \to 1$.

b. $f(-3) = \dfrac{1}{2}$ and $g(f(-3)) = g\left(\dfrac{1}{2}\right) = -3$.   ***Answer***

Notice that $f: -3 \to \dfrac{1}{2}$ and $g: \dfrac{1}{2} \to -3$.

Parts (a) and (b) suggest that the functions f and g "undo each other."
Parts (c) and (d) prove that this is so for any number x.

c. $f(g(x)) = f(2x - 4) = \dfrac{2x - 4 + 4}{2} = x$ ***Answer***

d. $g(f(x)) = g\left(\dfrac{x+4}{2}\right) = 2\left(\dfrac{x+4}{2}\right) - 4 = x$ ***Answer***

In multiplication, two numbers whose product is the identity 1, such as 2 and 2^{-1}, are called inverses. Similarly, two functions whose composite is the identity I, such as f and g in Example 2, are called *inverse functions*.

Inverse Functions

The functions f and g are **inverse functions** if

$$f(g(x)) = x \text{ for all } x \text{ in the domain of } g$$
and
$$g(f(x)) = x \text{ for all } x \text{ in the domain of } f.$$

The inverse of a function f is usually denoted f^{-1}, read "f inverse."

Caution: The superscript -1 in f^{-1} is *not* an exponent. The symbol $f^{-1}(x)$ denotes the value of f inverse at x; it does *not* mean $\dfrac{1}{f(x)}$.

Suppose that two functions f and g are inverses and that $f(a) = b$, so that (a, b) is on the graph of f. Then $g(b) = g(f(a)) = a$, and the point (b, a) must be on the graph of g. This means every point (a, b) on the graph of f corresponds to a point (b, a) on the graph of g. Therefore, the graphs are *mirror images* of each other with respect to the line $y = x$. You can verify this by drawing graphs of inverse functions on the same axes. A computer or a graphing calculator may be helpful. The diagram shows the inverse functions of Example 2.

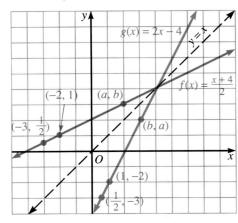

Some functions do not have inverse functions. If the reflection of the graph of a function f is itself to be the graph of a function, the graph of f must *not* contain two different points with the same y-coordinate. Therefore, a function has an inverse function if and only if it is one-to-one.

As you learned in Lesson 10-2, every horizontal line intersects the graph of a one-to-one function in at most one point. Therefore, you can use the *horizontal-line test* to tell whether a given function has an inverse function.

Horizontal-Line Test

A function has an inverse function if and only if every horizontal line intersects the graph of the function in *at most* one point.

If a function has an inverse function, you can find it by writing y for $f(x)$, interchanging x and y, and solving for y. Example 3 illustrates.

Example 3 Let $f(x) = x^3 - 1$.

    a. Graph f and determine whether f has an inverse function. If so, graph f^{-1} by reflecting f across the line $y = x$.

    b. Find $f^{-1}(x)$.

Solution

    a. The graph of f is shown in red at the right. The graph of f passes the horizontal-line test, so f has an inverse. The graph of f^{-1}, shown in blue, is the reflection of the graph of f across the line $y = x$.

    b. Replace $f(x)$ by y: $y = x^3 - 1$

        Interchange x and y: $x = y^3 - 1$

        Solve for y: $y^3 = x + 1$

                               $y = \sqrt[3]{x + 1}$

             $f^{-1}(x) = \sqrt[3]{x + 1}$ *Answer*

Oral Exercises

Suppose $f(x) = 3x$, $g(x) = x + 1$, and $h(x) = x^2 + 2$. Find the following.

1. a. $f(g(3))$ **b.** $f(g(0))$ **c.** $f(g(-6))$ **d.** $f(g(x))$

2. a. $g(f(4))$ **b.** $g(f(5))$ **c.** $g(f(-6))$ **d.** $g(f(x))$

3. a. $f(h(2))$ **b.** $h(f(2))$ **c.** $f(h(x))$ **d.** $h(f(x))$

4. a. $g(h(3))$ **b.** $h(g(3))$ **c.** $g(h(x))$ **d.** $h(g(x))$

5. Find $f^{-1}(x)$.   **6.** Find $g^{-1}(x)$.   **7.** Does h^{-1} exist? Why or why not?

Written Exercises

Suppose $f(x) = \frac{x}{2}$, $g(x) = x - 3$, and $h(x) = \sqrt{x}$. Find a real-number value or an expression in x for each of the following. If no real value can be found, say so.

A
1. **a.** $f(g(8))$ **b.** $f(g(-5))$ **c.** $f(g(0))$ **d.** $f(g(x))$
2. **a.** $g(f(8))$ **b.** $g(f(-5))$ **c.** $g(f(0))$ **d.** $g(f(x))$
3. **a.** $f(h(9))$ **b.** $f(h(4))$ **c.** $f(h(-4))$ **d.** $f(h(x))$
4. **a.** $h(f(32))$ **b.** $h(f(16))$ **c.** $h(f(x))$ **d.** $f(f(x))$
5. **a.** $h(g(12))$ **b.** $h(g(2))$ **c.** $h(g(x))$ **d.** $h(h(x))$
6. **a.** $g(h(9))$ **b.** $g(h(\sqrt{3}))$ **c.** $g(h(x))$ **d.** $g(g(x))$

Use the horizontal-line test to determine whether each function f has an inverse function. If so, draw a rough sketch of f^{-1} by reflecting f across $y = x$.

7.   **8.**   **9.**   **10.**

In Exercises 11–14, find $f^{-1}(x)$. Then graph f and f^{-1} in the same coordinate system. You may wish to verify your graphs on a computer or a graphing calculator.

11. $f(x) = 2x - 3$ **12.** $f(x) = \frac{x+6}{3}$ **13.** $f(x) = x^3$ **14.** $f(x) = \frac{12}{x}$

In Exercises 15–22, graph g and use the horizontal-line test to determine if g has an inverse function. If so, find $g^{-1}(x)$. If g has no inverse, say so. You may wish to verify your graphs on a computer or a graphing calculator.

15. $g(x) = \left(\frac{8}{x}\right)^3$ **16.** $g(x) = \sqrt[3]{2x}$ **17.** $g(x) = x^4$ **18.** $g(x) = |x|$

B
19. $g(x) = x^2 - x$ **20.** $g(x) = x^3 + 2$ **21.** $g(x) = \sqrt{x^2}$ **22.** $g(x) = (2x + 3)^5$

23. a. Draw the graph of $f(x) = 2^x$ by making a table of values and carefully plotting several points.

    b. Draw the graph of f^{-1} on the same coordinate system by reflecting the graph of f in the line $y = x$.

    c. Find $f^{-1}(2)$, $f^{-1}(4)$, $f^{-1}(8)$, and $f^{-1}(\frac{1}{2})$.

    d. Give the domain and range of f and f^{-1}.

24. The squaring function $f(x) = x^2$ has no inverse function. However, the square root function $g(x) = \sqrt{x}$ *does* have an inverse function. Find this inverse and explain why it is not the same as $f(x) = x^2$. [*Hint:* Compare the domains of f and g^{-1}.]

C **25.** Let $f(x) = mx + b$. For what values of m and b will $f(x) = f^{-1}(x)$ for all x?

26. Prove that the graphs of a linear function and its inverse function are never perpendicular.

Mixed Review Exercises

Simplify.

1. $25^{-3/2}$

2. $\sqrt{18} + \sqrt{32}$

3. $(2 - 3i)^2$

4. $(3^{\sqrt{2}})^{2\sqrt{2}}$

5. $\sqrt{-48} \cdot \sqrt{-12}$

6. $\dfrac{2^{\pi-2}}{2^{\pi+2}}$

7. $(-27)^{2/3}$

8. $\dfrac{\sqrt{2} - 3}{1 + \sqrt{2}}$

9. $\dfrac{5^2 - 1}{1 - 3^2}$

10. $16^{1.75}$

11. $\dfrac{4 + 3i}{2 - i}$

12. $4^{1.5} \cdot 2^{-2}$

Self-Test 1

Vocabulary   exponential form (p. 456)
                        exponential function (p. 460)
                        one-to-one function (p. 460)
                        exponential equation (p. 460)
                        composite (p. 463)

                composition (p. 463)
                identity function (p. 463)
                inverse functions (p. 464)
                horizontal-line test (p. 465)

1. Write in exponential form.   **Obj. 10-1, p. 455**

    a. $\sqrt[6]{64x^4y^{-2}}$ **b.** $\sqrt[3]{6^{1/2}} \cdot \sqrt[6]{6^{-3}}$

2. Write in simplest radical form.

    a. $\left(\dfrac{25}{\sqrt[3]{20}}\right)^{3/2}$ **b.** $\sqrt[3]{x^2y^4} \cdot \sqrt{x^3y}$

3. Solve $(x - 2)^{3/5} = 8$.

4. Simplify.   **Obj. 10-2, p. 459**

    a. $4^{3/\sqrt{2}} \div 16^{2\sqrt{2}}$ **b.** $\dfrac{16^{\sqrt{5}}}{4^{\sqrt{5}}} \cdot 32^{-2\sqrt{5}}$

5. Solve $7^{x-1} = 49^{4-x}$.

6. If $f(x) = 6x + 1$ and $g(x) = \sqrt{x}$, find the following.   **Obj. 10-3, p. 463**

    a. $f(g(4))$ **b.** $g(f(4))$ **c.** $f(g(x))$ **d.** $g(f(x))$

7. Show that $f(x) = 3x - 7$ and $g(x) = \dfrac{x + 7}{3}$ are inverse functions.

Logarithmic Functions

10-4 Definition of Logarithms

Objective   To define logarithmic functions and to learn how they are related to exponential functions.

The graph of $f(x) = 2^x$ is shown in Figure 1. Since the function passes the horizontal-line test, it has an inverse function. The graph of f^{-1} can be found by reflecting the graph of f across the line $y = x$, as shown in Figure 2.

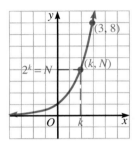

$f(x) = 2^x$

Figure 1

Figure 2

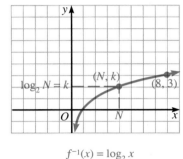

$f^{-1}(x) = \log_2 x$

Figure 3

The inverse of f, the exponential function with base 2, is called the **logarithmic function** with base 2, and is denoted by $f^{-1}(x) = \log_2 x$. (See Figure 3.)

$\log_2 x$ is read "the base 2 logarithm of x" or "log base 2 of x."

Notice that (3, 8) is on the graph of $f(x) = 2^x$ and (8, 3) is on the graph of $f^{-1}(x) = \log_2 x$. Since these functions are inverses, it follows that $2^3 = 8$ means $\log_2 8 = 3$. This and other examples are given below.

Exponential Form		*Logarithmic Form*
$2^3 = 8$	means	$\log_2 8 = 3$
$2^4 = 16$	means	$\log_2 16 = 4$
$2^0 = 1$	means	$\log_2 1 = 0$
$2^{-1} = \dfrac{1}{2}$	means	$\log_2 \dfrac{1}{2} = -1$
$2^k = N$	means	$\log_2 N = k$

A base other than 2 can be used with logarithmic functions. In fact the base can be any positive number except 1, since every power of 1 is 1.

Definition of Logarithm

If b and N are positive numbers ($b \neq 1$),

$$\log_b N = k \text{ if and only if } b^k = N.$$

Every positive number has a unique logarithm with base b. Since the base b exponential function is one-to-one, its inverse is one-to-one. Therefore,

$$\log_b M = \log_b N \text{ if and only if } M = N.$$

Several other properties of logarithms are easy to verify. Since the statements $\log_b N = k$ and $b^k = N$ are equivalent, by substitution

$$\log_b b^k = k \qquad \text{and} \qquad b^{\log_b N} = N.$$

Since $b = b^1$ and $1 = b^0$, it follows from the definition of a logarithm that

$$\log_b b = 1 \qquad \text{and} \qquad \log_b 1 = 0.$$

Example 1 Write each equation in exponential form.

    a. $\log_6 36 = 2$ **b.** $\log_2 2 = 1$ **c.** $\log_{10} (0.001) = -3$

Solution   **a.** $6^2 = 36$ **b.** $2^1 = 2$ **c.** $10^{-3} = 0.001$

Example 2 Write each equation in logarithmic form.

    a. $6^0 = 1$ **b.** $8^{-2/3} = \dfrac{1}{4}$ **c.** $5^{3/2} = 5\sqrt{5}$

Solution   **a.** $\log_6 1 = 0$ **b.** $\log_8 \dfrac{1}{4} = -\dfrac{2}{3}$ **c.** $\log_5 (5\sqrt{5}) = \dfrac{3}{2}$

Example 3 Simplify each logarithm.

    a. $\log_5 25$ **b.** $\log_2 8\sqrt{2}$ **c.** $2^{\log_2 7}$

Solution   **a.** Since $5^2 = 25$, $\log_5 25 = 2$.   *Answer*

           b. Let $\log_2 8\sqrt{2} = x$.

              Then: $2^x = 8\sqrt{2}$

                      $2^x = 2^3 \cdot 2^{1/2}$

                      $2^x = 2^{7/2}$ Set the exponents equal.

                         $x = \dfrac{7}{2}$

            $\therefore \log_2 8\sqrt{2} = \dfrac{7}{2}$ *Answer*

           c. Since $b^{\log_b N} = N$, $2^{\log_2 7} = 7$.   *Answer*

Example 4 Solve each equation.
  a. $\log_4 x = 3$
  b. $\log_x 81 = 4$

Solution   **a.** Rewrite in exponential form.

$$4^3 = x$$
$$64 = x$$

∴ the solution set is {64}.
Answer

b. Rewrite in exponential form.

$$x^4 = 81$$

Since $81 = 3^4$,

$$x^4 = 3^4$$
$$x = 3$$

∴ the solution set is {3}.   **Answer**

Oral Exercises

Express in exponential form.

1. $\log_2 32 = 5$ **2.** $\log_3 9 = 2$ **3.** $\log_7 \sqrt{7} = \frac{1}{2}$ **4.** $\log_3 \frac{1}{81} = -4$

Express in logarithmic form.

5. $4^3 = 64$ **6.** $9^{3/2} = 27$ **7.** $10^{-2} = 0.01$ **8.** $16^{-3/4} = \frac{1}{8}$

Simplify.

9. $\log_6 36$ **10.** $\log_2 16$ **11.** $\log_{10} 100$ **12.** $\log_3 \frac{1}{9}$

13. $\log_2 2\sqrt{2}$ **14.** $\log_7 1$ **15.** $4^{\log_4 16}$ **16.** $\log_6 (6^5)$

17. $\log_2 10$ lies between the consecutive integers __?__ and __?__ .

18. $\log_{10} 101$ lies between the consecutive integers __?__ and __?__ .

19. $\log_3 \frac{1}{50}$ lies between the consecutive integers __?__ and __?__ .

20. If you solve $\log_2 x^2 = \log_2 100$, then the solution set is {?, ?}.

Written Exercises

Simplify each logarithm.

A **1.** $\log_5 125$ **2.** $\log_4 16$ **3.** $\log_3 81$

  4. $\log_6 6$ **5.** $\log_3 1$ **6.** $\log_8 4$

  7. $\log_5 \frac{1}{25}$ **8.** $\log_2 \frac{1}{8}$ **9.** $\log_6 6\sqrt{6}$

  10. $\log_5 25\sqrt{5}$ **11.** $\log_4 \sqrt{2}$ **12.** $\log_{27} \sqrt{3}$

13. $\log_7 \sqrt[3]{49}$ **14.** $\log_3 \sqrt[5]{9}$ **15.** $\log_{1/2} 8$

16. $\log_{1/3} 27$ **17.** $\log_2 \sqrt[3]{\dfrac{1}{4}}$ **18.** $\log_{10} \dfrac{1}{\sqrt{1000}}$

Solve for x.

19. $\log_7 x = 2$ **20.** $\log_6 x = 3$ **21.** $\log_9 x = -\dfrac{1}{2}$

22. $\log_6 x = 2.5$ **23.** $\log_4 x = -\dfrac{3}{2}$ **24.** $\log_{1/9} x = -\dfrac{1}{2}$

B **25.** $\log_x 27 = \dfrac{3}{2}$ **26.** $\log_x 64 = 6$ **27.** $\log_x 7 = -\dfrac{1}{2}$

28. $\log_x 7 = 1$ **29.** $\log_x 1 = 0$ **30.** $\log_x 2 = 0$

31. a. Show that $\log_2 8 + \log_2 4 = \log_2 32$ by simplifying the three logarithms.

b. Show that $\log_9 3 + \log_9 27 = \log_9 81$ by simplifying the three logarithms.

c. State a generalization based on parts (a) and (b).

32. a. Simplify $\log_2 8$ and $\log_8 2$.

b. Simplify $\log_3 \sqrt{3}$ and $\log_{\sqrt{3}} 3$.

c. State a generalization based on parts (a) and (b).

33. a. If $f(x) = 6^x$, then $f^{-1}(x) = $ __?__ .

b. Find $f^{-1}(36)$ and $f^{-1}\left(\dfrac{1}{\sqrt{6}}\right)$.

c. Give the domain and range of f and f^{-1}.

34. a. If $g(x) = \log_4 x$, then $g^{-1}(x) = $ __?__ .

b. Find $g^{-1}(2)$ and $g^{-1}\left(-\dfrac{3}{2}\right)$.

c. Give the domain and range of g and g^{-1}.

Sketch the graph of each function and its inverse function in the same coordinate system. Label at least three points on each graph. You may wish to check your graphs on a computer or a graphing calculator.

35. $f(x) = 6^x$ **36.** $g(x) = \log_4 x$

37. $h(x) = \log_{10} x$ **38.** $k(x) = \left(\dfrac{1}{2}\right)^x$

Solve.

C **39.** $\log_5 (\log_3 x) = 0$ **40.** $\log_4 (\log_3 (\log_2 x)) = 0$

41. If $0 < b < 1$ and $0 < u < 1$, is $\log_b u$ positive or negative?

42. If $1 < a < b$, is $\log_b (\log_b a)$ positive or negative?

43. The decibel (dB) is used to measure the loudness of sound. Sound that is just barely audible has a decibel level of 0 and an intensity level of 10^{-12} W/m^2 (watts per square meter). Sound painful to the ear has a decibel level of 130 and an intensity level of 10 W/m^2. The formula

$D = 10 \log_{10} \left(\dfrac{I}{I_0} \right)$ gives the decibel level of a sound whose intensity is I,

where I_0 is the intensity of barely audible sound. Some decibel levels and intensities are shown in the table below. The intensity is given in watts per square meter.

	Intensity	Decibels
Barely audible sound	10^{-12}	0
Quiet conversation	10^{-9}	30
Heavy traffic	10^{-3}	90
Jet plane (at 20 m)	10^2	140

a. The intensity of thunder is 10^{12} times the intensity of barely audible sound. What is the decibel level of thunder?

b. The decibel levels of a subway train entering a station and normal conversation are 100 dB and 60 dB, respectively. How many times as intense as normal conversation is the noise of the subway train?

Historical Note / Logarithms

The late sixteenth century was a time of great mathematical activity, particularly in astronomy and navigation. Because of the involved computations that these fields required, there was a need for a method to simplify arithmetic calculation and make multiplication and division easier. The Scottish mathematician John Napier found a way to turn multiplication into addition and division into subtraction, and in 1614 he published a work demonstrating the use of logarithms.

In the nineteenth century, the great French mathematician Laplace said of Napier that by lessening the work of computation his invention doubled the life of the astronomer. Scientists and mathematicians continued to use logarithms until calculators and computers provided more efficient tools.

Napier's logarithms are related to what we today call *natural logarithms*. A year after Napier published his work, the English mathematician Henry Briggs became interested in Napier's invention. It was Briggs who suggested that 10 be the logarithmic base and who introduced the terms *characteristic* and *mantissa*. He compiled an extensive table of logarithms giving 14-place approximations for logarithms of numbers from 1 to 20,000 and 90,000 to 100,000.

10-5 Laws of Logarithms

Objective   To learn and apply the basic properties of logarithms.

The laws of exponents can be used to derive the *laws of logarithms*.

Laws of Logarithms

Let b be the base of a logarithmic function ($b > 0$, $b \neq 1$). Let M and N be positive numbers.

$$\textbf{1. } \log_b MN = \log_b M + \log_b N$$

$$\textbf{2. } \log_b \frac{M}{N} = \log_b M - \log_b N$$

$$\textbf{3. } \log_b M^k = k \log_b M$$

Laws 1 and 3 are proved below. The proof of Law 2 is left as Exercise 43.

Proof: Let $\log_b M = x$ and $\log_b N = y$. Therefore, $b^x = M$ and $b^y = N$.

<table>
<tr><td align="center">Law 1</td><td align="center">Law 3</td></tr>
<tr><td>$MN = b^x b^y = b^{x+y}$</td><td>$M^k = (b^x)^k = b^{kx}$</td></tr>
<tr><td>$\log_b MN = \log_b b^{x+y}$</td><td>$\log_b M^k = \log_b b^{kx}$</td></tr>
<tr><td>$\log_b MN = x + y$</td><td>$\log_b M^k = kx$</td></tr>
<tr><td>$\therefore \log_b MN = \log_b M + \log_b N$</td><td>$\therefore \log_b M^k = k \log_b M$</td></tr>
</table>

Example 1   Express $\log_6 M^2 N^3$ in terms of $\log_6 M$ and $\log_6 N$.

Solution

$$
\begin{aligned}
\log_6 M^2 N^3 &= \log_6 M^2 + \log_6 N^3 && \text{Use Law 1.}\\
&= 2 \log_6 M + 3 \log_6 N && \text{Use Law 3.}
\end{aligned}
$$

Example 2   Express $\log_2 \sqrt{\dfrac{M}{N^5}}$ in terms of $\log_2 M$ and $\log_2 N$.

Solution

$$
\begin{aligned}
\log_2 \sqrt{\frac{M}{N^5}} &= \log_2 \left(\frac{M}{N^5}\right)^{1/2}\\[2mm]
&= \frac{1}{2} \log_2 \left(\frac{M}{N^5}\right) && \text{Use Law 3.}\\[2mm]
&= \frac{1}{2}(\log_2 M - \log_2 N^5) && \text{Use Law 2.}\\[2mm]
&= \frac{1}{2}(\log_2 M - 5 \log_2 N) && \text{Use Law 3.}
\end{aligned}
$$

The process illustrated in Examples 1 and 2 can also be reversed.

Example 3   Express as a single logarithm.

     a. $\log_{10} p + 3 \log_{10} q$ **b.** $4 \log_{10} p - 2 \log_{10} q$

Solution   **a.** $\log_{10} p + 3 \log_{10} q = \log_{10} p + \log_{10} q^3$ Use Law 3.

$$= \log_{10} pq^3 \qquad \text{Use Law 1.}$$

     b. $4 \log_{10} p - 2 \log_{10} q = \log_{10} p^4 - \log_{10} q^2$ Use Law 3.

$$= \log_{10} \frac{p^4}{q^2} \qquad \text{Use Law 2.}$$

You can often use the laws of logarithms to compute the numerical value of a logarithm from given logarithms.

Example 4   If $\log_{10} 2 = 0.30$ and $\log_{10} 3 = 0.48$, find the following.

     a. $\log_{10} 18$ **b.** $\log_{10} \frac{20}{3}$ **c.** $\log_{10} 5$ **d.** $\log_{10} \left(\frac{1}{\sqrt[3]{2}} \right)$

Solution   **a.** Since $18 = 2 \cdot 3^2$,

$$\log_{10} 18 = \log_{10} 2 + 2 \log_{10} 3$$
$$= 0.30 + 2(0.48)$$
$$= 1.26 \quad \textbf{\textit{Answer}}$$

     b. Since $\dfrac{20}{3} = \dfrac{2 \cdot 10}{3}$,

$$\log_{10} \frac{20}{3} = \log_{10} 2 + \log_{10} 10 - \log_{10} 3$$
$$= 0.30 + 1 - 0.48$$
$$= 0.82 \quad \textbf{\textit{Answer}}$$

     c. Since $5 = \dfrac{10}{2}$,

$$\log_{10} 5 = \log_{10} 10 - \log_{10} 2$$
$$= 1 - 0.30$$
$$= 0.70 \quad \textbf{\textit{Answer}}$$

     d. Since $\dfrac{1}{\sqrt[3]{2}} = 2^{-1/3}$,

$$\log_{10} \left(\frac{1}{\sqrt[3]{2}} \right) = -\frac{1}{3} \log_{10} 2$$
$$= -\frac{1}{3}(0.30)$$
$$= -0.10 \quad \textbf{\textit{Answer}}$$

You can use the laws and properties of logarithms to solve logarithmic equations. *Because the logarithm of a variable expression is defined only if the expression is positive, be sure to check the answers you obtain.*

Example 5 Solve $\log_3 x + \log_3 (x - 6) = 3$ for x.

Solution

$$\log_3 x + \log_3 (x - 6) = 3$$
$$\log_3 x(x - 6) = 3 \qquad \text{Use Law 1.}$$
$$\log_3 (x^2 - 6x) = 3$$

Change to exponential form: $\qquad x^2 - 6x = 3^3$
$$x^2 - 6x - 27 = 0$$
$$(x + 3)(x - 9) = 0$$
$$x = -3 \qquad \text{or} \qquad x = 9$$

Check: If $x = -3$, $\log_3 x$ is not defined. So -3 is *not* a root.

If $x = 9$, $\log_3 x + \log_3 (x - 6) = \log_3 9 + \log_3 3$
$$= \log_3 (9 \cdot 3)$$
$$= \log_3 27$$
$$= 3 \quad \checkmark$$

$\therefore$ the solution set is $\{9\}$. *Answer*

Oral Exercises

Express each logarithm in terms of $\log_3 M$ and $\log_3 N$.

1. $\log_3 M^4$

2. $\log_3 N^6$

3. $\log_3 M^4 N$

4. $\log_3 \left(\dfrac{M}{N^3} \right)$

5. $\log_3 \left(\dfrac{1}{M} \right)$

6. $\log_3 \sqrt{M}$

7. $\log_3 \sqrt[3]{N^2}$

8. $\log_3 \left(\dfrac{1}{N\sqrt{N}} \right)$

Express as a logarithm of a single number or expression.

9. $\log_a 3 + \log_a 4$

10. $\log_a 7 - \log_a 5$

11. $4 \log_a 2$

12. $2 \log_a 9$

13. $\dfrac{1}{2} \log_a 36$

14. $-\log_a \dfrac{1}{6}$

15. $\log_b 3 + \log_b 5 + \log_b 2$

16. $\log_b 6 + \log_b 5 - \log_b 2$

17. $2 \log_b p + \log_b q$

18. $\log_b x - 3 \log_b y$

19. $\dfrac{1}{2} \log_b r + \dfrac{1}{2} \log_b s$

20. $\dfrac{1}{2} (\log_b x - \log_b y)$

Let $c = \log_3 10$ and $d = \log_3 5$. Express the following in terms of c and d.

21. $\log_3 50$

22. $\log_3 500$

23. $\log_3 250$

24. $\log_3 2$

Written Exercises

Express each logarithm in terms of $\log_2 M$ and $\log_2 N$.

A   **1.** $\log_2 M^6 N^3$ **2.** $\log_2 (MN)^4$ **3.** $\log_2 M\sqrt{N}$ **4.** $\log_2 \sqrt[3]{M^2 N}$

5. $\log_2 \dfrac{M^4}{N^3}$ **6.** $\log_2 \left(\dfrac{M}{N}\right)^7$ **7.** $\log_2 \sqrt{\dfrac{M}{N^3}}$ **8.** $\log_2 \dfrac{1}{MN}$

If $\log_{10} 9 = 0.95$ and $\log_{10} 2 = 0.30$ (accurate to two decimal places), find the following.

9. $\log_{10} 81$ **10.** $\log_{10} \dfrac{9}{2}$ **11.** $\log_{10} \sqrt{2}$ **12.** $\log_{10} 3$

13. $\log_{10} 8$ **14.** $\log_{10} 36$ **15.** $\log_{10} \dfrac{20}{9}$ **16.** $\log_{10} 900$

17. $\log_{10} \dfrac{1}{9}$ **18.** $\log_{10} \dfrac{1}{2000}$ **19.** $\log_{10} \sqrt[3]{\dfrac{2}{9}}$ **20.** $\log_{10} 162$

Express as a logarithm of a single number or expression.

21. $5 \log_4 p + \log_4 q$

22. $\log_{10} x - 4 \log_{10} y$

23. $4 \log_3 A - \dfrac{1}{2} \log_3 B$

24. $\log_5 M + \dfrac{1}{4} \log_5 N$

B   **25.** $\log_2 M + \log_2 N + 3$

26. $\log_5 x - \log_5 y + 2$

27. $1 - 3 \log_5 x$

28. $\dfrac{1 + \log_9 x}{2}$

Simplify.

29. $2 \log_{10} 5 + \log_{10} 4$

30. $2 \log_3 6 - \log_3 4$

31. $\log_4 40 - \log_4 5$

32. $\log_4 3 - \log_4 48$

Solve each equation.

33. $\log_a x = 2 \log_a 3 + \log_a 5$

34. $\log_a x = \dfrac{3}{2} \log_a 9 + \log_a 2$

35. $\log_b (x + 3) = \log_b 8 - \log_b 2$

36. $\log_b (x^2 + 7) = \dfrac{2}{3} \log_b 64$

37. $\log_a x - \log_a (x - 5) = \log_a 6$

38. $\log_a (3x + 5) - \log_a (x - 5) = \log_a 8$

39. $\log_2 (x^2 - 9) = 4$

40. $\log_3 (x + 2) + \log_3 6 = 3$

41. If $f(x) = \log_2 x$ and $g(x) = 4^x$, find: **a.** $f(g(3))$ **b.** $g\left(f\left(\dfrac{1}{2}\right)\right)$ **c.** $f(g^{-1}(16))$

42. If $f(x) = 3^x$ and $g(x) = \log_9 x$, find: **a.** $f(f(-1))$ **b.** $g\left(g\left(\dfrac{1}{81}\right)\right)$ **c.** $f^{-1}(g(9))$

43. Prove the second law of logarithms.

44. Simplify $4^{\log_2 (2^{\log_2 5})}$.

Solve each equation.

C **45.** $\log_5 (\log_3 x) = 0$ **46.** $\log_2 (\log_4 x) = 1$

      47. $\log_6 (x + 1) + \log_6 x = 1$ **48.** $\log_{10} (x + 6) + \log_{10} (x - 6) = 2$

      49. $\dfrac{1}{2} \log_a (x + 2) + \dfrac{1}{2} \log_a (x - 2) = \dfrac{2}{3} \log_a 27$

      50. $2 \log_3 x - \log_3 (x - 2) = 2$

      51. $\log_b (x - 1) + \log_b (x + 2) = \log_b (8 - 2x)$

Mixed Review Exercises

Solve.

1. $\log_2 x = -\dfrac{1}{2}$ **2.** $x^3 - 7x + 6 = 0$ **3.** $2^{x+3} = 4^{x-1}$

4. $\sqrt{x} + 2 = x$ **5.** $3^x = \dfrac{\sqrt{3}}{9}$ **6.** $3(2x - 1) = 5x + 4$

7. $\dfrac{x}{x + 1} - 1 = \dfrac{1}{x}$ **8.** $\log_x 8 = \dfrac{3}{2}$ **9.** $(x - 2)^2 = 5$

If $f(x) = x^2 - 1$ and $g(x) = \sqrt{x + 1}$, find each of the following.

10. $f(-2)$ **11.** $g(3)$ **12.** $f(g(1))$ **13.** $g(f(-1))$

Self-Test 2

Vocabulary   logarithmic function (p. 468)   laws of logarithms (p. 473)
                   logarithm (p. 469)

1. Write in exponential form.   **Obj. 10-4 p. 468**

     a. $\log_3 81 = 4$ **b.** $\log_6 216 = 3$

2. Write in logarithmic form.

     a. $5^4 = 625$ **b.** $25^{3/2} = 125$

3. Evaluate each expression.

     a. $\log_2 4^{3/2}$ **b.** $4^{\log_4 12}$

4. Solve $\log_b 27 = 3$.

5. Write $\log_2 (M^5 N^6)^{1/3}$ in terms of $\log_2 M$ and $\log_2 N$.   **Obj. 10-5 p. 473**

6. If $\log_{10} 5 = 0.70$, find $\log_{10} 0.04$.

7. Solve $\log_a x + \log_a (x - 2) = \log_a 3$.

Check your answers with those at the back of the book.

Applications

10-6 Applications of Logarithms

Objective   To use common logarithms to solve equations involving
powers and to evaluate logarithms with any given base.

Logarithms were invented to simplify difficult calculations. Because of the decimal nature of our number system, it is easiest to work with base 10 logarithms. These are called **common logarithms.** When common logarithms are used in calculations, the base 10 is usually not written. For example, log 6 means $\log_{10} 6$.

Most scientific calculators have a key marked "log" that gives the common logarithm of a number. You also can find the common logarithm of a number by using Table 3 on page 812.

Although calculators and computers have replaced logarithms for doing heavy computational work, logarithms still provide the best, or even the only, method of solution for many types of problems. Some of these will be shown later in this lesson and in the next.

If you have a calculator, you should practice finding the logarithms of a few numbers and then skip ahead to the paragraph preceding Example 2.

Without a calculator, you will need to use the table of logarithms (Table 3), a portion of which is shown below.

N	0	1	2	3	4	5	6	7	8	9
20	3010	3032	3054	3075	3096	3118	3139	3160	3181	3201
21	3222	3243	3263	3284	3304	3324	3345	3365	3385	3404
22	3424	3444	3464	3483	3502	3522	3541	3560	3579	3598
23	3617	3636	3655	3674	3692	3711	3729	3747	3766	3784
24	3802	3820	3838	3856	3874	3892	3909	3927	3945	3962

The table gives the logarithms of numbers between 1 and 10 rounded to four decimal places. The decimal point is omitted. To find an approximation for log x for $1 < x < 10$, find the first two digits of x in the column headed N and the third digit in the row to the right of N. For example, to find log 2.36, look for 23 under N and move across row 23 to the column headed 6, where you find 3729. Therefore, log $2.36 \approx 0.3729$.

To find an approximation for the logarithm of a positive number greater than 10 or less than 1, write the number in scientific notation (see page 221) and use Law 1: log MN = log M + log N. For example,

$$\begin{aligned}
\log 236 &= \log (2.36 \times 10^2) \\
&= \log 2.36 + \log 10^2 && \text{Use Law 1.} \\
&= \log 2.36 + 2 \\
&\approx 0.3729 + 2, \text{ or } 2.3729 && \text{Use Table 3.}
\end{aligned}$$

The example shows that the common logarithm of a number can be written as the sum of an integer, called the **characteristic,** and a nonnegative number less than 1, called the **mantissa,** which can be found in a table of logarithms.

Example 1   Find each logarithm. Use Table 3.
   a. log 3.8   **b.** log 97,500   **c.** log 0.000542

Solution   **a.** Find 38 under N; then read across to the column headed 0.
      $\therefore$ log 3.8 = 0.5798   *Answer*

   b. Write 97,500 in scientific notation and use laws of logarithms.
      log 97,500 = log (9.75×10^4) = log 9.75 + log 10^4
      The mantissa for 9.75 is 0.9890. The characteristic is 4.
      $\therefore$ log 97,500 = 0.9890 + 4 = 4.9890   *Answer*

   c. Write 0.000542 in scientific notation and use laws of logarithms.
      log 0.000542 = log (5.42×10^{-4}) = log 5.42 + log 10^{-4}
      The mantissa for 5.42 is 0.7340. The characteristic is -4.
      $\therefore$ log 0.000542 = 0.7340 + (-4) = -3.2660 *Answer*

The logarithms in the tables, and therefore the answers, are approximations. However, it is a common practice to use =, as in Example 1, rather than ≈.
   If log $y = a$, then the number y is sometimes called the **antilogarithm** of a. The value of y can be found by using a calculator or log tables.

Example 2   Find y to three significant digits if:
   a. log $y = 0.8995$ **b.** log $y = 2.4825$

Solution 1   Using a Calculator   On some calculators you find antilogarithms by using the inverse function key with the logarithmic function key. On many others, you can use the 10^x key as shown.
   a. If $\log_{10} y = 0.8995$, then $y = 10^{0.8995} = 7.93$.   *Answer*
   b. $y = 10^{2.4825} = 304$ *Answer*

Solution 2   Using Tables
   a. In the body of Table 3 find the value closest to 0.8995, that is, 0.8993. This is the entry in the row labeled 79 and in the column labeled 3.
      $\therefore y = 7.93$ *Answer*

   b. First find the number that has 0.4825 as its mantissa. The table entry with mantissa closest to 0.4825 is 3.04.
$$\log y = 2.4825 = 0.4825 + 2$$
$$= \log 3.04 + \log 10^2$$
$$= \log (3.04 \times 10^2)$$
$$\log y = \log 304$$
      $\therefore y = 304$ *Answer*

In the previous example the answers were rounded to three significant digits. For most problems the answers you find using Table 3 will be accurate to three significant digits.

Example 3 Find the value of each expression to three significant digits.
> **a.** $\sqrt[5]{493}$ **b.** $(0.173)^6$

Solution 1 Using a Calculator

a. Since $\sqrt[5]{493} = 493^{1/5} = 493^{0.2}$, use the power key, y^x.

$$y^x = 493^{0.2} = 3.46 \quad \textit{Answer}$$

b. $y^x = (0.173)^6 = 0.0000268 \quad \textit{Answer}$

Solution 2 Using Tables

a. Let $x = \sqrt[5]{493}$.

$$\log x = \frac{1}{5} \log 493$$

$$\log x = \frac{1}{5}(2.6928)$$

$$\log x = 0.5386$$

$$\therefore \quad x = 3.46 \quad \textit{Answer}$$

b. $(0.173)^6 = (1.73 \times 10^{-1})^6$
$$= (1.73)^6 \times 10^{-6}$$

Let $x = (1.73)^6$. Then:
$$\log x = 6 \log 1.73$$
$$\log x = 1.428$$
$$x = 26.8$$

$$\therefore (0.173)^6 = 26.8 \times 10^{-6}$$
$$= 0.0000268 \quad \textit{Answer}$$

Whether or not you use a calculator, you need the properties of logarithms to solve the exponential equation given in Example 4. The solution to Example 4 is given in *calculation-ready form*, where the next step is to obtain a decimal approximation using a calculator or a table.

Example 4 Solve $3^{2x} = 5$.
> **a.** Give the solution in calculation-ready form.
> **b.** Give the solution as a decimal with three significant digits.

Solution

a. $3^{2x} = 5$ Take logarithms of both sides.

$$\log 3^{2x} = \log 5$$

$$2x \log 3 = \log 5 \qquad \text{Use laws of logarithms to simplify.}$$

$$x = \frac{\log 5}{2 \log 3} \qquad \text{Solve for } x.$$

$\therefore$ the solution in calculation-ready form is $\dfrac{\log 5}{2 \log 3}$.   *Answer*

b. Find log 5 and log 3 with a calculator or table.

$$x = \frac{0.6990}{2(0.4771)} = 0.733$$

$\therefore$ to three significant digits, the solution is 0.733.   *Answer*

If you know the base b logarithm of a number and wish to find its base a logarithm, you can use the following formula:

Change-of-Base Formula

$$\log_a x = \frac{\log_b x}{\log_b a}$$

The proof of this formula is left as Exercise 42 on page 482.

Example 5 Find $\log_4 7$.

Solution Use the change-of-base formula letting $b = 10$:

$$\log_4 7 = \frac{\log_{10} 7}{\log_{10} 4} = \frac{0.8451}{0.6021} = 1.404 \quad \textit{Answer}$$

Oral Exercises

Use a calculator or Table 3 to find each logarithm.

1. log 4.05
2. log 40.5
3. log 405
4. log 0.405
5. log 8.36
6. log 83.6
7. log 0.836
8. log 0.0836

Use a calculator or Table 3 to find x to three significant digits.

9. $\log x = 0.6072$
10. $\log x = 0.9212$
11. $\log x = 1.9212$
12. $\log x = 0.9212 - 1$

Solve for x in calculation-ready form.

13. $2^x = 7$
14. $9^x = 8$
15. $8^{2x} = 3$
16. $3^{-x} = 7$

Written Exercises

Use a calculator or Table 3 to find the value of each expression to three significant digits.

A
1. $(1.06)^{10}$
2. $(10.6)^{10}$
3. $(0.38)^5$
4. $(347)^{1.5}$
5. $(12.7)^{5/2}$
6. $\sqrt[6]{786}$
7. $\sqrt[5]{(81.2)^4}$
8. $\sqrt[3]{(412)^2}$

Use a calculator or Table 3 to find x to three significant digits.

9. $\log x = 0.8531$
10. $\log x = 0.4065$
11. $\log x = 2.84$
12. $\log x = 1.605$
13. $\log x = -1.8$
14. $\log x = -2.91$

Solve each equation.

a. Give the solution in calculation-ready form.
b. Give the solution to three significant digits.

15. $3^x = 30$ **16.** $5^t = 10$ **17.** $5.6^x = 56$ **18.** $(1.02)^x = 2$

19. $30^{-x} = 5$ **20.** $12^{2x} = 1000$ **21.** $3.5^{2t} = 60$ **22.** $\dfrac{4^{2-t}}{3} = 7$

Solve each equation *without* using a calculator or logarithms. (See Example 2 of Lesson 10-2, page 461.)

23. $4^x = 8\sqrt{2}$ **24.** $3^x = \sqrt[5]{9}$ **25.** $125^x = 25\sqrt{5}$ **26.** $8^x = 16\sqrt[3]{2}$

Unlike Exercises 15–26, Exercises 27–34 are *not* exponential equations. Solve each equation using a calculator or logarithms. Give answers to three significant digits.

Sample   Solve $x^{2/3} = 12$.

Solution   1. Raise each side to the $\dfrac{3}{2}$ power. You find that $x = 12^{3/2}$.
      2. To simplify $12^{3/2}$.
         (a) Use a calculator and the y^x key to obtain $x = 41.6$, or
         (b) Use logarithms to obtain $\log x = \dfrac{3}{2} \log 12 = 1.619$. Then find the anti-
            logarithm: $x = 41.6$.   *Answer*

27. $x^{2/5} = 34$ **28.** $x^{2/3} = 50$ **29.** $\sqrt[3]{x^4} = 60$ **30.** $\sqrt[5]{x^3} = 900$

B **31.** $2x^5 = 100$ **32.** $\dfrac{\sqrt[5]{x}}{9} = 7$ **33.** $(3y - 1)^6 = 80$ **34.** $\sqrt[3]{4t + 3} = 8.15$

Find each logarithm. Use the change-of-base formula.

35. $\log_2 9$ **36.** $\log_6 8$ **37.** $\log_3 40$ **38.** $\log_7 \dfrac{1}{2}$

Solve for x to three significant digits.

39. $3^{2x} - 7 \cdot 3^x + 10 = 0$ **40.** $3^{2x} - 7 \cdot 3^x + 12 = 0$
41. a. Simplify $\log_7 49$ and $\log_{49} 7$.
    b. Simplify $\log_2 8$ and $\log_8 2$.
    c. How are $\log_b a$ and $\log_a b$ related? Give a convincing argument to justify your answer.

C **42.** Derive the change-of-base formula. (*Hint:* Let $\log_a x = y$, so that $x = a^y$.)

10-7 Problem Solving: Exponential Growth and Decay

Objective To use exponential and logarithmic functions to solve growth and decay problems.

Suppose an investment of P dollars earns 8% interest compounded annually. Then the value of the investment will be multiplied by 1.08 each year because:

Value at end of year = 100%(value at beginning of year) + interest earned

$\qquad\qquad\qquad$ = 1.00(value at beginning of year) + 0.08(value at beginning of year)

$\qquad\qquad\qquad$ = 1.08(value at beginning of year)

From this equation you can obtain the following table of values.

Time in years	0	1	2	3	$\cdots$	t
Value in dollars	P	$P(1.08)$	$P(1.08)^2$	$P(1.08)^3$	$\cdots$	$P(1.08)^t$

$\qquad\qquad\qquad\quad \times 1.08 \quad\; \times 1.08 \quad\; \times 1.08$

Because the value of the investment in t years is given by the exponential function $A = P(1.08)^t$, we say that the investment has **exponential growth** and that the annual *growth rate* is 8%.

If the 8% annual interest is compounded quarterly (that is, compounded four times per year), the quarterly growth rate is 2%. Then

$$\text{value at end of } q \text{ quarters} = P(1.02)^q$$

and $\qquad\qquad\qquad$ value at end of t years $= P(1.02)^{4t}.$

This result can be generalized as follows.

Compound Interest Formula

If an amount P (called the *principal*) is invested at an annual interest rate r compounded n times a year, then in t years the investment will grow to an amount A given by

$$A = P\left(1 + \frac{r}{n}\right)^{nt}$$

where r is expressed as a decimal.

Example 1 How long will it take an investment of $1000 to triple in value if it is invested at an annual rate of 12% compounded quarterly?

(Solution is on the next page.)

Solution   Use the compound interest formula. Let $P = 1000$, $A = 3000$, $r = 0.12$, and $n = 4$.

$$3000 = 1000\left(1 + \frac{0.12}{4}\right)^{4t}$$

$$3 = (1.03)^{4t}$$

$$\log 3 = \log (1.03)^{4t}$$

$$\log 3 = 4t \log 1.03$$

$$t = \frac{\log 3}{4 \log 1.03} = \frac{0.4771}{4(0.0128)}$$

$$t = 9.3 \text{ (years)}$$

Since the interest is compounded *quarterly*, you cannot give 9.3 as an answer. Since 9.3 is closer to $9\frac{1}{4}$ than $9\frac{1}{2}$, the investment will triple in approximately $9\frac{1}{4}$ years.   ***Answer***

Investing money is just one example of exponential growth. Another example is the growth of a population. If a population now has N_0 people and is growing at a rate of 1.4% per year, then in t years the population will have $N_0(1.014)^t$ people.

A growth rate of 1.4% per year approximately doubles the population every 50 years, since $N_0(1.014)^{50} \approx 2N_0$. From this equation you can see that $1.014 \approx 2^{1/50}$. Therefore, the number N of people in the population t years from now can be written in either of two ways:

$N = N_0(1.014)^t$ (The population grows 1.4% per year.)

$N = N_0 \cdot 2^{t/50}$ (The population doubles every 50 years.)

Doubling-Time Growth Formula

If a population of size N_0 doubles every d years (or hours, or days, or any other unit of time), then the number N in the population at time t is given by

$$N = N_0 \cdot 2^{t/d}.$$

Example 2   A certain bacteria population doubles in size every 12 hours. By how much will it grow in 2 days?

Solution   Use the doubling-time growth formula. Let $t = 48$ hours (2 days) and let the doubling time $d = 12$ hours.

$$N = N_0 \cdot 2^{t/d}$$

$$N = N_0 \cdot 2^{48/12} = N_0 \cdot 2^4 = 16N_0$$

$\therefore$ the population grows by a factor of 16 in 2 days.   ***Answer***

While investments and populations grow exponentially, the value of a car decreases exponentially, or experiences **exponential decay.** If a car loses 25% of its value each year, then its value N at the end of a year is

$$N = 0.75 \times \text{(value at beginning of year)}.$$

Therefore, after t years a car initially valued at N_0 dollars will have a value of

$$N = N_0(0.75)^t.$$

One frequently studied example of exponential decay is the decay of a radioactive substance. Over a period of time, some of the substance will change to a different element. The *half-life* of the radioactive substance is the amount of time it takes until exactly half of the original substance remains unchanged. For example, the half-life of radioactive radium is 1600 years. Therefore, every 1600 years half of the radium decays.

Time in years	0	1600	3200	4800	. . .	t
Amount left	N_0	$N_0 \cdot \dfrac{1}{2}$	$N_0 \cdot \left(\dfrac{1}{2}\right)^2$	$N_0 \cdot \left(\dfrac{1}{2}\right)^3$	. . .	$N_0 \cdot \left(\dfrac{1}{2}\right)^{t/1600}$

The half-life decay formula is very similar to the doubling-time growth formula.

Half-Life Decay Formula

If an amount N_0 has a half-life h, then the amount remaining at time t is

$$N = N_0\left(\frac{1}{2}\right)^{t/h}.$$

Example 3 The half-life of carbon-14 (C-14) is 5730 years. How much of a 10.0 mg sample will remain after 4500 years?

Solution Use the half-life decay formula. Let $N_0 = 10.0$, $h = 5730$, and $t = 4500$.

$$N = N_0\left(\frac{1}{2}\right)^{t/h}$$

$$N = 10.0\left(\frac{1}{2}\right)^{4500/5730}$$

Using a calculator: $N = 5.80$ (to three significant digits)

If a calculator is not available, take logarithms of both sides:

$$\log N = \log 10.0 + \frac{4500}{5730} \log 0.5$$

$$\log N = 1 + (0.7853)(-0.3010)$$

$$\log N = 0.7636$$

$$N = 5.80$$

∴ 5.80 mg of the original C-14 remains after 4500 years.   *Answer*

Oral Exercises

Complete each table.

1.

Item	Cost now	Annual growth rate	Cost in t years
Movie ticket	$5	10%	?
Sweater	$40	5%	?
Sneakers	?	?	$60(1.07)^t$

2.

Item	Cost now	Annual decay rate	Value in t years
Car	$10,000	30%	?
Bike	$300	20%	?
Skis	?	?	$250(0.95)^t$

3.

Population size now	Doubling time	Size in t years
1000	9 years	?
3 million	25 years	?
?	?	$7500 \cdot 2^{t/12}$

4.

Amount of radioactive substance now	Half-life in days	Amount of substance in t days
80	8	?
1200	30	?
?	?	$500\left(\dfrac{1}{2}\right)^{t/3}$

Problems

Solve. Give money answers in dollars and cents and all other answers to three significant digits. A calculator may be helpful.

A 1. One thousand dollars is invested at 12% interest compounded annually. Determine how much the investment is worth after:
   a. 1 year   **b.** 2 years   **c.** 3 years   **d.** 10 years

2. One hundred dollars is invested at 7.2% interest compounded annually. Determine how much the investment is worth after:
   a. 1 year   **b.** 5 years   **c.** 10 years   **d.** 20 years
   e. Use your answers to parts (a)–(d) to estimate the doubling time for the investment.

3. Redo Exercise 1 assuming that the interest is compounded quarterly.

4. Redo Exercise 1 assuming that the interest is compounded monthly.

5. The value of a new $12,500 auto-
mobile decreases 20% per year.
Find its value after:

   a. 1 year   **b.** 2 years
   c. 3 years   **d.** 10 years

6. The value of a new $3500 sail-
boat decreases 10% per year.
Find its value after:

   a. 1 year   **b.** 5 years
   c. 10 years   **d.** 20 years

7. A certain population of bacteria doubles every 3 weeks. The number of
bacteria in the population is now N_0. Find its size in:

   a. 6 weeks   **b.** 15 weeks   **c.** W weeks

8. A culture of yeast doubles in size every 20 min. The size of the culture is
now N_0. Find its size in:

   a. 1 hour   **b.** 12 hours   **c.** 1 day

9. The half-life of carbon-14 is approximately 6000 years. Determine how
much of 100 kg of this substance will remain after:

   a. 12,000 years   **b.** 24,000 years   **c.** y years

10. The radioactive gas radon has a half-life of approximately $3\frac{1}{2}$ days. About
how much of a 100 g sample will remain after 1 week?

B **11.** How long will it take you to double your money if you invest it at a rate
of 8% compounded annually?

12. How long will it take you to triple your money if you invest it at a rate of
6% compounded annually?

13. Savings institutions sometimes use the term "effective yield" to describe
the interest rate compounded *annually* (once a year) that is equivalent to a
given rate. The effective yield can be found by computing the value of $1
at the end of the year at the given rate. For example, at 8% compounded
quarterly, the value A of $1 at the end of a year is

$$A = 1\left(1 + \frac{0.08}{4}\right)^4 = (1.02)^4.$$

Evaluating A directly (without using logs), you find that $A = 1.0824$ to
four decimal places. Therefore, the original $1 has grown in value by
$0.0824, an increase of 8.24%. The effective yield is 8.24% per year.
Find the effective yield of 12% compounded quarterly.

14. Bank A offers 6% interest compounded monthly. Bank B offers 6.1% compounded quarterly. Which bank pays more interest per year?

15. One of the many by-products of uranium is radioactive plutonium, which
has a half-life of approximately 25,000 years. If some of this plutonium is
encased in concrete, what fraction of it will remain after 100 years? after a
million years?

16. Sugar is put into a large quantity of water and the mixture is stirred. After 2 minutes 50% of the sugar has dissolved. How much longer will it take until 90% of the sugar has dissolved? (Use the half-life decay formula.)

Sample   A baseball card increased in value from \$50 to \$500 in 15 years. Find its average annual rate of appreciation.

Solution   Let r be the annual rate of appreciation.

$$\text{Then} \quad 50(1 + r)^{15} = 500$$
$$(1 + r)^{15} = 10$$
$$1 + r = 10^{1/15} \approx 1.17$$
$$\therefore r \approx 0.17 = 17\%$$

17. A gold coin appreciated in value from \$100 to \$238 in 8 years. Find the average annual rate of appreciation.

18. Ten years ago Michael paid \$250 for a rare 1823 stamp. Its current value is \$1000. Find the average annual rate of growth.

19. A new car that cost \$12,000 decreased in value to \$4000 in 5 years. Find the average annual rate of depreciation.

20. A tractor that 4 years ago cost \$8000, now is worth only \$3200. Find the average annual rate of depreciation.

C **21.** One million dollars is invested at 6.4% interest. Find the value of the investment after one year if the interest is compounded:

a. quarterly   **b.** daily   **c.** hourly
d. Why do you think your answers to parts (a)–(c) are so close in value?

22. A colony of bacteria has 6.5×10^6 members at 8 A.M. and 9.75×10^6 members at 10:30 A.M. Find its population at noon.

Mixed Review Exercises

If the domain of each function is $D = \{1, 2, 4\}$, find the range.

1. $f(x) = \log_4 x$ **2.** $g(x) = |x - 3|$ **3.** $h(x) = (\sqrt{2})^x$

4. $F(x) = x^2 - 3x + 2$ **5.** $G(x) = \dfrac{x}{x + 2}$ **6.** $H(x) = x^{1/2}$

Simplify.

7. $\log_3 12 - 2 \log_3 2$ **8.** $(4^{1/3})^{3/2}$ **9.** $\dfrac{2^{-1} - 2^{-2}}{1 + 2^{-1}}$

10. $\dfrac{8^{0.5}}{2^{1.5}}$ **11.** $\sqrt{\left(\dfrac{1}{3}\right)^2 + \left(\dfrac{1}{4}\right)^2}$ **12.** $\log_6 24 + 2 \log_6 3$

10-8 The Natural Logarithm Function

Objective   To define and use the natural logarithm function.

In advanced work in science and mathematics, the most important logarithm function is the **natural logarithm function.** Its base is the irrational number e, which has the approximate value 2.71828. The **natural logarithm** of x is sometimes denoted by $\log_e x$, but more often by $\ln x$.

The number e is defined to be the limiting value of $\left(1 + \frac{1}{n}\right)^n$ as n becomes larger and larger. Using a calculator, you can make a table and show that as n gets very large, the value of this expression approaches a number a bit larger than 2.718.

n	100	1000	10,000	100,000
$\left(1 + \frac{1}{n}\right)^n$	2.70481	2.71692	2.71815	2.71827

The exercises of this lesson are just like exercises in previous lessons, except that the symbol $\ln x$ is used instead of $\log_b x$. Example 1 illustrates.

Example 1 *Working with base 2 logs*

1. If $\log_2 x = 5$, then $x = 2^5$.
2. If $2^x = 7$, then $x = \log_2 7$.
3. $\log_2 2^5 = 5$ and $2^{\log_2 7} = 7$

Working with base e logs

1. If $\ln x = 5$, then $x = e^5$.
2. If $e^x = 7$, then $x = \ln 7$.
3. $\ln e^5 = 5$ and $e^{\ln 7} = 7$

Example 2 **a.** Simplify $\ln \dfrac{1}{e^2}$.

b. Write as a single logarithm: $2 \ln 5 + \ln 4 - 3$.

Solution **a.** $\ln \dfrac{1}{e^2} = \ln e^{-2} = -2 \ln e = -2 \cdot 1 = -2$ *Answer*

b. $2 \ln 5 + \ln 4 - 3 = \ln 5^2 + \ln 4 - \ln e^3$

$$= \ln \frac{5^2 \cdot 4}{e^3} = \ln \frac{100}{e^3} \quad Answer$$

Example 3 Solve: **a.** $\ln x = 2$ **b.** $\ln \dfrac{1}{x} = 2$

Solution   **a.** $x = e^2$ *Answer*   **b.** $\dfrac{1}{x} = e^2$

$$x = \frac{1}{e^2}, \text{ or } e^{-2} \quad Answer$$

Example 4 Solve $e^{2x} = 9$.

Solution 1 $e^{2x} = 9$
$\qquad\qquad 2x = \ln 9$ $\qquad\qquad$ Rewrite in logarithmic form.

$\qquad\qquad\quad x = \dfrac{1}{2} \ln 9$

$\qquad\qquad\quad x = \ln 9^{1/2} = \ln 3$ $\quad$ *Answer*

Solution 2 $\qquad\quad e^{2x} = 9$ $\qquad\qquad$ $\Big\{$ Take the natural log of both sides
$\qquad\qquad\quad \ln e^{2x} = \ln 9$ $\qquad\qquad$ of the equation.

$\qquad\qquad 2x \cdot \ln e = \ln 9$

$\qquad\qquad\quad 2x \cdot 1 = \ln 9$

$\qquad\qquad\qquad\quad x = \dfrac{1}{2} \ln 9 = \ln 9^{1/2} = \ln 3$ $\quad$ *Answer*

Oral Exercises

Give each equation in exponential form.

1. $\ln 4 = 1.39$ $\qquad\qquad$ **2.** $\ln \dfrac{1}{4} = -1.39$ $\qquad\qquad$ **3.** $\ln e = 1$

Give each equation in logarithmic form.

4. $e^2 = 7.39$ $\qquad\qquad$ **5.** $e^{-2} = 0.14$ $\qquad\qquad$ **6.** $e^{1/5} = 1.22$

Simplify.

7. $\ln \dfrac{1}{e}$ $\qquad\qquad$ **8.** $\ln e^{12}$ $\qquad\qquad$ **9.** $\ln \sqrt{e}$

Solve.

10. $\ln x = 5$ $\qquad\qquad$ **11.** $\ln x = \dfrac{1}{3}$ $\qquad\qquad$ **12.** $e^x = 3$

13. $x = \ln e^{3/4}$ $\qquad\qquad$ **14.** $e^{\ln x} = 10$ $\qquad\qquad$ **15.** $\ln x = -\dfrac{1}{2}$

16. Approximate to three decimal places the value of $\left(1 + \dfrac{1}{5000}\right)^{5000}$.

Written Exercises

Write each equation in exponential form.

A **1.** $\ln 8 = 2.08$ $\qquad$ **2.** $\ln 100 = 4.61$ $\qquad$ **3.** $\ln \dfrac{1}{3} = -1.10$ $\qquad$ **4.** $\ln \dfrac{1}{e^2} = -2$

Write each equation in logarithmic form.

5. $e^3 = 20.1$ **6.** $e^7 = 1097$ **7.** $e^{1/2} = 1.65$ **8.** $\sqrt[3]{e} = 1.40$

Simplify. If the expression is undefined, say so.

9. $\ln e^2$ **10.** $\ln e^{10}$ **11.** $\ln \dfrac{1}{e^3}$ **12.** $\ln \dfrac{1}{\sqrt{e}}$

13. $\ln 1$ **14.** $\ln 0$ **15.** $e^{\ln 5}$ **16.** $e^{\ln 0.5}$

Write as a single logarithm.

17. $\ln 3 + \ln 4$ **18.** $\ln 8 - \ln 2$ **19.** $2 \ln 3 - \ln 5$

20. $\ln 7 + \dfrac{1}{2} \ln 9$ **21.** $\dfrac{1}{3} \ln 8 + \ln 5 + 3$ **22.** $4 \ln 2 - \ln 3 - 1$

Solve for x. Leave answers in terms of e.

23. $\ln x = 3$ **24.** $\ln \dfrac{1}{x} = 2$ **25.** $\ln (x - 4) = -1$

26. $\ln |x| = 1$ **27.** $\ln x^2 = 9$ **28.** $\ln \sqrt{x} = 3$

Solve for x. Leave answers in terms of natural logarithms.

29. $e^x = 2$ **30.** $e^{-x} = 3$ **31.** $e^{2x} = 25$

32. $e^{3x} = 8$ **33.** $e^{x-2} = 2$ **34.** $\dfrac{1}{e^x} = 7$

Solve. Leave answers in terms of e or natural logarithms.

35. $\sqrt{e^x} = 3$ **36.** $e^{-2x} = 0.2$ **37.** $(e^x)^5 = 1000$

38. $3e^{2x} + 2 = 50$ **39.** $\ln (\ln x) = 0$ **40.** $|\ln x| = 1$

41. $\ln x + \ln (x + 3) = \ln 10$ **42.** $2 \ln x = \ln (x + 1)$ **43.** $e^{2x} - 7e^x + 12 = 0$

Give the domain and range of each function.

B   **44.** $f(x) = \ln x$ **45.** $f(x) = \ln |x|$ **46.** $f(x) = \ln x^2$ **47.** $f(x) = \ln (x - 5)$

      48. Graph $y = \ln x$ and $y = e^x$ in the same coordinate system.

      49. Graph $y = 2^x$, $y = e^x$, and $y = 3^x$ in the same coordinate system.

C   **50.** Express in terms of e the approximate value of each expression when n is very large.

      a. $\left(1 + \dfrac{1}{n}\right)^{5n}$ **b.** $\left(1 + \dfrac{2}{n}\right)^n$ **c.** $\left(\dfrac{n}{n + 1}\right)^{2n}$

    51. Refer to the compound interest formula on page 483.

      a. Show that if interest is compounded daily, so that n is quite large, then $A \approx Pe^{rt}$.

      b. Use part (a) to find the amount of money that you would have after 1 year if you invest $1000 at 6% interest compounded daily.

If your calculator has an e^x or y^x key, or a natural logarithm ($\ln x$) and an inverse function key, you can evaluate powers of e.

1. If you invest P dollars at an annual interest rate r compounded daily, then the value of the investment t years later will be about Pe^{rt}.
   (Note that the rate is expressed as a decimal. Therefore, if the rate is 6%, then $r = 0.06$.) Find the value of a $1000 investment at 6% interest compounded daily after:

   a. 1 year   **b.** 2 years   **c.** 10 years

2. Which investment is worth more?

   Investment A: $1000 at 8% compounded daily after 10 years

   Investment B: $2000 at 8% compounded daily after 5 years

3. **a.** Complete the following table for $y = e^{-x^2}$.

x	0	0.1	0.2	0.3	0.4	0.5	0.6	0.7	0.8	0.9	1.0
y											

   b. Use the table to sketch the graph of $y = e^{-x^2}$. (*Hint:* The graph is symmetric about the y-axis.) This graph is closely related to the so-called bell-shaped curve so important in statistics.

4. **a.** Evaluate $\left(1 + \dfrac{1}{n}\right)^n$ when n is a million.

   b. Compare your answer with the following approximation for e:
      2.718281828.

5. The notation $n!$ (read "n factorial") represents the product of the integers from 1 to n.

$$n! = 1 \cdot 2 \cdot 3 \cdot \ldots \cdot n$$

A very important number in mathematics is approximated by the sum

$$1 + \frac{1}{1!} + \frac{1}{2!} + \frac{1}{3!} + \frac{1}{4!} + \frac{1}{5!} + \frac{1}{6!}.$$

Find this sum and tell what important number it approximates.

Challenge

A friend is thinking of an integer between 1 and 1,000,000. You are to guess the number by asking your friend questions that require only yes-or-no answers. What is the minimum number of questions that you must ask to be sure of guessing the number? How should you guess?

Self-Test 3

Vocabulary   common logarithms (p. 478)   compound interest formula (p. 483)
                characteristic (p. 479)   doubling-time formula (p. 484)
                mantissa (p. 479)   exponential decay (p. 485)
                antilogarithm (p. 479)   half-life (p. 485)
                calculation-ready form (p. 480)   half-life decay formula (p. 485)
                change-of-base formula (p. 481)   natural logarithm function (p. 489)
                exponential growth (p. 483)   natural logarithm (p. 489)

Solve. Give answers to three significant digits.

1. $x^{2/3} = 75$ **2.** $2^{x-2} = 14$ **3.** $10^{2x} = 5^{x+2}$ Obj. 10-6, p. 478

4. You invest $2000 in a bank offering 10% interest compounded   Obj. 10-7, p. 483
quarterly. Find the value of your investment after a total of five
years' growth.

5. If $e^{3x+2} = 5$, find x in terms of natural logarithms.   Obj. 10-8, p. 489

Check your answers with those at the back of the book.

Application / *Radiocarbon Dating*

When archaeologists uncover a piece of bone or wood from the site of a
dig, they are often able to date the time when the animal or tree was alive
by using radiocarbon dating. Most of the carbon in Earth's atmosphere is
the isotope C-12, but a small amount is the radioactive isotope C-14.

The C-14 decays with a half-life of 5730 years,
but the ratio of C-14 to C-12 in the atmosphere remains approximately constant because
the C-14 is restored by cosmic ray bombard-
ment of atoms in the atmosphere. The carbon
taken in and used by growing plants and animals contains the same fraction of C-14 as the
atmosphere. When an organism dies, the
amount of C-12 it contains remains the same,
but the C-14 decays. Archaeologists therefore
use the half-life decay formula,

$$N = N_0 \cdot \left(\frac{1}{2}\right)^{t/h}, \quad \text{or} \quad \frac{N}{N_0} = \left(\frac{1}{2}\right)^{t/h},$$

where N is the amount of C-14 in the sample
discovered, N_0 is the amount of C-14 in the
sample when it was alive, and h is the half-
life of C-14, 5730 years.

Exponential and Logarithmic Functions **493**

The quantity N is found by measuring the radioactivity of the sample. N_0 is not known, but it can be estimated by finding the total amount of carbon in the sample and assuming that the atmospheric C-14 to C-12 ratio was the same when the sample was alive as it is now, about 1 to 10^{12}. The age t of the sample can then be computed. For example, if the amount N of C-14 is $\frac{1}{8}$ of the estimated amount N_0 that was in the sample when it was alive, then

$$\frac{N}{N_0} = \frac{1}{8} = \left(\frac{1}{2}\right)^{t/5730}$$

$$2^{-3} = 2^{-t/5730}$$

$$-3 = -\frac{t}{5730}$$

$$t = 3 \cdot 5730 = 17{,}190$$

Therefore, the age of the sample is approximately 17,000 years.

The dates found by radiocarbon dating can often be checked against other dating methods. For example, *dendrochronology* (establishing a chronology by correlating sequences of growth rings in wood) has dated some very old samples of long-lived tree species. A piece of wood from a bristle-cone pine has been dated in this way to 6000 B.C. Radiocarbon dating of this sample gave a date of 5500 B.C., showing that 8000 years ago the atmospheric C-14 to C-12 ratio was greater than it is now. (Can you see why a greater ratio then gives a younger date?) Estimates of such atmospheric changes made by comparing tree-ring and radiocarbon dates can be used to obtain more accurate radiocarbon dates. With such corrections these dates are usually accurate to within two centuries for dates up to 10,000 years ago.

Exercises

1. The amount of C-14 in a sample of wood is 45% of the amount that would be found in a living sample with the same total carbon content. How many years have passed since the sample was part of a living tree? Give your answer to two significant digits.

2. A scientist finds that a bone sample contains C-14 and C-12 in the ratio 0.08 to 10^{12}.

   a. What is the ratio of C-14 in the sample now (N) to C-14 in the sample when it was part of a living animal (N_0)? Assume that the atmospheric C-14 fraction was the same then as it is now.

   b. How many years have passed since the animal was living? Give your answer to two significant digits.

3. Using tree-ring chronology it is found that 400 years have passed since a given wood sample was part of a living tree.

   a. What percent of the original C-14 can a scientist expect to find today?

   b. What percent would a scientist 200 years from now expect to find?

   Give your answers to the nearest tenth of a percent.

Chapter Summary

1. The meaning of rational exponents is defined as follows:

$$b^{p/q} = (\sqrt[q]{b})^p = \sqrt[q]{b^p}$$

   Rational exponents can be used to define powers with real exponents, and the laws of exponents hold for real exponents.

2. The function defined by $y = b^x$ is called the *exponential function* with base b, where $b > 0$ and $b \neq 1$.

3. The function whose value at x is $f(g(x))$ is called the *composite* of f and g. The operation that combines f and g to produce their composite is called *composition*.

4. *Inverse functions* are functions that "undo" one another. The functions f and g are inverse functions if

$$f(g(x)) = x \text{ for all } x \text{ in the domain of } g$$

   and $\quad g(f(x)) = x$ for all x in the domain of f.

5. The function $y = \log_b x$ is called the *logarithmic function* with base b, where $b > 0$ and $b \neq 1$. This function is the inverse of the exponential function with base b.

6. *Logarithms* are defined by the relationship

$$\log_b N = k \text{ if and only if } b^k = N,$$

   where b and N are positive numbers and $b \neq 1$.

7. In working with logarithms, you can use the *laws of logarithms*.

   Law 1 $\qquad \log_b MN = \log_b M + \log_b N$

   Law 2 $\qquad \log_b \dfrac{M}{N} = \log_b M - \log_b N$

   Law 3 $\qquad \log_b M^k = k \log_b M$

8. There are certain types of equations for which logarithms provide the best, or even the only, method of solution. Logarithms are used in studying exponential equations and their applications to such occurrences as compound interest, population growth, and radioactive decay. Here are some related formulas:

   Compound interest formula: $\quad A = P\left(1 + \dfrac{r}{n}\right)^{nt}$ $\quad$ (See page 483.)

   Doubling-time growth formula: $\quad N = N_0 \cdot 2^{t/d}$ $\quad$ (See page 484.)

   Half-life decay formula: $\quad N = N_0 \cdot \left(\dfrac{1}{2}\right)^{t/h}$ $\quad$ (See page 485.)

9. An important function used in science and mathematics is the *natural logarithm function* denoted by $\ln x$ or $\log_e x$. Its base is the irrational number e, which is approximately 2.71828.

Chapter Review

Write the letter of the correct answer.

1. For positive x and y, which expression is equivalent to $(x^{1/2} - y^{1/2})^2$?
   a. $x - y$ **b.** $x + 2\sqrt{xy} + y$ **c.** $x - 2\sqrt{xy} + y$

   10-1

2. Simplify $\dfrac{2}{\sqrt[6]{8}}$.

   a. $\dfrac{1}{\sqrt{2}}$ **b.** $\dfrac{1}{2}$ **c.** $\sqrt{2}$ **d.** $\sqrt[3]{4}$

3. Solve $25^{x+2} = 5^{3x-3}$.

   10-2

   a. 2   **b.** 3   **c.** 4   **d.** 7

4. Simplify $\left(\dfrac{2^{1+\sqrt{2}}}{2^{1-\sqrt{2}}}\right)^{\sqrt{2}}$.

   a. 4   **b.** $2^{\sqrt{2}}$ **c.** 8   **d.** 16

5. Which of the following could be the graph of $y = \left(\dfrac{1}{3}\right)^x$?

   a.   **b.**   **c.**   **d.**

6. If $f(x) = 3x - 2$ and $g(x) = 2x - 1$, then $g(f(x))$ is equal to:

   10-3

   a. $6x - 4$ **b.** $6x - 5$ **c.** $6x^2 - x - 2$ **d.** $6x - 3$

7. Are the functions f and g inverse functions if $f(x) = \dfrac{5}{3}x + 1$ and $g(x) = \dfrac{3(x-1)}{5}$?

   a. Yes   **b.** No

8. If $f(x) = 9x - 7$, find the function g that is the inverse of f.

   a. $g(x) = 7x + 9$ **b.** $g(x) = \dfrac{1}{9}x + 7$ **c.** $g(x) = \dfrac{x+7}{9}$

9. Simplify $\log_4 64$.

   10-4

   a. $\dfrac{1}{3}$ **b.** 3   **c.** 16   **d.** 4

10. Between which two numbers is $\log_5 150$?
    a. 0 and 1   **b.** 1 and 2   **c.** 2 and 3   **d.** 3 and 4

11. Simplify $\log_6\left(\dfrac{36}{6^{-10}}\right)$.

    10-5

    a. 12   **b.** -8 **c.** -20 **d.** -5

12. If $\log_b z = \dfrac{1}{3}\log_b x + \log_b y$, write z in terms of x and y.

    a. $y\sqrt[3]{x}$ **b.** $(x + y)^{1/3}$ **c.** $(xy)^{1/3}$ **d.** $\dfrac{x}{3} + y$

13. Solve $10^{5t} = 2$.

10-6

a. $\frac{2}{5}$ **b.** $\frac{\log_{10} 2}{\log_{10} 5}$ **c.** $5 \log_{10} 2$ **d.** $\frac{1}{5} \log_{10} 2$

14. If 40 mg of a radioactive substance decays to 5 mg in 12 min, find the half-life, in minutes, of the substance.

10-7

a. 2 min **b.** 4 min **c.** 6 min **d.** 8 min

15. Simplify $\ln \frac{1}{e^3}$.

10-8

a. 3 **b.** -3 **c.** 0 **d.** 1

16. Solve $e^{2x-1} = 3$.

a. $\ln 3 + 1$ **b.** $\frac{\ln 3}{2x - 1}$ **c.** $\frac{\ln 3 + 1}{2}$ **d.** $1 - \ln 3$

Chapter Test

1. Find the value of x.

10-1

a. $\left(\frac{1}{16}\right)^{-3/4} = x$ **b.** $27^x = 81$

2. Simplify.

a. $\sqrt[3]{\sqrt{125y^6}}$ **b.** $(64^{2/3} + 27^{2/3})^{3/2}$

3. Solve $4^{x-2} = 8^{\pi+1} \div 8^{\pi-1}$.

10-2

4. Suppose $f(x) = 2x - 1$ and $g(x) = x^2 + 4$. Find:

10-3

a. $f(g(-2))$ **b.** $g(f(x))$

5. Suppose that $f(x) = \sqrt[3]{x - 1}$ and $g(x) = x^3 + 1$. Show that f and g are inverse functions.

6. a. Write in logarithmic form using the base 2: $32^{3/5} = 8$.

10-4

b. Write in exponential form: $\log_{16}\left(\frac{1}{64}\right) = -\frac{3}{2}$.

7. Simplify: **a.** $6^{\log_6 3}$ **b.** $\log_3 27^{\sqrt{2}}$

8. Solve: **a.** $\log_3 x = \log_3 12 + \log_3 2 - \log_3 6$

10-5

b. $\log_4 (x - 6) + \log_4 x = 2$

9. If $\log_{10} 2 = 0.301$ and $\log_{10} 3 = 0.477$, find the following.

a. $\log_{10} 8$ **b.** $\log_{10} 12$ **c.** $\log_{10} 15$

10. Express t in terms of common logarithms: $5^{3t} = 2$.

10-6

11. The population of a certain colony of bacteria doubles every 5 hours. How long will it take for the population to triple? Give the answer to two significant digits.

10-7

12. Write x in terms of e: $\ln x^2 = 8$.

10-8

13. Write as a single natural logarithm: $\frac{1}{2} - \ln 7$.

Mixed Problem Solving

Solve each problem that has a solution. If a problem has no solution, explain why.

A

1. How much iodine should be added to 50 mL of a 10% iodine solution to obtain a 25% solution?

2. When the reciprocal of a real number is subtracted from 1, the difference is equal to the number. Find the number.

3. The number of representatives to Congress from California is equal to the sum of the numbers of representatives from New York and North Carolina. Together the three states have 90 representatives. The number from New York is 23 more than the number from North Carolina. How many representatives does each state have?

4. When two meshed gears revolve, their speeds vary inversely as the numbers of teeth they have. A gear with 36 teeth runs at 200 r/min (revolutions/min). Find the speed of a 60-tooth gear meshed with it.

5. The difference of two numbers is 6. Find their minimum product.

6. A rectangular garden has perimeter 44 m. A walkway 2 m wide surrounds the garden. The combined area of the garden and walkway is 224 m^2. Find the dimensions of the garden.

7. When Susan had annual incomes of $15,000 and $18,000, she saved $600 and $900, respectively. If the amount saved is a linear function of the amount earned, how much would she save when her income becomes $24,000?

8. The *eagle* is a 10-dollar gold United States coin minted from 1795 to 1933. Find all the possible ways a person could receive $25 using eagles and quarter-eagles.

B

9. The area of a circle is directly proportional to the square of its circumference. A circle with circumference 14π has area 49π. Find a general formula relating the circumference and the area of a circle.

10. Find three consecutive integers such that the square of the sum of the first two is 80 more than the sum of the squares of the last two.

11. Two water purification plants can process a day's water supply in 5 h. Operating alone, the smaller plant would need 4 h more than the larger one would. How long would each need to do the job alone?

12. The value of a certain car decreases at a rate of 12% annually. About how long does it take for the car to be worth half its original value?

13. If Rachel jogs for awhile at 8 km/h and then walks at 5 km/h, she travels 23 km. If she reverses the amount of time spent jogging and walking, she travels 6 km farther. Find her total traveling time.

Preparing for College Entrance Exams

Strategy for Success

If you skip a question, don't forget to leave the line for that answer blank on your answer sheet. To ensure an accurate pairing between questions and answers, every so often compare the numbering in your test booklet with the one on your answer sheet.

Decide which is the best of the choices given and write the corresponding letter on your answer sheet.

1. Which equation represents the set of all points that are equidistant from $(-7, 4)$ and $(3, 6)$?
   - **(A)** $x - 5y = -27$
   - **(B)** $5x - y = -15$
   - **(C)** $5x + y = -5$
   - **(D)** $5x - y = 15$
   - **(E)** $x + 5y = 23$

2. The graph of $9x^2 - 18 = y^2$ is
   - **(A)** a parabola
   - **(B)** a circle
   - **(C)** an ellipse
   - **(D)** a hyperbola
   - **(E)** a point

3. Which expression below is equivalent to $-3 \log_8 4$?
   - **(A)** $\dfrac{\sqrt[5]{32^4}}{32}$
   - **(B)** $\log_5 \sqrt{5} - \log_5 5\sqrt{5}$
   - **(C)** $\log 1 - \log 0.01$
   - **(D)** $\sqrt{-4}$
   - **(E)** $\ln \dfrac{1}{e^2}$

4. An ellipse with equation $\dfrac{x^2}{16} + \dfrac{y^2}{36} = 1$ has which point as a focus?
   - **(A)** $(0, -2\sqrt{3})$
   - **(B)** $(0, 2\sqrt{5})$
   - **(C)** $(-2\sqrt{5}, 0)$
   - **(D)** $(2\sqrt{13}, 0)$
   - **(E)** $(0, 6)$

5. If $f(x) = \dfrac{4}{x} - 1$, find $f^{-1}(7)$ if possible.
   - **(A)** $\dfrac{1}{2}$
   - **(B)** $-\dfrac{3}{7}$
   - **(C)** 4
   - **(D)** 0
   - **(E)** f^{-1} does not exist.

6. Solve $27^{2t-1} = 81^{t+2}$.
   - **(A)** 3
   - **(B)** -3
   - **(C)** $-\dfrac{1}{2}$
   - **(D)** -2
   - **(E)** $\dfrac{11}{2}$

7. Find the number of real solutions of this system: $\quad x^2 + 16y^2 = 25$
   $$xy - 3 = 0$$
   - **(A)** 0
   - **(B)** 1
   - **(C)** 2
   - **(D)** 3
   - **(E)** 4

8. Find the z-coordinate of the solution of this system: $\quad 5x - 3y + z = 5$
   $$4x + 3y - 2z = -4$$
   $$2x - 3y - 7z = 13$$
   - **(A)** 0
   - **(B)** 1
   - **(C)** -1
   - **(D)** 2
   - **(E)** -2

11 Sequences and Series

The spiral form of a nautilus shell can be approximated by drawing quarter circles within squares positioned in a spiral sequence.

Sequences

11-1 Types of Sequences

Objective   To determine whether a sequence is arithmetic, geometric, or neither and to supply missing terms of a sequence.

Sequences of numbers are very common in the world around us. Here are some examples:

(1)   Detroit's noontime temperatures
        (in °F) for a particularly cold week $18°, 15°, 17°, 14°, 0°, -2°, -5°$

(2)   Value (in dollars and cents) of an
        investment at three-month intervals $507.50, 515.11, 522.84, 530.68, \ldots$

(3)   Total distance (in miles) traveled
        by a jet after each hour of a trans-
        continental flight $500, 1100, 1700, 2300, 2900$

(4)   Number of bacteria at each stage in
        the growth of a colony $1, 2, 4, 8, 16, \ldots$

        Because the members of a sequence are arranged in a definite order, you can set up a correspondence between the members and their positions in the list. A correspondence for sequence (4) is shown in the mapping diagram at the right.

        A **sequence** can be defined as a function whose domain consists of consecutive positive integers. Each corresponding value is called a **term** of the sequence. A sequence is **finite** if it has a limited number of terms and **infinite** if it does not.

        The terms of sequence (1) above seem to have no pattern in their occurrence. On the other hand, there are definite patterns in sequences (2)–(4). If you know what the pattern is, then you can find terms of the sequence that are not shown. For example, in sequence (3), the *difference* between each term and the term before it is always 600. If the terms of this sequence were to continue, then the next term would be $2900 + 600$, or 3500. Sequence (3) is an example of an *arithmetic sequence*.

A sequence in which the difference between any two successive terms is constant is called an **arithmetic sequence,** or **arithmetic progression.** The constant difference is called the **common difference** and is usually denoted by d.

Example 1 For each arithmetic sequence, find the common difference and the next two terms of the sequence.

a. 4, 7, 10, 13, 16, . . .   b. 50, 45, 40, 35, 30, . . .

Solution a. The common difference is found by subtracting any term from the term that follows it.

$$d = 7 - 4 = 3$$

The next two terms are $16 + 3 = 19$ and $19 + 3 = 22$.   *Answer*

b. The common difference is

$$d = 45 - 50 = -5.$$

The next two terms are $30 + (-5) = 25$ and $25 + (-5) = 20$.   *Answer*

Look at the pattern in sequence (4): 1, 2, 4, 8, 16, ... The sequence is not arithmetic because the difference between successive terms is *not* constant. However, the *ratio* of successive terms *is* constant.

$$\frac{2}{1} = \frac{4}{2} = \frac{8}{4} = \frac{16}{8} = \cdots$$

Since the ratio is always 2, every term is twice the preceding term. So the next term of this sequence would be $16 \cdot 2$, or 32. Sequence (4) is an example of a *geometric sequence*.

A sequence in which the ratio of every pair of successive terms is constant is called a **geometric sequence,** or **geometric progression.** The constant ratio is called the **common ratio** and is usually denoted by r.

Notice that sequence (2) on the preceding page is almost a geometric sequence. Each pair of successive terms has a ratio approximately equal to 1.015. In fact, the sequence was obtained by rounding the terms of a true geometric sequence to the nearest hundredth.

Example 2 For each geometric sequence, find the common ratio and the next two terms of the sequence.

a. 2, 6, 18, 54, . . .   b. 80, −40, 20, −10, . . .

Solution a. Find the common ratio by dividing any term by the term before it:

$$r = \frac{6}{2} = 3$$

The next two terms are $54 \cdot 3 = 162$ and $162 \cdot 3 = 486$.   *Answer*

b. The common ratio is

$$r = \frac{-40}{80} = -\frac{1}{2}.$$

The next two terms are $-10 \cdot \left(-\frac{1}{2}\right) = 5$ and $5 \cdot \left(-\frac{1}{2}\right) = -\frac{5}{2}$.   *Answer*

The nth term of a sequence is usually denoted by the symbol t_n. For example, in the sequence 2, 6, 18, 54, . . . , the first term is $t_1 = 2$, the second term is $t_2 = 6$, and so on. If you know a formula for the nth term of a sequence in terms of n, then you can find *any* term of the sequence.

Example 3 Using the given formula for the nth term, find t_1, t_2, t_3, and t_4. Then tell whether the sequence is arithmetic, geometric, or neither.

a. $t_n = 5 + 4n$ **b.** $t_n = n^2$ **c.** $t_n = 3 \cdot 2^n$

Solution **a.**

n	$t_n = 5 + 4n$
1	$t_1 = 5 + 4(1) = 9$
2	$t_2 = 5 + 4(2) = 13$
3	$t_3 = 5 + 4(3) = 17$
4	$t_4 = 5 + 4(4) = 21$

The sequence 9, 13, 17, 21, . . . is arithmetic since the common difference is 4. *Answer*

b.

n	$t_n = n^2$
1	$t_1 = 1^2 = 1$
2	$t_2 = 2^2 = 4$
3	$t_3 = 3^2 = 9$
4	$t_4 = 4^2 = 16$

The sequence 1, 4, 9, 16, . . . is neither arithmetic nor geometric. *Answer*

c.

n	$t_n = 3 \cdot 2^n$
1	$t_1 = 3 \cdot 2^1 = 6$
2	$t_2 = 3 \cdot 2^2 = 12$
3	$t_3 = 3 \cdot 2^3 = 24$
4	$t_4 = 3 \cdot 2^4 = 48$

The sequence 6, 12, 24, 48, . . . is geometric since the common ratio is 2. *Answer*

Sometimes it is possible to find a formula for the nth term of a sequence if a sufficient number of terms are given to indicate a pattern. (This will be discussed in Lessons 11-2 and 11-3.) Of course, when you use a pattern to guess a formula for t_n, you are making the assumption that the pattern continues throughout the sequence.

When it is difficult to find a pattern in a sequence of numbers, you may be able to find a pattern in the differences between terms.

Example 4 Find the next term in the sequence 2, 6, 12, 20, 30, . . . by using the pattern in the differences between terms.

Solution

sequence   2   6   12   20   30   ?

differences   4   6   8   10   ?

Although the original sequence is neither arithmetic nor geometric, the *differences* form an arithmetic sequence with common difference 2. So the next difference will be 12, and the next term in the original sequence will be 30 + 12, or 42. *Answer*

Oral Exercises

Give the common difference and supply the missing terms for each arithmetic sequence.

1. 3, 7, 11, 15, __?__ , __?__

2. 21, 15, 9, 3, __?__ , __?__

3. 7, 10, __?__ , 16, 19, __?__

4. __?__ , __?__ , 25, 50, 75, 100

Give the common ratio and supply the missing terms for each geometric sequence.

5. 3, 6, 12, 24, __?__ , __?__

6. 1, -2, 4, -8, __?__ , __?__

7. $\frac{1}{100}$, $\frac{1}{10}$, __?__ , 10, 100, __?__

8. __?__ , __?__ , $\frac{1}{3}$, $\frac{1}{9}$, $\frac{1}{27}$, $\frac{1}{81}$

Tell the first four terms of the sequence with the given formula. Then tell whether the sequence is arithmetic, geometric, or neither.

9. $t_n = 1 - 2n$

10. $t_n = \dfrac{1}{n+1}$

11. $t_n = 3^n$

12. $t_n = n^2 - 1$

Give the next two terms of each sequence by using the pattern in the differences between terms.

13. 8, 9, 11, 14, __?__ , __?__

14. 5, 7, 11, 17, __?__ , __?__

Written Exercises

Tell whether each sequence is arithmetic, geometric, or neither. Then supply the missing terms of the sequence.

A

1. 20, 17, 14, 11, __?__ , __?__

2. 5, 9, 13, 17, __?__ , __?__

3. 1, 5, 25, 125, __?__ , __?__

4. 256, 64, 16, 4, __?__ , __?__

5. 18, 22, 26, __?__ , 34, __?__

6. 4, __?__ , -4, -8, -12, __?__

7. 1, $\frac{1}{4}$, $\frac{1}{9}$, $\frac{1}{16}$, __?__ , __?__

8. 32, -16, 8, -4, __?__ , __?__

9. $4^{1/2}$, $4^{3/2}$, $4^{5/2}$, $4^{7/2}$, __?__ , __?__

10. $\frac{1}{12}$, $\frac{2}{13}$, $\frac{3}{14}$, $\frac{4}{15}$, __?__ , __?__

Find the first four terms of the sequence with the given formula. Then tell whether the sequence is arithmetic, geometric, or neither.

11. $t_n = 4n + 3$

12. $t_n = 2n + 1$

13. $t_n = 3^{n-1}$

14. $t_n = 2 \cdot 3^n$

15. $t_n = \dfrac{(-2)^n}{8}$

16. $t_n = 13 - 4n$

17. $t_n = \log (n + 1)$

18. $t_n = \log 10^n$

19. a. What type of sequence is $-3, -1, 1, 3, \ldots$?

   b. What type of sequence is $2^{-3}, 2^{-1}, 2^1, 2^3, \ldots$?

20. a. What type of sequence is $1, 4, 16, 64, \ldots$?

   b. What type of sequence is $\log_2 1, \log_2 4, \log_2 16, \log_2 64, \ldots$?

Find the next two terms of each sequence by using the pattern in the differences between terms.

B **21.** $2, 4, 8, 14, 22, \ldots$ **22.** $-3, 1, 9, 21, 37, \ldots$

23. $60, 48, 38, 30, 24, \ldots$ **24.** $24, 23, 21, 17, 9, \ldots$

25. $1, 3, 7, 15, 31, \ldots$ **26.** $0, 1, 4, 13, 40, \ldots$

27. $1, 1, 2, 3, 5, 8, 13, \ldots$ **28.** $1, 2, 6, 15, 31, \ldots$

   (The well-known sequence in Exercise 27 is called the *Fibonacci sequence.* Each term is the sum of the two terms before it.)

29. $1, 3, 6, 11, 19, 31, \ldots$ **30.** $5, 7, 10, 16, 27, 45, \ldots$

[*Hint:* For Exercises 29 and 30, look at the *second* differences (that is, the differences of the differences between terms).]

31. A *triangular number* can be represented by dots that are arranged in the shape of an equilateral triangle, as shown below. The first four triangular numbers are given.

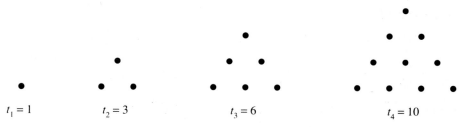

$t_1 = 1$ $t_2 = 3$ $t_3 = 6$ $t_4 = 10$

   a. Find the next two triangular numbers.

   b. Find the tenth triangular number without actually drawing a diagram.

32. A *pentagonal number* can be represented by dots that are arranged in the shape of a pentagon, as shown below. The first four pentagonal numbers are given.

$t_1 = 1$ $t_2 = 5$ $t_3 = 12$ $t_4 = 22$

   a. Find the next two pentagonal numbers.

   b. Find the tenth pentagonal number without drawing a diagram.

33. A diagonal of a polygon is a line segment joining any two nonadjacent vertices of the polygon. If n is the number of sides of a polygon ($n \geq 3$), then let t_n be the number of its diagonals.

$t_3 = 0$

$t_4 = 2$

$t_5 = 5$

a. Draw a hexagon and find t_6.   **b.** Find t_{10} without drawing a decagon.

34. If n chords of a circle are drawn, then let t_n be the *maximum* number of non-overlapping regions that can be formed within the circle, as shown.

1 chord
$t_1 = 2$ regions

2 chords
$t_2 = 4$ regions

3 chords
$t_3 = 7$ regions

a. Use drawings to find t_4 and t_5.   **b.** Find t_6 without using a drawing.

C **35.** Let t_n be the *maximum* number of non-overlapping regions that can be formed when n points of a circle are connected, as shown.

1 point
$t_1 = 1$ region

2 points
$t_2 = 2$ regions

3 points
$t_3 = 4$ regions

4 points
$t_4 = 8$ regions

a. Make careful drawings and determine t_5 and t_6 by counting the regions formed. Is the resulting sequence geometric?

b. Use the pattern in the *third* differences (that is, the differences of the differences of the differences between terms) to guess the value of t_7.

Mixed Review Exercises

Graph each equation.

1. $y = x^2 - 2x + 3$

2. $y = 2^{-x}$

3. $4x^2 + y^2 = 16$

4. $y = \log_4 x$

5. $x^2 + 4y^2 = 36$

6. $(x - 1)^2 + (y + 2)^2 = 9$

Solve each system.

7. $2x + y = 1$
   $x - y = 5$

8. $y = 2x^2 - 1$
   $x + y = 2$

9. $x^2 + y^2 = 5$
   $x^2 - y^2 = 3$

11-2 Arithmetic Sequences

Objective   To find a formula for the *n*th term of an arithmetic sequence and to find specified terms of arithmetic sequences.

How would you find the hundredth term (t_{100}) of the arithmetic sequence

$$3, 7, 11, 15, \ldots ?$$

One way is to consider the number of terms in the sequence and the number of differences between the terms.

As you can see from the diagram, the number of differences is always one less than the number of terms. Therefore, to find t_{100}, start with the first term 3 and add the common difference 4 ninety-nine times.

$$t_{100} = 3 + 99 \cdot 4$$
$$= 399$$

Example 1   Find a formula for the *n*th term of the sequence 5, 8, 11, 14, ...

Solution   The first term is $t_1 = 5$. The common difference is $d = 8 - 5 = 3$.
Since there are *n* terms, start with 5 and add the difference $n - 1$ times.
$$t_n = 5 + (n - 1)3$$
$$= 2 + 3n \quad \textbf{\textit{Answer}}$$

The method used in Example 1 can be generalized to *any* arithmetic sequence.

In an arithmetic sequence with first term t_1 and common difference *d*, the *n*th (or *general*) term is given by
$$t_n = t_1 + (n - 1)d.$$

Example 2   Find t_{17} for the following arithmetic sequence:
$$5, 8, 11, 14, \ldots$$

Solution   Use the formula $t_n = 2 + 3n$ from Example 1.
$$t_{17} = 2 + 3(17)$$
$$= 53 \quad \textbf{\textit{Answer}}$$

Example 3 Find t_{23} for the arithmetic sequence in which $t_2 = 4$ and $t_5 = 22$.

Solution Substitute 4 for t_2 and 22 for t_5 in the formula $t_n = t_1 + (n - 1)d$ to obtain a system of linear equations in t_1 and d.

$$t_2 = t_1 + (2 - 1)d \longrightarrow 4 = t_1 + d$$
$$t_5 = t_1 + (5 - 1)d \longrightarrow 22 = t_1 + 4d$$

Solve the first equation for t_1: $t_1 = 4 - d$.
Substitute $4 - d$ for t_1 in the second equation and solve for d:

$$22 = (4 - d) + 4d$$
$$22 = 4 + 3d$$
$$d = 6$$

Then $t_1 = 4 - d = -2$. Now using $t_1 = -2$ and $d = 6$ in the formula $t_n = t_1 + (n - 1)d$, find t_{23}:

$$t_{23} = t_1 + (23 - 1)d$$
$$= -2 + 22 \cdot 6$$
$$= 130 \quad \textbf{\textit{Answer}}$$

The terms between two given terms of an arithmetic sequence are called **arithmetic means** of the given terms.

10, 13, 16, 19, 22	10, 14, 18, 22	10, 16, 22
Three arithmetic means between 10 and 22	Two arithmetic means between 10 and 22	*The* arithmetic mean of 10 and 22

A single arithmetic mean between two numbers is called *the* **arithmetic mean** of the numbers. The arithmetic mean, or *average*, of two numbers a and b is $\dfrac{a + b}{2}$.

Example 4 **a.** Find the arithmetic mean of 4 and 9.
b. Insert four arithmetic means between 15 and 50.

Solution **a.** The arithmetic mean is the average of 4 and 9.

$$\frac{4 + 9}{2} = \frac{13}{2} = 6.5 \quad \textbf{\textit{Answer}}$$

b. Outline the sequence: 15, _?_, _?_, _?_, _?_, 50

In this sequence, 15 is the *first* term and 50 is the *sixth* term. So to find d substitute 15 for t_1 and 50 for t_6 in the formula $t_n = t_1 + (n - 1)d$.

$$50 = 15 + (6 - 1)d$$
$$35 = 5d$$
$$d = 7$$

The four means are obtained by adding 7 to successive terms:
15, 22, 29, 36, 43, 50. *Answer*

Give a formula for the *n*th term of each arithmetic sequence.

1. −1, 0, 1, 2, . . .   **2.** 2, 4, 6, 8, . . .   **3.** 0, 3, 6, 9, . . .

For each arithmetic sequence, find the value of t_{11}.

4. 4, 7, 10, 13, . . .   **5.** 10, 5, 0, −5, . . .   **6.** 3, 13, 23, 33, . . .

Written Exercises

Find a formula for the *n*th term of each arithmetic sequence.

A   **1.** 24, 32, 40, 48, . . .   **2.** 30, 20, 10, 0, . . .
   3. −3, −10, −17, −24, . . .   **4.** −6, −1, 4, 9, . . .
   5. 7, 11, 15, 19, . . .   **6.** 13, 4, −5, −14, . . .

Find the specified term of each arithmetic sequence.

   7. 4, 9, 14, 19, . . . ; t_{21} **8.** 3, 11, 19, . . . ; t_{31}
   9. 100, 98, 96, . . . ; t_{25} **10.** 3, 3.5, 4, 4.5, . . . ; t_{101}
   11. −2, −11, −20, . . . ; t_{101} **12.** 17, 7, −3, . . . ; t_{1000}
   13. $t_1 = 5$, $t_3 = 20$; t_{12} **14.** $t_2 = 7$, $t_4 = 8$; t_1
   15. $t_5 = 24$, $t_9 = 40$; t_1 **16.** $t_8 = 60$, $t_{12} = 48$; t_{40}
   17. $t_7 = -19$, $t_{10} = -28$; t_{21} **18.** $t_{10} = 41$, $t_{15} = 61$; t_3

Find the arithmetic mean of each pair of numbers.

   19. −3, 7   **20.** 2.3, 9.1   **21.** $\dfrac{4}{5}, \dfrac{11}{5}$ **22.** $-\sqrt{2}, 3\sqrt{2}$

Insert (a) two, (b) three, and (c) four arithmetic means between each pair.

B   **23.** −27, 33   **24.** 15, 45   **25.** 11, 35   **26.** 0, 20
   27. How many terms are in the sequence 18, 24, 30, . . . , 618?
   28. How many terms are in the sequence 44, 36, 28, . . . , −380?
   29. How many multiples of 3 are there between 100 and 1000?
   30. How many numbers between 50 and 500 are divisible by 7?

Find the position *n* of the term in red in each arithmetic sequence.

   31. 25, 33, 41, . . . , 145, . . .   **32.** 40, 37, 34, . . . , −29, . . .

C   **33.** Show that the sequence $a, \dfrac{a + b}{2}, b$ is arithmetic.

11-3 Geometric Sequences

Objective   To find a formula for the nth term of a geometric sequence and to find specified terms of geometric sequences.

The method for finding the nth term of a geometric sequence is similar to the method used for arithmetic sequences. For example, to find the tenth term of the geometric sequences 5, 10, 20, 40, . . . , look at the number of times the first term 5 is multiplied by the common ratio 2 to produce t_{10}.

As the diagram shows, the number of ratios is always one less than the number of terms. Therefore, the tenth term is the product of 5 and the ninth power of 2.

$$t_{10} = 5 \cdot 2^9$$
$$= 2560$$

Example 1   Find a formula for the nth term of the sequence 3, -12, 48, -192, ...

Solution   The first term is $t_1 = 3$. The common ratio is $r = \dfrac{-12}{3} = -4$.

Since there are n terms, start with 3 and multiply by the ratio $n - 1$ times:
$$t_n = 3(-4)^{n-1} \quad \textbf{\textit{Answer}}$$

The method used in Example 1 can be generalized to *any* geometric sequence.

In a geometric sequence with first term t_1 and common ratio r, the nth (or *general*) term is given by

$$t_n = t_1 \cdot r^{n-1}.$$

Example 2   Find t_6 for the following geometric sequence:
$$3, -12, 48, \ldots$$

Solution   Use the formula $t_n = 3(-4)^{n-1}$ from Example 1.
$$t_6 = 3(-4)^{6-1}$$
$$= 3(-4)^5$$
$$= 3(-1024) = -3072 \quad \textbf{\textit{Answer}}$$

Example 3 Find t_7 for the geometric sequence in which $t_2 = 24$ and $t_5 = 3$.

Solution Substitute 24 for t_2 and 3 for t_5 in the formula $t_n = t_1 \cdot r^{n-1}$ to obtain a system of equations in t_1 and r.

$$t_2 = t_1 \cdot r^{2-1} \longrightarrow 24 = t_1 \cdot r$$
$$t_5 = t_1 \cdot r^{5-1} \longrightarrow 3 = t_1 \cdot r^4$$

Solve the first equation for t_1: $t_1 = \dfrac{24}{r}$. Substitute this expression for t_1 in the second equation and solve for r:

$$3 = \frac{24}{r} \cdot r^4$$

$$\frac{1}{8} = r^3, \quad \text{so } r = \frac{1}{2}$$

Substitute $\frac{1}{2}$ for r in the first equation:

$$24 = t_1 \cdot \frac{1}{2}$$
$$48 = t_1$$

Using $t_1 = 48$ and $r = \frac{1}{2}$ in the formula $t_n = t_1 \cdot r^{n-1}$, find t_7:

$$t_7 = 48 \cdot \left(\frac{1}{2}\right)^{7-1}$$

$$= 48 \cdot \frac{1}{64} = \frac{3}{4} \quad \textbf{\textit{Answer}}$$

Geometric means are the terms between two given terms of a geometric sequence.

1, 2, 4, 8, 16	1, −2, 4, −8, 16	1, 4, 16
Three geometric means between 1 and 16	Three geometric means between 1 and 16	*The* geometric mean of 1 and 16

Because 1, 4, 16 and 1, −4, 16 are both geometric sequences, you might think that 4 or −4 is the geometric mean of 1 and 16. However, so that the geometric mean of two numbers will be unique, it is a common practice to consider *the* **geometric mean** of a and b to be $\sqrt{ab}$ if a and b are positive and $-\sqrt{ab}$ if a and b are negative.

Example 4 **a.** Find the geometric mean of 4 and 9.

b. Insert three geometric means between $\dfrac{1}{2}$ and $\dfrac{1}{162}$.

Solution **a.** The geometric mean of 4 and 9 is $\sqrt{4 \cdot 9} = \sqrt{36}$, or 6. **_Answer_**

b. Outline the sequence: $\dfrac{1}{2}, \underline{}, \underline{}, \underline{}, \dfrac{1}{162}$

(Solution continues on the next page.)

To find r, substitute $\frac{1}{162}$ for t_5 and $\frac{1}{2}$ for t_1 in the formula

$t_n = t \cdot r^{n-1}$:

$$\frac{1}{162} = \frac{1}{2} \cdot r^{5-1}$$

$$\frac{1}{81} = r^4$$

$$\pm \frac{1}{3} = r$$

If $r = \frac{1}{3}$, then the geometric sequence is $\frac{1}{2}, \frac{1}{6}, \frac{1}{18}, \frac{1}{54}, \frac{1}{162}$.

If $r = -\frac{1}{3}$, then the geometric sequence is $\frac{1}{2}, -\frac{1}{6}, \frac{1}{18}, -\frac{1}{54}, \frac{1}{162}$.

Answer

Example 5 George has taken a job with a starting salary of $20,000. Find his salary during his fourth year on the job if he receives annual raises of:

a. $1100   **b.** 5%

Solution **a.** With annual raises of $1100, George's yearly salaries form an *arithmetic* sequence with $t_1 = 20{,}000$ and $d = 1100$.
The fourth term in the sequence is:

$$t_4 = 20{,}000 + (4 - 1)1100$$
$$= 20{,}000 + 3300$$
$$= 23{,}300$$

∴ George's fourth-year salary will be $23,300.   *Answer*

b. With annual raises of 5%, George's yearly salaries form a *geometric* sequence with $t_1 = 20{,}000$ and $r = 1.05$. (Notice that the common ratio is 1.05 and not 0.05, because a 5% raise means that for a salary S one year, the salary will be $S + 0.05S$, or $1.05S$, the next year.)
The fourth term in the sequence is:

$$t_4 = 20{,}000(1.05)^{4-1}$$
$$= 20{,}000(1.157625)$$
$$= 23{,}152.50$$

∴ George's fourth-year salary will be $23,152.50.   *Answer*

Oral Exercises

Give a formula for the nth term of each geometric sequence.

1. 1000, 200, 40, 8, . . .   **2.** 3, −6, 12, −24, . . .

3. 4, −12, 36, −108, . . .   **4.** 125, 25, 5, 1, . . .

For each geometric sequence, find the value of t_{11} (in exponential form).

5. 1, 4, 16, 64, . . .   **6.** 5, 10, 20, 40, . . .   **7.** 18, −6, 2, −$\frac{2}{3}$, . . .

8. Give the geometric mean of each pair of numbers.

   a. 1, 4   **b.** −3, −27   **c.** 3, 12

Written Exercises

Find a formula for the nth term of each geometric sequence.

A **1.** 2, 6, 18, 54, . . .   **2.** 500, 100, 20, 4, . . .

   3. 1, $\sqrt{2}$, 2, 2$\sqrt{2}$, . . .   **4.** 8, 12, 18, 27, . . .

   5. 64, −48, 36, −27, . . .   **6.** −1, 0.1, −0.01, 0.001, . . .

Find the specified term of each geometric sequence.

   7. 2, 6, 18, 54, . . . ; t_{10} **8.** 5, 10, 20, 40, . . . ; t_{12}

   9. 320, 80, 20, 5, . . . ; t_8 **10.** 1, −3, 9, −27, . . . ; t_8

   11. 40, −20, 10, −5, . . . ; t_{11} **12.** −10, 50, −250, 1250, . . . ; t_9

   13. $t_2 = 18$, $t_3 = 12$; t_5 **14.** $t_3 = -12$, $t_6 = 96$; t_9

   15. $t_1 = 5$, $t_3 = 80$; t_6 **16.** $t_2 = 8$, $t_4 = 72$; t_1

   17. y, y^3, y^5, . . . ; t_{20} **18.** ab^2, a^2b^5, a^3b^8, . . . ; t_{25}

Find the geometric mean of each pair of numbers.

   19. 2, 8   **20.** $\frac{1}{12}$, $\frac{1}{18}$ **21.** $\sqrt{3}$, 3$\sqrt{3}$ **22.** −18, −36

Insert the given number of geometric means between the pairs of numbers.

   23. Three;   5, 80   **24.** Two;   −4, 108   **25.** Four;   1, 2   **26.** Three; $\frac{1}{5}$, $\frac{5}{4}$

Tell whether each sequence is arithmetic or geometric. Then find a formula for the nth term.

B **27.** The sequence of positive even integers   **28.** The sequence of odd integers greater than two

   29. 25, 33, 41, 49, . . .   **30.** −17, −11, −5, 1, . . .

   31. 200, −100, 50, −25   **32.** e^x, e^{2x}, e^{3x}, e^{4x}, . . .

   33. $2a + 1$, $3a + 3$, $4a + 5$, . . .   **34.** $\frac{a^2}{2}$, $\frac{a^4}{4}$, $\frac{a^6}{8}$, . . .

   35. The sequence of negative even integers that are multiples of 5   **36.** The sequence of positive integers that give a remainder of 1 when divided by 4

Find a formula for the *n*th term of each sequence. The sequences are neither arithmetic nor geometric. (*Hint:* Analyze the patterns in the numerators and denominators separately.)

37. $\dfrac{2}{1}, \dfrac{3}{4}, \dfrac{4}{9}, \dfrac{5}{16}, \ldots$

38. $\dfrac{1}{2}, \dfrac{2}{5}, \dfrac{3}{8}, \dfrac{4}{11}, \ldots$

C 39. Show that the sequence a, $\sqrt{ab}$, b (a, b positive) is geometric.

40. The first three numbers of the sequence 8, x, y, 36 form an arithmetic sequence, but the last three numbers form a geometric sequence. Find all possible values of x and y.

Problems

The following problems involve arithmetic and geometric sequences. A calculator may be helpful in solving these problems.

A 1. Allysa has taken a job with a starting salary of $17,600 and annual raises of $850. What will be her salary during her fifth year on the job?

2. Frank has taken a job with a starting salary of $15,000 and annual raises of 4%. What will be his salary during his third year on the job?

3. The cost of an annual subscription to a magazine is $20 this year. The cost is expected to rise by 10% each year. What will be the cost 6 years from now?

4. A new pair of running shoes costs $70 now. Assuming an annual 8% price increase, find the price 4 years from now.

5. A carpenter is building a staircase from the first floor to the second floor of a house. The distance between floors is 3.3 m. Each step rises 22 cm. Not counting either floor itself, how many steps will there be?

6. The width of a pyramid decreases by 1.57 m for each successive 1 m of height. If the width at a height of 1 m is 229.22 m, what is the width at a height of 86 m?

7. A culture of yeast doubles in size every 4 hours. If the yeast population is estimated to be 3 million now, what will it be one day from now?

8. An advertisement for a mutual fund claims that people who invested in the fund 5 years ago have doubled their money. If the fund's future performance is similar to its past performance, how much would a $2000 investment be worth in 40 years?

B **9.** A pile of bricks has 85 bricks in the bottom row, 79 bricks in the second row, 73 in the third row, and so on until there is only 1 brick in the top row.

   a. How many bricks are in the 12th row?

   b. How many rows are there in all?

10. A projectile fired vertically upward rises 15,840 ft the first second, 15,808 ft the following second, and 15,776 ft the third second.

   a. How many feet does it rise the 45th second?

   b. How many feet and in what direction does it move during the first second of the tenth minute after it has been fired?

11. A new $12,000 automobile decreases in value by 25% each year. What is its value 7 years from now?

12. A house purchased last year for $80,000 is now worth $96,000. Assuming that the value of the house continues to appreciate (increase) at the same rate each year, find the value 2 years from now.

13. Job A has a starting salary of $12,000 with annual increases of $800. Job B has a starting salary of $11,000 with annual increases of 10%. Which job will pay more after 3 years? after 5 years?

C **14.** There are 12 steps in the chromatic scale from the A below middle C to the next higher A. This scale can be played by playing the white and black keys of a piano in order from left to right. The frequency of a note is given in hertz (Hz). For example, the A below middle C has a frequency of 220 Hz, which means the piano string vibrates 220 times per second. Notice that the frequency of the A above middle C is twice as great, 440 Hz. If the frequencies of the notes in the chromatic scale form a geometric sequence, show that the common ratio is $2^{1/12}$ and then find the frequency of middle C.

Mixed Review Exercises

Solve each inequality and graph the solution set.

   1. $|2x - 5| > 9$ **2.** $x^2 + x \leq 6$ **3.** $2 \leq \frac{x}{2} + 4 < 5$

   4. $x^3 > 4x$ **5.** $5 - 3x \geq -1$ **6.** $|x + 3| < 2$

Find an equation for each figure described.

   7. The line containing $(-2, 3)$ and $(1, 0)$

   8. The parabola with focus $(0, -1)$ and directrix $y = 1$

   9. The ellipse with x-intercepts ± 2 and y-intercepts $\pm \sqrt{3}$

Self-Test 1

Vocabulary   sequence (p. 501)   the arithmetic mean (p. 508)
            terms of a sequence (p. 501)   average (p. 508)
            finite sequence (p. 501)   geometric sequence (p. 502)
            infinite sequence (p. 501)   geometric progression (p. 502)
            arithmetic sequence (p. 501)   common ratio (p. 502)
            arithmetic progression (p. 501)   geometric means (p. 511)
            common difference (p. 501)   the geometric mean (p. 511)
            arithmetic means (p. 508)

1. Tell whether each sequence is arithmetic, geometric, or   Obj. 11-1, p. 501
   neither. Then supply the missing terms of the sequence.

   a. $\frac{2}{3}$, 1, $\frac{3}{2}$, $\frac{9}{4}$, _?_, _?_

   b. 5, 1, −3, −7, _?_, _?_

   c. $\frac{1}{3}$, $\frac{2}{5}$, $\frac{3}{7}$, $\frac{4}{9}$, _?_, _?_

   d. 1, 1.3, 1.6, 1.9, _?_, _?_

2. For the arithmetic sequence 36, 29, 22, 15, . . . , find   Obj. 11-2, p. 507
   (a) a formula for the nth term and **(b)** t_{19}.

3. Insert **(a)** one and **(b)** two arithmetic means between
   −2 and 10.

4. For the geometric sequence 48, −24, 12, −6, . . . , find   Obj. 11-3, p. 510
   (a) a formula for the nth term and **(b)** t_{10}.

5. Insert **(a)** one and **(b)** two geometric means between
   2 and 16.

6. During a week-long sale, a store reduced its prices on sale
   items by 10% each day. A coat that was priced $80 on the
   first day of the sale was sold on the fourth day. At what
   price was the coat sold?

Check your answers with those at the back of the book.

Challenge

The two arithmetic sequences given below have infinitely many terms in common. Find three of these common terms.

$$2, 14, 26, . . .$$

$$\text{and} \quad 1, 8, 15, . . .$$

Graphing Sequences

The table and graphs below show that the arithmetic sequence $t_n = 2n - 1$ is related to the linear function $f(x) = 2x - 1$. The only difference is that $f(x)$ is defined for all real numbers x, while t_n is defined only for positive integers n.

n	1	2	3	4
$t_n = 2n - 1$	1	3	5	7

In general, the arithmetic sequence $t_n = t_1 + (n - 1)d$ can be rewritten as $t_n = dn + (t_1 - d)$. In this form you can see that the slope of the graph is the common difference of the sequence.

$$\left.\begin{array}{l} t_n = dn + (t_1 - d) \\ f(x) = mx + b \end{array}\right\} \quad \text{common difference } d = \text{slope } m$$

Just as arithmetic sequences and linear functions are related, geometric sequences and exponential functions are also related. A comparison of the sequence $t_n = 16\left(\frac{1}{2}\right)^n$ and the function $f(x) = 16\left(\frac{1}{2}\right)^x$ is shown below.

n	1	2	3	4
$t_n = 16\left(\frac{1}{2}\right)^n$	8	4	2	1

Exercises

Give the slope of the graph of each arithmetic sequence.

1. 3, 9, 15, . . .   **2.** 9, 5, 1, . . .   **3.** $t_n = 7 - 2n$ **4.** $t_n = 3 + 5(n - 1)$

Give the domain of the function $f(x)$ and of the sequence t_n. Sketch the graph of each.

5. $f(x) = 3x + 5$ **6.** $f(x) = 8 - 2x$ **7.** $f(x) = 2^x$ **8.** $f(x) = 9\left(\frac{2}{3}\right)^x$

$t_n = 3n + 5$ $t_n = 8 - 2n$ $t_n = 2^n$ $t_n = 9\left(\frac{2}{3}\right)^n$

Series

11-4 Series and Sigma Notation

Objective   To identify series and to use sigma notation.

When the terms of a sequence are added together, the resulting expression is called a **series.** Here are some examples:

Finite sequence	3, 7, 11, 15, 19
Related finite series	$3 + 7 + 11 + 15 + 19$
Infinite sequence	$\frac{1}{2}, \frac{1}{4}, \frac{1}{8}, \frac{1}{16}, \cdots$
Related infinite series	$\frac{1}{2} + \frac{1}{4} + \frac{1}{8} + \frac{1}{16} + \cdots$

Many of the words used to describe sequences are also used to describe series. For example, the finite series above is an arithmetic series with first term $t_1 = 3$, last term $t_5 = 19$, and general term $t_n = 4n - 1$. An **arithmetic series** is a series whose related sequence is arithmetic. Similarly, the infinite series above is a **geometric series** because its related sequence is geometric.

A series can be written in an abbreviated form by using the Greek letter Σ *(sigma)*, called the **summation sign.** For instance, to abbreviate the writing of the series

$$2 + 4 + 6 + 8 + \cdots + 100$$

first notice that the general term of the series is $2n$. The series begins with the term for $n = 1$ and ends with the term for $n = 50$. Using sigma notation you can write this series as $\sum\limits_{n=1}^{50} 2n$, which is read "the sum of $2n$ for values of n from 1 to 50."

$$\sum_{n=1}^{50} 2n = 2 \cdot 1 + 2 \cdot 2 + 2 \cdot 3 + 2 \cdot 4 + \cdots + 2 \cdot 50$$

$$= 2 + 4 + 6 + 8 + \cdots + 100$$

Similarly, the sigma expression below represents "the sum of n^2 for values of n from 1 to 10."

$$\sum_{n=1}^{10} n^2 = 1^2 + 2^2 + 3^2 + \cdots + 10^2$$

In sigma notation the general term, n^2, is called the **summand,** and the letter n is called the **index.**

Any letter can be used as the index. For example, replacing the index n by k in the series just given does not change the series:

$$\sum_{k=1}^{10} k^2 = 1^2 + 2^2 + 3^2 + \cdots + 10^2$$

Example 1 Write the series $\displaystyle\sum_{j=1}^{20} (-1)^j(j + 2)$ in expanded form.

Solution

$$\sum_{j=1}^{20} (-1)^j(j + 2) = (-1)^1(1 + 2) + (-1)^2(2 + 2) + (-1)^3(3 + 2) + \cdots + (-1)^{20}(20 + 2)$$

$$= -3 + 4 - 5 + \cdots + 22 \quad \textbf{\textit{Answer}}$$

The first and last values of the index are called the **limits of summation.** In Example 1, the *lower limit* is 1 and the *upper limit* is 20. If a series is infinite, then the symbol ∞ is used for the upper limit to indicate that the summation does not end. For example,

$$\sum_{k=1}^{\infty} \frac{1}{2^{k-1}}$$

is read "the sum of $\dfrac{1}{2^{k-1}}$ for values of k from 1 to infinity."

$$\sum_{k=1}^{\infty} \frac{1}{2^{k-1}} = \frac{1}{2^{1-1}} + \frac{1}{2^{2-1}} + \frac{1}{2^{3-1}} + \frac{1}{2^{4-1}} + \cdots$$

$$= 1 + \frac{1}{2} + \frac{1}{4} + \frac{1}{8} + \cdots$$

By changing the lower limit of summation from 1 to 0, this infinite geometric series can be rewritten with a simpler summand.

$$\sum_{k=0}^{\infty} \frac{1}{2^k} = \frac{1}{2^0} + \frac{1}{2^1} + \frac{1}{2^2} + \frac{1}{2^3} + \cdots$$

$$= 1 + \frac{1}{2} + \frac{1}{4} + \frac{1}{8} + \cdots$$

As you will see in Lesson 11-6, this infinite series has a finite sum of 2.

Example 2 Use sigma notation to write the series $10 + 15 + 20 + \cdots + 100$.

Solution 1 By inspection,
$$10 + 15 + 20 + \cdots + 100 = 5 \cdot 2 + 5 \cdot 3 + 5 \cdot 4 + \cdots + 5 \cdot 20.$$

$\therefore$ the series is $\displaystyle\sum_{k=2}^{20} 5k.$ $\quad$ **\textit{Answer}}**

Solution 2 Since the series is arithmetic with common difference 5, the nth term is:

$$t_n = 10 + (n - 1)5$$
$$= 5n + 5, \text{ or } 5(n + 1)$$

Now find n such that the last term is 100.

$$t_n = 5n + 5$$
$$100 = 5n + 5$$
$$n = 19$$

$\therefore$ the series is $\displaystyle\sum_{n=1}^{19} 5(n + 1)$. **Answer**

In the two solutions of Example 2, notice that the expressions $\displaystyle\sum_{k=2}^{20} 5k$ and $\displaystyle\sum_{n=1}^{19} 5(n + 1)$ represent the same series.

Example 3 Use sigma notation to write the series $\dfrac{5}{2} - \dfrac{5}{4} + \dfrac{5}{6} - \dfrac{5}{8} + \cdots$.

Solution Since the infinite series is neither arithmetic nor geometric, you need to look for patterns:

(1) The numerators are all equal to 5.

(2) Since the denominators are consecutive even integers, a general expression for the denominators is $2n$.

(3) To make the terms of a series alternate in sign, choose $n = 1$ for the lower limit of the summand and include in the expression for the summand one of the following factors:

$(-1)^n$ makes *odd-numbered* terms *negative*;
$(-1)^{n+1}$ makes *even-numbered* terms *negative*.

So the expression for the summand is $(-1)^{n+1}\left(\dfrac{5}{2n}\right)$.

$\therefore$ the series is $\displaystyle\sum_{n=1}^{\infty} (-1)^{n+1}\left(\dfrac{5}{2n}\right)$. **Answer**

Oral Exercises

Exercises 1–4 refer to the series $\displaystyle\sum_{m=2}^{5} (4m^2 - 3)$.

1. What is the index?

2. What is the summand?

3. What are the limits of summation?

4. What are the first and last terms?

For each series, read each symbol aloud and give the expanded form.

5. $\displaystyle\sum_{i=1}^{3} i$

6. $\displaystyle\sum_{j=2}^{4} (5 - j)$

7. $\displaystyle\sum_{n=1}^{5} 4n$

8. $\displaystyle\sum_{k=1}^{4} k^3$

9. $\displaystyle\sum_{z=5}^{\infty} \frac{1}{z - 1}$

10. $\displaystyle\sum_{j=0}^{6} (-1)^j$

Tell whether or not each pair of series is the same.

11. $\displaystyle\sum_{k=0}^{8} \frac{1}{k + 3}, \sum_{j=3}^{11} \frac{1}{j}$

12. $\displaystyle\sum_{n=0}^{4} (-1)^n(n + 1)^2, \sum_{j=1}^{5} (-1)^j(j^2)$

13. $\displaystyle\sum_{k=0}^{4} \frac{3k + 3}{k + 6}, \sum_{i=1}^{3} \frac{3i}{i + 5}$

14. $\displaystyle\sum_{k=0}^{\infty} \frac{k + 1}{3^k}, \sum_{n=1}^{\infty} \frac{n}{3^{n-1}}$

Written Exercises

Write each series in expanded form.

A

1. $\displaystyle\sum_{n=1}^{6} (n + 10)$

2. $\displaystyle\sum_{k=1}^{8} 3k$

3. $\displaystyle\sum_{n=1}^{6} 2^n$

4. $\displaystyle\sum_{n=4}^{10} (3n - 2)$

5. $\displaystyle\sum_{j=0}^{5} \frac{(-1)^j}{j + 1}$

6. $\displaystyle\sum_{k=0}^{3} 4^{-k}$

7. $\displaystyle\sum_{n=3}^{8} |5 - n|$

8. $\displaystyle\sum_{k=1}^{4} (-k)^{k+1}$

Write each series using sigma notation.

9. $2 + 4 + 6 + \cdots + 1000$

10. $5 + 10 + 15 + \cdots + 250$

11. $1^3 + 2^3 + 3^3 + \cdots + 20^3$

12. $3 \cdot 4 + 3 \cdot 4^2 + 3 \cdot 4^3 + \cdots + 3 \cdot 4^{24}$

13. $\frac{1}{2} + \frac{2}{3} + \frac{3}{4} + \cdots + \frac{99}{100}$

14. $\frac{1}{5^3} + \frac{1}{5^4} + \frac{1}{5^5} + \cdots + \frac{1}{5^{15}}$

15. $1 + 3 + 5 + \cdots + 199$

16. $3 + 7 + 11 + 15 + \cdots + 399$

17. $1 + 2 + 4 + 8 + \cdots + 64$

18. $\sqrt{7} + \sqrt{14} + \sqrt{21} + \cdots + \sqrt{77}$

19. $1 + \frac{1}{2} + \frac{1}{3} + \frac{1}{4} + \cdots$

20. $1 + \frac{1}{3} + \frac{1}{9} + \frac{1}{27} + \cdots$

B

21. $-9 + 3 - 1 + \frac{1}{3} + \cdots$

22. $8 - 4 + 2 - 1 + \cdots$

23. $6 - 12 + 24 - 48$

24. $2 + 5 + 10 + 17 + 26 + 37$

25. $1 + \frac{1}{4} + \frac{1}{9} + \frac{1}{16} + \cdots$

26. $1 - \frac{1}{3} + \frac{1}{5} - \frac{1}{7} + \cdots$

27. The series consisting of positive three-digit integers divisible by 5

28. The series consisting of positive two-digit integers ending in 2

29. An infinite geometric series with first term 1 and common ratio $\frac{1}{4}$

30. An arithmetic series consisting of eleven terms with first term 8 and last term 508

Determine the missing summand so that each statement is true.

31. $\displaystyle\sum_{k=5}^{8} \frac{k}{k+4} = \sum_{j=1}^{4} \underline{\ ?\ }$

32. $\displaystyle\sum_{k=6}^{9} \frac{k+1}{k-1} = \sum_{j=7}^{10} \underline{\ ?\ }$

C **33.** Show that $\displaystyle\sum_{k=1}^{4} k \log 5 = \log (5^{10})$.

34. Write $\log (1 \cdot 2 \cdot 3 \cdot 4 \cdot 5 \cdot 6)$ using sigma notation.

35. Find the value of $\displaystyle\sum_{k=1}^{6} \left(\sum_{j=1}^{k} 1\right)$.

Career Note / *Marine Biologist*

Although nearly seventy percent of the earth's surface is covered by water, we know much less about what life exists and how it exists in the sea than we do about life on land. Scientists who study the variety and activity of life in the sea are called marine biologists.

Some marine biologists engage in basic research. They seek and describe new species of plants and animals living in the sea, and they study the relationships between the sealife and its environment. Other marine biologists undertake applied research. They investigate how sealife can meet human needs, especially in providing new sources of food and pharmaceuticals.

Marine biologists are as committed to conserving the resources of the sea as to utilizing them. Conservation requires knowing a species' population, and this is done through random sampling. A sample of the population is caught, tagged, and then released. Later on when a second sample is taken, some of those caught will already be tagged. The ratio of those that are found tagged

to the size of the second sample should approximately equal the ratio of the size of the original sample to the entire population. From this proportion marine biologists can estimate the population of the species.

Extra / Induction

How could you prove that

$$\frac{1}{1 \cdot 2} + \frac{1}{2 \cdot 3} + \frac{1}{3 \cdot 4} + \cdots + \frac{1}{n(n+1)} = \frac{n}{n+1}$$

is true for all positive integers n? Since the series is neither arithmetic nor geometric, there is no formula that can be used to find the sum. Instead, you can think of the general statement as a *sequence of statements*.

(1) $$\frac{1}{1 \cdot 2} = \frac{1}{2}$$

(2) $$\frac{1}{1 \cdot 2} + \frac{1}{2 \cdot 3} = \frac{2}{3}$$

(3) $$\frac{1}{1 \cdot 2} + \frac{1}{2 \cdot 3} + \frac{1}{3 \cdot 4} = \frac{3}{4}$$

$\vdots$ $\qquad\qquad\qquad\qquad\qquad\qquad\qquad$ $\vdots$

(n) $$\frac{1}{1 \cdot 2} + \frac{1}{2 \cdot 3} + \cdots + \frac{1}{n(n+1)} = \frac{n}{n+1}$$

It is easy to verify that statements (1), (2), and (3) above are true. However, this does not *prove* that statement (*n*) is true for all positive integers *n*. One way to prove statement (*n*) is to use a method called the *principle of mathematical induction*.

Principle of Mathematical Induction

To show that a sequence of statements is true for all positive integers *n*:
1. Show that statement (1) is true.
2. Show that if statement (*k*) is true, then statement (*k* + 1) is also true.

Using mathematical induction to prove a sequence of statements can be compared to knocking down an infinite row of dominoes. The first part of such a proof shows that the first domino falls. The second part shows that if any one domino falls, then the next in line also falls. Taken together, the two parts of the proof guarantee that all the dominoes fall.

Example   Show that statement (*n*) above is true for all positive integers *n*.

Solution   1. Statement (1) is $\frac{1}{1 \cdot 2} = \frac{1}{2}$. This is obviously true.

2. Assume that statement (*k*) is true:

$$\frac{1}{1 \cdot 2} + \frac{1}{2 \cdot 3} + \cdots + \frac{1}{k(k+1)} = \frac{k}{k+1}$$

(Solution continues on next page.)

Prove that statement $(k + 1)$ is true:

$$\frac{1}{1 \cdot 2} + \frac{1}{2 \cdot 3} + \cdots + \frac{1}{k(k + 1)} + \frac{1}{(k + 1)(k + 2)} = \frac{k + 1}{k + 2}$$

Proof:

Assuming that

$$\frac{1}{1 \cdot 2} + \frac{1}{2 \cdot 3} + \cdots + \frac{1}{k(k + 1)} = \frac{k}{k + 1}$$

is true, add $\dfrac{1}{(k + 1)(k + 2)}$ to both sides and then simplify the right side of the equation:

$$\frac{1}{1 \cdot 2} + \frac{1}{2 \cdot 3} + \cdots + \frac{1}{k(k + 1)} + \frac{1}{(k + 1)(k + 2)} = \frac{k}{k + 1} + \frac{1}{(k + 1)(k + 2)}$$

$$= \frac{k(k + 2) + 1}{(k + 1)(k + 2)}$$

$$= \frac{k^2 + 2k + 1}{(k + 1)(k + 2)}$$

$$= \frac{(k + 1)^2}{(k + 1)(k + 2)}$$

$$= \frac{k + 1}{k + 2} \quad \checkmark$$

Exercises

Use the principle of mathematical induction to prove that each statement is true for all positive integers n.

1. $\dfrac{1}{1 \cdot 3} + \dfrac{1}{3 \cdot 5} + \dfrac{1}{5 \cdot 7} + \cdots + \dfrac{1}{(2n - 1)(2n + 1)} = \dfrac{n}{2n + 1}$

2. $\dfrac{1}{1 \cdot 4} + \dfrac{1}{4 \cdot 7} + \dfrac{1}{7 \cdot 10} + \cdots + \dfrac{1}{(3n - 2)(3n + 1)} = \dfrac{n}{3n + 1}$

3. $1^2 + 2^2 + 3^2 + \cdots + n^2 = \dfrac{n(n + 1)(2n + 1)}{6}$

4. $1 \cdot 2 + 2 \cdot 3 + 3 \cdot 4 + \cdots + n(n + 1) = \dfrac{n(n + 1)(n + 2)}{3}$

5. If, in a room of n people, every person shakes hands once with every other person, there will be $\dfrac{n^2 - n}{2}$ handshakes.

6. A convex polygon with n sides has $\dfrac{n^2 - 3n}{2}$ diagonals. (*Note:* The first step in the proof is to show that the statement is true for $n = 3$, since a polygon must have at least three sides.)

7. $n^3 + 2n$ is a multiple of 3. (This means $n^3 + 2n = 3m$ where m is some integer.)

8. $1^3 + 2^3 + 3^3 + \cdots + n^3 = (1 + 2 + 3 + \cdots + n)^2$

11-5 Sums of Arithmetic and Geometric Series

Objective   To find sums of finite arithmetic and geometric series.

In the preceding lesson you worked with series such as

$$2 + 4 + 6 + 8 + \cdots + 100$$

but did not compute the sums. If a series consists of just a few terms, then the sum is easy to compute. For example, the sum of the first three terms of the series above, which we denote by S_3, is

$$S_3 = 2 + 4 + 6$$
$$= 12$$

The sum of the first n terms of a series is denoted by S_n. If a series is arithmetic or geometric, there are ways to find its sum without actually adding all of the terms. Theorems 1 and 2 state methods of finding such sums.

Theorem 1   The sum of the first n terms of an arithmetic series is

$$S_n = \frac{n(t_1 + t_n)}{2}.$$

Proof   Write the series for S_n twice, the second time with the order reversed. Then add the two equations, term by term:

$$S_n = \quad t_1 + (t_1 + d) + (t_1 + 2d) + \cdots + (t_n - d) + t_n$$
$$S_n = \quad t_n + (t_n - d) + (t_n - 2d) + \cdots + (t_1 + d) + t_1$$

$$2S_n = \underbrace{(t_1 + t_n) + (t_1 + t_n) + (t_1 + t_n) + \cdots + (t_1 + t_n) + (t_1 + t_n)}_{n \text{ addends}}$$

$$2S_n = n(t_1 + t_n)$$
$$S_n = \frac{n(t_1 + t_n)}{2}$$

Example 1   Find the sum of the first 40 terms of the arithmetic series $2 + 5 + 8 + 11 + \cdots$.

Solution   First find the 40th term: $t_{40} = t_1 + (n - 1)d$
$$= 2 + (39)3 = 119$$

Then use Theorem 1: $S_n = \dfrac{n(t_1 + t_n)}{2}$

$$S_{40} = \frac{40(2 + 119)}{2} = 2420 \quad \textbf{\textit{Answer}}$$

Example 2 Evaluate $\displaystyle\sum_{k=1}^{20} (5k + 2)$.

Solution Evaluating $\displaystyle\sum_{k=1}^{20} (5k + 2)$ means finding S_{20} for this arithmetic series:

$$7 + 12 + 17 + \cdots + 102$$

$$\therefore \quad S_{20} = \frac{20(7 + 102)}{2} = 1090 \quad \textbf{\textit{Answer}}$$

Theorem 2 The sum of the first n terms of a geometric series with common ratio r is

$$S_n = \frac{t_1(1 - r^n)}{1 - r}. \quad (r \neq 1)$$

Proof Multiply the series for S_n by the common ratio r. Then subtract the new series from the original one, as shown below:

$$S_n = t_1 + t_1r + t_1r^2 + \cdots + t_1r^{n-2} + t_1r^{n-1}$$
$$rS_n = \qquad t_1r + t_1r^2 + \cdots + t_1r^{n-2} + t_1r^{n-1} + t_1r^n$$

$$S_n - rS_n = t_1 + 0 \ + 0 \ + \cdots + 0 \qquad + 0 \qquad - t_1r^n$$
$$S_n - rS_n = t_1 - t_1r^n$$
$$S_n(1 - r) = t_1(1 - r^n)$$

Since $r \neq 1$, divide both sides by $1 - r$:

$$S_n = \frac{t_1(1 - r^n)}{1 - r}$$

Note that the formula just given is not defined for $r = 1$. (If $r = 1$, the geometric series $t_1 + t_1r + t_1r^2 + \cdots + t_1r^{n-1}$ consists of n terms equal to t_1, and the sum is nt_1.)

Example 3 Evaluate $\displaystyle\sum_{n=1}^{10} 3(-2)^{n-1}$.

Solution You are finding S_{10} for the series $3 - 6 + 12 - 24 + \cdots$, whose common ratio is -2.

$$S_n = \frac{t_1(1 - r^n)}{1 - r}$$

$$S_{10} = \frac{3[1 - (-2)^{10}]}{1 - (-2)}$$

$$= \frac{3(1 - 1024)}{3} = -1023 \quad \textbf{\textit{Answer}}$$

Oral Exercises

Find S_5 for each series.

1. $8 + 12 + 16 + 20 + \cdots$

2. $3 - 6 + 12 - 24 + \cdots$

3. $1 + \dfrac{1}{2} + \dfrac{1}{4} + \dfrac{1}{8} + \cdots$

4. $7 + 5 + 3 + 1 + \cdots$

Written Exercises

Find the sum of each arithmetic series.

A

1. $n = 20$, $t_1 = 5$, $t_{20} = 62$

2. $n = 100$, $t_1 = 17$, $t_{100} = 215$

3. $n = 40$, $t_1 = -12$, $t_{40} = 183$

4. $n = 50$, $t_1 = 187$, $t_{50} = 40$

5. $\displaystyle\sum_{k=1}^{100} 5k$

6. $\displaystyle\sum_{n=1}^{24} (2n - 1)$

7. $\displaystyle\sum_{j=1}^{50} (3j + 2)$

8. $\displaystyle\sum_{m=10}^{20} (30 - m)$

9. The first 100 terms of the series
$4 + 7 + 10 + 13 + \cdots$

10. The first 100 terms of the series
$100 + 98 + 96 + 94 + \cdots$

11. $11 + 15 + 19 + \cdots + 83$

12. $50 + 48 + 46 + \cdots + 10$

Find the sum of each geometric series.

13. $n = 8$, $r = 2$, $t_1 = 1$

14. $n = 10$, $r = -2$, $t_1 = 1$

15. $n = 10$, $r = -3$, $t_1 = 2$

16. $n = 12$, $r = 3$, $t_1 = \dfrac{1}{9}$

17. $\displaystyle\sum_{k=1}^{12} 2^{-k}$

18. $\displaystyle\sum_{j=1}^{10} \left(-\dfrac{1}{2}\right)^j$

19. Find S_{10} if the series $24 + 12 + \cdots$ is **(a)** arithmetic; **(b)** geometric.

20. Find S_{20} if the series $1 + 1.1 + \cdots$ is **(a)** arithmetic; **(b)** geometric.

Find the sum of each of the following.

B

21. The first 20 positive integers ending in 3

22. The positive two-digit integers ending in 4

23. The positive three-digit odd integers

24. The positive three-digit integers divisible by 6

25. **a.** The first twenty multiples of 2, starting with 2
b. The first twenty powers of 2, starting with 2^1

26. **a.** The first ten powers of 5
b. The first ten powers of 5 that are perfect squares

Find the sum of each of the following.

27. The positive two-digit integers that are *not* divisible by 3

28. The positive two-digit integers that are *not* divisible by 5

29. a. Show that $\displaystyle\sum_{k=1}^{n} 2^{k-1} = 2^n - 1$.

 b. How many terms are needed before the sum exceeds one million?

30. If n is an even number, show that the sum of series (A) is three times the sum of series (B).

 (A) $1 + \dfrac{1}{2} + \dfrac{1}{4} + \cdots + \left(\dfrac{1}{2}\right)^{n-1}$ (B) $1 - \dfrac{1}{2} + \dfrac{1}{4} - \cdots + \left(-\dfrac{1}{2}\right)^{n-1}$

31. The sum of a series is given by $S_n = 8(3^n - 1)$. Find the first three terms of the related sequence. Then identify the series as arithmetic, geometric, or neither. (*Hint:* $S_1 = t_1$.)

32. If $S_n = 2n^2 + 5n$, find t_n.

33. Guess the approximate value of the following expressions. Check your guess with a calculator.

 a. $\left(\dfrac{1}{2}\right)^{20}$ **b.** $(0.9)^{100}$ **c.** $(-0.8)^{50}$

Use the results of Exercise 33 to estimate the answers for Exercise 34.

C **34.** If $|r| < 1$ and n is a very large integer, find the approximate value of:

 a. r^n **b.** S_n where $S_n = \dfrac{t_1(1 - r^n)}{1 - r}$

The following series are neither arithmetic nor geometric, but by analyzing their patterns, you can find their sums. Find the sum of each series.

35. $2 - 4 + 6 - 8 + 10 - \cdots - 100$

36. $1 + 2 + 4 + 5 + 7 + 8 + \cdots + 95 + 97 + 98$

37. $\displaystyle\sum_{k=1}^{20} (2^k + k)$ **38.** $\displaystyle\sum_{n=1}^{12} (2^n - 1)$

Problems

Solve. A calculator may be helpful.

A **1.** The front row of a theater has 25 seats. Each of the other rows has two more seats than the row before it. How many seats are there altogether in the first 20 rows?

 2. Kristen is given a test consisting of 15 questions. The first question is worth five points, and each question after the first is worth three points more than the question before it. What is the maximum score that Kristen can obtain?

3. Every hour a clock chimes as many times as the hour. How many times does it chime from 1 A.M. through midnight, inclusive?

4. A ship's clock strikes every half hour of a 4-hour period. After the first half hour it strikes "one bell," after the second half hour it strikes "two bells," and so on until it strikes "eight bells" at the end of the 4-hour period. It then begins a new 4-hour period. How many strikes of the bell occur in one day?

5. Kurt can trace his ancestors back through 10 generations. He counts his parents as the first generation back, his four grandparents as the second generation back, and so on. How many ancestors does he have in these 10 generations?

6. The Kemps have rented a house for the past six years. During their first year of renting, they paid $600 per month. If their rent was increased by 5% for each year after that, what is the total amount of money they have paid in rent over the past six years?

B 7. You have won a contest sponsored by a local radio station. If you are given the choice of the two payment plans listed below, which plan will pay you more? How much more?

Plan A: $1 on the first day, $2 on the second day, $3 on the third day, and so on for two weeks.

Plan B: 1¢ on the first day, 2¢ on the second day, 4¢ on the third day, and so on for two weeks.

8. Refer to Problem 7 and find how much each plan would pay you if the payments extended over three weeks instead of two.

9. Which of the two jobs described below will pay you the higher salary during the fifth year of employment? Which will pay you the greater total amount for all five years?

Job A: Make $20,000 the first year with annual raises of $1500.
Job B: Make $18,000 the first year with annual raises of 10%.

10. Refer to the two jobs described in Problem 9. Which job will pay you the greater total amount over a 10-year period? How much more?

11. On the first day of each year, the Ortegas invest $1000 at 6% interest compounded annually. Find the value of their investment after 10 years.

12. The 1560 members of the Western Environmental Society have a method of quickly notifying members. The president and treasurer each call 3 members (round 1), each of whom then calls 3 other members (round 2), each of whom then calls 3 more members (round 3), and so on. How many rounds of calls are needed before everyone is contacted?

13. The number $T_n = 1 + 2 + 3 + 4 + \cdots + n$ is called a *triangular number* because it is possible to represent the number by dots arranged as an equilateral triangle.

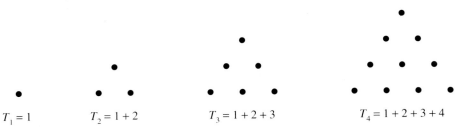

$T_1 = 1$ $\quad\quad T_2 = 1 + 2$ $\quad\quad T_3 = 1 + 2 + 3$ $\quad\quad T_4 = 1 + 2 + 3 + 4$

   a. Find a formula for T_n in terms of n.
   b. Write your formula using sigma notation.

C 14. Add any two consecutive triangular numbers. What special kind of number is obtained? Make a conjecture and then prove it.

15. Almost 2000 years ago, Plutarch observed that 1 more than eight times a triangular number always gives a special kind of number. What is this kind of number? Give a convincing argument to support your answer.

Mixed Review Exercises

Evaluate if $x = -2$ and $y = 8$.

1. $y^{1/x}$

2. $2x^2 - y$

3. $x^{-1} \cdot y^{2/3}$

4. $\sqrt{-xy}$

5. $\dfrac{y}{x^3}$

6. $\sqrt{x^2 + y^2}$

7. $|xy|$

8. $\log_4\left(-\dfrac{x}{y}\right)$

9. $\dfrac{y - x^2}{x}$

10. Find the slope of the line $3x - 2y = 6$.

11. Find the radius of the circle $x^2 + y^2 - 4x + 6y - 3 = 0$.

12. Find the x-intercepts of the hyperbola $4x^2 - 9y^2 = 36$.

Computer Exercises   *For students with some programming experience.*

1. Write a program that will find the sum of a series when the user enters the limits of summation and the general term of the series. (*Hint:* You can have the user enter the general term by retyping a numbered program line before each run.)

2. Use the program in Exercise 1 to find the sum of each series.

   a. $\displaystyle\sum_{k=10}^{20} (3k + 1)$

   b. $\displaystyle\sum_{k=1}^{30} \left(\frac{2}{3}\right)^k$

   c. $\displaystyle\sum_{k=2}^{100} \frac{1}{k^2 - 1}$

11-6 Infinite Geometric Series

Objective To find sums of infinite geometric series having ratios with absolute value less than one.

Consider the infinite geometric series

$$\frac{1}{2} + \frac{1}{4} + \frac{1}{8} + \frac{1}{16} + \cdots$$

and several of the following *partial sums* for this series.

$$S_1 = \frac{1}{2}$$

$$S_2 = \frac{1}{2} + \frac{1}{4} = \frac{3}{4}$$

$$S_3 = \frac{1}{2} + \frac{1}{4} + \frac{1}{8} = \frac{7}{8}$$

$$S_4 = \frac{1}{2} + \frac{1}{4} + \frac{1}{8} + \frac{1}{16} = \frac{15}{16}$$

$$\vdots$$

$$S_{10} = \frac{1}{2} + \frac{1}{4} + \frac{1}{8} + \frac{1}{16} + \cdots + \frac{1}{1024} = \frac{1023}{1024}$$

You can see that as more terms are added, the sum gets closer and closer to the value 1. In fact, this can be proved as follows:

1. First, find a formula for S_n, the sum of the first n terms.

$$S_n = \frac{t_1(1 - r^n)}{1 - r} = \frac{\frac{1}{2}[1 - (\frac{1}{2})^n]}{1 - \frac{1}{2}} = \frac{\frac{1}{2}[1 - (\frac{1}{2})^n]}{\frac{1}{2}} = 1 - \left(\frac{1}{2}\right)^n$$

This formula tells you that S_n is always less than 1 for any n.

2. Next, notice that as n becomes increasingly large, $\left(\frac{1}{2}\right)^n$ gets closer and closer to 0. This means S_n gets closer and closer to 1:

$$S_n = 1 - \left(\frac{1}{2}\right)^n \approx 1 - 0 = 1$$

If enough terms are added (making n sufficiently large), the sum will approximate 1 as closely as you may require. Therefore, we say that the sum of the infinite geometric series $\frac{1}{2} + \frac{1}{4} + \frac{1}{8} + \frac{1}{16} + \cdots$ is 1.

Some infinite geometric series do not have a sum. For example, the partial sums of the series

$$1 + 10 + 100 + 1000 + \cdots$$

continue to become larger and larger. The sums do not approach, or "home in" on, a fixed value.

In order for an infinite geometric series to have a sum, the common ratio r must be between -1 and 1. Then as n increases, r^n gets closer and closer to 0. The fact that r^n approximates 0 for large n is used to prove the following theorem.

Theorem

An infinite geometric series with common ratio r has a sum S if $|r| < 1$. This sum is

$$S = \frac{t_1}{1 - r}.$$

Proof

1. The sum of the first n terms is $S_n = \dfrac{t_1(1 - r^n)}{1 - r}$.

2. As n becomes larger and larger, r^n gets closer and closer to 0 since $|r| < 1$. Then

$$S_n = \frac{t_1(1 - r^n)}{1 - r} \approx \frac{t_1(1 - 0)}{1 - r} = \frac{t_1}{1 - r}.$$

Therefore, the sum S is given by

$$S = \frac{t_1}{1 - r}.$$

Example 1 Find the sum of each infinite geometric series. If the series has no sum, say so.

a. $8 - 4 + 2 - 1 + \cdots$ **b.** $8 + 12 + 18 + 27 + \cdots$

Solution **a.** Since $r = -\frac{1}{2}$, $|r| < 1$ and the series has a sum.

$$S = \frac{t_1}{1 - r} = \frac{8}{1 - (-\frac{1}{2})} = \frac{8}{\frac{3}{2}} = \frac{16}{3} \quad \textbf{\textit{Answer}}$$

b. Since $r = \frac{3}{2}$, $|r| \not< 1$. $\therefore$ the series has no sum. **_Answer_**

The next example shows that any infinite repeating decimal can be expressed as an infinite geometric series.

Example 2 Write $0.121212 \ldots$ as a common fraction.

Solution The infinite repeating decimal can be written as the infinite series

$$0.12 + 0.0012 + 0.000012 + \cdots.$$

Since $r = 0.01$, $|r| < 1$ and the series has a sum.

$$S = \frac{t_1}{1 - r} = \frac{0.12}{1 - 0.01} = \frac{0.12}{0.99} = \frac{4}{33} \quad \textbf{\textit{Answer}}$$

Give the common ratio of each geometric series and tell whether the series has a sum.

1. $27 + 9 + 3 + \cdots$

2. $27 - 9 + 3 - \cdots$

3. $\frac{2}{3} + 1 + \frac{3}{2} + \cdots$

4. $\frac{9}{4} + \frac{3}{2} + 1 + \cdots$

5. $1 + \sqrt{3} + 3 + \cdots$

6. $8^{-1} - 8^{-2} + 8^{-3} - \cdots$

7. $\sum_{n=0}^{\infty} 3\left(\frac{4}{5}\right)^n$

8. $\sum_{n=1}^{\infty} 2(1.1)^n$

9. $\sum_{n=1}^{\infty} (-0.35)^n$

10. $\sum_{n=1}^{\infty} \frac{1}{10^n}$

Written Exercises

For each geometric series, find the sum. If the series has no sum, say so.

A

1. $24 + 12 + 6 + 3 + \cdots$

2. $24 - 12 + 6 - 3 + \cdots$

3. $27 - 18 + 12 - 8 + \cdots$

4. $27 + 18 + 12 + 8 + \cdots$

5. $256 + 320 + 400 + 500 + \cdots$

6. $500 + 400 + 320 + 256 + \cdots$

7. $3 + 4 + 5\frac{1}{3} + 7\frac{1}{9} + \cdots$

8. $\frac{1}{2} - \frac{1}{3} + \frac{2}{9} - \frac{4}{27} + \cdots$

9. $3\sqrt{3} - 3 + \sqrt{3} - 1 + \cdots$

10. $4^{-1/2} + 4^{-3/2} + 4^{-5/2} + 4^{-7/2} + \cdots$

11. $\sum_{n=0}^{\infty} 3\left(\frac{1}{4}\right)^n$

12. $\sum_{n=1}^{\infty} \frac{2^n}{5^n}$

For each geometric series find S_1, S_2, S_3, S_4, and S_5. Use these sums to approximate S. Then use the formula to find S. Compare your approximation with the value obtained from the formula.

13. $8 + 2 + \frac{1}{2} + \frac{1}{8} + \cdots$

14. $9 + \frac{9}{10} + \frac{9}{100} + \frac{9}{1000} + \cdots$

15. $\frac{3}{4} - \frac{3}{8} + \frac{3}{16} - \frac{3}{32} + \cdots$

16. $\frac{1}{3} + \frac{1}{9} + \frac{1}{27} + \frac{1}{81} + \cdots$

Write each repeating decimal as a common fraction. Use the method of Example 2.

17. $0.3333\ldots$

18. $0.4444\ldots$

19. $3.12312312\ldots$

20. $0.363636\ldots$

21. $0.49999\ldots$

22. $1.045045045\ldots$

Write the first three terms of the infinite geometric series satisfying the given condition.

B **23.** $t_1 = 8$, $S = 12$ **24.** $t_1 = 40$, $S = 200$ **25.** $r = -\frac{1}{3}$, $S = 30$ **26.** $r = \frac{2}{5}$, $S = 125$

27. Explain why there is no infinite geometric series with first term 10 and sum 4.

28. There are two infinite geometric series with $t_1 = 100$ and $t_3 = 1$. Find the sum of each.

C **29. a.** Find the sum of the series $1 + \frac{2}{3} + \frac{4}{9} + \frac{8}{27} + \cdots$.

      b. How many terms of this series must be added before their sum is within 0.00001 of the sum found in part (a)? (*Hint:* Consider $S - S_n$.)

30. If $|x| < 1$, find the sum S of the series $1 + 2x + 3x^2 + 4x^3 + \cdots$. (*Hint:* Consider $S - xS$.)

Problems

Solve. A calculator may be helpful.

A **1.** A child on a swing is given a big push. She travels 12 ft on the first back-and-forth swing but only $\frac{5}{6}$ as far on each successive back-and-forth swing. How far does she travel before the swing stops?

2. Suppose an indecisive man starts out from home and walks 1 mi east, then $\frac{1}{2}$ mi west, then $\frac{1}{4}$ mi east, then $\frac{1}{8}$ mi west, and so on. Relative to his home, approximately where would he end up?

3. A side of a square is 12 cm. The midpoints of its sides are joined to form an inscribed square, and this process is continued as shown in the diagram. Find the sum of the perimeters of the squares if this process is continued without end.

4. A side of an equilateral triangle is 10 cm. The midpoints of its sides are joined to form an inscribed equilateral triangle, and this process is continued. Find the sum of the perimeters of the triangles if the process is continued without end.

5. Find the sum of the areas of the squares in Problem 3.

6. Find the sum of the areas of the triangles in Problem 4.

B **7.** The diagram shows a superball that rebounds 95% of the distance it falls. This ball is thrown 12 m in the air (so that the initial up-and-down distance traveled is 24 m). What is the total vertical distance traveled by the ball before it stops bouncing?

12 m

8. Refer to Problem 7 and suppose that the ball is dropped from a height of 10 m. Make a diagram of the bouncing ball's path and note that it differs from the diagram above because the ball does not *begin* at ground level. What is the total vertical distance traveled by the ball before it stops bouncing?

9. A morning glory vine grows more slowly as it gets longer. Suppose that its growth during the week after it emerges from the soil is 50 in. and that during each succeeding week it grows $\frac{4}{5}$ as much as it did the previous week.

   a. How much does the vine grow during its 26th week?
   b. What is its total growth for a 26-week growing period?
   c. Assuming that the vine could grow forever, what would be the vine's total growth?

   (*Note:* Since the answer to part (c) is easier to calculate than the answer to part (b) and since the two answers are nearly the same, the sum of a finite geometric series with many terms can be estimated by the sum of the corresponding infinite series.)

10. A piledriver pounds a steel column into the earth. On its first drive, the column penetrates 1.5 m into the earth and on each succeeding drive it moves 92% as far as it did on its previous drive.

    a. How far is the column driven on the 60th drive?
    b. What is the approximate total distance that the column moves in 60 drives? (See the note after Problem 9(c).)

C 11. The Koch curve is constructed as follows: Start with a square (see Figure 1). Construct a square on the middle third of each side of Figure 1 (see Figure 2). Then construct a square on the middle third of each side of Figure 2 (see Figure 3). Continue forming new figures by constructing a square on the middle third of each side of the previous figure.

Figure 1   Figure 2   Figure 3

   a. Complete the table below.

	Fig. 1	Fig. 2	Fig. 3	Fig. 4	. . .	Fig. n
Number of new squares	1	4	?	?	. . .	?
Area of each new square	1	$\frac{1}{9}$	?	?	. . .	?
Total new area	1	$\frac{4}{9}$	?	?	. . .	?

   b. Find the total area bounded by the Koch curve. (*Hint:* Add the entries in the last row of the table. Notice that the entries *after* the first one form a geometric series.)

12. Show that the Koch curve described in Problem 11 has no perimeter (that is, the perimeter is of unbounded length).

13. A snowflake curve is constructed as follows: Start with an equilateral triangle (see Figure 1). Construct an equilateral triangle on the middle third of each side of Figure 1 (see Figure 2). Then construct an equilateral triangle on the middle third of each side of Figure 2 (see Figure 3). Continue forming new figures by constructing an equilateral triangle on the middle third of each side of the previous figure. If the first equilateral triangle has sides of length 1, what is the total area bounded by the snowflake curve?

| Figure 1 | Figure 2 | Figure 3 |

14. Show that the snowflake curve described in Problem 13 has no perimeter (that is, the perimeter is of unbounded length).

Self-Test 2

Vocabulary   series (p. 518)   summation sign (p. 518)
          arithmetic series (p. 518)   summand (p. 518)
          geometric series (p. 518)   index (p. 518)
          sigma (p. 518)   limits of summation (p. 519)

1. Identify each series as arithmetic, geometric, or neither. Then rewrite the series using sigma notation.   Obj. 11-4, p. 518

    a. $27 - 18 + 12 - 8$ **b.** $\frac{1}{2} + \frac{2}{3} + \frac{3}{4} + \cdots + \frac{15}{16}$

    c. $20 + 17 + 14 + \cdots + 2$ **d.** $1 + 4 + 9 + 16 + \cdots$

2. Find the sum of each arithmetic series.   Obj. 11-5, p. 525

    a. $n = 30,\ t_1 = 7,\ t_{30} = 65$ **b.** $\sum_{k=1}^{15} (45 - 6k)$

3. Find the sum of each geometric series.

    a. $n = 6,\ r = 2,\ t_1 = 5$ **b.** $\sum_{k=1}^{4} \left(\frac{2}{3}\right)^k$

4. Find the sum of each infinite geometric series. If the series has no sum, say so.   Obj. 11-6, p. 531

    a. $\frac{25}{8} + \frac{5}{2} + 2 + \frac{8}{5} + \cdots$ **b.** $1 - 1.1 + 1.21 - 1.331 + \cdots$

Check your answers with those at the back of the book.

Binomial Expansions

11-7 Powers of Binomials

Objective   To expand powers of binomials.

When you multiply out a power of a binomial, the resulting sum of terms is called a **binomial expansion.** An interesting pattern is formed by the coefficients of the expansion of $(a + b)^n$ where n is a nonnegative integer.

$$(a + b)^0 = 1$$
$$(a + b)^1 = 1a^1 + 1b^1$$
$$(a + b)^2 = 1a^2 + 2ab + 1b^2$$
$$(a + b)^3 = 1a^3 + 3a^2b + 3ab^2 + 1b^3$$
$$(a + b)^4 = 1a^4 + 4a^3b + 6a^2b^2 + 4ab^3 + 1b^4$$
$$(a + b)^5 = 1a^5 + 5a^4b + 10a^3b^2 + 10a^2b^3 + 5ab^4 + 1b^5$$

The pattern of coefficients suggests the triangular array of numbers, shown at the right, called **Pascal's triangle.** The triangle has 1's at the beginning and end of each row. Each of the other numbers is the sum of the two numbers above it. For example, $15 = 10 + 5$

													Row
						1							0
					1		1						1
				1		2		1					2
			1		3		3		1				3
		1		4		6		4		1			4
	1		5		10		10		5		1		5
1		6		15		20		15		6		1	6

Because the top row of the triangle corresponds to $(a + b)^0$, it is convenient to call it row 0. The next row is row 1, followed by row 2, and so on. Then for any positive integer n, the coefficients of the expansion of $(a + b)^n$ are the elements of row n of Pascal's triangle.

Example 1   Expand $(a + b)^6$. Use row 6 of Pascal's triangle.

Solution   From row 6 of Pascal's triangle, you find that the coefficients are
$$1, \quad 6, \quad 15, \quad 20, \quad 15, \quad 6, \quad 1.$$
The powers of a are $a^6, \quad a^5, \quad a^4, \quad a^3, \quad a^2, \quad a^1, \quad a^0.$

The powers of b are $b^0, \quad b^1, \quad b^2, \quad b^3, \quad b^4, \quad b^5, \quad b^6.$

Combining this information, you get
$$(a + b)^6 = a^6 + 6a^5b + 15a^4b^2 + 20a^3b^3 + 15a^2b^4 + 6ab^5 + b^6.$$
Answer

Notice that the exponents of a in Example 1 decrease from 6 to 0 and the exponents of b increase from 0 to 6. There are seven terms, and the sum of the exponents in each term is equal to 6. When the binomial contains a negative term, the signs in the expansion will alternate, as Example 2 shows.

Example 2 Find the first four terms in the expansion of $(x - 2y)^7$.

Solution First find row 7 of Pascal's triangle:

	1	6	15	20	15	6	1		Row 6
1	7	21	35	35	21	7	1		Row 7

Now expand $(a + b)^7$.
$$(a + b)^7 = a^7 + 7a^6b + 21a^5b^2 + 35a^4b^3 + \cdots$$
Then replace a by x and b by $(-2y)$.
$$(x - 2y)^7 = x^7 + 7x^6(-2y) + 21x^5(-2y)^2 + 35x^4(-2y)^3 + \cdots$$
The first four terms are: $x^7 - 14x^6y + 84x^5y^2 - 280x^4y^3$. **Answer**

The pattern of coefficients given in Pascal's triangle is symmetric. For example, in the expansion of $(a + b)^5$ the coefficients of the last three terms are the same as the coefficients of the first three in reverse order. And in the expansion of $(a + b)^6$ the coefficients to the right of the middle term, $20a^3b^3$, are the same as those to the left. This symmetry is useful in writing expansions and finding particular terms, as shown in Example 3.

Example 3 The first three terms in the expansion of $(a + b)^{20}$ are $a^{20} + 20a^{19}b + 190a^{18}b^2$. Write the last three terms.

Solution $190a^2b^{18} + 20ab^{19} + b^{20}$ **Answer**

Oral Exercises

Row 8 of Pascal's triangle is 1, 8, 28, 56, 70, 56, 28, 8, 1. Use this row to complete Exercises 1 and 2.

1. Give the first four terms in the expansion of $(a + b)^8$.

2. What is row 9 of Pascal's triangle?

3. Give the number of terms in the expansion of each of the following.
   a. $(m - n)^1$ **b.** $(2x + y)^6$ **c.** $(3p^2 - 1)^9$

4. Give the first term in the expansion of each of the following.
   a. $(a + b)^{10}$ **b.** $(3m + n)^3$ **c.** $(x^3 - y^2)^5$

5. Give the last term in the expansion of each of the following.
   a. $(a - b)^7$ **b.** $(p^2 - 2)^4$ **c.** $(m + n^2)^8$

Written Exercises

Expand and simplify each expression.

A   **1.** $(x + y)^3$ **2.** $(x + y)^4$ **3.** $(c - d)^5$ **4.** $(p - q)^6$

      5. $(a + 1)^8$ **6.** $(x - 2)^3$ **7.** $(3t + 4)^5$ **8.** $(2y - 3)^4$

      9. $(x^2 - 1)^6$ **10.** $(\sqrt{x} + 3)^4$ **11.** $(p^2 + q^3)^3$ **12.** $\left(a + \dfrac{1}{a}\right)^5$

In Exercises 13–16, use the symmetry of the coefficients.

13. The first three terms of $(x + y)^{17}$ are $x^{17} + 17x^{16}y + 136x^{15}y^2$. Write the last three terms.

14. The first three terms of $(p - q)^{22}$ are $p^{22} - 22p^{21}q + 231p^{20}q^2$. Write the last three terms.

15. The tenth term of $(x + 1)^{20}$ is $167{,}960x^{11}$. Write the twelfth term. (*Hint:* The eleventh term is the middle term.)

16. The eighth term of $(p - q)^{15}$ is $-6435p^8q^7$. Write the ninth term.

Expand and simplify each expression.

B   **17.** $(a + b)^7 + (a - b)^7$ **18.** $(a + b)^8 - (a - b)^8$

      19. $(a^2 + 2ab + b^2)^3$ **20.** $(x - 1)^4(x + 1)^4$

21. Use the first three terms of a binomial expansion to approximate $(2.1)^7$. [*Hint:* $(2.1)^7 = (2 + 0.1)^7$.]

22. Use the first four terms of a binomial expansion to approximate $(9.8)^4$. [*Hint:* $(9.8)^4 = (10 - 0.2)^4$.]

C   **23.** **a.** Find the sum of the entries in each of the first 7 rows of Pascal's triangle.
           b. Guess a formula for the sum of the entries in the nth row.

24. Find a formula for the sum of *all* entries in rows 0 through n of Pascal's triangle.

Mixed Review Exercises

Tell whether each sequence is arithmetic, geometric, or neither. Then find a formula for the nth term of the sequence.

1. $\dfrac{2}{3}, \dfrac{4}{5}, \dfrac{6}{7}, \ldots$ **2.** $\dfrac{9}{16}, \dfrac{3}{4}, 1, \ldots$ **3.** $7, 3, -1, \ldots$ **4.** $-24, 12, -6, \ldots$

Find the sum of each series.

5. $5 + 9 + 13 + \cdots + 45$ **6.** $6 + \dfrac{9}{2} + \dfrac{27}{8} + \cdots$ **7.** $\displaystyle\sum_{n=1}^{6} 3 \cdot 2^n$

11-8 The General Binomial Expansion

Objective   To use the binomial theorem to find a particular term of a binomial expansion.

If you used Pascal's triangle to find the coefficients in the expansion of $(a + b)^{20}$, you would need to compute all the rows up through row 20. Fortunately, there is another way to find these coefficients. The following example shows how each coefficient depends on the power of the binomial and the position of the term in the expansion.

Example 1   The first four terms in the expansion of $(a + b)^{20}$ are given below. Find the next two terms.

$$(a + b)^{20} = a^{20} + \frac{20}{1}a^{19}b + \frac{20 \cdot 19}{1 \cdot 2}a^{18}b^2 + \frac{20 \cdot 19 \cdot 18}{1 \cdot 2 \cdot 3}a^{17}b^3 + \cdots$$

Solution   The coefficient of the term containing b^k has k factors in the numerator beginning with the power 20 and decreasing, and k factors in the denominator beginning with 1 and increasing. So the next two terms are

$$\frac{20 \cdot 19 \cdot 18 \cdot 17}{1 \cdot 2 \cdot 3 \cdot 4}a^{16}b^4 \quad \text{and} \quad \frac{20 \cdot 19 \cdot 18 \cdot 17 \cdot 16}{1 \cdot 2 \cdot 3 \cdot 4 \cdot 5}a^{15}b^5. \quad \textbf{\textit{Answer}}$$

Example 2   Find the first five terms in the expansion of $(a - b)^{12}$. Simplify each coefficient.

Solution   The terms containing odd powers of b will have a minus sign. The expansion begins

$$a^{12} - \frac{12}{1}a^{11}b + \frac{12 \cdot 11}{1 \cdot 2}a^{10}b^2 - \frac{12 \cdot 11 \cdot 10}{1 \cdot 2 \cdot 3}a^9b^3 + \frac{12 \cdot 11 \cdot 10 \cdot 9}{1 \cdot 2 \cdot 3 \cdot 4}a^8b^4 - \cdots.$$

When the fractions are simplified, each coefficient is an integer. The first five terms are

$$a^{12} - 12a^{11}b + 66a^{10}b^2 - 220a^9b^3 + 495a^8b^4. \quad \textbf{\textit{Answer}}$$

The coefficients in the expansions of $(a + b)^n$ can be written more simply using **factorial notation.** The symbol $r!$ (read "r factorial") is defined as follows:

$$r! = r(r - 1)(r - 2) \cdots \cdots 3 \cdot 2 \cdot 1 \text{ if } r \text{ is a positive integer}$$

$$0! = 1$$

For example, $4! = 4 \cdot 3 \cdot 2 \cdot 1 = 24$ and $5! = 5 \cdot 4 \cdot 3 \cdot 2 \cdot 1 = 120$.

Notice what happens when factorials are divided.

$$\frac{12!}{8!} = \frac{12 \cdot 11 \cdot 10 \cdot 9 \cdot 8 \cdot 7 \cdot 6 \cdot 5 \cdot 4 \cdot 3 \cdot 2 \cdot 1}{8 \cdot 7 \cdot 6 \cdot 5 \cdot 4 \cdot 3 \cdot 2 \cdot 1} = 12 \cdot 11 \cdot 10 \cdot 9$$

Therefore, you can write the fifth term of the expansion in Example 2 as follows:

$$\frac{12 \cdot 11 \cdot 10 \cdot 9}{1 \cdot 2 \cdot 3 \cdot 4} a^8 b^4 = \frac{12! \div 8!}{4!} a^8 b^4 = \frac{12!}{8!4!} a^8 b^4.$$

The terms found in Examples 1 and 2 illustrate the general formula for the expansion of $(a + b)^n$ given by the *binomial theorem,* which is proved in more advanced courses by using the principle of mathematical induction. (See the Extra on page 523.)

The Binomial Theorem

If n is a positive integer, then

$$(a + b)^n = a^n + \frac{n}{1} a^{n-1}b + \frac{n(n-1)}{1 \cdot 2} a^{n-2}b^2 + \frac{n(n-1)(n-2)}{1 \cdot 2 \cdot 3} a^{n-3}b^3 + \cdots + b^n$$

$$= a^n + \frac{n!}{(n-1)!1!} a^{n-1}b + \frac{n!}{(n-2)!2!} a^{n-2}b^2 + \frac{n!}{(n-3)!3!} a^{n-3}b^3 + \cdots + b^n$$

The $(k + 1)$st term of $(a + b)^n$ is $\dfrac{n!}{(n-k)!k!} a^{n-k}b^k$.

Example 3   Find and simplify the seventh term in the expansion of $(2x - y)^{10}$.

Solution   First find the seventh term in the expansion of $(a + b)^{10}$. This is the term containing b^6.

$$\frac{10!}{4!6!} a^4 b^6 = \frac{10 \cdot \overset{3}{\cancel{9}} \cdot \cancel{8} \cdot 7}{\cancel{4} \cdot \cancel{3} \cdot \cancel{2} \cdot 1} a^4 b^6 = 210 a^4 b^6$$

Now replace a by $2x$ and b by $-y$.

$$210(2x)^4(-y)^6 = 210 \cdot 16x^4 \cdot y^6$$
$$= 3360x^4 y^6 \quad \textbf{Answer}$$

Because $0!$ is defined as 1, the formula for the $(k + 1)$st term in the expansion of $(a + b)^n$ remains valid for the first and last terms.

First term $(k = 0)$: $\dfrac{n!}{(n-0)!0!} a^{n-0}b^0 = \dfrac{n!}{n!} a^n \cdot 1 = a^n$

Last term $(k = n)$: $\dfrac{n!}{(n-n)!n!} a^{n-n}b^n = \dfrac{n!}{n!} 1 \cdot b^n = b^n$

This means you can write the binomial theorem in a compact form using sigma notation:

$$(a + b)^n = \sum_{k=0}^{n} \frac{n!}{(n-k)!k!} a^{n-k} b^k$$

Oral Exercises

Evaluate.

1. 3!

2. 2!

3. $\frac{5!}{4!}$

4. $\frac{11!}{9!}$

5. $\frac{4!}{3!1!}$

6. $\frac{5!}{2!3!}$

7. $\frac{6!}{4!2!}$

8. $\frac{4!}{4!0!}$

9. Give the coefficient of the third term in the expansion of each of the following.

    a. $(a + b)^{10}$ **b.** $(c - d)^8$ **c.** $(x + 2y)^5$

10. Give **(a)** the power of a and **(b)** the power of b for the given term in the expansion of $(a + b)^{20}$.

    a. The fifth term   **b.** The eleventh term

Written Exercises

Evaluate.

A   **1.** 6!

2. 7!

3. $\frac{5!}{3!}$

4. $\frac{8!}{7!}$

5. $\frac{100!}{99!}$

6. $\frac{20!}{18!}$

7. $\frac{10!}{7!3!}$

8. $\frac{10!}{4!6!}$

9. $\frac{(n+1)!}{n!}$

10. $\frac{n!}{(n-2)!}$

11. $\frac{(n+1)!}{(n-1)!2!}$

12. $\frac{n!}{3!(n-3)!}$

Write the first four terms in the expansion of each of the following.

13. a. $(a + b)^{14}$ **14. a.** $(a + b)^{12}$ **15. a.** $(a + b)^{19}$ **16. a.** $(a + b)^{21}$
     b. $(a - b)^{14}$ **b.** $(a - b)^{12}$ **b.** $(x - 2y)^{19}$ **b.** $(c - d^2)^{21}$

Find and simplify the specified term in each expansion.

17. The term containing b^5 in $(a + b)^{20}$

18. The term containing b^{13} in $(a - b)^{15}$

19. The eleventh term of $(s - t)^{14}$

20. The eighteenth term of $(s + t)^{18}$

B   **21.** The term containing a^4 in $(a + 2b)^9$

22. The term containing x^2 in $(2x - y)^7$

23. The middle term of $(c^2 - 2d)^8$

24. The term containing b^9 in $(a + 3b^3)^5$

Decide whether each statement is true or false. If the statement is false, give a counterexample.

25. $(a \cdot b)! = a!b!$ **26.** $(n^2)! = (n!)^2$ **27.** $(2n)! = 2^n(n!)$

C   **28.** Find the term in the expansion of $(y - y^{-2})^{12}$ that does not contain y.

29. Assume that the binomial theorem continues to hold when n is a positive rational number that is not an integer. In general, the coefficients in the expansion will not be integers, and instead of terminating, the expansion will be an infinite series:

$$(a + b)^n = a^n + \frac{n}{1}a^{n-1}b + \frac{n(n-1)}{1 \cdot 2}a^{n-2}b^2 + \cdots$$

a. Write the first four terms in the expansion of $(1 + x)^{1/2}$.
b. Use your answer in (a) with $x = 1$ to find a numerical estimate of $\sqrt{2}$.

Calculator Key-In

1. Use the factorial key on your calculator to find 10!.

2. What is the largest positive integer n for which $n!$ can be displayed exactly (that is, every digit is shown) on your calculator?

3. If your calculator displays numbers in scientific notation, you will be able to find approximate values of $n!$ when the exact integer value of $n!$ is too long to be displayed.
a. Find 20! to four significant digits.
b. Find the coefficient of the middle term of $(a + b)^{36}$ to four significant digits.

4. What is the largest positive integer n for which $n!$ can be displayed in scientific notation on your calculator?

Self-Test 3

Vocabulary   binomial expansion (p. 537)   factorial notation (p. 540)
                 Pascal's triangle (p. 537)   binomial theorem (p. 541)

Expand and simplify each expression.

1. $(x - 2)^5$ **2.** $(4a + 3)^4$ **Obj. 11-7, p. 537**
3. Write the first three terms in the expansion of $(2x + 3y)^{10}$.   **Obj. 11-8, p. 540**

Find and simplify the specified term in each expansion.

4. The fourth term in $(a^2 - b)^7$
5. The term containing x^8 in $(x - 2y)^{11}$

Check your answers with those at the back of the book.

Alice Hamilton (1869–1970), Harvard's first female professor, thought factory workers should have a safe environment in which to work. As a physician and specialist in industrial toxicology, she was dismayed by the high death rates in American industry in the early part of this century. Before laws protecting workers could be passed, someone had to gather data about the number of deaths and fight for better working conditions. Hamilton performed this service for the pottery, lead, and munitions industries.

In her zeal to find poisonous dusts, she explored mine shafts and climbed dangerous catwalks. Her commitment to preventing lead and benzene poisoning led to some of the first industrial disease laws in the United States.

Chapter Summary

1. Sequences can be *arithmetic, geometric,* or neither. If there is a *common difference d* between any two successive terms, then the sequence is arithmetic. If the *ratio* of successive terms is constant, then the sequence is geometric.

2. In any arithmetic sequence with first term t_1 and common difference d, the nth term is given by $t_n = t_1 + (n - 1)d$.

   In any geometric sequence with first term t_1 and common ratio r, the nth term is given by $t_n = t_1 \cdot r^{n-1}$.

3. The *arithmetic mean* of two numbers a and b is $\dfrac{a + b}{2}$. The *geometric mean* is $\sqrt{ab}$ (if $a, b > 0$) or $-\sqrt{ab}$ (if $a, b < 0$).

4. A *series* is an expression in which the terms of a sequence are added together. If the related sequence is arithmetic or geometric, then the series is classified respectively as arithmetic or geometric.

   The Greek letter Σ (sigma) can be used to represent the sum of a series in an abbreviated form. The upper and lower limits of summation are the first and last values of the *index.* If a series is infinite, then the symbol ∞ is used for the upper limit.

5. The sum of the first n terms of an *arithmetic series* is given by

$$S_n = \frac{n(t_1 + t_n)}{2}.$$

6. The sum of the first n terms of a *geometric series* with common ratio r is given by

$$S_n = \frac{t_1(1 - r^n)}{1 - r}, \; r \neq 1.$$

7. An *infinite geometric series* with common ratio r has a sum if $|r| < 1$. This sum is

$$S = \frac{t_1}{1 - r}.$$

8. When $(a + b)^n$ is multiplied out, the resulting sum of terms is a *binomial expansion*. For any positive integer n, the coefficients of the expansion of $(a + b)^n$ are the elements of row n of Pascal's triangle. The *binomial theorem* can be used to expand $(a + b)^n$.

Chapter Review

Write the letter of the correct answer.

1. The sequence 56, 28, 14, 7, . . . is __?__.

   a. arithmetic   **b.** geometric   **c.** neither

   11-1

2. Find the next term in the sequence 8, 5, 2, -1, ...

   a. -2 **b.** -3 **c.** -4 **d.** -5

3. For the arithmetic sequence -4, -1.5, 1, 3.5, . . . , find t_{21}.

   a. 43.5   **b.** 46   **c.** 48.5   **d.** 51

   11-2

4. Insert two arithmetic means between -2 and 13.

   a. 1, 10   **b.** 2, 9   **c.** 3, 8   **d.** 4, 7

5. For the geometric sequence $\frac{81}{8}$, $\frac{27}{4}$, $\frac{9}{2}$, 3, . . . , find t_9.

   a. $\frac{8}{9}$ **b.** $\frac{16}{27}$ **c.** $\frac{32}{81}$ **d.** $\frac{64}{243}$

   11-3

6. Find the geometric mean of 10 and 40.

   a. 15   **b.** 20   **c.** 25   **d.** 30

7. Write the series $\sum_{k=0}^{3} 2^{k-1}$ in expanded form.

   11-4

   a. $\frac{1}{2} + 1 + 2 + 4$ **b.** $1 + 2 + 4 + 8$

   c. $-2 + 0 + 2 + 4$ **d.** $0 + 2 + 4 + 6$

8. Write the series $47 + 41 + 35 + \cdots + 5$ using sigma notation.

   a. $\sum_{n=0}^{8} (53 - 6n)$ **b.** $\sum_{n=1}^{8} (53 - 6n)$

   c. $\sum_{n=1}^{7} (47 - 6n)$ **d.** $\sum_{n=0}^{8} (47 - 6n)$

9. Find the sum of the arithmetic series $\sum\limits_{n=1}^{25} (3n + 2)$. 11-5

    a. 984   **b.** 1000   **c.** 1025   **d.** 2050

10. Find the sum of the geometric series with $n = 7$, $r = -2$, and $t_1 = 5$.

    a. -105 **b.** 215   **c.** 315   **d.** 635

11. Find the sum of the infinite geometric series $64 + 48 + 36 + 27 + \cdots$ if it has one. 11-6

    a. $\dfrac{256}{7}$ **b.** $\dfrac{256}{3}$ **c.** 256   **d.** no sum

12. Expand and simplify $(x^2 - 2)^4$. 11-7

    a. $x^8 + 8x^6 + 24x^4 + 32x^2 + 16$ **b.** $x^8 + 4x^6 + 6x^4 + 4x^2 + 1$

    c. $x^8 - 8x^6 + 24x^4 - 32x^2 + 16$ **d.** $x^4 - 8x^3 + 24x^2 - 32x + 16$

13. Find the fourth term in the expansion of $(3x - y^2)^6$. 11-8

    a. $-540x^3y^6$ **b.** $540x^3y^6$ **c.** $-540x^3y^3$ **d.** $-27x^3y^6$

Chapter Test

1. Tell whether the sequence $\dfrac{2}{3}, \dfrac{3}{5}, \dfrac{4}{7}, \dfrac{5}{9}, \ldots$ is arithmetic, geometric, or neither. Then find the next term in the sequence. 11-1

2. Find a formula for the nth term of the arithmetic sequence $-7, -3, 1, 5, \ldots$. 11-2

3. Insert three arithmetic means between 1 and 47.

4. Find the eighth term of the geometric sequence $11, -22, 44, -88, \ldots$. 11-3

5. Find the geometric mean of $\dfrac{3}{2}$ and $\dfrac{8}{3}$.

6. Write the series $\dfrac{1}{2} - \dfrac{1}{4} + \dfrac{1}{6} - \dfrac{1}{8} + \cdots$ using sigma notation. 11-4

7. Find the sum of the first 20 terms of the arithmetic series $53 + 46 + 39 + 32 + \cdots$. 11-5

8. Find the sum of the geometric series $\sum\limits_{n=1}^{5} 2 \cdot 3^{n-1}$.

9. Find the sum of the infinite geometric series $250 + 150 + 90 + 54 + \cdots$ if it has one. 11-6

10. Expand and simplify $(a^2 + b)^8$. 11-7

11. Find the 12th term in the expansion of $(2x - y)^{13}$. 11-8

Cumulative Review *(Chapters 8–11)*

1. Suppose z varies directly as x and inversely as the square of y. If $z = 2$ when $x = 36$ and $y = 3$, find z when $x = 24$ and $y = 2$.

2. Divide $8x^3 - 12x + 11$ by $2x + 3$.

3. Solve $x^4 - 2x^3 + 4x^2 + 30x - 13 = 0$ completely given that $2 + 3i$ is a root.

4. Find the center and radius of the circle $x^2 + y^2 - 6x + 8y + 21 = 0$.

5. Find and graph an equation of the parabola with vertex $(4, 0)$ and directrix $x = 0$.

6. Graph the hyperbola $\dfrac{x^2}{16} - \dfrac{y^2}{4} = 1$ and find the coordinates of its foci.

Solve each system of equations.

7. $2x^2 + y^2 = 6$
   $x - y = 3$

8. $y = x^2 - 1$
   $y^2 - x^2 = 5$

9. $x - y - z = 5$
   $2x - y + z = 3$
   $x + 2y - 2z = 4$

10. $5x - 3y + z = -6$
    $x + 3y - z = 0$
    $4x + y + z = 7$

11. Evaluate $\dfrac{64^{2/3}}{64^{-1/2}}$.

12. If $f(x) = x^2 + 2$ and $g(x) = 2x - 3$, find the following:
    a. $f(g(1))$
    b. $g(f(1))$

Solve.

13. $9^{x+2} = 27$

14. $4^{x-2} = \dfrac{1}{8}$

15. $\log_9 x = 1.5$

16. $\log_x 16 = -2$

17. Give the solution of $5^x = 40$ to three significant digits.

18. Evaluate $e^{2 \ln 3}$.

19. Tell whether the sequence $108, -72, 48, -32, \ldots$ is arithmetic, geometric, or neither. Then find a formula for the nth term.

20. Tell whether the series $\displaystyle\sum_{n=1}^{11} (35 - 3n)$ is arithmetic, geometric, or neither. Then find its sum.

21. Expand and simplify $(x - y^2)^4$.

22. Find and simplify the twelfth term in the expansion of $(2x + y)^{13}$.

12 Triangle Trigonometry

By finding angle measurements between our sun and a star on different occasions, an astronomer can use trigonometry to determine how far the star is from Earth.

Trigonometric Functions

12-1 Angles and Degree Measure

Objective   To use degrees to measure angles.

Trigonometry, which means "triangle measurement," is a branch of mathematics that was originally used for surveying. Today trigonometry is used in astronomy, navigation, architecture, and many other fields where measurement is important.

In this chapter angle measurements will be given in *degrees*. Recall that one **degree** (1°) is $\frac{1}{360}$ of a complete revolution. In geometry most angles you work with have measures between 0° and 180°. In trigonometry you will also use angles with measures greater than 180° and less than 0°.

You can generate any angle by using a rotation that moves one of the rays forming the angle, called the **initial side,** onto the other ray, called the **terminal side.** As shown in the figure below, *counterclockwise* rotations produce **positive angles** and *clockwise* rotations produce **negative angles.** An angle generated in this manner is called a **directed angle.**

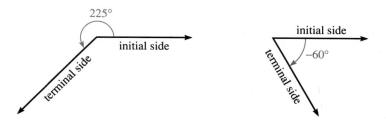

An angle is in **standard position** when its initial side coincides with the positive *x*-axis. As shown in the diagrams below, angles in standard position can be classified according to the quadrant in which their terminal side lies. If the terminal side lies on a coordinate axis, the angle is called a **quadrantal angle.**

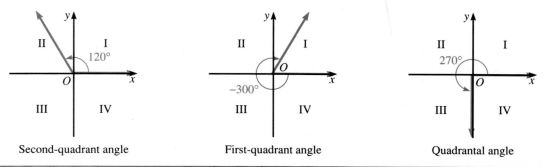

Second-quadrant angle   First-quadrant angle   Quadrantal angle

Two angles in standard position are **coterminal** if their terminal sides coincide. The diagrams below show several angles coterminal with a 150° angle.

150° + 0 • 360°

150° + 1 • 360°

150° + 2 • 360°

150° + (−1) • 360°

The measures of all angles coterminal with a 150° angle are given by the expression

$$150° + n \cdot 360°, \text{ where } n \text{ is an integer.}$$

In the diagrams above, n has the values 0, 1, 2, and −1, respectively.

Example 1 **a.** Write a formula for the measures of all angles coterminal with a 30° angle.

b. Use the formula to find two positive angles and two negative angles that are coterminal with a 30° angle.

Solution **a.** $30° + n \cdot 360°$, where n is an integer.

b. To find the measures of the angles, let n equal 1, 2, −1, and −2.

$30° + (1)360° = 390°$
$30° + (2)360° = 750°$
$30° + (−1)360° = −330°$
$30° + (−2)360° = −690°$

To measure angles more precisely than to the nearest degree, you can use either decimal degrees or degrees, minutes, and seconds. One **minute** (1′) is $\frac{1}{60}$ of 1°, and one **second** (1″) is $\frac{1}{60}$ of 1′.

Many calculators will convert decimal degrees to degrees, minutes, and seconds, and vice versa. (See the Calculator Key-In on page 554.) If you do not use a calculator, you can use the methods shown in Example 2.

Example 2 **a.** Express 14°36′54″ in decimal degrees.

b. Express 72.568° in degrees, minutes, and seconds to the nearest second.

Solution **a.** Use these facts: $1' = \left(\frac{1}{60}\right)°$ and $1'' = \left(\frac{1}{60}\right)' = \left(\frac{1}{3600}\right)°$.

$$14°36'54'' = 14° + \left(\frac{36}{60}\right)° + \left(\frac{54}{3600}\right)°$$

$$= 14° + 0.6° + 0.015°$$

$$= 14.615° \quad \textit{Answer}$$

b. $72.568° = 72° + (0.568 \times 60)'$

$$= 72° + 34.08'$$

$$= 72° + 34' + (0.08 \times 60)''$$

$$= 72° + 34' + 4.8''$$

$$\approx 72°34'5''$$

Answer

We often use capital letters, as in $\angle A$, or Greek letters such as θ (theta), ϕ (phi), and α (alpha) to name angles. (A list of Greek letters used in this book appears on page xvi.) To denote the measure of an angle, we may write, for example, $\theta = 20°$, $\angle A = 20°$, $m\angle A = 20°$, or $\angle A = 20$. In this book we will use $\theta = 20°$ or $\angle A = 20°$ to denote an angle with a measure of 20°. Also, when we say that 120° is a second-quadrant angle, we mean that a 120° *angle* in standard position has its terminal side in the second quadrant.

Oral Exercises

1. The measure of an angle in quadrant I is between 0° and 90°. What can you say about the measure of an angle in quadrant II? in quadrant III? in quadrant IV?

2. Give the measures of the quadrantal angles from 0° to 360°, inclusive.

Estimate the measure of angle θ.

3.

4.

5.

6.

7.

8.

9.

10.
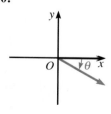

Name the quadrant of each angle.

11. $140°$

12. $215°$

13. $-45°$

14. $-150°$

15. $-315°$

16. $440°$

17. $-400°$

18. $500°$

Give two angles, one positive and one negative, that are coterminal with the given angle.

19. 50° **20.** 100° **21.** 120° **22.** 40°

23. −45° **24.** −210° **25.** 390° **26.** 480°

Written Exercises

Sketch each angle in standard position. Indicate its rotation by a curved arrow. Classify each angle by its quadrant. If the angle is a quadrantal angle, say so. (See the diagrams at the bottom of page 549.)

A **1. a.** 135° **b.** −135° **2. a.** 40° **b.** −40°

3. a. 300° **b.** −300° **4. a.** 240° **b.** −240°

5. −270° **6.** 315° **7.** 290° **8.** −90°

9. 495° **10.** 1080° **11.** −810° **12.** 750°

Sketch in standard position the angle described and then find its measure.

Sample 1 $\frac{2}{5}$ of a clockwise rotation

Solution $\frac{2}{5} \times (-360) = -144$

∴ the measure is −144°.

13. $\frac{2}{3}$ of a counterclockwise revolution

14. $\frac{3}{8}$ of a counterclockwise revolution

15. $\frac{3}{4}$ of a clockwise revolution **16.** $\frac{1}{6}$ of a clockwise revolution

17. $1\frac{3}{5}$ counterclockwise revolutions **18.** $2\frac{1}{3}$ counterclockwise revolutions

Sketch each angle in standard position when $n = 0$, $n = 1$, $n = 2$, and $n = -1$.

19. $45° + n \cdot 360°$ **20.** $-30° + n \cdot 360°$ **21.** $60° + n \cdot 180°$ **22.** $-30° + n \cdot 180°$

For Exercises 23–30:

a. Write a formula for the measures of all angles coterminal with the given angle.

b. Use the formula to find two angles, one positive and one negative, that are coterminal with the given angle.

23. 35° **24.** 140° **25.** −100° **26.** −210°

27. 520° **28.** 355° **29.** 1000° **30.** −3605°

Express in decimal degrees to the nearest tenth of a degree.

31. $15°30'$ **32.** $47°36'$ **33.** $72°50'$ **34.** $51°20'$

Express in decimal degrees to the nearest hundredth of a degree.

35. $25°45'$ **36.** $33°15'$ **37.** $45°18'20''$ **38.** $0°42'30''$

Express in degrees and minutes to the nearest minute.

39. $25.4°$ **40.** $63.6°$ **41.** $44.9°$ **42.** $27.1°$

Express in degrees, minutes, and seconds to the nearest second.

43. $34.41°$ **44.** $18.27°$ **45.** $23.67°$ **46.** $58.83°$

The terminal side of an angle in standard position passes through the given point. Draw the angle and use a protractor to estimate its measure.
$(\sqrt{3} \approx 1.73)$

Sample 2 $(-1, \sqrt{3})$

Solution

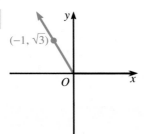

The measure is about $120°$. *Answer*

47. $(4, -4)$ **48.** $(3, 4)$ **49.** $(4, 3)$ **50.** $(-4, 4)$

51. $(\sqrt{3}, 1)$ **52.** $(-1, \sqrt{3})$ **53.** $(-1, -\sqrt{3})$ **54.** $(\sqrt{3}, -1)$

Find a first-quadrant angle θ, $0° < \theta < 90°$, for which an angle four times as large as θ will be in the given quadrant.

Sample 3   Find a first-quadrant angle θ, $0° < \theta < 90°$, for which an angle four times as large will be in the second quadrant.

Solution   You want 4θ to be in quadrant II: $90° < 4\theta < 180°$

$$\therefore 22.5° < \theta < 45°$$

Since any angle θ between $22.5°$ and $45°$ is in quadrant I, and 4θ is in quadrant II, you can choose, for example, $30°$. *Answer*

B   **55.** the first quadrant   **56.** the third quadrant   **57.** the fourth quadrant

Find a first-quadrant angle θ, $0° < \theta < 90°$, for which an angle six times as large as θ will be in the given quadrant.

58. the second quadrant   **59.** the third quadrant   **60.** the fourth quadrant

Find an angle θ for which an angle half as large as θ will be in the given quadrant.

61. the first quadrant   **62.** the second quadrant   **63.** the third quadrant

Find an angle θ for which an angle one fifth as large as θ will be in the given quadrant.

64. the first quadrant   **65.** the second quadrant   **66.** the third quadrant

---

Mixed Review Exercises

Find the third term in the expansion of each binomial.

1. $(x + 2y)^5$ **2.** $(a - 3b)^8$ **3.** $(m^2 + n^3)^{13}$

Simplify.

4. $\sqrt{-3} \cdot \sqrt{-12}$ **5.** $\log_2 14 - \log_2 7$ **6.** $(2^{3/4} - 2^{1/4})^2$

7. $\ln \sqrt[3]{e^2}$ **8.** $(3 - 2i)(4 + i)$ **9.** $\sqrt{20} + \sqrt{45} - \sqrt{125}$

10. $\dfrac{-6(3 - 5)}{(-7 + 9)^2}$ **11.** $\sqrt{\dfrac{98}{27}}$ **12.** $\log_3 3\sqrt{3}$

 Calculator Key-In

Many scientific calculators have keys that convert between degrees, minutes, and seconds and decimal degrees. On some calculators there is a conversion key (usually labeled ° ′ ″) that allows you to convert, say, 32°45′10″ to decimal degrees as follows: enter 32, press the key; enter 45, press the key again; enter 10, press the key a third time. The result is 32.7527778°.

To convert from decimal degrees to degrees, minutes, and seconds, you just enter the decimal degrees and press the inverse conversion key. (If there is no such key, press the inverse key and then the conversion key.)

On other calculators you may be required to enter 32°45′10″ differently and the keys may be labeled with "DMS" for degrees-minutes-seconds. Test your calculator using 32°45′10″ and its decimal equivalent 32.7527778° before doing the following exercises.

Convert each measure to degrees, minutes, and seconds or to decimal degrees.

1. 58.6°   **2.** 29.7°   **3.** 86.43°   **4.** 108.26°

5. 36°25′36″   **6.** 45°11′19″   **7.** 73°52′25″   **8.** 115°42′51″

12-2 Trigonometric Functions of Acute Angles

Objective   To define trigonometric functions of acute angles.

To define the trigonometric functions of an acute angle θ, first place θ in standard position as shown in the diagram below. Next, choose any point $P(x, y)$, other than the origin, on the terminal side of θ and let r be the distance OP. Then the following definitions can be made:

The **sine** of θ, written sin θ, is equal to $\dfrac{y}{r}$.

The **cosine** of θ, written cos θ, is equal to $\dfrac{x}{r}$.

The **tangent** of θ, written tan θ, is equal to $\dfrac{y}{x}$.

The ratios $\dfrac{y}{r}$, $\dfrac{x}{r}$, and $\dfrac{y}{x}$ depend only on θ and not on the choice of P. (See Exercise 32.) Therefore, the sine, cosine, and tangent are functions of θ. Notice that x, y, and r are the lengths of the legs and hypotenuse, respectively, of right triangle OMP. The definitions above can be restated in terms of these lengths, for any right triangle containing the acute angle θ.

$$\sin \theta = \frac{y}{r} = \frac{\text{length of side opposite } \theta}{\text{length of the hypotenuse}}$$

$$\cos \theta = \frac{x}{r} = \frac{\text{length of side adjacent to } \theta}{\text{length of the hypotenuse}}$$

$$\tan \theta = \frac{y}{x} = \frac{\text{length of side opposite } \theta}{\text{length of side adjacent to } \theta}$$

Example 1   An acute angle θ is in standard position and its terminal side passes through $P(4, 5)$. Find sin θ, cos θ, and tan θ.

Solution   Here $x = 4$ and $y = 5$. To find r, use the Pythagorean theorem (page 401):

$$r^2 = 4^2 + 5^2 = 16 + 25 = 41, \quad \text{so} \quad r = \sqrt{41}$$

$$\therefore \sin \theta = \frac{y}{r} = \frac{5}{\sqrt{41}} = \frac{5\sqrt{41}}{41}; \cos \theta = \frac{x}{r} = \frac{4}{\sqrt{41}} = \frac{4\sqrt{41}}{41}; \tan \theta = \frac{5}{4}$$

The three remaining trigonometric functions are **reciprocal functions** of those functions already defined.

The **cotangent** of θ, written cot θ, is equal to $\frac{x}{y}$, or $\frac{1}{\tan \theta}$.

The **secant** of θ, written sec θ, is equal to $\frac{r}{x}$, or $\frac{1}{\cos \theta}$.

The **cosecant** of θ, written csc θ, is equal to $\frac{r}{y}$, or $\frac{1}{\sin \theta}$.

Example 2 Find the values of the six trigonometric functions of an angle θ in standard position whose terminal side passes through (5, 12).

Solution Sketch θ in standard position.
By the Pythagorean theorem:

$$r^2 = x^2 + y^2 = 5^2 + 12^2 = 169$$
$$r = 13$$

$$\sin \theta = \frac{y}{r} = \frac{12}{13} \qquad \csc \theta = \frac{r}{y} = \frac{13}{12}$$

$$\cos \theta = \frac{x}{r} = \frac{5}{13} \qquad \sec \theta = \frac{r}{x} = \frac{13}{5}$$

$$\tan \theta = \frac{y}{x} = \frac{12}{5} \qquad \cot \theta = \frac{x}{y} = \frac{5}{12}$$

An equation involving trigonometric functions of an angle θ that is true for all values of θ is a *trigonometric identity*. Using the diagram above and the definitions of this lesson, notice that

$$\frac{\sin \theta}{\cos \theta} = \frac{\frac{y}{r}}{\frac{x}{r}} = \frac{y}{x} = \tan \theta \quad \text{and} \quad \frac{\cos \theta}{\sin \theta} = \frac{\frac{x}{r}}{\frac{y}{r}} = \frac{x}{y} = \cot \theta.$$

Therefore, the following equations are identities:

$$\tan \theta = \frac{\sin \theta}{\cos \theta} \qquad \cot \theta = \frac{\cos \theta}{\sin \theta}$$

The next identity is called a *Pythagorean identity*. Notice that we write $\sin^2 \theta$ for $(\sin \theta)^2$ and $\cos^2 \theta$ for $(\cos \theta)^2$.

$$\sin^2 \theta + \cos^2 \theta = 1$$

To prove this identity, you use the Pythagorean relationship $x^2 + y^2 = r^2$.

$$\sin^2 \theta + \cos^2 \theta = \left(\frac{y}{r}\right)^2 + \left(\frac{x}{r}\right)^2 = \frac{y^2}{r^2} + \frac{x^2}{r^2} = \frac{y^2 + x^2}{r^2} = \frac{r^2}{r^2} = 1$$

Example 3 Find $\cos \theta$ and $\tan \theta$ if θ is an acute angle and $\sin \theta = \frac{1}{3}$.

Solution 1 Use the identity $\sin^2 \theta + \cos^2 \theta = 1$.

$$\left(\frac{1}{3}\right)^2 + \cos^2 \theta = 1$$

$$\cos^2 \theta = 1 - \left(\frac{1}{3}\right)^2 = 1 - \frac{1}{9} = \frac{8}{9}$$

$$\cos \theta = \sqrt{\frac{8}{9}} = \frac{2\sqrt{2}}{3} \quad \textit{Answer}$$

$$\tan \theta = \frac{\sin \theta}{\cos \theta} = \frac{\frac{1}{3}}{\frac{2\sqrt{2}}{3}} = \frac{1}{2\sqrt{2}} = \frac{\sqrt{2}}{4} \quad \textit{Answer}$$

Solution 2 Make a diagram showing

$$\sin \theta = \frac{\text{length of side opposite } \theta}{\text{length of the hypotenuse}} = \frac{1}{3}.$$

Then find x: $\quad x^2 = 3^2 - 1^2 = 8$

$$x = 2\sqrt{2}$$

Then $\cos \theta = \frac{2\sqrt{2}}{3}$ and $\tan \theta = \frac{1}{2\sqrt{2}} = \frac{\sqrt{2}}{4}$. *Answer*

The sine and cosine are called **cofunctions.** Other pairs of cofunctions are the tangent and cotangent and the secant and cosecant. There are six cofunction identities that can be derived from right triangle ABC.

$$\sin A = \frac{a}{c} = \cos B \qquad \cos A = \frac{b}{c} = \sin B$$

$$\tan A = \frac{a}{b} = \cot B \qquad \cot A = \frac{b}{a} = \tan B$$

$$\sec A = \frac{c}{b} = \csc B \qquad \csc A = \frac{c}{a} = \sec B$$

Notice that $\angle A$ and $\angle B$ are complementary; that is, $m\angle A + m\angle B = 90°$. Notice also that:

$$\sin \theta = \cos (90° - \theta) \qquad \cos \theta = \sin (90° - \theta)$$

$$\tan \theta = \cot (90° - \theta) \qquad \cot \theta = \tan (90° - \theta)$$

$$\sec \theta = \csc (90° - \theta) \qquad \csc \theta = \sec (90° - \theta)$$

Thus, any trigonometric function of an acute angle is equal to the cofunction of the complement of the angle. Since 30° angles and 60° angles are complementary, they illustrate the cofunction relationship. For example, $\tan 60° = \cot 30°$.

You can read the trigonometric functions of 30°, 45°, and 60° from the triangles shown below. To recall these values, you may find it easier to draw the triangles rather than to memorize the table.

θ	$\sin \theta$	$\cos \theta$	$\tan \theta$	$\csc \theta$	$\sec \theta$	$\cot \theta$
30°	$\frac{1}{2}$	$\frac{\sqrt{3}}{2}$	$\frac{\sqrt{3}}{3}$	2	$\frac{2\sqrt{3}}{3}$	$\sqrt{3}$
45°	$\frac{\sqrt{2}}{2}$	$\frac{\sqrt{2}}{2}$	1	$\sqrt{2}$	$\sqrt{2}$	1
60°	$\frac{\sqrt{3}}{2}$	$\frac{1}{2}$	$\sqrt{3}$	$\frac{2\sqrt{3}}{3}$	2	$\frac{\sqrt{3}}{3}$

Example 4 Use the diagram at the right. Find the lengths of side $\overline{BC}$ and side $\overline{AB}$.

Solution
$$\tan 30° = \frac{BC}{AC} \qquad \cos 30° = \frac{AC}{AB}$$
$$\frac{\sqrt{3}}{3} = \frac{a}{12} \qquad \frac{\sqrt{3}}{2} = \frac{12}{c}$$
$$3a = 12\sqrt{3} \qquad \sqrt{3}c = 24$$
$$a = 4\sqrt{3} \qquad c = \frac{24}{\sqrt{3}} = 8\sqrt{3}$$

Example 5 Use the diagram at the right to find the measure of $\angle A$.

Solution
$$\cos A = \frac{AC}{AB} = \frac{3}{6} = \frac{1}{2}$$
$$\cos 60° = \frac{1}{2}$$
$$\therefore \angle A = 60°. \quad \textbf{\textit{Answer}}$$

Oral Exercises

Give the values of the six trigonometric functions of θ.

1.

2.

3.

4.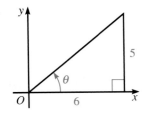

Use the cofunction identities to find the measure of the acute angle θ.

5. $\sin \theta = \cos 25°$　　　　　　　　　　　　**6.** $\tan \theta = \cot 70°$

7. $\sec \theta = \csc 15°$　　　　　　　　　　　　**8.** $\cos \theta = \sin 45°$

The length of one side of a right triangle is given. Give the lengths of the other two sides.

9.

10.

11.
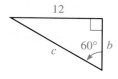

12. If $\sin \theta = \dfrac{2}{3}$ for an acute angle θ, state two ways to find $\cos \theta$ and $\tan \theta$.

Written Exercises

Find the values of the six trigonometric functions of angle θ.

A　**1.**

2.

3.

4.

Find the values of the six trigonometric functions of an angle θ in standard position whose terminal side passes through point P.

5. $P(8, 15)$　　　　**6.** $P(3, 4)$　　　　**7.** $P(3, 1)$　　　　**8.** $P(5, 5)$

Complete the table. In each case, θ is an acute angle.

	9.	10.	11.	12.	13.	14.	15.
$\sin \theta$	$\dfrac{3}{5}$	?	?	$\dfrac{\sqrt{2}}{3}$	$\dfrac{\sqrt{161}}{15}$	?	?
$\cos \theta$	?	$\dfrac{1}{2}$	$\dfrac{\sqrt{3}}{2}$	?	?	$\dfrac{5}{13}$	$\dfrac{1}{4}$
$\tan \theta$	?	?	?	?	?	?	?

Use the cofunction identities to find the measure of the acute angle φ.

16. $\sin \phi = \cos 50°$　　**17.** $\cos \phi = \sin 40°$　　**18.** $\tan \phi = \cot 17°$　　**19.** $\sec \phi = \csc 80°$

In Exercises 20–25, use the diagram at the
right. Find the lengths of the sides and the
measures of the angles that are not given.
Leave your answers in simplest radical form.

20. $a = 6$, $\angle A = 30°$ **21.** $b = 2$, $\angle A = 45°$

22. $c = 10$, $\angle A = 45°$ **23.** $c = 20$, $\angle A = 60°$

24. $a = 3$, $c = 6$ **25.** $a = 4$, $b = 4$

In Exercises 26–29, find the length of x. Leave your answers in simplest
radical form.

B   26.

27.

28.

29.

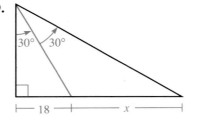

30. Prove that $\cot \theta = \dfrac{\cos \theta}{\sin \theta}$.

31. The three *Pythagorean identities* in trigonometry are as follows:

$$\sin^2 \theta + \cos^2 \theta = 1 \qquad \tan^2 \theta + 1 = \sec^2 \theta \qquad 1 + \cot^2 \theta = \csc^2 \theta$$

We have shown the first of these to be true for every acute angle θ. Use
this first Pythagorean identity to prove that the other two identities are true
for every acute angle θ.

32. $P(x, y)$ and $Q(x', y')$ are two points on the terminal side of an acute
angle θ in standard position. Let $OP = r$ and $OQ = s$. Use similar triangles
to show that $\dfrac{y}{r} = \dfrac{y'}{s}$, $\dfrac{x}{r} = \dfrac{x'}{s}$, and $\dfrac{y}{x} = \dfrac{y'}{x'}$. (This shows that using point Q
to define the trigonometric functions θ will yield the same values as using
point P.)

C   33. Angle θ is acute and $\sin \theta = u$. Express the other five trigonometric functions of θ in terms of u. (*Hint:* Recall that $\sin^2 \theta + \cos^2 \theta = 1$.)

34. Angle ϕ is acute and $\cos \phi = v$. Express the other five trigonometric functions of ϕ in terms of v.

12-3 Trigonometric Functions of General Angles

Objective   To define trigonometric functions of general angles.

In Lesson 12-2, you learned the definitions for the six trigonometric functions for acute angles. Now you'll extend these definitions to angles of any measure.

As you did with acute angles, place the angle θ in standard position, as shown in each diagram below. Choose a point $P(x, y)$ on the terminal side of θ in each quadrant and let r be the distance OP.

  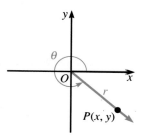

For any angle θ and any point (x, y) on the terminal side:

$$\sin \theta = \frac{y}{r} \qquad \cos \theta = \frac{x}{r} \qquad \tan \theta = \frac{y}{x}, \text{ if } x \neq 0$$

$$\csc \theta = \frac{r}{y}, \text{ if } y \neq 0 \qquad \sec \theta = \frac{r}{x}, \text{ if } x \neq 0 \qquad \cot \theta = \frac{x}{y}, \text{ if } y \neq 0$$

Example 1   Find the values of the six trigonometric functions of an angle θ in standard position whose terminal side passes through $(8, -15)$.

Solution   First make a sketch similar to the third one above.
Here $x = 8$ and $y = -15$. Since r is the distance between $(0, 0)$ and $(8, -15)$, you can find r by using the distance formula (page 402).

$$r = \sqrt{8^2 + (-15)^2} = \sqrt{64 + 225} = \sqrt{289} = 17$$

$$\sin \theta = \frac{y}{r} = -\frac{15}{17} \qquad \cos \theta = \frac{x}{r} = \frac{8}{17} \qquad \tan \theta = \frac{y}{x} = -\frac{15}{8}$$

$$\csc \theta = \frac{r}{y} = -\frac{17}{15} \qquad \sec \theta = \frac{r}{x} = \frac{17}{8} \qquad \cot \theta = \frac{x}{y} = -\frac{8}{15}$$

In the definitions above, r is always positive. Therefore, the signs of the functions of θ are determined by the signs of x and y, and these signs depend only on the quadrant of the terminal side of θ, as shown on the next page.

Function value	Quadrant of θ			
	I	II	III	IV
sin θ csc θ	+	+	−	−
cos θ sec θ	+	−	−	+
tan θ cot θ	+	−	+	−

The sine and cosine functions are defined for all angles, but the other four functions are undefined for certain quadrantal angles.

Example 2 Determine which functions are defined for a 180° angle and find their values.

Solution Place the angle in standard position. Choose a point on the terminal side, such as $P(-1, 0)$. Then $x = -1$, $y = 0$, and $r = 1$.

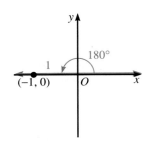

$$\sin 180° = \frac{y}{r} = \frac{0}{1} = 0$$

$$\cos 180° = \frac{x}{r} = \frac{-1}{1} = -1$$

$$\tan 180° = \frac{y}{x} = \frac{0}{-1} = 0$$

$$\csc 180° = \frac{r}{y} \text{ is undefined since } y = 0.$$

$$\sec 180° = \frac{r}{x} = \frac{1}{-1} = -1$$

$$\cot 180° = \frac{x}{y} \text{ is undefined since } y = 0.$$

If θ is not a quadrantal angle, there is a unique acute angle α, corresponding to θ, formed by the terminal side of θ and the positive or negative x-axis. When θ is in standard position, we call α the **reference angle** of θ. Diagrams like those below are helpful in finding reference angles.

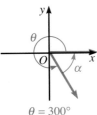

$\theta = 300°$
$\alpha = 60°$

$\theta = 225°$
$\alpha = 45°$

$\theta = -210°$
$\alpha = 30°$

$\theta = 420°$
$\alpha = 60°$

Example 3 Find the measure of the reference angle α for each given angle θ.

        a. $\theta = 140°$ **b.** $\theta = 300°10'$ **c.** $\theta = -135°$

Solution   **a.** Since θ is in quadrant II: $\alpha = 180° - \theta = 180° - 140° = 40°$

        b. Since θ is in quadrant IV: $\alpha = 360° - \theta$

$$= 360° - 300°10'$$
$$= 359°60' - 300°10' = 59°50'$$

        c. Find the positive angle that is coterminal with $-135°$:

$$-135° + 360° = 225°$$

Since $225°$ is in quadrant III: $\alpha = \theta - 180° = 225° - 180° = 45°$

     You can use the reference angle to draw a **reference triangle** for any angle θ and then use the reference triangle to find values for the trigonometric functions of θ. For example, from right triangle OMP shown at the right, you see that $\cos \alpha = \dfrac{|x|}{r}$.

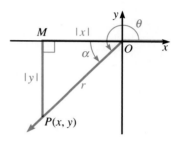

Since $\cos \theta = \dfrac{x}{r}$, you have

$$\cos \theta = \frac{x}{r} = \pm \frac{|x|}{r} = \pm \cos \alpha.$$

     Similar results hold for the other trigonometric functions and for angles in other quadrants. Therefore, each trigonometric function of θ is either equal to the same function of α or equal to its opposite. That is, $\sin \theta = \pm \sin \alpha$, $\cos \theta = \pm \cos \alpha$, and so forth, where α is the reference angle of θ. The correct sign is determined by the quadrant of θ.

Example 4 Write $\cos 200°$ as a function of an acute angle.

Solution   The reference angle of $200°$ is $200° - 180° = 20°$.
Since $200°$ is a third-quadrant angle, its cosine is negative.
$\therefore \cos 200° = -\cos 20°$.   *Answer*

Example 5 Find the exact value of the following.

        a. $\tan 330°$ **b.** $\csc (-225°)$

Solution   **a.** The reference angle of $330°$ is $30°$.

          Since $330°$ is a fourth-quadrant angle, its tangent is negative.

          $\therefore \tan 330° = -\tan 30° = -\dfrac{\sqrt{3}}{3}$ *Answer*

        b. The reference angle of $-225°$ is $45°$.

          Since $-225°$ is a second-quadrant angle, its cosecant is positive.

          $\therefore \csc (-225°) = \csc 45° = \sqrt{2}$ *Answer*

The identities proved for acute angles in Lesson 12-2 are true for general angles θ as well. If $\cos \theta \neq 0$, you have

$$\frac{\sin \theta}{\cos \theta} = \frac{\frac{y}{r}}{\frac{x}{r}} = \frac{y}{x} = \tan \theta \quad \text{and} \quad \frac{\cos \theta}{\sin \theta} = \frac{\frac{x}{r}}{\frac{y}{r}} = \frac{x}{y} = \cot \theta.$$

And, if $P(x, y)$ is at a distance r from the origin on the terminal side of θ, the distance formula gives

$$r = \sqrt{x^2 + y^2} \quad \text{or} \quad r^2 = x^2 + y^2.$$

Therefore, for any angle θ,

$$\sin^2 \theta + \cos^2 \theta = \frac{y^2 + x^2}{r^2} = 1.$$

Example 6 Find the five other trigonometric functions of θ if $\cos \theta = -\frac{2}{5}$ and $180° < \theta < 360°$.

Solution 1 Use $\sin^2 \theta + \cos^2 \theta = 1$.

$$\sin^2 \theta + \left(-\frac{2}{5}\right)^2 = 1$$

$$\sin^2 \theta = 1 - \frac{4}{25} = \frac{21}{25}$$

$$\sin \theta = \pm\sqrt{\frac{21}{25}}$$

$$\sin \theta = \frac{\sqrt{21}}{5} \quad \text{or} \quad \sin \theta = -\frac{\sqrt{21}}{5}$$

Since $\cos \theta$ is negative and $180° < \theta < 360°$, θ is a third-quadrant angle and $\sin \theta$ is negative.

$$\sin \theta = -\frac{\sqrt{21}}{5}$$

$$\tan \theta = \frac{\sin \theta}{\cos \theta} = \frac{-\frac{\sqrt{21}}{5}}{-\frac{2}{5}} = \frac{\sqrt{21}}{2}$$

The three remaining functions are reciprocals of those already known:

$$\sec \theta = -\frac{5}{2} \qquad \csc \theta = -\frac{5}{\sqrt{21}} = -\frac{5\sqrt{21}}{21} \qquad \cot \theta = \frac{2}{\sqrt{21}} = \frac{2\sqrt{21}}{21}$$

Solution 2 If $180° < \theta < 360°$ and $\cos \theta$ is negative, θ must be a third-quadrant angle.

Make a sketch as shown.

The Pythagorean theorem gives:
$y^2 = 5^2 - (-2)^2 = 21 \quad \text{or} \quad y = -\sqrt{21}$

Thus, $\sin \theta = -\frac{\sqrt{21}}{5}$ and $\tan \theta = \frac{-\sqrt{21}}{-2} = \frac{\sqrt{21}}{2}$.

The remaining functions are obtained as in Solution 1.

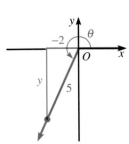

Oral Exercises

State the measure of the reference angle α for each angle θ.

1.

$\theta = 135°$

2.

$\theta = -120°$

3.

$\theta = 330°$

4.

$\theta = 405°$

For each angle θ, name its quadrant and its reference angle.

5. $\theta = 170°$ **6.** $\theta = 250°$ **7.** $\theta = 210°$ **8.** $\theta = -70°$

9. $\theta = -305°$ **10.** $\theta = -200°$ **11.** $\theta = 460°$ **12.** $\theta = -400°$

Give the values of the six trigonometric functions of θ. If any value is undefined, say so.

13.

$P(-4, 3)$

14.

$P(3, -4)$

15.

$P(1, 1)$

16.

$P(-1, -1)$

17.

$P(0, 1)$

18.

$P(1, 0)$

19.

$P(-3, -2)$

20.

$P(-2, 1)$

In Exercises 21–28, name the quadrant of θ. (Use the chart on page 562.)

21. $\sin \theta < 0$, $\cos \theta > 0$ **22.** $\sin \theta < 0$, $\cos \theta < 0$

23. $\sin \theta > 0$, $\cos \theta > 0$ **24.** $\sin \theta > 0$, $\cos \theta < 0$

25. $\sec \theta > 0$, $\tan \theta < 0$ **26.** $\csc \theta < 0$, $\cos \theta < 0$

27. $\cos \theta < 0$, $180° < \theta < 360°$ **28.** $\tan \theta > 0$, $90° < \theta < 270°$

29. If $\cos \theta = \dfrac{1}{4}$, $\sec \theta = $ _?_   **30.** If $\sin \theta = -\dfrac{3}{5}$, $\csc \theta = $ _?_

31. If $\cot \theta = -3$, $\tan \theta = $ _?_   **32.** If $\sec \theta = 1.5$, $\cos \theta = $ _?_

Written Exercises

Find the values of the six trigonometric functions of an angle θ in standard position whose terminal side passes through point P.

A 1. $P(3, -4)$ 2. $P(-5, 12)$ 3. $P(-7, 24)$ 4. $P(-8, -15)$

Complete the table. If any value is undefined, say so.

		$\sin \theta$	$\cos \theta$	$\tan \theta$	$\sec \theta$	$\csc \theta$	$\cot \theta$
5.	$0°$	?	?	?	?	?	?
6.	$90°$	?	?	?	?	?	?
7.	$180°$	?	?	?	?	?	?
8.	$270°$	?	?	?	?	?	?

Find the measure of the reference angle α of the given angle θ.

9. $\theta = 233°$ 10. $\theta = 126°$ 11. $\theta = -205°$ 12. $\theta = -112°$

13. $\theta = 512°$ 14. $\theta = 659°$ 15. $\theta = -725°$ 16. $\theta = -611°$

17. $\theta = 96.4°$ 18. $\theta = 134.7°$ 19. $\theta = -184.1°$ 20. $\theta = -344.2°$

21. $\theta = 156°20'$ 22. $\theta = 213°40'$ 23. $\theta = 152°30'$ 24. $\theta = 312°50'$

Write each of the following as a function of an acute angle.

25. $\cos 216°$ 26. $\tan 334°$ 27. $\sin (-17°)$

28. $\sec (-106°)$ 29. $\cot 287.1°$ 30. $\csc 117.8°$

31. $\cos (-221.9°)$ 32. $\sin (-46.6°)$ 33. $\tan 265°20'$

34. $\cot 137°10'$ 35. $\sin 212°40'$ 36. $\sin 324°30'$

Find the exact value of the six trigonometric functions of each angle.

37. $330°$ 38. $240°$ 39. $135°$ 40. $-60°$

41. $315°$ 42. $300°$ 43. $-150°$ 44. $480°$

First give the quadrant of angle θ. Then find the five other trigonometric functions of θ. Give answers involving radicals in simplest radical form.

B 45. $\cos \theta = -\dfrac{8}{17}$, $0° < \theta < 180°$ 46. $\sin \theta = -\dfrac{4}{5}$, $90° < \theta < 270°$

47. $\sin \theta = -\dfrac{5}{13}$, $\cos \theta > 0$ 48. $\cos \theta = -\dfrac{3}{5}$, $\sin \theta > 0$

49. $\cos \theta = \dfrac{2}{3}$, $0° < \theta < 270°$ 50. $\sin \theta = -\dfrac{2}{5}$, $-90° < \theta < 180°$

51. $\tan \theta = -\dfrac{3}{4}$, $\cos \theta < 0$ 52. $\sec \theta = \dfrac{13}{5}$, $\sin \theta < 0$

Name all angles θ, $0° \le \theta < 360°$, that make the statement true.

53. $\sin \theta = 0$

54. $\cos \theta = 0$

55. $\tan \theta = 0$

56. $\sin \theta = 1$

57. $\tan \theta = 1$

58. $\cos \theta = -1$

59. $\sin \theta = -1$

60. $\sin \theta = \frac{1}{2}$

61. $\cos \theta = -\frac{\sqrt{3}}{2}$

62. $\sin \theta = \sin 300°$

63. $\tan \theta = \tan 15°$

64. $\sin \theta = \sin 40°$

65. $\cos \theta = \cos 70°$

66. $\cos \theta = \cos 100°$

67. $\tan \theta = \tan 20°$

Write each of the following in terms only of $\sin \theta$ and $\cos \theta$. (*Hint:* A sketch will be useful.)

C   **68.** $\sin (-\theta)$

69. $\cos (-\theta)$

70. $\tan (-\theta)$

71. $\sin (180° + \theta)$

72. $\sin (180° - \theta)$

73. $\cos (180° + \theta)$

74. $\cos (180° - \theta)$

75. $\tan (180° + \theta)$

76. $\tan (180° - \theta)$

Mixed Review Exercises

Evaluate.

1. $\displaystyle\sum_{k=1}^{8} 5 \cdot 2^{k-1}$

2. $\displaystyle\sum_{k=1}^{20} (3k + 1)$

3. $\displaystyle\sum_{k=1}^{\infty} 10\left(\frac{3}{5}\right)^{k-1}$

Find the zeros of each function. If the function has no zeros, say so.

4. $f(x) = 3x + 5$

5. $g(x) = \log (x - 1)$

6. $h(x) = 2x^2 + x - 3$

7. $F(x) = \dfrac{x^3 - 4x}{x^2 + 1}$

8. $G(x) = x^2 - 4x + 5$

9. $H(x) = \left(\dfrac{1}{2}\right)^x$

 Career Note / *Flight Engineer*

Modern aircraft are operated with the aid of many instruments. A flight engineer assists the pilot in checking the plane before take-off, checks many of the instruments in flight, and records performance levels and fuel usage. Flight engineers report any mechanical problems to the pilot and sometimes make repairs. Flight engineers are licensed by the Federal Aviation Administration.

Most airlines look for candidates with a high school education and at least two years of college. Courses in mathematics and science are important.

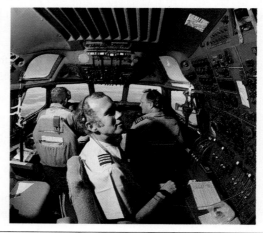

12-4 Values of Trigonometric Functions

Objective To use a calculator or trigonometric tables to find values of trigonometric functions.

It is not possible to give *exact* trigonometric function values for most angles. You can, however, get good approximations using a scientific calculator. For example, to find cos 52°, enter 52 and press the cos key to obtain 0.6156615.

If a calculator is not available, you can use Table 4 or Table 5 on pages 814–826 to approximate the functions of an acute angle θ as follows:

1. If $0° \le \theta \le 45°$, find the function name in the *top* row and look *down* to the entry opposite the degree measure of θ at the extreme *left*.
2. If $45° \le \theta \le 90°$, find the function name in the *bottom* row and look *up* to the entry opposite the degree measure of θ at the extreme *right*.

Example 1 Find each function value to four significant digits.
  a. sec 34.7°   **b.** cos 84°20′

Solution 1 Using a Calculator

  a. Many scientific calculators have keys labeled *sin, cos,* and *tan,* but not *sec, csc,* and *cot.* So to find values for secant, cosecant, and cotangent, you may need to use reciprocal functions (page 556).

$$\sec 34.7° = \frac{1}{\cos 34.7°} = \frac{1}{0.822144} = 1.2163319$$

  ∴ to four significant digits, sec 34.7° = 1.216. *Answer*

  b. Some calculators operate only with decimal degrees. First you may need to change 84°20′ to decimal degrees by dividing 20′ by 60. (60′ = 1°)

$$\cos 84°20′ = \cos 84.333333° = 0.0987408$$

  ∴ to four significant digits, cos 84°20′ = 0.0987. *Answer*

Solution 2 Using Tables

  a. In Table 4 look *down* the column under "sec θ" until you see the entry opposite 34.7° at the left. You should find "1.216."

  ∴ sec 34.7° = 1.216 *Answer*

  b. In Table 5 look *up* the column over "Cos" until you see the entry opposite 84°20′ at the right. You should find "0.0987."

  ∴ cos 84°20′ = 0.0987 *Answer*

Even though the values given by Table 4 and Table 5 are approximate, it's a common practice to use =, as in Example 1, rather than ≈.

Using a calculator you can easily find function values for angles with measures like 64.72° and 26°27′. However, these measures fall between consecutive entries in the tables. To use the tables, you may either use the table entry nearest the given measure or use *linear interpolation* (Lesson 8-9).

Example 2 Find cos 26°27′

Solution 1 Using a Calculator

You may need to change 26°27′ to decimal degrees by dividing 27′ by 60.

∴ cos 26°27′ = cos 26.45° = 0.8953234. *Answer*

Solution 2 Using Tables

Notice that cos 26°20′ > cos 26°27′ > cos 26°30′. Using a vertical arrangement for the linear interpolation, you can write:

$$10'\left[7'\left[\begin{array}{c|c} \theta & \cos\theta \\ \hline 26°20' & 0.8962 \\ 26°27' & \underline{\quad ? \quad} \\ 26°30' & 0.8949 \end{array}\right]d\right]-0.0013 \left\{\begin{array}{l}\text{A negative number}\\ \text{because the function}\\ \text{value decreases.}\end{array}\right.$$

$$\frac{7}{10} = \frac{d}{-0.0013}; \ d = \frac{7}{10}(-0.0013) = -0.0009$$

∴ cos 26°27′ = 0.8962 + (−0.0009) = 0.8953 *Answer*

Example 3 Find the measure of the acute angle θ to the nearest tenth of a degree when sin θ = 0.8700.

Solution 1 Using a Calculator

To find an acute angle when given one of its function values, you use the inverse function keys ($\sin^{-1}$, $\cos^{-1}$, $\tan^{-1}$, or inv sin, inv cos, inv tan).

If sin θ = 0.8700, then

$$\theta = \sin^{-1} 0.8700 = 60.458639.$$

∴ to the nearest tenth of a degree, θ = 60.5°. *Answer*

Caution: $\sin^{-1} x$ is *a number whose sine is x*. The notation $\sin^{-1} x$ is not exponential notation.

It does *not* mean $\dfrac{1}{\sin x}$.

Solution 2 Using Tables

Reverse the process described in Example 1. Look in the sine columns of Table 4 until you find the entry nearest 0.8700. This entry is 0.8704, and it is opposite 60.5° on the right.

∴ to the nearest tenth of a degree, θ = 60.5°. *Answer*

Sometimes you may be asked to find solutions to the nearest minute instead of to the nearest tenth of a degree. You can use a calculator or tables to do this. To use the tables, you again use linear interpolation.

Example 4 Find the measure of the acute angle θ to the nearest minute when $\cot \theta = 0.0782$.

Solution 1 Using a Calculator

Since $\cot \theta = 0.0782$, $\tan \theta = \dfrac{1}{0.0782} = 12.787724$. Therefore,

$$\theta = \tan^{-1} 12.787724 = 85.52857°.$$

To convert $0.52857°$ to minutes, multiply by 60 to obtain $31.714191'$.
$\therefore$ to the nearest minute, $\theta = 85°32'$. **Answer**

Solution 2 Using Tables

First locate in Table 5 the nearest values for $\cot \theta$ that are above and below 0.0782. Then arrange the values as follows.

$$10'\left[d'\begin{bmatrix} \dfrac{\theta}{85°40'} & \dfrac{\cot \theta}{0.0758} \\ \dfrac{?}{85°30'} & \dfrac{0.0782}{0.0787} \end{bmatrix}-0.0005\right]-0.0029 \begin{cases} \text{A negative number} \\ \text{because} \\ \cot 85°40' < \cot 85°30'. \end{cases}$$

$\dfrac{d}{10} = \dfrac{-0.0005}{-0.0029}$; $d = \dfrac{5}{29}(10) = 2$

$\therefore$ to the nearest minute, $\theta = 85°30' + 2'$, or $85°32'$. **Answer**

The trigonometric functions of *any* angle can be found with the help of a calculator or with reference angles.

Example 5 Find $\sin 147.8°$ and $\cos 147.8°$.

Solution 1 Using a Calculator
You can obtain the values directly.

$$\sin 147.8° = 0.5329$$
$$\cos 147.8° = -0.8462$$

Solution 2 Using Tables
The reference angle for $147.8°$ is $180° - 147.8° = 32.2°$. Since $147.8°$ is a second-quadrant angle, its sine is positive and its cosine is negative. Therefore:

$$\sin 147.8° = \sin 32.2° = 0.5329$$
$$\cos 147.8° = -\cos 32.2° = -0.8462$$

Example 6 Find to the nearest tenth of a degree the measures of two angles satisfying $\cos \theta = -0.7455$ and $0° < \theta < 360°$.

Solution 1 Using a Calculator

You can follow the method of Solution 2 of Example 5 and first find the reference angle α of θ. Or you can work with θ directly:

$$\theta = \cos^{-1}(-0.7455) = 138.20206.$$

Therefore, one angle is $138.2°$. (This is a second-quadrant angle.)
Since $\cos \theta < 0$, θ is a second- or third-quadrant angle. To find the other angle, first subtract $138.2°$ from $180°$ to obtain a reference angle of $41.8°$. Then the third-quadrant angle is $\theta = 180° + 41.8° = 221.8°$.
∴ the angles are $138.2°$ and $221.8°$. *Answer*

Solution 2 Using Tables

First find the reference angle α of θ.
Since $\cos \alpha = 0.7455$, $\alpha = 41.8°$.
Because $\cos \theta < 0$, θ is a second- or third-quadrant angle.
Therefore,
$\theta_1 = 180° - \alpha = 180° - 41.8° = 138.2°$ and
$\theta_2 = 180° + \alpha = 180° + 41.8° = 221.8°$. *Answer*

Oral Exercises

Use a calculator, Table 4, or Table 5 to find each function value to four significant digits.

1. $\sin 15°$
2. $\cos 42°$
3. $\tan 22.6°$
4. $\cot 34.7°$
5. $\cos 85.9°$
6. $\sin 58.2°$
7. $\csc 27°40'$
8. $\sec 49°20'$
9. $\cot 73°50'$
10. $\tan 56°10'$
11. $\sec 12°30'$
12. $\csc 47°20'$

Use a calculator or Table 4 to find the measure of acute angle θ to the nearest tenth of a degree.

13. $\cos \theta = 0.8960$
14. $\sin \theta = 0.6160$
15. $\cot \theta = 7.596$
16. $\tan \theta = 1.632$
17. $\sec \theta = 2.089$
18. $\csc \theta = 1.095$

Use a calculator or Table 5 to find the measure of acute angle θ to the nearest ten minutes.

19. $\sin \theta = 0.5324$
20. $\cos \theta = 0.8511$
21. $\tan \theta = 1.455$
22. $\cot \theta = 0.4950$
23. $\csc \theta = 1.046$
24. $\sec \theta = 5.164$

25. Explain how you would find the measures of two angles θ between $0°$ and $360°$ such that $\sin \theta = -0.5000$.

Written Exercises

Find each function value to four significant digits.

A
1. tan 15.2°
2. cos 38.7°
3. sin 65.2°
4. cot 81.1°

5. cos 31°10'
6. sin 46°30'
7. sec 54°40'
8. csc 23°20'

9. csc 7.44°
10. sec 13.22°
11. cos 32.43°
12. sin 26.57°

13. cos 52°43'
14. tan 29°36'
15. cot 78°15'
16. sin 87°53'

17. sec 111.3°
18. csc 163.4°
19. sin 268.5°
20. cos 328.1°

21. cot 143°30'
22. tan 173°20'
23. cos 231°30'
24. sin 312°40'

Find the measure of the acute angle θ to the nearest tenth of a degree.

25. $\sin \theta = 0.3400$
26. $\cos \theta = 0.8400$
27. $\cot \theta = 1.700$

28. $\tan \theta = 1.325$
29. $\sec \theta = 3.555$
30. $\csc \theta = 3.000$

Find the measure of the acute angle θ to the nearest minute.

31. $\cos \theta = 0.8621$
32. $\sin \theta = 0.2654$
33. $\tan \theta = 0.1482$

34. $\cos \theta = 0.2715$
35. $\sin \theta = 0.7321$
36. $\tan \theta = 2.550$

Find the measures of two angles between 0° and 360° with the given function value. Give answers to the nearest tenth of a degree.

B
37. $\sin \theta = 0.4875$
38. $\cos \theta = 0.7851$
39. $\tan \theta = -1.752$

40. $\sin \theta = -0.8300$
41. $\cos \theta = 0.2524$
42. $\tan \theta = 0.6182$

Find the measure of an angle θ between 0° and 360° that satisfies the stated conditions. Give answers to the nearest tenth of a degree.

43. $\cos \theta = 0.4275$, $180° < \theta < 360°$
44. $\sin \theta = -0.5212$, $90° < \theta < 270°$

45. $\sin \theta = -0.6118$, $\cos \theta > 0$
46. $\cos \theta = 0.7815$, $\tan \theta < 0$

Self-Test 1

Vocabulary degree (p. 549)
initial side (p. 549)
terminal side (p. 549)
positive angle (p. 549)
negative angle (p. 549)
directed angle (p. 549)
standard position (p. 549)
quadrantal angle (p. 549)
coterminal angles (p. 550)
minute, second (p. 550)

sine (p. 555)
cosine (p. 555)
tangent (p. 555)
reciprocal functions (p. 556)
cotangent (p. 556)
secant (p. 556)
cosecant (p. 556)
identity (p. 556)
cofunction (p. 557)
reference angle (p. 562)

1. Name two angles, one positive and one negative, that are coterminal with the given angle.
   a. 250°   **b.** −300°

   Obj. 12-1, p. 549

2. Express 32°35′ in decimal degrees to the nearest hundredth of a degree.

3. Express 74.26′ in degrees, minutes, and seconds to the nearest second.

4. When angle θ is in standard position, its terminal side passes through ($\sqrt{7}$, 3). Find the values of the six trigonometric functions of θ.

   Obj. 12-2, p. 555

5. In the right triangle shown at the right, find a, b, and the measure of $\angle B$.

6. Find the reference angle α for an angle with a measure of 300°. Find the exact values of the six trigonometric functions of 300°, using radicals when necessary.

   Obj. 12-3, p. 561

7. Find the following to four significant digits:
   a. cos 36°42′   **b.** csc 64.33°

   Obj. 12-4, p. 568

8. Find the measure of angle θ to the nearest minute if θ is an acute angle and sin θ = 0.8760.

Check your answers with those at the back of the book.

 Historical Note / *Tables of Sines*

Among the first historical appearances of trigonometry were "tables of chords," recognizable as sine tables. Values for these early tables, such as the one devised by the Greek astronomer Hipparchus around 140 B.C., were found by using geometric methods to measure lengths of chords of a circle.

Modern trigonometric tables (and also the values found by a calculator or a computer) are computed by using terms of an infinite series. One such series is the *Taylor series* for sin x, where x is given in radian measure (see Lesson 13-1):

$$\sin x = x - \frac{x^3}{3!} + \frac{x^5}{5!} - \frac{x^7}{7!} + \cdots$$

By adding up the terms of this series you can approximate the sine of a given angle to any desired accuracy. For example, using the four terms shown, you find that

$$\sin 1 \approx 1 - \frac{1}{3!} + \frac{1}{5!} - \frac{1}{7!} \approx 0.84147,$$

a value that is correct to five decimal places.

Triangle Trigonometry

12-5 Solving Right Triangles

Objective   To find the sides and angles of a right triangle.

Any triangle ABC has six measurements associated with it:
the lengths of the three sides, denoted by a, b, and c, and the
measures of the three angles, denoted by $\angle A$, $\angle B$, and $\angle C$.
(Note that the lower-case letters denote the lengths of sides
opposite angles labeled by the corresponding capital letters.)
If any three of these measurements (other than the three angle
measures) are given, then the other three can be found.

    Finding measurements for all the sides and angles of a
triangle is called **solving the triangle.**

    In the case of a right triangle ABC, one angle is 90°. You
can solve the triangle if the lengths of two sides, or the length
of one side and the measurement of one acute angle, are known.

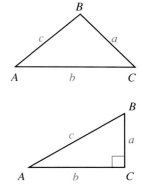

Example 1   Solve the right triangle shown above if $\angle A = 36°$ and $b = 50$.

Solution 1   Since $\angle A + \angle B = 90°$, $\angle B = 90° - \angle A = 90° - 36° = 54°$.

$$\cot A = \frac{b}{a} \qquad\qquad \cos A = \frac{b}{c}$$

$$\cot 36° = \frac{50}{a} \qquad\qquad \cos 36° = \frac{50}{c}$$

$$a = \frac{50}{\cot 36°} \qquad\qquad c = \frac{50}{\cos 36°}$$

$$= \frac{50}{1.376} \qquad\qquad\quad = \frac{50}{0.8090}$$

$$= 36.3 \qquad\qquad\qquad = 61.8$$

$$\therefore \ \angle B = 54°, \ a = 36.3, \text{ and } c = 61.8 \quad \textit{Answer}$$

Solution 2   Again $\angle B = 90° - 36° = 54°$.

$$\tan A = \frac{a}{b} \qquad\qquad\qquad \sec A = \frac{c}{b}$$

$$\tan 36° = \frac{a}{50} \qquad\qquad\quad \sec 36° = \frac{c}{50}$$

$$a = 50(\tan 36°) \qquad\qquad b = 50(\sec 36°)$$

$$= 50(0.7265) \qquad\qquad = 50(1.236)$$

$$= 36.3 \qquad\qquad\qquad\ = 61.8$$

$$\therefore \ \angle B = 54°, \ a = 36.3, \text{ and } c = 61.8 \quad \textit{Answer}$$

In Example 1, Solution 1 or Solution 2 are both well suited for calculator use. However, if tables are used, Solution 2 would be a better choice because long division is avoided.

Measurements of lengths and angles are approximations, as are the values given by tables and calculators. The following is a guide to the corresponding accuracies of length and angle measurements.

An angle measured to		A length measured to
1°		2 significant digits.
0.1° or 10'	corresponds to	3 significant digits.
0.01° or 1'		4 significant digits.

The figure at the right shows a helicopter and a pickup truck. To see the helicopter, the driver's line of sight must be raised, or *elevated*, at an angle θ above the horizontal. This angle θ is called the **angle of elevation** of the helicopter. Similarly, the angle ϕ is called the **angle of depression** of the truck from the helicopter. The angle of elevation and the corresponding angle of depression always have the same measure.

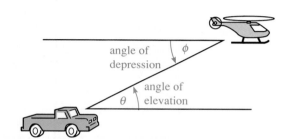

Example 2 A research ship finds that the angle of elevation of a volcanic island peak is 25.6°. After the ship has moved 1050 m closer to the island, the angle of elevation is 31.2°. What is the height of the peak above sea level? (Note that the sketch shown at the right is not drawn to scale.)

Solution

$$\cot 31.2° = \frac{x}{y} \qquad \text{and} \qquad \cot 25.6° = \frac{1050 + x}{y}$$

$$x = y \cot 31.2° \qquad\qquad y \cot 25.6° = 1050 + x$$

$$x = 1.651y \qquad\qquad 2.087y = 1050 + x$$

Substitute $1.651y$ for x in the second equation.

$$2.087y = 1050 + 1.651y$$
$$0.436y = 1050$$
$$y = \frac{1050}{0.436} = 2410 \text{ (to three significant digits)}$$

∴ the volcanic peak is 2410 m above sea level. ***Answer***

The altitude drawn in the diagram at the right below divides isosceles triangle *ABC* into two congruent right triangles. Constructing such an altitude will help you to solve any isosceles triangle.

Example 3   Solve isosceles triangle *ABC* if $a = 31.0$ and $c = 42.5$.

Solution   Make a sketch as shown.

In the right triangle on the left in the diagram, the side adjacent to $\angle B$ has length $\frac{1}{2}a$, or 15.5. So

$$\cos B = \frac{15.5}{42.5} = 0.3647 \text{ and } \angle B = 68.6°.$$

Since $\triangle ABC$ is isosceles,

$$\angle C = \angle B = 68.6° \text{ and } b = c = 42.5$$
$$\angle A = 180° - (\angle B + \angle C)$$
$$= 180° - 137.2° = 42.8°$$

$\therefore \ \angle B = 68.6°, \ \angle C = 68.6°, \ \angle A = 42.8°, \text{ and } b = 42.5$ ***Answer***

Oral Exercises

Give an equation that can be used to find the value of *x*.

1.

2.

3.

4.

5.

6.

7. The angle of depression of *Q* from *P* measures 25°. Therefore, the angle of elevation of *P* from *Q* measures __?__.

8. What fact from geometry did you use in answering the question in Exercise 7?

9. Give each angle to the nearest 10′.
   a. 42°13′21″   **b.** 140°9′   **c.** 0°11′59″

10. Give each number to three significant digits.
   a. 6.758   **b.** 1284.56   **c.** 0.39975

Written Exercises

Give lengths to three significant digits and angle measures to the nearest tenth of a degree or nearest ten minutes. You may wish to use a calculator.

Solve each right triangle ABC.

A
1. $\angle A = 36.2°$, $c = 68$
2. $\angle B = 15.8°$, $c = 12.2$
3. $\angle B = 65.4°$, $a = 2.35$
4. $\angle A = 82.1°$, $b = 246$
5. $\angle B = 48.3°$, $b = 74.7$ 6. $\angle A = 24.0°$, $a = 5.25$
7. $a = 230$, $c = 320$ 8. $a = 52.5$, $b = 28.0$
9. $a = 0.123$, $b = 0.315$ 10. $b = 3.90$, $c = 42.5$
11. $\angle B = 58°10'$, $c = 420$ 12. $\angle A = 38°40'$, $c = 42.5$
13. $\angle A = 15°30'$, $a = 4.50$ 14. $\angle B = 67°20'$, $a = 450$
15. $\angle A = 30°50'$, $b = 53.5$ 16. $\angle B = 85°10'$, $b = 0.620$

B 17. The radius of circle O is 13 cm and the length of $\overline{AB}$ is 10 cm. Find the measure of $\angle AOB$.

Ex. 17

Ex. 18

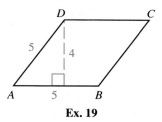

Ex. 19

18. The height of an isosceles trapezoid is 6 units and the bases have lengths 4 units and 20 units. Find the measures of the angles.

19. A rhombus has sides 5 units long and its height is 4 units. Find its angles.

20. **a.** Find the perimeter of a regular pentagon inscribed in a unit circle.
  b. Find the perimeter of a regular pentagon circumscribed about a unit circle.

(*Hint:* Use the shaded right triangles in the figures below.)

21. Repeat Exercise 20 for a regular polygon having 10 sides.

22. Repeat Exercise 20 for a regular polygon having 20 sides.

23. Explain why the number 2π lies between the answers to the (a) and (b) parts of Exercises 20, 21, and 22.

C 24. a. Repeat Exercise 20 for a polygon having n sides.

    b. Use the results of part (a) to tell what number is approached by $n \sin\left(\dfrac{180}{n}\right)^{\circ}$ as n gets larger and larger. What number is approached by $n \tan\left(\dfrac{180}{n}\right)^{\circ}$ as n gets larger and larger?

Problems

Give lengths to three significant digits and angle measures to the nearest tenth of a degree. You may wish to use a calculator.

A 1. What is the angle of elevation of the sun when a tree 6.25 m tall casts a shadow 10.1 m long?

2. A boy flying a kite is standing 30 ft from a point directly under the kite. If the string to the kite is 50 ft long, find the angle of elevation of the kite.

3. A cable 4 m long is attached to a pole. The cable is staked to the ground 1.75 m from the base of the pole. Find the angle that the cable makes with the ground.

4. How far from the base of a building is the bottom of a 30 ft ladder that makes an angle of 75° with the ground?

5. The angle of elevation of the summit of a mountain from the bottom of a ski lift is 33°. A skier rides 1000 ft on this ski lift to get to the summit. Find the vertical distance between the bottom of the ski lift and the summit.

6. The approach pattern to an airport requires pilots to set an 11° angle of descent toward the runway. If a plane is flying at an altitude of 9500 m, at what distance (measured along the ground) from the airport must the pilot start the descent?

7. The distance from the point directly under a hot air balloon to the point where the balloon is staked to the ground with a rope is 285 ft. The angle of elevation up the rope to the balloon is 48°. Find the height of the balloon.

8. Opposite corners of a small rectangular park are joined by diagonal paths, each 360 m long. What are the dimensions of the park if the paths intersect at a 65° angle?

B **9.** A pendulum in a grandfather clock is 160 cm long. The horizontal distance between the farthest points in a complete swing is 65 cm. Through what angle does the pendulum swing?

10. A camping tent is supported by a rope stretched between two trees at a height of 210 cm. If the sides of the tent make an angle of 55° with the level ground, how wide is the tent at the bottom?

11. From the top of a 135 ft observation tower, a park ranger sights two forest fires on opposite sides of the tower. If their angles of depression are 42.5° and 32.6°, how far apart are the fires?

12. From a point 250 m from the base of a vertical cliff, the angles of elevation to the top and bottom of a radio tower on top of the cliff are 62.2° and 59.5°. How tall is the tower?

Ex. 13

13. The side of the beach shelter shown at the right is made up of four 30°–60°–90° triangles. Find the dimension marked *x*.

14. A pendulum 50 cm long is moved 26° from the vertical. How much is the lower end of the pendulum raised?

C **15.** Two observers 1600 m apart on a straight, flat road measure the angles of elevation of a helicopter hovering over the road between them. If these angles are 32.0° and 50.5°, how high is the helicopter?

Ex. 14

16. The pilot of a hot air balloon sees a field straight ahead that he knows is 1000 ft long. The angles of depression to the ends of the field are 25.4° and 34.7°. What is the height of the balloon and its horizontal distance from the nearer end of the field?

17. In a molecule of carbon tetrachloride, the four chlorine atoms are at the vertices of a regular tetrahedron, with the carbon atom in the center. What is the angle between two of the carbon-chlorine bonds (shown in red in the figure)?

Ex. 17

Mixed Review Exercises

Give the exact values of the six trigonometric functions of each angle. If any function is not defined for the angle, say so.

1. 450°   **2.** −135°   **3.** 330°   **4.** −240°

Write in simplest form without negative exponents.

5. $\dfrac{x^{-1} - 1}{x - x^{-1}}$ **6.** $\sqrt{50m^4}$ **7.** $\dfrac{1}{y + 2} + \dfrac{4}{y^2 - 4}$

8. $(-5a^3b)^2(2a^2b)^3$ **9.** $(x^{2/3})^{-3/4}$ **10.** $(2u - 3)(u^2 + u - 2)$

12-6 The Law of Cosines

Objective   To use the law of cosines to find sides and angles of triangles.

If you solve the equations

$$\sin \theta = \frac{y}{r} \quad \text{and} \quad \cos \theta = \frac{x}{r}$$

for x and y you obtain the following useful fact.

If θ is an angle in standard position and P is a point on its terminal side, then the coordinates of P are $(r \cos \theta, r \sin \theta)$, where $r = OP$.

This fact is used to prove the *law of cosines*.

The Law of Cosines

In any triangle ABC,

$$c^2 = a^2 + b^2 - 2ab \cos C$$
$$b^2 = a^2 + c^2 - 2ac \cos B$$
$$a^2 = b^2 + c^2 - 2bc \cos A$$

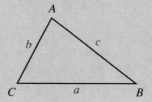

To prove the first form of the law of cosines, draw a coordinate system with $\angle C$ in standard position. (The diagrams below show an acute $\angle C$ and an obtuse $\angle C$ on such a coordinate system.) Then apply the distance formula to AB.

$$
\begin{aligned}
c^2 &= (b \cos C - a)^2 + (b \sin C - 0)^2 \\
&= b^2 \cos^2 C - 2ab \cos C + a^2 + b^2 \sin^2 C \\
&= a^2 + b^2(\cos^2 C + \sin^2 C) - 2ab \cos C
\end{aligned}
$$

But, since $\cos^2 C + \sin^2 C = 1$, the equation becomes

$$c^2 = a^2 + b^2 - 2ab \cos C.$$

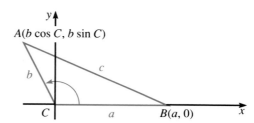

The other two forms of the law of cosines can be proved by repeating the process just shown with $\angle A$ and $\angle B$ in turn put into standard position. Notice that if $\angle C$ is a right angle, $\cos C = 0$, and the law of cosines reduces to the Pythagorean theorem.

Example 1 In $\triangle ABC$, $a = 10$, $b = 13$, and $\angle C = 70°$. Find c to three significant digits.

Solution

$c^2 = a^2 + b^2 - 2ab \cos C$

$c^2 = 10^2 + 13^2 - 2(10)(13) \cos 70°$

$\quad = 100 + 169 - 260(0.3420)$

$\quad = 269 - 88.92$

$\quad = 180.08$

$c = \sqrt{180.08} = 13.42$

$\therefore$ to three significant digits, $c = 13.4$. **Answer**

By rewriting the equation

$$c^2 = a^2 + b^2 - 2ab \cos C$$

in the form

$$\cos C = \frac{a^2 + b^2 - c^2}{2ab},$$

you can use the law of cosines to find the measure of an angle of a triangle when you know the lengths of the sides.

Example 2 A triangular-shaped lot has sides of length 50 m, 120 m, and 150 m. Find the largest angle of the lot to the nearest tenth of a degree.

Solution The largest angle of a triangle is opposite the longest side. Make a sketch. Let $a = 50$, $b = 120$, and $c = 150$. Substitute in this formula:

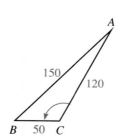

$\cos C = \dfrac{a^2 + b^2 - c^2}{2ab}$

$\cos C = \dfrac{50^2 + 120^2 - 150^2}{2(50)(120)}$

$\quad = \dfrac{2500 + 14{,}400 - 22{,}500}{12{,}000}$

$\quad = \dfrac{-5600}{12{,}000}$

$\quad = -0.4667$

$\therefore \angle C = 117.8°$. **Answer**

You may find it helpful to use a calculator to solve problems like the ones illustrated in Examples 1 and 2.

Oral Exercises

Use the law of cosines to give an equation involving the side or angle labeled x.

1.

2.

3.

4.

5.

6.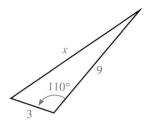

Written Exercises

Find lengths to three significant digits and the measures of the angles to the nearest tenth of a degree. You may wish to use a calculator.

In Exercises 1–12, find the indicated part of $\triangle ABC$.

A

1. $a = 6$, $b = 7$, $\angle C = 20°$, $c = $ ___?___   **2.** $b = 12$, $c = 17$, $\angle A = 74°$, $a = $ ___?___

3. $c = 15$, $a = 13$, $\angle B = 83°$, $b = $ ___?___   **4.** $b = 3$, $a = 4$, $\angle C = 40°$, $c = $ ___?___

5. $c = 15$, $b = 30$, $\angle A = 140°$, $a = $ ___?___   **6.** $a = 100$, $c = 200$, $\angle B = 150°$, $b = $ ___?___

7. $a = 8$, $b = 10$, $c = 12$, $\angle B = $ ___?___   **8.** $a = 9$, $b = 10$, $c = 15$, $\angle C = $ ___?___

9. $a = 13$, $b = 30$, $c = 40$, smallest angle = ___?___

10. $a = 30$, $b = 20$, $c = 40$, largest angle = ___?___

11. $a = 1.6$, $b = 0.9$, $c = 1.8$, largest angle = ___?___

12. $a = 1.2$, $b = 2.4$, $c = 2.0$, smallest angle = ___?___

B

13. Given a circle O, chord $AB = 10.1$, chord $BC = 15.5$ and $\angle ABC = 26°10'$. Find the length of chord AC.

14. A parallelogram has sides 6 cm and 8 cm and a 65° angle. Find the lengths of the diagonals. (Recall that adjacent angles of a parallelogram are supplementary.)

15. Three circles with radii 25 in., 36 in., and 30 in. are externally tangent to each other. Find the angles of the triangle formed by joining their centers.

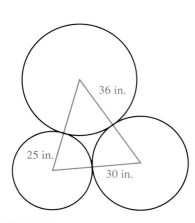

16. Find the lengths of the sides of a parallelogram whose diagonals intersect at a 35° angle and have lengths 6 and 10. (Recall that the diagonals of a parallelogram bisect each other.)

17. In a parallelogram having sides of lengths a and b, let the diagonals have lengths d_1 and d_2. Show that $d_1^2 + d_2^2 = 2(a^2 + b^2)$. (*Hint:* Apply the law of cosines twice and use the fact that $\cos(180° - \theta) = -\cos \theta$.)

C **18.** In $\triangle ABC$, $a = 2$, $b = 4$, and $c = 3$. Find the length of the median from A to $\overline{BC}$.

19. Prove that $\cos 2\theta = 1 - 2 \sin^2 \theta$ for $0° < \theta < 90°$ as follows. First, using the triangle shown, express b^2 in two ways: (1) by using the law of cosines and (2) by first finding $\dfrac{b}{2}$ from the shaded right triangle. Then set the two expressions equal.

Problems

Find lengths to three significant digits and measures of angles to the nearest tenth of a degree. You may wish to use a calculator.

A **1.** A ranger in an observation tower can sight the north end of a lake 15 km away and the south end of the same lake 19 km away. The angle between these two lines of sight is 104°. How long is the lake?

2. Two planes leave an airport at the same time, one flying due west at 500 km/h and the other flying due southeast at 300 km/h. What is the distance between the planes two hours later?

3. A triangular-shaped lot of land has sides of length 130 m, 150 m, and 80 m. What are the measures of the angles?

4. Two streets meet at an angle of 52°. If a triangular lot has frontages of 60 m and 65 m on the two streets, what is the perimeter of the lot?

5. Newtown is 8 mi east of Oldtown and Littleton is 10 mi northwest of Oldtown. How far is Newtown from Littleton?

6. An oil tanker and a cruise ship leave port at the same time and travel straight-line courses at 10 mi/h and 25 mi/h, respectively. Two hours later they are 40 mi apart. What is the angle between their courses?

7. A baseball diamond is a square 90 ft on a side. The pitcher's mound is 60.5 feet from home plate. How far is it from the mound to first base?

8. A water molecule consists of two hydrogen atoms and one oxygen atom joined as in the diagram. The distance from the nucleus of each hydrogen atom to the nucleus of the oxygen atom is 9.58×10^{-9} cm, and the bond angle θ is 104.8°. How far are the nuclei of the hydrogen atoms from each other?

B **9.** A large park in the shape of a parallelogram has diagonal paths that meet at a 60° angle. If the diagonal paths are 12 km and 20 km long, find the perimeter and the area of the park.

10. A flagpole 4 m tall stands on a sloping roof. A support wire 5 m long joins the top of the pole to a point on the roof 6 m up from the bottom of the pole. At what angle is the roof inclined to the horizontal?

11. A vertical pole 20 m tall standing on a 15° slope is braced by two cables extending from the top of the pole to two points on the ground, 30 m up the slope and 30 m down the slope. How long are the cables?

C **12.** The measures of two sides of a parallelogram are 50 cm and 80 cm, and one diagonal is 90 cm long. How long is the other diagonal?

Computer Exercises
For students with some programming experience.

1. Given a right triangle ABC with $\angle C = 90°$ and the lengths of two sides, write a program that will find the length of the other side and the degree measures of the two acute angles. (*Hint:* You can find the measure of the acute angles by using a ratio of the lengths of the legs as the argument for the computer's built-in inverse tangent function ATN. This function gives an angle measure in radians (see Lesson 13-1). To convert radian measure R to degree measure D you can use the formula D = R*180/3.14159.)

2. Run the program in Exercise 1 to find the length of the remaining side and the measures of the remaining angles in each right triangle.

    a. $a = 52$, $b = 38$ **b.** $a = 29.6$, $c = 47.4$ **c.** $b = 5.93$, $c = 9.07$

3. Given a right triangle ABC with $\angle C = 90°$ and the length of one side and the measure of one acute angle, write a program that will find the lengths of the other two sides and the measure of the third angle. (*Hint:* Remember that the computer's functions SIN(X), COS(X), and TAN(X) require an argument X in radians. To convert degree measure D to radian measure R you can use the formula R = D*3.14159/180.)

4. Run the program in Exercise 3 to find the lengths of the remaining sides and the measure of the remaining angle in each triangle.

    a. $\angle A = 32°$, $a = 37$ **b.** $\angle B = 15°$, $b = 17.6$ **c.** $\angle B = 19.8°$, $c = 9.89$

5. Given the degree measure of two angles and the lengths of two sides of a triangle, write a program that will find the degree measure of the third angle and the length of the other side.

6. Run the program in Exercise 5 to find the measures of the remaining sides and angle in each triangle.

    a. $\angle A = 68°$, $\angle B = 72°$, $c = 13.91$, $a = 20.06$
    b. $\angle A = 51.3°$, $\angle B = 102.8°$, $c = 27.05$, $b = 60.39$
    c. $\angle B = 115.6°$, $\angle C = 12.4°$, $a = 7.385$, $c = 2.012$

Reading Algebra / Making a Sketch

Read the following problem and try to visualize how the points, angles, and distances are related.

> Two points, X and Y, are separated by a swamp. To find the distance between them, Kelly walks 150 m from X to another point, Z. She estimates $\angle ZXY$ to be 45° and $\angle XZY$ to be 60°. Find the approximate distance from X to Y.

Can you picture the relative positions of X, Y, and Z? Many problems dealing with distance, angles, and geometric shapes are difficult to think about if you have only words to look at. Making a sketch as you read can help you understand the given information and see what is being asked for. For example, a sketch like the one below shows all the essential information in the problem stated above.

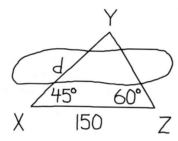

Notice that your sketch for a problem can be useful even if it is not to scale. For example, in the sketch above, there is no need to draw the angles exactly to scale, though we have shown them both as acute, with the larger one at Z.

Most diagrams in your textbook are drawn to scale. When a diagram is drawn for your book, the artist begins with complete information about dimensions. But you usually won't know the exact proportions of the shapes in your sketch until you have solved the problem. So don't wait until your solution is complete to make a drawing. Start right away with a rough sketch; you can improve it later as you discover more information.

Exercises

For each problem listed below, draw a sketch showing the essential information, labeling segments and angles with their given measures, and using variables of your choice to represent unknowns.

1. Problem 5, page 589

2. Problem 10, page 590

3. Problem 1, page 595

4. Problem 12, page 596

12-7 The Law of Sines

Objective   To use the law of sines to find sides and angles of triangles.

The *law of sines*, like the law of cosines, allows you to use given information about three measurements of a triangle to determine the other measurements. For example, if you know the lengths of two sides and the measure of the included angle, you can find the length of the third side by using the law of cosines. But if the given angle is not the included angle, the law of cosines cannot be used and you need to use the law of sines.

The Law of Sines

In any triangle ABC,

$$\frac{\sin A}{a} = \frac{\sin B}{b} = \frac{\sin C}{c}$$

To prove the law of sines, you first find a formula for the area, K, of $\triangle ABC$. Draw a coordinate system with $\angle A$ in standard position. ($\angle A$ may be acute or obtuse.) Let h be the height of the triangle measured from B.

Since $K = \frac{1}{2}$(base)(height), you can write

$$K = \frac{1}{2}bh.$$

Since the y-coordinate of B equals h, you know $h = c \sin A$. Therefore,

$$K = \frac{1}{2}bc \sin A.$$

You can find two other formulas for the area K of $\triangle ABC$ by replacing $\angle A$ in turn with $\angle B$ and $\angle C$:

$$K = \frac{1}{2}ac \sin B \qquad K = \frac{1}{2}ab \sin C$$

Therefore, by substitution,

$$\frac{1}{2}bc \sin A = \frac{1}{2}ac \sin B = \frac{1}{2}ab \sin C.$$

By dividing each expression by $\frac{1}{2}abc$, you obtain the law of sines shown above.

Example 1 In $\triangle ABC$, $\angle A = 40°$ and $a = 15$.

  a. Find $\angle B$ if $b = 20$. **b.** Find $\angle B$ if $b = 8$.

Solution Use the formula $\dfrac{\sin B}{b} = \dfrac{\sin A}{a}$.

a. $\dfrac{\sin B}{20} = \dfrac{\sin 40°}{15}$

$\sin B = \dfrac{20 \sin 40°}{15}$

$= \dfrac{20(0.6428)}{15}$

$= 0.8571$

$\angle B = 59°$ or $\angle B = 121°$

Since $40° + 59° = 99°$
and $40° + 121° = 161°$,
and both $99°$ and $161°$
are less than $180°$, there
are two solutions.

$\therefore \angle B = 59°$ or
$\angle B = 121°$. **Answer**

b. $\dfrac{\sin B}{8} = \dfrac{\sin 40°}{15}$

$\sin B = \dfrac{8 \sin 40°}{15}$

$= \dfrac{8(0.6428)}{15}$

$= 0.3428$

$\angle B = 20°$ or $\angle B = 160°$

$160°$ is not a solution because
$\angle A + \angle B = 40° + 160° = 200°$,
which is greater than $180°$.

$\therefore \angle B = 20°$. **Answer**

If you are given two sides of a triangle and the angle opposite one of the sides, there may be two solutions, one solution, or no solution. You'll learn why in Lesson 12-8.

Example 2 A 123 ft support wire for a transmitting tower makes an angle of $61°$ with the ground. This wire is to be replaced by a new wire whose angle with the ground is $46°$. How long will the new wire be?

Solution Let $\angle A = 46°$, and let $c =$ the length of the new wire.
$\angle ACB = 180° - 61° = 119°$, and $a = 123$ (ft).

$\dfrac{\sin A}{a} = \dfrac{\sin ACB}{c}$

$\dfrac{\sin 46°}{123} = \dfrac{\sin 119°}{c}$

$c = \dfrac{123 \sin 119°}{\sin 46°}$

$= \dfrac{123(0.8746)}{0.7193}$

$= 149.6$

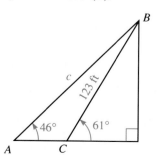

$\therefore$ the new wire will be 149.6 ft long. **Answer**

Given triangle *ABC* with sides *a*, *b*, and *c*, use the law of sines to find an expression equivalent to the given one.

1. $\dfrac{b}{a}$

2. $\dfrac{a}{\sin A}$

3. $\dfrac{\sin A}{\sin B}$

4. $\dfrac{a \sin B}{b \sin A}$

5. $\dfrac{a \sin B}{b}$

6. $\dfrac{\cos A}{\sin A}$ if $\angle C = 90°$

Written Exercises

Find the indicated part of $\triangle ABC$ to three significant digits or to the nearest tenth of a degree. If there are two solutions, give both. You may wish to use a calculator.

A

1. $a = 14$, $\angle A = 25°$, $\angle B = 75°$, $b = \underline{\ ?\ }$

2. $c = 12$, $\angle A = 42°$, $\angle C = 69°$, $a = \underline{\ ?\ }$

3. $b = 3.40$, $\angle A = 110°$, $\angle C = 50°$, $a = \underline{\ ?\ }$

4. $a = 2.60$, $\angle B = 60°$, $\angle C = 100°$, $c = \underline{\ ?\ }$

5. $c = 35$, $\angle A = 38°$, $\angle C = 102°$, $b = \underline{\ ?\ }$

6. $b = 130$, $\angle B = 95°$, $\angle C = 35°$, $a = \underline{\ ?\ }$

7. $a = 4$, $b = 3$, $\angle A = 40°$, $\angle B = \underline{\ ?\ }$

8. $a = 4.5$, $b = 6.0$, $\angle B = 35°$, $\angle A = \underline{\ ?\ }$

9. $a = 4.0$, $c = 6.4$, $\angle C = 125°$, $\angle B = \underline{\ ?\ }$

10. $a = 18$, $b = 12$, $\angle A = 110°$, $\angle C = \underline{\ ?\ }$

11. $a = 5$, $c = 7$, $\angle A = 42°$, $\angle C = \underline{\ ?\ }$

12. $b = 15$, $c = 11$, $\angle C = 40°$, $\angle B = \underline{\ ?\ }$

In Exercises 13–18, find the exact value of $\dfrac{a}{b}$ in $\triangle ABC$ without using tables or calculators. (Recall that $\sin^2 \theta + \cos^2 \theta = 1$.)

Sample $\sin A = \dfrac{3}{4}$, $\cos B = \dfrac{2}{3}$

Solution The law of sines can be written in the form $\dfrac{a}{b} = \dfrac{\sin A}{\sin B}$.

You need to find $\sin B$. Since $\sin^2 B + \cos^2 B = 1$, you have

$\sin^2 B = 1 - \cos^2 B = 1 - \left(\dfrac{2}{3}\right)^2 = \dfrac{5}{9}$. Therefore, $\sin B = \dfrac{\sqrt{5}}{3}$.

By substitution, $\dfrac{a}{b} = \dfrac{\sin A}{\sin B} = \dfrac{\dfrac{3}{4}}{\dfrac{\sqrt{5}}{3}} = \dfrac{9}{4\sqrt{5}} = \dfrac{9\sqrt{5}}{20}$.

13. $\sin A = \frac{2}{3}$, $\cos B = \frac{4}{5}$

14. $\cos A = \frac{3}{5}$, $\sin B = \frac{4}{5}$

15. $\cos A = \frac{5}{13}$, $\cos B = \frac{3}{5}$

16. $\cos A = \frac{8}{17}$, $\cos B = \frac{15}{17}$

17. $\cos A = \frac{\sqrt{3}}{2}$, $\cos B = \frac{\sqrt{2}}{2}$

18. $\cos A = \frac{1}{2}$, $\tan B = 1$

Show that the following formulas are true for any triangle ABC.

B 19. $\dfrac{\sin A + \sin B}{\sin B} = \dfrac{a + b}{b}$

20. $\dfrac{\sin A - \sin B}{\sin B} = \dfrac{a - b}{b}$

21. Prove the formula $\sin 2\theta = 2 \sin \theta \cos \theta$ for $0° < \theta < 90°$. Use the law of sines and the fact that $\sin (90° - \theta) = \cos \theta$.

22. Use the law of sines to show that the bisector of an angle of a triangle divides the opposite side in the same ratio as the two adjacent sides; that is,

$$\frac{a}{b} = \frac{x}{y}.$$

(*Hint:* Use the law of sines in $\triangle BNC$ and $\triangle ANC$ and the fact that $\sin (180° - B) = \sin B$.)

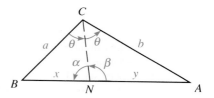

Problems

Give answers to three significant digits. You may wish to use a calculator.

A 1. Two angles of a triangle measure $32°$ and $53°$. The longest side is 55 cm. Find the length of the shortest side.

2. Two angles of a triangle measure $75°$ and $51°$. The side opposite the $75°$ angle is 25 in. How long is the shortest side?

3. How long is the base of an isosceles triangle if each leg is 27 cm and each base angle measures $23°$?

4. A college football pennant is in the shape of an isosceles triangle. The base is 16 in. long. The sides meet at an angle of $35°$. How long are the sides?

5. A fire is sighted from two ranger stations that are 5000 m apart. The angles of observation to the fire measure $52°$ from one station and $41°$ from the other station. Find the distance along the line of sight to the fire from the closer of the two stations.

6. Two surveyors are on opposite sides of a swamp. To find the distance between them, one surveyor locates a point T that is 180 m from her location at point P. The angles of sight from T to the other surveyor's position, R, measure $72°$ for $\angle RPT$ and $63°$ for $\angle PTR$. How far apart are the surveyors?

7. Two markers are located at points A and B on opposite sides of a lake. To find the distance between the markers, a surveyor laid off a base line, $\overline{AC}$, 25 m long and found that $\angle BAC = 85°$ and $\angle BCA = 66°$. Find AB.

8. From a hang glider approaching a 5000 ft clearing the angles of depression of the opposite ends of the field measure 24° and 30°. How far is the hang glider from the nearer end of the field?

B **9.** A loading ramp 5 m long makes a 25° angle with the level ground beneath it. The ramp is replaced by another ramp 15 m long. Find the angle that the new ramp makes with the ground.

10. From the top of an office building 72 ft high, the angle of elevation to the top of an apartment building across the street is 31°. From the base of the office building, the angle of elevation is 46°. How tall is the apartment building?

11. The captain of a freighter 6 km from the nearer of two unloading docks on the shore finds that the angle between the lines of sight to the two docks is 35°. If the docks are 10 km apart, how far is the tanker from the farther dock?

12. Two wires bracing a transmission tower are attached to the same location on the ground. One wire is attached to the tower 10 m above the other. The longer wire is 30 m long and the angle of inclination of the shorter wire is 53°. To the nearest tenth of a degree, what angle does the longer wire make with the ground?

C **13.** To find the height of a tree across a river, Carlos laid off a base line 70 m long and measured the angles shown. He found the angle of elevation of the top of the tree from A to be 10°. How tall is the tree?

Mixed Review Exercises

Give the following function values to four significant digits.

1. cos 78.2° **2.** tan $(-43°10')$ **3.** sin 213.7°

4. csc 305°45' **5.** cot $(-101.9°)$ **6.** sec 488°34'

Write an equation for each figure described.

7. The circle with center $(-1, 4)$ and radius 3.

8. The line through $(-3, 5)$ and $(2, 0)$.

9. The parabola with focus $(0, -2)$ and directrix $y = 2$.

12-8 Solving General Triangles

Objective   To solve any given triangle.

The problem of solving triangles can be divided into four cases:

SSS: Given three sides.

SAS: Given two sides and the included angle.

SSA: Given two sides and the angle opposite one of them.

ASA and **AAS:** Given two angles and one side.

You may remember that SSS, SAS, ASA, and AAS are used in geometry to prove triangles congruent. However, knowing two sides and a nonincluded angle (the SSA case) is not enough to prove two triangles congruent. You will see later in this lesson that when you try to solve a triangle given these measurements, they may not determine a triangle.

In the examples that follow, lengths will be found to three significant digits and angle measures to the nearest tenth of a degree.

Example 1   (SSS case) Solve $\triangle ABC$ if $a = 4$, $b = 6$, and $c = 5$.

Solution   Use the law of cosines to find one of the angles, say $\angle A$.

$$\cos A = \frac{b^2 + c^2 - a^2}{2bc}$$

$$\cos A = \frac{6^2 + 5^2 - 4^2}{2 \cdot 6 \cdot 5}$$

$$= \frac{36 + 25 - 16}{60}$$

$$= \frac{45}{60} = 0.7500$$

$\therefore \angle A = 41.4°$

To find another angle, it is usually easier to use the law of sines.

$$\frac{\sin C}{c} = \frac{\sin A}{a}$$

$$\sin C = \frac{c \sin A}{a}$$

$$\sin C = \frac{5 \sin 41.4°}{4} = \frac{5(0.6613)}{4} = 0.8266$$

Since $\angle C$ is not the largest angle, it must be acute.

$\therefore \angle C = 55.8°$

Since $\angle A + \angle B + \angle C = 180°$, you know $\angle B = 180° - \angle A - \angle C$.
So by substitution, $\angle B = 180° - 41.4° - 55.8° = 82.8°$.

$\therefore \angle A = 41.4°$, $\angle B = 82.8°$, and $\angle C = 55.8°$.   **Answer**

Example 2 (SAS case) Solve $\triangle ABC$ if $a = 8$, $c = 7$, and $\angle B = 31.8°$.

Solution Use the law of cosines to find the third side, b.

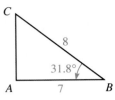

$$b^2 = a^2 + c^2 - 2ac \cos B$$
$$= 8^2 + 7^2 - 2(8)(7) \cos 31.8°$$
$$= 64 + 49 - 112(0.8499)$$
$$= 17.8$$
$$b = \sqrt{17.8} = 4.2$$

Use the law of sines to find the measure of $\angle C$, the smaller of the two remaining angles.

$$\frac{\sin C}{c} = \frac{\sin B}{b}$$
$$\sin C = \frac{c \sin B}{b}$$
$$= \frac{7 \sin 31.8°}{4.2}$$
$$= \frac{7(0.5270)}{4.2} = 0.8783$$

Since $\angle C$ is not the largest angle, it must be acute.
$$\therefore \angle C = 61.4°$$
$$\angle A = 180° - 31.8° - 61.4°$$
$$= 86.8°$$
$$\therefore \angle A = 86.8°, \ b = 4.2, \text{ and } \angle C = 61.4°. \quad \textbf{\textit{Answer}}$$

The SSA case is often called the *ambiguous case* because there are six possible outcomes, four if $\angle A$ is acute and two if $\angle A$ is right or obtuse. These possibilities are illustrated below and on the opposite page.

$$\angle A < 90°$$

$\angle A \geq 90°$

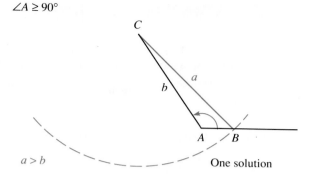

$a \leq b$
No solution

$a > b$
One solution

Example 3 (SSA case) Solve $\triangle ABC$ if $b = 22$, $c = 30$, and $\angle B = 30°$.

Solution

$\dfrac{\sin C}{c} = \dfrac{\sin B}{b}$

$\sin C = \dfrac{c \sin B}{b}$

$\qquad = \dfrac{30 \sin 30°}{22}$

$\qquad = \dfrac{30(0.5000)}{22} = 0.6818$

$\angle C = 43.0°$ or $\angle C = 180° - 43.0° = 137.0°$

$\angle A = 180° - 30° - 43.0° = 107.0°$ $\qquad \angle A = 180° - 30° - 137.0° = 13.0°$

$a = \dfrac{b \sin A}{\sin B}$ $\qquad\qquad\qquad a = \dfrac{b \sin A}{\sin b}$

$\quad = \dfrac{22 \sin 107.0°}{\sin 30°}$ $\qquad\qquad \quad = \dfrac{22 \sin 13.0°}{\sin 30°}$

$\quad = \dfrac{22(0.9563)}{0.5000}$ $\qquad\qquad \quad = \dfrac{22(0.2250)}{0.5000}$

$a = 42.1$ $\qquad\qquad\qquad\qquad a = 9.90$

$\therefore \angle A = 107.0°$, $\angle C = 43.0°$, and $a = 42.1$
or $\angle A = 13.0°$, $\angle C = 137.0°$, and $a = 9.90$ **Answer**

Example 4 (AAS case) Solve $\triangle ABC$ if $a = 40$, $\angle A = 45°$, and $\angle C = 55°$.

Solution

$\angle B = 180° - \angle A - \angle C = 180° - 45° - 55° = 80°$

$b = \dfrac{a \sin B}{\sin A}$ $\qquad\qquad c = \dfrac{a \sin C}{\sin A}$

$\quad = \dfrac{40 \sin 80°}{\sin 45°}$ $\qquad\qquad \quad = \dfrac{40 \sin 55°}{\sin 45°}$

$\quad = \dfrac{40(0.9848)}{0.7071}$ $\qquad\qquad \quad = \dfrac{40(0.8192)}{0.7071}$

$b = 55.7$ $\qquad\qquad\qquad\qquad c = 46.3$

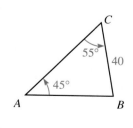

$\therefore b = 55.7$, $c = 46.3$, and $\angle B = 80°$. **Answer**

Oral Exercises

Given the following measurements, tell which law you would use first to solve the triangle.

1. SSS   **2.** SAS   **3.** SSA   **4.** AAS   **5.** ASA

Outline a strategy for solving each triangle.

Sample $a = 20, c = 30, \angle B = 40°$

Solution   1. Use the law of cosines
                      to find b.
                 2. Use the law of sines
                      to find $\angle A$.
                 3. Use $\angle C = 180° - \angle A - \angle B$
                      to find $\angle C$.

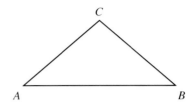

6. $a = 17, \angle B = 20°, \angle C = 60°$ **7.** $a = 8, c = 13, \angle B = 150°$

8. $a = 5, b = 7, c = 11$ **9.** $b = 15, \angle B = 100°, \angle C = 25°$

10. $b = 10, c = 13, \angle A = 120°$ **11.** $a = 10, b = 25, c = 30$

12. $a = 30, b = 20, \angle A = 130°$ **13.** $b = 20, c = 18, \angle B = 120°$

14. $a = 9, b = 10, \angle B = 40°$ **15.** $a = 12, b = 11, \angle A = 35°$

16. $b = 14, c = 18, \angle B = 50°$ **17.** $b = 15, c = 14.5, \angle C = 70°$

18. In $\triangle ABC$, if $\angle A$ is an obtuse angle, what can you conclude about $\angle C$?

19. In $\triangle ABC$, if b is the longest side, what can you conclude about $\angle A$?

20. Explain why the SSA case is called the ambiguous case.

Written Exercises

Give lengths to three significant digits and angle measures to the nearest tenth of a degree. You may wish to use a calculator.

A   **1–12.** Solve the triangles given in Oral Exercises 6–17. If there are two solutions, find both. If there are no solutions, say so.

In Exercises 13 and 14, recall that a median is the segment from a vertex to the midpoint of the opposite side.

B   **13.** In $\triangle ABC$, $a = 4$, $b = 5$, and $\angle C = 110°$. Find the length of the median to the longest side.

14. In $\triangle ABC$, $c = 10$ and $\angle A = \angle B = 40°$. Find the length of the median to $\overline{AC}$.

Solve Exercises 15 and 16 by using both the law of sines and the law of cosines. You may wish to use a calculator.

15. In quadrilateral $ABCD$, $\angle 1 = 30°$, $\angle 2 = 80°$, $\angle B = 80°$, $AD = 8$, and $AB = 10$. Find the length of $\overline{CD}$.

16. In quadrilateral $ABCD$, $\angle B = 110°$, $\angle D = 75°$, $\angle 3 = 35°$, $AB = 5$, and $BC = 6$. Find the length of $\overline{CD}$.

17. Find the lengths of the diagonals of the trapezoid shown below.

18. Find the lengths of the diagonals of the quadrilateral shown below.

C 19. The angles of a triangle measure 57°, 60°, and 63°. The longest side is 5 units longer than the shortest side. Find the lengths of the three sides.

20. One angle of a triangle has measure 65°. The side opposite this angle has length 10. If the area of the triangle is 20, find the perimeter.

Problems

Give lengths to three significant digits and angle measures to the nearest tenth of a degree. You may wish to use a calculator.

A 1. Two planes leave an airport at the same time, one flying at 300 km/h and the other at 420 km/h. The angle between their flight paths is 75°. After three hours, how far apart are they?

2. Jan is flying a plane on a triangular course at 320 mi/h. She flies due east for two hours and then turns right through a 65° angle. How long after turning will she be exactly southeast of where she started?

3. Two cables of length 300 m and 270 m extend from the top of a television antenna to the level ground on opposite sides of the antenna. The longer cable makes an angle of 48° with the ground. Find the acute angle that the shorter cable makes with the ground and the distance between the cables along the ground.

4. A vertical television mast is mounted on the roof of a building. From a point 750 ft from the base of the building, the angles of elevation to the bottom and top of the mast measure 34° and 50° respectively. How tall is the mast?

5. Jim is flying a plane from Midville to Vista, a distance of 500 km. Because of a thunderstorm, he had to fly 17.5° off course for 300 km. How far is he now from Vista, and through what angle should he turn to fly directly there?

6. A monument consists of a flagpole 15 m tall standing on a mound in the shape of a cone with vertex angle 140°. How long a shadow does the pole cast on the cone when the angle of elevation of the sun is 62°?

B 7. From the top of a building 10 m tall, the angle of elevation to the top of a flagpole is 11°. At the base of the building, the angle of elevation to the top of the flagpole is 39°. Find the height of the flagpole.

8. A communication satellite is in orbit 35,800 km above the equator. It completes one orbit every 24 hours, so that from Earth it appears to be stationary above a point on the equator. If this point has the same longitude as Houston, find the measure of θ, the satellite's angle of elevation from Houston. The latitude of Houston is 29.7°N. The radius of Earth is 6400 km.

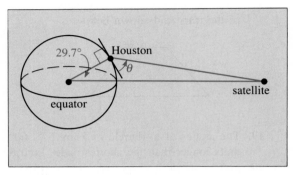

9. In Problem 8, what is the greatest latitude from which a signal can travel to the satellite in a straight line?

10. From the top of an observation post that is 90 m high, a ranger sights a campsite at an angle of depression of 10°. Turning in a different direction, the ranger sees another campsite at an angle of depression of 13°. The angle between these two lines of sight is 35°. How far apart are the campsites?

C 11. In quadrilateral $ABCD$, $AB = 3$, $BC = 4$, $CD = 5$, and $DA = 6$. The length of diagonal $\overline{BD}$ is 7. Find the length of the other diagonal.

12. From Base Camp, located 2000 m above sea level, the angle of elevation of Camp A is 25° and the angle of elevation of the summit of Mount Snow is 41°. From Camp A the angle of elevation of the summit is 53°. If Camp A is 3 km closer to the summit (air distance) than Base Camp is, how high is the summit above sea level?

13. In quadrilateral $ABCD$ shown at the left below, find the length of $\overline{CD}$.

14. Two balloons are moored directly over a straight, level road. The diagram on the right above shows the angle of elevation of the balloons from two observers on the road one kilometer apart. How far apart are the balloons? Which balloon is higher, and by how many meters?

12-9 Areas of Triangles

Objective   To apply triangle area formulas.

You know that the area of a triangle can be found by using the formula
$$K = \frac{1}{2}(\text{base})(\text{height}).$$
Here are some other area formulas for triangles.

The area K of $\triangle ABC$ is given by each of the formulas listed below.

$$K = \frac{1}{2}bc \sin A \qquad K = \frac{1}{2}ac \sin B \qquad K = \frac{1}{2}ab \sin C$$

$$K = \frac{1}{2}a^2 \frac{\sin B \sin C}{\sin A} \qquad K = \frac{1}{2}b^2 \frac{\sin A \sin C}{\sin B} \qquad K = \frac{1}{2}c^2 \frac{\sin A \sin B}{\sin C}$$

$$K = \sqrt{s(s-a)(s-b)(s-c)}, \text{ where } s = \frac{1}{2}(a+b+c)$$

The first three formulas are useful for finding areas of triangles in the SAS case. They were derived in Lesson 12-7. The formulas in the second row above are useful in the AAS case. You can prove them using the law of sines in the form $b = \frac{a \sin B}{\sin A}$. By substituting $\frac{a \sin B}{\sin A}$ for b in $K = \frac{1}{2}ab \sin C$, the equation becomes $K = \frac{1}{2}a^2 \frac{\sin B \sin C}{\sin A}$. The other formulas on the second row are proved similarly. The formula on the third row is called **Hero's formula** and is used to find areas in the SSS case. Its proof is outlined in Exercises 17–19 on pages 599–600.

Example 1   (SAS case) Find the area of $\triangle ABC$ if $b = 32$, $c = 27$, and $\angle A = 108°$.

Solution $K = \frac{1}{2}bc \sin A = \frac{1}{2}(32)(27) \sin 108°$

$$= 432 \,(0.9511)$$
$$= 411$$

∴ the area is 411 square units.   *Answer*

Example 2 (ASA case) Find the area of the triangle shown at the right.

Solution Let $\angle A = 100°$ and $\angle B = 65°$.
Then $\angle C = 180° - 100° - 65° = 15°$.

Use the formula $K = \frac{1}{2}c^2 \frac{\sin A \sin B}{\sin C}$.

$K = \frac{1}{2}(2.2)^2 \frac{\sin 100° \sin 65°}{\sin 15°}$

$= 2.42 \frac{(0.9848)(0.9063)}{(0.2588)}$

$= 8.35$

65°

100°

2.2 m

$\therefore$ the area is 8.35 m². **Answer**

Example 3 (SSS case) A triangular city lot has sides of lengths 50 ft, 60 ft, and 80 ft. Find its area.

Solution Use Hero's formula: $K = \sqrt{s(s-a)(s-b)(s-c)}$, where

$s = \frac{1}{2}(a + b + c)$

$= \frac{1}{2}(50 + 60 + 80) = 95.$

Then: $s - a = 95 - 50 = 45$

$s - b = 95 - 60 = 35$

$s - c = 95 - 80 = 15$

Now substitute: $K = \sqrt{s(s-a)(s-b)(s-c)}$

$= \sqrt{95(45)(35)(15)} = 75\sqrt{399} = 1498$

$\therefore$ to three significant digits, the area is 1500 ft². **Answer**

In the ambiguous case (SSA), the triangle must be partially solved before the area, if any, can be found.

Example 4 (SSA case) Find the area of $\triangle ABC$ if $b = 22$, $c = 30$ and $\angle B = 30°$.

Solution In Example 3, page 593, two triangles fit the given data, one with $\angle A = 107.0°$ and the other with $\angle A = 13.0°$. Therefore:

$\angle A = 107.0°$ or $\angle A = 13.0°$

$K = \frac{1}{2}bc \sin A$ $K = \frac{1}{2}bc \sin A$

$= \frac{1}{2}(22)(30) \sin 107.0°$ $= \frac{1}{2}(22)(30) \sin 13.0°$

$= 330(0.9563)$ $= 330(0.2250)$

$= 316$ $= 74.3$

$\therefore$ the area is 316 square units or 74.3 square units. **Answer**

Oral Exercises

Three parts of $\triangle ABC$ are given. Tell which formula you would use to find the area of the triangle.

1. $\angle A = 25°$, $\angle B = 50°$, $b = 30$

2. $a = 18$, $b = 10$, $\angle C = 45°$

3. $a = 6$, $c = 14$, $\angle B = 62°$

4. $\angle A = 70°$, $b = 15$, $c = 36$

5. $a = 12$, $b = 8$, $c = 12$

6. $\angle B = 20°$, $\angle C = 115°$, $c = 16$

7. $c = 15$, $b = 22$, $\angle A = 150°$

8. $a = 7$, $b = 12$, $c = 8$

9. Suppose $a = 12$, $b = 11$, and $\angle B = 60°$. Explain why this is an example of the ambiguous case.

Written Exercises

Give lengths to three significant digits and angle measures to the nearest tenth of a degree. You may wish to use a calculator.

A   **1–8.** Find the areas of each triangle described in Oral Exercises 1–8.

    9. Find the area of a parallelogram that has a 45° angle and sides with lengths 10 and 18.

   10. Find the area of a rhombus that has perimeter 48 and an angle of 55°.

B   **11.** The area of $\triangle ABC$ is 36 square units. If $\angle B = 30°$ and $c = 8$, find the value of a.

   12. The area of $\triangle ABC$ is 20 square units. If $\angle A = 130°$ and $b = 6$, find the value of c.

   13. A triangle has area 48 cm² and its shorter sides have lengths 9 cm and 12 cm. Find the largest angle of the triangle.

   14. The angles of a triangle are 25°, 45°, and 110°. What is the length of the longest side if the area of the triangle is 75 square units?

   15. Find the area of a regular octagon whose sides are 10 cm long.

   16. Find the area of a regular octagon inscribed in a unit circle.

Exercises 17–19 make up a proof of Hero's formula.
Exercises 17 and 18 are used in Exercise 19 on the following page.

   17. Show that $K^2 = \frac{1}{4}a^2b^2 \sin^2 C$.

   18. Recall that $s = \frac{1}{2}(a + b + c)$, so that $2s = a + b + c$.

Therefore, $2(s - a) = 2s - 2a = a + b + c - 2a = -a + b + c$.
Show that $2(s - b) = a - b + c$ and $2(s - c) = a + b - c$.

C **19.** Justify each of the statements from (a)–(i) below. Use the results of Exercises 17 and 18 where necessary.

a. $a^2 + b^2 - c^2 = 2ab \cos C$

b. $(a^2 + b^2 - c^2)^2 = 4a^2b^2 \cos^2 C = 4a^2b^2(1 - \sin^2 C)$

c. $(a^2 + b^2 - c^2)^2 = 4a^2b^2 - 4a^2b^2 \sin^2 C = 4a^2b^2 - 16K^2$

d. $16K^2 = 4a^2b^2 - (a^2 + b^2 - c^2)^2$

e. $16K^2 = [2ab + (a^2 + b^2 - c^2)][2ab - (a^2 + b^2 - c^2)]$

f. $16K^2 = [(a + b)^2 - c^2][c^2 - (a - b)^2]$

g. $16K^2 = [(a + b) + c][(a + b) - c][c + (a - b)][c - (a - b)]$

h. $16K^2 = [2s][2(s - c)][2(s - b)][2(s - a)]$

i. $K = \sqrt{s(s - a)(s - b)(s - c)}$

Mixed Review Exercises

Find the indicated part of $\triangle ABC$ to three significant digits or to the nearest tenth of a degree.

1. $a = 8$, $b = 5$, $\angle C = 62°$, $c = $ ___?___

2. $\angle A = 100°$, $\angle B = 30°$, $b = 15$, $a = $ ___?___

3. $a = 6$, $b = 4$, $c = 9$, $\angle C = $ ___?___

4. $a = 20$, $\angle B = 76°$, $\angle C = 48°$, $b = $ ___?___

Find the five other trigonometric functions of θ.

5. $\sin \theta = \frac{1}{2}$, $90° < \theta < 180°$

6. $\cos \theta = -\frac{3}{4}$, $180° < \theta < 270°$

7. $\tan \theta = -1$, $270° < \theta < 360°$

8. $\sin \theta = \frac{5}{13}$, $0° < \theta < 90°$

Self-Test 2

Vocabulary   solving a triangle (p. 574)   law of cosines (p. 580)
            angle of elevation (p. 575)   law of sines (p. 586)
            angle of depression (p. 575)   ambiguous case (p. 592)

Give lengths to three significant digits and angle measures to the nearest tenth of a degree.

1. In $\triangle ABC$, $\angle C = 90°$, $\angle B = 27.3°$, and $a = 30$. Find $\angle A$ and   Obj. 12-5, p. 574
sides b and c.

2. In $\triangle DEF$, $d = 18$, $e = 24$ and $\angle F = 42°$. Find side f.   Obj. 12-6, p. 580

3. In $\triangle XYZ$, $\angle X = 36°$, $x = 14$, and $z = 23.5$. Find $\angle Z$.   Obj. 12-7, p. 586

4. Solve $\triangle ABC$ if $a = 15$, $b = 12$, and $c = 26$.   Obj. 12-8, p. 591

5. Find the area of $\triangle RST$ if $r = 9$, $s = 12$, and $\angle T = 53.7°$.   Obj. 12-9, p. 597

Check your answers with those at the back of the book.

A comet usually is visible for only a few weeks or months on its path through the solar system. This fact makes its motion difficult to analyze and predict. Early astronomers debated whether comets moved in straight paths or along some regular curve. The Englishman Edmund Halley (1656–1742) solved this problem by describing comets' orbits as ellipses with predictable paths. He claimed that the comets of 1531, 1607, and 1682 were in fact the same comet and that this comet could be seen from Earth approximately every seventy-five years. Later Halley suggested that the comets reported in 1305, 1380, and 1456 were also sightings of this single comet.

Halley predicted the comet's next return for December, 1758. Off schedule by just a few days, the comet appeared in the same part of the sky he had predicted. Although Halley did not live to see this proof of his theory, the scientific world recognized his achievement by naming the comet after him. Halley's comet was last visible from Earth in 1986 and is not due to be seen again until 2061.

Mathematical problems, as well as the positions of the planets, the size of the universe, and stellar motion, also intrigued

Halley. He published papers on topics ranging from higher geometry and the roots of equations to the computation of logarithms and trigonometric functions. One result of his interest in social statistics was the first explanation of how mortality tables could be used to calculate life insurance premiums. Halley was appointed as astronomer royal in 1720. He continued his work well into his old age.

Challenge

One day a visitor to a distant land stood at a fork in the road. The visitor knew that each path led to a different village. The inhabitants of one village always told the truth. The inhabitants of the other village always lied. Luckily, a local citizen happened to be along and agreed to answer one and only one question. The visitor thought for awhile and finally asked a simple question that was guaranteed to indicate the right path to the village of truth tellers, no matter which village the person was from.

What was the question? Explain the visitor's reasoning for asking that particular question.

Chapter Summary

1. Angles may be measured in *degrees,* where one degree is $\frac{1}{360}$ of a complete revolution. The measure is *positive* if the rotation of the angle's *initial side* onto its *terminal side* is counterclockwise. The measure is *negative* if the rotation is clockwise.

2. Let $P(x, y)$ be a point other than the origin on the terminal side of an angle θ in standard position, and let $r = OP$. Then the *trigonometric functions* of θ are:

$$\sin \theta = \frac{y}{r} \qquad \cos \theta = \frac{x}{r} \qquad \tan \theta = \frac{y}{x}$$

$$\csc \theta = \frac{r}{y} \qquad \sec \theta = \frac{r}{x} \qquad \cot \theta = \frac{x}{y}$$

   If θ is acute, the numbers x, y, and r can be thought of as being the lengths of the legs and hypotenuse, respectively, of a right triangle.

3. Scientific calculators give good approximations to the values of the trigonometric functions of any angle. Tables giving values of the functions of acute angles can also be used because each angle θ has an acute *reference angle* α such that $\sin \theta = \pm \sin \alpha$, $\cos \theta = \pm \cos \alpha$, and so forth.

4. To *solve a triangle* is to find the measures of all of its sides and angles when three of the measures (including at least one side) are known. Right triangles can be solved using the definitions of the trigonometric functions. *Any* triangle ABC can be solved with the help of the law of cosines (see page 580) and the law of sines (page 586). Depending on the information given, there may be one solution, two solutions, or no solution.

5. The area K of $\triangle ABC$ can be found by using one of the formulas given on page 597.

Chapter Review

Give the letter of the correct answer.

1. Which of the following is coterminal with $-210°$?   12-1
   a. $30°$ **b.** $150°$ **c.** $-150°$ **d.** $210°$

2. Express $45°14'42''$ in decimal degrees.
   a. $45.452°$ **b.** $45.542°$ **c.** $45.245°$ **d.** $45.45°$

3. Find the product of all six trigonometric functions of an angle θ in   12-2
   standard position whose terminal side passes through $(-\sqrt{3}, 1)$.

   a. -1 **b.** 1 **c.** $\sqrt{3}$ **d.** $\frac{\sqrt{3}}{2}$

4. In $\triangle ABC$, $\angle A = 60°$, $b = 10$, and $\angle B = 90°$. Find c.
　　a. $5\sqrt{3}$　　　　　**b.** 5　　　　　　**c.** 10　　　　　**d.** 20

5. Find x in the diagram at the right.
　　a. $16\sqrt{2}$　　　　**b.** $6\sqrt{2}$
　　c. $4\sqrt{3}$　　　　　**d.** $8\sqrt{6}$

6. Express $\cos 230°$ as a function of an acute angle.　　　　　　　　　　12-3
　　a. $-\cos 40°$　　　**b.** $-\sin 40°$　　　**c.** $-\cos 50°$　　　**d.** $\cos 50°$

7. Find $\tan \theta$ if $\cos \theta = \dfrac{8}{17}$ and $\sin \theta > 0$.

　　a. $\dfrac{15}{8}$　　　　**b.** $-\dfrac{15}{17}$　　　**c.** $-\dfrac{8}{15}$　　　**d.** $-\dfrac{15}{8}$

8. Give the exact value of $\sin 300°$.
　　a. $-\dfrac{1}{2}$　　　　**b.** $\dfrac{1}{2}$　　　　**c.** $-\dfrac{\sqrt{3}}{2}$　　　**d.** $\dfrac{\sqrt{3}}{2}$

9. Find $\sec 116.7°$ to four decimal places.　　　　　　　　　　　　　12-4
　　a. 1.119　　　　**b.** 2.226　　　　**c.** -1.119　　　**d.** -2.226

10. Find θ to the nearest ten minutes if $\cot \theta = -0.7046$ and $0° < \theta < 180°$.
　　a. $54°50'$　　　**b.** $125°10'$　　　**c.** $144°50'$　　　**d.** $-56°50'$

11. In $\triangle DEF$, $\angle E = 90°$, $e = 10$, and $d = 8$. Find $\angle F$ to the nearest tenth of　　12-5
a degree.
　　a. $36.9°$　　　**b.** $53.1°$　　　**c.** $38.7°$　　　**d.** $44.2°$

12. In $\triangle ABC$, $\angle B = 90°$, $\angle C = 55°$, and $c = 12$. Find b to the nearest tenth.
　　a. 14.6　　　　**b.** 6.9　　　　**c.** 20.9°　　　**d.** 9.8

13. The base angles of an isosceles trapezoid are $45°$ and the bases have
lengths 15 and 25 units. Find the height of the trapezoid.
　　a. 5　　　　　**b.** 5.2　　　　**c.** 10.2　　　　**d.** 5.3

14. In $\triangle ABC$, $a = 6$, $b = 10$, and $c = 7$. Find the measure of the largest angle　12-6
to the nearest tenth of a degree.
　　a. $43.5°$　　　**b.** $100.3°$　　　**c.** $36.2°$　　　**d.** $46.5°$

15. Two planes leave the airport at noon, one traveling east at 300 km/h and
the other traveling northwest at 450 km/h. How far apart are they at
2 PM?
　　a. 637 km　　　**b.** 821 km　　　**c.** 1390 km　　　**d.** 1080 km

16. In $\triangle ABC$, find $\angle A$ to the nearest tenth of a degree if $\angle B = 95°$, $b = 15.4$,　12-7
and $a = 8.0$.
　　a. $32.1°$　　　**b.** $56.8°$　　　**c.** $31.2°$　　　**d.** $92.3°$

17. Wires of lengths 20 m and 30 m extend from the top of a tower to the
ground on the same side of the tower. The shorter wire makes an angle
of $42°$ with the ground. What angle do the wires make with each other?
　　a. $15.5°$　　　**b.** $63.5°$　　　**c.** $26.5°$　　　**d.** $48.2°$

(Chapter Review continues on the next page.)

18. In $\triangle ABC$, $\angle B = 30°$ and $c = 20$. For what value(s) of b will the triangle 12-8
have two solutions?
   a. $0 < b < 10$ **b.** $10 < b < 20$ **c.** $b < 20$ **d.** $b = 10$

19. Find x in the diagram at the right.
   a. 15.8   **b.** 18.4
   c. 19.8   **d.** 20.7

20. Find the area of $\triangle ABC$ if $a = 8$, $b = 15$, and $\angle C = 40°$. 12-9
   a. 77.1   **b.** 60.0   **c.** 45.9   **d.** 38.6

Chapter Test

1. Express $13.24°$ in degrees, minutes, and seconds. 12-1

2. Find two angles, one positive and one negative, that are coterminal with $285°$.

3. Find x in the diagram at the right. 12-2

4. If $\cos \theta = -\dfrac{3}{4}$ and $\sin \theta < 0$, find $\tan \theta$. 12-3

5. Give the exact values of the six trigonometric functions of $240°$.

6. **a.** Find sec $145.8°$. 12-4
   b. If $\cos \theta = 0.5606$ and $90° < \theta < 360°$, find θ to the nearest tenth of a degree.

7. The angle of elevation from an observer on the street to the top 12-5
of a building is $55.6°$. If the observer is 150 ft from the base
of the building, how tall is the building?

8. A compass with legs 3 in. long is opened to measure the diameter of 12-6
a circle. If the diameter is 5 in., what is the angle between the legs
of the compass?

Give lengths to three significant digits and angle measures to the nearest tenth of a degree.

9. Solve $\triangle DEF$ if $\angle D = 32°$, $\angle E = 108°$, and $f = 12$. 12-7

10. Solve $\triangle ABC$ if $\angle A = 40°$, $a = 6$, and $b = 8$. 12-8

11. Find the area of a triangular plot of land if the sides have length 12-9
200 m, 150 m, and 100 m.

Preparing for College Entrance Exams

Decide which is the best of the choices given and write the corresponding letter on your answer sheet.

1. Find the next term of this sequence: 1, 3, 7, 13, 21, . . .
   - **(A)** 27
   - **(B)** 37
   - **(C)** 31
   - **(D)** 33
   - **(E)** 29

2. Evaluate $64 \sum_{k=1}^{6} 2^{-k}$.
   - **(A)** 64
   - **(B)** 128
   - **(C)** 63
   - **(D)** 1
   - **(E)** $\frac{31}{32}$

3. The infinite series $1 - \frac{5}{4} + \frac{25}{16} - \frac{125}{64} + \cdots$
   - **(A)** has the common difference $\frac{5}{4}$
   - **(B)** has the sum $\frac{4}{9}$
   - **(C)** has no sum
   - **(D)** has the sum -4
   - **(E)** is neither arithmetic nor geometric

4. Which function(s) is (are) not defined for $\phi = -90°$?
   - I. $\cos \phi$
   - II. $\sec \phi$
   - III. $\tan \phi$
   - **(A)** II only
   - **(B)** I and II only
   - **(C)** II and III only
   - **(D)** I and III only
   - **(E)** I, II, and III

5. If $\csc \phi = 3$ and $90° < \phi < 270°$, find $\cot \phi$.
   - **(A)** $-2\sqrt{2}$
   - **(B)** $\frac{\sqrt{2}}{4}$
   - **(C)** $-\frac{\sqrt{2}}{4}$
   - **(D)** -8
   - **(E)** $-\frac{\sqrt{2}}{3}$

6. In $\triangle XYZ$, $\angle X = 45°$, $\angle Z = 30°$, and $z = 8$. Find x.
   - **(A)** $\frac{8}{3}\sqrt{6}$
   - **(B)** $8\sqrt{2}$
   - **(C)** $4\sqrt{2}$
   - **(D)** $4\sqrt{6}$
   - **(E)** $32\sqrt{2}$

7. Find the area of a triangle with sides of lengths 5, 6, and 7.
   - **(A)** $2\sqrt{6}$
   - **(B)** $3\sqrt{15}$
   - **(C)** $2\sqrt{14}$
   - **(D)** $6\sqrt{6}$
   - **(E)** $\frac{21\sqrt{2}}{2}$

8. In $\triangle RST$, $\angle R = 137.4°$, $t = 15$, and $s = 12$. Find r to the nearest integer.
   - **(A)** 10
   - **(B)** 17
   - **(C)** 21
   - **(D)** 23
   - **(E)** 25

The upper graph shows the sine waves for three guitar strings vibrating at different frequencies. The lower graph shows the sum of the two solid sine waves.

Circular Functions and Their Graphs

13-1 Radian Measure

Objective   To use radians to measure angles.

In advanced mathematics angles are usually measured in *radians* rather than degrees. To define the radian measure of an angle θ, let O be a circle of radius r centered at the vertex of θ, as shown at the right. Let s be the length of the arc of O intercepted by θ. The arc length s is considered to be positive if θ is a positive angle and negative if θ is a negative angle.

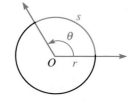

The **radian measure** of θ is defined to be the ratio of arc length s to radius r:

$$\theta = \frac{s}{r}$$

The symbol θ here stands for the *measure* of angle θ. When no unit of angle measure is specified, you can assume that radian measure is being used. Therefore, "$\theta = 2$" means "the measure of θ is 2 radians," and the length of the arc of O intercepted by θ is twice the length of the radius r of circle O.

If θ is a complete revolution, $s = 2\pi r$ and $\theta = \frac{2\pi r}{r} = 2\pi$. If θ is a complete revolution, the degree measure of θ is 360°. Likewise, an angle that is half a revolution has degree measure 180° and radian measure π. The conversion formula

$$180° = \pi \text{ radians}$$

can be used to convert from one system of measure to the other:

$$1° = \frac{\pi}{180} \text{ radians} \qquad 1 \text{ radian} = \frac{180°}{\pi} \approx 57.3°$$

Example 1   **a.** Express 85° in radians.   **b.** Express $\frac{\pi}{3}$ radians in degrees.

Solution   **a.** $85° = 85 \cdot \dfrac{\pi}{180}$ radians

$\qquad\qquad\quad = \dfrac{17\pi}{36}$ radians

b. $\dfrac{\pi}{3}$ radians $= \dfrac{\pi}{3} \cdot \dfrac{180°}{\pi} = 60°$

$\left(\text{or } \dfrac{\pi}{3} \text{ radians} = \dfrac{1}{3} \cdot \pi \text{ radians}\right.$

$\qquad\qquad = \dfrac{1}{3} \cdot 180° = 60°\big)$

Here are the degree and radian measures of some frequently used angles.

Degree measure	0°	30°	45°	60°	90°	120°	135°	150°	180°
Radian measure	0	$\dfrac{\pi}{6}$	$\dfrac{\pi}{4}$	$\dfrac{\pi}{3}$	$\dfrac{\pi}{2}$	$\dfrac{2\pi}{3}$	$\dfrac{3\pi}{4}$	$\dfrac{5\pi}{6}$	π

Scientific calculators accept both degree and radian measurements in decimal form. You can convert between degrees and radians on a calculator by using the π key or a decimal approximation, such as 3.1416, for π.

Example 2 **a.** Express 35° in radians to the nearest hundredth of a radian.

b. Express 2 radians in degrees to the nearest tenth of a degree.

Solution **a.** $35° = 35 \cdot \dfrac{\pi}{180}$ radians

$\approx \dfrac{35(3.1416)}{180}$ radians

≈ 0.61 radians *Answer*

b. 2 radians $= 2 \cdot \dfrac{180°}{\pi}$

$\approx \dfrac{2 \cdot 180°}{3.1416}$

$\approx 114.6°$ *Answer*

Using radian measure you can write simple formulas for finding the length of an arc and the area of a sector, as shown below. In the formulas, θ denotes radian measure.

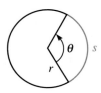

Arc length

$s = r\theta$

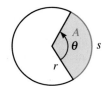

Area of a sector

$A = \dfrac{1}{2}r^2\theta$ or $A = \dfrac{1}{2}rs$

The formula $s = r\theta$ is equivalent to $\theta = \dfrac{s}{r}$, which was used to define radian measure. The area formulas follow from the fact that the areas of sectors are proportional to the measures of their central angles.

$$\dfrac{\text{area of sector}}{\text{area of circle}} = \dfrac{\text{measure of central angle of sector}}{\text{measure of central angle of circle}}$$

or

$$\dfrac{A}{\pi r^2} = \dfrac{\theta}{2\pi}.$$

Solving for A gives

$$A = (\pi r^2)\left(\dfrac{\theta}{2\pi}\right) = \dfrac{1}{2}r^2\theta.$$

Using $s = r\theta$ gives

$$A = \dfrac{1}{2}r^2\theta = \dfrac{1}{2}r(r\theta) = \dfrac{1}{2}rs.$$

Example 3   A central angle of a circle of radius 3 cm measures 1.5 radians. Find
(a) the length of the intercepted arc and **(b)** the area of the related sector.

Solution   **a.** $s = r\theta$ **b.** $A = \frac{1}{2}r^2\theta$

$\qquad = (3)(1.5) = 4.5$ (cm)

$\qquad\qquad\qquad = \frac{1}{2}(3^2)(1.5) = 6.75$ (cm^2)

Suppose that a particle moves with a constant speed v around a circle of
radius r cm. If the particle travels s centimeters in t seconds, its speed is

$$v = \frac{s}{t} \text{ (cm/s)}.$$

Similarly, if the particle moves through an angle of θ radians in t seconds, its
angular speed, denoted by ω, is

$$\omega = \frac{\theta}{t} \text{ (radians/s)}.$$

Since arc length s and radian measure θ are related by the formula $s = r\theta$, the
speed v and angular speed ω of the particle are related by the formula

$$v = \frac{s}{t} = r\frac{\theta}{t} = r\omega.$$

Therefore, the particle's speed is r times its angular speed.

Example 4   A Ferris wheel 60 ft in diameter makes one revolution in
3 min. Find **(a)** the speed of
a seat on the rim and **(b)** the
angular speed.

Solution   In 1 revolution the seat travels
the distance of the circumfer-
ence of the wheel ($s = 2\pi r$) and
through 360°, or 2π radians.

a. $v = \frac{s}{t} = \frac{\pi \cdot 60}{3}$

$\qquad = 20\pi \approx 62.8$ (ft/min)

b. $\omega = \frac{\theta}{t} = \frac{2\pi}{3}$

$\qquad \approx 2.1$ (radians/min)

Oral Exercises

Express each degree measure in radians. Leave your answers in terms of π.

1. 180°   **2.** 360°   **3.** −90°   **4.** −720°

Express each radian measure in degrees.

5. $\dfrac{\pi}{2}$
6. -2π
7. $-\pi$
8. 3π

Give the radian measure of ϕ and θ.

9.

10.
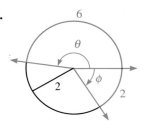

Give the arc length s and the area A of the sector. Angle measures are given in radians.

11.

12.

13. If a circle has radius r, how long is the arc that is intercepted by an angle with radian measure 1?

Written Exercises

Express each degree measure in radians. Leave your answer in terms of π.

A
1. $45°$
2. $30°$
3. $60°$
4. $270°$

5. $-120°$
6. $135°$
7. $150°$
8. $-180°$

9. $-330°$
10. $240°$
11. $-315°$
12. $495°$

Express each radian measure in degrees. In Exercises 25–28, leave your answers in terms of π.

13. $\dfrac{\pi}{6}$
14. $-\dfrac{\pi}{4}$
15. $\dfrac{\pi}{3}$
16. $-\dfrac{\pi}{2}$

17. $\dfrac{4\pi}{3}$
18. $\dfrac{5\pi}{6}$
19. $-\dfrac{7\pi}{6}$
20. $\dfrac{3\pi}{4}$

21. 3π
22. $-\dfrac{7\pi}{4}$
23. $-\dfrac{7\pi}{2}$
24. $\dfrac{9\pi}{4}$

25. 4
26. 3
27. -2
28. $-\dfrac{1}{2}$

Express each degree measure in radians. Give answers to the nearest hundredth of a radian.

29. $10°$ **30.** $-50°$ **31.** $80°$ **32.** $300°$

33. $48°$ **34.** $265°$ **35.** $-174°$ **36.** $255°$

Express each radian measure in degrees. Give answers to the nearest tenth of a degree.

37. 3 **38.** -2.5 **39.** 0.4 **40.** -1.6

41. -1.5 **42.** 5.5 **43.** 15 **44.** 8

In Exercises 45–56, r, s, θ, and area A are as shown in the adjacent figure. Angle θ is measured in radians. Find the missing measures.

45. $r = 4$, $\theta = 1$, $s = \underline{\ ?\ }$, $A = \underline{\ ?\ }$
46. $r = 5$, $\theta = 2.5$, $s = \underline{\ ?\ }$, $A = \underline{\ ?\ }$
47. $r = 4$, $s = 12$, $\theta = \underline{\ ?\ }$, $A = \underline{\ ?\ }$
48. $r = 5$, $s = 30$, $\theta = \underline{\ ?\ }$, $A = \underline{\ ?\ }$

B **49.** $r = 5$, $A = 15$, $\theta = \underline{\ ?\ }$, $s = \underline{\ ?\ }$ **50.** $r = 2$, $A = 6$, $\theta = \underline{\ ?\ }$, $s = \underline{\ ?\ }$
51. $s = 10$, $\theta = 2.5$, $r = \underline{\ ?\ }$, $A = \underline{\ ?\ }$ **52.** $s = 1.2$, $\theta = 0.5$, $r = \underline{\ ?\ }$, $A = \underline{\ ?\ }$
53. $r = 2$, $A = 3$, $s = \underline{\ ?\ }$, $\theta = \underline{\ ?\ }$ **54.** $r = 8$, $A = 6$, $s = \underline{\ ?\ }$, $\theta = \underline{\ ?\ }$
55. $A = 8$, $\theta = 4$, $r = \underline{\ ?\ }$, $s = \underline{\ ?\ }$ **56.** $A = 6$, $\theta = 3$, $r = \underline{\ ?\ }$, $s = \underline{\ ?\ }$

The angle α has the given radian measure. Find, to the nearest hundredth of a radian, the measure of the angle between 0 and 2π that is coterminal with α.

57. $\alpha = 8.12$ **58.** $\alpha = 6.55$ **59.** $\alpha = -3$ **60.** $\alpha = -20$

Problems

Solve. Give answers to two significant digits.

The *latitude* of point P shown in the drawing of Earth at the right is the measure of $\angle EOP$. The radius of Earth is about 4000 miles.

A **1.** The latitude of Houston is $29.8°$N. How far is it from the equator?

2. The latitude of New York is $40.7°$N. How far is it from the North Pole?

3. A phonograph record with diameter 12 in. rotates at 45 rev./min. Find the distance that a point on the rim travels in one minute.

4. An automobile tire with diameter 2 ft turns 10 times per second. How far does a point on the tread travel in one minute?

Trigonometric Graphs and Identities   **611**

5. A Ferris wheel with diameter 50 ft makes one revolution every two minutes. What is the speed of a seat on the rim of the wheel?

6. A merry-go-round 40 ft in diameter makes four revolutions every minute. What is the speed of a seat on the rim?

7. From an observation point on Earth, the angle subtended by the sun 150,000,000 km away is about 0.0093 radians. Approximate the diameter of the sun by finding the length of the red arc.

8. The diameter of the moon is approximately 2200 mi. How far is the moon from the Earth if the angle it subtends from an observation point on Earth measures about 0.518°? (Use the diagram in Problem 7.)

B 9. The angular speed of the larger pulley wheel shown at the right is 50 rev./min. Find the angular speed and rim speed of the smaller wheel.

6 cm

15 cm

10. A clock pendulum swings back and forth once every two seconds. If its length is one meter and the greatest angle that it makes with the vertical is 12°, how many kilometers does the bottom end of the pendulum travel in one day?

11. What is the angular speed in radians per second of the hour hand of a clock? If the hand is 5 cm long, how fast is the tip moving in millimeters per second?

12. Each point on Earth's surface (except the poles) moves in a circle as Earth rotates on its axis. What is the angular speed in radians per second of the point where you are located?

C 13. A cylindrical water tank 4 ft in diameter is lying on its side as shown. What percent of its capacity is used when it contains water to a depth of one foot?

Mixed Review Exercises

Solve each triangle. Give lengths to three significant digits and angle measures to the nearest tenth of a degree.

1. $a = 8$, $b = 5$, $\angle C = 42°$
2. $\angle A = 75°$, $\angle B = 30°$, $b = 12$
3. $b = 10$, $c = 7$, $\angle B = 104°$
4. $a = 5$, $b = 6$, $c = 9$

5–8. Find the area of each triangle in Exercises 1–4. Give answers to three significant digits.

13-2 Circular Functions

Objective   To define the circular functions.

There are many nongeometric uses of the trigonometric functions. For example, the first equation shown below describes the shape of an ocean wave (height y and horizontal distance x in meters), and the second equation gives the line voltage V of ordinary alternating current (time t in seconds). In neither case do angles play any part.

$$y = 2.5 \sin (0.03x) \qquad V = 170 \cos (120\pi t)$$

These examples suggest that $\sin s$ and $\cos s$ can be defined with s representing a *number* rather than an angle.

To develop these definitions, let O be the unit circle $x^2 + y^2 = 1$, and let A be the point $(1, 0)$, as shown at the right. Given any real number s, start at A and measure $|s|$ units around O in a counterclockwise direction if $s \geq 0$, and in a clockwise direction if $s < 0$, arriving at a point $P(x, y)$. The sine of s and the cosine of s are then defined by the coordinates of point P. The tangent, cotangent, secant, and cosecant functions are defined in terms of sine and cosine.

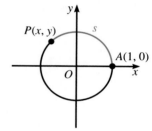

$$\sin s = y \qquad\qquad \cos s = x$$

$$\tan s = \frac{\sin s}{\cos s} \text{ if } \cos s \neq 0 \qquad \cot s = \frac{\cos s}{\sin s} \text{ if } \sin s \neq 0$$

$$\sec s = \frac{1}{\cos s} \text{ if } \cos s \neq 0 \qquad \csc s = \frac{1}{\sin s} \text{ if } \sin s \neq 0$$

These functions are sometimes called *circular functions* because a circle is used in their definition.

Example 1   For some number s, the point $P\left(\frac{2}{3}, -\frac{\sqrt{5}}{3}\right)$ is s units from $A(1, 0)$ along a unit circle. Find the exact values of the six circular functions of s.

Solution $\sin s = -\dfrac{\sqrt{5}}{3}$ $\qquad\qquad\qquad$ $\cos s = \dfrac{2}{3}$

$\tan s = \dfrac{-\dfrac{\sqrt{5}}{3}}{\dfrac{2}{3}} = -\dfrac{\sqrt{5}}{2}$ $\qquad$ $\cot s = \dfrac{\dfrac{2}{3}}{-\dfrac{\sqrt{5}}{3}} = -\dfrac{2}{\sqrt{5}} = -\dfrac{2\sqrt{5}}{5}$

$\sec s = \dfrac{1}{\dfrac{2}{3}} = \dfrac{3}{2}$ $\qquad\qquad$ $\csc s = \dfrac{1}{-\dfrac{\sqrt{5}}{3}} = -\dfrac{3}{\sqrt{5}} = -\dfrac{3\sqrt{5}}{5}$

There is a close relationship between the circular functions and the corresponding trigonometric functions. The diagram at the right shows a central angle θ, in standard position, that intercepts $\overarc{AP}$. Notice that $\sin \theta = \dfrac{y}{1} = y = \sin s$. Also, the radian measure

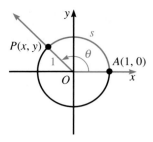

of θ is $\dfrac{s}{1}$, or s. Therefore, $\sin s$ equals the sine of any angle having radian measure s. Similarly, the other circular functions of the number s are equal to the corresponding trigonometric functions of any angle having radian measure s.

You can use a calculator or a table to find values of circular functions. Scientific calculators can be set to work with radians, and Table 6 gives values of trigonometric functions for angles with radian measure between 0 and $\dfrac{\pi}{2}$.

Example 2 Find $\sin 1.44$ to four decimal places.

Solution 1 Using a Calculator

Set the calculator in radian mode.

$\sin 1.44 = 0.9915$ to four decimal places. **Answer**

Solution 2 Using Tables

Use Table 6 to find the sine of an angle having radian measure 1.44.

$\sin 1.44 = 0.9915$ **Answer**

When multiples of $\dfrac{\pi}{6}$ and $\dfrac{\pi}{4}$ are involved, use exact values of the trigonometric functions unless you are asked for a decimal approximation.

Example 3 Find the exact values of the six trigonometric functions of $\dfrac{\pi}{6}$.

Solution Since $\dfrac{\pi}{6}$ radians $= 30°$, use values from a 30°-60°-90° triangle (page 558).

$$\sin \frac{\pi}{6} = \frac{1}{2} \qquad \cos \frac{\pi}{6} = \frac{\sqrt{3}}{2} \qquad \tan \frac{\pi}{6} = \frac{\sqrt{3}}{3}$$

$$\csc \frac{\pi}{6} = 2 \qquad \sec \frac{\pi}{6} = \frac{2\sqrt{3}}{3} \qquad \cot \frac{\pi}{6} = \sqrt{3}$$

If you are given the value of a trigonometric function of x, you can use a calculator or Table 6 to find the value of x to two decimal places when x is between 0 and $\dfrac{\pi}{2}$.

Example 4 Find x to two decimal places if sec $x = 1.975$ and $0 \le x \le \dfrac{\pi}{2}$.

Solution 1 Using a Calculator

Since sec $x = 1.975$, cos $x = \dfrac{1}{1.975} = 0.5063291$. Set the calculator in radian mode and use the inverse function key $\cos^{-1}$ (or inv cos):

$$x = \cos^{-1} 0.5063291 = 1.0398738.$$

$\therefore x = 1.04$ to two decimal places. *Answer*

Solution 2 Using Tables

Use Table 6 and read across from 1.975 in the sec θ column. You should find that sec $1.04 = 1.975$.

$\therefore x = 1.04$ *Answer*

For some given function values, you may be able to express x exactly in terms of π.

Example 5 Find the exact value of x if cos $x = 0.5$ and $0 \le x \le \dfrac{\pi}{2}$.

Solution Recall from the geometry of a 30°-60°-90° triangle that cos $60° = 0.5$. An angle with degree measure 60° has radian measure $\dfrac{\pi}{3}$, and $0 \le \dfrac{\pi}{3} \le \dfrac{\pi}{2}$.

$\therefore x = \dfrac{\pi}{3}$ *Answer*

The equation of the unit circle centered at the origin is $x^2 + y^2 = 1$. By definition, $x = \cos s$ and $y = \sin s$. Therefore, by substitution,

$$\cos^2 s + \sin^2 s = 1.$$

This is the same identity that was proved on page 556 for functions of angles.

For the rest of this book we will follow the usual practice of referring to the circular functions as trigonometric functions. It will be clear from the context whether we are talking about functions of angles or functions of numbers.

Oral Exercises

Sketch a circle with equation $x^2 + y^2 = 1$.

1. How many points on the circle have the property that cos $s = 0$?

2. How many points on the circle have the property that cos $s = 1$?

3. Explain why the value of sin s is always between -1 and 1, inclusive.

Give the exact values of the six trigonometric functions of the number s.
Each circle is a unit circle centered at the origin.

4.

$\left(\dfrac{3}{5}, \dfrac{4}{5}\right)$

5.

$\left(-\dfrac{4}{5}, \dfrac{3}{5}\right)$

6.

$\left(-\dfrac{5}{13}, -\dfrac{12}{13}\right)$

7.

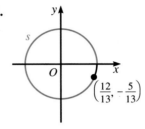

$\left(\dfrac{12}{13}, -\dfrac{5}{13}\right)$

Complete the table by using *positive* or *negative* for each entry.

	8.	**9.**	**10.**	**11.**	**12.**	**13.**
	$\sin s$	$\cos s$	$\tan s$	$\cot s$	$\sec s$	$\csc s$
$0 < s < \dfrac{\pi}{2}$	?	?	?	?	?	?
$\dfrac{\pi}{2} < s < \pi$	?	?	?	?	?	?
$\pi < s < \dfrac{3\pi}{2}$	?	?	?	?	?	?
$\dfrac{3\pi}{2} < s < 2\pi$	?	?	?	?	?	?

14. Complete the following statements showing that
$\cos (s + \pi) = -\cos s$ and $\sin (s + \pi) = -\sin s$:
The length of $\overset{\frown}{APQ}$ of the unit circle shown at
the right is __?__. Therefore, in terms of s, the
coordinates of Q are (__?__, __?__). Therefore,
__?__ $= -x = -\cos s$, and __?__ $= -y = -\sin s$.

$P(\cos s, \sin s)$

(x, y) s

$A(1, 0)$

$(-x, -y)$

$Q(?, ?)$

Tell the value of each of the following functions.
If the function is undefined for the given
number, say so.

15. $\sin \dfrac{\pi}{2}$

16. $\tan \dfrac{3\pi}{2}$

17. $\cos (-\pi)$

18. $\cot \left(-\dfrac{\pi}{2}\right)$

Written Exercises

For some number s, the point P is s units from $A(1, 0)$ along the unit circle $x^2 + y^2 = 1$. Find the exact values of the six trigonometric functions of s.

A **1.** $P\left(\frac{3}{5}, -\frac{4}{5}\right)$ **2.** $P\left(-\frac{5}{13}, \frac{12}{13}\right)$ **3.** $P\left(\frac{\sqrt{7}}{4}, \frac{3}{4}\right)$ **4.** $P\left(-\frac{2}{5}, -\frac{\sqrt{21}}{5}\right)$

Find the sine, cosine, and tangent of each number to four decimal places.

5. 1.25 **6.** 1.4 **7.** 0.25 **8.** 0.7

Find the exact values of the six trigonometric functions of the given number. If a function is undefined for the number, say so.

9. $\frac{\pi}{3}$ **10.** $\frac{\pi}{4}$ **11.** $\frac{3\pi}{2}$ **12.** $\frac{3\pi}{4}$ **13.** $\frac{11\pi}{6}$

14. π **15.** $\frac{8\pi}{3}$ **16.** $\frac{15\pi}{4}$ **17.** $-\frac{5\pi}{6}$ **18.** $-\frac{3\pi}{4}$

Find the number x if $0 \le x \le \frac{\pi}{2}$ and x has the given function value.

B **19.** $\cos x = 0.7248$ **20.** $\csc x = 1.018$ **21.** $\sin x = \frac{\sqrt{3}}{2}$ **22.** $\tan x = 1$

Verify the identity $\cos^2 s + \sin^2 s = 1$ for the given value of s.

23. $\frac{3\pi}{4}$ **24.** $\frac{7\pi}{6}$ **25.** 0.7 **26.** 1

27. Explain how to use Table 6 to find an approximate value for sin 7.

In Exercises 28–31, use the diagram at the right.

28. Show that if $0 < s < \frac{\pi}{2}$, then $\tan s = AQ$.
(*Hint:* $\triangle OAQ \sim \triangle OMP$.)

29. Show that if $0 < s < \frac{\pi}{2}$, then $\sec s = OQ$.
(See the hint for Exercise 28.)

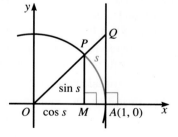

C **30.** Show that if $0 < s < \frac{\pi}{2}$, then $\sin s < s$.

31. Show that if $0 < s < \frac{\pi}{2}$, then $s < \tan s$.
(*Hint:* Compare the areas of sector OAP and $\triangle OAQ$.)

32. Show that if $0 < s < \frac{\pi}{2}$, then $1 < \frac{s}{\sin s} < \frac{1}{\cos s}$.
(*Hint:* Start by combining the inequalities in Exercises 30 and 31.)

Maria Goeppert Mayer (1906–1972) was the first woman to receive the Nobel Prize for theoretical physics. She began her career at the University of Göttingen as a mathematics student, but she soon started working on quantum mechanics.

In the same year that she married the American chemist Joseph Mayer, she finished her Ph.D. dissertation on the theory of light quanta.

After the Mayers moved to Baltimore, Maria was unable to get a university appointment. Instead, she decided to help other scientists without pay.

In early 1946 the Mayers accepted appointments at the University of Chicago. There Maria was surrounded by some of the greatest names in physics, and she set out to learn nuclear theory and study the origin of elements.

Along with the physicist Hans Jensen she wrote the book *Elementary Theory of Nuclear Shell Structure*.

One of the most famous questions in the history of mathematics, asked by the Greeks about 300 B.C., was the following: Given an arbitrary angle, can this angle be *trisected* (that is, subdivided into three equal angles) with only a straightedge and a compass? This question remained unanswered until the 19th century, when the following apparently unrelated theorem was proved:

> From a given unit length, a segment whose length is a root of a cubic equation with no rational root cannot be constructed.

By a trigonometric identity: $\cos 3\theta = 4 \cos^3 \theta - 3 \cos \theta$.
If $\theta = 20°$, then $3\theta = 60°$ and $\cos 3\theta = \frac{1}{2}$. Therefore,
$$\tfrac{1}{2} = 4 \cos^3 20° - 3 \cos 20°.$$
If you let $x = \cos 20°$, this equation becomes
$$0 = 8x^3 - 6x - 1.$$

But by applying the rational root theorem (page 382), you find that this equation has no rational root. Therefore, by the theorem above, a segment whose length is a root of this equation, such as $\cos 20°$, cannot be constructed. If you could trisect a 60° angle, you could construct a 20° angle, and by dropping a perpendicular to one side of this angle from a point on the other side a unit distance from the vertex, you could construct a segment of length $\cos 20°$. It follows that a 60° angle cannot be trisected.

13-3 Periodicity and Symmetry

Objective   To use periodicity and symmetry in graphing functions.

Because the circumference of the unit circle is 2π, the numbers s and $s + 2\pi$ are associated with the same point $P(x, y)$ on the unit circle, as shown in the diagram at the right. Therefore,

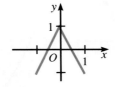

$$\sin (s + 2\pi) = \sin s \quad \text{and} \quad \cos (s + 2\pi) = \cos s.$$

These identities tell you that the sine and cosine functions have *period* 2π.

In general, a function is **periodic** if for some positive constant p,

$$f(x + p) = f(x)$$

for every x in the domain of f. The smallest such p is the **period** of f.

Example 1   Part of the graph of a function f having period 2 is shown. Graph f in the interval $-5 \le x \le 5$.

Solution   Since the given part of the graph covers one period, repeat it to the right until you reach 5 and to the left until you reach -5.

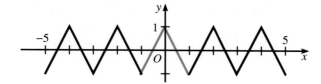

From the diagram at the right you can see that

$$\sin (-s) = -y = -\sin s$$

and $$\cos (-s) = x = \cos s.$$

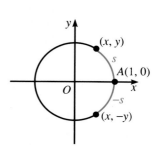

These identities tell you that the sine is an *odd* function and that the cosine is an *even* function.

In general,

f is an **odd function** if $f(-x) = -f(x)$

and f is an **even function** if $f(-x) = f(x)$

for every x in the domain of f. Unlike an integer, a function does not have to be even or odd; it can be neither.

Example 2 Determine whether each function is even, odd, or neither.

a. $f(x) = \dfrac{x^2}{x^4 + 1}$ **b.** $g(x) = x^3 - 4x$ **c.** $h(x) = x^2 + x$

Solution **a.** $f(-x) = \dfrac{(-x)^2}{(-x)^4 + 1} = \dfrac{x^2}{x^4 + 1} = f(x)$

∴ f is even. **Answer**

b. $g(-x) = (-x)^3 - 4(-x) = -x^3 + 4x = -(x^3 - 4x) = -g(x)$

∴ g is odd. **Answer**

c. $h(-x) = (-x)^2 + (-x) = x^2 - x$

Notice that $x^2 - x \neq x^2 + x$ and $x^2 - x \neq -(x^2 + x)$.

∴ h is neither even nor odd. **Answer**

The diagrams below illustrate the effect of evenness or oddness on the graph of a function.

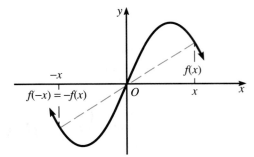

When f is even, the graph is symmetric with respect to the y-axis. That is, if the point (x, y) is on the graph of f, then so is the point $(-x, y)$.

When f is odd, the graph is symmetric with respect to the origin. That is, if the point (x, y) is on the graph of f, then so is the point $(-x, -y)$.

Example 3 Part of the graph of a function f is shown at the right. Graph f in the interval $-3 \leq x \leq 3$ assuming that f is **(a)** even and **(b)** odd.

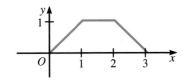

Solution **a.** Reflect the given part of the graph in the y-axis.

b. Reflect the given part of the graph in the origin.

Oral Exercises

Tell the period of the periodic function whose graph is shown. Is the function even, odd, or neither?

1.

2.

3.

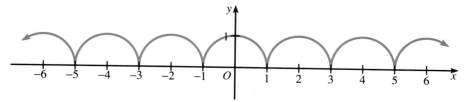

Tell whether each function is even, odd, or neither.

4. $g(x) = 2x^3 - 3x$

5. $h(x) = 2x^4 - 3x^2 + 5$

6. $f(x) = 4x^3 - 2x^2$

7. $t(x) = |x|$

8. $s(x) = \dfrac{x}{x^2 + 1}$

9. $r(x) = x^4 - 2x^3 + x^2 - 2$

Written Exercises

Part of the graph of a function f having the given period p is shown. Graph f in the indicated interval and tell whether f is even, odd, or neither.

A **1.**

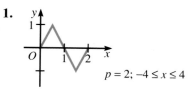

$p = 2; -4 \le x \le 4$

2.

$p = 2; -4 \le x \le 4$

3.

$p = 3; -6 \le x \le 6$

4.

$p = 4; -8 \le x \le 8$

5.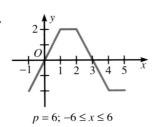

$p = 6; -6 \le x \le 6$

6.

$p = 5; -10 \le x \le 10$

Part of the graph of a function f is given. Graph f in the interval $-3 \le x \le 3$ assuming f is (a) even and (b) odd.

7.

8.

9.

10.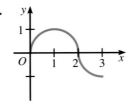

Determine whether each function f is even, odd, or neither.

11. $f(x) = x^3 + x$

12. $f(x) = x^3 + 5x + 3$

13. $f(x) = x^4 + x^3$

14. $f(x) = x^4 + 2x^2 + 1$

15. $f(x) = \dfrac{x^2}{x^2 + 4}$

16. $f(x) = x\sqrt{x^2 + 1}$

B 17. $f(x) = x \sin x$

18. $f(x) = x \cos x$

19. $f(x) = \sin x \cos x$

20. $f(x) = \sin x + \cos x$

21. Show that the product of two odd functions is even.

22. Show that the product of an even function and an odd function is odd.

23. Show that if two functions have period p then the function that is their sum has a period that divides p. (*Hint:* Let $h(x) = f(x) + g(x)$ where f and g have period p.)

24. Show that the sum of an even function and an odd function is neither even nor odd.

25. Show that $\sin 2x$ has period π. (*Hint:* Find the smallest positive p such that $\sin 2(x + p) = \sin 2x$.)

26. Show that $\cos 2x$ has period π. (See the hint for Exercise 25.)

27. Show that $\cos \frac{1}{2}x$ has period 4π.   28. Show that $\sin \frac{1}{2}x$ has period 4π.

C 29. Show that if $f(x)$ has period p, then $f(kx)$ has period $\frac{p}{k}$ for any $k > 0$.

30. For any function $f(x)$, define two functions $O(x)$ and $E(x)$ as follows:

$$O(x) = \frac{f(x) - f(-x)}{2} \quad \text{and} \quad E(x) = \frac{f(x) + f(-x)}{2}$$

  a. Show that $O(x)$ is odd and $E(x)$ is even.
  b. Show that $O(x) + E(x) = f(x)$. (Any function can be written as the sum of an odd function and an even function in this way.)
  c. Find $O(x)$ and $E(x)$ if $f(x) = x^3 - 3x^2 + 2x + 1$.
  d. Find $O(x)$ and $E(x)$ if $f(x) = e^x$. (These functions are called the hyperbolic sine and cosine of x and are written $\sinh x$ and $\cosh x$.)

Mixed Review Exercises

Graph each equation.

1. $y = x^2 + 6x + 11$ 2. $y = 2^x - 1$ 3. $4x - 3y = 12$
4. $y = \log_{1/2} x$ 5. $x^2 + y^2 + 2y = 8$ 6. $9x^2 - y^2 = 9$

Solve.

7. $\sqrt{7x + 4} = x + 2$ 8. $2x + 7 = 1 - x$ 9. $x^3 - 2x^2 + 2x - 4 = 0$
10. $\frac{x^2}{12} + 1 = \frac{2x}{3}$ 11. $2x^2 + x - 10 = 0$ 12. $\frac{1}{x - 1} + \frac{1}{x} = \frac{3}{x^2 - x}$

Challenge

The function $f(x) = \sin \frac{\pi}{2}x$ has period 4, and the function $g(x) = \cos \frac{\pi}{3}x$ has period 6. Is the function $h(x) = f(x) + g(x)$ periodic? If so, what is its period?

13-4 Graphs of the Sine and Cosine

Objective   To graph the sine, cosine, and related functions.

To obtain the graphs of $y = \sin x$ and $y = \cos x$, you can use a computer or a graphing calculator. Otherwise a table of values for $0 \le x \le 2\pi$ is needed.

x	0	$\frac{\pi}{6}$	$\frac{\pi}{3}$	$\frac{\pi}{2}$	$\frac{2\pi}{3}$	$\frac{5\pi}{6}$	π	$\frac{7\pi}{6}$	$\frac{4\pi}{3}$	$\frac{3\pi}{2}$	$\frac{5\pi}{3}$	$\frac{11\pi}{6}$	2π
	0	0.52	1.05	1.57	2.09	2.62	3.14	3.67	4.19	4.71	5.24	5.76	6.28
$\sin x$	0	0.50	0.87	1	0.87	0.50	0	−0.50	−0.87	−1	−0.87	−0.50	0
$\cos x$	1	0.87	0.50	0	−0.50	−0.87	−1	−0.87	−0.50	0	0.50	0.87	1

The graphs shown below are obtained by plotting the ordered pairs $(x, \sin x)$ and $(x, \cos x)$ from the table above and by joining the points with smooth curves.

Since both the sine and cosine functions have period 2π, their complete graphs consist of the above graphs repeated over and over, as shown below.

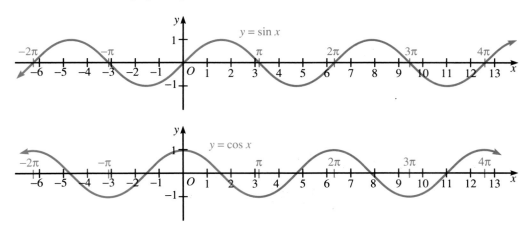

Notice that the sine curve is symmetric with respect to the origin, since sine is an odd function. The cosine curve is symmetric with respect to the y-axis, since cosine is an even function. Notice also that if the cosine curve were shifted $\frac{\pi}{2}$ units to the right, it would coincide with the sine curve.

If you have a computer or a graphing calculator, you can investigate the graphs of functions of the form $y = c + \sin x$, such as those shown at the right. Notice that the graph of the sine function is shifted upward c units if $c > 0$, and it is shifted downward $|c|$ units if $c < 0$. (This also applies to functions of the form $y = c + \cos x$.)

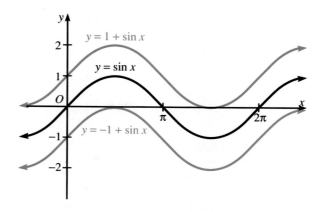

The diagram at the right shows the graphs of three functions of the form $y = a \sin x (a > 0)$. In each case the value of the function varies from a minimum value of $-a$ to a maximum value of a. (This also applies to functions with the form $y = a \cos x$.) The positive number a is called the **amplitude** of $a \sin x$ (or $a \cos x$). Therefore, the amplitude of $2 \sin x$ is 2, of $\sin x$ is 1, and of $\frac{1}{2} \sin x$ is $\frac{1}{2}$.

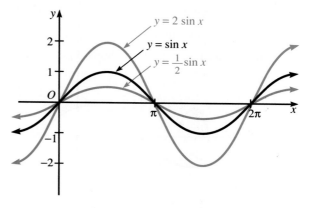

Example 1 Graph $y = 2 + 3 \sin x$.

Solution The sine curve is shifted upward 2 units, and its amplitude is 3.

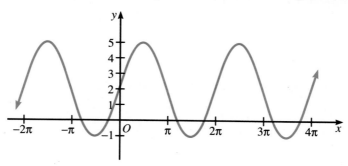

Notice in the solution of Example 1 that the function has a maximum value of 5 and a minimum value of -1. In general, any function of the form $y = c + a \sin x$ (or $y = c + a \cos x$) has maximum value $M = c + a$ when $\sin x = 1$ (or $\cos x = 1$) and minimum value $m = c - a$ when $\sin x = -1$ (or $\cos x = -1$).

Since

$$\frac{M + m}{2} = \frac{(c + a) + (c - a)}{2} = \frac{2c}{2} = c,$$

the average of the extreme values is the vertical shift in the sine (or cosine) curve. Also, since

$$\frac{M - m}{2} = \frac{(c + a) - (c - a)}{2} = \frac{2a}{2} = a,$$

half the difference between the extreme values is the amplitude of the curve.

While the a in $y = a \sin x$ causes a *vertical* stretch or shrinking in the graph of $y = \sin x$, the b in $y = \sin bx$ causes a *horizontal* stretch or shrinking. In other words, the coefficient b affects the period.

For example, to find the period of $f(x) = \sin 3x$, look at $f(x + p)$: $f(x + p) = \sin 3(x + p) = \sin (3x + 3p)$. Then $f(x + p) = f(x)$ if and only if $\sin (3x + 3p) = \sin 3x$. This equation holds if $3p = 2\pi$. Therefore, $p = \frac{2\pi}{3}$.

The graph of $y = \sin 3x$ is shown below with the graph of $y = \sin x$.

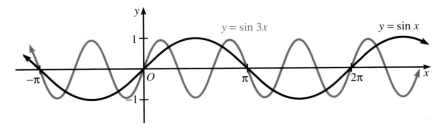

For positive real numbers a and b, the functions $f(x) = a \sin bx$ and $f(x) = a \cos bx$ have amplitude a and period $\frac{2\pi}{b}$.

Example 2 Graph $y = 2 \cos \frac{\pi}{2}x$.

Solution The amplitude is 2, and the period is $\frac{2\pi}{\frac{\pi}{2}}$, or 4.

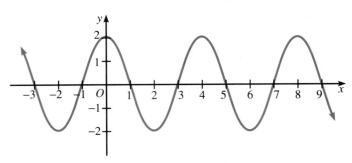

Example 3 Each graph shown below has an equation of the form $y = c + a \sin bx$ or $y = c + a \cos bx$. Find an equation of each graph.

a.

b.

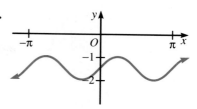

Solution **a.** The graph is the same as the one in Example 2 except that it has been shifted upward 1 unit.

∴ the equation is $y = 1 + 2 \cos \dfrac{\pi}{2}x.$ **Answer**

b. Since the graph shows a function with period π, $\dfrac{2\pi}{b} = \pi$ and $b = 2$. Since the maximum value M is -1 and the minimum value m is -2,

$$c = \frac{M + m}{2} = \frac{-1 + (-2)}{2} = -\frac{3}{2},$$

and

$$a = \frac{M - m}{2} = \frac{-1 - (-2)}{2} = \frac{1}{2}.$$

Finally, since the graph is halfway between the maximum and minimum values of the function at $x = 0$, the graph is a sine curve.

∴ the equation is $y = -\dfrac{3}{2} + \dfrac{1}{2} \sin 2x.$ **Answer**

Oral Exercises

Identify each curve as sine or cosine. Then give its amplitude and period.

1.

2.

3.

4.

5.

6.

7.

8.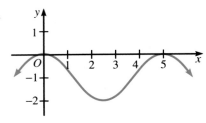

9. Explain why the b in $y = \sin bx$ has no effect on the maximum and minimum values of the sine function.

In Exercises 10–13, match each description with the letter of the function that fits the description.

10. The amplitude is 3.

11. The period is $\frac{\pi}{2}$.

12. The minimum value is 1.

13. The period is 2.

a. $y = 3 + \sin 2\pi x$

b. $y = 2 \cos 4x$

c. $y = 2 + \sin 3x$

d. $y = 3 \cos 2x$

e. $y = 1 + \sin \pi x$

Written Exercises

In Exercises 1–12:
a. Find the amplitude of each function.
b. Find the maximum and minimum values.
c. Find the period.

A　**1.** $y = 2 \cos x$

2. $y = \frac{1}{2} \sin x$

3. $y = \sin 4x$

4. $y = \cos 2x$

5. $y = 3 \sin 3x$

6. $y = \frac{1}{2} \sin 2x$

7. $y = \frac{1}{2} \sin 2\pi x$

8. $y = 2 \cos \pi x$

9. $y = 2 \cos \frac{\pi}{2} x$

10. $y = 3 \sin \frac{\pi}{4} x$

11. $y = \cos 2\pi x - 1$

12. $y = 2 \sin \pi x + 3$

13–20. Find an equation of each curve pictured in Oral Exercises 1–8.

Find an equation of the form $y = c + a \sin bx$ that satisfies the given conditions. (M is the maximum value of the function y, and m is the minimum value.)

Sample $M = 5$, $m = 1$, period π

Solution   1. $a = \dfrac{M - m}{2} = \dfrac{5 - 1}{2} = 2$

2. $c = \dfrac{M + m}{2} = \dfrac{5 + 1}{2} = 3$

3. $\dfrac{2\pi}{b} = \pi$, so $b = 2$

$\therefore y = 3 + 2 \sin 2x$.   *Answer*

21. $M = 3$, $m = -3$, period 2π **22.** $M = 1$, $m = -1$, period 8π

23. $M = 2$, $m = -2$, period 2 **24.** $M = 2$, $m = -2$, period 3

25. $M = 7$, $m = 1$, period π **26.** $M = 0$, $m = -6$, period 4

27–32. Graph each function in Exercises 1–6. Show at least two periods. You may wish to verify your graphs on a computer or a graphing calculator.

B   **33–38.** Graph each function in Exercises 7–12. Show at least two periods. You may wish to verify your graphs on a computer or a graphing calculator.

In a function of the form $y = \sin (x - d)$ or $y = \cos (x - d)$, the number d determines the *phase shift* of the graph. In general, for any function f, the graph of $y = f(x - d)$ is obtained by shifting the graph of $y = f(x)$ to the right d units if $d > 0$ or to the left $|d|$ units if $d < 0$.

The graph of $y = -f(x)$ is obtained by reflecting the graph of $y = f(x)$ in the x-axis.

Use the information given in the preceding paragraphs to sketch the graph of each of the following equations. Show at least two periods.

39. $y = \cos \left(x - \dfrac{\pi}{4} \right) - 1$ **40.** $y = -2 \sin 2x$ **41.** $y = -\cos 2 \left(x + \dfrac{\pi}{3} \right)$

C   **42.** $y = -\sin (2x - \pi) + 1$ **43.** $y = -2 \cos (4x + \pi) + \dfrac{1}{2}$ **44.** $y = -2 \cos (2\pi x - \pi) + 4$

45. Show that the graph of $y = \cos \left(x - \dfrac{\pi}{2} \right)$ is a sine curve.

46. Explain how the graphs of $y = \sin x$ and $y = \cos \left(x + \dfrac{\pi}{2} \right)$ are related.

47. Use the equation $\sin 2x = 2 \sin x \cos x$ to show that the graph of $y = \sin x \cos x$ is a sine curve. Sketch the graph.

48. Use the equation $\cos 2x = 1 - 2 \sin^2 x$ to show that the graph of $y = \sin^2 x$ is a cosine curve. Sketch the graph.

13-5 Graphs of the Other Functions

Objective   To graph the tangent, cotangent, secant, cosecant, and related functions.

The tangent function is odd, has period π, and has asymptotes. These properties are investigated below and will be useful in graphing the function.

Example 1   Determine whether the tangent function is even, odd, or neither.

Solution   Since sine is an odd function and cosine is even,

$$\tan (-x) = \frac{\sin (-x)}{\cos (-x)} = \frac{-\sin x}{\cos x} = -\tan x.$$

$\therefore$ tangent is an odd function.   *Answer*

In addition to being odd, tangent has period π. To see why, use the fact that $\sin (x + \pi) = -\sin x$ and $\cos (x + \pi) = -\cos x$ (see Oral Exercise 14, page 616):

$$\tan (x + \pi) = \frac{\sin (x + \pi)}{\cos (x + \pi)} = \frac{-\sin x}{-\cos x} = \tan x$$

Another property of tangent comes from the fact that $\tan x$ is undefined when $\cos x = 0$ $\left(\text{that is, when } x = \frac{k\pi}{2} \text{ where } k \text{ is}\right.$ an odd integer$\left.\right)$. To see what happens to $\tan x$ as x *approaches* $\frac{\pi}{2}$, for example, look at the diagram. For $0 < x < \frac{\pi}{2}$, you know $\tan x = AQ$ (see Exercise 28 on page 617). The nearer x is to $\frac{\pi}{2}$, the closer P is to B, and the larger AQ is. Therefore,

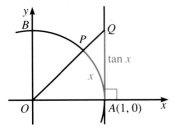

as x approaches $\frac{\pi}{2}$, $\tan x$ increases without bound.

A short table of values is used to start the graph of $y = \tan x$ in the figure below. The dashed line is an *asymptote* of the graph (recall Lesson 9-5).

x	$\tan x$
0	0
$\frac{\pi}{6} \approx 0.52$	0.58
$\frac{\pi}{4} \approx 0.79$	1.00
$\frac{\pi}{3} \approx 1.05$	1.73
$\frac{\pi}{2} \approx 1.57$	—

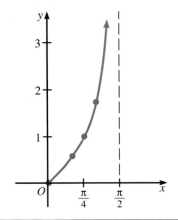

Since tangent is odd, the graph on the previous page can be reflected in the origin to obtain the graph of one period of the function, as shown in Figure 1. The complete graph of the tangent function, shown in Figure 2, is obtained by using the fact that the period of the tangent function is π.

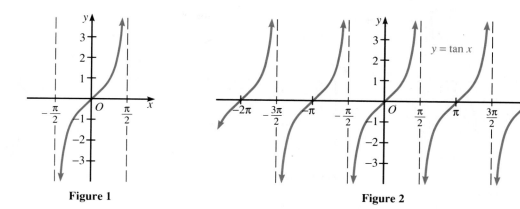

Figure 1   Figure 2

You can use a computer or a graphing calculator to explore the graphs of functions of the form $y = c + a \tan bx$, for various values of a, b, and c. Remember (see Lesson 13-4) that the number c produces a vertical shift in the curve, a produces a vertical stretching or shrinking, and b produces a change in the period.

Example 2   Graph one period of the function $y = 2 \tan \frac{1}{2}x$.

Solution   1. Since $b = \frac{1}{2}$, the period of the function is $\frac{\pi}{\frac{1}{2}} = 2\pi$.

Asymptotes occur when $\frac{1}{2}x = \frac{\pi}{2}$ and $\frac{1}{2}x = -\frac{\pi}{2}$. Solving these equations gives asymptotes at $x = \pi$ and $x = -\pi$.

2. Since $a = 2$, there is a vertical stretch of the tangent curve. For example, since $\tan x = 1$ when $x = \frac{\pi}{4}$, $2 \tan \frac{1}{2}x = 2$ when $\frac{1}{2}x = \frac{\pi}{4}$ (that is, when $x = \frac{\pi}{2}$). Similarly, $2 \tan \frac{1}{2}x = -2$ when $x = -\frac{\pi}{2}$.

The graph of one period of the function is shown at the right.

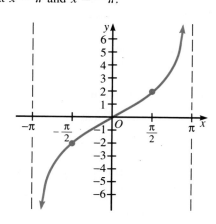

To graph the cotangent, secant, and cosecant functions or variations of them, you can use methods similar to those discussed in this lesson for tangent. The graphs of $y = \cot x$, $y = \sec x$, and $y = \csc x$ are shown below in red along with the graphs of their reciprocal functions in black.

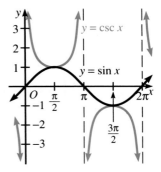

Notice from the graphs that cotangent has period π and that secant and cosecant have period 2π.

Oral Exercises

Give the period of each function.

Sample $y = \csc 4x$ **Solution**   Since $b = 4$, the period is $\frac{2\pi}{4} = \frac{\pi}{2}$.

1. $y = \sec 2x$ **2.** $y = 3 \csc x$ **3.** $y = \frac{1}{2} \cot x$ **4.** $y = \tan \frac{1}{3}x$

5. $y = 2 \csc \pi x$ **6.** $y = \cot 2x + 1$ **7.** $y = \sec \frac{1}{2}x - 1$ **8.** $y = 3 \tan \frac{\pi}{4}x$

For which numbers x, $0 \le x < 2\pi$, are the following functions *not* defined?

9. $\tan x$ **10.** $\sec x$ **11.** $\cot x$ **12.** $\csc x$

For each description, give the letter of the function that has a graph fitting that description.

13. The graph has $x = \frac{\pi}{2}$ as an asymptote and passes through the origin.

14. The graph is symmetric in the y-axis and passes through the point $(0, 3)$.

15. The graph does not intersect either axis.

16. The graph has the y-axis and the line $x = \frac{\pi}{2}$ as asymptotes.

a. $y = \cot 2x$

b. $y = \tan \frac{x}{2}$

c. $y = 3 \sec x$

d. $y = 3 \csc x$

e. $y = 2 \tan x$

f. $y = 2 \cot \pi x$

Written Exercises

Use the fact that the cosine is even and the sine is odd to determine if the following are even, odd, or neither.

A **1.** $\cot x$ **2.** $\sec x$ **3.** $\csc x$ **4.** $\tan x + \sec x$

5. $\sec x - \csc x$ **6.** $x \csc x$ **7.** $x^2 \cot x$ **8.** $\sec x \tan x$

9–16. Sketch the graph of at least one period of the functions given in Oral Exercises 1–8.

17. Does the tangent function have an amplitude? Justify your answer.

18. Does the secant function have an amplitude? Justify your answer.

Graph each pair of functions in the same coordinate plane for $-\frac{\pi}{2} < x < \frac{\pi}{2}$. You may wish to verify your graphs on a computer or a graphing calculator.

B **19.** $y = \tan x$; $y = -\tan x$ **20.** $y = \sec x$; $y = -\sec x$

21. Graph $x = \tan y$ for $-\frac{\pi}{2} < y < \frac{\pi}{2}$.   **22.** Graph $x = \cot y$ for $0 < y < \pi$.

Graph each of the following over the interval $0 \le x \le 2\pi$. (*Hint:* See Exercises 39–44 on page 629.)

23. $y = \tan\left(x - \frac{\pi}{6}\right)$ **24.** $y = \cot\left(x - \frac{\pi}{4}\right)$ **25.** $y = \csc\left(x + \frac{\pi}{2}\right)$

26. $y = \sec\left(x - \frac{\pi}{4}\right)$ **27.** $y = \cot\left(x - \frac{\pi}{4}\right) + 1$ **28.** $y = \tan\left(x + \frac{\pi}{4}\right) - 1$

C **29.** Graph $y = \sin x$, $y = x$, and $y = \tan x \left(0 < x < \frac{\pi}{2}\right)$ in the same coordinate plane. Explain why the resulting figure illustrates the inequality $\sin x < x < \tan x$. (See also Exercises 30 and 31 on page 617.)

Mixed Review Exercises

Give the exact values of the six trigonometric functions of each number. If a function is not defined for the number, say so.

1. $\frac{5\pi}{2}$ **2.** $-\frac{4\pi}{3}$ **3.** $\frac{19\pi}{6}$ **4.** $-\frac{5\pi}{4}$

Evaluate if $a = -2$ and $b = 4$.

5. $\frac{a - b}{a + b}$ **6.** $\log_b{(-a)}$ **7.** $\sum_{n=1}^{5} b \cdot a^{n-1}$ **8.** $b^{1/a}$

Self-Test 1

Vocabulary   radian measure (p. 607)   periodic, period (p. 619)
          arc length (p. 608)   odd function (p. 619)
          area of a sector (p. 608)   even function (p. 619)
          circular functions (p. 613)   amplitude (p. 625)

1. Express in radians.   Obj. 13-1, p. 607
   a. 165° (Leave your answer in terms of π.)
   b. 256° (Give your answer to the nearest hundredth of a radian.)

2. Express in degrees. Give your answers to the nearest tenth of a degree.
   a. $-\dfrac{5\pi}{6}$ **b.** 2.5

3. Find the exact values of the six trigonometric functions of $\dfrac{3\pi}{4}$.   Obj. 13-2, p. 613

4. Find the sine, cosine, and tangent of 1.44 to four decimal places.

5. Determine whether each function is even, odd, or neither.   Obj. 13-3, p. 619
   a. $f(x) = x^4 - 3x^2 + 5$ **b.** $g(x) = 2x^2 + 6x + 4$

6. Find the amplitude, the maximum and minimum values, and   Obj. 13-4, p. 624
   the period of $y = 1 + 3 \sin 2x$. Then sketch the graph, showing at least two periods.

7. Sketch the graph of $y = \dfrac{3}{2} \cot \dfrac{\pi}{2} x$ over the interval $0 < x < 2$.   Obj. 13-5, p. 630

Check your answers with those at the back of the book.

Application / *Frequencies in Music*

The musical tone produced by a vibrating string, a guitar string for example, can be represented mathematically by a graph. The y-coordinates of the graph represent the displacement of a fixed point in the middle of the string at time t. If you ignore the complicated secondary motions of the string, this graph will be a *sinusoid,* a graph resembling a sine or cosine curve.

The fundamental period of the tone graphed at the top of the next page is $\frac{1}{440}$ s, and its *frequency,* defined as the reciprocal of the period, is therefore 440 Hz (hertz, or cycles per second).

The frequency of a vibrating string is determined by its tension, which the guitarist changes when tuning the instrument, and by its length, which the guitarist controls while playing by pressing the string against one of the metal ridges, or *frets,* on the neck of the guitar.

The C major scale beginning at middle C ends on C′, the note one octave above. As the graphs of C and C′ indicate, the fundamental period of C′ is exactly half that of C, and thus its frequency (528 Hz) is exactly twice that of C (264 Hz). The table below shows other relationships among the frequencies of the C major scale. On the left are listed the names of each pair of notes, or *interval.*

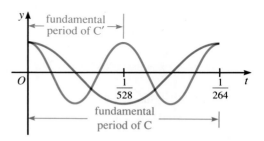

Name of note	C	D	E	F	G	A	B	C′	Ratio of frequencies
Frequency (Hz)	264	297	330	352	396	440	495	528	
Octave									2:1
Fifth									3:2
Fourth									4:3
Major third									5:4
Minor third									6:5

Exercises

1. Suppose a string tuned to F vibrates with an amplitude of 0.5 mm. Write an equation of the form $y = a \cos bt$, where y is in millimeters and t is in seconds, for the motion of a point in the middle of the string.

2. The frequency of vibration of a string with a given tension is inversely proportional to its length. For example, if you halve the length of the string, its frequency doubles and its pitch goes up by one octave.
   a. A guitar string tuned to G (396 Hz) is shortened to $\frac{3}{4}$ of its original length. What will its frequency be? What note will it play?
   b. A guitar string 30 cm long is tuned to E (330 Hz). How long is the vibrating part of the string when the guitarist plays G on this string?

Trigonometric Identities

13-6 The Fundamental Identities

Objective   To simplify trigonometric expressions and to prove identities.

You learned earlier that a trigonometric identity is an equation such as
$$\sin^2 \alpha + \cos^2 \alpha = 1$$
that holds for *all* values of α, except those for which either side of the equation is undefined. This is true of all trigonometric identities. You are already familiar with most of the *fundamental identities* listed below.

The Reciprocal Identities

$$\sin \alpha = \frac{1}{\csc \alpha} \qquad\qquad \sin \alpha \csc \alpha = 1 \qquad\qquad \csc \alpha = \frac{1}{\sin \alpha}$$

$$\cos \alpha = \frac{1}{\sec \alpha} \qquad\qquad \cos \alpha \sec \alpha = 1 \qquad\qquad \sec \alpha = \frac{1}{\cos \alpha}$$

$$\tan \alpha = \begin{cases} \dfrac{1}{\cot \alpha} \\[2mm] \dfrac{\sin \alpha}{\cos \alpha} \end{cases} \qquad \tan \alpha \cot \alpha = 1 \qquad \cot \alpha = \begin{cases} \dfrac{1}{\tan \alpha} \\[2mm] \dfrac{\cos \alpha}{\sin \alpha} \end{cases}$$

The Cofunction Identities

$$\sin \theta = \cos (90° - \theta) \qquad\qquad \cos \theta = \sin (90° - \theta)$$
$$\tan \theta = \cot (90° - \theta) \qquad\qquad \cot \theta = \tan (90° - \theta)$$
$$\sec \theta = \csc (90° - \theta) \qquad\qquad \csc \theta = \sec (90° - \theta)$$

The Pythagorean Identities

$$\sin^2 \alpha + \cos^2 \alpha = 1 \qquad\qquad 1 + \tan^2 \alpha = \sec^2 \alpha \qquad\qquad 1 + \cot^2 \alpha = \csc^2 \alpha$$

The reciprocal, cofunction, and Pythagorean identities can be used to simplify trigonometric expressions and to prove other identities.

Example 1   Simplify $\cos \alpha + \tan \alpha \sin \alpha$.

Solution   Write the functions in terms of sines and cosines. Then simplify.

$$\cos \alpha + \tan \alpha \sin \alpha = \cos \alpha + \frac{\sin \alpha}{\cos \alpha} \sin \alpha = \frac{\cos^2 \alpha + \sin^2 \alpha}{\cos \alpha} = \frac{1}{\cos \alpha} = \sec \alpha$$

Because $\tan \alpha$ is not defined if $\cos \alpha = 0$, the trigonometric identity $\cos \alpha + \tan \alpha \sin \alpha = \sec \alpha$ is valid only if $\cos \alpha \neq 0$, that is, only if $\alpha \neq 90° + n \cdot 180°$, where n is an integer. In general, such restrictions are understood to apply even though they are not specifically mentioned.

Example 2 Write $\tan x$ in terms of $\sin x$.

Solution You know $\tan x = \dfrac{\sin x}{\cos x}$. You must now find $\cos x$ in terms of $\sin x$. Use the first Pythagorean identity.

$$\sin^2 x + \cos^2 x = 1$$
$$\cos^2 x = 1 - \sin^2 x$$
$$\cos x = \pm \sqrt{1 - \sin^2 x}$$

$$\therefore \tan x = \frac{\sin x}{\cos x} = \pm \frac{\sin x}{\sqrt{1 - \sin^2 x}}. \quad \textbf{\textit{Answer}}$$

The value of x determines which sign is used in the answer to Example 2. If x is the measure of a first- or fourth-quadrant angle, $\sin x$ and $\tan x$ have the same sign and the $+$ is used. If x is the measure of a second- or third-quadrant angle, $\sin x$ and $\tan x$ have opposite signs and the $-$ is used.

You can usually use one of two general strategies in proving identities.

General Strategies for Proving Identities

1. Simplify the more complicated side of the identity until it is identical to the other side.
2. Transform both sides of the identity into the same expression.

Caution: It is incorrect to "work across the $=$ sign." For example, do not cross multiply. To do so assumes that what is being proved is true.

Here are some special strategies that can be helpful in proving identities.

Special Strategies for Proving Identities

1. Express functions in terms of sines and cosines (see Example 1).
2. Look for expressions to which the Pythagorean identities can be applied. For example: $\tan^2 x + 1$ $(= \sec^2 x)$, and $1 - \sin^2 x$ $(= \cos^2 x)$.
3. Use factoring. For example: $\sin^2 x = 1 - \cos^2 x = (1 + \cos x)(1 - \cos x)$.
4. Combine terms on each side of the identity into a single fraction.
5. Multiply one side of the equation by an expression equal to 1 (see Example 4).

Example 3   Prove that $\tan^2 \alpha + \csc^2 \alpha - \cot^2 \alpha = \sec^2 \alpha$.

Solution   Use general strategy 1.

$$\tan^2 \alpha + \csc^2 \alpha - \cot^2 \alpha \overset{?}{=} \sec^2 \alpha$$

$\tan^2 \alpha + \dfrac{1}{\sin^2 \alpha} - \dfrac{\cos^2 \alpha}{\sin^2 \alpha}$	Use the reciprocal identities.
$\tan^2 \alpha + \dfrac{1 - \cos^2 \alpha}{\sin^2 \alpha}$	Use special strategy 4.
$\tan^2 \alpha + \dfrac{\sin^2 \alpha}{\sin^2 \alpha}$	Use special strategy 2.
$\tan^2 \alpha + 1$	Simplify.
$\sec^2 \alpha \qquad = \sec^2 \alpha$	Use special strategy 2.

Example 4   Prove that $\dfrac{1 - \cos t}{\sin t} = \dfrac{\sin t}{1 + \cos t}$.

Solution   Remember, you cannot cross multiply. Use special strategy 5 and multiply the left side by 1 in the form $\dfrac{1 + \cos t}{1 + \cos t}$.

$$\dfrac{1 - \cos t}{\sin t} \overset{?}{=} \dfrac{\sin t}{1 + \cos t}$$

$\dfrac{(1 - \cos t)(1 + \cos t)}{\sin t(1 + \cos t)}$	Use special strategy 5.
$\dfrac{1 - \cos^2 t}{\sin t(1 + \cos t)}$	Use FOIL.
$\dfrac{\sin^2 t}{\sin t(1 + \cos t)}$	Use special strategy 2.
$\dfrac{\sin t}{1 + \cos t} = \dfrac{\sin t}{1 + \cos t}$	Simplify.

Example 5   Prove that $(\csc x - 1)(\sin x + 1) = \cos x \cot x$.

Solution   Use general strategy 2.

$$(\csc x - 1)(\sin x + 1) \overset{?}{=} \cos x \cot x$$

$\csc x \sin x + \csc x - \sin x - 1$	$\cos x \cdot \dfrac{\cos x}{\sin x}$
$1 \quad + \csc x - \sin x - 1$	$\dfrac{\cos^2 x}{\sin x}$
$1 + \dfrac{1}{\sin x} - \sin x - 1$	$\dfrac{1 - \sin^2 x}{\sin x}$
$\dfrac{1}{\sin x} - \sin x = \dfrac{1}{\sin x} - \sin x$	

Since $(\csc x - 1)(\sin x + 1) = \dfrac{1}{\sin x} - \sin x$ and $\cos x \cot x = \dfrac{1}{\sin x} - \sin x$,

you have proved the identity.

There are often several ways of proving an identity. For instance, look back to the first step in Example 3. There you could use a Pythagorean identity and replace $\csc^2 \alpha - \cot^2 \alpha$ by the number 1, since $1 + \cot^2 \alpha = \csc^2 \alpha$. Then you would immediately obtain $\tan^2 \alpha + 1 = \sec^2 \alpha$.

Oral Exercises

Simplify.

1. $1 - \cos^2 \phi$

2. $\csc^2 \theta - 1$

3. $\sin^2 \alpha - 1$

4. $1 - \sec^2 x$

5. $\tan \alpha \cos \alpha$

6. $\cot \theta \sin \theta$

7. $\dfrac{\cos x}{\sec x}$

8. $\dfrac{\csc x}{\sin x}$

9. $\dfrac{\cot x}{\tan x}$

Express the first function in terms of the second.

10. $\sin t$; $\cos t$

11. $\sec \theta$; $\tan \theta$

12. $\tan x$; $\sec x$

13. $\cos x$; $\sin x$

14. Explain why it is incorrect to prove the following identity by dividing both sides of the equation by $\cos x$.

$$\frac{\cos x}{\sin^2 x} = \cos x \csc^2 x$$

Written Exercises

Simplify.

A **1.** $\sin \alpha \sec \alpha$

2. $\cos \theta \csc \theta$

3. $\dfrac{\tan x}{\sin x}$

4. $\dfrac{\cos t}{\cot t}$

5. $(1 - \sin \theta)(1 + \sin \theta)$

6. $(\sec \phi - 1)(\sec \phi + 1)$

7. $\cos \alpha \csc \alpha \tan \alpha$

8. $\cot t \sec t \sin t$

9. $\dfrac{\sin x}{\csc x} + \dfrac{\cos x}{\sec x}$

10. $\dfrac{\csc x}{\sin x} - \dfrac{\cot x}{\tan x}$

11. $\dfrac{\sec \theta}{\cos \theta} - 1$

12. $\dfrac{\sec t}{\cos t} - \sec t \cos t$

13. $\dfrac{1 + \tan^2 \alpha}{\tan^2 \alpha}$

14. $\dfrac{1 + \tan^2 \theta}{1 + \cot^2 \theta}$

15. $\cos^2 50° + \sin^2 50°$

16. $1 + \tan^2 50°$

17. $\csc^2 t - \cot^2 t$

18. $\sec^2 350° - \tan^2 350°$

Trigonometric Graphs and Identities **639**

19. $\dfrac{\sin \alpha \cos \alpha}{1 - \sin^2 \alpha}$

20. $\dfrac{1 - \cos^2 \theta}{\sin \theta \cos \theta}$

21. $\sec x - \sin x \tan x$

22. $\cos^2 t (\cot^2 t + 1)$

23. $\dfrac{1 - \sin^2 \phi}{1 - \sin \phi} - 1$

24. $\dfrac{\sec^2 \alpha - 1}{\sec \alpha + 1} + 1$

Write the first function in terms of the second.

25. $\sec x$; $\sin x$

26. $\csc x$; $\cos x$

27. $\tan t$; $\sin t$

28. $\tan t$; $\cos t$

29. $\tan \alpha \sec \alpha$; $\sin \alpha$

30. $\tan \alpha \sec \alpha$; $\cos \alpha$

Prove each identity.

31. $\sin^2 x (1 + \cot^2 x) = 1$

32. $\sin^2 \alpha (\csc^2 \alpha + \sec^2 \alpha) = \sec^2 \alpha$

33. $\cos^2 \theta - \sin^2 \theta = 1 - 2 \sin^2 \theta$

34. $\cos^2 t - \sin^2 t = 2 \cos^2 t - 1$

35. $\cot \phi + \tan \phi = \csc \phi \sec \phi$

36. $\sec^2 \alpha + \csc^2 \alpha = \sec^2 \alpha \csc^2 \alpha$

37. $\cos^4 x - \sin^4 x = \cos^2 x - \sin^2 x$

38. $\cos^2 \theta \tan^2 \theta + \sin^2 \theta \tan^2 \theta + 1 = \sec^2 \theta$

Simplify.

B **39.** $\dfrac{\tan t + \cot t}{\sec^2 t}$

40. $\dfrac{\sec \alpha - \cos \alpha}{\tan^2 \alpha}$

41. $\cot \theta (\cos \theta \tan \theta + \sin \theta)$

42. $\sin \phi + \cos \phi \cot \phi$

43. $\dfrac{\tan^2 x}{\sec x + 1} + 1$

44. $\dfrac{\sec t + \csc t}{1 + \tan t}$

45. $\dfrac{\tan \theta}{1 + \sec \theta} + \dfrac{1 + \sec \theta}{\tan \theta}$

46. $\dfrac{\sin \alpha}{1 + \cos \alpha} + \dfrac{1 + \cos \alpha}{\sin \alpha}$

47. $(\cos t + \sin t)^2 + (\cos t - \sin t)^2$

48. $(1 + \tan x)^2 + (1 - \tan x)^2$

Prove each identity.

49. $\dfrac{1 + \cos x}{1 - \cos x} = \dfrac{\sec x + 1}{\sec x - 1}$

50. $\dfrac{\csc \phi}{\csc \phi - 1} + \dfrac{\csc \phi}{\csc \phi + 1} = 2 \sec^2 \phi$

51. $(\cot \alpha + \tan \alpha)^2 = \csc^2 \alpha \sec^2 \alpha$

52. $\sec t + \csc t = (\tan t + \cot t)(\cos t + \sin t)$

53. $\dfrac{\sin x}{1 + \cos x} = \csc x - \cot x$

54. $\dfrac{\sec y - \tan y}{1 - \sin y} = \sec y$

55. $\dfrac{\cos t}{1 - \sin t} = \dfrac{1}{\sec t - \tan t}$

56. $\dfrac{\sec \theta - 1}{\tan \theta} = \dfrac{\tan \theta}{\sec \theta + 1}$

C **57.** $\dfrac{\sin^2 \phi + 2 \cos \phi - 1}{\sin^2 \phi + 3 \cos \phi - 3} = \dfrac{1}{1 - \sec \phi}$

58. $\sqrt{\dfrac{1 - \cos x}{1 + \cos x}} = |\csc x - \cot x|$

13-7 Trigonometric Addition Formulas

Objective   To use formulas for the sine and cosine of a sum or difference.

If you were asked to evaluate sin 75°, you could use a calculator or tables to find an approximate answer. To four decimal places, sin 75° = 0.9659. But an exact answer is sometimes needed. Notice that sin 75° can be expressed as follows:

$$\sin 75° = \sin (45° + 30°).$$

Therefore, if sin (45° + 30°) can be evaluated in terms of trigonometric functions of 45° and 30°, you can find an exact expression for sin 75°.
    The following identities are called *trigonometric addition formulas*.

Addition Formulas for the Sine and Cosine

$\sin (\alpha + \beta) = \sin \alpha \cos \beta + \cos \alpha \sin \beta$

$\sin (\alpha - \beta) = \sin \alpha \cos \beta - \cos \alpha \sin \beta$

$\cos (\alpha + \beta) = \cos \alpha \cos \beta - \sin \alpha \sin \beta$

$\cos (\alpha - \beta) = \cos \alpha \cos \beta + \sin \alpha \sin \beta$

Caution: $\sin (\alpha + \beta) \neq \sin \alpha + \sin \beta$. Here is a counterexample:

$$\sin 45° + \sin 30° = \frac{\sqrt{2}}{2} + \frac{1}{2} = 1.207,$$

but $\qquad \sin (45° + 30°) = \sin 75° = 0.9659.$

Example 1   Find the exact value of each of the following.
   a. sin 75°   **b.** cos 195°

Solution   **a.** $\sin 75° = \sin (45° + 30°)$

$\qquad\qquad = \sin 45° \cos 30° + \cos 45° \sin 30°$

$\qquad\qquad = \dfrac{\sqrt{2}}{2} \cdot \dfrac{\sqrt{3}}{2} + \dfrac{\sqrt{2}}{2} \cdot \dfrac{1}{2}$

$\qquad\qquad = \dfrac{\sqrt{6} + \sqrt{2}}{4}$ ***Answer***

b. $\cos 195° = \cos (240° - 45°)$

$\qquad\qquad = \cos 240° \cos 45° + \sin 240° \sin 45°$

$\qquad\qquad = \left(-\dfrac{1}{2}\right) \cdot \dfrac{\sqrt{2}}{2} + \left(-\dfrac{\sqrt{3}}{2}\right) \cdot \dfrac{\sqrt{2}}{2}$

$\qquad\qquad = \dfrac{-\sqrt{2} - \sqrt{6}}{4}$, or $-\dfrac{\sqrt{2} + \sqrt{6}}{4}$ ***Answer***

Example 2 Prove the identity $\cos\left(\frac{3\pi}{2} - \theta\right) = -\sin\theta$.

Solution $\cos\left(\frac{3\pi}{2} - \theta\right) = \cos\frac{3\pi}{2}\cos\theta + \sin\frac{3\pi}{2}\sin\theta$

$$= 0 \cdot \cos\theta + (-1)\sin\theta$$

$$= -\sin\theta \quad \checkmark$$

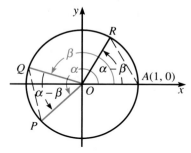

To prove the formula for $\cos(\alpha - \beta)$, place the angles α, β, and $\alpha - \beta$ in standard position and let P, Q, and R be, respectively, the points where their terminal sides intersect the unit circle. The coordinates of the points shown in the diagram are

$P(\cos\alpha, \sin\alpha)$, $\quad Q(\cos\beta, \sin\beta)$,
$R(\cos(\alpha - \beta), \quad \sin(\alpha - \beta))$, $\quad$ and $A(1, 0)$.

Since $\overset{\frown}{QP}$ and $\overset{\frown}{AR}$ both have central angles of measure $\alpha - \beta$, they are congruent. Therefore chords $\overline{QP}$ and $\overline{AR}$ are congruent and $(QP)^2 = (AR)^2$. Now use the distance formula (page 402) and simplify the result using the Pythagorean identity $\sin^2\theta + \cos^2\theta = 1$.

$$(QP)^2 = (\cos\alpha - \cos\beta)^2 + (\sin\alpha - \sin\beta)^2$$
$$= \cos^2\alpha - 2\cos\alpha\cos\beta + \cos^2\beta + \sin^2\alpha - 2\sin\alpha\sin\beta + \sin^2\beta$$
$$= 2 - 2(\cos\alpha\cos\beta + \sin\alpha\sin\beta)$$
$$(AR)^2 = [\cos(\alpha - \beta) - 1]^2 + [\sin(\alpha - \beta) - 0]^2$$
$$= \cos^2(\alpha - \beta) - 2\cos(\alpha - \beta) + 1 + \sin^2(\alpha - \beta)$$
$$= 2 - 2\cos(\alpha - \beta)$$

Since $(QP)^2 = (AR)^2$,

$$2 - 2(\cos\alpha\cos\beta + \sin\alpha\sin\beta) = 2 - 2\cos(\alpha - \beta),$$

or $\quad\quad\quad\quad \cos(\alpha - \beta) = \cos\alpha\cos\beta + \sin\alpha\sin\beta$.

To derive the formula for $\cos(\alpha + \beta)$, use the formula for $\cos(\alpha - \beta)$ and the fact that $\cos(-\beta) = \cos\beta$ and $\sin(-\beta) = -\sin\beta$.

$$\cos(\alpha + \beta) = \cos[\alpha - (-\beta)]$$
$$= \cos\alpha\cos(-\beta) + \sin\alpha\sin(-\beta)$$
$$= \cos\alpha\cos\beta - \sin\alpha\sin\beta$$

To prove the formula for $\sin(\alpha + \beta)$, use these cofunction identities:

$$\cos(90° - \theta) = \sin\theta \quad\quad \text{and} \quad\quad \sin(90° - \phi) = \cos\phi.$$

The proof is obtained by writing $\sin(\alpha + \beta)$ as a cosine and simplifying.

$$\sin(\alpha + \beta) = \cos[90° - (\alpha + \beta)] = \cos[(90° - \alpha) - \beta]$$
$$= \cos(90° - \alpha)\cos\beta + \sin(90° - \alpha)\sin\beta$$
$$= \sin\alpha\cos\beta + \cos\alpha\sin\beta.$$

The proof of the formula for $\sin(\alpha - \beta)$ is left as Exercise 33.

Oral Exercises

Use a trigonometric addition formula to expand each expression. Do not evaluate.

Sample 1 $\cos (45° - 30°) = \cos 45° \cos 30° + \sin 45° \sin 30°$

1. $\cos (60° - 45°)$

2. $\sin (30° + 45°)$

3. $\sin (45° - 30°)$

4. $\cos (60° + 45°)$

5. $\cos \left(\dfrac{\pi}{4} + \dfrac{\pi}{3} \right)$

6. $\sin \left(\dfrac{\pi}{2} - \dfrac{\pi}{6} \right)$

Give each angle as a sum or difference of multiples of 30°, 45°, or 60°.

7. $105°$

8. $15°$

9. $165°$

10. $285°$

Simplify to a trigonometric function of a single angle. Then give the exact value of the function.

Sample 2 $\cos 100° \cos 40° + \sin 100° \sin 40° = \cos (100° - 40°) = \cos 60° = \dfrac{1}{2}$

11. $\cos 20° \cos 70° - \sin 20° \sin 70°$

12. $\sin 100° \cos 10° - \cos 100° \sin 10°$

13. $\sin \dfrac{3\pi}{4} \cos \dfrac{\pi}{4} + \cos \dfrac{3\pi}{4} \sin \dfrac{\pi}{4}$

14. $\cos \dfrac{\pi}{2} \cos \dfrac{\pi}{6} + \sin \dfrac{\pi}{2} \sin \dfrac{\pi}{6}$

Simplify to a trigonometric function of a single angle.

15. $\cos 2\theta \cos \theta + \sin 2\theta \sin \theta$

16. $\sin 3\phi \cos 2\phi - \cos 3\phi \sin 2\phi$

Written Exercises

Find the exact value of each of the following.

A

1. $\cos 105°$

2. $\sin 15°$

3. $\sin 165°$

4. $\cos 285°$

5. $\cos 195°$

6. $\sin 255°$

Simplify to a trigonometric function of a single angle. Then give the exact value if possible. (See Sample 2 above.)

7. $\cos 20° \cos 40° - \sin 20° \sin 40°$

8. $\sin 50° \cos 5° - \cos 50° \sin 5°$

9. $\sin 130° \cos 80° + \cos 130° \sin 80°$

10. $\cos 50° \cos 140° + \sin 50° \sin 140°$

11. $\cos \dfrac{\pi}{4} \cos \dfrac{\pi}{12} + \sin \dfrac{\pi}{4} \sin \dfrac{\pi}{12}$

12. $\sin \dfrac{\pi}{3} \cos \dfrac{\pi}{12} - \cos \dfrac{\pi}{3} \sin \dfrac{\pi}{12}$

13. $\sin 3\theta \cos \theta - \cos 3\theta \sin \theta$

14. $\cos 2\phi \cos \phi - \sin 2\phi \sin \phi$

Prove each identity.

15. $\sin\left(\dfrac{\pi}{2} + \theta\right) = \cos\theta$

16. $\cos\left(\dfrac{\pi}{2} - \theta\right) = \sin\theta$

17. $\cos(\pi - x) = -\cos x$

18. $\sin(\pi + x) = -\sin x$

19. $\sin\left(\dfrac{3\pi}{2} - x\right) = -\cos x$

20. $\cos\left(\dfrac{3\pi}{2} + x\right) = \sin x$

Simplify to a trigonometric function of a single angle.

B **21.** $\sin\left(\dfrac{\pi}{6} + \theta\right) + \sin\left(\dfrac{\pi}{6} - \theta\right)$

22. $\cos\left(\dfrac{\pi}{3} + \theta\right) + \cos\left(\dfrac{\pi}{3} - \theta\right)$

23. $\cos(x - y)\cos y - \sin(x - y)\sin y$

24. $\sin(x + y)\cos y - \cos(x + y)\sin y$

Prove each identity.

25. $\dfrac{\sin(\alpha + \beta)}{\cos\alpha\cos\beta} = \tan\alpha + \tan\beta$

26. $\dfrac{\sin(\alpha + \beta)}{\sin\alpha\sin\beta} = \cot\alpha + \cot\beta$

27. $\cos\theta\cos\phi(\tan\theta + \tan\phi) = \sin(\theta + \phi)$

28. $\sin\theta\sin\phi(\cot\theta\cot\phi - 1) = \cos(\theta + \phi)$

29. $2\sin\alpha\cos\beta = \sin(\alpha - \beta) + \sin(\alpha + \beta)$

30. $2\cos\alpha\cos\beta = \cos(\alpha - \beta) + \cos(\alpha + \beta)$

**In Exercises 31 and 32, $\sin\alpha = -\dfrac{3}{5}$, $\cos\beta = \dfrac{8}{17}$, and α and β are third-
and fourth-quadrant angles, respectively. (*Hint:* First find $\cos\alpha$ and $\sin\beta$.)**

31. Find **(a)** $\sin(\alpha + \beta)$, **(b)** $\cos(\alpha + \beta)$, and **(c)** the quadrant of $\alpha + \beta$.

32. Find **(a)** $\sin(\alpha - \beta)$, **(b)** $\cos(\alpha - \beta)$, and **(c)** the quadrant of $\alpha - \beta$.

33. Show that $\sin(\alpha - \beta) = \sin\alpha\cos\beta - \cos\alpha\sin\beta$.
[*Hint:* $\alpha - \beta = \alpha + (-\beta)$.]

34. Use a trigonometric addition formula to derive a formula for $\sin 2\theta$.

35. Prove that $\tan(\pi - \alpha) = -\tan\alpha$. $\left(\textit{Hint:} \tan(\pi - \alpha) = \dfrac{\sin(\pi - \alpha)}{\cos(\pi - \alpha)}\right)$.

C **36.** Show that in the figure at the left below, $h = \dfrac{d\sin\alpha\sin\beta}{\sin(\alpha + \beta)}$.

37. Show that in the figure at the right above, $h = \dfrac{d\sin\alpha\sin\beta}{\sin(\alpha + \beta)}$.
(*Hint:* See Exercise 36.)

Mixed Review Exercises

Graph each function. Show at least two periods.

1. $y = \frac{1}{2} \sin 2x$

2. $y = \tan \frac{\pi}{2} x$

3. $y = 3 \cos \pi x$

Simplify.

4. $\sin^2 x (1 + \cot^2 x)$

5. $x^6 (x^{n-2})^3$

6. $\sqrt[3]{16x^6}$

7. $\dfrac{\log_2 x}{\log_2 \frac{1}{x}}$

8. $\dfrac{1 + \sqrt{3}}{1 - \sqrt{3}}$

9. $\sqrt{\dfrac{1 - \left(-\frac{5}{13}\right)}{2}}$

10. $\left(\dfrac{x^{2/3}}{x^{1/2}}\right)^6$

11. $\dfrac{1}{x} + \dfrac{1}{x^2 - x}$

12. $\csc x - \cos x \cot x$

Computer Exercises

For students with some programming experience.

1. Using only the fact that sin 5° = 0.087155743 and the addition formula for sine on page 641, write a program that will display a table of degree measures and their sines in 5° increments for angles from 5° to 90°. Note that sin 10° = sin (5° + 5°) = sin 5° cos 5° + cos 5° sin 5°. Do *not* use the computer's SIN (X) or COS (X) built-in functions. To find cosines use the identity

$$\cos \theta = \sqrt{1 - \sin^2 \theta}, \ 0° \leq \theta \leq 90°.$$

Since the computer will produce sine values having 8 or 9 digits, you can improve the appearance of the table by rounding these values to the nearest ten thousandth using the computer's greatest-integer function INT:

100 LET X=INT(10000*X+0.5)/10000

Note that, without the 0.5 being added before INT is applied, the computer would merely truncate, or cut off, the value of X after four decimal places.

2. Modify the program in Exercise 1 to include cosine and tangent values. For the tangent, use the fact that

$$\tan \theta = \frac{\sin \theta}{\cos \theta}, \ 0° \leq \theta < 90°.$$

Run this program to display a table of sines, cosines, and tangents for angles from 5° to 85°. Compare your values with those in Table 6.

3. Given the *sine* of an angle, write a program that will compute the sine of any given multiple of that angle that is less than 90°. For example, if sin θ = 0.224935 is entered, the program should be able to find, say, sin 6θ. Again, do *not* use any of the computer's built-in functions.

4. Run the program in Exercise 3 to find the following.
   a. sin 7θ if sin θ = 0.156422
   b. sin 15θ if sin θ = 0.104528

13-8 Double-Angle and Half-Angle Formulas

Objective   To use the double-angle and half-angle formulas for the sine and cosine.

You can obtain formulas for the sine and cosine of twice an angle by substituting α for β in the formulas for $\sin(\alpha + \beta)$ and $\cos(\alpha + \beta)$:

$$\begin{aligned}\sin 2\alpha &= \sin(\alpha + \alpha) \\ &= \sin\alpha\cos\alpha + \cos\alpha\sin\alpha \\ &= 2\sin\alpha\cos\alpha\end{aligned} \qquad \begin{aligned}\cos 2\alpha &= \cos(\alpha + \alpha) \\ &= \cos\alpha\cos\alpha - \sin\alpha\sin\alpha \\ &= \cos^2\alpha - \sin^2\alpha\end{aligned}$$

These formulas are called the *double-angle formulas* for sine and cosine.

Two other useful formulas for $\cos 2\alpha$ can be obtained by using the Pythagorean identity $\sin^2\alpha + \cos^2\alpha = 1$.

$$\begin{aligned}\cos 2\alpha &= \cos^2\alpha - \sin^2\alpha \\ &= (1 - \sin^2\alpha) - \sin^2\alpha \\ &= 1 - 2\sin^2\alpha\end{aligned} \qquad \begin{aligned}\cos 2\alpha &= \cos^2\alpha - \sin^2\alpha \\ &= \cos^2\alpha - (1 - \cos^2\alpha) \\ &= 2\cos^2\alpha - 1\end{aligned}$$

Double-Angle Formulas for Sine and Cosine

$$\sin 2\alpha = 2\sin\alpha\cos\alpha \qquad \cos 2\alpha = \cos^2\alpha - \sin^2\alpha$$
$$\cos 2\alpha = 1 - 2\sin^2\alpha$$
$$\cos 2\alpha = 2\cos^2\alpha - 1$$

Example 1   If $\sin\alpha = \frac{3}{5}$ and α is in the second quadrant, find $\sin 2\alpha$ and $\cos 2\alpha$.

Solution   Make a sketch. You can see that if $\sin\alpha = \frac{3}{5}$, then $\cos\alpha = -\frac{4}{5}$.

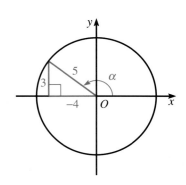

$$\sin 2\alpha = 2\sin\alpha\cos\alpha$$
$$= 2\left(\frac{3}{5}\right)\left(-\frac{4}{5}\right) = -\frac{24}{25}$$

$$\cos 2\alpha = \cos^2\alpha - \sin^2\alpha$$
$$= \left(-\frac{4}{5}\right)^2 - \left(\frac{3}{5}\right)^2 = \frac{7}{25}$$

$\therefore \sin 2\alpha = -\frac{24}{25}$ and $\cos 2\alpha = \frac{7}{25}$.   *Answer*

Formulas for the sine and cosine of half an angle can be derived by replacing α by $\frac{\theta}{2}$ in two of the formulas for $\cos 2\alpha$.

$$\cos 2\alpha = 1 - 2 \sin^2 \alpha \qquad\qquad\qquad \cos 2\alpha = 2 \cos^2 \alpha - 1$$

$$\cos 2\left(\frac{\theta}{2}\right) = 1 - 2 \sin^2\left(\frac{\theta}{2}\right) \qquad\qquad \cos 2\left(\frac{\theta}{2}\right) = 2 \cos^2\left(\frac{\theta}{2}\right) - 1$$

$$\cos \theta = 1 - 2 \sin^2 \frac{\theta}{2} \qquad\qquad\qquad \cos \theta = 2 \cos^2 \frac{\theta}{2} - 1$$

$$\sin^2 \frac{\theta}{2} = \frac{1 - \cos \theta}{2} \qquad\qquad\qquad \cos^2 \frac{\theta}{2} = \frac{1 + \cos \theta}{2}$$

Now take the square root of both sides of these equations to obtain the *half-angle formulas*.

Half-Angle Formulas for Sine and Cosine

$$\sin \frac{\theta}{2} = \pm \sqrt{\frac{1 - \cos \theta}{2}} \qquad\qquad \cos \frac{\theta}{2} = \pm \sqrt{\frac{1 + \cos \theta}{2}}$$

Notice that both of the half-angle formulas above contain a plus-or-minus sign. You must decide which sign to use by determining in which quadrant $\frac{\theta}{2}$ is located. Example 2 illustrates.

Example 2 Find the exact value of: **a.** $\sin \frac{7\pi}{8}$ **b.** $\cos 165°$

Solution Since $\frac{7\pi}{8}$ is a second-quadrant angle, $\sin \frac{7\pi}{8}$ is positive. Since $165°$ is a second-quadrant angle, $\cos 165°$ is negative.

a. $\dfrac{7\pi}{8} = \dfrac{1}{2}\left(\dfrac{7\pi}{4}\right)$ **b.** $165° = \dfrac{1}{2}(330°)$

$$\sin \frac{7\pi}{8} = \sin \frac{1}{2}\left(\frac{7\pi}{4}\right) \qquad\qquad \cos 165° = \cos \frac{1}{2}(330°)$$

$$= +\sqrt{\frac{1 - \cos \frac{7\pi}{4}}{2}} \qquad\qquad = -\sqrt{\frac{1 + \cos 330°}{2}}$$

$$= \sqrt{\frac{1 - \frac{\sqrt{2}}{2}}{2}} \qquad\qquad = -\sqrt{\frac{1 + \frac{\sqrt{3}}{2}}{2}}$$

$$= \sqrt{\frac{2 - \sqrt{2}}{4}} \qquad\qquad = -\sqrt{\frac{2 + \sqrt{3}}{4}}$$

$$= \frac{1}{2}\sqrt{2 - \sqrt{2}} \quad \textit{Answer} \qquad\qquad = -\frac{1}{2}\sqrt{2 + \sqrt{3}} \quad \textit{Answer}$$

Example 3 If $180° < \alpha < 360°$, and $\cos \alpha = -\dfrac{5}{13}$, find the exact value of each of the following.

a. $\sin \dfrac{\alpha}{2}$ $\qquad\qquad$ **b.** $\cos \dfrac{\alpha}{2}$

Solution Since $180° < \alpha < 360°$, $\quad 90° < \dfrac{\alpha}{2} < 180°$. Therefore, $\dfrac{\alpha}{2}$ is a second-quadrant angle and $\sin \dfrac{\alpha}{2}$ is positive and $\cos \dfrac{\alpha}{2}$ is negative.

a. $\sin \dfrac{\alpha}{2} = +\sqrt{\dfrac{1 - \cos \alpha}{2}} = \sqrt{\dfrac{1 - \left(-\dfrac{5}{13}\right)}{2}} = \sqrt{\dfrac{9}{13}} = \dfrac{3\sqrt{13}}{13}$

b. $\cos \dfrac{\alpha}{2} = -\sqrt{\dfrac{1 + \cos \alpha}{2}} = -\sqrt{\dfrac{1 + \left(-\dfrac{5}{13}\right)}{2}} = -\sqrt{\dfrac{4}{13}} = -\dfrac{2\sqrt{13}}{13}$

Example 4 Prove the following identity:
$$\cos \theta \csc \theta - \sin \theta \sec \theta = 2 \cot 2\theta.$$

Solution $\cos \theta \csc \theta - \sin \theta \sec \theta \overset{?}{=} 2 \cot 2\theta$

$$\dfrac{\cos \theta}{\sin \theta} - \dfrac{\sin \theta}{\cos \theta}$$

$$\dfrac{\cos^2 \theta - \sin^2 \theta}{\sin \theta \cos \theta}$$

$$2\,\dfrac{\cos^2 \theta - \sin^2 \theta}{2 \sin \theta \cos \theta}$$

$$2\,\dfrac{\cos 2\theta}{\sin 2\theta}$$

$$2 \cot 2\theta = 2 \cot 2\theta \quad \checkmark$$

Oral Exercises

Simplify to a trigonometric function of a single angle. Do not evaluate.

1. $2 \sin \theta \cos \theta$ $\qquad\qquad$ **2.** $2 \cos^2 \theta - 1$

3. $\cos^2 x - \sin^2 x$ $\qquad\qquad$ **4.** $1 - \cos^2 x$

5. $\sin \phi \cos \phi$ $\qquad\qquad$ **6.** $2 \sin^2 \phi - 1$

7. $2 \sin 35° \cos 35°$ $\qquad\qquad$ **8.** $1 - 2 \sin^2 22.5°$

9. $\sqrt{\dfrac{1 + \cos 80°}{2}}$ $\qquad\qquad$ **10.** $\sqrt{\dfrac{1 - \cos \pi}{2}}$

11. $2 \sin 5x \cos 5x$ $\qquad\qquad$ **12.** $\cos^2 4x - \sin^2 4x$

Written Exercises

Simplify to a trigonometric function of a single angle. Do not evaluate.

A   **1.** $\cos^2 10° - \sin^2 10°$

    2. $2 \sin 105° \cos 105°$

    3. $1 - 2 \sin^2 4°$

    4. $2 \cos^2 \pi - 1$

    5. $\sqrt{\dfrac{1 - \cos \alpha}{2}}, \; 0° < \alpha < 180°$

    6. $\sqrt{\dfrac{1 + \cos \alpha}{2}}, \; 0° < \alpha < 180°$

    7. $1 - 2 \sin^2 \dfrac{\theta}{2}$

    8. $2 \cos 2\alpha \sin 2\alpha$

    9. $\cos^2 2t - \sin^2 2t$

    10. $2 \cos^2 \dfrac{x}{2} - 1$

    11. $\sqrt{\dfrac{1 - \cos 2\alpha}{2}}, \; 0 < \alpha < \pi$

    12. $\sqrt{\dfrac{1 + \cos 2\alpha}{2}}, \; -\dfrac{\pi}{2} < \alpha < \dfrac{\pi}{2}$

Find the exact value of each of the following.

    13. $2 \sin 15° \cos 15°$

    14. $2 \cos^2 22.5° - 1$

    15. $\cos^2 \dfrac{\pi}{12} - \sin^2 \dfrac{\pi}{12}$

    16. $\cos^2 \dfrac{\pi}{8} + \sin^2 \dfrac{\pi}{8}$

    17. $1 - 2 \sin^2 105°$

    18. $2 \sin 165° \cos 165°$

    19. $\cos 22.5°$

    20. $\sin 75°$

    21. $\sin \dfrac{7\pi}{12}$

    22. $\cos \dfrac{5\pi}{8}$

    23. $\sin 112.5°$

    24. $\cos 202.5°$

In Exercises 25–28, $0 < \alpha < 180°$ and $\cos \alpha = -\dfrac{4}{5}$. Find the following.

    25. $\cos 2\alpha$

    26. $\sin 2\alpha$

    27. $\sin \dfrac{\alpha}{2}$

    28. $\cos \dfrac{\alpha}{2}$

In Exercises 29–32, $180° < \alpha < 360°$ and $\cos \alpha = \dfrac{8}{17}$. Find the following.

    29. $\sin 2\alpha$

    30. $\cos 2\alpha$

    31. $\cos \dfrac{\alpha}{2}$

    32. $\sin \dfrac{\alpha}{2}$

Prove each identity.

B   **33.** $\cos^4 \theta - \sin^4 \theta = \cos 2\theta$

    34. $(\sin \theta + \cos \theta)^2 = 1 + \sin 2\theta$

    35. $\sin 4x = 4 \sin x \cos x \cos 2x$

    36. $\sin x + \sin x \cos 2x = \sin 2x \cos x$

    37. $\dfrac{\sin 2x}{1 + \cos 2x} = \tan x$

    38. $\dfrac{\sin 2x}{1 - \cos 2x} = \cot x$

    39. $\cot \alpha + \tan \alpha = 2 \csc 2\alpha$

    40. $\csc 2\alpha + \cot 2\alpha = \cot \alpha$

    41. Graph $y = \cos^2 x - \sin^2 x$. Give the period and amplitude of the graph.

    42. Graph $y = \sin x \cos x$. Give the period and amplitude of the graph.

C   **43.** Express $\cos 4\theta$ in terms only of $\cos \theta$.

13-9 Formulas for the Tangent

Objective   To use addition, double-angle, and half-angle formulas for the tangent.

An addition formula for the tangent can be obtained by using the addition formulas for the sine and cosine. If $\tan(\alpha + \beta)$, $\tan\alpha$, and $\tan\beta$ are defined, then:

$$\tan(\alpha + \beta) = \frac{\sin(\alpha + \beta)}{\cos(\alpha + \beta)}$$

$$= \frac{\sin\alpha\cos\beta + \cos\alpha\sin\beta}{\cos\alpha\cos\beta - \sin\alpha\sin\beta}$$

$$= \frac{\dfrac{\sin\alpha\cos\beta}{\cos\alpha\cos\beta} + \dfrac{\cos\alpha\sin\beta}{\cos\alpha\cos\beta}}{\dfrac{\cos\alpha\cos\beta}{\cos\alpha\cos\beta} - \dfrac{\sin\alpha\sin\beta}{\cos\alpha\cos\beta}}$$
$\left\{\begin{array}{l}\text{Divide numerator}\\\text{and denominator}\\\text{by } \cos\alpha\cos\beta.\end{array}\right.$

$$= \frac{\tan\alpha + \tan\beta}{1 - \tan\alpha\tan\beta}$$

To obtain a formula for $\tan(\alpha - \beta)$, use the formula for $\tan(\alpha + \beta)$ and the fact that $\tan(-\beta) = -\tan\beta$.

$$\tan(\alpha - \beta) = \tan[\alpha + (-\beta)] = \frac{\tan\alpha + \tan(-\beta)}{1 - \tan\alpha\tan(-\beta)} = \frac{\tan\alpha - \tan\beta}{1 + \tan\alpha\tan\beta}$$

Addition Formulas for the Tangent

$$\tan(\alpha + \beta) = \frac{\tan\alpha + \tan\beta}{1 - \tan\alpha\tan\beta}$$

$$\tan(\alpha - \beta) = \frac{\tan\alpha - \tan\beta}{1 + \tan\alpha\tan\beta}$$

Example 1 Find the exact value of $\tan 285°$.

Solution $\tan 285° = \tan(240° + 45°) = \dfrac{\tan 240° + \tan 45°}{1 - \tan 240° \tan 45°}$

$$= \frac{\sqrt{3} + 1}{1 - (\sqrt{3})(1)}$$

$$= \frac{\sqrt{3} + 1}{1 - \sqrt{3}} \cdot \frac{1 + \sqrt{3}}{1 + \sqrt{3}}$$

$$= \frac{\sqrt{3} + 3 + 1 + \sqrt{3}}{1 - 3}$$

$$= \frac{4 + 2\sqrt{3}}{-2} = -2 - \sqrt{3} \quad \textit{Answer}$$

A double-angle formula for the tangent can be obtained by substituting α for β in the formula $\tan (\alpha + \beta)$.

$$\tan 2\alpha = \tan (\alpha + \alpha) = \frac{\tan \alpha + \tan \alpha}{1 - \tan \alpha \tan \alpha} = \frac{2 \tan \alpha}{1 - \tan^2 \alpha}$$

Double-Angle Formula for the Tangent

$$\tan 2\alpha = \frac{2 \tan \alpha}{1 - \tan^2 \alpha}$$

A formula for $\tan \frac{\theta}{2}$ can be obtained by using the half-angle formulas for sine and cosine.

$$\tan \frac{\theta}{2} = \frac{\sin \dfrac{\theta}{2}}{\cos \dfrac{\theta}{2}} = \frac{\pm \sqrt{\dfrac{1 - \cos \theta}{2}}}{\pm \sqrt{\dfrac{1 + \cos \theta}{2}}} = \pm \sqrt{\frac{1 - \cos \theta}{1 + \cos \theta}}$$

A second formula for $\tan \frac{\theta}{2}$ can be derived by using the fact that $2 \cos^2 \alpha = 1 + \cos 2\alpha$.

$$\tan \alpha = \frac{\sin \alpha}{\cos \alpha} = \frac{2 \sin \alpha \cos \alpha}{2 \cos \alpha \cos \alpha} = \frac{2 \sin \alpha \cos \alpha}{2 \cos^2 \alpha} = \frac{\sin 2\alpha}{1 + \cos 2\alpha}$$

By substituting $\frac{\theta}{2}$ for α you will obtain the second formula given in the chart below. The derivation of the third formula is left as Exercise 32.

Half-Angle Formulas for the Tangent

$$\tan \frac{\theta}{2} = \pm \sqrt{\frac{1 - \cos \theta}{1 + \cos \theta}} \qquad \tan \frac{\theta}{2} = \frac{\sin \theta}{1 + \cos \theta} \qquad \tan \frac{\theta}{2} = \frac{1 - \cos \theta}{\sin \theta}$$

Example 2 Find the exact value of $\tan \frac{5\pi}{12}$.

Solution Use the formula $\tan \frac{\theta}{2} = \frac{1 - \cos \theta}{\sin \theta}$ with $\theta = \frac{5\pi}{6}$.

$$\tan \frac{5\pi}{12} = \tan \frac{1}{2}\left(\frac{5\pi}{6}\right) = \frac{1 - \cos \dfrac{5\pi}{6}}{\sin \dfrac{5\pi}{6}} = \frac{1 - \left(-\dfrac{\sqrt{3}}{2}\right)}{\dfrac{1}{2}} = 2 + \sqrt{3} \quad \textbf{\textit{Answer}}$$

Oral Exercises

Simplify to a trigonometric function of a single angle.

1. $\dfrac{\tan 3\theta + \tan \theta}{1 - \tan 3\theta \tan \theta}$

2. $\dfrac{\tan 3\theta - \tan \theta}{1 + \tan 3\theta \tan \theta}$

3. $\dfrac{2 \tan x}{1 - \tan^2 x}$

4. $\dfrac{2 \tan 4x}{1 - \tan^2 4x}$

5. $\dfrac{\sin \dfrac{\alpha}{2}}{1 + \cos \dfrac{\alpha}{2}}$

6. $\dfrac{1 - \cos 2\alpha}{\sin 2\alpha}$

7. $\dfrac{\sin x}{1 - \cos x}$

8. $\dfrac{1 + \cos x}{\sin x}$

9. $\dfrac{2 \sin x \cos x}{1 - 2 \sin^2 x}$

10. Explain how you would find the exact value of $\tan 22.5°$.

11. Explain how you would find the exact value of $\tan \dfrac{\pi}{12}$.

Written Exercises

Simplify to a trigonometric function of a single angle. Do not evaluate.

A 1. $\dfrac{\tan 80° - \tan 20°}{1 + \tan 80° \tan 20°}$

2. $\dfrac{\tan 15° + \tan 30°}{1 - \tan 15° \tan 30°}$

3. $\dfrac{\tan 140° + \tan 40°}{1 - \tan 140° \tan 40°}$

4. $\dfrac{\tan 250° - \tan 25°}{1 + \tan 250° \tan 25°}$

5. $\dfrac{\tan \dfrac{7\pi}{12} + \tan \dfrac{\pi}{4}}{1 - \tan \dfrac{7\pi}{12} \tan \dfrac{\pi}{4}}$

6. $\dfrac{\tan \dfrac{7\pi}{6} - \tan \dfrac{\pi}{3}}{1 + \tan \dfrac{7\pi}{6} \tan \dfrac{\pi}{3}}$

7. $\dfrac{2 \tan 22.5°}{1 - \tan^2 22.5°}$

8. $\dfrac{2 \tan 67.5°}{1 - \tan^2 67.5°}$

Find the exact value of each of the following.

9. $\tan 75°$ **10.** $\tan 15°$ **11.** $\tan 165°$ **12.** $\tan 105°$

13. $\tan \dfrac{11\pi}{12}$ **14.** $\tan \dfrac{3\pi}{8}$ **15.** $\tan 22.5°$ **16.** $\tan 112.5°$

17. $\tan \dfrac{11\pi}{8}$ **18.** $\tan \dfrac{5\pi}{8}$ **19.** $\cot 195°$ **20.** $\cot 255°$

In Exercises 21–24, $\tan \alpha = -\dfrac{8}{15}$, $\tan \beta = \dfrac{3}{4}$, and α is a second-quadrant angle. Find the value of each expression.

21. $\tan (\alpha + \beta)$ **22.** $\tan (\alpha - \beta)$ **23.** $\tan 2\alpha$ **24.** $\tan \dfrac{\alpha}{2}$

Prove each identity.

B **25.** $\tan (\alpha + \pi) = \tan \alpha$ **26.** $\tan (\alpha - \pi) = \tan \alpha$

27. $\tan \left(\dfrac{\pi}{4} - \alpha \right) = \dfrac{1 - \tan \alpha}{1 + \tan \alpha}$ **28.** $\tan \left(\dfrac{\pi}{4} + \alpha \right) = \dfrac{1 + \tan \alpha}{1 - \tan \alpha}$

29. Express $\cot (\alpha + \beta)$ in terms of $\cot \alpha$ and $\cot \beta$.

30. Express $\cot (\alpha - \beta)$ in terms of $\cot \alpha$ and $\cot \beta$.

31. Use the formula derived in Exercise 29 to find a formula for $\cot 2\alpha$ in terms of $\cot \alpha$.

32. Prove the following identity:

$$\tan \frac{\theta}{2} = \frac{1 - \cos \theta}{\sin \theta}$$

C **33. a.** Find $\sin \dfrac{7\pi}{8}$ and $\cos \dfrac{7\pi}{8}$ using half-angle formulas for sine and cosine.

       b. Using the answers found in part (a), show that

$$\tan \frac{7\pi}{8} = -\sqrt{3 - 2\sqrt{2}}.$$

       c. Using a half-angle formula for tangent, show that

$$\tan \frac{7\pi}{8} = 1 - \sqrt{2}.$$

       d. Show that the answers to parts (b) and (c) are equal.

Prove each identity.

34. $\dfrac{2 \tan \dfrac{x}{2}}{1 + \tan^2 \dfrac{x}{2}} = \sin x$ **35.** $\tan 3\theta = \dfrac{3 \tan \theta - \tan^3 \theta}{1 - 3 \tan^2 \theta}$

Mixed Review Exercises

Prove each identity.

1. $(\sin x - \cos x)^2 = 1 - \sin 2x$ **2.** $\cos \left(\dfrac{3\pi}{2} - x \right) = -\sin x$

3. $\sec x - \sin x \tan x = \cos x$ **4.** $1 - \sin 2x \tan x = \cos 2x$

Solve each inequality and graph the solution set.

5. $|x - 3| < 2$ **6.** $x^2 > 2x + 3$ **7.** $2 \le 3x + 5 \le 14$

8. $x^3 < 3x^2$ **9.** $|2x + 1| \ge 3$ **10.** $7 - 4x > -1$

11. $3x - 5 > 13$ **12.** $x^2 - 2x \le 3$ **13.** $|2x| \ge -1$

Self-Test 2

1. Simplify to a trigonometric function of a single angle:

$$\cos x + \tan x \sin x$$

Obj. 13-6, p. 636

2. Prove the following identity: $\sec \theta - \tan \theta \sin \theta = \cos \theta$.

3. Write $\cos 310° \cos 260° - \sin 310° \sin 260°$ as a trigonometric function of a single angle. Then give the exact value of the function.

Obj. 13-7, p. 641

4. Simplify to a trigonometric function of a single angle:

$$2 \sin 5\alpha \cos 5\alpha$$

Obj. 13-8, p. 646

5. If $180° < \theta < 360°$, and $\cos \theta = -\dfrac{21}{29}$, find the following:

    a. $\cos 2\theta$ **b.** $\sin 2\theta$ **c.** $\cos \dfrac{\theta}{2}$

Simplify to a trigonometric function of a single angle. Do not evaluate.

6. $\dfrac{\tan 50° - \tan 40°}{1 + \tan 50° \tan 40°}$ **7.** $\dfrac{2 \tan 35°}{1 - \tan^2 35°}$

Obj. 13-9, p. 650

8. $\dfrac{1 - \cos 130°}{\sin 130°}$

Check your answers with those at the back of the book.

Chapter Summary

1. The *radian measure* of an angle θ is $\dfrac{s}{r}$, where r is the radius of any circle centered at the vertex of θ and s is the length of the arc intercepted by θ. To convert from one system of measure to another, you use the conversion formula $180° = \pi$ radians.

   Circular arc length is given by

   $$s = r\theta,$$

   and sector area by $\qquad\qquad A = \tfrac{1}{2}r^2\theta,$

   where θ denotes the radian measure of the central angle.

2. *Circular functions* of the number s are equal to the corresponding trigonometric functions of an angle having radian measure s. If angle θ has radian measure s, then $\sin \theta = \sin s$, $\cos \theta = \cos s$, and so on. Therefore, the values of circular functions can be found by using a calculator or tables.

3. A function f has *period p* ($p > 0$) if $f(x + p) = f(x)$ for every x in the domain of f. The sine, cosine, secant, and cosecant have period 2π; the tangent and cotangent have period π.

4. For every x in the domain of f, a function f is *even* if $f(-x) = f(x)$, and *odd* if $f(-x) = -f(x)$. The sine, cosecant, tangent, and cotangent are odd functions; the cosine and secant are even functions.

5. The graphs of $y = a \sin bx$ and $y = a \cos bx$ each have amplitude a and period $\dfrac{2\pi}{b}$.

6. The fundamental *trigonometric identities* (page 636), as well as the *addition, double-angle,* and *half-angle formulas* are used to simplify trigonometric expressions and to prove other identities.

Chapter Review

Write the letter of the correct answer.

1. Express $\dfrac{13\pi}{4}$ radians in degrees.

   a. 495°   **b.** 225°   **c.** 585°   **d.** 135°

   13-1

2. Express 150° in radians.

   a. $\dfrac{7\pi}{3}$ **b.** $\dfrac{5\pi}{6}$ **c.** $\dfrac{7\pi}{6}$ **d.** $\dfrac{7\pi}{12}$

3. Find the exact value of $\csc\left(-\dfrac{7\pi}{6}\right)$.

   13-2

   a. $\dfrac{\sqrt{3}}{2}$ **b.** $-\dfrac{\sqrt{3}}{2}$ **c.** -2 **d.** 2

4. If $\tan s = \dfrac{\sqrt{3}}{3}$ and $-\pi < s < 0$, then $s = \underline{\ ?\ }$.

   a. $-\dfrac{4\pi}{3}$ **b.** $-\dfrac{2\pi}{3}$ **c.** $-\dfrac{5\pi}{6}$ **d.** $-\dfrac{7\pi}{6}$

5. If $f(x) = \dfrac{x}{x^2 - 9}$, then f is $\underline{\ ?\ }$.

   13-3

   a. odd   **b.** even   **c.** neither odd nor even   **d.** both even and odd

6. If $f(x)$ has period 8, then $f(2x)$ has period $\underline{\ ?\ }$.

   a. 16   **b.** 8   **c.** 4   **d.** 2

7. Which of the following is an equation of the cosine curve shown at the right?

   13-4

   a. $y = 2 \cos \dfrac{2x}{3}$ **b.** $y = 2 \cos \dfrac{3\pi x}{2}$

   c. $y = 2 \cos \dfrac{2\pi x}{3}$ **d.** $y = 2 \cos 3\pi x$

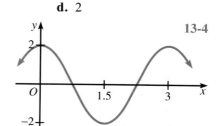

8. Which of the following is an equation of a sine curve with maximum value 3, minimum value -1 and period $\frac{\pi}{3}$?

   a. $y = 2 \sin 6x + 1$ **b.** $y = 2 \sin 6\pi x + 1$

   c. $y = 2 \sin 3x + 1$ **d.** $y = 2 \sin 3x$

9. Find the period of the function defined by $y = 2 \sec \pi x$.   13-5

   a. π **b.** 2π **c.** 1   **d.** 2

10. Which of the following statements is *not* true?

   a. $\cot x$ is odd   **b.** $\sec x$ is even   **c.** $\tan x$ is even   **d.** $\csc x$ is odd

11. Simplify $\sec \theta - \tan \theta \sin \theta$.   13-6

   a. $\sin^2 \theta$ **b.** $\cos^2 \theta$ **c.** $\sec \theta$ **d.** $\cos \theta$

12. If $\sin t = u$, then $\tan t = \underline{}$.

   a. $\pm \dfrac{u}{\sqrt{1 - u^2}}$ **b.** $\pm \dfrac{1}{\sqrt{1 - u^2}}$ **c.** $\pm \dfrac{1}{1 - u^2}$ **d.** $\pm \dfrac{u}{\sqrt{u^2 - 1}}$

13. Write $\sin 2\pi \cos \frac{\pi}{6} - \cos 2\pi \sin \frac{\pi}{6}$ as a trigonometric function of a single angle and evaluate.   13-7

   a. $\dfrac{\sqrt{3}}{2}$ **b.** $-\dfrac{\sqrt{3}}{2}$ **c.** $\dfrac{1}{2}$ **d.** $-\dfrac{1}{2}$

14. Find the exact value of $\sin 165°$.

   a. $\dfrac{\sqrt{6} + \sqrt{2}}{4}$ **b.** $\dfrac{\sqrt{6} - \sqrt{2}}{4}$ **c.** $\dfrac{\sqrt{2} - \sqrt{6}}{4}$ **d.** $-\dfrac{\sqrt{2} + \sqrt{6}}{4}$

15. If $\sin \theta = \frac{2}{3}$, then $\cos 2\theta = \underline{}$.   13-8

   a. $\dfrac{1}{3}$ **b.** $\dfrac{5}{9}$ **c.** $-\dfrac{1}{9}$ **d.** $\dfrac{1}{9}$

16. Evaluate $\cos \frac{7\pi}{8}$.

   a. $-\dfrac{\sqrt{2 - \sqrt{2}}}{2}$ **b.** $\dfrac{\sqrt{2 + \sqrt{2}}}{2}$ **c.** $\dfrac{\sqrt{2 - \sqrt{2}}}{2}$ **d.** $-\dfrac{\sqrt{2 + \sqrt{2}}}{2}$

17. Simplify $\tan \left(\alpha - \frac{\pi}{4} \right) \tan \left(\alpha + \frac{\pi}{4} \right)$.   13-9

   a. 1   **b.** -1 **c.** $\tan 2\alpha$ **d.** $\dfrac{1}{2} \tan 2\alpha$

18. If θ is a fourth-quadrant angle and $\cos \theta = \frac{8}{17}$, find $\tan \frac{\theta}{2}$.

   a. $-\dfrac{3}{5}$ **b.** $\dfrac{3}{5}$ **c.** $-\dfrac{15}{8}$ **d.** $\dfrac{8}{15}$

Chapter Test

1. Express $-120°$ in radians. 13-1

2. Express $\frac{7\pi}{10}$ in degrees.

3. The central angle of a sector of a circle having diameter 8 in. measures $65°$. Find the following:
   a. the length, s, of the intercepted arc.
   b. the area, A, of the sector.

4. Find the exact values of the six trigonometric functions of $\frac{4\pi}{3}$. 13-2

5. Determine if $f(x) = x \sin x - \tan x$ is even, odd, or neither. 13-3

6. Find the amplitude, the maximum and minimum values, and the period of the function $y = 3 \cos 4x$. 13-4

7. Find an equation of the sine curve shown at the right.

8. Sketch one period of the graph of $y = 2 \cot \pi x$. 13-5

9. Prove the following identity: 13-6
$$\csc \theta - \cot \theta \cos \theta = \sin \theta$$

10. Prove the following identity:
$$\tan x + \cot x = \sec x \csc x$$

11. Simplify the following expression: 13-7
$$\cos (\alpha + 45°) + \sin (\alpha - 45°)$$

12. If $\sin \alpha = \frac{8}{17}$ and $\cos \alpha < 0$, find the following: 13-8
    a. $\sin 2\alpha$ b. $\cos 2\alpha$ c. $\sin \frac{\alpha}{2}$

13. If $\sin \theta = -\frac{5}{13}$ and $\cos \theta > 0$, find the following: 13-9
    a. $\tan 2\theta$ b. $\tan \frac{\theta}{2}$

14 Trigonometric Applications

A ship's radar uses a polar co-ordinate system. The location of a point is determined by its direction and distance from the center (O), called the pole.

Vectors

14-1 Vector Operations

Objective   To define vector operations and apply the resultant of two vectors.

The *velocity* of an airplane is described by giving both its *speed* and its *direction*. For example, you might say its velocity is 600 km/h northeast. Any quantity that has both magnitude (size) and direction is called a **vector quantity.** You can represent a vector quantity by an arrow, called a **vector,** whose length is proportional to the magnitude of the quantity, and whose direction gives the direction of the vector quantity (see Figure 1).

A vector is often labeled with the magnitude of the vector quantity it represents or the magnitude is indicated by giving a scale. For example, the arrow on the diagram at the left in Figure 1 represents a velocity of 600 km/h northeast. The arrow on the diagram at the right is $4\frac{1}{2}$ times as long as the given 100-m scale and therefore represents a displacement of 450 m due east.

A velocity of
600 km/h northeast

Figure 1

A displacement of
450 m due east

The symbol $\overrightarrow{AB}$ (read "vector AB") denotes the vector extending from point A, the **initial point,** to point B, the **terminal point.** Boldface letters such as **u** and **v** are also used to denote vectors. (When handwriting vectors, use letters with bars or small arrows over them: $\bar{u}, \bar{v},$ or $\vec{u}, \vec{v},$ in place of boldface.) Two vectors that have the same magnitude and same direction, such as $\overrightarrow{AB}$ and $\overrightarrow{CD}$ in the diagram, are called **equivalent vectors,** and you write $\overrightarrow{AB} = \overrightarrow{CD}$.

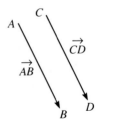

In the vector diagram at the right, a plane flies from A to B and then from B to C. Since the plane could get to the same final position by flying directly from A to C, $\overrightarrow{AC}$ is called the *sum,* or *resultant,* of $\overrightarrow{AB}$ and $\overrightarrow{BC}$. That is, $$\overrightarrow{AC} = \overrightarrow{AB} + \overrightarrow{BC}.$$

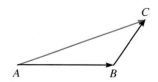

To find the resultant of any two vectors, you use the following fact: For any vector **v** and point P, there is exactly one vector with initial point P that has the same direction and length as **v**. We will often treat vectors as movable arrows and label this vector **v** as well. Therefore, we can *place the initial point of* **v** *at P* by constructing the unique vector with initial point P that is equivalent to **v**. In general, you have the following methods for finding the resultant.

Figure 2

Vector Addition

Given two vectors **u** and **v,** you can find their **sum,** or **resultant,** by using either of the following two methods. These methods show that vector addition is commutative: **u + v = v + u.**

The Triangle Method

Place the initial point of **v** at the terminal point of **u.** Then **u + v** is the vector extending from the initial point of **u** to the terminal point of **v.**

The Parallelogram Method

Form a parallelogram with **u** and **v** as adjacent sides starting from a common point. Then **u + v** extends from that point to the opposite vertex of the parallelogram.

Example 1   For the following two vector quantities, make a scale drawing showing the vectors and their sum:

A 10 km trip due west followed by a 15 km trip southeast.

Solution   Let **u** be the vector representing a 10 km trip west and **v** be the vector representing a 15 km trip southeast. Their sum or resultant is shown below.

The chart at the top of the next page tells you how to multiply a vector by a real number, or *scalar.*

To multiply the vector **v** by the real number t, multiply the length of **v** by $|t|$ and reverse the direction if $t < 0$.

This rule is illustrated in Figure 3. Notice that $(-1)\mathbf{v}$ is written as $-\mathbf{v}.$

Figure 3   **Figure 4**

The difference $\mathbf{u} - \mathbf{v}$ is defined to be $\mathbf{u} + (-\mathbf{v})$. (See Figure 4.)

You can think of a point as a vector of length 0. This zero vector is denoted by **0.** For any vector **u,** you have $\mathbf{u} - \mathbf{u} = \mathbf{0}.$

The length or magnitude of a vector **v** is also called its **norm** and is denoted by $\|\mathbf{v}\|$. For example, if **v** is an airplane velocity vector, then $\|\mathbf{v}\|$ is the speed of the airplane. A vector with a norm of 1 is called a **unit vector.**

The direction of a vector **v** is often given by its *bearing*.

The **bearing** of a vector **v** is the angle measured clockwise from due north around to **v.**

Some terms used in navigation are illustrated in Figure 5. You should be aware that a craft may not travel in the direction in which it is headed because of the effect of winds and currents. For example, if you row with a heading of 90° (due east) across a river flowing from north to south, the bearing of your *true course* will be somewhere between 90° and 180°.

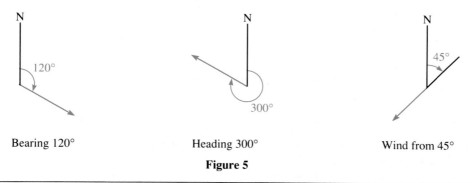

Bearing 120°   Heading 300°   Wind from 45°

Figure 5

Example 2 Use vectors **u** and **v** as shown below. Vector **u** has magnitude 4 and bearing 90° and vector **v** has magnitude 8 and bearing 225°. Sketch (a) **u** + **v** and (b) **u** − **v**.

Solution **a.** **u** + **v**

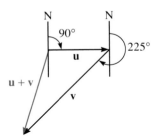

b. Use the fact that **u** − **v** = **u** + (−**v**).

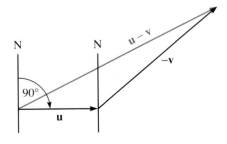

Example 3 The magnitude and direction of vectors **u** and **v** are given below. If **w** = **u** + **v**, find the magnitude of **w** to three significant digits and its bearing to the nearest tenth of a degree.

$$\text{\textbf{u}: magnitude 7, bearing } 135°$$
$$\text{\textbf{v}: magnitude 5, bearing } 245°$$

Solution The figure at the right shows the vector quantities involved. The vector **w** is the resultant vector. Since $65° + \alpha + 45° = 180°$, $\alpha = 70°$. Use the law of cosines to find $\|\mathbf{w}\|$, the magnitude of **w**.

$$\|\mathbf{w}\|^2 = 7^2 + 5^2 - 2(7)(5) \cos 70°$$
$$= 49 + 25 - (70)(0.342)$$
$$= 50.06$$
$$\|\mathbf{w}\| = 7.08 \text{ (to three significant digits)}$$

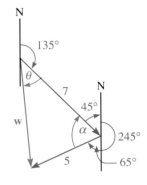

To find the bearing of vector **w**, first use the law of sines to find θ:

$$\frac{\sin \theta}{5} = \frac{\sin 70°}{7.08}$$

$$\sin \theta = \frac{5 \sin 70°}{7.08} = \frac{5(0.940)}{7.08} = 0.6640$$

$$\theta = 41.6°$$

Then the bearing of **w** is $135° + \theta = 135° + 41.6° = 176.6°$.

∴ the magnitude of **w** is 7.08 and its bearing is 176.6°. ***Answer***

Example 4 The air speed (speed in still air) of a light plane is 220 km/h, and its heading is 150°. A wind of 45 km/h is blowing from 40°. Find the plane's ground speed (the speed relative to the ground) and the bearing of its true course.

Solution The figure at the right shows the velocity vectors involved. The vector **v** is the true-course velocity. To find the ground speed ||**v**||, use the law of cosines. Since $\alpha + 30° + 40° = 180°$, $\alpha = 110°$.

$$\|\mathbf{v}\|^2 = 220^2 + 45^2 - (2)(220)(45)\cos 110°$$

$$= 48{,}400 + 2025 + 6772$$

$$= 57{,}197$$

$$\|\mathbf{v}\| = 239 \quad \text{(km/h)}$$

To find the bearing of **v**, the true course, first use the law of sines to find θ.

$$\frac{\sin \theta}{45} = \frac{\sin 110°}{239}$$

$$\sin \theta = \frac{45(0.9397)}{239} = 0.177$$

$$\theta = 10.2°$$

Then the bearing of **v** is $150° + \theta = 150° + 10.2° = 160.2°$.

∴ the ground speed of the plane is 239 km/h and the bearing of the true course is 160.2°. **Answer**

Oral Exercises

Exercises 1–16 refer to parallelogram *ABCD*. In Exercises 1–10, determine whether the statement is true or false.

1. $\overrightarrow{DC} + \overrightarrow{CX} = \overrightarrow{DX}$

2. $\overrightarrow{XB} = \overrightarrow{XD}$

3. $\overrightarrow{XD} + \overrightarrow{XB} = 0$

4. $2\overrightarrow{XB} = \overrightarrow{DB}$

5. $\overrightarrow{BA} + \overrightarrow{AD} = \overrightarrow{BD}$

6. $\overrightarrow{DC} + \overrightarrow{CB} = \overrightarrow{DB}$

7. $\overrightarrow{AX} + \overrightarrow{XB} + \overrightarrow{DA} = 0$

8. $\overrightarrow{BD} - \overrightarrow{AD} = \overrightarrow{BA}$

9. $2\overrightarrow{DX} + \overrightarrow{BC} = \overrightarrow{DC}$

10. $\overrightarrow{BA} + \overrightarrow{DA} = \overrightarrow{CA}$

Find a vector joining two lettered points in the diagram that is equivalent to the given vector.

11. $2\overrightarrow{XD}$

12. $-\overrightarrow{AC}$

13. $\overrightarrow{CD} - \overrightarrow{DA}$

14. $\overrightarrow{AD} + \overrightarrow{DC} + \overrightarrow{CX} + \overrightarrow{XB}$

15. $\overrightarrow{AD} + \overrightarrow{CX}$

16. $\overrightarrow{DC} - \overrightarrow{CB}$

Written Exercises

In Exercises 1–6, two vector quantities are given. Make a scale drawing showing the two vectors and their sum.

A

1. A 9 km trip east followed by a 4 km trip southwest.

2. A 5 km trip southeast followed by a 7 km trip southwest.

3. The heading of a plane is northwest, and its speed is 350 km/h. A wind of 50 km/h is blowing from the west.

4. The heading of a fishing boat is due west, and its speed is 18 km/h. The current is flowing at 6 km/h due south.

5. A small boat sails for 15 km bearing 60° and then sails for 10 km bearing 120°.

6. The air speed of a plane is 390 km/h and its heading is 230°. A wind of 35 km/h is blowing from 150°.

In the diagram at the right, vector u has magnitude 3 and bearing 315°, vector v has magnitude 2 and bearing 180°, and vector w has magnitude 3 and bearing 90°. Use these vectors to sketch the following vectors.

7. $\mathbf{u} + \mathbf{w}$

8. $\mathbf{u} + \mathbf{v}$

9. $\mathbf{u} - \mathbf{v}$

10. $\mathbf{u} - \mathbf{w}$

11. $\mathbf{u} + 2\mathbf{v}$

12. $2\mathbf{u} + \mathbf{w}$

13. $\mathbf{u} + 2\mathbf{v} - \mathbf{w}$

14. $\mathbf{w} - \mathbf{u} - 2\mathbf{v}$

In Exercises 15–22, the magnitude and direction of vectors u and v are given. Find the magnitude of w to three significant digits and its bearing to the nearest tenth of a degree.

15. **u:** magnitude 117, bearing 130°
    v: magnitude 102, bearing 220°
    w: $\mathbf{u} + \mathbf{v}$

16. **u:** magnitude 218, bearing 22°
    v: magnitude 170, bearing 112°
    w: $\mathbf{u} - \mathbf{v}$

B

17. **u:** magnitude 136, bearing 220°
    v: magnitude 197, bearing 300°
    w: $2\mathbf{u} - \mathbf{v}$

18. **u:** magnitude 1850, bearing 125°
    v: magnitude 2960, bearing 25°
    w: $\mathbf{u} + 2\mathbf{v}$

19. **u:** magnitude 460, bearing 0°
    v: magnitude 712, bearing 130°
    w: $\mathbf{u} + \mathbf{v}$

20. **u:** magnitude 23.0, bearing 215°
    v: magnitude 14.5, bearing 105°
    w: $\mathbf{u} + \mathbf{v}$

21. **u:** magnitude 3.62, bearing 25°
    v: magnitude 14.5, bearing 105°
    w: $\mathbf{u} + \mathbf{v}$

22. **u:** magnitude 621, bearing 305°
    v: magnitude 336, bearing 15°
    Find **w** if $3(\mathbf{w} + \mathbf{u}) = 2(\mathbf{w} - \mathbf{v})$.

Problems

Solve. Find magnitudes to three significant digits and bearings to the nearest tenth of a degree.

A   **1.** A plane flies 285 km due west and then 320 km due north. What are its distance and bearing from the starting point?

2. A ship leaves port and sails 175 km due southeast and then 168 km due southwest. What are the distance and bearing of the port from the ship?

3. To fly due north at 580 km/h when there is a 45 km/h wind blowing from due east, with what heading and speed should a plane travel?

4. Repeat Problem 3 assuming that the wind is blowing from 130°.

5. A plane's air speed is 270 km/h, and its heading is 70°. Find its ground speed and true-course bearing if a wind of 50 km/h is blowing from 330°.

6. Repeat Problem 5 assuming that the wind is blowing from due north.

7. Two ships leave port at noon. One ship's speed is 15 km/h, and its heading is 75°. The other ship's speed is 24 km/h, and its heading is 155°. How far apart are the ships at 3 P.M. and what is the bearing of each from the other?

B   **8.** A plane leaves Mapleton bound for Logan Beach, which is 600 km away at a bearing of 25° from Mapleton. To avoid a storm, the pilot first flies 140 km on a 325° course. From there the pilot can fly the plane directly to Logan Beach in still air. What heading should be used and how far must the plane fly?

9. Los Angeles Airport bears 140° from San Francisco Airport and is 540 km away. A pilot is planning a direct flight from San Francisco Airport to Los Angeles Airport to leave at 2 P.M. The plane's air speed will be 640 km/h, and there will be a 60 km/h wind blowing from 290°. What should the compass heading be, and what is the plane's estimated time of arrival (ETA) to the nearest minute?

Mixed Review Exercises

Determine whether the given function is odd, even, or neither.

1. $f(x) = \dfrac{x + 2}{x - 3}$

2. $y = \cos x - \sin x$

3. $f(x) = \dfrac{6x^3 - x}{x^2 + 1}$

Solve.

4. $x^2 + 6x = 8$

5. $|3z - 7| = 5$

6. $(3x - 2)^2 = 36$

7. $\log_9 x = \dfrac{3}{2}$

8. $\sqrt{4t + 9} = t - 3$

9. $p^2 + \dfrac{11p}{2} = 3$

14-2 Vectors in the Plane

Objective   To find vectors in component form and to apply th dot product.

In a coordinate plane we denote by **i** the vector having initial point (0, 0) and terminal point (1, 0). We denote by **j** the vector having initial point (0, 0) and terminal point (0, 1). If point P has coordinates (a, b), then $\overrightarrow{OP}$ is the sum or resultant vector of $a\mathbf{i} + b\mathbf{j}$ and $\overrightarrow{OP} = a\mathbf{i} + b\mathbf{j}$ (see the diagram). In fact, every vector in the plane can be expressed in terms of **i** and **j**.

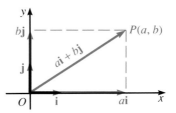

> If $\mathbf{u} = a\mathbf{i} + b\mathbf{j}$, the number a is the **x-component** of **u** and the number b is the **y-component** of **u**. When **u** is written in the form $a\mathbf{i} + b\mathbf{j}$, vector **u** is said to be in **component form**.

Example 1   Given the points $A(-3, 4)$ and $B(2, 1)$, express $\overrightarrow{AB}$ in component form.

Solution 1   Draw an equivalent vector, say $\overrightarrow{OP}$, with its initial point at the origin, as shown below. Then P has coordinates $(5, -3)$, and $\overrightarrow{AB} = \overrightarrow{OP} = 5\mathbf{i} - 3\mathbf{j}$.

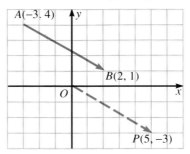

Solution 2   Notice that the x-component of $\overrightarrow{AB}$ is the difference of the x-coordinates of B and A. The y-component is the difference in the y-coordinates of B and A. $\therefore \overrightarrow{AB} = [2 - (-3)]\mathbf{i} + [1 - 4]\mathbf{j} = 5\mathbf{i} - 3\mathbf{j}$.

Example 2   Find the coordinates of B if A has coordinates $(3, -1)$ and $\overrightarrow{AB} = -\mathbf{i} + 2\mathbf{j}$.

Solution   Draw the vector $\overrightarrow{OP} = -\mathbf{i} + 2\mathbf{j}$.
Point P is 1 unit to the left and 2 units above O. If $\overrightarrow{AB} = \overrightarrow{OP}$, then B must be 1 unit to the left and 2 units above A.

$\therefore B$ has coordinates $(2, 1)$.   **Answer**

The concepts of Lesson 14-1 can also be expressed in terms of components.

If $\mathbf{u} = a\mathbf{i} + b\mathbf{j}$, $\mathbf{v} = c\mathbf{i} + d\mathbf{j}$ and t is a scalar, then:
$$\mathbf{u} = \mathbf{v} \text{ if and only if } a = c \text{ and } b = d$$
$$\mathbf{u} + \mathbf{v} = (a + c)\mathbf{i} + (b + d)\mathbf{j}$$
$$t\mathbf{u} = ta\mathbf{i} + tb\mathbf{j}$$
$$\|\mathbf{u}\| = \sqrt{a^2 + b^2}$$

The last statement, $\|\mathbf{u}\| = \sqrt{a^2 + b^2}$, results from applying the Pythagorean theorem to the vector diagram showing $\mathbf{u} = a\mathbf{i} + b\mathbf{j}$ (see the diagram below). The vectors $a\mathbf{i}$ and $b\mathbf{j}$ are perpendicular, and since $\mathbf{i}$ and $\mathbf{j}$ are unit vectors, $\|a\mathbf{i}\| = |a|$ and $\|b\mathbf{j}\| = |b|$.

Example 3 Let $\mathbf{u} = \mathbf{i} + 2\mathbf{j}$, $\mathbf{v} = 4\mathbf{i} - 2\mathbf{j}$, and $\mathbf{w} = 2\mathbf{u} - \mathbf{v}$.
a. Find $\mathbf{w}$ in component form.
b. Draw a diagram showing $\mathbf{u}$, $\mathbf{v}$, and $\mathbf{w}$.
c. Find $\|\mathbf{w}\|$.
d. Find the angle γ that $\mathbf{w}$ makes with the positive x-axis.

Solution **a.** $\mathbf{w} = 2\mathbf{u} - \mathbf{v} = 2(\mathbf{i} + 2\mathbf{j}) - (4\mathbf{i} - 2\mathbf{j})$ **b.**
$$= 2\mathbf{i} + 4\mathbf{j} - 4\mathbf{i} + 2\mathbf{j}$$
$$= -2\mathbf{i} + 6\mathbf{j}$$
c. $\|\mathbf{w}\| = \sqrt{(-2)^2 + 6^2} = \sqrt{40} = 2\sqrt{10}$
d. From part (a), $\mathbf{w}$ has x-component -2 and y-component 6. So

$$\tan \gamma = \frac{6}{-2} = -3$$
and
$$\gamma = 108.4°.$$

Example 4 The vector $\mathbf{u}$ makes an angle of $143°$ with the positive x-axis and $\|\mathbf{u}\| = 5$. Express $\mathbf{u}$ in component form.

Solution Let $\mathbf{u} = a\mathbf{i} + b\mathbf{j}$. If the initial point of $\mathbf{u}$ is placed at the origin, the terminal point will have coordinates (a, b). From the definitions of the trigonometric functions, you have

$$\cos 143° = \frac{a}{\|\mathbf{u}\|} \quad \text{and} \quad \sin 143° = \frac{b}{\|\mathbf{u}\|}.$$

(Solution continues on the next page.)

Solve: $\cos 143° = \dfrac{a}{\|\mathbf{u}\|}$ $\sin 143° = \dfrac{b}{\|\mathbf{u}\|}$

$$a = \|\mathbf{u}\| \cdot \cos 143°   \qquad b = \|\mathbf{u}\| \cdot \sin 143°$$
$$= 5(-0.7986)   \qquad\qquad = 5(0.6018)$$
$$= -3.99   \qquad\qquad\quad = 3.01$$

Therefore, $\mathbf{u} = -3.99\mathbf{i} + 3.01\mathbf{j}.$ *Answer*

The angle between two nonzero vectors is the angle θ, $0° \le \theta \le 180°$, that the vectors determine when their initial points are placed together. This angle is used to define the *dot product* of two vectors.

The **dot product** of two nonzero vectors $\mathbf{u}$ and $\mathbf{v}$ is defined to be

$$\mathbf{u} \cdot \mathbf{v} = \|\mathbf{u}\|\,\|\mathbf{v}\| \cos \theta$$

where θ is the angle between $\mathbf{u}$ and $\mathbf{v}.$

Caution: The dot product is a real number, not a vector.

The following useful theorem is proved in Exercise 31, page 670.

Theorem

If $\mathbf{u} = a\mathbf{i} + b\mathbf{j}$ and $\mathbf{v} = c\mathbf{i} + d\mathbf{j},$ then

$$\mathbf{u} \cdot \mathbf{v} = ac + bd.$$

Example 5   Use the dot product to find the angle between $\mathbf{u}$ and $\mathbf{v}$ if $\mathbf{u} = -2\mathbf{i} + \mathbf{j}$ and $\mathbf{v} = 4\mathbf{i} + 3\mathbf{j}.$

Solution   By the definition of dot product, $\mathbf{u} \cdot \mathbf{v} = \|\mathbf{u}\| \cdot \|\mathbf{v}\| \cos \theta.$ Solving for $\cos \theta$ gives

$$\cos \theta = \frac{\mathbf{u} \cdot \mathbf{v}}{\|\mathbf{u}\| \cdot \|\mathbf{v}\|} = \frac{(-2) \cdot 4 + 1 \cdot 3}{\sqrt{(-2)^2 + 1^2}\sqrt{4^2 + 3^2}}$$

$$= \frac{-5}{5\sqrt{5}} = \frac{-\sqrt{5}}{5} = -0.4472.$$

Therefore, $\theta = 116.6°.$ *Answer*

Two vectors are called **orthogonal** if either vector is $\mathbf{0}$ or if they are perpendicular. Since $\cos 90° = 0,$ the dot product of two vectors will be 0 whenever they are orthogonal.

Vectors $\mathbf{u}$ and $\mathbf{v}$ are orthogonal if and only if $\mathbf{u} \cdot \mathbf{v} = 0.$

Example 6   Find a unit vector orthogonal to $\mathbf{u} = 4\mathbf{i} - 3\mathbf{j}$.

Solution   Let $\mathbf{v} = x\mathbf{i} + y\mathbf{j}$. Then $\mathbf{u}$ and $\mathbf{v}$ are orthogonal if and only if

$$\mathbf{u} \cdot \mathbf{v} = 4x - 3y = 0.$$

One solution of this equation is $(3, 4)$, since $4(3) - 3(4) = 0$. So the vector $\mathbf{v} = 3\mathbf{i} + 4\mathbf{j}$ is orthogonal to $\mathbf{u}$. To find a unit vector orthogonal to $\mathbf{u}$, use the fact that any nonzero vector divided by its norm has length 1. So,

$$\frac{\mathbf{v}}{\|\mathbf{v}\|} = \frac{3\mathbf{i} + 4\mathbf{j}}{\sqrt{3^2 + 4^2}} = \frac{3\mathbf{i} + 4\mathbf{j}}{5} = \frac{3}{5}\mathbf{i} + \frac{4}{5}\mathbf{j}$$

is a unit vector orthogonal to $\mathbf{u}$.   *Answer*

Oral Exercises

Find s and t, given the following.

1. $s\mathbf{i} + t\mathbf{j} = 5\mathbf{i} - 2\mathbf{j}$

2. $-7\mathbf{i} + t\mathbf{j} = s\mathbf{i} + 7\mathbf{j}$

3. $(s + 2)\mathbf{i} + \mathbf{j} = 6\mathbf{i} + t\mathbf{j}$

4. $4s\mathbf{i} - t\mathbf{j} = 8\mathbf{j}$

If $\mathbf{u} = 3\mathbf{i} - 2\mathbf{j}$ and $\mathbf{v} = 4\mathbf{i} + 5\mathbf{j}$, find the following.

5. $\mathbf{u} + \mathbf{v}$

6. $\mathbf{v} - \mathbf{u}$

7. $2\mathbf{v}$

8. $2\mathbf{v} + \mathbf{u}$

9. $\|\mathbf{u}\|$

10. $\|\mathbf{v}\|$

11. $\mathbf{u} \cdot \mathbf{v}$

12. $\mathbf{v} \cdot \mathbf{u}$

Written Exercises

Refer to the diagram at the right and express each vector in component form.

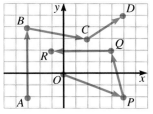

A   **1.** $\overrightarrow{OP}$

2. $\overrightarrow{PQ}$

3. $\overrightarrow{QR}$

4. $\overrightarrow{AB}$

5. $\overrightarrow{BC}$

6. $\overrightarrow{CD}$

Find the coordinates of B, given the following.

7. $A(4, -3)$, $\overrightarrow{AB} = 2\mathbf{i} - 4\mathbf{j}$ **8.** $A(-5, 1)$, $\overrightarrow{AB} = 2\mathbf{i} + \mathbf{j}$

Find the coordinates of A, given the following.

9. $B(4, 5)$, $\overrightarrow{AB} = 2\mathbf{i} - \mathbf{j}$ **10.** $B(0, -4)$, $\overrightarrow{AB} = 6\mathbf{i} - \mathbf{j}$

Find s and t, given that

11. $(s + 1)\mathbf{i} + 5\mathbf{j} = 2\mathbf{i} + (t - 2)\mathbf{j}$

12. $(3s - 2)\mathbf{i} + t\mathbf{j} = s\mathbf{i} + (2t - 3)\mathbf{j}$

13. $(s + t)\mathbf{i} + (s - t)\mathbf{j} = 5\mathbf{i} - \mathbf{j}$

14. $(s + 2t)\mathbf{i} + (s - t)\mathbf{j} = \mathbf{i}$

In Exercises 15–18:

a. Find **w** in component form.
b. Draw a diagram showing the vectors **u**, **v**, and **w**.
c. Find ‖**w**‖.
d. Find the angle γ, $0° \le \gamma \le 360°$, that **w** makes with the positive x-axis.

15. $\mathbf{u} = 4\mathbf{i} + 3\mathbf{j}$, $\mathbf{v} = 2\mathbf{i} - \mathbf{j}$; $\mathbf{w} = \mathbf{u} + \mathbf{v}$

16. $\mathbf{u} = \mathbf{i} - \mathbf{j}$, $\mathbf{v} = 5\mathbf{i} + 2\mathbf{j}$; $\mathbf{w} = \mathbf{u} - \mathbf{v}$

17. $\mathbf{u} = 4\mathbf{i} + 3\mathbf{j}$, $\mathbf{v} = 2\mathbf{i} - \mathbf{j}$; $\mathbf{w} = 2\mathbf{u} - \mathbf{v}$

18. $\mathbf{u} = -4\mathbf{i} + 3\mathbf{j}$, $\mathbf{v} = \mathbf{i} + 2\mathbf{j}$; $\mathbf{w} = 2\mathbf{u} + 3\mathbf{v}$

Use the dot product to find the angle between u and v.

19. $\mathbf{u} = 4\mathbf{i} + 3\mathbf{j}$, $\mathbf{v} = 2\mathbf{i} - \mathbf{j}$ **20.** $\mathbf{u} = 5\mathbf{i} - 3\mathbf{j}$, $\mathbf{v} = 2\mathbf{i} + 4\mathbf{j}$

21. $\mathbf{u} = 2\mathbf{i} - \mathbf{j}$, $\mathbf{v} = \mathbf{i} + 3\mathbf{j}$ **22.** $\mathbf{u} = 2\mathbf{i} + 3\mathbf{j}$, $\mathbf{v} = 6\mathbf{i} - 4\mathbf{j}$

Find a unit vector orthogonal to v.

23. $\mathbf{v} = \mathbf{i} + \mathbf{j}$ **24.** $\mathbf{v} = 5\mathbf{i} - 12\mathbf{j}$

25. $\mathbf{v} = 3\mathbf{i} - 4\mathbf{j}$ **26.** $\mathbf{v} = 2\mathbf{i} + 4\mathbf{j}$

In Exercises 27–30, α is the angle that vector **u** makes with the positive x-axis, and β is the angle that vector **v** makes with the positive x-axis.

a. Express **u** and **v** in component form.
b. Find $\mathbf{u} \cdot \mathbf{v}$ using the definition of dot product.
c. Find $\mathbf{u} \cdot \mathbf{v}$ using the theorem on page 668.

Give answers to three significant digits.

B **27.** ‖**u**‖ = 8, $\alpha = 20°$; ‖**v**‖ = 17, $\beta = 80°$

28. ‖**u**‖ = 6, $\alpha = 15°$; ‖**v**‖ = 10, $\beta = 65°$

29. ‖**u**‖ = 12.5, $\alpha = 68°$; ‖**v**‖ = 18.0, $\beta = 116°$

30. ‖**u**‖ = 8.70, $\alpha = 57°$; ‖**v**‖ = 6.60, $\beta = -28°$

C **31.** Let $\mathbf{u} = a\mathbf{i} + b\mathbf{j}$ and $\mathbf{v} = c\mathbf{i} + d\mathbf{j}$. Prove that $\mathbf{u} \cdot \mathbf{v} = ac + bd$ by applying the law of cosines to the triangle shown. [*Hint:* $\|\mathbf{u} - \mathbf{v}\|^2 = (a - c)^2 + (b - d)^2$.]

32. Prove that for any vector **u**, $\mathbf{u} = (\mathbf{u} \cdot \mathbf{i})\mathbf{i} + (\mathbf{u} \cdot \mathbf{j})\mathbf{j}$.

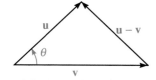

Computer Exercises   *For students with some programming experience.*

1. Write a program to find the magnitude of a vector given in component form, and the angle in degrees (measured counterclockwise) that it makes with the positive x-axis. Be aware that the function ATN(Z) returns the angle in *radians* in Quadrant I or IV whose tangent is Z. If the vector is vertical, you cannot use the ATN function to find the direction angle.

2. Run the program in Exercise 1 to find the magnitude of each vector and the angle it makes with the positive x-axis.

a. $7\mathbf{i} - 6\mathbf{j}$ **b.** $-15\mathbf{i} + 8\mathbf{j}$ **c.** $-57.3\mathbf{i} - 28.9\mathbf{j}$

3. Write a program to find the resultant of two vectors. Given the magnitudes of two vectors $\mathbf{u}$ and $\mathbf{v}$ and the angle in degrees that each makes with the positive x-axis, the program should find the magnitude of $\mathbf{u} + \mathbf{v}$ and the angle that it makes with the positive x-axis.
(*Hint:* Convert the given vectors to component form and add them. Then use the program of Exercise 1.)

4. Run the program in Exercise 3 to find the sum of each pair $\mathbf{u}$ and $\mathbf{v}$. α and β are the angles that $\mathbf{u}$ and $\mathbf{v}$ make with the positive x-axis.

a. $\|\mathbf{u}\| = 12.5,$ $\alpha = 26°;$ $\|\mathbf{v}\| = 17.4,$ $\beta = 115°$
b. $\|\mathbf{u}\| = 7.03,$ $\alpha = -15°;$ $\|\mathbf{v}\| = 9.22,$ $\beta = 78°$
c. $\|\mathbf{u}\| = 52.8,$ $\alpha = 129°;$ $\|\mathbf{v}\| = 43.3,$ $\beta = 216°$
d. $\|\mathbf{u}\| = 431,$ $\alpha = 65°;$ $\|\mathbf{v}\| = 256,$ $\beta = 190°$
e. $\|\mathbf{u}\| = 30,$ $\alpha = 210°;$ $\|\mathbf{v}\| = 40,$ $\beta = 300°$

Self-Test 1

Vocabulary
vector quantity (p. 659)
vector (p. 659)
initial point (p. 659)
terminal point (p. 659)
equivalent vectors (p. 659)
sum of vectors (p. 660)
resultant (p. 660)
parallelogram rule (p. 660)
difference of vectors (p. 661)

norm (p. 661)
unit vector (p. 661)
bearing (p. 661)
x-component (p. 666)
y-component (p. 666)
component form (p. 666)
dot product (p. 668)
orthogonal (p. 668)

1. If $\mathbf{u}$ has bearing $37°$ and magnitude 5, and $\mathbf{v}$ has bearing $125°$ and magnitude 6, find the magnitude of $2\mathbf{u} + \mathbf{v}$ and find its bearing to the nearest degree.   Obj. 14-1, p. 659

2. An ocean current flows due south at 15 km/h. If a ship is going to travel through this current on a course with heading $130°$ at a speed of 20 km/h, find its true-course bearing.

3. Find $\mathbf{w}$ in component form if $\mathbf{u} = 4\mathbf{i} + 3\mathbf{j},$ $\mathbf{v} = -2\mathbf{i} + 7\mathbf{j},$ and $\mathbf{w} = 2\mathbf{u} + 3\mathbf{v}.$ Obj. 14-2, p. 666

4. Use the dot product to find the angle between $\mathbf{u} = 5\mathbf{i} + 2\mathbf{j}$ and $\mathbf{v} = 8\mathbf{i} - 3\mathbf{j}$ to the nearest tenth of a degree.

Check your answers with those at the back of the book.

Application / Force, Work, and Energy

In the metric system the basic unit of the vector quantity *force* is the **newton** (N). Near the surface of Earth, the force exerted by gravity on a mass of m kg has a magnitude of approximately $9.8m$ newtons, directed downward. When you work with forces, you often use the following principle:

> When an object is at rest or moving with constant velocity, the sum of the forces acting on it is **0**. (If the sum of the forces is not **0**, the object will accelerate, and its velocity will vary.)

Example 1 A loading ramp makes an angle of 25° with the horizontal. A 120 kg crate slides down the ramp with constant velocity. What is the frictional force acting on the crate?

Solution The figure at the left above shows the three forces acting on the crate. The gravitational force **G** has magnitude

$$\|\mathbf{G}\| = 9.8 \times 120 = 1176 \text{ N.}$$

The forces **F** and **H** are respectively parallel and perpendicular to the ramp. **F** is the frictional force you need to find. Since the crate is moving with constant velocity, the sum of **G, H,** and **F** is **0**, as indicated in the vector diagram at the right above. From the diagram,

$$\frac{\|\mathbf{F}\|}{\|\mathbf{G}\|} = \sin 25°.$$

If you solve this equation for $\|\mathbf{F}\|$, you obtain

$$\|\mathbf{F}\| = \|\mathbf{G}\| \sin 25° = (1176)(0.4226) = 497.$$

∴ the frictional force is 497 N up the ramp. *Answer*

When you exert a force **F** on an object and move it from A to B, you do *work* and expend *energy*. If **F** has the same direction as $\mathbf{d} = \overrightarrow{AB}$, the **work** W done is

$$W = \|\mathbf{F}\| \, \|\mathbf{d}\|.$$

This is also the **energy** expended in moving the object from A to B.

The basic unit of work and energy is the *joule* (J). A **joule** is the work done when a force of one newton moves an object one meter in the direction of the force. Another unit of work and energy is the *kilowatt-hour* (kW · h). One **kilowatt-hour** is 3.6×10^6 J.

Example 2 How much energy in kilowatt-hours is needed to lift a 2500 kg elevator 80 m?

Solution The motor lifting the elevator must overcome the force of gravity and therefore must exert an upward force **F** of magnitude

$$2500 \times 9.8 = 2.45 \times 10^4 \text{ N.}$$

The displacement **d** is also upward and has norm 80 m. Therefore, the work done is

$$W = \|\mathbf{F}\| \, \|\mathbf{d}\|$$
$$= (2.45 \times 10^4) \times 80$$
$$= 1.96 \times 10^6 \text{ J.}$$

∴ the energy needed is

$$\frac{1.96 \times 10^6}{3.6 \times 10^6} = 0.544 \text{ kW · h.} \quad \textbf{\textit{Answer}}$$

Suppose now that a constant force **F** moves an object from A to B but makes an angle θ with $\mathbf{d} = \overrightarrow{AB}$ (see the diagram). In this case only the part of **F** in the direction of **d** is effective in doing work. The magnitude of this part is $\|\mathbf{F}\| \cos \theta$, and the work done by **F** is therefore

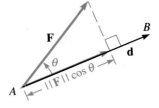

$$\|\mathbf{F}\| \cos \theta \times \|\mathbf{d}\|,$$

or

$$\|\mathbf{F}\| \, \|\mathbf{d}\| \cos \theta.$$

Since $\|\mathbf{F}\| \, \|\mathbf{d}\| \cos \theta = \mathbf{F} \cdot \mathbf{d},$ the work is given by

$$W = \mathbf{F} \cdot \mathbf{d}.$$

Example 3 Find the work done by the force $\mathbf{F} = 6\mathbf{i} - 3\mathbf{j}$ in moving an object from $A(0, 1)$ to $B(5, 2)$. The force is in newtons and the distance is in meters.

Solution $\mathbf{d} = \overrightarrow{AB} = (5 - 0)\mathbf{i} + (2 - 1)\mathbf{j} = 5\mathbf{i} + \mathbf{j}$

∴ $W = \mathbf{F} \cdot \mathbf{d} = (6\mathbf{i} - 3\mathbf{j}) \cdot (5\mathbf{i} + \mathbf{j})$
$= 6 \cdot 5 - 3 \cdot 1 = 27 \text{ J} \quad \textbf{\textit{Answer}}$

Exercises

Forces are in newtons and distances are in meters. Give work in joules unless kilowatt-hours are called for.

In Exercises 1 and 2, the two given forces act on an object. What additional force is needed to keep the object at rest?

A
1. $\mathbf{F}_1 = 3\mathbf{i} - 2\mathbf{j}$
   $\mathbf{F}_2 = -\mathbf{i} + 3\mathbf{j}$

2. $\mathbf{F}_1 = 5\mathbf{i} - \mathbf{j}$
   $\mathbf{F}_2 = -2\mathbf{i} + \mathbf{j}$

In Exercises 3 and 4, find the work done by the force F in moving an object from _A_ to _B_.

3. $\mathbf{F} = 3\mathbf{i} - \mathbf{j}$
   $A(-2, 0)$, $B(2, 3)$

4. $\mathbf{F} = 5\mathbf{i}$
   $A(-3, 2)$, $B(5, -1)$

5. How much work is done by the force $\mathbf{F} = 3\mathbf{i} - \mathbf{j}$ in moving an object (a) from $A(-2, 1)$ to $B(3, 3)$ to $C(5, 0)$? (b) from $A(-2, 1)$ to $C(5, 0)$?

6. What is the combined work done by the forces $\mathbf{F} = 2\mathbf{i} + 5\mathbf{j}$ and $\mathbf{G} = 4\mathbf{i} - 2\mathbf{j}$ in moving an object from $A(-1, 0)$ to $B(4, 1)$? What is the work done by $\mathbf{F} + \mathbf{G}$ in moving the object from _A_ to _B_?

In Exercises 7 and 8, the given forces act on an object. Find the magnitude and bearing of the additional force F that will keep the object at rest.

7. $\mathbf{F}_1$: 10 N, bearing 120°
   $\mathbf{F}_2$: 5 N, bearing 180°

8. $\mathbf{F}_1$: 6 N, bearing 300°
   $\mathbf{F}_2$: 15 N, bearing 30°

9. The mass of a loaded helicopter is 1500 kg. How much energy does its engine expend in ascending vertically for 100 m?

10. A ramp makes a 20° angle with the horizontal. A 50 kg crate slides down the ramp with constant velocity. Find the frictional force acting on the crate.

11. How much energy does a 75 kg man expend in climbing up a 12 m ladder that makes a 78° angle with the horizontal?

B
12. A conveyer belt carries coal up a 22° slope through a distance of 5 km measured along the slope. How many kilowatt-hours of energy are used in transporting 120 metric tons of coal? (1 metric ton = 1000 kg.)

13. A sled is pulled for 200 m up a 23° incline by a force of 120 N making a 40° angle with the horizontal. Find the work done.

In Exercises 14 and 15, use the fact that the tension in a wire, cable, or rope is the magnitude of the force it exerts.

14. A 20 kg mirror is hung from the middle of a wire as shown. Find the tension in the wire if each half of it makes a 30° angle with the horizontal.

15. A 200 kg load is suspended from two cables that make angles of 40° and 60° with the horizontal. Find the tension in each cable.

Polar Coordinates and Complex Numbers

14-3 Polar Coordinates

Objective   To define polar coordinates and graph polar equations.

If you know how far away an object is and in what direction, you can locate the object. This is the principle of the *polar coordinate system,* which is the coordinate system used to show the position of stars on a map such as the one shown at the right.

A **polar coordinate system** consists of a point O called the **pole,** and a ray, called the **polar axis,** having O as its endpoint (see Figure 1). The **polar coordinates** of a point P are an ordered pair (r, θ), where $r = OP$ and θ is the measure of an angle from the polar axis to the segment $\overline{OP}$.

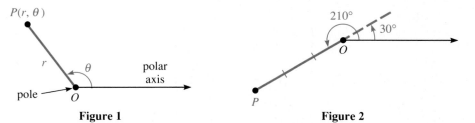

Figure 1   **Figure 2**

Each point has many pairs of polar coordinates. For example, $(3, 210°)$, $(3, -150°)$, and $(3, 570°)$ are all polar coordinates of the point P in Figure 2. *Negative* values of r are used to indicate that P is on the ray *opposite* to the terminal side of θ. Therefore, $(-3, 30°)$ and $(-3, 390°)$ are also polar coordinates of P in Figure 2.

Example 1   Graph these points in the same polar coordinate system:

$A(3, 120°)$
$B(-3, 45°)$
$C(4, -30°)$

Solution

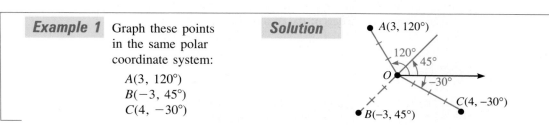

When polar coordinates and rectangular coordinates are used together, the polar axis is taken to coincide with the nonnegative x-axis, as shown in the diagram. The following formulas, which can be derived using the diagram, will help you to convert from one coordinate system to the other.

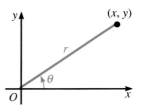

Coordinate-System Conversion Formulas

From polar to rectangular

$$x = r \cos \theta$$

$$y = r \sin \theta$$

From rectangular to polar

$$r = \pm\sqrt{x^2 + y^2}$$

$$\cos \theta = \frac{x}{r}, \quad \sin \theta = \frac{y}{r}$$

Example 2
a. Convert $A(-3, 30°)$ to rectangular coordinates.
b. Convert $B(4, -4)$ to polar coordinates.

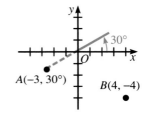

Solution
a. $x = r \cos \theta = -3 \cos 30° = -3\left(\dfrac{\sqrt{3}}{2}\right) = -\dfrac{3\sqrt{3}}{2}$

$y = r \sin \theta = -3 \sin 30° = -3\left(\dfrac{1}{2}\right) = -\dfrac{3}{2}$

rectangular coordinates: $\left(-\dfrac{3\sqrt{3}}{2}, -\dfrac{3}{2}\right)$

b. $r = \pm\sqrt{x^2 + y^2} = \pm\sqrt{(4)^2 + (-4)^2} = \pm 4\sqrt{2}$.

If you choose the positive value of r, then the terminal side of θ passes through B and B is a fourth quadrant angle. Since

$$\sin \theta = \frac{-4}{4\sqrt{2}} = -\frac{\sqrt{2}}{2},$$

you can use 315° for θ.

polar coordinates: $(4\sqrt{2}, 315°)$

Choosing the negative value of r would result in answers like $(-4\sqrt{2}, 135°)$ or $(-4\sqrt{2}, -225°)$, which are also correct.

Just like with graphs of equations in x and y, the graph of a polar equation in r and θ is the set of all points (r, θ) that satisfy the equation. To graph polar equations, you use your knowledge of the trigonometric functions.

Example 3 Graph $r = 3 \cos 3\theta$.

Solution To graph an equation in polar coordinate form, a polar coordinate graph should be used. Construct a table for θ.

θ	3θ	$\cos 3\theta$	$3 \cos 3\theta$
0°	0°	1.0000	3.00
15°	45°	0.7071	2.12
30°	90°	0.0000	0.00
45°	135°	−0.7071	−2.12
60°	180°	−1.0000	−3.00
75°	225°	−0.7071	−2.12
90°	270°	0.0000	0.00
105°	315°	0.7071	2.12
120°	360°	1.0000	3.00
135°	405°	0.7071	2.12
150°	450°	0.0000	0.00
165°	495°	−0.7071	−2.12
180°	540°	−1.0000	−3.00

Because the equation involves 3θ it is helpful to consider values of θ at intervals of 15°.

As θ increases from 180° to 360° the graph is repeated. The complete graph, called a *three-leafed rose,* is shown at the right.

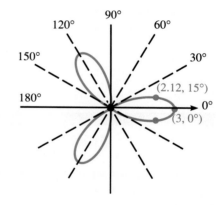

Sometimes a polar equation can be graphed more easily by using its rectangular form. The conversion formulas given on the preceding page can be used to transform equations from one system to another.

Example 4 **a.** Transform $r(1 + \cos \theta) = 3$ to a rectangular-coordinate equation.

b. Transform $x^2 + y^2 = 3y$ to a polar-coordinate equation.

Solution **a.** $r(1 + \cos \theta) = 3$

$r + r \cos \theta = 3$

$r = 3 - r \cos \theta$

$\sqrt{x^2 + y^2} = 3 - x$

$x^2 + y^2 = 9 - 6x + x^2$

$y^2 = 9 - 6x$ ***Answer***

b. $x^2 + y^2 = 3y$

$r^2 = 3r \sin \theta$

$r = 3 \sin \theta$

Answer

Oral Exercises

Give a pair of polar coordinates for each pair of rectangular coordinates.

1. $(-3, 0)$ **2.** $(0, -5)$ **3.** $(-5, 5)$ **4.** $(0, 0)$

Give a pair of rectangular coordinates for each pair of polar coordinates.

5. $(5, 180°)$ **6.** $(-3, 90°)$ **7.** $(\sqrt{2}, 45°)$ **8.** $(6, -135°)$

Give three pairs of polar coordinates (r, θ) for each point, including one with negative r and one with negative θ.

9. A **10.** B

11. C **12.** D

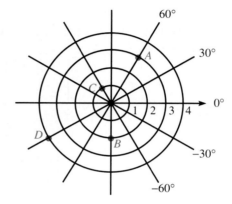

Written Exercises

Plot the point whose polar coordinates are given and find its rectangular coordinates. Give answers in simplest radical form.

A **1.** $(4, 30°)$ **2.** $(2, 45°)$ **3.** $(-3, 120°)$ **4.** $(-5, 135°)$

    5. $(7, -60°)$ **6.** $(6, -45°)$ **7.** $(-6, -150°)$ **8.** $(-3, -120°)$

Find a pair of polar coordinates for each pair of rectangular coordinates.

    9. $(4, 0)$ **10.** $(0, -3)$ **11.** $(-2, 2)$ **12.** $(\sqrt{3}, -1)$

    13. $(-1, \sqrt{3})$ **14.** $(-\sqrt{2}, -\sqrt{2})$ **15.** $(-\sqrt{5}, \sqrt{5})$ **16.** $(-\sqrt{2}, \sqrt{6})$

Find a polar equation for each rectangular-coordinate equation.

    17. $x = 5$ **18.** $y = -3$ **19.** $y = x$

    20. $x^2 + y^2 = 9$ **21.** $x^2 + y^2 = 6y$ **22.** $x^2 + y^2 + 8x = 0$

Find a rectangular-coordinate equation for each polar equation.

    23. $r = 2$ **24.** $\theta = 120°$ **25.** $r \sin \theta = 2$ **26.** $r \cos \theta = -1$

B **27.** $r = 2 \sin \theta$ **28.** $r + 2 \cos \theta = 0$ **29.** $r(1 - \cos \theta) = 2$ **30.** $r(1 - \sin \theta) = 1$

Graph using the method of Example 3.

31. $r = 2 \cos 2\theta$ \quad (Four-leafed rose)

32. $r = 4 \sin 3\theta$ \quad (Three-leafed rose)

33. $r = 1 + \sin \theta$ $\left.\begin{array}{c} \\ \\ \end{array}\right\}$ \quad (These heart-shaped curves are called *cardioids*.)

34. $r = 1 + \cos \theta$

C **35.** $r^2 = 2 \cos 2\theta$ $\left.\begin{array}{c} \\ \\ \end{array}\right\}$ \quad (These curves are called *lemniscates*.)

36. $r^2 = 4 \sin 2\theta$

Mixed Review Exercises

Find the quotient and the remainder.

1. $\dfrac{x^4 - 2x^2 + 3}{x + 2}$

2. $\dfrac{3x^3 + x^2 - 3x - 1}{3x + 1}$

3. $\dfrac{6x^2 - 4x + 7}{2x + 2}$

4. $\dfrac{1 - 2x + 3x^2 - 4x^3}{x - 2}$

Graph each equation.

5. $x^2 + 6x - y + 7 = 0$

6. $x^2 + y^2 - 4x + 2y - 4 = 0$

7. $7x + 5y = 10$

8. $y = \log_2 x$

9. $x = -3$

10. $3x^2 + y^2 = 27$

Extra / *Spirals*

Some polar equations have graphs that are spirals. In these equations the polar angle θ is usually measured in radians. For example, the *spiral of Archimedes* has an equation of the form $r = c\theta$, c a constant, θ in radians.

The *logarithmic spiral* has the equation $r = \frac{c}{\theta}$. And the *equiangular spiral* has an equation of the form $r = cb^\theta$, c and b constants.

Sketch the graph of each polar equation below for the given values of θ, using radian measure for θ. First determine what the largest value of r will be, so that you can use a convenient scale for r.

1. $r = \theta,\ 0 \le \theta \le 6\pi$

2. $r = \frac{1}{\theta},\ 0.2 \le \theta \le 4\pi$

3. $r = 2^\theta,\ -\pi \le \theta \le \pi$ (*Hint:* Choose values of θ for which 2^θ is easy to compute, such as $\theta = -3, -2, \ldots , 2, 3$. Use the fact that 3 radians is a bit less than half a complete rotation to draw the corresponding angles.)

14-4 The Geometry of Complex Numbers

Objective To plot complex numbers in the complex plane and to use the polar form of complex numbers.

You know that every complex number can be written in the form $x + yi$, where x and y are real numbers, and $i^2 = -1$. You can represent complex numbers in a coordinate plane by letting $x + yi$ correspond to the point (x, y), as shown in the diagram.

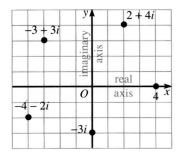

When the plane is used for this purpose, it is called the **complex plane.** The horizontal axis is called the **real axis,** and the vertical axis is called the **imaginary axis.**

Let $z = x + yi$ be any complex number. Geometrically, the **absolute value** of z, $|z| = \sqrt{x^2 + y^2}$, is the distance from the origin O to the point z; that is, $|z|$ is the magnitude of the vector $\overrightarrow{Oz}$. This is illustrated in Figure 1 below, which also shows the **conjugate** of z, $\bar{z} = x - yi$, and the negative of z, $-z = -x - yi$.

Figure 1

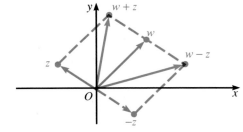

Figure 2

The sum and difference of $w = u + vi$ and $z = x + yi$ can be constructed by adding and subtracting the vectors $\overrightarrow{Ow}$ and $\overrightarrow{Oz}$, as shown in Figure 2.

Example 1 Plot $w = -2 + 4i$, $z = 3 - i$, $w + z$, and $w - z$ in the complex plane.

Solution

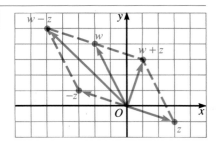

By introducing polar coordinates into the complex plane, you can express the real and imaginary parts of the complex number $z = x + yi$ in terms of its distance and direction from the origin. If the vector $\overrightarrow{Oz}$ has length r and makes an angle θ with the positive real axis, as shown at the right, then

$$x = r \cos \theta \quad \text{and} \quad y = r \sin \theta.$$

Therefore, the complex number $z = x + yi$ can be written as

$$z = (r \cos \theta) + (r \sin \theta)i = r(\cos \theta + i \sin \theta).$$

The **polar,** or **trigonometric, form** of a nonzero complex number $z = x + yi$ is given by

$$z = r(\cos \theta + i \sin \theta),$$

where $r = \sqrt{x^2 + y^2}$, and θ is an angle such that $\cos \theta = \dfrac{x}{r}$ and $\sin \theta = \dfrac{y}{r}$.

The angle θ is called the **amplitude,** or **argument,** of z. In this book, angle θ will be chosen such that $0° \le \theta < 360°$. The absolute value, $r = |z|$, is sometimes called the **modulus** of z.

Example 2 **a.** Write $-4 + 5i$ in polar form with absolute value in simplest radical form and amplitude to the nearest tenth of a degree.

b. Write -6 in polar form.

c. Write $4(\cos 300° + i \sin 300°)$ in $x + yi$ form using radicals.

d. Write $3(\cos 20° + i \sin 20°)$ in $x + yi$ form to four significant digits.

Solution **a.** $r = \sqrt{(-4)^2 + 5^2} = \sqrt{41}$

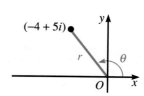

$$\sin \theta = \frac{5}{\sqrt{41}} = 0.7809$$

So the reference angle of θ is $51.3°$.
Since θ is a second-quadrant angle,
$\theta = 180° - 51.3° = 128.7°$.
$\therefore -4 + 5i = \sqrt{41}\,(\cos 128.7° + i \sin 128.7°)$.

b. $r = |-6| = 6$ and $\theta = 180°$.
$\therefore -6 = 6(\cos 180° + i \sin 180°)$.

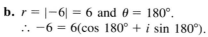

c. $\cos 300° = \dfrac{1}{2}$ and $\sin 300° = -\dfrac{\sqrt{3}}{2}$

$\therefore 4(\cos 300° + i \sin 300°) = 4\left(\dfrac{1}{2} - \dfrac{\sqrt{3}}{2}i\right) = 2 - 2i\sqrt{3}$.

d. $\cos 20° = 0.9397$ and $\sin 20° = 0.3420$
$\therefore 3(\cos 20° + i \sin 20°) = 3(0.9397 + 0.3420i) = 2.819 + 1.026i$.

The first part of the following theorem tells you that to multiply two complex numbers, you multiply their absolute values and add their amplitudes. The second part tells you that to divide two complex numbers, you divide their absolute values and subtract their amplitudes.

If $w = a(\cos \alpha + i \sin \alpha)$ and $z = b(\cos \beta + i \sin \beta)$, then

(1) $wz = ab[\cos (\alpha + \beta) + i \sin (\alpha + \beta)]$

(2) $\dfrac{w}{z} = \dfrac{a}{b} [\cos (\alpha - \beta) + i \sin (\alpha - \beta)]$.

Proof   Here is a proof of part (1). A proof of part (2) is asked for in Exercise 36.

$$wz = a(\cos \alpha + i \sin \alpha) \cdot b(\cos \beta + i \sin \beta)$$
$$= ab(\cos \alpha \cos \beta + i \cos \alpha \sin \beta + i \sin \alpha \cos \beta - \sin \alpha \sin \beta)$$
$$= ab[\cos \alpha \cos \beta - \sin \alpha \sin \beta + i(\sin \alpha \cos \beta + \cos \alpha \sin \beta)]$$
$$= ab[\cos (\alpha + \beta) + i \sin (\alpha + \beta)] \quad \checkmark$$

Example 3   Let $w = 3.1 (\cos 35° + i \sin 35°)$ and $z = 2.4(\cos 140° + i \sin 140°)$. Find wz and $\dfrac{w}{z}$ in polar form with $0° \le \theta < 360°$.

Solution $wz = (3.1)(2.4)[\cos (35° + 140°) + i \sin (35° + 140°)]$
$$= 7.44[\cos 175° + i \sin 175°] \quad \textit{Answer}$$

$$\dfrac{w}{z} = \dfrac{3.1}{2.4} [\cos (35° - 140°) + i \sin (35° - 140°)]$$

$$= 1.29[\cos (-105°) + i \sin (-105°)]$$

$$= 1.29[\cos 255° + i \sin 255°] \quad \textit{Answer}$$

Oral Exercises

Give each complex number in $x + yi$ form.

1. $5(\cos 90° + i \sin 90°)$ **2.** $4(\cos 180° + i \sin 180°)$

3. $\sqrt{2} (\cos 45° + i \sin 45°)$ **4.** $\sqrt{6} (\cos 135° + i \sin 135°)$

Give the polar form of each complex number.

5. 3   **6.** $5i$ **7.** $\sqrt{3} - i$ **8.** $-1 - i$

Tell what happens to the absolute value and the amplitude of a complex number when it is multiplied by each of the following numbers.

9. i **10.** -1 **11.** -2 **12.** $-i$

Written Exercises

Plot w, z, $w + z$, and $w - z$ in the complex plane.

A
1. $w = 1 - i$, $z = 4$
2. $w = -2 - 3i$, $z = -2 + 3i$
3. $w = 3 + i$, $z = 3 - i$
4. $w = 2i$, $z = 2 + i$
5. $w = 4 - 3i$, $z = -2 + 5i$
6. $w = -1 + 3i$, $z = 4 - i$

Find wz and $\dfrac{w}{z}$ in polar form with $0° \leq \theta < 360°$.

7. $w = 5(\cos 30° + i \sin 30°)$, $z = 2(\cos 80° + i \sin 80°)$
8. $w = 4(\cos 0° + i \sin 0°)$, $z = 3(\cos 130° + i \sin 130°)$
9. $w = 4.5(\cos 150° + i \sin 150°)$, $z = 1.2(\cos 315° + i \sin 315°)$
10. $w = 6.3(\cos 160° + i \sin 160°)$, $z = 2.0(\cos 210° + i \sin 210°)$

Write in $x + yi$ form using radicals.

11. $3(\cos 30° + i \sin 30°)$
12. $2(\cos 120° + i \sin 120°)$
13. $4.5(\cos 120° + i \sin 120°)$
14. $4.0(\cos 240° + i \sin 240°)$

Write in $x + yi$ form to four significant digits. You may wish to use a calculator.

15. $5(\cos 36° + i \sin 36°)$
16. $3(\cos 20° + i \sin 20°)$
17. $2.2(\cos 150° + i \sin 150°)$
18. $0.8(\cos 250° + i \sin 250°)$

Write in polar form with absolute value in simplest radical form and amplitude to the nearest tenth of a degree.

19. $1 - i\sqrt{3}$
20. $2\sqrt{2}(1 - i)$
21. $2\sqrt{3}(-1 + i)$
22. $12 + 5i$
23. $-3 - 4i$
24. $5 - i$
25. $3 - 2i$
26. $-\sqrt{5} + i$

In Exercises 27–30:

a. Use the theorem on page 682 to find wz.

b. Convert the answer in part (a) to $x + yi$ form and simplify.

c. Convert w and z to $x + yi$ form.

d. Multiply w and z in part (c) and simplify.

(Leave your answers to parts (b) and (d) in simplest radical form. They should agree.)

B
27. $w = 3(\cos 120° + i \sin 120°)$, $z = 6(\cos 150° + i \sin 150°)$
28. $w = 2(\cos 60° + i \sin 60°)$, $z = 7(\cos 30° + i \sin 30°)$
29. $w = 2(\cos 30° + i \sin 30°)$, $z = 4(\cos 120° + i \sin 120°)$
30. $w = 8(\cos 150° + i \sin 150°)$, $z = 4(\cos 60° + i \sin 60°)$

In Exercises 31 and 32, let w and z be any two nonzero complex numbers.

31. Show that $|wz| = |w| \cdot |z|$. (*Hint:* Use polar form.)

32. Use geometry to show that $|w + z| \le |w| + |z|$.

In Exercises 33–36, let w and z be any two complex numbers.

33. Show that $z\bar{z} = |z|^2$. (*Hint:* Use $z = x + yi$.)

34. Show that $\dfrac{1}{z} = \dfrac{\bar{z}}{|z|^2}$. (*Hint:* Use Exercise 33.)

C **35.** Show that if $z = b(\cos \beta + i \sin \beta)$, then

$$\frac{1}{z} = \frac{1}{b}(\cos \beta - i \sin \beta).$$

(*Hint:* Use Exercise 34.)

36. Prove the formula for $\dfrac{w}{z}$ stated on page 682.

(*Hint:* $\dfrac{w}{z} = w \cdot \dfrac{1}{z}$. Use Exercise 35 and the product formula.)

37. Explain why the shaded triangles in the diagram at the right are similar.

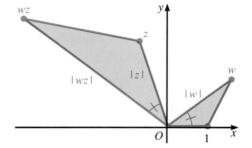

Biographical Note / Henri Poincaré

Henri Poincaré (1854–1912), whose broad vision encompassed all branches of mathematics, was interested also in the psychology of mathematical discovery and the role of beauty and harmony in mathematical thought.

Poincaré was born in Nancy, France. His mathematical genius was recognized while he was still in school. His ability to solve complex problems without hesitation was coupled with a physical clumsiness that limited him in sports and geometry. His drawings were in fact so bad that he would have to label them, "this is a horse, this is a tree. . . . "

During his lifetime Poincaré published more than 500 papers on new mathematics as well as more than thirty books. This outpouring revolutionized practically all branches of mathematical physics,

theoretical physics, and astronomy as they existed at that time.

14-5 De Moivre's Theorem

Objective   To use De Moivre's theorem.

You can use the product formula (formula (1), page 682) to find powers of complex numbers in polar form. For example, if $z = r(\cos \theta + i \sin \theta)$, then

$$z^2 = z \cdot z = r(\cos \theta + i \sin \theta) \cdot r(\cos \theta + i \sin \theta)$$
$$= r \cdot r[\cos (\theta + \theta) + i \sin (\theta + \theta)]$$
$$\therefore z^2 = r^2 (\cos 2\theta + i \sin 2\theta)$$

Also, $$z^3 = z \cdot z^2 = [r(\cos \theta + i \sin \theta)][r^2 (\cos 2\theta + i \sin 2\theta)]$$
$$= r \cdot r^2[\cos (\theta + 2\theta) + i \sin (\theta + 2\theta)]$$
$$\therefore z^3 = r^3(\cos 3\theta + i \sin 3\theta)$$

These results can be generalized.

De Moivre's Theorem

If $z = r(\cos \theta + i \sin \theta)$, and n is a positive integer, then

$$z^n = r^n(\cos n\theta + i \sin n\theta).$$

Example 1   Use De Moivre's Theorem to find $(-1 + i\sqrt{3})^8$. Give the answer in $x + yi$ form.

Solution   Let $z = -1 + i\sqrt{3}$. Then write z in polar form.

$$|z| = 2, \text{ so } z = 2\left(-\frac{1}{2} + i \frac{\sqrt{3}}{2}\right).$$

Since $-\frac{1}{2} = \cos 120°$ and $\frac{\sqrt{3}}{2} = \sin 120°$, you have

$$z = 2(\cos 120° + i \sin 120°).$$

Now use De Moivre's Theorem.

$$z^8 = 2^8[\cos (8 \cdot 120°) + i \sin (8 \cdot 120°)]$$
$$= 256(\cos 960° + i \sin 960°)$$
$$= 256[\cos (240° + 2 \cdot 360°) + i \sin (240° + 2 \cdot 360°)]$$
$$= 256(\cos 240° + i \sin 240°) \qquad \text{Simplify.}$$
$$= 256\left(-\frac{1}{2} - i\frac{\sqrt{3}}{2}\right) \qquad \text{Convert to } x + yi \text{ form.}$$
$$= -128 - 128i\sqrt{3}$$
$$\therefore (-1 + i\sqrt{3})^8 = -128 - 128i\sqrt{3} \quad \textbf{\textit{Answer}}$$

Example 2 Find the cube roots of $2 - 2i$
a. in polar form;
b. in $x + yi$ form, with x and y in simplest radical form.

Solution **a.** Write $2 - 2i$ in polar form:
$$2 - 2i = 2\sqrt{2}\left(\frac{\sqrt{2}}{2} - i\frac{\sqrt{2}}{2}\right) = 2\sqrt{2}(\cos 315° + i \sin 315°)$$

Let $z = r(\cos \theta + i \sin \theta)$ be a cube root of $2 - 2i$.
Then $z^3 = 2 - 2i$ and by De Moivre's Theorem
$$z^3 = r^3(\cos 3\theta + i \sin 3\theta) = 2\sqrt{2}(\cos 315° + i \sin 315°).$$

From this equation,
$$r^3 = 2\sqrt{2} = 2^{3/2}, \text{ so } r = 2^{1/2} = \sqrt{2}.$$
Also, $\cos 3\theta = \cos 315°$ and $\sin 3\theta = \sin 315°$.

Therefore, 3θ differs from $315°$ by an integral multiple of $360°$:
$$3\theta = 315° + k \cdot 360° \text{ (k is an integer)}$$
$$\theta = 105° + k \cdot 120° \text{ (k is an integer)}$$
Substituting 0, 1, and 2 for k gives
$$\theta = 105°, \ 225°, \ 345°.$$

(Other values of k give angles coterminal with one of these angles.)

$\therefore 2 - 2i$ has these three cube roots:
$$\sqrt{2}(\cos 105° + i \sin 105°)$$
$$\sqrt{2}(\cos 225° + i \sin 225°)$$
$$\sqrt{2}(\cos 345° + i \sin 345°)$$

b. The root $\sqrt{2}(\cos 225° + i \sin 225°) = \sqrt{2}\left(-\frac{\sqrt{2}}{2} - i\frac{\sqrt{2}}{2}\right)$
$$= -1 - i.$$

To write the other two roots in simplest radical form, you can use the methods of Lesson 13-7 to obtain:
$$\sqrt{2}(\cos 105° + i \sin 105°) = \left(\frac{1 - \sqrt{3}}{2}\right) + i\left(\frac{1 + \sqrt{3}}{2}\right)$$
$$\sqrt{2}(\cos 345° + i \sin 345°) = \left(\frac{1 + \sqrt{3}}{2}\right) - i\left(\frac{1 - \sqrt{3}}{2}\right)$$

The nth roots of the number 1 are usually called the **nth roots of unity**. By following the pattern of Example 2, we can show that there are n nth roots of unity, given by this formula:
$$\cos \frac{k \cdot 360°}{n} + i \sin \frac{k \cdot 360°}{n} \quad (k = 0, 1, 2, \ldots, n - 1)$$

Example 3 Find the fifth roots of unity in polar form.

Solution Let $n = 5$ in the formula at the bottom of the preceding page. Then the five fifth roots of unity are as follows:

$$1$$
$$\cos 72° + i \sin 72°$$
$$\cos 144° + i \sin 144°$$
$$\cos 216° + i \sin 216°$$
$$\cos 288° + i \sin 288°$$

As shown in the diagram, the fifth roots of unity are evenly spaced around the unit circle. Therefore, they are the vertices of a regular pentagon inscribed in the circle.

Oral Exercises

1. If $z = 2(\cos 40° + i \sin 40°)$, give the polar form of z^2 and z^5.
2. Give the polar form of each complex number.
   a. i **b.** i^6 **c.** The two square roots of i
3. Let $z = 1 + i$. Give each complex number in $x + yi$ form.
   a. z^2 **b.** z^4 **c.** z^8 **d.** z^9

Written Exercises

Use De Moivre's theorem to find each power. Give answers in $x + yi$ form.

A
1. $(\sqrt{3} + i)^4$ 2. $(-1 - i)^6$ 3. $(-1 + i)^9$ 4. $(1 - i)^7$
5. $(1 - i\sqrt{3})^7$ 6. $(1 + i\sqrt{3})^{10}$ 7. $(-\sqrt{3} - i)^8$ 8. $(-\sqrt{3} + i)^9$

Find the required roots in polar form.

9. The ninth roots of unity   10. The 10th roots of unity
11. The cube roots of $4\sqrt{3} - 4i$ 12. The cube roots of $-4 + 4i\sqrt{3}$

Find the required roots in $x + yi$ form.

B
13. The cube roots of unity   14. The fourth roots of unity
15. The sixth roots of unity   16. The eighth roots of unity
17. The cube roots of -1 18. The cube roots of i
19. The fourth roots of $-6 + 6i\sqrt{3}$ 20. The fourth roots of $-6 - 6i\sqrt{3}$
21. Plot the complex number $z = 1 + i$. Then plot the following.
   a. z^2 **b.** z^3 **c.** z^4 **d.** z^5 **e.** z^6 **f.** z^7

C **22.** Let $w = \cos 60° + i \sin 60°$. **(a)** Show that w is a sixth root of unity.
(b) Show that the other sixth roots of unity are: w^2, w^3, w^4, w^5, w^6.

23. An nth root of unity, such as w in Exercise 22, whose powers give all of the other nth roots of unity, is said to be **primitive**. Show that w^5 is another primitive sixth root of unity. (*Hint:* Show that the six powers of w^5, namely $(w^5)^1$, $(w^5)^2$, . . . , $(w^5)^6$, simplify to the sixth roots of unity. For example, $(w^5)^2 = w^{10} = w^6 \cdot w^4 = 1 \cdot w^4 = w^4$.)

Mixed Review Exercises

Let $\mathbf{u} = 3\mathbf{i} + \mathbf{j}$, $\mathbf{v} = 2\mathbf{i} - 4\mathbf{j}$, and $\mathbf{w} = \mathbf{u} + \mathbf{v}$.

1. Find $\mathbf{w}$ in component form.

2. Draw a diagram showing $\mathbf{u}$, $\mathbf{v}$, and $\mathbf{w}$.

3. Find $\|\mathbf{w}\|$.

4. Find the angle γ that $\mathbf{w}$ makes with the positive x-axis.

Plot each pair of polar coordinates and find its rectangular coordinates.

5. $(4, 30°)$ **6.** $(3, -120°)$ **7.** $(-2, 150°)$

Self-Test 2

Vocabulary
pole (p. 675)
polar axis (p. 675)
polar coordinates (p. 675)
complex plane (p. 680)
real axis (p. 680)
imaginary axis (p. 680)
absolute value (p. 680)

conjugate (p. 680)
polar form (p. 681)
trigonometric form (p. 681)
amplitude (p. 681)
argument (p. 681)
modulus (p. 681)
nth roots of unity (p. 686)

1. Find the rectangular coordinates of $(4, -60°)$. Give your answer in simplest radical form.

Obj. 14-3, p. 675

2. Find a pair of polar coordinates (r, θ) for the point with rectangular coordinates $(-12, 5)$. Give r to the nearest tenth and θ to the nearest tenth of a degree.

3. Plot $w = 4 + 5i$, $z = -6 + 2i$, $w + z$, and $w - z$ in the complex plane.

Obj. 14-4, p. 680

4. Find wz and $\dfrac{w}{z}$ in polar form if $w = 6(\cos 35° + i \sin 35°)$ and $z = 3(\cos 70° + i \sin 70°)$.

5. Write $3 - 2i$ in polar form with absolute value to three significant digits and amplitude to the nearest tenth of a degree.

6. Use De Moivre's Theorem to find $(-\sqrt{3} + i)^6$ in $x + yi$ form.

Obj. 14-5, p. 685

Inverse Functions

14-6 The Inverse Cosine and Inverse Sine

Objective   To evaluate expressions involving the inverse cosine and inverse sine.

The equation $\cos x = y$ assigns many values of x to each value of y. For example, $\cos x = \frac{1}{2}$ if x has any of the values $\frac{\pi}{3}$, $-\frac{\pi}{3}$, or any number differing from one of these by an integral multiple of 2π (see below).

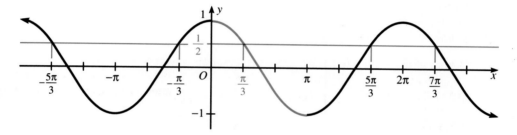

If, however, the domain of the cosine function is restricted to the interval $0 \le x \le \pi$, then only one value of x corresponds to each value of y. This new function, called Cosine, is a *one-to-one function* (recall Lesson 10-2).

By studying the graph of $y = \text{Cos } x$ at the right, you can see that for each number u such that $-1 \le u \le 1$, there is a unique number v $(0 \le v \le \pi)$ such that $\text{Cos } v = u$. For example, if $u = \frac{1}{2}$, then $v = \frac{\pi}{3}$. The number v is denoted by $\text{Cos}^{-1} u$ and is called the **inverse cosine** of u. $\left(\text{Cos}^{-1} u \text{ does } not \text{ mean } \frac{1}{\text{Cos } u}.\right)$ The inverse cosine is also called the **Arc cosine**, denoted **Arccos**.

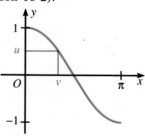

Example 1   Find the value of each of the following.

a. $\text{Cos}^{-1}(-1)$ **b.** $\text{Cos}^{-1} 0$ **c.** $\text{Cos}^{-1}\left(-\frac{\sqrt{2}}{2}\right)$

Solution   **a.** Let $v = \text{Cos}^{-1}(-1)$.
     Then $\cos v = -1$ and $0 \le v \le \pi$. So $v = \pi$ and therefore
$$\text{Cos}^{-1}(-1) = \pi. \quad \textit{Answer}$$

(Solution continues on the next page.)

b. Let $v = \text{Cos}^{-1}\ 0$. Then $\cos v = 0$ and $0 \le v \le \pi$.

So $v = \dfrac{\pi}{2}$ and therefore $\text{Cos}^{-1}\ 0 = \dfrac{\pi}{2}$.   *Answer*

c. Let $v = \text{Cos}^{-1}\left(-\dfrac{\sqrt{2}}{2}\right)$. Then $\cos v = -\dfrac{\sqrt{2}}{2}$ and $0 \le v \le \pi$.

So $v = \dfrac{3\pi}{4}$ and therefore $\text{Cos}^{-1}\left(-\dfrac{\sqrt{2}}{2}\right) = \dfrac{3\pi}{4}$.   *Answer*

The graph of $y = \text{Cos}\ x$, which is the red curve in the diagram at the right, has as its inverse the graph of

$$y = \text{Cos}^{-1}\ x,$$

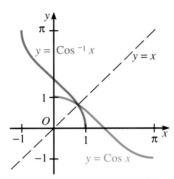

which is the blue curve in the diagram. Notice that the blue curve is the reflection across the line $y = x$ of the red curve.

The definition of the inverse cosine can be restated as follows.

$$y = \text{Cos}^{-1}\ x$$
if and only if
$$\cos y = x \quad \text{and} \quad 0 \le y \le \pi.$$

In defining the **inverse sine** function, denoted **$\text{Sin}^{-1}\ x$,** we reason as above. However, to obtain a *one-to-one function* Sine, we restrict the domain of the sine function to the interval $-\dfrac{\pi}{2} \le x \le \dfrac{\pi}{2}$.

$$y = \text{Sin}^{-1}\ x$$
if and only if
$$\sin y = x \quad \text{and} \quad -\dfrac{\pi}{2} \le y \le \dfrac{\pi}{2}.$$

The diagram at the right shows the graph of $y = \text{Sin}^{-1}\ x$ in blue and of $y = \text{Sin}\ x$ in red. The inverse sine function is also called the **Arc sine,** denoted **Arcsin.**

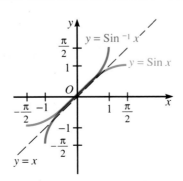

Example 2 Find the value of each of the following.

        a. $\text{Sin}^{-1}\left(\cos\frac{5\pi}{3}\right)$ **b.** $\text{Cos}^{-1}\left[\cos\left(\frac{7\pi}{6}\right)\right]$

        c. $\sin\left(\text{Sin}^{-1}\frac{1}{9}\right)$ **d.** $\cos\left[\text{Sin}^{-1}\left(-\frac{1}{5}\right)\right]$

Solution   **a.** $\text{Sin}^{-1}\left(\cos\frac{5\pi}{3}\right) = \text{Sin}^{-1}\left(\frac{1}{2}\right) = \frac{\pi}{6}$ *Answer*

        b. $\text{Cos}^{-1}\left[\cos\left(\frac{7\pi}{6}\right)\right] = \text{Cos}^{-1}\left(-\frac{\sqrt{3}}{2}\right) = \frac{5\pi}{6}$ *Answer*

        c. Let $v = \text{Sin}^{-1}\frac{1}{9}$.

            Then by the definition of the inverse sine, $\sin v = \frac{1}{9}$.

            $\therefore \sin\left(\text{Sin}^{-1}\frac{1}{9}\right) = \frac{1}{9}$ *Answer*

        d. Let $v = \text{Sin}^{-1}\left(-\frac{1}{5}\right)$. Then $\sin v = -\frac{1}{5}$, and $-\frac{\pi}{2} \leq v \leq \frac{\pi}{2}$.
            You want to find $\cos v$.

$$\cos v = \pm\sqrt{1 - \sin^2 v} = \pm\sqrt{1 - \left(-\frac{1}{5}\right)^2} = \pm\frac{2\sqrt{6}}{5}$$

            Since $\sin v$ is negative and $-\frac{\pi}{2} \leq v \leq \frac{\pi}{2}$, $\cos v$ is positive.

            $\therefore \cos\left[\text{Sin}^{-1}\left(-\frac{1}{5}\right)\right] = \frac{2\sqrt{6}}{5}$ *Answer*

    Although the values of the inverse sine and inverse cosine functions are real numbers, it is often convenient to think of them as measures of angles in radians. For example, to find $\cos\left[\text{Sin}^{-1}\left(-\frac{1}{5}\right)\right]$, you might make a sketch showing $v = \text{Sin}^{-1}\left(-\frac{1}{5}\right)$ as an angle with a sine of $-\frac{1}{5}$. Since $-\frac{\pi}{2} \leq v \leq \frac{\pi}{2}$, you can represent v as a fourth-quadrant

angle. Since $\sin v = -\frac{1}{5}$, the terminal side of v passes through a point 5 units from O with y-coordinate -1. By the Pythagorean theorem, the x-coordinate of this point is $2\sqrt{6}$, and so $\cos v = \frac{2\sqrt{6}}{5}$. Therefore, $\cos\left[\text{Sin}^{-1}\left(-\frac{1}{5}\right)\right] = \frac{2\sqrt{6}}{5}$.

    In some applications the values of inverse trigonometric functions may be given in degrees. For example, you might think of $\text{Cos}^{-1}\left(-\frac{1}{2}\right)$ as "the angle between $0°$ and $180°$ whose cosine is $-\frac{1}{2}$" and give its value as $120°$.

Oral Exercises

Give the value of each of the following.

1. $\text{Sin}^{-1}(-1)$ 2. $\text{Cos}^{-1} 1$ 3. $\text{Sin}^{-1} 0$ 4. $\text{Sin}^{-1}\left(-\frac{1}{2}\right)$

5. $\text{Cos}^{-1}\left(\frac{\sqrt{3}}{2}\right)$ 6. $\text{Cos}^{-1}\left(\frac{\sqrt{2}}{2}\right)$ 7. $\text{Sin}^{-1} 1$ 8. $\text{Sin}^{-1}\frac{\sqrt{2}}{2}$

Written Exercises

Find the value of each of the following.

A   1. $\text{Sin}^{-1}\frac{\sqrt{3}}{2}$ 2. $\text{Cos}^{-1}\left(-\frac{\sqrt{2}}{2}\right)$ 3. $\text{Cos}^{-1}\left(-\frac{\sqrt{3}}{2}\right)$

4. $\text{Sin}^{-1}\left[\sin\left(\frac{5\pi}{6}\right)\right]$ 5. $\text{Cos}^{-1}\left(\cos\frac{2\pi}{3}\right)$ 6. $\text{Sin}^{-1}\left(\sin\frac{3\pi}{2}\right)$

7. $\text{Cos}^{-1}\left(\cos\frac{5\pi}{3}\right)$ 8. $\text{Sin}^{-1}\left(\cos\frac{3\pi}{4}\right)$ 9. $\text{Cos}^{-1}\left[\sin\left(-\frac{\pi}{3}\right)\right]$

10. $\cos\left[\text{Sin}^{-1}\left(-\frac{\sqrt{3}}{2}\right)\right]$ 11. $\sin\left[\text{Cos}^{-1}\left(-\frac{\sqrt{3}}{2}\right)\right]$ 12. $\sin\left[\text{Cos}^{-1}\left(-\frac{4}{5}\right)\right]$

13. $\cos\left[\text{Sin}^{-1}\left(-\frac{5}{13}\right)\right]$ 14. $\cos\left(\text{Sin}^{-1}\frac{3}{4}\right)$ 15. $\sin\left[\text{Cos}^{-1}\left(-\frac{2}{5}\right)\right]$

B   16. $\cos\left(\text{Sin}^{-1}\frac{1}{2} + \text{Cos}^{-1}\frac{1}{2}\right)$ 17. $\sin\left[\text{Cos}^{-1}\left(-\frac{1}{2}\right) - \text{Sin}^{-1}\frac{1}{2}\right]$

Express without using trigonometric or inverse trigonometric functions.

Sample $\cos(2\,\text{Sin}^{-1} x)$

Solution   Let $y = \text{Sin}^{-1} x$. Then $\sin y = x$, where $-\frac{\pi}{2} \le y \le \frac{\pi}{2}$.

So $\cos y = \sqrt{1 - \sin^2 y} = \sqrt{1 - x^2}$.

$\therefore \cos(2\,\text{Sin}^{-1} x) = \cos 2y = \cos^2 y - \sin^2 y = 1 - 2x^2$ *Answer*

18. $\sin(\text{Cos}^{-1} x)$ 19. $\cos(\text{Sin}^{-1} x)$

20. $\cos(2\,\text{Cos}^{-1} x)$ 21. $\sin(2\,\text{Sin}^{-1} x)$

22. $\sin(\text{Sin}^{-1} u + \text{Sin}^{-1} v)$ 23. $\cos(\text{Sin}^{-1} 1 - \text{Sin}^{-1} v)$

Prove the following.

C   24. $\text{Cos}^{-1}\frac{3}{5} - \text{Cos}^{-1}\frac{4}{5} = \text{Sin}^{-1}\frac{7}{25}$ 25. $\text{Cos}^{-1}\frac{5}{13} - \text{Sin}^{-1}\frac{4}{5} = \text{Cos}^{-1}\frac{63}{65}$

14-7 Other Inverse Functions

Objective   To evaluate expressions involving the inverse trigonometric functions.

Figures 1 and 2 below suggest the following definitions of the **inverse tangent** function (denoted **Tan⁻¹ x**) and the **inverse cotangent** function (denoted **Cot⁻¹ x**).

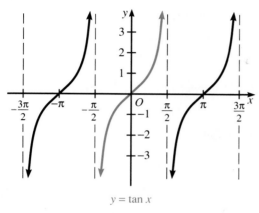

$y = \tan x$

Figure 1

$y = \cot x$

Figure 2

$y = \text{Tan}^{-1} x$
if and only if
$\tan y = x$ and $-\dfrac{\pi}{2} < y < \dfrac{\pi}{2}$.

$y = \text{Cot}^{-1} x$
if and only if
$\cot y = x$ and $0 < y < \pi$.

The graphs of these functions, shown below, are the reflections of the red curves of Figures 1 and 2 across the line $y = x$.

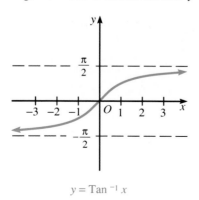

$y = \text{Tan}^{-1} x$

Figure 3

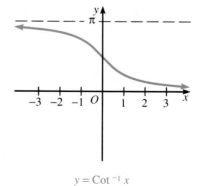

$y = \text{Cot}^{-1} x$

Figure 4

Example 1 Find the value of each of the following.

a. $\text{Tan}^{-1}(-1)$ **b.** $\text{Cot}^{-1} 0$ **c.** $\text{Tan}^{-1}\left(-\frac{\sqrt{3}}{3}\right)$

Solution

a. Let $y = \text{Tan}^{-1}(-1)$. Then $\tan y = -1$ and $-\frac{\pi}{2} < y < \frac{\pi}{2}$.

So $y = -\frac{\pi}{4}$ and therefore $\text{Tan}^{-1}(-1) = -\frac{\pi}{4}$. **Answer**

b. Let $y = \text{Cot}^{-1} 0$. Then $\cot y = 0$ and $0 < y < \pi$.

So $y = \frac{\pi}{2}$ and therefore $\text{Cot}^{-1} 0 = \frac{\pi}{2}$. **Answer**

c. Let $y = \text{Tan}^{-1}\left(-\frac{\sqrt{3}}{3}\right)$. Then $\tan y = -\frac{\sqrt{3}}{3}$ and $-\frac{\pi}{2} < y < \frac{\pi}{2}$.

So $y = -\frac{\pi}{6}$ and therefore $\text{Tan}^{-1}\left(-\frac{\sqrt{3}}{3}\right) = -\frac{\pi}{6}$. **Answer**

Example 2 Show that if $x > 0$, then $\text{Cot}^{-1} x = \text{Tan}^{-1} \frac{1}{x}$.

Solution Let $y = \text{Cot}^{-1} x$.

Then $\cot y = x$, and since x is positive, $0 < y < \frac{\pi}{2}$.

So $\tan y = \frac{1}{\cot y} = \frac{1}{x}$ and $0 < y < \frac{\pi}{2}$.

Since $\tan y = \frac{1}{x}$, $y = \text{Tan}^{-1} \frac{1}{x}$.

Therefore, by substitution, $\text{Cot}^{-1} x = \text{Tan}^{-1} \frac{1}{x}$.

The result of Example 2 uses the fact that the tangent and cotangent are reciprocal functions. This suggests that we define the **inverse secant** function (denoted **Sec**$^{-1}$ x) in terms of the inverse cosine and the **inverse cosecant** function (**Csc**$^{-1}$ x) in terms of the inverse sine.

If $|x| \geq 1$,

$$\text{Sec}^{-1} x = \text{Cos}^{-1} \frac{1}{x} \qquad \text{and} \qquad \text{Csc}^{-1} x = \text{Sin}^{-1} \frac{1}{x}.$$

Example 3 Find the value of $\sin\left[\text{Sec}^{-1}\left(\frac{4}{3}\right)\right]$.

Solution By the definition above, $\text{Sec}^{-1}\left(\frac{4}{3}\right) = \text{Cos}^{-1}\left(\frac{3}{4}\right)$.

Let $y = \text{Cos}^{-1}\left(\frac{3}{4}\right)$. Then $\cos y = \frac{3}{4}$ and $0 < y < \pi$.

So $\sin y = \sqrt{1 - \left(\frac{3}{4}\right)^2} = \sqrt{\frac{7}{16}} = \frac{\sqrt{7}}{4}$. Since $\cos y$ is positive and $0 < y < \pi$, $\sin y$ is positive.

$\therefore \sin\left[\text{Sec}^{-1}\left(\frac{4}{3}\right)\right] = \frac{\sqrt{7}}{4}$.

Example 4 Find the value of $\cot\left[\text{Cot}^{-1}\frac{4}{5} + \text{Sin}^{-1}\left(-\frac{4}{5}\right)\right]$.

Solution Let $a = \text{Cot}^{-1}\frac{4}{5}$ and $b = \text{Sin}^{-1}\left(-\frac{4}{5}\right)$. Since

$$\cot(a + b) = \frac{1}{\tan(a+b)} = \frac{1 - \tan a \tan b}{\tan a + \tan b},$$

you need to find $\tan a$ and $\tan b$. Since $a = \text{Cot}^{-1}\frac{4}{5}$, $\cot a = \frac{4}{5}$ and so $\tan a = \frac{5}{4}$. An easy way to find $\tan b$ is to think of b as the measure of an angle between $-\frac{\pi}{2}$ and $\frac{\pi}{2}$ whose sine is $-\frac{4}{5}$. Then make a sketch like the one at the right. Since $z = 3$, you have $\tan b = -\frac{4}{3}$.

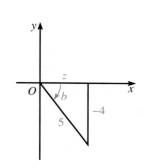

$$\therefore \cot\left[\text{Cot}^{-1}\frac{4}{5} + \text{Sin}^{-1}\left(-\frac{4}{5}\right)\right] = \frac{1 - \left(\frac{5}{4}\right)\left(-\frac{4}{3}\right)}{\frac{5}{4} + \left(-\frac{4}{3}\right)} = -32. \quad \textbf{\textit{Answer}}$$

Oral Exercises

Give the value of each of the following.

1. $\text{Cot}^{-1}(-1)$ 2. $\text{Tan}^{-1}(\sqrt{3})$ 3. $\text{Cot}^{-1}0$ 4. $\text{Csc}^{-1}(-1)$

5. $\text{Tan}^{-1}\frac{\sqrt{3}}{3}$ 6. $\text{Cot}^{-1}\frac{\sqrt{3}}{3}$ 7. $\text{Csc}^{-1}(-2)$ 8. $\text{Sec}^{-1}2$

Written Exercises

Find the value of each of the following.

A   1. $\text{Csc}^{-1}2$ 2. $\text{Sec}^{-1}(-1)$ 3. $\text{Cot}^{-1}\left(\cot\frac{5\pi}{6}\right)$

4. $\text{Sec}^{-1}\left[\sec\left(-\frac{\pi}{4}\right)\right]$ 5. $\text{Cot}^{-1}\left[\cot\left(-\frac{3\pi}{2}\right)\right]$ 6. $\text{Tan}^{-1}\left[\tan\frac{11\pi}{6}\right]$

7. $\text{Cot}^{-1}\left(\tan\frac{3\pi}{4}\right)$ **8.** $\text{Tan}^{-1}\left[\cot\left(-\frac{5\pi}{6}\right)\right]$ **9.** $\tan\left(\text{Cot}^{-1}2\right)$

10. $\sin\left(\text{Csc}^{-1}3\right)$ **11.** $\tan\left[\text{Sin}^{-1}\left(-\frac{3}{5}\right)\right]$ **12.** $\cot\left[\text{Sin}^{-1}\frac{12}{13}\right]$

13. $\sin\left(\text{Tan}^{-1}2\right)$ **14.** $\cos\left(\text{Cot}^{-1}\frac{1}{2}\right)$

B   **15.** $\tan\left(\text{Sin}^{-1}\frac{5}{13}+\text{Tan}^{-1}\frac{8}{15}\right)$ **16.** $\cos\left(\text{Tan}^{-1}\frac{4}{3}-\text{Sin}^{-1}\frac{5}{13}\right)$

     17. $\sin\left(2\,\text{Cot}^{-1}2\right)$ **18.** $\tan\left[2\,\text{Cos}^{-1}\left(-\frac{3}{5}\right)\right]$

Express without using trigonometric or inverse trigonometric functions.
(*Hint:* See the Sample on page 692.)

19. $\csc\left(\text{Sin}^{-1}x\right)$ **20.** $\cot\left(\text{Tan}^{-1}x\right)$

21. $\cos\left(\text{Tan}^{-1}x\right)$ **22.** $\sin\left(\text{Cot}^{-1}x\right)$

23. $\sin\left(2\,\text{Tan}^{-1}x\right)$ **24.** $\cos\left(2\,\text{Cot}^{-1}x\right)$

25. $\tan\left(\text{Cot}^{-1}x+\text{Cot}^{-1}y\right)$ **26.** $\cot\left(\text{Tan}^{-1}x-\text{Tan}^{-1}y\right)$

C   **27.** Graph $y=\text{Sec}^{-1}x$.   **28.** Graph $y=\text{Csc}^{-1}x$.

     29. In Example 2 we showed that

$$\text{Cot}^{-1}x=\text{Tan}^{-1}\frac{1}{x}\text{ if }x>0.$$

If $x<0$, then

$$\text{Cot}^{-1}x=\pi+\text{Tan}^{-1}\frac{1}{x}.$$

A proof of this identity starts like this: Let $y=\text{Cot}^{-1}x$.

Then $\cot y=x$ and, since $x<0$, $\frac{\pi}{2}<y<\pi$.

Let $z=y-\pi$.

Then $\cot(z+\pi)=x$ and $-\frac{\pi}{2}<z<0$.

Complete the proof using the fact that $\cot(z+\pi)=\cot z$.

Mixed Review Exercises

Determine whether each sequence is arithmetic or geometric. Then find the specified term of the sequence.

1. $7, 12, 17, \ldots\;;\; t_{21}$ **2.** $-3, -6, -12, \ldots\;;\; t_8$

3. $9, 18, 36, \ldots\;;\; t_{11}$ **4.** $5, -1, -7, \ldots\;;\; t_{15}$

Write each series using sigma notation. Then find the sum.

5. $3+12+48+\cdots+3072$ **6.** $5+8+11+\cdots+302$

7. $-5-1+3+\cdots+71$ **8.** $100+20+4+\cdots$

14-8 Trigonometric Equations

Objective   To solve trigonometric equations.

To solve equations involving trigonometric functions you usually use algebraic transformations and trigonometric identities. The strategy used in Examples 1 and 2 is to transform the equation so that it involves only one trigonometric function.

Example 1   Solve $\cos 2x = \sin x$, $0 \le x < 2\pi$.

Solution   Of the three formulas for $\cos 2x$, it is best to use $\cos 2x = 1 - 2 \sin^2 x$ so that $\sin x$ will be the only function in the transformed equation.

$$\cos 2x = \sin x$$
$$1 - 2 \sin^2 x = \sin x \qquad \text{Make one side equal 0.}$$
$$2 \sin^2 x + \sin x - 1 = 0 \qquad \text{Factor.}$$
$$(2 \sin x - 1)(\sin x + 1) = 0 \qquad \text{Solve.}$$

$$2 \sin x - 1 = 0 \qquad \text{or} \qquad \sin x + 1 = 0$$

$$\sin x = \frac{1}{2} \qquad\qquad\qquad \sin x = -1$$
$$x = \frac{\pi}{6}, \frac{5\pi}{6} \qquad\qquad\qquad x = \frac{3\pi}{2}$$

$\therefore$ the solution set over the interval $0 \le x < 2\pi$ is:

$$\left\{ \frac{\pi}{6}, \frac{5\pi}{6}, \frac{3\pi}{2} \right\} \quad \textbf{\textit{Answer}}$$

Example 2   Solve $\sin^2 \theta = \cos^2 \theta$. Find **(a)** the solutions in the interval $0° \le \theta < 360°$ and **(b)** the formulas giving general solutions.

Solution   **a.** $\sin^2 \theta = \cos^2 \theta$ Divide both sides by $\cos^2 \theta$ ($\cos \theta \neq 0$).

$$\frac{\sin^2 \theta}{\cos^2 \theta} = 1 \qquad \text{Use the identity } \frac{\sin \theta}{\cos \theta} = \tan \theta.$$

$$\tan^2 \theta = 1 \qquad \text{Take the square root of both sides.}$$

$$\tan \theta = \pm 1 \qquad \text{Solve.}$$

$$\tan \theta = 1 \qquad\qquad \text{or} \qquad\qquad \tan \theta = -1$$
$$\theta = 45°, 225° \qquad\qquad\qquad\qquad \theta = 135°, 315°$$

$\therefore$ the solution set over the interval $0° \le \theta < 360°$ is:

$$\{45°, 135°, 225°, 315°\} \quad \textbf{\textit{Answer}}$$

Notice in the first step of part (a) that you may divide by $\cos^2 \theta$ because there is no solution of the equation for which $\cos \theta = 0$.

b. To find formulas giving general solutions, you should recognize that all other solutions differ from 45° and 135° by integral multiples of 180°, the period of the tangent function. Therefore, all solutions are of this form:

$$\theta = 45° + k \cdot 180°, \ k \text{ is an integer}$$
$$\theta = 135° + k \cdot 180°, \ k \text{ is an integer}$$

∴ the set of general solutions is:

$$\{\theta\colon \theta = 45° + k \cdot 180° \quad \text{or} \quad \theta = 135° + k \cdot 180°\}$$
Answer

Solutions in a specified interval, usually $0 \le x < 2\pi$ or $0° \le \theta \le 360°$ as in Examples 1 and 2, are sometimes called **primary solutions.** Formulas that name all possible solutions, such as the ones in the second part of Example 2, give the **general solutions.**

Sometimes a calculator or tables must be used to solve a trigonometric equation.

Example 3 Solve $3 \cos^2 \theta - 8 \cos \theta + 4 = 0$ for θ, $0° \le \theta < 360°$.

Solution

| $3 \cos^2 \theta - 8 \cos \theta + 4 = 0$ | | Factor. |
| $(\cos \theta - 2)(3 \cos \theta - 2) = 0$ | | Solve. |

$\cos \theta - 2 = 0$ or $3 \cos \theta - 2 = 0$

$\cos \theta = 2$ $\qquad\qquad$ $\cos \theta = \dfrac{2}{3} = 0.6667$

No solution, since $-1 \le \cos \theta \le 1$.

To the nearest tenth of a degree, a calculator or table gives

$$\theta = 48.2°.$$

Another solution in the given interval is $\theta = 311.8°$.

∴ the solution set over the interval $0° \le \theta < 360°$ is:

$$\{48.2°, 311.8°\} \quad \textit{Answer}$$

Example 4 Solve $\sin (3\theta - 20°) = \dfrac{1}{2}$ for θ, $0° \le \theta < 360°$.

Solution Let $a = 3\theta - 20°$. Then $\sin a = \dfrac{1}{2}$, so

$a = 30° + k \cdot 360°$ or $a = 150° + k \cdot 360°$, where k is an integer.

Therefore, by substitution,

$3\theta - 20° = 30° + k \cdot 360°$	or	$3\theta - 20° = 150° + k \cdot 360°$
$3\theta = 50° + k \cdot 360°$		$3\theta = 170° + k \cdot 360°$
$\theta = 16.7° + k \cdot 120°$		$\theta = 56.7° + k \cdot 120°$

Substituting 0, 1, and 2 for k gives

$$\theta = 16.7°, \ 136.7°, \ 256.7° \qquad \text{or} \qquad \theta = 56.7°, \ 176.7°, \ 296.7°$$

$\therefore$ the solution set over the interval $0° \leq \theta < 360°$ is:

$$\{16.7°, \ 56.7°, \ 136.7°, \ 176.7°, \ 256.7°, \ 296.7°\}. \quad \textbf{\textit{Answer}}$$

Oral Exercises

How many solutions does each equation have in the interval $0 \leq x < 2\pi$?

1. $\cos x = \dfrac{4}{5}$

2. $\sin x = -1$

3. $\sin^2 x = \dfrac{1}{9}$

4. $\cos x(\tan x - 2) = 0$

5. $\cos x(\sin x - 2) = 0$

6. $\cos 2x = \dfrac{1}{2}$

7. $\sin \dfrac{x}{9} = \dfrac{1}{3}$

8. $\cos \dfrac{x}{9} = \dfrac{1}{3}$

9. In solving $2 \sin 2x = 1$, would you use a double-angle formula? If not, what *would* you do?

10. In solving $2 \cos \left(x - \dfrac{\pi}{4}\right) = 1$, would you use an addition formula? If not, what *would* you do?

Outline a strategy for solving each of the following equations.

11. $\cos \theta = \sin \theta$

12. $\tan x = \cos x$

13. $\sin 2x = \cos x$

14. $2 \cos (\theta - 25°) = 1$

15. $\sin 3\theta = 1$

16. $\sin 4x = \cos 2x$

Written Exercises

Find the primary solution of each equation for $0° \leq \theta < 360°$ or $0 \leq x < 2\pi$. Then find formulas giving the general solution.

Give your answer to the nearest tenth of a degree if the variable is θ. Give your answer to the nearest hundredth of a radian if the variable is x, unless you can express the answer in terms of π.

A

1. $4 \sin \theta = 3$

2. $7 \cos \theta = 2$

3. $2 \cos^2 x = 1$

4. $4 \sin^2 x = 3$

5. $\sin \theta + \cos \theta = 0$

6. $\csc \theta = \sec \theta$

7. $\csc \theta = 16 \sin \theta$

8. $5 \sin x + 2 \cos x = 0$

9. $\cot \left(x - \dfrac{\pi}{8}\right) = 1$

10. $2 \sin \left(x + \dfrac{\pi}{4}\right) = 1$

11. $\sin (\theta + 25°) = 0$

12. $\tan (\theta - 20°) = \sqrt{3}$

Trigonometric Applications **699**

Solve for θ, $0° \le \theta < 360°$, or for x, $0 \le x < 2\pi$.

13. $4 \cos \theta = 3 \sec \theta$

14. $3 \tan \theta = \cot \theta$

15. $3 \sec x = \csc x$

16. $2 \cos 2\theta = 1$

17. $\sec^2 \theta = 2$

18. $2 \cot^2 \theta = 1$

19. $\sin 2x - \cos x = 0$

20. $\sin 2\theta = 2 \sin \theta$

21. $\cos 2\theta = \cos \theta$

22. $\cos 2x = \sin x$

23. $2 \sin 2\theta = \sqrt{3}$

24. $2 \cos x = \csc x$

25. $2 \tan^2 \theta + \tan \theta - 3 = 0$

26. $6 \sin^2 \theta = \sin \theta + 1$

27. $\cos 2x = 11 \sin x + 6$

28. $\cos 2x + 7 \sin x + 3 = 0$

B   **29.** $\tan^2 \left(x - \dfrac{\pi}{3} \right) = 1$

30. $2 \cos^2 \left(x + \dfrac{\pi}{6} \right) = 1$

31. $\sin 3x + \cos 3x = 0$

32. $\tan^2 3x = 3$

33. $\sin 4\theta = \cos 2\theta$

34. $\cos 3x + \cos x = 0$

35. $\sin 2x = \cot x$

36. $2 \sin 2\theta = \tan \theta$

37. $\cos 2\theta = 2 \sin^2 \theta$

38. $\cos 2\theta + 2 \cos^2 \theta = 0$

39. $\cos (2\theta + 10°) = -\dfrac{1}{2}$

40. $\sin (3\theta - 25°) = \dfrac{\sqrt{3}}{2}$

41. $\sin (3x + \pi) = -\dfrac{\sqrt{2}}{2}$

42. $\tan \left(2x - \dfrac{\pi}{3} \right) = 1$

C   **43.** $\tan (\theta + 32°) = \cot (\theta - 20°)$

44. $\sin (\theta - 10°) = \cos \theta$

45. $\cos 2\theta = \sin \theta - 1$

46. $\cot^2 x - \cos^2 x = \cot^2 x \cos^2 x$

Self-Test 3

Vocabulary   one-to-one (p. 689)
inverse cosine, $\text{Cos}^{-1} x$ (p. 689)
inverse sine, $\text{Sin}^{-1} x$ (p. 690)
inverse tangent, $\text{Tan}^{-1} x$ (p. 693)
inverse cotangent, $\text{Cot}^{-1} x$ (p. 693)

inverse secant, $\text{Sec}^{-1} x$ (p. 694)
inverse cosecant, $\text{Csc}^{-1} x$ (p. 694)
primary solution (p. 698)
general solution (p. 698)

Find the value of each of the following.

1. $\text{Sin}^{-1} \left(-\dfrac{\sqrt{3}}{2} \right)$

2. $\text{Sin}^{-1} \left(\sin \dfrac{7\pi}{6} \right)$ Obj. 14-6, p. 689

3. $\text{Cos}^{-1} \left[\sin \left(-\dfrac{\pi}{3} \right) \right]$

4. $\cos \left[\text{Sin}^{-1} \left(-\dfrac{4}{5} \right) \right]$

5. $\text{Cot}^{-1} \left(-\dfrac{\sqrt{3}}{3} \right)$

6. $\sec \left(\text{Tan}^{-1} \dfrac{8}{15} \right)$ Obj. 14-7, p. 693

7. $\text{Csc}^{-1} \dfrac{2\sqrt{3}}{3}$

8. $\cos \left[\text{Sec}^{-1} (-5) \right]$

9. Solve the equation $3 \sin^2 \theta - \cos^2 \theta = 0$, $0° \le \theta < 360°$.   Obj. 14-8, p. 697
Give θ to the nearest tenth of a degree.

Computer Key-In

Mathematicians have used the inverse tangent function and the infinite series for $\text{Tan}^{-1} x$ shown below to approximate the irrational number π.

$$\text{Tan}^{-1} x = x - \frac{x^3}{3} + \frac{x^5}{5} - \frac{x^7}{7} + \cdots + (-1)^{n-1} \frac{x^{2n-1}}{2n-1} + \cdots \qquad \textbf{(1)}$$

This series, which was obtained by the Scottish mathematician James Gregory in 1671, gives the value of $\text{Tan}^{-1} x$ for any real number x with $|x| \leq 1$. If $x = 1$, equation (1) becomes

$$\frac{\pi}{4} = \text{Tan}^{-1} 1 = 1 - \frac{1}{3} + \frac{1}{5} - \frac{1}{7} + \cdots$$

By summing a large number of terms on the right and multiplying the sum by 4, you can approximate π. If you use 1000 terms, however, the approximation is accurate to only three decimal places.

You can use equation (1) to obtain a more efficient method for approximating π by applying the following formula. Let a and b be real numbers with $|a| < 1$ and $|b| < 1$. Then

$$\text{Tan}^{-1} a + \text{Tan}^{-1} b = \text{Tan}^{-1} \left(\frac{a+b}{1-ab} \right).$$

If a and b are chosen so that $\frac{a+b}{1-ab} = 1$, then

$$\text{Tan}^{-1} a + \text{Tan}^{-1} b = \frac{\pi}{4}.$$

Since a and b are less than 1 in absolute value, you can compute $\text{Tan}^{-1} a$ and $\text{Tan}^{-1} b$ using equation (1) and then add these values and multiply by 4 to find π. For example, you can use $a = \frac{1}{2}$ and $b = \frac{1}{3}$ since $\dfrac{\frac{1}{2} + \frac{1}{3}}{1 - \frac{1}{2} \cdot \frac{1}{3}} = 1$.

Using 15 terms in equation (1), you obtain for π the value

$$4(0.463647609 + 0.321750551) = 3.141592640,$$

which is correct to seven decimal places. (The value of π to ten places is 3.1415926536.)

The computer program which follows prints an approximation for π when you enter a value of N (the number of terms used in equation (1)) and values of a and b.

```
10 INPUT "ENTER VALUES FOR N, A, B: "; N, A, B
20 LET X=A
30 GOSUB 110
40 LET P=S
50 LET X=B
60 GOSUB 110
70 LET P=P+S
80 PRINT "AN APPROXIMATE VALUE FOR PI IS: ";
```

(Program continues on the next page.)

```
90 PRINT 4*P
100 END
110 LET S=X
120 FOR J=2 TO N
130 LET T=(−1)∧(J−1)*X∧(2*J−1)/(2*J−1)
140 LET S=S+T
150 NEXT J
160 RETURN
```

Exercises

1. For each pair (a, b) given below, run the program for $N = 5$, $N = 10$, and $N = 15$, and record the approximations for π. (Note that you must enter the fractions as decimals; for the repeating decimals, enter as many places as your computer will accept.)

a. $\left(\dfrac{1}{4}, \dfrac{3}{5}\right)$ **b.** $\left(\dfrac{6}{11}, \dfrac{5}{17}\right)$ **c.** $\left(\dfrac{5}{6}, \dfrac{1}{11}\right)$ **d.** $\left(\dfrac{2}{5}, \dfrac{3}{7}\right)$

 e. For a given value of N, which pair above gives the best approximation?

2. a. Find another pair (a, b) that meets the requirements. (*Hint:* Choose an a and solve for b.)

 b. Run the program for your pair, using $N = 15$. Record the approximation.

3. Prove the formula for $\text{Tan}^{-1} a + \text{Tan}^{-1} b$. *Hint:* Let $x = \text{Tan}^{-1} a$ and $y = \text{Tan}^{-1} b$, so that $\tan x = a$ and $\tan y = b$. Use the addition formula for the tangent (page 650) to find $\tan(x + y)$. Then use the fact that

$$\text{Tan}^{-1}(\tan(x + y)) = x + y \text{ if } -\frac{\pi}{2} < x + y < \frac{\pi}{2}.$$

Chapter Summary

1. *Vector quantities* have both magnitude and direction. A given vector quantity can be represented by an arrow called a *vector*. Vectors can be added, subtracted, and multiplied by real numbers. Two vectors are *equivalent* if they have the same magnitude and direction. If two vectors are perpendicular or if either one is **0**, then they are *orthogonal*.

2. A vector can be expressed in component form, $\mathbf{u} = a\mathbf{i} + b\mathbf{j}$, where the numbers a and b are the *x-* and *y-components*, respectively, of $\mathbf{u}$. The *dot product* of two vectors can be used to find the angle between them.

3. The *polar coordinates* of a point P are (r, θ), where r is the distance from point O, called the pole, to point P, and θ is the measure of the angle that $\overrightarrow{OP}$ makes with the *polar axis*.

4. The complex number $z = x + yi$ can be represented by the point (x, y) in the *complex plane*. The *polar*, or *trigonometric*, *form* of z is

$$z = r(\cos \theta + i \sin \theta),$$

where (r, θ) are the polar coordinates of the point z.

5. Powers and roots of complex numbers can be found with the help of *De Moivre's theorem:*

$$z^n = r^n (\cos n\theta + i \sin n\theta).$$

6. When the domains of the trigonometric functions are restricted to certain intervals, they become one-to-one, and the *inverse trigonometric functions* can be defined. For example:

$y = \mathrm{Sin}^{-1} x$ if and only if $\sin y = x$ and $-\dfrac{\pi}{2} \le y \le \dfrac{\pi}{2}$;

$y = \mathrm{Cos}^{-1} x$ if and only if $\cos y = x$ and $0 \le y \le \pi$;

$y = \mathrm{Tan}^{-1} x$ if and only if $\tan y = x$ and $-\dfrac{\pi}{2} < y < \dfrac{\pi}{2}$.

7. A *trigonometric equation* can be solved by using algebraic transformations and trigonometric identities.

Chapter Review

Write the letter of the correct answer.

1. A ship leaves port on a heading of 172° with a speed of 25 km/h. The current is flowing 6 km/h at a heading of 200°. What is the actual heading of the ship to the nearest tenth of a degree?
   a. 5.3° **b.** 166.7° **c.** 177.3° **d.** 194.7°

   14-1

2. Vector **u** has magnitude 6 and bearing 40°. Vector **v** has magnitude 10 and bearing 160°. Find $\|\mathbf{v} - \mathbf{u}\|$ to the nearest hundredth.
   a. 4.00 **b.** 8.72 **c.** 14.00 **d.** 11.66

3. Find the coordinates of A, given $B(4, 3)$ and $\overrightarrow{AB} = -5 + 4i$.
   a. $(9, -1)$ **b.** $(-1, 7)$ **c.** $(9, -7)$ **d.** $(1, -1)$

   14-2

4. Find a unit vector orthogonal to $9\mathbf{i} - 12\mathbf{j}$.
   a. $\dfrac{\sqrt{2}}{2}\mathbf{i} + \dfrac{\sqrt{2}}{2}\mathbf{j}$ **b.** $4\mathbf{i} + 3\mathbf{j}$ **c.** $-\dfrac{3}{5}\mathbf{i} + \dfrac{4}{5}\mathbf{j}$ **d.** $-\dfrac{4}{5}\mathbf{i} - \dfrac{3}{5}\mathbf{j}$

5. Find a polar equation of the curve whose rectangular-coordinate equation is $x + y - 2 = 0$.
   a. $r = 2 \sin \theta$ **b.** $r = 2 \cos \theta$ **c.** $r = \dfrac{2}{\cos \theta + \sin \theta}$ **d.** $r = \dfrac{-2}{\cos \theta + \sin \theta}$

   14-3

6. Find the rectangular coordinates of the point $(5, -60°)$.
   a. $\left(\dfrac{5}{2}, \dfrac{-5\sqrt{3}}{2}\right)$ **b.** $\left(\dfrac{5\sqrt{3}}{2}, -\dfrac{5}{2}\right)$ **c.** $\left(-\dfrac{5}{2}, -\dfrac{5\sqrt{3}}{2}\right)$ **d.** $\left(-\dfrac{5\sqrt{3}}{2}, -\dfrac{5}{2}\right)$

7. Express $2 - 2i\sqrt{3}$ in polar form.
   a. $2 (\cos 300° + i \sin 300°)$ **b.** $2 (\cos 120° + i \sin 120°)$
   c. $4 (\cos 60° + i \sin 60°)$ **d.** $4 (\cos 300° + i \sin 300°)$

   14-4

8. Find $\dfrac{w}{z}$ if $w = 10 (\cos 83° + i \sin 83°)$ and $z = 5 (\cos 113° + i \sin 113°)$.
   a. $2i$ **b.** $\sqrt{3} - i$ **c.** $2\sqrt{3} - 2i$ **d.** $1 - i\sqrt{3}$

9. Express $\left(-\dfrac{\sqrt{2}}{2} + \dfrac{\sqrt{2}}{2}i\right)^{10}$ in $x + yi$ form.

14-5

   a. 1   **b.** -1 **c.** i **d.** $-i$

10. Which of the following is not a cube root of $-i$?

   a. i **b.** $-i$ **c.** $\dfrac{\sqrt{3}}{2} - \dfrac{1}{2}i$ **d.** $-\dfrac{\sqrt{3}}{2} - \dfrac{1}{2}i$

11. Find the value of $\mathrm{Sin}^{-1}\left[\sin\left(-\dfrac{4\pi}{3}\right)\right]$.

14-6

   a. $\dfrac{\pi}{3}$ **b.** $-\dfrac{\pi}{3}$ **c.** $\dfrac{2\pi}{3}$ **d.** $\dfrac{5\pi}{6}$

12. Express $\cos(2\,\mathrm{Sin}^{-1} x)$ without using trigonometric or inverse trigonometric functions.

   a. $2x\sqrt{1 - x^2}$ **b.** $1 - 2x^2$ **c.** $2x^2 - 1$ **d.** $\sqrt{1 - x^2} - x$

13. Find the value of $\mathrm{Csc}^{-1}\left(-\dfrac{2\sqrt{3}}{3}\right)$.

14-7

   a. $\dfrac{2\pi}{3}$ **b.** $\dfrac{\pi}{3}$ **c.** $-\dfrac{\pi}{3}$ **d.** does not exist

14. Find the value of $\sin[2\,\mathrm{Tan}^{-1}(-2)]$.

   a. $\dfrac{4}{5}$ **b.** $-\dfrac{4}{5}$ **c.** $\dfrac{\sqrt{5}}{5}$ **d.** $-\dfrac{4\sqrt{5}}{5}$

15. Which of the following is the solution set of $2\sin 2\theta = 1$ over the interval $0° \le \theta < 360°$?

14-8

   a. $\{30°, 150°\}$ **b.** $\{15°, 30°, 75°, 150°\}$

   c. $\{15°, 75°\}$ **d.** $\{15°, 75°, 195°, 255°\}$

16. How many solutions of $4\cos^2(2x + \pi) = 3$ are there in the interval $0 \le x < 2\pi$?

   a. 2   **b.** 4   **c.** 8   **d.** 16

Chapter Test

1. Two planes leave the airport at 7:30 A.M., each traveling at 250 km/h. Their headings are 165° and 220°. How far apart are they at 9:30 A.M. (to the nearest integer)? What is the bearing of each from the other to the nearest degree?

14-1

2. Let $\mathbf{u} = 4\mathbf{i} + 3\mathbf{j}$ and $\mathbf{v} = -12\mathbf{i} + 5\mathbf{j}$.

14-2

   a. Find $\|3\mathbf{u} + \mathbf{v}\|$.   **b.** Find the angle between $\mathbf{u}$ and $\mathbf{v}$.

3. Copy and complete the table.

14-3

Polar coordinates	(3, 120°)	?	(−6, 135°)	?
Rectangular coordinates	?	(0, −4)	?	$(-\sqrt{5}, \sqrt{5})$

4. Sketch the polar curve $r = 2\sin 2\theta$.

5. Let $w = 9 (\cos 125° + i \sin 125°)$ and $z = 3 (\cos 230° + i \sin 230°)$.   14-4

Find and simplify wz and $\frac{w}{z}$.

6. Find the five fifth roots of $-i$ in polar form.   14-5

7. Find the value of $\cos \left[2 \operatorname{Sin}^{-1} \left(-\frac{24}{25} \right) \right]$.   14-6

8. Find the value of $\sec [\operatorname{Cot}^{-1} (-6)]$.   14-7

9. Find the general solution of $\sin 2x = \tan x$.   14-8

10. Solve $\sin 2\theta = 2 \sin^2 \theta,\ 0° \leq \theta < 360°$.

Mixed Problem Solving

Solve. If a problem has no solution, explain why.

A

1. A jewelry store manager bought some necklaces for $6250 and sold all but 40 for a total of $7130. If the purchase price was $3 less than the selling price, find the number sold.

2. Rico won a prize of $5 one week, $10 the second week, $15 the third week, and so on for 52 weeks. How much did he win in all?

3. Three oxygen atoms and two nitrogens atoms contain 38 protons. Five oxygen atoms and three nitrogen atoms contain 61 protons. Find the number of protons in one nitrogen atom.

4. Find the maximum possible area of a rectangular garden enclosed with 36 m of fencing.

5. A snail moves 2 cm in 1 min. In each succeeding minute it travels half as far as in the previous minute. How far would it travel (a) during the eighth minute? (b) if it could travel forever?

6. A mixture of raisins and peanuts is worth $56. Raisins and peanuts are worth $3/kg and $5/kg, respectively. Find all integral possibilities for the number of kilograms of each used.

7. It takes Rachel 7 h to split a cord of wood. Meg can do the job in 5 h. At 9:00 A.M. Rachel began to work. At 11:00 A.M. Meg joined her. At what time was the job finished?

B

8. A small plane made a round trip of 900 km in 6 h with a 30 km/h head and tail wind. Find the plane's air speed to the nearest whole number.

9. The area of an equilateral triangle varies directly as the square of its perimeter. An equilateral triangle with perimeter 18 has area $9\sqrt{3}$. Find the perimeter of a triangle with area $3\sqrt{3}$.

10. The half-life of Carbon-11 is 20 min. How many minutes does it take for a 50 g sample to decay to 5 g?

11. The top of a vertical tree broken by the wind hit the ground 10 ft from the base of the tree. The angle formed by the ground and the treetop was 25°. Find the original height of the tree.

Preparing for College Entrance Exams

Questions 1–6 each consist of two quantities, one in Column A and one in Column B. Write the letter A if the quantity in Column A is greater; B if the quantity in Column B is greater; C if the two quantities are equal; and D if the relationship cannot be determined from the information given.

	Column A		Column B
1.	x		$|x|$
2.	$|a - b|$		$|b - a|$
3.	$\csc \dfrac{10\pi}{3}$		$\sec \dfrac{10\pi}{3}$
4.	$\tan(-x)$		$-\tan x$
5.	$\sqrt{1 + \tan^2 \theta}$		$\sqrt{1 - \cos^2 \theta}$
6.	$\sin(180° + \theta)$		$\cos(90° + \theta)$

Decide which is the best of the choices given and write the corresponding letter on your answer sheet.

7. Find a unit vector orthogonal to $\mathbf{u} = 6\mathbf{i} - 8\mathbf{j}$.

(A) $\frac{3}{5}\mathbf{i} + \frac{4}{5}\mathbf{j}$ (B) $\frac{3}{5}\mathbf{i} - \frac{4}{5}\mathbf{j}$ (C) $4\mathbf{i} - 3\mathbf{j}$ (D) $\frac{4}{5}\mathbf{i} + \frac{3}{5}\mathbf{j}$ (E) $\frac{4}{5}\mathbf{i} - \frac{3}{5}\mathbf{j}$

8. Express $\sin(2 \operatorname{Tan}^{-1} x)$ without using trigonometric or inverse trigonometric functions.

(A) $\dfrac{x}{\sqrt{x^2 + 1}}$ (B) $\dfrac{2x}{\sqrt{1 - x^2}}$ (C) $\dfrac{x^2 - 1}{x^2 + 1}$ (D) $\dfrac{2x\sqrt{1 - x^2}}{1 - 2x^2}$ (E) $\dfrac{2x}{x^2 + 1}$

9. How many solutions does $\cot x + \tan x = -2$ have in the interval $0 \le x < 2\pi$?

(A) 0 (B) 1 (C) 2 (D) 3 (E) 4

10. Which polar equation is that of a circle with center $(0, 1)$ and radius 1?

(A) $r \cos \theta = 2$ (B) $r = \cos 2\theta$ (C) $r = 2 \sin \theta$ (D) $r = 1 \csc \theta$ (E) $r^2 = 2 \sin 2\theta$

Cumulative Review (Chapters 12–14)

1. Name two angles, one positive and one negative, that are coterminal with 105°.

2. Find the values of the six trigonometric functions of an angle θ in standard position whose terminal side passes through the point $(-2, 5)$.

3. A truck has an 8-foot loading ramp inclined at a 20° angle with respect to the ground. How many feet above the ground is the top of the ramp? Give your answer to three significant digits.

4. In $\triangle ABC$, $a = 5$, $b = 9$, and $c = 6$. Find $\angle A$, $\angle B$, and $\angle C$ to the nearest tenth of a degree.

5. Find the area of $\triangle ABC$ in Exercise 4.

6. Write $-200°$ in radians.

7. Find the exact values of the six trigonometric functions of $-\dfrac{13\pi}{6}$.

8. Find the amplitude, the maximum and minimum values, and the period of
$y = 5 \cos \dfrac{\pi}{3}x - 2$.

9. Write $\dfrac{\cos \theta}{\sin^3 \theta}$ in terms of $\cot \theta$.

10. Prove: **a.** $\sin x = \dfrac{\sec x}{\cot x + \tan x}$ **b.** $\cot (\alpha + \beta) = \dfrac{\cot \alpha \cot \beta - 1}{\cot \alpha + \cot \beta}$

11. If $\sin \phi = -\dfrac{3}{5}$ and $90° < \phi < 270°$, find:

    a. $\cos \dfrac{\phi}{2}$ **b.** $\cos 2\phi$ **c.** $\tan (180° - \phi)$ **d.** $\sin (90° + \phi)$

12. The vector **u** has magnitude 10 and bearing 200°. Vector **v** has magnitude 8 and bearing 35°. Find the magnitude of **u** − **v** to three significant digits and its bearing to the nearest tenth of a degree.

13. Find a unit vector orthogonal to $\mathbf{v} = 2\mathbf{i} - \mathbf{j}$.

14. Use DeMoivre's theorem to express $(2 - 2i\sqrt{3})^5$ in $x + yi$ form.

15. Find the cube roots of $-8i$ in $x + yi$ form.

Find the value of each of the following.

16. $\mathrm{Cos}^{-1} \left(\tan \dfrac{3\pi}{4} \right)$ 17. $\sin \left(\mathrm{Cos}^{-1} \dfrac{1}{2} - \mathrm{Cot}^{-1} \dfrac{4}{3} \right)$ 18. $\tan \left[2 \mathrm{Sin}^{-1} \left(-\dfrac{5}{13} \right) \right]$

Solve for x, $0 \le x < 2\pi$, or for θ, $0° \le \theta < 360°$.

19. $2 \csc^2 x + 3 \csc x = 2$ 20. $\sin \left(x + \dfrac{\pi}{3} \right) = -\dfrac{1}{2}$

21. $\cos 2\theta = 3 \cos \theta + 1$ 22. $3 \tan^2 \theta - 1 = 0$

15 Statistics and Probability

A more appropriate name for a combination lock might be "permutation lock." To open such a lock, you need to know the order *in which* to dial the numbers.

Statistics

15-1 Presenting Statistical Data

Objective   To display data using frequency distributions, histograms, and stem-and-leaf plots, and to compute measures of central tendency.

The results of an algebra test are shown in the table at the right. This table is called a **frequency distribution** because it shows how many times a given score was attained.

A **histogram,** shown below, can be used to display a frequency distribution. To draw a histogram, you first group the data into convenient intervals. (The intervals used for the test scores, for example, are 50–59, 60–69, 70–79, and so on.) For each interval you then construct a rectangle having a width equal to the interval width and a height equal to the frequency of the data falling within the interval.

Score	Number of Students
99	2
98	1
95	3
94	1
93	1
90	2
88	2
87	3
84	5
83	3
82	1
80	2
78	1
75	2
74	1
73	2
68	1
65	1
56	1

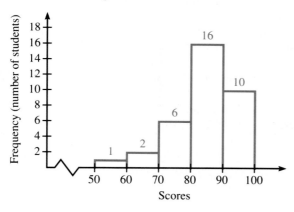

Example 1   Use the histogram above to answer the following questions.
 a. Which interval contains the most test scores?
 b. Which contains the fewest test scores?
 c. How many test scores are 80 or above?
 d. How many test scores are below 70?

Solution   **a.** The interval 80–89, with 16 scores, contains the most.
 b. The interval 50–59, with 1 score, contains the least.
 c. Add the frequencies for the intervals 80–89 and 90–99: 16 + 10 = 26
 d. Add the frequencies for the intervals 50–59 and 60–69: 1 + 2 = 3

A **stem-and-leaf plot,** which is another way of displaying data, actually includes the data in the display.

Example 2 Draw a stem-and-leaf plot for the test scores on the previous page.

Solution You first obtain the *stems* by using only the tens' digit of each score. These are written, in order, to the left of a vertical line. Then, for each score, you record the *leaf,* or units' digit, to the right of its stem.

Stem	Leaf
5	6
6	5, 8
7	3, 3, 4, 5, 5, 8
8	0, 0, 2, 3, 3, 3, 4, 4, 4, 4, 4, 7, 7, 7, 8, 8
9	0, 0, 3, 4, 5, 5, 5, 8, 9, 9

Notice that a stem-and-leaf plot has the shape of a histogram drawn horizontally.

Numbers used to describe a set of data are called **statistics.** Three different statistics are often used to measure the *central tendency* of a distribution: the *mode,* the *median,* and the *mean.*

The **mode** of a distribution is the number that occurs most frequently. From the stem-and-leaf plot of Example 2 you can see that the mode of the test scores is 84.

The **median** of a distribution is the middle number: There are as many numbers greater than the median as there are less than it. Since the stem-and-leaf plot of Example 2 contains 35 scores, the median is the 18th score counting from either the top or the bottom of the distribution. The median of the test scores is therefore 84.

The **mean** of a distribution is the *arithmetic average* of the numbers. Since the sum of the 35 scores in the stem-and-leaf plot of Example 2 is 2933, the mean is $\frac{2933}{35}$, or 83.8.

Computers and calculators, especially graphing calculators, are very useful when working with statistics.

Example 3 Use the distribution of ages given in the stem-and-leaf plot at the right to find **(a)** the mode, **(b)** the median, and **(c)** the mean (*M*).

1	8, 9, 9
2	2, 2, 2, 5
3	1, 1, 1, 6, 6, 6, 6
4	0, 0, 3, 3, 3

Solution **a.** The age that occurs most often is 36. Therefore, the mode is 36.

b. There are 19 ages, so the 10th age from the top (or bottom) is the median. Therefore the median is 31.

c. $M = \dfrac{18 + 2(19) + 3(22) + 25 + 3(31) + 4(36) + 2(40) + 3(43)}{19}$

≈ 31.2

Finding the mode and median of a distribution is not always as straight-forward as finding the mean. When a distribution has two or more numbers that occur most often with equal frequency, each of these numbers is a mode of the distribution. Also, when the count of the numbers in a distribution is even, there are two "middle" numbers, and their average is the median of the distribution.

Example 4 Find **(a)** the mode and **(b)** the median of the following distribution:

5, 6, 6, 6, 8, 10, 13, 13, 13, 16, 18, 20

Solution **a.** Since both 6 and 13 occur three times, each is a mode.

b. The two middle scores are 10 and 13.

∴ the median is $\dfrac{10 + 13}{2}$, or 11.5.

In Exercises 1–4, use the histogram at the right.

1. Find the mode.

2. How many scores are 28 or above?

3. How many scores are there altogether?

4. Find the median.

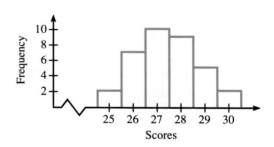

In Exercises 5–8, use the stem-and-leaf plot at the right below.

5. Find the mode.

6. Find the median.

7. In computing the mean, by what number do you divide?

8. Suppose another 32 was in the distribution. Then what would the mode be?

1	2, 2, 7, 8
2	3, 4, 5, 5, 9
3	2, 2, 2, 6, 8
4	0, 0, 0, 0, 6
5	0

Written Exercises

In Exercises 1–4, draw a stem-and-leaf plot for the given distribution.

A **1.** 16, 54, 23, 38, 22, 22, 40, 46, 52, 19, 20

2. 61, 38, 55, 65, 66, 42, 61, 48, 50, 39, 62, 61

3. 2, 5, 13, 28, 61, 9, 18, 10, 52, 34, 28, 42, 19, 28, 7

4. 123, 129, 132, 135, 140, 151, 152, 160, 166, 168, 168, 169

5–8. Find **(a)** the mode, **(b)** the median, and **(c)** the mean for each distribution given in Exercises 1–4. A calculator may be helpful.

9. The frequency distribution of the heights of the members of a high school basketball team is shown at the right. Find **(a)** the mode, **(b)** the median, and **(c)** the mean.

Height (cm)	Frequency
175	1
178	1
180	2
181	1
184	3
185	2
188	1
192	1

10. The average monthly temperatures, in degrees Celsius, for Indianapolis during a recent year were -0.05, 2.00, 6.28, 12.72, 18.61, 24.33, 27.44, 26.39, 22.00, 15.89, 8.94, and 2.17. For this distribution find **(a)** the median and **(b)** the mean.

11. Draw a histogram for the frequency distribution of quiz scores shown at the right.

12. For the frequency distribution shown at the right, find **(a)** the mode, **(b)** the median, and **(c)** the mean.

Score	Frequency
10	2
15	5
20	8
25	15
30	10

B **13.** James has test scores of 82, 73, 76, and 92. What must he score on a fifth test if his average test score is to be 82?

14. The mean of 12 numbers is 15. What is the sum of the numbers?

15. If each score in a set of scores were increased by 5 points, how would this affect the mode, median, and mean of these scores?

16. If each score in a set of scores were reduced by 40%, how would this affect the mode, median, and mean of these scores?

C **17.** At high school A, the mean score of 50 students on a science test is 75. At high school B, the mean score of 40 students on the same test is 80. What is the mean score of the 90 students?

18. The mean of m scores is M_1. The mean of n scores is M_2.
a. What is the mean of $m + n$ scores?
b. Under what conditions is the combined mean equal to $\dfrac{M_1 + M_2}{2}$?

Mixed Review Exercises

Convert the given rectangular coordinates to polar coordinates.

1. $(-3, 3\sqrt{3})$　　　　　**2.** $(\sqrt{6}, \sqrt{2})$　　　　　**3.** $(0, 3)$

Write in simplest form without negative exponents.

4. $\left(\dfrac{m^2 n^{-3}}{m^{-4} n}\right)^{-2}\left(\dfrac{mn^{-1}}{m^3 n}\right)$　　　　**5.** $\dfrac{w^2 - 3w - 4}{3w^2 - 10w - 8}$　　　　**6.** $8^{\sqrt{2}} \cdot 2^{\sqrt{8}}$

15-2 Analyzing Statistical Data

Objective   To compute measures of dispersion and, together with measures of central tendency, to describe and compare distributions using these statistics.

In Lesson 15-1 you learned that the median of a distribution divides the data into two halves. Finding the median of each half of the data therefore divides the data into quarters. The median of the *lower half* of the data is called the **first quartile** of the distribution. The median of the *upper half* of the data is called the **third quartile.** Thus one fourth of the numbers in the distribution are less than the first quartile, and three fourths are less than the third quartile.

Example 1   For the distribution of test scores shown in the stem-and-leaf plot at the right find **(a)** the median, **(b)** the first quartile, and **(c)** the third quartile.

6	4, 6, 6
7	1, 3, 8, 8
8	0, 5, 7, 9
9	2, 2, 8
10	0, 0, 0

Solution   **a.** With 17 scores in the distribution, the median is the 9th score counting from either end: 85.

b. With 9 scores (including the median) in the lower half of the distribution, the first quartile is the 5th score counting from the low end: 73.

c. With 9 scores (including the median) in the upper half of the distribution, the third quartile is the 5th score counting from the high end: 92.

In some distributions the numbers are clumped together, while in others the numbers are spread out. One way to measure the *dispersion* of a set of data is to find the **range,** which is the difference between the largest and smallest numbers in the set. For example, the range of the distribution in Example 1 is 100 − 64, or 36.

A diagram known as a **box-and-whisker plot** can be used to show the median, the first and third quartiles, and the range of a distribution.

Example 2   Draw a box-and-whisker plot for the distribution in Example 1.

Solution   1. First show the median, the first and third quartiles, and the lowest and highest scores as dots below a number line.

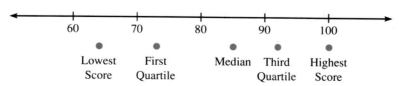

(Solution continues on the next page.)

2. Next draw a narrow rectangular box with its shorter sides containing the two quartile dots. Then draw a line segment through the median dot parallel to the shorter sides. Finally, draw line segments, or "whiskers," from the first- and third-quartile dots to the lowest- and highest-score dots, respectively.

In a box-and-whisker plot, the box encloses the middle half of the distribution and the whiskers indicate the spread of data through the lowest and highest fourths of the distribution. The distance between the two extreme points represents the range of the distribution.

By analyzing the box-and-whisker plots for two distributions you can easily make comparisons between them.

Example 3 Two classes took the same algebra test. The results are shown in the box-and-whisker plots below.

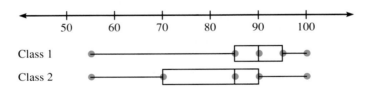

a. Which class has the higher median?

b. Which class has the smaller range?

c. For which class are the scores in the middle half closer together?

d. Which class has the better set of scores?

Solution a. Class 1: The median of Class 1, which is 90, is greater than the median of Class 2, which is 85.

b. Both classes have the same range: $100 - 55$, or 45.

c. Class 1: The box for Class 1 is shorter than the box for Class 2. (Note that by ignoring the extreme scores, you may get a better picture than the range provides of the "typical" spread in the scores.)

d. Class 1: Three fourths of the scores for Class 1 are at or above the median for Class 2.

Besides the range, two other statistics are often used to measure the dispersion of a distribution. They are called the *variance* and the *standard deviation* of the distribution.

The results of a history test are shown in the table at the right. Since the sum of the 8 scores is 632, the mean of the scores is $632 \div 8$, or 79. The table also shows how much each score differs from the mean. These differences are called *deviations* from the mean.

To get an idea of how the scores are generally scattered about the mean, you can make the deviations positive and find their average. (To make the deviations positive, mathematicians choose to square them rather than take their absolute values.) The result of computing the average of the squares of the deviations is called the *variance* of the distribution.

Score	Deviation from mean
94	+15
89	+10
86	+7
82	+3
75	−4
72	−7
71	−8
63	−16

If $x_1, x_2, \ldots, x_n$ are n numbers and M is their mean, then the **variance** of the distribution is

$$\frac{(x_1 - M)^2 + (x_2 - M)^2 + \cdots + (x_n - M)^2}{n}.$$

Example 4 Find the variance of the distribution of the test scores shown above.

Solution Variance $= \dfrac{15^2 + 10^2 + 7^2 + 3^2 + (-4)^2 + (-7)^2 + (-8)^2 + (-16)^2}{8}$

$= \dfrac{768}{8} = 96$ **Answer**

The principal square root of the variance is called the **standard deviation.** It is denoted by the Greek letter sigma (σ):

$$\sigma = \sqrt{\frac{\text{sum of the squares of the deviations from the mean}}{\text{number of elements in the distribution}}} = \sqrt{\text{variance}}$$

From Example 4 you can see that the standard deviation of the history test scores is $\sqrt{96}$, or 9.8 to the nearest tenth.

The mean, the median, and the mode each indicate where the center of a distribution is. They are called **measures of central tendency.** The range, the first and third quartiles, the variance, and the standard deviation each indicate how scattered a distribution is. They are called **measures of dispersion.**

A useful way to characterize a distribution is to give a measure of central tendency and a measure of dispersion for it. The mean and standard deviation are the most common pair of measures used.

Example 5 The stem-and-leaf plot at the right shows the distribution of the heights, in centimeters, of the members of a high school basketball team. Find the mean and standard deviation for the distribution.

17	5, 8
18	0, 0, 1, 4, 4, 4, 5, 5, 8
19	2

(Solution is on the next page.)

Solution   Use a table like the one shown below to organize your computations.

Height	Frequency	Height × Frequency	Deviation	(Deviation)^2	(Deviation)^2 × Frequency
175	1	175	−8	64	64
178	1	178	−5	25	25
180	2	360	−3	9	18
181	1	181	−2	4	4
184	3	552	+1	1	3
185	2	370	+2	4	8
188	1	188	+5	25	25
192	1	192	+9	81	81
Sums	12	2196			228

After completing the first three columns, you can compute the mean.

$$\therefore M = \frac{2196}{12} = 183 \quad \textbf{\textit{Answer}}$$

After completing the last three columns, you can compute the standard deviation.

$$\therefore \sigma = \sqrt{\frac{228}{12}} \approx 4.4 \quad \textbf{\textit{Answer}}$$

Oral Exercises

In Exercises 1–4, use the stem-and-leaf plot at the right.

```
1 | 0, 0, 5, 6
2 | 0, 4, 5, 5, 5, 5, 7, 9
3 | 7, 8, 9, 9, 9
```

1. Find the median.

2. Find the first quartile.

3. Find the third quartile.

4. Find the range.

5. Use the box-and-whisker plot at the right to state what each of the following points represents and to give its value.

    a. *A*   **b.** *B*   **c.** *C*   **d.** *D*   **e.** *E*

In Exercises 6–13, use the box-and-whisker plots below.

Which class has:

6. the highest score?

7. the lowest score?

8. the smaller range?

9. the higher median?

10. the higher first quartile?

11. the higher third quartile?

12. scores in the middle half closer together?

13. the better set of scores?

14. What can you say about distributions A and B if they have the same mean but the standard deviation of A is greater than that of B?

Written Exercises

For each stem-and-leaf plot find (a) the median, (b) the first quartile, (c) the third quartile, and (d) the range.

A **1.**

0	2, 3, 7
1	0, 4, 4, 8
2	2, 5, 5, 5, 9, 9
3	0, 4, 4, 8, 8
4	0, 1, 7

2.

5	9
6	3, 3, 7
7	1, 6, 6, 6
8	0, 0, 4, 4, 8
9	2, 5, 5

3–4. Make a box-and-whisker plot for each of the distributions in Exercises 1 and 2.

In Exercises 5–8, use the box-and-whisker plots at the right and justify your answers.

5. Which class has the highest median?

6. Which class has the smallest range?

7. For which class are the scores in the middle half closest together?

8. Which class has the best set of scores?

For the distributions in Exercises 9–14 find (a) the mean, (b) the variance, and (c) the standard deviation to the nearest tenth.

9. 1, 4, 6, 6, 7, 8, 8, 8

10. 3, 4, 5, 5, 5, 5, 6, 9

11. 34, 42, 44, 70, 73, 79

12. 8, 15, 38, 64, 85, 102

13. 42, 46, 50, 50, 52, 54, 56

14. 37, 38, 41, 45, 45, 47, 48

15. In a golf tournament the 18-hole totals for the top nine golfers were 67, 69, 70, 70, 71, 72, 73, 73, and 74. Find **(a)** the mean, **(b)** the variance, and **(c)** the standard deviation of the scores to the nearest tenth of a stroke.

16. The highest temperatures for the first eleven days of June in St. Louis were 70, 68, 63, 66, 70, 70, 73, 87, 70, 68, and 65. Find **(a)** the mean, **(b)** the variance, and **(c)** the standard deviation of the temperatures to the nearest tenth of a degree.

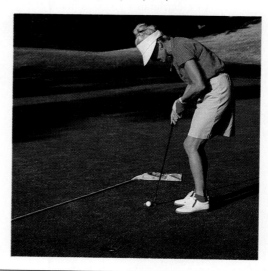

B 17. If each score in a set of scores was increased by 10 points, how would this affect the mean and the standard deviation?

18. If each score in a set of scores was multiplied by 2, how would this affect the mean and the standard deviation?

19. Make up frequency distributions of four scores between 0 and 100 that have the given mean and standard deviation.
   a. $M = 80$; $\sigma < 3$ **b.** $M = 80$; $\sigma > 20$ **c.** $M = 80$; $\sigma = 10$

C 20. The means and standard deviations of an algebra test given to four different groups of 100 students each are listed in the table at the right. You would need more information to answer the following questions with certainty. However, based on the data given, respond to the questions and tell why you gave your response.

Group	Mean	σ
1	60	16
2	70	12
3	70	14
4	80	6

   a. Which group probably has the lowest individual score?
   b. Which group probably has the smallest range?
   c. Which group most closely resembles the total population of 400 students?

Career Note / Electrician

Most people tend to take electricity for granted. When you turn on a light or use an electrical appliance, you usually don't think about the power that flows through the wires hidden in the walls of your home. Electricians, however, are very much concerned with getting that power safely to where it is needed.

An electrician's work generally involves either installations or maintenance. When wiring is placed in a building under construction, an electrician must install outlets and switches in each room, run wires through metal or plastic tubing (called conduit) in the walls, and connect the wires to circuit boxes. Once the wiring is installed, it must be maintained through periodic inspection and repair. An electrician deals with problems ranging from blown fuses to short circuits.

A knowledge of mathematics is important to electricians. For example, to increase the current in a circuit, an electrician must increase voltage or

decrease resistance. This fact is expressed in Ohm's Law:

$$I = \frac{E}{R}$$

where I measures current (in amperes), E measures voltage (in volts), and R measures resistance (in ohms).

15-3 The Normal Distribution

Objective   To recognize and analyze normal distributions.

The histogram at the right shows the annual rainfall in a Nebraska county over a period of 78 years. Each bar in the histogram shows the number of years in which the annual rainfall was within a given range. For example, the leftmost bar shows that for one year out of the 78 the rainfall was between 11 and 14 inches.

The two histograms below give the weights of 100 pennies and the lifetimes of 433 light bulbs.

Figure 1

Figure 2

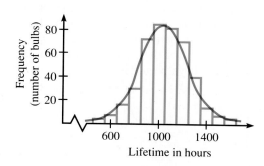

Figure 3

When each of these histograms is approximated by drawing a smooth curve (shown in blue), the resulting curves resemble each other. Such "bell-shaped" curves often appear when various sets of data are plotted. For example, all of the following frequency distributions might be represented by bell-shaped curves: the heights of all eleventh-grade girls in a large city, the acidity levels of soil samples from all farms in a rural county, and the achievement test scores for all high school students in a state.

Certain distributions represented by bell-shaped curves are known as *normal distributions*. A normal distribution with mean equal to zero and standard deviation equal to one is called the **standard normal distribution.** The bell-shaped curve that represents a standard normal distribution is called the **standard normal curve.**

The figure at the top of the next page shows the standard normal curve. Note that the *x*-coordinates represent the number of standard deviations from the mean.

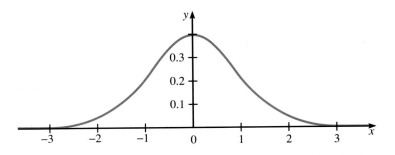

The standard normal curve has the following properties:
1. It is symmetric with respect to the y-axis.
2. It approaches the x-axis asymptotically as |x| increases.
3. The total area under the curve and above the x-axis is equal to 1.

For any set of data that are approximately normally distributed, the standard normal curve can be used to estimate the fraction of the data falling between two given values, a and b. The fraction is simply the shaded area shown in the figure at the right.

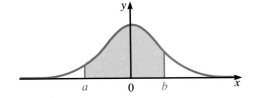

To find this area, extensive tables of area measures for the standard normal curve have been developed. One such table, shown below, gives area measures from 0 to x, at intervals of 0.2, for $0 \leq x \leq 4.0$.

Area under the Standard Normal Curve for $0 \leq x \leq 4.0$

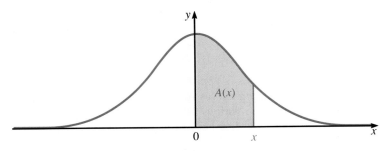

x	Area, $A(x)$	x	Area, $A(x)$	x	Area, $A(x)$
0.0	0.0000	1.4	0.4192	2.8	0.4974
0.2	0.0793	1.6	0.4452	3.0	0.4987
0.4	0.1554	1.8	0.4641	3.2	0.4993
0.6	0.2257	2.0	0.4772	3.4	0.4997
0.8	0.2881	2.2	0.4861	3.6	0.4998
1.0	0.3413	2.4	0.4918	3.8	0.4999
1.2	0.3849	2.6	0.4953	4.0	0.5000

Example 1 In a standard normal distribution, what fraction of the data is between 2.6 and 2.8?

Solution The fraction desired is equal to the area under the standard normal curve bounded by $x = 2.6$ and $x = 2.8$. This is the difference between the area from 0 to 2.8 and the area from 0 to 2.6. Therefore:

$$A(2.6 \le x \le 2.8) = A(2.8) - A(2.6)$$
$$= 0.4974 - 0.4953$$
$$= 0.0021 \quad \textit{Answer}$$

Example 2 In a standard normal distribution, what percent of the data is within two standard deviations from the mean?

Solution The mean is 0 and the standard deviation is 1. Since the standard normal curve is symmetric with respect to the y-axis, the area under the curve between -2 and 2 is twice the area between 0 and 2. Therefore:

$$A(-2 \le x \le 2) = 2 \cdot A(2)$$
$$= 2(0.4772)$$
$$= 0.9544, \text{ or } 95.44\% \quad \textit{Answer}$$

Notice in Example 2 that slightly more than 95% of the data in a standard normal distribution are within two standard deviations of the mean. Likewise you can show that slightly more than 68% of the data are within one standard deviation of the mean.

Because all normal distributions are similar to the standard normal distribution, you can use the standard normal curve to answer questions about any normal distribution. Suppose a normal distribution has mean M and standard deviation σ. Then the percent of data falling between $M + a\sigma$ and $M + b\sigma$ is just the area between $x = a$ and $x = b$ under the standard normal curve.

To obtain a from the number $M + a\sigma$, you would first subtract M and then divide by σ. In general, if z is a given value from a normal distribution with mean M and standard deviation σ, then the corresponding value from the standard normal distribution is x, where

$$x = \frac{z - M}{\sigma}.$$

The number x is called the **standardized value** corresponding to z. It gives the number of standard deviations that z is from the mean of the distribution.

Example 3 The heights of a certain group of adults are normally distributed with a mean of 180 cm and a standard deviation of 8 cm.

a. Find the percent of the group having a height greater than 196 cm.
b. Find the percent of the group having heights between 172 cm and 180 cm.

(Solution is on the next page.)

Solution   **a.** The standardized value corresponding to 196 is:

$$x = \frac{196 - M}{\sigma} = \frac{196 - 180}{8} = 2$$

The area under the standard normal curve to the right of $x = 2$ is $0.5 - 0.4772$, or 0.0228.

∴ about 2.3% of the group has height greater than 196 cm.

b. The standardized values corresponding to 172 and 180 are:

$$\frac{172 - 180}{8} = -1 \quad \text{and} \quad \frac{180 - 180}{8} = 0$$

The area under the standard normal curve between -1 and 0 is the same as the area between 0 and 1, which is 0.3413.

∴ about 34% of the group has height between 172 cm and 180 cm.

Oral Exercises

In Exercises 1–3, use the histograms on page 719.

1. In Figure 1, about what fraction of the 78 years had rainfall between 32 and 38 inches?

2. In Figure 2, what percent of the 100 pennies had a weight of 3.11 g?

3. In Figure 3, about what fraction of the 433 light bulbs had lifetimes of more than 1000 h?

4. What is the area under the standard normal curve to the left of the y-axis?

5. If the area under the standard normal curve between $x = 0$ and $x = 0.5$ is 0.1915, what percent of the data in a standard normal distribution is within one-half standard deviation from the mean?

6. For a normal distribution with mean 12 and standard deviation 4, what is the standardized value corresponding to 20?

Written Exercises

In Exercises 1–9, use the table on page 720.

A   1. In a standard normal distribution, what percent of the data is between the mean and 0.4?

2. In a standard normal distribution, what percent of the data is between -0.2 and 0.2?

3. In a standard normal distribution, what percent of the data is between three and four standard deviations below the mean?

4. In a standard normal distribution, what percent of the data is within three standard deviations from the mean?

5. The mean weight of a loaf of bread was found by sampling to be 455 g, with a standard deviation of 5 g. Assuming a normal distribution, find the percent of loaves with weights that are:
   a. less than 450 g
   b. greater than 445 g
   c. greater than 470 g
   d. between 450 g and 460 g

6. A college aptitude test is scaled so that its scores approximate a normal distribution with a mean of 500 and a standard deviation of 100. Find the percent of the students taking the test who are expected to score:
   a. above 800 points
   b. less than 400 points
   c. between 700 and 900 points
   d. between 800 and 820 points

7. The mean life of a certain kind of light bulb is 900 h with a standard deviation of 30 h. Assuming the lives of the light bulbs are normally distributed, find the percent of the light bulbs that will last:
   a. less than 900 h
   b. more than 984 h

8. A type of coin in circulation for the past 10 years has a mean weight of 0.22 oz and a standard deviation of 0.01 oz. Assuming a normal distribution, find the percent of these coins with weights that are:
   a. greater than 0.24 oz
   b. less than 0.206 oz

B 9. Assuming a standard normal distribution, find a number k such that the percent of data less than k is:
   a. 50%
   b. 100%
   c. 15.87%
   d. 84.13%

10. Assuming a standard normal distribution, explain why the fraction of data greater than a constant k equals 1 minus the fraction of data less than k.

C 11. Assuming a standard normal distribution, explain why the fraction of data having absolute value greater than a nonnegative constant k equals twice the difference between 1 and the fraction of data less than k.

12. The equation of the standard normal curve is

$$y = \frac{1}{\sqrt{2\pi}} e^{(-1/2)x^2}.$$

The base of the natural logarithm, e, is equal to 2.71828 . . . . Use a calculator or a table of logarithms to compute the y-coordinate of the point on the standard normal curve for:
   a. $x = 0$
   b. $x = 1$
   c. $x = 2$

Mixed Review Exercises

For the line containing the given points, find (a) the slope and (b) an equation in standard form.

1. $(4, -7)$, $(0, 5)$
2. $(3, 2)$, $(7, 5)$
3. $(0, 0)$, $(-2, 1)$
4. $(-2, 6)$, $(1, 8)$
5. $(-5, -1)$, $(3, -1)$
6. $(-3, -2)$, $(-1, -4)$

15-4 Correlation

Objective   To draw a scatter plot, determine the correlation coefficient, and use the regression line for a set of ordered pairs of data.

The director of sales for a national distributor of prerecorded videocassettes wants to predict future sales. Although such a prediction could be based on past sales alone, the sales director believes that sales are tied to the number of households that have a VCR.

The table at the right gives the nationwide data on VCR households and videocassette sales for five consecutive years. To determine whether a mathematical relationship exists between these two sets of data, you can treat the data as ordered pairs and plot them to obtain a graph called a **scatter plot.**

VCR Households (in millions)	Prerecorded Videocassette Sales (in millions)
8.3	9.5
15.0	22.0
23.5	52.0
32.5	84.0
45.8	110.0

Example 1   Draw a scatter plot of the data in the table above.

Solution   Plot the ordered pairs (8.3, 9.5), (15.0, 22.0), (23.5, 52.0), (32.5, 84.0), and (45.8, 110.0) in a coordinate plane, as shown below.

Notice that the points in the scatter plot of Example 1 are very close to being collinear. A statistic called the *correlation coefficient* is used to characterize how closely the points in a scatter plot cluster about a line.

Given a set of ordered pairs (x, y), the **correlation coefficient,** denoted by r_{xy} or merely r, of the ordered pairs is

$$r = \frac{M_{xy} - M_x \cdot M_y}{\sigma_x \cdot \sigma_y},$$

where M_x and σ_x are the mean and standard deviation of the x-values,
M_y and σ_y are the mean and standard deviation of the y-values,
and M_{xy} is the mean of the products of the ordered pairs.

Because of the way in which the correlation coefficient is defined, its value for any set of ordered pairs is always between -1 and 1, inclusive. The more the ordered pairs cluster about a line with a positive slope, the closer r is to 1. Similarly, the more the ordered pairs cluster about a line with a negative slope, the closer r is to -1. If the ordered pairs tend not to be collinear at all, the correlation coefficient is close to 0. These three situations are illustrated in the figures below.

  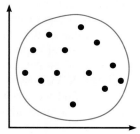

High positive correlation: r **is close to 1.**   **High negative correlation:** r **is close to -1.**   **Relatively no correlation:** r **is close to 0.**

Example 2   **a.** Using the scatter plot shown in Example 1, describe the nature of the correlation.

b. Suppose x represents the number of VCR households and y represents the number of videocassette sales. Determine the correlation coefficient of the ordered pairs from the table on the previous page given that $M_{xy} \approx 1879.77$, $M_x \approx 25.02$, $M_y = 55.5$, $\sigma_x \approx 13.19$, and $\sigma_y \approx 37.50$.

Solution   **a.** Since the ordered pairs tend to cluster about a line with a positive slope, there is a *high positive correlation* between VCR households and videocassette sales.

b. The correlation coefficient of the ordered pairs is

$$r = \frac{M_{xy} - M_x \cdot M_y}{\sigma_x \cdot \sigma_y} \approx \frac{1879.77 - (25.02)(55.5)}{(13.19)(37.50)} \approx 0.99.$$

A few words of caution are in order here: Although the correlation coefficient measures the strength of the linear relationship between two variables, a high correlation does *not* necessarily imply a cause-and-effect relationship. For instance, research may show a high positive correlation between the body weights and achievement test scores among children. Even if these two variables are linearly related, the correlation between them does not suggest that children should overeat to improve their achievement scores.

A more reasonable conclusion is that body weights and achievement scores are tied to a third variable, such as age. Therefore, as age increases, so do body weights and achievement scores.

When the correlation between two variables is high, you can draw a line, called the **regression line,** that best fits the known values of the variables. The figure at the right, for instance, shows the regression line drawn for the scatter plot of Example 1.

Since the regression line stays reasonably close to the points in a scatter plot, its equation can be used to estimate the value of one variable for a given value of the other. For example, the number of expected videocassette sales could be determined for any given number of VCR households.

To find the equation of a regression line, you use the following fact, proved in a more advanced study of statistics.

For a set of ordered pairs (x, y), the regression line relating x and y contains the point (M_x, M_y) and has as its slope $r\left(\dfrac{\sigma_y}{\sigma_x}\right)$.

Example 3   **a.** Using part (b) of Example 2, determine an equation for the regression line relating x and y.
   b. Find the expected number of videocassette sales when the number of VCR households is 60 million.

Solution   **a.** The regression line contains the point $(M_x, M_y) = (25.02, 55.5)$.

The slope of the regression line is: $r\left(\dfrac{\sigma_y}{\sigma_x}\right) \approx 0.99\left(\dfrac{37.50}{13.19}\right) \approx 2.81$.

To obtain the equation of the regression line, use the point-slope form of the equation of a line:
$$y - 55.5 = 2.81(x - 25.02)$$
$$y = 2.81x - 14.81 \quad \textit{Answer}$$

b. Substituting 60 for x in the equation from part (a), you get a value of $2.81(60) - 14.81$, or 153.79, for y.

∴ the number of expected videocassette sales is about 154 million.

Answer

Note that the answer to part (b) of Example 3 is only approximate. In a more advanced study of statistics, you will learn how to measure the accuracy of an approximation obtained from a regression line. Generally speaking, the closer $|r|$ is to 1, the greater the accuracy of the approximation.

Oral Exercises

1. a. When two variables have a high __?__ correlation, one variable increases as the other variable increases.
   b. When two variables have a high __?__ correlation, one variable decreases as the other variable increases.

2. For each value of the correlation coefficient, tell whether the linear relationship between x and y is strong or weak.
   a. $r_{xy} = 0.05$ **b.** $r_{xy} = -0.95$ **c.** $r_{xy} = -0.10$ **d.** $r_{xy} = 0.90$

3. For each pair of variables, tell whether you think the correlation is positive, negative, or close to zero.
   a. A high school student's height and weight
   b. A car's age and its value
   c. A team's standing in its conference and the attendance at its games
   d. A state's monthly temperature averages and precipitation totals
   e. A company's advertising budget and its volume of sales

Written Exercises

In Exercises 1–4, draw a scatter plot. Then describe the nature of the correlation.

A   1.

Congressional Elections (for 6 consecutive elections)	
Registered to Vote (percent)	Actually Voted (percent)
70	55
68	55
62	45
63	46
64	49
64	46

2.

Sales of Single-Family Homes (for 6 consecutive months)	
Median Price (in thousands of $)	Number Sold (in thousands)
99	710
95	740
98	720
97	730
105	640
108	650

3.

Moviegoing (for 7 consecutive years)	
Average Ticket Price (in dollars)	Admissions (in billions)
2.80	1.06
2.90	1.18
3.20	1.20
3.40	1.20
3.60	1.06
3.70	1.02
3.90	1.09

4.

Price of Precious Metals (for 7 consecutive months)	
Gold (in $ per troy oz)	Platinum (in $ per oz)
400	520
410	530
440	590
460	610
450	570
450	580
460	610

5. In Exercise 1, suppose x represents the percent registered to vote and y represents the percent that actually voted. Given that $M_{xy} \approx 3226.33$, $M_x \approx 65.17$, $M_y \approx 49.33$, $\sigma_x \approx 2.85$, and $\sigma_y \approx 4.19$, find **(a)** the value of the correlation coefficient for the ordered pairs and **(b)** an equation of the regression line relating x and y.

6. In Exercise 2, suppose x represents the median price of single-family homes and y represents the number sold. Given that $M_{xy} \approx 69{,}893.33$, $M_x \approx 100.33$, $M_y \approx 698.33$, $\sigma_x \approx 4.61$, and $\sigma_y \approx 38.91$, find **(a)** the value of the correlation coefficient for the ordered pairs and **(b)** an equation of the regression line relating x and y.

7. The regression equation for two chemistry tests is $y = 0.7x + 15$, where x represents a student's score on the first test and y represents the student's score on the second test. What is the predicted score on the second test for a student who missed that test but scored 90 on the first test?

8. The regression equation for people on a weight-reducing diet is $y = 1.8x + 1.6$, where x is the number of weeks on the diet and y is the number of pounds of weight lost. If a person wants to lose 10 lb, how long should that person stay on the diet?

B   9. The correlation between two tests given to 100 students in the junior class is 0.70. The mean for the first test was 125 with standard deviation 20. The mean of the second test is 80 with standard deviation 10. A student took the first test and scored 100. What is your best estimate of this student's score on the second test?

10. The correlation between the values and the ages of 50 automobiles is -0.85. The average value of these automobiles is $8000 with standard deviation $1000. The average age is 4 years with standard deviation 2 years. If you had no specific information about an automobile except that its age was 2 years, what is your best estimate of its value?

11. State-by-state data on annual per capita income and annual spending on education per student correlate above 0.80 for all the states in the United States. On the basis of this data it is argued that a state should raise educational spending to increase per capita income. Is this a valid argument? Explain your answer.

12. The correlation between mental age and height was 0.08 for a large sample of boys of the same age. For a large sample of boys ranging in age from 6 to 15 years old, this correlation was 0.78. How do you explain the difference in these results? How would you answer the question, "Are height and mental age related?"

13. Using two real-world variables of your choosing, form a hypothesis concerning their relationship. Test the hypothesis by gathering data for the two variables and calculating the correlation coefficient. Then write a paragraph indicating whether or not the data supported your hypothesis.

Self-Test 1

Vocabulary   frequency distribution (p. 709)
histogram (p. 709)
stem-and-leaf plot (p. 710)
statistics (p. 710)
mode, median, mean (p. 710)
first, third quartile (p. 713)
range (p. 713)
box-and-whisker plot (p. 713)
variance (p. 715)
standard deviation (p. 715)

measures of central tendency (p. 715)
measures of dispersion (p. 715)
standard normal distribution (p. 719)
standard normal curve (p. 719)
standardized value (p. 721)
scatter plot (p. 724)
correlation coefficient (p. 725)
regression line (p. 726)

1. To the nearest whole number, find the mode, median, and mean for the distribution of numbers given in the stem-and-leaf plot below.

$$\begin{array}{c|l} 1 & 4,\ 7,\ 7,\ 8 \\ 2 & 1,\ 2,\ 2,\ 5,\ 6,\ 8,\ 8,\ 8 \\ 3 & 2,\ 3,\ 4,\ 6,\ 6,\ 7 \end{array}$$

Obj. 15-1, p. 709

2. Find the first and third quartile scores for the distribution in Exercise 1.

Obj. 15-2, p. 713

3. To the nearest whole number, find the variance and standard deviation for the distribution in Exercise 1.

4. In a standard normal distribution, what fraction of the data is between $x = -2.0$ and $x = 1.0$? (Use the table on page 720.)

Obj. 15-3, p. 719

5. The mean weight of the lobsters in a certain bay is 3 kg with a standard deviation of 0.5 kg. What percent of the lobsters weigh more than 4.1 kg? (Use the table on page 720.)

6. Draw a scatter plot for the ordered pairs given in the table at the right. Then describe the nature of the correlation between x and y.

x	y
3	12
5	10
8	7
9	5

Obj. 15-4, p. 724

Counting

15-5 Fundamental Counting Principles

Objective   To apply fundamental counting principles.

A local moped dealer sells 6 different models of mopeds. Each model is available in 3 colors. Therefore, there are

$$6 \cdot 3, \text{ or } 18,$$

different combinations of model and color that can be ordered. This illustrates a *fundamental counting principle*.

If one selection can be made in m ways, and for each of these a second selection can be made in n ways, then the number of ways the two selections can be made is $m \times n$.

Example 1   How many *odd* 2-digit whole numbers less than 70 are there?

Solution   A diagram such as ☐☐ is useful to help analyze the problem.

There are *six* possible selections for the tens' digit: 1, 2, 3, 4, 5, and 6. Write 6 in the first box: $\boxed{6}$☐.

There are *five* possible selections for the units' digit: 1, 3, 5, 7, and 9. Write 5 in the second box: $\boxed{6}\boxed{5}$.

∴ by the fundamental counting principle stated above, there are $6 \cdot 5$, or 30, odd 2-digit whole numbers less than 70.   *Answer*

To get to school, Rita can either walk or take a bus. These are *mutually exclusive* choices. That is, she can either ride or walk to school, but not both. If she chooses to ride, there are two possible bus routes she can use, and if she walks, there are three routes she can take. Thus there are $2 + 3$, or 5, possible routes for Rita's trip to school. This illustrates another useful counting principle, the additive rule for mutually exclusive possibilities.

If the possibilities being counted can be grouped into *mutually exclusive* cases, then the total number of possibilities is the *sum* of the number of possibilities in each case.

Example 2 How many positive integers less than 100 can be written using the digits 6, 7, 8, and 9?

Solution Consider two mutually exclusive cases: (1) the 1-digit integers and (2) the 2-digit integers.

	T	U	
1-digit integers	—	4	4
2-digit integers	4	4	$4 \cdot 4 = \underline{16}$
			Total $= 20$

∴ there are 20 positive integers less than 100 that can be written using the digits 6, 7, 8, and 9. **Answer**

Example 3 How many license plates of 3 symbols (letters and digits) can be made using at least one letter in each?

Solution There are three mutually exclusive cases: license plates with one letter, two letters, or three letters.

1-letter case: There are 26 possibilities for the letter and 10 for each digit. Thus there are $26 \cdot 10 \cdot 10$, or 2600, letter-digit-digit combinations. Since the letter can be in any of three positions on the plate (left, middle, or right), there are $3 \cdot 2600$, or 7800, possible 1-letter plates.

2-letter case: There are $26 \cdot 26 \cdot 10$, or 6760, letter-letter-digit combinations. Since the digit can be in any of three positions, there are $3 \cdot 6760$, or 20,280, possible 2-letter plates.

3-letter case: There are $26 \cdot 26 \cdot 26$, or 17,576, possible 3-letter plates.

∴ the number of plates is the sum of the three mutually exclusive possibilities: $7800 + 20,280 + 17,576$, or 45,656. **Answer**

Oral Exercises

1. Elena can wear one of 2 blouses and one of 5 scarves. How many blouse-scarf combinations are available to her?

2. There are 3 trails on the north face of Mount Ezra and 2 trails on the south face of Mount Ezra. How many routes are there going up the north face and down the south face?

3. Kelly must buy hamburger rolls for a cookout. She can buy them in one of 4 supermarkets or one of 3 bakery shops. In how many ways can Kelly run her errand?

4. George can choose among 15 different flavors of ice cream, 6 different flavors of sherbet, and 5 different flavors of frozen yogurt. In how many ways can he choose a single dessert?

Written Exercises

A **1.** How many even 2-digit positive integers less than 50 are there?

2. How many odd 2-digit positive integers greater than 20 are there?

3. How many odd 3-digit positive integers can be written using the digits 2, 3, 4, 5, and 6?

4. How many even 3-digit positive integers can be written using the digits 1, 2, 4, 7, and 8?

5. In how many ways can you select one algebra book, one geometry book, and one calculus book from a collection of 8 different algebra books, 5 different geometry books, and 3 different calculus books?

6. A student council has 5 seniors, 4 juniors, 3 sophomores, and 2 freshmen as members. In how many ways can a 4-member council committee be formed that includes one member of each class?

7. In how many different ways can a 10-question true-false test be answered if every question must be answered?

8. In how many different ways can a 10-question true-false test be answered if it is all right to leave questions unanswered?

9. How many ways are there to select 3 cards, one after the other, from a deck of 52 cards if the cards are not returned to the deck after being selected?

10. How many ways are there to write a 3-digit positive integer using the digits 1, 3, 5, 7, and 9 if no digit is used more than once?

B **11.** How many 7-digit telephone numbers can be created if the first digit must be 8, the second must be 5, and the third must be 2 or 3?

12. How many positive odd integers less than 10,000 can be written using the digits 3, 4, 6, 8, and 0?

13. How many license plates of 3 symbols (letters and digits) can be made using at least 2 letters for each?

14. How many license plates of 4 symbols can be made using 2 letters and 2 digits?

15. How many multiples of 3 less than 100 can be formed from the digits 1, 4, 5, 7, and 8? (*Hint:* The sum of the digits of any multiple of 3 is also a multiple of 3.)

16. How many multiples of 3 less than 1000 can be formed from the digits 2, 5, and 9?

C **17.** Suppose you have totally forgotten the combination to your locker. There are three numbers in the combination, and you're sure each number is different. The numbers on the lock's dial range from 0 to 35. If you test one combination every 12 seconds, how long will it take to test all possible combinations?

18. DNA (deoxyribonucleic acid) molecules include the base units adenine, thymine, cytosine, and guanine (A, T, C, and G). The sequence of base units along a strand of DNA encodes genetic information. In how many different sequences can A, T, C, and G be arranged along a short strand of DNA that has only 8 base units?

19. Protein molecules are made up of many amino acid residues joined end-to-end. Proteins have different properties, depending on the sequence of amino acid residues in the molecules. If there are 20 naturally occurring amino acids, how many different sequences of amino acid residues can occur in a 6-residue-long fragment of a protein molecule?

20. How many 3-letter code words can be formed if at least one of the letters is to be chosen from the vowels *a, e, i, o,* and *u*?

Mixed Review Exercises

Evaluate if $x = -2$ and $y = 8$.

1. $\dfrac{y^2}{x^5}$ **2.** $|x - y|$ **3.** y^x **4.** $\sqrt{xy}$

Use the given root to solve each equation completely.

5. $x^3 + x^2 - 2x - 2 = 0; \ \sqrt{2}$ **6.** $x^3 - 4x^2 + 3x - 12 = 0; \ -i\sqrt{3}$

7. $2x^3 - 11x^2 + 4x + 5 = 0; \ 5$ **8.** $x^3 + 6x^2 + 16x + 96 = 0; \ 4i$

Challenge

If you have two colors of paint available, in how many ways can you paint the faces of a cube so that each face has one of the two colors? (Two colorings are the same if one can be turned so that it matches the other exactly. You don't have to use both colors.)

15-6 Permutations

Objective   To find the number of permutations of the elements of a set.

The letters *a*, *b*, and *c* can be arranged in six different ways:

abc	*bac*	*cab*
acb	*bca*	*cba*

Each of these arrangements is called a *permutation* of the letters *a*, *b*, and *c*. When the elements of a set are arranged in a definite order, the arrangement is called a **permutation** of the elements.

To determine the number of permutations of the letters *a*, *b*, and *c* without actually listing them, you can use the first fundamental counting principle stated in Lesson 15-5. Since any one of the three letters may be written first, a 3 appears in the first box of the following diagram:

$$\boxed{3}\ \boxed{}\ \boxed{}$$

After the first letter has been selected, the second must be selected from the remaining two letters:

$$\boxed{3}\ \boxed{2}\ \boxed{}$$

Only one selection remains for the last letter:

$$\boxed{3}\ \boxed{2}\ \boxed{1}$$

Therefore the number of permutations of the letters *a*, *b*, and *c* is

$$3 \cdot 2 \cdot 1, \text{ or } 6.$$

Example 1   Find the number of permutations of the four letters *p*, *q*, *r*, and *s*.

Solution   Complete a diagram: $\boxed{4}\ \boxed{3}\ \boxed{2}\ \boxed{1}$

∴ the number of permutations is $4 \cdot 3 \cdot 2 \cdot 1$, or 24. ***Answer***

Recall from Lesson 11-8 that for any positive integer *n*, the product

$$n \cdot (n - 1) \cdot (n - 2) \cdot \ \dots \ \cdot 3 \cdot 2 \cdot 1$$

can be written as *n*! (read ''*n* factorial''). As you have seen, there are 3! permutations of 3 objects and 4! permutations of 4 objects. In general:

The number of permutations of *n* objects is *n*!.

The next example illustrates what happens when not all the elements of a set are used to form permutations.

Example 2 In how many ways can the letters in the word JUSTICE be arranged using only 5 letters at a time?

Solution The first choice can be any one of the 7 letters, the second can be any one of the 6 remaining letters, and so on:

$$\boxed{7}\,\boxed{6}\,\boxed{5}\,\boxed{4}\,\boxed{3}$$

∴ there are $7 \cdot 6 \cdot 5 \cdot 4 \cdot 3$, or 2520, permutations of the 7 letters in JUSTICE when the letters are taken 5 at a time. *Answer*

The symbol $_nP_r$ is used to indicate the number of permutations of n objects taken r at a time. From Example 2 you can see that $_7P_5 = 2520$. This result was obtained by filling five boxes with a decreasing sequence of consecutive integers, starting with 7. When you find $_nP_r$, the diagram has r boxes to be filled:

$$\boxed{n}\,\boxed{n-1}\,\boxed{n-2}\,\boxed{\cdots}\,\boxed{n-(r-1)}$$

Therefore

$$_nP_r = \underbrace{n(n-1)(n-2)\cdots[n-(r-1)]}_{r \text{ factors}}$$

Notice that when $r = n$, the final factor in the equation above, $n - (r - 1)$, becomes $n - (n - 1)$, or 1. Therefore,

$$_nP_n = n!$$

Example 3 From a set of 9 different books, 4 are to be selected and arranged on a shelf. How many arrangements are possible?

Solution Find the number of permutations of 9 books taken 4 at a time:

$$_9P_4 = 9 \cdot 8 \cdot 7 \cdot 6 = 3024 \quad \textbf{\textit{Answer}}$$

In Example 3, $_9P_4$ can be rewritten as

$$\frac{9 \cdot 8 \cdot 7 \cdot 6 \cdot 5 \cdot 4 \cdot 3 \cdot 2 \cdot 1}{5 \cdot 4 \cdot 3 \cdot 2 \cdot 1} = \frac{9!}{5!}.$$

In general, when the expression $_nP_r$ is multiplied by $\frac{(n-r)\dots 2\cdot 1}{(n-r)\dots 2\cdot 1}$, it becomes

$$\frac{n(n-1)\dots[n-(r-1)](n-r)[n-(r+1)]\dots 2\cdot 1}{(n-r)[n-(r+1)]\dots 2\cdot 1}.$$

Therefore, you have the following formula.

$$_nP_r = \frac{n!}{(n-r)!}$$

If you consider the letters a and a to be distinct, the number of permutations of a, a, and b taken 3 at a time is 3!, or 6. These 6 permutations are listed below:

$$aab \qquad aba \qquad baa$$
$$aab \qquad aba \qquad baa$$

In contrast, the letters a, a, and b, taken 3 at a time, give only 3 distinguishable permutations:

$$aab \qquad aba \qquad baa$$

Each of these permutations corresponds to two of the original 6 permutations, because there are 2!, or 2, permutations of a, a. Therefore, if P is the number of permutations of a, a, and b,

$$P \cdot 2! = 3!, \quad \text{so} \quad P = \frac{3!}{2!} = 3.$$

In general, when n elements of a set are taken n at a time and n_1 of the elements are alike, the number of permutations, P, is $\frac{n!}{n_1!}$. A similar relationship holds when two or more repeated elements occur.

If a set of n elements has n_1 elements of one kind alike, n_2 of another kind alike, and so on, then the number of permutations, P, of the n elements taken n at a time is given by

$$P = \frac{n!}{n_1! n_2! \cdots}.$$

Example 4 Find the number of ways the letters in the word HUBBUB can be arranged.

Solution Since there are 6 letters, 2 of which are U's and 3 are B's,

$$P = \frac{n!}{n_1! n_2!} = \frac{6!}{2! 3!} = 60. \quad \textbf{\textit{Answer}}$$

Oral Exercises

Evaluate.

1. 5!

2. $\frac{10!}{9!}$

3. $\frac{6!}{(6-3)!}$

4. $\frac{4!}{2!2!}$

5. $_4P_4$

6. $_{10}P_3$

7. $_{100}P_2$

8. $_7P_0$

Tell the number of ways the letters in each word can be arranged.

9. FOR

10. ALL

11. SOME

12. NONE

Written Exercises

Evaluate each of the following.

A **1.** $3!5!$
  2. $3(5!)$
  3. $\dfrac{8!}{(8-3)!}$
  4. $\dfrac{8!}{3!5!}$

Find $_nP_r$ for the given values of n and r.

  5. $n = 7, r = 7$
  6. $n = 5, r = 2$
  7. $n = 6, r = 1$
  8. $n = 8, r = 4$

  9. In how many ways can 6 different books be arranged on a shelf?

  10. In how many ways can 8 people be lined up in a row for a photograph?

  11. In how many ways can 3 cards from a deck of 52 cards be laid in a row face up?

  12. In how many ways can 4 of 7 different kinds of bushes be planted along one side of a house?

  13. In how many ways can the letters of the word MONDAY be arranged using all 6 letters?

  14. In how many ways can the letters of the word TODAY be arranged using only 3 of the letters at a time?

Find the number of ways the letters of each word can be arranged.

B **15.** ADDEND
  16. BEEKEEPER
  17. ROTOR

  18. DEEMED
  19. MISSISSIPPI
  20. ALBUQUERQUE

  21. How many different signals can be made by displaying five flags all at one time on a flagpole? The flags differ only in color: two are red, two are white, and one is blue.

  22. How many different signals could be made if the flags of Exercise 21 were of the following colors: three red, one white, and one blue?

  23. In how many ways can 3 identical emeralds, 2 identical diamonds, and 2 different opals be arranged in a row in a display case?

  24. In how many ways can 3 red, 4 blue, and 2 green pens be distributed to 9 students seated in a row if each student receives one pen?

  25. Show that $_6P_4 = 6(_5P_3)$.
  26. Show that $_5P_r = 5(_4P_{r-1})$.

  27. Show that $_nP_5 - {_nP_4} = (n-5)_nP_4$.

  28. Show that $_nP_r - {_nP_{r-1}} = (n-r)_nP_{r-1}$.

C **29.** Solve for n: $_nP_5 = 14(_nP_4)$.
  30. Solve for n: $_nP_3 = 17(_nP_2)$.

  31. Find the number of 6-letter permutations that can be formed from the letters in the word SHOPPER.

15-7 Combinations

Objective   To find the combinations of a set of elements.

Set B is a **subset** of set A if each member of B is also a member of A. Therefore, every set is a subset of itself. The empty set, $\emptyset$, is considered to be a subset of every set.

Example 1   For the 3-letter set {P, Q, R}, find:

        a. all the subsets.   **b.** the 2-letter subsets.

Solution   **a.** {P, Q, R}, {P, Q}, {P, R}, {Q, R}, {P}, {Q}, {R}, $\emptyset$ *Answer*

             b. {P, Q}, {P, R}, {Q, R}   *Answer*

Consider the 2-letter subsets of the 4-letter set {S, T, U, V}:

$$\{S, T\}, \{S, U\}, \{S, V\}, \{T, U\}, \{T, V\}, \text{ and } \{U, V\}.$$

These subsets are also known as the *combinations* of the letters S, T, U, and V taken two at a time. An *r*-element subset of a set of *n* elements is called a **combination** of *n* elements taken *r* at a time.

The symbol $_nC_r$ denotes the number of combinations of *n* elements taken *r* at a time. For example, since there are six 2-letter subsets of the 4-letter set {S, T, U, V}, $_4C_2 = 6$.

Recall that the *order* of the elements matters in a permutation: ST and TS, for example, are different permutations. Order is not important in a combination, however: {S, T} and {T, S} are the same combination.

For a given combination there may be a number of different permutations. For example, there are $_2P_2$, or $2!$, permutations possible for each of the 2-letter combinations of the letters S, T, U, and V. So the total number of permutations of the 4 letters taken 2 at a time is just the number of combinations multiplied by the number of permutations per combination:

$$_4P_2 = (_4C_2)(_2P_2).$$

In general,

$$_nP_r = (_nC_r)(_rP_r),$$

and therefore

$$_nC_r = \frac{_nP_r}{_rP_r}.$$

Recall from Lesson 15-6 that $_nP_r = \dfrac{n!}{(n-r)!}$ and $_rP_r = r!$. Therefore:

$$_nC_r = \frac{_nP_r}{_rP_r} = \frac{n!}{r!\,(n-r)!}$$

Example 2 Find the number of combinations of the letters in the word SOLVE, taking them **(a)** 5 at a time and **(b)** 2 at a time. List each combination.

Solution **a.** $_5C_5 = \dfrac{5!}{5!0!} = 1$ {S, O, L, V, E}

b. $_5C_2 = \dfrac{5!}{2!3!} = 10$ {S, O} {S, L} {S, V} {S, E} {O, L}
{O, V} {O, E} {L, V} {L, E} {V, E}

Example 3 In how many ways can a committee of 6 be chosen from 5 teachers and 4 students if:
a. all are equally eligible?
b. the committee must include 3 teachers and 3 students?

Solution **a.** There are 9 people eligible for the committee:

$$_9C_6 = \frac{9!}{6!3!} = 84 \quad \textbf{\textit{Answer}}$$

b. There are $_5C_3$ ways to choose 3 teachers and $_4C_3$ ways to choose 4 students. Use the first fundamental counting principle stated on page 730 to find the number of ways of selecting the committee:

$$_5C_3 \cdot {}_4C_3 = \frac{5!}{3!2!} \cdot \frac{4!}{3!1!} = 10 \cdot 4 = 40 \quad \textbf{\textit{Answer}}$$

Example 4 A standard deck of cards consists of 4 suits (clubs, diamonds, hearts, and spades) of 13 cards each. How many 5-card hands can be dealt that include 4 cards from the same suit and one card from a different suit?

Solution There are 4 suits with $_{13}C_4$ possible combinations from any one suit:

$$4(_{13}C_4) = 4\left(\frac{13!}{4!9!}\right) = 4(715) = 2860$$

The remaining card must come from the other 3 suits having $52 - 13$, or 39, cards:

$$_{39}C_1 = \frac{39!}{1!38!} = 39$$

$\therefore$ the number of 5-card hands that include exactly 4 cards from any one suit is $2860 \cdot 39$, or 111,540.   **_Answer_**

Oral Exercises

Tell whether each of the following is a combination or a permutation.

1. Nine books placed in a row on a shelf.
2. Three books selected from a collection of 20 books.
3. An arrangement of the letters in the word BOOK.

Evaluate.

4. $\dfrac{5!}{4!1!}$ **5.** $\dfrac{7!}{3!4!}$ **6.** $_4C_2$ **7.** $_6C_4$

Written Exercises

A

1. For the 2-letter set {J, K}, find:
   a. all the subsets.
   b. the subsets containing fewer than 2 letters.

2. For the 4-digit set {1, 3, 5, 7}, find:
   a. the 3-digit subsets.
   b. the subsets in which the sum of the digits is at least 9.

Evaluate.

3. $_5C_3$ **4.** $_6C_1$ **5.** $_8C_6$ **6.** $_7C_4$

7. $_{10}C_8$ **8.** $_9C_2$ **9.** $_{12}C_5$ **10.** $_{100}C_2$

11. How many combinations can be formed from the letters in EIGHT, taking them:
   a. 4 at a time?   **b.** 3 at a time?   **c.** 2 at a time?

12. How many combinations can be formed from the letters in HEXAGON, taking them:
   a. 6 at a time?   **b.** 4 at a time?   **c.** 2 at a time?

13. A volleyball team has 12 members, one coach, and 2 managers. How many different combinations of 7 people can be chosen to kneel in the front row of the team picture?

14. A sample of 4 mousetraps taken from a batch of 100 mousetraps is to be inspected. How many different samples could be selected?

15. In a group of 10 people, each person shakes hands with everyone else once. How many handshakes are there?

16. You can order a hamburger with cheese, onion, pickle, relish, mustard, lettuce, tomato, or mayonnaise. How many different combinations of the "extras" can you order, choosing any four of them?

B **17.** Seven points lie on a circle, as shown in the figure at the right. How many inscribed triangles can be constructed having any three of these points as vertices?

18. Ten points lie on the circumference of a circle. How many inscribed quadrilaterals can be drawn having these points as vertices?

19. A school club has 15 boys and 16 girls as members. How many different 6-person committees can be selected from the membership if equal numbers of boys and girls are to be selected?

20. The junior and senior class councils each have 10 members. In how many ways can a prom committee be formed if it is to consist of 3 seniors and 2 juniors selected from the two class councils?

Exercises 21–24 refer to a standard deck of cards as described in Example 4, page 739.

21. How many 13-card hands having exactly 11 diamonds can be dealt?

22. How many 13-card hands having exactly 11 cards from any suit can be dealt?

23. How many 5-card hands having exactly 3 aces and 2 other cards can be dealt?

24. How many 7-card hands having exactly 3 spades, 3 clubs, and 1 heart can be dealt?

25. In how many ways can 4 or more students be selected from 8 students?

26. In how many ways can up to 4 students be selected from 6 girls and 5 boys if each selection must have an equal number of girls and boys?

C **27.** Prove that $_nC_r = {_nC_{n-r}}$ by using the formula $_nC_r = \dfrac{n!}{r!(n-r)!}$.

28. Prove that $_nC_r = {_{n-1}C_{r-1}} + {_{n-1}C_r}$. (*Hint:* $n! = n \cdot (n-1)!$)

29. a. Refer to the binomial theorem on page 541. What is the relationship between the binomial coefficients in the expansion of $(a + b)^n$ and the values of $_nC_r$ for $r = 0, 1, 2, \ldots , n$?
b. Based on the results of part (a), rewrite the binomial theorem using sigma notation (see Lesson 11-4) and $_nC_r$.

30. Show that the total number of subsets of a set having n elements is 2^n. (*Hint:* Each member of the set is or is not selected in forming a subset. Recall that the empty set is defined to be a subset of every set.)

Mixed Review Exercises

Find the value of each function if $x = 4$.

1. $f(x) = \dfrac{x + 6}{2x - 3}$ **2.** $g(x) = |1 - x|$ **3.** $h(x) = (x + 2)^2$

4. $F(x) = -3$ **5.** $G(x) = -2x + 1$ **6.** $H(x) = x^4 + 5x^3 - 12x^2 + 6$

Simplify.

7. $\sqrt{72} - \sqrt{18} + \sqrt{8}$ **8.** $\log_6 2 + \log_6 3$ **9.** $\dfrac{1}{m + 2} + \dfrac{6}{m^2 - 2m - 8}$

10. $(4 - 3i)(2 + i)$ **11.** $\dfrac{2t^2 + 3t - 9}{t^2 - 9}$ **12.** $\left(\dfrac{\sqrt{13} + 2}{3}\right)\left(\dfrac{\sqrt{13} - 2}{3}\right)$

Computer Exercises

For students with some programming experience.

1. Write a program that will calculate $n!$ for an input value of n.

2. Run the program in Exercise 1 to find $n!$ for each given value of n.
   a. 10   b. 13   c. 16   d. 20

3. Use the program in Exercise 1 as a subroutine in a program that will compute the value of $_nC_r$, for input values of n and r. Have the program check first to make sure n and r are positive integers with $n \geq r$.

4. Run the program in Exercise 3 to find the value of $_nC_r$ for each given pair of values.
   a. $n = 12, r = 3$ b. $n = 18, r = 9$

5. Modify the program in Exercise 3 so that you can compute $_{52}C_r$ for values of r up to 15. (Note that 52! is too large a number for most microcomputers to handle. You can compute $52 \cdot 51 \cdot \ldots \cdot [52 - (r - 1)]$ and then divide by $r!$.) Use your program to compute the following numbers.
   a. $_{52}C_5$, the number of different 5-card hands that can be dealt from a 52-card deck.
   b. $_{52}C_{13}$, the number of possible bridge hands.
   c. Considering your answer to part (b), do you think it is likely that there is some possible bridge hand that has *never* been dealt?

Self-Test 2

Vocabulary   fundamental counting principles (p. 730)   permutation (p. 734)

               mutually exclusive (p. 730)   subset (p. 738)

                                                    combination (p. 738)

1. How many even 3-digit positive integers can be written using the digits 2, 4, 5, 6, and 8?   Obj. 15-5, p. 730

2. A menu lists 4 appetizers, 7 main courses, and 5 desserts. How many complete meals can be chosen from this menu?

3. In how many ways can 5 toy blocks of different color be stacked?   Obj. 15-6, p. 734

4. How many different permutations can be formed using all the letters in the word MEMBER?

5. In how many ways can a 3-person committee be selected from a group of 8 people?   Obj. 15-7, p. 738

6. How many combinations can be formed from the letters in LEAST, taking them 2 at a time?

Compare your answers with those at the back of the book.

Probability

15-8 Sample Spaces and Events

Objective   To specify sample spaces and events for random experiments.

Suppose you toss a coin and it lands "heads up." You repeat the "experiment" and the coin lands "tails up." An experiment in which you do not necessarily get the same outcome when you repeat it under the same conditions is called a **random experiment.**

The set of all possible outcomes of a random experiment is known as the **sample space** for the experiment. When you toss a coin, there are only two possible outcomes: heads (H) or tails (T). The sample space for the coin tossing experiment is therefore $\{H, T\}$.

Any subset of possible outcomes for an experiment is known as an **event.** When an event is a single element of the sample space, it is often called a **simple event.** $\{H\}$ and $\{T\}$ are simple events for the coin tossing experiment.

Example 1   For the rolling of a single die, specify:
   a. the sample space for the experiment.
   b. the event that a number greater than 2 results.
   c. the event that an odd number results.

Solution   **a.** $\{1, 2, 3, 4, 5, 6\}$ **b.** $\{3, 4, 5, 6\}$ **c.** $\{1, 3, 5\}$

Suppose you roll two dice, one red and one blue. An outcome in this experiment can be represented by the ordered pair (r, b), where r is the number showing on the red die and b is the number on the blue die.

The sample space for the two-dice experiment, shown as a graph in the figure at the right, contains $6 \cdot 6$, or 36, elements. Point A, with coordinates $(2, 3)$, represents the outcome "red die shows 2 and blue die shows 3."

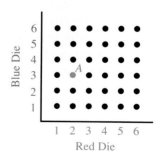

Example 2   For the two-dice experiment, specify each event.
   a. The sum of the numbers showing on the two dice is 7.
   b. The red die shows 5.

Solution   **a.** $\{(1, 6), (2, 5), (3, 4), (4, 3), (5, 2), (6, 1)\}$
   b. $\{(5, 1), (5, 2), (5, 3), (5, 4), (5, 5), (5, 6)\}$

---

1. Each of the letters A, B, C, D, E, F, and G is written on a separate card. The cards are then shuffled, and one card is drawn.
   a. Specify the sample space for this experiment.
   b. Specify the event that a vowel is drawn.

2. A penny and a quarter are tossed. Let (p, q) represent each outcome.
   a. Specify the sample space for this experiment.
   b. Specify the event that exactly one head is up.

Written Exercises

For Exercises 1–6, refer to the two-dice experiment described in this lesson. Specify each event.

A 1. Both dice show the same number.

2. The sum of the numbers showing on the two dice is 6.

3. The product of the numbers showing on the two dice is 12.

4. The sum of the numbers showing on the two dice is greater than 8.

5. The product of the numbers showing on the two dice is less than 10.

6. The number showing on the red die is greater than the number showing on the blue die.

B 7. Two bags contain marbles. The first bag contains a red and a blue marble. The second bag contains a red, a blue, a yellow, and a green marble. One marble is drawn from each bag.
   a. Specify the sample space.
   b. Specify the event that at least one red marble is drawn.
   c. Specify the event that neither marble drawn is blue.

8. A red, a blue, and a green die are tossed. Let (r, b, g) represent each outcome.
   a. How many elements are in the sample space?
   b. Specify the event that all three dice show the same number.
   c. Specify the event that the sum of the numbers showing on the red and blue dice is less than the number showing on the green die.

For Exercises 9–16, refer again to the two-dice experiment. The ordered pair (r, b) represents an element in the sample space. Specify each event.

9. $r + b = 9$ 10. $b > r$ 11. $r = 2b$ 12. $r - b = 2$

C 13. $b > 4$ or $r < 3$ 14. $r + b = 4$ or $r = b$

15. $r + b$ is prime, and b is a perfect square.   16. r is a multiple of 2, and $r - b$ is a multiple of 2.

15-9 Probability

Objective   To find the probability that an event will occur.

Consider the toss of a single coin. The sample space contains two elements, H and T. If the coin is fair, these two outcomes are *equally likely* to occur. That is, if you repeated the coin toss a great many times, you would expect heads about half of the time and tails about half of the time. Therefore, the probability of the event $\{H\}$, denoted by $P(H)$, equals the probability of the event $\{T\}$, denoted by $P(T)$:

$$P(H) = P(T) = \frac{1}{2}.$$

Now consider the rolling of a die. There are six possible outcomes, and the sample space is $\{1, 2, 3, 4, 5, 6\}$. If the die is fair, the six simple events $\{1\}, \{2\}, \{3\}, \{4\}, \{5\},$ and $\{6\}$ are equally likely. Therefore:

$$P(1) = P(2) = P(3) = P(4) = P(5) = P(6) = \frac{1}{6}.$$

If $\{a_1, a_2, a_3, \ldots, a_n\}$ is a sample space containing n equally likely outcomes, then the probability of each simple event is $\frac{1}{n}$:

$$P(a_1) = P(a_2) = P(a_3) = \cdots = P(a_n) = \frac{1}{n}.$$

Returning to the roll of a die, suppose A is the event that the outcome is an odd number. Event A is therefore the subset $\{1, 3, 5\}$ of the sample space $\{1, 2, 3, 4, 5, 6\}$. The probability of A is defined as the sum of the probabilities of the simple events that make up A. Therefore:

$$P(A) = P(1) + P(3) + P(5) = \frac{1}{6} + \frac{1}{6} + \frac{1}{6} = \frac{3}{6} = \frac{1}{2}.$$

If the sample space for an experiment consists of n equally likely outcomes, and if k of them are in event E, then:

$$P(E) = \underbrace{\frac{1}{n} + \frac{1}{n} + \cdots + \frac{1}{n}}_{k \text{ addends}} = \frac{k}{n}.$$

If the event E contains all elements of the sample space, then $P(E) = \frac{n}{n} = 1$; that is, the event is certain to occur. If event E contains no elements of the sample space, then $P(E) = \frac{0}{n} = 0$; that is, the event is certain *not* to occur.

Example 1 A die is rolled. Find the probability of each event.

 a. Event A: The number showing is less than 5.
 b. Event B: The number showing is between 2 and 6.

Solution The sample space is $\{1, 2, 3, 4, 5, 6\}$. Since the six outcomes are equally likely, the probability of each simple event is $\frac{1}{6}$.

 a. Event $A = \{1, 2, 3, 4\}$ **b.** Event $B = \{3, 4, 5\}$

 $\therefore P(A) = \frac{4}{6} = \frac{2}{3}$ *Answer* $\therefore P(B) = \frac{3}{6} = \frac{1}{2}$ *Answer*

Example 2 Two dice are rolled and the numbers are noted. Find the probability of each event.

 a. Event A: The sum of the numbers is less than 5.
 b. Event B: The sum of the numbers is 4 or 5.

Solution Look at the figure on page 743. Notice that there are 36 possible outcomes when two dice are rolled.

 a. Event $A = \{(1, 1), (1, 2), (1, 3), (2, 1), (2, 2), (3, 1)\}$

 $\therefore P(A) = \frac{6}{36} = \frac{1}{6}$ *Answer*

 b. Event $B = \{(1, 3), (1, 4), (2, 2), (2, 3), (3, 1), (3, 2), (4, 1)\}$

 $\therefore P(B) = \frac{7}{36}$ *Answer*

Example 3 There are 12 tulip bulbs in a package. Nine will yield yellow tulips and three will yield red tulips. If two tulip bulbs are selected at random, find the probability of each event.

 a. Event Q: Both tulips will be red.
 b. Event R: One tulip will be yellow and the other red.

Solution **a.** Since the bulbs are selected at random, all possible pairs are equally likely. Thus:

$$P(Q) = \frac{\text{number of ways to pick a pair of red bulbs}}{\text{number of ways to pick any pair of bulbs}}$$

There are 3 red bulbs. The number of ways of picking 2 of them is:

$$_3C_2 = \frac{3!}{2!1!} = 3.$$

There are 12 bulbs in all. The number of ways of picking 2 of them is:

$$_{12}C_2 = \frac{12!}{2!10!} = 66.$$

 $\therefore P(Q) = \frac{3}{66} = \frac{1}{22}$ *Answer*

b. The number of ways of picking one yellow bulb and one red bulb is:

$$_9C_1 \cdot {}_3C_1 = \frac{9!}{1!8!} \cdot \frac{3!}{1!2!} = 9 \cdot 3 = 27.$$

$$\therefore P(R) = \frac{27}{66} = \frac{9}{22} \quad \textbf{\textit{Answer}}$$

Consider a randomly chosen element of a normal distribution with mean M and standard deviation σ. The probability of this element falling between the values $M + a\sigma$ and $M + b\sigma$ is just the area under the standard normal curve between $x = a$ and $x = b$ (see Lesson 15-3).

Example 4 The lifetimes of 60-watt light bulbs are normally distributed with mean 1000 hours and standard deviation 100 hours. What is the probability that a bulb just put in a lamp will last more than 1100 hours?

Solution The diagram at the right shows the hours h of light bulb life and the standardized values x that correspond to them. The probability that $h > 1100$ is the same as the area under the standard normal curve to the right of $x = 1$. Therefore:

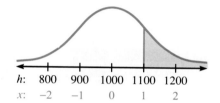

| h: | 800 | 900 | 1000 | 1100 | 1200 |
| x: | -2 | -1 | 0 | 1 | 2 |

$$P(h > 1100) = P(x > 1)$$
$$= 0.5 - A(1)$$
$$= 0.5 - 0.3413 \qquad \text{Use the table on page 720.}$$
$$= 0.1587$$

$\therefore$ the probability that the bulb lasts more than 1100 hours is about 0.16.

Example 5 For a certain airline, the trip times for daily flights between two cities are normally distributed. The mean trip time is 144 minutes with a standard deviation of 15 minutes. What is the probability that a flight takes less than 2 hours?

Solution The standardized value corresponding to 2 hours, or 120 minutes, is:

$$x = \frac{120 - M}{\sigma} = \frac{120 - 144}{15} = -\frac{24}{15} = -1.6.$$

Therefore, for a random trip time t:

$$P(t < 120) = P(x < -1.6)$$
$$= P(x > 1.6) \qquad \left\{ \begin{array}{l} \text{The standard normal curve is sym-} \\ \text{metric with respect to the } y\text{-axis.} \end{array} \right.$$
$$= 0.5 - A(1.6)$$
$$= 0.5 - 0.4452 \qquad \text{Use the table on page 720.}$$
$$= 0.0548$$

$\therefore$ the probability that a flight takes less than 2 hours is about 0.05.

Oral Exercises

A number wheel is divided into 16 congruent sectors, numbered 1–16. You spin the wheel and record the number. Find the probability of each event.

1. It is a 7.

2. It is an odd number.

3. It is a 4 or a 9.

4. It is a 1, 2, or 3.

5. It is between 0 and 17.

6. It is less than 13.

7. It is less than 3 or greater than 12.

8. It is a 19.

Two dice are rolled. Find the probability of each event. (Use the figure on page 743 for the sample space of this experiment.)

9. The numbers showing on the two dice are equal.

10. The sum of the numbers showing is 7.

11. The sum of the numbers showing is 13.

Written Exercises

A

1. A box contains 10 slips of paper numbered 1–10. A slip of paper is drawn at random from the box and the number is noted. Find the probability of each event.

   a. It is a 2.

   b. It is an odd number.

   c. It is less than 4.

   d. It is less than 11.

   e. It is greater than 8.

   f. It is between 3 and 4.

2. A letter is selected at random from those in the word TRIANGLE. Find the probability of each event.

   a. It is a vowel.

   b. It is a consonant.

   c. It is from the first half of the alphabet.

   d. It is between F and Q in the alphabet.

3. One marble is drawn at random from a bag containing 4 white, 6 red, and 6 green marbles. Find the probability of each event.

   a. It is white.

   b. It is white, red, or green.

   c. It is red or green.

   d. It is not red.

4. One card is drawn at random from a 52-card deck. Find the probability of each event.

   a. It is an ace.

   b. It is a diamond.

   c. It is black.

   d. It is the king of clubs.

   e. It is a red queen.

   f. It is a black heart.

5. Two coins are tossed. Find the probability of each event.

   a. Both come up tails.

   b. At least one coin comes up heads.

   c. The coins match.

   d. The coins don't match.

6. A 10-speed bicycle is given as a door prize. A total of 220 tickets numbered 1–220 are sold. If the winning number is chosen at random, find the probability of each event.
   a. The winning number is less than 101.
   b. The winning number is greater than 50.
   c. The winning number is between 10 and 21.

7. Of 1,000,000 income tax returns received, all are checked for arithmetic accuracy and 9000 returns, selected at random, are checked thoroughly. For a given return, find the probability of each event.
   a. It is checked for arithmetic accuracy.
   b. It is checked thoroughly.
   c. It is not checked thoroughly.

8. Two dice are rolled. Find the probability of each event.
   a. The sum of the numbers showing on the dice is 10.
   b. The sum is at least 10.
   c. Exactly one die shows a 4.
   d. At least one die shows a 4.

B 9. Three coins are tossed. Find the probability of each event.
   a. All come up heads.
   b. All come up tails.
   c. At least one comes up tails.
   d. Exactly two come up heads.

10. Evangeline makes necklaces from beads of four different colors: red, white, yellow, and turquoise. There are twice as many red beads as white beads, and twice as many white beads as yellow beads or turquoise beads. If Evangeline chooses a bead at random from a box containing all the beads, what is the probability she chooses:
   a. a red bead?
   b. a white bead?
   c. either a yellow bead or a turquoise bead?

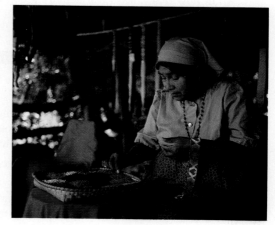

11. A bag contains 2 red, 4 yellow, and 6 blue marbles. Two marbles are drawn at random. Find the probability of each event.
   a. Both are red.   b. Both are yellow.
   c. Both are blue.   d. One is red and one is yellow.
   e. Neither is red.   f. Neither is blue.

12. Two cards are drawn at random from a 52-card deck. Find the probability of each event.
   a. Both are hearts.   b. Both are jacks.
   c. Neither is red.   d. Neither is a spade.

In Exercises 13 and 14, use the table on page 720.

13. For a certain math class, the mean time required to complete homework assignments is 36 minutes with a standard deviation of 10 minutes. Assuming that the completion times are normally distributed, what is the probability that the time required to complete a given homework assignment in this class will be:
    a. less than 20 minutes?   b. more than one hour?

14. A certain brand of cereal claims that the mean number of raisins in each box is 74 with a standard deviation of 5. Assuming that the raisins are normally distributed, what is the probability that a given box of this cereal will contain:
    a. fewer than 65 raisins?   b. more than 80 raisins?

Mixed Review Exercises

Solve each inequality and graph the solution set.

1. $|t - 2| > 4$ 2. $x^2 - x \le 2$ 3. $4 - 3n < -2$

4. $-3 \le 2x - 3 \le 5$ 5. $4y^2 > 16$ 6. $|3k + 1| < 4$

7. How many permutations are there of all the letters in PAPER?

8. In how many ways can a study group of four be selected from a class of 24 students?

Computer Exercises   *For students with some programming experience.*

1. Write a program that will use the computer's random number generator (RND) to simulate the tossing of a coin a given number of times. Have the computer print the number of times that heads come up.

2. Run the program in Exercise 1 for each number of tosses and record the number of heads.   **a.** 20   **b.** 80   **c.** 200   **d.** 800

3. When a coin is tossed a number of times, the *expected* number of heads is half the number of tosses. Which of the runs in Exercise 2 comes closest, in relative terms, to the expected number of heads?

4. Write a program that will use the RND function to simulate rolling two dice a given number of times. Have the computer print the number of "doubles" (that is, when the numbers showing on the dice are the same).

5. Run the program in Exercise 4 for each number of rolls and record the number of doubles.   **a.** 12   **b.** 72   **c.** 120   **d.** 720

6. When two dice are rolled a number of times, the *expected* number of doubles is one sixth the number of rolls. Which of the runs in Exercise 5 comes closest, in relative terms, to the expected number of doubles?

Reading Algebra / Probability

When English words are used as technical terms in algebra they often have a meaning different from their usual meaning in the language. It is particularly important in reading about probability and statistics to distinguish the mathematical usage of a term from its common use, since many of the terms have mathematical meanings that are close to, but not identical to, their meanings in common speech. In the brief lesson that follows, you will be introduced to the concept of *expectation*. As you read the lesson and do the exercises, notice how the meaning of expectation in the study of probability is related to its ordinary meaning.

Often each simple event in a sample space can be measured by a number or score. For example, if you toss two coins you can count the number of heads as the score; it will be 0, 1, or 2. The *expectation* of the score is found by multiplying the score of each simple event by the probability of that event and adding up these products. Here is how the computation looks for the coin tossing experiment:

Event		Score	Probability	Product
Coin 1	Coin 2			
H	H	2	$\frac{1}{4}$	$\frac{1}{2}$
H	T	1	$\frac{1}{4}$	$\frac{1}{4}$
T	H	1	$\frac{1}{4}$	$\frac{1}{4}$
T	T	0	$\frac{1}{4}$	0

The expectation is $\frac{1}{2} + \frac{1}{4} + \frac{1}{4} + 0$, or 1. Therefore, the *expected* number of heads is 1.

Exercises

1. If you toss four coins, what is the expected number of heads?

2. What is the probability that you get two heads on a toss of four coins?

3. If you repeat the experiment of tossing four coins many times, will you score two heads more than half the time or less than half the time?

4. When you roll a die with faces numbered 1 to 6, what is your expected score? Will you ever get this score?

5. Suppose you roll the die many times and find the average (mean) of all your scores. Can you predict what this average will approximately be?

6. In what sense does the expectation tell you what to expect?

Application / *Sampling*

A factory produces 200,000 light bulbs during each 8-hour shift. The factory managers want to know how many defective bulbs are being produced. If the number is too high, production methods must be changed to improve the output. One way to determine the number of defective bulbs is to test every one produced. Although this would give the managers the information they need, it would also be too expensive and time-consuming.

Another method is to test only a *sample* of the bulbs to get some idea of what the entire *population* of light bulbs is like. Testing only a relatively small sample is certainly less expensive than testing the entire population, but can the managers of the factory have confidence in the results? That is, do the results in a sample provide a good estimate of the results in the population as a whole? Studies have shown that the answer to this question is: It depends on how samples are drawn from the population.

The best estimates are obtained by drawing random samples. A sample from a population is a *random sample* provided:

1. Each element in the population is equally likely to be in the sample.
2. The selection of each element is independent of the selection of every other element.

It is not always easy to obtain a random sample, however. For example, if each element in a population is to be selected on an "equally likely" basis, then all elements must be known and available. This is usually not the case with large populations, such as all the eligible voters in the United States.

Because random samples are difficult to obtain, nonrandom samples are often used to characterize populations. Here are a few alternative sampling procedures.

Convenience sampling:	Pick the elements in the population that are readily at hand.
Self-selected sampling:	Allow elements in the population to volunteer to be in the sample.
Judgment sampling:	Have an expert choose a sample believed to be representative of the population.

Compared to random sampling, these and other sampling procedures often yield less dependable results. They may, however, be sufficient for whatever purpose the population is being sampled.

Exercises

Which of the following samples is random? If nonrandom, explain why.

1. A random-number generator activates a device that tests light bulbs coming off the assembly line.
   a. Testing is done throughout the 8-hour shift each day.
   b. Testing is done only during the first 2 hours of production each day.

2. A 20-student committee is to be selected from a student body of 500.

    a. Twenty names are drawn from a box containing all the students' names, each written on a separate slip of paper.

    b. One name is selected at random from a roster of all the students. The remaining selections are made by picking every 25th name on the roster following the first one selected.

3. A TV-talk-show host lets members of the audience who raise their hands voice their opinions on an issue.

4. A newspaper reporter asks passersby their opinions on an issue.

Describe how samples could be obtained in each of the following situations. Also describe any limitations of the sampling techniques that you propose.

5. The editorial staff of a magazine needs to know how the subscribers would react to a change in the layout of the magazine.

6. A national news organization wants to know how the current national economy is affecting typical American families.

7. A researcher wants to investigate whether optimists are generally more healthy than pessimists.

8. A school superintendent wants to know whether the voters in the school district will support a bond issue to build a new school.

Extra / *Random Numbers*

Computers use arithmetic to generate sequences of "random" numbers that can be used to simulate experiments such as the tossing of a coin. Since these sequences are completely determined by the first number chosen, the numbers are not truly random (in fact, they are often called *pseudo*random numbers), but they can still be used to program realistic simulations.

One common method for generating random numbers is illustrated at the right. First choose a multiplier, 23, and a starting value, 809. Then repeat the following steps: Multiply by 23, save the last three digits, and record the hundreds' digit. The sequence of random numbers is 6, 9, 1, 3, 4, ...

Notice that the rightmost two digits must be discarded, since the units' digit repeats the pattern 7, 1, 3, 9, and the tens' digit is always even. Since the last three digits are saved from each product, the sequence of "random" digits must repeat after at most 1000 steps. Actual computer routines save many more digits, and they provide a sequence of 7- or 8-digit pseudorandom numbers that will repeat only after millions of numbers have been drawn.

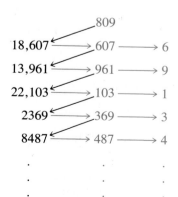

15-10 Mutually Exclusive and Independent Events

Objective To identify mutually exclusive and independent events and find the probability of such events.

Consider the sets A and B where

$A = \{1, 2, 3, 4\}$ and $B = \{3, 4, 5, 6\}$.

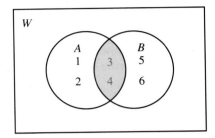

Notice that these sets have two elements in common, namely 3 and 4. This fact can be illustrated through the use of a diagram like the one at the right. (Such diagrams are called *Venn diagrams* in honor of the English mathematician John Venn, who used them extensively to picture sets and set relations.)

In the Venn diagram, sets A and B are shown as subsets of the set W of whole numbers. Notice that the region where sets A and B overlap is shaded. The shaded region represents $\{3, 4\}$, the set of elements that belong to *both A and B*. This set is called the *intersection* of A and B. You can use the symbol $\cap$ to indicate an intersection of sets:

$$\{1, 2, 3, 4\} \cap \{3, 4, 5, 6\} = \{3, 4\}.$$

In general, if A and B are any sets, then the set whose members are the elements belonging to both A and B is called the **intersection** of A and B and is denoted by $A \cap B$.

The sets $\{1, 2, 3, 4\}$ and $\{5, 6, 7, 8\}$ have no elements in common. That is,

$$\{1, 2, 3, 4\} \cap \{5, 6, 7, 8\} = \emptyset.$$

Sets that have no elements in common are called **disjoint** sets.

In the Venn diagram at the right, notice that the shading extends throughout sets A and B. The shaded region represents $\{1, 2, 3, 4, 5, 6\}$, the set containing all the elements of A together with all the elements of B. In other words, the shaded region represents the set consisting of all the elements in *at least one* of the sets A and B. This set is called the *union* of A and B. You can use the symbol $\cup$ to indicate a union of sets:

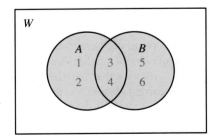

$$\{1, 2, 3, 4\} \cup \{3, 4, 5, 6\} = \{1, 2, 3, 4, 5, 6\}.$$

In general, if A and B are any sets, then the set whose members are the elements belonging to either A or B (or both) is called the **union** of A and B and is denoted by $A \cup B$.

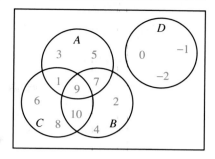

Example 1 Specify each of the following sets by listing its elements. Use the Venn diagram at the right.

    a. $A \cap B$ **b.** $A \cup B$
    c. $B \cap C$ **d.** $B \cup C$
    e. $A \cap D$ **f.** $(A \cup B) \cap C$

Solution

    a. $\{7, 9\}$
    b. $\{1, 2, 3, 4, 5, 7, 9, 10\}$
    c. $\{9, 10\}$
    d. $\{1, 2, 4, 6, 7, 8, 9, 10\}$
    e. $\emptyset$
    f. $\{1, 9, 10\}$

The Venn diagram below shows the sample space S for the experiment of drawing a number at random from $\{1, 2, 3, 4, 5, 6\}$. Every subset of S is an event, including S itself and the empty set, $\emptyset$. These two events have probabilities 1 and 0, respectively.

For any sample space S,

$$P(S) = 1 \quad \text{and} \quad P(\emptyset) = 0.$$

The Venn diagram also shows events A and B in S. Event A, $\{1, 2\}$, is the drawing of a number less than 3. Event B, $\{1, 3, 5\}$, is the drawing of an odd number. Note that

$$P(A) = \tfrac{2}{6} \text{ and } P(B) = \tfrac{3}{6}.$$

The probability that either A or B occurs is $P(A \cup B)$. Since $A \cup B = \{1, 2, 3, 5\}$,

$$P(A \cup B) = \tfrac{4}{6}.$$

To see how $P(A \cup B)$, $P(A)$, and $P(B)$ are related, first note that $P(A \cap B) = \tfrac{1}{6}$, since $A \cap B = \{1\}$. Then:

$$P(A \cup B) = \frac{4}{6} = \frac{2}{6} + \frac{3}{6} - \frac{1}{6} = P(A) + P(B) - P(A \cap B).$$

For any two events A and B in a sample space,

$$P(A \cup B) = P(A) + P(B) - P(A \cap B).$$

If two events have no elements in common (that is, if their intersection is empty), the events are called **mutually exclusive.** For such events, $P(A \cap B) = P(\emptyset) = 0$. Thus, $P(A \cup B) = P(A) + P(B) - 0 = P(A) + P(B)$.

If A and B are mutually exclusive events,

$$P(A \cup B) = P(A) + P(B).$$

Example 2 From a group of 10 seniors and 8 juniors, 3 students are to be selected at random to form a committee. What is the probability that at least 2 seniors are selected?

Solution The selection will include at least 2 seniors if either exactly 2 seniors and 1 junior are selected (event M) *or* 3 seniors are selected (event N). Since the two events are mutually exclusive, $P(M \cup N) = P(M) + P(N)$.

$$P(M) = \frac{{}_{10}C_2 \cdot {}_8C_1}{{}_{18}C_3} = \frac{45 \cdot 8}{816} = \frac{360}{816} = \frac{15}{34}$$

$$P(N) = \frac{{}_{10}C_3}{{}_{18}C_3} = \frac{120}{816} = \frac{5}{34}$$

$$P(M \cup N) = \frac{15}{34} + \frac{5}{34} = \frac{20}{34} = \frac{10}{17}$$

$\therefore$ the probability of at least 2 seniors being selected is $\frac{10}{17}$. **Answer**

Consider the experiment of tossing a coin twice. The sample space for the experiment is $\{(H, H), (H, T), (T, H), (T, T)\}$. Let A be the event that a head is obtained on the first toss. Let B be the event that a head is obtained on the second toss. Then:

$$A = \{(H, H), (H, T)\} \text{ and } B = \{(H, H), (T, H)\}.$$

Since the experiment has four possible outcomes, the probabilities of the events are:

$$P(A) = \frac{2}{4} = \frac{1}{2} \text{ and } P(B) = \frac{2}{4} = \frac{1}{2}.$$

Notice that A and B are not mutually exclusive, since $A \cap B = \{(H, H)\}$. However, the two events are *independent,* because obtaining a head on the first toss has nothing to do with obtaining a head on the second toss. In general, if the occurrence of one event has no effect on the probability of another event, the two events are said to be **independent.**

Since $P(A \cap B) = \frac{1}{4}$, you can see the following pattern:

$$P(A) \cdot P(B) = P(A \cap B)$$

$$\frac{1}{2} \cdot \frac{1}{2} = \frac{1}{4}$$

Two events A and B are independent if and only if:

$$P(A \cap B) = P(A) \cdot P(B).$$

Example 3 A bag contains 4 green and 4 blue buttons. Two buttons are drawn at random from the bag as follows:

a. The first button drawn is put back into the bag before the second button is drawn.

b. The first button drawn is *not* put back into the bag before the second button is drawn.

Let A be the event that the first button drawn is green.
Let B be the event that the second button drawn is blue.
Are the events in each case dependent or independent?

Solution **a.** $P(A) = \dfrac{4}{8} = \dfrac{1}{2}$

$P(B) = \dfrac{4}{8} = \dfrac{1}{2}$

$P(A \cap B) = \dfrac{{}_4C_1 \cdot {}_4C_1}{{}_8C_1 \cdot {}_8C_1} = \dfrac{4 \cdot 4}{8 \cdot 8} = \dfrac{16}{64} = \dfrac{1}{4}$

$P(A) \cdot P(B) = \dfrac{1}{2} \cdot \dfrac{1}{2} = \dfrac{1}{4} = P(A \cap B)$

$\therefore$ events A and B are independent. (Since the first button drawn is put back, the result of the first draw does *not* affect the outcome of the second draw.) **Answer**

b. $P(A) = \dfrac{1}{2}$

Event B can come about in two mutually exclusive ways: Either you draw a green button first and then draw one of the four blue buttons among the seven remaining, or you draw a blue button first and then draw one of the other three blue buttons among the seven remaining. Therefore:

$P(B) = P(\text{first green, then blue}) + P(\text{first blue, then blue})$

$\qquad = \qquad \dfrac{4}{8} \cdot \dfrac{4}{7} \qquad + \qquad \dfrac{4}{8} \cdot \dfrac{3}{7} \qquad = \dfrac{28}{56} = \dfrac{1}{2}$

The event $A \cap B$ is equivalent to drawing a green button first and then a blue one. That is,

$P(A \cap B) = P(\text{first green, then blue}) = \dfrac{4}{8} \cdot \dfrac{4}{7} = \dfrac{2}{7}.$

$P(A) \cdot P(B) = \dfrac{1}{2} \cdot \dfrac{1}{2} = \dfrac{1}{4} \neq P(A \cap B)$

$\therefore$ events A and B are dependent. (In this case, the result of the first draw does affect the second draw.) **Answer**

Suppose an event A is a subset of a sample space S. It is sometimes useful to consider the **complement** of A, which consists of the elements in S that are *not* members of A. The symbol for the complement of an event A is $\overline{A}$. Recall that $P(S) = 1$. Since $A \cup \overline{A} = S$, $P(A \cup \overline{A}) = 1$. Since A and $\overline{A}$ are mutually exclusive, $P(A \cup \overline{A}) = P(A) + P(\overline{A})$. Therefore, $P(A) + P(\overline{A}) = 1$ and $P(\overline{A}) = 1 - P(A)$.

Example 4 Based on past performances in the school band, the probability that Ryan will be selected to perform in the county band is $\frac{3}{4}$, that Faye will be selected is $\frac{2}{3}$, and that Chung will be selected is $\frac{1}{2}$. Suppose that the selection of one student does not affect another student's chances. Find the probabilities of each of the following events.

a. Ryan and Chung are selected, but Faye is not.

b. At least one of the three is selected.

Solution Let R be the event that Ryan is selected for the county band, F the event that Faye is selected, and C the event that Chung is selected. Then:

$$P(R) = \frac{3}{4}, \; P(F) = \frac{2}{3}, \text{ and } P(C) = \frac{1}{2}.$$

a. The probability that Faye is *not* selected is:

$$P(\overline{F}) = 1 - P(F) = 1 - \frac{2}{3} = \frac{1}{3}.$$

$$\therefore P(R) \cdot P(C) \cdot P(\overline{F}) = \frac{3}{4} \cdot \frac{1}{2} \cdot \frac{1}{3} = \frac{1}{8} \quad \textit{Answer}$$

b. Let D be the event that *at least one* of the three students is selected, and let E be the event that *none* of them is selected. Note that $D = \overline{E}$.

$$P(D) = P(\overline{E})$$
$$= 1 - P(E)$$
$$= 1 - P(\overline{R}) \cdot P(\overline{F}) \cdot P(\overline{C}) \qquad \begin{cases} E \text{ is equivalent to } \overline{R}, \overline{F}, \\ \text{and } \overline{C} \text{ occurring together.} \end{cases}$$
$$= 1 - \frac{1}{4} \cdot \frac{1}{3} \cdot \frac{1}{2}$$
$$= 1 - \frac{1}{24} = \frac{23}{24}$$

$\therefore$ the probability of at least one being selected is $\frac{23}{24}$. *Answer*

Oral Exercises

In Exercises 1–3, use the Venn diagram at the right.

1. Specify each of the following sets by listing its elements.

  a. $C \cup D$ **b.** $B \cap D$

  c. $B \cap C$ **d.** $(B \cup C) \cap A$

2. Find the probability of each event.

  a. B **b.** $\overline{A}$

  c. $C \cap D$ **d.** $A \cup B$

3. Tell whether the events are mutually exclusive.

  a. A and B **b.** B and C **c.** B and D

  d. $(B \cup D)$ and A **e.** $(A \cup B)$ and D **f.** $\overline{A}$ and B

4. Use the given probabilities to tell whether A and B are dependent or independent events.

  a. $P(A) = \frac{1}{16}$; $P(B) = \frac{1}{4}$; $P(A \cap B) = \frac{1}{4}$

  b. $P(A) = \frac{1}{3}$; $P(B) = \frac{3}{4}$; $P(A \cap B) = \frac{1}{4}$

  c. $P(A) = 0.3$; $P(B) = 0.7$; $P(A \cap B) = 0.05$

  d. $P(A) = 0.4$; $P(B) = 0.5$; $P(A \cap B) = 0.2$

5. Identify the events in each problem as independent or dependent, and then find the probability requested.

  a. In a box there are 3 pink and 3 green erasers. You pick one at random, replace it, and select another at random. What is the probability that both erasers are green?

  b. There are two 60-watt light bulbs and two 75-watt light bulbs in a box. You pick one at random and place it in your study lamp. Then you select another at random. What is the probability that you first picked a 75-watt and then a 60-watt light bulb?

Written Exercises

A 1. Tanya randomly guesses a whole number from 1 to 10. Find the probability of each event.

  a. She guesses a number less than 6.

  b. She guesses an odd number.

  c. She guesses an odd number less than 6.

2. A single marble is drawn from a bag containing 3 red, 5 white, and 2 blue marbles. Find the probability of each event.

  a. A red or blue marble is drawn.

  b. A blue or white marble is drawn.

  c. A red, white, or blue marble is drawn.

3. The names of 3 seniors, 4 juniors, and 5 sophomores are placed in a bowl. One name is drawn at random, set aside, and a second name is then drawn at random. Find the probability of each event.

  a. The first name drawn is a junior and the second is a senior.

  b. Both names drawn are sophomores.

4. There are 3 red, 2 blue, and 3 yellow crayons in a box. Jeff randomly selects one, returns it to the box, and then randomly selects another. Find the probability of each event.

  a. The first crayon selected is blue and the second is yellow.

  b. Both crayons selected are red.

5. A red and a green die are rolled. Let A be the event that the red die shows 2 and B be the event that the sum of the numbers showing is 8.

  a. Find the probability of A, B, $A \cup B$, and $A \cap B$.

  b. Are A and B independent events?

6. A die and a coin are tossed. Let *A* be the event that the die shows a 5 or 6, and let *B* be the event that the coin shows heads.

    a. Specify the sample space for the experiment.
    b. Specify the simple events in *A*, *B*, *A* ∪ *B*, and *A* ∩ *B*.
    c. Find the probability of *A*, *B*, *A* ∪ *B*, and *A* ∩ *B*.
    d. Are *A* and *B* mutually exclusive? Are they independent?

7. Two dice are rolled. Find the probability of each event.

    a. The sum of the numbers showing is 10.
    b. Either the sum of the numbers showing is 4 or both numbers are 4.

8. Two cards are drawn at the same time from a 52-card deck. Find the probability of each event.

    a. Both cards are jacks.
    b. Both cards are sixes.
    c. Either both cards are jacks or both are sixes.

9. A coin is tossed three times. Find the probability of each event.

    a. At least two tosses come up tails.
    b. At least one toss comes up heads.

10. From a 52-card deck a card is drawn and then replaced. After the deck is shuffled, a second card is drawn. Find the probability of each event.

    a. The first card is a 5 and the second is a 6.
    b. Both cards are clubs.

B **11.** If a six-volume set of books is placed on a shelf at random, what is the probability that the books will be arranged in either correct or reverse order?

12. When Carlos shoots a basketball, the probability of a basket is 0.4. When Brad shoots, the probability of a basket is 0.45. What is the probability that at least one basket is made if they each take one shot?

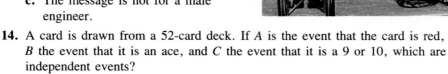

13. Four managers, five engineers, and one lawyer are in a meeting when a message arrives for one of them. Of these people, three managers, two engineers, and the lawyer are women. Find the probability of each event.

    a. The message is for an engineer.
    b. The message is for a man.
    c. The message is not for a male engineer.

14. A card is drawn from a 52-card deck. If *A* is the event that the card is red, *B* the event that it is an ace, and *C* the event that it is a 9 or 10, which are independent events?

    a. *A* and *B*   **b.** *A* and *C*   **c.** *B* and *C*

15. The probability that Karla will ask Frank to be her tennis partner is $\frac{1}{4}$, that Juan will ask Frank is $\frac{1}{3}$, and that Roger will ask Frank is $\frac{3}{4}$. Find the probability of each event.

    a. Karla and Juan ask him.

    b. Juan and Roger ask him, but Karla doesn't.

    c. At least two of the three ask him.

    d. At least one of the three asks him.

16. The probability of rain on a certain day is 65% in Yellow Falls and 40% in Copper Creek. Find the probability of each event.

    a. It will rain in Yellow Falls but not in Copper Creek.

    b. It will rain in both towns.

    c. It will rain in neither town.

    d. It will rain in at least one of the towns.

C **17.** When it is not snowing, the probability of the Colorado Drifters winning a football game is $\frac{7}{10}$. When it is snowing, the probability of the Drifters winning is $\frac{2}{5}$. The probability of snow on any day in January is $\frac{1}{2}$.

    a. What is the probability of the Drifters winning on January 17?

    b. If the Drifters won the game on January 17, what is the probability that the team played in snow?

Self-Test 3

Vocabulary   random experiment (p. 743)   disjoint (p. 754)

                 sample space (p. 743)   union (p. 754)

                 event (p. 743)   mutually exclusive events (p. 755)

                 simple event (p. 743)   independent events (p. 756)

                 equally likely events (p. 745)   complement (p. 757)

                 intersection (p. 754)

Solve.

1. Each of the letters in the word SQUARE is written on a different   **Obj. 15-8, p. 743** card, and the six cards are shuffled. One card is drawn at random.

    a. Specify the sample space for the drawing.

    b. Specify the event that a vowel is drawn.

2. One chip is drawn at random from a box containing 3 green,   **Obj. 15-9, p. 745** 5 blue, 2 white, and 8 red chips. What is the probability that the chip drawn is white?

3. Two dice are rolled.   **Obj. 15-10, p. 754**

    a. Find the probability that either or both numbers showing are 6's or their sum is greater than 8.

    b. Are the two events independent?

    c. Are the two events mutually exclusive?

Check your answers with those at the back of the book.

Computers are often used to simulate real-world events. A typical simulation involves waiting lines, such as those found in banks and grocery stores.

The program below simulates a single waiting line. The user controls four variables: the duration of the simulation (DUR), the maximum length of the line (MAX), the probability that a customer arrives at any given moment (PROB), and the average service time for a customer (AST).

During the course of the simulation, the computer keeps a running count of the number of customers in line (CIL), the number served (CS), and the number lost because the line was too long (CL). The final status of these counts is reported at the end of the simulation.

```
10 PRINT "THIS PROGRAM SIMULATES A WAITING LINE."
20 INPUT "ENTER THE DURATION (IN MINUTES) OF THE
   SIMULATION: "; DUR
30 INPUT "ENTER THE MAXIMUM LENGTH OF THE WAITING
   LINE: "; MAX
40 INPUT "ENTER THE PROBABILITY THAT A CUSTOMER WILL
   ARRIVE AT ANY GIVEN MINUTE: "; PROB
50 INPUT "ENTER THE AVERAGE SERVICE TIME (IN MINUTES)
   FOR A CUSTOMER: "; AST
60 LET CIL=0
70 LET CS=0
80 LET CL=0
90 LET ST=0
100 FOR T=1 TO DUR
110 IF RND(1)>PROB THEN 140
120 IF CIL=MAX THEN LET CL=CL+1
130 IF CIL<MAX THEN LET CIL=CIL+1
140 IF CIL>0 AND ST=0 THEN LET ST=AST
150 IF ST>0 THEN LET ST=ST-1
160 IF CIL>0 AND ST=0 THEN LET CS=CS+1
170 IF CIL>0 AND ST=0 THEN LET CIL=CIL-1
180 NEXT T
190 PRINT "NUMBER OF CUSTOMERS SERVED: "; CS
200 PRINT "NUMBER OF CUSTOMERS LOST: "; CL
210 PRINT "NUMBER LEFT IN LINE: "; CIL
220 END
```

Exercises

1. Holding DUR, MAX, and AST constant, run the simulation for the following values of PROB. Record the results.

   a. 0.1   **b.** 0.3   **c.** 0.5   **d.** 0.7   **e.** 0.9

2. Based on the results of Exercise 1, what conclusion can you draw?

Chapter Summary

1. *Frequency distributions, histograms, stem-and-leaf plots,* and *box-and-whisker plots* are four ways of displaying statistical data.

2. The *mode,* the *median,* and the *mean* are used to describe the *central tendency* of a frequency distribution.

3. The *range,* the *variance,* and the *standard deviation* are used to describe the *dispersion* of a frequency distribution.

4. The *variance* of a distribution is defined as follows:

$$\text{variance} = \frac{(x_1 - M)^2 + (x_2 - M)^2 + \cdots + (x_n - M)^2}{n}$$

   where $x_1, x_2, \ldots, x_n$ are n numbers and M is their mean.

5. The *standard deviation,* σ, is the principal square root of the variance:

$$\sigma = \sqrt{\text{variance}}.$$

6. The *standard normal curve* is symmetric with respect to the y-axis and approaches the x-axis asymptotically as $|x|$ increases. The total area under the curve and above the x-axis equals 1. For a normal distribution with mean M and standard deviation σ, the fraction of the data falling between $M + a\sigma$ and $M + b\sigma$ is the area under the standard normal curve between $x = a$ and $x = b$.

7. The *correlation coefficient, r,* between two variables x and y is a measure of how closely the ordered pairs (x, y) cluster about a line called the *regression line.* The closer the points (x, y) are to being collinear, the closer $|r|$ is to 1.

8. The number of *permutations* of n things taken r at a time is given by:

$$_nP_r = n(n - 1)(n - 2) \cdots [n - (r - 1)] = \frac{n!}{(n - r)!}$$

9. The number of *combinations* of n things taken r at a time is given by:

$$_nC_r = \frac{n!}{r!(n - r)!}$$

10. The set of possible outcomes of a random experiment is called the *sample space* of the experiment. Any subset of the sample space is an *event.* An event consisting of a single outcome is a *simple event.*

11. If the sample space for an experiment consists of n equally likely outcomes, the probability of each simple event is $\frac{1}{n}$. If the event E contains k elements of this event space,

$$P(E) = \frac{1}{n} + \frac{1}{n} + \cdots + \frac{1}{n} = \frac{k}{n}.$$

12. Two events, A and B, are *independent* if and only if

$$P(A \cap B) = P(A) \cdot P(B).$$

13. In general, for any two events A and B in a sample space,
$$P(A \cup B) = P(A) + P(B) - P(A \cap B).$$
If the events have no simple events in common (their intersection is empty), the events are said to be *mutually exclusive*. For mutually exclusive events,
$$P(A \cup B) = P(A) + P(B).$$

Chapter Review

Write the letter of the correct answer.

1. Nine quiz scores were 6, 7, 7, 8, 8, 9, 9, 10, and 10. What is the median of these scores?   15-1
   a. 7   **b.** 8   **c.** 9   **d.** 10

2. On four tests, Jean got the following scores: 100, 79, 86, and 91. What is the variance for this distribution?   15-2
   a. 89   **b.** 15.3   **c.** 58.5   **d.** 234

3. Under the standard normal curve, the area bounded by the x- and y-axes and the line $x = 3$ is 0.4987. In a standard normal distribution, what percent of the data is between -3 and 3?   15-3
   a. 49.87%   **b.** 24.94%   **c.** 0%   **d.** 99.74%

4. The regression line for two variables x and y is $y = 0.45x - 32$. Predict the value of x that is associated with a value of 13 for y.   15-4
   a. 10   **b.** 100   **c.** -13 **d.** 553

5. How many positive integers less than 25 can be formed by using the digits 1, 2, and 3?   15-5
   a. 3   **b.** 6   **c.** 9   **d.** 12

6. In how many ways can 3 cards be selected from a 12-card deck if the selected cards are not returned to the deck?
   a. 220   **b.** 1728   **c.** 1440   **d.** 1320

7. How many different permutations can be made using all the letters in the word ATLANTA?   15-6
   a. 420   **b.** 5040   **c.** 840   **d.** 2520

8. In how many ways can 5 players be chosen from 9 players?   15-7
   a. 3024   **b.** 36   **c.** 126   **d.** 15,120

9. If two coins are tossed, what does $\{(H, H), (H, T), (T, H), (T, T)\}$ represent?   15-8
   a. the event that at least one head turns up
   b. the event that exactly one tail turns up
   c. the sample space for the experiment

10. Three coins are tossed. What is the probability that at least one tail turns up?   15-9
    a. $\frac{1}{8}$ **b.** $\frac{3}{8}$ **c.** $\frac{7}{8}$ **d.** 1

11. If A and B are mutually exclusive events, which of the following relationships applies?

15-10

a. $P(A \cup B) = P(A) \cdot P(B)$ **b.** $P(A \cap B) \neq 0$

c. $P(A \cap B) = P(A) \cdot P(B)$ **d.** $P(A \cup B) = P(A) + P(B)$

12. A dime and a penny are tossed. Let A be the event that the dime turns up heads, and let B be the event that the penny turns up tails. What is $P(A \cap B)$?

a. 1   **b.** $\frac{3}{4}$ **c.** $\frac{1}{2}$ **d.** $\frac{1}{4}$

Chapter Test

1. Draw a stem-and-leaf plot for the set of data below. Then find the mode, median, and mean to the nearest whole number.

15-1

13, 25, 17, 16, 31, 28, 25, 17, 25, 32, 33

2. Draw a box-and-whisker plot for the data in Exercise 1. Then compute the variance and standard deviation to the nearest whole number.

15-2

3. The mean score on a history test was 74.3 and the standard deviation was 5.7. Assuming the scores were normally distributed, what fraction of the scores were above 80?

15-3

4. Draw a scatter plot for the data shown at the right. Then state whether there is a high positive correlation, a high negative correlation, or relatively no correlation between price and demand.

15-4

Price	Demand
12	260
15	250
19	210
24	180

5. How many whole numbers less than 500 can be formed using the digits 1, 2, 4, and 5?

15-5

6. In how many ways can you arrange 5 pictures in one row on a wall?

15-6

7. How many different signals can be made by displaying four pennants, all at one time, on a vertical flagpole? The pennants are identical except for color: three are blue and one is red.

8. How many 4-person committees can be formed from a group of 8 people?

15-7

9. Two dice are rolled. Specify each of the following events:

15-8

a. The dice show the same number.

b. The sum of the numbers showing is 4.

10. One card is drawn at random from a 52-card deck. Find the probability of each event.

15-9

a. It is a spade.   **b.** It is an ace.   **c.** It is a 9 or 10.

11. There are 3 red, 3 white, and 3 green marbles in a bag. Two are drawn at random. Let A be the event that at least one marble is red. Let B be the event that both marbles are the same color.

15-10

a. Find $P(A)$, $P(B)$, $P(A \cap B)$, and $P(A \cup B)$.

b. Are A and B mutually exclusive events? Are they independent?

16 Matrices and Determinants

In today's movies artificial landscapes may be created by computers using fractals. Rotations of graphic images in 3-D use matrix transformations.

Matrices

16-1 Definition of Terms

Objective   To learn and apply matrix terminology.

A **matrix** (plural, matrices) is a rectangular array of numbers enclosed by brackets. Here are some examples of matrices:

$$\begin{bmatrix} 2 & 0 \\ 7 & 1\frac{1}{2} \\ -3 & 19 \end{bmatrix} \qquad \begin{bmatrix} 3 & 0 & 9 \\ 0 & -2 & 0 \end{bmatrix} \qquad \begin{bmatrix} \frac{4}{5} \\ 16 \end{bmatrix} \qquad \begin{bmatrix} \sqrt{2} & 5 & -0.34 \end{bmatrix}$$

The numbers in a matrix are called the **elements** of the matrix. The number of **rows** (horizontal) and the number of **columns** (vertical) determine the **dimensions** of the matrix.

When giving the dimensions of a matrix, you always write the number of rows first. The matrices above are, in order, a 3×2 (read "three by two") matrix, a 2×3 matrix, a 2×1 matrix, and a 1×3 matrix.

Some special matrices are illustrated below.

row matrix:	**column matrix:**	**square matrix:**
exactly one row	exactly one column	an $n \times n$ matrix
$\begin{bmatrix} 3 & -10 & \frac{1}{2} \end{bmatrix}$	$\begin{bmatrix} 6 \\ -2 \end{bmatrix}$	$\begin{bmatrix} 1 & -1 \\ 3 & 0 \end{bmatrix}$

In a square matrix the number of rows and the number of columns are equal.

Capital letters are used to name matrices. Subscripts are used to indicate dimensions. For example, $A_{2\times 5}$ denotes a 2×5 matrix named A. If all the elements of a matrix are zeros, the matrix is called a **zero matrix.** Therefore, the matrix at the right is a 3×3 zero matrix, denoted by $O_{3\times 3}$.

$$\begin{bmatrix} 0 & 0 & 0 \\ 0 & 0 & 0 \\ 0 & 0 & 0 \end{bmatrix}$$

Example 1   Write the zero matrix denoted by $O_{2\times 4}$.

Solution $O_{2\times 4}$ has 2 rows and 4 columns.
Each element is zero.

$$\therefore O_{2\times 4} = \begin{bmatrix} 0 & 0 & 0 & 0 \\ 0 & 0 & 0 & 0 \end{bmatrix} \quad \textit{Answer}$$

Two matrices are equal if and only if they have the same dimensions and the elements in all corresponding positions are equal. For example,

$$\begin{bmatrix} 3 & 2 & \frac{5}{5} \\ -5 & 0 & \frac{4}{6} \end{bmatrix} = \begin{bmatrix} 3 & 2 & 1 \\ -5 & 0 & \frac{2}{3} \end{bmatrix} \quad \text{but} \quad \begin{bmatrix} 6 & 2 \\ \sqrt{2} & -4 \end{bmatrix} \neq \begin{bmatrix} 6 & 2 \\ -4 & \sqrt{2} \end{bmatrix}$$

Example 2 Find the value of each variable.

$$\begin{bmatrix} x + 5 & -1 \\ 4 & 6 \end{bmatrix} = \begin{bmatrix} 2 & -1 \\ 4 & 3y \end{bmatrix}$$

Solution Since the matrices are equal, elements in corresponding positions are equal:

$$x + 5 = 2 \quad \text{and} \quad 3y = 6$$
$$\therefore \quad x = -3 \quad \text{and} \quad y = 2 \quad \textbf{\textit{Answer}}$$

Oral Exercises

Let $A = \begin{bmatrix} 3 & -4 & \frac{1}{2} & 0 \\ 9 & 0 & -3 & 5 \\ 2 & 6 & 1 & -6 \end{bmatrix}$. **Give each of the following.**

1. The elements in row 2.
2. The elements in column 1.
3. The elements in column 3.
4. The elements in row 3.
5. The number of rows in A.
6. The number of columns in A.

Give the dimensions of each matrix.

7. $\begin{bmatrix} 9 & 2 & -4 \\ 3 & \frac{1}{2} & 6 \end{bmatrix}$
8. $\begin{bmatrix} 4 & 1 \\ 6 & 0 \\ 9 & -4 \end{bmatrix}$
9. $\begin{bmatrix} 12 \end{bmatrix}$
10. $\begin{bmatrix} 6 \\ 0 \\ -8 \\ 1 \end{bmatrix}$

11. $\begin{bmatrix} a & b \\ c & d \end{bmatrix}$
12. $\begin{bmatrix} 4 & 9 & -3 & 0 & 2 \end{bmatrix}$
13. $\begin{bmatrix} 4 \\ -5 \\ 3 \\ 14 \end{bmatrix}$
14. $\begin{bmatrix} 0 & 0 & 0 & 0 \\ 0 & 0 & 0 & 0 \\ 0 & 0 & 0 & 0 \end{bmatrix}$

List the matrices in Exercises 7–14 above that can be described as follows.

15. a row matrix
16. a column matrix
17. a square matrix
18. a zero matrix

Written Exercises

Write the zero matrix denoted by each of the following.

A **1.** $O_{1 \times 2}$ **2.** $O_{3 \times 1}$ **3.** $O_{2 \times 5}$ **4.** $O_{4 \times 3}$

Find the value of each variable.

5. $O_{2 \times 2} = \begin{bmatrix} x & y \\ z & 0 \end{bmatrix}$ **6.** $O_{3 \times 2} = \begin{bmatrix} 0 & y \\ x & 0 \\ 0 & z \end{bmatrix}$ **7.** $O_{2 \times 3} = \begin{bmatrix} x + 4 & 5y & 10 - z \\ 0 & 0 & 0 \end{bmatrix}$

8. $\begin{bmatrix} 5 - x \\ y + 6 \\ -3z \end{bmatrix} = O_{3 \times 1}$ **9.** $\begin{bmatrix} 1 \\ 0 \\ x \end{bmatrix} = \begin{bmatrix} y \\ 0 \\ -4 \end{bmatrix}$ **10.** $\begin{bmatrix} 2 & x & 0 \end{bmatrix} = \begin{bmatrix} y & 4 & 0 \end{bmatrix}$

11. $\begin{bmatrix} 3 & 2 \\ -1 & 0 \\ x & y \end{bmatrix} = \begin{bmatrix} 3 & 2 \\ -1 & 0 \\ 8 & -5 \end{bmatrix}$ **12.** $\begin{bmatrix} 4 & x & -3 & y \\ 19 & -1 & 5 & 0 \end{bmatrix} = \begin{bmatrix} z & -1 & -3 & 6 \\ 19 & -1 & 5 & 0 \end{bmatrix}$

B **13.** $\begin{bmatrix} x - 3 \\ 12 \\ 0 \end{bmatrix} = \begin{bmatrix} 8 \\ y \\ z + 4 \end{bmatrix}$ **14.** $\begin{bmatrix} 2x & 4 \\ -1 & 8 \end{bmatrix} = \begin{bmatrix} -6 & y + 2 \\ z & 8 \end{bmatrix}$

15. $\begin{bmatrix} x + y & 1 \\ 0 & x - y \end{bmatrix} = \begin{bmatrix} 2 & 1 \\ 0 & 8 \end{bmatrix}$ **16.** $\begin{bmatrix} 8 - y & 0 \\ x + 2z & x + z \end{bmatrix} = \begin{bmatrix} -3y & 0 \\ -5 & -1 \end{bmatrix}$

17. $\begin{bmatrix} 2x + 3y \\ x - y \end{bmatrix} = \begin{bmatrix} 3 \\ 4 \end{bmatrix}$ **18.** $\begin{bmatrix} 2x - y \\ 3x + 5y \end{bmatrix} = \begin{bmatrix} 6 \\ 22 \end{bmatrix}$

Let a, b, c, and d be nonzero real numbers. Find the values of x and y in each of the equations below.

19. $\begin{bmatrix} ax \\ by \end{bmatrix} = \begin{bmatrix} 1 \\ 0 \end{bmatrix}$ **20.** $\begin{bmatrix} ax + y \\ bx + y \end{bmatrix} = \begin{bmatrix} 1 \\ 0 \end{bmatrix}$

21. $\begin{bmatrix} ax + b & cy + d \end{bmatrix} = \begin{bmatrix} 1 & 0 \end{bmatrix}$ **22.** $\begin{bmatrix} ax + by & cx + dy \end{bmatrix} = \begin{bmatrix} 1 & 1 \end{bmatrix}$

Mixed Review Exercises

Solve each triangle. Give lengths to three significant digits and angle measures to the nearest tenth of a degree.

1. $a = 7$, $b = 10$, $\angle C = 49°$ **2.** $c = 8$, $\angle C = 90°$, $\angle B = 30°$

3. $a = \dfrac{3\sqrt{2}}{2}$, $b = \dfrac{3\sqrt{2}}{2}$, $\angle C = 90°$ **4.** $a = 24$, $b = 17$, $\angle A = 72°$

5–8. Find the area of each triangle in Exercises 1–4 above. Give answers to three significant digits.

16-2 Addition and Scalar Multiplication

Objective   To find sums and differences of matrices and products of a scalar and a matrix.

The **sum of matrices** having the same dimensions is the matrix whose elements are the sums of the corresponding elements of the matrices being added.

Example 1   Simplify: $\begin{bmatrix} 4 & 7 \\ 0 & -2 \\ 1 & -6 \end{bmatrix} + \begin{bmatrix} 2 & -5 \\ -3 & 2 \\ 0 & -4 \end{bmatrix}$

Solution $\begin{bmatrix} 4 & 7 \\ 0 & -2 \\ 1 & -6 \end{bmatrix} + \begin{bmatrix} 2 & -5 \\ -3 & 2 \\ 0 & -4 \end{bmatrix} = \begin{bmatrix} 4+2 & 7+(-5) \\ 0+(-3) & (-2)+2 \\ 1+0 & (-6)+(-4) \end{bmatrix}$

$$= \begin{bmatrix} 6 & 2 \\ -3 & 0 \\ 1 & -10 \end{bmatrix} \quad \textbf{Answer}$$

Since matrices with different dimensions do not have corresponding elements, *addition of matrices of different dimensions is not defined.* In each set of $m \times n$ matrices, addition is both commutative and associative. (Exercises 21 and 22 ask you to prove these facts for the set of 2×2 matrices.)

For each set of $m \times n$ matrices, $O_{m\times n}$ is the *identity* for addition. For example, if $A_{2\times 2} = \begin{bmatrix} a & b \\ c & d \end{bmatrix}$, then

$$A_{2\times 2} + O_{2\times 2} = \begin{bmatrix} a & b \\ c & d \end{bmatrix} + \begin{bmatrix} 0 & 0 \\ 0 & 0 \end{bmatrix} = \begin{bmatrix} a+0 & b+0 \\ c+0 & d+0 \end{bmatrix} = \begin{bmatrix} a & b \\ c & d \end{bmatrix} = A_{2\times 2}.$$

For each set of $m \times n$ matrices, $O_{m\times n}$ is the **identity** for addition.
$$A_{m\times n} + O_{m\times n} = O_{m\times n} + A_{m\times n} = A_{m\times n}.$$

The *additive inverse of matrix A* is the matrix $-A$. Each element of $-A$ is the opposite of its corresponding element in A. For example,

$$\text{if } A = \begin{bmatrix} 2 & -1 \\ 3 & 0 \end{bmatrix}, \text{ then } -A = \begin{bmatrix} -2 & 1 \\ -3 & 0 \end{bmatrix}.$$

For each set of $m \times n$ matrices, the **additive inverse** of A is the matrix $-A$.
$$A_{m\times n} + (-A_{m\times n}) = -A_{m\times n} + A_{m\times n} = O_{m\times n}.$$

As with real numbers, *subtraction of matrices* is defined in terms of addition.

For each set of $m \times n$ matrices, **subtraction** is defined as follows.
$$A_{m \times n} - B_{m \times n} = A_{m \times n} + (-B_{m \times n}).$$

Example 2 Let $A = \begin{bmatrix} 6 & -2 \\ 5 & 4 \end{bmatrix}$ and $B = \begin{bmatrix} 4 & 3 \\ 5 & -2 \end{bmatrix}$. Find $A - B$.

Solution Apply the definition of subtraction: $A - B = A + (-B)$.

$$\begin{bmatrix} 6 & -2 \\ 5 & 4 \end{bmatrix} - \begin{bmatrix} 4 & 3 \\ 5 & -2 \end{bmatrix} = \begin{bmatrix} 6 & -2 \\ 5 & 4 \end{bmatrix} + \begin{bmatrix} -4 & -3 \\ -5 & 2 \end{bmatrix} = \begin{bmatrix} 2 & -5 \\ 0 & 6 \end{bmatrix}$$
Answer

The properties of addition of matrices are summarized below.

Properties of Addition of Matrices

Let A, B, and C be $m \times n$ matrices. Let $O_{m \times n}$ be the $m \times n$ zero matrix.

1. Closure Property   **$A + B$ is an $m \times n$ matrix.**
2. Commutative Property   **$A + B = B + A$**
3. Associative Property   **$(A + B) + C = A + (B + C)$**
4. Identity Property   **$A + O_{m \times n} = O_{m \times n} + A = A$**
5. Inverse Property   **$A + (-A) = -A + A = O_{m \times n}$**

In matrix algebra a real number is called a **scalar.** The **scalar product** of a real number r and a matrix A is the matrix rA. Each element of rA is r times its corresponding element in A. In the expression $2A$, the number 2 is a scalar and $2A$ (which is another matrix) is the scalar product of 2 and A.

Example 3 Let $A = \begin{bmatrix} 2 & 0 \\ -2 & 3 \end{bmatrix}$. Find $3A$ and rA.

Solution
$$3A = 3 \begin{bmatrix} 2 & 0 \\ -2 & 3 \end{bmatrix} = \begin{bmatrix} 3 \cdot 2 & 3 \cdot 0 \\ 3(-2) & 3 \cdot 3 \end{bmatrix} = \begin{bmatrix} 6 & 0 \\ -6 & 9 \end{bmatrix} \quad \textit{Answer}$$

$$rA = r \begin{bmatrix} 2 & 0 \\ -2 & 3 \end{bmatrix} = \begin{bmatrix} r \cdot 2 & r \cdot 0 \\ r(-2) & r \cdot 3 \end{bmatrix} = \begin{bmatrix} 2r & 0 \\ -2r & 3r \end{bmatrix} \quad \textit{Answer}$$

The properties of scalar multiplication are summarized in the chart at the top of the next page.

An equation in which the variable stands for a matrix is called a **matrix equation.** You can solve some simple matrix equations by using matrix addition and scalar multiplication.

Example 4 Solve for the matrix X: $2X + 2\begin{bmatrix} 1 & -2 \\ 0 & 3 \end{bmatrix} = 8\begin{bmatrix} \frac{1}{2} & 0 \\ -1 & \frac{1}{4} \end{bmatrix}$

Solution   Find the scalar products:

$$2X + \begin{bmatrix} 2 & -4 \\ 0 & 6 \end{bmatrix} = \begin{bmatrix} 4 & 0 \\ -8 & 2 \end{bmatrix}$$

Add $\begin{bmatrix} -2 & 4 \\ 0 & -6 \end{bmatrix}$, the inverse of $\begin{bmatrix} 2 & -4 \\ 0 & 6 \end{bmatrix}$, to each side.

$$2X + \begin{bmatrix} 2 & -4 \\ 0 & 6 \end{bmatrix} + \begin{bmatrix} -2 & 4 \\ 0 & -6 \end{bmatrix} = \begin{bmatrix} 4 & 0 \\ -8 & 2 \end{bmatrix} + \begin{bmatrix} -2 & 4 \\ 0 & -6 \end{bmatrix}$$

$$2X + O_{2 \times 2} = \begin{bmatrix} 2 & 4 \\ -8 & -4 \end{bmatrix}$$

$$X = \frac{1}{2}\begin{bmatrix} 2 & 4 \\ -8 & -4 \end{bmatrix} = \begin{bmatrix} 1 & 2 \\ -4 & -2 \end{bmatrix}$$

The check is left for you.   ***Answer***

Oral Exercises

Give the value of each element of the sum or difference.

$$\begin{bmatrix} 4 & 0 \\ 2 & 1 \end{bmatrix} + \begin{bmatrix} 12 & -4 \\ 9 & 3 \end{bmatrix} = \begin{bmatrix} a & b \\ c & d \end{bmatrix} \qquad \begin{bmatrix} 9 & 0 \\ -2 & 7 \end{bmatrix} - \begin{bmatrix} 3 & 7 \\ 8 & -5 \end{bmatrix} = \begin{bmatrix} e & f \\ g & h \end{bmatrix}$$

   1. a **2.** b **3.** c **4.** d **5.** e **6.** f **7.** g **8.** h

Give the value of each element of the scalar product.

$$-2\begin{bmatrix} 15 & -2 & 0 & 8 & \sqrt{3} \end{bmatrix} = \begin{bmatrix} p & q & r & s & t \end{bmatrix}$$

9. p **10.** q **11.** r **12.** s **13.** t

Written Exercises

Simplify.

A   **1.** $\begin{bmatrix} 4 & 1 \\ -2 & 0 \end{bmatrix} + \begin{bmatrix} 0 & 3 \\ 2 & 3 \end{bmatrix}$ **2.** $\begin{bmatrix} 2 & -3 \\ -2 & 10 \end{bmatrix} + \begin{bmatrix} -2 & 3 \\ 2 & -8 \end{bmatrix}$

3. $\begin{bmatrix} 8 & 0 \\ 4 & 10 \end{bmatrix} + \begin{bmatrix} 3 & 4 \\ -1 & 6 \end{bmatrix}$ **4.** $\begin{bmatrix} -8 & 5 \\ 3 & 0 \end{bmatrix} + \begin{bmatrix} 4 & -2 \\ -2 & 9 \end{bmatrix}$

5. $\begin{bmatrix} 4 & 7 \\ 2 & -1 \\ 0 & 5 \end{bmatrix} - \begin{bmatrix} 0 & 5 \\ 2 & 4 \\ 3 & -2 \end{bmatrix}$ **6.** $\begin{bmatrix} 5 & 8 \\ -2 & 14 \\ 0 & -6 \end{bmatrix} - \begin{bmatrix} 5 & 4 \\ 4 & -2 \\ -5 & -5 \end{bmatrix}$

7. $\begin{bmatrix} 0 & -2 & 4 \\ 3 & -2 & -6 \end{bmatrix} - \begin{bmatrix} 1 & -5 & 0 \\ 3 & -2 & -6 \end{bmatrix}$ **8.** $\begin{bmatrix} 4 & 6 & -8 \\ -2 & 5 & 0 \end{bmatrix} - \begin{bmatrix} 18 & 3 & 12 \\ 1 & 0 & 5 \end{bmatrix}$

9. $6\begin{bmatrix} 7 & 8 \\ -4 & 0 \end{bmatrix}$ **10.** $12\begin{bmatrix} 3 & 0 \\ -5 & 2 \end{bmatrix}$

11. $O_{2\times3} + \begin{bmatrix} 3 & 5 & 2 \\ -2 & 8 & 0 \end{bmatrix}$ **12.** $O_{2\times3} - 5\begin{bmatrix} 3 & 0 & -1 \\ 6 & 2 & 4 \end{bmatrix}$

13. $2\begin{bmatrix} 5 & -2 \\ -3 & 4 \\ 0 & 6 \end{bmatrix} + \begin{bmatrix} 1 & 7 \\ 0 & -4 \\ 6 & 5 \end{bmatrix}$ **14.** $\begin{bmatrix} 4 & 2 \\ 1 & -3 \end{bmatrix} + \begin{bmatrix} -4 & -2 \\ -1 & 3 \end{bmatrix}$

B   **15.** $3\begin{bmatrix} 2 \\ 2 \\ 5 \end{bmatrix} + 5\begin{bmatrix} 0 \\ 6 \\ 3 \end{bmatrix}$ **16.** $\begin{bmatrix} 3 & 0 \\ 5 & -4 \\ 0 & -3 \end{bmatrix} - 4\begin{bmatrix} 10 & -2 \\ 2 & 4 \\ -7 & 0 \end{bmatrix}$

Solve each equation for matrix X.

17. $X + \begin{bmatrix} 3 & 2 \\ 1 & 0 \end{bmatrix} = \begin{bmatrix} 6 & 3 \\ 7 & -1 \end{bmatrix}$ **18.** $X + \begin{bmatrix} 0 & 4 \\ 9 & -1 \end{bmatrix} = \begin{bmatrix} 2 & 0 \\ -1 & 2 \end{bmatrix}$

19. $X + 3\begin{bmatrix} -3 & 2 \\ 0 & -1 \end{bmatrix} = \begin{bmatrix} -4 & 10 \\ 12 & 0 \end{bmatrix}$ **20.** $X - \begin{bmatrix} 0 & 5 \\ 1 & -2 \end{bmatrix} = 3\begin{bmatrix} 2 & -3 \\ 3 & 1 \end{bmatrix}$

Let p and q be scalars and let A, B, and C be 2×2 matrices. Prove these properties for all 2×2 matrices.

C   **21.** $A + B = B + A$ **22.** $(A + B) + C = A + (B + C)$

23. $(-1)A = -A$ **24.** $pA = Ap$

25. $(p + q)A = pA + qA$ **26.** $(pq)B = p(qB)$

27. $p \cdot O_{2\times2} = O_{2\times2}$ **28.** $0 \cdot A = O_{2\times2}$

16-3 Matrix Multiplication

Objective   To find the product of two matrices.

The method of multiplying two matrices is illustrated by this example.

A discount department store sells jeans, sweaters, and shirts. The prices in dollars are shown in the 1×3 matrix, A, below.

$$A = \begin{bmatrix} \overset{\text{Jeans}}{22} & \overset{\text{Sweaters}}{15} & \overset{\text{Shirts}}{6} \end{bmatrix}$$

In June, sales of jeans, sweaters, and shirts numbered in order 20, 30, and 48. In July, sales were 15, 20, and 72. Sales are displayed in the 3×2 matrix, B, below.

$$B = \begin{bmatrix} \overset{\text{June}}{20} & \overset{\text{July}}{15} \\ 30 & 20 \\ 48 & 72 \end{bmatrix}$$

In June, income from sales of jeans was $22 \cdot 20$; from sweaters, $15 \cdot 30$; and from shirts, $6 \cdot 48$. Income from sales in June and July is shown below:

$$\text{June:} \quad 22 \cdot 20 \;+\; 15 \cdot 30 \;+\; 6 \cdot 48$$
$$\text{July:} \quad 22 \cdot 15 \;+\; 15 \cdot 20 \;+\; 6 \cdot 72$$

The monthly income from these sales is shown in matrix S.

$$S = \begin{bmatrix} \overset{\text{June}}{(440 + 450 + 288)} & \overset{\text{July}}{(330 + 300 + 432)} \end{bmatrix} = \begin{bmatrix} 1178 & 1062 \end{bmatrix}$$

Matrix S is the *product* of matrices A and B:

$$\begin{aligned}
S = A \times B &= \begin{bmatrix} 22 & 15 & 6 \end{bmatrix} \begin{bmatrix} 20 & 15 \\ 30 & 20 \\ 48 & 72 \end{bmatrix} \\
&= \begin{bmatrix} (22 \cdot 20 + 15 \cdot 30 + 6 \cdot 48) & (22 \cdot 15 + 15 \cdot 20 + 6 \cdot 72) \end{bmatrix} \\
&= \begin{bmatrix} (440 + 450 + 288) & (330 + 300 + 432) \end{bmatrix} \\
&= \begin{bmatrix} 1178 & 1062 \end{bmatrix}
\end{aligned}$$

Each element in S is obtained by multiplying elements of the *row* in A with elements of a *column* in B. This example suggests the following definition.

The **product of matrices** $A_{m \times n}$ and $B_{n \times p}$ is the $m \times p$ matrix whose element in the ath row and bth column is the sum of the products of corresponding elements of the ath row of A and the bth column of B.

Two matrices can be multiplied only if the second dimension of the first matrix equals the first dimension of the second matrix. For example,

if $A_{1\times2} = \begin{bmatrix} 3 & -1 \end{bmatrix}$ and $B_{2\times3} = \begin{bmatrix} 1 & 0 & 2 \\ 0 & 1 & -1 \end{bmatrix}$,

then $A_{1\times2} \cdot B_{2\times3} = \begin{bmatrix} 3 & -1 \end{bmatrix} \begin{bmatrix} 1 & 0 & 2 \\ 0 & 1 & -1 \end{bmatrix} = \begin{bmatrix} 3 & -1 & 7 \end{bmatrix}$;

but $B_{2\times3} \cdot A_{1\times2} = \begin{bmatrix} 1 & 0 & 2 \\ 0 & 1 & -1 \end{bmatrix} \begin{bmatrix} 3 & -1 \end{bmatrix}$ is not defined.

The product $A_{1\times2} \cdot B_{2\times3}$ is a 1×3 matrix. However, matrix multiplication is not defined for $B_{2\times3} \cdot A_{1\times2}$ because there are more elements in each row of B (three elements) than there are in each column of A (one element).

Example 1 Multiply: $\begin{bmatrix} 3 & 1 & 0 \\ -1 & 4 & 2 \end{bmatrix} \begin{bmatrix} -1 & 0 \\ 2 & 3 \\ 5 & 0 \end{bmatrix}$

Solution A 2×3 matrix times a 3×2 matrix gives a 2×2 matrix.

$\begin{bmatrix} 3 & 1 & 0 \\ -1 & 4 & 2 \end{bmatrix} \begin{bmatrix} -1 & 0 \\ 2 & 3 \\ 5 & 0 \end{bmatrix} = \begin{bmatrix} -1 & \\ & \end{bmatrix}$

row 1 × column 1:
$3(-1) + 1 \cdot 2 + 0 \cdot 5 = -1$

$\begin{bmatrix} 3 & 1 & 0 \\ -1 & 4 & 2 \end{bmatrix} \begin{bmatrix} -1 & 0 \\ 2 & 3 \\ 5 & 0 \end{bmatrix} = \begin{bmatrix} -1 & 3 \\ & \end{bmatrix}$

row 1 × column 2:
$3 \cdot 0 + 1 \cdot 3 + 0 \cdot 0 = 3$

$\begin{bmatrix} 3 & 1 & 0 \\ -1 & 4 & 2 \end{bmatrix} \begin{bmatrix} -1 & 0 \\ 2 & 3 \\ 5 & 0 \end{bmatrix} = \begin{bmatrix} -1 & 3 \\ 19 & \end{bmatrix}$

row 2 × column 1:
$(-1)(-1) + 4 \cdot 2 + 2 \cdot 5 = 19$

$\begin{bmatrix} 3 & 1 & 0 \\ -1 & 4 & 2 \end{bmatrix} \begin{bmatrix} -1 & 0 \\ 2 & 3 \\ 5 & 0 \end{bmatrix} = \begin{bmatrix} -1 & 3 \\ 19 & 12 \end{bmatrix}$

row 2 × column 2:
$(-1)0 + 4 \cdot 3 + 2 \cdot 0 = 12$

$\therefore$ the product is $\begin{bmatrix} -1 & 3 \\ 19 & 12 \end{bmatrix}$. **Answer**

Powers of square matrices are defined just like powers of real numbers.

$$A^2 = A \cdot A, \qquad A^3 = A \cdot A \cdot A, \qquad A^n = \underbrace{A \cdot A \cdot A \cdot \ldots \cdot A}_{n \text{ factors}}$$

An $n \times n$ matrix whose *main diagonal* from upper left to lower right has all elements 1, while all other elements are 0, is called an **identity matrix** and is denoted by $I_{n\times n}$. Here are the 2×2 and 3×3 identity matrices:

$$I_{2\times2} = \begin{bmatrix} 1 & 0 \\ 0 & 1 \end{bmatrix} \text{ and } I_{3\times3} = \begin{bmatrix} 1 & 0 & 0 \\ 0 & 1 & 0 \\ 0 & 0 & 1 \end{bmatrix}$$

You can show (see Exercise 27) that for any $n \times n$ square matrix A,

$$I_{n \times n} \cdot A = A \cdot I_{n \times n} = A.$$

Multiplication of a matrix by a zero matrix yields a zero matrix. In particular, if A is any $n \times n$ square matrix and $O_{n \times n}$ is the zero matrix,

$$O_{n \times n} \cdot A = A \cdot O_{n \times n} = O_{n \times n}.$$

A major difference between real number multiplication and matrix multiplication is the fact that $CD = O_{n \times n}$ does *not* imply that either $C = O_{n \times n}$ or $D = O_{n \times n}$. For example,

$$\text{if } C = \begin{bmatrix} -3 & 1 \\ 6 & -2 \end{bmatrix} \text{ and } D = \begin{bmatrix} 2 & 1 \\ 6 & 3 \end{bmatrix},$$

$$\text{then } CD = \begin{bmatrix} -3 & 1 \\ 6 & -2 \end{bmatrix} \begin{bmatrix} 2 & 1 \\ 6 & 3 \end{bmatrix} = \begin{bmatrix} 0 & 0 \\ 0 & 0 \end{bmatrix} = O_{2 \times 2}.$$

Caution: Unlike real number multiplication, matrix multiplication is *not* in general commutative. Here is a counterexample:

$$\text{Let } A = \begin{bmatrix} 0 & 3 \\ 1 & -1 \end{bmatrix} \text{ and } B = \begin{bmatrix} 2 & 0 \\ 1 & 4 \end{bmatrix}.$$

$$\text{Then } AB = \begin{bmatrix} 0 & 3 \\ 1 & -1 \end{bmatrix} \begin{bmatrix} 2 & 0 \\ 1 & 4 \end{bmatrix}$$

$$= \begin{bmatrix} 3 & 12 \\ 1 & -4 \end{bmatrix}.$$

$$\text{But } BA = \begin{bmatrix} 2 & 0 \\ 1 & 4 \end{bmatrix} \begin{bmatrix} 0 & 3 \\ 1 & -1 \end{bmatrix}$$

$$= \begin{bmatrix} 0 & 6 \\ 4 & -1 \end{bmatrix}.$$

$$\therefore AB \ne BA$$

Be sure to compute a product in the order in which it is written.

Although the commutative property does not apply to matrix multiplication, the associative and distributive properties do hold. These and other properties for multiplication of square matrices are summarized below.

Properties of Matrix Multiplication

Let A, B, and C be $n \times n$ matrices. Let $I_{n \times n}$ be the identity matrix and $O_{n \times n}$ be the zero matrix.

1. Associative Property $(AB)C = A(BC)$
2. Distributive Property $A(B + C) = AB + AC$
                                        $(B + C)A = BA + CA$
3. Identity Property $I_{n \times n} \cdot A = A \cdot I_{n \times n} = A$
4. Multiplicative Property of $O_{n \times n}$ $O_{n \times n} \cdot A = A \cdot O_{n \times n} = O_{n \times n}$

Oral Exercises

Give the dimensions of each matrix and tell whether the matrices can be multiplied.

1. $\begin{bmatrix} 3 & 5 \\ 0 & 4 \end{bmatrix}\begin{bmatrix} -1 & 3 \\ 5 & 7 \end{bmatrix}$

2. $\begin{bmatrix} 6 & 2 & 0 \\ 1 & 0 & 8 \\ 4 & -2 & 5 \end{bmatrix}\begin{bmatrix} 2 & 0 \\ 1 & -3 \\ 0 & 4 \end{bmatrix}$

3. $\begin{bmatrix} 2 \\ 7 \\ -4 \end{bmatrix}\begin{bmatrix} 1 & 0 & 6 & 4 \end{bmatrix}$

4. $\begin{bmatrix} 8 & 1 & 0 \end{bmatrix}\begin{bmatrix} 5 & 7 & -2 \\ 0 & 4 & 1 \end{bmatrix}$

5. $\begin{bmatrix} 0 & 0 \\ 0 & 0 \\ 0 & 0 \\ 0 & 0 \end{bmatrix}\begin{bmatrix} 3 & 5 & 9 & 0 \end{bmatrix}$

6. $\begin{bmatrix} -2 & 3 & 0 & 1 \end{bmatrix}\begin{bmatrix} 1 & 0 & 0 \\ 0 & 1 & 0 \\ 0 & 0 & 1 \end{bmatrix}$

Give the dimensions of the product of the matrices.

7. $\begin{bmatrix} 3 & 2 & 1 & 8 \end{bmatrix}\begin{bmatrix} 4 \\ 0 \\ 3 \\ -5 \end{bmatrix}$

8. $\begin{bmatrix} 6 & 2 & 8 \\ -1 & 0 & 3 \end{bmatrix}\begin{bmatrix} 9 & 0 \\ 3 & 0 \\ 1 & -4 \end{bmatrix}$

9. $\begin{bmatrix} \frac{2}{3} & 0 \\ \frac{1}{5} & \frac{1}{2} \end{bmatrix}\begin{bmatrix} \frac{1}{6} & \frac{1}{10} \\ \frac{1}{3} & 0 \end{bmatrix}$

10. $\begin{bmatrix} 0 \\ 3 \\ 5 \end{bmatrix}\begin{bmatrix} 2 & -2 & 0 & 8 & 3 \end{bmatrix}$

Give the value of each variable shown in the computation of the product.

$$\begin{bmatrix} -4 & 0 & 2 \\ 5 & 3 & -2 \end{bmatrix}\begin{bmatrix} -1 \\ 5 \\ 1 \end{bmatrix} = \begin{bmatrix} (-4)(-1) + a \cdot 5 + 2c \\ 5(-1) + b \cdot 5 + (-2)d \end{bmatrix} = \begin{bmatrix} e \\ f \end{bmatrix}$$

11. a **12.** b **13.** c **14.** d **15.** e **16.** f

Written Exercises

Multiply.

A

1. $\begin{bmatrix} 3 & 1 \end{bmatrix}\begin{bmatrix} 4 \\ 6 \end{bmatrix}$

2. $\begin{bmatrix} 0 & -3 & 4 \end{bmatrix}\begin{bmatrix} 1 \\ 0 \\ -4 \end{bmatrix}$

3. $\begin{bmatrix} 4 \\ -2 \end{bmatrix}\begin{bmatrix} 3 & 0 & -1 & 5 \end{bmatrix}$

4. $\begin{bmatrix} 0 & 3 \\ 5 & -1 \end{bmatrix}\begin{bmatrix} -1 \\ 3 \end{bmatrix}$

5. $\begin{bmatrix} 3 & 0 \\ 1 & 2 \end{bmatrix}\begin{bmatrix} -1 & 8 \\ 0 & 3 \end{bmatrix}$

6. $\begin{bmatrix} 0 & 8 \\ 3 & 1 \\ -1 & 5 \end{bmatrix}\begin{bmatrix} 3 & 1 & -2 \\ 0 & 8 & -5 \end{bmatrix}$

7. $\begin{bmatrix} 0 & 2 & -1 \\ 4 & 1 & 0 \\ 0 & -1 & 2 \end{bmatrix} \begin{bmatrix} 4 & 3 & 0 \\ -1 & 0 & 2 \\ 1 & 0 & -2 \end{bmatrix}$
   8. $\begin{bmatrix} 1 & 0 & 0 \\ 2 & 0 & -1 \\ 0 & 1 & 0 \end{bmatrix} \begin{bmatrix} 1 & 0 & 2 \\ 0 & -3 & 1 \\ 0 & 1 & 0 \end{bmatrix}$

9. $\begin{bmatrix} 1 & 2 \\ 3 & -1 \\ 0 & 4 \end{bmatrix} \begin{bmatrix} 1 & 0 & 0 & -1 \\ -1 & 1 & 1 & 0 \end{bmatrix}$
   10. $\begin{bmatrix} 0 & 2 & 0 \\ 0 & 1 & 1 \\ 1 & -1 & 0 \end{bmatrix} \begin{bmatrix} 1 & 1 & 0 \\ 0 & 1 & 0 \\ -1 & 0 & 1 \end{bmatrix}$

11. $\begin{bmatrix} 1 & 0 & 2 \\ -2 & 1 & 0 \\ 0 & 1 & 0 \end{bmatrix} \begin{bmatrix} a & b & c \\ d & e & f \\ g & h & i \end{bmatrix}$
   12. $\begin{bmatrix} 2 & 0 & 1 \\ 5 & -1 & 0 \\ 0 & -2 & 4 \end{bmatrix} \begin{bmatrix} r & s & t \\ u & v & w \\ x & y & z \end{bmatrix}$

In Exercises 13–20, let $A = \begin{bmatrix} 1 & 2 \\ 0 & -1 \end{bmatrix}$, $B = \begin{bmatrix} 0 & -1 \\ 1 & 1 \end{bmatrix}$, $C = \begin{bmatrix} -2 & 0 \\ 0 & 1 \end{bmatrix}$.

B **13.** Find AB and BA.
   14. Find BC and CB.

15. Find $(AB)C$ and $A(BC)$.
   16. Find $AC + BC$ and $(A + B)C$.

17. Find $(A + B)(A - B)$ and $A^2 - B^2$.
   18. Find $(A + B)^2$ and $A^2 + 2AB + B^2$.

19. Find $O_{2 \times 2} \cdot A$ and $A \cdot O_{2 \times 2}$.
   20. Find $I_{2 \times 2} \cdot C$ and $C \cdot I_{2 \times 2}$.

Let $A = \begin{bmatrix} 2 & -3 \\ 2 & 1 \end{bmatrix}$, $B = \begin{bmatrix} \frac{1}{8} & \frac{3}{8} \\ -\frac{1}{4} & \frac{1}{4} \end{bmatrix}$, **and** $D = \begin{bmatrix} 1 & 0 & -1 \\ 0 & 2 & -1 \\ 1 & 0 & 0 \end{bmatrix}$. **Find each product.**

21. AB
   22. BA
   23. D^2

24. D^3
   25. $(-D)^2$
   26. $(-D)^3$

Prove the following properties for all 2×2 matrices.

C **27.** $A \cdot I_{2 \times 2} = I_{2 \times 2} \cdot A = A$
   28. $A \cdot O_{2 \times 2} = O_{2 \times 2} \cdot A = O_{2 \times 2}$

29. $(AB)C = A(BC)$
   30. $A(B + C) = AB + AC$

Mixed Review Exercises

Find the real solutions of each system.

1. $2x^2 + y^2 = 5$
   $x^2 = 1 - y^2$

2. $3s + 4t = 2$
   $s - 2t = 4$

3. $y^2 - x^2 = 24$
   $2x - 3 = y$

4. $4m + 2n = -8$
   $3m - n = -11$

5. $xy - 15 = 0$
   $y^2 - x^2 = -16$

6. $a + b - c = -2$
   $2a - b + c = 5$
   $a + 3b + 2c = 2$

Solve.

7. $\sqrt{2x + 3} = x$

8. $\dfrac{m^2 + 2m}{3} = \dfrac{1 - m}{2}$

9. $3x + 8 = 3 - 2x$

16-4 Applications of Matrices

Objective   To solve problems using matrices.

Matrices are used to solve a variety of problems in the social and physical sciences. Two applications of matrices are introduced in this lesson, *communication networks* and *dominance relations*.

Consider a network of four computers as shown in Figure 1. The arrows indicate the directions in which data can be transmitted and received.

A Communication Network

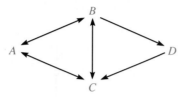

Figure 1

Computer *A*, for example, can transmit directly to computers *B* and *C*, but not to *D*. Computer *A* can receive directly from computers *B* and *C*, but not from *D*. Matrix *X* below illustrates this network. A "1" indicates that direct transmission is possible; a "0" that direct transmission is not possible.

$$
\text{From Computer} \quad
\begin{array}{c}
 \\ A \\ B \\ C \\ D
\end{array}
\overset{\displaystyle \text{To Computer}}{
\overset{\displaystyle A \quad B \quad C \quad D}{
\begin{bmatrix}
0 & 1 & 1 & 0 \\
1 & 0 & 1 & 1 \\
1 & 1 & 0 & 0 \\
0 & 0 & 1 & 0
\end{bmatrix}}} = X
$$

For example, the "1" in the first row and second column indicates that computer *A* can transmit data directly to computer *B*. The "0" in the third row and fourth column indicates that computer *C* can't transmit data directly to computer *D*.

In general, rows of matrix *X* represent the directions data can be transmitted; columns represent the sources from which data may be received. Therefore, computer *C* can receive data from computers *A, B,* and *D* (third column).

Compare each row and column of matrix *X* with the corresponding point of Figure 1. Matrix *X* is called a **communication matrix.**

In addition to sending data directly, it is possible to send data from one point to another via a relay point. For example, computer *A* can send data to computer *D* via computer *B*. It can be proved that matrix X^2 represents the number of routes one computer can send data to a second computer in the network through exactly one relay point.

Compare matrix X^2 below with Figure 1. Trace the possible routes for sending data from computer B to computer C via one relay point. There are two possible routes:

(1) from B to A, and then to C;

(2) from B to D, and then to C.

The element in the second row and third column of X^2 is "2," the number of routes data can be sent from B to C using exactly one relay point. The "1" written in the fourth row and first column of X^2 indicates there is exactly one route from D to A using exactly one relay point. Verify this by checking Figure 1.

$$\text{To Computer}$$

$$\text{From Computer} \quad \begin{array}{c} \\ A \\ B \\ C \\ D \end{array} \overset{\begin{array}{cccc} A & B & C & D \end{array}}{\begin{bmatrix} 2 & 1 & 1 & 1 \\ 1 & 2 & 2 & 0 \\ 1 & 1 & 2 & 1 \\ 1 & 1 & 0 & 0 \end{bmatrix}} = X^2$$

Notice that some of the elements on the main diagonal of X^2 are not zero. For example, the "2" in the first row and first column indicates that computer A can send data to itself by two different routes using exactly one relay point.

Matrix X represents direct communication and matrix X^2 represents communication using 2 steps (with exactly one relay). It can be proved that X^3 represents the number of routes data can be sent from one point to another using 3 steps (with exactly two relay points).

The matrix $X + X^2$ represents the total number of routes data can be sent from one point to another in the network using no more than two steps. The matrix $X + X^2 + X^3$ shows the total number of routes data can be sent using no more than three steps.

Example 1 Use Figure 1 and matrices X and X^2 to solve these problems.

 a. Which computer can send data directly to the most points?

 b. Which computer can receive data directly from the most points?

 c. Find the matrix that represents the total number of routes data can be sent from one computer to another using no more than 2 steps (that is, one relay point).

Solution a. The row entries in matrix X represent routes on which data can be transmitted directly. Therefore computer B (with three routes) can send data to the most points. *Answer*

 b. The column entries in matrix X represent routes on which data can be received directly. Therefore computer C (with three routes) can receive data from the most points. *Answer*

c. X represents the number of direct data communication routes. X^2 represents the number of routes using 2 steps. So $X + X^2$ represents the number of routes using no more than 2 steps.

$$X + X^2 = \begin{bmatrix} 0 & 1 & 1 & 0 \\ 1 & 0 & 1 & 1 \\ 1 & 1 & 0 & 0 \\ 0 & 0 & 1 & 0 \end{bmatrix} + \begin{bmatrix} 2 & 1 & 1 & 1 \\ 1 & 2 & 2 & 0 \\ 1 & 1 & 2 & 1 \\ 1 & 1 & 0 & 0 \end{bmatrix} = \begin{bmatrix} 2 & 2 & 2 & 1 \\ 2 & 2 & 3 & 1 \\ 2 & 2 & 2 & 1 \\ 1 & 1 & 1 & 0 \end{bmatrix}$$

A **dominance relation** exists between members of a group when, between any two members, one dominates the other.

Consider a four-team volleyball tournament. The Broncos beat the Mavericks and the Knights, but lost to the Raiders. The Mavericks beat only the Knights. The Raiders beat the Broncos and the Mavericks. The Knights beat only the Raiders.

Matrix M characterizes these dominances. A "1" indicates that the team named in the corresponding row beat the team in the corresponding column. The number of games won is indicated in the rows. The number of games lost is indicated in the columns.

	Broncos	Knights	Mavericks	Raiders
Broncos	0	1	1	0
Knights	0	0	0	1
Mavericks	0	1	0	0
Raiders	1	0	1	0

$= M$

For example, the Broncos won two (row one) and lost one (column one). Notice that the teams with the best records are the Broncos and Raiders, both of which won two games and lost one. To pick a tournament winner, you must look beyond these direct dominances because the Broncos and Raiders are tied at that level.

A member x is said to have a second-stage dominance over a member z when x dominates y and y dominates z. Second-stage dominances are shown in matrix M^2.

	Broncos	Knights	Mavericks	Raiders
Broncos	0	1	0	1
Knights	1	0	1	0
Mavericks	0	0	0	1
Raiders	0	2	1	0

$= M^2$

The fourth row shows that the Raiders have a second-stage dominance over the Mavericks and a pair of second-stage dominances over the Knights. In other words, the Raiders beat two other teams that, in turn, beat the Knights.

From the sum of M and M^2, the winner of the tournament can be decided.

$$
\begin{array}{c}
 \\
\text{Broncos} \\
\text{Knights} \\
\text{Mavericks} \\
\text{Raiders}
\end{array}
\begin{array}{cccc}
\text{Broncos} & \text{Knights} & \text{Mavericks} & \text{Raiders} \\
\left[\begin{array}{cccc}
0 & 2 & 1 & 1 \\
1 & 0 & 1 & 1 \\
0 & 1 & 0 & 1 \\
1 & 2 & 2 & 0
\end{array}\right]
\end{array} = M + M^2
$$

The row sums of $M + M^2$ show that the Raiders, with a total of 5 direct and second-stage dominances, are the tournament winners.

Oral Exercises

Matrix Y illustrates a communication network. Use matrix Y to answer the questions below.

$$
\begin{array}{c}
 \\
 \\
\text{From Point}
\end{array}
\begin{array}{cc}
 & \text{To Point} \\
\begin{array}{c}
 \\
A \\
B \\
C \\
D
\end{array}
&
\begin{array}{cccc}
A & B & C & D \\
\left[\begin{array}{cccc}
0 & 0 & 0 & 1 \\
1 & 0 & 0 & 1 \\
1 & 1 & 0 & 0 \\
1 & 0 & 1 & 0
\end{array}\right]
\end{array}
\end{array} = Y
$$

1. Name the destinations to which each point can send data.
   a. Point A **b.** Point B **c.** Point C **d.** Point D

2. Which point(s) can send data to the most points?

3. Name the sources from which each point can receive data.
   a. Point A **b.** Point B **c.** Point C **d.** Point D

4. Which point(s) can receive data from the most points?

Written Exercises

Use the radio network shown below for Exercises 1–3.

A 1. Write the matrix that illustrates this network.

2. Find the matrix that represents the number of ways a message can be sent from one station to another using exactly one relay.

3. Find the matrix that represents the number of routes a message can be sent from one station to another using at most one relay.

Use the computer network described below for Exercises 4–10.

Computer A sends data to computer B. Computer B sends and receives data from computers A, C, and D. Computer E receives data from computers C and D.

4. Draw this communication network.

5. Write the matrix that illustrates this network.

6. Which computer(s), if any, cannot send data to other computers in the network?

7. Find the matrix that represents the number of routes data can be sent from one computer to another using exactly one relay.

8. Name the computer(s) that can receive data from the most points via one relay.

9. Name the computer(s) that can send data to the most points via one relay.

10. Find the matrix that represents the number of routes data can be sent from one computer to another using at most one relay.

Consider the situation described below for Exercises 11–12.

In a tennis tournament Jane beats Cindy, but loses to Yoko and Anna. Cindy beats Yoko and Anna. Yoko beats Jane and Anna. Anna beats only Jane.

B 11. Write the dominance matrix for the above tournament results and find the matrix that shows the second-stage dominances for this tournament.

12. Find the matrix that shows the tournament winner using both direct and second-stage dominances. Name the winner.

Use the communication network at the right for Exercises 13–16.

13. Write the matrix for the communication network shown and find the matrix that represents the number of ways a message can be sent from one point to another via one relay.

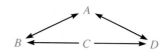

14. Find the matrix that represents the number of ways a message can be sent from one point to another using at most one relay.

15. Find the matrix that represents the number of ways a message can be sent from one point to another using exactly two relays.

16. Find the matrix that represents the number of ways a message can be sent from one point to another using at most two relays.

17. In the communication network at the right, a "hook" indicates that two line segments do not intersect. Find the number of ways each point can send a message back to itself using no more than two relays.

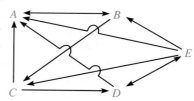

18. Matrix operations used to analyze communication networks and dominance relations are similar. For each concept listed below name the comparable concept from the other application area.

Communication Networks	Dominance Relations
a. *A* can send data to *B*.	___?___
b. ___?___	*A* is dominated by *B*.
c. *B* is a relay point from *A* to *C*	___?___
d. ___?___	*B* wins more games in a tournament than any other team.
e. ___?___	*A* has a second-stage dominance over *B*.
f. *C* can send messages to more stations than any other station in the network using at most one relay.	___?___

Self-Test 1

Vocabulary

matrix (p. 767)
elements of a matrix (p. 767)
row of a matrix (p. 767)
column of a matrix (p. 767)
dimensions of a matrix (p. 767)
row matrix (p. 767)
column matrix (p. 767)
square matrix (p. 767)
zero matrix (p. 767)

sum of matrices (p. 770)
additive inverse of a matrix (p. 770)
subtraction of matrices (p. 771)
scalar, scalar product (p. 771)
matrix equation (p. 772)
product of matrices (p. 774)
identity matrix (p. 775)
communication matrix (p. 779)
dominance relations (p. 781)

Find the value of each variable in Exercises 1 and 2.

1. $O_{2 \times 2} = \begin{bmatrix} 0 & y - 5 \\ x + 1 & z^2 \end{bmatrix}$
2. $\begin{bmatrix} 3x + 2 \\ 10 \\ 0 \end{bmatrix} = \begin{bmatrix} 8 \\ -y \\ z - 4 \end{bmatrix}$

Obj. 16-1, p. 767

Let $A = \begin{bmatrix} 1 & 4 \\ 0 & 3 \end{bmatrix}$ and $B = \begin{bmatrix} -1 & 2 \\ 1 & 0 \end{bmatrix}$. **Find the following.**

3. $A + B$	**4.** $A - B$	**5.** $3B$ Obj. 16-2, p. 770
6. AB	**7.** A^2	**8.** $A - 2B$ Obj. 16-3, p. 774

9. a. Write the communication matrix for the network shown.
    b. Use matrices to find the number of routes a message can be sent from point *A* to *C* using at most one relay point.

Obj. 16-4, p. 779

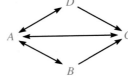

Extra / Augmented Matrices

When solving systems of linear equations in three variables, you learned a method called Gaussian elimination (see Lesson 9-9). This method involves transforming a given system into an equivalent triangular system.

Although you learned to do Gaussian elimination by working with the complete equations of a system, you can also solve the system by working only with the numbers in the equations. The coefficients and constants are put in a matrix called the *augmented matrix* for the system. For example, the augmented matrix for the system

$$x + y - 2z = 7$$
$$-x + 4y + 3z = 2$$
$$2x - 3y + 2z = -2$$

is

$$\begin{bmatrix} 1 & 1 & -2 & 7 \\ -1 & 4 & 3 & 2 \\ 2 & -3 & 2 & -2 \end{bmatrix}$$

Recall that as you transform a system into triangular form, you use linear combinations of the system's equations. That is, you multiply one equation by a nonzero constant and add the result to another equation. This technique can also be applied to an augmented matrix.

> To obtain an augmented matrix in triangular form, you replace any row with the sum of that row and any other row multiplied by a nonzero constant.

Example 1 Solve this system (see Example 2, page 445):

$$x + y - 2z = 7$$
$$-x + 4y + 3z = 2$$
$$2x - 3y + 2z = -2$$

Solution

1. Write the augmented matrix. The goal is to obtain a matrix in triangular form (that is, with zeros in the lower left corner).

$$\begin{bmatrix} 1 & 1 & -2 & 7 \\ -1 & 4 & 3 & 2 \\ 2 & -3 & 2 & -2 \end{bmatrix}$$

2. Replace the second row with the sum of the first and the second rows.

$$\begin{bmatrix} 1 & 1 & -2 & 7 \\ 0 & 5 & 1 & 9 \\ 2 & -3 & 2 & -2 \end{bmatrix}$$

3. Replace the third row with the sum of the third row and -2 times the first row.

$$\begin{bmatrix} 1 & 1 & -2 & 7 \\ 0 & 5 & 1 & 9 \\ 0 & -5 & 6 & -16 \end{bmatrix}$$

(Solution continues on the next page.)

4. Replace the third row with the sum of the second and third rows.

$$\begin{bmatrix} 1 & 1 & -2 & 7 \\ 0 & 5 & 1 & 9 \\ 0 & 0 & 7 & -7 \end{bmatrix}$$

The third row tells you that $7z = -7$, so $z = -1$.

The second row tells you that $5y + z = 9$. Since $z = -1$, you have $5y + (-1) = 9$, or $y = 2$.

The first row tells you that $x + y - 2z = 7$. Since $y = 2$ and $z = -1$, you have $x + 2 - 2(-1) = 7$, or $x = 3$.

∴ the solution of the system is $(3, 2, -1)$. **Answer**

Example 2 Solve this system: $\begin{aligned} 2x - y + 3z &= 5 \\ 4x + y &= 13 \\ 3x \quad + 4z &= 4 \end{aligned}$

Solution

1. Write the augmented matrix.

$$\begin{bmatrix} 2 & -1 & 3 & 5 \\ 4 & 1 & 0 & 13 \\ 3 & 0 & 4 & 4 \end{bmatrix}$$

2. Replace the second row with the sum of the second row and -2 times the first row.

$$\begin{bmatrix} 2 & -1 & 3 & 5 \\ 0 & 3 & -6 & 3 \\ 3 & 0 & 4 & 4 \end{bmatrix}$$

3. Replace the third row with the sum of the third row and $-\frac{3}{2}$ times the first row.

$$\begin{bmatrix} 2 & -1 & 3 & 5 \\ 0 & 3 & -6 & 3 \\ 0 & \frac{3}{2} & -\frac{1}{2} & -\frac{7}{2} \end{bmatrix}$$

4. Replace the third row with the sum of the third row and $-\frac{1}{2}$ times the second row.

$$\begin{bmatrix} 2 & -1 & 3 & 5 \\ 0 & 3 & -6 & 3 \\ 0 & 0 & \frac{5}{2} & -5 \end{bmatrix}$$

From the third row you get $z = -2$. Then from the second row you get $y = -3$. Finally, from the first row you get $x = 4$.

∴ the solution of the system is $(4, -3, -2)$. **Answer**

Computer software can be very useful in reducing augmented matrices to triangular form.

Exercises

Solve each system by using an augmented matrix.

1. $\begin{aligned} x + y + z &= 0 \\ x - y + z &= 2 \\ x - y - z &= 10 \end{aligned}$

2. $\begin{aligned} x + 2y - z &= 0 \\ -x + y &= 5 \\ x \quad + 2z &= 6 \end{aligned}$

3. $\begin{aligned} x - 2y + z &= 9 \\ 3x + y &= 1 \\ -2x - 3y - z &= 0 \end{aligned}$

4. $\begin{aligned} x + 2y + 3z &= 3 \\ 2x + 3y + 4z &= 2 \\ -3x - 5y + 2z &= 4 \end{aligned}$

5. $\begin{aligned} 2x + y + z &= 3 \\ 4x \quad + 3z &= 5 \\ 3x + 2y &= 1 \end{aligned}$

6. $\begin{aligned} 2x - y - 2z &= 5 \\ 3x - y &= 1 \\ 5x \quad + 4z &= -2 \end{aligned}$

Inverses of Matrices

16-5 Determinants

Objective   To find the determinant of a 2 × 2 or 3 × 3 matrix.

A real number called the *determinant* is associated with each *square* matrix.
The determinant of a matrix is usually displayed in the same form as the
matrix, but with vertical bars rather than brackets enclosing the elements.
The number of elements in any row or column is called the **order** of the
determinant.

Determinant of a 2 × 2 matrix

Let $A = \begin{bmatrix} a & b \\ c & d \end{bmatrix}$. Then the determinant of A, denoted by det A, is defined as
follows:

$$\det A = \begin{vmatrix} a & b \\ c & d \end{vmatrix} = ad - bc$$

Since the determinant of A has 2 rows and 2 columns, det A is of order 2.

Example 1   Evaluate $\begin{vmatrix} 5 & 2 \\ 4 & -1 \end{vmatrix}$.

Solution $\begin{vmatrix} 5 & 2 \\ 4 & -1 \end{vmatrix} = 5(-1) - (2)(4) = -13$ ***Answer***

The determinant of a 3 × 3 matrix is defined below.

Determinant of a 3 × 3 matrix

Let $B = \begin{bmatrix} a_1 & b_1 & c_1 \\ a_2 & b_2 & c_2 \\ a_3 & b_3 & c_3 \end{bmatrix}$. Then det B is defined as follows:

$$\det B = \begin{vmatrix} a_1 & b_1 & c_1 \\ a_2 & b_2 & c_2 \\ a_3 & b_3 & c_3 \end{vmatrix}$$

$$= a_1 b_2 c_3 + a_2 b_3 c_1 + a_3 b_1 c_2 - a_3 b_2 c_1 - a_2 b_1 c_3 - a_1 b_3 c_2$$

A convenient method for finding the six terms needed to evaluate a 3×3 determinant is shown below.

1. Copy the first two columns of the matrix in order to the right of the third column.

$$\begin{array}{ccc|cc} a_1 & b_1 & c_1 & a_1 & b_1 \\ a_2 & b_2 & c_2 & a_2 & b_2 \\ a_3 & b_3 & c_3 & a_3 & b_3 \end{array}$$

2. Multiply each element in the first row of the original matrix by the other two elements on the left-to-right downward diagonal. These products are the first three terms of the determinant.

$$\begin{array}{ccc|cc} a_1 & b_1 & c_1 & a_1 & b_1 \\ a_2 & b_2 & c_2 & a_2 & b_2 \\ a_3 & b_3 & c_3 & a_3 & b_3 \end{array}$$

$$a_1b_2c_3 + a_3b_1c_2 + a_2b_3c_1$$

3. Multiply each element in the last row of the original matrix by the other two elements on the left-to-right upward diagonal. The opposites of these products are the last three terms of the determinant.

$$\begin{array}{ccc|cc} a_1 & b_1 & c_1 & a_1 & b_1 \\ a_2 & b_2 & c_2 & a_2 & b_2 \\ a_3 & b_3 & c_3 & a_3 & b_3 \end{array}$$

$$- a_3b_2c_1 - a_1b_3c_2 - a_2b_1c_3$$

Caution: This method works *only* for 3×3 determinants.

Example 2 Evaluate $\det A = \begin{vmatrix} 2 & -1 & 1 \\ 5 & 3 & -1 \\ 0 & 4 & 2 \end{vmatrix}$.

Solution

$$\begin{array}{ccc|cc} 2 & -1 & 1 & 2 & -1 \\ 5 & 3 & -1 & 5 & 3 \\ 0 & 4 & 2 & 0 & 4 \end{array}$$

$$\det A = 12 + 0 + 20 - 0 - (-8) - (-10) = 50 \quad \textit{Answer}$$

Recall that the *main diagonal* of a square matrix runs from the upper left to the lower right. A matrix whose only nonzero elements are on the main diagonal is called a **diagonal matrix.** As you can easily verify for 2×2 and 3×3 diagonal matrices, the determinant of a diagonal matrix is the product of the elements on the main diagonal.

Oral Exercises

Evaluate.

1. $\begin{vmatrix} 2 & 0 \\ 0 & 1 \end{vmatrix}$

2. $\begin{vmatrix} 0 & 0 \\ 0 & 0 \end{vmatrix}$

3. $\begin{vmatrix} 1 & 2 \\ 2 & 5 \end{vmatrix}$

4. $\begin{vmatrix} 2 & 1 \\ -2 & 1 \end{vmatrix}$

5. $\begin{vmatrix} 8 & -6 \\ 5 & 3 \end{vmatrix}$

6. $\begin{vmatrix} 2 & 0 \\ 0 & 2 \end{vmatrix} + \begin{vmatrix} 0 & 1 \\ -1 & 0 \end{vmatrix}$

7. $\begin{vmatrix} 0 & 5 \\ -1 & 0 \end{vmatrix} - \begin{vmatrix} 2 & 1 \\ 1 & 2 \end{vmatrix}$

8. $\begin{vmatrix} 3 & 0 \\ 0 & -2 \end{vmatrix} \div \begin{vmatrix} 1 & 0 \\ 0 & 2 \end{vmatrix}$

Written Exercises

Evaluate.

A **1.** $\begin{vmatrix} 3 & 1 \\ 4 & 3 \end{vmatrix}$
   2. $\begin{vmatrix} 2 & 10 \\ 0 & 3 \end{vmatrix}$
   3. $\begin{vmatrix} 5 & -3 \\ 3 & 1 \end{vmatrix}$

4. $\begin{vmatrix} 15 & 0 \\ 0 & 4 \end{vmatrix}$
   5. $\begin{vmatrix} 0 & 8 \\ -4 & 0 \end{vmatrix}$
   6. $\begin{vmatrix} 3 & 9 \\ 2 & 6 \end{vmatrix}$

7. $\begin{vmatrix} 2 & -5 & 3 \\ 0 & 8 & 1 \\ -5 & 4 & 0 \end{vmatrix}$
   8. $\begin{vmatrix} 3 & 0 & 1 \\ 5 & 2 & 2 \\ -2 & 0 & 4 \end{vmatrix}$
   9. $\begin{vmatrix} 12 & -9 & 13 \\ 0 & 0 & 8 \\ -9 & 2 & 1 \end{vmatrix}$

10. $\begin{vmatrix} -10 & 0 & 0 \\ 0 & 4 & 0 \\ 0 & 0 & -12 \end{vmatrix}$
   11. $\begin{vmatrix} 5 & 0 & 0 \\ 0 & -6 & 0 \\ 0 & 0 & 2 \end{vmatrix}$
   12. $\begin{vmatrix} 0 & 2 & -3 \\ 3 & 5 & -3 \\ 1 & 2 & 0 \end{vmatrix}$

B **13.** $\begin{vmatrix} 8 & -1 & -5 \\ 0 & 9 & 6 \\ -2 & 0 & 3 \end{vmatrix}$
   14. $\begin{vmatrix} 2 & 3 & -1 \\ -3 & 0 & -8 \\ 11 & -4 & 6 \end{vmatrix}$

15. Show that for any 2×2 matrix, A, if all the elements in one row or one column are zeros, then det $A = 0$.

16. Show that for any 3×3 matrix, B, if all the elements in one row or one column are zeros, then det $B = 0$.

17. Let A be any 2×2 matrix. Show that if the two rows are the same, then det $A = 0$.

18. Let A be any 2×2 matrix. Show that if any row or column of A is multiplied by a scalar r, then det A is multiplied by r.

C **19.** Let r be a scalar, and let A be any 2×2 matrix. Show that det $(rA) = r^2$ det A.

20. Let A and B be 2×2 matrices. Show that det $AB =$ det $A \cdot$ det B.

Mixed Review Exercises

Write in simplest form without zero or negative exponents.

1. $\dfrac{t^2 - 3t + 2}{6 - 5t + t^2}$
   2. $\dfrac{a^2 b^3}{a^3 b^7}$
   3. $\dfrac{3}{s + 1} + \dfrac{4}{3s - 3}$

4. $\sqrt[3]{a^2} \cdot \sqrt[3]{a^4}$
   5. $\dfrac{\dfrac{1}{c} - 1}{c - \dfrac{1}{c}}$
   6. $\dfrac{y + 5}{y - 4} \div \dfrac{y^2 - 25}{y^2 - 6y + 8}$

7. $(-9pr^4 q^3)(4p^2 r^{-4} q^3)$
   8. $\dfrac{x^3 - 1}{x^2 + 3x - 4}$
   9. $(2^\pi)^{-1/\pi}$

16-6 Inverses of Matrices

Objective   To solve systems of equations using inverses of matrices.

The product of the two matrices below is the identity matrix, $I_{2\times2}$:

$$\begin{bmatrix} 2 & 1 \\ 3 & -2 \end{bmatrix}\begin{bmatrix} \frac{2}{7} & \frac{1}{7} \\ \frac{3}{7} & -\frac{2}{7} \end{bmatrix} = \begin{bmatrix} 1 & 0 \\ 0 & 1 \end{bmatrix}$$

For any two matrices A and B, if $AB = BA = I$, then A and B are called **inverse matrices.** The symbol A^{-1} denotes the inverse of matrix A. Therefore, $B = A^{-1}$ and $A = B^{-1}$.

Does every nonzero 2×2 matrix, A, have an inverse, A^{-1}?

Let $A = \begin{bmatrix} a & b \\ c & d \end{bmatrix}$. Find $A^{-1} = \begin{bmatrix} x & u \\ y & v \end{bmatrix}$ such that $AA^{-1} = I_{2\times2}$.

$$\begin{bmatrix} a & b \\ c & d \end{bmatrix}\begin{bmatrix} x & u \\ y & v \end{bmatrix} = \begin{bmatrix} ax + by & au + bv \\ cx + dy & cu + dv \end{bmatrix} = \begin{bmatrix} 1 & 0 \\ 0 & 1 \end{bmatrix}$$

The equation above is true if and only if

$$ax + by = 1 \qquad\qquad au + bv = 0$$
$$cx + dy = 0 \qquad\qquad cu + dv = 1.$$

If $ad - bc \neq 0$, these systems may be solved to obtain the following:

$$x = \frac{d}{ad - bc} \qquad\qquad u = \frac{-b}{ad - bc}$$

$$y = \frac{-c}{ad - bc} \qquad\qquad v = \frac{a}{ad - bc}$$

Substitute these values in the matrix A^{-1}.

$$A^{-1} = \begin{bmatrix} \dfrac{d}{ad - bc} & \dfrac{-b}{ad - bc} \\ \dfrac{-c}{ad - bc} & \dfrac{a}{ad - bc} \end{bmatrix} = \frac{1}{ad - bc}\begin{bmatrix} d & -b \\ -c & a \end{bmatrix}, \quad ad - bc \neq 0.$$

You can verify that $AA^{-1} = A^{-1}A = I_{2\times2}$. Since $ad - bc$ is the determinant of A, you have this general result.

If $A = \begin{bmatrix} a & b \\ c & d \end{bmatrix}$ and det $A \neq 0$, then

$$A^{-1} = \frac{1}{\det A}\begin{bmatrix} d & -b \\ -c & a \end{bmatrix}.$$

If det A = 0, then A has no inverse.

Example 1 Find A^{-1} if $A = \begin{bmatrix} 6 & 2 \\ -3 & 1 \end{bmatrix}$.

Solution Since det $A = 12 \neq 0$, A has an inverse. Here is what the general result says to do to find A^{-1}: Interchange a and d, replace both b and c with their opposites, and then multiply the resulting matrix by $\dfrac{1}{\det A}$.

$$A^{-1} = \frac{1}{12}\begin{bmatrix} 1 & -2 \\ 3 & 6 \end{bmatrix} = \begin{bmatrix} \frac{1}{12} & -\frac{1}{6} \\ \frac{1}{4} & \frac{1}{2} \end{bmatrix} \quad \textit{Answer}$$

You can use the inverse of a matrix to solve a system of linear equations

$$ax + by = e$$
$$cx + dy = f$$

where a, b, c, d, e, and f are real numbers. You can write this system using the definitions of matrix multiplication and equality of matrices.

$$\begin{bmatrix} a & b \\ c & d \end{bmatrix}\begin{bmatrix} x \\ y \end{bmatrix} = \begin{bmatrix} e \\ f \end{bmatrix}$$

This equation is in the form $AX = B$ where the *matrix of coefficients*

$$A = \begin{bmatrix} a & b \\ c & d \end{bmatrix}, \quad X = \begin{bmatrix} x \\ y \end{bmatrix}, \quad \text{and} \quad B = \begin{bmatrix} e \\ f \end{bmatrix}.$$

Therefore, if A^{-1} exists you have: $A^{-1}(AX) = A^{-1}(B)$
$$(A^{-1}A)X = A^{-1}B$$
$$IX = A^{-1}B$$
$$X = A^{-1}B$$

Thus if det $A \neq 0$, so that A^{-1} exists, the system has a unique solution, $X = A^{-1}B$.

Example 2 Use matrices to solve this system: $2x - 3y = 1$
$$x + 4y = 6$$

Solution Write the matrix equation $AX = B$: $\begin{bmatrix} 2 & -3 \\ 1 & 4 \end{bmatrix}\begin{bmatrix} x \\ y \end{bmatrix} = \begin{bmatrix} 1 \\ 6 \end{bmatrix}$

Since det $A = 11 \neq 0$, A^{-1} exists, and the solution of the equation $AX = B$ is $X = A^{-1}B$. So find A^{-1}, the inverse of the matrix of coefficients.

$$A^{-1} = \frac{1}{11}\begin{bmatrix} 4 & 3 \\ -1 & 2 \end{bmatrix}$$

Then $\begin{bmatrix} x \\ y \end{bmatrix} = \dfrac{1}{11}\begin{bmatrix} 4 & 3 \\ -1 & 2 \end{bmatrix}\begin{bmatrix} 1 \\ 6 \end{bmatrix} = \dfrac{1}{11}\begin{bmatrix} 22 \\ 11 \end{bmatrix} = \begin{bmatrix} 2 \\ 1 \end{bmatrix}.$

∴ the solution of the system is (2, 1). *Answer*

Oral Exercises

Tell whether each matrix has an inverse.

1. $\begin{bmatrix} 4 & -10 \\ -1 & 3 \end{bmatrix}$

2. $\begin{bmatrix} 6 & -8 \\ 3 & -4 \end{bmatrix}$

3. $\begin{bmatrix} 10 & 5 \\ 9 & 4 \end{bmatrix}$

4. $\begin{bmatrix} 0 & 16 \\ 0 & -12 \end{bmatrix}$

5. $\begin{bmatrix} 6 & -3 \\ -10 & 5 \end{bmatrix}$

6. $\begin{bmatrix} 8 & -5 \\ 3 & -2 \end{bmatrix}$

Tell whether each system has a unique solution.

7. $5x + y = 1$
$4x + 2y = 3$

8. $4x + 2y = -6$
$6x + 3y = -9$

9. $3x - 2y = 2$
$3x + 2y = -9$

10. $6x = 3$
$2x + y = 5$

11. $2x + 4y = 12$
$3x + 6y = 20$

12. $8x - y = 10$
$y = 5$

13. If a system of equations does not have a unique solution, can you conclude that the system has no solution? Explain.

Written Exercises

Find the inverse of each matrix. If the matrix has no inverse, say so.

A

1. $\begin{bmatrix} 2 & 1 \\ -1 & 2 \end{bmatrix}$

2. $\begin{bmatrix} 5 & 3 \\ 2 & 1 \end{bmatrix}$

3. $\begin{bmatrix} 4 & 6 \\ 2 & 3 \end{bmatrix}$

4. $\begin{bmatrix} -3 & 1 \\ 4 & -2 \end{bmatrix}$

5. $\begin{bmatrix} 1 & 0 \\ 0 & 1 \end{bmatrix}$

6. $\begin{bmatrix} 0 & 9 \\ -1 & 3 \end{bmatrix}$

7. $\begin{bmatrix} 3 & -1 \\ 6 & 2 \end{bmatrix}$

8. $\begin{bmatrix} 0 & 0 \\ 5 & -7 \end{bmatrix}$

Use matrices to find the solution of each system of equations. If a system has no unique solution, say so.

9. $2x - y = 6$
$3x + 2y = -19$

10. $3x + 2y = 5$
$3x - y = 2$

11. $2x + 2y = 16$
$x - 3y = -4$

12. $4x - 6y = 8$
$-6x + 9y = -12$

13. $x + y = 6$
$2x = 4$

14. $4x + 3y = 1$
$4x + y = -5$

15. $x + y = 1$
$2x + 2y = 5$

16. $x = 3$
$-x + y = 2$

Solve each equation for matrix X. (*Hint:* These equations are in the form $AX = B$.)

B

17. $\begin{bmatrix} 1 & 3 \\ 2 & 7 \end{bmatrix} X = \begin{bmatrix} 4 & 1 \\ -2 & 3 \end{bmatrix}$

18. $\begin{bmatrix} 4 & 4 \\ 3 & 4 \end{bmatrix} X = \begin{bmatrix} 1 & 0 \\ 0 & 1 \end{bmatrix}$

19. $\begin{bmatrix} 4 & 2 \\ 1 & 1 \end{bmatrix} X - \begin{bmatrix} 0 & \frac{1}{2} \\ -1 & \frac{3}{2} \end{bmatrix} = \begin{bmatrix} 2 & 3 \\ 0 & 1 \end{bmatrix}$

20. $\begin{bmatrix} 5 & 3 \\ 3 & 2 \end{bmatrix} X - \begin{bmatrix} 1 & 3 \\ 2 & 1 \end{bmatrix} = \begin{bmatrix} 1 & 2 \\ -1 & 3 \end{bmatrix}$

Self-Test 2

Vocabulary   determinant (p. 787)   diagonal matrix (p. 788)
                 order of a determinant (p. 787)   inverse matrices (p. 790)

In Exercises 1 and 2 evaluate each determinant.

1. $\begin{vmatrix} 3 & 0 \\ 5 & -1 \end{vmatrix}$
            2. $\begin{vmatrix} 2 & -2 & 0 \\ 1 & 1 & -2 \\ 0 & -3 & 1 \end{vmatrix}$
              Obj. 16-5, p. 787

3. Find A^{-1} if $A = \begin{bmatrix} 1 & -1 \\ 0 & 3 \end{bmatrix}$.
                                       Obj. 16-6, p. 790

4. Solve for matrix A.

$$\begin{bmatrix} 2 & 1 \\ -1 & 4 \end{bmatrix} A = \begin{bmatrix} 3 & -1 \\ -6 & 5 \end{bmatrix}.$$

Check your answers with those at the back of the book.

Historical Note / *Matrices*

Matrices are a relatively new concept in mathematics. They were not devised until 1857 when the British mathematician Arthur Cayley (1821–1895) began to use them in the study of linear transformations.

    Cayley was working on linear transformations of the type

$$x_1 = ax + by$$
$$y_1 = cx + dy$$

which map an ordered pair (x, y) onto another ordered pair (x_1, y_1). In this transformation a, b, c, and d are real numbers. Cayley noticed that the entire transformation was determined by these four real numbers, which could be arranged in the square array

$$\begin{bmatrix} a & b \\ c & d \end{bmatrix}.$$

    When the product of two such arrays is defined as in this chapter, the matrix corresponding to a *sequence* of transformations acting one after the other is the *product* of their matrices. Further analysis led Cayley to formulate most of the rules of the algebra of matrices.

    A linear transformation maps any point (x, y) in the plane onto an image point (x_1, y_1), and Cayley saw that the matrix of a transformation gives geometric information about this mapping. For example, the image of a polygon is another polygon, generally with different sides and angles.

    However, the area of the image is always equal to the area of the original figure multiplied by the absolute value of the determinant of the matrix of the transformation.

Working with Determinants

16-7 Expansion of Determinants by Minors

Objective   To evaluate third-order determinants using expansion by minors.

It is important to distinguish between matrices and determinants. A matrix is a rectangular array of numbers. A determinant, often denoted by a square array of numbers, is in fact a real number associated with a square matrix. The process of evaluating a determinant is called "expanding the determinant."

You have already learned to evaluate a determinant of order 2 or 3. In this lesson you will learn a method of expansion that is applicable to determinants of order n when $n \geq 3$. This method is called "expansion by minors."

The **minor** of an element in a determinant is the determinant resulting from the deletion of the row and column containing the element. Therefore in the determinant

$$\begin{vmatrix} 3 & -1 & 0 \\ 8 & 4 & 2 \\ 0 & -2 & 4 \end{vmatrix},$$

the minor of the element 3 is

$$\begin{vmatrix} 3 & -1 & 0 \\ 8 & 4 & 2 \\ 0 & -2 & 4 \end{vmatrix}, \quad \text{or} \quad \begin{vmatrix} 4 & 2 \\ -2 & 4 \end{vmatrix}.$$

The minor of the element 8 is

$$\begin{vmatrix} 3 & -1 & 0 \\ 8 & 4 & 2 \\ 0 & -2 & 4 \end{vmatrix}, \quad \text{or} \quad \begin{vmatrix} -1 & 0 \\ -2 & 4 \end{vmatrix}.$$

By definition (see page 787), the value of a third-order determinant is given by

$$\begin{vmatrix} a_1 & b_1 & c_1 \\ a_2 & b_2 & c_2 \\ a_3 & b_3 & c_3 \end{vmatrix} = a_1b_2c_3 + a_2b_3c_1 + a_3b_1c_2 - a_3b_2c_1 - a_2b_1c_3 - a_1b_3c_2$$

$$= a_1(b_2c_3 - b_3c_2) - b_1(a_2c_3 - a_3c_2) + c_1(a_2b_3 - a_3b_2) \qquad \left\{ \begin{array}{l} \text{Rearranging} \\ \text{terms and} \\ \text{factoring} \end{array} \right.$$

$$= a_1\begin{vmatrix} b_2 & c_2 \\ b_3 & c_3 \end{vmatrix} - b_1\begin{vmatrix} a_2 & c_2 \\ a_3 & c_3 \end{vmatrix} + c_1\begin{vmatrix} a_2 & b_2 \\ a_3 & b_3 \end{vmatrix}.$$

Notice that the three determinants are the minors of a_1, b_1, and c_1, named below by A_1, B_1, and C_1, respectively. Therefore,

$$\begin{vmatrix} a_1 & b_1 & c_1 \\ a_2 & b_2 & c_2 \\ a_3 & b_3 & c_3 \end{vmatrix} = a_1 A_1 - b_1 B_1 + c_1 C_1.$$

The right-hand expression is the expansion of this determinant by minors *about the first row*.

A determinant may be expanded by minors about any row or column.

To Expand a Determinant by Minors

1. Multiply each element of a given row or column by its minor.
2. Add the number of the row and the number of the column for each element. If the sum is odd, multiply the product obtained in Step 1 by -1.
3. Add the products to obtain the value of the determinant.

Example 1 Expand $\begin{vmatrix} 2 & -6 & 3 \\ 4 & 5 & -2 \\ 3 & -1 & 0 \end{vmatrix}$ by the minors of the first row.

Solution Multiply each element of the first row by its minor. Also multiply the product for the element in the first row, second column by -1, since the sum $1 + 2$, or 3, is odd. Then add the products.

$$\begin{vmatrix} 2 & -6 & 3 \\ 4 & 5 & -2 \\ 3 & -1 & 0 \end{vmatrix} = 2 \begin{vmatrix} 5 & -2 \\ -1 & 0 \end{vmatrix} + (-1)(-6) \begin{vmatrix} 4 & -2 \\ 3 & 0 \end{vmatrix} + 3 \begin{vmatrix} 4 & 5 \\ 3 & -1 \end{vmatrix}$$

$$= 2(0 - 2) + 6(0 + 6) + 3(-4 - 15)$$

$$= -4 + 36 + (-57) = -25 \quad \textbf{\textit{Answer}}$$

You can shorten the work of expanding a determinant by minors by choosing a row or column having the greatest number of zeros as elements.

Example 2 Expand $\begin{vmatrix} 1 & 3 & 0 \\ -1 & 4 & 2 \\ 2 & 1 & 0 \end{vmatrix}$ by the minors of the third column.

Solution $\begin{vmatrix} 1 & 3 & 0 \\ -1 & 4 & 2 \\ 2 & 1 & 0 \end{vmatrix} = 0(C_1) + (-1)2 \begin{vmatrix} 1 & 3 \\ 2 & 1 \end{vmatrix} + 0(C_3)$

$$= 0 + (-2)(1 - 6) + 0$$

$$= (-2)(-5) = 10 \quad \textbf{\textit{Answer}}$$

The method of expansion by minors is used to define determinants of order $n \geq 3$. For example, consider the matrix

$$M = \begin{bmatrix} a_1 & b_1 & c_1 & d_1 \\ a_2 & b_2 & c_2 & d_2 \\ a_3 & b_3 & c_3 & d_3 \\ a_4 & b_4 & c_4 & d_4 \end{bmatrix}.$$

Let

$$A_1 = \begin{vmatrix} b_2 & c_2 & d_2 \\ b_3 & c_3 & d_3 \\ b_4 & c_4 & d_4 \end{vmatrix}, \qquad B_1 = \begin{vmatrix} a_2 & c_2 & d_2 \\ a_3 & c_3 & d_3 \\ a_4 & c_4 & d_4 \end{vmatrix},$$

$$C_1 = \begin{vmatrix} a_2 & b_2 & d_2 \\ a_3 & b_3 & d_3 \\ a_4 & b_4 & d_4 \end{vmatrix}, \quad \text{and} \quad D_1 = \begin{vmatrix} a_2 & b_2 & c_2 \\ a_3 & b_3 & c_3 \\ a_4 & b_4 & c_4 \end{vmatrix}.$$

The determinant of M is defined as

$$\det M = a_1 A_1 - b_1 B_1 + c_1 C_1 - d_1 D_1.$$

Note that A_1, B_1, C_1, and D_1 are the minors of a_1, b_1, c_1, and d_1, respectively. Thus they are minors of the first row of this 4×4 determinant. As with third-order determinants, determinants of order four may be expanded by minors about any row or any column. The minors of a fourth-order determinant are each third-order determinants. In general, minors of a determinant of order n are determinants of order $n - 1$. Since the expansion of higher order determinants can be tedious, computers are often used to evaluate them.

Oral Exercises

Give the requested element or determinant using the given expansion by minors.

$$\begin{vmatrix} 0 & 1 & 2 \\ 1 & -1 & -3 \\ -2 & 4 & 5 \end{vmatrix} = a_1 \begin{vmatrix} -1 & -3 \\ b_3 & 5 \end{vmatrix} - b_1 \begin{vmatrix} a_2 & -3 \\ -2 & 5 \end{vmatrix} + 2C_1$$

1. a_1 **2.** b_3 **3.** b_1 **4.** a_2 **5.** C_1

6. Complete the computation to evaluate the determinant given above.

Written Exercises

Expand each determinant about the given row or column.

A   **1.** $\begin{vmatrix} 2 & 1 & 3 \\ -2 & 1 & 4 \\ 1 & 2 & 5 \end{vmatrix}$;   row one   **2.** $\begin{vmatrix} 2 & 3 & 1 \\ 0 & -1 & -4 \\ 5 & 0 & 0 \end{vmatrix}$;   row three

3. $\begin{vmatrix} -2 & -2 & 0 \\ 0 & 2 & 3 \\ 1 & -1 & 1 \end{vmatrix}$; column one

4. $\begin{vmatrix} 4 & 0 & -5 \\ 9 & 1 & 11 \\ -4 & 0 & 3 \end{vmatrix}$; column two

5. $\begin{vmatrix} -1 & 3 & 4 \\ 0 & 1 & 0 \\ 3 & 6 & -2 \end{vmatrix}$; row two

6. $\begin{vmatrix} 2 & 0 & 0 \\ 10 & 1 & -1 \\ 3 & 0 & 4 \end{vmatrix}$; row one

7. $\begin{vmatrix} 1 & -1 & 0 \\ 2 & 3 & -1 \\ 10 & -4 & 0 \end{vmatrix}$; column three

8. $\begin{vmatrix} 0 & -2 & 4 \\ 1 & 3 & -1 \\ 0 & 0 & 5 \end{vmatrix}$; column one

Expand each determinant about any row or column.

9. $\begin{vmatrix} 1 & 0 & -1 \\ 0 & 1 & 2 \\ -1 & 2 & 1 \end{vmatrix}$
10. $\begin{vmatrix} 0 & 2 & 1 \\ 4 & 0 & -2 \\ -1 & 3 & 1 \end{vmatrix}$
11. $\begin{vmatrix} 0 & 3 & 1 \\ 4 & -1 & 2 \\ 0 & 2 & 0 \end{vmatrix}$
12. $\begin{vmatrix} 1 & -1 & 2 \\ 3 & 5 & 1 \\ 4 & 2 & -2 \end{vmatrix}$

Solve each equation for x.

B **13.** $\begin{vmatrix} 2 & 1 & 1 \\ 1 & 2 & 1 \\ x & 1 & 2 \end{vmatrix} = 4$

14. $\begin{vmatrix} 1 & 3 & 4 \\ 5 & 15 & 10 \\ -1 & x & 2 \end{vmatrix} = 80$

15. $\begin{vmatrix} 12 & -7 & 19 \\ 0 & x & 5 \\ -9 & 3 & 43 \end{vmatrix} = 135$

16. $\begin{vmatrix} 4 & -3 & 9 \\ 8 & x & -4 \\ 1 & 5 & 3 \end{vmatrix} = 545$

Find the value of each determinant.

17. $\begin{vmatrix} 5 & 0 & 0 & 0 \\ 3 & -3 & 0 & 0 \\ 8 & 2 & 2 & 0 \\ -4 & 10 & 3 & 2 \end{vmatrix}$

18. $\begin{vmatrix} 1 & 2 & 3 & 4 \\ 5 & 6 & 7 & 8 \\ 9 & 10 & 11 & 12 \\ 13 & 14 & 15 & 16 \end{vmatrix}$

C **19.** Let A be a 3×3 matrix. Show that if any two rows of A are alike, then det $A = 0$.

Mixed Review Exercises

Simplify.

1. $125^{-2/3}$

2. $\cos \dfrac{3\pi}{4}$

3. $\dfrac{8!}{3!5!}$

4. $\sqrt{-2} \cdot \sqrt{-8}$

5. $\log_3 9\sqrt{3}$

6. $\sec^2 20° - \tan^2 20°$

16-8 Properties of Determinants

Objective   To use the properties of determinants to simplify the expansion of determinants by minors.

Determinants have properties that you can use to simplify their evaluation. Although third-order determinants are used as examples in this lesson, the properties apply to determinants of any order.

Property 1.   If each element in any row (or any column) is 0, then the determinant is equal to 0. For example,

$$\begin{vmatrix} 3 & 5 & -2 \\ 0 & 0 & 0 \\ -1 & 2 & 4 \end{vmatrix} = 0 \begin{vmatrix} 5 & -2 \\ 2 & 4 \end{vmatrix} - 0 \begin{vmatrix} 3 & -2 \\ -1 & 4 \end{vmatrix} + 0 \begin{vmatrix} 3 & 5 \\ -1 & 2 \end{vmatrix}$$

$$= 0 + 0 + 0 = 0.$$

Property 2.   If two rows (or two columns) of a determinant have corresponding elements that are equal, then the determinant is equal to 0. For example,

$$\begin{vmatrix} 8 & -4 & 3 \\ 6 & 2 & 1 \\ 6 & 2 & 1 \end{vmatrix} = 8 \begin{vmatrix} 2 & 1 \\ 2 & 1 \end{vmatrix} - (-4) \begin{vmatrix} 6 & 1 \\ 6 & 1 \end{vmatrix} + 3 \begin{vmatrix} 6 & 2 \\ 6 & 2 \end{vmatrix}$$

$$= 8(0) + 4(0) + 3(0) = 0.$$

Property 3.   If two rows (or two columns) of a determinant are interchanged, then the resulting determinant is the opposite of the original determinant. For example, expand these determinants by minors about the second row:

$$\begin{vmatrix} 1 & 2 & 4 \\ 3 & 1 & 0 \\ -2 & -1 & 2 \end{vmatrix} = (-1)3 \begin{vmatrix} 2 & 4 \\ -1 & 2 \end{vmatrix} + 1 \begin{vmatrix} 1 & 4 \\ -2 & 2 \end{vmatrix} - 0 \begin{vmatrix} 1 & 2 \\ -2 & -1 \end{vmatrix}$$

$$= -3(8) + 10 + 0 = -14;$$

$$\begin{vmatrix} 2 & 1 & 4 \\ 1 & 3 & 0 \\ -1 & -2 & 2 \end{vmatrix} = (-1)(1) \begin{vmatrix} 1 & 4 \\ -2 & 2 \end{vmatrix} + 3 \begin{vmatrix} 2 & 4 \\ -1 & 2 \end{vmatrix} - 0 \begin{vmatrix} 2 & 1 \\ -1 & -2 \end{vmatrix}$$

$$= (-1)(10) + 3(8) + 0 = 14.$$

Property 4.   If each element in one row (or one column) of a determinant is multiplied by a real number k, then the determinant is multiplied by k. For example, expand these determinants by minors about the first row:

$$\begin{vmatrix} 1 & 2 & 1 \\ -1 & 4 & -2 \\ 2 & -1 & 1 \end{vmatrix} = 1 \begin{vmatrix} 4 & -2 \\ -1 & 1 \end{vmatrix} - 2 \begin{vmatrix} -1 & -2 \\ 2 & 1 \end{vmatrix} + 1 \begin{vmatrix} -1 & 4 \\ 2 & -1 \end{vmatrix}$$

$$= 2 - 2(3) - 7 = -11;$$

$$\begin{vmatrix} 5\cdot 1 & 5\cdot 2 & 5\cdot 1 \\ -1 & 4 & -2 \\ 2 & -1 & 1 \end{vmatrix} = 5\cdot 1 \begin{vmatrix} 4 & -2 \\ -1 & 1 \end{vmatrix} - 5\cdot 2 \begin{vmatrix} -1 & -2 \\ 2 & 1 \end{vmatrix} + 5\cdot 1 \begin{vmatrix} -1 & 4 \\ 2 & -1 \end{vmatrix}$$

$$= 5\left(1 \begin{vmatrix} 4 & -2 \\ -1 & 1 \end{vmatrix} - 2 \begin{vmatrix} -1 & -2 \\ 2 & 1 \end{vmatrix} + 1 \begin{vmatrix} -1 & 4 \\ 2 & -1 \end{vmatrix} \right)$$

$$= 5(-11) = -55.$$

Property 5. If each element of one row (or one column) is multiplied by a real number k and if the resulting products are then added to the corresponding elements of another row (or another column), then the resulting determinant equals the original one. For example, expand these determinants by minors about the first column:

$$\begin{vmatrix} -4 & -1 & 1 \\ 6 & -3 & 2 \\ -6 & 3 & -1 \end{vmatrix} = -4 \begin{vmatrix} -3 & 2 \\ 3 & -1 \end{vmatrix} - 6 \begin{vmatrix} -1 & 1 \\ 3 & -1 \end{vmatrix} - 6 \begin{vmatrix} -1 & 1 \\ -3 & 2 \end{vmatrix}$$

$$= -4(-3) - 6(-2) - 6(1) = 18;$$

$$\begin{vmatrix} -4 + 2(-1) & -1 & 1 \\ 6 + 2(-3) & -3 & 2 \\ -6 + 2(3) & 3 & -1 \end{vmatrix} = \begin{vmatrix} -6 & -1 & 1 \\ 0 & -3 & 2 \\ 0 & 3 & -1 \end{vmatrix} = -6 \begin{vmatrix} -3 & 2 \\ 3 & -1 \end{vmatrix} - 0 \begin{vmatrix} -1 & 1 \\ 3 & -1 \end{vmatrix} + 0 \begin{vmatrix} -1 & 1 \\ -3 & 2 \end{vmatrix}$$

$$= -6(-3) + 0 + 0 = 18.$$

Properties 1–5 can be used to simplify the expansion of determinants. Notice how transforming the first column in the example for Property 5 simplified the expansion. A similar approach is taken in the example below, where the first row is reduced to $\begin{matrix} 1 & 0 & 0 \end{matrix}$ to simplify the expansion.

Example 1 Evaluate $\begin{vmatrix} 15 & 0 & 2 \\ 45 & -1 & 3 \\ 30 & 2 & 8 \end{vmatrix}$.

Solution Factor 15 from the elements in the first column. (Property 4)

$$\begin{vmatrix} 15 & 0 & 2 \\ 45 & -1 & 3 \\ 30 & 2 & 8 \end{vmatrix} = \begin{vmatrix} 15(1) & 0 & 2 \\ 15(3) & -1 & 3 \\ 15(2) & 2 & 8 \end{vmatrix} = 15 \begin{vmatrix} 1 & 0 & 2 \\ 3 & -1 & 3 \\ 2 & 2 & 8 \end{vmatrix}$$

Multiply the first column by -2 and add the results to the third column. (Property 5)

$$15 \begin{vmatrix} 1 & 0 & 2 \\ 3 & -1 & 3 \\ 2 & 2 & 8 \end{vmatrix} = 15 \begin{vmatrix} 1 & 0 & 2 + (-2) \\ 3 & -1 & 3 + (-6) \\ 2 & 2 & 8 + (-4) \end{vmatrix} = 15 \begin{vmatrix} 1 & 0 & 0 \\ 3 & -1 & -3 \\ 2 & 2 & 4 \end{vmatrix}$$

(Solution continues on the next page.)

Expand by minors about the first row.

$$15\begin{vmatrix} 1 & 0 & 0 \\ 3 & -1 & -3 \\ 2 & 2 & 4 \end{vmatrix} = 15\left(1\begin{vmatrix} -1 & -3 \\ 2 & 4 \end{vmatrix} - 0 + 0\right)$$

$$= 15(2 - 0 + 0) = 30 \quad \textbf{\textit{Answer}}$$

Oral Exercises

Use the properties of determinants to find the value of each determinant below, given that

$$\begin{vmatrix} 0 & 3 & 1 \\ 1 & 2 & 0 \\ -1 & 2 & 1 \end{vmatrix} = 1.$$

1. $\begin{vmatrix} 1 & 2 & 0 \\ 0 & 3 & 1 \\ -1 & 2 & 1 \end{vmatrix}$ **2.** $\begin{vmatrix} 0 & 3 & 3 \\ 1 & 2 & 0 \\ -1 & 2 & 3 \end{vmatrix}$ **3.** $\begin{vmatrix} 0 & 3 & 1 \\ 1 & 2 & 0 \\ -1 & 5 & 2 \end{vmatrix}$ **4.** $\begin{vmatrix} 0 & 3 & 1 \\ p & 2 & 0 \\ -p & 5 & 2 \end{vmatrix}$

Written Exercises

Evaluate each determinant. Use Properties 1–5 to simplify the work.

A **1.** $\begin{vmatrix} 41 & 13 & -5 \\ 0 & 5 & 9 \\ 41 & 13 & -5 \end{vmatrix}$ **2.** $\begin{vmatrix} -2 & 0 & 12 \\ 4 & 0 & 0 \\ 3 & 0 & 6 \end{vmatrix}$ **3.** $\begin{vmatrix} 2 & -1 & 1 \\ 1 & 3 & -3 \\ 7 & -8 & 8 \end{vmatrix}$

4. $\begin{vmatrix} 5 & 0 & 0 \\ 0 & 5 & 0 \\ 0 & 0 & 5 \end{vmatrix}$ **5.** $\begin{vmatrix} 2 & -4 & 6 \\ 5 & 0 & -8 \\ 3 & -6 & 9 \end{vmatrix}$ **6.** $\begin{vmatrix} -1 & 4 & 12 \\ 1 & -2 & -12 \\ 8 & 16 & 6 \end{vmatrix}$

7. $\begin{vmatrix} 1 & 5 & 10 \\ 0 & -5 & -8 \\ 1 & 5 & -2 \end{vmatrix}$ **8.** $\begin{vmatrix} 26 & 29 & 29 \\ 25 & 28 & 26 \\ 25 & 30 & 27 \end{vmatrix}$ **9.** $\begin{vmatrix} 5 & 1 & 1 \\ 1 & 5 & 1 \\ 1 & 1 & 5 \end{vmatrix}$

B **10.** $\begin{vmatrix} 4 & 0 & 0 & 0 \\ 0 & 4 & 0 & 0 \\ 0 & 0 & 4 & 0 \\ 0 & 0 & 0 & 4 \end{vmatrix}$ **11.** $\begin{vmatrix} 1 & -1 & -1 & 1 \\ 1 & 1 & 1 & -1 \\ -1 & -1 & 1 & 1 \\ -1 & 1 & -1 & -1 \end{vmatrix}$ **12.** $\begin{vmatrix} 1 & 1 & 1 & -1 \\ 2 & 2 & -2 & 2 \\ -3 & -3 & 3 & 3 \\ -4 & 4 & 4 & 4 \end{vmatrix}$

C **13.** Explain how Properties 1 and 5 can be used to justify Property 2.

   14. Prove Property 2 for third-order determinants.

   15. Prove Property 5 for second-order determinants.

16-9 Cramer's Rule

Objective   To solve systems of equations using determinants.

Determinants may be used to solve systems of linear equations. Consider this system of two linear equations in two variables:

$$ax + by = e$$
$$cx + dy = f$$

Using the linear-combination method of Lesson 3-5, you can verify that

$$x = \frac{de - bf}{ad - bc} \quad \text{and} \quad y = \frac{af - ce}{ad - bc}$$

provided that $ad - bc \neq 0$. Note that the denominators are equal to the determinant of coefficients,

$$D = \begin{vmatrix} a & b \\ c & d \end{vmatrix}.$$

The numerators are equal to the determinants D_x and D_y where

$$D_x = \begin{vmatrix} e & b \\ f & d \end{vmatrix} \quad \text{and} \quad D_y = \begin{vmatrix} a & e \\ c & f \end{vmatrix}.$$

Note that D_x is formed by replacing the column of coefficients of x in D with the column of constants. Similarly, D_y is formed by replacing the column of coefficients of y in D with the column of constants.

By substitution,

$$x = \frac{\begin{vmatrix} e & b \\ f & d \end{vmatrix}}{\begin{vmatrix} a & b \\ c & d \end{vmatrix}} = \frac{D_x}{D} \quad \text{and} \quad y = \frac{\begin{vmatrix} a & e \\ c & f \end{vmatrix}}{\begin{vmatrix} a & b \\ c & d \end{vmatrix}} = \frac{D_y}{D}.$$

Example 1   Use determinants to solve this system: $2x - y = 6$
$3x + 5y = 22$

Solution $x = \dfrac{D_x}{D} = \dfrac{\begin{vmatrix} 6 & -1 \\ 22 & 5 \end{vmatrix}}{\begin{vmatrix} 2 & -1 \\ 3 & 5 \end{vmatrix}} = \dfrac{30 + 22}{10 + 3} = \dfrac{52}{13} = 4$

$y = \dfrac{D_y}{D} = \dfrac{\begin{vmatrix} 2 & 6 \\ 3 & 22 \end{vmatrix}}{\begin{vmatrix} 2 & -1 \\ 3 & 5 \end{vmatrix}} = \dfrac{44 - 18}{10 + 3} = \dfrac{26}{13} = 2$

$\therefore$ the solution of the system is (4, 2).   *Answer*

Similarly, determinants can be used to solve a system of three linear equations in three variables:

$$a_1 x + b_1 y + c_1 z = d_1$$
$$a_2 x + b_2 y + c_2 z = d_2$$
$$a_3 x + b_3 y + c_3 z = d_3$$

Let

$$D = \begin{vmatrix} a_1 & b_1 & c_1 \\ a_2 & b_2 & c_2 \\ a_3 & b_3 & c_3 \end{vmatrix} \neq 0, \qquad D_x = \begin{vmatrix} d_1 & b_1 & c_1 \\ d_2 & b_2 & c_2 \\ d_3 & b_3 & c_3 \end{vmatrix},$$

$$D_y = \begin{vmatrix} a_1 & d_1 & c_1 \\ a_2 & d_2 & c_2 \\ a_3 & d_3 & c_3 \end{vmatrix}, \qquad \text{and} \qquad D_z = \begin{vmatrix} a_1 & b_1 & d_1 \\ a_2 & b_2 & d_2 \\ a_3 & b_3 & d_3 \end{vmatrix}.$$

Then $x = \dfrac{D_x}{D}$, $y = \dfrac{D_y}{D}$, and $z = \dfrac{D_z}{D}$.

This method can be generalized, as indicated below. Known as Cramer's Rule, it was named for the Swiss mathematician Gabriel Cramer (1704–1752).

Cramer's Rule

The solution of a system of n linear equations in n variables is given by

$$x = \frac{D_x}{D}, \; y = \frac{D_y}{D}, \ldots,$$

where D is the determinant of the matrix of coefficients of the variables ($D \neq 0$) and D_x, D_y, $\ldots$, are derived from D by replacing the coefficients of x, y, $\ldots$, respectively, by the constants.

Example 2 Use Cramer's rule to solve:
$$x - 3y - 2z = 9$$
$$3x + 2y + 6z = 20$$
$$4x - y + 3z = 25$$

Solution

$$D = \begin{vmatrix} 1 & -3 & -2 \\ 3 & 2 & 6 \\ 4 & -1 & 3 \end{vmatrix} = 1(6 + 6) + 3(9 - 24) - 2(-3 - 8) = -11$$

$$D_x = \begin{vmatrix} 9 & -3 & -2 \\ 20 & 2 & 6 \\ 25 & -1 & 3 \end{vmatrix} = 9(6 + 6) + 3(60 - 150) - 2(-20 - 50) = -22$$

$$D_y = \begin{vmatrix} 1 & 9 & -2 \\ 3 & 20 & 6 \\ 4 & 25 & 3 \end{vmatrix} = 1(60 - 150) - 9(9 - 24) - 2(75 - 80) = 55$$

$$D_z = \begin{vmatrix} 1 & -3 & 9 \\ 3 & 2 & 20 \\ 4 & -1 & 25 \end{vmatrix} = 1(50 + 20) + 3(75 - 80) + 9(-3 - 8) = -44$$

$$x = \frac{D_x}{D} = \frac{-22}{-11} = 2 \qquad y = \frac{D_y}{D} = \frac{55}{-11} = -5 \qquad z = \frac{D_z}{D} = \frac{-44}{-11} = 4$$

$\therefore$ the solution of the system is $(2, -5, 4)$. **Answer**

Cramer's rule provides a method for finding a solution of a system of n linear equations in n variables. Recall that such a system has a unique solution if and only if $D \neq 0$. When $D = 0$ the system may have no solution or it may have infinitely many solutions.

To distinguish between inconsistent and dependent equations, you must consider the value of D_y. If $D = 0$ and $D_y \neq 0$, the equations are inconsistent and their graphs are parallel. If $D = 0$ and $D_y = 0$, the equations are dependent and their graphs coincide. The examples below illustrate systems with consistent equations, inconsistent equations, and dependent equations.

$x + y = 1$
$2x + y = 4$

$2x - y = -1$
$2x - y = 4$

$3x + 2y = 2$
$6x + 4y = 4$

$D \neq 0$

Consistent equations;
graphs intersect.

$D = 0$
$D_y \neq 0$

Inconsistent equations;
graphs are parallel.

$D = 0$
$D_y = 0$

Dependent equations;
graphs coincide.

Example 3 Determine the number of solutions for each system.

  a. $2x + y = -6$
       $x - 3y = 9$

  b. $x - 2y = 3$
       $-2x + 4y = 1$

Solution **a.** $D = \begin{vmatrix} 2 & 1 \\ 1 & -3 \end{vmatrix} = -6 - 1 = -7 \neq 0$

Since $D \neq 0$, the equations are consistent,
$\therefore$ the system has one solution. **Answer**

b. $D = \begin{vmatrix} 1 & -2 \\ -2 & 4 \end{vmatrix} = 4 - 4 = 0 \qquad D_y = \begin{vmatrix} 1 & 3 \\ -2 & 1 \end{vmatrix} = 1 - (-6) = 7$

Since $D = 0$ and $D_y \neq 0$, the equations are inconsistent.
$\therefore$ the system has no solution. **Answer**

For each system give the entries for columns one, two, and three of the determinant D.

1. $x + y + z = 2$
$x + 2y + 2z = -4$
$2x + 2y - z = 1$

2. $-x + 3y + 2z = 4$
$2x - y + z = 2$
$4x + 2y - 2z = 0$

3. $2x \qquad - 3z = 0$
$3x + y \qquad = 15$
$4y + z = -8$

4. $5x - y \qquad = -17$
$-2x + 3y + 3z = 12$
$z = 0$

Give the elements in columns one, two, and three of the indicated determinants.

5. Exercise 1, D_x
6. Exercise 1, D_y
7. Exercise 1, D_z

8. Exercise 2, D_x
9. Exercise 2, D_y
10. Exercise 2, D_z

11. Exercise 3, D_x
12. Exercise 3, D_y
13. Exercise 3, D_z

14. Exercise 4, D_x
15. Exercise 4, D_y
16. Exercise 4, D_z

Written Exercises

Use Cramer's rule to solve each system.

A

1. $x - 5y = 2$
$2x + y = 4$

2. $2x + 3y = 7$
$3x + 4y = 10$

3. $2x + y - z = 2$
$x + y + z = 7$
$x + 2y + z = 4$

4. $x - 3y - 2z = 9$
$3x + 2y + 6z = 20$
$4x - y + 3z = 25$

5. $x + y - z = 0$
$2x + y - 2z = 1$
$3x \qquad - 4z = 1$

6. $x + y + z = -2$
$2x - y - z = -1$
$3x + 2y + 4z = -15$

7. $x - y + z = 3$
$y + z = 3$
$-y + z = 1$

8. $x + 3y + z = 0$
$x \qquad + 4z = -2$
$-6y + z = 1$

9. $2x + y - z = 3$
$4x - y + 4z = 0$
$-3y + 2z = 6$

10. $x - 2y \qquad = 8$
$x \qquad + 2z = 3$
$-y - 2z = 2$

Determine whether each system has no solution, one solution, or more than one solution.

11. $2x - y = 5$
$-2x + y = -1$

12. $3x + 9y = 14$
$6x + 18y = 28$

13. $4x + 3y = 16$
$8x + 6y = 10$

14. $5x - 9y = 15$
$-2x + 5y = 7$

Use Cramer's rule to solve each system.

B **15.** $2a - 3b + c - 3d = -1$
$a + 2b + 3c - d = 1$
$3a + 5b + 6c = 4$
$3a - b - 2d = 6$

16. $p - 2q + r = -4$
$2p - r + s = -3$
$q + 2r - 3s = -14$
$3p - 3q + 2r = -13$

Prove the following statements about this system: $y = m_1x + b_1$
$y = m_2x + b_2$

C **17.** $D \neq 0$ if and only if $m_1 \neq m_2$.

18. $D = 0$ and $D_y \neq 0$ if and only if $m_1 = m_2 \neq 0$ and $b_1 \neq b_2$.

19. $D = 0$ and $D_y = 0$ if and only if $m_1 = m_2$ and $b_1 = b_2$.

Mixed Review Exercises

Perform the indicated matrix operations.

1. $\begin{bmatrix} 5 & 2 \\ 3 & -1 \end{bmatrix} + \begin{bmatrix} 1 & -2 \\ 0 & 5 \end{bmatrix}$

2. $4\begin{bmatrix} 2 & -1 \\ 3 & 0 \end{bmatrix}$

3. $\begin{bmatrix} -1 & 2 \\ -2 & 3 \end{bmatrix}\begin{bmatrix} 0 & 1 \\ 4 & 1 \end{bmatrix}$

Graph each function.

4. $y = \frac{1}{2} \sin 4x$

5. $y = \frac{2}{3}x - 4$

6. $y = -2x^2 + 8x - 7$

Self-Test 3

Vocabulary   minor of an element (p. 794)   expansion of a determinant (p. 795)

1. Expand by minors: $\begin{vmatrix} 3 & 0 & 0 \\ 0 & 2 & 1 \\ -2 & 4 & -1 \end{vmatrix}$ Obj. 16-7, p. 794

2. Evaluate: $\begin{vmatrix} 2 & 1 & 0 & -2 \\ -1 & 0 & 0 & 4 \\ 2 & 3 & 6 & -1 \\ 0 & 1 & 2 & 2 \end{vmatrix}$ Obj. 16-8, p. 798

3. Use Cramer's rule to solve: $2x + y - z = 1$
$x + 3y + z = 10$
$3x - 2y - 8z = -3$ Obj. 16-9, p. 801

Chapter Summary

1. Two *matrices* are equal if and only if they have the same *dimensions* and all corresponding *elements* are equal.

2. Matrices having the same dimensions can be added (or subtracted) by adding (or subtracting) corresponding elements. The properties of addition of matrices are listed on page 771.

3. In matrix algebra, a real number is called a *scalar*. The *scalar product* of a real number r and a matrix A is the matrix rA, each of whose elements is r times the corresponding element of A. The properties of scalar multiplication are listed on page 772.

4. The product of matrices $A_{m \times n}$ and $B_{n \times p}$ is the $m \times p$ matrix whose element in the ath row and bth column is the sum of the products of corresponding elements of the ath row of A and the bth column of B. Unlike multiplication of real numbers, matrix multiplication is not commutative. Moreover, $A \times B = O$ does not imply that $A = O$ or $B = O$. The properties of matrix multiplication are listed on page 776.

5. Problems in the physical and social sciences can be solved using matrices.

6. For each $n \times n$ matrix A there is a real number called the *determinant* of A. For 2×2 matrices of the form $\begin{bmatrix} a & b \\ c & d \end{bmatrix}$, the determinant is $ad - bc$. Determinants for higher-order matrices can be defined and expanded by minors about the elements of any row or column. Determinants have properties that can simplify expansion by minors. These properties are listed on pages 798 and 799.

7. If $AB = BA = I$, matrices A and B are said to be inverses of each other. If $A = \begin{bmatrix} a & b \\ c & d \end{bmatrix}$ and det $A \neq 0$, then $A^{-1} = \dfrac{1}{\det A} \begin{bmatrix} d & -b \\ -c & a \end{bmatrix}$.

8. The solution of a system of n linear equations in n variables is given by $x = \dfrac{D_x}{D}$, $y = \dfrac{D_y}{D}$, . . . , where D is the determinant of the matrix of coefficients of the variables $(D \neq 0)$ and D_x, D_y, . . . , are derived from D by replacing the coefficients of x, y, . . . , respectively, by the constants. This method of solution is called *Cramer's rule*.

Chapter Review

Write the letter of the correct answer.

1. Solve for x and y: $\begin{bmatrix} 4 - x & 0 & 3y \end{bmatrix} = O_{1 \times 3}$ 16-1
   - **a.** $x = 4$ $y = -3$
   - **b.** $x = -4$ $y = 0$
   - **c.** $x = -4$ $y = 3$
   - **d.** $x = 4$ $y = 0$

In Exercises 2 and 3, let $A = \begin{bmatrix} -2 & 3 \\ -1 & 6 \end{bmatrix}$ **and** $B = \begin{bmatrix} 3 & -2 \\ -1 & 4 \end{bmatrix}$.

2. Find $B - 2A$.

16-2

    a. $\begin{bmatrix} 5 & -5 \\ 0 & -2 \end{bmatrix}$
    b. $\begin{bmatrix} 7 & -8 \\ 1 & -8 \end{bmatrix}$
    c. $\begin{bmatrix} 7 & -3 \\ 1 & -8 \end{bmatrix}$
    d. $\begin{bmatrix} 1 & 3 \\ -3 & 8 \end{bmatrix}$

3. Find BA.

16-3

    a. $\begin{bmatrix} -9 & -16 \\ 19 & -22 \end{bmatrix}$
    b. $\begin{bmatrix} -3 & 8 \\ -9 & -22 \end{bmatrix}$
    c. $\begin{bmatrix} -4 & -3 \\ -2 & 21 \end{bmatrix}$
    d. $\begin{bmatrix} -9 & -16 \\ -9 & -26 \end{bmatrix}$

4. Find the matrix that shows the number of ways a message can be sent from one point to another using exactly one relay.

16-4

    a. $\begin{bmatrix} 1 & 0 & 1 \\ 1 & 1 & 0 \\ 0 & 1 & 0 \end{bmatrix}$
    b. $\begin{bmatrix} 1 & 1 & 0 \\ 0 & 1 & 1 \\ 1 & 0 & 0 \end{bmatrix}$
    c. $\begin{bmatrix} 0 & 1 & 0 \\ 1 & 0 & 1 \\ 1 & 0 & 0 \end{bmatrix}$
    d. $\begin{bmatrix} 0 & 0 & 1 \\ 1 & 1 & 0 \\ 0 & 1 & 1 \end{bmatrix}$

5. Evaluate $\begin{vmatrix} 4 & 2 \\ -1 & 3 \end{vmatrix}$.

16-5

    a. -14
    b. 10
    c. -10
    d. 14

6. Evaluate $\begin{vmatrix} -1 & 1 & 0 \\ 1 & 2 & 2 \\ 1 & 0 & 0 \end{vmatrix}$.

    a. -1
    b. 2
    c. 3
    d. -2

7. Find the inverse of $\begin{bmatrix} -2 & 1 \\ 0 & 6 \end{bmatrix}$.

16-6

    a. $\begin{bmatrix} -\frac{1}{2} & \frac{1}{12} \\ 0 & \frac{1}{6} \end{bmatrix}$
    b. $\begin{bmatrix} \frac{1}{2} & \frac{1}{12} \\ 0 & \frac{1}{6} \end{bmatrix}$
    c. $\begin{bmatrix} 6 & 0 \\ -1 & -2 \end{bmatrix}$
    d. $\begin{bmatrix} \frac{1}{2} & -\frac{1}{12} \\ 0 & -\frac{1}{6} \end{bmatrix}$

8. Expand this determinant by minors about the first row: $\begin{vmatrix} -1 & 0 & 1 \\ 2 & 1 & 0 \\ 1 & -1 & 3 \end{vmatrix}$

16-7

    a. 0
    b. 12
    c. -6
    d. 6

9. Evaluate the determinant $\begin{vmatrix} 3 & 2 & -2 \\ 1 & 0 & 2 \\ -12 & 3 & 8 \end{vmatrix}$.

16-8

    a. 40
    b. -40
    c. -44
    d. -88

10. Use Cramer's rule to solve this system:
$$\begin{aligned} 2x + y + 3z &= -2 \\ 5x + 2y &= 5 \\ 2y + 3z &= -13 \end{aligned}$$

16-9

    a. $\{(3, -5, -1)\}$
    b. $\{(3, 5, -1)\}$
    c. $\{(3, 5, 1)\}$
    d. $\{(-3, -5, 1)\}$

Chapter Test

1. Find the value x and y: $O_{1\times3} = \begin{bmatrix} 2x - 6 \\ 0 \\ y + 4 \end{bmatrix}$

16-1

In Exercises 2–7, let $A = \begin{bmatrix} 2 & 1 & 3 \\ 0 & -1 & 1 \\ 4 & -1 & 0 \end{bmatrix}$ **and** $B = \begin{bmatrix} -2 & 1 & 5 \\ 5 & -5 & 1 \\ 0 & 3 & -1 \end{bmatrix}$.

Express the following as a single matrix.

2. $A + B$ **3.** $B - A$ 16-2

4. $4A$ **5.** $A - 2B$

6. AB **7.** B^2 16-3

8. Find the number of routes that a message might take from point A to point B using at most one relay point.

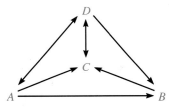

16-4

Evaluate these determinants.

9. $\begin{vmatrix} 6 & 3 \\ -2 & -2 \end{vmatrix}$ **10.** $\begin{vmatrix} 1 & 0 & -1 \\ 3 & 1 & 1 \\ 0 & 2 & 2 \end{vmatrix}$ 16-5

11. Solve this equation for A: $\begin{bmatrix} 2 & 3 \\ -1 & 0 \end{bmatrix} A = \begin{bmatrix} 7 & 3 \\ 4 & 0 \end{bmatrix}$ 16-6

12. Expand this determinant by minors about the third row: $\begin{vmatrix} 3 & 0 & -1 \\ 2 & -1 & 1 \\ 1 & -1 & 0 \end{vmatrix}$ 16-7

Use the properties of determinants to evaluate these determinants.

13. $\begin{vmatrix} 10 & 5 & -5 \\ 16 & 8 & -8 \\ 3 & 0 & -4 \end{vmatrix}$ **14.** $\begin{vmatrix} 4 & 3 & -1 \\ -3 & -3 & 5 \\ 5 & 5 & 3 \end{vmatrix}$ 16-8

15. Use Cramer's rule to solve this system:
$$\begin{aligned} 4x - 3y &= 1 \\ 6x \quad\quad - 8z &= 1 \\ y - 2z &= 0 \end{aligned}$$

16-9

Preparing for College Entrance Exams

Strategy for Success

It usually takes less time to solve an equation than to substitute all of the possible solutions. However, if you can eliminate possibilities that have the wrong sign, that are obviously too large or too small, or that do not meet the stated or implied conditions of the problem, then it may be quicker to substitute the remaining choices in the given equation to find the correct solution.

Decide which is the best of the choices given and write the corresponding letter on your answer sheet.

1. If $A = \begin{bmatrix} 1 & -1 \\ 2 & 4 \end{bmatrix}$, for which matrix X is $AX \neq XA$?

(A) $\begin{bmatrix} 1 & -1 \\ 2 & 4 \end{bmatrix}$　　　　**(B)** $\begin{bmatrix} 1 & 0 \\ 0 & 1 \end{bmatrix}$　　　　**(C)** $\begin{bmatrix} 0 & 0 \\ 0 & 0 \end{bmatrix}$

(D) $\begin{bmatrix} 4 & 2 \\ -1 & 1 \end{bmatrix}$　　　　**(E)** $\begin{bmatrix} \frac{2}{3} & \frac{1}{6} \\ -\frac{1}{3} & \frac{1}{6} \end{bmatrix}$

2. Find the inverse of $\begin{bmatrix} 5 & -6 \\ 2 & 4 \end{bmatrix}$ if it exists.

(A) $\begin{bmatrix} \frac{5}{32} & \frac{1}{16} \\ -\frac{3}{16} & \frac{1}{8} \end{bmatrix}$　　　　**(B)** $\begin{bmatrix} \frac{1}{8} & \frac{1}{16} \\ -\frac{3}{16} & \frac{5}{32} \end{bmatrix}$　　　　**(C)** $\begin{bmatrix} \frac{1}{8} & \frac{3}{16} \\ -\frac{1}{16} & \frac{5}{32} \end{bmatrix}$

(D) $\begin{bmatrix} \frac{1}{8} & -\frac{3}{16} \\ \frac{1}{16} & \frac{5}{32} \end{bmatrix}$　　　　**(E)** does not exist

3. In how many ways can the letters in the word INITIAL be arranged?

(A) 120　　　　**(B)** 840　　　　**(C)** 1680　　　　**(D)** 2520　　　　**(E)** 5040

4. Two dice are tossed. Find the probability that both the sum and the product of the numbers are less than six.

(A) $\frac{5}{18}$　　　　**(B)** $\frac{2}{9}$　　　　**(C)** $\frac{1}{3}$　　　　**(D)** $\frac{1}{6}$　　　　**(E)** $\frac{1}{4}$

5. There are 80 sophomores, 68 juniors, and 72 seniors enrolled at a high school. In how many ways can a 7-member student council be formed if two representatives are chosen from each grade and then the president is elected from the entire student body?

(A) $_{80}C_2 + _{68}C_2 + _{72}C_2 + _{220}C_1$　　　　**(B)** $_{80}C_2 \cdot _{68}C_2 \cdot _{72}C_2 \cdot _{220}C_1$

(C) $_{80}P_2 + _{68}P_2 + _{72}P_2 + _{214}P_1$　　　　**(D)** $_{80}P_2 \cdot _{68}P_2 \cdot _{72}P_2 \cdot _{214}P_1$

(E) $_{80}C_2 \cdot _{68}C_2 \cdot _{72}C_2 \cdot _{214}C_1$

Table 1 / *Squares and Square Roots*

N	N²	√N	√10N	N	N²	√N	√10N
1.0	1.00	1.000	3.162	5.5	30.25	2.345	7.416
1.1	1.21	1.049	3.317	5.6	31.36	2.366	7.483
1.2	1.44	1.095	3.464	5.7	32.49	2.387	7.550
1.3	1.69	1.140	3.606	5.8	33.64	2.408	7.616
1.4	1.96	1.183	3.742	5.9	34.81	2.429	7.681
1.5	2.25	1.225	3.873	6.0	36.00	2.449	7.746
1.6	2.56	1.265	4.000	6.1	37.21	2.470	7.810
1.7	2.89	1.304	4.123	6.2	38.44	2.490	7.874
1.8	3.24	1.342	4.243	6.3	39.69	2.510	7.937
1.9	3.61	1.378	4.359	6.4	40.96	2.530	8.000
2.0	4.00	1.414	4.472	6.5	42.25	2.550	8.062
2.1	4.41	1.449	4.583	6.6	43.56	2.569	8.124
2.2	4.84	1.483	4.690	6.7	44.89	2.588	8.185
2.3	5.29	1.517	4.796	6.8	46.24	2.608	8.246
2.4	5.76	1.549	4.899	6.9	47.61	2.627	8.307
2.5	6.25	1.581	5.000	7.0	49.00	2.646	8.367
2.6	6.76	1.612	5.099	7.1	50.41	2.665	8.426
2.7	7.29	1.643	5.196	7.2	51.84	2.683	8.485
2.8	7.84	1.673	5.292	7.3	53.29	2.702	8.544
2.9	8.41	1.703	5.385	7.4	54.76	2.720	8.602
3.0	9.00	1.732	5.477	7.5	56.25	2.739	8.660
3.1	9.61	1.761	5.568	7.6	57.76	2.757	8.718
3.2	10.24	1.789	5.657	7.7	59.29	2.775	8.775
3.3	10.89	1.817	5.745	7.8	60.84	2.793	8.832
3.4	11.56	1.844	5.831	7.9	62.41	2.811	8.888
3.5	12.25	1.871	5.916	8.0	64.00	2.828	8.944
3.6	12.96	1.897	6.000	8.1	65.61	2.846	9.000
3.7	13.69	1.924	6.083	8.2	67.24	2.864	9.055
3.8	14.44	1.949	6.164	8.3	68.89	2.881	9.110
3.9	15.21	1.975	6.245	8.4	70.56	2.898	9.165
4.0	16.00	2.000	6.325	8.5	72.25	2.915	9.220
4.1	16.81	2.025	6.403	8.6	73.96	2.933	9.274
4.2	17.64	2.049	6.481	8.7	75.69	2.950	9.327
4.3	18.49	2.074	6.557	8.8	77.44	2.966	9.381
4.4	19.36	2.098	6.633	8.9	79.21	2.983	9.434
4.5	20.25	2.121	6.708	9.0	81.00	3.000	9.487
4.6	21.16	2.145	6.782	9.1	82.81	3.017	9.539
4.7	22.09	2.168	6.856	9.2	84.64	3.033	9.592
4.8	23.04	2.191	6.928	9.3	86.49	3.050	9.644
4.9	24.01	2.214	7.000	9.4	88.36	3.066	9.695
5.0	25.00	2.236	7.071	9.5	90.25	3.082	9.747
5.1	26.01	2.258	7.141	9.6	92.16	3.098	9.798
5.2	27.04	2.280	7.211	9.7	94.09	3.114	9.849
5.3	28.09	2.302	7.280	9.8	96.04	3.130	9.899
5.4	29.16	2.324	7.348	9.9	98.01	3.146	9.950
5.5	30.25	2.345	7.416	10	100.00	3.162	10.000

Table 2 / Cubes and Cube Roots

N	N^3	$\sqrt[3]{N}$	$\sqrt[3]{10N}$	$\sqrt[3]{100N}$	N	N^3	$\sqrt[3]{N}$	$\sqrt[3]{10N}$	$\sqrt[3]{100N}$
1.0	1.000	1.000	2.154	4.642	5.5	166.375	1.765	3.803	8.193
1.1	1.331	1.032	2.224	4.791	5.6	175.616	1.776	3.826	8.243
1.2	1.728	1.063	2.289	4.932	5.7	185.193	1.786	3.849	8.291
1.3	2.197	1.091	2.351	5.066	5.8	195.112	1.797	3.871	8.340
1.4	2.744	1.119	2.410	5.192	5.9	205.379	1.807	3.893	8.387
1.5	3.375	1.145	2.466	5.313	6.0	216.000	1.817	3.915	8.434
1.6	4.096	1.170	2.520	5.429	6.1	226.981	1.827	3.936	8.481
1.7	4.913	1.193	2.571	5.540	6.2	238.328	1.837	3.958	8.527
1.8	5.832	1.216	2.621	5.646	6.3	250.047	1.847	3.979	8.573
1.9	6.859	1.239	2.668	5.749	6.4	262.144	1.857	4.000	8.618
2.0	8.000	1.260	2.714	5.848	6.5	274.625	1.866	4.021	8.662
2.1	9.261	1.281	2.759	5.944	6.6	287.496	1.876	4.041	8.707
2.2	10.648	1.301	2.802	6.037	6.7	300.763	1.885	4.062	8.750
2.3	12.167	1.320	2.844	6.127	6.8	314.432	1.895	4.082	8.794
2.4	13.824	1.339	2.884	6.214	6.9	328.509	1.904	4.102	8.837
2.5	15.625	1.357	2.924	6.300	7.0	343.000	1.913	4.121	8.879
2.6	17.576	1.375	2.962	6.383	7.1	357.911	1.922	4.141	8.921
2.7	19.683	1.392	3.000	6.463	7.2	373.248	1.931	4.160	8.963
2.8	21.952	1.409	3.037	6.542	7.3	389.017	1.940	4.179	9.004
2.9	24.389	1.426	3.072	6.619	7.4	405.224	1.949	4.198	9.045
3.0	27.000	1.442	3.107	6.694	7.5	421.875	1.957	4.217	9.086
3.1	29.791	1.458	3.141	6.768	7.6	438.976	1.966	4.236	9.126
3.2	32.768	1.474	3.175	6.840	7.7	456.533	1.975	4.254	9.166
3.3	35.937	1.489	3.208	6.910	7.8	474.552	1.983	4.273	9.205
3.4	39.304	1.504	3.240	6.980	7.9	493.039	1.992	4.291	9.244
3.5	42.875	1.518	3.271	7.047	8.0	512.000	2.000	4.309	9.283
3.6	46.656	1.533	3.302	7.114	8.1	531.441	2.008	4.327	9.322
3.7	50.653	1.547	3.332	7.179	8.2	551.368	2.017	4.344	9.360
3.8	54.872	1.560	3.362	7.243	8.3	571.787	2.025	4.362	9.398
3.9	59.319	1.574	3.391	7.306	8.4	592.704	2.033	4.380	9.435
4.0	64.000	1.587	3.420	7.368	8.5	614.125	2.041	4.397	9.473
4.1	68.921	1.601	3.448	7.429	8.6	636.056	2.049	4.414	9.510
4.2	74.088	1.613	3.476	7.489	8.7	658.503	2.057	4.431	9.546
4.3	79.507	1.626	3.503	7.548	8.8	681.472	2.065	4.448	9.583
4.4	85.184	1.639	3.530	7.606	8.9	704.969	2.072	4.465	9.619
4.5	91.125	1.651	3.557	7.663	9.0	729.000	2.080	4.481	9.655
4.6	97.336	1.663	3.583	7.719	9.1	753.571	2.088	4.498	9.691
4.7	103.823	1.675	3.609	7.775	9.2	778.688	2.095	4.514	9.726
4.8	110.592	1.687	3.634	7.830	9.3	804.357	2.103	4.531	9.761
4.9	117.649	1.698	3.659	7.884	9.4	830.584	2.110	4.547	9.796
5.0	125.000	1.710	3.684	7.937	9.5	857.375	2.118	4.563	9.830
5.1	132.651	1.721	3.708	7.990	9.6	884.736	2.125	4.579	9.865
5.2	140.608	1.732	3.733	8.041	9.7	912.673	2.133	4.595	9.899
5.3	148.877	1.744	3.756	8.093	9.8	941.192	2.140	4.610	9.933
5.4	157.464	1.754	3.780	8.143	9.9	970.299	2.147	4.626	9.967
5.5	166.375	1.765	3.803	8.193	10	1000.000	2.154	4.642	10.000

Table 3 / *Common Logarithms of Numbers*

N	0	1	2	3	4	5	6	7	8	9
10	0000	0043	0086	0128	0170	0212	0253	0294	0334	0374
11	0414	0453	0492	0531	0569	0607	0645	0682	0719	0755
12	0792	0828	0864	0899	0934	0969	1004	1038	1072	1106
13	1139	1173	1206	1239	1271	1303	1335	1367	1399	1430
14	1461	1492	1523	1553	1584	1614	1644	1673	1703	1732
15	1761	1790	1818	1847	1875	1903	1931	1959	1987	2014
16	2041	2068	2095	2122	2148	2175	2201	2227	2253	2279
17	2304	2330	2355	2380	2405	2430	2455	2480	2504	2529
18	2553	2577	2601	2625	2648	2672	2695	2718	2742	2765
19	2788	2810	2833	2856	2878	2900	2923	2945	2967	2989
20	3010	3032	3054	3075	3096	3118	3139	3160	3181	3201
21	3222	3243	3263	3284	3304	3324	3345	3365	3385	3404
22	3424	3444	3464	3483	3502	3522	3541	3560	3579	3598
23	3617	3636	3655	3674	3692	3711	3729	3747	3766	3784
24	3802	3820	3838	3856	3874	3892	3909	3927	3945	3962
25	3979	3997	4014	4031	4048	4065	4082	4099	4116	4133
26	4150	4166	4183	4200	4216	4232	4249	4265	4281	4298
27	4314	4330	4346	4362	4378	4393	4409	4425	4440	4456
28	4472	4487	4502	4518	4533	4548	4564	4579	4594	4609
29	4624	4639	4654	4669	4683	4698	4713	4728	4742	4757
30	4771	4786	4800	4814	4829	4843	4857	4871	4886	4900
31	4914	4928	4942	4955	4969	4983	4997	5011	5024	5038
32	5051	5065	5079	5092	5105	5119	5132	5145	5159	5172
33	5185	5198	5211	5224	5237	5250	5263	5276	5289	5302
34	5315	5328	5340	5353	5366	5378	5391	5403	5416	5428
35	5441	5453	5465	5478	5490	5502	5514	5527	5539	5551
36	5563	5575	5587	5599	5611	5623	5635	5647	5658	5670
37	5682	5694	5705	5717	5729	5740	5752	5763	5775	5786
38	5798	5809	5821	5832	5843	5855	5866	5877	5888	5899
39	5911	5922	5933	5944	5955	5966	5977	5988	5999	6010
40	6021	6031	6042	6053	6064	6075	6085	6096	6107	6117
41	6128	6138	6149	6160	6170	6180	6191	6201	6212	6222
42	6232	6243	6253	6263	6274	6284	6294	6304	6314	6325
43	6335	6345	6355	6365	6375	6385	6395	6405	6415	6425
44	6435	6444	6454	6464	6474	6484	6493	6503	6513	6522
45	6532	6542	6551	6561	6571	6580	6590	6599	6609	6618
46	6628	6637	6646	6656	6665	6675	6684	6693	6702	6712
47	6721	6730	6739	6749	6758	6767	6776	6785	6794	6803
48	6812	6821	6830	6839	6848	6857	6866	6875	6884	6893
49	6902	6911	6920	6928	6937	6946	6955	6964	6972	6981
50	6990	6998	7007	7016	7024	7033	7042	7050	7059	7067
51	7076	7084	7093	7101	7110	7118	7126	7135	7143	7152
52	7160	7168	7177	7185	7193	7202	7210	7218	7226	7235
53	7243	7251	7259	7267	7275	7284	7292	7300	7308	7316
54	7324	7332	7340	7348	7356	7364	7372	7380	7388	7396

Table 3 / Common Logarithms of Numbers

N	0	1	2	3	4	5	6	7	8	9
55	7404	7412	7419	7427	7435	7443	7451	7459	7466	7474
56	7482	7490	7497	7505	7513	7520	7528	7536	7543	7551
57	7559	7566	7574	7582	7589	7597	7604	7612	7619	7627
58	7634	7642	7649	7657	7664	7672	7679	7686	7694	7701
59	7709	7716	7723	7731	7738	7745	7752	7760	7767	7774
60	7782	7789	7796	7803	7810	7818	7825	7832	7839	7846
61	7853	7860	7868	7875	7882	7889	7896	7903	7910	7917
62	7924	7931	7938	7945	7952	7959	7966	7973	7980	7987
63	7993	8000	8007	8014	8021	8028	8035	8041	8048	8055
64	8062	8069	8075	8082	8089	8096	8102	8109	8116	8122
65	8129	8136	8142	8149	8156	8162	8169	8176	8182	8189
66	8195	8202	8209	8215	8222	8228	8235	8241	8248	8254
67	8261	8267	8274	8280	8287	8293	8299	8306	8312	8319
68	8325	8331	8338	8344	8351	8357	8363	8370	8376	8382
69	8388	8395	8401	8407	8414	8420	8426	8432	8439	8445
70	8451	8457	8463	8470	8476	8482	8488	8494	8500	8506
71	8513	8519	8525	8531	8537	8543	8549	8555	8561	8567
72	8573	8579	8585	8591	8597	8603	8609	8615	8621	8627
73	8633	8639	8645	8651	8657	8663	8669	8675	8681	8686
74	8692	8698	8704	8710	8716	8722	8727	8733	8739	8745
75	8751	8756	8762	8768	8774	8779	8785	8791	8797	8802
76	8808	8814	8820	8825	8831	8837	8842	8848	8854	8859
77	8865	8871	8876	8882	8887	8893	8899	8904	8910	8915
78	8921	8927	8932	8938	8943	8949	8954	8960	8965	8971
79	8976	8982	8987	8993	8998	9004	9009	9015	9020	9025
80	9031	9036	9042	9047	9053	9058	9063	9069	9074	9079
81	9085	9090	9096	9101	9106	9112	9117	9122	9128	9133
82	9138	9143	9149	9154	9159	9165	9170	9175	9180	9186
83	9191	9196	9201	9206	9212	9217	9222	9227	9232	9238
84	9243	9248	9253	9258	9263	9269	9274	9279	9284	9289
85	9294	9299	9304	9309	9315	9320	9325	9330	9335	9340
86	9345	9350	9355	9360	9365	9370	9375	9380	9385	9390
87	9395	9400	9405	9410	9415	9420	9425	9430	9435	9440
88	9445	9450	9455	9460	9465	9469	9474	9479	9484	9489
89	9494	9499	9504	9509	9513	9518	9523	9528	9533	9538
90	9542	9547	9552	9557	9562	9566	9571	9576	9581	9586
91	9590	9595	9600	9605	9609	9614	9619	9624	9628	9633
92	9638	9643	9647	9652	9657	9661	9666	9671	9675	9680
93	9685	9689	9694	9699	9703	9708	9713	9717	9722	9727
94	9731	9736	9741	9745	9750	9754	9759	9763	9768	9773
95	9777	9782	9786	9791	9795	9800	9805	9809	9814	9818
96	9823	9827	9832	9836	9841	9845	9850	9854	9859	9863
97	9868	9872	9877	9881	9886	9890	9894	9899	9903	9908
98	9912	9917	9921	9926	9930	9934	9939	9943	9948	9952
99	9956	9961	9965	9969	9974	9978	9983	9987	9991	9996

Table 4 / *Trigonometric Functions of θ*

(θ in decimal degrees)

θ Degrees	θ Radians	sin θ	cos θ	tan θ	cot θ	sec θ	csc θ		
0.0	.0000	.0000	1.0000	.0000	undefined	1.000	undefined	1.5708	**90.0**
0.1	.0017	.0017	1.0000	.0017	573.0	1.000	573.0	1.5691	89.9
0.2	.0035	.0035	1.0000	.0035	286.5	1.000	286.5	1.5673	89.8
0.3	.0052	.0052	1.0000	.0052	191.0	1.000	191.0	1.5656	89.7
0.4	.0070	.0070	1.0000	.0070	143.2	1.000	143.2	1.5638	89.6
0.5	.0087	.0087	1.0000	.0087	114.6	1.000	114.6	1.5621	89.5
0.6	.0105	.0105	.9999	.0105	95.49	1.000	95.49	1.5603	89.4
0.7	.0122	.0122	.9999	.0122	81.85	1.000	81.85	1.5586	89.3
0.8	.0140	.0140	.9999	.0140	71.62	1.000	71.62	1.5568	89.2
0.9	.0157	.0157	.9999	.0157	63.66	1.000	63.66	1.5551	89.1
1.0	.0175	.0175	.9998	.0175	57.29	1.000	57.30	1.5533	**89.0**
1.1	.0192	.0192	.9998	.0192	52.08	1.000	52.09	1.5516	88.9
1.2	.0209	.0209	.9998	.0209	47.74	1.000	47.75	1.5499	88.8
1.3	.0227	.0227	.9997	.0227	44.07	1.000	44.08	1.5481	88.7
1.4	.0244	.0244	.9997	.0244	40.92	1.000	40.93	1.5464	88.6
1.5	.0262	.0262	.9997	.0262	38.19	1.000	38.20	1.5446	88.5
1.6	.0279	.0279	.9996	.0279	35.80	1.000	35.81	1.5429	88.4
1.7	.0297	.0297	.9996	.0297	33.69	1.000	33.71	1.5411	88.3
1.8	.0314	.0314	.9995	.0314	31.82	1.000	31.84	1.5394	88.2
1.9	.0332	.0332	.9995	.0332	30.14	1.001	30.16	1.5376	88.1
2.0	.0349	.0349	.9994	.0349	28.64	1.001	28.65	1.5359	**88.0**
2.1	.0367	.0366	.9993	.0367	27.27	1.001	27.29	1.5341	87.9
2.2	.0384	.0384	.9993	.0384	26.03	1.001	26.05	1.5324	87.8
2.3	.0401	.0401	.9992	.0402	24.90	1.001	24.92	1.5307	87.7
2.4	.0419	.0419	.9991	.0419	23.86	1.001	23.88	1.5289	87.6
2.5	.0436	.0436	.9990	.0437	22.90	1.001	22.93	1.5272	87.5
2.6	.0454	.0454	.9990	.0454	22.02	1.001	22.04	1.5254	87.4
2.7	.0471	.0471	.9989	.0472	21.20	1.001	21.23	1.5237	87.3
2.8	.0489	.0488	.9988	.0489	20.45	1.001	20.47	1.5219	87.2
2.9	.0506	.0506	.9987	.0507	19.74	1.001	19.77	1.5202	87.1
3.0	.0524	.0523	.9986	.0524	19.08	1.001	19.11	1.5184	**87.0**
3.1	.0541	.0541	.9985	.0542	18.46	1.001	18.49	1.5167	86.9
3.2	.0559	.0558	.9984	.0559	17.89	1.002	17.91	1.5149	86.8
3.3	.0576	.0576	.9983	.0577	17.34	1.002	17.37	1.5132	86.7
3.4	.0593	.0593	.9982	.0594	16.83	1.002	16.86	1.5115	86.6
3.5	.0611	.0610	.9981	.0612	16.35	1.002	16.38	1.5097	86.5
3.6	.0628	.0628	.9980	.0629	15.89	1.002	15.93	1.5080	86.4
3.7	.0646	.0645	.9979	.0647	15.46	1.002	15.50	1.5062	86.3
3.8	.0663	.0663	.9978	.0664	15.06	1.002	15.09	1.5045	86.2
3.9	.0681	.0680	.9977	.0682	14.67	1.002	14.70	1.5027	86.1
4.0	.0698	.0698	.9976	.0699	14.30	1.002	14.34	1.5010	**86.0**
4.1	.0716	.0715	.9974	.0717	13.95	1.003	13.99	1.4992	85.9
4.2	.0733	.0732	.9973	.0734	13.62	1.003	13.65	1.4975	85.8
4.3	.0750	.0750	.9972	.0752	13.30	1.003	13.34	1.4957	85.7
4.4	.0768	.0767	.9971	.0769	13.00	1.003	13.03	1.4940	85.6
4.5	.0785	.0785	.9969	.0787	12.71	1.003	12.75	1.4923	85.5
4.6	.0803	.0802	.9968	.0805	12.43	1.003	12.47	1.4905	85.4
4.7	.0820	.0819	.9966	.0822	12.16	1.003	12.20	1.4888	85.3
4.8	.0838	.0837	.9965	.0840	11.91	1.004	11.95	1.4870	85.2
4.9	.0855	.0854	.9963	.0857	11.66	1.004	11.71	1.4853	85.1
5.0	.0873	.0872	.9962	.0875	11.43	1.004	11.47	1.4835	**85.0**
5.1	.0890	.0889	.9960	.0892	11.20	1.004	11.25	1.4818	84.9
5.2	.0908	.0906	.9959	.0910	10.99	1.004	11.03	1.4800	84.8
5.3	.0925	.0924	.9957	.0928	10.78	1.004	10.83	1.4783	84.7
5.4	.0942	.0941	.9956	.0945	10.58	1.004	10.63	1.4765	84.6
5.5	.0960	.0958	.9954	.0963	10.39	1.005	10.43	1.4748	84.5
5.6	.0977	.0976	.9952	.0981	10.20	1.005	10.25	1.4731	84.4
5.7	.0995	.0993	.9951	.0998	10.02	1.005	10.07	1.4713	84.3
5.8	.1012	.1011	.9949	.1016	9.845	1.005	9.895	1.4696	84.2
5.9	.1030	.1028	.9947	.1033	9.677	1.005	9.728	1.4678	84.1
6.0	.1047	.1045	.9945	.1051	9.514	1.006	9.567	1.4661	**84.0**
		cos θ	sin θ	cot θ	tan θ	csc θ	sec θ	θ Radians	θ Degrees

Table 4 / *Trigonometric Functions of θ*
(θ in decimal degrees)

θ Degrees	θ Radians	sin θ	cos θ	tan θ	cot θ	sec θ	csc θ		
6.0	.1047	.1045	.9945	.1051	9.514	1.006	9.567	1.4661	**84.0**
6.1	.1065	.1063	.9943	.1069	9.357	1.006	9.411	1.4643	83.9
6.2	.1082	.1080	.9942	.1086	9.205	1.006	9.259	1.4626	83.8
6.3	.1100	.1097	.9940	.1104	9.058	1.006	9.113	1.4608	83.7
6.4	.1117	.1115	.9938	.1122	8.915	1.006	8.971	1.4591	83.6
6.5	.1134	.1132	.9936	.1139	8.777	1.006	8.834	1.4574	83.5
6.6	.1152	.1149	.9934	.1157	8.643	1.007	8.700	1.4556	83.4
6.7	.1169	.1167	.9932	.1175	8.513	1.007	8.571	1.4539	83.3
6.8	.1187	.1184	.9930	.1192	8.386	1.007	8.446	1.4521	83.2
6.9	.1204	.1201	.9928	.1210	8.264	1.007	8.324	1.4504	83.1
7.0	.1222	.1219	.9925	.1228	8.144	1.008	8.206	1.4486	**83.0**
7.1	.1239	.1236	.9923	.1246	8.028	1.008	8.091	1.4469	82.9
7.2	.1257	.1253	.9921	.1263	7.916	1.008	7.979	1.4451	82.8
7.3	.1274	.1271	.9919	.1281	7.806	1.008	7.870	1.4434	82.7
7.4	.1292	.1288	.9917	.1299	7.700	1.008	7.764	1.4416	82.6
7.5	.1309	.1305	.9914	.1317	7.596	1.009	7.661	1.4399	82.5
7.6	.1326	.1323	.9912	.1334	7.495	1.009	7.561	1.4382	82.4
7.7	.1344	.1340	.9910	.1352	7.396	1.009	7.463	1.4364	82.3
7.8	.1361	.1357	.9907	.1370	7.300	1.009	7.368	1.4347	82.2
7.9	.1379	.1374	.9905	.1388	7.207	1.010	7.276	1.4329	82.1
8.0	.1396	.1392	.9903	.1405	7.115	1.010	7.185	1.4312	**82.0**
8.1	.1414	.1409	.9900	.1423	7.026	1.010	7.097	1.4294	81.9
8.2	.1431	.1426	.9898	.1441	6.940	1.010	7.011	1.4277	81.8
8.3	.1449	.1444	.9895	.1459	6.855	1.011	6.927	1.4259	81.7
8.4	.1466	.1461	.9893	.1477	6.772	1.011	6.845	1.4242	81.6
8.5	.1484	.1478	.9890	.1495	6.691	1.011	6.765	1.4224	81.5
8.6	.1501	.1495	.9888	.1512	6.612	1.011	6.687	1.4207	81.4
8.7	.1518	.1513	.9885	.1530	6.535	1.012	6.611	1.4190	81.3
8.8	.1536	.1530	.9882	.1548	6.460	1.012	6.537	1.4172	81.2
8.9	.1553	.1547	.9880	.1566	6.386	1.012	6.464	1.4155	81.1
9.0	.1571	.1564	.9877	.1584	6.314	1.012	6.392	1.4137	**81.0**
9.1	.1588	.1582	.9874	.1602	6.243	1.013	6.323	1.4120	80.9
9.2	.1606	.1599	.9871	.1620	6.174	1.013	6.255	1.4102	80.8
9.3	.1623	.1616	.9869	.1638	6.107	1.013	6.188	1.4085	80.7
9.4	.1641	.1633	.9866	.1655	6.041	1.014	6.123	1.4067	80.6
9.5	.1658	.1650	.9863	.1673	5.976	1.014	6.059	1.4050	80.5
9.6	.1676	.1668	.9860	.1691	5.912	1.014	5.996	1.4032	80.4
9.7	.1693	.1685	.9857	.1709	5.850	1.015	5.935	1.4015	80.3
9.8	.1710	.1702	.9854	.1727	5.789	1.015	5.875	1.3998	80.2
9.9	.1728	.1719	.9851	.1745	5.730	1.015	5.816	1.3980	80.1
10.0	.1745	.1736	.9848	.1763	5.671	1.015	5.759	1.3963	**80.0**
10.1	.1763	.1754	.9845	.1781	5.614	1.016	5.702	1.3945	79.9
10.2	.1780	.1771	.9842	.1799	5.558	1.016	5.647	1.3928	79.8
10.3	.1798	.1788	.9839	.1817	5.503	1.016	5.593	1.3910	79.7
10.4	.1815	.1805	.9836	.1835	5.449	1.017	5.540	1.3893	79.6
10.5	.1833	.1822	.9833	.1853	5.396	1.017	5.487	1.3875	79.5
10.6	.1850	.1840	.9829	.1871	5.343	1.017	5.436	1.3858	79.4
10.7	.1868	.1857	.9826	.1890	5.292	1.018	5.386	1.3840	79.3
10.8	.1885	.1874	.9823	.1908	5.242	1.018	5.337	1.3823	79.2
10.9	.1902	.1891	.9820	.1926	5.193	1.018	5.288	1.3806	79.1
11.0	.1920	.1908	.9816	.1944	5.145	1.019	5.241	1.3788	**79.0**
11.1	.1937	.1925	.9813	.1962	5.097	1.019	5.194	1.3771	78.9
11.2	.1955	.1942	.9810	.1980	5.050	1.019	5.148	1.3753	78.8
11.3	.1972	.1959	.9806	.1998	5.005	1.020	5.103	1.3736	78.7
11.4	.1990	.1977	.9803	.2016	4.959	1.020	5.059	1.3718	78.6
11.5	.2007	.1994	.9799	.2035	4.915	1.020	5.016	1.3701	78.5
11.6	.2025	.2011	.9796	.2053	4.872	1.021	4.973	1.3683	78.4
11.7	.2042	.2028	.9792	.2071	4.829	1.021	4.931	1.3666	78.3
11.8	.2059	.2045	.9789	.2089	4.787	1.022	4.890	1.3648	78.2
11.9	.2077	.2062	.9785	.2107	4.745	1.022	4.850	1.3631	78.1
12.0	.2094	.2079	.9781	.2126	4.705	1.022	4.810	1.3614	**78.0**
		cos θ	sin θ	cot θ	tan θ	csc θ	sec θ	θ Radians	θ Degrees

Table 4 / *Trigonometric Functions of θ*

(θ in decimal degrees)

θ Degrees	θ Radians	sin θ	cos θ	tan θ	cot θ	sec θ	csc θ		
12.0	.2094	.2079	.9781	.2126	4.705	1.022	4.810	1.3614	**78.0**
12.1	.2112	.2096	.9778	.2144	4.665	1.023	4.771	1.3596	77.9
12.2	.2129	.2113	.9774	.2162	4.625	1.023	4.732	1.3579	77.8
12.3	.2147	.2130	.9770	.2180	4.586	1.023	4.694	1.3561	77.7
12.4	.2164	.2147	.9767	.2199	4.548	1.024	4.657	1.3544	77.6
12.5	.2182	.2164	.9763	.2217	4.511	1.024	4.620	1.3526	77.5
12.6	.2199	.2181	.9759	.2235	4.474	1.025	4.584	1.3509	77.4
12.7	.2217	.2198	.9755	.2254	4.437	1.025	4.549	1.3491	77.3
12.8	.2234	.2215	.9751	.2272	4.402	1.025	4.514	1.3474	77.2
12.9	.2251	.2233	.9748	.2290	4.366	1.026	4.479	1.3456	77.1
13.0	.2269	.2250	.9744	.2309	4.331	1.026	4.445	1.3439	**77.0**
13.1	.2286	.2267	.9740	.2327	4.297	1.027	4.412	1.3422	76.9
13.2	.2304	.2284	.9736	.2345	4.264	1.027	4.379	1.3404	76.8
13.3	.2321	.2300	.9732	.2364	4.230	1.028	4.347	1.3387	76.7
13.4	.2339	.2317	.9728	.2382	4.198	1.028	4.315	1.3369	76.6
13.5	.2356	.2334	.9724	.2401	4.165	1.028	4.284	1.3352	76.5
13.6	.2374	.2351	.9720	.2419	4.134	1.029	4.253	1.3334	76.4
13.7	.2391	.2368	.9715	.2438	4.102	1.029	4.222	1.3317	76.3
13.8	.2409	.2385	.9711	.2456	4.071	1.030	4.192	1.3299	76.2
13.9	.2426	.2402	.9707	.2475	4.041	1.030	4.163	1.3282	76.1
14.0	.2443	.2419	.9703	.2493	4.011	1.031	4.134	1.3265	**76.0**
14.1	.2461	.2436	.9699	.2512	3.981	1.031	4.105	1.3247	75.9
14.2	.2478	.2453	.9694	.2530	3.952	1.032	4.077	1.3230	75.8
14.3	.2496	.2470	.9690	.2549	3.923	1.032	4.049	1.3212	75.7
14.4	.2513	.2487	.9686	.2568	3.895	1.032	4.021	1.3195	75.6
14.5	.2531	.2504	.9681	.2586	3.867	1.033	3.994	1.3177	75.5
14.6	.2548	.2521	.9677	.2605	3.839	1.033	3.967	1.3160	75.4
14.7	.2566	.2538	.9673	.2623	3.812	1.034	3.941	1.3142	75.3
14.8	.2583	.2554	.9668	.2642	3.785	1.034	3.915	1.3125	75.2
14.9	.2601	.2571	.9664	.2661	3.758	1.035	3.889	1.3107	75.1
15.0	.2618	.2588	.9659	.2679	3.732	1.035	3.864	1.3090	**75.0**
15.1	.2635	.2605	.9655	.2698	3.706	1.036	3.839	1.3073	74.9
15.2	.2653	.2622	.9650	.2717	3.681	1.036	3.814	1.3055	74.8
15.3	.2670	.2639	.9646	.2736	3.655	1.037	3.790	1.3038	74.7
15.4	.2688	.2656	.9641	.2754	3.630	1.037	3.766	1.3020	74.6
15.5	.2705	.2672	.9636	.2773	3.606	1.038	3.742	1.3003	74.5
15.6	.2723	.2689	.9632	.2792	3.582	1.038	3.719	1.2985	74.4
15.7	.2740	.2706	.9627	.2811	3.558	1.039	3.695	1.2968	74.3
15.8	.2758	.2723	.9622	.2830	3.534	1.039	3.673	1.2950	74.2
15.9	.2775	.2740	.9617	.2849	3.511	1.040	3.650	1.2933	74.1
16.0	.2793	.2756	.9613	.2867	3.487	1.040	3.628	1.2915	**74.0**
16.1	.2810	.2773	.9608	.2886	3.465	1.041	3.606	1.2898	73.9
16.2	.2827	.2790	.9603	.2905	3.442	1.041	3.584	1.2881	73.8
16.3	.2845	.2807	.9598	.2924	3.420	1.042	3.563	1.2863	73.7
16.4	.2862	.2823	.9593	.2943	3.398	1.042	3.542	1.2846	73.6
16.5	.2880	.2840	.9588	.2962	3.376	1.043	3.521	1.2828	73.5
16.6	.2897	.2857	.9583	.2981	3.354	1.043	3.500	1.2811	73.4
16.7	.2915	.2874	.9578	.3000	3.333	1.044	3.480	1.2793	73.3
16.8	.2932	.2890	.9573	.3019	3.312	1.045	3.460	1.2776	73.2
16.9	.2950	.2907	.9568	.3038	3.291	1.045	3.440	1.2758	73.1
17.0	.2967	.2924	.9563	.3057	3.271	1.046	3.420	1.2741	**73.0**
17.1	.2985	.2940	.9558	.3076	3.251	1.046	3.401	1.2723	72.9
17.2	.3002	.2957	.9553	.3096	3.230	1.047	3.382	1.2706	72.8
17.3	.3019	.2974	.9548	.3115	3.211	1.047	3.363	1.2689	72.7
17.4	.3037	.2990	.9542	.3134	3.191	1.048	3.344	1.2671	72.6
17.5	.3054	.3007	.9537	.3153	3.172	1.049	3.326	1.2654	72.5
17.6	.3072	.3024	.9532	.3172	3.152	1.049	3.307	1.2636	72.4
17.7	.3089	.3040	.9527	.3191	3.133	1.050	3.289	1.2619	72.3
17.8	.3107	.3057	.9521	.3211	3.115	1.050	3.271	1.2601	72.2
17.9	.3124	.3074	.9516	.3230	3.096	1.051	3.254	1.2584	72.1
18.0	.3142	.3090	.9511	.3249	3.078	1.051	3.236	1.2566	**72.0**
		cos θ	sin θ	cot θ	tan θ	csc θ	sec θ	θ Radians	θ Degrees

Table 4 / *Trigonometric Functions of θ*

(θ in decimal degrees)

θ Degrees	θ Radians	sin θ	cos θ	tan θ	cot θ	sec θ	csc θ		
18.0	.3142	.3090	.9511	.3249	3.078	1.051	3.236	1.2566	**72.0**
18.1	.3159	.3107	.9505	.3269	3.060	1.052	3.219	1.2549	71.9
18.2	.3177	.3123	.9500	.3288	3.042	1.053	3.202	1.2531	71.8
18.3	.3194	.3140	.9494	.3307	3.024	1.053	3.185	1.2514	71.7
18.4	.3211	.3156	.9489	.3327	3.006	1.054	3.168	1.2497	71.6
18.5	.3229	.3173	.9483	.3346	2.989	1.054	3.152	1.2479	71.5
18.6	.3246	.3190	.9478	.3365	2.971	1.055	3.135	1.2462	71.4
18.7	.3264	.3206	.9472	.3385	2.954	1.056	3.119	1.2444	71.3
18.8	.3281	.3223	.9466	.3404	2.937	1.056	3.103	1.2427	71.2
18.9	.3299	.3239	.9461	.3424	2.921	1.057	3.087	1.2409	71.1
19.0	.3316	.3256	.9455	.3443	2.904	1.058	3.072	1.2392	**71.0**
19.1	.3334	.3272	.9449	.3463	2.888	1.058	3.056	1.2374	70.9
19.2	.3351	.3289	.9444	.3482	2.872	1.059	3.041	1.2357	70.8
19.3	.3368	.3305	.9438	.3502	2.856	1.060	3.026	1.2339	70.7
19.4	.3386	.3322	.9432	.3522	2.840	1.060	3.011	1.2322	70.6
19.5	.3403	.3338	.9426	.3541	2.824	1.061	2.996	1.2305	70.5
19.6	.3421	.3355	.9421	.3561	2.808	1.062	2.981	1.2287	70.4
19.7	.3438	.3371	.9415	.3581	2.793	1.062	2.967	1.2270	70.3
19.8	.3456	.3387	.9409	.3600	2.778	1.063	2.952	1.2252	70.2
19.9	.3473	.3404	.9403	.3620	2.762	1.064	2.938	1.2235	70.1
20.0	.3491	.3420	.9397	.3640	2.747	1.064	2.924	1.2217	**70.0**
20.1	.3508	.3437	.9391	.3659	2.733	1.065	2.910	1.2200	69.9
20.2	.3526	.3453	.9385	.3679	2.718	1.066	2.896	1.2182	69.8
20.3	.3543	.3469	.9379	.3699	2.703	1.066	2.882	1.2165	69.7
20.4	.3560	.3486	.9373	.3719	2.689	1.067	2.869	1.2147	69.6
20.5	.3578	.3502	.9367	.3739	2.675	1.068	2.855	1.2130	69.5
20.6	.3595	.3518	.9361	.3759	2.660	1.068	2.842	1.2113	69.4
20.7	.3613	.3535	.9354	.3779	2.646	1.069	2.829	1.2095	69.3
20.8	.3630	.3551	.9348	.3799	2.633	1.070	2.816	1.2078	69.2
20.9	.3648	.3567	.9342	.3819	2.619	1.070	2.803	1.2060	69.1
21.0	.3665	.3584	.9336	.3839	2.605	1.071	2.790	1.2043	**69.0**
21.1	.3683	.3600	.9330	.3859	2.592	1.072	2.778	1.2025	68.9
21.2	.3700	.3616	.9323	.3879	2.578	1.073	2.765	1.2008	68.8
21.3	.3718	.3633	.9317	.3899	2.565	1.073	2.753	1.1991	68.7
21.4	.3735	.3649	.9311	.3919	2.552	1.074	2.741	1.1973	68.6
21.5	.3752	.3665	.9304	.3939	2.539	1.075	2.729	1.1956	68.5
21.6	.3770	.3681	.9298	.3959	2.526	1.076	2.716	1.1938	68.4
21.7	.3787	.3697	.9291	.3979	2.513	1.076	2.705	1.1921	68.3
21.8	.3805	.3714	.9285	.4000	2.500	1.077	2.693	1.1903	68.2
21.9	.3822	.3730	.9278	.4020	2.488	1.078	2.681	1.1886	68.1
22.0	.3840	.3746	.9272	.4040	2.475	1.079	2.669	1.1868	**68.0**
22.1	.3857	.3762	.9265	.4061	2.463	1.079	2.658	1.1851	67.9
22.2	.3875	.3778	.9259	.4081	2.450	1.080	2.647	1.1833	67.8
22.3	.3892	.3795	.9252	.4101	2.438	1.081	2.635	1.1816	67.7
22.4	.3910	.3811	.9245	.4122	2.426	1.082	2.624	1.1798	67.6
22.5	.3927	.3827	.9239	.4142	2.414	1.082	2.613	1.1781	67.5
22.6	.3944	.3843	.9232	.4163	2.402	1.083	2.602	1.1764	67.4
22.7	.3962	.3859	.9225	.4183	2.391	1.084	2.591	1.1746	67.3
22.8	.3979	.3875	.9219	.4204	2.379	1.085	2.581	1.1729	67.2
22.9	.3997	.3891	.9212	.4224	2.367	1.086	2.570	1.1711	67.1
23.0	.4014	.3907	.9205	.4245	2.356	1.086	2.559	1.1694	**67.0**
23.1	.4032	.3923	.9198	.4265	2.344	1.087	2.549	1.1676	66.9
23.2	.4049	.3939	.9191	.4286	2.333	1.088	2.538	1.1659	66.8
23.3	.4067	.3955	.9184	.4307	2.322	1.089	2.528	1.1641	66.7
23.4	.4084	.3971	.9178	.4327	2.311	1.090	2.518	1.1624	66.6
23.5	.4102	.3987	.9171	.4348	2.300	1.090	2.508	1.1606	66.5
23.6	.4119	.4003	.9164	.4369	2.289	1.091	2.498	1.1589	66.4
23.7	.4136	.4019	.9157	.4390	2.278	1.092	2.488	1.1572	66.3
23.8	.4154	.4035	.9150	.4411	2.267	1.093	2.478	1.1554	66.2
23.9	.4171	.4051	.9143	.4431	2.257	1.094	2.468	1.1537	66.1
24.0	.4189	.4067	.9135	.4452	2.246	1.095	2.459	1.1519	**66.0**
		cos θ	sin θ	cot θ	tan θ	csc θ	sec θ	θ Radians	θ Degrees

Table 4 / *Trigonometric Functions of* θ

(θ in decimal degrees)

θ Degrees	θ Radians	sin θ	cos θ	tan θ	cot θ	sec θ	csc θ		
24.0	.4189	.4067	.9135	.4452	2.246	1.095	2.459	1.1519	**66.0**
24.1	.4206	.4083	.9128	.4473	2.236	1.095	2.449	1.1502	65.9
24.2	.4224	.4099	.9121	.4494	2.225	1.096	2.439	1.1484	65.8
24.3	.4241	.4115	.9114	.4515	2.215	1.097	2.430	1.1467	65.7
24.4	.4259	.4131	.9107	.4536	2.204	1.098	2.421	1.1449	65.6
24.5	.4276	.4147	.9100	.4557	2.194	1.099	2.411	1.1432	65.5
24.6	.4294	.4163	.9092	.4578	2.184	1.100	2.402	1.1414	65.4
24.7	.4311	.4179	.9085	.4599	2.174	1.101	2.393	1.1397	65.3
24.8	.4328	.4195	.9078	.4621	2.164	1.102	2.384	1.1380	65.2
24.9	.4346	.4210	.9070	.4642	2.154	1.102	2.375	1.1362	65.1
25.0	.4363	.4226	.9063	.4663	2.145	1.103	2.366	1.1345	**65.0**
25.1	.4381	.4242	.9056	.4684	2.135	1.104	2.357	1.1327	64.9
25.2	.4398	.4258	.9048	.4706	2.125	1.105	2.349	1.1310	64.8
25.3	.4416	.4274	.9041	.4727	2.116	1.106	2.340	1.1292	64.7
25.4	.4433	.4289	.9033	.4748	2.106	1.107	2.331	1.1275	64.6
25.5	.4451	.4305	.9026	.4770	2.097	1.108	2.323	1.1257	64.5
25.6	.4468	.4321	.9018	.4791	2.087	1.109	2.314	1.1240	64.4
25.7	.4485	.4337	.9011	.4813	2.078	1.110	2.306	1.1222	64.3
25.8	.4503	.4352	.9003	.4834	2.069	1.111	2.298	1.1205	64.2
25.9	.4520	.4368	.8996	.4856	2.059	1.112	2.289	1.1188	64.1
26.0	.4538	.4384	.8988	.4877	2.050	1.113	2.281	1.1170	**64.0**
26.1	.4555	.4399	.8980	.4899	2.041	1.114	2.273	1.1153	63.9
26.2	.4573	.4415	.8973	.4921	2.032	1.115	2.265	1.1135	63.8
26.3	.4590	.4431	.8965	.4942	2.023	1.115	2.257	1.1118	63.7
26.4	.4608	.4446	.8957	.4964	2.014	1.116	2.249	1.1100	63.6
26.5	.4625	.4462	.8949	.4986	2.006	1.117	2.241	1.1083	63.5
26.6	.4643	.4478	.8942	.5008	1.997	1.118	2.233	1.1065	63.4
26.7	.4660	.4493	.8934	.5029	1.988	1.119	2.226	1.1048	63.3
26.8	.4677	.4509	.8926	.5051	1.980	1.120	2.218	1.1030	63.2
26.9	.4695	.4524	.8918	.5073	1.971	1.121	2.210	1.1013	63.1
27.0	.4712	.4540	.8910	.5095	1.963	1.122	2.203	1.0996	**63.0**
27.1	.4730	.4555	.8902	.5117	1.954	1.123	2.195	1.0978	62.9
27.2	.4747	.4571	.8894	.5139	1.946	1.124	2.188	1.0961	62.8
27.3	.4765	.4586	.8886	.5161	1.937	1.125	2.180	1.0943	62.7
27.4	.4782	.4602	.8878	.5184	1.929	1.126	2.173	1.0926	62.6
27.5	.4800	.4617	.8870	.5206	1.921	1.127	2.166	1.0908	62.5
27.6	.4817	.4633	.8862	.5228	1.913	1.128	2.158	1.0891	62.4
27.7	.4835	.4648	.8854	.5250	1.905	1.129	2.151	1.0873	62.3
27.8	.4852	.4664	.8846	.5272	1.897	1.130	2.144	1.0856	62.2
27.9	.4869	.4679	.8838	.5295	1.889	1.132	2.137	1.0838	62.1
28.0	.4887	.4695	.8829	.5317	1.881	1.133	2.130	1.0821	**62.0**
28.1	.4904	.4710	.8821	.5340	1.873	1.134	2.123	1.0804	61.9
28.2	.4922	.4726	.8813	.5362	1.865	1.135	2.116	1.0786	61.8
28.3	.4939	.4741	.8805	.5384	1.857	1.136	2.109	1.0769	61.7
28.4	.4957	.4756	.8796	.5407	1.849	1.137	2.103	1.0751	61.6
28.5	.4974	.4772	.8788	.5430	1.842	1.138	2.096	1.0734	61.5
28.6	.4992	.4787	.8780	.5452	1.834	1.139	2.089	1.0716	61.4
28.7	.5009	.4802	.8771	.5475	1.827	1.140	2.082	1.0699	61.3
28.8	.5027	.4818	.8763	.5498	1.819	1.141	2.076	1.0681	61.2
28.9	.5044	.4833	.8755	.5520	1.811	1.142	2.069	1.0664	61.1
29.0	.5061	.4848	.8746	.5543	1.804	1.143	2.063	1.0647	**61.0**
29.1	.5079	.4863	.8738	.5566	1.797	1.144	2.056	1.0629	60.9
29.2	.5096	.4879	.8729	.5589	1.789	1.146	2.050	1.0612	60.8
29.3	.5114	.4894	.8721	.5612	1.782	1.147	2.043	1.0594	60.7
29.4	.5131	.4909	.8712	.5635	1.775	1.148	2.037	1.0577	60.6
29.5	.5149	.4924	.8704	.5658	1.767	1.149	2.031	1.0559	60.5
29.6	.5166	.4939	.8695	.5681	1.760	1.150	2.025	1.0542	60.4
29.7	.5184	.4955	.8686	.5704	1.753	1.151	2.018	1.0524	60.3
29.8	.5201	.4970	.8678	.5727	1.746	1.152	2.012	1.0507	60.2
29.9	.5219	.4985	.8669	.5750	1.739	1.154	2.006	1.0489	60.1
30.0	.5236	.5000	.8660	.5774	1.732	1.155	2.000	1.0472	**60.0**
		cos θ	sin θ	cot θ	tan θ	csc θ	sec θ	θ Radians	θ Degrees

Table 4 / *Trigonometric Functions of θ*

(θ in decimal degrees)

θ Degrees	θ Radians	sin θ	cos θ	tan θ	cot θ	sec θ	csc θ		
30.0	.5236	.5000	.8660	.5774	1.732	1.155	2.000	1.0472	**60.0**
30.1	.5253	.5015	.8652	.5797	1.725	1.156	1.994	1.0455	59.9
30.2	.5271	.5030	.8643	.5820	1.718	1.157	1.988	1.0437	59.8
30.3	.5288	.5045	.8634	.5844	1.711	1.158	1.982	1.0420	59.7
30.4	.5306	.5060	.8625	.5867	1.704	1.159	1.976	1.0402	59.6
30.5	.5323	.5075	.8616	.5890	1.698	1.161	1.970	1.0385	59.5
30.6	.5341	.5090	.8607	.5914	1.691	1.162	1.964	1.0367	59.4
30.7	.5358	.5105	.8599	.5938	1.684	1.163	1.959	1.0350	59.3
30.8	.5376	.5120	.8590	.5961	1.678	1.164	1.953	1.0332	59.2
30.9	.5393	.5135	.8581	.5985	1.671	1.165	1.947	1.0315	59.1
31.0	.5411	.5150	.8572	.6009	1.664	1.167	1.942	1.0297	**59.0**
31.1	.5428	.5165	.8563	.6032	1.658	1.168	1.936	1.0280	58.9
31.2	.5445	.5180	.8554	.6056	1.651	1.169	1.930	1.0263	58.8
31.3	.5463	.5195	.8545	.6080	1.645	1.170	1.925	1.0245	58.7
31.4	.5480	.5210	.8535	.6104	1.638	1.172	1.919	1.0228	58.6
31.5	.5498	.5225	.8526	.6128	1.632	1.173	1.914	1.0210	58.5
31.6	.5515	.5240	.8517	.6152	1.625	1.174	1.908	1.0193	58.4
31.7	.5533	.5255	.8508	.6176	1.619	1.175	1.903	1.0175	58.3
31.8	.5550	.5270	.8499	.6200	1.613	1.177	1.898	1.0158	58.2
31.9	.5568	.5284	.8490	.6224	1.607	1.178	1.892	1.0140	58.1
32.0	.5585	.5299	.8480	.6249	1.600	1.179	1.887	1.0123	**58.0**
32.1	.5603	.5314	.8471	.6273	1.594	1.180	1.882	1.0105	57.9
32.2	.5620	.5329	.8462	.6297	1.588	1.182	1.877	1.0088	57.8
32.3	.5637	.5344	.8453	.6322	1.582	1.183	1.871	1.0071	57.7
32.4	.5655	.5358	.8443	.6346	1.576	1.184	1.866	1.0053	57.6
32.5	.5672	.5373	.8434	.6371	1.570	1.186	1.861	1.0036	57.5
32.6	.5690	.5388	.8425	.6395	1.564	1.187	1.856	1.0018	57.4
32.7	.5707	.5402	.8415	.6420	1.558	1.188	1.851	1.0001	57.3
32.8	.5725	.5417	.8406	.6445	1.552	1.190	1.846	.9983	57.2
32.9	.5742	.5432	.8396	.6469	1.546	1.191	1.841	.9966	57.1
33.0	.5760	.5446	.8387	.6494	1.540	1.192	1.836	.9948	**57.0**
33.1	.5777	.5461	.8377	.6519	1.534	1.194	1.831	.9931	56.9
33.2	.5794	.5476	.8368	.6544	1.528	1.195	1.826	.9913	56.8
33.3	.5812	.5490	.8358	.6569	1.522	1.196	1.821	.9896	56.7
33.4	.5829	.5505	.8348	.6594	1.517	1.198	1.817	.9879	56.6
33.5	.5847	.5519	.8339	.6619	1.511	1.199	1.812	.9861	56.5
33.6	.5864	.5534	.8329	.6644	1.505	1.201	1.807	.9844	56.4
33.7	.5882	.5548	.8320	.6669	1.499	1.202	1.802	.9826	56.3
33.8	.5899	.5563	.8310	.6694	1.494	1.203	1.798	.9809	56.2
33.9	.5917	.5577	.8300	.6720	1.488	1.205	1.793	.9791	56.1
34.0	.5934	.5592	.8290	.6745	1.483	1.206	1.788	.9774	**56.0**
34.1	.5952	.5606	.8281	.6771	1.477	1.208	1.784	.9756	55.9
34.2	.5969	.5621	.8271	.6796	1.471	1.209	1.779	.9739	55.8
34.3	.5986	.5635	.8261	.6822	1.466	1.211	1.775	.9721	55.7
34.4	.6004	.5650	.8251	.6847	1.460	1.212	1.770	.9704	55.6
34.5	.6021	.5664	.8241	.6873	1.455	1.213	1.766	.9687	55.5
34.6	.6039	.5678	.8231	.6899	1.450	1.215	1.761	.9669	55.4
34.7	.6056	.5693	.8221	.6924	1.444	1.216	1.757	.9652	55.3
34.8	.6074	.5707	.8211	.6950	1.439	1.218	1.752	.9634	55.2
34.9	.6091	.5721	.8202	.6976	1.433	1.219	1.748	.9617	55.1
35.0	.6109	.5736	.8192	.7002	1.428	1.221	1.743	.9599	**55.0**
35.1	.6126	.5750	.8181	.7028	1.423	1.222	1.739	.9582	54.9
35.2	.6144	.5764	.8171	.7054	1.418	1.224	1.735	.9564	54.8
35.3	.6161	.5779	.8161	.7080	1.412	1.225	1.731	.9547	54.7
35.4	.6178	.5793	.8151	.7107	1.407	1.227	1.726	.9529	54.6
35.5	.6196	.5807	.8141	.7133	1.402	1.228	1.722	.9512	54.5
35.6	.6213	.5821	.8131	.7159	1.397	1.230	1.718	.9495	54.4
35.7	.6231	.5835	.8121	.7186	1.392	1.231	1.714	.9477	54.3
35.8	.6248	.5850	.8111	.7212	1.387	1.233	1.710	.9460	54.2
35.9	.6266	.5864	.8100	.7239	1.381	1.235	1.705	.9442	54.1
36.0	.6283	.5878	.8090	.7265	1.376	1.236	1.701	.9425	**54.0**
		cos θ	sin θ	cot θ	tan θ	csc θ	sec θ	θ Radians	θ Degrees

Table 4 / *Trigonometric Functions of θ*

(θ in decimal degrees)

θ Degrees	θ Radians	sin θ	cos θ	tan θ	cot θ	sec θ	csc θ		
36.0	.6283	.5878	.8090	.7265	1.376	1.236	1.701	.9425	**54.0**
36.1	.6301	.5892	.8080	.7292	1.371	1.238	1.697	.9407	53.9
36.2	.6318	.5906	.8070	.7319	1.366	1.239	1.693	.9390	53.8
36.3	.6336	.5920	.8059	.7346	1.361	1.241	1.689	.9372	53.7
36.4	.6353	.5934	.8049	.7373	1.356	1.242	1.685	.9355	53.6
36.5	.6370	.5948	.8039	.7400	1.351	1.244	1.681	.9338	53.5
36.6	.6388	.5962	.8028	.7427	1.347	1.246	1.677	.9320	53.4
36.7	.6405	.5976	.8018	.7454	1.342	1.247	1.673	.9303	53.3
36.8	.6423	.5990	.8007	.7481	1.337	1.249	1.669	.9285	53.2
36.9	.6440	.6004	.7997	.7508	1.332	1.250	1.666	.9268	53.1
37.0	.6458	.6018	.7986	.7536	1.327	1.252	1.662	.9250	**53.0**
37.1	.6475	.6032	.7976	.7563	1.322	1.254	1.658	.9233	52.9
37.2	.6493	.6046	.7965	.7590	1.317	1.255	1.654	.9215	52.8
37.3	.6510	.6060	.7955	.7618	1.313	1.257	1.650	.9198	52.7
37.4	.6528	.6074	.7944	.7646	1.308	1.259	1.646	.9180	52.6
37.5	.6545	.6088	.7934	.7673	1.303	1.260	1.643	.9163	52.5
37.6	.6562	.6101	.7923	.7701	1.299	1.262	1.639	.9146	52.4
37.7	.6580	.6115	.7912	.7729	1.294	1.264	1.635	.9128	52.3
37.8	.6597	.6129	.7902	.7757	1.289	1.266	1.632	.9111	52.2
37.9	.6615	.6143	.7891	.7785	1.285	1.267	1.628	.9093	52.1
38.0	.6632	.6157	.7880	.7813	1.280	1.269	1.624	.9076	**52.0**
38.1	.6650	.6170	.7869	.7841	1.275	1.271	1.621	.9058	51.9
38.2	.6667	.6184	.7859	.7869	1.271	1.272	1.617	.9041	51.8
38.3	.6685	.6198	.7848	.7898	1.266	1.274	1.613	.9023	51.7
38.4	.6702	.6211	.7837	.7926	1.262	1.276	1.610	.9006	51.6
38.5	.6720	.6225	.7826	.7954	1.257	1.278	1.606	.8988	51.5
38.6	.6737	.6239	.7815	.7983	1.253	1.280	1.603	.8971	51.4
38.7	.6754	.6252	.7804	.8012	1.248	1.281	1.599	.8954	51.3
38.8	.6772	.6266	.7793	.8040	1.244	1.283	1.596	.8936	51.2
38.9	.6789	.6280	.7782	.8069	1.239	1.285	1.592	.8919	51.1
39.0	.6807	.6293	.7771	.8098	1.235	1.287	1.589	.8901	**51.0**
39.1	.6824	.6307	.7760	.8127	1.230	1.289	1.586	.8884	50.9
39.2	.6842	.6320	.7749	.8156	1.226	1.290	1.582	.8866	50.8
39.3	.6859	.6334	.7738	.8185	1.222	1.292	1.579	.8849	50.7
39.4	.6877	.6347	.7727	.8214	1.217	1.294	1.575	.8831	50.6
39.5	.6894	.6361	.7716	.8243	1.213	1.296	1.572	.8814	50.5
39.6	.6912	.6374	.7705	.8273	1.209	1.298	1.569	.8796	50.4
39.7	.6929	.6388	.7694	.8302	1.205	1.300	1.566	.8779	50.3
39.8	.6946	.6401	.7683	.8332	1.200	1.302	1.562	.8762	50.2
39.9	.6964	.6414	.7672	.8361	1.196	1.304	1.559	.8744	50.1
40.0	.6981	.6428	.7660	.8391	1.192	1.305	1.556	.8727	**50.0**
40.1	.6999	.6441	.7649	.8421	1.188	1.307	1.552	.8709	49.9
40.2	.7016	.6455	.7638	.8451	1.183	1.309	1.549	.8692	49.8
40.3	.7034	.6468	.7627	.8481	1.179	1.311	1.546	.8674	49.7
40.4	.7051	.6481	.7615	.8511	1.175	1.313	1.543	.8657	49.6
40.5	.7069	.6494	.7604	.8541	1.171	1.315	1.540	.8639	49.5
40.6	.7086	.6508	.7593	.8571	1.167	1.317	1.537	.8622	49.4
40.7	.7103	.6521	.7581	.8601	1.163	1.319	1.534	.8604	49.3
40.8	.7121	.6534	.7570	.8632	1.159	1.321	1.530	.8587	49.2
40.9	.7138	.6547	.7559	.8662	1.154	1.323	1.527	.8570	49.1
41.0	.7156	.6561	.7547	.8693	1.150	1.325	1.524	.8552	**49.0**
41.1	.7173	.6574	.7536	.8724	1.146	1.327	1.521	.8535	48.9
41.2	.7191	.6587	.7524	.8754	1.142	1.329	1.518	.8517	48.8
41.3	.7208	.6600	.7513	.8785	1.138	1.331	1.515	.8500	48.7
41.4	.7226	.6613	.7501	.8816	1.134	1.333	1.512	.8482	48.6
41.5	.7243	.6626	.7490	.8847	1.130	1.335	1.509	.8465	48.5
41.6	.7261	.6639	.7478	.8878	1.126	1.337	1.506	.8447	48.4
41.7	.7278	.6652	.7466	.8910	1.122	1.339	1.503	.8430	48.3
41.8	.7295	.6665	.7455	.8941	1.118	1.341	1.500	.8412	48.2
41.9	.7313	.6678	.7443	.8972	1.115	1.344	1.497	.8395	48.1
42.0	.7330	.6691	.7431	.9004	1.111	1.346	1.494	.8378	**48.0**
		cos θ	sin θ	cot θ	tan θ	csc θ	sec θ	θ Radians	θ Degrees

Table 4 / *Trigonometric Functions of θ*

(θ in decimal degrees)

θ Degrees	θ Radians	sin θ	cos θ	tan θ	cot θ	sec θ	csc θ		
42.0	.7330	.6691	.7431	.9004	1.111	1.346	1.494	.8378	**48.0**
42.1	.7348	.6704	.7420	.9036	1.107	1.348	1.492	.8360	47.9
42.2	.7365	.6717	.7408	.9067	1.103	1.350	1.489	.8343	47.8
42.3	.7383	.6730	.7396	.9099	1.099	1.352	1.486	.8325	47.7
42.4	.7400	.6743	.7385	.9131	1.095	1.354	1.483	.8308	47.6
42.5	.7418	.6756	.7373	.9163	1.091	1.356	1.480	.8290	47.5
42.6	.7435	.6769	.7361	.9195	1.087	1.359	1.477	.8273	47.4
42.7	.7453	.6782	.7349	.9228	1.084	1.361	1.475	.8255	47.3
42.8	.7470	.6794	.7337	.9260	1.080	1.363	1.472	.8238	47.2
42.9	.7487	.6807	.7325	.9293	1.076	1.365	1.469	.8221	47.1
43.0	.7505	.6820	.7314	.9325	1.072	1.367	1.466	.8203	**47.0**
43.1	.7522	.6833	.7302	.9358	1.069	1.370	1.464	.8186	46.9
43.2	.7540	.6845	.7290	.9391	1.065	1.372	1.461	.8168	46.8
43.3	.7557	.6858	.7278	.9424	1.061	1.374	1.458	.8151	46.7
43.4	.7575	.6871	.7266	.9457	1.057	1.376	1.455	.8133	46.6
43.5	.7592	.6884	.7254	.9490	1.054	1.379	1.453	.8116	46.5
43.6	.7610	.6896	.7242	.9523	1.050	1.381	1.450	.8098	46.4
43.7	.7627	.6909	.7230	.9556	1.046	1.383	1.447	.8081	46.3
43.8	.7645	.6921	.7218	.9590	1.043	1.386	1.445	.8063	46.2
43.9	.7662	.6934	.7206	.9623	1.039	1.388	1.442	.8046	46.1
44.0	.7679	.6947	.7193	.9657	1.036	1.390	1.440	.8029	**46.0**
44.1	.7697	.6959	.7181	.9691	1.032	1.393	1.437	.8011	45.9
44.2	.7714	.6972	.7169	.9725	1.028	1.395	1.434	.7994	45.8
44.3	.7732	.6984	.7157	.9759	1.025	1.397	1.432	.7976	45.7
44.4	.7749	.6997	.7145	.9793	1.021	1.400	1.429	.7959	45.6
44.5	.7767	.7009	.7133	.9827	1.018	1.402	1.427	.7941	45.5
44.6	.7784	.7022	.7120	.9861	1.014	1.404	1.424	.7924	45.4
44.7	.7802	.7034	.7108	.9896	1.011	1.407	1.422	.7906	45.3
44.8	.7819	.7046	.7096	.9930	1.007	1.409	1.419	.7889	45.2
44.9	.7837	.7059	.7083	.9965	1.003	1.412	1.417	.7871	45.1
45.0	.7854	.7071	.7071	1.0000	1.000	1.414	1.414	.7854	**45.0**
		cos θ	sin θ	cot θ	tan θ	csc θ	sec θ	θ Radians	θ Degrees

Table 5 / *Trigonometric Functions of θ*

(θ in degrees and minutes)

Angle	Sin	Cos	Tan	Cot	Sec	Csc	
0° 00′	.0000	1.0000	.0000	- - - -	1.000	- - - -	90° 00′
10′	.0029	1.0000	.0029	343.8	1.000	343.8	50′
20′	.0058	1.0000	.0058	171.9	1.000	171.9	40′
30′	.0087	1.0000	.0087	114.6	1.000	114.6	30′
40′	.0116	.9999	.0116	85.94	1.000	85.95	20′
50′	.0145	.9999	.0145	68.75	1.000	68.76	10′
1° 00′	.0175	.9998	.0175	57.29	1.000	57.30	89° 00′
10′	.0204	.9998	.0204	49.10	1.000	49.11	50′
20′	.0233	.9997	.0233	42.96	1.000	42.98	40′
30′	.0262	.9997	.0262	38.19	1.000	38.20	30′
40′	.0291	.9996	.0291	34.37	1.000	34.38	20′
50′	.0320	.9995	.0320	31.24	1.001	31.26	10′
2° 00′	.0349	.9994	.0349	28.64	1.001	28.65	88° 00′
10′	.0378	.9993	.0378	26.43	1.001	26.45	50′
20′	.0407	.9992	.0407	24.54	1.001	24.56	40′
30′	.0436	.9990	.0437	22.90	1.001	22.93	30′
40′	.0465	.9989	.0466	21.47	1.001	21.49	20′
50′	.0494	.9988	.0495	20.21	1.001	20.23	10′
3° 00′	.0523	.9986	.0524	19.08	1.001	19.11	87° 00′
10′	.0552	.9985	.0553	18.07	1.002	18.10	50′
20′	.0581	.9983	.0582	17.17	1.002	17.20	40′
30′	.0610	.9981	.0612	16.35	1.002	16.38	30′
40′	.0640	.9980	.0641	15.60	1.002	15.64	20′
50′	.0669	.9978	.0670	14.92	1.002	14.96	10′
4° 00′	.0698	.9976	.0699	14.30	1.002	14.34	86° 00′
10′	.0727	.9974	.0729	13.73	1.003	13.76	50′
20′	.0756	.9971	.0758	13.20	1.003	13.23	40′
30′	.0785	.9969	.0787	12.71	1.003	12.75	30′
40′	.0814	.9967	.0816	12.25	1.003	12.29	20′
50′	.0843	.9964	.0846	11.83	1.004	11.87	10′
5° 00′	.0872	.9962	.0875	11.43	1.004	11.47	85° 00′
10′	.0901	.9959	.0904	11.06	1.004	11.10	50′
20′	.0929	.9957	.0934	10.71	1.004	10.76	40′
30′	.0958	.9954	.0963	10.39	1.005	10.43	30′
40′	.0987	.9951	.0992	10.08	1.005	10.13	20′
50′	.1016	.9948	.1022	9.788	1.005	9.839	10′
6° 00′	.1045	.9945	.1051	9.514	1.006	9.567	84° 00′
10′	.1074	.9942	.1080	9.255	1.006	9.309	50′
20′	.1103	.9939	.1110	9.010	1.006	9.065	40′
30′	.1132	.9936	.1139	8.777	1.006	8.834	30′
40′	.1161	.9932	.1169	8.556	1.007	8.614	20′
50′	.1190	.9929	.1198	8.345	1.007	8.405	10′
7° 00′	.1219	.9925	.1228	8.144	1.008	8.206	83° 00′
10′	.1248	.9922	.1257	7.953	1.008	8.016	50′
20′	.1276	.9918	.1287	7.770	1.008	7.834	40′
30′	.1305	.9914	.1317	7.596	1.009	7.661	30′
40′	.1334	.9911	.1346	7.429	1.009	7.496	20′
50′	.1363	.9907	.1376	7.269	1.009	7.337	10′
8° 00′	.1392	.9903	.1405	7.115	1.010	7.185	82° 00′
10′	.1421	.9899	.1435	6.968	1.010	7.040	50′
20′	.1449	.9894	.1465	6.827	1.011	6.900	40′
30′	.1478	.9890	.1495	6.691	1.011	6.765	30′
40′	.1507	.9886	.1524	6.561	1.012	6.636	20′
50′	.1536	.9881	.1554	6.435	1.012	6.512	10′
9° 00′	.1564	.9877	.1584	6.314	1.012	6.392	81° 00′
	Cos	Sin	Cot	Tan	Csc	Sec	Angle

Table 5 / *Trigonometric Functions of θ*

(θ in degrees and minutes)

Angle	Sin	Cos	Tan	Cot	Sec	Csc	
9° 00′	.1564	.9877	.1584	6.314	1.012	6.392	**81° 00′**
10′	.1593	.9872	.1614	6.197	1.013	6.277	**50′**
20′	.1622	.9868	.1644	6.084	1.013	6.166	**40′**
30′	.1650	.9863	.1673	5.976	1.014	6.059	**30′**
40′	.1679	.9858	.1703	5.871	1.014	5.955	**20′**
50′	.1708	.9853	.1733	5.769	1.015	5.855	**10′**
10° 00′	.1736	.9848	.1763	5.671	1.015	5.759	**80° 00′**
10′	.1765	.9843	.1793	5.576	1.016	5.665	**50′**
20′	.1794	.9838	.1823	5.485	1.016	5.575	**40′**
30′	.1822	.9833	.1853	5.396	1.017	5.487	**30′**
40′	.1851	.9827	.1883	5.309	1.018	5.403	**20′**
50′	.1880	.9822	.1914	5.226	1.018	5.320	**10′**
11° 00′	.1908	.9816	.1944	5.145	1.019	5.241	**79° 00′**
10′	.1937	.9811	.1974	5.066	1.019	5.164	**50′**
20′	.1965	.9805	.2004	4.989	1.020	5.089	**40′**
30′	.1994	.9799	.2035	4.915	1.020	5.016	**30′**
40′	.2022	.9793	.2065	4.843	1.021	4.945	**20′**
50′	.2051	.9787	.2095	4.773	1.022	4.876	**10′**
12° 00′	.2079	.9781	.2126	4.705	1.022	4.810	**78° 00′**
10′	.2108	.9775	.2156	4.638	1.023	4.745	**50′**
20′	.2136	.9769	.2186	4.574	1.024	4.682	**40′**
30′	.2164	.9763	.2217	4.511	1.024	4.620	**30′**
40′	.2193	.9757	.2247	4.449	1.025	4.560	**20′**
50′	.2221	.9750	.2278	4.390	1.026	4.502	**10′**
13° 00′	.2250	.9744	.2309	4.331	1.026	4.445	**77° 00′**
10′	.2278	.9737	.2339	4.275	1.027	4.390	**50′**
20′	.2306	.9730	.2370	4.219	1.028	4.336	**40′**
30′	.2334	.9724	.2401	4.165	1.028	4.284	**30′**
40′	.2363	.9717	.2432	4.113	1.029	4.232	**20′**
50′	.2391	.9710	.2462	4.061	1.030	4.182	**10′**
14° 00′	.2419	.9703	.2493	4.011	1.031	4.134	**76° 00′**
10′	.2447	.9696	.2524	3.962	1.031	4.086	**50′**
20′	.2476	.9689	.2555	3.914	1.032	4.039	**40′**
30′	.2504	.9681	.2586	3.867	1.033	3.994	**30′**
40′	.2532	.9674	.2617	3.821	1.034	3.950	**20′**
50′	.2560	.9667	.2648	3.776	1.034	3.906	**10′**
15° 00′	.2588	.9659	.2679	3.732	1.035	3.864	**75° 00′**
10′	.2616	.9652	.2711	3.689	1.036	3.822	**50′**
20′	.2644	.9644	.2742	3.647	1.037	3.782	**40′**
30′	.2672	.9636	.2773	3.606	1.038	3.742	**30′**
40′	.2700	.9628	.2805	3.566	1.039	3.703	**20′**
50′	.2728	.9621	.2836	3.526	1.039	3.665	**10′**
16° 00′	.2756	.9613	.2867	3.487	1.040	3.628	**74° 00′**
10′	.2784	.9605	.2899	3.450	1.041	3.592	**50′**
20′	.2812	.9596	.2931	3.412	1.042	3.556	**40′**
30′	.2840	.9588	.2962	3.376	1.043	3.521	**30′**
40′	.2868	.9580	.2994	3.340	1.044	3.487	**20′**
50′	.2896	.9572	.3026	3.305	1.045	3.453	**10′**
17° 00′	.2924	.9563	.3057	3.271	1.046	3.420	**73° 00′**
10′	.2952	.9555	.3089	3.237	1.047	3.388	**50′**
20′	.2979	.9546	.3121	3.204	1.048	3.356	**40′**
30′	.3007	.9537	.3153	3.172	1.049	3.326	**30′**
40′	.3035	.9528	.3185	3.140	1.049	3.295	**20′**
50′	.3062	.9520	.3217	3.108	1.050	3.265	**10′**
18° 00′	.3090	.9511	.3249	3.078	1.051	3.236	**72° 00′**
	Cos	Sin	Cot	Tan	Csc	Sec	Angle

Table 5 / *Trigonometric Functions of θ*

(θ in degrees and minutes)

Angle	Sin	Cos	Tan	Cot	Sec	Csc	
18° 00'	.3090	.9511	.3249	3.078	1.051	3.236	**72° 00'**
10'	.3118	.9502	.3281	3.047	1.052	3.207	**50'**
20'	.3145	.9492	.3314	3.018	1.053	3.179	**40'**
30'	.3173	.9483	.3346	2.989	1.054	3.152	**30'**
40'	.3201	.9474	.3378	2.960	1.056	3.124	**20'**
50'	.3228	.9465	.3411	2.932	1.057	3.098	**10'**
19° 00'	.3256	.9455	.3443	2.904	1.058	3.072	**71° 00'**
10'	.3283	.9446	.3476	2.877	1.059	3.046	**50'**
20'	.3311	.9436	.3508	2.850	1.060	3.021	**40'**
30'	.3338	.9426	.3541	2.824	1.061	2.996	**30'**
40'	.3365	.9417	.3574	2.798	1.062	2.971	**20'**
50'	.3393	.9407	.3607	2.773	1.063	2.947	**10'**
20° 00'	.3420	.9397	.3640	2.747	1.064	2.924	**70° 00'**
10'	.3448	.9387	.3673	2.723	1.065	2.901	**50'**
20'	.3475	.9377	.3706	2.699	1.066	2.878	**40'**
30'	.3502	.9367	.3739	2.675	1.068	2.855	**30'**
40'	.3529	.9356	.3772	2.651	1.069	2.833	**20'**
50'	.3557	.9346	.3805	2.628	1.070	2.812	**10'**
21° 00'	.3584	.9336	.3839	2.605	1.071	2.790	**69° 00'**
10'	.3611	.9325	.3872	2.583	1.072	2.769	**50'**
20'	.3638	.9315	.3906	2.560	1.074	2.749	**40'**
30'	.3665	.9304	.3939	2.539	1.075	2.729	**30'**
40'	.3692	.9293	.3973	2.517	1.076	2.709	**20'**
50'	.3719	.9283	.4006	2.496	1.077	2.689	**10'**
22° 00'	.3746	.9272	.4040	2.475	1.079	2.669	**68° 00'**
10'	.3773	.9261	.4074	2.455	1.080	2.650	**50'**
20'	.3800	.9250	.4108	2.434	1.081	2.632	**40'**
30'	.3827	.9239	.4142	2.414	1.082	2.613	**30'**
40'	.3854	.9228	.4176	2.394	1.084	2.595	**20'**
50'	.3881	.9216	.4210	2.375	1.085	2.577	**10'**
23° 00'	.3907	.9205	.4245	2.356	1.086	2.559	**67° 00'**
10'	.3934	.9194	.4279	2.337	1.088	2.542	**50'**
20'	.3961	.9182	.4314	2.318	1.089	2.525	**40'**
30'	.3987	.9171	.4348	2.300	1.090	2.508	**30'**
40'	.4014	.9159	.4383	2.282	1.092	2.491	**20'**
50'	.4041	.9147	.4417	2.264	1.093	2.475	**10'**
24° 00'	.4067	.9135	.4452	2.246	1.095	2.459	**66° 00'**
10'	.4094	.9124	.4487	2.229	1.096	2.443	**50'**
20'	.4120	.9112	.4522	2.211	1.097	2.427	**40'**
30'	.4147	.9100	.4557	2.194	1.099	2.411	**30'**
40'	.4173	.9088	.4592	2.177	1.100	2.396	**20'**
50'	.4200	.9075	.4628	2.161	1.102	2.381	**10'**
25° 00'	.4226	.9063	.4663	2.145	1.103	2.366	**65° 00'**
10'	.4253	.9051	.4699	2.128	1.105	2.352	**50'**
20'	.4279	.9038	.4734	2.112	1.106	2.337	**40'**
30'	.4305	.9026	.4770	2.097	1.108	2.323	**30'**
40'	.4331	.9013	.4806	2.081	1.109	2.309	**20'**
50'	.4358	.9001	.4841	2.066	1.111	2.295	**10'**
26° 00'	.4384	.8988	.4877	2.050	1.113	2.281	**64° 00'**
10'	.4410	.8975	.4913	2.035	1.114	2.268	**50'**
20'	.4436	.8962	.4950	2.020	1.116	2.254	**40'**
30'	.4462	.8949	.4986	2.006	1.117	2.241	**30'**
40'	.4488	.8936	.5022	1.991	1.119	2.228	**20'**
50'	.4514	.8923	.5059	1.977	1.121	2.215	**10'**
27° 00'	.4540	.8910	.5095	1.963	1.122	2.203	**63° 00'**
	Cos	Sin	Cot	Tan	Csc	Sec	Angle

Table 5 / *Trigonometric Functions of θ*

(θ in degrees and minutes)

Angle	Sin	Cos	Tan	Cot	Sec	Csc	
27° 00′	.4540	.8910	.5095	1.963	1.122	2.203	63° 00′
10′	.4566	.8897	.5132	1.949	1.124	2.190	50′
20′	.4592	.8884	.5169	1.935	1.126	2.178	40′
30′	.4617	.8870	.5206	1.921	1.127	2.166	30′
40′	.4643	.8857	.5243	1.907	1.129	2.154	20′
50′	.4669	.8843	.5280	1.894	1.131	2.142	10′
28° 00′	.4695	.8829	.5317	1.881	1.133	2.130	62° 00′
10′	.4720	.8816	.5354	1.868	1.134	2.118	50′
20′	.4746	.8802	.5392	1.855	1.136	2.107	40′
30′	.4772	.8788	.5430	1.842	1.138	2.096	30′
40′	.4797	.8774	.5467	1.829	1.140	2.085	20′
50′	.4823	.8760	.5505	1.816	1.142	2.074	10′
29° 00′	.4848	.8746	.5543	1.804	1.143	2.063	61° 00′
10′	.4874	.8732	.5581	1.792	1.145	2.052	50′
20′	.4899	.8718	.5619	1.780	1.147	2.041	40′
30′	.4924	.8704	.5658	1.767	1.149	2.031	30′
40′	.4950	.8689	.5696	1.756	1.151	2.020	20′
50′	.4975	.8675	.5735	1.744	1.153	2.010	10′
30° 00′	.5000	.8660	.5774	1.732	1.155	2.000	60° 00′
10′	.5025	.8646	.5812	1.720	1.157	1.990	50′
20′	.5050	.8631	.5851	1.709	1.159	1.980	40′
30′	.5075	.8616	.5890	1.698	1.161	1.970	30′
40′	.5100	.8601	.5930	1.686	1.163	1.961	20′
50′	.5125	.8587	.5969	1.675	1.165	1.951	10′
31° 00′	.5150	.8572	.6009	1.664	1.167	1.942	59° 00′
10′	.5175	.8557	.6048	1.653	1.169	1.932	50′
20′	.5200	.8542	.6088	1.643	1.171	1.923	40′
30′	.5225	.8526	.6128	1.632	1.173	1.914	30′
40′	.5250	.8511	.6168	1.621	1.175	1.905	20′
50′	.5275	.8496	.6208	1.611	1.177	1.896	10′
32° 00′	.5299	.8480	.6249	1.600	1.179	1.887	58° 00′
10′	.5324	.8465	.6289	1.590	1.181	1.878	50′
20′	.5348	.8450	.6330	1.580	1.184	1.870	40′
30′	.5373	.8434	.6371	1.570	1.186	1.861	30′
40′	.5398	.8418	.6412	1.560	1.188	1.853	20′
50′	.5422	.8403	.6453	1.550	1.190	1.844	10′
33° 00′	.5446	.8387	.6494	1.540	1.192	1.836	57° 00′
10′	.5471	.8371	.6536	1.530	1.195	1.828	50′
20′	.5495	.8355	.6577	1.520	1.197	1.820	40′
30′	.5519	.8339	.6619	1.511	1.199	1.812	30′
40′	.5544	.8323	.6661	1.501	1.202	1.804	20′
50′	.5568	.8307	.6703	1.492	1.204	1.796	10′
34° 00′	.5592	.8290	.6745	1.483	1.206	1.788	56° 00′
10′	.5616	.8274	.6787	1.473	1.209	1.781	50′
20′	.5640	.8258	.6830	1.464	1.211	1.773	40′
30′	.5664	.8241	.6873	1.455	1.213	1.766	30′
40′	.5688	.8225	.6916	1.446	1.216	1.758	20′
50′	.5712	.8208	.6959	1.437	1.218	1.751	10′
35° 00′	.5736	.8192	.7002	1.428	1.221	1.743	55° 00′
10′	.5760	.8175	.7046	1.419	1.223	1.736	50′
20′	.5783	.8158	.7089	1.411	1.226	1.729	40′
30′	.5807	.8141	.7133	1.402	1.228	1.722	30′
40′	.5831	.8124	.7177	1.393	1.231	1.715	20′
50′	.5854	.8107	.7221	1.385	1.233	1.708	10′
36° 00′	.5878	.8090	.7265	1.376	1.236	1.701	54° 00′
	Cos	Sin	Cot	Tan	Csc	Sec	Angle

Table 5 / *Trigonometric Functions of θ*

(θ in degrees and minutes)

Angle	Sin	Cos	Tan	Cot	Sec	Csc	
36° 00′	.5878	.8090	.7265	1.376	1.236	1.701	**54° 00′**
10′	.5901	.8073	.7310	1.368	1.239	1.695	50′
20′	.5925	.8056	.7355	1.360	1.241	1.688	40′
30′	.5948	.8039	.7400	1.351	1.244	1.681	30′
40′	.5972	.8021	.7445	1.343	1.247	1.675	20′
50′	.5995	.8004	.7490	1.335	1.249	1.668	10′
37° 00′	.6018	.7986	.7536	1.327	1.252	1.662	**53° 00′**
10′	.6041	.7969	.7581	1.319	1.255	1.655	50′
20′	.6065	.7951	.7627	1.311	1.258	1.649	40′
30′	.6088	.7934	.7673	1.303	1.260	1.643	30′
40′	.6111	.7916	.7720	1.295	1.263	1.636	20′
50′	.6134	.7898	.7766	1.288	1.266	1.630	10′
38° 00′	.6157	.7880	.7813	1.280	1.269	1.624	**52° 00′**
10′	.6180	.7862	.7860	1.272	1.272	1.618	50′
20′	.6202	.7844	.7907	1.265	1.275	1.612	40′
30′	.6225	.7826	.7954	1.257	1.278	1.606	30′
40′	.6248	.7808	.8002	1.250	1.281	1.601	20′
50′	.6271	.7790	.8050	1.242	1.284	1.595	10′
39° 00′	.6293	.7771	.8098	1.235	1.287	1.589	**51° 00′**
10′	.6316	.7753	.8146	1.228	1.290	1.583	50′
20′	.6338	.7735	.8195	1.220	1.293	1.578	40′
30′	.6361	.7716	.8243	1.213	1.296	1.572	30′
40′	.6383	.7698	.8292	1.206	1.299	1.567	20′
50′	.6406	.7679	.8342	1.199	1.302	1.561	10′
40° 00′	.6428	.7660	.8391	1.192	1.305	1.556	**50° 00′**
10′	.6450	.7642	.8441	1.185	1.309	1.550	50′
20′	.6472	.7623	.8491	1.178	1.312	1.545	40′
30′	.6494	.7604	.8541	1.171	1.315	1.540	30′
40′	.6517	.7585	.8591	1.164	1.318	1.535	20′
50′	.6539	.7566	.8642	1.157	1.322	1.529	10′
41° 00′	.6561	.7547	.8693	1.150	1.325	1.524	**49° 00′**
10′	.6583	.7528	.8744	1.144	1.328	1.519	50′
20′	.6604	.7509	.8796	1.137	1.332	1.514	40′
30′	.6626	.7490	.8847	1.130	1.335	1.509	30′
40′	.6648	.7470	.8899	1.124	1.339	1.504	20′
50′	.6670	.7451	.8952	1.117	1.342	1.499	10′
42° 00′	.6691	.7431	.9004	1.111	1.346	1.494	**48° 00′**
10′	.6713	.7412	.9057	1.104	1.349	1.490	50′
20′	.6734	.7392	.9110	1.098	1.353	1.485	40′
30′	.6756	.7373	.9163	1.091	1.356	1.480	30′
40′	.6777	.7353	.9217	1.085	1.360	1.476	20′
50′	.6799	.7333	.9271	1.079	1.364	1.471	10′
43° 00′	.6820	.7314	.9325	1.072	1.367	1.466	**47° 00′**
10′	.6841	.7294	.9380	1.066	1.371	1.462	50′
20′	.6862	.7274	.9435	1.060	1.375	1.457	40′
30′	.6884	.7254	.9490	1.054	1.379	1.453	30′
40′	.6905	.7234	.9545	1.048	1.382	1.448	20′
50′	.6926	.7214	.9601	1.042	1.386	1.444	10′
44° 00′	.6947	.7193	.9657	1.036	1.390	1.440	**46° 00′**
10′	.6967	.7173	.9713	1.030	1.394	1.435	50′
20′	.6988	.7153	.9770	1.024	1.398	1.431	40′
30′	.7009	.7133	.9827	1.018	1.402	1.427	30′
40′	.7030	.7112	.9884	1.012	1.406	1.423	20′
50′	.7050	.7092	.9942	1.006	1.410	1.418	10′
45° 00′	.7071	.7071	1.000	1.000	1.414	1.414	**45° 00′**
	Cos	Sin	Cot	Tan	Csc	Sec	Angle

Table 6 / *Trigonometric Functions of θ*

(θ in radians)

θ Radians	θ Degrees	sin θ	cos θ	tan θ	cot θ	sec θ	csc θ
0.00	0° 00′	0.0000	1.000	0.0000	Undefined	1.000	Undefined
.01	0° 34′	.0100	1.000	.0100	100.0	1.000	100.0
.02	1° 09′	.0200	0.9998	.0200	49.99	1.000	50.00
.03	1° 43′	.0300	0.9996	.0300	33.32	1.000	33.34
.04	2° 18′	.0400	0.9992	.0400	24.99	1.001	25.01
0.05	2° 52′	0.0500	0.9988	0.0500	19.98	1.001	20.01
.06	3° 26′	.0600	.9982	.0601	16.65	1.002	16.68
.07	4° 01′	.0699	.9976	.0701	14.26	1.002	14.30
.08	4° 35′	.0799	.9968	.0802	12.47	1.003	12.51
.09	5° 09′	.0899	.9960	.0902	11.08	1.004	11.13
0.10	5° 44′	0.0998	0.9950	0.1003	9.967	1.005	10.02
.11	6° 18′	.1098	.9940	.1104	9.054	1.006	9.109
.12	6° 53′	.1197	.9928	.1206	8.293	1.007	8.353
.13	7° 27′	.1296	.9916	.1307	7.649	1.009	7.714
.14	8° 01′	.1395	.9902	.1409	7.096	1.010	7.166
0.15	8° 36′	0.1494	0.9888	0.1511	6.617	1.011	6.692
.16	9° 10′	.1593	.9872	.1614	6.197	1.013	6.277
.17	9° 44′	.1692	.9856	.1717	5.826	1.015	5.911
.18	10° 19′	.1790	.9838	.1820	5.495	1.016	5.586
.19	10° 53′	.1889	.9820	.1923	5.200	1.018	5.295
0.20	11° 28′	0.1987	0.9801	0.2027	4.933	1.020	5.033
.21	12° 02′	.2085	.9780	.2131	4.692	1.022	4.797
.22	12° 36′	.2182	.9759	.2236	4.472	1.025	4.582
.23	13° 11′	.2280	.9737	.2341	4.271	1.027	4.386
.24	13° 45′	.2377	.9713	.2447	4.086	1.030	4.207
0.25	14° 19′	0.2474	0.9689	0.2553	3.916	1.032	4.042
.26	14° 54′	.2571	.9664	.2660	3.759	1.035	3.890
.27	15° 28′	.2667	.9638	.2768	3.613	1.038	3.749
.28	16° 03′	.2764	.9611	.2876	3.478	1.041	3.619
.29	16° 37′	.2860	.9582	.2984	3.351	1.044	3.497
0.30	17° 11′	0.2955	0.9553	0.3093	3.233	1.047	3.384
.31	17° 46′	.3051	.9523	.3203	3.122	1.050	3.278
.32	18° 20′	.3146	.9492	.3314	3.018	1.053	3.179
.33	18° 55′	.3240	.9460	.3425	2.920	1.057	3.086
.34	19° 29′	.3335	.9428	.3537	2.827	1.061	2.999
0.35	20° 03′	0.3429	0.9394	0.3650	2.740	1.065	2.916
.36	20° 38′	.3523	.9359	.3764	2.657	1.068	2.839
.37	21° 12′	.3616	.9323	.3879	2.578	1.073	2.765
.38	21° 46′	.3709	.9287	.3994	2.504	1.077	2.696
.39	22° 21′	.3802	.9249	.4111	2.433	1.081	2.630
0.40	22° 55′	0.3894	0.9211	0.4228	2.365	1.086	2.568
.41	23° 30′	.3986	.9171	.4346	2.301	1.090	2.509
.42	24° 04′	.4078	.9131	.4466	2.239	1.095	2.452
.43	24° 38′	.4169	.9090	.4586	2.180	1.100	2.399
.44	25° 13′	.4259	.9048	.4708	2.124	1.105	2.348
0.45	25° 47′	0.4350	0.9004	0.4831	2.070	1.111	2.299
.46	26° 21′	.4439	.8961	.4954	2.018	1.116	2.253
.47	26° 56′	.4529	.8916	.5080	1.969	1.122	2.208
.48	27° 30′	.4618	.8870	.5206	1.921	1.127	2.166
.49	28° 05′	.4706	.8823	.5334	1.875	1.133	2.125

Table 6 / _Trigonometric Functions of_ θ

(θ in radians)

θ Radians	θ Degrees	sin θ	cos θ	tan θ	cot θ	sec θ	csc θ
0.50	28° 39′	0.4794	0.8776	0.5463	1.830	1.139	2.086
.51	29° 13′	.4882	.8727	.5594	1.788	1.146	2.048
.52	29° 48′	.4969	.8678	.5726	1.747	1.152	2.013
.53	30° 22′	.5055	.8628	.5859	1.707	1.159	1.978
.54	30° 56′	.5141	.8577	.5994	1.668	1.166	1.945
0.55	31° 31′	0.5227	0.8525	0.6131	1.631	1.173	1.913
.56	32° 05′	.5312	.8473	.6269	1.595	1.180	1.883
.57	32° 40′	.5396	.8419	.6410	1.560	1.188	1.853
.58	33° 14′	.5480	.8365	.6552	1.526	1.196	1.825
.59	33° 48′	.5564	.8309	.6696	1.494	1.203	1.797
0.60	34° 23′	0.5646	0.8253	0.6841	1.462	1.212	1.771
.61	34° 57′	.5729	.8196	.6989	1.431	1.220	1.746
.62	35° 31′	.5810	.8139	.7139	1.401	1.229	1.721
.63	36° 06′	.5891	.8080	.7291	1.372	1.238	1.697
.64	36° 40′	.5972	.8021	.7445	1.343	1.247	1.674
0.65	37° 15′	0.6052	0.7961	0.7602	1.315	1.256	1.652
.66	37° 49′	.6131	.7900	.7761	1.288	1.266	1.631
.67	38° 23′	.6210	.7838	.7923	1.262	1.276	1.610
.68	38° 58′	.6288	.7776	.8087	1.237	1.286	1.590
.69	39° 32′	.6365	.7712	.8253	1.212	1.297	1.571
0.70	40° 06′	0.6442	0.7648	0.8423	1.187	1.307	1.552
.71	40° 41′	.6518	.7584	.8595	1.163	1.319	1.534
.72	41° 15′	.6594	.7518	.8771	1.140	1.330	1.517
.73	41° 50′	.6669	.7452	.8949	1.117	1.342	1.500
.74	42° 24′	.6743	.7385	.9131	1.095	1.354	1.483
0.75	42° 58′	0.6816	0.7317	0.9316	1.073	1.367	1.467
.76	43° 33′	.6889	.7248	.9505	1.052	1.380	1.452
.77	44° 07′	.6961	.7179	.9697	1.031	1.393	1.437
.78	44° 41′	.7033	.7109	.9893	1.011	1.407	1.422
.79	45° 16′	.7104	.7038	1.009	.9908	1.421	1.408
0.80	45° 50′	0.7174	0.6967	1.030	0.9712	1.435	1.394
.81	46° 25′	.7243	.6895	1.050	.9520	1.450	1.381
.82	46° 59′	.7311	.6822	1.072	.9331	1.466	1.368
.83	47° 33′	.7379	.6749	1.093	.9146	1.482	1.355
.84	48° 08′	.7446	.6675	1.116	.8964	1.498	1.343
0.85	48° 42′	0.7513	0.6600	1.138	0.8785	1.515	1.331
.86	49° 17′	.7578	.6524	1.162	.8609	1.533	1.320
.87	49° 51′	.7643	.6448	1.185	.8437	1.551	1.308
.88	50° 25′	.7707	.6372	1.210	.8267	1.569	1.297
.89	51° 00′	.7771	.6294	1.235	.8100	1.589	1.287
0.90	51° 34′	0.7833	0.6216	1.260	0.7936	1.609	1.277
.91	52° 08′	.7895	.6137	1.286	.7774	1.629	1.267
.92	52° 43′	.7956	.6058	1.313	.7615	1.651	1.257
.93	53° 17′	.8016	.5978	1.341	.7458	1.673	1.247
.94	53° 52′	.8076	.5898	1.369	.7303	1.696	1.238
0.95	54° 26′	0.8134	0.5817	1.398	0.7151	1.719	1.229
.96	55° 00′	.8192	.5735	1.428	.7001	1.744	1.221
.97	55° 35′	.8249	.5653	1.459	.6853	1.769	1.212
.98	56° 09′	.8305	.5570	1.491	.6707	1.795	1.204
.99	56° 43′	.8360	.5487	1.524	.6563	1.823	1.196

$\frac{\pi}{6}$ → .53

$\frac{\pi}{4}$ → .79

Table 6 / *Trigonometric Functions of θ*
(θ in radians)

θ Radians	θ Degrees	sin θ	cos θ	tan θ	cot θ	sec θ	csc θ
1.00	57° 18′	0.8415	0.5403	1.557	0.6421	1.851	1.188
1.01	57° 52′	.8468	.5319	1.592	.6281	1.880	1.181
1.02	58° 27′	.8521	.5234	1.628	.6142	1.911	1.174
1.03	59° 01′	.8573	.5148	1.665	.6005	1.942	1.166
1.04	59° 35′	.8624	.5062	1.704	.5870	1.975	1.160
1.05	60° 10′	0.8674	0.4976	1.743	0.5736	2.010	1.153
1.06	60° 44′	.8724	.4889	1.784	.5604	2.046	1.146
1.07	61° 18′	.8772	.4801	1.827	.5473	2.083	1.140
1.08	61° 53′	.8820	.4713	1.871	.5344	2.122	1.134
1.09	62° 27′	.8866	.4625	1.917	.5216	2.162	1.128
1.10	63° 02′	0.8912	0.4536	1.965	0.5090	2.205	1.122
1.11	63° 36′	.8957	.4447	2.014	.4964	2.249	1.116
1.12	64° 10′	.9001	.4357	2.066	.4840	2.295	1.111
1.13	64° 45′	.9044	.4267	2.120	.4718	2.344	1.106
1.14	65° 19′	.9086	.4176	2.176	.4596	2.395	1.101
1.15	65° 53′	0.9128	0.4085	2.234	0.4475	2.448	1.096
1.16	66° 28′	.9168	.3993	2.296	.4356	2.504	1.091
1.17	67° 02′	.9208	.3902	2.360	.4237	2.563	1.086
1.18	67° 37′	.9246	.3809	2.428	.4120	2.625	1.082
1.19	68° 11′	.9284	.3717	2.498	.4003	2.691	1.077
1.20	68° 45′	0.9320	0.3624	2.572	0.3888	2.760	1.073
1.21	69° 20′	.9356	.3530	2.650	.3773	2.833	1.069
1.22	69° 54′	.9391	.3436	2.733	.3659	2.910	1.065
1.23	70° 28′	.9425	.3342	2.820	.3546	2.992	1.061
1.24	71° 03′	.9458	.3248	2.912	.3434	3.079	1.057
1.25	71° 37′	0.9490	0.3153	3.010	0.3323	3.171	1.054
1.26	72° 12′	.9521	.3058	3.113	.3212	3.270	1.050
1.27	72° 46′	.9551	.2963	3.224	.3102	3.375	1.047
1.28	73° 20′	.9580	.2867	3.341	.2993	3.488	1.044
1.29	73° 55′	.9608	.2771	3.467	.2884	3.609	1.041
1.30	74° 29′	0.9636	0.2675	3.602	0.2776	3.738	1.038
1.31	75° 03′	.9662	.2579	3.747	.2669	3.878	1.035
1.32	75° 38′	.9687	.2482	3.903	.2562	4.029	1.032
1.33	76° 12′	.9711	.2385	4.072	.2456	4.193	1.030
1.34	76° 47′	.9735	.2288	4.256	.2350	4.372	1.027
1.35	77° 21′	0.9757	0.2190	4.455	0.2245	4.566	1.025
1.36	77° 55′	.9779	.2092	4.673	.2140	4.779	1.023
1.37	78° 30′	.9799	.1994	4.913	.2035	5.014	1.021
1.38	79° 04′	.9819	.1896	5.177	.1931	5.273	1.018
1.39	79° 39′	.9837	.1798	5.471	.1828	5.561	1.017
1.40	80° 13′	0.9854	0.1700	5.798	0.1725	5.883	1.015
1.41	80° 47′	.9871	.1601	6.165	.1622	6.246	1.013
1.42	81° 22′	.9887	.1502	6.581	.1519	6.657	1.011
1.43	81° 56′	.9901	.1403	7.055	.1417	7.126	1.010
1.44	82° 30′	.9915	.1304	7.602	.1315	7.667	1.009
1.45	83° 05′	0.9927	0.1205	8.238	0.1214	8.299	1.007
1.46	83° 39′	.9939	.1106	8.989	.1113	9.044	1.006
1.47	84° 14′	.9949	.1006	9.887	.1011	9.938	1.005
1.48	84° 48′	.9959	.0907	10.98	.0911	11.03	1.004
1.49	85° 22′	.9967	.0807	12.35	.0810	12.39	1.003

$\dfrac{\pi}{3}$ →

Table 6 / *Trigonometric Functions of θ*
(θ in radians)

θ Radians	θ Degrees	sin θ	cos θ	tan θ	cot θ	sec θ	csc θ
1.50	85° 57′	0.9975	0.0707	14.10	0.0709	14.14	1.003
1.51	86° 31′	.9982	.0608	16.43	.0609	16.46	1.002
1.52	87° 05′	.9987	.0508	19.67	.0508	19.70	1.001
1.53	87° 40′	.9992	.0408	24.50	.0408	24.52	1.001
1.54	88° 14′	.9995	.0308	32.46	.0308	32.48	1.000
1.55	88° 49′	0.9998	0.0208	48.08	0.0208	48.09	1.000
1.56	89° 23′	.9999	.0108	92.62	.0108	92.63	1.000
1.57	89° 57′	1.000	.0008	1256	.0008	1256	1.000

$\dfrac{\pi}{2} \longrightarrow$

Acknowledgments

Book designed by Ligature, Inc.
Technical Art by Precision Graphics
Illustrations by Gary Torrisi
Cover design by Ligature, Inc.
Cover and chapter opener photos by John Payne Photo, Ltd.;
    Chin Wai-Lam/Image Bank/Chicago (background photo)

Photos

xiv	Peter Chapman	356	Index Stock
xv	Peter Chapman	362	John Payne
14	Cleo Freelance Photography	422	Hubertus Kanus/Shostal Associates
25	NASA/Peter Arnold, Inc.	423	Steve Dunwell
39	NASA	425	NASA
43	Friend/Denny/Stock Shop	436	Index Stock
51	Philip Jon Bailey/Picture Cube	472	Arjen Nerkaik/Stock Market
64	Michael L. Abramson/Woodfin Camp	487	Jerome Wexler/Uniphoto
65	Nancy Sheehan	493	Fred Whitehead/Nawrocki
116	Russ Kinne/Comstock	514	FPG
133	Norman Owen Tomalin/Bruce Coleman	535	David Witbeck/Picture Cube
141	Gabe Palmer/After Image	567	Jordan Coonrad/Light Images
151	Greg Mancuso/After Image	578	Philip Jon Bailey/Stock Boston
159	John Marshall	609	Bob Daemmrich/Stock Boston
183	Bob McKeever/Tom Stack and Associates	634	Eric Carle/Shostal Associates
200	Shostal Associates	673	Ron Jantz/Folio, Inc.
221	Index Stock	675	© Rand McNally, and Co., R.L.
243	Dave Bartuff/FPG	717	Nancy A. Santullo/Stock Market
251	David Stoecklin/Stock Market	718	John Scheiber/Stock Market
280	James F. Palka/Nawrocki	733	Tom Stack and Associates
292	Manley Features/Shostal Associates	749	Shostal Associates
304	Coco McCoy/Rainbow	760	Marty Heitner/Taurus Photo
337	Norman Prince	774	Dan McCoy/Rainbow
344	Jeffrey W. Myers/FPG	781	Bob Daemmrich

Introducing Explorations

The following sixteen pages provide you with activities for exploring various concepts of algebra. The activities give you a chance to discover for yourself some of the ideas presented in this textbook. They carefully lead you to interesting conclusions and applications; they will make some of the abstract concepts of algebra easier to understand.

Some of the questions in these activities are open-ended. They ask you to describe, explain, analyze, design, summarize, write, predict, check, generalize, and recognize patterns. Often there is more than one correct way to answer a question. You can work on these activities by yourself or in small groups. You will need to use the materials listed below.

With these explorations we hope you enjoy exploring algebra!

Use With Lessons	Titles	Materials	Pages
1-1	Exploring Irrational Numbers	compass, ruler, graph paper, string, scissors	832
2-1	Exploring Inequalities	calculator	833
3-8	Exploring Functions	graph paper	834
4-5, 4-6	Exploring Polynomial Factors	algebra tiles	835
5-7	Exploring Continued Fractions	calculator	836
6-2	Exploring Radicals	calculator	837
7-5	Exploring Quadratic Equations	computer or graphing calculator	838
8-1	Exploring Direct Variation	springs and weights	839
9-6	Exploring Circles and Ellipses	computer	840
10-2	Exploring Powers and Roots	computer or graphing calculator	841
11-7	Exploring Pascal's Triangle	none needed	842
12-2	Exploring Trigonometric Ratios	protractor, straightedge, metric ruler, scientific calculator	843
13-4	Exploring Sine Curves	ruler, graph paper	844
14-3	Exploring Polar Coordinate Equations	computer or graphing calculator	845
15-9	Exploring Probability with Experiments	coins, computer	846
16-3	Exploring Matrices in Geometry	graph paper	847

Explorations

Exploring Irrational Numbers

Use before Lesson 1-1

In this activity, you will explore locating irrational numbers on a number line.

Explore Square Roots

1. a. Draw a large number line on graph paper. Label point A at 0 and point B at 1. Draw $\overline{BC}$ perpendicular to the number line at B and having length 1. Draw $\overline{AC}$. What kind of triangle have you constructed?
  b. Find the length of the legs and hypotenuse of $\triangle ABC$.

2. a. Use a compass to locate point D on the number line with $AD = AC$. Point D should be to the right of point B. Draw $\overline{DE}$ perpendicular to the number line at D and having length 1. Draw $\overline{AE}$. Find the lengths of the legs and hypotenuse of $\triangle ADE$.
  b. What is the coordinate of point D? Write it on your number line.

3. a. Use a compass to locate point F on the number line with $AF = AE$. Point F should be to the right of point D. Draw $\overline{FG}$ perpendicular to the number line at F and having length 1. Draw $\overline{AG}$. Find the lengths of the legs and hypotenuse of $\triangle AFG$.
  b. What is the coordinate of point F? Write it on your number line.

4. Continue this pattern, drawing three new triangles and locating square roots on the number line. Which irrational numbers have you located?

Explore π

5. a. Draw a large number line on graph paper. Cut a string to form a circle with diameter 1 unit, as shown at the right. Use the formula $C = \pi d$ to find the circumference of the circle.

  b. Now place the string on your number line, with one end at zero. At what point on the number line does the other end of the string lie?

Use What You Have Observed

6. Describe how you can locate $\sqrt{8}$ and $-\sqrt{5}$ on the number line by using geometric constructions.

7. Using the method of Exercise **5**, how can you locate 3π and $\frac{\pi}{2}$ on the number line?

Exploring Inequalities

Use with Lesson 2-1

In this activity, you will explore statements about inequalities. These statements may be false in general and yet be true for certain replacements of the variables.

Explore Multiplication

1. Investigate this statement: If $a < b$, then $ac < bc$. A calculator will be useful.
   a. First choose a value for c such that $c > 0$. Choose various values for a and b so that you consider all possible cases. The table below will help you to organize your work. Make a table and fill in the values you choose.

a	b	c	ac	bc	$ac < bc$ true?
choose $a > 0$	choose $b > 0$				
choose $a < 0$	choose $b > 0$				
choose $a < 0$	choose $b < 0$				
$a = 0$					
	$b = 0$				

   b. Next choose a value for c such that $c < 0$, and make a new table.
   c. Now let $c = 0$. Is $ac < bc$ true? Why?
   d. For which case(s) is the statement true? For which case(s) is it false?

2. Investigate this statement: If $a > b$ and $c > d$, then $ac > bd$.
   a. Do you think this statement is always true?
   b. Choose values for a, b, c, and d until you find a case for which the statement is false.

Explore Reciprocals

3. Investigate this statement: If $a > b$, then $\frac{1}{a} < \frac{1}{b}$.

   a. Choose $a > 0$ and $b > 0$. Does the statement appear to be always true?
   b. Choose $a < 0$ and $b < 0$. Does the statement appear to be always true?

   c. Find a case for which the statement is *not* true.

Use What You Have Observed

4. Decide whether this generalization is true or false: If two numbers are larger than two other numbers, then the product of the larger numbers is greater than the product of the smaller numbers. Explain your answer.

5. What is the error in this solution? $\frac{1}{x+2} > \frac{1}{2}$
$$x + 2 < 2$$
$$x < 0$$

Exploring Functions

Use with Lesson 3-8

In this activity, you will explore properties and graphs of two special functions.

Explore a Real-Life Function

An overnight carrier service will deliver a package weighing 1 lb or less for $4. The delivery charge for a package weighing more than 1 lb but not more than 2 lb is $8. The charge for a package weighing more than 2 lb but not more than 3 lb is $12.

1. If the weight of a package is given in ounces, how can you determine the delivery charge for the package?

2. Give the delivery charge for packages of the following weights.

   a. 1 oz; 2 oz; 3 oz; $5\frac{1}{2}$ oz; 8 oz; 14 oz; 16 oz

   b. 17 oz; 20 oz; $26\frac{1}{2}$ oz; 30 oz; $31\frac{1}{2}$ oz; 32 oz

   c. 33 oz; 40 oz; 42 oz; 45 oz; 46 oz; 48 oz

3. Draw a graph with weights shown on the horizontal axis and delivery charges shown on the vertical axis for all packages weighing from 0 lb through 3 lb.

4. Based on the results of Exercises **2** and **3,** is every weight assigned to exactly one delivery charge? Is the delivery charge a *function* of the weight of the package?

Explore a Rounding Function

5. The following rule describes a particular function.

   > Given any number x,
   > $f(x) = x$ if x is an integer
   > $f(x) =$ the next integer less than x if x is not an integer

   Find $f(x)$ for each of the following values of x.

   a. 2   **b.** $3\frac{1}{2}$ **c.** $3.\overline{6}$ **d.** $\sqrt{2}$

6. Graph the function described in Exercise **5** for all x such that $0 \le x < 4$.

Use What You Have Observed

7. Why might the two functions considered here be called *step* functions?

8. Refer to the rule for the rounding-down function given in Exercise **5.** Describe a similar rounding-up function and graph it for $0 < x \le 4$.

Exploring Polynomial Factors

Use before Lessons 4-5 and 4-6

You can use square and rectangular tiles like those shown below to represent algebraic quantities. In this activity, you will use them to factor polynomials.

x-by-x tile

Area = x^2

1-by-x tile

Area = x

1-by-1 tile

Area = 1

We will let blue tiles represent positive quantities (x^2, x, and 1), and let red tiles represent negative quantities ($-x^2$, $-x$, and -1).

Explore Finding Factors

1. Represent $3x^2 + 6x$ with three blue x-by-x tiles and six blue 1-by-x tiles as shown at the right.

   a. What are the lengths of the sides of the rectangle?
   b. How would you express $3x^2 + 6x$ in factored form?

2. Use tiles to factor $4x^2 + 8x + 3$.

   a. What tiles can you use to represent the given polynomial?
   b. Arrange the tiles to form a rectangle as shown.
   c. What are the lengths of the sides of the rectangle?
   d. What are the factors of $4x^2 + 8x + 3$?

3. Use tiles to factor $x^2 + x - 12$.
   a. What tiles can you use to represent the given polynomial?
   b. Try to form a rectangle using only these tiles. Is it possible?
   c. Arrange your tiles as shown at the right. To complete the rectangle, add 3 blue 1-by-x tiles along the top and 3 red 1-by-x tiles along the left side. Why can we do this without changing the value of the polynomial?

   d. What are the lengths of the sides of the rectangle?
   e. What are the factors of $x^2 + x - 12$?

Use What You Have Observed

4. Use tiles to factor each polynomial.
   a. $5x^2 - 10x$
   b. $x^2 + 9x + 14$
   c. $x^2 - 5x + 6$
   d. $3x^2 + 7x + 2$
   e. $x^2 - 16$
   f. $x^2 - 2x + 1$

Exploring Continued Fractions

Use with Lesson 5-7

The two expressions below are continued fractions. Example A is finite, or terminating; Example B is infinite, or nonterminating. Both are *simple* continued fractions because each numerator is 1. Nonsimple continued fractions have at least one numerator that is not 1. We will explore simple continued fractions.

Example A $2 + \cfrac{1}{3 + \cfrac{1}{4 + \cfrac{1}{2}}}$

Example B $1 + \cfrac{1}{2 + \cfrac{1}{2 + \cfrac{1}{2 + \cdots}}}$

Explore Finite Continued Fractions

1. Evaluate Example A above to find a simple fraction.

2. Find a simple fraction for each continued fraction below.

   a. $2 + \cfrac{1}{3 + \cfrac{1}{4}}$ **b.** $3 + \cfrac{1}{2 + \cfrac{1}{1 + \cfrac{1}{6 + \cfrac{1}{2}}}}$

3. A simple continued fraction can be represented as $[a_1, a_2, a_3, \ldots, a_n]$, where a_1 through a_{n-1} are the whole numbers preceding the plus signs and a_n is the last denominator. Example A above is $[2, 3, 4, 2]$. Write the two continued fractions in Exercise **2** in this form.

4. Write each simple continued fraction below in expanded form and evaluate it.
   a. $[2, 1, 3, 4]$ **b.** $[4, 2, 3]$

Explore Infinite Continued Fractions

5. Example B can be expressed as $[1, \overline{2}]$, which means $[1, 2, 2, 2, \ldots]$. Using a calculator, evaluate the following. Each is a number of terms of $[1, \overline{2}]$.
   a. $[1, 2]$ **b.** $[1, 2, 2]$ **c.** $[1, 2, 2, 2]$
   d. $[1, 2, 2, 2, 2]$ **e.** $[1, 2, 2, 2, 2, 2]$
   f. Analyze your answers in parts **a–e**. Each successive value is a closer approximation of what irrational number (square root)?

6. Evaluate the first several terms of each continued fraction below to find an irrational number the fraction appears to represent.
   a. $[1, \overline{1, 2}]$ **b.** $[2, \overline{2, 4}]$

Use What You Have Observed

7. Based on these explorations, make a conjecture about the kind of numbers represented by finite continued fractions and infinite continued fractions.

Exploring Radicals

Use before Lesson 6-2

In this activity, you will use a calculator to explore some properties of radicals.

Explore Addition and Subtraction

1. Use a calculator to evaluate each pair of expressions.

   a. $\sqrt{64} + \sqrt{36}$ and $\sqrt{64 + 36}$
        b. $\sqrt{15} - \sqrt{8}$ and $\sqrt{15 - 8}$

   c. $\sqrt[3]{27} + \sqrt[3]{125}$ and $\sqrt[3]{27 + 125}$
        d. $\sqrt[3]{27} - \sqrt[3]{64}$ and $\sqrt[3]{27 - 64}$

2. Based on the results of Exercise **1**, does each of the following generalizations appear to be true if $\sqrt[n]{a}$ and $\sqrt[n]{b}$ are any real numbers?

   a. $\sqrt[n]{a} + \sqrt[n]{b} = \sqrt[n]{a + b}$
        b. $\sqrt[n]{a} - \sqrt[n]{b} = \sqrt[n]{a - b}$

Explore Multiplication and Division

3. Use a calculator to evaluate each pair of expressions.

   a. $\sqrt{64} \cdot \sqrt{9}$ and $\sqrt{64 \cdot 9}$
        b. $\dfrac{\sqrt{24}}{\sqrt{2}}$ and $\sqrt{\dfrac{24}{2}}$

   c. $\sqrt[3]{7} \cdot \sqrt[3]{2}$ and $\sqrt[3]{7 \cdot 2}$
        d. $\dfrac{\sqrt[3]{27}}{\sqrt[3]{8}}$ and $\sqrt[3]{\dfrac{27}{8}}$

4. Based on the results of Exercise **3**, does each of the following generalizations appear to be true if $\sqrt[n]{a}$ and $\sqrt[n]{b}$ are any real numbers?

   a. $\sqrt[n]{a} \cdot \sqrt[n]{b} = \sqrt[n]{ab}$
        b. $\dfrac{\sqrt[n]{a}}{\sqrt[n]{b}}$ and $\sqrt[n]{\dfrac{a}{b}}$ if $b \neq 0$

Explore Roots and Powers

5. Use a calculator to evaluate each pair of expressions.

   a. $\sqrt[3 \cdot 2]{64}$ and $\sqrt[3]{\sqrt{64}}$
     b. $\sqrt[2 \cdot 3]{18}$ and $\sqrt{\sqrt[3]{18}}$
     c. $\sqrt[3]{8^2}$ and $(\sqrt[3]{8})^2$
     d. $\sqrt{2^3}$ and $(\sqrt{2})^3$

6. Based on the results of Exercise **5**, does each of the following generalizations appear to be true if each radical represents a real number?

   a. $\sqrt[nq]{b} = \sqrt[n]{\sqrt[q]{b}}$
        b. $\sqrt[n]{b^m} = (\sqrt[n]{b})^m$

Use What You Have Observed

7. Mentally evaluate each of the following.

   a. $\sqrt{9 + 16}$
        b. $\sqrt{100 - 64}$
        c. $\sqrt{16 \cdot 36}$
        d. $\sqrt[3]{\dfrac{125}{64}}$

Exploring Quadratic Equations

Use before Lesson 7-5

In this activity, you will use a graphing calculator or a computer with graphing software to explore equations of the form
$$y - k = a(x - h)^2.$$
The form above is equivalent to
$$y = a(x - h)^2 + k.$$
If you let $a = 1$, $h = 0$, and $k = 0$, the equation becomes
$$y = x^2.$$
The graph of this equation is shown at the right.

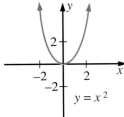

Explore Changes in the Value of a

1. Use the software to graph $y = ax^2$ for several values of a, with $a > 0$. For example, let $a = \frac{1}{4}, \frac{1}{3}, \frac{1}{2}, 2, 3$, and 4. What do you observe?

2. Now graph $y = ax^2$ for several values of a, with $a < 0$.
   a. In which direction does the graph of $y = ax^2$ open if $a > 0$? If $a < 0$?
   b. What can you conclude about the shape of the graph from the value of $|a|$?

Explore Changes in the Value of h

3. **a.** Graph $y = 2(x - h)^2$ for $h = 0$.
   b. Increase the value of h by 1. Graph this equation.
   c. Increase the value of h by 1 several more times. Graph each equation. Describe what you observe.

4. Repeat Exercise **3**, but decrease the value of h. What do you observe?

Explore Changes in the Value of k

5. **a.** Graph $y = 2(x - h)^2 + k$ for $h = 0$ and $k = 0$.
   b. Increase the value of k by 1 several times. Graph each equation.
   c. Decrease the value of k by 1 several times. Graph each equation.
   d. Describe what you observe.

Use What You Have Observed

6. How can you obtain the graph of each equation from the graph of $y = 2x^2$?
   a. $y = 2(x - 3)^2 + 1$ **b.** $y = 2(x + 7)^2 - 3$

7. How would you expect the graph of each of the following to differ from the graph of $y = 2(x - 3)^2 + 1$?
   a. $y = 4(x - 3)^2 + 1$ **b.** $y = -\frac{1}{2}(x - 3)^2 + 1$

Exploring Direct Variation

Use before Lesson 8-1

In this activity, you will explore a special type of linear function, called a direct variation, by using springs and weights.

Explore with a Spring

1. Hang the first spring from the stand. Find its length to the nearest 0.1 cm.

2. Attach a weight to the spring. Record the weight to the nearest 0.01 kg. Measure and record the new length of the spring. Compute and record the difference between this length and the length you found in step 1. (This is called the *stretch* of the spring.)

3. Repeat step 2 for two different weights.

Weight (kg)	Original Length (cm)	New Length (cm)	Stretch (cm)	Stretch ÷ Weight

4. Complete the table by finding stretch ÷ weight.

5. Write a statement about stretch ÷ weight for the spring.

Explore with More Springs

6. Repeat Exercises 1–5 for a second spring. Does your last answer change?

7. Repeat Exercises 1–5 for a third spring. Does your last answer change?

Use What You Have Observed

8. Use the data you recorded in the first table to predict the stretch for the first spring for a weight $1\frac{1}{2}$ times that of the heaviest weight used. Check by measuring.

9. Exercise 8 required you to think of stretch (of the spring) as a function of weight. This function can be described by the equation $y = kx$. Refer to the table for the questions below.
   a. In which column are the values of x (the independent variable) located?
   b. In which column are the corresponding values of y (the dependent variable) located?
   c. Where in the table was the value of k determined?

Exploring Circles and Ellipses

Use before Lesson 9-6

You have already seen in Lesson 9-2 that the following equation represents a circle with center (h, k) and radius r.

$$(x - h)^2 + (y - k)^2 = r^2$$

Graphing software makes it easy to explore such graphs. But you might have to graph the equation in the following two parts.

$$y = \sqrt{r^2 - (x - h)^2} + k \qquad\qquad y = -\sqrt{r^2 - (x - h)^2} + k$$

Explore Circles

1. Let $(h, k) = (3, 1)$ and $r = 2$. Using graphing software, graph the equation above. What do you observe?

2. Choose several different sets of values for (h, k) and r and graph each equation. What do you observe?

Explore Ellipses

You have already seen that the following equation represents an ellipse with center $(0, 0)$, horizontal axis of length $2a$, and vertical axis of length $2b$.

$$\frac{x^2}{a^2} + \frac{y^2}{b^2} = 1$$

Now you will graph equations of this form.

$$\frac{(x - h)^2}{a^2} + \frac{(y - k)^2}{b^2} = 1$$

You might have to graph this equation in the following two parts.

$$y = b\sqrt{1 - \frac{(x - h)^2}{a^2}} + k \qquad\qquad y = -b\sqrt{1 - \frac{(x - h)^2}{a^2}} + k$$

3. Let $(h, k) = (5, 2)$, $a = 3$, and $b = 4$. Graph the equation. What do you observe?

4. Choose several different sets of values of (h, k), a, and b and graph each equation. What do you observe?

Use What You Have Observed

5. From the graph of the unit circle $x^2 + y^2 = 1$, describe how you can obtain the graph of each equation below.

   a. $x^2 + y^2 = 4$ **b.** $x^2 + y^2 = \frac{1}{16}$ **c.** $(x - (-2))^2 + (y - 4)^2 = 9$

   d. $\frac{x^2}{4} + \frac{y^2}{9} = 1$ **e.** $\frac{(x - 1)^2}{4} + \frac{(y - 3)^2}{9} = 1$ **f.** $\frac{(x - 2)^2}{\frac{1}{4}} + \frac{(y - (-3))^2}{16} = 1$

Exploring Powers and Roots

Use before Lesson 10-2

In this activity, you will see how graphs can be used to explore and investigate powers and roots. You will need a graphing calculator or graphing software.

Explore Powers

1. Use the software to graph $y = x$, $y = x^3$, and $y = x^5$. If the graphing software you are using can display more than one graph at the same time, your combined graph will look like the figure at the right.
   a. In which quadrants do all three graphs lie?
   b. Which equation's graph is closest to the y-axis for $0 < x < 1$?
   c. Which equation's graph is closest to the y-axis for $x > 1$?

2. Think about the graphs $y = x^{13}$ and $y = x^{15}$. (Do not graph them yet.)
   a. In which quadrants will these graphs lie?
   b. Which equation's graph will be closer to the y-axis for $0 < x < 1$? for $x > 1$?
   c. Graph the equations to see whether your answers to part **b** were correct.
   d. Change the scale to examine the curves for $0 < x < 1$ and $0 < y < 0.001$. Then examine your curves for $1 < x < 1.001$ and $1 < y < 1.01$. Are your conclusions in part **b** verified in these magnified views?

3. Graph the three equations $y = x^2$, $y = x^4$, and $y = x^6$.
   a. In which quadrants do these graphs lie?
   b. Which equation's graph is closest to the y-axis for $0 < x < 1$? for $x > 1$?
   c. Which is larger, $(0.9)^2$ or $(0.9)^6$? Which is larger, $(1.01)^2$ or $(1.01)^6$?

Explore Roots

4. Graph the equation $y = x^2 - 3$.
   a. Between what pairs of integers does this graph cross the x-axis?
   b. What are the roots of the equation $x^2 - 3 = 0$?
   c. What is your best (decimal) estimate of $\sqrt{3}$? (Hint: Estimate where the graph crosses the positive x-axis.)
   d. Obtain a more precise estimate of $\sqrt{3}$ by using the zoom feature.

Use What You Have Observed

5. For $a > 0$ where n and m are positive integers, what conditions are necessary for $a^m > a^n$?

6. Use graphing software to find the root(s) of $x^2 + 4x + 1 = 0$ to three decimal places.

Exploring Pascal's Triangle

Use before Lesson 11-7

The following triangular array of numbers is called Pascal's triangle. Notice that, except for the 1's, each number in the triangle is the sum of the two numbers above it. For example, $15 = 10 + 5$.

```
            1                        Row 0
          1   1                          1
        1   2   1                        2
      1   3   3   1                      3
    1   4   6   4   1                    4
  1   5  10  10   5   1                  5
1   6  15  20  15   6   1                6
         ⋮
```

Explore Sets of Numbers

1. Complete Rows 7 and 8 of Pascal's triangle.

2. Find a diagonal in Pascal's triangle that consists of the set of positive integers.

3. Find a diagonal that consists of the set of triangular numbers.

4. Rearrange Pascal's triangle so the rows align on the left. Then find the sum along each diagonal as shown at right. What sequence of numbers is formed?

5. Find the sums along each row of Pascal's triangle. What sequence of numbers is formed? Express each number as a power of 2.

Explore Patterns

6. Alternate – and + signs between elements in any row (except row 0). What is the sum?

7. Find the sum of the squares of all the elements in any row (except row 0). Can you find the resulting sum elsewhere in Pascal's triangle?

Use What You Have Observed

8. Many patterns like those above have been discovered about Pascal's triangle, and new ones are still being discovered. Explore the triangle further to discover a pattern on your own.

Exploring Trigonometric Ratios

Use before Lesson 12-2

In this activity, you will explore the trigonometric ratios tangent, sine, and cosine, abbreviated tan, sin, and cos. Each ratio is defined below in terms of the lengths of the sides of a right triangle.

$$\tan x = \frac{\text{opp.}}{\text{adj.}}$$

$$\sin x = \frac{\text{opp.}}{\text{hyp.}}$$

$$\cos x = \frac{\text{adj.}}{\text{hyp.}}$$

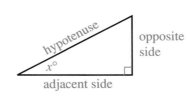

Explore the Ratios in Right Triangles

1. Make a chart similar to the one below to record your results.

Right Triangle	opp. 30°	adj. 30°	hyp.	tan 30°	sin 30°	cos 30°
ABC						
ADE						
AFG						

2. Use a protractor and a straightedge to draw a 30° angle, keeping one side horizontal.

3. Create three right triangles so that the right angle is formed with the horizontal side, as shown. Make $\overline{AB}$ at least 15 cm long.

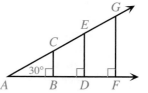

4. For each right triangle, measure to the nearest mm the side opposite the 30° angle, the side adjacent to the 30° angle, and the hypotenuse. Record the measurements in your chart.

5. Use a calculator to compute to the nearest hundredth the ratios for the last three columns of your chart. Record the results.

6. Use the trigonometric function keys on your calculator to find tan 30°, sin 30°, and cos 30°. Compare the values with those in your table.

7. Repeat Exercise **1–6** for angles of 45° and 60°.

Use What You Have Observed

8. For each of the last three columns in each chart, how do the calculated ratios compare? Why does this make sense?

Exploring Sine Curves

Use with Lesson 13-4

Light and sound travel in waves that can be represented by sine curves. When waves of the same type travel in the same plane at the same time, the resultant light or sound is the "sum" of the waves. In this activity, you will explore finding the sum of two sine curves by using graphs and investigate the connection to light and sound.

Explore Addition

Figure 1

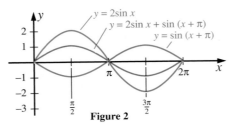

Figure 2

1. Using a ruler, measure the y-value at $x = \frac{\pi}{2}$ for $y = \sin x$ in Figure 1. By sliding the ruler upward add this y-value to the y-value for $y = 2 \sin x$. Notice that the resulting sum is the y-value for $y = 2 \sin x + \sin x$.

2. For several values of x between 0 and π in Figure 1, repeat the addition procedure described in Exercise **1**.

3. For values of x between π and 2π in Figure 1, how will the graphical addition procedure be different from that described in Exercise **1**? Why?

4. For several values of x in Figure 2, verify that the y-value of $y = 2 \sin x + \sin (x + \pi)$ is the sum of the y-values of $y = 2 \sin x$ and $y = \sin (x + \pi)$.

Use What You Have Observed

5. In each of the following, graph both equations (for $0 < x \leq 2\pi$) on the same set of axes. Then graphically find the sine curve that is the sum of the two curves.
   a. $y = \sin x$; $y = \sin 2x$ **b.** $y = \sin x$; $y = \sin (x + \frac{\pi}{2})$

6. When light or sound waves are in phase, as are $y = 2 \sin x$ and $y = \sin x$ in Figure 1, the result is called *constructive interference*. Constructive interference increases the volume of sound or the brightness of light. In Figure 2 the waves represented by $y = \sin (x + \pi)$ and $y = 2 \sin x$ are out of phase. What do you think this condition is called and what might its effect be on light or sound?

Exploring Polar Coordinate Equations

Use with Lesson 14-3

In this activity, you will use a graphing calculator or a computer with graphing software to explore various types of polar equations.

Explore Equations of Lines

1. Graph $r \cos \theta = a$ for several positive values of a, for example, $a = \frac{1}{4}, \frac{1}{2}$, 1, 2, 4. The graph for $a = 1$ is shown below. What do you observe?

2. Graph $r \cos \theta = a$ for several negative values of a. What do you observe?

3. Graph $r \sin \theta = a$ for several nonzero values of a. What do you observe?

4. Graph $r = k$ for the following values of k.
   a. $k = 1, 2, 3$ **b.** $k = -1, -2, -3$
   c. Compare the graphs in part **a** with the graphs in part **b**. How are they related?

5. Graph $p = r \cos (\theta - k)$ for several values of p and k, for example, $(p, k) =$ (1, 60°), (1, 45°), (1, 30°), (-2, 45°), and (-2, 30°). What do you observe?

Explore Other Polar Equations

6. Graph $r = 2a \cos \theta$ for several values of a. What do you observe?

7. Graph $r = 2a \sin \theta$ for several values of a. What do you observe?

8. Graph $r = 2 \cos b\theta$ for $b = 2$ and $b = 3$. What do you observe?

Use What You Have Observed

9. Based on your results in Exercises **6–8**, what would you predict to be the graphs for $r = 2 \sin b\theta$ where $b = 2$ and $b = 3$?

Exploring Probability With Experiments

Use after Lesson 15-9

In theory, if a coin is tossed, the probability of "heads" is $\frac{1}{2}$. What happens in practice? You can actually toss a coin a number of times and record the results, or you can simulate a coin toss by using random digits.

Explore Tossing a Coin

1. **a.** Toss a coin 20 times and record your results for the number of heads and the number of tails. Based on your experiment, what is the probability of "heads"? Of "tails"?
   b. Compare your results with a classmate. Are they exactly the same? Are they approximately the same? Should they be the same?
   c. Combine your results with your classmate's results. Find the probability of "heads" based on these 40 coin tosses. Is it close to $\frac{1}{2}$?
   d. Combine the results of all the students in your class, and find the probability of "heads" based on these results. Is it even closer to $\frac{1}{2}$?
   e. Compare experimental probability with theoretical probability and try to answer questions like these. Which is more accurate? more realistic? more reliable? more valid? more believable?

Explore Simulating a Coin Toss

2. Use the random number function RND on a computer to generate a list of 100 random digits 0–9. Let each even digit represent "heads" and each odd digit represent "tails." How many heads resulted from this simulation? Is the probability of "heads" approximately $\frac{1}{2}$?

3. Repeat Exercise **2** by generating a new list of 100 random digits. Are the results exactly the same? Is the probability approximately $\frac{1}{2}$?

Use What You Have Observed

If a baseball player has a batting average of .300, in theory, the player should get 3 hits in every 10 at bats. The player's probability of a hit is $\frac{300}{1000} = \frac{3}{10}$.

4. Design and describe a simulation, using random digits, for the next 100 at bats of a player whose batting average is .300.

5. Carry out the simulation you designed in Exercise **4**. What is the probability of a hit?

Exploring Matrices in Geometry

Use with Lesson 16-3

In this activity, you will use matrix multiplication to explore geometric transformations. You can use matrices to represent geometric figures such as points and triangles.

Point $(-1, 2)$
Matrix
$\begin{bmatrix} -1 \\ 2 \end{bmatrix}$

Triangle with vertices at $(-1, 2)$, $(0, 0)$, and $(2, 1)$
Matrix
$\begin{bmatrix} -1 & 0 & 2 \\ 2 & 0 & 1 \end{bmatrix}$

Explore Transforming Figures

1. Let $B = \begin{bmatrix} -1 & 0 & 2 \\ 2 & 0 & 1 \end{bmatrix}$ represent a triangle and $A = \begin{bmatrix} 1 & 0 \\ 0 & -1 \end{bmatrix}$ represent a geometric transformation. Draw B on graph paper. Find AB and draw the resulting triangle on the same axes. How are the two triangles related?

2. Find AB for a few more triangles, for example,
$$B = \begin{bmatrix} 2 & 3 & 1 \\ 3 & 5 & 2 \end{bmatrix}, B = \begin{bmatrix} -2 & -3 & -1 \\ -3 & -5 & -2 \end{bmatrix}, \text{ and } B = \begin{bmatrix} 0 & 0 & 3 \\ 2 & -1 & 1 \end{bmatrix}.$$
Does multiplying by A flip (reflect) each triangle over the x-axis?

3. Now let $A = \begin{bmatrix} 0 & -1 \\ 1 & 0 \end{bmatrix}$ and let B represent matrices for triangles as before. Find AB for several triangles. Graph the triangles represented by B and AB. What kind of geometric transformation does A seem to be?

4. Next we will represent figures and transformations in a slightly different way.
Let $A = \begin{bmatrix} 1 & 0 & 2 \\ 0 & 1 & -1 \\ 0 & 0 & 1 \end{bmatrix}$ represent a transformation and let $B = \begin{bmatrix} -1 & 0 & 2 \\ 2 & 0 & 1 \\ 1 & 1 & 1 \end{bmatrix}$ represent the same triangle as in Exercise 1. (In both of these matrices, the third row is included only to make matrix multiplication possible.) Find AB. Graph B and AB. What kind of geometric transformation does A seem to be?

Use What You Have Observed

5. Describe the transformation each matrix below represents.
    a. $A = \begin{bmatrix} -1 & 0 \\ 0 & 1 \end{bmatrix}$
    b. $A = \begin{bmatrix} -1 & 0 \\ 0 & -1 \end{bmatrix}$

6. Write a transformation matrix for a slide of 2 units to the left and 3 units up.

Portfolio Projects

To make a portfolio, an artist selects a variety of original work to represent the range of his or her skills. Each of the following projects will give you a chance to create a finished product that you will be proud to add to your algebra portfolio.

The projects will help you develop your ability to present and communicate your ideas. They will also help you develop your problem-solving and reasoning abilities as you make connections between what you know and what is new. Your individual insight and creativity will help shape the mathematics you discover.

Let these projects be springboards for further exploration. Feel free to expand them to include new questions or areas of interest that arise. Most of all, have fun!

Efficiency Expert *(Chapter 1)*

Two warehouse workers do the tasks outlined in the table to assemble and pack a chest of drawers. Certain tasks must be done before others: A before B; B before C; D before E; C and E before I; C, F, G, and H before J; and K, of course, must be done last.

	Task	Time
A.	Install drawer supports on sides.	10 min
B.	Attach sides to top and bottom.	15 min
C.	Attach back panel to assembly.	8 min
D.	Assemble door.	8 min
E.	Attach hinges to door.	9 min
F.	Assemble top drawer.	12 min
G.	Assemble middle drawer.	12 min
H.	Assemble bottom drawer.	12 min
I.	Attach door.	6 min
J.	Insert and adjust drawers.	10 min
K.	Pack the completed chest.	6 min

1. Set up a work schedule that takes the least time. Use the following work rules to assign the tasks to the two workers:

   Rule 1. Work on only one task at a time.

   Rule 2. Take breaks only between tasks.

   Rule 3. Do not work together on any task.

2. Alter the rules so that the job takes less time. Develop a new work schedule.

Swing Time *(Chapter 2)*

In major league baseball, a fastball travels the distance (60 ft 6 in.) from the pitcher to the catcher in 0.41 s. On average, a batter connects with the ball 0.28 s after the start of the swing. Using these figures, estimate how long the ball is in range to be hit by the batter. In other words, by how much time can the batter be off in the start of the swing and still hit the ball? Draw a diagram and explain your reasoning.

Parametric Graphs *(Chapter 3)*

The graph shows the path of an object moving at a constant speed in a straight line. Each point on the graph represents the position of the object at a different time. Notice that the *x*-coordinate of the starting point is 3 and that the object moves 2 units horizontally in each unit of time. The *parametric equation* $x = 3 + 2t$ gives the object's *x*-coordinate in terms of time *t*, which is called a *parameter*.

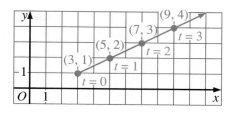

1. What is the parametric equation for *y*?

2. How would the parametric equations change if the object moved twice as fast? Write the new equations.

Another object moving at a constant speed in a straight line has the parametric equations $x = 1 + t$ and $y = 10 - t$.

3. What is an equation of the path in terms of *x* and *y*?

4. Do the paths of the two objects ever cross? Do the objects ever meet? Explain.

5. To the nearest unit of time, determine when the two objects are closest together.

45° Paths *(Chapter 4)*

Materials: *Graph paper, straightedge*

Interesting patterns result from the study of *45° paths*. A 45° path originates at a vertex on the boundary of an *m* by *n* grid. The path continues in a straight line along diagonals until it hits the grid boundary. Then it "bounces off" at the same angle (45°) and continues until it either reaches a corner of the grid or returns to its starting point, whichever occurs first. A path that begins at a corner and ends at a corner is called a *corner-to-corner path*. A path that begins at a vertex that is not at a corner and then returns to its starting point is called a *closed path*. The length of any 45° path is the total number of grid squares through which it passes.

Corner-to-Corner Path

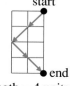

length = 4 units

Closed Path

length = 8 units

1. **a.** Draw all corner-to-corner and all closed paths for each of the following grids: 3 by 6, 4 by 6, 4 by 8, 6 by 8, 6 by 9.
   b. For each grid, record the lengths of all corner-to-corner paths and all closed paths.

2. Look for a pattern in your answers to Exercise 1.
   a. Predict the lengths of all the corner-to-corner paths for any *m* by *n* grid.
   b. Predict which *m* by *n* grids have closed paths and what the lengths of the paths are.

Travel Safety (Chapter 5)

Materials: *Almanac*

Which do you think is safer, traveling by car or by plane? Travel safety is calculated in two ways, as deaths per *passenger-mile* and deaths per *passenger-hour*. A **passenger-mile** is equal to one person traveling one mile. A **passenger-hour** is equal to one person traveling for one hour. For example, suppose a family of four travels 120 miles. This represents 4 • 120 = 480 passenger-miles. By car, at an average speed of 40 mi/h, the family's trip takes 3 h. By plane, at an average speed of 400 mi/h, it takes only 0.3 h. By car the trip represents 4 • 3 = 12 passenger-hours, but by plane it represents only 4 • 0.3 = 1.2 passenger-hours.

1. Use an almanac to find out how many deaths resulted from car accidents and how many from plane crashes in the United States in a recent year.
   a. Calculate the number of deaths per billion (10^9) passenger-miles for cars and for planes.
   b. Using the average speeds given above, calculate the number of deaths per billion passenger-hours by car and by air.
   c. Which method of travel seems to be safer? Explain.
   d. The airline industry claims that statistics show that air travel is much safer than driving. Do you agree? Which statistic do you think the industry uses?

2. Find recent statistics for travel by car, plane, bus, and train. Calculate the safety of each using passenger-miles and passenger-hours. Which seems to be the safest way to travel? Explain.

Square Roots of Complex Numbers (Chapter 6)

You can use a scientific calculator to find the square root of a real number. To find the square root of a complex number, you can use algebra. For example, to find the square root of $3 - 4i$, you will have to find $z = a + bi$ such that $z^2 = 3 - 4i$.

$$z^2 = 3 - 4i$$
$$(a + bi)^2 = 3 - 4i$$
$$a^2 - b^2 + 2abi = 3 - 4i$$
$$a^2 - b^2 = 3, \ 2ab = -4$$

1. Solve the pair of simultaneous equations $a^2 - b^2 = 3$ and $2ab = -4$. Use the results to find all values of z.

2. Write four different complex numbers. Use the demonstrated technique to find all their square roots.

3. Using your results in Exercise 2, draw a conclusion about how many square roots a complex number has and how the roots are related. How does this compare to real numbers?

Population Growth (Chapter 7)

Materials: *Calculator, graph paper*

Biologists who study how populations grow over time sometimes use mathematical models to help them. One such model is based on this quadratic equation:

$$y = x + rx(1 - x/C)$$

In the model, x represents the current size of the population, y represents the size of the next generation, and r and C are constants. While r is a growth rate (which measures the population's reproductive ability), C is the environment's carrying capacity (which is the maximum number of individuals that the population's environment can sustain). To see how the model works, let $r = 1$ and $C = 5000$. If the size of the population is $x = 1000$, then the size of the first generation is:

$$y = 1000 + 1 \cdot 1000 \cdot (1 - 1000/5000) = 1800$$

For the second generation, let $x = 1800$ and recalculate y:

$$y = 1800 + 1 \cdot 1800 \cdot (1 - 1800/5000) = 2952$$

The table and graph below show what happens to the population over several generations. Notice how the population approaches the carrying capacity.

Using a calculator, explore what happens to a population for other values of r, such as 1.5, 2, and 3. Always let $x = 1000$ initially, and keep $C = 5000$. Write a report (including tables and graphs) of your findings.

Generation	Population
0	1000
1	1800
2	2952
3	4161
4	4859
5	4996

Fitting a Function (Chapter 8)

Using a simple "trick," you can find a polynomial function whose graph goes through any given set of points. For example, the graph of the following function goes through the points (1, 3), (2, 5), and (3, –1):

$$f(x) = 3 \cdot \frac{(x-2)(x-3)}{(1-2)(1-3)} + 5 \cdot \frac{(x-1)(x-3)}{(2-1)(2-3)} + (-1) \cdot \frac{(x-1)(x-2)}{(3-1)(3-2)}$$

1. Simplify $f(x)$. Verify by substitution that $f(1) = 3$, $f(2) = 5$, and $f(3) = -1$.
2. Explain why this function works.
3. Find a polynomial function g that satisfies $g(-2) = 1$, $g(0) = 3$, $g(1) = -2$, and $g(4) = 3$. Show your calculations and simplify the polynomial.
4. What conditions must be met by a set of points to make this "trick" work?

Paper Folding a Parabola *(Chapter 9)*

Materials: *Wax paper*

1. **a.** Draw a horizontal line *l* near the bottom of a large piece of wax paper and mark a point *P* several inches above *l*.

   b. Fold the paper so that *P* coincides with a point *P′* near the right end of *l*. Open the paper and notice the crease that has been formed.

   c. Fold the paper so that *P* coincides with another point on *l*, this time a little farther to the left than before. Continue folding the paper in this way until *P* has traveled all the way across *l*.

   (a)

   (b)

   (c)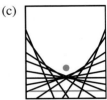

2. The shape outlined by the creases is a parabola, with each crease being tangent to the parabola. Referring to the diagram for step (b) of Exercise 1, describe how to locate the point on the crease that lies on the parabola with focus *P* and directrix *l*. Explain your reasoning.

3. Predict what will happen to the shape of the parabola if you increase the distance between *P* and *l*. Then, using another piece of wax paper, form a second parabola to confirm your prediction.

A Model of Exponential Decay *(Chapter 10)*

Materials: *Quarter-cup measure, beaker or clear jar, medicine dropper, food coloring*

A factory releases a pollutant into a pond. Suppose that each day one fourth of the pond water is replaced by fresh water from a stream that enters the pond on one side and leaves the pond on another side. Gradually, the pollutant is diluted and carried away. Use the following steps to simulate this situation.

Step 1 Start with one cup of water (representing the pond) in a beaker. Stir in one drop of food coloring (representing the pollutant).

Step 2 Remove one-quarter cup of the mixture and stir in one-quarter cup of clean water. How much of the food coloring is left?

Step 3 Remove another quarter cup of mixture and stir in another quarter cup of clean water. How much of the food coloring is left?

Step 4 Continue this process until the food coloring is no longer visible. How many steps does this take?

Step 1

Step 2

1. Interpret each step of the simulation in terms of the pollutant in the pond. Is the pollutant ever completely removed from the pond?

2. Suppose the factory releases the same amount of pollutant into the pond every day. What effect do you think this will have on the amount of pollutant present in the pond over time? Modify the simulation to test your conjecture.

The Weekly Allowance (Chapter 11)

Katerina has been negotiating with her parents for a raise in her weekly allowance. They offer her the following options:

(1) Beginning now, she can have a one-cent raise in her allowance each week.
(2) After one year, and at the end of each year thereafter, she can have a one-dollar raise in her weekly allowance.

Katerina must stick with the option she chooses for at least five years.

1. Which option should Katerina choose? Explain your reasoning.

2. Katerina bargains for a two-cent raise every week. If her parents agree, should she choose this option? Explain your reasoning.

Indirect Measurement (Chapter 12)

Materials: *Protractor, navigation compass*

Some measurements, such as the width of a pond or the height of a tree, are not easy to find with a ruler or tape measure. Since ancient times people have used trigonometry to find hard-to-obtain measurements *indirectly*.

Choose a local landmark, such as a pond, a tree, a tall statue, or a building, and then devise a method to measure its width or height using trigonometry. If possible, measure your chosen landmark in several different ways or from several different spots and compare the results that you get.

You may find some of the following measuring techniques helpful:

- To measure a distance over flat ground, determine the length of your stride and then pace off the distance you want to know.

- To find an angle of elevation, attach a weighted string to a protractor. Use the edge of the protractor to sight the top of the landmark. Be sure you understand that the angle of elevation is equal to the angle between the 90° mark on the protractor and the point where the weighted string crosses the protractor.

- To find an angle formed by your line of sight to two points on the same level, hold a compass in your hand while looking at one of the points. Note the compass bearing (the degree measure of your line of sight from north). Then turn to sight the other point and again note the compass bearing. Use the two bearings to find the angle through which you turned.

In an illustrated report, describe your method and present your data, your calculations, your results, and any conclusions you draw about their accuracy.

Get the Beat (Chapter 13)

A musical note can be modeled by a sine curve. The reciprocal of the curve's period is the note's *frequency* measured in *hertz* (Hz), the number of cycles per second.

When two notes with very nearly the same frequency are played together, the resulting tone seems to pulse, or *beat*, with the tone's volume repeatedly rising and falling. For example, beats occur if notes of frequency 8 Hz and 10 Hz are played together. The equation $y = \sin 16\pi x + \sin 20\pi x$ (x is in seconds) describes the resulting tone. (Note: Although humans cannot hear tones of such low frequency, these low frequencies will be easier to work with as an example.)

1. How are the numbers 16π and 20π related to the frequencies 8 Hz and 10 Hz?

   You can use a graphing calculator or computer with graphing software to see the beats in the graph of the equation given above. Be sure to use radian mode.

2. Set the viewing window for $-0.25 \le x \le 1.25$ and $-2.5 \le y \le 2.5$. Graph the equation $y = \sin 16\pi x + \sin 20\pi x$. What do you think the graph's changing amplitude has to do with the tone's beat?

3. How many beats do you see in the graph of $y = \sin 16\pi x + \sin 20\pi x$ for $-0.25 \le x \le 1.25$? What is the frequency of the beats? (*Hint*: How many seconds are represented by this interval of the x-axis?) Make a conjecture about the relationship of the beat frequency and the frequencies of the two notes.

4. Use graphs to test your conjecture with other pairs of notes having frequencies that are close together, such as 12 Hz and 15 Hz, 20 Hz and 21 Hz, and 35 Hz and 39 Hz. (You may want to use an interval of one second on the x-axis.)

The Stunt (Chapter 14)

A stuntman on a motorcycle will attempt to launch himself off an angled ramp, pass over 20 cars parked side-by-side, and then land on a ramp on the other side. The x- and y-components of the launch vector (shown in blue) give the stuntman his horizontal and vertical motion. His vertical position, however, is also affected by gravity. If v is the stuntman's launch speed (in ft/s) and θ is the angle the ramp makes with the horizontal, then the stuntman's horizontal and vertical positions (in ft) t seconds after launch are:

$$x = (v \cos \theta)t \qquad \text{and} \qquad y = \underbrace{(v \sin \theta)t}_{\text{vertical motion from launch vector}} - 16t^2 \longleftarrow \text{effect of gravity on vertical motion}$$

The stuntman must decide at what angle to build the ramp and what launch speed to use. Find some combinations of ramp angles and launch speeds that allow the stuntman to complete his jump successfully. Assume each car is 6 ft wide.

Simpson's Paradox *(Chapter 15)*

A baseball player's batting average is given by the ratio $\dfrac{\text{number of hits}}{\text{number of times at bat}}$.
The result is expressed as a decimal rounded to three places, although the decimal point is often omitted. For example, a ball player who had 64 hits in 400 times at bat has a batting average of $\dfrac{64}{400} = .160$, or 160.

Here are statistics for two baseball players for the months of May and June:

Player	May	June
Kapinsky	7 hits; 20 times at bat	28 hits; 100 times at bat
Mendoza	32 hits; 100 times at bat	5 hits; 20 times at bat

1. **a.** Find the batting average of each player for May and for June separately. Based on these averages, which player is the better hitter?
   b. Find the batting average of each player for the months of May and June combined. Based on these averages, which player is the better hitter?

2. The results of Exercise 1 are an example of a statistical phenomenon called *Simpson's paradox*. Find another set of statistics that displays the same behavior. Explain what it is about the data that causes Simpson's paradox.

Transition Matrices *(Chapter 16)*

In a certain metropolitan area, 300,000 people live in the city and 450,000 people live in the suburbs. Census data from the last 20 years show that each year 4% of those who live in the city move to the suburbs, while 2% of those who live in the suburbs move into the city. The two matrices shown below display this information. The right matrix, which shows the population shifts, is a *transition matrix*.

$$\text{Population} \begin{array}{cc} \text{City} & \text{Suburb} \\ [\,300{,}000 & 450{,}000\,] \end{array} = P_0$$

$$\text{From} \begin{array}{c} \\ \text{City} \\ \text{Suburb} \end{array} \overset{\displaystyle \begin{array}{cc} \text{To} \\ \text{City} \quad \text{Suburb} \end{array}}{\begin{bmatrix} 0.96 & 0.04 \\ 0.02 & 0.98 \end{bmatrix}} = T$$

2% of suburbanites move to the city.　　98% of suburbanites stay in the suburbs.

1. Find the matrix product $P_1 = P_0 T$, and verify that P_1 gives the city and suburban populations one year from now.

2. **a.** Find $P_2 = P_1 T$, $P_3 = P_2 T$, and $P_4 = P_3 T$. What does each matrix represent?
   b. Use a calculator or computer with matrix capabilities to find $P_n = P_{n-1} T$ for $n = 5, 6, 7, \ldots$. What happens? Interpret the results.

3. Suppose 3% of the suburban dwellers move to the city each year. What effect does this have on the results of Exercise 2?

Appendix

A-1 Common Logarithms: Notation and Interpolation

Objective   To find logarithms and antilogarithms using interpolation.

You can use common logarithms to find approximations for complex algebraic expressions such as

$$\frac{\sqrt{93.8}}{0.0592 \times 3.67}$$

You may recall from Lesson 10-6 that every common logarithm has two parts, a *characteristic* and a *mantissa* (see page 479). To use Table 3 as an aid in computation, the mantissa must remain positive. Consider

$$\log 0.0217 = \log (2.17 \times 10^{-2})$$
$$= \log 2.17 + \log 10^{-2}$$
$$= 0.3365 - 2.$$

To retain the positive mantissa, the characteristic -2 is usually written as $8.0000 - 10$, giving

$$\log 0.0217 = 8.3365 - 10.$$

In some cases it may be more convenient to use other differences such as $28.0000 - 30$ or $98.0000 - 100$ to represent -2. Thus, $\log 0.0217$ might be written as $28.3365 - 30$ or as $98.3365 - 100$.

Table 3 gives logarithms of numbers with three significant digits. By interpolating (see Lesson 8-9) you can find better approximations for (1) the logarithms of numbers with four significant digits and (2) the antilogarithms of numbers that fall between entries in the body of the table.

Example   Find: **a.** $\log 26.83$ **b.** antilog $(6.7382 - 10)$

Solution   **a.** Since $\log 26.83 = \log 2.683 + 1$, interpolate using Table 3 to find $\log 2.683$.

	x	$\log x$	
	2.690	0.4298	
0.010 ⎡ 0.003 ⎡	2.683	log 2.683 ⎤ d	⎤ 0.0017
	2.680	0.4281	

$$\frac{d}{0.0017} = \frac{0.003}{0.010}$$

$$d = \frac{3}{10} \times 0.0017 \approx 0.0005$$

The difference d is rounded to four decimal places because mantissas in Table 3 are accurate to only four places.

Thus, log 2.683 ≈ 0.4281 + 0.0005

$$= 0.4286$$

∴ log 26.83 ≈ 1.4286   *Answer*

b. To find antilog (6.7382 − 10), locate in the body of Table 3 the consecutive entries 7380 and 7390, between which the mantissa 7382 lies. Make note of the corresponding four-digit sequences 5.470 and 5.480 and interpolate as follows:

	x	$\log x$	
	5.480	0.7388	
0.010 $\left[\,c\left[\vphantom{}\right.\right.$ antilog 0.7382	0.7382	$\left.\right]$ 0.0002 $\left.\right]$ 0.0008	
	5.470	0.7380	

$$\frac{c}{0.010} = \frac{0.0002}{0.0008}$$

$$c = 0.0025 \approx 0.003$$

antilog 0.7382 ≈ 5.470 + 0.003

$$= 5.473$$

∴ antilog (6.7382 − 10) ≈ 5.473 × 10^{-4}

$$= 0.0005473 \quad \textit{Answer}$$

Note that the value of c was rounded to one significant figure because reverse interpolation in a four-place table yields at most four significant digits for the number whose logarithm is given. With practice, much of the interpolation can be performed mentally.

Written Exercises

Find each logarithm using interpolation.

A

1. log 3.475	**2.** log 8.612	**3.** log 77.68
4. log 239.6	**5.** log 0.5624	**6.** log 0.03857
7. log 68244	**8.** log 138760	**9.** log 0.4962
10. log 0.001634	**11.** log 48560	**12.** log 79820

Find each antilogarithm to four significant digits using interpolation.

13. antilog 0.6968	**14.** antilog 0.7728	**15.** antilog 1.3843
16. antilog 3.8402	**17.** antilog 8.9531 − 10	**18.** antilog 5.1322 − 10
19. antilog 9.9296 − 10	**20.** antilog 7.2388 − 10	**21.** antilog 6.8041 − 10

A-2 Common Logarithms: Computation

Objective   To use common logarithms to calculate products, quotients, powers, and roots.

For common logarithms, the first two laws stated on page 473 become

$$(1)\ \log p \times q = \log p + \log q \quad \text{and}$$
$$(2)\ \log p \div q = \log p - \log q.$$

When using logarithms in computations, you can use these laws to replace the operations of multiplication and division with addition and subtraction.

You can avoid making mistakes when computing with logarithms by arranging your work according to these rules:

1. Estimate the answer.
2. List numbers to be added or subtracted in vertical columns.
3. Align equality symbols and decimal points vertically.
4. Write all characteristics before looking up the mantissas.
5. Indicate the operations $(+, -, \times)$ to be performed.
6. Label each step so that if you must check back, you will know what you are checking.

The purpose of making a preliminary estimate of the answer is to avoid the common error of misplacing the decimal point.

Example 1 below does not require interpolation. In general, examples and exercises *do* require interpolation and will be worked to four-digit accuracy.

Example 1   Simplify 0.0192×4370.

Solution
1. Let $N = 0.0192 \times 4370$.
2. Estimate the value of N.

   $0.0192 \approx 0.02$
   $4370 \approx 4000$

   $\therefore N \approx 0.02 \times 4000 = 80$ (estimate)

3. $\log N = \log (0.0192 \times 4370)$
            $= \log 0.0192 + \log 4370$

   $\log 0.0192 = \quad 8.2833 - 10$
   $\underline{\log 4370 = \quad 3.6405 \qquad\qquad +}$

   $\quad \log N = 11.9238 - 10$
   $\qquad\quad N = \text{antilog } 1.9238 = 83.9$

   $\therefore 0.0192 \times 4370 = 83.9$ ***Answer***

Note that the answer and the estimate agree to one significant digit.

Only common logarithms of numbers that are powers of 10 can be stated exactly. Thus, although 0.0192×4370 is only approximately equal to 83.9, the "=" sign will be used for simplicity.

Example 2 Simplify $\dfrac{-7.394}{91.47}$.

Solution

1. Since you know the answer will be negative, let $N = \dfrac{7.394}{91.47}$ to make it easier to compute with logarithms.

2. Estimate the value of N.

$$\frac{7.394}{91.47} \approx \frac{7}{9 \times 10^1} \approx 0.7 \times 10^{-1} = 0.07$$

3. $\quad \log N = \log \dfrac{7.394}{91.47} = \log 7.394 - \log 91.47$

$\quad \underline{\log 7.394 = 0.8688 \qquad \text{(by interpolation)}}$
$\quad \log 91.47 = 1.9613 \quad - \quad \text{(by interpolation)}$

To obtain a difference in which the mantissa is positive, write 0.8688 as $10.8688 - 10$.

$\quad \log 7.394 = 10.8688 - 10$
$\quad \underline{\log 91.47 = 1.9613 -}$
$\qquad\qquad \log N = 8.9075 - 10$
$\qquad\qquad\qquad N = 0.08082 \quad \text{(by interpolation)}$

$\therefore \dfrac{-7.394}{91.47} = -0.08082$ ***Answer***

Check the answer against the estimate.

You can use the third law of logarithms, $\log p^n = n \log p$, to compute powers and roots.

Example 3 Simplify $(0.239)^7$.

Solution

1. Let $N = (0.239)^7$.

2. To estimate, round 3 factors up and 4 factors down.
$\quad N \approx (0.2)^4(0.3)^3 = (0.0016)(0.027) = 0.0000432 \quad \text{(estimate)}$

3. $\quad \log N = 7 \log 0.239$
$\quad \underline{\log 0.239 = 9.3784 - 10}$
$\qquad\qquad\qquad\qquad\qquad 7 \quad \times$
$\quad \log N = 65.6488 - 70$
$\qquad\quad = 5.6488 - 10$
$\qquad\qquad N = \text{antilog } (5.6488 - 10) = 0.00004454$
$\therefore (0.239)^7 = 0.00004454$ ***Answer***

Example 4 Simplify $\sqrt[3]{0.482}$.

Solution

1. Let $N = \sqrt[3]{0.482} = (0.482)^{1/3}$.

2. Estimate the value of N.
   $(0.7)^3 = 0.343;$ $(0.8)^3 = 0.512$
   $\therefore 0.7 < N < 0.8$

3. $\log N = \frac{1}{3}(0.482)$
   $\log 0.482 = 9.6830 - 10$

   To ensure that the negative part of the characteristic remains an integer after it is multiplied by $\frac{1}{3}$, write $9.6830 - 10$ as $29.6830 - 30$.

   $\log 0.482 = 29.6830 - 30$

   $\underline{} \quad \frac{1}{3} \quad \times$

   $\log N = 9.8943 -10$

   $N = \text{antilog } (9.8943 - 10) = 0.7841$

   $\therefore \sqrt[3]{0.482} = 0.7841$ **Answer**

Example 5 applies the laws of logarithms to several operations at once.

Example 5 Simplify $\sqrt[5]{\dfrac{(81.2)^2}{0.59 \times 367}}$. Give the answer to four significant digits.

Solution

1. Estimate: $\sqrt[5]{\dfrac{(81.2)^2}{0.59 \times 367}} \approx \sqrt[5]{\dfrac{81^2}{6 \times 36}} = \sqrt[5]{\dfrac{3^8}{3^3 \cdot 2^3}}$

   $\approx \sqrt[5]{\dfrac{3^5}{2^3}} \approx \sqrt[5]{\dfrac{243}{8}} \approx \sqrt[5]{30} \approx \sqrt[5]{32} = 2$

2. Let $H = (81.2)^2$, $K = 0.59 \times 367$, and $N = \left(\dfrac{H}{K}\right)^{1/5}$.

   Then $\log N = \frac{1}{5}(\log H - \log K)$.

$\log H = 2 \log 81.2$	$\log K = \log 0.59 + \log 367$
$\log 81.2 = 1.9096$	$\log 0.59 = 9.7709 - 10$
$\underline{} \quad 2 \quad \times$	$\underline{\log 367 = 2.5647} \quad\quad +$
$\log H = 3.8192$	$\log K = 2.3356$

   $\log H = 3.8192$
   $\underline{\log K = 2.3356} \quad -$
   $\log H - \log K = 1.4836$

   $\underline{} \quad \frac{1}{5} \quad \times$

   $\log N = 0.2967$
   $\text{antilog } 0.2967 = 1.980$

   $\therefore \sqrt[5]{\dfrac{(81.2)^2}{0.59 \times 367}} = 1.980$ **Answer**

Written Exercises

Given that log A = 3.6100, log B = 8.1234 − 10, and log C = 1.2000, find the following.

A

1. log AC
2. log BC
3. log $\dfrac{A}{B}$
4. log $\dfrac{A}{C}$

5. log $100B$
6. log $0.01A$
7. log $\dfrac{1000}{C}$
8. log $\dfrac{0.001}{B}$

Given that log A = 13.600, log B = 8.1310 − 10, and log C = 3.2148, find the following.

9. log B^2
10. log C^{10}
11. log $\sqrt{C}$
12. log $\sqrt[3]{A}$

13. log $C^{3/4}$
14. log $\sqrt[3]{B^2}$
15. log $\dfrac{\sqrt{A}}{\sqrt[3]{C}}$
16. log $\dfrac{B}{\sqrt{A}}$

Use Table 3 to evaluate each expression to three significant digits. Do not interpolate.

17. 281×0.94
18. 0.749×0.562
19. 943×804
20. 0.321×8.35

21. $\dfrac{117}{3.26}$
22. $\dfrac{4960}{54,600}$
23. $\dfrac{0.013}{427}$
24. $\dfrac{0.526}{0.049}$

25. $(2.81)^{10}$
26. $(90.3)^5$
27. $(0.395)^7$
28. $(0.0431)^2$

29. $(21.4)^{1/3}$
30. $(8.26)^{2/3}$
31. $\sqrt{12.4}$
32. $\sqrt[3]{1.36}$

Use Table 3 to evaluate each expression to four significant digits. Interpolate when necessary.

B

33. 0.572×6370
34. 23.64×1.47
35. 0.8215×0.051

36. $\dfrac{0.1496}{0.582}$
37. $\dfrac{137}{5242}$
38. $\dfrac{0.05274}{0.9412}$

39. $2.16 \times 46.73 \times 134.5$
40. $94 \times 0.05 \times 1.728$
41. $1.8 \times 32 \times 0.01347$

42. $(12.4)^3(3.86)^2$
43. $(37.4)^5(812)^2$
44. $(42.3)^4(0.0016)^2$

45. $\sqrt{\dfrac{427}{0.592}}$
46. $\sqrt[3]{\dfrac{(2.37)^2}{1.15}}$
47. $\sqrt{\dfrac{(2.03)^3}{97.5 \times 1.98}}$

C

48. $\sqrt[5]{\dfrac{(527.4)^2}{1542 \times (0.2592)^2}}$
49. $\sqrt{\dfrac{12,560^3}{79.21 \times 8004}}$

50. $92.37^{0.562}$
51. $0.3142^{1.273}$

52. Using $\pi \approx 3.142$, find V when $V = \dfrac{4}{3}\pi r^3$ and $r = 0.1526$.

53. Determine the number of digits in 3^{51} without using a calculator.

A-3 Discrete Mathematics

Objective   To explore some topics in discrete mathematics.

We frequently associate the study of mathematics with people who intend to pursue careers in mathematics, engineering, science, and teaching. People in other disciplines such as social science and business management are now turning to mathematical models to solve problems that arise in their respective careers. Many of these problems involve counting a finite number of objects, completing a process in a finite number of steps, drawing flow charts or diagrams to represent choices that must be made, or determining by a logical method the most efficient solution of a problem. These and other skills have been grouped under the name *discrete mathematics*.

You have already studied some discrete mathematics topics, for example, functions and relations (Lessons 3-8 and 3-10), counting techniques (Lessons 15-5 through 15-7), and algorithm analysis (page 215).

Mathematical induction, a method of proof by which statements involving the set of natural numbers can be established, is an important topic in discrete mathematics (refer to page 523 for an explanation of mathematical induction). To illustrate, we pose and solve the chessboard problem. Is it possible to cover with triominoes a chessboard having 2^n squares on a side and one square removed? A triomino is a set of three squares in the shape of an "el." We will show that it is possible and that the number of triominoes needed is $\frac{1}{3}(2^{2n} - 1)$. Note that if it is possible to cover the board, then $2^n \cdot 2^n - 1$ squares must be covered by 3-square triominoes. That is, it would take exactly $\frac{1}{3}(2^{2n} - 1)$ triominoes to do it.

We will prove what is said above by mathematical induction. To begin, let $n = 1$. We have a 2×2 chessboard that can be covered by 1 triomino as shown in the figure at the right. Assume that it is possible to cover a board with dimensions $2^k \times 2^k$. We must prove that the number of triominoes needed for a board with dimensions $2^{k+1} \times 2^{k+1}$ is $\frac{1}{3}(2^{2(k+1)} - 1)$.

We divide the board into four congruent squares, each with dimensions $2^k \times 2^k$. We place 1 triomino in the center of the board as shown. Each of the four squares is now missing one square. Thus, by the induction hypothesis, the total number of triominoes needed is:

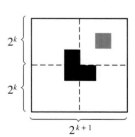

$$
\begin{aligned}
4[\tfrac{1}{3}(2^{2k} - 1)] + 1 &= \tfrac{1}{3}[4 \cdot 2^{2k} - 4] + 1 \\
&= \tfrac{1}{3}[(2^2 \cdot 2^{2k} - 1) - 3] + 1 \\
&= \tfrac{1}{3}[2^{2k+2} - 1] - 1 + 1 \\
&= \tfrac{1}{3}[2^{2(k+1)} - 1]
\end{aligned}
$$

We have shown that the statement is true for $n = 1$ and that if it is true for a natural number k, it is also true for $k + 1$. Thus, by the principle of mathematical induction, the statement is true for all natural numbers.

Another interesting topic in discrete mathematics is *graph theory*. The map in Figure 1 shows seven bridges across the River Pregel in the 18th century city of Konigsberg. Citizens wondered if it was possible to take a walk in such a way as to cross each bridge exactly once.

Figure 1

Figure 2

In 1736 the Swiss mathematician Leonard Euler visited the city and showed that such a walk was not possible. He replaced the actual configuration with a graph as in Figure 2. The land masses A, B, C, and D were represented by vertices a, b, c, and d. The seven bridges were represented by the seven edges. The question then became: Is there a path P in the graph that traverses each edge exactly once?

We can show by contradiction that there is no such P. To do so, we call a vertex even (odd) if an even (odd) number of edges meet there. Suppose that such a path P exists and let v be any vertex at which P neither begins nor ends. Whenever a point traversing P comes into v, it must also leave. Thus v must be an even vertex. If P begins at one vertex and ends at a different one, then we see that there are exactly two odd vertices. If P begins and ends at the same vertex, then there are no odd vertices. Figure 2 shows four odd vertices. This contradiction shows that no such P exists. You cannot walk through the city of Konigsberg and cross each bridge exactly once.

Graphs such as the one in Figure 2 are used to illustrate such things as crystals, networks, circuits, and organizational management.

A-4 Preparing for College Entrance Exams

If you plan to attend college, you will most likely be required to take college entrance examinations. Some of these exams attempt to measure the extent to which your verbal and mathematical reasoning skills have been developed. Others test your knowledge of specific subject areas. Usually the best preparation for college entrance examinations is to follow a strong academic program in high school, to study, and to read as extensively as possible.

The following test-taking strategies may prove useful:

- Familiarize yourself with the test you will be taking well in advance of the test date. Sample tests, with accompanying explanatory material, are available for many standardized tests. By working through this sample material, you become comfortable with the types of questions and directions that will appear on the test and you develop a feeling for the pace at which you must work in order to complete the test.

- Find out how the test is scored so that you know whether it is advantageous to guess.

- Skim sections of the test before starting to answer the questions, to get an overview of the questions. You may wish to answer the easiest questions first. In any case, do not waste time on questions you do not understand; go on to those that you do.

- Mark your answer sheet carefully, checking the numbering on the answer sheet about every five questions to avoid errors caused by misplaced answer markings.

- Write in the test booklet if it is helpful; for example, cross out incorrect alternatives and do mathematical calculations.

- Work carefully, but do not take time to double-check your answers unless you finish before the deadline and have extra time.

- Arrive at the test center early and come well prepared with any necessary supplies such as sharpened pencils and a watch.

College entrance examinations that test general reasoning abilities, such as the Scholastic Aptitude Test, usually include questions dealing with basic algebraic concepts and skills. The College Board Achievement Tests in mathematics (Level I and Level II) include many questions on algebra. The following second-year algebra topics often appear on these exams. For each of the topics listed on pages 865 – 867, a page reference to the place in your textbook where this topic is discussed has been provided. As you prepare for college entrance exams, you may wish to review the topics on these pages.

Types of Numbers

Integers (p. 1)
Rational, irrational, and real numbers (pp. 1, 273, 283–285)
Imaginary and complex numbers (pp. 288–290, 292–294)
Polar form of complex numbers (pp. 680–682, 685–687)

Properties of Real Numbers

Properties of equality (p. 14)
Properties of addition and multiplication (pp. 15, 27)
Properties of order (p. 59)
Proving properties of real numbers (pp. 81–84, 88–89)

Sets and Logic

Set notation (pp. 1, 37, 61, 153)
Quantifiers (p. 20)
Logic (pp. 95–97)
Union and intersection, Venn diagrams (pp. 754–755)

Solving Equations

Linear equations (pp. 37–39)
Linear systems in two variables (pp. 124–128, 801, 803)
Linear systems in three variables (pp. 444–447, 802–803)
Equations with fractional coefficients (pp. 242–244)
Fractional equations (pp. 247–249)
Radical equations (pp. 277–279)
Quadratic equations, completing the square (pp. 307–309)
Quadratic equations, the quadratic formula (pp. 311–313, 317–319)
Equations in quadratic form (pp. 322–323)
Polynomial equations (pp. 194–196, 377–379)
Finding rational roots (pp. 382–383)
Approximating irrational roots (pp. 386–388)
Quadratic systems (pp. 439–441)
Exponential equations (pp. 460–461, 480)
Logarithmic equations (pp. 475, 479)
Trigonometric equations (pp. 697–699)

Solving Inequalities

Linear inequalities (pp. 59–61)
Combined inequalities (pp. 65–66)
Inequalities involving absolute value (pp. 73–74, 76–77)
Systems of inequalities (pp. 135–137)
Polynomial inequalities (pp. 202–203)

Factoring

Prime factorization of integers (pp. 179–180)
Perfect square trinomials, difference of squares (pp. 183–184)
Sum and difference of cubes (p. 184)
Quadratic polynomials (pp. 188–189)
Using the factor theorem (pp. 372–374)

Graphing

The number line (pp. 1–2)
Inequalities (pp. 60, 65–66, 73, 76–77)
Using a sign graph (p. 203)
The coordinate plane (pp. 107–108)
Linear equations (pp. 108–110)
Slope (pp. 112–114)
Intercepts, parallel and perpendicular lines (pp. 119–120)
Systems of linear equations (p. 124)
Linear inequalities and systems of linear inequalities (pp. 135–137)
Parabolas (pp. 326–330, 412–415)
Polynomial equations (pp. 386–388)
Circles (pp. 407–409)
Ellipses (pp. 418–421)
Hyperbolas (pp. 426–429)
Conics with center not at the origin (pp. 432–433)
Quadratic systems (pp. 436–437, 439–441)
Exponential functions (pp. 459–460)
Graphing the inverse of a function (pp. 464–465)
Logarithmic functions (p. 468)
Periodicity and symmetry (pp. 619–620)
Sine and cosine, period and amplitude (pp. 624–627)
Tangent, cotangent, secant, cosecant (pp. 630–632)
Inverse trigonometric functions (pp. 690, 693)

Functions

Functions and relations, domain and range (pp. 141–143, 153–155)
Linear functions (pp. 146–148)
Quadratic functions (pp. 333–335)
Polynomial functions (pp. 167–169, 386–388)
Division of polynomials (pp. 340–341, 344–345)
Interpolation (pp. 391–393)
Composition, inverses (pp. 463–465)
Trigonometric functions (pp. 555–558, 561–564)
Circular functions (pp. 613–615)
Trigonometric identities (pp. 598–601)
Inverse trigonometric functions (pp. 689–691, 693–695)

Exponents and Logarithms

Laws of exponents (pp. 171, 212, 455, 459–460)
Laws of logarithms (pp. 473–475)
Change of base (p. 481)

Sequences and Series

Arithmetic and geometric sequences (pp. 501–503, 507–508, 510–512)
Series, sigma notation (pp. 518–520)
Sums of finite arithmetic and geometric series (pp. 525–526)
Sums of infinite geometric series (pp. 531–532)
Expansion of binomials, the binomial theorem (pp. 537–538, 540–542)

Vectors

Operations (pp. 659–661)
The norm (p. 661)
Components (pp. 666–668)
The dot product (p. 668)

Statistics, Counting, Probability

Measures of central tendency, measures of dispersion
   (pp. 710–711, 713–716)
Counting principles (pp. 730–731)
Permutations (pp. 734–736)
Combinations (pp. 738–739)
Computing probabilities (pp. 745–747, 754–758)

Solving Problems

Translating words into symbols (pp. 43–45)
Using linear equations (pp. 49–52)
Using inequalities (pp. 69–70)
Using systems (pp. 131–132)
Using linear functions (pp. 147–148)
Using polynomial equations (pp. 198–199)
Using rational expressions (pp. 243–244, 248–249)
Maximum and minimum problems (pp. 340–341)
Using direct variation and proportions (pp. 351–353)
Using inverse and joint variation (pp. 358–360)
Problems on exponential growth and decay (pp. 483–485)
Finding distances and angles using right triangles (pp. 574–576)
Using the law of cosines and the law of sines (pp. 580–581, 586–587, 591–593)
Using vectors (pp. 659–663, 672–673)
Using the normal distribution (pp. 719–722, 747)
Using matrices (pp. 779–782)

Types of Questions

The types of questions you can expect may include five-choice *Multiple–Choice* questions, four-choice *Quantitative Comparison* questions, or *Grid–in* questions. Here is an example of a **Multiple–Choice** question:

If $4(x - 1) = 20$, what is the value of $x + 3$?
(A) 6
(B) 7
(C) 8
(D) 9   *the correct answer*
(E) 10

Quantitative Comparison questions give you two quantities and ask you to compare them. Here is an example of a Quantitative Comparison question:

Column A	Column B
4^{10}	$4^{11} - 4^{10}$

You must choose whether the quantity in Column A is greater, the quantity in column B is greater, the two quantities are equal, or if you cannot determine which is greater from the information given. For this example, the quantity in Column A is greater.

Grid–in questions emphasize active problem solving and critical thinking by asking you to grid the answer directly on the answer sheet rather than recognize it from among the choices. Using the same example as the Multiple–Choice question, the answer is still 9, but you would have to grid 9 on the answer sheet rather than choose (D).

Calculator Use

Some college entrance exams allow students to use calculators. If calculator use is permitted, bring a familiar calculator with you to the test center, but don't plan to use it for every problem. First decide how to solve the problem, and then decide if a calculator will help with the computation. For example, suppose you are given that $P(x) = x^2 - 2x - 3$. If you are asked to find $P(x + 1)$, a calculator will not help. You should substitute $(x + 1)$ for x in $P(x)$ to get $P(x + 1) = (x + 1)^2 - 2(x + 1) - 3 = x^2 - 4$. If you are asked to find $P(-4)$, however, using a calculator may help. Substitute (-4) into the expression $x^2 - 2x - 3$ to get $P(-4) = (-4)^2 - 2(-4) - 3 = 21$. In general, calculators may help solve problems involving computations, data, number patterns, or guess-and-check problem solving.

Glossary

abscissa (p. 107): The first coordinate in an ordered pair of real numbers associated with a point in the coordinate plane; also called *x-coordinate*.

absolute value of a complex number (pp. 298, 680): The absolute value of the complex number $z = x + yi$ is defined as $|z| = \sqrt{x^2 + y^2}$, the distance from z to the origin in the complex plane.

absolute value of a number (p. 3): For each real number a, $|a| = a$ if $a > 0$, $|a| = -a$ if $a < 0$. On a number line, $|a|$ is the distance between the graph of the number a and the origin.

additive inverse of a matrix (p. 770): The additive inverse of matrix A is matrix $-A$. Each element of $-A$ is the opposite of its corresponding element in A.

amplitude of a complex number (p. 681): For a nonzero complex number in the form $z = n(\cos \theta + i \sin \theta)$, the angle θ is the amplitude of z; also called *argument* of z.

amplitude of a periodic function (p. 625): For a function of the form $y = a \sin x$ ($a > 0$) or $y = a \cos x$, the positive number a is called the amplitude of the function.

angle (p. 549): A figure formed by two rays that have the same endpoint.

angle of elevation [depression] (p. 575): A measure of the deviation of one's line of sight from the horizontal.

antilogarithm (p. 479): If $\log x = a$, then the number x is called the antilogarithm of a.

arithmetic means (pp. 287, 508): The terms between two given terms of an arithmetic sequence. A single arithmetic mean between two numbers is called *the arithmetic mean*, or the *average*, of the two numbers.

arithmetic sequence (p. 501): A sequence in which the difference between any two successive terms (called the *common difference*) is constant. Also called *arithmetic progression*.

arithmetic series (p. 518): *See under* series.

associated equation (p. 135): The equation from which an inequality is derived by replacing the equals sign by an inequality symbol.

axes (p. 107): Two number lines, one horizontal and one vertical, intersecting at a point O.

axiom (p. 81): A statement that is assumed to be true; also called *postulate*.

base of a power (p. 7): The repeated factor of the power. In a^5, a is the base.

bearing of a vector (p. 661): The angle measured clockwise from due north around to the vector.

binomial (p. 174): A polynomial that has two terms.

binomial expansion (p. 537): The sum of terms that results from multiplying out a power of a binomial.

Boolean algebra (p. 95): The algebra of logic.

box-and-whisker plot (p. 713): A diagram that shows the median, quartiles, and range of a distribution.

Cartesian coordinate system (p. 107): A rectangular coordinate system.

central conics (p. 432): Circles, ellipses, and hyperbolas.

characteristic (p. 479): The integer part of a logarithm.

circle (p. 407): The set of all points in a plane that are a fixed distance (the *radius*) from a fixed point (the *center*).

coefficient (p. 167): The constant (or numerical) factor in a variable.

column matrix (p. 767): A matrix consisting of only one column.

combination (p. 738): An *r*-element subset of a set of *n* elements is a combination of *n* elements taken *r* at a time.

communication matrix (p. 779): A matrix representing routes by which data can be transmitted and received.

complement of an event (p. 757): If an event *A* is a subset of a sample space *S*, the complement of *A* consists of the elements of *S* that are not members of *A*.

completing the square (p. 308): Changing the form of a quadratic equation to make the left side a trinomial square.

complex fraction (p. 238): A fraction whose numerator or denominator (or both) has one or more fractions or powers with negative exponents.

complex number (p. 292): A number of the form $a + bi$, where *a* and *b* are real. The number *a* is called the *real part* of $a + bi$; *b* is called the *imaginary part*.

complex plane (p. 680): The rectangular coordinate system when it is used to represent complex numbers.

component form of a vector (p. 666): When the vector **u** is written in the form $a\mathbf{i} + b\mathbf{j}$, it is in component form. The number *a* is the *x*-component, and *b* is the *y*-component.

composite of two functions (p. 463): The function whose value at *x* is $f(g(x))$ is called the composite of the functions *f* and *g*. The operation that combines *f* and *g* to produce their composite is called *composition*.

conics, or **conic sections** (p. 407): Plane curves having second-degree equations.

conjugate of a complex number (p. 680): The conjugate of the number $z = x + yi$ is $z = x - yi$.

conjugates (p. 274, 294): Expressions of the form $a\sqrt{b} + c\sqrt{d}$ and $a\sqrt{b} - c\sqrt{d}$. The numbers $a + bi$ and $a - bi$ are *complex conjugates*.

conjunction (p. 65): A sentence formed by joining two sentences with the word *and*.

consistent equations (p. 128): The equations in a system that has at least one solution.

constant (p. 39): A number.

constant of variation (pp. 351, 358): In a direct variation, specified by $y = mx$ ($m \neq 0$), *m* is the constant of variation. In an inverse variation, specified by $xy = k$ ($k \neq 0$), *k* is the constant of variation. Also called *constant of proportionality*.

converse (p. 82): The statement that results from interchanging the hypothesis and the conclusion of an if-then statement.

coordinate of a point (p. 1): The real number paired with that point on a number line.

coordinates (p. 107): The unique ordered pair of real numbers associated with each point in the plane.

corollary (p. 81): A theorem that can be proved easily from another theorem.

correlation coefficient (pp. 724–725): A statistic used to determine how closely the points in a scatter plot cluster about a line.

cosecant (p. 556): The cosecant of an angle θ is the reciprocal of $\sin \theta$.

cosine (p. 555): For an acute angle θ in standard position, with $P(x, y)$ a point (not the origin) on the terminal side and $r =$ the distance OP,

$$\cos \theta = \frac{x}{r}.$$

cotangent (p. 556): The cotangent of an angle θ is the reciprocal of $\tan \theta$.

coterminal angles (p. 550): Two angles whose terminal sides coincide when the angles are in standard position.

counterexample (p. 82): In algebra, a single numerical example that makes a statement false.

cube root (p. 260): A cube root of a number *b* is a solution of the equation $x^3 = b$.

degree (p. 549): $1° = \frac{1}{360}$ of a complete revolution.

degree of a monomial (p. 167): The sum of the degrees of the variables in the monomial.

degree of a polynomial (p. 167): The greatest of the degrees of its terms after it has been simplified.

degree of a polynomial equation (p. 377): The degree of a polynomial equation of the form $P(x) = 0$ is the degree of $P(x)$.

degree of a variable in a monomial (p. 167): The number of times the variable occurs as a factor in the monomial.

dense (p. 285): A set S of real numbers is dense if between any two numbers in the set there is a member of S.

dependent equations (p. 128): The equations in a consistent system that has infinitely many solutions.

determinant (p. 787): A real number associated with a square matrix and usually displayed in the same form.

diagonal matrix (p. 788): A matrix whose only nonzero elements are on the main diagonal.

difference of vectors (p. 661): The difference $\mathbf{u} - \mathbf{v}$ is defined to be $\mathbf{u} + (-\mathbf{v})$.

dimensions of a matrix (p. 767): The number of rows (horizontal) and the number of columns (vertical).

direct variation (p. 351): A linear function defined by an equation of the form $y = mx$ ($m \neq 0$). The number m is called the *constant of variation*.

directed angle (p. 549): An angle generated by the rotation of a ray (the *initial side*) onto another ray (the *terminal side*).

discriminant (p. 318): The expression $b^2 - 4ac$ is called the discriminant of the quadratic equation $ax^2 + bx + c = 0$.

disjoint sets (p. 754): Sets that have no elements in common.

disjunction (p. 65): A sentence formed by joining two sentences with the word *or*.

division algorithm (p. 364): Dividend = Quotient $\times$ Divisor + Remainder

domain of a function (p. 141): *See function.*

domain of a variable (p. 9): The set whose members may serve as replacements for the variable; also called *replacement set*.

dominance relation (p. 781): A dominance relation exists between members of a group when, between any two members, one dominates the other.

dot product (p. 668): The dot product of two nonzero vectors $\mathbf{u}$ and $\mathbf{v}$ is defined to be $\mathbf{u} \cdot \mathbf{v} = \|\mathbf{u}\| \|\mathbf{v}\| \cos \theta$, where θ is the angle between $\mathbf{u}$ and $\mathbf{v}$.

ellipse (pp. 418–419): The set of all points P in the plane such that the sum of the distances from P to two fixed points (the *foci*) is a given constant.

empty set (p. 38): The set with no members; also called the *null set*.

equation (p. 6): A sentence formed by placing an equals sign between two expressions, called *sides* of the equation.

equivalent equations (p. 37): Equations that have the same solution set over a given domain.

equivalent expressions (p. 16): Two expressions that are equal for every value of each variable they contain.

equivalent inequalities (p. 60): Inequalities with the same solution set.

equivalent systems (p. 124): Systems of equations that have the same solution set.

equivalent vectors (p. 659): Two vectors that have the same magnitude and the same direction.

even function (pp. 337, 619): A function f is even if $f(-x) = f(x)$ for every x in the domain of f.

event (p. 743): Any subset of possible outcomes for an experiment.

exponent (p. 7): A positive exponent tells how many times the base occurs as a factor of a power.

exponential equation (p. 460): An equation in which a variable appears in an exponent.

exponential form (p. 456): A radical expression is in exponential form when it is expressed as a power or a product of powers.

exponential function (p. 460): If $b > 0$ and $b \neq 1$, the function defined by $y = b^x$ is called the exponential function with base b.

extraneous root (p. 247): A root of a transformed equation that is not a root of the original equation.

factor a polynomial (p. 183): To express the polynomial as a product of other polynomials taken from a specified factor set.

factor set (pp. 179, 183): The set from which numbers or polynomials are chosen as factors.

factorial notation (p. 540): The symbol $r!$, read ''r factorial,'' is defined as $r! = r(r - 1)(r - 2) \cdot \ldots \cdot 3 \cdot 2 \cdot 1$ if r is a positive integer, and as 1 if $r = 0$.

field (pp. 14–15): A set of numbers having the associative, closure, commutative, distributive, identity, and inverse properties.

finite sequence (p. 501): A sequence that has a limited number of terms.

fractional equation (p. 247): An equation in which a variable occurs in a denominator.

frequency distribution (p. 709): A display of data that shows the number of occurrences of each item.

function (pp. 141, 153): A correspondence between two sets, D and R, that assigns to each member of D exactly one element of R. The *domain* of the function is D. The *range* is the set of all members of R assigned to at least one member of D.

geometric means (pp. 287, 511): The terms between two given terms of a geometric sequence. *The geometric mean* of a and b is generally considered as $\sqrt{ab}$ if a and b are positive and $-\sqrt{ab}$ if a and b are negative.

geometric sequence (p. 502): A sequence in which the ratio of every pair of successive terms (called the *common ratio*) is constant. Also called *geometric progression*.

geometric series (p. 518): *See under* series.

graph of a function (p. 142): The set of all points (x, y) such that x is in the domain of the function and the rule of the function assigns y to x.

graph of an open sentence in two variables (p. 108): The set of all points in the coordinate plane whose coordinates satisfy the sentence.

graph of an ordered pair (p. 107): The point in an xy-coordinate plane associated with an ordered pair of real numbers.

graph of a real number (p. 1): The point on a number line paired with that number.

graph of a system of inequalities (p. 137): Points satisfying all the inequalities in the system.

greatest common factor (GCF) (pp. 179, 180): The GCF of two or more integers is the greatest integer that is a factor of each. The GCF of two or more monomials is the common factor that has the greatest degree and the greatest numerical coefficient.

greatest common factor (GCF) of two or more polynomials (p. 190): The common factor having the greatest degree and the greatest constant factor.

half-plane (p. 135): The set of all points on one side of a given line (the *boundary*).

histogram (p. 709): A bar graph displaying a frequency distribution.

hyperbola (p. 426): The set of all points P in the plane such that the difference between the distances from P to two fixed points (the *foci*) is a given constant.

identity (p. 38): An equation that is satisfied by all values of the variable.

identity function (p. 463): The function $I(x) = x$.

identity matrix (p. 775): An $n \times m$ matrix whose main diagonal from upper left to lower right has all elements 1, while all other elements are 0.

imaginary axis (p. 680): The vertical axis in the complex plane.

imaginary number (p. 288): The number i, defined as $\sqrt{-1}$.

imaginary numbers (p. 288): All numbers of the form $a + bi$, $b \neq 0$. Numbers of the form bi, $b \neq 0$, are called *pure imaginary numbers*.

inconsistent equations (p. 128): The equations in a system that has no solution.

independent events (p. 756): Two events are independent if the occurrence of one event does not affect the probability of the other.

inequality (p. 6): A sentence formed by placing an inequality symbol between two expressions.

inequality symbol (p. 6): One of the symbols $<$, $>$, $\leq$, $\geq$, and $\neq$.

infinite sequence (p. 501): A sequence that has an unlimited number of terms.

initial point (p. 659): The starting point of a vector.

integers (p. 1): Members of the set $\{ \dots, -3, -2, -1, 0, 1, 2, 3, \dots \}$.

intersection of sets (p. 754): The intersection of sets A and B is the set whose members are elements of both A and B.

inverse functions (p. 464): The functions f and g are inverse functions if $f(g(x)) = x$ for all x in the domain of g and $g(f(x)) = x$ for all x in the domain of f.

inverse matrices (p. 790): For any two matrices A and B, if $AB = BA =$ the identity matrix, then A and B are inverse matrices.

inverse variation (p. 358): A function defined by an equation of the form
$$xy = k \text{ or } y = \frac{k}{x} \ (x \neq 0, \ k \neq 0).$$
The number k is called the *constant of variation*.

irrational numbers (p. 1): Real numbers that are not rational, such as $\sqrt{7}$ and π.

irreducible polynomial (p. 189): A polynomial that has more than one term and cannot be expressed as a product of polynomials of lower degree taken from a given factor set.

joint variation (p. 359): When a quantity varies directly as the product of two or more other quantities, the variation is called joint variation.

least common multiple (LCM) (pp. 179, 180): The LCM of two or more integers is the least positive integer having each as a factor. The LCM of two or more monomials is the common multiple having the least degree and the least positive numerical coefficient.

least common multiple (LCM) of two or more polynomials (p. 190): The common multiple having least degree and least positive constant factor.

like radicals (p. 270): Two radicals with the same index and radicand.

linear combination (p. 125): The result of adding two linear equations.

linear equation in two variables (p. 109): Any equation that can be expressed in the form $Ax + By = C$ (A and B not both 0).

linear function (p. 146): A function f that can be defined by $f(x) = mx + b$, where x, m, and b are any real numbers. If $m = 0$, f is a *constant function*.

linear inequality (p. 135): A sentence obtained by replacing the equals sign in a linear equation by an inequality symbol.

linear system (p. 124): A set of linear equations in the same two variables.

logarithm (p. 469): If b and N are positive numbers ($b \neq 1$), then $\log_b N = k$ if and only if $b^k = N$.

logarithmic function with base 2 (p. 468): The inverse of $f(x) = 2^x$, $b > 0$ and $b \neq 1$.

mantissa (p. 479): The decimal or fractional part of a logarithm.

mapping diagram (p. 141): A diagram that pictures a correspondence between two sets.

mathematical model (p. 198): An equation that represents a real-life problem.

matrix (pp. 447, 767): A rectangular array of numbers, enclosed by brackets. The numbers in the matrix are its *elements*.

mean (p. 710): The arithmetic average of the numbers in a distribution.

measures of central tendency (p. 715): The mean, median, and mode of a distribution.

measures of dispersion (p. 715): The range, first and third quartiles, variance, and standard deviation of a distribution.

median (p. 710): The middle number in a distribution or the average of the two middle numbers.

minor (p. 794): The minor of an element in a determinant is the determinant resulting from the deletion of the row and column containing the element.

mode (p. 710): The number that occurs most frequently in a distribution.

monomial (p. 167): A constant, a variable, or a product of a constant and one or more variables.

multiple roots, multiple zeros (p. 196): Roots of equations or zeros of functions that arise from repeated factors.

mutually exclusive events (p. 755): Events that have no elements in common.

natural logarithm function (p. 489): The logarithm function with base e; $e \approx 2.71828$.

natural numbers (p. 1): Members of the set $\{1, 2, 3, \ldots\}$.

negative angle (p. 549): An angle generated by a clockwise rotation.

negative numbers (p. 2): Numbers less than zero.

norm (p. 661): The length or magnitude of a vector.

nth root (p. 260): An nth root of b is a solution of the equation $x^n = b$, where n is a positive integer.

nth roots of unity (p. 686): The nth roots of the number 1.

null set (p. 38): *See* empty set.

numerical expression (p. 6): A symbol or group of symbols used to represent a number; also called *numeral*.

objective function (p. 159): In linear programming, the expression to be maximized or minimized.

odd function (pp. 337, 619): A function f is odd if $f(-x) = -f(x)$ for every x in the domain of f.

one-to-one function (p. 460): A function f is called one-to-one if for every p and q in the domain of f, $f(p) = f(q)$ implies $p = q$.

open sentence (pp. 37, 101): An equation or inequality that contains one or more variables.

opposites (p. 3): On a number line, numbers that are the same distance from the origin but on opposite sides of it.

ordered pair (p. 101): A pair of numbers having a definite order.

ordinate (p. 107): The second coordinate in an ordered pair of real numbers associated with a point in the coordinate plane; also called *y-coordinate*.

origin (pp. 2, 107): On the number line, the graph of 0. In a plane rectangular coordinate system, the meeting point of the axes; the graph of (0, 0).

orthogonal vectors (p. 668): Vectors **u** and **v** are orthogonal if and only if $\mathbf{u} \cdot \mathbf{v} = 0$.

parabola (p. 412): The set of all points equidistant from a fixed line (the *directrix*) and a fixed point not on the line (the *focus*).

Pascal's triangle (p. 537): A triangular array of numbers representing the coefficients of the expansion of $(a + b)^n$, where n is a nonnegative integer.

periodic function (p. 619): A function is periodic if, for some positive constant p, $f(x + p) = f(x)$ for every x in the domain of f. The smallest such p is the *period* of f.

permutation (p. 734): An arrangement of the elements of a set in a definite order.

plane rectangular coordinate system (p. 107): A system for locating the point associated with any ordered pair by reference to two number lines intersecting at right angles.

polar coordinate system (p. 675): A system that consists of a point O, called the *pole*, and a ray, called the *polar axis*, having O as its endpoint.

polar coordinates (p. 675): The polar coordinates of a point P are an ordered pair (r, θ), where $r = OP$ and θ is the measure of an angle from the polar axis to $\overline{OP}$.

polar form of a complex number (p. 681): The polar form of a nonzero complex number $z = x + yi$ is $z = r(\cos \theta + i \sin \theta)$, where $r = \sqrt{x^2 + y^2}$ and θ is an angle such that $\cos \theta = \frac{x}{r}$ and $\sin \theta = \frac{y}{r}$.

polynomial (p. 167): A monomial or a sum of monomials. The monomials are called the *terms* of the polynomial.

polynomial equation (p. 194): An equation that is equivalent to one with a polynomial as one side and 0 as the other side.

polynomial inequality (p. 202): An inequality equivalent to an inequality with a polynomial as one side and 0 as the other side.

positive angle (p. 549): An angle generated by a counterclockwise rotation.

positive numbers (p. 2): Numbers greater than zero.

postulate (p. 81): *See* axiom.

power (p. 7): A product of equal factors.

prime factorization (p. 179): The process of writing a positive integer as a product of primes.

prime number, or **prime** (p. 179): An integer greater than 1 whose only positive factors are itself and 1.

prime polynomial (p. 189): An irreducible polynomial with integral coefficients is prime if the greatest common factor of its coefficients is 1.

product of matrices (p. 774): The product of matrices $A_{m \times n}$ and $B_{n \times p}$ is the $m \times p$ matrix whose element in the ath row and bth column is the sum of the products of corresponding elements of the ath row of A and the bth column of B.

proof (p. 81): Logical reasoning from hypothesis to conclusion.

proportion (p. 352): An equality of ratios. In the proportion $\frac{y_1}{x_1} = \frac{y_2}{x_2}$, the numbers x_1 and y_2 are the *means*; y_1 and x_2 are the *extremes*.

quadrant (p. 107): One of the four regions into which a plane rectangular coordinate system is divided by its axes.

quadrantal angle (p. 549): An angle whose terminal side lies on a coordinate axis.

quadratic equation in one variable (p. 307): A polynomial equation in the form $ax^2 + bx + c = 0$.

quadratic form (p. 322): An equation is in quadratic form if it can be written as $a[f(x)]^2 + b[f(x)] + c = 0$.

quadratic formula (p. 311): Solutions of the quadratic equation $ax^2 + bx + c = 0$ $(a \neq 0)$ are given by the formula $x = \dfrac{-b \pm \sqrt{b^2 - 4ac}}{2a}$.

quadratic function (p. 333): A function that can be written in either of the forms $f(x) = ax^2 + bx + c$ $(a \neq 0)$ or $f(x) = a(x - h)^2 + k$ $(a \neq 0)$.

quadratic polynomial, or **second-degree polynomial** (p. 188): A polynomial of the form $a^2 + bx + c$ $(a \neq 0)$. The *quadratic term* is ax^2, the *linear term* is bx, the *constant term* is c.

quadratic system (p. 436): A system containing only quadratic equations or a combination of linear and quadratic equations in the same two variables.

quadratic trinomial (p. 188): A quadratic polynomial in which a, b, and c are all nonzero integers.

quartile (p. 713): The median of the lower half of a set of data is the first quartile; the median of the upper half is the third quartile.

quotient (p. 7): The result of dividing one number by another.

radian (p. 607): An angular measure. In a circle of radius r, centered at the vertex of an angle θ, the radian measure of θ is defined as the ratio of the length of s, the arc intercepted by θ, to the radius r: $\theta = \dfrac{s}{r}$.

radical (p. 261): The symbol $\sqrt[n]{b}$ is called a radical; n is the *index*, and b is the *radicand*.

radical equation (p. 277): An equation that contains a radical with a variable in the radicand.

range of a function (p. 141): *See* function.

range of a set of data (p. 713): The difference between the largest and smallest numbers in the set.

rational algebraic expression (p. 227): An algebraic expression that can be expressed as a quotient of polynomials. Also called *rational expression.*

rational function (p. 228): A function defined by a simplified rational expression in one variable.

rational numbers (p. 1): Numbers that are the result of dividing an integer by a nonzero integer.

real axis (p. 680): The horizontal axis in the complex plane.

real numbers (p. 1): The set consisting of all the rational and irrational numbers.

reciprocal (p. 15): Each nonzero real number has a reciprocal, or *multiplicative inverse,* such that the product of the number and its reciprocal is 1.

reference angle (p. 562): If θ is not a quadrantal angle, there is a unique angle α corresponding to θ such that either $\theta + \alpha$ or $\theta - \alpha$ is an integral multiple of $180°$; α is called the *reference angle* of θ.

regression line (p. 726): A line that relates two variables having a high correlation.

relation (p. 153): Any set of ordered pairs.

repeating decimal (p. 284): When the division process is used to express a rational number as a decimal, the result is a repeating decimal if a digit or block of digits repeats endlessly as the remainder.

resultant (pp. 659–660): The sum of two vectors.

root (solution) of a polynomial equation (p. 194): A value of the variable that satisfies the equation.

row matrix (p. 767): A matrix consisting of only one row.

sample space (p. 743): The set of all possible outcomes of a random experiment.

scalar (p. 771): In matrix algebra, a real number.

scalar product (p. 771): The scalar product of a real number r and a matrix A is the matrix rA, in which each element is r times its corresponding element in A.

scatter plot (p. 724): A graph of ordered pairs used to determine whether there is a mathematical relationship between two sets of data.

scientific notation (p. 221): A method of representing a number in the form $m \times 10^n$, where $1 \le m < 10$ and n is an integer.

secant (p. 556): The secant of an angle θ is the reciprocal of cos θ.

sequence (p. 501): A function whose domain consists of consecutive positive integers. Each corresponding value is a *term* of the sequence.

series (p. 518): The indicated sum of the terms of a sequence. An *arithmetic series* is a series whose related sequence is arithmetic. A *geometric series* is a series whose related sequence is geometric.

sides of an equation (p. 6): The two expressions joined by the equals sign.

significant digit (p. 221): A significant digit of a number written in decimal form is any nonzero digit or any zero that has a purpose other than placing the decimal point.

similar (like) monomials (p. 167): Monomials that are identical or that differ only in their coefficients.

similar terms (p. 23): Terms that contain the same variable factor; also called *like terms*.

simple event (p. 743): A single element of a sample space.

simplified polynomial (p. 167): A polynomial in which no two terms are similar.

sine (p. 555): For an acute angle θ in standard position, with $P(x, y)$ a point (not the origin) on the terminal side and $r = $ the distance OP, $\sin \theta = \dfrac{y}{r}$.

slope (p. 112): For any two distinct points (x_1, y_1) and (x_2, y_2) on line L, the slope of L is $\dfrac{y_2 - y_1}{x_2 - x_1}$.

solution set (pp. 37, 101): The set of all solutions of an open sentence that belong to a given domain of the variable. For an open sentence in two variables, the set of all ordered pairs satisfying the sentence.

square matrix (p. 767): A matrix with the same number of rows and columns.

square root (p. 259): A square root of a number b is a solution of the equation $x^2 = b$. The *principal square root* of b is the positive square root, $\sqrt{b}$.

standard deviation (p. 715): The principal square root of the variance of a distribution, denoted by σ.

standard normal curve (p. 719): The bell-shaped curve that represents a standard normal distribution.

standard normal distribution (p. 719): A normal distribution with mean equal to 0 and standard deviation equal to 1.

standard position (p. 549): An angle is in standard position when its initial side coincides with the positive x-axis.

statistics (p. 710): Numbers used to describe a set of data.

stem-and-leaf plot (p. 710): A way of displaying the data in a frequency distribution.

subset (p. 738): Set B is a subset of set A if each member of B is also a member of A.

subtraction of matrices (p. 771): For each set of $m \times n$ matrices, $A_{m \times n} - B_{m \times n} = A_{m \times n} + (-B_{m \times n})$.

sum of matrices (p. 770): The sum of matrices having the same dimensions is the matrix whose elements are the sums of the corresponding elements of the matrices being added.

sum of vectors (p. 660): For vectors $\mathbf{u}$ and $\mathbf{v}$, the sum $\mathbf{u} + \mathbf{v}$ is the vector extending from the initial point of $\mathbf{u}$ to the terminal point of $\mathbf{v}$.

summand (p. 518): In a series written in sigma notation, for example $\sum_{n=1}^{10} n^2$, the general term n^2 is called the summand. The letter n is called the *index*. The first and last values of n are called the *limits of summation*.

summation sign (p. 518): The Greek letter Σ (sigma), used to write a series in abbreviated form.

synthetic division (p. 368): A method that uses only coefficients to display the process of dividing a polynomial in x by $x - c$. Also called *synthetic substitution*.

tangent (p. 555): For an acute angle θ in standard position, with $P(x, y)$ a point (not the origin) on the terminal side and $r =$ the distance OP, $\tan \theta = \frac{y}{x}$.

tautology (p. 97): A compound statement that is true for all truth values of its component statements.

terminal point (p. 659): The ending point of a vector.

terminating decimal (p. 284): When the division process is used to express a rational number as a decimal, the result is a terminating decimal if the division ends with no remainder. Also called *finite decimal*.

theorem (p. 81): A statement that can be proved.

translation (p. 408): Sliding a graph to a new position in the coordinate plane without changing its shape.

trigonometric form of a complex number (p. 681): *See* polar form.

trinomial (p. 174): A polynomial that has three terms.

uniform motion (p. 44): Motion at a constant speed.

union of sets (p. 754): The union of sets A and B is the set whose members are the elements belonging to either A or B (or both).

unique (p. 14): Identified with one and only one member of the set under consideration.

unit vector (p. 661): A vector with a norm of 1.

value of a numerical expression (p. 6): The number represented by the expression.

values of a function (p. 142): Members of the range of the function.

values of a variable (p. 9): The members of the domain of the variable.

variable (p. 9): A symbol used to represent any member of a given set.

variance of a distribution (p. 715): The average of the squares of the deviations from the mean of the distribution.

vector (p. 659): An arrow representing a vector quantity.

vector quantity (p. 659): Any quantity that has both magnitude and direction.

vertex of a parabola (p. 326): The point where the parabola crosses its axis.

whole numbers (p. 1): Members of the set $\{0, 1, 2, 3, \ldots\}$.

***x*-axis, *y*-axis** (p. 107): A horizontal and a vertical number line meeting at right angles, used in graphing points in the plane.

***x*-coordinate** (p. 107): *See* abscissa.

***x*-intercept** (pp. 119, 329): The x-coordinate of the point where a line or curve intersects the x-axis.

***y*-coordinate** (p. 107): *See* ordinate.

***y*-intercept** (pp. 119, 329): The y-coordinate of the point where a line or curve intersects the y-axis.

zero of a function (p. 195): A number r is a zero of a function f if $f(r) = 0$.

zero matrix (p. 767): A matrix whose elements are all zeros.

Index

Abel, Niels Henrik, 385
Abscissa, 107
Absolute value, 3
  of complex numbers, 294, 298–300, 680
  definition, 9, 89
  in open sentences, 73–79
  theorems involving, 89–91
Accuracy in measurement, 221–222
Addition
  cancellation property, 82
  of fractions, 235–237
  of matrices, 770–771
  of polynomials, 168
  of radicals, 270–273
  of real numbers, 21–25
  of vectors, 660
Addition formulas, trigonometric, 641–644, 650
Addition properties, 15
  of equality, 14
  of matrices, 771
  of order, 59
Additive inverse, 15
  of a matrix, 770
Air speed, 131, 662
Algebraic expressions(s), 9
  equivalent, 16
  rational, 227–229
Algorithm, division, 364, 367
Amplitude
  of a complex number, 681
  of sine and cosine functions, 625
Analytic geometry, 401–453
Angle(s)
  coterminal, 550, 551
  of depression, 575
  directed, 549
  of elevation, 575
  initial side, 549
  naming, 551
  negative, 549
  positive, 549
  quadrantal, 549
  reference, 562–563
  in standard position, 549
  terminal side, 549
Angle measure
  conversion equations for, 607
  decimal degrees, 550, 554

  degrees, minutes, seconds, 550–551, 554
  radians, 607–612
  relation to arc length, 607, 608
Antilogarithm, 479
Applications
  astronomy, 224–225, 350, 424–425, 436, 548
  architecture, 100
  communication networks, 779–781
  compound interest, 483–484
  dominance relations, 781–782
  electrical circuits, 253–254, 258
  exponential growth and decay, 454, 483–488
  force, work, and energy, 672–674
  geometry, 39, 44, 50–51, 198–199, 312–313. *See also* Problems.
  graphics, 766
  light, 166, 210, 258, 306
  light and sound waves, 844
  linear programming, 159–161
  logarithms, 478–482, 489–491
  LORAN, 426
  navigation, 426, 658, 661–663
  music, 606, 634–635
  patterns in nature, 500
  planetary orbits, 424–425
  radiocarbon dating, 493–494
  sampling, 752–753
  trigonometric, 659–705
  See also Modeling, Problems, *and* Formulas.
Approximations
  decimal, of powers having irrational exponents, 459
  irrational roots, 386–389
  length and angle measurements, 575
  by linear interpolation, 391–395
  of measurements, 221–222
  of π, 701–702
  using regression line, 726–727
Arc cosine, 689
Arc length, 607, 608, 609
Arc sine, 690
Arc tangent, 693 (not so named)
Archimedes, 431

Areas
  parabolic section, 431
  sector, 608–609
  triangle, 586, 597–600
  under standard normal curve, 720–722
Argument of a complex number, 681
Associative properties, 15
Astronomical unit, 224–225
Asymptote(s), 230, 231, 630
Average, 508, 710
Axiom, 81
  See also Properties of real numbers.
Axis (Axes), 107
  of ellipse, 419
  imaginary, 680
  polar, 675, 676
  real, 680
  of symmetry of a parabola, 326, 412

Base
  of a power, 7
  of logarithms, 468, 481, 482
BASIC, 5
Bearing, 661
between, meaning of, 65
Binomial(s), 174
  containing radicals, 274–276
  powers of, 537–539
Binomial expansion, 537
  general, 540–543
Binomial theorem, 541
Biographical notes
  Ch'in Chiu-Shao, 220
  Halley, Edmund, 601
  Hamilton, Alice, 544
  Just, Ernest Everett, 177
  Kovalevski, Sonya, 32
  Mayer, Maria Goeppert, 618
  Poincaré, Henri, 684
  Steinmetz, Charles, 316
Bombelli, Rafael, 297
Boole, George, 95
Boolean algebra, 95–97
Boundary, 135
Box-and-whisker plot, 713–714
Briggs, Henry, 472

Calculator Key-In
  base *e* exponentials, 492
  conversion: degrees/minutes/
    seconds to decimal degrees, 554
  factorials, 543
  finding roots, 263, 269
  order of operations, 12
  scientific notation, 225–226
Calculators
  graphing, 114, 119, 138, 231,
    327, 336, 386, 387, 410, 416,
    421, 423, 430, 436, 441, 464,
    471, 624, 625, 631, 710, 838,
    840, 841, 845
  suggestions for when to use, 10,
    285, 419, 427, 437, 459, 478–
    480, 485, 489, 550, 568–571,
    608, 614, 615, 698, 710
  use in exercises, 24, 30, 35,
    224, 240, 268, 314, 316, 324,
    481, 482, 486–488, 514, 528,
    534, 571, 572, 577, 578, 582,
    583, 588, 589, 594, 595, 599,
    683, 833, 836, 837, 843
Cancellation property
  of addition, 82
  of multiplication, 82
Cancellation rule. *See under*
  Fractions.
Cardano, Girolamo, 297
Cardioid, 679 (Exs. 33, 34)
Cayley, Arthur, 793
Career notes
  automotive engineer, 64
  computer graphics artist, 423
  electrician, 718
  flight engineer, 567
  marine biologist, 522
  statistician, 337
Cartesian coordinate system, 107
Challenge (problems), 48, 140,
  158, 325, 363, 411, 492, 516,
  601, 623, 733
Chapter reviews, 55–56, 93–94,
  162–163, 207–208, 255–256,
  302–303, 346–347, 397–398,
  451–453, 496–497, 545–546,
  602–604, 655–656, 703–704,
  764–765, 806–807
Chapter summaries, 55, 92, 161–
  162, 206, 254–255, 301, 346,
  396–397, 450–451, 495, 544–
  545, 602, 654–655, 702–703,
  763–764, 806
Chapter tests, 57, 94, 164, 208,
  257, 303, 348, 398, 453, 497,
  546, 604, 657, 704, 765, 808

**Characteristic of common
  logarithm,** 479
Checking solutions, 38, 247, 278
  graphing, 109, 137, 203
  polynomial division, 365
  radical equations, 278
  word problems, 50, 51, 69, 70,
    198, 313, 371
Circles, 407–411
  center, 407
  equations, 407–409
  exploring graphs of, 840
  radius, 407
  translating, 408
  unit, 613
Closed
  under addition, 19
  under division, 35 (Exs. 31–36)
  under multiplication, 19
  under subtraction, 25 (Exs. 51–
    56)
**Closure properties of real
  numbers,** 15, 19
Coefficient(s), 113, 167
  fractional, 242–246
  and roots of quadratic equations,
    338–341
Cofunctions, 557
**College Entrance Exams
  (preparations),** 99, 209, 305,
  399, 499, 605, 706, 809, 864–
  867
Combinations, 738–741
Common difference, in arithmetic
  sequence, 501
Common ratio, in geometric
  sequence, 502
Communication networks, 779–
  781
Commutative properties, 15
Complement of an event, 757
Completeness, property of, 283
Complex fractions, 238–241
Complex numbers, 292–296
  absolute value, 294, 298–300,
    680
  addition and subtraction of, 293,
    680
  amplitude, or argument, 681
  conjugate, 294, 298, 299, 680
  equality of, 293
  geometry of, 680–684
  historical note, 297
  operations with, 293–294, 680,
    682
  polar/trigonometric form, 681
  powers and roots of, 685–688

  product, definition of, 293
  real and imaginary parts, 292,
    681
  square roots of, 442
  sum, definition of, 293
Complex plane, 680
Components, *x*- and *y*-, 666
Composite of two functions, 463
Composition of functions, 463–464
Computer Exercises, 5, 32, 64,
  75, 122, 187, 321, 395, 406,
  530, 584, 645, 670, 742, 750
**Computer graphing ideas
  (techniques),** 114, 119, 160,
  327, 386, 387, 436, 460, 464,
  624, 625, 631, 838, 840, 841,
  845
  See also Calculators, graphing.
Computer Key-In
  approximating π, 701–702
  the Euclidean algorithm, 215
  factoring trinomials, 193
  locating zeros, 389–390
  simulating an experiment, 762
  solving an inequality, 80
Conclusion, 81
Conditional, 95, 96
Conic sections, 407
  circles, 407–411
  ellipses, 418–423, 424–425, 432
  hyperbolas, 426–431, 432–433
  parabolas, 412–416, 431
Conics
  central, 432
  with center not at the origin, 432–
    435
  See also Conic sections.
Conjugate hyperbolas, 430
Conjugates, 274, 294, 298, 680
Conjunction, 65, 66
  and absolute value, 73
  symbol, 95
Consecutive integers, 45
Consecutive numbers, 45
Constant, 39, 167
  degree of, 167
  of proportionality, 352, 358
  of variation, 351, 358
Constant term, 188
Constraint, 159
Converse, 82
Coordinate(s)
  of point on number line, 1, 2,
  polar, 675–676
  x-, *y*-, 107
Coordinate system
  Cartesian, 107

plane rectangular, 107
polar, 675–676
Corollary, 81–82
Correlation coefficient, 724–726
Correspondence, one-to-one, 107
Cosecant function, 556, 561, 613
graph, 632
inverse, 694
Cosine function, 555, 561
as function of real numbers, 613
graph, 624–629, 689
inverse, 689–690, 691
Cosines, law of, 580
Cotangent function, 556, 561, 613
graph, 632, 693
inverse, 693
Counterexample, 82
Counting, fundamental principles, 730–733
Cramer, Gabriel, 802
Cramer's rule, 801–805
Critical thinking skills. *See* Thinking skills.
Cube root(s), 260
approximating, 394–395 (linear interpolation)
table, 811
Cubes, sum and difference of, 184
Cumulative reviews, 165, 348–349, 547, 707
Curve
area under, 720, 721
bell-shaped, 719
standard normal, 719–722

Data analysis. *See under* Statistics.
Decimals
repeating, 284
terminating, 284
converting rational numbers to, 283–284
finite, 284
infinite, 284, 285
nonrepeating, 285
Degree
of a monomial, 167
of a polynomial, 167, 168
of a variable in a monomial, 167
Degree in angle measure, 549, 550, 554
De Moivre's theorem, 685–688
Denominator, rationalizing, 265–266, 274–275

Density, 285
Descartes, René, 107
Descartes' rule of signs, 378, 379
Determinant(s), 787–789
element, 787
expansion by minors, 794–800
order, 787
properties, 787, 798–799
of 2×2 matrix, 787
of 3×3 matrix, 787
use in solving systems, 801–805
Diagonal, main, in matrix, 775, 788
Diagonal of polygon, 506
Difference(s), 7
of cubes, 184
of matrices, 771
of rational expressions, 235–237
of squares, 183
of vectors, 661
Digits, significant, 221–222
Directrix, 412
Dirichlet, 152
Discrete mathematics, 862–863
See also Algorithm, division; Counting, fundamental principles; Functions; Induction, mathematical; Matrix; Probability; Relations; Sets; *and* Statistics.
Discriminant, 317–318
Disjunction, 65, 66
and absolute value, 73, 74
symbol, 95
Dispersion of a set of data, 713
Distance formula, 401–402
Distribution, frequency, 709–729
normal, 719–723
Distributive properties, 15, 23, 34
Division
checking, 365
of complex numbers, 682
definition of, 33
fractions, rule for dividing, 232
long, 364–365
missing terms in, 365
of polynomials, 364–367
quotient in, 7, 33, 232
of rational expressions, 232–234
of real numbers, 33–35
synthetic, 368–370
Domain
of a function, 141
of a relation, 153
of a variable, 9
Dominance relations, 781–782
Dot product, 668

Double-angle formulas
sine and cosine, 646–647
tangent, 651

e, 489
Eccentricity, 422, 424
Electrical circuits, 253–254
Ellipse, 418–423, 424–425
center, 419
equations, 418–421, 432
focal radii, 418
focus (foci), 418
graph, 419–420, 840
extent, 419
intercepts, 419
symmetry, 419
major axis, 419
minor axis, 419
Energy, 672, 673
Enrichment. *See* Applications, Biographical notes, Career notes, Challenge, Explorations, Extras, Historical notes.
Equality of real numbers, 14
Equation(s), 6
associated, 135
of circles, 407–409
cubic, 307
depressed, 374
of ellipses, 418–421, 432
equivalent, 37
exponential, 460–462, 480
fractional, 247–252
with fractional coefficients, 242–246
of hyperbolas, 426–429, 432
logarithmic, 475
matrix, 772–773
in one variable, 37–42, 307
of parabolas, 326–330, 412–415
quadratic, 307–321
in quadratic form, 322–325
radical, 277–281
sides of, 6
transforming, 37–38
trigonometric, 697–700
See also Linear equations, Polynomial equations, Quadratic equations, Systems of equations.
Equivalence, 95, 96–97
Error analysis (cautions), 34, 142, 247, 259, 289, 319, 365, 637, 668, 788
Estimation
of angle measures, 551, 553

in problem solving, 371
by sampling, 752–753
using scientific notation, 222–224
Euclid, 182
Euclidean algorithm, 215
Euler, Leonhard, 297
Evaluating an expression, 9–10
Event(s), 743
complement of, 757
equally likely, 745
independent, 756–757
mutually exclusive, 755–756
simple, 743
Expectation, 751
Explorations
circles and ellipses, 840
continued fractions, 836
direct variation, 839
functions, 834
inequalities, 833
irrational numbers, 832
matrices in geometry, 847
Pascal's triangle, 842
polar coordinate equations, 845
polynomial factors, 835
powers and roots, 841
probability, 846
quadratic equations, 838
radicals, 837
sine curves, 844
trigonometric ratios, 843
Exponential functions, 455–467
applications of, 483–488
Exponent(s), 7
irrational, 459
laws of, 171, 212, 217
negative, 216, 217, 218
rational, 455–458
real number, 459–462
zero, 216
Expression
algebraic, 9
in exponential form, 456
numerical, 6
rational, 227
Extrapolation, linear, 395
Extras
augmented matrices, 785–786
conjugates and absolute value, 298–300
graphing rational functions, 230–231
graphing sequences, 517
growth of functions, 462
induction, 523–524
irrationality of $\sqrt{2}$, 273

logical symbols: quantifiers, 20
random numbers, 753
spirals, 679
symbolic logic: Boolean algebra, 95–97

Factor set, 179, 183
Factor(s), 7
exploring polynomial, 835
greatest common (GCF), 179, 180, 190
greatest monomial, 183
integral, 179
linear, 194
Factor theorem, 373
Factorization
complete, 189–190
prime, 179–181
Factorial notation, 540–541, 734
Fibonacci, 42
Field, 14–15
Focal radii, 418, 426
Focus (foci)
of an ellipse, 418
of a hyperbola, 426
of a parabola, 412
FOIL method of multiplying binomials, 174–175
Force, 672, 673
Formula(s), 39
area of triangle, 586, 597, 599
change of base in logarithms, 481
circular-arc length, 608
circular-sector area, 608
combinations, 738
compound interest, 483–484, 487, 491 (Ex. 51)
coordinate system conversion, 676
distance, 402
double-angle, 646, 651
doubling-time growth, 484
geometric, 39, 40, 41, 47, 830–831
half-angle, 647, 651
half-life decay, 485, 493
Hero's, 597, 599–600
midpoint, 402
multiplying and dividing complex numbers, 682
nth term of sequence, 507, 510
Ohm's law, 253, 718
permutations, 735, 736
probability, 745, 755, 756

projectile motion, 39, 199, 201
quadratic, 311
rate of change, 147
from science, 39, 40, 41, 134, 199, 201, 253, 351, 353, 359, 472 (Ex. 43), 484, 485, 493, 609, 672
solving, 39, 41–42
trigonometric addition, 641, 650
uniform motion, 44
Fractals, 535, 536, 766
Fractional coefficients, 242–246
Fractions
adding and subtracting, 235–237
cancellation rule, 211
complex, 238–241
continued, 240, 836
division rule for, 232
in equations, 242–252
least common denominator, 235
multiplying and dividing, 211, 232–234
simplifying, 211–214, 227–229
Frequency, 634
Frequency distribution, 709
Functions, 141–152, 153–154
circular, 613–617
composite, 463
composition of, 463–464
constant, 146
domain of, 141
double zero of, 196
exploring, 834
exponential, 455–467
graphs. See Graphs of functions.
greatest-integer, 145
growth of, 462
historical note, 152
identity, 463–464
inverse, 464–465, 689–696
linear, 146–152
logarithmic, 468–477
natural logarithm, 489–492
notation for, 142
objective, 159
odd and even, 619–620
one-to-one, 460, 465, 689
periodic, 619–623
quadratic, 333–337, 340–341
range of, 141
rational, 228, 230–231
signum, 145
trigonometric, 549–574
values of, 142, 334, 335
zero(s) of, 195, 196, 386
Fundamental theorem of algebra, 377

Gauss, Karl Friedrich, 297, 377, 445

Gaussian elimination, 445–446, 785–786

Geometry
analytic, 401–453
of complex numbers, 680–684
exploring transformations, 847
formulas, xviii, 39, 40, 41, 47, 330–331
problems. *See under* Problems.
of quadratic systems, 436–438

Golden rectangle, 315

Graphs
checking, 109
circles, 407–411, 840
conjunction of inequalities, 65–66
disjunction of inequalities, 65–66
ellipses, 419–420, 840
exploring, 834, 838, 840, 841, 844, 845
of functions. *See* Graphs of functions.
hyperbolas, 427–429
linear equations in two variables, 107–111
linear inequalities, 135–139
linear systems, 124, 803
on number line, 1–3, 59–61
open sentence involving absolute values, 73–74, 76–77
open sentence in two variables, 108
of opposites, 2–3
ordered pairs, 107–108
parabolas, 326–337, 412–416
polar equations, 676–679, 845
quadratic systems, 436–438, 439–440
of relations, 153, 154, 155
of sequences, 517
sign, 203
solution set of inequality, 59, 61, 65–66, 202–203

Graphs of functions, 142, 143, 146
amplitude, 625
cotangent, secant, and cosecant, 632
exponential, 460, 468
inverse of a function, 464–465
inverse trigonometric, 689–690, 693
logarithmic, 468
odd and even, 620
periodic, 619–620

phase shift, 629
polynomial, 386–388
quadratic, 333–335, 340
rational, 230–231
sine and cosine, 624–629, 689, 690
symmetric, 620
tangent, 630–631, 693

Greatest common factor (GCF), 179, 180, 190

Greatest monomial factor, 183

Greek letters, xvi, 551

Ground speed, 131, 662

Growth and decay, exponential, 483–488

Half-angle formulas
sine and cosine, 647–648
tangent, 651

Half-life, 485

Half-plane, 135
closed, 136
open, 135

Harmonic mean, 252

Head wind, 131

Hero's formula, 597, 599–600

Hertz (Hz), 634

Hipparchus, 573

Histogram, 709, 710

Historical notes
area of a parabolic section, 431
complex numbers, 297
functions, 152
linkages, 68
logarithms, 472
matrices, 793
prime numbers, 182
roots of polynomial equations, 385
tables of sines, 573
a trigonometric identity, 618
word problems, 42

Horizontal-line test, 465

Hyperbola(s), 426–431, 432–433
asymptotes, 427
center, 428
conjugate, 430
equations, 426–429, 432
focal radii, 426
focus (foci), 426
graphs, 427–429

Hypothesis, 81

i, 288

Identities, trigonometric, 556, 564
addition formulas, 641–644

cofunction, 636
fundamental, 636–640
proving, 637–639
Pythagorean, 556, 560, 636
reciprocal, 636

Identity, 38

Identity properties of real numbers, 15

Imaginary numbers, 288–291

Independent events, 756–757

Index of radical, 261

Index of summation, 518–519

Induction, mathematical, 523–524

Inequalities
combined, 65–68
equivalent, 60
exploring, 833
with fractional coefficients, 242–246
graphs of, 59–61, 65–67, 73–74, 76–77, 135–137, 202–203
linear in two variables, 135–139
in one variable, 59–64, 242, 244, 245
polynomial, 202–205
reversing direction, 60
symbols for, 2, 6
systems of, 137, 139
theorems involving, 88–89
transforming, 60
word problems, 69–72

Inequality, 6
sides of, 6

Integers, 1
congruent, 178
even and odd, 45
perfect, 182
relatively prime, 182

Intercepts
of an ellipse, 49
of a line, 119
of a parabola, 329

Intermediate-value theorem, 387

Interpolation
inverse, 391
linear, 391–395
in log tables, 856–857

Inverse functions, 689–696
cosecant (Csc^{-1}), 694
cosine (Cos^{-1}), 689–690, 691
cotangent (Cot^{-1}), 693
secant (Sec^{-1}), 694
sine (Sin^{-1}), 690–691
tangent (Tan^{-1}), 693

Inverse properties of real numbers, 15

Inverse variation, 358–362, 429
Inverses of matrices, 787–792
Irrational numbers, 1, 273, 283, 285, 832

Joint variation, 359–362
Joule (J), 673

Kepler, Johann, 424
Kilowatt-hour, 673

Latitude, 611
Latus rectum, 416
Law(s)
  of cosines, 580
  of exponents, 171, 212, 217
  of logarithms, 473–477
  of sines, 586
Least common denominator (LCD), 235
Least common multiple (LCM), 179, 180, 190
Leibniz, 42, 152
Lemniscates, 679 (Exs. 35, 36)
Light-year, 224–225 (Exs. 9, 10)
Line(s)
  finding equation, 118–122
  horizontal, 112
  parallel, 119, 124
  perpendicular, 119–120
  regression, 726–727
  slope of, 112–117
  vertical, 112
Linear combination, 125, 785
Linear equation(s)
  consistent, 128, 803
  Cramer's rule, 802
  dependent, 128, 446, 803
  inconsistent, 128, 446–447, 803
  intercept form, 122
  one variable, 49–54, 307
  point-slope form, 118
  simultaneous solution, 124
  slope-intercept form, 119
  standard form, 118
  systems in three variables, 444–449, 785–786, 802–803
  systems in two variables, 124–134, 791–792, 801
  two-point form, 122
  in two variables, 107–111
    graphs, 107–111
    solution set, 101

Linear function, 146
Linear programming, 159–161
Linear term, 188, 189
Linkage, 68
Logarithm(s), 468–470, 856–861
  change of base, 481, 482
  characteristic of, 479
  common, 478, 856–861
  computation with, 478–482, 858–861
  laws of, 473–477
  mantissa of, 479
  Napier's, 472
  natural, 472
  table of, 812–813
Logarithmic functions, 468–477
  applications of, 478–488
Logic, 20, 95–97
Logically equivalent statements, 96–97

Mantissa, 479
Mapping diagram, 141, 153
Matrix (matrices), 447, 767–808
  addition of, 770, 771
  additive inverse, 770
  applications, 779–784
  augmented, 785–786
  column, 767
  communication, 779–781
  determinant of, 787–788
  diagonal, 788
  dimensions, 767
  dominance relations, 779
  elements of, 767
  equal, 768
  exploring, 847
  history, 793
  identity, 770, 775
  inverses of, 787–792
  main diagonal, 775, 788
  powers of, 775–776
  product of, 774–776
  row, 767
  scalar multiplication, 771–772
  square, 767, 775–776
  subtraction of, 771
  sum of, 770
  zero, 767, 776
Matrix equation(s), 772–773
Maximum and minimum values, 335, 340–341
Mean(s)
  arithmetic, 287, 508
  deviations from, 715
  of a distribution, 710

geometric, 287, 352, 511
harmonic, 252
Mean proportional, 355
Measure(s)
  of central tendency, 710–711, 715
  degree, 550–551
  of dispersion, 715–716
  radian, 607–608
Median of a distribution, 710, 711, 713
Members of a set, 1
Midpoint formula, 402–403
Minor, 794
Minute, 550
Mixed Problem Solving, 98, 304, 498, 705
Mixed Review Exercises, *at end of each odd-numbered lesson.*
Mode of a distribution, 710, 711
Modeling, mathematical, 147–148, 150–152, 159–161, 183, 198–199, 340–341, 483–485, 779–782
Modulus, 681
Monomial(s), 167
  degree of, 167
  factoring, 180
  quotient of, 212
  similar, or like, 167
Multiple
  integral, 179
  least common, 179, 180, 190
Multiple-choice exercises
  See College Entrance Exams, Chapter reviews.
Multiple roots, 196
Multiplication
  cancellation property, 82
  of complex numbers, 682
  of fractions, 211, 232–234
  matrix, 774–776, 847
  of matrix by scalar, 771–772
  polynomial by monomial, 172, 173
  polynomial by polynomial, 174–176
  of rational expressions, 232–234
  of real numbers, 27–31
  using laws of exponents, 171–173
  of vector by scalar, 660–661
Multiplication property of equality, 14
Multiplicative inverse, 15
  of a matrix, 790
Multiplicative properties of 0 and −1, 27

Napier, John, 472
Natural logarithms, 472
Negation, 95, 96
Negative exponents, 216, 217, 218
Newton (N), 672
Norm, 661
Normal distribution, 719–723
Notation
    factorial, 540–541, 734
    functional, 142
    scientific, 218, 221–225
    set, 153
    sigma, 518–520
    for solution set of inequality, 61
    for a vector, 659
nth root(s), 260
    principal, 260
    of unity, 686–687
Number line, 1
    graphing on, 1–5, 202–203
Numbers
    absolute value, 3
    comparing, 2
    complex, 292–296, 680–684
    consecutive, 45
    integers, 1, 45
    irrational, 1, 283, 285, 832
    natural, 1
    opposite of, 2–3
    pentagonal, 505
    prime, 179, 182
    rational, 1, 283–284
    real, 1–2. See also Real
        numbers.
    triangular, 505, 530
    whole, 1
Numerical expression(s), 6
    simplifying, 7–8
    value of, 6

Ohm's law, 253, 718
Open sentence(s), 37
    involving absolute values, 73–79
    solution, or root, 37
    solving, 37, 101
    in two variables, 101–111
    See also Equations, Inequalities.
Opposite of a product, property
    of, 28
Opposite of a sum, property of,
    29
Opposites, properties of, 15
Opposites of real numbers, 2–3
Order
    of operations, 8, 233
    properties of, 59

in the real numbers, 2
    theorems involving, 88–89
Ordered pair, 101, 107, 157
    graphing, 107–108
Ordered triple, 444
Ordinate, 107
Origin, 1, 107

Parabola, 326
    area of section, 431
    axis, 326, 412
    directrix, 412
    equations, 326–330, 412–415
    focus, 412
    graphs, 326–337, 412–415
    intercepts, 329
    latus rectum, 416
    vertex, 326, 334, 412
Parallel lines, 119, 124
Parallelogram method, 660
Parsec, 224–225 (Exs. 9, 10)
Pascal's triangle, 537–538, 842
Patterns
    in multiplying and factoring
        polynomials, 175, 183, 184
    in sequences, 501, 502, 503
Peaucellier, 68
Percent, 243
Period, 619
    frequency, 634, 635
Permutations, 734–737
    with repeated elements, 736
Perpendicular lines, 119–120, 403
Pi (π), 607, 608, 701–702, 832
Plane
    complex, 680
    xy-coordinate, 107
Plotting a point, 107
Point(s)
    coordinates of, 107, 675
    distance between, 401–402
    lattice, 411
    plotting, 107
Polar coordinates, 675–676
    exploring equations with, 845
Pole, 675
Polynomial equations, 194–201,
    372–389
    bounds on roots, 389 (Exs. 27,
        28)
    degree of, 377
    double root, 196
    irrational roots, 386–389
    multiple roots, 196, 374, 377
    in problem solving, 198–201

rational roots, 382–385
    root, or solution, 194
    solution by factoring, 194–195,
        374
Polynomial inequality, 202
    graph of solution set, 202–203
    solution, 202–205
Polynomial(s), 167
    adding, 168
    degree of, 167, 168
    division, 364–367
    factoring, 183–192, 835
    irreducible, 189
    multiplying by monomial, 172
    multiplying by polynomial, 174
    prime, 189
    quadratic, 188–189
    simplified, 167
    simplifying, 168, 169
    special products of, 175
    subtracting, 168–169
    terms of, 167, 168, 188
Portfolio Projects, 848–855
Postulates, 81
    See also under Properties.
Power(s), 7
    of a complex number, 685
    exploring, 837, 841
    with irrational exponent, 459
    of a power, 171
    of a product, 171
Preparing for College Entrance
    Exams. See College Entrance
    Exams.
Prime, 179, 182
Prime factorization, 179–181
Prime polynomial, 189
Probability, 743–750
    expectation, 751
    exploring, 846
Problem solving strategies
    checking solution, 50, 51, 69,
        70, 199, 313, 371
    estimating answer, 371
    insufficient information, 51, 53
    making a table, 49, 51–52, 132,
        244, 248–249
    making a sketch, 50, 51–52,
        198, 312, 576, 585
    plan, 49–50, 371
    recognizing a pattern, 175, 183,
        184, 501, 502, 503
    unneeded information, 51–52, 53
Problems
    aircraft, 131–132, 133–134, 248–
        249, 662–663, 665
    areas of triangles, 597–599

checking answers, 50, 51, 69, 70, 199, 313, 371
communication networks, 779–781, 782–784
digit, 103, 106
distance/rate/time, 44, 47, 48, 51–52, 53, 246, 248–249, 250–251
dominance relations, 781–782, 783, 784
electrical circuits, 253–254
force, 672–674
geometric, 44, 46–48, 52–53, 71–72, 198–199, 200–201, 357, 362, 443–444, 449, 506, 534–536
growth and decay, 483–488, 493–494
historical note, 42
latitude, 611
maximum and minimum, 159–161, 340–341, 343–345
mixture, 244, 245, 246
money, 102–103, 105–106, 448, 449
music, 634–635
navigation, 661–663, 665
plan for solving, 49–50
probability, 746–750, 759–761
science, 151, 199–201, 224–225, 344, 351–354, 356–357, 359–362, 392, 394, 443, 472, 487, 579, 733
speed, 609, 611–612
three-dimensional, 39, 47, 444, 579, 611, 612
using fractional equations, 242–252
using inequalities, 69–72
using polynomial equations, 198–202
using systems of equations, 131–134, 442–443, 448–449
trigonometric, 578–579, 583–584, 589–590, 595–596, 599, 611–612
work, 243, 245, 246, 248, 250, 251
See also Applications, Challenge (problems).
Product(s), 7
of complex numbers, 293
dot, 668
estimating, 222
factors of, 7
polynomial and monomial, 172, 173

polynomial and polynomial, 174–176
power of, 171
of powers, 171
of rational expressions, 232–234
of real numbers, 27–31
scalar, 771
Progressions, 501–517
arithmetic, 501
geometric, 502
Proof, 16, 18–19, 81–91
by analytic geometry, 403, 405–406, 411
indirect, 273
informal (convincing argument), 62, 63, 87, 117, 157, 181, 182, 201, 204, 287, 321, 367, 379, 389, 482, 530, 600, 699, 723, 729
Properties of matrices and determinants
addition of matrices, 771
determinants, 798–799
matrix multiplication, 776
scalar multiplication, 772
See also Properties of real numbers.
Properties of real numbers
associative, 15
cancellation, 82, 87
closure, 15, 19
commutative, 15
completeness, 283
density, 285
distributive, 15, 23, 34, 86
of equality
addition, 14
multiplication, 14
reflexive, 14
symmetric, 14
transitive, 14
identity, 15
inverse, 15
multiplicative, of -1, 27, 84
multiplicative of zero, 27, 82
opposite of a product, 28, 83, 85
opposite of a sum, 29, 84
of opposites, 15, 85
of order
addition, 59
comparison, 59
multiplication, 59, 88–89
transitive, 59
of radicals, 261, 264
of reciprocals, 15
zero-product, 83, 86, 194
Proportion(s), 352

extremes, 352
means, 352
See also Variation.
Pythagoras, 273
Pythagorean theorem, 401

Quadrants, 107
Quadratic equation(s), 307
completing the square, 307–310, 319
discriminant, 317–318
double root, 317, 318
exploring graphs of, 838
quadratic formula, 311–315, 319
roots, nature of, 317–318
roots and coefficients, 338–341
solving, 307–321
systems of, 436–438, 439–443
See also Polynomial equations.
Quadratic formula, 311
Quadratic functions, 333–337, 340–341
Quadratic term, 188
Quantifiers, 20
Quartiles, 713–714
Quotient(s), 7, 33
estimating, 222
of monomials, 211–214
of rational expressions, 232–234
simplified, 212

Radian, 607
related to degree, 607–608
Radical(s), 261
binomials containing, 274–276
exploring, 837
properties of, 261, 264, 837
simplifying, 264–269
sums of, 270–273
equations containing, 277–281
Radical equations, 277–281
Radical sign, 259, 261
Radicand, 261
Radiocarbon dating, 493–494
Radius (radii)
of a circle, 407
focal, of ellipse, 418
Random experiment, 743
Random numbers, 753
Random sample, 752
Range
of a frequency distribution, 713, 714
of a function, 141
of a relation, 153

Rate of change, 147
Rational expressions, 227–241
  products and quotients, 232–234
  in simplest form, 227
  sums and differences, 235–237
Rational functions, 228
  graphing, 230–231
Rational numbers. *See under* Numbers.
Rational root theorem, 382, 385
Reading algebra, xiv–xv
  independent study, 26
  making a sketch, 585
  probability, 751
  problem solving, 371
  symbols, xvi, 178
Real numbers, 1–2
  decimal representations of, 283–285
  equality of, 14
  graphs of, 1–3
  irrational, 285, 832
  rational, 283–284
  roots of, 259–263
  set of, 1, 283
  See also Properties of real numbers.
Reciprocal(s), 15, 33
  property of, 15
Recorde, Robert, 42
Reflexive property, 14
Region, feasible, 160
Regression line, 726–727
Relations, 153–157
  domain, 153
  dominance, 781–782
  graphs, 153, 154, 155
  range, 153
Remainder theorem, 372
Replacement set, 9
Resultant, 659–660
Reviews. *See* Chapter reviews, Chapter tests, College Entrance Exams, Mixed Problem Solving, Mixed Review Exercises, Self-Tests.
Rhind Papyrus, 42
Root(s)
  approximating irrational, 386–389, 836, 841
  bounds on, 389 (Exs. 27, 28)
  and coefficients, 338–341
  conjugate imaginary, 318
  cube, 260, 394–395
  double, 196, 317, 318
  extraneous, 247, 278
  finding rational, 318

  multiple, 196
  *n*th, 260
  of open sentence, 37
  of polynomial equation, 194, 377–379, 382–383, 385, 386–388
  and radicals, 259–282
  square, 259, 288, 394–395
  theorems regarding, 266
Roots of unity, 687–688
  primitive, 688 (Ex. 22)
Rose
  three-leafed, 677, 679 (Ex. 32)
  four-leafed, 679 (Ex. 31)

Sample space, 743
Sampling, 752–753
Scalar, 660, 661, 771
Scalar product, 660–661, 771–772
Scatter plot, 724
Science. *See* Applications, Formulas, Problems.
Scientific notation, 218, 221–225, 478
Secant function, 556, 561, 613
  graph, 632
  inverse, 694
Second, 550
Self-Tests
  Chap. 1: 13, 36, 54
  Chap. 2: 72, 79, 92
  Chap. 3: 123, 140, 158
  Chap. 4: 177, 192, 205
  Chap. 5: 226, 241, 252
  Chap. 6: 282, 297
  Chap. 7: 316, 325, 345
  Chap. 8: 363, 381, 396
  Chap. 9: 417, 435, 450
  Chap. 10: 467, 477, 493
  Chap. 11: 516, 536, 543
  Chap. 12: 572, 600
  Chap. 13: 634, 654
  Chap. 14: 671, 688, 700
  Chap. 15: 729, 742, 761
  Chap. 16: 784, 793, 805
Sequence(s), 501–517
  arithmetic, 501, 507–509
  Fibonacci, 505
  finite, 501
  formula for *n*th term, 503, 507–508, 510
  geometric, 502, 510–514
  graphing, 517
  infinite, 501
  term of, 501

Series, 518–536
  arithmetic, 518
  geometric, 518
  infinite geometric, 531–536
  sums of, 525–536
  Taylor, 573
Set(s)
  dense, 285
  disjoint, 754
  empty, 38, 738
  factor, 179, 183
  intersection of, 754
  null, 38
  of primes, 179
  replacement, 9
  solution, 37, 61
  union of, 754
Set notation, 153
Sigma notation, 518–520
Signs, Descartes' rule of, 378, 379
Sign graph, 203
Significant digit(s), 221–222
Simplest radical form, 266
Simplifying an expression, 7–8
Simultaneous solution, 124
Sine function(s), 555, 561
  exploring addition of, 844
  as function of real numbers, 613
  graph, 624–629, 690
  inverse, 690–691
Sines, law of, 586
Sinusoid, 634
Slope, 112
Solution(s)
  of conjunction, 65, 66
  of disjunction, 65, 66
  of equations and inequalities having fractional coefficients, 242–244
  of equations containing radicals, 277–279
  of exponential equations, 460–461, 480
  of linear equations in two variables, 108–110
  of linear systems, 124–130, 444–449, 785–786
  of open sentences, 37, 101–103
  of open sentences involving absolute values, 73–74, 76–77
  of polynomial inequalities, 202–203
  primary and general, 698
  of quadratic equations, 307–321
  of quadratic systems, 436–437, 439–441

of trigonometric equations, 697–700

Speed, 609
angular, 609
Spirals, 679
Square(s)
completing the, 307–310, 319
difference of, 183
trinomial, 183
Square root(s), 259
approximating, 394–395
exploring, 832
of negative numbers, 259
principal, 259
table, 810
Standard deviation, 714, 715
Standard normal curve, 719–722
Standardized value in normal distribution, 721
Statistics
analyzing data, 391–395, 713–718
correlation, 724–729
gathering data, 752–753
normal distribution, 719–723
presenting data, 709–712
Stem-and-leaf plot, 710
Subset, 1, 738
Substitution, 8
synthetic, 373
transformation by, 125
Substitution principle, 8
Subtraction
definition, 22
of matrices, 771
of polynomials, 168–169
of rational expressions, 235–237
of real numbers, 22–25
Sum(s), 7, 19
of complex numbers, 293
of cubes, 184
of first *n* terms
arithmetic series, 525–526
geometric series, 526
of infinite geometric series, 531–536
opposite of, 29
of radicals, 270–273
of rational expressions, 235–237
of roots, 338–340
of two matrices, 770
of vectors, 659, 660
Summand, 518
Summation
limits of, 519
sign for, 518
Symbolic logic, 95–97

Symbols
absolute value, 3
division, 7
equality, 6
greater than, less than, 2
grouping, 7
history of, 42
inequality, 2, 6
logical, 20
multiplication, 7, 8
table of, xvi
words into, 43–48
Symmetric property, 14
Symmetry, axis of, 326, 412
Synthetic division, 368–370
Synthetic substitution, 373
Systems of equations
consistent, 128, 803
dependent, 128, 446, 803
equivalent, 124–125
homogeneous, 448
inconsistent, 128, 446–447, 803
infinitely many solutions, 127–128
linear in three variables, 444–449, 785–786, 802–803
linear in two variables, 124–134, 791–792, 801
quadratic, 436–438, 439–443
triangular form, 445–446, 785
used to solve word problems, 131–134, 442–443, 448–449
Systems of inequalities, 137, 139

Tables
common logarithms, 812–813
cubes and cube roots, 811
of measurement, xvii
squares and square roots, 810
in statistics, 709, 715, 716, 720, 724
symbols, xvi
trigonometric functions, 814–830
truth, 95–97
Tail wind, 131
Tangent function, 555, 561
as function of real numbers, 613
graph, 630–631, 693
inverse, 693–694
Tautology, 97
Technology. *See* Calculator Key-In, Calculators, Computer Exercises, Computer graphing ideas, Computer Key-In, *and* Science.

Term(s), 7
constant, in quadratic polynomial, 188
like, 23
linear, in quadratic polynomial, 188, 189
of polynomial, 167, 168
quadratic, in quadratic polynomial, 188
of sequence, 501
similar, 23
Tests. *See* Chapter tests, Self-Tests, College Entrance Exams.
Theorem(s), 81
about absolute value, 89
binomial, 541
about complex numbers, 682
conjugate root, 377, 381
De Moivre's, 685
Descartes' rule of signs, 378
dot product of vectors, 668
equation of line, 114
factor, 373
fundamental, of algebra, 377
about graphs and related equations, 108
intermediate-value, 387
about order, 89
about parallel and perpendicular lines, 119
proving, 81–90
Pythagorean, 401
converse of, 401
about radicals, 266
rational root, 382, 385
remainder, 372
roots of polynomial equation, 377–379
roots of quadratic equation, 338, 339
slope of line, 113
sum of infinite geometric series, 532
Thinking skills
Analysis. *See* Data analysis, Error analysis, Patterns, *and* Problem solving strategies.
Applying concepts. *See* Applications, Enrichment, Formulas, *and* Modeling.
Interpreting. *See* Patterns, Problem solving strategies, Reading Algebra, *and* Statistics.
Reasoning and inferencing. *See* Hypothesis, Induction, Logic, Probability, Proof, *and* Set(s).

Recall and transfer. *See*
Applications, Enrichment,
Formulas, Modeling, Mixed
Problem Solving, *and*
Problems.
Spatial perception. *See*
Geometry, Modeling, *and*
Problems, three-dimensional.
Synthesis. *See* Algorithm,
division; Conjunction;
Disjunction; Enrichment;
Euclidean Algorithm; Mixed
Problem Solving; Problems;
and Proof.
See also Explorations.
Transformations
exploring geometric, 847
that produce equivalent
equations, 37
that produce equivalent
inequalities, 60
that produce equivalent systems,
125
Transitive property, 14
Translating
a graph, 327–328, 408, 432
words into symbols, 43–48, 70
Triangle(s)
ambiguous case, 592–593
areas of, 586, 597–600
reference, 563
solving general, 591–596
solving right, 574–579
Trigonometry, triangle, 574–600
Trigonometric functions
of acute angles, 555–560, 843
cofunction relationships, 557–
558
of general angles, 561–567
identities, 636–654
inverses, 689–696
of real numbers, 613–617
tables, 568–571 (use of)
30°, 60°, 45° angles, 557–558
values of, 568–572

Trinomial(s), 174
factoring, 188–189
perfect square, 183
quadratic, 188
Truth table, 95–97

Union of sets, 754
unique, meaning of, 14
Unit circle, 613
Unit vector, 661

Value(s)
of function, 142, 334, 335
of numerical expression, 6
of variable, 9
Variable, 9
domain of, 9
replacement set of, 9
values of, 9
Variance, 714, 715
Variation
constant of, 351, 358
direct, 351–357, 839
inverse, 358–362, 429
joint, 359–362
Vertical line test, 154, 433
Vector(s), 659
addition, 660
in component form, 666
difference of, 661
dot product of, 668
equivalent, 659
initial and terminal points, 659,
666
norm of, 661
notation for, 659
orthogonal, 668–669
in the plane, 666–671
resultant of, 659–660
scalar multiplication, 660–661
sum, 659, 660

unit, 661
zero, 661
Vector quantity, 659
Venn, John, 754
Venn diagrams, 754–755
Vertex, 326, 412
Vertical-line test for function, 154

Watt, James, 68
Wind speed, 131
Word problems. *See* Problems.
Work, 672–673

***x*-axis,** 107
***x*-coordinate,** 107
***x*-intercept(s)**
of a graph of a polynomial
function, 386
of a line, 119
of a parabola, 329

***y*-axis,** 107
***y*-coordinate,** 107
***y*-intercept**
of a line, 119
of a parabola, 329

Zero(s)
absolute value of, 3
degree of, 167
division by, 33
double, 196
as exponent, 216
of a function, 195, 386, 392
graph as origin, 1
multiplicative property, 27, 82
multiple, 196
Zero-product property, 83, 194

Answers to Selected Exercises

Chapter 1 Basic Concepts of Algebra

Written Exercises, pages 4–5 **1.** -6 **3.** 0
5. -1 **7.** -4 **9.** $-\frac{1}{2}$ **11.** $0 > -6$
13. $-3 < -1$ **15.** $6 > -5$
17. $0 > -4$

19. $\frac{1}{2} > -\frac{3}{2}$

21. $0.5 > -1.5$

23. -5 **25.** 3 **27.** $-5, -1, 0, 2, 4$
29. $-2, -\frac{3}{2}, -\frac{1}{2}, 2, \frac{5}{2}$
31. $-2.6, -1.8, -1.6, -0.6$ **33.** -1
35. -3.6 **37.** $\frac{1}{3}$ **39.** $-1, 7$

Mixed Review Exercises, page 5 **1.** False
2. False **3.** False **4.** True **5.** False
6. False **7.** True **8.** True **9.** True **10.** True
11. True **12.** False

Computer Exercises, page 5 **2. a.** $-8 < -7$
b. $0.1 > 0.01$ **c.** $0 = 0$ **d.** $3.65 > 3.56$
e. $4 > -4$ **f.** $800 < 8000$ **4. a.** 5.5 **b.** 0
c. 1.565 **d.** -5.16 **e.** 9.055 **f.** 1

Written Exercises, pages 10–12 **1.** $=$ **3.** $>$
5. $=$ **7.** $<$ **9.** $=$ **11. a.** 11 **b.** 1 **c.** 5
13. a. 44 **b.** 180 **c.** 84 **15.** 2 **17.** 3
19. $\frac{2}{3}$ **21.** 28 **23.** 3 **25.** 19 **27.** 343
29. 100 **31.** $\frac{7}{2}$ **33.** 625 **35.** $\frac{1}{3}$ **37.** 14
39. 32 **41.** 8 **43.** 90
45. $18 \div 2 - 3 \cdot (2 + 1) = 0$
47. $[6 - (5 - 3)] \cdot 2 = 8$
49. $(3^2 - 2)^2 - 4 \cdot (3 + 3) = 25$
51. a. 9, 36, 100, 225 **b.** 9, 36, 100, 225
53. Answers may vary; $3 \cdot 3 + (3 + 3) \div 3$

Note that for all Calculator Key-Ins your calculator may give answers to more places or fewer places than shown in these Answers to Selected Exercises. Also, some calculators round the last displayed digit, while others simply discard digits that cannot be displayed. Answers given here are rounded to the number of places shown.

Calculator Key-In, page 12 **1.** 45 **3.** 15
5. 4.25 **7.** 3.965517 **9.** 0.7777778
11. 22.7375

Self-Test 1, page 13
1.

2. $-2 < -\frac{1}{2}$ **3.** False **4.** True **5.** True
6. True **7.** 12 **8.** 4 **9.** 16 **10.** 3

Written Exercises, pages 17–19 **1.** 150 **3.** 0
5. $4z + 1$ **7.** 0 **9.** $2a$ **11.** True **13.** False
15. True **17. a.** Dist. prop. **b.** Assoc. prop.
of mult. **c.** Prop. of reciprocals **d.** Ident.
prop. of mult. **19. a.** Comm. prop. of add.
b. Assoc. prop. of add. **c.** Prop. of opposites
d. Ident. prop. of add. **e.** Prop. of reciprocals
21. a. Assoc. prop. of add. **b.** Ident. prop. of
mult. **c.** Dist. prop. **d.** Ident. prop. of mult.
e. Dist. prop. **23. a.** Add. prop. of $=$
b. Assoc. prop. of add. **c.** Prop. of opposites
d. Ident. prop. of add. **e.** Mult. prop. of $=$
f. Assoc. prop. of mult. **g.** Prop. of
reciprocals **h.** Ident. prop. of mult.
25. a. Closed under addition **b.** Closed under
multiplication **27. a.** Closed under addition
b. Closed under multiplication **29. a.** Closed
under addition **b.** Closed under multiplication
31. a. Closed under addition **b.** Closed under
multiplication **33. a.** Not closed under addition
$\left(\frac{2}{3} + \frac{2}{3} = \frac{4}{3}\right)$ **b.** Closed under multiplication
35. No; Prop. of reciprocals **37.** No; Ident.
prop. of add.

Mixed Review Exercises, page 20 **1.** $>$
2. $<$ **3.** $<$ **4.** $=$ **5.** $<$ **6.** $<$

Extra, page 20 **1.** $\forall_x \ 2x = x + x$
3. $\forall_x \forall_y \ x + y = y + x$ **5.** False **7.** True

Written Exercises, pages 24–25 **1.** -21
3. 34 **5.** 25 **7.** 7.4 **9.** -23 **11.** 30.7
13. -2 **15.** 8 **17.** 19 **19.** -9 **21.** -21
23. $4p - 8q$ **25.** $a - 2b + 3$ **27.** $10c + 6$
29. $11x - 2$ **31.** $-3p - q$ **33.** $-5s + 9$
35. 9 **37.** -3 **39.** 4 **41.** 10.1 **43.** 152.7°C
45. 3822 ft below sea level **47.** $74\frac{5}{8}$
49. Answers may vary; $7 - 2 \neq 2 - 7$ **51.** No;
$0 - 1 = -1$ **53.** No; $1 - 2 = -1$ **55.** Yes

Reading Algebra, page 26 **1.** Solve certain
equations in one variable. **3.** An equation or
inequality containing a variable; equations having
the same solution set over a given domain
5. {3} **7.** Table 7, p. 830

Written Exercises, pages 30–31 **1.** -210
3. 12 **5.** -2.4 **7.** $60xy$ **9.** $-6abc$ **11.** 10
13. 96 **15.** 96 **17.** $-2 + 4x + 6x^2$
19. $-3z^2 + 2z - 1$ **21.** $-9xy$ **23.** $-5k - 24$
25. $2x - 2y$ **27.** $-18a^3 - 16$
29. $2tw - 13t - 14$ **31. a.** 3 **b.** -3 **c.** 0
33. a. 0 **b.** 0 **c.** 4 **35.** 1. Ident. prop. of
mult. 2. Prop. of opp. of a prod. 3. Comm.
prop. of mult. 4. Trans. prop. of = 5. Symm.
prop. of =

Mixed Review Exercises, page 31 **1.** 0 **2.** 1
3. 7 **4.** -1 **5.** -4 **6.** 14 **7.** Comm. prop.
of add. **8.** Ident. prop. of add. **9.** Symm.
prop. of = **10.** Prop. of reciprocals
11. Assoc. prop. of mult. **12.** Dist. prop.

Computer Exercises, page 32 **2. a.** 3
b. 380.585 **c.** -127.375 **d.** -212.28352
4. a. 5.897 **b.** -186.542125 **c.** 29.921848
d. 392.66634

Written Exercises, page 35 **1.** 3 **3.** 8
5. -18 **7.** 4 **9.** 48 **11.** $-\frac{3}{2}$ **13.** 8
15. $-\frac{2}{5}$ **17.** -54 **19.** $-3x^2 - 9$ **21.** $n^2 - 1$
23. $6c^2 - 4c - 1$ **25. a.** 2 **b.** 2 **c.** 0
27. a. -1 **b.** $\frac{3}{2}$ **c.** -15 **29.** Answers may
vary; $4 \div 2 \neq 2 \div 4$ **31.** Yes **33.** Yes
35. Yes

Self-Test 2, page 36 **1. a.** Comm. prop. of
add. **b.** Assoc. prop. of add. **c.** Prop. of opposites **d.** Indent. prop. of add. **e.** Prop. of
reciprocals **2.** 2 **3.** -10 **4.** 8 **5.** -2.8
6. -9 **7.** $3cd$ **8.** 25 **9.** 2 **10.** -2
11. $3x^2 - 5x + 1$ **12.** -1

Written Exercises, pages 40–42 **1.** {3}
3. {12} **5.** {6} **7.** {2} **9.** {2} **11.** Ø **13.** {3}
15. {−7.5} **17.** {−4} **19.** {3}
21. {real numbers} **23.** {4} **25.** Yes; No
27. Yes; Yes **29.** No; Yes **31.** $x = \frac{5}{2}y + 5$
33. $p = \frac{I}{rt}$ **35.** $x = \frac{y - b}{m}$ **37.** $w = \frac{1}{2}p - l$
39. $x = 2b + \frac{c}{a}$ **41.** $v = \frac{s}{t} + \frac{1}{2}gt$ **43.** $h = \frac{2}{\pi}$
45. $t = 1.5$ **47.** $b_1 = 28$ **49.** $s = 15$
51. $h = \frac{A}{2\pi r} - r$ **53.** $a = \frac{S(1 - r)}{1 - r^n}$

Mixed Review Exercises, page 42 **1.** 8
2. -6 **3.** 25 **4.** 4 **5.** -12 **6.** -2
7. $x - 8$ **8.** $4ab$ **9.** $-x^2y^3$ **10.** $-3c + 4d$
11. 0 **12.** $m - 1$

Written Exercises, pages 46–48
1. $(60 - w)$ ft **3.** $\frac{1}{2}x + 2$ **5. a.** $l(l - 2)$ cm^2
b. $(4l - 4)$ cm **7.** $9r$ mi **9.** $\left(120 - \frac{a}{3}\right)°$
11. $\frac{1}{3}x(2x - 3)^2$ **13.** $(4t - 250)$ dollars
15. $(0.15q + 1.80)$ dollars **17–29.** Answers
may vary; examples are given. **17.** Let $x =$ the
length of shortest side;
$x + (x + 2) + (x + 4) + (x + 6) = 60$ **19.** Let
$x =$ the regular price; $x + 2(x - 2) = 41$
21. Let $g =$ the amount of money that Greg has;
$g + (g + 12) + 2(g + 12) = 124$ **23.** Let $c =$
the car's speed; $2c + 2\left(\frac{2}{3}c\right) = 140$ **25.** Let $b =$
the measure of $\angle B$; $(b + 20) + b + 2b + 4b = 360$ **27.** Let $c =$ the number of kilograms of
cashews; $7c + 9(20 - c) = 7.80(20)$ **29.** Let
$r =$ Kevin's speed going; $\frac{320}{r} + \frac{1}{3} = \frac{320}{r - 4}$

Challenge, page 48 Answers may vary;
$23 = 2^3 + 2^3 + 1^3 + 1^3 + 1^3 + 1^3 + 1^3 + 1^3 + 1^3$

Problems, pages 52–54 **1.** Amy: $11;
Maria: $19 **3.** 180 **5.** 84 ft × 49 ft **7.** 51°
9. 12:15 P.M. **11.** 34 dimes, 6 quarters
13. not enough information **15.** no solution
17. $5\frac{1}{2}$% and $8\frac{1}{2}$%; extra information
19. 2:30 P.M. **21.** 80 ppm/h **23.** winner: 728
votes; loser: 673 votes

Mixed Review Exercises, page 54 **1.** {−2}
2. {3} **3.** {−1} **4.** -20 **5.** 100 **6.** 26
7. -1 **8.** -3 **9.** 2

Self-Test 3, page 54　**1.** {5}　**2.** {2}
3. $b = 2m - a$　**4.** $2(n + n^2)$　**5.** $4x + 10$
6. $2\frac{1}{2}$ h

Chapter Review, pages 55–56　**1.** c　**3.** d
5. a　**7.** c　**9.** d　**11.** c　**13.** b　**15.** c　**17.** a

Chapter 2 Working with Inequalities

Written Exercises, pages 62–63
　1. {x: x > 2}

　3. {t: t < 3}

　5. {x: x > −2}

　7. {t: t < −3}

　9. {s: s > −1}

　11. {y: y > 4}

　13. {h: h > −1}

　15. {u: u > −3}

　17. Ø　**19.** {y: y > −1}

21. {k: k < 1}　**23.** {real numbers}
25. True　**27.** False; $-1 < 0$, but $(-1)^2 > 0^2$.
29. True　**31.** True　**33.** If you multiply any
number between 0 and 1 by itself, the result is
smaller than the original number.

Mixed Review Exercises, page 63　**1.** $16t - 10$
2. -60　**3.** $-16p^3$　**4.** 12　**5.** $-4cd + 3$
6. 2　**7.** 14　**8.** 54　**9.** $a^2 - 5a + 7$
10. $8x - 5y$　**11.** -5　**12.** $11 - 6m$

Computer Exercises, page 64
2. a. {1, 2, 3, 4, 5, 6}　**b.** {5, 6, 7}　**c.** Ø
4. a. No solutions　**b.** {−10, −9, −8, −7,
−6, −5, −4, −3, −2, −1, 0, 1}　**c.** {1}

Written Exercises, pages 67–68
　1. {x: 3 ≤ x < 5}

　3. {t: t < 1 or t ≥ 3}

　5. {y: y ≥ 3}

　7. {real numbers}

　9. {x: 2 ≤ x < 5}

　11. {r: −2 < r < 2}

　13. {z: z ≤ 3 or z > 5}

　15. {t: t ≥ 3 or t < 2}

　17. Ø
　19. {k: −1 < k < 3}

　21. {d: 9 ≤ d < 15}

　23. {real numbers}

　25. {x: −1 < x < 8}

　27. {real numbers}

29. $\left\{t: 0 < t \le \frac{9}{2}\right\}$ **31.** $\left\{t: t > -\frac{4}{3}\right\}$

33. $\{x: -2 \le x \le 4\}$

35. $\{y: -2 < y < -1 \text{ or } 3 < y < 4\}$

Problems, pages 71–72 **1.** At most 64
3. 9 cm **5.** $\{5, 7, 9\}, \{7, 9, 11\}$ **7.** 93 or
more **9.** At most 8 cm **11.** At least 160 min
13. At most $2\frac{1}{2}$ min **15.** Width: at least 13 m;
length: at least 18 m

Mixed Review Exercises, page 72
1. $\{x: x \ge -2\}$

2. $\{t: t \le -4 \text{ or } t \ge 1\}$

3. $\{m: m < -5\}$

4. $\emptyset$

5. $\{d: d > 1\}$

6. $\{p: 0 < p < 3\}$

7. 4 **8.** 10 **9.** 21 **10.** 21

Self-Test 1, page 72
1. $\{m: m \le 3\}$

2. $\{y: y > -2\}$

3. $\{x: x < 5\}$

4. $\{n: n \le -1\}$

5. $\{a: -1 < a < 1\}$

6. $\{c: c < -3 \text{ or } c \ge 3\}$

7. 83 or more

Written Exercises, page 75
1.

3.

5.

7.

9. $\{t: -4 < t < -1\}$

11. $\left\{\frac{5}{2}\right\}$

13. $\{x: x \le 1 \text{ or } x \ge 5\}$

15. $\{\text{real numbers}\}$

17. $\{t: 0 \le t \le 4\}$

19. $\{-1, -9\}$ **21.** $\{u: -1 \le u \le 2\}$

23. $\left\{d: \frac{3}{2} \le d \le 2\right\}$ **25.** $\left\{t: t < \frac{4}{3} \text{ or } t > 2\right\}$

27. $\emptyset$

29.

31.

33. $\{x: -3 \le x \le 1\}$

Computer Exercises, page 75 **2. a.** $x = -3$,
$-2, -1, 0, 1, 2, 15, 16, 17, 18, 19$
b. $x = -12, -11, -10, -9, -8, -7, -6, -5,$
$-4, -3, -2$

Written Exercises, pages 78–79 **1.** $\{-4, 4\}$
3. $\{t: -3 < t < 3\}$ **5.** $\{u: u \le -2 \text{ or } u \ge 2\}$
7. $\{y: -1 < y < 5\}$ **9.** $\{t: -1 \le t \le 4\}$
11. $\{r: r < -7 \text{ or } r > 3\}$ **13.** $\{-1, 2\}$
15. $\{p: p \le -4 \text{ or } p \ge -1\}$ **17.** $\{t: 1 \le t \le 5\}$
19. $\left\{x: \frac{1}{2} < x < \frac{5}{2}\right\}$ **21.** $\left\{f: -\frac{13}{3} < f < -\frac{5}{3}\right\}$

23. $\{t: -8 < t < 2\}$
25. $\{x: x \le a - c \text{ or } x \ge a + c\}$
27. $\{x: 0 < x < 2c\}$
29. $\left\{x: x \le \dfrac{-a-c}{b} \text{ or } x \ge \dfrac{-a+c}{b}\right\}$
31. $\left\{x: x < -\dfrac{a-c}{b} \text{ or } x > \dfrac{a-c}{b}\right\}$
33. $\left\{x: -a - \dfrac{c}{b} < x < -a + \dfrac{c}{b}\right\}$

Mixed Review Exercises, page 79
1. $\{x: -1 \le x \le 3\}$ **2.** $\{1, 4\}$ **3.** $\{c: c > 3\}$
4. $\{y: y \ne -1\}$ **5.** $\{w: w \le -2 \text{ or } w \ge 2\}$
6. $\{p: p < -3 \text{ or } p \ge 2\}$ **7.** $\{z: z \le -2\}$
8. $\{-4\}$ **9.** $\{m: -2 < m < 4\}$

Self-Test 2, page 79 **1.** $|3x + 2| \le 4$
2. $\{x: x < -2 \text{ or } x > 1\}$

3. $\{y: 2 \le y \le 4\}$

4. $|x + 2| \ge 5$ **5.** $\{m: 2 \le m \le 6\}$
6. $\{n: n < -2 \text{ or } n > -1\}$

Computer Key-In, page 80 **1.** $-1.5 < x < 4$
3. $x < -0.2 \text{ or } x > 0.6$ **5.** $-8 < x < 5$
7. No solution

Written Exercises, pages 85–87 $(-1)^2 = 1^2$,
but $-1 \ne 1$ **3.** $|0 - 1| \ne |0| - |1|$
5. (1) Comm. prop. of mult.; (2) Proved in Example 4; (3) Comm. prop. of mult. **7.** (1) Def. of subtr.; (2) Assoc. prop. of add.; (3) Prop. of opp.; (4) Ident. prop. of add. **9.** (1) Given; (2) Def. of u^2; (3) Prop. of recip.; (4) Mult. prop. of eq.; (5) Assoc. prop. of mult.; (6) Prop. of recip.; (7) Ident. prop. of mult.
11. (1) Given; (2) Prop. of recip.; (3) Mult. prop. of eq.; (4) Mult. prop. of 0; (5) Assoc. prop. of mult.; (6) Prop. of recip.; (7) Ident. prop. of mult.; (8) Steps 2–7 **13.** (1) Given, prop. of recip.; (2) Def. of div.; (3) Dist. prop. of mult. with respect to subtr.; (4) Def. of div.
15. (1) $a \ne 0$ (Given); (2) $a \cdot \dfrac{1}{a} = 1$ (Prop. of recip.); (3) $\dfrac{1}{a} \cdot \dfrac{1}{\frac{1}{a}} = 1$ (Prop. of recip.);
(4) $\therefore a = \dfrac{1}{\frac{1}{a}}$ (Steps 2 and 3 and prop. of recip. uniqueness) **17.** (1) $b \ne 0$ (Given);

(2) $(-b)\left(-\dfrac{1}{b}\right) = \dfrac{1}{b} \cdot b$ (Prod. of opp.)
(3) $(-b)\left(-\dfrac{1}{b}\right) = 1$ (Prop. of recip.);
(4) $\therefore -\dfrac{1}{b} = \dfrac{1}{-b}$ (Prop. of recip. uniqueness)
19. (1) $b \ne 0$ (Given); (2) $\dfrac{-a}{-b} = (-a)\left(\dfrac{1}{-b}\right)$
(Def. of div.); (3) $\dfrac{-a}{-b} = (-a)\left(-\dfrac{1}{b}\right)$ (Proved in Exercise 17); (4) $\dfrac{-a}{-b} = a \cdot \dfrac{1}{b}$ (Product of opp.
(Ex. 6)); (5) $\therefore \dfrac{-a}{-b} = \dfrac{a}{b}$ (Def. of div.)
21. (1) $b \ne 0,\ d \ne 0$ (Given);
(2) $\dfrac{a}{b} \cdot \dfrac{c}{d} = \left(a \cdot \dfrac{1}{b}\right) \cdot \left(c \cdot \dfrac{1}{d}\right)$ (Def. of div.);
(3) $\dfrac{a}{b} \cdot \dfrac{c}{d} = (ac)\left(\dfrac{1}{b} \cdot \dfrac{1}{d}\right)$ (Assoc. and Comm.
prop. of mult.); (4) $\dfrac{a}{b} \cdot \dfrac{c}{d} = (a \cdot c)\left(\dfrac{1}{bd}\right)$ (Proved in Ex. 20); (5) $\therefore \dfrac{a}{b} \cdot \dfrac{c}{d} = \dfrac{ac}{bd}$ (Def. of div.)
23. (1) $c \ne 0,\ d \ne 0$ (Given);
(2) $\dfrac{a}{b} \div \dfrac{c}{d} = \dfrac{a}{b} \cdot \dfrac{1}{\frac{c}{d}}$ (Def. of div.);
(3) $\dfrac{a}{b} \div \dfrac{c}{d} = \dfrac{a}{b} \cdot \dfrac{d}{c}$ (Proved in Ex. 22);
(4) $\therefore \dfrac{a}{b} \div \dfrac{c}{d} = \dfrac{ad}{bc}$ (Proved in Ex. 21) **25.** Let
the numbers be a and b. If $ab = 0$, then at least one of a and b must be 0 by the zero-product property. But a and b are given to be nonzero.

Written Exercises, pages 90–91 **1.** (1) Given;
(2) Second mult. prop. of order; (3) Mult. prop. of -1 **3.** (1) Given; (2) Add. prop. of order; (3) Given; (4) Add. prop. of order; (5) Trans. prop. of order (Steps 2 and 4) **5.** (1) Given; (2) Second mult. prop. of order; (3) Given; (4) Second mult. prop. of order; (5) Trans. prop. of order (Steps 2 and 4); (6) Def. of a^2 and b^2
7. (1) $a > 0$ (Given); (2) $a + (-a) > 0 + (-a)$ (Add. prop. of order); (3) $0 > 0 + (-a)$ (Prop. of opp.); (4) $\therefore 0 > -a$, or $-a < 0$ (Ident. prop. of add.) **9.** (1) $a > 0,\ b > 0$ (Given);
(2) $ab > 0 \cdot b$ (First mult. prop. of order);
(3) $\therefore ab > 0$ (Mult. prop. of 0) **11.** (1) $a > 0$,
$b > 0$ (Given); (2) $a \cdot \dfrac{1}{b} > 0 \cdot \dfrac{1}{b}$ (First mult.
prop. of order (assuming $\dfrac{1}{b} > 0$)); (3) $a \cdot \dfrac{1}{b} > 0$
(Mult. prop. of 0); (4) $\therefore \dfrac{a}{b} > 0$ (Def. of div.)

13. (1) $a > 0$, $b < 0$ (Given); (2) $a \cdot \frac{1}{b} < 0 \cdot \frac{1}{b}$ (Second mult. prop. of order (assuming $\frac{1}{b} < 0$)); (3) $a \cdot \frac{1}{b} < 0$ (Mult. prop. of 0); (4) $\therefore \frac{a}{b} < 0$ (Def. of div.) **15.** (1) $a > 0$, $b > 0$, $\frac{1}{a} > \frac{1}{b}$ (Given); (2) $ab > 0$ (Exercise 9); (3) $ab \cdot \frac{1}{a} > ab \cdot \frac{1}{b}$ (First mult. prop. of order); (4) $\left(a \cdot \frac{1}{a}\right)b > a\left(b \cdot \frac{1}{b}\right)$ (Assoc. and Comm. prop. of mult.); (5) $1 \cdot b > a \cdot 1$ (Prop. of recip.); (6) $\therefore b > a$, or $a < b$ (Ident. prop. of mult.). **17.** (1) $a < b$ (Given); (2) $a + a < a + b$ and $a + b < b + b$ (Add. prop. of order); (3) $2a < a + b$ and $a + b < 2b$ (Simplification); (4) $\therefore a < \frac{a + b}{2}$ and $\frac{a + b}{2} < b$, or $a < \frac{a + b}{2} < b$ (First mult. prop. of order (using the fact $\frac{1}{2} > 0$))

Mixed Review Exercises, page 91 **1.** True **2.** True **3.** True
4. $\{2, 4\}$

5. $\{d: d \geq -1\}$

6. $\{y: -1 < y < 3\}$

7. $\{2\}$

8. $\varnothing$ **9.** $\{k: k < -4 \text{ or } k \geq 4\}$

Self-Test 3, page 92 **1.** (1) Given; (2) Prop. of opp.; (3) Add. prop. of eq.; (4) Assoc. prop. of add.; (5) Prop. of opp.; (6) Ident. prop. of add.; (7) Def. of subtr. **2.** (1) $ab = c$, $b \neq 0$ (Given); (2) $\frac{1}{b}$ is a real number (Prop. of recip.); (3) $ab \cdot \frac{1}{b} = c \cdot \frac{1}{b}$ (Mult. prop. of eq.); (4) $a\left(b \cdot \frac{1}{b}\right) = c \cdot \frac{1}{b}$ (Assoc. prop. of mult.);

(5) $a \cdot 1 = c \cdot \frac{1}{b}$ (Prop. of recip.); (6) $a = c \cdot \frac{1}{b}$ (Ident. prop. of mult.); (7) $\therefore a = \frac{c}{b}$ (Def. of div.). **3.** (1) $a > 1$ (Given); (2) $a > 0$ (Trans. prop. of order (using the fact $1 > 0$)); (3) $a \cdot a > a \cdot 1$ (First mult. prop. of order); (4) $a \cdot a > a$ (Ident. prop. of mult.); (5) $\therefore a^2 > a$ (Def. of a^2)

Chapter Review, pages 93–94 **1.** b **3.** a **5.** b **7.** c **9.** b **11.** d **13.** c

Extra, page 97 **1.** true **3.** true **5.** true **7.** true **9.** true **11.** p is true and $\sim q$ is true; $p \wedge \sim q$ is true. **13.** Logically equivalent **15.** Not a tautology **17.** Tautology

Mixed Problem Solving, page 98 **1.** At least 14 quarters **3.** 32 h **5.** $1600 at 5%; $2400 at 8% **7.** $\frac{32}{d}$ **9.** 60° **11.** 3 lb
13. 7 cm $\times$ 12 cm

Preparing for College Entrance Exams, page 99 **1.** C **3.** C **5.** D **7.** A

Chapter 3 Linear Equations and Functions

Written Exercises, pages 104–105
1. $\left\{(-1, 3), \left(0, \frac{7}{3}\right), (2, 1)\right\}$
3. $\left\{\left(-1, \frac{1}{2}\right), (0, 0), (2, -1)\right\}$ **5.** $\left\{(-1, -1), \left(0, -\frac{5}{9}\right), \left(2, \frac{1}{3}\right)\right\}$
7. $\left\{\left(-2, \frac{11}{3}\right), \left(1, \frac{5}{3}\right), \left(3, \frac{1}{3}\right)\right\}$
9. $\left\{(-2, 1), \left(1, -\frac{1}{2}\right), \left(3, -\frac{3}{2}\right)\right\}$
11. $\left\{\left(-2, -\frac{13}{9}\right), \left(1, -\frac{1}{9}\right), \left(3, \frac{7}{9}\right)\right\}$
13. 6, 4, 3 **15.** $-\frac{7}{2}, \frac{7}{5}, -11$
17. $-\frac{1}{2}, -1, 2$ **19.** $-\frac{1}{4}, 1, 2$ **21.** 5
23. -2 **25.** -3
27. $\{(0, 4), (1, 3), (2, 2), (3, 1), (4, 0)\}$
29. $\{(0, 15), (1, 11), (2, 7), (3, 3)\}$
31. $\{(0, 6), (3, 4), (6, 2), (9, 0)\}$
33. $\{(1, 1), (1, 2), (1, 3), (2, 1), (2, 2), (3, 1)\}$
35. $\{(1, 1), (1, 2), (1, 3), (2, 1), (2, 2), (3, 1), (3, 2), (4, 1)\}$ **37.** $\{(1, 3), (6, 2), (9, 1)\}$

39. Let $N = 10t + u$. Then $K = 10u + t$.
$N - K = (10t + u) - (10u + t) =$
$10t - t + u - 10u = 9t - 9u = 9(t - u)$.
So $N - K$ is an integral multiple of 9.

Problems, pages 105–106   1. 3 \$5 and 3 \$20;
or 7 \$5 and 2 \$20; or 11 \$5 and 1 \$20; or 15 \$5
and 0 \$20   **3.** 7 dimes and 1 quarter; or 2 dimes
and 3 quarters   **5.** 4 cm, 4 cm, 4 cm and 7 cm;
or 5 cm, 5 cm, 5 cm and 4 cm; or 6 cm, 6 cm,
6 cm and 1 cm   **7.** 34 nickels, 3 dimes, and
0 quarters; 27 nickels, 4 dimes, and 1 quarter; or
20 nickels, 5 dimes, and 2 quarters; or 13 nickels, 6 dimes, and 3 quarters; or 6 nickels,
7 dimes and 4 quarters   **9.** 41, 51, 61, 63, 71,
73, 81, 83, 85, 91, 93, 95   **11.** 10, 11, 12, 13,
20, 21   **13.** 4, 15 and 8; 5, 12 and 10; 6, 9 and
12; 7, 6, and 14

Mixed Review Exercises, page 106   1. -26
2. -36 **3.** 7 **4.** -25 **5.** -2 **6.** 12 **7.** 9
8. -7
  9. $\{y: -1 < y \le 4\}$

10. $\{m: m < 2 \text{ or } m > 4\}$

11. $\{n: n \ge 4\}$

Written Exercises, page 111

1.

3.

5.

7.

9.

11.

13.

15.

17.

19.

21.

23. 2   **25.** 0
27. $(5, -2)$

29. $(-3, 5)$

31.

33.

33.

35.

35.

37.

37. 60 ft **39.** 6 **41.** $\frac{3}{2}$ **43.** 1 **45.** 2

47. -4 **49.** $y - y_1 = m(x - x_1)$;
$y - y_1 = mx - mx_1$; $mx - y = mx_1 - y_1$; $A = m$,
$B = -1$, $C = mx_1 - y_1$

51. slope $= -\dfrac{A}{B} = -\dfrac{m}{-1} = m$ **53.** The coordinates of $Q(x', y')$ satisfy the equation of L.

Mixed Review Exercises, page 117 **1.** -3
2. 1 **3.** 2 **4.** -2

5.

6.

39.

7.

8.

Written Exercises, pages 116–117 **1.** 2

3. $-\frac{2}{3}$ **5.** vertical **7.** 4 **9.** $-\frac{4}{5}$ **11.** -1

13. -1 **15.** $-\frac{1}{2}$ **17.** 1 **19.** $\frac{1}{3}$ **21.** $-\frac{3}{2}$

23. 6 **25–35.** Points chosen may vary.
Examples are given.

25.

27.

9.

10.

Written Exercises, pages 121–122
1. $x - y = -1$ **3.** $2x + y = 10$
5. $x - 2y = 1$ **7.** $x - 5y = 19$ **9.** $y = -1$
11. $2x - 5y = -24$ **13.** $x + y = 2$
15. $x - 2y = -3$ **17.** $6x - 5y = 3$
19. $2x + 5y = 0$ **21.** $x + y = 1$

23. $2x + 3y = 0$ **25.** $x = -2$ **27.** $y = \frac{1}{2}$

29. $6x - 6y = 7$ **31. a.** $x + y = 3$
b. $x - y = -3$ **33. a.** $x - 2y = 8$
b. $2x + y = -4$ **35. a.** $x + 2y = 2$
b. $2x - y = 4$ **37. a.** $y = 1$ **b.** $x = -4$

29.

31.

39. $y = 4$ **41.** $x = 0$ **43.** $y = 6$
45. $3x + 4y = 12$ **47.** $x + y = -1$
49. Parallelogram, not a rectangle
51. Parallelogram, rectangle **53.** Not a parallelogram

Computer Exercises, pages 122–123
2. a. 4.5 **b.** -2.28571429 **c.** vertical
4. $y = 1.72x + 17.2$ **6. a.** $y = 4.5x - 14.5$
b. $y = -2.28571429x - 4.42857143$ **c.** $x = 7$

Self-Test 1, page 123
1. $\left\{\left(-3, -\frac{21}{2}\right), \left(-1, -\frac{13}{2}\right), \left(2, -\frac{1}{2}\right)\right\}$
2. 10 jazz, 1 classical; 7 jazz, 3 classical; 4 jazz, 5 classical; 1 jazz, 7 classical

3.   **4.**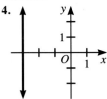

5. $\frac{3}{2}$ **6.** $-\frac{5}{4}$ **7.** 0 **8.** $y - 4 = \frac{1}{6}(x - 2)$ or
$y - 3 = \frac{1}{6}(x + 4)$ **9.** $2x - y = -8$ **10.** $y = 1$

Written Exercises, pages 129–130 **1.** $(-1, 3)$
3. $(3, 1)$ **5.** $(3, 1)$ **7.** $(-2, -2)$ **9.** $(2, 4)$
11. $\left(-\frac{9}{2}, 5\right)$
13. $(3.5, -0.5)$ **15.** $(2, 8)$

  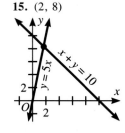

17. $\left(\frac{27}{11}, -\frac{5}{11}\right)$ **19.** $(0, 2)$ **21.** $\left(\frac{23}{4}, -\frac{1}{2}\right)$
23. $\{(x, y): 2x - y = 1\}$; examples: $(0, -1)$,
$\left(\frac{1}{2}, 0\right)$, $(2, 3)$ **25.** $\left(-\frac{2}{3}, -\frac{1}{3}\right)$
27. No solution **29.** inconsistent
31. consistent **33.** $(2, -3)$ **35.** $(-2, 3)$
37. $\left(\frac{1}{2}, 2\right)$ **39.** $(-1, a + b)$; $a \neq b$
41. $\left(\dfrac{de - bf}{ad - bc}, \dfrac{af - ce}{ad - bc}\right)$; $ad \neq bc$

43. $\left(\dfrac{b_2 - b_1}{m_1 - m_2}, \dfrac{m_1 b_2 - m_2 b_1}{m_1 - m_2}\right)$ is a common
solution, and it is made up of real numbers since $m_1 \neq m_2$ and hence $m_1 - m_2 \neq 0$.

Mixed Review Exercises, page 130 **1. a.** $\frac{1}{2}$
b. $x - 2y = -2$ **2. a.** 3 **b.** $3x - y = 5$
3. a. $\frac{2}{3}$ **b.** $2x - 3y = 0$ **4. a.** 0 **b.** $y = -1$
5. a. -1 **b.** $x + y = -1$ **6. a.** $-\frac{4}{3}$
b. $4x + 3y = 0$ **7.** $m = 1$, $b = -4$
8. $m = -\frac{5}{3}$, $b = 2$ **9.** $m = 2$, $b = 1$
10. $m = 0$, $b = 5$ **11.** $m = 3$, $b = 2$
12. $m = -\frac{1}{6}$, $b = \frac{2}{3}$

Problems, pages 132–134 **1.** 12 \$20 bills,
3 \$50 bills **3.** 55°, 55°, 70° **5.** 9 pairs of
jeans, 12 shirts **7.** air speed, 165 mi/h; wind
speed, 15 mi/h **9.** first minute, \$2.80; add.
minute, \$1.20 **11.** \$5000 at 15%; \$3000 at 6%
13. air speed, 60 km/h; distance 105 km
15. $v_0 = 8$ m/s; $a = 5$ m/s^2 **17.** weekly charge,
\$120; charge per mile, \$.25
19. a.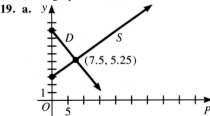

b. \$7.50 **c.** Supply is greater than demand.

Written Exercises, pages 138–139
1.   **3.**

5.   **7.**

9.

11.

33.

35.

13.

15.

37.

39.

17.

19.

41.

43.

21.

23.

45.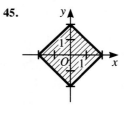

Mixed Review Exercises, page 139   **1.** $(-2, 1)$
2. $(0, -4)$ **3.** $(3, -4)$
4. $x - y = -4$ **5.** $x = 2$

25.

27.

6. $4x + y = 0$ **7.** $2x - 3y = -6$

29.

31.

8. $y = 7$

9. $x + 2y = -6$

21.

23.

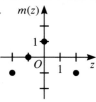

25. 2 **27.** 2 **29.** {real numbers}
31. {real numbers}
33. $\{x: x \neq -2 \text{ and } x \neq -3\}$ **35.** 0, 1
37. 8, -5 **39.** 8, -1 **41.** $-\dfrac{24}{25}$ **43.** -2
45. $2x + h$
47. a. $f(0) = f(0 + 0) = f(0) + f(0)$. Since
$f(0) = f(0) + f(0)$, $f(0)$ is additive identity, i.e.
$f(0) = 0$
b. $f(2a) = f(a + a) = f(a) + f(a) = 2f(a)$
c. $f(a + (-a)) = f(a) + f(-a)$;
$f(0) = f(a) + f(-a)$; from part **(a)**, $f(0) = 0$;
$0 = f(a) + f(-a)$; $f(-a) = -f(a)$
49. $s(0) = 0$, but $f(0)$ is undefined

Self-Test 2, p. 140 **1.** (6, 12) **2.** No solution
3. $\{(x, y): 4x + 3y = 10\}$; examples: $\left(0, \dfrac{10}{3}\right)$,
$\left(\dfrac{5}{2}, 0\right)$, (1, 2) **4.** $(-2, 7)$ **5.** wind speed,
15 mi/h; air speed, 105 mi/h **6.** Bret, \$1760;
Sandra, \$1800

7.

8.

9.

10.

Written Exercises, pages 144–145
1. $\{-3, -1, 1, 3\}$ **3.** $\{-2, 2\}$ **5.** $\{-2, 0, 4\}$
7. $\{0, 1, 4\}$ **9.** $\{0, 12\}$ **11.** $\{-1, 0, 1\}$
13. $F(x)$

15. $g(x)$

17. $f(x)$

19. $G(x)$

Written Exercises, pages 149–150
1. $f(x) = 2x + 3$ **3.** $f(x) = 3x + 1$
5. $f(x) = \dfrac{1}{2}x - 2$ **7.** $f(x) = 2x + 1$
9. $f(x) = 2x + 3$ **11.** $f(x) = -\dfrac{3}{2}x + 5$
13. $f(x) = 2x + 1$ **15.** $f(x) = -3$
17. $f(x) = 3x - 1$ **19.** $f(x) = -\dfrac{3}{2}x$
21. $f(x) = -\dfrac{3}{2}x - \dfrac{1}{2}$ **23.** $g(0) = -5$;
$g(-1) = -8$ **25.** $g(0) = 2$; $g(3) = \dfrac{1}{2}$
27. $f(-3) = -8$; $f(10) = \dfrac{41}{3}$ **29.** $f(10) = -10$;
$f(20) = -\dfrac{70}{3}$ **31.** m

Problems, pages 150–152 **1. a.** \$3150
b. 3 years, 8 months **3.** \$235 **5.** \$275
7. a. 179 lb **b.** 70 days
9. a. $c(f) = \dfrac{5}{9}f - \dfrac{160}{9}$ **b.** 37°C **c.** $-40°$
11. a. $C(n) = 4.5 + 0.062n$
b. $C(n) = 10.1 + 0.055n$

11. c. $C(n)$

(1500, 92.6)
(800, 54.1)
(0, 4.5)
20
200

1. not a function   **3.** function

5. not a function   **7.** $D = \{-1, 0, 1\}$; no

9. $D = \{0, 1, 2, 3\}$; no
11. $D = \{-2, -1, 0, 1, 2\}$; no

Ex. 9   Ex. 11

13. $D = \{-2, -1, 0, 1, 2\}$; yes
15. $D = \{-2, -1, 0, 1, 2\}$; no

Mixed Review Exercises, page 152

1.

2.

3.

4.

5.

6.

Ex. 13   Ex. 15

17. $D = \{-2, -1, 1, 2\}$; no
19. $D = \{-2, -1, 1, 2\}$; yes

7.

8.

9.

10. $\{-4, -1, 2\}$ **11.** $\{0, 1\}$ **12.** $\{5, 6, 7\}$
13. $\{-7, -2, 3\}$

Ex. 17   Ex. 19

21. $f \neq g$ **23.** $f = g$
25. yes   **27.** yes

29. yes   **31.** no

33. no   **35.** no

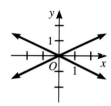

37. a. $\{\{5, 3\}, 5\}, \{5, \{5, 3\}\}, \{5, \{3, 5\}\}$
b. $(4, 1)$ **c.** $\{a, \{\{a, b\}, \{a, b, c\}\}\}$

Self-Test 3, page 158 **1.** $R = \{1, 3, 7\}$
2. $D = \left\{x: x \neq \dfrac{5}{2}\right\}$

3.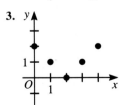

4. $f(x) = -\dfrac{2}{3}x + 1$

5. $f(x) = -2x + 5$

6. $f(x) = 0.2x + 8.5$

7. $\$710$

8. no
9. $D = \{-2, -1, 0, 1, 2\}$; yes

Ex. 8   Ex. 9

Application, pages 159–161 **1.** max.
value = 470; min. value = 0 **3.** max.
value = 121; min. value = 15 **5.** 12 packages
of Trailblazer mix and 24 packages of Frontier
mix

Chapter Review, pages 162–163 **1.** c **3.** d
5. b **7.** a **9.** c **11.** d **13.** b **15.** d **17.** c

Cumulative Review (Chapters 1–3), page 165
1. -3 **3.** $5x - y$ **5.** 4 h
7. $\{y: y \leq -3\}$

9. $\{-4, 1\}$

11. $\{n: -2 \leq n \leq 2\}$

13. $\left\{(-1, -2), \left(0, -\dfrac{1}{2}\right), (1, 1)\right\}$ **15.** 0
17. $7x - 2y = -22$ **19.** $5x - 4y = 3$
21. $(2, -1)$ **23.** $\left(-\dfrac{1}{2}, \dfrac{1}{2}\right)$ **25.** $\$8$ **27.** -13

Chapter 4 Products and Factors of Polynomials

Written Exercises, page 170 **1.** $x^2 - 2x + 2$; 2
3. $3x^3 - x$; 3 **5.** $9x^2y^2 - 3x^2 + 5xy^2$; 4
7. $5x^3y^2z^2 + 6x^2yz^3 - xyz$; 7 **9. a.** $7m - 1$
b. $3m - 7$ **11. a.** $2t^2 - 3t - 13$ **b.** $-13t - 1$
13. a. $5v^3 + v^2 - 1$ **b.** $5v^3 - v^2 - 4v + 3$
15. a. $5x^2 - 2xy + 7y^2$ **b.** $x^2 - 2xy + y^2$
17. $13x^2 - 6x - 2$ **19.** $m^2 + 21$ **21.** $7ax$
23. $12p^2 - 21pq - 20q^2$ **25.** $a = -2, b = 4,$
$c = 3, d = 2$ **27.** $a = -1, b = 3, c = 2$

Mixed Review Exercises, page 170
1. $f(x) = -2x + 2$ **2.** $f(x) = x + 5$
3. $f(x) = -\dfrac{1}{2}x + \dfrac{5}{2}$ **4. a.** -1 **b.** $x + y = 4$
5. a. $\dfrac{1}{4}$ **b.** $x - 4y = -7$ **6. a.** $-\dfrac{2}{3}$
b. $2x + 3y = 0$ **7. a.** $-\dfrac{3}{4}$ **b.** $3x + 4y = -2$
8. a. $\dfrac{2}{5}$ **b.** $2x - 5y = -43$ **9. a.** 0
b. $y = 5$

Written Exercises, page 173 **1.** $6z^5$ **3.** $-t^{12}$
5. $3x^3y^3$ **7.** $2u^5v^5$ **9.** $16a^6b^4$ **11.** $-27p^3q^{12}r^6$
13. z^6 **15.** s^8t^9 **17.** $3y^4 - 6y^3 + 9y$
19. $r^3s^2 - 2r^2s^3 - rs^4$ **21.** z^{2n} **23.** x^{2m}
25. r^{3h} **27.** $t^n + t^{n+1} + t^{n+2}$ **29.** $p^{m+1} + p^m$
31. $z^{2m} + z^m - 1$ **33.** t^{n^2} **35.** $\{5\}$ **37.** $\{2\}$

Written Exercises, pages 175–176
1. $6v^2 - 13v - 5$ **3.** $12z^2 - 7z - 12$
5. $9x^2 + 60x + 100$ **7.** $25y^2 - 4$
9. $14t^2 - 3t - 2$ **11.** $1 - 81t^2$
13. $3x^2 - 2xy - 8y^2$ **15.** $6p^2 + 5pq - 6q^2$
17. $x^4 - 9$ **19.** $s^6 + 2s^3t^3 + t^6$
21. $t^3 - t^2 - 2t$ **23.** $x^3y - 2x^2y^2 + xy^3$
25. $2c^3 - 5c^2 + c + 2$ **27.** $x^3 + 5x^2 + x - 10$
29. $y^6 - 5y^4 + 7y^2 - 2$ **31.** $x^4 + 3x - 2$
33. $a^4 - 4a^2b^2 - ab^3 - 2b^4$ **35.** $p^{2n} - 2p^n + 1$
37. $r^{2n} + r^ns^n - 2s^{2n}$
39. $a^3 - 3a^2b + 3ab^2 - b^3$ **41.** $a^3 + b^3$
43. $a^4 - b^4$ **45.** $x^4 - y^4$ **47.** $x^4 + 4$ **49.** 4
51. 4 **53.** -2 **55.** $4, -4$

Mixed Review Exercises, page 176
1. $2x^2 - 11x + 10$ **2.** a^6b^9 **3.** $-12m^3n^4$
4. $7y - 11$ **5.** $5cd - 6c + 12d$
6. $8p^3 - 4p^2 + 20p$ **7.** $-u^{11}$
8. $9z^3 + 5z^2 - 4z - 8$ **9.** $-10y$ **10.** $2x^6$

Self-Test 1, page 177 **1.** $x^2 + x - 1$ **2.** $t - 6$
3. $12p^4q^4$ **4.** t^{2n} **5.** $8a^2 + 2ab - 15b^2$
6. $2y^3 + 3y^2 + 4y - 3$ **7.** $9x^2 - 12x + 4$
8. $16c^2 - 9$

Reading Algebra, page 178 **1. a.** True
b. True **c.** False **d.** True **e.** False
3. a. True **b.** True **c.** False **5.** No, they
might be opposites. **7.** $5|(19 - 4)$ **9. a.** True
b. False **c.** True **d.** True **11.** odd integers

Written Exercises, pages 181–182 **1.** $2^2 \cdot 5 \cdot 7$
3. prime **5.** $3^2 \cdot 7^2$ **7.** $2^2 \cdot 7^2 \cdot 13$ **9.** 5; 140
11. 12; 432 **13.** 28; 840 **15.** 1; 315 **17.** $3p^2$;
$45p^3q$ **19.** $17y^2z$; $204xy^2z^2$ **21.** $22h^2k^2r$;
$440h^3k^2r^2$ **23.** 7; $42abc$ **25.** $13p^2q^2r^2$; $78p^3q^3r^3$
27. 126
29. $1 + 2 + 4 + 8 + 16 + 31 + 62 + 124 +$
$248 = 496$ **31. a.** $1, p, q, pq$ **b.** $1, p, q, p^2,$
$q^2, pq, p^2q, pq^2, p^2q^2$ **c.** $1, p, q, p^2, q^2, pq,$
$p^3, p^2q, pq^2, q^3, p^3q, p^2q^2, pq^3, p^3q^2, p^2q^3, p^3q^3$

Written Exercises, pages 185–186
1. $16x^2(x - 4)$ **3.** $(t + 9)^2$ **5.** $(4k - 1)(4k + 1)$
7. $(2y + 5)^2$ **9.** $(4x - 5)(4x + 5)$
11. $(11s - 3t)^2$ **13.** $(6p - 7q)(6p + 7q)$

15. $s(t - 1)(t + 1)$ **17.** $(t - 3)(t^2 + 3t + 9)$
19. $2rs(2r + s)(4r^2 - 2rs + s^2)$
21. $(x + 2)(y - 3)$ **23.** $(x - 2)(y - 3)$
25. $(q + 2)(p - 2)$ **27.** $(a - 2)(b + 1)$
29. $(x - 3 + 2y)(x - 3 - 2y)$
31. $(u - v + 1)(u + v - 1)$ **33.** $(x^2 - y)^2$
35. $(a^2 + b)(a^4 - a^2b + b^2)$
37. $(2s - 3)(2s + 3)(4s^2 + 9)$
39. $(x - y)(x^2 + xy + y^2)(x + y)(x^2 - xy + y^2)$
41. $(u + v)(u - v - 2)$ **43.** $2q(3p^2 + q^2)$
45. $(s + t)(s^2 + t^2)$ **47.** $4ab(3a^2 + b^2)(a^2 + 3b^2)$
49. $(x^n - 1)(x^n + 1)$ **51.** $(x^n + y^n)^2$
53. $x^{2n}(x^n - y^n)(x^n + y^n)$
55. $(x^2 - x + 1)(x^2 + x + 1)$
57. $(x^2 + y^2 - xy)(x^2 + y^2 + xy)$

Mixed Review Exercises, page 187
1. $x^2 - 6x + 9$ **2.** $2y^2 - 5y - 12$
3. $2m^3n - 3m^2n^2$ **4.** $4c - 9$
5. $a^3 - 6a^2 + 2a - 12$ **6.** $8w^5$ **7.** $6x$; $72x^3$
8. $15a^2b^3$; $150a^3b^4$ **9.** $21mp^2$; $126mp^3$

Computer Exercises, page 187
2. a. $(x - 28y)(x - 28y)$
b. $(3x + 24y)(3x + 24y)$
c. $64x^2 + 56xy + 49y^2$; not a perfect square
d. $(25x - 21y)(25x - 21y)$
4. a. $(12x - 2y)(12x - 2y)$
b. $(12x + 2y)(12x - 2y)$
c. $(12x + 2y)(12x + 2y)$ **d.** $144x^2 + 0xy + 4y^2$;
not a perfect square or diff. of squares

Written Exercises, pages 191–192
1. $(x - 8)(x - 1)$ **3.** $(z - 9)(z - 2)$
5. $(r + 2)(r + 10)$ **7.** prime
9. $(s - 2)(s - 18)$ **11.** $(x + 4)(x - 3)$
13. $(t - 7)(t + 5)$ **15.** $(3z + 1)(z + 1)$
17. $-(s - 4)(s + 2)$ **19.** $(x - 6y)(x + 5y)$
21. prime **23.** $(2t - 1)(t + 3)$
25. $(p - 3)(3p + 2)$ **27.** $(2x - 3y)(3x + y)$
29. $(2h - 3k)(h + 5k)$ **31.** $(6x - 5)(x + 2)$
33. prime **35.** prime **37.** $x(2x - y)(2x + 5y)$
39. $4pq(q - 2)(q^2 + 2q + 4)$
41. $(r - 2s)(r + 2s)(r^2 + 4s^2)$
43. $(x - 2)(x + 2)(x^2 + 1)$ **45.** $x - 2$;
$(x - 1)(x - 2)^2$ **47.** $t(t + 1)$;
$t(t + 1)(t + 4)(t - 3)$ **49.** $p - q$;
$(p - q)(p^2 + pq + q^2)$
53. $(a^n - b^n)(a^n + b^n)(a^{2n} + b^{2n})$
55. $(x^n - 2)(x^n + 2)(x^n - 1)(x^n + 1)$
57. $4(x + 2)(x^2 - 3x + 1)$ **59.** $x^2 - x + 1$

Self-Test 2, page 192 **1.** $2 \cdot 3^2 \cdot 5 \cdot 11$
2. $2^3 \cdot 3 \cdot 5^3$ **3.** $18ab$; $108a^2b^3c^2$ **4.** $2p^2q^2$;
$24p^3q^3r^2$ **5.** $3x(3x - 2)(3x + 2)$ **6.** $(3t + 1)^2$

Answers to Selected Exercises **14**

7. $(5s - 2t)(25s^2 + 10st + 4t^2)$
8. $(a + 1)(b - 1)$ **9.** $2(3t - 1)(t + 1)$
10. $(4z - 7)(z + 2)$

Computer Key-In, page 193
1. $(3x + 8)(4x - 15)$ **3.** $(x - 8)(6x - 7)$
5. $(x - 1)(2x + 10)$

Written Exercises, page 196 **1.** $\{1, 4\}$
3. $\{0, -1, 2\}$ **5.** $\{1 \text{ (d.r.)}, 3 \text{ (d.r.)}\}$ **7.** $\{1, 3\}$
9. $\{-1, 0, 1\}$ **11.** $\{0, 2 \text{ (d.r.)}\}$ **13.** $\left\{\frac{1}{3}, 1\right\}$
15. $\left\{-2, \frac{3}{2}\right\}$ **17.** $\left\{-3, \frac{2}{3}\right\}$ **19.** $\{-6, 12\}$
21. $\{2, 8\}$ **23.** $\left\{-1, \frac{4}{3}\right\}$ **25.** $\{-2, 1 \text{ (d.r.)}\}$
27. $\{-3, 0 \text{ (d.r.)}, 3 \text{ (d.r.)}\}$
29. $\{-1 \text{ (d.r.)}, 1 \text{ (d.r.)}\}$
31. $\{-3, -1, 0 \text{ (d.r.)}, 1, 3\}$ **33.** $\{0, 3 \text{ (d.r.)}\}$
35. $\left\{-1, \frac{1}{3}\right\}$ **37.** $\{1, -4\}$ **39.** $\{1 \text{ (d.r.)}, 2\}$
41. $\{-2 \text{ (t.r.)}, 2 \text{ (t.r.)}\}$ **43.** $(-1 \text{ (d.r.)}, 1 \text{ (d.r.)})$
45. 2, 4, 6 **47.** 0 (d.z.), $\frac{2}{3}$ (d.z.)
49. 1 (t.z.), 5 **51.** $\left\{-\frac{b}{a}, \frac{b}{a}\right\}$ **53.** $\{-a, 1, -1\}$
55. $x = 0$ or $y = 0$

Mixed Review Exercises, page 197 **1.** $(x - 3)^2$
2. prime **3.** $(2n + 3)(4m - 5)$
4. $(4a - 5b)(4a + 5b)$ **5.** $2x(4x - 3)(3x - 2)$
6. $(2u + 1)(4u^2 - 2u + 1)$ **7.** $z^3 - z^2$
8. $16a^2 + 40a + 25$ **9.** $5m + 1$
10. $p^4 - p^2 - 6$ **11.** -1 **12.** $-27c^6d^3$

Problems, pages 199–201 **1.** -9 or 8
3. -13 and -11 or 11 and 13 **5.** 9 cm by
13 cm **7.** 8 m and 11 m **9.** 9 ft **11.** 15 m
and 20 m **13.** 11 cm by 14 cm **15.** 490 m;
20 s **17.** 40 s **19.** 3062.5 m **21.** No
23. 343 cm^3 **25.** 6, 8, 10 **27.** 18 m by 38 m
29. 2.5 m

Written Exercises, pages 204–205
1. $\{x: 2 < x < 5\}$ **3.** $\{x: -2 \le x \le 2\}$
5. $\{z: z < 0 \text{ or } z > 4\}$ **7.** $\{y: y \le -3 \text{ or } y \ge 3\}$
9. $\{x: 1 < x < 4\}$ **11.** $\{t: t < 3 \text{ or } t > 6\}$
13. $\left\{z: -\frac{3}{2} \le z \le \frac{5}{2}\right\}$ **15.** {real numbers}
17. $\{s: -3 \le s \le 4\}$ **19.** $\{x: -4 < x < 0 \text{ or }$
$x > 4\}$ **21.** $\{y: y < -3 \text{ or } 0 < y < 2\}$
23. $\{x: -2 < x < 0 \text{ or } 2 < x < 4\}$
25. $\{x: x < -2 \text{ or } x > 2\}$ **27.** $\{x: -3 \le x \le -1$
or $1 \le x \le 3\}$ **29.** $\{x: x < -2 \text{ or } x > 3\}$
31. $\{0, 2\}$ **33.** $p \neq q$

Mixed Review Exercises, page 205
1. $\left\{-\frac{3}{2}, 1\right\}$ **2.** $\{5\}$ **3.** $\{-4, 4\}$ **4.** $\{0\}$
5. $\{1, 4\}$ **6.** $\left\{-\frac{1}{2}, 4\right\}$ **7.** $\{8\}$ **8.** $\{0, 2\}$
9. $\left\{-\frac{1}{3}\right\}$ **10.** $\{0, 1\}$ **11.** $\{0, 4\}$ **12.** $\{11\}$

Self-Test 3, page 205 **1.** $\{-3, 14\}$
2. $\left\{-1, \frac{2}{3}\right\}$ **3.** 3 m **4.** $\left\{x: x < -5 \text{ or } x > \frac{3}{2}\right\}$
5. $\{x: 1 \le x \le 4\}$

Chapter Review, pages 207–208 **1.** c **3.** d
5. a **7.** c **9.** d **11.** b **13.** b **15.** c **17.** a
19. a

Preparing for College Entrance Exams,
page 209 **1.** B **3.** C **5.** A **7.** D

Chapter 5 Rational Expressions

Written Exercises, pages 213–214 **1.** $3x^2$
3. $-\frac{3p}{q}$ **5.** $\frac{3uv}{5}$ **7.** $\frac{27r^3}{s^6}$ **9.** $\frac{3s^3}{t^3}$ **11.** $\frac{3x}{2y}$
13. $\frac{r^2}{r^2}$ **15.** $\frac{9v}{u^2}$ **17.** s^2 **19.** $\frac{z^2}{x^2}$ **21.** $\frac{4x}{3}$
23. $-\frac{c^9}{d^5h^3}$ **25.** b **27.** p^{n-1} **29.** t^n
31. $(p - q) + (r - s) = (p + r) - (q + s)$; yes
33. 1. Def. of division; 2. Comm. prop. and
Assoc. prop. of mult.; 3. Prop. of the reciprocal
of a product; 4. Def. of division

Mixed Review Exercises, page 215
1. $\{x: x < -4 \text{ or } x > 2\}$ **2.** $\{y: y \le -2\}$
3. $\{w: -2 < w < 2\}$ **4.** $\{p: p < 4\}$
5. $\{t: -4 \le t \le -1\}$ **6.** $\left\{a: -1 < a < \frac{3}{2}\right\}$
7. $\{c: -1 < c < 3\}$ **8.** $\{m: m \le -1 \text{ or } m \ge 2\}$
9. $\{d: d < -3 \text{ or } d > -1\}$
10. $\{g: g \le 0 \text{ or } g \ge 1\}$ **11.** $\{n: 1 \le n \le 2\}$
12. $\{u: u \le 2 \text{ or } u > 3\}$

Computer Key-In, page 215 **1.** **a.** 48 **b.** 2
c. 37 **d.** 6

Written Exercises, pages 218–220 **1.** $\frac{3}{5}$ **3.** 9
5. 12 **7.** $\frac{25}{2}$ **9.** 7×10^{-4} **11.** $6x^2y^{-3}$
13. 5.96 **15.** 0.0072 **17.** -0.216
19. 0.5625 **21.** $\frac{3y}{x^2}$ **23.** $\frac{1}{st^3}$ **25.** u^2v **27.** $\frac{x^4}{4y^4}$
29. $\frac{y^2}{3}$ **31.** $\frac{x^2}{8y^2}$ **33.** $\frac{1}{x^5}$ **35.** $\frac{p}{q}$ **37.** $\frac{y^2}{x^2}$

39. $\dfrac{1}{u^3 v^6}$ **41.** $\dfrac{5x^3}{y^6}$ **43 and 45.** Answers may vary. Examples are given. **43.** $(1+2)^{-1} = 3^{-1} = \dfrac{1}{3}$; $1^{-1} + 2^{-1} = 1 + \dfrac{1}{2} = \dfrac{3}{2}$; $\dfrac{1}{3} \neq \dfrac{3}{2}$

45. $2 \cdot 3^{-1} = \dfrac{2}{3}$; $\dfrac{1}{2 \cdot 3} = \dfrac{1}{6}$; $\dfrac{2}{3} \neq \dfrac{1}{6}$

47. $x^2 - 4x + 2$ **49.** $4x^2 - 5x + 1$

51. $x^2 - 4x + 4$

Written Exercises, page 223 **1.** 7.5×10^3
3. 6.08×10^{-1} **5.** 1.005×10^1
7. 3.20×10^{-2} **9.** 6.55×10^5
11. 5.60×10^{-4} **13.** 5000 **15.** 0.0043
17. 67,500 **19.** 0.00750 **21.** > **23.** >
25. 100 **27.** 0.03 **29.** 4×10^2 **31.** 2.6×10^3

Problems, pages 224–225 **1.** U.S.A.:
244,000,000; 9,360,000; China: 1,060,000,000;
9,600,000; Italy: 57,400,000; 301,000; World:
5,030,000,000; 149,000,000 **3. a.** 4.85%
b. 21.1% **c.** 1.14% **5.** $10,200 **7.** no
greater than 96 cm **9.** 1 AU $\approx 4.85 \times 10^{-6}$
parsecs; 1 parsec $\approx 2.06 \times 10^5$ AU or 3.26 light
years; 1 light year $\approx 6.33 \times 10^4$ AU
11. 0.389 AU; 30.3 AU **13.** 1.9×10^{22}

Mixed Review Exercises, page 225 **1.** $\dfrac{2x}{3y^3}$

2. $\dfrac{9a^2}{4b^2}$ **3.** $\dfrac{4m^3}{5n^2}$ **4.** $\dfrac{c^3}{d^3}$ **5.** $\dfrac{q^2}{p}$ **6.** -1 **7.** $\dfrac{7}{5}$

8. $\dfrac{v^3}{u^2}$ **9.** $\dfrac{x^3}{y^2}$ **10.** $c^2 d^2$ **11.** $\dfrac{3}{2}$ **12.** $\dfrac{b^4}{a^6}$

Calculator Key-In, page 226 **1.** 1×10^9
3. 1×10^{-10} **5.** 1.081×10^{-4} **7.** 9.517×10^0

Self-Test 1, page 226 **1.** $\dfrac{3s^2}{2t}$ **2.** $\dfrac{4y^8}{wx^2}$ **3.** $\dfrac{q^3 r^2}{p^2}$

4. $\dfrac{y^4 z^5}{x^2}$ **5.** $x^2 - 4$ **6.** 4.8260×10^2
7. 2.10×10^{-4} **8.** 3609 **9.** 0.05400
10. 90,000

Written Exercises, pages 228–229 **1.** $\dfrac{x-3}{2x}$

3. $\dfrac{u-2}{u}$ **5.** -1 **7.** $\dfrac{s+t}{s-t}$ **9.** $\dfrac{x-2}{x-4}$

11. $\dfrac{1-2y}{1+2y}$ **13.** $\dfrac{r-1}{r-4}$ **15.** -1 **17.** $\dfrac{z^2+1}{z^2}$

19. $(x+a)^2$ **21.** reals except 0 and 9; ± 3
23. reals except 1; ± 2 **25.** reals except -2, 0,
and 2; -3, $\dfrac{3}{2}$ **27.** reals except -1 and 1; 2

29. $\dfrac{x+1}{x-1}$ **31.** $\dfrac{1}{x+y}$ **33.** $\dfrac{s^2+t^2}{(s+t)(s-t)}$

35. $\dfrac{x^2+xy+y^2}{x^2+y^2}$ **37.** $\dfrac{x+y+2}{x+y-2}$ **39.** $\dfrac{x+y+z}{x+y-z}$

41. $\dfrac{x^2+xy+y^2}{x+y}$ **43.** $\dfrac{x^n-y^n}{x^n+4y^n}$

Extra, page 231

1.

3.

5.

7.

9.

11.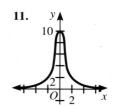

Written Exercises, page 234 **1.** $-\dfrac{8}{9}$ **3.** x

5. $12t$ **7.** $3y$ **9.** $\dfrac{r}{3}$ **11.** $\dfrac{x}{(x-2)(x-1)}$

13. $\dfrac{2u+1}{u+2}$ **15.** $\dfrac{x+2}{x-2}$ **17.** $\dfrac{p-q}{p+q}$ **19.** $\dfrac{u}{v-u}$

21. $\dfrac{x+y}{x+3y}$ **23.** $\dfrac{a^2+b^2}{a+b}$

Mixed Review Exercises, page 234
1. 5.4×10^{-4} **2.** 6.34×10^7 **3.** 1×10^{-1}
4. 3.281×10^3 **5.** $\dfrac{x+3}{x+1}$ **6.** $\dfrac{3}{3y-2}$ **7.** $\dfrac{2c^2}{3d^3}$

8. $\dfrac{a-1}{a+1}$ **9.** $\dfrac{3}{2}$ **10.** $\dfrac{9}{16z^5}$ **11.** $\dfrac{t^2+1}{2(t+1)(t-1)}$

12. $u^2 - u + 1$

Written Exercises, page 237 **1.** $-\dfrac{1}{4}$ **3.** $\dfrac{3}{7}$

5. $\dfrac{5}{9}$ **7.** $\dfrac{t}{2}$ **9.** $\dfrac{z^2+1}{z^2}$ **11.** $\dfrac{1}{6}$ **13.** $\dfrac{p^2+4q^2}{2p^3 q^4}$

15. $\dfrac{(a+b)(a-b)}{ab}$ **17.** $\dfrac{a^2+b^2+c^2}{abc}$

19. $\dfrac{2(y-z)}{yz}$ **21.** $\dfrac{8}{(z+4)(z-4)}$

23. $\dfrac{2}{(t+1)(t-1)}$ **25.** $-\dfrac{2}{(x+1)(x-1)^2}$

27. $-\dfrac{2}{(s+1)^2(s-1)}$ **29.** $\dfrac{2b}{a-b}$

31. $\dfrac{2y}{(x+y)(x-y)}$ **33.** $\dfrac{5x}{(x-2)(x+2)(x-3)}$

35. $\dfrac{3u}{(2u-v)(2u+v)(u-v)}$ **37.** $\dfrac{2a}{x+a}$

41. $A=2;\ B=-1$

Written Exercises, pages 239–241 **1.** 2 **3.** 4

5. x **7.** $-ab$ **9.** $\dfrac{v+u}{uv}$ **11.** $\dfrac{y-x}{y+x}$

13. $\dfrac{h^2-h+1}{h}$ **15.** $\dfrac{st+1}{t}$ **17.** $-\dfrac{y}{y+3}$

19. $-a$ **21.** $\dfrac{x+1}{x}$ **23.** $\dfrac{a-b}{a+b}$ **25.** $\dfrac{x-1}{x}$

27. a. 1.500 **b.** 1.400 **c.** 1.417 **d.** 1.414

29. $-\dfrac{1}{(x+1)(x+h+1)}$ **31.** $-\dfrac{2x+h}{x^2(x+h)^2}$

33. $\dfrac{x}{2x+1}$ **35.** $\dfrac{x-1}{x}$

Mixed Review Exercises, page 241 **1.** $\dfrac{y+x}{x^2y^2}$

2. $\dfrac{4(a+3)}{3a^2}$ **3.** $\dfrac{8v^2}{15u^2}$ **4.** $\dfrac{t^2+t+2}{(t+2)(t-2)}$ **5.** $-\dfrac{b}{2a}$

6. $\dfrac{x^2+y^2}{2xy}$ **7.** $\dfrac{3x-6}{8}$ **8.** -1 **9.** $\dfrac{2(a-2)}{3b}$

10. $(3,-2)$ **11.** $(1,3)$ **12.** $\left(-2,\dfrac{1}{2}\right)$

Self-Test 2, page 241 **1.** $\dfrac{t+1}{t}$ **2.** $\dfrac{x-1}{x}$

3. $\dfrac{p^7r^5}{9s^5}$ **4.** $\dfrac{u+2v}{u-2v}$ **5.** $\dfrac{(b-a)^2}{a^2b^2}$

6. $\dfrac{2}{(2x+1)(2x-1)^2}$ **7.** $\dfrac{1}{2}$ **8.** h

Written Exercises, page 245 **1.** $\left\{\dfrac{9}{2}\right\}$ **3.** $\{12\}$

5. $\{z:z\le 3\}$ **7.** $\{r:r>-2\}$ **9.** $\{2,4\}$

11. $\left\{-\dfrac{3}{2},2\right\}$ **13.** $\left\{-\dfrac{3}{2},\dfrac{3}{5}\right\}$ **15.** $\left\{\dfrac{2}{3},-\dfrac{1}{2}\right\}$

17. $\left\{\dfrac{3}{2},-\dfrac{5}{3}\right\}$ **19.** $\left\{\dfrac{2}{3},\dfrac{3}{2}\right\}$

21. $\{y:-3<y<1\}$ **23.** $\left\{t:t\le-2 \text{ or } t\ge\dfrac{5}{2}\right\}$

Problems, pages 245–246 **1.** 60 **3.** 500

5. $13\dfrac{1}{3}$ h **7.** 6 L **9.** 36 km **11.** $400 at

4.5%, $1800 at 7% **13.** 0.8 L of 7.5%; 1 L of

12% **15.** 135 h **17.** 75 km at 45 km/h; 75 km

at 75 km/h **19.** $2\dfrac{5}{8}$ h **21.** 160 km

Written Exercises, pages 249–250 **1.** $\{4\}$

3. $\{-3\}$ **5.** $\{9\}$ **7.** $\{2\}$ **9.** $\{-7\}$ **11.** $\{6\}$

13. no solution **15.** $\{4\}$ **17.** $\{2,3\}$ **19.** $\{1,3\}$

21. $\{0,3\}$ **23.** $\{-4\}$ **25.** no solution

27. $\{1,-3\}$ **29.** $\left\{-2,\dfrac{1}{2}\right\}$

Problems, pages 250–252 **1.** 4 and 12 **3.** 5

5. 20 h **7.** 35 mi/gal **9.** 2.5 km/h **11.** 40

13. 250 km/h **15.** 48 mi/h **17.** 75 km/h

19. 5 h **21.** 20

Mixed Review Exercises, page 252 **1.** $-x-1$

2. $\dfrac{2t-1}{2}$ **3.** $\dfrac{4m^3}{3n}$ **4.** $\dfrac{1}{8z^2}$ **5.** $\dfrac{a(b^2+a)}{b(b+a^2)}$

6. $\dfrac{c+3}{c+2}$ **7.** $\{4\}$ **8.** $\{x:x\ge-2\}$ **9.** $\{-3,5\}$

10. $\{x:x>-6\}$ **11.** $\{4\}$ **12.** $\left\{x:-6<x<\dfrac{4}{3}\right\}$

Self-Test 3, page 252 **1.** $\{4\}$ **2.** $\left\{-\dfrac{6}{5},1\right\}$

3. $3\dfrac{1}{3}$ h **4.** no solution **5.** $\{-4\}$ **6.** 120 km/h

Application, page 254 **1.** 1.5 A **3.** 10 Ω

Chapter Review, pages 255–256 **1.** b **3.** a

5. b **7.** d **9.** a **11.** b **13.** d **15.** c **17.** b

Chapter 6 Irrational and Complex Numbers

Written Exercises, pages 262–263 **1. a.** 4

b. -4 **c.** Not real **d.** 2 **3. a.** 9 **b.** -9

c. Not real **d.** 3 **5. a.** 0.1 **b.** Not real

c. 0.1 **d.** -0.1 **7. a.** 7 **b.** 7 **c.** 7 **d.** -7

9. a. $\dfrac{1}{8}$ **b.** $\dfrac{1}{8}$ **c.** $-\dfrac{1}{4}$ **d.** $-\dfrac{1}{4}$ **11. a.** 10

b. 10^2 **c.** 10^3 **d.** 10^{10} **13. a.** $|a|$ **b.** a^2

c. a^2 **d.** $|a|$ **15.** ± 12 **17.** None **19.** $\pm\dfrac{2}{3}$

21. $\pm\dfrac{5}{4}$ **23.** $\pm\dfrac{1}{2}$ **25.** ± 3 **27.** $x\ge-5$

29. All reals **31.** All reals **33. a.** $x\ge-1$

b. $x\ge 1$ **c.** All reals **d.** $x\le-1$ or $x\ge 1$

35. $-3\le x\le 0$ or $x\ge 3$ **37.** $0\le x\le 1$

Mixed Review Exercises, page 263 **1.** $\{0,4\}$

2. $\{-3,0\}$ **3.** $\{3\}$ **4.** $\{2,3\}$ **5.** $\{-3,2\}$

6. $\left\{-\dfrac{1}{2},1\right\}$ **7.** $\{5\}$ **8.** $\left\{-4,\dfrac{3}{2}\right\}$ **9.** No solu-

tion **10.** No solution

Calculator Key-In, page 263 **1.** 105 **3.** 0.99

5. 2×10^5 **7.** Both expressions equal

0.51763809 on a calculator.

Written Exercises, pages 267–269 1. $2\sqrt{13}$
3. $9\sqrt{2}$ 5. 14 7. $\frac{2\sqrt{2}}{3}$ 9. $\frac{2\sqrt{3}}{3}$ 11. $2\sqrt{2}$
13. $3\sqrt{5}$ 15. $6\sqrt{35}$ 17. 2 19. $5\sqrt[3]{2}$
21. $\frac{\sqrt[3]{10}}{2}$ 23. 3 25. 28 27. $3\sqrt[3]{20}$
29. $\frac{\sqrt[3]{45}}{3}$ 31. a. $4\sqrt{2}$ b. $2\sqrt[3]{4}$ c. $2\sqrt[4]{2}$
d. 2 33. 6.24 35. 7.66 37. 3.46
39. $3|x|\sqrt{2}$ 41. $5a\sqrt[3]{3a^2}$ 43. $\frac{|x|\sqrt{y}}{y^2}$
45. $\frac{3\sqrt[3]{2ab^2}}{2b^2}$ 47. $4\sqrt{a+b}$ 49. $|a+1|\sqrt{2}$
51. $\frac{1}{6}$ 53. $\frac{\sqrt[3]{3}}{2}$ 55. -1 57. $12\sqrt[3]{12}$
59. $\frac{3x\sqrt{3x}}{|y|}$ 61. $\frac{3\sqrt[3]{5a}}{a}$ 63. $\frac{3\sqrt[3]{2y}}{y}$ 65. $\frac{1}{2x}$

Calculator Key-In, page 269 1. 5.47 3. 0.46
5. No, 10 is not a power of 2. 7. 7.83
9. 107.81

Written Exercises, pages 272–273 1. $8\sqrt{2}$
3. $2\sqrt{3}$ 5. Not possible 7. $6+7\sqrt{6}$
9. $\sqrt{2}+3\sqrt{7}$ 11. $5\sqrt[3]{2}+2\sqrt[3]{5}$ 13. $\frac{2\sqrt{15}}{5}$
15. $\frac{5\sqrt{6}}{6}$ 17. $\frac{3\sqrt[3]{4}}{2}$ 19. $4+2\sqrt{5}$
21. $3\sqrt{5}+10\sqrt{3}$ 23. -36 25. $-\sqrt{3}$
27. 17 29. $3-\sqrt{3}$ 31. $10-2\sqrt[3]{10}$
33. $\sqrt[3]{6}+3\sqrt[3]{18}$ 35. $-x\sqrt{2x}$ 37. $(p+r)\sqrt{pr}$
39. $\frac{7\sqrt{10a}}{10}$ 41. $3w\sqrt{2}+2w^2\sqrt{3}$ 43. $\sqrt{102}$

Mixed Review Exercises, page 273 1. $2|x|\sqrt{7x}$
2. $\frac{9}{7}$ 3. $\frac{a}{xy}$ 4. $\frac{3\sqrt{10x}}{8y^2}$ 5. $3-x$ 6. $-3\sqrt[3]{5}$
7. $9x^2-1$ 8. $\frac{\sqrt{x}}{|x|}$ 9. $2xy^2\sqrt[3]{x}$

Written Exercises, pages 275–276 1. 2
3. $8+2\sqrt{7}$ 5. $5+4\sqrt{2}$ 7. $\frac{4+\sqrt{3}}{13}$
9. $9-2\sqrt{14}$ 11. $-6+5\sqrt{3}$ 13. $\sqrt{5}-\sqrt{2}$
15. 4 17. $34-2\sqrt{3}$ 19. $\frac{5\sqrt{3}-3\sqrt{5}}{2}$
21. $27+4\sqrt{35}$ 23. 7 25. $21-6\sqrt{10}$
27. $\frac{1}{2}$ 29. $60-60\sqrt{2}+12\sqrt{3}-12\sqrt{6}$
31. $-2-\sqrt{5}$ 33. $-\frac{\sqrt{2}}{2}$ 37. a. $2\sqrt{5}+3\sqrt{2}$
b. $\frac{2\sqrt{5}-3\sqrt{2}}{2}$ c. $\frac{2\sqrt{5}-3\sqrt{2}}{2}$ 39. 6 41. 1
43. $\frac{2w}{w-1}$ 45. $\sqrt{a}$ 47. It equals $2a^3+6ab$,
which is rational.

Written Exercises, pages 280–281 1. {7}
3. {36} 5. {-4, 4} 7. {21} 9. {-4}
11. {0, 8} 13. {2} 15. {3} 17. {2}
19. a. {4} b. {2√5} 21. a. {1} b. {√2}
23. a. {9} b. {-3 - 3√2} 25. {4} 27. No
solution 29. {22} 31. {0} 33. a. 30 mi
b. $h=\frac{2}{3}d^2$ 35. a. $10-5\sqrt{2}$ b. $2\sqrt{6}$
37. b. 12 cm 39. b. $3\sqrt[6]{2}$

Mixed Review Exercises, page 282
1. $\sqrt{5}+\sqrt{2}$ 2. $6\sqrt{2}-6\sqrt{3}$
3. a^2+ab+b^2 4. $\frac{4+\sqrt{6}}{5}$ 5. m^8n^9
6. $15+10\sqrt{2}$ 7. $\frac{1}{x-1}$ 8. $6\sqrt{3}$
9. $18-36\sqrt{2}+2\sqrt{3}-4\sqrt{6}$ 10. $\frac{2\sqrt{6}}{3}$
11. $5y^3-13y^2+14y+8$ 12. $4\sqrt[3]{5}$

Self-Test 1, page 282 1. $\frac{7}{11}$ 2. -0.5 3. 13
4. $x\le4$ 5. $5\sqrt{6}$ 6. $\frac{2a^2\sqrt{b}}{3b^2}$ 7. $|x-3|\sqrt{2}$
8. $3\sqrt[3]{2}$ 9. $\frac{3}{2}$ 10. $9+14\sqrt{3}$ 11. $6\sqrt{2}+8$
12. {9} 13. {6}

Written Exercises, pages 286–287 1. a. R
b. Ir 3. a. Ir b. R 5. a. R b. R c. Ir
7. 0.625 9. $1.\overline{857142}$ 11. $\frac{253}{50}$ 13. $\frac{118}{25}$
15. $\frac{4}{9}$ 17. $\frac{5}{6}$ 19. $\frac{26}{11}$ 21. $\frac{1010}{333}$
23–33. Answers may vary. Examples are given.
23. a. 0.15 b. $\frac{\sqrt{2}}{10}$ 25. a. 2.5 b. $\sqrt{6.5}$
27. a. 3.9 b. $\sqrt{15.5}$ 29. a. $\frac{127}{144}$
b. 0.87515115111 . . . 31. a. 0.45 b. $\frac{\sqrt{5}}{5}$
33. a. $x=\sqrt{2}$, $y=-\sqrt{2}$ b. $x=\sqrt{2}$,
$y=\sqrt{3}$ 35. (1) and (4) 37. Answers may
vary. For example: $\frac{1}{10}$, $\frac{1}{16}$, $\frac{1}{20}$, $\frac{1}{50}$

Written Exercises, pages 290–291 1. $9i$
3. $-24i$ 5. $2i\sqrt{5}$ 7. $6i\sqrt{2}$ 9. -6 11. $7i$
13. $-5\sqrt{2}$ 15. -49 17. -1 19. -2
21. -3 23. $2i$ 25. $-\frac{i\sqrt{5}}{5}$ 27. $-\frac{i\sqrt{3}}{2}$
29. $-2i$ 31. $\{\pm12i\}$ 33. $\{\pm7i\}$ 35. $\{\pm2i\sqrt{3}\}$
37. a. $11i$ b. -30 39. a. $-2i\sqrt{2}$ b. 30
41. a. $5i\sqrt{2}$ b. -12 43. $-6a$ 45. -2
47. $4ic\sqrt{3}-3ic\sqrt{5}$ 49. $-6ri\sqrt{r}$
51. $5x^2i-4x^2i\sqrt{x}$ 53. $-1, -i, 1, i$: these 4
values repeat in order.

Mixed Review Exercises, page 291 1. {14}
2. {−5, 3} **3.** {2} **4.** {13} **5.** No solution
6. {−8} **7.** {2, 3} **8.** {−2, 2} **9.** {5} **10.** R
11. R **12.** Ir **13.** Ir

Written Exercises, pages 295–296 1. $10 - 5i$
3. $-3 - 9i$ **5.** $-18 + 11i$ **7.** $-4 + 3i$
9. $4 + 8i$ **11.** 10 **13.** 58 **15.** $-37 - 12i$
17. $-17 - i$ **19.** 21 **21.** $-12 - 16i$
23. $-2 - 2i\sqrt{3}$ **25.** 169 **27.** 7 **29.** $\frac{3}{5} - \frac{4}{5}i$
31. $\frac{3}{5} + \frac{1}{5}i$ **33.** $-\frac{3}{5} + \frac{4}{5}i$ **35.** $\frac{17}{19} - \frac{6\sqrt{2}}{19}i$
37. $\frac{2}{13} - \frac{3}{13}i$ **39.** $-\frac{\sqrt{3}}{9} - \frac{\sqrt{6}}{9}i$
41. $\frac{11}{10} + \frac{27}{10}i$ **43.** $\frac{3}{7} + \frac{2\sqrt{3}}{7}i$
45. $(2 + i)^2 - 4(2 + i) + 5 =$
$3 + 4i - 8 - 4i + 5 = 0$ **47. b.** $-\frac{\sqrt{2}}{2} - \frac{\sqrt{2}}{2}i$

Self-Test 2, page 297 1. $0.\overline{384615}$ **2.** $\frac{19}{110}$
3. Answers may vary. Examples: 10.02;
10.010110111 . . . **4.** -9 **5.** $5i\sqrt{7}$
6. $-2\sqrt{15}$ **7.** $\frac{i}{2}$ **8.** $\{\pm 3i\sqrt{5}\}$ **9.** $3 + 4i$
10. $-3 - 4i$ **11.** $-14 + 13i\sqrt{3}$
12. $-\frac{10}{17} + \frac{11}{17}i$

Extra, page 300 1. 5 **3.** 2 **5.** 1
7. $-\frac{4}{25} - \frac{3}{25}i$ **9.** $-\frac{i}{2}$ **11.** $\frac{\sqrt{2}}{2} - \frac{\sqrt{2}}{2}i$
15. a. $\frac{(\bar{z} - z)i}{2}$

Chapter Review, pages 302–303 1. c **3.** b
5. a **7.** c **9.** d **11.** b **13.** a **15.** b

Mixed Problem Solving, page 304 1. 3 lemons, 6 limes; 8 lemons, 2 limes **3.** 27
5. $2000 at 6% and $5000 at 10%. **7.** $\frac{27}{12}$
9. length 6 cm; width 3 cm **11.** 11:36 A.M.

**Preparing for College Entrance Exams,
page 305 1.** B **3.** D **5.** B **7.** C

Chapter 7 Quadratic Equations and Functions

Written Exercises, pages 309–310
1. a. $\{\pm\sqrt{3}\}$ **b.** $\{1 \pm \sqrt{3}\}$ **c.** $\left\{\frac{1 \pm \sqrt{3}}{2}\right\}$
3. a. $\{\pm 4\}$ **b.** $\{-3, -11\}$ **c.** $\left\{-1, -\frac{11}{3}\right\}$

5. a. $\{\pm 2i\}$ **b.** $\{-7 \pm 2i\}$ **c.** $\left\{-\frac{7}{2} \pm i\right\}$
7. $\{7 \pm 2\sqrt{3}\}$ **9.** $\{-6 \pm 9\sqrt{2}\}$ **11.** $\{7 \pm 2i\}$
13. $\{1 \pm \sqrt{6}\}$ **15.** $\{-3 \pm \sqrt{11}\}$
17. $\{-10 \pm 10i\}$ **19.** $\{2 \pm \sqrt{5}\}$
21. $\left\{-1 \pm \frac{\sqrt{2}}{2}\right\}$ **23.** $\{3 \pm i\sqrt{11}\}$
25. $\left\{\frac{1}{2} \pm \frac{\sqrt{5}}{2}\right\}$ **27.** $\left\{-\frac{2}{3}, -1\right\}$ **29.** $\{1 \pm i\sqrt{3}\}$
31. $\left\{2 \pm \frac{\sqrt{6}}{3}\right\}$ **33.** $\left\{-\frac{1}{2}, 3\right\}$ **35.** $\{-3 \pm \sqrt{5}\}$
37. $\{3\}$ **39.** $\{8\}$ **41.** $\left\{-\frac{b}{2} \pm \frac{\sqrt{b^2 - 4c}}{2}\right\}$

Mixed Review Exercises, page 310 1. $5i\sqrt{2}$
2. $14 - 5i$ **3.** $54\sqrt{2}$ **4.** $5\sqrt{2}$ **5.** -24
6. $12 - i$ **7.** $\frac{7\sqrt{10}}{10}$ **8.** $-\frac{x}{3y}$ **9.** $\frac{x - 2}{x + 3}$

Written Exercises, pages 313–314
1. $\{-3 \pm \sqrt{5}\}$ **3.** $\{2 \pm 3i\}$ **5.** $\left\{\frac{2}{5}, -1\right\}$
7. $\left\{-\frac{6}{5} \pm \frac{2}{5}i\right\}$ **9.** $\left\{\frac{-1 \pm \sqrt{13}}{6}\right\}$
11. $\left\{\frac{-1 \pm \sqrt{15}}{2}\right\}$ **13.** $\left\{\frac{4 \pm \sqrt{10}}{3}\right\}$
15. $\left\{\frac{2 \pm \sqrt{10}}{2}\right\}$ **17.** $\left\{\frac{5}{2} \pm \frac{\sqrt{7}}{2}i\right\}$
19. $\{-1.24, 3.24\}$ **21.** $\{-0.83, 2.83\}$
23. $\{-1.41, -0.59\}$ **25.** $\{\pm 3\}$ **27.** $\{1\}$
29. $\left\{\frac{\sqrt{2} \pm \sqrt{6}}{2}\right\}$ **31.** $\{\sqrt{2} \pm 1\}$
33. $\left\{-2\sqrt{2}, -\frac{\sqrt{2}}{2}\right\}$ **35.** $\{-2i, i\}$
37. $\{2 + i, 1 + i\}$ **39.** $\{-i, 7i\}$

Problems, pages 314–315 1. 1.66 **3.** 1.62
5. 50 m by 100 m **7.** 1.68 m **9.** 18.94 mm
11. 1.39 **13.** 5.37
15. Going out, 45 km/h; returning, 30 km/h

Self-Test 1, page 316 1. $\{3 \pm \sqrt{7}\}$
2. $\left\{\frac{-9 \pm \sqrt{57}}{6}\right\}$ **3.** $\left\{\frac{5 \pm \sqrt{37}}{2}\right\}$
4. $\left\{-\frac{1}{4} \pm \frac{\sqrt{3}}{4}i\right\}$ **5.** 0.70 cm by 4.30 cm
6. 3 ft

Written Exercises, pages 320–321 1. two different irrational real roots **3.** two conjugate
imaginary roots **5.** two different irrational real
roots **7.** two conjugate imaginary roots **9.** two
different rational real roots **11.** two different
irrational real roots **13.** {1, 5} **15.** $\left\{\frac{2}{3}\right\}$
17. $\{-7\}$ **19.** $\{-9, 11\}$ **21.** $\{2 \pm \sqrt{6}\}$

23. $\left\{-\frac{5}{2}, 3\right\}$ **25.** $\{2 \pm \sqrt{2}\}$ **27.** $\left\{\frac{3}{2} \pm \frac{3\sqrt{3}}{2}i\right\}$

29. $\{3\}$ **31. a.** $k = 2$ **b.** $k < 2$ **c.** $k > 2$

33. a. $k = \pm 2$ **b.** $-2 < k < 2$ **c.** $k < -2$ or $k > 2$ **35. a.** $k = \pm 2$ **b.** $-2 < k < 2$

c. $k < -2$ or $k > 2$ **37.** $\frac{1}{2}$

Mixed Review Exercises, page 321

1. $\{2 \pm 3\sqrt{2}\}$ **2.** $\{-1, 4\}$ **3.** $\left\{\frac{3 \pm \sqrt{7}}{2}\right\}$

4. $\left\{\frac{1}{2} \pm \frac{\sqrt{19}}{2}i\right\}$ **5.** $\left\{-\frac{4}{3}, \frac{7}{2}\right\}$ **6.** $\left\{\frac{-2 \pm \sqrt{7}}{3}\right\}$

7. x^6 **8.** $6x\sqrt{2x}$ **9.** $\frac{1}{x^2}$ **10.** x^3y^4 **11.** $\frac{\sqrt{x}}{x^2}$

12. $\frac{y^2}{x}$ **13.** $5x$ **14.** $\frac{y^6}{x^4}$

Computer Exercises, page 321 **2. a.** real; irrational **b.** real, rational **c.** imaginary **4. a.** 5, -2.33333334 **b.** 0.192582404, -5.1925824 **c.** 5.10977223, -4.10977223 **6. a.** $-1 \pm 2.64575131i$ **b.** 2.39718086, -1.14718086 **c.** $1.5 \pm 1.11803399i$

Written Exercises, pages 324–325

1. a. $\{1, -2\}$ **b.** $\left\{1, \frac{5}{2}\right\}$ **c.** $\{\pm 1, \pm 2\}$

3. a. $\left\{-\frac{1}{6}, 1\right\}$ **b.** $\left\{\pm\frac{3\sqrt{2}}{2}, \pm 1\right\}$ **c.** $\left\{-\frac{1}{3}, 2\right\}$

5. a. $\{\pm i, \pm 2\}$ **b.** $\{16\}$ **7. a.** $\{\pm 3i, \pm 2\}$

b. $\left\{\pm\frac{1}{3}i, \pm\frac{1}{2}\right\}$ **9.** $\{25, 36\}$ **11.** $\{\pm 2, \pm 3\}$

13. $\{-7, 11\}$ **15.** $\{4\}$ **17.** $\{25\}$ **19.** $\{4\}$

21. $\{\pm 2.91\}$ **23.** $\{0.08\}$

25. $\{-1.19, 1, 2, 4.19\}$

27. $144x^4 - 73x^2 + 4 = 0$ **29.** $\{10, 250\}$

31. $\{\pm 2, \pm 3, \pm i\}$

Self-Test 2, page 325 **1.** Imaginary, conjugate **2.** Real, unequal, rational **3.** Real, unequal, irrational **4.** Real, unequal, rational **5.** $\{2 \pm \sqrt{5}\}$ **6.** $\{-1, 4\}$ **7.** $\{4 \pm \sqrt{10}\}$ **8.** $k > \frac{4}{3}$ **9.** $\{\pm 2i, \pm\sqrt{3}\}$ **10.** $\left\{\frac{4}{3}\right\}$

Written Exercises, pages 331–332

1.

3.

5.

7.

9.

11.

13.

15.

17.

19. $y + 3 = \frac{1}{2}(x - 4)^2$ **21.** $y = \frac{1}{3}x^2$

23. $y - 5 = -\frac{1}{3}(x - 3)^2$ **25.** $y - 2 = -2(x - 4)^2$ **27.** 5 **29.** 2 **31.** $a = 2, k = 3$

Mixed Review Exercises, page 332
1. Imaginary, conjugate **2.** Real, double, rational **3.** Real, unequal, irrational

4. $\{\pm 1, \pm 2\}$ **5.** $\left\{\frac{1}{2} \pm i\right\}$ **6.** $\{4\}$

Written Exercises, pages 336–337
1. **3.**

5. **7.**

9. **11.**

13. **15.**

17. 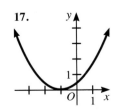 **19.** min., -8; $(-2, -8)$
21. min., $-\frac{5}{2}$; $\left(\frac{3}{2}, -\frac{5}{2}\right)$
23. max., $\frac{49}{4}$; $\left(-\frac{3}{2}, \frac{49}{4}\right)$

25. a. $(2, -7)$ **b.** {real numbers}
c. $\{y: y \geq -7\}$ **d.** $2 \pm \sqrt{7}$

27. a. $(-1, 9)$ **b.** {real numbers} **c.** $\{y: y \leq 9\}$
d. $-4, 2$ **29. a.** $(1, -6)$ **b.** {real numbers}
c. $\{y: y \geq -6\}$ **d.** $1 \pm \sqrt{6}$ **31. a.** $(1, -72)$
b. {real numbers} **c.** $\{y: y \geq -72\}$ **d.** $-5, 7$
33. a. $(-6, 9)$ **b.** {real numbers}
c. $\{y: y \leq 9\}$ **d.** $-9, -3$
35. a. $(-1, 2)$ **b.** {real numbers}
c. $\{y: y \leq 2\}$ **d.** $-3, 1$ **37. a.** neg. **b.** pos.
c. pos. **d.** zero **39. a.** zero **b.** pos.
c. pos. **d.** pos. **41.** $y - \frac{4c - b^2}{4} = \left(x - \frac{-b}{2}\right)^2$

Written Exercises, pages 342–343
1. $\left\{\frac{3}{2} \pm \frac{\sqrt{15}}{2}i\right\}$ **3.** $\left\{-\frac{3}{2}, 2\right\}$ **5.** $\left\{\frac{3}{4} \pm \frac{\sqrt{7}}{4}i\right\}$
7. $x^2 - 7x + 10 = 0$ **9.** $2x^2 - x - 10 = 0$
11. $x^2 - 3 = 0$ **13.** $x^2 - 2x - 2 = 0$
15. $9x^2 - 6x - 1 = 0$ **17.** $x^2 + 5 = 0$
19. $x^2 - 6x + 10 = 0$ **21.** $x^2 - 10x + 27 = 0$
23. $8x^2 - 4x + 3 = 0$ **25.** $f(x) = -10x^2 +$
$40x - 30$ **27.** $f(x) = \frac{1}{2}x^2 - 4x$ **29.** $f(x) =$
$-\frac{1}{3}x^2 + \frac{4}{3}x + \frac{32}{3}$ **31.** $f(x) = \frac{2}{3}x^2 - \frac{8}{3}x - \frac{10}{3}$
33. $f(x) = -\frac{2}{3}x^2 - \frac{8}{3}x + \frac{10}{3}$ **35.** $f(x) = \frac{2}{5}x^2 -$
$\frac{12}{5}x + 2$ **37. a.** $(3 + i\sqrt{2})^2 - 6(3 + i\sqrt{2}) +$
$11 = 7 + 6i\sqrt{2} - 18 - 6i\sqrt{2} + 11 = 0$
b. $3 - i\sqrt{2}$

Problems, pages 343–345 **1.** $20 - x$;
$-x^2 + 20x$; 100 **3.** 400 **5.** 625 cm^2
7. a. \$210; \$227.50; $(20 + n)(8 - 0.10n)$
b. 30 **9.** 25 passengers **11.** 3.75 amps; 225
watts **13.** $26,666\frac{2}{3}$ m^2 **15. a.** 40 ft **b.** Yes

Mixed Review Exercises, page 345
1. **3.**

5. **7.** {real numbers}; ± 3
8. {real numbers}; $\frac{8}{3}$
9. {real numbers};
$-3 \pm \sqrt{11}$

21 *Answers to Selected Exercises*

10. {real numbers}; no zeros 11. {real numbers}; no real zeros 12. $\left\{x: x \neq \frac{5}{2}\right\}; -\frac{1}{4}$

Self-Test 3, page 345 1. $(-3, 4); x = -3$
2. $(0, -6), (6, 0), (2, 0)$ 3. $y - 1 = 3(x + 2)^2$
4. vertex $(-3, -1)$; axis $x = -3$
5. $D = \{$real numbers$\}; R = \{y: y \geq -6\}$; zeros: $3 \pm \sqrt{6}$ 6. Answers may vary.
$2x^2 - 2x - 1 = 0$ 7. 36 ft

Chapter Review, pages 346–347 1. c 3. a
5. a 7. a 9. c 11. b 13. a 15. d

Cumulative Review (Chapters 4–7), pages 348–349 1. $x^2 - 9x + 8$ 3. $x^8y^4z^{12}$
5. $(2x - 3)(x - 4)$ 7. $(4a - 3b)(4a + 3b)$
9. $\{-1, 4\}$ 11. $\{-3, 0, 3\}$ 13. 9 and 11 or
-9 and -11 15. $\frac{3x}{4y^3}$ 17. $\frac{a}{a + 2}$ 19. $\frac{1}{2y}$
21. $\{2\}$ 23. \$1200 at 5%; \$2800 at 8%
25. $\frac{3\sqrt{6}}{2}$ 27. $\frac{\sqrt{7} + \sqrt{3}}{2}$ 29. $\frac{5}{13} - \frac{12}{13}i$
31. $\left\{\frac{3 \pm \sqrt{2}}{2}\right\}$ 33. $k > 3$
35. vertex $(3, 1)$; axis $x = 3$ 37. -9

Chapter 8 Variation and Polynomial Equations

Written Exercises, pages 354–355 1. 10
3. 20 5. 75 7. 50 9. 6 11. $6\sqrt{2}$

Problems, pages 356–357 1. 15 m 3. \$6525
5. 175 ft 7. 12,500 9. 25 kg 11. 115 L
13. 13 cm 15. 96 ft 17. 2 amperes
19. $\frac{\pi}{6}; V = \frac{\pi}{6}d^3$

Mixed Review Exercises, page 357
1. $\left\{\frac{5 \pm \sqrt{13}}{6}\right\}$ 2. $\{4\}$ 3. $\left\{0, \frac{5}{2}\right\}$ 4. $\{-8, 4\}$
5. $\left\{\frac{1}{2}\right\}$ 6. $\left\{-4, \frac{1}{2}\right\}$ 7. $\{-1 \pm \sqrt{2}\}$ 8. $\{6\}$
9. $\left\{\frac{1 \pm 2\sqrt{3}}{3}\right\}$

Written Exercises, pages 360–361 1. 1 3. 12
5. 36 7. 8 9. 0.5

Problems, pages 361–362 1. 750 kHz
3. 864 BTU 5. 6 lux 7. 576 r/m; about
61.7 r/m 9. 1.0125 mm 11. 378 kg
13. About 35,800 km

Self-Test 1, page 363 1. a. $\frac{1}{7}$ b. 3 2. 216
3. 68.6 kPa 4. a. 15 b. $\frac{1}{2}$ 5. 9 6. 90
7. 800 r/min

Written Exercises, pages 366–367
1. $x + 1 + \frac{-6}{x + 2}$ 3. $x + 4 + \frac{-2}{2 - x}$
5. $2t - 3 + \frac{4}{2t + 1}$ 7. $x^2 + 2x - 4 + \frac{-2}{x - 3}$
9. $2s^2 - 8s + 3 + \frac{1}{s + 4}$
11. $2t^2 - t + 3 + \frac{4}{3t + 2}$ 13. $5z + 3$
15. $2x^2 - x + 3$ 17. $3u^2 - 2 + \frac{8u + 8}{3u^2 + 2u + 2}$
19. $x^2 + 2ax + \frac{a^3}{x + 2a}$
21. $3t^2 - ct - c^2 + \frac{c^3t + 2c^4}{2t^2 + ct + c^2}$
23. $x^2 + x + 1 + \frac{2x^3}{x^4 - x^2 + 1}$
25. $x^2 - a^2 + \frac{2a^4}{x^2 + a^2}$ 27. $x^4 - ax^3 + a^3x - a^4$
29. $x^3 - x^2 - 10$ 31. $-\frac{9}{2}$ 33. $c = 1; k = -2$

Mixed Review Exercises, page 367 1. 12
2. 16 3. $3\sqrt{3}$ 4. 25.6 5. $\frac{16}{3}$ 6. 1 7. $\sqrt{3}$
8. 0.4

Written Exercises, page 370
1. $3x^2 + x + 3 + \frac{4}{x - 2}$ 3. $x^2 - 2$
5. $t^3 - 2 + \frac{3}{t + 5}$ 7. $2s^3 - s^2 - 3s - 2$
9. $x^4 + x^3 + x^2 + x + 1$ 11. $2x^3 - x^2 + x - 2$
13. $x^2 - 2x + 3 + \frac{-5}{2x + 1}$
15. $2t^3 + 3t^2 + 2t - 2$
17. $z^2 + (-2 + 2i)z - 4i + \frac{3}{z - 2i}$
19. $Q(x) = 2x^2 - x + 3; R = -5$
21. $Q(z) = 2z^2 - (3 + 4i)z + 6i; R = 2$ 23. 4

Reading Algebra, page 371 3. \$126.50; 7%
account 5. a. 6 h b. less than; Tom's time is
greater than Sue's c. 7.5 h

Written Exercises, pages 375–376 1. 5
3. -10 5. 13 7. 10 9. Yes 11. No
13. Yes 15. Yes 17. $\{-3, \pm\sqrt{3}\}$
19. $\{-4, 2 \pm i\}$ 21. $x^3 - 7x + 6 = 0$
23. $x^4 + 2x^3 - 5x^2 - 6x = 0$
25. $2x^3 - 3x^2 - 11x + 6 = 0$
27. $x^4 - x^3 - 12x^2 - 4x + 16 = 0$

29. $\{-1, 4, -2, 2\}$ **31.** $\left\{1, -\frac{2}{3}, -1 \pm \sqrt{2}\right\}$

35. 3 **37.** 2

Mixed Review Exercises, page 376

1. $x + 1 + \frac{-3}{x + 2}$ **2.** $y - 1$ **3.** $2u - 3$

4. $3w^2 + 2 + \frac{-1}{w^2 - 2}$ **5.** $a^4 - a^3 - 3a^2 + 3a$

6. $c + 2$ **7.** Imag. **8.** Rat. **9.** Irrat.

Written Exercises, pages 380–381

1. $x^3 + x^2 + 25x + 25 = 0$

3. $x^3 + 4x^2 + 6x + 4 = 0$ **5.** $-2i$ **7.** $1 + 2i$

9. $\{-1 \pm 2i, 2\}$ **11.** $\{3 \pm i, \pm\sqrt{10}\}$

13–19. Answers are given in this order: pos. real, neg. real, imaginary. **13.** 1, 1, 2

15. 0, 2, 2; or 0, 0, 4 **17.** 1, 2, 2; or 1, 0, 4

19. 3, 2, 0; 1, 2, 2; 3, 0, 2; or 1, 0, 4

21. $x^4 - 2x^3 + 6x^2 - 8x + 8 = 0$

Self-Test 2, page 381 **1.** $x^2 - 2x + 3$

2. $2x^3 + x^2 + 2x - 1 + \frac{2}{x - 2}$ **3.** 16

4. Yes **5.** $2x^3 - 5x^2 - 4x + 3 = 0$

6. a. 3 or 1 pos.; 1 neg. **b.** $\left\{2 \pm i, 1, -\frac{1}{2}\right\}$

7. $x^3 + 6x + 20 = 0$

Written Exercises, pages 384–385

1. $\{-3, 1, 2\}$ **3.** none **5.** $\{-1, 3, -1 \pm 2i\}$

7. $\left\{\frac{1}{2}, -2 \pm i\right\}$ **9.** $\left\{\frac{3}{2}, 2 \pm \sqrt{2}\right\}$

11. $\left\{-2, \frac{2}{3}, \pm i\right\}$ **23.** $\left\{-\frac{3}{2}, \pm i\sqrt{2}\right\}$

25. $\left\{\frac{5}{2}, \pm 2i\right\}$

Mixed Review Exercises, page 385 **1.** $\{3\}$

2. $\left\{\frac{2}{3} \pm \frac{\sqrt{2}}{3}i\right\}$ **3.** $\{4\}$ **4.** $\{-1\}$ **5.** $\left\{-\frac{2}{3}, \frac{1}{4}\right\}$

6. $\{\pm 2, \pm i\sqrt{2}\}$ **7.** $\{\pm i, 2\}$ **8.** $\{-3, 1 \pm \sqrt{2}\}$

9. $\{2 \pm i, -1\}$ **10.** $\{1 \pm i, -2, 3\}$

Written Exercises, pages 388–389

1. 2.5 **3.** -1.5 **5.** 2 **7.** $-1.5, -0.5, 2$

9. 1.5, 2.5 **11.** 2.5 **13.** -1.5 **15.** 2.1

17. $-1.5, -0.3, 1.9$ **19.** 1.5, 2.8 **21.** $c = 3$

23. $c = \frac{1}{2}$ **25.** 1.53 **29.** 4; -2

Computer Key-In, page 390 **1. a.** $x \approx -0.7$, 2.0, 2.7 **b.** $x \approx 2.73$ **3.** 1.149

Written Exercises, pages 394–395

1. 99 million **3.** 198 million **5.** 1916

7. 1957 **9.** 1.090 kg/m³ **11.** 1.179 kg/m³

13. 216 m **15.** 2070 m

Computer Exercises, page 395 **3.** 60, 68, 251, 259 **5. a.** $f(2.8) = 23.2$; exact value = 21.952

b. $f(7) = 14$; exact value = 14

c. $f(13.71) = 0.08145$; exact value = 0.0729394603

d. $f(14.3) = 606292.576$; exact value = 599967.756

Self-Test 3, page 396 **1.** The only possible rational roots, ± 1 and ± 3, do not satisfy the equation. **2.** $\left\{-\frac{1}{2}, 1, \pm 2i\right\}$ **3.** -1.7

4. 2.315 **5. a.** 33.2°C **b.** 0.15 min

Chapter Review, pages 397–398 **1.** c **3.** d

5. b **7.** c **9.** c **11.** c **13.** c

Preparing for College Entrance Exams, page 399 **1.** B **3.** E **5.** B **7.** C **9.** E

Chapter 9 Analytic Geometry

Written Exercises, pages 404–405 **1.** 13;

$\left(\frac{13}{2}, 6\right)$ **3.** $\sqrt{74}$; $\left(-\frac{5}{2}, \frac{5}{2}\right)$ **5.** $2\sqrt{5}$; (4, 4)

7. $\frac{13}{3}$; $\left(\frac{7}{6}, 0\right)$ **9.** $11\sqrt{2}$; $\left(\frac{11}{2}, \frac{11}{2}\right)$ **11.** 3;

$\left(0, \frac{1}{2}\right)$ **13.** 6; $(1, \sqrt{3})$ **15.** $|a|$; $\left(\frac{a}{2}, b\right)$

17. $2\sqrt{a^2 + b^2}$; (b, a) **19.** (6, 10) **21.** (10, 6)

23. $(-h, -k)$ **25.** yes; yes; $\frac{17}{2}$ **27.** yes; no

29. collinear **31.** not collinear

Problems, pages 405–406 **1.** $3x + 5y - 8 = 0$

3. $5x - 4y - 11 = 0$ **5.** $\left(\frac{8}{3}, 0\right), \left(0, \frac{8}{5}\right)$

Mixed Review Exercises, page 406

1. Graph is a line passing through $(0, -2)$ and $(4, 0)$. **2.** Graph is a parabola with vertex $(0, 3)$ and axis $x = 0$. **3.** Graph is a parabola with vertex $(-2, 1)$ and axis $x = -2$. **4.** 8

5. $\frac{1}{2}$ **6.** -4

Computer Exercises, page 406
2. a. 7.07106782 **b.** 18.4390889
c. 294.49618 **4. a.** 22.471793 **b.** 150.512173

Written Exercises, pages 410–411
1. $(x - 3)^2 + y^2 = 9$
3. $(x - 2)^2 + (y + 5)^2 = 64$ **5.** $x^2 + y^2 = 144$

23 *Answers to Selected Exercises*

7. $(x - 6)^2 + (y - 1)^2 = 2$

9.

11.

13.

15. $(0, 0)$; 4
17. $(0, -4)$; 4
19. $(2, -1)$; 3
21. no graph
23. $\left(-\frac{3}{2}, 2\right)$; $\frac{5}{2}$

25.

27.

29.

31. $(2, 3)$; 2 **33.** $\left(1, -\frac{1}{4}\right)$; $\frac{\sqrt{17}}{4}$
35. $x^2 + (y - 5)^2 = 25$
37. $(x - 1)^2 + (y - 4)^2 = 2$
39. $(x + 2)^2 + (y - 4)^2 = 16$
41. $(x - 5)^2 + (y + 4)^2 = 16$

Written Exercises, pages 415–416 **1.** $F(4, 7)$
3. $D: y = 4$ **5.** $D: y = -8$ **7.** $y - 2 = -\frac{1}{8}x^2$
9. $y = -\frac{1}{16}x^2$ **11.** $x = \frac{1}{4}y^2$ **13.** $x - 1 =$
$-\frac{1}{4}(y - 2)^2$ **15.** $y - 2 = \frac{1}{8}(x - 3)^2$ **17.** $(0, 0)$;
$\left(0, \frac{3}{2}\right)$; $y = -\frac{3}{2}$; $x = 0$ **19.** $(-1, 2)$; $(0, 2)$;
$x = -2$; $y = 2$ **21.** $(-2, 1)$; $(-2, -1)$; $y = 3$;
$x = -2$ **23.** $\left(-\frac{3}{2}, 3\right)$; $\left(\frac{1}{2}, 3\right)$; $x = -\frac{7}{2}$; $y = 3$
25. $(-5, -2)$; $\left(-5, -\frac{3}{2}\right)$; $y = -\frac{5}{2}$; $x = -5$

27.

29.

31.

33.

Mixed Review Exercises, page 417 **1.** 0; 8;
$(1, 2)$ **2.** 4; $2\sqrt{17}$; $(2, -2)$ **3.** $-\frac{1}{7}$; $5\sqrt{2}$;
$\left(\frac{1}{2}, \frac{7}{2}\right)$ **4.** $2x - 3y = 5$ **5.** $x + y = -1$
6. $(x + 1)^2 + (y - 3)^2 = 16$

Self-Test 1, page 417 **1.** $\sqrt{65}$ **2.** $\left(\frac{5}{2}, 8\right)$
3. not collinear **4.** $(x - 5)^2 + (y + 4)^2 = 49$
5. $(-6, 2)$; $2\sqrt{2}$ **6.** $x + 4 = \frac{1}{16}(y - 3)^2$
7. $(-5, 2)$; $(-5, -2)$; $y = 6$; $x = -5$
8. vertex $(-1, -3)$; focus $(1, -3)$; directrix
$x = -3$; axis $y = -3$

Written Exercises, pages 421–423
1. $(-\sqrt{5}, 0)$, $(\sqrt{5}, 0)$ **3.** $(-4\sqrt{2}, 0)$,
$(4\sqrt{2}, 0)$ **5.** $(0, -\sqrt{6})$, $(0, \sqrt{6})$
7. $(0, -2\sqrt{5})$, $(0, 2\sqrt{5})$ **9.** $(0, -2\sqrt{2})$,
$(0, 2\sqrt{2})$ **11.** $(0, -2)$, $(0, 2)$
13. $\left(-\frac{2\sqrt{2}}{3}, 0\right)$, $\left(\frac{2\sqrt{2}}{3}, 0\right)$
15. $\frac{x^2}{25} + \frac{y^2}{4} = 1$ **17.** $\frac{x^2}{4} + \frac{y^2}{2} = 1$
19. $\frac{x^2}{81} + \frac{y^2}{45} = 1$ **21.** $\frac{x^2}{128} + \frac{y^2}{144} = 1$
25. a. 0 **b.** 0 **c.** At the center **d.** A circle
27. $\frac{4\sqrt{5}}{3}$ m (≈ 3 m)

Application, page 425 **1.** max:
1.017 AU $\approx$ 153 million km; min:
0.983 AU $\approx$ 147 million km **3.** $\approx 1.0001:1$
5. $\approx 1.78:1$; since the ratio is much larger, the
orbit is much more elongated than that of Earth.

Written Exercises, pages 430–431

1. a.

b. $(-\sqrt{41}, 0)$, $(\sqrt{41}, 0)$

3. a.

b. $(0, -\sqrt{26})$, $(0, \sqrt{26})$

5. a.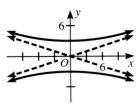

b. $(-\sqrt{29}, 0)$, $(\sqrt{29}, 0)$

7. a.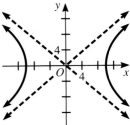

b. $(0, -3\sqrt{10})$, $(0, 3\sqrt{10})$

9. a.

b. $(-5\sqrt{7}, 0)$, $(5\sqrt{7}, 0)$

11. a.

b. $\left(0, -\dfrac{\sqrt{5}}{2}\right)$, $\left(0, \dfrac{\sqrt{5}}{2}\right)$

13. $\dfrac{y^2}{25} - \dfrac{x^2}{39} = 1$ **15.** $\dfrac{y^2}{9} - \dfrac{x^2}{4} = 1$

17. $\dfrac{y^2}{9} - \dfrac{x^2}{1} = 1$

Mixed Review Exercises, page 431

1.

3.

5. yes **6.** yes **7.** no **8.** yes **9.** yes **10.** no

Written Exercises, pages 434–435

1. $\dfrac{(x-6)^2}{16} + \dfrac{(y-3)^2}{25} = 1$

3. $\dfrac{(x+3)^2}{7} + \dfrac{y^2}{16} = 1$

5. $\dfrac{(x-2)^2}{25} + \dfrac{(y+3)^2}{9} = 1$

7. $\dfrac{(x-4)^2}{1} - \dfrac{(y+2)^2}{15} = 1$

9. $\dfrac{(y+5)^2}{4} - \dfrac{(x-3)^2}{5} = 1$

11. $\dfrac{(y+5)^2}{9} - \dfrac{(x-5)^2}{7} = 1$ **13.** hyperbola; $(1, -3)$; $(1 + \sqrt{5}, -3)$, $(1 - \sqrt{5}, -3)$
15. circle; $(3, 8)$; radius 4 **17.** ellipse; $(-2, 3)$; $(-6, 3)$, $(2, 3)$
19. $20x^2 + 20y^2 - 32xy - 9 = 0$
21. a. parabola **b.** circle **c.** ellipse
d. hyperbola **e.** circle: $A = B$; parabola: $AB = 0$; ellipse: $AB > 0$ and $A \neq B$; hyperbola: $AB < 0$

Self-Test 2, page 435

1.

2. $\dfrac{x^2}{16} + \dfrac{y^2}{7} = 1$

3.

4. $\dfrac{y^2}{4} - \dfrac{x^2}{12} = 1$

5. hyperbola; see figure at right.

6. $\dfrac{x^2}{36} + \dfrac{(y-2)^2}{11} = 1$

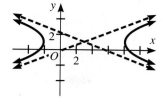

Written Exercises, page 438

1. 2 **3.** 0 **5.** 1 **7.** 4 **9.** 3
11. (2.5, 1.5) **13.** (4.5, 2), (−4.5, −2),
(2, 4.5), (−2, −4.5) **15.** (−2, 0), (2.5, 1),
(2.5, −1)

Mixed Review Exercises, page 438

1. $y - 1 = \dfrac{1}{8}(x - 2)^2$ **2.** $\dfrac{x^2}{6} + \dfrac{y^2}{4} = 1$

3. $\dfrac{y^2}{4} - \dfrac{x^2}{5} = 1$ **4.** $(x + 3)^2 + (y - 4)^2 = 25$

5. {(1, 5)} **6.** {(2, −1)} **7.** {(3, −3)}

Written Exercises, pages 441–442

1. {(−4, 11), (2, −1)} **3.** {($\sqrt{3}$, 3),
(−$\sqrt{3}$, 3)} **5.** {(8, −7)} **7.** {(2, 4), (4, 2)}
9. {(−43, 9), (5, −3)} **11.** {(4, 3), (4, −3),
(−4, 3), (−4, −3)} **13.** {($\sqrt{21}$, 2),
($\sqrt{21}$, −2), (−$\sqrt{21}$, 2), (−$\sqrt{21}$, −2)} **15.** No
real solution **17.** {(3, −2), (−3, 2), (2, −3),
(−2, 3)} **19.** {(−6, −1), (2, 3), (−3, −2)}
21. $2 + i$, $−2 − i$ **23.** $3 + 2i$, $−3 − 2i$
25. $3 + 4i$, $−3 − 4i$

Problems, pages 442–443
1. 5 and 11
3. 35 m by 43 m **5.** 5 ft by 12 ft **7.** 57
9. 30 m by 45 m **11.** 3.7 km/h and 4.7 km/h
13. $\left(\dfrac{4}{5}, \dfrac{3}{5}\right)$

Written Exercises, pages 447–448

1. {(−10, 14, −2)} **3.** {(3, −2, −4)}
5. {(−4, −2, 1)} **7.** {(−1, 0, −4)}
9. {(−1, −3, 5)} **11.** {(1, 2, 2)}
13. {(1, 3, 2)} **15.** {(−1, 2, 3)}
17. $\left\{\left(1, -1, -\dfrac{1}{2}\right)\right\}$ **19. a.** {(5 + z, 2 + 2z, z)}
b. Examples: (5, 2, 0), (8, 8, 3), (3, −2, −2)
23. {(2z, 5z, z)} **25.** {(0, 2z, z)}

Problems, pages 448–449
1. 6 nickels, 8
dimes, 5 quarters **3.** 2 carats, 4 carats, 8 carats
5. 30°, 70°, 80° **7.** 48 patrons, 103 sponsors,
175 donors **9.** $y = 3x^2 - 2x - 1$
11. $x^2 + y^2 - 4x + 6y + 9 = 0$

Mixed Review Exercises, page 449
1. {($\sqrt{2}$, 1), ($\sqrt{2}$, −1), (−$\sqrt{2}$, 1),
(−$\sqrt{2}$, −1)} **2.** {(2, 5)} **3.** $\left\{\left(\dfrac{5}{4}, \dfrac{3}{2}\right), (3, -2)\right\}$

4. $\dfrac{x^2}{9y^4}$ **5.** $12x\sqrt{x}$ **6.** $−12x^5y^5$ **7.** $12x^2$

8. $\dfrac{5x^2}{3y^5}$ **9.** $\dfrac{3x}{5y^3}$ **10.** $\dfrac{3x^2\sqrt{3}}{2}$ **11.** $\dfrac{y^4}{x^5}$

Self-Test 3, page 450
1. b. 4 **c.** {(4.5, 2), (4.5, −2), (−4.5, 2),
(−4.5, −2)} **2. b.** 1 **c.** {(3, 2.5)} **3. b.** 2
c. {(3, 0), (−3, 0)}
4. {(0, 5), (0, −5)} **5.** {(−9, −11), (1, −1)}
6. {(2, −1, 3)} **7.** {(2, −3, 4)}

Chapter Review, pages 451–453 **1.** d **3.** c
5. a **7.** c **9.** c **11.** a **13.** c **15.** d **17.** d
19. c

Chapter 10 Exponential and Logarithmic Functions

Written Exercises, page 458 **1.** 9 **3.** $\dfrac{1}{7}$ **5.** 8

7. 8 **9.** $−\dfrac{1}{5}$ **11.** $\dfrac{1}{2}$ **13.** −4 **15.** $\dfrac{1}{5}$ **17.** $\dfrac{1}{2}$
19. 49 **21.** $x^{3/2}y^{5/2}$ **23.** $a^{-1}b^{3/2}$ **25.** $a^{-5}b^{5/2}$
27. $8a^{-1/2}b^{-3/2}$ **29.** $2\sqrt[3]{2}$ **31.** $\sqrt{2}$ **33.** $\sqrt[4]{2}$
35. 3 **37.** x **39.** $x^{1/12}$ **41.** $a^2 - 2a$
43. a. {16} **b.** {5} **45. a.** $\left\{\dfrac{1}{25}\right\}$ **b.** $\left\{\dfrac{1}{200}\right\}$
47. {−56} **49.** {±11}

Mixed Review Exercises, page 458 **1.** $\dfrac{x+2}{x+1}$

2. $\dfrac{2\sqrt{2xy}}{y^2}$ **3.** $2a^2 - a + 1$ **4.** $−8c^4$ **5.** ab^3
6. $2n^2 - 5n - 12$

Written Exercises, pages 461–462 **1. a.** $9^{\sqrt{2}}$
b. $9^{\sqrt{2}}$ **c.** 9 **d.** 81 **3.** 100^π **5.** 6^π
7. $\dfrac{1}{10,000}$ **9.** 1 **11.** $\dfrac{1}{2}$ **13.** 1 **15.** $\dfrac{2 - \sqrt{3}}{2}$
17. $\dfrac{1}{3^\pi}$ **19.** {−3} **21.** $\left\{-\dfrac{5}{3}\right\}$ **23.** $\left\{\dfrac{2}{3}\right\}$
25. No solution **27.** {2} **29.** {3} **31.** {3}
33.

35.

37. {6} **39.** {0}

Written Exercises, pages 466–467 **1. a.** $\frac{5}{2}$

b. -4 **c.** $-\frac{3}{2}$ **d.** $\frac{x-3}{2}$ **3. a.** $\frac{3}{2}$ **b.** 1

c. not real **d.** $\frac{\sqrt{x}}{2}$ **5. a.** 3 **b.** not real

c. $\sqrt{x-3}$ **d.** $\sqrt[4]{x}$ **7.** no **9.** no

11. $f^{-1}(x) = \frac{x+3}{2}$ **13.** $f^{-1}(x) = \sqrt[3]{x}$

 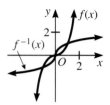

15. $g^{-1}(x) = \frac{8}{\sqrt[3]{x}}$ **17.** no inverse **19.** no

inverse **21.** no inverse **23. c.** $1; 2; 3; -1$
d. $f: D = \{\text{reals}\}, R = \{y: y > 0\}; f^{-1}: D = \{x: x > 0\}, R = \{\text{reals}\}$ **25.** $m = 1, b = 0$;
or $m = -1$

Mixed Review Exercises, page 467 **1.** $\frac{1}{125}$

2. $7\sqrt{2}$ **3.** $-5 - 12i$ **4.** 81 **5.** -24 **6.** $\frac{1}{16}$
7. 9 **8.** $5 - 4\sqrt{2}$ **9.** -3 **10.** 128
11. $1 + 2i$ **12.** 2

Self-Test 1, page 467 **1. a.** $2x^{2/3}y^{-1/3}$
b. $6^{-1/3}$ **2. a.** $\frac{25\sqrt{5}}{2}$ **b.** $x^2y\sqrt[6]{xy^5}$ **3.** $\{34\}$
4. a. $2^{-5\sqrt{2}}$ **b.** $2^{-8\sqrt{5}}$ **5.** $\{3\}$ **6. a.** 13
b. 5 **c.** $6\sqrt{x} + 1$ **d.** $\sqrt{6x + 1}$
7. $f(g(x)) = 3\left(\frac{x+7}{3}\right) - 7 = x + 7 - 7 = x$;
$g(f(x)) = \frac{(3x-7)+7}{3} = \frac{3x}{3} = x$

Written Exercises, pages 470–472 **1.** 3 **3.** 4
5. 0 **7.** -2 **9.** $\frac{3}{2}$ **11.** $\frac{1}{4}$ **13.** $\frac{2}{3}$ **15.** -3
17. $-\frac{2}{3}$ **19.** $\{49\}$ **21.** $\left\{\frac{1}{3}\right\}$ **23.** $\left\{\frac{1}{8}\right\}$ **25.** $\{9\}$
27. $\left\{\frac{1}{49}\right\}$ **29.** $\{x: x > 0 \text{ and } x \neq 1\}$
31. a. $3 + 2 = 5$ **b.** $\frac{1}{2} + \frac{3}{2} = 2$
c. $\log_b M + \log_b N = \log_b MN$ **33. a.** $\log_6 x$
b. $2; -\frac{1}{2}$ **c.** $f: D = \{\text{reals}\}, R = \{y: y > 0\}$;
$f^{-1}: D = \{x: x > 0\}, R = \{\text{reals}\}$

35.

37.

39. $\{3\}$
41. positive
43. a. 120 dB
b. 10^4

Written Exercises, pages 476–477
1. $6 \log_2 M + 3 \log_2 N$ **3.** $\log_2 M + \frac{1}{2} \log_2 N$
5. $4 \log_2 M - 3 \log_2 N$
7. $\frac{1}{2} \log_2 M - \frac{3}{2} \log_2 N$ **9.** 1.90 **11.** 0.15
13. 0.90 **15.** 0.35 **17.** -0.95 **19.** -0.22
21. $\log_4 p^5 q$ **23.** $\log_3 \frac{A^4}{\sqrt{B}}$ **25.** $\log_2 8MN$
27. $\log_5 \frac{5}{x^3}$ **29.** 2 **31.** $\frac{3}{2}$ **33.** $\{45\}$ **35.** $\{1\}$
37. $\{6\}$ **39.** $\{\pm 5\}$ **41. a.** 6 **b.** $\frac{1}{4}$ **c.** 1
45. $\{3\}$ **47.** $\{2\}$ **49.** $\{\sqrt{85}\}$ **51.** $\{2\}$

Mixed Review Exercises, page 477 **1.** $\left\{\frac{\sqrt{2}}{2}\right\}$

2. $\{-3, 1, 2\}$ **3.** $\{5\}$ **4.** $\{4\}$ **5.** $\left\{-\frac{3}{2}\right\}$ **6.** $\{7\}$
7. $\left\{-\frac{1}{2}\right\}$ **8.** $\{4\}$ **9.** $\{2 \pm \sqrt{5}\}$ **10.** 3 **11.** 2
12. 1 **13.** 1

Self-Test 2, page 477 **1. a.** $3^4 = 81$
b. $6^3 = 216$ **2. a.** $\log_5 625 = 4$
b. $\log_{25} 125 = \frac{3}{2}$ **3. a.** 3 **b.** 12 **4.** $\{3\}$
5. $\frac{5}{3} \log_2 M + 2 \log_2 N$ **6.** -1.40 **7.** $\{3\}$

Written Exercises, pages 481–482 **1.** 1.79
3. 0.00792 **5.** 575 **7.** 33.7 **9.** 7.13
11. 692 **13.** 0.0158 **15. a.** $\frac{\log 30}{\log 3}$ **b.** 3.10

17. a. $\dfrac{\log 56}{\log 5.6}$ **b.** 2.34 **19. a.** $-\dfrac{\log 5}{\log 30}$

b. -0.473 **21. a.** $\dfrac{\log 60}{2\log 3.5}$ **b.** 1.63

23. $\left\{\dfrac{7}{4}\right\}$ **25.** $\left\{\dfrac{5}{6}\right\}$ **27.** $\{6740\}$ **29.** $\{21.6\}$

31. $\{2.19\}$ **33.** $\{1.03\}$ **35.** 3.17 **37.** 3.36

39. $\{0.631, 1.46\}$ **41. a.** $2; \dfrac{1}{2}$ **b.** $3; \dfrac{1}{3}$

c. They are reciprocals. $\log_b a = \dfrac{\log_a a}{\log_a b} = \dfrac{1}{\log_a b}$

Problems, pages 486–488 **1. a.** $1120
b. $1254.40 **c.** $1404.93 **d.** $3105.85
3. a. $1125.51 **b.** $1266.77 **c.** $1425.76
d. $3262.04 **5. a.** $10,000 **b.** $8000
c. $6400 **d.** $1342.18 **7. a.** $4N_0$ **b.** $32N_0$
c. $(2^{w/3})N_0$ **9. a.** 25 kg **b.** 6.25 kg
c. $100\left(\dfrac{1}{2}\right)^{y/6000}$ kg **11.** about 9 years
13. 12.6% **15.** 0.997; 9.09×10^{-13}
17. 11.4% **19.** 19.7% **21. a.** $1,065,552.45
b. $1,066,086.39 **c.** $1,066,091.81

Mixed Review Exercises, page 488

1. $\left\{0, \dfrac{1}{2}, 1\right\}$ **2.** $\{1, 2\}$ **3.** $\{\sqrt{2}, 2, 4\}$

4. $\{0, 6\}$ **5.** $\left\{\dfrac{1}{3}, \dfrac{1}{2}, \dfrac{2}{3}\right\}$ **6.** $\{1, \sqrt{2}, 2\}$ **7.** 1

8. 2 **9.** $\dfrac{1}{6}$ **10.** 1 **11.** $\dfrac{5}{12}$ **12.** 3

Written Exercises, pages 490–491 **1.** $e^{2.08} = 8$

3. $e^{-1.10} = \dfrac{1}{3}$ **5.** $\ln 20.1 = 3$ **7.** $\ln 1.65 = \dfrac{1}{2}$

9. 2 **11.** -3 **13.** 0 **15.** 5 **17.** $\ln 12$

19. $\ln \dfrac{9}{5}$ **21.** $\ln 10e^3$ **23.** $\{e^3\}$ **25.** $\{4 + e^{-1}\}$

27. $\{\pm e^{9/2}\}$ **29.** $\{\ln 2\}$ **31.** $\{\ln 5\}$

33. $\{2 + \ln 2\}$ **35.** $\{\ln 9\}$ **37.** $\left\{\dfrac{3}{5}\ln 10\right\}$

39. $\{e\}$ **41.** $\{2\}$ **43.** $\{\ln 3, \ln 4\}$
45. $D = \{x: x \neq 0\}$; $R = \{\text{reals}\}$
47. $D = \{x: x > 5\}$; $R = \{\text{reals}\}$

Calculator Key-In, page 492 **1. a.** $1061.84
b. $1127.50 **c.** $1822.12 **3. a.** 1; 0.99; 0.96;
0.91; 0.85; 0.78; 0.70; 0.61; 0.53; 0.44; 0.37
5. 2.718056; e

Self-Test 3, page 493 **1.** $\{650\}$ **2.** $\{5.81\}$

3. $\{1.07\}$ **4.** $3277.23 **5.** $\ln\left(\dfrac{5}{e^2}\right)^{1/3}$

$\left(\text{or } \dfrac{\ln 5 - 2}{3}\right)$

Application, pages 493–494 **1.** 6600
3. a. 95.3% **b.** 93.0%

Chapter Review, pages 496–497 **1.** c **3.** d
5. b **7.** a **9.** b **11.** a **13.** d **15.** b

Mixed Problem Solving, page 498 **1.** 10 mL
3. CA:45; NY:34; NC:11 **5.** -9 **7.** $1500
9. $A = \dfrac{C^2}{4\pi}$ **11.** Larger plant: 8.4 h; smaller:
12.4 h **13.** 4 h

**Preparing for College Entrance Exams,
page 499** **1.** C **3.** E **5.** A **7.** E

Chapter 11 Sequences and Series

Written Exercises, pages 504–506 **1.** A; 8, 5
3. G; 625, 3125 **5.** A; 30, 38 **7.** N; $\dfrac{1}{25}, \dfrac{1}{36}$
9. G; $4^{9/2}, 4^{11/2}$ **11.** 7, 11, 15, 19; A **13.** 1,
3, 9, 27; G **15.** $-\dfrac{1}{4}, \dfrac{1}{2}, -1, 2$; G **17.** log 2,
log 3, log 4, log 5; N **19. a.** A **b.** G
21. 32, 44 **23.** 20, 18 **25.** 63; 127 **27.** 21,
34 **29.** 48, 71 **31. a.** 15, 21 **b.** 55
33. a. 9 **b.** 35 **35. a.** 16, 31; No **b.** 57

Mixed Review Exercises, page 506

1.   **2.**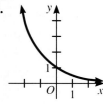

7. $(2, -3)$ **8.** $\left(-\dfrac{3}{2}, \dfrac{7}{2}\right)$, $(1, 1)$ **9.** $(2, 1)$,
$(-2, 1)$, $(2, -1)$, $(-2, -1)$

Written Exercises, page 509 **1.** $t_n = 8n + 16$
3. $t_n = 4 - 7n$ **5.** $t_n = 4n + 3$ **7.** 104 **9.** 52
11. -902 **13.** 87.5 **15.** 8 **17.** -61 **19.** 2
21. $\dfrac{3}{2}$ **23. a.** $-7, 13$ **b.** $-12, 3, 18$
c. $-15, -3, 9, 21$ **25. a.** 19, 27 **b.** 17, 23,
29 **c.** 15.8, 20.6, 25.4, 30.2 **27.** 101
29. 300 **31.** 16
33. $\dfrac{a + b}{2} - a = b - \dfrac{a + b}{2} = \dfrac{b - a}{2}$

Written Exercises, pages 513–514
1. $t_n = 2 \cdot 3^{n-1}$ **3.** $t_n = (\sqrt{2})^{n-1}$
5. $t_n = 64\left(-\frac{3}{4}\right)^{n-1}$ **7.** 39,366 **9.** $\frac{5}{256}$
11. $\frac{5}{128}$ **13.** $\frac{16}{3}$ **15.** 5120 **17.** y^{39} **19.** 4
21. 3 **23.** 10; 20; 40 **25.** $\sqrt[5]{2}, \sqrt[5]{4}, \sqrt[5]{8},$
$\sqrt[5]{16}$ **27.** A; $t_n = 2n$ **29.** A;
$t_n = 25 + 8(n-1)$ **31.** G; $t_n = 200\left(-\frac{1}{2}\right)^{n-1}$
33. A; $t_n = (n+1)a + (2n-1)$ **35.** A;
$t_n = -10n$ **37.** $t_n = \frac{n+1}{n^2}$
39. $\frac{\sqrt{ab}}{a} = \frac{b}{\sqrt{ab}} = \sqrt{\frac{b}{a}}$

Problems, pages 514–515 **1.** $21,000
3. $35.43 **5.** 14 **7.** 192 million **9. a.** 19
b. 15 **11.** $1601.87 **13.** B; B

Mixed Review Exercises, page 515
1. $\{x: x < -2 \text{ or } x > 7\}$ **2.** $\{x: -3 \le x \le 2\}$
3. $\{x: -4 \le x < 2\}$
4. $\{x: -2 < x < 0 \text{ or } x > 2\}$ **5.** $\{x: x \le 2\}$
6. $\{x: -5 < x < -1\}$ **7.** $x + y = 1$
8. $y = -\frac{1}{4}x^2$ **9.** $\frac{x^2}{4} + \frac{y^2}{3} = 1$

Self-Test 1, page 516 **1. a.** G; $\frac{27}{8}, \frac{81}{16}$ **b.** A;
$-11, -15$ **c.** N; $\frac{5}{11}, \frac{6}{13}$ **d.** A; 2.2, 2.5
2. a. $t_n = 43 - 7n$ **b.** -90 **3. a.** 4 **b.** 2, 6
4. a. $t_n = 48\left(-\frac{1}{2}\right)^{n-1}$ **b.** $-\frac{3}{32}$ **5. a.** $4\sqrt{2}$
b. 4, 8 **6.** $58.32

Extra, page 517 **1.** 6 **3.** -2
5. {real numbers}; {positive integers}
7. {real numbers}; {positive integers}

Written Exercises, pages 521–522
1. $11 + 12 + 13 + 14 + 15 + 16$
3. $2 + 4 + 8 + 16 + 32 + 64$
5. $1 - \frac{1}{2} + \frac{1}{3} - \frac{1}{4} + \frac{1}{5} - \frac{1}{6}$
7. $2 + 1 + 0 + 1 + 2 + 3$ **9.** $\sum_{n=1}^{500} 2n$
11. $\sum_{n=1}^{20} n^3$ **13.** $\sum_{n=1}^{99} \frac{n}{n+1}$ **15.** $\sum_{n=1}^{100} (2n-1)$
17. $\sum_{n=0}^{6} 2^n$ **19.** $\sum_{n=1}^{\infty} \frac{1}{n}$ **21.** $\sum_{n=1}^{\infty} 27\left(-\frac{1}{3}\right)^n$

23. $\sum_{n=0}^{3} 6(-2)^n$ **25.** $\sum_{n=1}^{\infty} \frac{1}{n^2}$ **27.** $\sum_{n=20}^{199} 5n$
29. $\sum_{n=0}^{\infty} \left(\frac{1}{4}\right)^n$ **31.** $\frac{j+4}{j+8}$ **33.** $\sum_{k=1}^{4} k \log 5 =$
$\log 5 + 2 \log 5 + 3 \log 5 + 4 \log 5 =$
$10 \log 5 = \log 5^{10}$ **35.** 21

Written Exercises, pages 527–528 **1.** 670
3. 3420 **5.** 25,250 **7.** 3925 **9.** 15,250
11. 893 **13.** 255 **15.** $-29,524$ **17.** $\frac{4095}{4096}$
19. a. -300 **b.** $\frac{3069}{64}$ **21.** 1960 **23.** 247,500
25. a. 420 **b.** 2,097,150 **27.** 3240
29. a. $\sum_{k=1}^{n} 2^{k-1} = \frac{2^0(1 - 2^n)}{1 - 2} = \frac{1 - 2^n}{-1} = 2^n - 1$
b. 20 **31.** 16, 48, 144; geometric
33. Answers will vary. **a.** 0.00000095
b. 0.00002656 **c.** 0.00001427 **35.** -50
37. 2,097,360

Problems, pages 528–530 **1.** 880 **3.** 156
5. 2046 **7.** Plan B; $58.83 more **9.** B; A
11. $13,971.64 **13. a.** $T_n = \frac{n(n+1)}{2}$
b. $T_n = \sum_{k=1}^{n} k$ **15.** the square of an odd number

Mixed Review Exercises, page 530 **1.** $\frac{\sqrt{2}}{4}$
2. 0 **3.** -2 **4.** 4 **5.** -1 **6.** $2\sqrt{17}$ **7.** 16
8. -1 **9.** -2 **10.** $\frac{3}{2}$ **11.** 4 **12.** (3, 0) and
$(-3, 0)$

Computer Exercises, page 530 **2. a.** 506
b. 1.99998957 **c.** 0.740049503

Written Exercises, pages 533–534 **1.** 48
3. 16.2 **5.** no sum **7.** no sum
9. $\frac{9}{2}(\sqrt{3} - 1)$ **11.** 4 **13.** 8, 10, $\frac{21}{2}, \frac{85}{8}, \frac{341}{32}$;
$S \approx 11$; $S = \frac{32}{3}$ **15.** $\frac{3}{4}, \frac{3}{8}, \frac{9}{16}, \frac{15}{32}, \frac{33}{64}$; $S \approx \frac{1}{2}$;
$S = \frac{1}{2}$ **17.** $\frac{1}{3}$ **19.** $\frac{1040}{333}$ **21.** $\frac{1}{2}$ **23.** 8, $\frac{8}{3}, \frac{8}{9}$
25. 40, $-\frac{40}{3}, \frac{40}{9}$ **29. a.** 3 **b.** 32

Problems, pages 534–536 **1.** 72 ft
3. $48(2 + \sqrt{2})$ cm **5.** 288 cm^2 **7.** 480 m
9. a. 0.18889 in. **b.** 249.24 in. **c.** 250 in.

11. a. $20, 100, 4 \cdot 5^{n-2}$; $\dfrac{1}{81}, \dfrac{1}{729}, \left(\dfrac{1}{9}\right)^{n-1}$; $\dfrac{20}{81}$,
$\dfrac{100}{729}, \dfrac{4 \cdot 5^{n-2}}{9^{n-1}} = \dfrac{4}{9}\left(\dfrac{5}{9}\right)^{n-2}$ **b.** 2 **13.** $\dfrac{2\sqrt{3}}{5}$

Self-Test 2, page 536 1. a. G; $\displaystyle\sum_{n=0}^{3} 27\left(-\dfrac{2}{3}\right)^n$
b. N; $\displaystyle\sum_{n=1}^{15} \dfrac{n}{n+1}$ **c.** A; $\displaystyle\sum_{n=1}^{7} (23 - 3n)$ **d.** N;
$\displaystyle\sum_{n=1}^{\infty} n^2$ **2. a.** 1080 **b.** -45 **3. a.** 315
b. $\dfrac{130}{81}$ **4. a.** $\dfrac{125}{8}$ **b.** no sum

Written Exercises, page 539
1. $x^3 + 3x^2y + 3xy^2 + y^3$
3. $c^5 - 5c^4d + 10c^3d^2 - 10c^2d^3 + 5cd^4 - d^5$
5. $a^8 + 8a^7 + 28a^6 + 56a^5 + 70a^4 + 56a^3 +$
$28a^2 + 8a + 1$ **7.** $243t^5 + 1620t^4 +$
$4320t^3 + 5760t^2 + 3840t + 1024$
9. $x^{12} - 6x^{10} + 15x^8 - 20x^6 + 15x^4 - 6x^2 + 1$
11. $p^6 + 3p^4q^3 + 3p^2q^6 + q^9$
13. $136x^2y^{15} + 17xy^{16} + y^{17}$ **15.** $167,960x^9$
17. $2a^7 + 42a^5b^2 + 70a^3b^4 + 14ab^6$
19. $a^6 + 6a^5b + 15a^4b^2 + 20a^3b^3 + 15a^2b^4 +$
$6ab^5 + b^6$ **21.** 179.52 **23. a.** 1, 2, 4, 8, 16,
32, 64 **b.** 2^n

Mixed Review Exercises, page 539 1. N;
$\dfrac{2n}{2n+1}$ **2.** G; $\dfrac{9}{16}\left(\dfrac{4}{3}\right)^{n-1}$ **3.** A; $11 - 4n$ **4.** G;
$-24\left(-\dfrac{1}{2}\right)^{n-1}$ **5.** 275 **6.** 24 **7.** 378

Written Exercises, pages 542–543 1. 720
3. 20 **5.** 100 **7.** 120 **9.** $n+1$ **11.** $\dfrac{n^2+n}{2}$
13. a. $a^{14} + 14a^{13}b + 91a^{12}b^2 + 364a^{11}b^3$
b. $a^{14} - 14a^{13}b + 91a^{12}b^2 - 364a^{11}b^3$
15. a. $a^{19} + 19a^{18}b + 171a^{17}b^2 + 969a^{16}b^3$
b. $x^{19} - 38x^{18}y + 684x^{17}y^2 - 7752x^{16}y^3$
17. $15,504a^{15}b^5$ **19.** $1001s^4t^{10}$ **21.** $4032a^4b^5$
23. $1120c^8d^4$ **25.** False; $a = 2, b = 3$ for example **27.** False; $n = 3$ for example
29. a. $1 + \dfrac{1}{2}x - \dfrac{1}{8}x^2 + \dfrac{1}{16}x^3$ **b.** 1.4375

Calculator Key-In, page 543 1. 3,628,800
3. a. 2.433×10^{18} **b.** 9.075×10^9

Self-Test 3, page 543
1. $x^5 - 10x^4 + 40x^3 - 80x^2 + 80x - 32$
2. $256a^4 + 768a^3 + 864a^2 + 432a + 81$
3. $1024x^{10} + 15,360x^9y + 103,680x^8y^2$
4. $-35a^8b^3$ **5.** $-1320x^8y^3$

Chapter Review, pages 545–546 1. b **3.** b
5. c **7.** a **9.** c **11.** c **13.** a

Cumulative Review, page 547 1. 3
3. $\{2 \pm 3i, -1 \pm \sqrt{2}\}$ **5.** $x = \dfrac{1}{16}y^2 + 4$
7. $(1, -2)$ **9.** $(2, -1, -2)$ **11.** 128
13. $\left\{-\dfrac{1}{2}\right\}$ **15.** $\{27\}$ **17.** 2.29 **19.** G;
$t_n = 108\left(-\dfrac{2}{3}\right)^{n-1}$
21. $x^4 - 4x^3y^2 + 6x^2y^4 - 4xy^6 + y^8$

Chapter 12 Triangle Trigonometry

Written Exercises, pages 552–554
1. a. second quadrant **b.** third quadrant

3. a. fourth quadrant **b.** first quadrant
5. quadrantal **7.** fourth quadrant **9.** second
quadrant **11.** quadrantal **13.** 240° **15.** $-270°$
17. 576° **19.** 45°, 405°, 765°, $-315°$
21. 60°, 240°, 420°, $-120°$
23–29. Answers may vary. Examples are
given. **23. a.** $35° + n \cdot 360°$ **b.** 395°, $-325°$
25. a. $-100° + n \cdot 360°$ **b.** 260°, $-460°$
27. a. $160° + n \cdot 360°$ **b.** 160°, $-200°$
29. a. $280° + n \cdot 360°$ **b.** 280°, $-80°$
31. 15.5° **33.** 72.8° **35.** 25.75° **37.** 45.31°
39. 25°24′ **41.** 44°54′ **43.** 34°24′36″
45. 23°40′12″ **47.** 315° **49.** 37° **51.** 30°
53. 240° **55.** Any angle between 0° and 22.5°.
57. Any angle between 67.5° and 90°. **59.** Any
angle between 30° and 45°. **61.** Any angle between 0° and 180°. **63.** Any angle between
360° and 540°. **65.** Any angle between 450°
and 900°.

Mixed Review Exercises, page 554 1. $40x^3y^2$
2. $252a^6b^2$ **3.** $78m^{22}n^6$ **4.** -6 **5.** 1
6. $3\sqrt{2} - 4$ **7.** $\dfrac{2}{3}$ **8.** $14 - 5i$ **9.** 0 **10.** 3
11. $\dfrac{7\sqrt{6}}{9}$ **12.** $\dfrac{3}{2}$

Calculator Key-In, page 554 1. 58°36′
3. 86°25′48″ **5.** 36.4266667° **7.** 73.8736111°

Written Exercises, pages 559–560

1. $\sin \theta = \frac{4}{5}$; $\cos \theta = \frac{3}{5}$; $\tan \theta = \frac{4}{3}$; $\cot \theta = \frac{3}{4}$; $\sec \theta = \frac{5}{3}$; $\csc \theta = \frac{5}{4}$ **3.** $\sin \theta = \frac{\sqrt{2}}{2}$; $\cos \theta = \frac{\sqrt{2}}{2}$; $\tan \theta = 1$; $\cot \theta = 1$; $\sec \theta = \sqrt{2}$; $\csc \theta = \sqrt{2}$ **5.** $\sin \theta = \frac{15}{17}$; $\cos \theta = \frac{8}{17}$; $\tan \theta = \frac{15}{8}$; $\cot \theta = \frac{8}{15}$; $\sec \theta = \frac{17}{8}$; $\csc \theta = \frac{17}{15}$

7. $\sin \theta = \frac{\sqrt{10}}{10}$; $\cos \theta = \frac{3\sqrt{10}}{10}$; $\tan \theta = \frac{1}{3}$; $\cot \theta = 3$; $\sec \theta = \frac{\sqrt{10}}{3}$; $\csc \theta = \sqrt{10}$ **9.** $\frac{4}{5}$; $\frac{3}{4}$

11. $\frac{1}{2}$; $\frac{\sqrt{3}}{3}$ **13.** $\frac{8}{15}$; $\frac{\sqrt{161}}{8}$ **15.** $\frac{\sqrt{15}}{4}$; $\sqrt{15}$
17. $50°$ **19.** $10°$ **21.** $\angle B = 45°$, $a = 2$, $c = 2\sqrt{2}$ **23.** $\angle B = 30°$, $b = 10$, $a = 10\sqrt{3}$
25. $\angle A = 45°$, $\angle B = 45°$, $c = 4\sqrt{2}$
27. $6\sqrt{3} + 6$ **29.** 36 **33.** $\cos \theta = \sqrt{1 - u^2}$, $\tan \theta = \frac{u\sqrt{1-u^2}}{1-u^2}$, $\cot \theta = \frac{\sqrt{1-u^2}}{u}$, $\sec \theta = \frac{\sqrt{1-u^2}}{1-u^2}$, $\csc \theta = \frac{1}{u}$

Written Exercises, pages 566–567

1. $\sin \theta = -\frac{4}{5}$, $\cos \theta = \frac{3}{5}$, $\tan \theta = -\frac{4}{3}$, $\cot \theta = -\frac{3}{4}$, $\sec \theta = \frac{5}{3}$, $\csc \theta = -\frac{5}{4}$
3. $\sin \theta = \frac{24}{25}$, $\cos \theta = -\frac{7}{25}$, $\tan \theta = -\frac{24}{7}$, $\cot \theta = -\frac{7}{24}$, $\sec \theta = -\frac{25}{7}$, $\csc \theta = \frac{25}{24}$ **5.** 0;
1; 0; 1; undef.; undef. **7.** 0; -1; 0; -1;
undef.; undef. **9.** $53°$ **11.** $25°$ **13.** $28°$
15. $5°$ **17.** $83.6°$ **19.** $4.1°$ **21.** $23°40'$
23. $27°30'$ **25.** $-\cos 36°$ **27.** $-\sin 17°$
29. $-\cot 72.9°$ **31.** $-\cos 41.9°$
33. $\tan 85°20'$ **35.** $-\sin 32°40'$
37. $\sin 330° = -\frac{1}{2}$, $\cos 330° = \frac{\sqrt{3}}{2}$,
$\tan 330° = -\frac{\sqrt{3}}{3}$, $\cot 330° = -\sqrt{3}$,
$\sec 330° = \frac{2\sqrt{3}}{3}$, $\csc 330° = -2$
39. $\sin 135° = \frac{\sqrt{2}}{2}$, $\cos 135° = -\frac{\sqrt{2}}{2}$,
$\tan 135° = -1$, $\cot 135° = -1$,
$\sec 135° = -\sqrt{2}$, $\csc 135° = \sqrt{2}$
41. $\sin 315° = -\frac{\sqrt{2}}{2}$, $\cos 315° = \frac{\sqrt{2}}{2}$,
$\tan 315° = -1$, $\cot 315° = -1$, $\sec 315° = \sqrt{2}$,
$\csc 315° = -\sqrt{2}$ **43.** $\sin (-150°) = -\frac{1}{2}$,

$\cos (-150°) = -\frac{\sqrt{3}}{2}$, $\tan (-150°) = \frac{\sqrt{3}}{3}$,
$\cot (-150°) = \sqrt{3}$, $\sec (-150°) = -\frac{2\sqrt{3}}{3}$,
$\csc (-150°) = -2$ **45.** second; $\sin \theta = \frac{15}{17}$,
$\tan \theta = -\frac{15}{8}$, $\cot \theta = -\frac{8}{15}$, $\sec \theta = -\frac{17}{8}$,
$\csc \theta = \frac{17}{15}$ **47.** fourth; $\cos \theta = \frac{12}{13}$,
$\tan \theta = -\frac{5}{12}$, $\cot \theta = -\frac{12}{5}$, $\sec \theta = \frac{13}{12}$,
$\csc \theta = -\frac{13}{5}$ **49.** first; $\sin \theta = \frac{\sqrt{5}}{3}$,
$\tan \theta = \frac{\sqrt{5}}{2}$, $\cot \theta = \frac{2\sqrt{5}}{5}$, $\sec \theta = \frac{3}{2}$,
$\csc \theta = \frac{3\sqrt{5}}{5}$ **51.** second; $\sin \theta = \frac{3}{5}$,
$\cos \theta = -\frac{4}{5}$, $\cot \theta = -\frac{4}{3}$, $\sec \theta = -\frac{5}{4}$,
$\csc \theta = \frac{5}{3}$ **53.** $0°$, $180°$ **55.** $0°$, $180°$
57. $45°$, $225°$ **59.** $270°$ **61.** $150°$, $210°$
63. $15°$, $195°$ **65.** $70°$, $290°$ **67.** $20°$, $200°$
69. $\cos \theta$ **71.** $-\sin \theta$ **73.** $-\cos \theta$ **75.** $\frac{\sin \theta}{\cos \theta}$

Mixed Review Exercises, page 567 **1.** 1275
2. 650 **3.** 25 **4.** $-\frac{5}{3}$ **5.** 2 **6.** $-\frac{3}{2}$, 1 **7.** 0,
2, -2 **8.** no real zeros **9.** no real zeros
10. no real zeros

Written Exercises, page 572 **1.** 0.2717
3. 0.9078 **5.** 0.8557 **7.** 1.729 **9.** 7.723
11. 0.8440 **13.** 0.6058 **15.** 0.2080
17. -2.753 **19.** -0.9997 **21.** -1.351
23. -0.6225 **25.** $19.9°$ **27.** $30.5°$ **29.** $73.7°$
31. $30°27'$ **33.** $8°26'$ **35.** $47°4'$ **37.** $29.2°$,
$150.8°$ **39.** $299.7°$, $119.7°$ **41.** $75.4°$, $284.6°$
43. $295.3°$ **45.** $322.3°$

Self-Test 1, pages 572–573 **1.** Answers may
vary. **a.** $610°$, $-110°$ **b.** $60°$, $-660°$
2. $32.58°$ **3.** $74°15'36''$ **4.** $\sin \theta = \frac{3}{4}$,
$\cos \theta = \frac{\sqrt{7}}{4}$, $\tan \theta = \frac{3\sqrt{7}}{7}$, $\cot \theta = \frac{\sqrt{7}}{3}$,
$\sec \theta = \frac{4\sqrt{7}}{7}$, $\csc \theta = \frac{4}{3}$ **5.** $a = 12$,
$b = 12\sqrt{3}$, $\angle B = 60°$ **6.** $60°$; $\sin 300° =$
$-\frac{\sqrt{3}}{2}$, $\cos 300° = \frac{1}{2}$, $\tan 300° = -\sqrt{3}$, $\cot 300°$
$= -\frac{\sqrt{3}}{3}$, $\sec 300° = 2$, $\csc 300° = -\frac{2\sqrt{3}}{3}$
7. **a.** 0.8018 **b.** 1.110 **8.** $61°10'$

Lessons 12-5 to 12-9: Note that because of roundoff error, answers obtained by using the trigonometric tables occasionally differ slightly from the answers given here, which are correct to the number of significant digits shown.

Written Exercises, pages 577–578
1. $\angle B = 53.8°$; $a = 40.2$; $b = 54.9$
3. $\angle A = 24.6°$; $b = 5.13$; $c = 5.65$
5. $\angle A = 41.7°$; $a = 66.6$; $c = 100$ **7.** $b = 222$;
$\angle A = 46.0°$; $\angle B = 44.0°$ **9.** $c = 0.338$;
$\angle A = 21.3°$; $\angle B = 68.7°$ **11.** $\angle A = 31°50'$;
$a = 222$; $b = 357$ **13.** $\angle B = 74°30'$; $b = 16.2$;
$c = 16.8$ **15.** $\angle B = 59°10'$; $a = 31.9$; $c = 62.3$
17. $45.2°$ **19.** $53.1°$, $126.9°$ **21. a.** 6.18
b. 6.50

Problems, pages 578–579 **1.** $31.7°$ **3.** $64.1°$
5. 545 ft **7.** 317 ft **9.** $23.4°$ **11.** 358 ft
13. 1.13 m **15.** 660 m **17.** $109.4°$

Mixed Review Exercises, page 579
1. $\sin 450° = 1$, $\cos 450° = 0$, $\tan 450°$ and
$\sec 450°$ undef., $\csc 450° = 1$, $\cot 450° = 0$

2. $\sin(-135°) = \cos(-135°) = -\dfrac{\sqrt{2}}{2}$,
$\tan(-135°) = \cot(-135°) = 1$,
$\sec(-135°) = \csc(-135°) = -\sqrt{2}$

3. $\sin 330° = -\dfrac{1}{2}$, $\cos 330° = \dfrac{\sqrt{3}}{2}$,
$\tan 330° = -\dfrac{\sqrt{3}}{3}$, $\cot 330° = -\sqrt{3}$,
$\sec 330° = \dfrac{2\sqrt{3}}{3}$, $\csc 330° = -2$

4. $\sin(-240°) = \dfrac{\sqrt{3}}{2}$, $\cos(-240°) = -\dfrac{1}{2}$,
$\tan(-240°) = -\sqrt{3}$, $\cot(-240°) = -\dfrac{\sqrt{3}}{3}$,
$\sec(-240°) = -2$, $\csc(-240°) = \dfrac{2\sqrt{3}}{3}$

5. $-\dfrac{1}{x+1}$ **6.** $5m^2\sqrt{2}$ **7.** $\dfrac{1}{y-2}$ **8.** $200a^{12}b^5$
9. $\dfrac{1}{x^{1/2}}$ **10.** $2u^3 - u^2 - 7u + 6$

Written Exercises, pages 582–583 **1.** 2.46
3. 18.6 **5.** 42.6 **7.** $55.8°$ **9.** $13.8°$
11. $87.4°$ **13.** 7.83 **15.** $69.2°$, $51.1°$, $59.7°$

Problems, pages 583–584 **1.** 26.9 km
3. $32.2°$, $87.8°$, $60.0°$ **5.** 16.6 mi **7.** 63.7 ft
9. perimeter = 45.4 km, area = 103.8 km^2
11. 31.5 m, 40.1 m

Computer Exercises, page 584
2. a. $\angle A = 53.8°$; $\angle B = 36.2°$; $c = 64.4$
b. $\angle A = 38.6°$; $\angle B = 51.4$; $b = 37.0$
c. $\angle A = 49.2°$; $\angle B = 40.8°$; $a = 6.86$
4. a. $\angle B = 58°$; $c = 69.8$; $b = 59.2$
b. $\angle A = 75°$; $c = 68.0$; $a = 65.7$
c. $\angle A = 70.2°$; $b = 3.35$; $a = 9.31$
6. a. $\angle C = 40°$; $b = 20.58$ **b.** $\angle C = 25.9°$;
$a = 48.33$ **c.** $\angle A = 52.0°$; $b = 8.451$

Written Exercises, pages 588–589 **1.** 32.0
3. 9.34 **5.** 23.0 **7.** $28.8°$ **9.** $24.2°$
11. $69.5°$ or $110.5°$ **13.** $\dfrac{10}{9}$ **15.** $\dfrac{15}{13}$ **17.** $\dfrac{\sqrt{2}}{2}$

Problems, pages 589–590 **1.** 29.3 cm
3. 49.7 cm **5.** 3280 m **7.** 47.1 m **9.** $8.1°$
11. 14.3 km **13.** 13.1 m

Mixed Review Exercises, page 590 **1.** 0.2045
2. -0.9380 **3.** -0.5548 **4.** -1.232
5. 0.2107 **6.** -1.604
7. $(x + 1)^2 + (y - 4)^2 = 9$ **8.** $x + y = 2$
9. $y = -\dfrac{1}{8}x^2$

Written Exercises, pages 594–595
1. $\angle A = 100°$, $b = 5.9$, $c = 14.9$
3. $\angle A = 19.7°$, $\angle B = 28.2°$, $\angle C = 132.1°$
5. $a = 20.0$, $\angle B = 25.7°$, $\angle C = 34.3°$
7. $\angle B = 30.7°$, $\angle C = 19.3°$, $c = 12.9$
9. $\angle A = 35.3°$, $\angle C = 104.7°$, $c = 15.1$
11. $\angle C = 80.0°$, $\angle A = 50.0°$, $a = 14.0$; or
$\angle C = 100.0°$, $\angle A = 30.0°$, $a = 9.14$ **13.** 2.61
15. 12.0 **17.** 4.33, 3.86 **19.** 80.1, 82.7, 85.1

Problems, pages 595–596 **1.** 1350 km
3. $55.7°$, 353 m **5.** 232 km, $40.4°$
7. 13.2 m **9.** $81.3°$ N or $81.3°$ S **11.** 5.51
13. 60.1 in.

Written Exercises, pages 599–600 **1.** 240
3. 37.1 **5.** 45.3 **7.** 82.5 **9.** 127 **11.** 18
13. $117.3°$ **15.** 483 cm^2
17. $K^2 = \left(\dfrac{1}{2}ab \sin C\right)^2 = \dfrac{1}{4}a^2b^2 \sin^2 C$
19. a. law of cosines **b.** $\sin^2 C + \cos^2 C = 1$
c. Ex. 17 **d.** Add. prop. of eq.
e. $x^2 - y^2 = (x + y)(x - y)$ **f.** Assoc. prop. of
add., factoring **g.** see **(e)** **h.** Ex. 18 **i.** Div.
prop. of eq., prop. of square roots

Mixed Review Exercises, page 600 **1.** 7.17
2. 29.5 **3.** $127.2°$ **4.** 23.4

5. $\cos \theta = -\dfrac{\sqrt{3}}{2}$, $\tan \theta = -\dfrac{\sqrt{3}}{3}$,

$\cot \theta = -\sqrt{3}$, $\sec \theta = -\dfrac{2\sqrt{3}}{3}$, $\csc \theta = 2$

6. $\sin \theta = -\dfrac{\sqrt{7}}{4}$, $\tan \theta = \dfrac{\sqrt{7}}{3}$, $\cot \theta = \dfrac{3\sqrt{7}}{7}$,

$\sec \theta = -\dfrac{4}{3}$, $\csc \theta = -\dfrac{4\sqrt{7}}{7}$ 7. $\sin \theta = -\dfrac{\sqrt{2}}{2}$,

$\cos \theta = \dfrac{\sqrt{2}}{2}$, $\cot \theta = -1$, $\sec \theta = \sqrt{2}$,

$\csc \theta = -\sqrt{2}$ 8. $\cos \theta = \dfrac{12}{13}$, $\tan \theta = \dfrac{5}{12}$,

$\cot \theta = \dfrac{12}{5}$, $\sec \theta = \dfrac{13}{12}$, $\csc \theta = \dfrac{13}{5}$

Self-Test 2, page 600 1. $\angle A = 62.7°$,
$b = 15.5$, $c = 33.8$ **2.** 16.1 **3.** $\angle Z = 80.6°$ or
99.4° **4.** $\angle A = 17.5°$, $\angle B = 13.9°$,
$\angle C = 148.6°$ **5.** 43.5

Chapter Review, pages 602–604 1. b **3.** b
5. b **7.** a **9.** d **11.** a **13.** a **15.** c **17.** a
19. d

Preparing for College Entrance Exams,
page 605 1. C **3.** C **5.** A **7.** D

Chapter 13 Trigonometric Graphs; Identities

Written Exercises, pages 610–611 1. $\dfrac{\pi}{4}$

3. $\dfrac{\pi}{3}$ **5.** $-\dfrac{2\pi}{3}$ **7.** $\dfrac{5\pi}{6}$ **9.** $-\dfrac{11\pi}{6}$ **11.** $-\dfrac{7\pi}{4}$
13. 30° **15.** 60° **17.** 240° **19.** −210°
21. 540° **23.** −630° **25.** $\dfrac{720°}{\pi}$ **27.** $-\dfrac{360°}{\pi}$
29. 0.17 **31.** 1.40 **33.** 0.84 **35.** −3.04
37. 171.9° **39.** 22.9° **41.** −85.9° **43.** 859.4°
45. 4; 8 **47.** 3; 24 **49.** 1.2; 6 **51.** 4; 20
53. 3; $\dfrac{3}{2}$ **55.** 2; 8 **57.** 1.84 **59.** 3.28

Problems, pages 611–612 1. 2100 mi
3. 1700 in. **5.** 79 ft/min **7.** 1,400,000 km
9. 250π radians/min; 4700 cm/min
11. 0.00015 radians/s; 0.0073 mm/s
13. Approximately 20%

Mixed Review Exercises, page 612
1. $c = 5.44$, $\angle A = 100.0°$, $\angle B = 38.0°$
2. $a = 23.2$, $c = 23.2$, $\angle C = 75°$ **3.** $a = 5.64$,
$\angle A = 33.2°$, $\angle C = 42.8°$ **4.** $\angle A = 31.6°$,
$\angle B = 39.0°$, $\angle C = 109.4°$ **5.** 13.4 **6.** 134
7. 19.2 **8.** 14.1

Written Exercises, page 617 1. $\sin s = -\dfrac{4}{5}$,

$\cos s = \dfrac{3}{5}$, $\tan s = -\dfrac{4}{3}$, $\cot s = -\dfrac{3}{4}$, $\sec s = \dfrac{5}{3}$,

$\csc s = -\dfrac{5}{4}$ **3.** $\sin s = \dfrac{3}{4}$, $\cos s = \dfrac{\sqrt{7}}{4}$,

$\tan s = \dfrac{3\sqrt{7}}{7}$, $\cot s = \dfrac{\sqrt{7}}{3}$, $\sec s = \dfrac{4\sqrt{7}}{7}$,

$\csc s = \dfrac{4}{3}$ **5.** 0.9490; 0.3153; 3.010

7. 0.2474; 0.9690; 0.2553 **9.** $\sin \dfrac{\pi}{3} = \dfrac{\sqrt{3}}{2}$,

$\cos \dfrac{\pi}{3} = \dfrac{1}{2}$, $\tan \dfrac{\pi}{3} = \sqrt{3}$, $\cot \dfrac{\pi}{3} = \dfrac{\sqrt{3}}{3}$,

$\sec \dfrac{\pi}{3} = 2$, $\csc \dfrac{\pi}{3} = \dfrac{2\sqrt{3}}{3}$ **11.** $\sin \dfrac{3\pi}{2} = -1$,

$\cos \dfrac{3\pi}{2} = 0$, $\tan \dfrac{3\pi}{2}$ and $\sec \dfrac{3\pi}{2}$ undef.,

$\cot \dfrac{3\pi}{2} = 0$, $\csc \dfrac{3\pi}{2} = -1$ **13.** $\sin \dfrac{11\pi}{6} = -\dfrac{1}{2}$,

$\cos \dfrac{11\pi}{6} = \dfrac{\sqrt{3}}{2}$, $\tan \dfrac{11\pi}{6} = -\dfrac{\sqrt{3}}{3}$,

$\cot \dfrac{11\pi}{6} = -\sqrt{3}$, $\sec \dfrac{11\pi}{6} = \dfrac{2\sqrt{3}}{3}$,

$\csc \dfrac{11\pi}{6} = -2$ **15.** $\sin \dfrac{8\pi}{3} = \dfrac{\sqrt{3}}{2}$,

$\cos \dfrac{8\pi}{3} = -\dfrac{1}{2}$, $\tan \dfrac{8\pi}{3} = -\sqrt{3}$,

$\cot \dfrac{8\pi}{3} = -\dfrac{\sqrt{3}}{3}$, $\sec \dfrac{8\pi}{3} = -2$, $\csc \dfrac{8\pi}{3} = \dfrac{2\sqrt{3}}{3}$

17. $\sin \left(-\dfrac{5\pi}{6}\right) = -\dfrac{1}{2}$, $\cos \left(-\dfrac{5\pi}{6}\right) = -\dfrac{\sqrt{3}}{2}$,

$\tan \left(-\dfrac{5\pi}{6}\right) = \dfrac{\sqrt{3}}{3}$, $\cot \left(-\dfrac{5\pi}{6}\right) = \sqrt{3}$,

$\sec \left(-\dfrac{5\pi}{6}\right) = -\dfrac{2\sqrt{3}}{3}$, $\csc \left(-\dfrac{5\pi}{6}\right) = -2$

19. 0.76 **21.** $\dfrac{\pi}{3}$

23. $\left(\dfrac{\sqrt{2}}{2}\right)^2 + \left(-\dfrac{\sqrt{2}}{2}\right)^2 = \dfrac{1}{2} + \dfrac{1}{2} = 1$
25. $(0.6442)^2 + (0.7648)^2 \approx 1$

Written Exercises, pages 621–623
1. odd

3. neither

5. odd

7. a.

b.

9. a.

b.

11. odd **13.** neither **15.** even **17.** even
19. odd

Mixed Review Exercises, page 623

1.

3.

5.

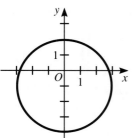

7. $\{0, 3\}$
8. $\{-2\}$
9. $\{2, \pm i\sqrt{2}\}$
10. $\{2, 6\}$
11. $\left\{-\dfrac{5}{2}, 2\right\}$
12. $\{2\}$

Written Exercises, pages 628–629 **1. a.** 2
b. $2, -2$ **c.** 2π **3. a.** 1 **b.** $1, -1$ **c.** $\dfrac{\pi}{2}$

5. a. 3 **b.** $3, -3$ **c.** $\dfrac{2\pi}{3}$ **7. a.** $\dfrac{1}{2}$ **b.** $\dfrac{1}{2}$,
$-\dfrac{1}{2}$ **c.** 1 **9. a.** 2 **b.** $2, -2$ **c.** 4 **11. a.** 1
b. $0, -2$ **c.** 1 **13.** $y = 2 \sin x$
15. $y = \dfrac{1}{2} \cos \dfrac{1}{2}x$ **17.** $y = \sin \pi x$
19. $y = 3 + 3 \cos \dfrac{\pi}{4}x$ **21.** $y = 3 \sin x$
23. $y = 2 \sin \pi x$ **25.** $y = 4 + 3 \sin 2x$
27.

29.

31.

33.

35.

37.

39.

41.

Written Exercises, page 633

1. odd

3. odd

5. neither

7. odd

9.

11.

13.

15.

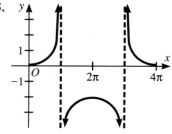

17. No; tangent has no maximum or minimum value.

19.

21.

23.

25.

27.

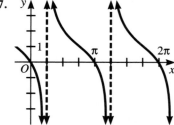

Mixed Review Exercises, page 633

1. $\sin \frac{5\pi}{2} = 1$; $\cos \frac{5\pi}{2} = 0$; $\tan \frac{5\pi}{2}$ and $\sec \frac{5\pi}{2}$ undef.; $\cot \frac{5\pi}{2} = 0$; $\csc \frac{5\pi}{2} = 1$

2. $\sin\left(-\frac{4\pi}{3}\right) = \frac{\sqrt{3}}{2}$; $\cos\left(-\frac{4\pi}{3}\right) = -\frac{1}{2}$; $\tan\left(-\frac{4\pi}{3}\right) = -\sqrt{3}$; $\cot\left(-\frac{4\pi}{3}\right) = -\frac{\sqrt{3}}{3}$; $\sec\left(-\frac{4\pi}{3}\right) = -2$; $\csc\left(-\frac{4\pi}{3}\right) = \frac{2\sqrt{3}}{3}$

35   *Answers to Selected Exercises*

3. $\sin \dfrac{19\pi}{6} = -\dfrac{1}{2}$; $\cos \dfrac{19\pi}{6} = -\dfrac{\sqrt{3}}{2}$;

$\tan \dfrac{19\pi}{6} = \dfrac{\sqrt{3}}{3}$; $\cot \dfrac{19\pi}{6} = \sqrt{3}$;

$\sec \dfrac{19\pi}{6} = -\dfrac{2\sqrt{3}}{3}$; $\csc \dfrac{19\pi}{6} = -2$

4. $\sin \left(-\dfrac{5\pi}{4}\right) = \dfrac{\sqrt{2}}{2}$; $\cos \left(-\dfrac{5\pi}{4}\right) = -\dfrac{\sqrt{2}}{2}$;

$\tan \left(-\dfrac{5\pi}{4}\right) = -1$; $\cot \left(-\dfrac{5\pi}{4}\right) = -1$;

$\sec \left(-\dfrac{5\pi}{4}\right) = -\sqrt{2}$; $\csc \left(-\dfrac{5\pi}{4}\right) = \sqrt{2}$ **5.** -3

6. $\dfrac{1}{2}$ **7.** 44 **8.** $\dfrac{1}{2}$

Self-Test 1, page 634 **1. a.** $\dfrac{11\pi}{12}$ **b.** 4.47

2. a. $-150°$ **b.** $143.2°$ **3.** $\sin \dfrac{3\pi}{4} = \dfrac{\sqrt{2}}{2}$,

$\cos \dfrac{3\pi}{4} = -\dfrac{\sqrt{2}}{2}$, $\tan \dfrac{3\pi}{4} = -1$, $\cot \dfrac{3\pi}{4} = -1$,

$\sec \dfrac{3\pi}{4} = -\sqrt{2}$, $\csc \dfrac{3\pi}{4} = \sqrt{2}$

4. 0.9915, 0.1304, 7.6018 **5. a.** even **b.** neither **6.** 3; 4, -2; π

7.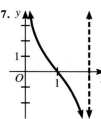

Application, page 635 **1.** $y = \dfrac{1}{2} \cos 704\pi t$

2. a. 528; C' **b.** 25 cm

Written Exercises, pages 639–640 **1.** $\tan \alpha$
3. $\sec x$ **5.** $\cos^2 \theta$ **7.** 1 **9.** 1 **11.** $\tan^2 \theta$
13. $\csc^2 \alpha$ **15.** 1 **17.** 1 **19.** $\tan \alpha$ **21.** $\cos x$

23. $\sin \phi$ **25.** $\pm \dfrac{1}{\sqrt{1 - \sin^2 x}}$

27. $\pm \dfrac{\sin t}{\sqrt{1 - \sin^2 t}}$ **29.** $\dfrac{\sin \alpha}{1 - \sin^2 \alpha}$ **39.** $\cot t$
41. $2 \cos \theta$ **43.** $\sec x$ **45.** $2 \csc \theta$ **47.** 2

Written Exercises, pages 643–644
1. $\dfrac{\sqrt{2} - \sqrt{6}}{4}$ **3.** $\dfrac{\sqrt{6} - \sqrt{2}}{4}$ **5.** $-\dfrac{\sqrt{6} + \sqrt{2}}{4}$

7. $\cos 60°$; $\dfrac{1}{2}$ **9.** $\sin 210°$; $-\dfrac{1}{2}$ **11.** $\cos \dfrac{\pi}{6}$;

$\dfrac{\sqrt{3}}{2}$ **13.** $\sin 2\theta$ **21.** $\cos \theta$ **23.** $\cos x$

31. a. $\dfrac{36}{85}$ **b.** $-\dfrac{77}{85}$ **c.** 2nd

Mixed Review Exercises, page 645
4. 1 **5.** x^{3n} **6.** $2x^2 \sqrt[3]{2}$ **7.** -1 **8.** $-2 - \sqrt{3}$
9. $\dfrac{3\sqrt{13}}{13}$ **10.** x **11.** $\dfrac{1}{x - 1}$ **12.** $\sin x$

Written Exercises, page 649 **1.** $\cos 20°$
3. $\cos 8°$ **5.** $\sin \dfrac{\alpha}{2}$ **7.** $\cos \theta$ **9.** $\cos 4t$

11. $\sin \alpha$ **13.** $\dfrac{1}{2}$ **15.** $\dfrac{\sqrt{3}}{2}$ **17.** $-\dfrac{\sqrt{3}}{2}$

19. $\dfrac{1}{2}\sqrt{2 + \sqrt{2}}$ **21.** $\dfrac{1}{2}\sqrt{2 + \sqrt{3}}$

23. $\dfrac{1}{2}\sqrt{2 + \sqrt{2}}$ **25.** $\dfrac{7}{25}$ **27.** $\dfrac{3\sqrt{10}}{10}$

29. $-\dfrac{240}{289}$ **31.** $-\dfrac{5\sqrt{34}}{34}$ **41.** period π, amp. 1
43. $8 \cos^4 \theta - 8 \cos^2 \theta + 1$

Written Exercises, pages 652–653 **1.** $\tan 60°$
3. $\tan 180°$ **5.** $\tan \dfrac{5\pi}{6}$ **7.** $\tan 45°$ **9.** $2 + \sqrt{3}$
11. $-2 + \sqrt{3}$ **13.** $-2 + \sqrt{3}$ **15.** $\sqrt{2} - 1$
17. $\sqrt{2} + 1$ **19.** $2 + \sqrt{3}$ **21.** $\dfrac{13}{84}$ **23.** $-\dfrac{240}{161}$

29. $\dfrac{\cot \alpha \cot \beta - 1}{\cot \alpha + \cot \beta}$ **31.** $\dfrac{\cot^2 \alpha - 1}{2 \cot \alpha}$

33. a. $\dfrac{1}{2}\sqrt{2 - \sqrt{2}}$; $-\dfrac{1}{2}\sqrt{2 + \sqrt{2}}$

Mixed Review Exercises, page 653
5. $\{x: 1 < x < 5\}$ **6.** $\{x: x < -1 \text{ or } x > 3\}$
7. $\{x: -1 \le x \le 3\}$ **8.** $\{x: x < 3, x \ne 0\}$
9. $\{x: x \le -2 \text{ or } x \ge 1\}$ **10.** $\{x: x < 2\}$
11. $\{x: x > 6\}$ **12.** $\{x: -1 \le x \le 3\}$
13. {real numbers}

Self-Test 2, page 654 **1.** $\sec x$ **3.** $\cos 570°$;
$-\dfrac{\sqrt{3}}{2}$ **4.** $\sin 10\alpha$ **5. a.** $\dfrac{41}{841}$ **b.** $\dfrac{840}{841}$
c. $-\dfrac{2\sqrt{29}}{29}$ **6.** $\tan 10°$ **7.** $\tan 70°$ **8.** $\tan 65°$

Chapter Review, pages 655–656 **1.** c **3.** d
5. a **7.** c **9.** d **11.** d **13.** d **15.** d **17.** b

Chapter 14 Trigonometric Applications

Written Exercises, page 664
1.

3.

5. 15 10

7.
w
u   u + w

9.
−v   u − v
u

11.
u
2v
u + 2v

13.
2v   u
u + 2v − w
−w

15. 155; 171.2°
17. 307; 180.8°
19. 545; 90°
21. 15.5; 92.1°

Problems, page 665 **1.** 429 km; 318.3°
3. 4.4°; 582 km/h   **5.** 283 km/h; 80.0°
7. 78 km; 2nd to 1st, 9.6°; 1st to 2nd, 189.6°
9. 142.7°; ETA 2:47 P.M.

Mixed Review Exercises, page 665 **1.** neither
2. neither   **3.** odd   **4.** $\{-3 \pm \sqrt{17}\}$ **5.** $\left\{\frac{2}{3}, 4\right\}$
6. $\left\{-\frac{4}{3}, \frac{8}{3}\right\}$ **7.** $\{27\}$ **8.** $\{10\}$ **9.** $\left\{-6, \frac{1}{2}\right\}$

Written Exercises, pages 669–670 **1.** $5\mathbf{i} - 2\mathbf{j}$
3. $-5\mathbf{i}$ **5.** $5\mathbf{i} - \mathbf{j}$ **7.** $(6, -7)$ **9.** $(2, 6)$
11. $s = 1, t = 7$ **13.** $s = 2, t = 3$
15. a. $6\mathbf{i} + 2\mathbf{j}$ **c.** $2\sqrt{10}$ **d.** $18.4°$
17. a. $6\mathbf{i} + 7\mathbf{j}$ **c.** $\sqrt{85}$ **d.** $49.4°$ **19.** $63.4°$
21. $98.1°$ **23.** $\frac{\sqrt{2}}{2}\mathbf{i} - \frac{\sqrt{2}}{2}\mathbf{j}$ **25.** $\frac{4}{5}\mathbf{i} + \frac{3}{5}\mathbf{j}$
27. a. $\mathbf{u} = 7.52\mathbf{i} + 2.74\mathbf{j}$; $\mathbf{v} = 2.95\mathbf{i} + 16.7\mathbf{j}$
b. 68.0   **c.** 67.9   **29. a.** $\mathbf{u} = 4.68\mathbf{i} + 11.6\mathbf{j}$;
$\mathbf{v} = -7.89\mathbf{i} + 16.2\mathbf{j}$ **b.** 151   **c.** 151

Self-Test 1, page 671 **1.** 12; 67°   **2.** 151.2°
3. $2\mathbf{i} + 27\mathbf{j}$ **4.** 42.4°

Application, page 674 **1.** $-2\mathbf{i} - \mathbf{j}$ **3.** 9 J
5. a. 22 J   **b.** 22 J   **7.** 13.2 N; 319°
9. 1.47×10^6 J   **11.** 8.63×10^3 J
13. 2.30×10^4 J   **15.** 995 N; 1525 N

Written Exercises, pages 678–679
1. $(2\sqrt{3}, 2)$ **3.** $\left(\frac{3}{2}, -\frac{3\sqrt{3}}{2}\right)$
5. $\left(\frac{7}{2}, -\frac{7\sqrt{3}}{2}\right)$ **7.** $(3\sqrt{3}, 3)$
9. $(4, 0°)$ **11.** $(2\sqrt{2}, 135°)$ **13.** $(2, 120°)$
15. $(\sqrt{10}, 135°)$ **17.** $r \cos \theta = 5$
19. $\sin \theta = \cos \theta$ **21.** $r = 6 \sin \theta$
23. $x^2 + y^2 = 4$ **25.** $y = 2$ **27.** $x^2 + y^2 = 2y$
29. $y^2 = 4 + 4x$
31.   **33.**

Mixed Review Exercises, page 679
1. $x^3 - 2x^2 + 2x - 4 + \frac{11}{x + 2}$ **2.** $x^2 - 1$
3. $3x - 5 + \frac{17}{2x + 2}$
4. $-4x^2 - 5x - 12 + \frac{-23}{x - 12}$
5. [graph]   **7.** [graph]

Written Exercises, pages 683–684
1.   **5.**
3. [graph]

7. $10(\cos 110° + i \sin 110°)$; $2.5(\cos 310° + i \sin 310°)$
9. $5.4(\cos 465° + i \sin 465°)$; $3.75(\cos 195° + i \sin 195°)$ **11.** $\dfrac{3\sqrt{3}}{2} + \dfrac{3}{2}i$
13. $-2.25 + 2.25i\sqrt{3}$ **15.** $4.045 + 2.939i$
17. $-1.905 + 1.1i$ **19.** $2(\cos 300° + i \sin 300°)$
21. $2\sqrt{6}(\cos 135° + i \sin 135°)$
23. $5(\cos 233.1° + i \sin 233.1°)$
25. $\sqrt{13}(\cos 326.3° + i \sin 326.3°)$
27. a. $18(\cos 270° + i \sin 270°)$ **b.** $-18i$
c. $w = -\dfrac{3}{2} + \dfrac{3\sqrt{3}}{2}i$, $z = -3\sqrt{3} + 3i$ **d.** $-18i$
29. a. $8(\cos 150° + i \sin 150°)$ **b.** $-4\sqrt{3} + 4i$
c. $w = \sqrt{3} + i$, $z = -2 + 2i\sqrt{3}$
d. $-4\sqrt{3} + 4i$

Written Exercises, pages 687–688
1. $-8 + 8i\sqrt{3}$ **3.** $-16 + 16i$ **5.** $64 - 64i\sqrt{3}$
7. $-128 - 128i\sqrt{3}$ **9.** $\cos 0° + i \sin 0°$;
$\cos 40° + i \sin 40°$; $\cos 80° + i \sin 80°$;
$\cos 120° + i \sin 120°$; $\cos 160° + i \sin 160°$;
$\cos 200° + i \sin 200°$; $\cos 240° + i \sin 240°$;
$\cos 280° + i \sin 280°$; $\cos 320° + i \sin 320°$
11. $r_1 = 2(\cos 110° + i \sin 110°)$;
$r_2 = 2(\cos 230° + i \sin 230°)$;
$r_3 = 2(\cos 350° + i \sin 350°)$ **13.** 1,
$-\dfrac{1}{2} \pm \dfrac{i\sqrt{3}}{2}$ **15.** ± 1, $\dfrac{1}{2} \pm \dfrac{i\sqrt{3}}{2}$, $-\dfrac{1}{2} \pm \dfrac{i\sqrt{3}}{2}$
17. -1, $\dfrac{1}{2} \pm \dfrac{i\sqrt{3}}{2}$ **19.** $\dfrac{\sqrt[4]{108}}{2} + \dfrac{i\sqrt[4]{12}}{2}$,
$-\dfrac{\sqrt[4]{108}}{2} - \dfrac{i\sqrt[4]{12}}{2}$, $-\dfrac{\sqrt[4]{12}}{2} + \dfrac{i\sqrt[4]{108}}{2}$,
$\dfrac{\sqrt[4]{12}}{2} - \dfrac{i\sqrt[4]{108}}{2}$

Mixed Review Exercises, page 688 **1.** $5i - 3j$
3. $\sqrt{34}$ **4.** $329.0°$ **5.** $(2\sqrt{3}, 2)$
6. $\left(-\dfrac{3}{2}, -\dfrac{3\sqrt{3}}{2}\right)$ **7.** $(\sqrt{3}, -1)$

Self-Test 2, page 688 **1.** $(2, -2\sqrt{3})$
2. $(13.0, 157.4°)$
3.

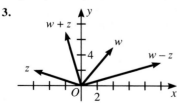

4. $wz = 18(\cos 105° + i \sin 105°)$,
$\dfrac{w}{z} = 2(\cos 325° + i \sin 325°)$
5. $3.61(\cos 326.3° + i \sin 326.3°)$ **6.** -64

Written Exercises, page 692 **1.** $\dfrac{\pi}{3}$ **3.** $\dfrac{5\pi}{6}$
5. $\dfrac{2\pi}{3}$ **7.** $\dfrac{\pi}{3}$ **9.** $\dfrac{5\pi}{6}$ **11.** $\dfrac{1}{2}$ **13.** $\dfrac{12}{13}$
15. $\dfrac{\sqrt{21}}{5}$ **17.** 1 **19.** $\sqrt{1 - x^2}$
21. $2x\sqrt{1 - x^2}$ **23.** v

Written Exercises, pages 695–696 **1.** $\dfrac{\pi}{6}$
3. $\dfrac{5\pi}{6}$ **5.** $\dfrac{\pi}{2}$ **7.** $\dfrac{3\pi}{4}$ **9.** $\dfrac{1}{2}$ **11.** $-\dfrac{3}{4}$
13. $\dfrac{2\sqrt{5}}{5}$ **15.** $\dfrac{171}{140}$ **17.** $\dfrac{4}{5}$ **19.** $\dfrac{1}{x}$
21. $\dfrac{\sqrt{x^2 + 1}}{x^2 + 1}$ **23.** $\dfrac{2x}{x^2 + 1}$ **25.** $\dfrac{x + y}{xy - 1}$

Mixed Review Exercises, page 696
1. Arithmetic; 107 **2.** Geometric; -384
3. Geometric; 9216 **4.** Arithmetic; -79
5. $\displaystyle\sum_{n=0}^{5} 3(4^n)$; 4095 **6.** $\displaystyle\sum_{n=1}^{100} (2 + 3n) = 15,350$
7. $\displaystyle\sum_{n=0}^{19} (-5 + 4n)$; 660 **8.** $\displaystyle\sum_{k=0}^{\infty} 100(5^{-k}) = 125$

Written Exercises, pages 699–700 **1.** $\{48.6°,$
$131.4°\}$ **3.** $\left\{\dfrac{\pi}{4}, \dfrac{3\pi}{4}, \dfrac{5\pi}{4}, \dfrac{7\pi}{4}\right\}$ **5.** $\{135°, 315°\}$
7. $\{14.5°, 165.5°, 194.5°, 345.5°\}$
9. $\left\{\dfrac{3\pi}{8}, \dfrac{11\pi}{8}\right\}$ **11.** $\{155°, 335°\}$
13. $\{30°, 150°, 210°, 330°\}$ **15.** $\{0.32, 3.46\}$
17. $\{45°, 135°, 225°, 315°\}$
19. $\left\{\dfrac{\pi}{6}, \dfrac{\pi}{2}, \dfrac{5\pi}{6}, \dfrac{3\pi}{2}\right\}$ **21.** $\{0°, 120°, 240°\}$
23. $\{30°, 60°, 210°, 240°\}$
25. $\{45°, 123.7°, 225°, 303.7°\}$ **27.** $\left\{\dfrac{7\pi}{6}, \dfrac{11\pi}{6}\right\}$
29. $\left\{\dfrac{\pi}{12}, \dfrac{7\pi}{12}, \dfrac{13\pi}{12}, \dfrac{19\pi}{12}\right\}$
31. $\left\{\dfrac{\pi}{4}, \dfrac{7\pi}{12}, \dfrac{11\pi}{12}, \dfrac{5\pi}{4}, \dfrac{19\pi}{12}, \dfrac{23\pi}{12}\right\}$
33. $\{15°, 45°, 75°, 135°, 195°, 225°, 255°, 315°\}$
35. $\left\{\dfrac{\pi}{4}, \dfrac{\pi}{2}, \dfrac{3\pi}{4}, \dfrac{5\pi}{4}, \dfrac{3\pi}{2}, \dfrac{7\pi}{4}\right\}$
37. $\{30°, 150°, 210°, 330°\}$
39. $\{55°, 115°, 235°, 295°\}$
41. $\left\{\dfrac{\pi}{12}, \dfrac{\pi}{4}, \dfrac{3\pi}{4}, \dfrac{11\pi}{12}, \dfrac{17\pi}{12}, \dfrac{19\pi}{12}\right\}$
43. $\{39°, 129°, 219°, 309°\}$ **45.** $\{51.3°, 128.7°\}$

Self-Test 3, page 700 **1.** $-\dfrac{\pi}{3}$ **2.** $-\dfrac{\pi}{6}$ **3.** $\dfrac{5\pi}{6}$
4. $\dfrac{3}{5}$ **5.** $\dfrac{2\pi}{3}$ **6.** $\dfrac{17}{15}$ **7.** $\dfrac{\pi}{3}$ **8.** $-\dfrac{1}{5}$
9. $\{30°, 150°, 210°, 330°\}$

Chapter Review, pages 703–704   **1.** c   **3.** a
5. c   **7.** d   **9.** d   **11.** a   **13.** c   **15.** d

Mixed Problem Solving, page 705   **1.** 460

3. 7   **5. a.** $\frac{1}{64}$ cm   **b.** 4 cm   **7.** 1:05 P.M.

9. $6\sqrt{3}$ **11.** 15.7 ft

Preparing for College Entrance Exams,
page 706   **1.** D   **3.** A   **5.** A   **7.** D   **9.** C

Cumulative Review, page 707   **1.** $465°, -255°$
for example   **3.** 2.74 ft   **5.** 14.1

7. $\sin\left(-\frac{13\pi}{6}\right) = -\frac{1}{2}$, $\cos\left(-\frac{13\pi}{6}\right) = \frac{\sqrt{3}}{2}$,
$\tan\left(-\frac{13\pi}{6}\right) = -\frac{\sqrt{3}}{3}$, $\cot\left(-\frac{13\pi}{6}\right) = -\sqrt{3}$,
$\sec\left(-\frac{13\pi}{6}\right) = \frac{2\sqrt{3}}{3}$, $\csc\left(-\frac{13\pi}{6}\right) = -2$

9. $\cot\theta + \cot^3\theta$ **11. a.** $-\frac{\sqrt{10}}{10}$ **b.** $\frac{7}{25}$

c. $-\frac{3}{4}$ **d.** $-\frac{4}{5}$ **13.** $\frac{\sqrt{5}}{5}\mathbf{i} + \frac{2\sqrt{5}}{5}\mathbf{j}$ **15.** $2i$,
$-\sqrt{3} - i$, $\sqrt{3} - i$ **17.** $\frac{4\sqrt{3} - 3}{10}$

19. $\left\{\frac{7\pi}{6}, \frac{11\pi}{6}\right\}$ **21.** $\{120°, 240°\}$

Chapter 15 Statistics and Probability

Written Exercises, pages 711–712

1. 1	6, 9
2	0, 2, 2, 3
3	8
4	0, 6
5	2, 4

3. 0	2, 5, 7, 9
1	0, 3, 8, 9
2	8, 8, 8
3	4
4	2
5	2
6	1

5. a. 22   **b.** 23   **c.** 32   **7. a.** 28   **b.** 19
c. 23.7   **9. a.** 184   **b.** 184   **c.** 183

11.

13. 87   **15.** Each would be increased by 5.
17. 77.2

Mixed Review Exercises, page 712
1. $(6, 120°)$ **2.** $(2\sqrt{2}, 30°)$ **3.** $(3, 90°)$

4. $\frac{n^6}{m^{14}}$ **5.** $\frac{w+1}{3w+2}$ **6.** $2^{5\sqrt{2}}$

Written Exercises, pages 717–718   **1. a.** 25
b. 14   **c.** 34   **d.** 45

3.

5. 2   **7.** 2   **9. a.** 6   **b.** 5.3   **c.** 2.3
11. a. 57   **b.** 305.3   **c.** 17.5   **13. a.** 50
b. 19.4   **c.** 4.4   **15. a.** 71   **b.** 4.4   **c.** 2.1
17. M is increased by 10; σ is unchanged.
19. Answers may vary.   **a.** 78; 80; 80; 82
b. 20; 100; 100; 100   **c.** 70; 70; 90; 90

Written Exercises, pages 722–723   **1.** 15.54%
3. 0.13%   **5. a.** 15.87%   **b.** 97.72%
c. 0.13%   **d.** 68.26%   **7. a.** 50%   **b.** 0.26%
9. a. 0   **b.** 4   **c.** -1 **d.** 1

Mixed Review Exercises, page 723   **1. a.** -3
b. $3x + y = 5$ **2. a.** $\frac{3}{4}$ **b.** $3x - 4y = 1$

3. a. $-\frac{1}{2}$ **b.** $x + 2y = 0$ **4. a.** $\frac{2}{3}$
b. $2x - 3y = -22$ **5. a.** 0   **b.** $y = -1$
6. a. -1 **b.** $x + y = -5$

Written Exercises, pages 727–728
1. High positive   **3.** No correlation
5. a. 0.96   **b.** $y = 1.41x - 42.56$ **7.** 78
9. 71   **11.** No; there may be no cause-and-
effect relationship.

Self-Test 1, page 729   **1.** 28; 27; 26   **2.** 21; 33
3. 51; 7   **4.** 81.85%   **5.** 1.39%   **6.** High negative

Written Exercises, pages 732–733   **1.** 20
3. 50   **5.** 120   **7.** 1024   **9.** 132,600
11. 20,000   **13.** 37,856   **15.** 12
17. 8568 min, or about 6 days   **19.** 64,000,000

Mixed Review Exercises, page 733   **1.** -2
2. 10   **3.** $\frac{1}{64}$ **4.** $4i$ **5.** $\{\pm\sqrt{2}, -1\}$

6. $\{\pm i\sqrt{3}, 4\}$ **7.** $\left\{5, 1, -\frac{1}{2}\right\}$ **8.** $\{\pm 4i, -6\}$

Written Exercises, page 737   **1.** 720   **3.** 336
5. 5040   **7.** 6   **9.** 720   **11.** 132,600   **13.** 720

15. 120 **17.** 30 **19.** 34,650 **21.** 30 **23.** 420
29. 18 **31.** 2520

Written Exercises, pages 740–741
1. a. {J, K}, {J}, {K}, Ø **b.** {J}, {K}, Ø **3.** 10
5. 28 **7.** 45 **9.** 792 **11. a.** 5 **b.** 10 **c.** 10
13. 6435 **15.** 45 **17.** 35 **19.** 254,800
21. 57,798 **23.** 4512 **25.** 163 **29. a.** $_nC_r$ is
the coefficient of $a^{n-r}b^r$.

b. $(a + b)^n = \sum\limits_{r=0}^{n} {_nC_r}a^{n-r}b^r$

Mixed Review Exercises, page 741 **1.** 2 **2.** 3
3. 36 **4.** −3 **5.** −7 **6.** 390 **7.** $5\sqrt{2}$ **8.** 1
9. $\dfrac{1}{m-4}$ **10.** $11 - 2i$ **11.** $\dfrac{2t-3}{t-3}$ **12.** 1

Self-Test 2, page 742 **1.** 100 **2.** 140 **3.** 120
4. 180 **5.** 56 **6.** 10

Written Exercises, page 744 **1.** {(1, 1), (2, 2),
(3, 3), (4, 4), (5, 5), (6, 6)} **3.** {(2, 6), (3, 4),
(4, 3), (6, 2)} **5.** {(1, 1), (1, 2), (1, 3), (1, 4),
(1, 5), (1, 6), (2, 1), (2, 2), (2, 3), (2, 4),
(3, 1), (3, 2), (3, 3), (4, 1), (4, 2), (5, 1),
(6, 1)} **7. a.** {(R, R), (R, B), (R, Y), (R, G),
(B, R), (B, B), (B, Y), (B, G)} **b.** {(R, R),
(R, B), (R, Y), (R, G), (B, R)} **c.** {(R, R),
(R, Y), (R, G)} **9.** {(3, 6), (4, 5), (5, 4),
(6, 3)} **11.** {(2, 1), (4, 2), (6, 3)} **13.** {(1, 1),
(1, 2), (1, 3), (1, 4), (1, 5), (1, 6), (2, 1), (2, 2),
(2, 3), (2, 4), (2, 5), (2, 6), (3, 5), (3, 6),
(4, 5), (4, 6), (5, 5), (5, 6), (6, 5), (6, 6)}
15. {1, 1), (2, 1), (4, 1), (6, 1), (1, 4), (3, 4)}

Written Exercises, pages 748–750 **1. a.** $\dfrac{1}{10}$
b. $\dfrac{1}{2}$ **c.** $\dfrac{3}{10}$ **d.** 1 **e.** $\dfrac{1}{5}$ **f.** 0 **3. a.** $\dfrac{1}{4}$ **b.** 1
c. $\dfrac{3}{4}$ **d.** $\dfrac{5}{8}$ **5. a.** $\dfrac{1}{4}$ **b.** $\dfrac{3}{4}$ **c.** $\dfrac{1}{2}$ **d.** $\dfrac{1}{2}$
7. a. 1 **b.** $\dfrac{9}{1000}$ **c.** $\dfrac{991}{1000}$ **9. a.** $\dfrac{1}{8}$ **b.** $\dfrac{1}{8}$
c. $\dfrac{7}{8}$ **d.** $\dfrac{3}{8}$ **11. a.** $\dfrac{1}{66}$ **b.** $\dfrac{1}{11}$ **c.** $\dfrac{5}{22}$ **d.** $\dfrac{4}{33}$
e. $\dfrac{15}{22}$ **f.** $\dfrac{5}{22}$ **13. a.** 0.0548 **b.** 0.0082

Mixed Review Exercises, page 750
1. {t: t < −2 or t > 6} **2.** {x: −1 ≤ x ≤ 2}
3. {n: n > 2} **4.** {x: 0 ≤ x ≤ 4}
5. {y: y < −2 or y > 2} **6.** $\left\{ k: -\dfrac{5}{3} < k < 1 \right\}$
7. 60 **8.** 10,626

Application, pages 752–753 **1. a.** Random
b. Not random **3.** Not random **5–8.** Answers
will vary.

Written Exercises, pages 759–761 **1. a.** $\dfrac{1}{2}$
b. $\dfrac{1}{2}$ **c.** $\dfrac{3}{10}$ **3. a.** $\dfrac{1}{11}$ **b.** $\dfrac{5}{33}$ **5. a.** $\dfrac{1}{6}$, $\dfrac{5}{36}$,
$\dfrac{5}{18}$, $\dfrac{1}{36}$ **b.** No **7. a.** $\dfrac{1}{12}$ **b.** $\dfrac{1}{9}$ **9. a.** $\dfrac{1}{2}$
b. $\dfrac{7}{8}$ **11.** $\dfrac{1}{360}$ **13. a.** $\dfrac{1}{2}$ **b.** $\dfrac{2}{5}$ **c.** $\dfrac{7}{10}$
15. a. $\dfrac{1}{12}$ **b.** $\dfrac{3}{16}$ **c.** $\dfrac{19}{48}$ **d.** $\dfrac{7}{8}$ **17. a.** $\dfrac{11}{20}$
b. $\dfrac{4}{11}$

Self-Test 3, page 761 **1. a.** {S, Q, U, A, R, E}
b. {U, A, E} **2.** $\dfrac{1}{9}$ **3. a.** $\dfrac{7}{18}$ **b.** No **c.** No

Chapter Review, pages 764–765 **1.** b **3.** d
5. c **7.** a **9.** c **11.** d

Chapter 16 Matrices and Determinants

Written Exercises, page 769
1. $\begin{bmatrix} 0 & 0 \end{bmatrix}$ **3.** $\begin{bmatrix} 0 & 0 & 0 & 0 & 0 \\ 0 & 0 & 0 & 0 & 0 \end{bmatrix}$
5. $x = y = z = 0$ **7.** $x = -4$, $y = 0$, $z = 10$
9. $x = -4$, $y = 1$ **11.** $x = 8$, $y = -5$
13. $x = 11$, $y = 12$, $z = -4$ **15.** $x = 5$,
$y = -3$ **17.** $x = 3$, $y = -1$ **19.** $x = \dfrac{1}{a}$, $y = 0$
21. $x = \dfrac{1-b}{a}$, $y = -\dfrac{d}{c}$

Mixed Review Exercises, page 769
1. $\angle A = 44.3°$, $\angle B = 86.7°$, $c = 7.56$
2. $\angle A = 60°$, $a = 4\sqrt{3}$, $b = 4$ **3.** $\angle A = 45°$,
$\angle B = 45°$, $c = 3$ **4.** $\angle B = 42.4°$, $\angle C = 65.6°$,
$c = 23.0$ **5.** 26.4 **6.** 13.8 **7.** 2.25 **8.** 186.0

Written Exercises, page 773
1. $\begin{bmatrix} 4 & 4 \\ 0 & 3 \end{bmatrix}$ **3.** $\begin{bmatrix} 11 & 4 \\ 3 & 16 \end{bmatrix}$ **5.** $\begin{bmatrix} 4 & 2 \\ 0 & -5 \\ -3 & 7 \end{bmatrix}$
7. $\begin{bmatrix} -1 & 3 & 4 \\ 0 & 0 & 0 \end{bmatrix}$ **9.** $\begin{bmatrix} 42 & 48 \\ -24 & 0 \end{bmatrix}$
11. $\begin{bmatrix} 3 & 5 & 2 \\ -2 & 8 & 0 \end{bmatrix}$ **13.** $\begin{bmatrix} 11 & 3 \\ -6 & 4 \\ 6 & 17 \end{bmatrix}$ **15.** $\begin{bmatrix} 6 \\ 36 \\ 30 \end{bmatrix}$

17. $\begin{bmatrix} 3 & 1 \\ 6 & -1 \end{bmatrix}$ **19.** $\begin{bmatrix} 5 & 4 \\ 12 & 3 \end{bmatrix}$

Written Exercises, pages 777–778

1. $\begin{bmatrix} 18 \end{bmatrix}$ **3.** $\begin{bmatrix} 12 & 0 & -4 & 20 \\ -6 & 0 & 2 & -10 \end{bmatrix}$

5. $\begin{bmatrix} -3 & 24 \\ -1 & 14 \end{bmatrix}$ **7.** $\begin{bmatrix} -3 & 0 & 6 \\ 15 & 12 & 2 \\ 3 & 0 & -6 \end{bmatrix}$

9. $\begin{bmatrix} -1 & 2 & 2 & -1 \\ 4 & -1 & -1 & -3 \\ -4 & 4 & 4 & 0 \end{bmatrix}$

11. $\begin{bmatrix} a+2g & b+2h & c+2i \\ -2a+d & -2b+e & -2c+f \\ d & e & f \end{bmatrix}$

13. $AB = \begin{bmatrix} 2 & 1 \\ -1 & -1 \end{bmatrix}$, $BA = \begin{bmatrix} 0 & 1 \\ 1 & 1 \end{bmatrix}$

15. $(AB)C = \begin{bmatrix} -4 & 1 \\ 2 & -1 \end{bmatrix}$, $A(BC) = \begin{bmatrix} -4 & 1 \\ 2 & -1 \end{bmatrix}$

17. $(A+B)(A-B) = \begin{bmatrix} 0 & 1 \\ 1 & 3 \end{bmatrix}$, $A^2 - B^2 = \begin{bmatrix} 2 & 1 \\ -1 & 1 \end{bmatrix}$

19. $O_{2\times2} \cdot A = A \cdot O_{2\times2} = \begin{bmatrix} 0 & 0 \\ 0 & 0 \end{bmatrix}$ **21.** $\begin{bmatrix} 1 & 0 \\ 0 & 1 \end{bmatrix}$

23. $\begin{bmatrix} 0 & 0 & -1 \\ -1 & 4 & -2 \\ 1 & 0 & -1 \end{bmatrix}$ **25.** $\begin{bmatrix} 0 & 0 & -1 \\ -1 & 4 & -2 \\ 1 & 0 & -1 \end{bmatrix}$

Mixed Review Exercises, page 778 **1.** No solution **2.** $(2, -1)$ **3.** $(5, 7), (-1, -5)$
4. $(-3, 2)$ **5.** $(5, 3), (-5, -3)$ **6.** $(1, -1, 2)$
7. $\{3\}$ **8.** $\left\{ \dfrac{-7 \pm \sqrt{73}}{4} \right\}$ **9.** $\{-1\}$

Written Exercises, pages 782–784

1.

	A	B	C
A	0	1	1
B	1	0	0
C	1	1	0

3.

	A	B	C
A	2	2	1
B	1	1	1
C	2	2	1

5.

	A	B	C	D	E
A	0	1	0	0	0
B	1	0	1	1	0
C	0	1	0	0	1
D	0	1	0	0	1
E	0	0	0	0	0

7. $\begin{bmatrix} 1 & 0 & 1 & 1 & 0 \\ 0 & 3 & 0 & 0 & 2 \\ 1 & 0 & 1 & 1 & 0 \\ 1 & 0 & 1 & 1 & 0 \\ 0 & 0 & 0 & 0 & 0 \end{bmatrix}$

9. B

11.

	J	C	Y	A
J	0	1	0	0
C	0	0	1	1
Y	1	0	0	1
A	1	0	0	0

; $\begin{bmatrix} 0 & 0 & 1 & 1 \\ 2 & 0 & 0 & 1 \\ 1 & 1 & 0 & 0 \\ 0 & 1 & 0 & 0 \end{bmatrix}$

13.

	A	B	C	D
A	0	1	0	1
B	1	0	0	0
C	0	1	0	1
D	1	0	0	0

; $\begin{bmatrix} 2 & 0 & 0 & 0 \\ 0 & 1 & 0 & 1 \\ 2 & 0 & 0 & 0 \\ 0 & 1 & 0 & 1 \end{bmatrix}$

15. $\begin{bmatrix} 0 & 2 & 0 & 2 \\ 2 & 0 & 0 & 0 \\ 0 & 2 & 0 & 2 \\ 2 & 0 & 0 & 0 \end{bmatrix}$

17. 2

Self-Test 1, page 784 **1.** $x = -1$, $y = 5$, $z = 0$
2. $x = 2$, $y = -10$, $z = 4$

3. $\begin{bmatrix} 0 & 6 \\ 1 & 3 \end{bmatrix}$ **4.** $\begin{bmatrix} 2 & 2 \\ -1 & 3 \end{bmatrix}$ **5.** $\begin{bmatrix} -3 & 6 \\ 3 & 0 \end{bmatrix}$

6. $\begin{bmatrix} 3 & 2 \\ 3 & 0 \end{bmatrix}$ **7.** $\begin{bmatrix} 1 & 16 \\ 0 & 9 \end{bmatrix}$ **8.** $\begin{bmatrix} 3 & 0 \\ -2 & 3 \end{bmatrix}$

9. a.

	A	B	C	D
A	0	1	1	1
B	1	0	1	0
C	1	0	0	0
D	1	0	1	0

b. 3

Extra, page 786 **1.** $(5, -1, -4)$
3. $(1, -2, 4)$ **5.** $(-1, 2, 3)$

Written Exercises, page 789 **1.** 5 **3.** 14
5. 32 **7.** 137 **9.** 456 **11.** −60 **13.** 138

Mixed Review Exercises, page 789 **1.** $\frac{t-1}{t-3}$

2. $\frac{1}{ab^4}$ **3.** $\frac{13s-5}{3(s+1)(s-1)}$ **4.** a^2 **5.** $-\frac{1}{c+1}$

6. $\frac{y-2}{y-5}$ **7.** $-36p^3q^6$ **8.** $\frac{x^2+x+1}{x+4}$ **9.** $\frac{1}{2}$

Written Exercises, page 792

1. $\begin{bmatrix} \frac{2}{5} & -\frac{1}{5} \\ \frac{1}{5} & \frac{2}{5} \end{bmatrix}$ **3.** No inverse

5. $\begin{bmatrix} 1 & 0 \\ 0 & 1 \end{bmatrix}$ **7.** $\begin{bmatrix} \frac{1}{6} & \frac{1}{12} \\ -\frac{1}{2} & \frac{1}{4} \end{bmatrix}$

9. (−1, −8) **11.** (5, 3) **13.** (2, 4) **15.** No
unique solution

17. $\begin{bmatrix} 34 & -2 \\ -10 & 1 \end{bmatrix}$ **19.** $\begin{bmatrix} 2 & -\frac{3}{4} \\ -3 & \frac{13}{4} \end{bmatrix}$

Self-Test 2, page 793 **1.** −3 **2.** −8

3. $\begin{bmatrix} 1 & \frac{1}{3} \\ 0 & \frac{1}{3} \end{bmatrix}$ **4.** $\begin{bmatrix} 2 & -1 \\ -1 & 1 \end{bmatrix}$

Written Exercises, pages 796–797 **1.** −7
3. −16 **5.** −10 **7.** 6 **9.** −4 **11.** 8 **13.** 1
15. 0 **17.** −60

Mixed Review Exercises, page 797 **1.** $\frac{1}{25}$

2. $-\frac{\sqrt{2}}{2}$ **3.** 56 **4.** −4 **5.** $\frac{5}{2}$ **6.** 1

Written Exercises, page 800 **1.** 0 **3.** 0 **5.** 0
7. 60 **9.** 112 **11.** 0

Written Exercises, pages 804–805 **1.** (2, 0)
3. (5, −3, 5) **5.** (3, −1, 2) **7.** (2, 1, 2)
9. (2, −4, −3) **11.** No solution **13.** No
solution **15.** (2, 2, −2, −1)

Mixed Review Exercises, page 805

1. $\begin{bmatrix} 6 & 0 \\ 3 & 4 \end{bmatrix}$ **2.** $\begin{bmatrix} 8 & -4 \\ 12 & 0 \end{bmatrix}$ **3.** $\begin{bmatrix} 8 & 1 \\ 12 & 1 \end{bmatrix}$

4.

5.

6.

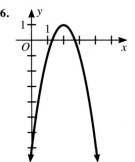

Self-Test 3 **1.** −18 **2.** −2 **3.** (−3, 5, −2)

Chapter Review, pages 806–807 **1.** d **3.** c
5. d **7.** a **9.** d

**Preparing for College Entrance Exams,
page 809** **1.** D **3.** B **5.** E

Explorations

Page 832 *(1-1)* **1. a.** isosceles right triangle
b. $AB = 1$, $BC = 1$, $AC = \sqrt{2}$ **3. a.** $AF = \sqrt{3}$, $FG = 1$, $AG = 2$ **b.** $\sqrt{3}$ **5. a.** π
b. $\pi \approx 3.14$ **7.** Start with a circle with
diameter 3 units to locate 3π, with diameter
$\frac{1}{2}$ unit to find $\frac{\pi}{2}$.

Page 833 *(2-1)* **1. c.** No; multiplying by 0
results in a product of 0. **d.** If $a < b$ and
$c > 0$, then $ac < bc$ is true. If $a < b$ and $c \le 0$,
then $ac < bc$ is false. **3. a.** yes **b.** yes

c. $3 > -2$ and $\frac{1}{3} \not< -\frac{1}{2}$ **5.** Invalid

transformation. See Exercise 3c.

Page 834 *(3-8)* **1.** Divide the weight by 16,
and find which whole numbers of pounds the
weight is between. **5. a.** 2 **b.** 3 **c.** 3 **d.** 1
7. The graphs of these functions look like steps
rather than points, lines, or curves.

Page 835 *(4-5, 4-6)* **1. a.** $3x$, $x + 2$
b. $3x(x + 2)$ **3. a.** 1 blue x-by-x tile, 1 blue
1-by-x tile, and 12 red 1-by-1 tiles **b.** no
c. because adding $3x$ and $-3x$ to the
polynomial is equivalent to adding zero
d. $x + 4$, $x - 3$ **e.** $x + 4$, $x - 3$

Page 836 *(5-7)* **1.** $\frac{67}{29}$ **3.** [2, 3, 4];
[3, 2, 1, 6, 2] **5. a.** 1.5 **b.** 1.4 **c.** $1.41\overline{6}$
d. ≈ 1.4138 **e.** ≈ 1.4143 **f.** $\sqrt{2}$ **7.** finite:
rational numbers; infinite: irrational numbers

Page 837 *(6-2)* **1. a.** 14; 10 **b.** 1.045; 2.646
c. 8; 5.337 **d.** -1; -3.332 **3. a.** 24; 24
b. 3.464; 3.464 **c.** 2.410; 2.410 **d.** 1.5; 1.5
5. a. 2; 2 **b.** 1.619; 1.619 **c.** 4; 4
d. 2.828; 2.828 **7. a.** 5 **b.** 6 **c.** 24
d. $\frac{5}{4} = 1.25$

Page 838 *(7-5)* **1.** For larger values of a, the
graph is narrower. **3. c.** The new graph is the
previous graph slid 1 unit to the right.
5. d. Increasing the value of k slides the graph
up, decreasing it slides the graph down.
7. a. It would be narrower. **b.** It would be
wider, and it would open downward instead
of upward.

Page 839 *(8-1)* **5.** It is a constant. **7.** No; a
different quotient may result but the quotient is
still a constant in each case. **9. a.** first column
b. fourth column **c.** fifth column

Page 840 *(9-6)* **1.** The graph is a circle with
center (3, 1) and radius 2. **3.** The graph is an
ellipse with center (5, 2), horizontal minor axis
of length 6, and vertical major axis of length 8.
5. a. double the radius **b.** divide the radius
by 4 **c.** triple the radius and slide the circle 2

units left and 4 units up **d.** stretch the circle
horizontally by a factor of 2 and vertically by a
factor of 3 **e.** place center at (1, 3) and stretch
as in part (d) **f.** place center at (2, -3), shrink
horizontally by a factor of $\frac{1}{2}$, and stretch
vertically by a factor of 4

Page 841 *(10-2)* **1. a.** first and third
b. $y = x$ **c.** $y = x^5$ **3. a.** first and second
b. $y = x^2$; $y = x^6$ **c.** $(0.9)^2$; $(1.01)^6$ **5.** $a > 1$
and $m > n$, or $0 < a < 1$ and $m < n$

Page 842 *(11-7)* **1.** Row 7: 1, 7, 21, 35, 35,
21, 7, 1; Row 8: 1, 8, 28, 56, 70, 56, 28, 8, 1
3. third diagonal from the left or right (1, 3,
6, 10, 15, ...) **5.** 2^0, 2^1, 2^2, 2^3, 2^4, ..., 2^n
7. Yes; it will always appear as the middle
element of a later row.

Page 844 *(13-4)* **3.** Slide the ruler down
instead of up; a negative y-value is being added.

Page 845 *(14-3)* **1.** Each graph is a line
perpendicular to the polar axis. **3.** Each graph
is a line parallel to the polar axis, above it if
$a > 0$ and below it if $a < 0$. **5.** Each graph is
a line that passes through (p, k) but not through
the pole. **7.** Each graph is a circle that passes
through the pole, has radius $|a|$, and has its
center on the 90° ray or its extension. **9.** Same
as for Ex. 8, but rotated 90° counterclockwise.

Page 846 *(15-9)* **1. c.** yes **d.** yes **3.** no;
yes **5.** Approx. $\frac{3}{10}$

Page 847 *(16-3)* **1.** AB is the result of flipping
(reflecting) B over the x-axis. **3.** a turn
(rotation) of 90° counterclockwise **5. a.** a flip
(reflection) over the y-axis **b.** a turn (rotation)
of 180°